Special Triangles

Name	*Characteristic*	*Examples*
Right Triangle	Triangle has a right angle.	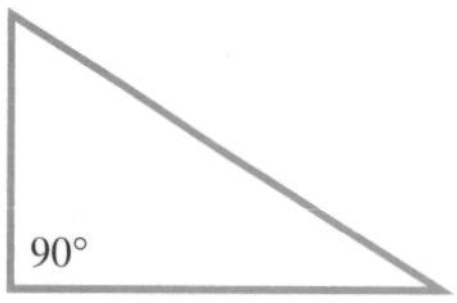
Isosceles Triangle	Triangle has two equal sides.	$AB = BC$ 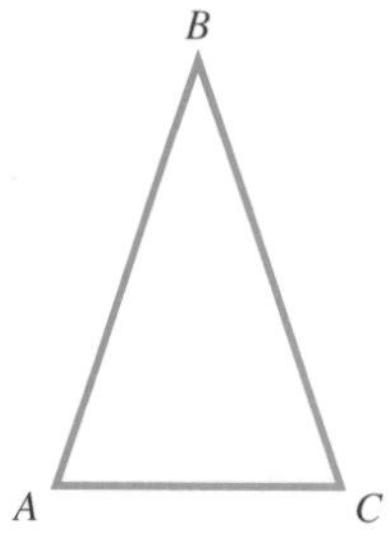
Equilateral Triangle	Triangle has three equal sides.	$AB = BC = CA$ 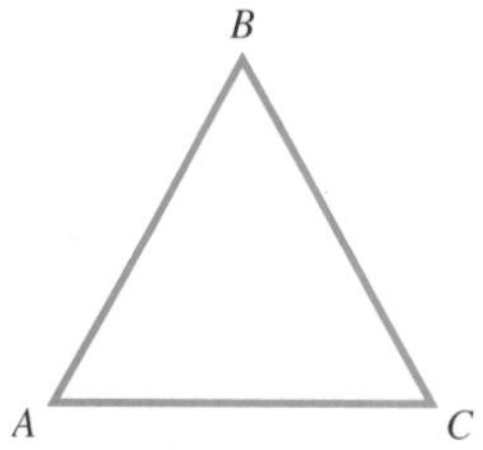
Similar Triangles	Corresponding angles are equal; corresponding sides are proportional.	$A = D, B = E, C = F$ $\frac{AB}{DE} = \frac{AC}{DF} = \frac{BC}{EF}$ 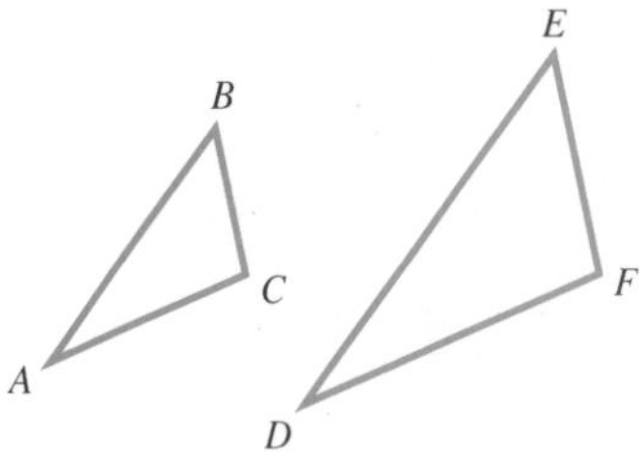

BEGINNING ALGEBRA

EIGHTH EDITION

ANNOTATED INSTRUCTOR'S EDITION

BEGINNING ALGEBRA

EIGHTH EDITION

ANNOTATED INSTRUCTOR'S EDITION

Margaret L. Lial
American River College

John Hornsby
University of New Orleans

ADDISON-WESLEY

An Imprint of Addison Wesley Longman, Inc.

Reading, Massachusetts • Menlo Park, California • New York • Harlow, England
Don Mills, Ontario • Sydney • Mexico City • Madrid • Amsterdam

Publisher: Jason Jordan

Acquisitions Editor: Jennifer Crum

Project Manager: Kari Heen

Developmental Editor: Terry McGinnis

Managing Editor: Ron Hampton

Production Supervisor: Kathleen A. Manley

Production Services: Elm Street Publishing Services, Inc.

Compositor: Typo-Graphics

Art Editor: Jennifer Bagdigian

Art Development: Meredith Nightingale

Artists: Precision Graphics, Jim Bryant, and Darwen and Vally Hennings

Marketing Manager: Craig Bleyer

Prepress Buyer: Caroline Fell

Manufacturing Coordinator: Evelyn Beaton

Text and Cover Designer: Susan Carsten

Cover Illustration: © Peter Siu/SIS

Library of Congress Cataloging-in-Publication Data

Lial, Margaret L.
Beginning algebra.—8th ed./Margaret L. Lial, John Hornsby.
p. cm.
Includes index.
ISBN 0-321-03644-1
1. Algebra. I. Hornsby, E. John. II. Title.
QA152.2.L5 1999
512.9—DC21 99-25871
CIP

Annotated Instructor's Edition ISBN: 0-321-04128-3

Printed in the U.S.A.

23456789-VH-02 01

Contents

CHAPTER 3 Linear Equations in Two Variables 175

CHAPTER 4 Polynomials and Exponents 223

CHAPTER 5 Factoring and Applications 288

CHAPTER 6 Rational Expressions 353

CHAPTER 7 Equations of Lines, Inequalities, and Functions 428

CHAPTER 8 Linear Systems 467

Preface

The eighth edition of *Beginning Algebra* is designed for college students who have never studied algebra or who want to review the basic concepts of algebra before taking additional courses in mathematics, science, business, nursing, or computer science. The primary objective of this text is to familiarize students with mathematical symbols and operations in order to solve first- and second-degree equations and applications that lead to these equations.

This revision of *Beginning Algebra* reflects our ongoing commitment to creating the best possible text and supplements package using the most up-to-date strategies for helping students succeed. One of these strategies, evident in our new Table of Contents and consistent with current teaching practices, involves the early introduction of functions and graphing lines in a rectangular coordinate system. We believe that this pedagogy has a great deal of merit as it provides students with the important "input-output" concept that will be an integral part of later mathematics courses. This organization also allows an early treatment of interesting interpretations of data in the form of line and bar graphs—two pictorial representations that students already see on a daily basis in magazines and newspapers. Chapter 3 introduces three sections on linear equations in two variables, ordered pairs, graphing, and slope, with a gentle introduction to the function concept in the form of input-output relationships. This allows students to read graphs in the chapters that immediately follow and to slowly develop an understanding of the basic idea of a function. Chapter 7 introduces the more involved concepts of forms of equations of lines and inequalities in two variables. The function concept is addressed again here, this time with a discussion of domain, range, and function notation.

If for any reason you choose not to cover these topics as our new edition suggests, it will not be difficult to defer Chapter 3 and combine it with Chapter 7 as in previous editions. You will need to skip the last objective in Section 4.1 (Graphing Simple Polynomials). Also, one or two applied problems in an example or exercise in Chapters 4–6 may refer to the function concept, but even those problems can be used without actually working through Chapter 3.

Other up-to-date pedagogical strategies to foster student success include a strong emphasis on vocabulary and problem solving, an increased number of real world applications in both examples and exercises, and a focus on relevant industry themes throughout the text.

Another strategy for student success, an exciting new CD-ROM called "Pass the Test," debuts with this edition of *Beginning Algebra.* Directly correlated to the text's content, "Pass the Test" helps students master concepts by providing interactive pretests, chapter tests, section reviews, and InterAct Math tutorial exercises. To support an increased emphasis on graphical manipulation, the CD-ROM also includes a graphing tool that can be used for open-ended, student-directed exploration of number lines and coordinate graphs, as well as for exercises relevant to the graphing content throughout the book.

The *Student's Study Guide and Journal,* redesigned and enhanced with an optional journal feature for those who would like to incorporate more writing in their mathematics curriculum, provides an additional strategy for student success.

Although *Beginning Algebra,* Eighth Edition, integrates many new elements, it also retains the time-tested features of previous editions: learning objectives for each section, careful exposition, fully developed examples, Cautions and Notes, and design features that highlight important definitions, rules, and procedures. Since the hallmark of any mathematics text is the quality of its exercise sets, we have carefully developed exercise sets that provide ample opportunity for drill and, at the same time, test conceptual understanding. In preparing this edition we have also addressed the standards of the National Council of Teachers of Mathematics and the American Mathematical Association of Two-Year Colleges, incorporating many new exercises focusing on concepts, writing, graph interpretation, technology use, collaborative work, and analysis of data from a wide variety of sources in the world around us.

CONTENT CHANGES

We have fine-tuned and polished presentations of topics throughout the text based on user and reviewer feedback. Some of the content changes you may notice include the following:

- Operations with real numbers are consolidated from four sections to two sections in Chapter 1.
- We consistently emphasize problem solving using a six-step problem-solving strategy, first introduced in Section 2.3 and continually reinforced in examples and the exercise sets throughout the text. Section 2.6 contains a comprehensive discussion of problem solving.
- New Chapter 3 introduces graphing and slope, along with an intuitive introduction to functions using input-output. Equations with two variables are presented earlier so that the applications used in later chapters (for example, with polynomials and rational expressions) can be more realistic and relevant. Both graphing and functions are continued in later chapters.
- New material is included on graphing parabolas in Section 4.1.
- Division of polynomials is consolidated in Section 4.6.
- Solution set and interval notation are now introduced in *Beginning Algebra,* consistent with the approach in *Intermediate Algebra.*

NEW FEATURES

We believe students and instructors will welcome the following new features:

Industry Themes To help motivate the material, each chapter features a particular industry that is presented in the chapter opener and revisited in examples and exercises in the chapter. Identified by special icons, these examples and exercises incorporate sourced data, often in the form of graphs and tables. Featured industries include business, health care, entertainment, sports, transportation, and others. (See pages 1, 88, 175, and 428.)

New Examples and Exercises We have added 25% more real application problems with data sources. These examples and exercises often relate to the industry themes. They are designed to show students how algebra is used to describe and interpret data in everyday life. (See pages 9, 112, 132, 434, and 492.)

The Olympic Committee has come to rely more and more on television rights and major corporate sponsors to finance the games. The pie charts show the funding plans for the first Olympics in Athens and the 1996 Olympics in Atlanta. Use proportions and the figures to answer the questions in Exercises 35 and 36.

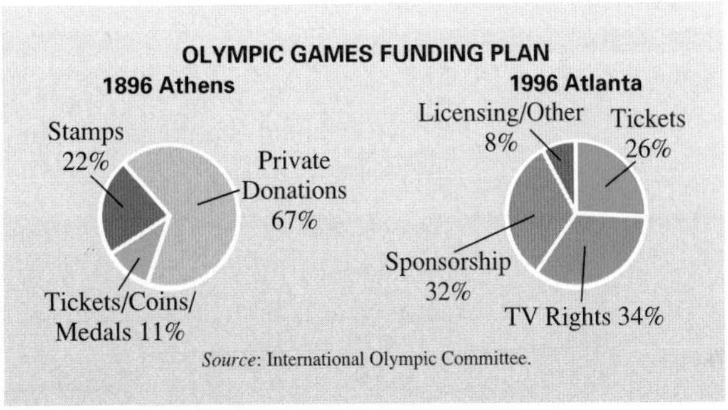

Source: International Olympic Committee.

35. In the 1996 Olympics, total revenue of $350 million was raised. There were 10 major sponsors.
  (a) Write a proportion to find the amount of revenue provided by tickets. Solve it.
  (b) What amount was provided by sponsors? Assuming the sponsors contributed equally, how much was provided per sponsor?
  (c) What amount was raised by TV rights?

36. Suppose the amount of revenue raised in the 1896 Olympics was equivalent to the $350 million in 1996.
  (a) Write a proportion for the amount of revenue provided by stamps and solve it.
  (b) What amount (in dollars) would have been provided by private donations?
  (c) In the 1988 Olympics, there were 9 major sponsors, and the total revenue was $95 million. What is the ratio of major sponsors in 1988 to those in 1996? What is the ratio of revenue in 1988 to revenue in 1996?

Technology Insights Exercises Technology is part of our lives, and we assume that all students of this text have access to scientific calculators. *While graphing calculators are not required for this text,* it is likely that students will go on to courses that use them. For this reason, we have included Technology Insights exercises in selected exercise sets. These exercises illustrate the power of graphing calculators and provide an opportunity for students to interpret typical results seen on graphing calculator screens. (See pages 212, 273, 324, and 506.)

Mathematical Journal Exercises While we continue to include conceptual and writing exercises that require short written answers, new journal exercises have been added that ask students to fully explain terminology, procedures, and methods, document their understanding using examples, or make connections between concepts. Instructors who wish to incorporate a journal component in their classes will find these exercises especially useful. For the greatest possible flexibility, both writing exercises and journal exercises are indicated with icons in the Annotated Instructor's Edition, but not in the Student Edition. (See pages 35, 103, 247, and 415.)

Group Activities Appearing at the end of each chapter, these activities allow students to apply the industry theme of the chapter to its mathematical content in a collaborative setting. (See pages 162, 458, and 506.)

Test Your Word Power To help students understand and master mathematical vocabulary, this new feature has been incorporated at the end of each chapter. Key terms from the chapter are presented with four possible definitions in multiple-choice format. Answers and examples illustrating each term are provided at the bottom of the appropriate page. (See pages 163, 277, and 343.)

HALLMARK FEATURES

We have retained the popular features of previous editions of the text. Some of these features are as follows:

Learning Objectives Each section begins with clearly stated numbered objectives, and material in the section is keyed to these objectives. In this way students know exactly what is being covered in each section.

OBJECTIVES

1. Learn the definition of *factor.*
2. Write fractions in lowest terms.
3. Multiply and divide fractions.
4. Add and subtract fractions.
5. Solve applied problems that involve fractions.
6. Interpret data in a circle graph.

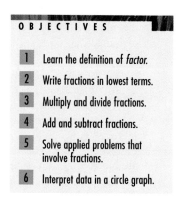

Cautions and Notes We often give students warnings of common errors and emphasize important ideas in Cautions and Notes that appear throughout the exposition.

Connections Retained from the previous edition, Connections boxes have been streamlined and now often appear at the beginning or the end of the exposition in selected sections. They continue to provide connections to the real world or to other mathematical concepts, historical background, and thought-provoking questions for writing or class discussion. (See pages 105, 201, 226, and 322.)

Problem Solving Increased emphasis has been given to our six-step problem-solving method to aid students in solving application problems. This method is continually reinforced in examples and exercises throughout the text. (See pages 105, 107, 330, 409, and 497.)

Ample and Varied Exercise Sets Students in beginning algebra require a large number and variety of practice exercises to master the material. This text contains approximately 5800 exercises, including about 1600 review exercises, plus numerous conceptual and writing exercises, journal exercises, and challenging exercises that go beyond the examples. More illustrations, diagrams, graphs, and tables now accompany exercises. Multiple-choice, matching, true/false, and completion exercises help to provide variety. Exercises suitable for calculator use are marked with a calculator icon in both the Student Edition and the Annotated Instructor's Edition. (See pages 27, 143, 347, and 438.)

Relating Concepts Previously titled Mathematical Connections, these sets of exercises often appear near the end of selected sections. They tie together topics and highlight the relationships among various concepts and skills. For example, they may show how algebra and geometry are related, or how a graph of a linear equation in two variables is related to the solution of the corresponding linear equation in one variable. Instructors have told us that these sets of exercises make great collaborative activities for small groups of students. (See pages 69, 210, 232, 381, and 490.)

Ample Opportunity for Review Each chapter concludes with a Chapter Summary that features Key Terms and Symbols, Test Your Word Power, and a Quick Review of each section's content. Chapter Review Exercises keyed to individual sections are included as well as mixed review exercises and a Chapter Test. Following every chapter after Chapter 1, there is a set of Cumulative Review Exercises that covers material going back to the first chapter. Students always have an opportunity to review material that appears earlier in the text, and this provides an excellent way to prepare for the final examination in the course. (See pages 214–222 and 459–466.)

SUPPLEMENTS

Our extensive supplements package includes the Annotated Instructor's Edition, testing materials, study guides, solutions manuals, CD-ROM software, videotapes, and a Web site. For more information on these and other helpful supplements, contact your Addison Wesley Longman sales representative.

FOR THE INSTRUCTOR

Annotated Instructor's Edition (ISBN 0-321-04128-3)
For immediate access, the Annotated Instructor's Edition provides answers to all text exercises and Group Activities in color in the margin or next to the corresponding exercise, as well as Chalkboard Examples and Teaching Tips. To assist instructors in assigning homework, additional icons not shown in the Student Edition indicate journal exercises, writing exercises, and challenging exercises ▲.

CHALKBOARD EXAMPLE

Write 90 as the product of prime factors.

Answer: $2 \cdot 3^2 \cdot 5$

TEACHING TIP The term *fraction bar* may be unfamiliar to some students.

Exercises designed for calculator use are indicated in both the Student Edition and the Annotated Instructor's Edition.

Instructor's Solutions Manual (ISBN 0-321-06193-4)
The *Instructor's Solutions Manual* provides solutions to all even-numbered exercises, including answer art, and lists of all writing, journal, challenging, Relating Concepts, and calculator exercises.

Answer Book (ISBN 0-321-06194-2)
The *Answer Book* contains answers to all exercises and lists of all writing, journal, challenging, Relating Concepts, and calculator exercises. Instructors may ask the bookstore to order multiple copies of the *Answer Book* for students to purchase.

Printed Test Bank (ISBN 0-321-06192-6)
The *Printed Test Bank* contains short answer and multiple-choice versions of a placement test and final exam; six forms of chapter tests for each chapter, including four open-response (short answer) and two multiple-choice forms; 10 to 20 additional exercises per objective for instructors to use for extra practice, quizzes, or tests; answer keys to all of the above listed tests and exercises; and lists of all writing, journal, challenging, Relating Concepts, and calculator exercises.

TestGen-EQ with QuizMaster EQ (ISBN 0-321-06132-2)
This fully networkable software presents a friendly graphical interface which enables professors to build, edit, view, print and administer tests. Tests can be printed or easily exported to HTML so they can be posted to the Web for student practice.

For the Student

Student's Study Guide and Journal (ISBN 0-321-06196-9)
The *Student's Study Guide and Journal* contains a "Chart Your Progress" feature for students to track their scores on homework assignments, quizzes, and tests, additional practice for each learning objective, section summary outlines that give students additional writing opportunities and help with test preparation, and self-tests with answers at the end of each chapter. A manual icon at the beginning of each section in the Student Edition identifies section coverage.

Student's Solutions Manual (ISBN 0-321-06195-0)
The *Student's Solutions Manual* provides solutions to all odd-numbered exercises (journal and writing exercises excepted). A manual icon at the beginning of each section in the Student Edition identifies section coverage.

InterAct Math Tutorial Software (ISBN 0-321-06140-3 (Student Version))
This tutorial software correlates with every odd-numbered exercise in the text. The program is highly interactive with sample problems and interactive guided solutions accompanying every exercise. The program recognizes common student errors and provides customized feedback with sophisticated answer recognition capabilities. The management system (InterAct Math Plus) allows instructors to create, administer, and track tests, and to monitor student performance during practice sessions.

"Real to Reel" Videotapes (0-321-05659-0)
This videotape series provides separate lessons for each section in the book. A videotape icon at the beginning of each section identifies section coverage. All objectives, topics, and problem-solving techniques are covered and content is specific to *Beginning Algebra,* Eighth Edition.

"Pass the Test" Interactive CD-ROM (ISBN 0-321-06204-3)
This CD helps students to master the course content by providing interactive pre-tests, chapter tests, section reviews, and InterAct tutorial exercises. After studying a chapter in class, students take a pre-test to determine what areas in that chapter need additional work. They are then directed to section reviews and tutorial exercises for continued practice. Students continue to take chapter tests and practice their skills until they have mastered the chapter. A unique graphing tool is provided for exploring the relationship between graphs and their algebraic representation.

World Wide Web Supplement (www.LialAlgebra.com)
Students can visit the Web site to explore additional real world applications related to the chapter themes, look up words in a complete glossary, and work through graphing calculator tutorials. The tutorials consist of step-by-step procedures as well as practice exercises for mastering basic graphing calculator skills.

MathXL (http://www.mathxl.com)
Available on-line with a pre-assigned ID and password by ordering a new copy of *Beginning Algebra,* Eighth Edition, with ISBN 0-201-68155-2, MathXL helps students prepare for tests by allowing them to take practice tests that are similar to the chapter tests in their text. Students also get a personalized study plan that identifies strengths and pinpoints topics where more review is needed. For more information on subscriptions, contact your Addison Wesley Longman sales representative.

Math Tutor Center
The Addison Wesley Longman Math Tutor Center is staffed by qualified mathematics instructors who provide students with tutoring on text examples, exercises, and problems. Tutoring assistance is provided by telephone, fax, and e-mail and is available five days a week, seven hours a day. A registration number for the tutoring service may be obtained by calling our toll-free customer service number, 1-800-922-0579 and requesting ISBN 0-201-44461-5. Registration for the service is active for one or more of the following time periods depending on the course duration: Fall (8/31–1/31), Spring (1/1–6/30), or Summer (5/1–8/31). The Math Tutor Center service is also available for other Addison Wesley Longman textbooks in developmental math, precalculus math, liberal arts math, applied math, applied calculus, calculus, and introductory statistics. For more information, please contact your Addison Wesley Longman sales representative.

Spanish Glossary (ISBN 0-321-01647-5)
This book includes math terms that would be encountered in Basic Math through College Algebra.

Acknowledgments

For a textbook to last through eight editions, it is necessary for the authors to rely on comments, criticisms, and suggestions of users, nonusers, instructors, and students. We are grateful for the many responses that we have received over the years. We wish to thank the following individuals who reviewed this edition of the text:

Josette Ahlering, *Central Missouri State*
Vickie Aldrich, *Dona Ana Branch Community College*
Lisa Anderson, *Ventura College*
Robert B. Baer, *Miami University–Hamilton*
Julie R. Bonds, *Sonoma State University*
Beverly R. Broomell, *Suffolk Community College*
Cheryl V. Cantwell, *Seminole Community College*
Stanley Carter, *Central Missouri State*
Jeff Clark, *Santa Rosa Junior College*
Ted Corley, *Glendale Community College*
Lisa Delong Cuneo, *Penn State–Dubois Campus*
Marlene Demerjian, *College of the Canyons*
Richard N. Dodge, *Jackson Community College*
Linda Franko, *Cuyahoga Community College*
Linda L. Galloway, *Macon State College*
Theresa A. Geiger, *St. Petersburg Junior College*
Martha Haehl, *Maple Woods Community College*
Melissa Harper, *Embry Riddle Aeronautical University*
W. Hildebrand, *Montgomery College*
Matthew Hudock, *St. Philips College*
Dale W. Hughes, *Johnson County Community College*
Linda Hurst, *Central Texas College*
Nancy Johnson, *Broward Community College–North*
Robert Kaiden, *Lorain County Community College*
Michael Karelius, *American River College*
Margaret Kimbell, *Texas State Technical College*
Linda Kodama, *Kapiolani Community College*
Jeff A. Koleno, *Lorain County Community College*
William R. Livingston, *Missouri Southern State College*
Doug Martin, *Mt. San Antonio College*
Larry Mills, *Johnson County Community College*
Mary Ann Misko, *Gadsden State Community College*
Elsie Newman, *Owens Community College*
Joanne V. Peeples, *El Paso Community College*
Janice Rech, *University of Nebraska–Omaha*
Joyce Saxon, *Morehead University*
Richard Semmler, *Northern Virginia Community College*
LeeAnn Spahr, *Durham Technical Community College*

No author can complete a project of this magnitude without the help of many other individuals. Our sincere thanks go to Jenny Crum of Addison Wesley Longman who coordinated the package of texts of which this book is a part. Other dedicated staff at Addison Wesley Longman who worked long and hard to make this revision a success include Jason Jordan, Kari Heen, Susan Carsten, Meredith Nightingale, and Kathy Manley.

While Terry McGinnis has assisted us for many years "behind the scenes" in producing our texts, she has contributed far more to these revisions than ever. There is no question that these books are improved because of her attention to detail and consistency, and we are most grateful for her work above and beyond the call of duty. Kitty Pellissier continues to do an outstanding job in checking the answers to exercises. Many thanks to Jenny Bagdigian who coordinated the art programs for the books.

Cathy Wacaser of Elm Street Publishing Services provided her usual excellent production work. She is indeed one of the best in the business. As usual, Paul Van Erden created an accurate, useful index. Becky Troutman prepared the Index of Applications. We are also grateful to Tommy Thompson who made suggestions for the feature "For the Student: 10 Ways to Succeed with Algebra," to Vickie Aldrich and Lucy Gurrola who wrote the Group Activity features, and Janis Cimperman of St. Cloud University and Steve C. Ouellette of the Walpole Massachusetts State Public Schools.

To these individuals and all the others who have worked on these books for 30 years, remember that we could not have done it without you. We hope that you share with us our pride in these books.

Margaret L. Lial
John Hornsby

An Introduction to Calculators

There is little doubt that the appearance of handheld calculators nearly three decades ago and the later development of scientific and graphing calculators have changed the methods of learning and studying mathematics forever. Where the study of computations with tables of logarithms and slide rules made up an important part of mathematics courses prior to 1970, today the widespread availability of calculators make their study a topic only of historical significance.

Most consumer models of calculators are inexpensive. At first, however, they were costly. One of the first consumer models available was the Texas Instruments SR-10, which sold for about $150 in 1973. It could perform the four operations of arithmetic and take square roots, but could do very little more.

Today calculators come in a large array of different types, sizes, and prices. *For the course for which this textbook is intended, the most appropriate type is the scientific calculator,* which costs $10–$20.

In this introduction, we explain some of the features of scientific and graphing calculators. However, remember that calculators vary among manufacturers and models, and that while the methods explained here apply to many of them, they may not apply to your specific calculator. For this reason, it is important to remember that *this introduction is only a guide, and is not intended to take the place of your owner's manual.* Always refer to the manual in the event you need an explanation of how to perform a particular operation.

Scientific Calculators

Scientific calculators are capable of much more than the typical four-function calculator that you might use for balancing your checkbook. Most scientific calculators use *algebraic logic.* (Models sold by Texas Instruments, Sharp, Casio, and Radio Shack, for example, use algebraic logic.) A notable exception is Hewlett Packard, a company whose calculators use *Reverse Polish Notation* (RPN). In this introduction, we explain the use of calculators with algebraic logic.

Arithmetic Operations

To perform an operation of arithmetic, simply enter the first number, press the operation key ([+], [−], [×], or [÷]), enter the second number, and then press the [=] key. For example, to add 4 and 3, use the following keystrokes.

[4] [+] [3] [=]   7

Change Sign Key

The key marked [±] allows you to change the sign of a display. This is particularly useful when you wish to enter a negative number. For example, to enter -3, use the following keystrokes.

[3] [±]   −3

Memory Key

Scientific calculators can hold a number in memory for later use. The label of the memory key varies among models; two of these are [M] and [STO]. [M+] and [M−] allow you to add to or subtract from the value currently in memory. The memory recall key, labeled [MR], [RM], or [RCL], allows you to retrieve the value stored in memory.

Suppose that you wish to store the number 5 in memory. Enter 5, then press the key for memory. You can then perform other calculations. When you need to retrieve the 5, press the key for memory recall.

If a calculator has a constant memory feature, the value in memory will be retained even after the power is turned off. Some advanced calculators have more than one memory. It is best to read the owner's manual for your model to see exactly how memory is activated.

Clearing/Clear Entry Keys

These keys allow you to clear the display or clear the last entry entered into the display. They are usually marked [C] and [CE]. In some models, pressing the [C] key once will clear the last entry, while pressing it twice will clear the entire operation in progress.

Second Function Key

This key is used in conjunction with another key to activate a function that is printed *above* an operation key (and not on the key itself). It is usually marked [2nd]. For example, suppose you wish to find the square of a number, and the squaring function (explained in more detail later) is printed above another key. You would need to press [2nd] before the desired squaring function can be activated.

Square Root Key

Pressing the square root key, [$\sqrt{x}$], will give the square root (or an approximation of the square root) of the number in the display. For example, to find the square root of 36, use the following keystrokes.

[3] [6] [$\sqrt{x}$] | 6 |

The square root of 2 is an example of an irrational number (Chapter 9). The calculator will give an approximation of its value, since the decimal for $\sqrt{2}$ never terminates and never repeats. The number of digits shown will vary among models. To find an approximation of $\sqrt{2}$, use the following keystrokes.

[2] [$\sqrt{x}$] | 1.4142136 | An approximation

Squaring Key

This key, [x^2], allows you to square the entry in the display. For example, to square 35.7, use the following keystrokes.

[3] [5] [.] [7] [x^2] | 1274.49 |

The squaring key and the square root key are often found on the same key, with one of them being a second function (that is, activated by the second function key, described above).

Reciprocal Key

The key marked [1/x] is the reciprocal key. (When two numbers have a product of 1, they are called *reciprocals*. See Chapter 1.) Suppose that you wish to find the reciprocal of 5. Use the following keystrokes.

[5] [1/x] | 0.2 |

Inverse Key

Some calculators have an inverse key, marked [INV]. Inverse operations are operations that "undo" each other. For example, the operations of squaring and taking the square root are inverse operations. The use of the [INV] key varies among different models of calculators, so read your owner's manual carefully.

Exponential Key

The key marked [x^y] or [y^x] allows you to raise a number to a power. For example, if you wish to raise 4 to the fifth power (that is, find 4^5, as explained in Chapter 4), use the following keystrokes.

[4] [x^y] [5] [=] [1024]

Root Key

Some calculators have this key specifically marked [$\sqrt[x]{y}$] or [$\sqrt[y]{x}$]; with others, the operation of taking roots is accomplished by using the inverse key in conjunction with the exponential key. Suppose, for example, your calculator is of the latter type and you wish to find the fifth root of 1024. Use the following keystrokes.

[1] [0] [2] [4] [INV] [x^y] [5] [=] [4]

Notice how this "undoes" the operation explained in the exponential key discussion above.

Pi Key

The number π is an important number in mathematics. It occurs, for example, in the area and circumference formulas for a circle. By pressing the [π] key, you can display the first few digits of π. (Because π is irrational, the display shows only an approximation.) One popular model gives the following display when the [π] key is pressed: [3.1415927].

Methods of Display

When decimal approximations are shown on scientific calculators, they are either *truncated* or *rounded*. To see how a particular model is programmed, evaluate 1/18 as an example. If the display shows .0555555 (last digit 5), it truncates the display. If it shows .0555556 (last digit 6), it rounds off the display.

When very large or very small numbers are obtained as answers, scientific calculators often express these numbers in scientific notation (Chapter 4). For example, if you multiply 6,265,804 by 8,980,591, the display might look like this:

[5.6270623 13].

The "13" at the far right means that the number on the left is multiplied by 10^{13}. This means that the decimal point must be moved 13 places to the right if the answer is to be expressed in its usual form. Even then, the value obtained will only be an approximation: 56,270,623,000,000.

Graphing Calculators

Graphing calculators are becoming increasingly popular in mathematics classrooms. While you are not expected to have a graphing calculator to study from this book, we do include a feature in many exercise sets called *Technology Insights* that asks you to interpret typical graphing calculator screens. These exercises can help to prepare you for future courses where graphing calculators may be recommended or even required.

Basic Features

Graphing calculators provide many features beyond those found on scientific calculators. In addition to the typical keys found on scientific calculators, they have keys that can be used to create graphs, make tables, analyze data, and change settings. One of the major differences between graphing and scientific calculators is that a graphing calculator has a larger viewing screen with graphing capabilities. The screens below illustrate the graphs of $y = x$ and $y = x^2$.

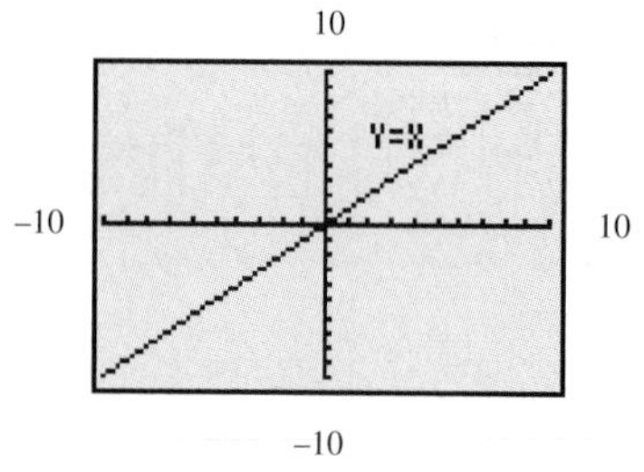

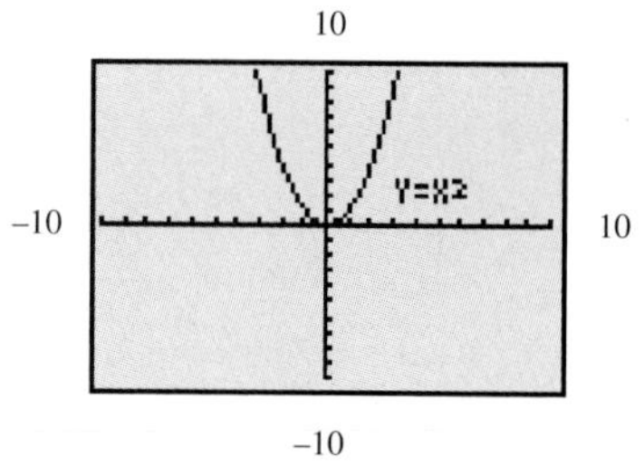

If you look closely at the screens, you will see that the graphs appear to be "jagged" rather than smooth, as they should be. The reason for this is that graphing calculators have much lower resolution than a computer screen. Because of this, graphs generated by graphing calculators must be interpreted carefully.

Editing Input

The screen of a graphing calculator can display several lines of text at a time. This feature allows you to view both previous and current expressions. If an incorrect expression is entered, an error message is displayed. The erroneous expression can be viewed and corrected by using various editing keys, much like a word-processing program. You do not need to enter the entire expression again. Many graphing calculators can also recall past expressions for editing or updating. The screen on the left below shows how two expressions are evaluated. The final line is entered incorrectly, and the resulting error message is shown in the screen on the right.

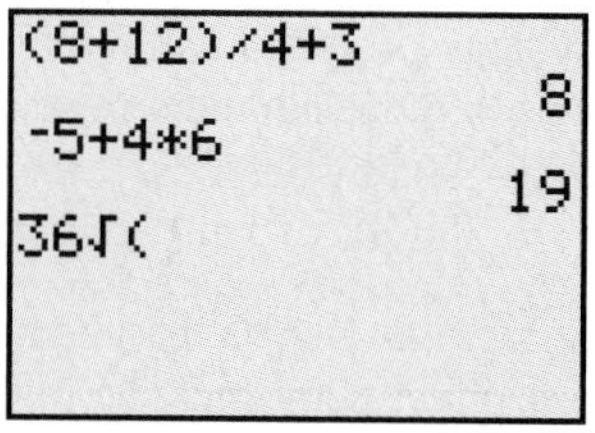

Order of Operations

Arithmetic operations on graphing calculators are usually entered as they are written in mathematical equations. For example, to evaluate $\sqrt{36}$ on a typical scientific calculator, you would first enter 36 and then press the square root key. As seen above, this is not the correct syntax for a graphing calculator. To find this root, you would first press the square root key, and then enter 36. See the screen on the left at the top of the next page. The order of operations on a graphing calculator is also important, and current models assist the user by inserting parentheses when typical errors might occur. The open parenthesis that follows the square root symbol is automatically entered by the calculator, so that an expression such as $\sqrt{2 \times 8}$ will not be calculated incorrectly as $\sqrt{2} \times 8$. Compare the two entries and their results in the screen on the right.

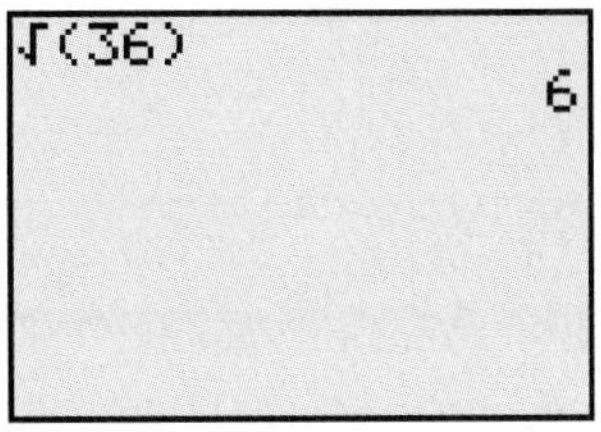

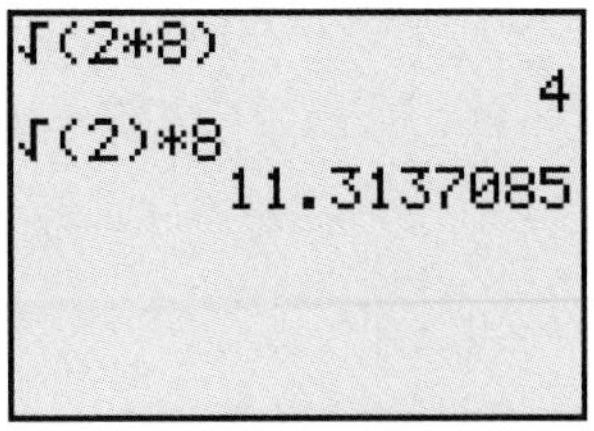

Viewing Windows

The viewing window for a graphing calculator is similar to the viewfinder in a camera. A camera usually cannot take a photograph of an entire view of a scene. The camera must be centered on some object and can only capture a portion of the available scenery. A camera with a zoom lens can photograph different views of the same scene by zooming in and out. Graphing calculators have similar capabilities. The *xy*-coordinate plane is infinite. The calculator screen can only show a finite, rectangular region in the plane, and it must be specified before the graph can be drawn. This is done by setting both minimum and maximum values for the *x*- and *y*-axes. The scale (distance between tick marks) is usually specified as well. Determining an appropriate viewing window for a graph is often a challenge, and many times it will take a few attempts before a satisfactory window is found.

The screen on the left shows a "standard" viewing window, and the graph of $y = 2x + 1$ is shown on the right. Using a different window would give a different view of the line.

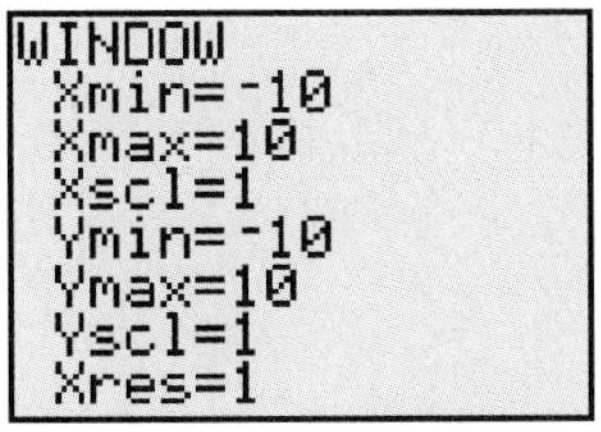

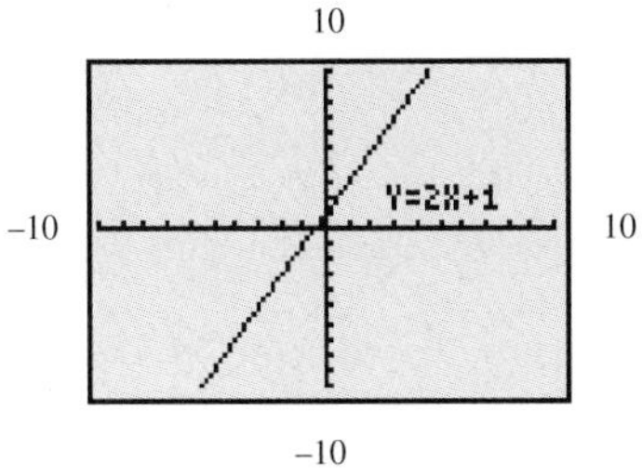

Locating Points on a Graph: Tracing and Tables

Graphing calculators allow you to trace along the graph of an equation, and, while doing this, display the coordinates of points on the graph. See the screen on the left at the top of the next page, which indicates that the point (2, 5) lies on the graph of $y = 2x + 1$. Tables for equations can also be displayed. The screen on the right shows a partial table for this same equation. Note the middle of the screen, which indicates that when $x = 2$, $y = 5$.

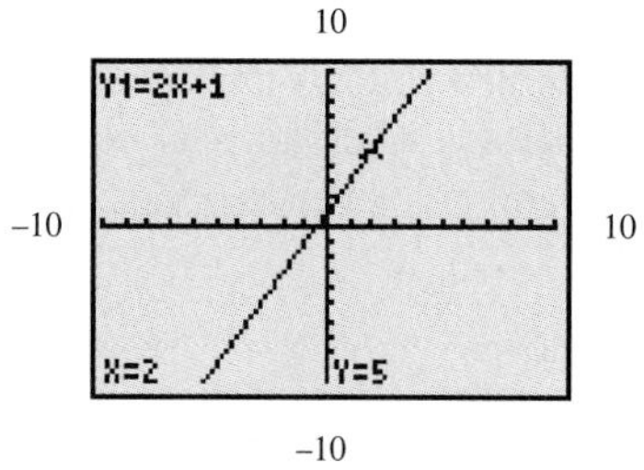

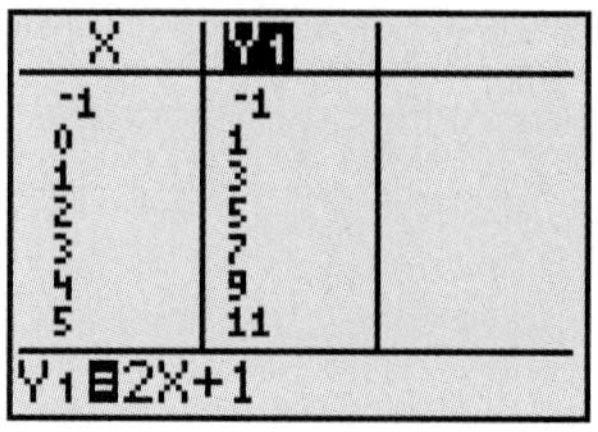

ADDITIONAL FEATURES

There are many features of graphing calculators that go far beyond the scope of this book. These calculators can be programmed, much like computers. Many of them can solve equations at the stroke of a key, analyze statistical data, and perform symbolic algebraic manipulations. Mathematicians from the past would have been amazed by today's calculators. Many important equations in mathematics cannot be solved by hand. However, their solutions can often be approximated using a calculator. Calculators also provide the opportunity to ask "What if . . . ?" more easily. Values in algebraic expressions can be altered and conjectures tested quickly.

FINAL COMMENTS

Despite the power of today's calculators, they cannot replace human thought. **In the entire problem-solving process, your brain is the most important component.** Calculators are only tools, and like any tool, they must be used appropriately in order to enhance our ability to understand mathematics. Mathematical insight may often be the quickest and easiest way to solve a problem; a calculator may neither be needed nor appropriate. By applying mathematical concepts, you can make the decision whether or not to use a calculator.

BEGINNING ALGEBRA

EIGHTH EDITION

ANNOTATED INSTRUCTOR'S EDITION

The Real Number System 1

Growth in productivity—the amount of goods and services produced for each hour of work—is the most important factor in improving living standards of our country's population. In the latter part of the 1990s, the U.S. construction industry finally began to revive after a long period of decline. Having survived many years of downsizing and layoffs, manufacturing-related industries are again contributing to economic growth. This bar graph shows the increase in construction spending in the latter half of 1996 and all of 1997. During which month and year was construction spending highest? Several exercises in Section 1.2 use this graph. Throughout this chapter, we will see other specific examples and exercises that relate to construction and manufacturing.

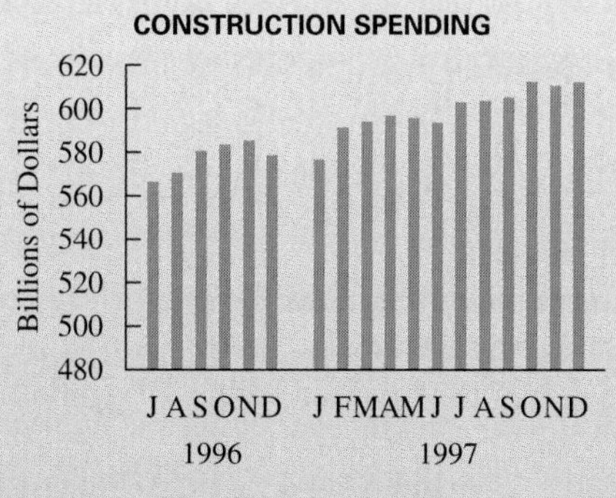

Source: Commerce Department.

Construction/
Manufacturing

Visit our Web site at www.LialAlgebra.com

1.1 Fractions

OBJECTIVES

1. Learn the definition of *factor.*
2. Write fractions in lowest terms.
3. Multiply and divide fractions.
4. Add and subtract fractions.
5. Solve applied problems that involve fractions.
6. Interpret data in a circle graph.

FOR EXTRA HELP

SSG Sec. 1.1
SSM Sec. 1.1
Pass the Test Software
InterAct Math Tutorial Software
Video 1

TEACHING TIP The term *fraction bar* may be unfamiliar to some students.

■ Answers will vary.

As preparation for the study of algebra, this section begins with a brief review of arithmetic. In everyday life the numbers seen most often are the **natural numbers,**

$$1, 2, 3, 4, \ldots,$$

the **whole numbers,**

$$0, 1, 2, 3, 4, \ldots,$$

and **fractions,** such as

$$\frac{1}{2}, \quad \frac{2}{3}, \quad \text{and} \quad \frac{15}{7}.$$

The parts of a fraction are named as follows.

$$\text{Fraction bar} \rightarrow \frac{4}{7} \quad \left(\frac{a}{b} = a \div b\right) \quad \begin{matrix}\leftarrow \text{Numerator} \\ \leftarrow \text{Denominator}\end{matrix}$$

As we will see later, the fraction bar represents division $\left(\frac{a}{b} = a \div b\right)$ and also serves as a grouping symbol.

CONNECTIONS

A common use of fractions is to measure dimensions of tools, amounts of building materials, and so on. We need to add, subtract, multiply, and divide fractions in order to solve many types of measurement problems.

FOR DISCUSSION OR WRITING

Discuss some situations in your experience where you have needed to perform the operations of addition, subtraction, multiplication, or division on fractions. (*Hint:* To get you started, think of art projects, carpentry projects, adjusting recipes, and working on cars.)

OBJECTIVE 1 Learn the definition of *factor.* In the statement $2 \times 9 = 18$, the numbers 2 and 9 are called **factors** of 18. Other factors of 18 include 1, 3, 6, and 18. The result of the multiplication, 18, is called the **product.**

The number 18 is **factored** by writing it as the product of two or more numbers. For example, 18 can be factored in several ways, as $6 \cdot 3$, or $18 \cdot 1$, or $9 \cdot 2$, or $3 \cdot 3 \cdot 2$. In algebra, a raised dot $\cdot$ is often used instead of the $\times$ symbol to indicate multiplication.

TEACHING TIP Mention that if 1 were not excluded from the prime numbers, then *composite numbers* (those that are composed of prime factors) could be factored into primes in more than one way. This would violate the important fact that there is only one way to factor a composite number into primes.

A natural number (except 1) is **prime** if it has only itself and 1 as factors. "Factors" are understood here to mean natural number factors. (By agreement, the number 1 is not a prime number.) The first dozen primes are

$$2, 3, 5, 7, 11, 13, 17, 19, 23, 29, 31, 37.$$

A natural number (except 1) that is not prime is a **composite** number.

It is often useful to find all the **prime factors** of a number—those factors that are prime numbers. For example, the only prime factors of 18 are 2 and 3.

INSTRUCTOR'S RESOURCES

PTB Sec. 1.1 ISM Sec. 1.1
AB Sec. 1.1

TEST GENERATOR

WORLD WIDE WEB
www.LialAlgebra.com

CHALKBOARD EXAMPLE

Write 90 as the product of prime factors.

Answer: $2 \cdot 3^2 \cdot 5$

TEACHING TIP You might wish to show your students alternate methods of prime factorization, as shown here.

$$\begin{array}{r} 3 \\ 2\overline{)6} \\ 2\overline{)12} \\ 2\overline{)24} \end{array}$$

$24 = 2 \cdot 2 \cdot 2 \cdot 3$ (True)

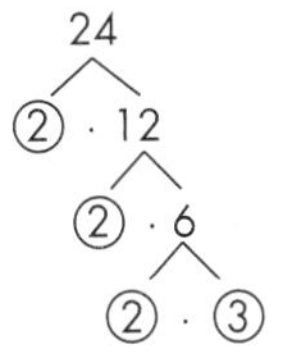

EXAMPLE 1 Factoring Numbers

Write the number as the product of prime factors.

(a) 35

Write 35 as the product of the prime factors 5 and 7, or as

$$35 = 5 \cdot 7.$$

(b) 24

One way to begin is to divide by the smallest prime, 2, to get

$$24 = 2 \cdot 12.$$

Now divide 12 by 2 to find factors of 12.

$$24 = 2 \cdot 2 \cdot 6$$

Since 6 can be written as $2 \cdot 3$,

$$24 = 2 \cdot 2 \cdot 2 \cdot 3,$$

where all factors are prime.

NOTE It is not necessary to start with the smallest prime factor, as shown in Example 1(b). In fact, no matter which prime factor we start with, we will *always* obtain the same prime factorization.

OBJECTIVE 2 **Write fractions in lowest terms.** We use prime numbers to write fractions in *lowest terms.* A fraction is in **lowest terms** when the numerator and denominator have no factors in common (other than 1). By the **basic principle of fractions,** if the numerator and denominator of a fraction are multiplied or divided by the *same* nonzero number, the value of the fraction is unchanged. To write a fraction in lowest terms, use these steps.

Writing a Fraction in Lowest Terms

Step 1 Write the numerator and the denominator as the product of prime factors.

Step 2 Divide the numerator and the denominator by the **greatest common factor,** the product of all factors common to both.

CHALKBOARD EXAMPLE

Write $\frac{12}{20}$ in lowest terms.

Answer: $\frac{3}{5}$

EXAMPLE 2 Writing Fractions in Lowest Terms

Write the fraction in lowest terms.

(a) $\dfrac{10}{15} = \dfrac{2 \cdot 5}{3 \cdot 5} = \dfrac{2 \cdot 1}{3 \cdot 1} = \dfrac{2}{3}$

Since 5 is the greatest common factor of 10 and 15, dividing both numerator and denominator by 5 gives the fraction in lowest terms.

(b) $\dfrac{15}{45} = \dfrac{3 \cdot 5}{3 \cdot 3 \cdot 5} = \dfrac{1 \cdot 1}{3 \cdot 1 \cdot 1} = \dfrac{1}{3}$

The factored form shows that 3 and 5 are the common factors of both 15 and 45. Dividing both 15 and 45 by $3 \cdot 5 = 15$ gives $\frac{15}{45}$ in lowest terms as $\frac{1}{3}$.

TEACHING TIP Students may ask, "Where does the 1 come from?" Remind them that when the numerator and the denominator are both divided by 3 (or 5), the result is 1. The phrase "cancel out" can be misleading.

We can simplify this process by finding the greatest common factor in the numerator and denominator by inspection. For instance, in Example 2(b), we can use 15 rather than $3 \cdot 5$.

$$\frac{15}{45} = \frac{\mathbf{15}}{3 \cdot \mathbf{15}} = \frac{\mathbf{1}}{3 \cdot \mathbf{1}} = \frac{1}{3}$$

CAUTION

Errors may occur when writing fractions in lowest terms if the factor 1 is not included. To see this, refer to Example 2(b). In the equation

$$\frac{3 \cdot 5}{3 \cdot 3 \cdot 5} = \frac{?}{3},$$

if 1 is not written in the numerator when dividing common factors, you may make an error. The **?** should be replaced by **1**.

OBJECTIVE 3 Multiply and divide fractions. The basic operations on whole numbers, addition, subtraction, multiplication, and division, also apply to fractions. We multiply two fractions by first multiplying their numerators and then multiplying their denominators. This rule is written in symbols as follows.

Multiplying Fractions

If $\frac{a}{b}$ and $\frac{c}{d}$ are fractions, then $\frac{a}{b} \cdot \frac{c}{d} = \frac{a \cdot c}{b \cdot d}$.

CHALKBOARD EXAMPLE

Find the product and write the answer in lowest terms.

$$\frac{7}{9} \cdot \frac{12}{14}$$

Answer: $\frac{2}{3}$

EXAMPLE 3 Multiplying Fractions

Find the product of $\frac{3}{8}$ and $\frac{4}{9}$, and write it in lowest terms.

First, multiply $\frac{3}{8}$ and $\frac{4}{9}$.

$$\frac{3}{8} \cdot \frac{4}{9} = \frac{\mathbf{3 \cdot 4}}{\mathbf{8 \cdot 9}} \quad \text{Multiply numerators; multiply denominators.}$$

It is easiest to write a fraction in lowest terms while the product is in factored form. Factor 8 and 9 and then divide out common factors in the numerator and denominator.

$$\frac{3 \cdot 4}{8 \cdot 9} = \frac{\mathbf{3 \cdot 4}}{\mathbf{2 \cdot 4 \cdot 3 \cdot 3}} = \frac{\mathbf{1} \cdot 3 \cdot 4}{2 \cdot 4 \cdot 3 \cdot 3} \quad \text{Factor. Introduce a factor of 1.}$$

$$= \frac{1}{2 \cdot 3} \quad \text{3 and 4 are common factors.}$$

$$= \frac{1}{6} \quad \text{Lowest terms}$$

TEACHING TIP Mention that the reciprocal of 6 is $\frac{1}{6}$.

Two fractions are **reciprocals** of each other if their product is 1. For example, $\frac{3}{4}$ and $\frac{4}{3}$ are reciprocals since

$$\frac{3}{4} \cdot \frac{4}{3} = \frac{12}{12} = 1.$$

Also, $\frac{7}{11}$ and $\frac{11}{7}$ are reciprocals of each other. We use the reciprocal to divide fractions. To *divide* two fractions, multiply the first fraction by the reciprocal of the second fraction.

Dividing Fractions

For the fractions $\frac{a}{b}$ and $\frac{c}{d}$, $\frac{a}{b} \div \frac{c}{d} = \frac{a}{b} \cdot \frac{d}{c}$.

(To divide by a fraction, multiply by its reciprocal.)

The reason this method works will be explained in Chapter 6. The answer to a division problem is called a **quotient.** For example, the quotient of 20 and 10 is 2, since $20 \div 10 = 2$.

CHALKBOARD EXAMPLE

Find the quotient and write the answer in lowest terms.

$$\frac{9}{10} \div \frac{3}{5}$$

Answer: $\frac{3}{2}$

EXAMPLE 4 Dividing Fractions

Find the following quotients, and write them in lowest terms.

(a) $\frac{3}{4} \div \frac{8}{5} = \frac{3}{4} \cdot \frac{5}{8} = \frac{3 \cdot 5}{4 \cdot 8} = \frac{15}{32}$ Multiply by the reciprocal of $\frac{8}{5}$.

(b) $\frac{3}{4} \div \frac{5}{8} = \frac{3}{4} \cdot \frac{8}{5} = \frac{3 \cdot 8}{4 \cdot 5} = \frac{3 \cdot 4 \cdot 2}{4 \cdot 5} = \frac{6}{5}$

OBJECTIVE 4 Add and subtract fractions. To find the **sum** of two fractions having the same denominator, add the numerators and keep the same denominator.

Adding Fractions

If $\frac{a}{b}$ and $\frac{c}{b}$ are fractions, then $\frac{a}{b} + \frac{c}{b} = \frac{a + c}{b}$.

CHALKBOARD EXAMPLE

Add $\frac{1}{9} + \frac{5}{9}$.

Answer: $\frac{2}{3}$

EXAMPLE 5 Adding Fractions with the Same Denominator

Add.

(a) $\frac{3}{7} + \frac{2}{7} = \frac{3 + 2}{7} = \frac{5}{7}$ Add numerators and keep the same denominator.

(b) $\frac{2}{10} + \frac{3}{10} = \frac{2 + 3}{10} = \frac{5}{10} = \frac{1}{2}$

If the fractions to be added do not have the same denominators, the procedure above can still be used, but only *after* the fractions are rewritten with a common denominator. For example, to rewrite $\frac{3}{4}$ as a fraction with a denominator of 32,

$$\frac{3}{4} = \frac{?}{32},$$

find the number that can be multiplied by 4 to give 32. Since $4 \cdot 8 = 32$, use the number 8. By the basic principle, we can multiply the numerator and the denominator by 8.

$$\frac{3}{4} = \frac{3 \cdot 8}{4 \cdot 8} = \frac{24}{32}$$

Finding the Least Common Denominator

To add or subtract fractions with different denominators, find the **least common denominator (LCD)** as follows.

Step 1 Factor both denominators.

Step 2 For the LCD, use every factor that appears in any factored form. If a factor is repeated, use the largest number of repeats in the LCD.

The next example shows this procedure.

CHALKBOARD EXAMPLE

Add $\frac{7}{30} + \frac{2}{45}$.

Answer: $\frac{5}{18}$

EXAMPLE 6 Adding Fractions with Different Denominators

Add the following fractions.

(a) $\frac{4}{15} + \frac{5}{9}$

To find the least common denominator, first factor both denominators.

$$15 = 5 \cdot 3 \qquad \text{and} \qquad 9 = 3 \cdot 3$$

Since 5 and 3 appear as factors, and 3 is a factor of 9 twice, the LCD is

$$5 \cdot 3 \cdot 3 \qquad \text{or} \qquad 45.$$

Write each fraction with 45 as denominator.

$$\frac{4}{15} = \frac{4 \cdot 3}{15 \cdot 3} = \frac{12}{45} \qquad \text{and} \qquad \frac{5}{9} = \frac{5 \cdot 5}{9 \cdot 5} = \frac{25}{45}$$

TEACHING TIP Warn students that $\frac{4}{15} + \frac{5}{9}$ is NOT EQUAL TO $\frac{9}{24}$.

Now add the two equivalent fractions.

$$\frac{4}{15} + \frac{5}{9} = \frac{12}{45} + \frac{25}{45} = \frac{37}{45}$$

(b) $3\frac{1}{2} + 2\frac{3}{4}$

These numbers are called mixed numbers. A **mixed number** is understood to be the sum of a whole number and a fraction. We can add mixed numbers using either of two methods.

Method 1

Rewrite both numbers as follows.

$$3\frac{1}{2} = 3 + \frac{1}{2} = \frac{3}{1} + \frac{1}{2} = \frac{6}{2} + \frac{1}{2} = \frac{6 + 1}{2} = \frac{7}{2}$$

$$2\frac{3}{4} = 2 + \frac{3}{4} = \frac{8}{4} + \frac{3}{4} = \frac{8 + 3}{4} = \frac{11}{4}$$

Now add. The common denominator is 4.

$$3\frac{1}{2} + 2\frac{3}{4} = \frac{7}{2} + \frac{11}{4} = \frac{14}{4} + \frac{11}{4} = \frac{25}{4} \quad \text{or} \quad 6\frac{1}{4}$$

Method 2

Write $3\frac{1}{2}$ as $3\frac{2}{4}$. Then add vertically.

$$\begin{array}{r} 3\frac{1}{2} \\ +\ 2\frac{3}{4} \\ \hline \end{array} \quad \rightarrow \quad \begin{array}{r} 3\frac{2}{4} \\ +\ 2\frac{3}{4} \\ \hline 5\frac{5}{4} \end{array}$$

Since $\frac{5}{4} = 1\frac{1}{4}$,

$$5\frac{5}{4} = 5 + 1\frac{1}{4} = 6\frac{1}{4}, \quad \text{or} \quad \frac{25}{4}.$$

NOTE

To multiply and divide mixed numbers, follow the same general procedure shown in Example 6(b), Method 1. First change to fractions, then perform the operation, and then convert back to a mixed number if desired. For example,

$$3\frac{1}{2} \cdot 2\frac{3}{4} = \frac{7}{2} \cdot \frac{11}{4} = \frac{77}{8} \quad \text{or} \quad 9\frac{5}{8}$$

$$3\frac{1}{2} \div 2\frac{3}{4} = \frac{7}{2} \div \frac{11}{4} = \frac{7}{2} \cdot \frac{4}{11} = \frac{14}{11} \quad \text{or} \quad 1\frac{3}{11}.$$

The **difference** between two numbers is found by subtraction. For example, $9 - 5 = 4$ so the difference between 9 and 5 is 4. Subtraction of fractions is similar to addition. Just subtract the numerators instead of adding them; again, keep the same denominator.

Subtracting Fractions

$$\frac{a}{b} - \frac{c}{b} = \frac{a - c}{b}$$

CHALKBOARD EXAMPLE

Subtract $\frac{3}{10} - \frac{1}{4}$.

Answer: $\frac{1}{20}$

EXAMPLE 7 **Subtracting Fractions**

Subtract. Write the differences in lowest terms.

(a) $\frac{15}{8} - \frac{3}{8} = \frac{15 - 3}{8}$ Subtract numerators; keep the same denominator.

$= \frac{12}{8} = \frac{3}{2}$ Lowest terms

(b) $\frac{7}{18} - \frac{4}{15}$

Here, $18 = 2 \cdot 3 \cdot 3$ and $15 = 3 \cdot 5$, so the LCD is $2 \cdot 3 \cdot 3 \cdot 5 = 90$.

$$\frac{7}{18} - \frac{4}{15} = \frac{7 \cdot 5}{2 \cdot 3 \cdot 3 \cdot 5} - \frac{4 \cdot 2 \cdot 3}{2 \cdot 3 \cdot 3 \cdot 5} = \frac{35}{90} - \frac{24}{90} = \frac{11}{90}$$

(c) $\dfrac{15}{32} - \dfrac{11}{45}$

Since $32 = 2 \cdot 2 \cdot 2 \cdot 2 \cdot 2$ and $45 = 3 \cdot 3 \cdot 5$, there are no common factors, and the LCD is $32 \cdot 45 = 1440$.

$$\frac{15}{32} - \frac{11}{45} = \frac{15 \cdot 45}{32 \cdot 45} - \frac{11 \cdot 32}{45 \cdot 32} \quad \text{Get a common denominator.}$$
$$= \frac{675}{1440} - \frac{352}{1440}$$
$$= \frac{323}{1440} \quad \text{Subtract.}$$

OBJECTIVE 5 Solve applied problems that involve fractions. Applied problems often require work with fractions. For example, when a carpenter reads diagrams and plans, he or she often must work with fractions whose denominators are 2, 4, 8, 16, or 32, as shown in the next example.

CHALKBOARD EXAMPLE

If an upholsterer needs $2\frac{1}{4}$ yards of fabric to recover a chair, how many chairs can be recovered with $23\frac{2}{3}$ yards of fabric?

Answer: 10 chairs (Some fabric will be left over.)

TEACHING TIP At times the fractions in an application can be a distraction. It is helpful for some students to read problems through, omitting the fractional parts of the numbers, and decide which operations are appropriate to the solution. Then they can do the work using the original quantities.

EXAMPLE 8 Adding Fractions to Solve a Manufacturing (Woodworking) Problem

The diagram in Figure 1 appears in the book *Woodworker's 39 Sure-Fire Projects.* It is the front view of a corner bookcase/desk. Add the fractions shown in the diagram to find the height of the bookcase/desk.

We must add the following measures (in inches):

$$\frac{3}{4},\quad 4\frac{1}{2},\quad 9\frac{1}{2},\quad \frac{3}{4},\quad 9\frac{1}{2},\quad \frac{3}{4},\quad 4\frac{1}{2}.$$

Begin by changing $4\frac{1}{2}$ to $4\frac{2}{4}$ and $9\frac{1}{2}$ to $9\frac{2}{4}$, since the common denominator is 4. Then, use Method 2 from Example 6(b).

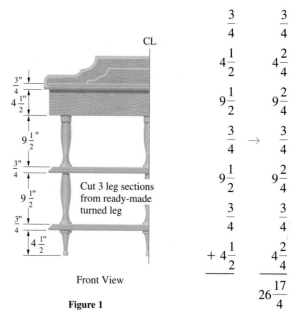

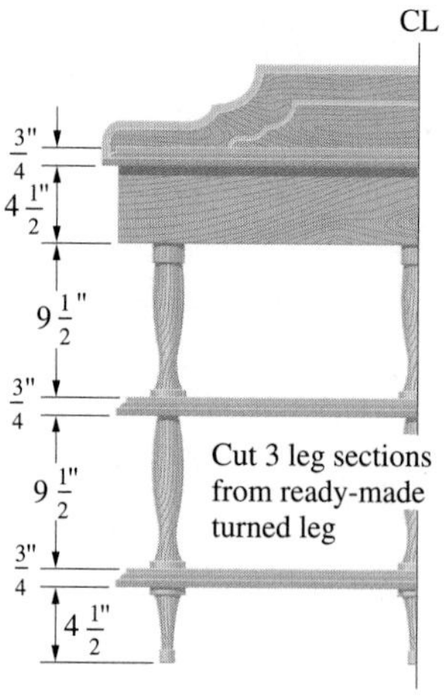

Front View

Figure 1

$$\begin{array}{r} \frac{3}{4} \\ 4\frac{1}{2} \\ 9\frac{1}{2} \\ \frac{3}{4} \\ 9\frac{1}{2} \\ \frac{3}{4} \\ +\,4\frac{1}{2} \\ \hline \end{array} \quad \rightarrow \quad \begin{array}{r} \frac{3}{4} \\ 4\frac{2}{4} \\ 9\frac{2}{4} \\ \frac{3}{4} \\ 9\frac{2}{4} \\ \frac{3}{4} \\ 4\frac{2}{4} \\ \hline 26\frac{17}{4} \end{array}$$

Since $\frac{17}{4} = 4\frac{1}{4}$, $26\frac{17}{4} = 26 + 4\frac{1}{4} = 30\frac{1}{4}$. The height is $30\frac{1}{4}$ inches. It is best to give answers as mixed numbers in applications like this.

OBJECTIVE 6 Interpret data in a circle graph. A **circle graph** or **pie chart** is often used to give a pictorial representation of data. A circle is used to indicate the total of all the categories represented. The circle is divided into sectors, or wedges (like pieces of pie) whose sizes show the relative magnitudes of the categories. The sum of all the fractional parts must be 1 (for 1 whole circle).

CHALKBOARD EXAMPLE

Based on Figure 2, how many people in a group of 200,000 such employees would we expect to be employed in farming/forestry/fishing?

Answer: 4000

EXAMPLE 9 Using a Pie Chart to Interpret Information

The pie chart in Figure 2 shows the job categories of African Americans employed in 1993 (age 16 or older).

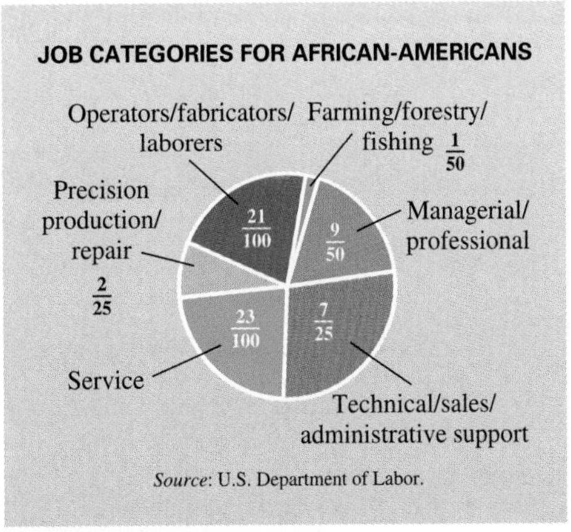

Figure 2

(a) In a group of 150,000 such employees, about how many would we expect to be employed in precision production/repair?

To find the answer, we multiply the fraction indicated in the chart for the category $\left(\frac{2}{25}\right)$ by the number of people in the group (150,000):

$$\frac{2}{25} \cdot 150{,}000 = \frac{300{,}000}{25} = 12{,}000.$$

About 12,000 people in the group are employed in precision production/repair.

(b) *Estimate* the number employed in service.

The fraction $\frac{23}{100}$ is approximately $\frac{1}{4}$. Therefore, a good estimate for this number is

$$\frac{1}{4} \cdot 150{,}000 = 37{,}500.$$

1.1 EXERCISES

1. true 2. true 3. false; The fraction $\frac{17}{51}$ can be reduced to $\frac{1}{3}$.

Decide whether each statement is true or false. If it is false, say why.

1. In the fraction $\frac{3}{7}$, 3 is the numerator and 7 is the denominator.

2. The mixed number equivalent of $\frac{41}{5}$ is $8\frac{1}{5}$.

3. The fraction $\frac{17}{51}$ is in lowest terms.

4. false; The reciprocal of $\frac{8}{2} = 4$ is $\frac{2}{8} = \frac{1}{4}$ 5. false; *Product* refers to multiplication, so the product of 8 and 2 is 16. 6. false; *Difference* refers to subtraction, so the difference between 12 and 2 is 10. 7. prime 8. prime 9. composite; $2 \cdot 2 \cdot 2 \cdot 2 \cdot 2 \cdot 2$ 10. composite; $3 \cdot 3 \cdot 11$ 11. composite; $2 \cdot 7 \cdot 13 \cdot 19$ 12. composite; $5 \cdot 5 \cdot 41$ 13. neither 14. neither 15. composite; $2 \cdot 3 \cdot 5$ 16. composite; $2 \cdot 2 \cdot 2 \cdot 5$ 17. composite $2 \cdot 2 \cdot 5 \cdot 5 \cdot 5$ 18. composite; $2 \cdot 2 \cdot 5 \cdot 5 \cdot 7$ 19. composite; $2 \cdot 2 \cdot 31$ 20. composite; $2 \cdot 2 \cdot 2 \cdot 3 \cdot 5$ 21. prime 22. prime 23. $\frac{1}{2}$ 24. $\frac{1}{3}$ 25. $\frac{5}{6}$ 26. $\frac{4}{5}$ 27. $\frac{1}{3}$ 28. $\frac{1}{4}$ 29. $\frac{6}{5}$ 30. $\frac{12}{7}$ 31. (c) 32. (a) 33. $\frac{24}{35}$ 34. $\frac{50}{63}$ 35. $\frac{6}{25}$ 36. $\frac{4}{11}$ 37. $\frac{6}{5}$ or $1\frac{1}{5}$ 38. $\frac{3}{2}$ or $1\frac{1}{2}$ 39. $\frac{232}{15}$ or $15\frac{7}{15}$ 40. $\frac{129}{5}$ or $25\frac{4}{5}$ 41. $\frac{10}{3}$ or $3\frac{1}{3}$ 42. $\frac{35}{27}$ or $1\frac{8}{27}$ 43. 12 44. 12 45. $\frac{1}{16}$ 46. $\frac{1}{75}$ 47. $\frac{84}{47}$ or $1\frac{37}{47}$ 48. $\frac{23}{78}$ 49. To multiply two fractions, multiply their numerators to get the numerator of the product and multiply their denominators to get the denominator of the product. For example, $\frac{2}{3} \cdot \frac{8}{5} = \frac{2 \cdot 8}{3 \cdot 5} = \frac{16}{15}$. To divide two fractions, replace the divisor with its reciprocal and then multiply. For example, $\frac{2}{5} \div \frac{7}{9} = \frac{2}{5} \cdot \frac{9}{7} = \frac{2 \cdot 9}{5 \cdot 7} = \frac{18}{35}$.

4. The reciprocal of $\frac{8}{2}$ is $\frac{4}{1}$.

5. The product of 8 and 2 is 10.

6. The difference between 12 and 2 is 6.

Identify each number as prime, composite, or neither. If the number is composite, write it as the product of prime factors. See Example 1.

7. 19 8. 31 9. 64 10. 99

11. 3458 12. 1025 13. 1 14. 0

15. 30 16. 40 17. 500 18. 700

19. 124 20. 120 21. 29 22. 83

Write each fraction in lowest terms. See Example 2.

23. $\frac{8}{16}$ 24. $\frac{4}{12}$ 25. $\frac{15}{18}$ 26. $\frac{16}{20}$

27. $\frac{15}{45}$ 28. $\frac{16}{64}$ 29. $\frac{144}{120}$ 30. $\frac{132}{77}$

31. One of the following is the correct way to write $\frac{16}{24}$ in lowest terms. Which one is it?

(a) $\frac{16}{24} = \frac{8+8}{8+16} = \frac{8}{16} = \frac{1}{2}$ (b) $\frac{16}{24} = \frac{4 \cdot 4}{4 \cdot 6} = \frac{4}{6}$

(c) $\frac{16}{24} = \frac{8 \cdot 2}{8 \cdot 3} = \frac{2}{3}$ (d) $\frac{16}{24} = \frac{14+2}{21+3} = \frac{2}{3}$

32. For the fractions $\frac{p}{q}$ and $\frac{r}{s}$, which one of the following can serve as a common denominator?

(a) $q \cdot s$ (b) $q + s$ (c) $p \cdot r$ (d) $p + r$

Find each product or quotient, and write it in lowest terms. See Examples 3 and 4.

33. $\frac{4}{5} \cdot \frac{6}{7}$ 34. $\frac{5}{9} \cdot \frac{10}{7}$ 35. $\frac{1}{10} \cdot \frac{12}{5}$ 36. $\frac{6}{11} \cdot \frac{2}{3}$

37. $\frac{15}{4} \cdot \frac{8}{25}$ 38. $\frac{4}{7} \cdot \frac{21}{8}$ 39. $2\frac{2}{3} \cdot 5\frac{4}{5}$ 40. $3\frac{3}{5} \cdot 7\frac{1}{6}$

41. $\frac{5}{4} \div \frac{3}{8}$ 42. $\frac{7}{6} \div \frac{9}{10}$ 43. $\frac{32}{5} \div \frac{8}{15}$ 44. $\frac{24}{7} \div \frac{6}{21}$

45. $\frac{3}{4} \div 12$ 46. $\frac{2}{5} \div 30$ 47. $2\frac{5}{8} \div 1\frac{15}{32}$ 48. $2\frac{3}{10} \div 7\frac{4}{5}$

49. Write a summary explaining how to multiply and divide two fractions. Give examples.

50. Write a summary explaining how to add and subtract two fractions. Give examples.

Find each sum or difference, and write it in lowest terms. See Examples 5–7.

51. $\frac{7}{12} + \frac{1}{12}$ 52. $\frac{3}{16} + \frac{5}{16}$ 53. $\frac{5}{9} + \frac{1}{3}$ 54. $\frac{4}{15} + \frac{1}{5}$

55. $3\frac{1}{8} + \frac{1}{4}$ 56. $5\frac{3}{4} + \frac{2}{3}$ 57. $\frac{7}{12} - \frac{1}{9}$ 58. $\frac{11}{16} - \frac{1}{12}$

59. $6\frac{1}{4} - 5\frac{1}{3}$ 60. $8\frac{4}{5} - 7\frac{4}{9}$ 61. $\frac{5}{3} + \frac{1}{6} - \frac{1}{2}$ 62. $\frac{7}{15} + \frac{1}{6} - \frac{1}{10}$

50. To add or subtract two fractions that have the same denominator, add or subtract the numerators and keep the same denominator. For example, $\frac{7}{8}+\frac{2}{8}=\frac{7+2}{8}=\frac{9}{8}$ and $\frac{7}{8}-\frac{2}{8}=\frac{7-2}{8}=\frac{5}{8}$. To add or subtract fractions that have different denominators, write both fractions with a common denominator, and then follow the earlier procedure. For example, $\frac{1}{2}+\frac{1}{3}=\frac{3}{6}+\frac{2}{6}=\frac{3+2}{6}=\frac{5}{6}$ and $\frac{1}{2}-\frac{1}{3}=\frac{3}{6}-\frac{2}{6}=\frac{3-2}{6}=\frac{1}{6}$. **51.** $\frac{2}{3}$ **52.** $\frac{1}{2}$ **53.** $\frac{8}{9}$ **54.** $\frac{7}{15}$ **55.** $\frac{27}{8}$ or $3\frac{3}{8}$ **56.** $\frac{77}{12}$ or $6\frac{5}{12}$ **57.** $\frac{17}{36}$ **58.** $\frac{29}{48}$ **59.** $\frac{11}{12}$ **60.** $\frac{61}{45}$ or $1\frac{16}{45}$ **61.** $\frac{4}{3}$ or $1\frac{1}{3}$ **62.** $\frac{8}{15}$ **63.** 6 cups **64.** $\frac{3}{8}$ tsp **65.** 34 dollars **66.** $47\frac{1}{16}$ dollars **67.** $\frac{9}{16}$ inch **68.** $\frac{3}{16}$ inch **69.** $618\frac{3}{4}$ feet

The following chart appears on a package of Quaker Quick Grits.

	Microwave	Stove Top		
Servings	1	1	4	6
Water	$\frac{3}{4}$ cup	1 cup	3 cups	4 cups
Grits	3 Tbsp	3 Tbsp	$\frac{3}{4}$ cup	1 cup
Salt (optional)	dash	dash	$\frac{1}{4}$ tsp	$\frac{1}{2}$ tsp

Use the chart to answer the questions in Exercises 63 and 64.

63. How many cups of water would be needed for 8 microwave servings?

64. How many tsp of salt would be needed for 5 stove top servings? (*Hint:* 5 is halfway between 4 and 6.)

Work each problem. See Example 8.

65. On Tuesday, February 10, 1998, Earthlink stock on the NASDAQ exchange closed the day at $4\frac{5}{8}$ (dollars) ahead of where it had opened. It closed at $38\frac{5}{8}$ (dollars). What was its opening price?

66. A report in *USA Today* on February 10, 1998, stated that Teva Pharmaceutical skidded $9\frac{9}{16}$ (dollars) to $37\frac{1}{2}$ (dollars) after the Israeli drugmaker said fourth-quarter net income was likely to be below expectations. What was its price before the skid?

67. A hardware store sells a 40-piece socket wrench set. The measure of the largest socket is $\frac{3}{4}$ inch, while the measure of the smallest socket is $\frac{3}{16}$ inch. What is the difference between these measures?

68. Two sockets in a socket wrench set have measures of $\frac{9}{16}$ inch and $\frac{3}{8}$ inch. What is the difference between these two measures?

69. A motel owner has decided to expand his business by buying a piece of property next to the motel. The property has an irregular shape, with five sides as shown in the figure. Find the total distance around the piece of property. (This is called the *perimeter* of the figure.)

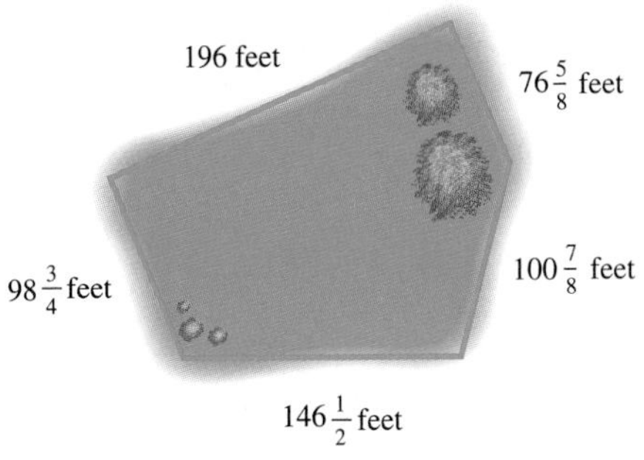

70. $22\frac{7}{8}$ feet
71. $5\frac{5}{24}$ inches 72. $\frac{15}{16}$ cup
73. $\frac{1}{3}$ cup 74. 8 cakes (There will be some sugar left over.)
75. 650 76. 225

70. Find the perimeter of the triangle in the figure.

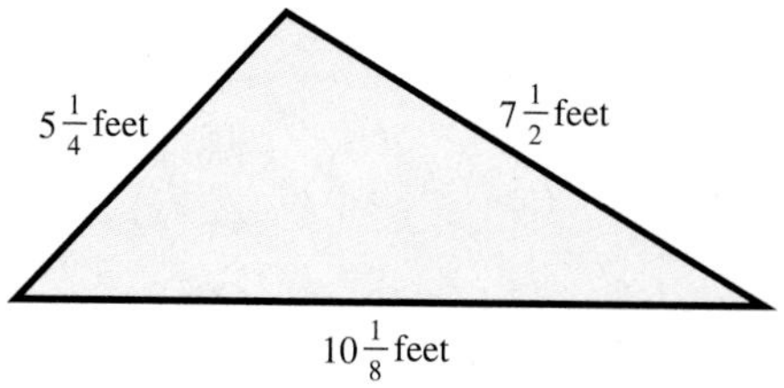

71. A piece of board is $15\frac{5}{8}$ inches long. If it must be divided into 3 pieces of equal length, how long must each piece be?

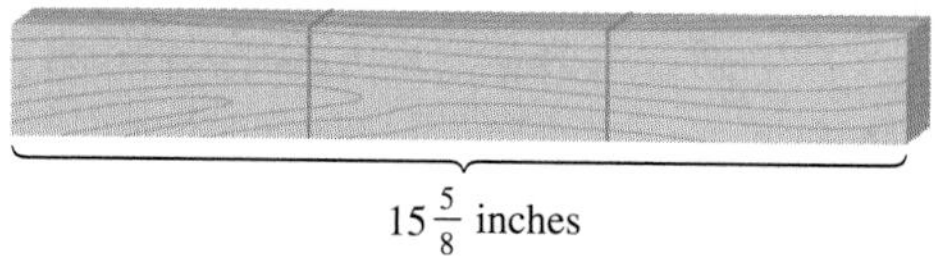

72. If one serving of a macaroni and cheese meal requires $\frac{1}{8}$ cup of chopped onions, how many cups of onions will $7\frac{1}{2}$ servings require?

73. Tex's favorite recipe for barbecue sauce calls for $2\frac{1}{3}$ cups of tomato sauce. The recipe makes enough barbecue sauce to serve 7 people. How much tomato sauce is needed for 1 serving?

74. A cake recipe calls for $1\frac{3}{4}$ cups of sugar. A caterer has $15\frac{1}{2}$ cups of sugar on hand. How many cakes can he make?

The pie chart gives fractional job categories of white employees in the workforce in 1993 (age 16 or older). Use the chart to answer the questions in Exercises 75 and 76. See Example 9.

75. In a random sample of 5000 such employees, about how many would we expect to be employed in service occupations?

76. In a random sample of 7500 such employees, about how many would we expect to be employed in farming/forestry/fishing occupations?

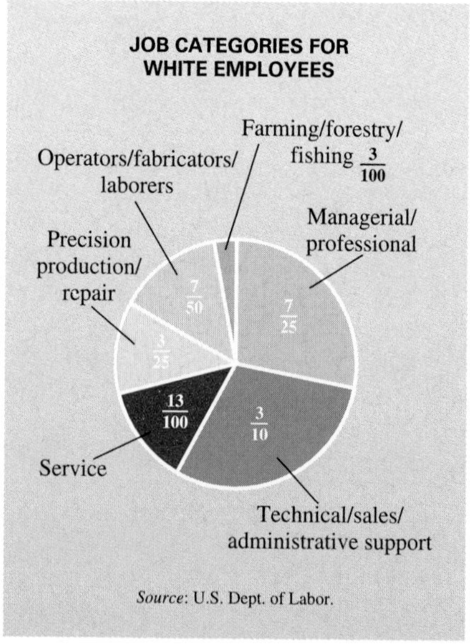

77. (a) Crum (b) Jordan (c) Jordan (d) Baldock (e) Tobin and Perry; $\frac{1}{2}$

78. (a) $\frac{1}{2}$ (b) $\frac{1}{4}$ (c) $\frac{1}{3}$ (d) $\frac{1}{6}$ 79. (b) 80. He is correct, because $\frac{8}{20} = \frac{12}{30} = \frac{2}{5}$.

77. At the conclusion of the AWL softball league season, batting statistics for five players were as follows.

Player	*At-bats*	*Hits*	*Home Runs*
Stephanie Baldock	40	9	2
Jennifer Crum	36	12	3
Jason Jordan	11	5	1
Greg Tobin	16	8	0
David Perry	20	10	2

Answer each of the following, using estimation skills as necessary.

(a) Which player got a hit in exactly $\frac{1}{3}$ of his or her at-bats?

(b) Which player got a hit in just less than $\frac{1}{2}$ of his or her at-bats?

(c) Which player got a home run in just less than $\frac{1}{10}$ of his or her at-bats?

(d) Which player got a hit in just less than $\frac{1}{4}$ of his or her at-bats?

(e) Which two players got hits in exactly the same fractional parts of their at-bats? What was the fractional part, expressed in lowest terms?

78. For each of the following, write a fraction in lowest terms that represents the region described.

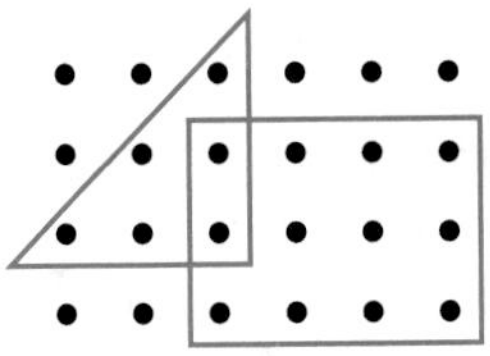

(a) the dots in the rectangle as a part of the dots in the entire figure

(b) the dots in the triangle as a part of the dots in the entire figure

(c) the dots in the overlapping region of the triangle and the rectangle as a part of the dots in the triangle alone

(d) the dots in the overlapping region of the triangle and the rectangle as a part of the dots in the rectangle alone

79. Estimate the best approximation for the following sum:

$$\frac{14}{26} + \frac{98}{99} + \frac{100}{51} + \frac{90}{31} + \frac{13}{27}.$$

(a) 6 **(b)** 7 **(c)** 5 **(d)** 8

80. In the local softball league, the first five games produced the following results: David Horwitz got 8 hits in 20 at-bats, and Chalon Bridges got 12 hits in 30 at-bats. David claims that he did just as well as Chalon. Is he correct? Why or why not?

1.2 Exponents, Order of Operations, and Inequality

OBJECTIVES

1. Use exponents.
2. Use the order of operations rules.
3. Use more than one grouping symbol.
4. Know the meanings of $\neq$, $<$, $>$, $\leq$, and $\geq$.
5. Translate word statements to symbols.
6. Write statements that change the direction of inequality symbols.
7. Interpret data in a bar graph.

FOR EXTRA HELP

SSG Sec. 1.2
SSM Sec. 1.2
Pass the Test Software
InterAct Math Tutorial Software
Video 1

CHALKBOARD EXAMPLE

Find the value of the exponential expression.

$$\left(\frac{1}{2}\right)^4$$

Answer: $\frac{1}{16}$

OBJECTIVE 1 Use exponents. In a multiplication problem, the same factor may appear several times. For example, in the product

$$3 \cdot 3 \cdot 3 \cdot 3 = 81,$$

the factor 3 appears four times. In algebra, repeated factors are written with an *exponent.* For example, in $3 \cdot 3 \cdot 3 \cdot 3$, the number 3 appears as a factor four times, so the product is written as 3^4, and is read "3 to the fourth power."

$$3 \cdot 3 \cdot 3 \cdot 3 = 3^4$$

The number 4 is the **exponent** or **power** and 3 is the **base** in the **exponential expression** 3^4. A natural number exponent, then, tells how many times the base is used as a factor. A number raised to the first power is simply that number. For example, $5^1 = 5$ and $\left(\frac{1}{2}\right)^1 = \frac{1}{2}$.

EXAMPLE 1 Evaluating an Exponential Expression

Find the values of the following.

(a) 5^2 $\quad 5 \cdot 5 = 25$ — 5 is used as a factor 2 times.

Read 5^2 as "5 squared."

(b) 6^3 $\quad 6 \cdot 6 \cdot 6 = 216$ — 6 is used as a factor 3 times.

Read 6^3 as "6 cubed."

(c) $2^5 = 2 \cdot 2 \cdot 2 \cdot 2 \cdot 2 = 32$ — 2 is used as a factor 5 times.

Read 2^5 as "2 to the fifth power."

(d) $\left(\frac{2}{3}\right)^3 = \frac{2}{3} \cdot \frac{2}{3} \cdot \frac{2}{3} = \frac{8}{27}$ — $\frac{2}{3}$ is used as a factor 3 times.

TEACHING TIP See Example 1(c). Students will often work 2^5 as $2 \cdot 5 = 10$. Illustrate the difference between these with several examples.

OBJECTIVE 2 Use the order of operations rules. Many problems involve more than one operation. To indicate the order in which the operations should be performed, we often use *grouping symbols.* If no grouping symbols are used, we apply the order of operations rules discussed below.

Consider the expression $5 + 2 \cdot 3$. To show that the multiplication should be performed before the addition, parentheses can be used to write

$$5 + (2 \cdot 3) = 5 + 6 = 11.$$

If addition is to be performed first, the parentheses should group $5 + 2$ as follows.

$$(5 + 2) \cdot 3 = 7 \cdot 3 = 21$$

Other grouping symbols used in more complicated expressions are brackets [ ], braces { }, and fraction bars. (For example, in $\frac{8 - 2}{3}$, the expression $8 - 2$ is considered to be grouped in the numerator.)

To work problems with more than one operation, use the following **order of operations.** This order is used by most calculators and computers.

INSTRUCTOR'S RESOURCES

PTB Sec. 1.2 **ISM** Sec. 1.2 **AB** Sec. 1.2 | **TEST GENERATOR** | **WORLD WIDE WEB** www.LialAlgebra.com

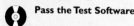

TEACHING TIP Using a memory device such as "**P**lease **E**xcuse **M**y **D**ear **A**unt **S**ally" helps students remember the rules for order of operations: **P**arentheses, **E**xponents, **M**ultiply, **D**ivide, **A**dd, **S**ubtract.

Order of Operations

If grouping symbols are present, simplify within them, innermost first (and above and below fraction bars separately), in the following order.

Step 1 Apply all exponents.

Step 2 Do any multiplications or divisions in the order in which they occur, working from left to right.

Step 3 Do any additions or subtractions in the order in which they occur, working from left to right.

If no grouping symbols are present, start with Step 1.

A dot has been used to show multiplication; another way to show multiplication is with parentheses. For example, 3(7), (3)7, and (3)(7) each mean $3 \cdot 7$ or 21. The next example shows the use of parentheses for multiplication.

CHALKBOARD EXAMPLE

Find the value of the expression.

$$7 \cdot 6 - 3(8 + 1)$$

Answer: 15

EXAMPLE 2 Using the Order of Operations

Find the values of the following.

(a) $9(6 + 11)$

Using the order of operations given above, work first inside the parentheses.

$$9(6 + 11) = 9(17) \quad \text{Work inside parentheses.}$$
$$= 153 \quad \text{Multiply.}$$

(b) $6 \cdot 8 + 5 \cdot 2$

Do any multiplications, working from left to right, and then add.

$$6 \cdot 8 + 5 \cdot 2 = 48 + 10 \quad \text{Multiply.}$$
$$= 58 \quad \text{Add.}$$

(c) $2(5 + 6) + 7 \cdot 3 = 2(11) + 7 \cdot 3$ Work inside parentheses.

$$= 22 + 21 \quad \text{Multiply.}$$
$$= 43 \quad \text{Add.}$$

(d) $9 - 2^3 + 5$

Find 2^3 first.

$$9 - 2^3 + 5 = 9 - 2 \cdot 2 \cdot 2 + 5 \quad \text{Use the exponent.}$$
$$= 9 - 8 + 5 \quad \text{Multiply.}$$
$$= 1 + 5 \quad \text{Subtract.}$$
$$= 6 \quad \text{Add.}$$

(e) $72 \div 2 \cdot 3 + 4 \cdot 2^3$

$$72 \div 2 \cdot 3 + 4 \cdot 2^3 = 72 \div 2 \cdot 3 + 4 \cdot 8 \quad \text{Use the exponent.}$$
$$= 36 \cdot 3 + 4 \cdot 8 \quad \text{Perform the division.}$$
$$= 108 + 32 \quad \text{Perform the multiplications.}$$
$$= 140 \quad \text{Add.}$$

Notice that the multiplications and divisions are performed from left to right *as they appear;* then the additions and subtractions should be done from left to right, *as they appear.*

OBJECTIVE 3 **Use more than one grouping symbol.** An expression with double (or *nested*) parentheses, such as $2(8 + 3(6 + 5))$, can be confusing. For clarity, square brackets, [ ], often are used in place of one pair of parentheses. Fraction bars also act as grouping symbols. The next example explains these situations.

CHALKBOARD EXAMPLE

Simplify $\frac{2(7 + 8) + 2}{3 \cdot 5 + 1}$.

Answer: 2

EXAMPLE 3 **Using Brackets and Fraction Bars as Grouping Symbols**

Simplify each expression.

(a) $2[8 + 3(6 + 5)]$

Work first within the parentheses, and then simplify inside the brackets until a single number remains.

$$\begin{aligned} 2[8 + 3(6 + 5)] &= 2[8 + 3(11)] \\ &= 2[8 + 33] \\ &= 2[41] \\ &= 82 \end{aligned}$$

(b) $\frac{4(5 + 3) + 3}{2(3) - 1}$

The expression can be written as the quotient

$$[4(5 + 3) + 3] \div [2(3) - 1],$$

which shows that the fraction bar groups the numerator and denominator separately. Simplify both numerator and denominator, then divide, if possible.

$$\begin{aligned} \frac{4(5 + 3) + 3}{2(3) - 1} &= \frac{4(8) + 3}{2(3) - 1} && \text{Work inside parentheses.} \\ &= \frac{32 + 3}{6 - 1} && \text{Multiply.} \\ &= \frac{35}{5} && \text{Add and subtract.} \\ &= 7 && \text{Divide.} \end{aligned}$$

NOTE Parentheses and fraction bars are used as grouping symbols to indicate an expression that represents a single number. That is why we must first simplify within parentheses and above and below fraction bars.

OBJECTIVE 4 **Know the meanings of $\neq$, $<$, $>$, $\leq$, and $\geq$.** So far, we have used the symbols for the operations of arithmetic and the symbol for equality (=). The equality symbol with a slash through it, $\neq$, means "is not equal to." For example,

$$7 \neq 8$$

indicates that 7 is not equal to 8.

If two numbers are not equal, then one of the numbers must be less than the other. The symbol $<$ represents "is less than," so that "7 is less than 8" is written

$$7 < 8.$$

Also, write "6 is less than 9" as $6 < 9$.

The symbol $>$ means "is greater than." Write "8 is greater than 2" as

$$8 > 2.$$

The statement "17 is greater than 11" becomes $17 > 11$.

TEACHING TIP Emphasize that the "arrowhead" always points to the smaller number.

Keep the meanings of the symbols $<$ and $>$ clear by remembering that the symbol always *points to the smaller number.* For example, write "8 is less than 15" by pointing the symbol toward the 8:

Smaller number → $8 < 15.$

Two other symbols, $\leq$ and $\geq$, also represent the idea of inequality. The symbol $\leq$ means "is less than or equal to," so that

$$5 \leq 9$$

means "5 is less than or equal to 9." This statement is true, since $5 < 9$ is true. **If either the $<$ part or the $=$ part is true, then the inequality $\leq$ is true.**

The symbol $\geq$ means "is greater than or equal to." Again,

$$9 \geq 5$$

is true because $9 > 5$ is true. Also, $8 \leq 8$ is true since $8 = 8$ is true. But it is not true that $13 \leq 9$ because neither $13 < 9$ nor $13 = 9$ is true.

CHALKBOARD EXAMPLE

Determine whether the statement is true or false.

$$12 > 6$$

Answer: true

TEACHING TIP The / symbol can also be used to indicate "not" with inequality symbols, e.g., $\not<$ (is not less than).

EXAMPLE 4 Using Inequality Symbols

Determine whether each statement is true or false.

(a) $6 \neq 6$
The statement is false because 6 *is equal to* 6.

(b) $5 < 19$
Since 5 represents a number that is indeed less than 19, this statement is true.

(c) $15 \leq 20$
The statement $15 \leq 20$ is true, since $15 < 20$.

(d) $25 \geq 30$
Both $25 > 30$ and $25 = 30$ are false. Because of this, $25 \geq 30$ is false.

(e) $12 \geq 12$
Since $12 = 12$, this statement is true.

OBJECTIVE 5 Translate word statements to symbols. An important part of algebra deals with translating words into algebraic notation.

PROBLEM SOLVING

As we will see throughout this book, the ability to solve problems using mathematics is based on translating the words of the problem into symbols. The next example is the first of many that illustrate such translations.

CHALKBOARD EXAMPLE

Write in symbols: Fourteen is greater than twelve.

Answer: $14 > 12$

EXAMPLE 5 Translating from Words to Symbols

Write each word statement in symbols.

(a) Twelve **equals** ten **plus** two.

$$12 = 10 + 2$$

(b) Nine **is less than** ten.

$$9 < 10$$

(c) Fifteen **is not equal to** eighteen.

$$15 \neq 18$$

(d) Seven **is greater than** four.

$$7 > 4$$

(e) Thirteen **is less than or equal to** forty.

$$13 \leq 40$$

(f) Eleven **is greater than or equal to** eleven.

$$11 \geq 11$$

OBJECTIVE 6 **Write statements that change the direction of inequality symbols.** Any statement with $<$ can be converted to one with $>$, and any statement with $>$ can be converted to one with $<$. We do this by reversing the order of the numbers and the direction of the symbol. For example, the statement $6 < 10$ can be written with $>$ as $10 > 6$. Similarly, the statement $4 \leq 10$ can be changed to $10 \geq 4$.

CHALKBOARD EXAMPLE

Write the statement with the inequality symbol reversed.

$$9 \leq 15$$

Answer: $15 \geq 9$

EXAMPLE 6 Converting between Inequality Symbols

The following examples show the same statement written in two equally correct ways.

(a) $9 < 16$ $\quad 16 > 9$

(b) $5 > 2$ $\quad 2 < 5$

(c) $3 \leq 8$ $\quad 8 \geq 3$

(d) $12 \geq 5$ $\quad 5 \leq 12$

Note that in each pair of inequalities, the point of the inequality symbol points toward the smaller number.

Here is a summary of the symbols discussed in this section.

Symbols of Equality and Inequality

$=$	is equal to	$\neq$	is not equal to
$<$	is less than	$>$	is greater than
$\leq$	is less than or equal to	$\geq$	is greater than or equal to

CAUTION The equality and inequality symbols are used to write mathematical *sentences* that describe the relationship between two numbers. On the other hand, the symbols for operations $(+, -, \times, \div)$ are used to write mathematical *expressions* that represent a single number. For example, compare the sentence $4 < 10$ with the expression $4 + 10$, which represents the number 14.

OBJECTIVE 7 **Interpret data in a bar graph.** **Bar graphs** are often used to summarize data in a concise manner.

EXAMPLE 7 Interpreting Inequality Concepts Using a Bar Graph

Figure 3 shows a bar graph that depicts the percentage growth of the ten fastest-growing U.S. manufacturing industries from 1993 to 1994.

CHALKBOARD EXAMPLE

In Figure 3, which industries showed growth less than or equal to 6.4%?

Answer: Mattresses and bedsprings, Leather tanning and finishing, Analytical instruments

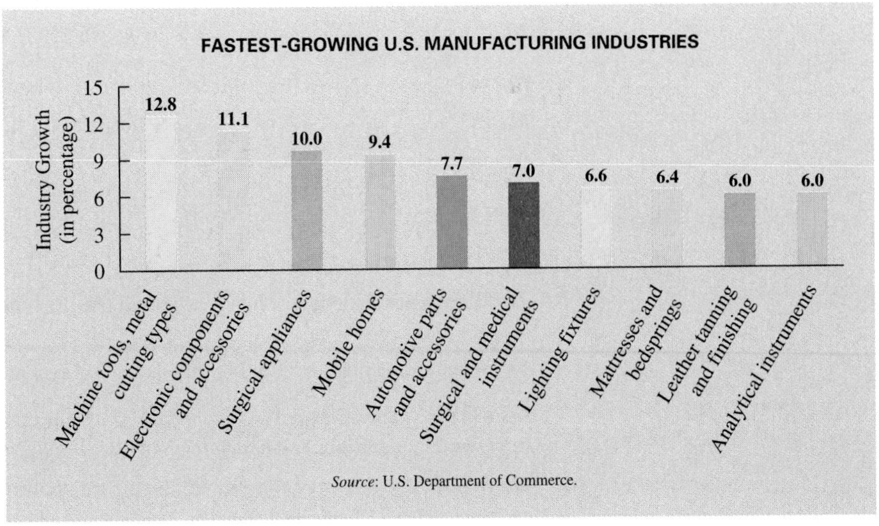

Figure 3

(a) Which industries showed growth greater than 9.4%?

Machine tools, metal cutting types (at 12.8%), electronic components and accessories (at 11.1%), and surgical appliances (at 10.0%) all showed growth greater than 9.4%.

(b) Which industries showed growth greater than or equal to 9.4%?

The three industries listed in part (a) and the mobile home industry, at 9.4%, showed growth greater than or equal to 9.4%. (These are industries that showed *at least* 9.4% growth.)

1.2 EXERCISES

1. false; $4 + 3(8 - 2) = 4 + 3 \cdot 6 = 4 + 18 = 22$. The common error leading to 42 is adding 4 to 3 and then multiplying by 6. One must follow the rules for order of operations. **2.** false; $3^3 = 3 \cdot 3 \cdot 3 = 27$ **3.** false; The correct interpretation is $4 = 16 - 12$. **4.** false; The correct interpretation is $6 = 10 - 4$. **5.** 49 **6.** 16 **7.** 144 **8.** 196 **9.** 64 **10.** 125 **11.** 1000 **12.** 1331 **13.** 81 **14.** 1296 **15.** 1024 **16.** 243 **17.** $\frac{16}{81}$ **18.** $\frac{27}{64}$ **19.** .000064 **20.** .00000625 **21.** Write the base as a factor the number of times indicated by the exponent. For example, $6^3 = 6 \cdot 6 \cdot 6 = 216$. **22.** For *any* number of factors of 1, the product must be 1. **23.** 32 **24.** 31 **25.** $\frac{49}{30}$ or $1\frac{19}{30}$

Decide whether each statement is true or false. If it is false, explain why.

1. When evaluated, $4 + 3(8 - 2)$ is equal to 42.

2. $3^3 = 9$

3. The statement "4 is 12 less than 16" is interpreted $4 = 12 - 16$.

4. The statement "6 is 4 less than 10" is interpreted $6 < 10 - 4$.

Find the value of each exponential expression. See Example 1.

5. 7^2 **6.** 4^2 **7.** 12^2 **8.** 14^2

9. 4^3 **10.** 5^3 **11.** 10^3 **12.** 11^3

13. 3^4 **14.** 6^4 **15.** 4^5 **16.** 3^5

17. $\left(\frac{2}{3}\right)^4$ **18.** $\left(\frac{3}{4}\right)^3$ **19.** $(.04)^3$ **20.** $(.05)^4$

21. Explain in your own words how to evaluate a power of a number, such as 6^3.

22. Explain why any power of 1 must be equal to 1.

Find the value of each expression. See Examples 2 and 3.

23. $9 \cdot 5 - 13$ **24.** $7 \cdot 6 - 11$ **25.** $\frac{1}{4} \cdot \frac{2}{3} + \frac{2}{5} \cdot \frac{11}{3}$

26. $\frac{17}{6}$ or $2\frac{5}{6}$ 27. 12 28. 74 29. 23.01 30. 36.14 31. 95 32. 1308 33. 90 34. 144 35. 14 36. 64 37. 9 38. 9 39. Begin by squaring 2. Then subtract 1, to get a result of $4 - 1 = 3$ within the parentheses. Next, raise 3 to the third power to get $3^3 = 27$. Multiply this result by 3 to obtain 81. Finally, add this result to 4 to get the final answer, 85. 40. 4 41. true 42. true 43. false 44. false 45. true 46. true 47. true 48. true 49. false 50. false 51. false 52. true 53. true 54. false 55. $15 = 5 + 10$ 56. $12 = 20 - 8$ 57. $9 > 5 - 4$ 58. $10 > 6 + 1$ 59. $16 \neq 19$ 60. $3 \neq 4$ 61. $2 \leq 3$ 62. $5 \leq 9$ 63. Seven is less than nineteen. True 64. Nine is less than ten. True 65. Three is not equal to six. True 66. Nine is not equal to thirteen. True 67. Eight is greater than or equal to eleven. False 68. Four is less than or equal to two. False 69. Answers will vary. One example is $5 + 3 \geq 2 \cdot 2$. 70. Answers will vary. One example is $5 - 2 \leq 6 \div 3$. It is false because it says that $3 \leq 2$, and actually $3 > 2$. By changing the 3 on the right side to 2, the statement becomes $3 \leq 3$, which is true. 71. $30 > 5$ 72. $4 < 8$ 73. $3 \leq 12$ 74. $41 \geq 25$ 75. is younger than 76. is taller than

26. $\frac{9}{4} \cdot \frac{2}{3} + \frac{4}{5} \cdot \frac{5}{3}$ **27.** $9 \cdot 4 - 8 \cdot 3$ **28.** $11 \cdot 4 + 10 \cdot 3$

29. $(4.3)(1.2) + (2.1)(8.5)$ **30.** $(2.5)(1.9) + (4.3)(7.3)$ **31.** $5[3 + 4(2^2)]$

32. $6[2 + 8(3^3)]$ **33.** $3^2[(11 + 3) - 4]$ **34.** $4^2[(13 + 4) - 8]$

35. $\dfrac{6(3^2 - 1) + 8}{3 \cdot 2 - 2}$ **36.** $\dfrac{2(8^2 - 4) + 8}{4 \cdot 3 - 10}$

37. $\dfrac{4(6 + 2) + 8(8 - 3)}{6(4 - 2) - 2^2}$ **38.** $\dfrac{6(5 + 1) - 9(1 + 1)}{5(8 - 6) - 2^3}$

39. Write an explanation of how you would use the order of operations to simplify $4 + 3(2^2 - 1)^3$.

40. When evaluating $(4^2 + 3^3)^4$, what is the *last* exponent that would be applied?

Tell whether each statement is true or false. In Exercises 45–54, first simplify the expression involving an operation. See Examples 2–4.

41. $5 < 6$ **42.** $3 < 7$ **43.** $8 \geq 17$

44. $10 \geq 41$ **45.** $17 \leq 18 - 1$ **46.** $12 \geq 10 + 2$

47. $6 \cdot 8 + 6 \cdot 6 \geq 0$ **48.** $4 \cdot 20 - 16 \cdot 5 \geq 0$

49. $6[5 + 3(4 + 2)] \leq 70$ **50.** $6[2 + 3(2 + 5)] \leq 135$

51. $\dfrac{9(7 - 1) - 8 \cdot 2}{4(6 - 1)} > 3$ **52.** $\dfrac{2(5 + 3) + 2 \cdot 2}{2(4 - 1)} > 1$

53. $8 \leq 4^2 - 2^2$ **54.** $10^2 - 8^2 > 6^2$

Write each word statement in symbols. See Example 5.

55. Fifteen is equal to five plus ten.

56. Twelve is equal to twenty minus eight.

57. Nine is greater than five minus four.

58. Ten is greater than six plus one.

59. Sixteen is not equal to nineteen.

60. Three is not equal to four.

61. Two is less than or equal to three.

62. Five is less than or equal to nine.

Write each statement in words and decide whether it is true or false.

63. $7 < 19$ **64.** $9 < 10$ **65.** $3 \neq 6$

66. $9 \neq 13$ **67.** $8 \geq 11$ **68.** $4 \leq 2$

69. Construct a true statement that involves an addition on the left side, the symbol $\geq$, and a multiplication on the right side.

70. Construct a false statement that involves subtraction on the left side, the symbol $\leq$, and a division on the right side. Then tell why the statement is false and how it could be changed to become true.

Write each statement with the inequality symbol reversed while keeping the same meaning. See Example 6.

71. $5 < 30$ **72.** $8 > 4$ **73.** $12 \geq 3$ **74.** $25 \leq 41$

75. What English-language phrase is used to express the fact that one person's age *is less than* another person's age?

76. What English-language phrase is used to express the fact that one person's height *is greater than* another person's height?

77. The inequality symbol $\geq$ implies a true statement if 12 equals 12 *or* if 12 is greater than 12.
78. is not greater than; no; is not less than
79. December 1996, January 1997, May 1997, June 1997, November 1997
80. July 1997 **81.** (a) .7 (b) 1.3% **82.** \$611.8 billion

77. $12 \geq 12$ is a true statement. Suppose that someone tells you the following: "$12 \geq 12$ is false, because even though 12 is equal to 12, 12 is not greater than 12." How would you respond to this?

78. The symbol $\neq$ means "is not equal to." How do you think we read the symbol $\not>$? Is this equivalent to $<$? How do you think we read $\not<$?

The bar graph shows total construction spending in billions of dollars at seasonally adjusted rates. Use the graph to answer the questions in Exercises 79 and 80. See Example 7.

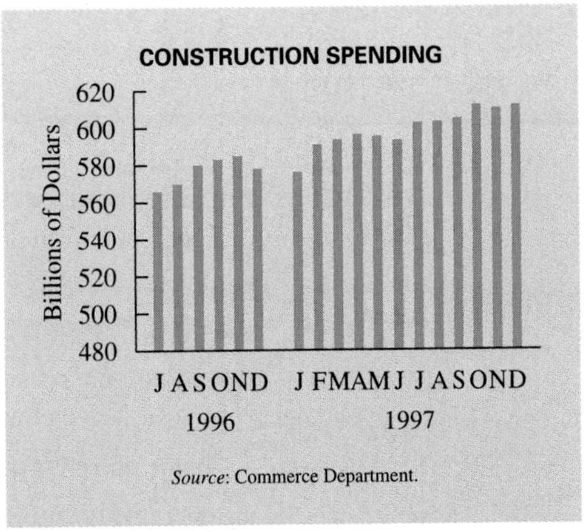

79. In five of the months depicted, construction spending decreased from the previous month. What were these five months?

80. What was the first month shown in which construction spending was greater than \$600 billion?

81. According to the Tuesday, February 3, 1998, issue of *The New York Times,* the National Association of Purchasing Management's index declined to 52.4 in January, 1998, from its level of 53.1 the previous month.
(a) By how much did the index decline?
(b) To find the percent of decline, divide the amount from part (a) by 53.1 and convert to a percent. What was the percent of decline?

82. According to the same article cited in Exercise 81, construction spending rose .1% in 1997. If construction spending the previous year was \$611.2 billion, what was the amount spent in 1997? (*Hint:* Find .1% of 611.2 and add it to 611.2 to get the amount in billions of dollars.)

1.3 Variables, Expressions, and Equations

OBJECTIVES

1. Define *variable,* and find the value of an algebraic expression, given the values of the variables.
2. Convert phrases from words to algebraic expressions.
3. Identify solutions of equations.

OBJECTIVE 1 Define *variable,* and find the value of an algebraic expression, given the values of the variables. A **variable** is a symbol, usually a letter, such as x, y, or z, used to represent any unknown number. An **algebraic expression** is a collection of numbers, variables, operation symbols, and grouping symbols (such as parentheses). For example,

$$6(x + 5), \quad 2m - 9, \quad \text{and} \quad 8p^2 + 6p + 2$$

are all algebraic expressions. In the algebraic expression $2m - 9$, the expression $2m$ indicates the product of 2 and m, just as $8p^2$ shows the product of 8 and p^2. Also, $6(x + 5)$ means the product of 6 and $x + 5$. An algebraic expression has different numerical values for different values of the variable.

INSTRUCTOR'S RESOURCES

PTB Sec. 1.3 ISM Sec. 1.3 AB Sec. 1.3 | TEST GENERATOR | WORLD WIDE WEB www.LialAlgebra.com

4 Identify solutions of equations from a set of numbers.

5 Distinguish between an *expression* and an *equation.*

FOR EXTRA HELP

SSG Sec. 1.3
SSM Sec. 1.3
Pass the Test Software
InterAct Math Tutorial Software
Video 1

For Example 1

CHALKBOARD EXAMPLE

Find the value of the expression if $p = 3$.

$$2p^3$$

Answer: 54

TEACHING TIP Mention that a variable in mathematics is analogous to a pronoun in English. For example, in "He went shopping," *he* can refer to any male shopper, while in $x + 3$, x can be replaced by any number.

TEACHING TIP The distinction between terms such as $3m^2$ and $(3m)^2$ in Example 1(b) will need to be reinforced several times.

For Example 2

CHALKBOARD EXAMPLE

Find the value of the expression if $x = 6$ and $y = 9$.

$$\frac{4x - 2y}{x + 1}$$

Answer: $\frac{6}{7}$

EXAMPLE 1 Evaluating Expressions

Find the numerical values of the following algebraic expressions when $m = 5$.

(a) $8m$

Replace m with 5, to get

$$8m = 8 \cdot 5 = 40.$$

(b) $3m^2$

For $m = 5$,

$$3m^2 = 3 \cdot 5^2 = 3 \cdot 25 = 75.$$

CAUTION In Example 1(b), notice that $3m^2$ means $3 \cdot m^2$; it *does not* mean $3m \cdot 3m$. The product $3m \cdot 3m$ is indicated by $(3m)^2$.

EXAMPLE 2 Evaluating Expressions

Find the value of each expression when $x = 5$ and $y = 3$.

(a) $2x + 7y$

Replace x with 5 and y with 3. Follow the order of operations; multiply first, then add.

$$\begin{aligned} 2x + 7y &= 2 \cdot 5 + 7 \cdot 3 && \text{Let } x = 5 \text{ and } y = 3. \\ &= 10 + 21 && \text{Multiply.} \\ &= 31 && \text{Add.} \end{aligned}$$

(b) $\dfrac{9x - 8y}{2x - y}$

Replace x with 5 and y with 3.

$$\begin{aligned} \frac{9x - 8y}{2x - y} &= \frac{9 \cdot 5 - 8 \cdot 3}{2 \cdot 5 - 3} && \text{Let } x = 5 \text{ and } y = 3. \\ &= \frac{45 - 24}{10 - 3} && \text{Multiply.} \\ &= \frac{21}{7} && \text{Subtract.} \\ &= 3 && \text{Divide.} \end{aligned}$$

(c)
$$\begin{aligned} x^2 - 2y^2 &= 5^2 - 2 \cdot 3^2 && \text{Let } x = 5 \text{ and } y = 3. \\ &= 25 - 2 \cdot 9 && \text{Use the exponents.} \\ &= 25 - 18 && \text{Multiply.} \\ &= 7 && \text{Subtract.} \end{aligned}$$

OBJECTIVE 2 Convert phrases from words to algebraic expressions. In Section 1.2 we saw how to translate from words to symbols.

PROBLEM SOLVING

Sometimes variables must be used to change word phrases into algebraic expressions. The next example illustrates this. Such translations are used in problem solving.

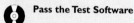

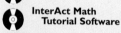

CHALKBOARD EXAMPLE

Write as an algebraic expression using x as the variable.

9 multiplied by the sum of a number and 5

Answer: $9(x + 5)$

EXAMPLE 3 Changing Word Phrases to Algebraic Expressions

Change the following word phrases to algebraic expressions. Use x as the variable.

(a) The **sum** of a number and 9

"**Sum**" is the answer to an addition problem. This phrase translates as

$$x + 9 \qquad \text{or} \qquad 9 + x.$$

(b) 7 **minus** a number

"**Minus**" indicates subtraction, so the answer is $7 - x$.

CAUTION

Here $x - 7$ would *not* be correct; this statement translates as "a number minus 7," not "7 minus a number." The expressions $7 - x$ and $x - 7$ are rarely equal. For example, if $x = 10$, $10 - 7 \neq 7 - 10$. ($7 - 10$ is a *negative number,* discussed in Section 1.4.)

TEACHING TIP The quantity that follows the words "less than" will go in front in a subtraction problem.

(c) 7 **less than** a number

Write 7 **less than** a number as $x - 7$. In this case $7 - x$ would not be correct, because "less than" means "subtracted from."

(d) The product of 11 and a number

$$11 \cdot x \qquad \text{or} \qquad 11x$$

As mentioned earlier, $11x$ means 11 times x. No symbol is needed to indicate the product of a number and a variable.

(e) 5 **divided by** a number

This translates as $\frac{5}{x}$. The expression $\frac{x}{5}$ would *not* be correct here.

(f) The **product** of 2, and the **sum** of a number and 8

We are multiplying 2 times another number. This number is the sum of x and 8, written $x + 8$. Using parentheses for this sum, the final expression is

$$2(x + 8).$$

OBJECTIVE 3 Identify solutions of equations. An **equation** is a statement that two algebraic expressions are equal. Therefore, an equation *always* includes the equality symbol, =. Examples of equations are

$$x + 4 = 11, \qquad 2y = 16, \qquad \text{and} \qquad 4p + 1 = 25 - p.$$

Solving an Equation

To **solve** an equation means to find the values of the variable that make the equation true. Such values of the variable are called the **solutions** of the equation.

CHALKBOARD EXAMPLE

Decide whether the given number is a solution of the equation.

$8p - 11 = 5;\quad 2$

Answer: yes

EXAMPLE 4 Deciding whether a Number Is a Solution

Decide whether the given number is a solution of the equation.

(a) $5p + 1 = 36;\quad 7$

Replace p with 7.

$$\begin{aligned} 5p + 1 &= 36 & & \\ 5 \cdot 7 + 1 &= 36 & ? \quad & \text{Let } p = 7. \\ 35 + 1 &= 36 & ? \quad & \\ 36 &= 36 & & \text{True} \end{aligned}$$

The number 7 is a solution of the equation.

(b) $9m - 6 = 32$; 4

$$\begin{aligned} 9m - 6 &= 32 & & \\ 9 \cdot 4 - 6 &= 32 & ? \quad & \text{Let } m = 4. \\ 36 - 6 &= 32 & ? \quad & \\ 30 &= 32 & & \text{False} \end{aligned}$$

The number 4 is not a solution of the equation.

TEACHING TIP Remind students that the rules for order of operations continue to apply here.

OBJECTIVE 4 Identify solutions of equations from a set of numbers. A **set** is a collection of objects. In mathematics, these objects are most often numbers. The objects that belong to the set, called **elements** of the set, are written between **set braces.** For example, the set containing the numbers 1, 2, 3, 4, and 5 is written as

$$\{1, 2, 3, 4, 5\}.$$

For more information about sets, see Appendix B at the back of this book.

In some cases, the set of numbers from which the solutions of an equation must be chosen is specifically stated. One way of determining solutions is direct substitution of all possible replacements. The ones that lead to a true statement are solutions.

CHALKBOARD EXAMPLE

Change the statement to an equation. Find all solutions from the set {0, 2, 4, 6, 8, 10}.

Three times a number is subtracted from 21, giving 15.

Answer: $21 - 3x = 15$; 2

EXAMPLE 5 Finding a Solution from a Given Set

Change each word statement to an equation. Use x as the variable. Then find all solutions of the equation from the set

$$\{0, 2, 4, 6, 8, 10\}.$$

(a) The sum of a number and four is six.

The word "is" suggests "equals." If x represents the unknown number, then translate as follows.

The sum of a number and four	is	six.
↓	↓	↓
$x + 4$	$=$	6

Try each number from the given set {0, 2, 4, 6, 8, 10}, in turn, to see that 2 is the only solution of $x + 4 = 6$.

(b) 9 more than five times a number is 49.

Use x to represent the unknown number. We start with $5x$ and then add 9 to it. The word "is" translates as =.

$$5x + 9 = 49$$

Try each number from {0, 2, 4, 6, 8, 10}. The solution is 8, since $5 \cdot 8 + 9 = 49$.

(c) The sum of a number and 12 is equal to four times the number.

If x represents the number, "the sum of a number and 12" is represented by $x + 12$. The translation is

$$x + 12 = 4x.$$

Trying each replacement leads to a true statement when $x = 4$, since $4 + 12 = 4(4) = 16$.

OBJECTIVE 5 **Distinguish between an *expression* and an *equation*.** Students often have trouble distinguishing between equations and expressions. Remember that an equation is a sentence; an expression is a phrase.

$4x + 5 = 9$	$4x + 5$
↑	↑
Equation (to solve)	Expression (to simplify or evaluate)

TEACHING TIP Point out that the words *equation* and *equal* are similar.

CHALKBOARD EXAMPLE

Decide whether the following is an equation or an expression.

$$\frac{3x - 1}{5}$$

Answer: expression

EXAMPLE 6 **Distinguishing between Equations and Expressions**

Decide whether each of the following is an equation or an expression.

(a) $2x - 5y$

There is no equals sign, so this is an expression.

(b) $2x = 5y$

Because of the equals sign, this is an equation.

1.3 EXERCISES

1. 10 **2.** 8 **3.** $12 + x$; 21 **4.** 4 **5.** no **6.** expression; equation **7.** $2x^3 = 2 \cdot x \cdot x \cdot x$, while $2x \cdot 2x \cdot 2x = (2x)^3$. **8.** The first is an expression indicating subtraction, $x - 5$, while the second is a statement relating 5 and x, $5 < x$. **9.** The exponent 2 applies only to its base, which is x. (The expression $(4x)^2$ would require multiplying 4 by $x = 3$ first.) **10.** 3; To equal 9, $2x$ must equal 6 and thus x must equal 3. **11.** (Answers will vary.) Two such pairs are $x = 0, y = 6$ and $x = 1, y = 4$. To determine them, choose a value for x, substitute it into the expression $2x + y$, and then subtract the value of $2x$ from 6. **12.** The value for y is 3. If x is 4, then $3x = 12$, and 3 from 12 equals 9.

Fill in each blank with the correct response.

1. If $x = 3$, then the value of $x + 7$ is ______.
2. If $x = 1$ and $y = 2$, then the value of $4xy$ is ______.
3. The sum of 12 and x is represented by the expression ______. If $x = 9$, the value of that expression is ______.
4. If x can be chosen from the set {0, 1, 2, 3, 4, 5}, the only solution of $x + 5 = 9$ is ______.
5. Will the equation $x = x + 4$ ever have a solution? ______.
6. $2x + 3$ is a(n) ________________ (equation/expression), while $2x + 3 = 8$ is a(n) ________________ (equation/expression).

Exercises 7–12 cover some of the concepts introduced in this section. Give a short explanation for each.

7. Why is $2x^3$ not the same as $2x \cdot 2x \cdot 2x$? Explain.
8. Why are "5 less than a number" and "5 is less than a number" translated differently?
9. Explain in your own words why, when evaluating the expression $4x^2$ for $x = 3$, 3 must be squared *before* multiplying by 4.
10. What value of x would cause the expression $2x + 3$ to equal 9? Explain your reasoning.
11. There are many pairs of values of x and y for which $2x + y$ will equal 6. Name two such pairs and describe how you determined them.
12. Suppose that for the equation $3x - y = 9$, the value of x is given as 4. What would be the corresponding value of y? How do you know this?

13. (a) 13 (b) 15 **14.** (a) 3 (b) 5 **15.** (a) 20 (b) 30 **16.** (a) 28 (b) 42 **17.** (a) 64 (b) 144 **18.** (a) 80 (b) 180 **19.** (a) $\frac{5}{3}$ (b) $\frac{7}{3}$ **20.** (a) $\frac{2}{5}$ (b) $\frac{4}{5}$ **21.** (a) $\frac{7}{8}$ (b) $\frac{13}{12}$ **22.** (a) $\frac{5}{4}$ (b) $\frac{23}{18}$ **23.** (a) 52 (b) 114 **24.** (a) 24 (b) 48 **25.** (a) 25.836 (b) 38.754 **26.** (a) 11.84 (b) 26.64 **27.** (a) 24 (b) 28 **28.** (a) 17 (b) 21 **29.** (a) 12 (b) 33 **30.** (a) 10 (b) 14 **31.** (a) 6 (b) $\frac{9}{5}$ **32.** (a) 5 (b) 13 **33.** (a) $\frac{4}{3}$ (b) $\frac{13}{6}$ **34.** (a) $\frac{13}{20}$ (b) $\frac{29}{20}$ **35.** (a) $\frac{2}{7}$ (b) $\frac{16}{27}$ **36.** (a) 5 (b) 18 **37.** (a) 12 (b) 55 **38.** (a) 28 (b) 26 **39.** (a) 1 (b) $\frac{28}{17}$ **40.** (a) $\frac{5}{13}$ (b) $\frac{2}{29}$ **41.** (a) 3.684 (b) 8.841 **42.** (a) 3.964 (b) 5.941 **43.** $12x$ **44.** $9x$ **45.** $x + 7$ **46.** $x + 13$ **47.** $x - 2$ **48.** $x - 8$ **49.** $7 - x$ **50.** $14 - x$ **51.** $x - 6$ **52.** $6 - x$ **53.** $\frac{12}{x}$ **54.** $\frac{x}{12}$ **55.** $6(x - 4)$ **56.** $9(x + 5)$ **57.** No, it is a connective word that joins the two factors: the number and 6. **58.** "Please excuse me, but I would like to point out that one *solves* an equation but *simplifies* an expression. You might change 'Solve' to 'Simplify.'" **59.** yes **60.** yes **61.** no **62.** no **63.** yes **64.** yes **65.** yes **66.** yes **67.** yes **68.** yes **69.** $x + 8 = 18$; 10 **70.** $x - 3 = 1$; 4 **71.** $16 - \frac{3}{4}x = 13$; 4 **72.** $\frac{6}{5}x + 2 = 14$; 10 **73.** $2x + 1 = 5$; 2 **74.** $3x = 6$; 2 **75.** $3x = 2x + 8$; 8 **76.** $\frac{12}{x} = \frac{1}{3}x$; 6

Find the numerical value **(a)** *if* $x = 4$ *and* **(b)** *if* $x = 6$. *See Example 1.*

13. $x + 9$ **14.** $x - 1$ **15.** $5x$ **16.** $7x$ **17.** $4x^2$

18. $5x^2$ **19.** $\frac{x + 1}{3}$ **20.** $\frac{x - 2}{5}$ **21.** $\frac{3x - 5}{2x}$ **22.** $\frac{4x - 1}{3x}$

23. $3x^2 + x$ **24.** $2x + x^2$ **25.** $6.459x$ **26.** $.74x^2$

Find the numerical value if **(a)** $x = 2$ *and* $y = 1$ *and* **(b)** $x = 1$ *and* $y = 5$. *See Example 2.*

27. $8x + 3y + 5$ **28.** $4x + 2y + 7$ **29.** $3(x + 2y)$ **30.** $2(2x + y)$

31. $x + \frac{4}{y}$ **32.** $y + \frac{8}{x}$ **33.** $\frac{x}{2} + \frac{y}{3}$ **34.** $\frac{x}{5} + \frac{y}{4}$

35. $\frac{2x + 4y - 6}{5y + 2}$ **36.** $\frac{4x + 3y - 1}{x}$ **37.** $2y^2 + 5x$ **38.** $6x^2 + 4y$

39. $\frac{3x + y^2}{2x + 3y}$ **40.** $\frac{x^2 + 1}{4x + 5y}$ **41.** $.841x^2 + .32y^2$ **42.** $.941x^2 + .2y^2$

Change each word phrase to an algebraic expression. Use x as the variable to represent the number. See Example 3.

43. Twelve times a number

44. Nine times a number

45. Seven added to a number

46. Thirteen added to a number

47. Two subtracted from a number

48. Eight subtracted from a number

49. A number subtracted from seven

50. A number subtracted from fourteen

51. The difference between a number and 6

52. The difference between 6 and a number

53. 12 divided by a number

54. A number divided by 12

55. The product of 6 and four less than a number

56. The product of 9 and five more than a number

57. In the phrase "Four more than the product of a number and 6," does the word *and* signify the operation of addition? Explain.

58. Suppose that the directions on a test read "Solve the following expressions." How would you politely correct the person who wrote these directions? What alternative directions might you suggest?

Decide whether each given number is a solution of the equation. See Example 4.

59. $5m + 2 = 7$; 1

60. $3r + 5 = 8$; 1

61. $2y + 3(y - 2) = 14$; 3

62. $6a + 2(a + 3) = 14$; 2

63. $6p + 4p + 9 = 11$; $\frac{1}{5}$

64. $2x + 3x + 8 = 20$; $\frac{12}{5}$

65. $3r^2 - 2 = 46$; 4

66. $2x^2 + 1 = 19$; 3

67. $\frac{z + 4}{2 - z} = \frac{13}{5}$; $\frac{1}{3}$

68. $\frac{x + 6}{x - 2} = \frac{37}{5}$; $\frac{13}{4}$

Change each word statement to an equation. Use x as the variable. Find all solutions from the set {2, 4, 6, 8, 10}. *See Example 5.*

69. The sum of a number and 8 is 18.

70. A number minus three equals 1.

71. Sixteen minus three-fourths of a number is 13.

72. The sum of six-fifths of a number and 2 is 14.

73. One more than twice a number is 5.

74. The product of a number and 3 is 6.

75. Three times a number is equal to 8 more than twice the number.

76. Twelve divided by a number equals $\frac{1}{3}$ times that number.

77. expression 78. expression
79. equation 80. equation
81. equation 82. expression
83. \$10.50; less by \$.33
84. \$11.78; less by \$.29
85. \$12.10; less by \$.27
86. \$12.41; less by \$.37

Identify as an expression *or an* equation. *See Example 6.*

77. $3x + 2(x - 4)$

78. $5y - (3y + 6)$

79. $7t + 2(t + 1) = 4$

80. $9r + 3(r - 4) = 2$

81. $x + y = 3$

82. $x + y - 3$

Mathematicians who study statistics have developed methods of determining mathematical models. Loosely speaking, a **mathematical model** is an equation that can be used to determine unknown quantities. We cannot always expect a model to give us an accurate answer, but at least we can obtain a rough estimate. For example, based on data from the U.S. Bureau of Labor Statistics, average hourly earnings of production workers in manufacturing industries in the United States during the period from 1990 through 1996 can be approximated by the equation $y = .319x - 624.31$, where x represents the year and y represents the hourly earnings in dollars.

Use this equation to approximate the average hourly earnings during the following years. Do this by replacing x in the equation by the year, and then simplifying the right side of the equation to find y, the hourly earnings. Then compare your answer to the actual earnings (given in parentheses) and tell whether the approximation is greater than or less than the actual earnings, and by how much.

83. 1990 (\$10.83)

84. 1994 (\$12.07)

85. 1995 (\$12.37)

86. 1996 (\$12.78)

1.4 Real Numbers and the Number Line

OBJECTIVES

1. Set up number lines.
2. Classify numbers.
3. Tell which of two different real numbers is smaller.
4. Find additive inverses and absolute values of real numbers.
5. Interpret real number meanings from a table of data.

FOR EXTRA HELP

SSG Sec. 1.4
SSM Sec. 1.4
Pass the Test Software
InterAct Math Tutorial Software
Video I

TEACHING TIP The "arrowhead" is used to indicate the positive direction on a number line.

OBJECTIVE 1 Set up number lines. In Section 1.1 we introduced two important sets of numbers, the *natural numbers* and the *whole numbers.*

Natural Numbers

$\{1, 2, 3, 4, \ldots\}$ is the set of **natural numbers.**

Whole Numbers

$\{0, 1, 2, 3, \ldots\}$ is the set of **whole numbers.**

NOTE The three dots show that the list of numbers continues in the same way indefinitely.

These numbers, along with many others, can be represented on **number lines** like the one pictured in Figure 4. We draw a number line by choosing any point on the line and calling it 0. Choose any point to the right of 0 and call it 1. The distance between 0 and 1 gives a unit of measure used to locate other points, as shown in Figure 4. The points labeled in Figure 4 and those continuing in the same way to the right correspond to the set of whole numbers.

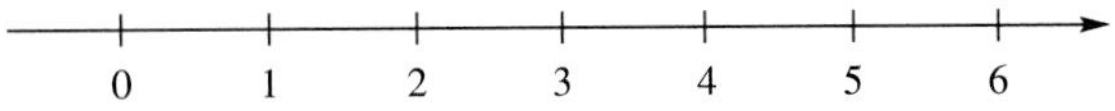

Figure 4

INSTRUCTOR'S RESOURCES

PTB Sec. 1.4 **ISM** Sec. 1.4
AB Sec. 1.4

TEST GENERATOR

WORLD WIDE WEB
www.LialAlgebra.com

All the whole numbers starting with 1 are located to the right of 0 on the number line. But numbers may also be placed to the left of 0. These numbers, written -1, -2, -3, and so on, shown in Figure 5, are called **negative numbers.** (The minus sign is used to show that these numbers are located to the *left* of 0.) The numbers to the *right* of 0 are **positive numbers.** The number 0 itself is neither positive nor negative. Positive numbers and negative numbers are called **signed numbers.**

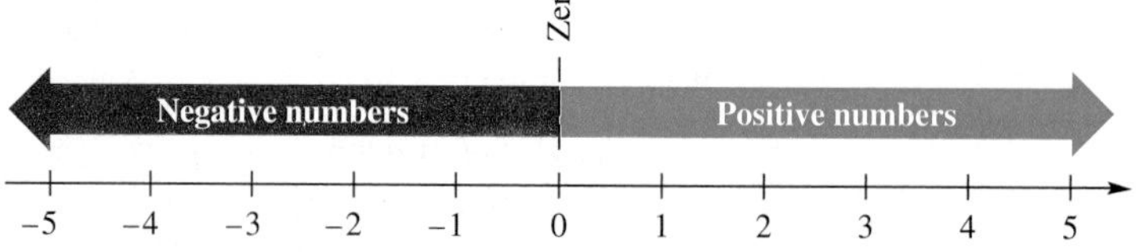

Figure 5

CONNECTIONS

There are many practical applications of negative numbers. For example, temperatures sometimes fall below zero. The lowest temperature ever recorded in meteorological records was $-128.6°$F at Vostok, Antarctica, on July 22, 1983. A business that spends more than it takes in has a negative "profit." Altitudes below sea level can be represented by negative numbers. The shore surrounding the Dead Sea is 1312 feet below sea level; this can be represented as -1312 feet. (*Source: The World Almanac and Book of Facts,* 1995.)

OBJECTIVE 2 Classify numbers. The set of numbers marked on the number line in Figure 5, including positive and negative numbers and zero, is part of the set of *integers.*

Integers

$\{\ldots, -3, -2, -1, 0, 1, 2, 3, \ldots\}$ is the set of **integers.**

Not all numbers are integers. For example, $\frac{1}{2}$ is not; it is a number halfway between the integers 0 and 1. Also, $3\frac{1}{4}$ is not an integer. Several numbers that are not integers are *graphed* in Figure 6. The **graph** of a number is a point on the number line. The number is called the **coordinate** of the point. Think of the graph of a set of numbers as a picture of the set. All the numbers in Figure 6 can be written as quotients of integers. These numbers are examples of *rational numbers.*

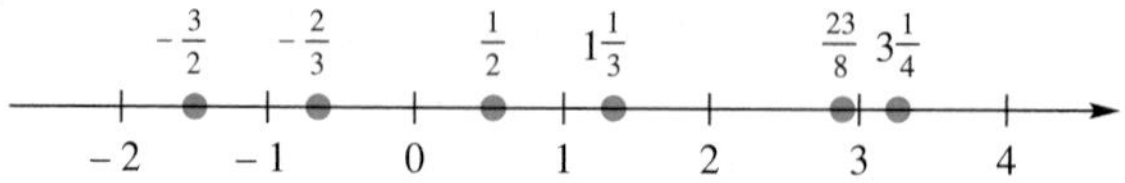

Figure 6

TEACHING TIP A rational number can be written as a ratio (or quotient) of two integers.

Rational Numbers

$\{x \mid x$ is a quotient of two integers, with denominator not $0\}$ is the set of **rational numbers.**

(Read the part in the braces as "the set of all numbers x such that x is a quotient of two integers, with denominator not 0.")

NOTE The set symbolism used in the definition of rational numbers,

$$\{x \mid x \text{ has a certain property}\},$$

is called **set-builder notation.** This notation is convenient to use when it is not possible to list all the elements of the set.

Since any integer can be written as the quotient of itself and 1, all integers also are rational numbers.

Although many numbers are rational, not all are. For example, a square that measures one unit on a side has a diagonal whose length is the square root of 2, written $\sqrt{2}$. See Figure 7. It can be shown that $\sqrt{2}$ cannot be written as a quotient of integers. Because of this, $\sqrt{2}$ is not rational; it is *irrational.* Other examples of irrational numbers are $\sqrt{3}$, $\sqrt{7}$, $-\sqrt{10}$, and π (the ratio of the *circumference* of a circle to its diameter).

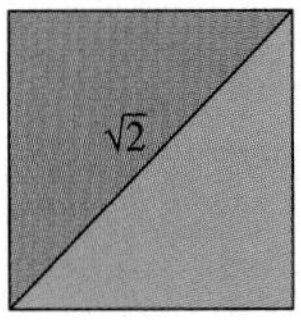

Figure 7

All numbers, both rational and irrational, can be represented by points on the number line and are called *real numbers.*

Real Numbers

$\{x \mid x$ is a rational or an irrational number$\}$ is the set of **real numbers.**

Real numbers can be written as decimals. Any rational number will have a decimal that either comes to an end (terminates) or repeats in a fixed "block" of digits. For example, $\frac{2}{5} = .4$ and $\frac{27}{100} = .27$ are rational numbers with terminating decimals; $\frac{1}{3} = .3333...$ and $\frac{3}{11} = .27272727...$ are repeating decimals. The decimal representation of an irrational number will neither terminate nor repeat. (A review of operations with decimals can be found in Appendix A.)

An example of a number that is not a real number is the square root of a negative number like $\sqrt{-5}$. These numbers are discussed in a later chapter.

Two ways to represent the relationships among the various types of numbers are shown in Figure 8. Part (a) also gives some examples. Notice that every real number is either a rational number or an irrational number.

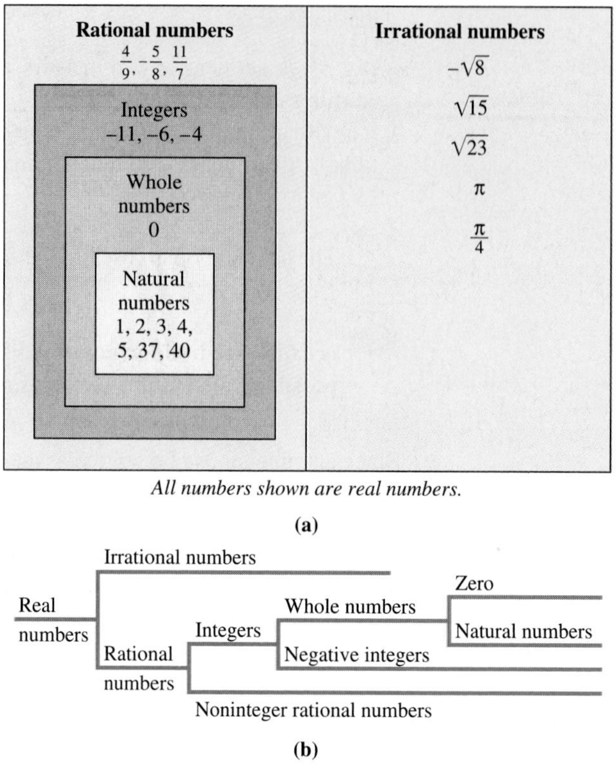

Figure 8

CHALKBOARD EXAMPLE

Identify each real number in the set below as rational or irrational.

$$\left\{\frac{5}{8}, -7, -1\frac{3}{5}, 0, \sqrt{11}, -\pi\right\}$$

Answer: $\frac{5}{8}$, $-7\left(\text{or } -\frac{7}{1}\right)$, $-1\frac{3}{5}\left(\text{or } -\frac{8}{5}\right)$, and $0\left(\text{or } \frac{0}{1}\right)$ are rational; $\sqrt{11}$ and $-\pi$ are irrational.

EXAMPLE 1 Determining whether a Number Belongs to a Set

List the numbers in the set

$$\left\{-5,\quad -\frac{2}{3},\quad 0,\quad \sqrt{2},\quad 3\frac{1}{4},\quad 5,\quad 5.8\right\}$$

that belong to each of the following sets of numbers.

(a) Natural numbers
The only natural number in the set is 5.

(b) Whole numbers
The whole numbers consist of the natural numbers and 0. So the elements of the set that are whole numbers are 0 and 5.

(c) Integers
The integers in the set are -5, 0, and 5.

(d) Rational numbers
The rational numbers are -5, $-\frac{2}{3}$, 0, $3\frac{1}{4}$, 5, 5.8, since each of these numbers *can* be written as the quotient of two integers. For example, $5.8 = \frac{58}{10}$.

(e) Irrational numbers
The only irrational number in the set is $\sqrt{2}$.

(f) Real numbers
All the numbers in the set are real numbers.

OBJECTIVE 3 **Tell which of two different real numbers is smaller.** Given any two whole numbers, you probably can tell which number is smaller. But what happens with negative numbers, as in the set of integers? Positive numbers decrease as the corresponding points on the number line go to the left. For example, $8 < 12$, and 8 is to the left of 12 on the number line. This ordering is extended to all real numbers by definition.

The Ordering of Real Numbers

For any two real numbers a and b, ***a* is less than *b*** if a is to the left of b on the number line.

This means that any negative number is smaller than 0, and any negative number is smaller than any positive number. Also, 0 is smaller than any positive number.

CHALKBOARD EXAMPLE

Determine whether the statement is true or false.

$$-4 \geq -1$$

Answer: false

EXAMPLE 2 **Determining the Order of Real Numbers**

Is it true that $-3 < -1$?

To decide whether the statement is true, locate both numbers, -3 and -1, on a number line, as shown in Figure 9. Since -3 is to the left of -1 on the number line, -3 is smaller than -1. The statement $-3 < -1$ is true.

Figure 9

NOTE In Section 1.2 we saw how it is possible to rewrite a statement involving $<$ as an equivalent statement involving $>$. The question in Example 2 can also be worded as follows: Is it true that $-1 > -3$? This is, of course, also a true statement.

We can say that for any two real numbers a and b, ***a* is greater than *b*** if a is to the right of b on the number line.

OBJECTIVE 4 **Find additive inverses and absolute values of real numbers.** By a property of the real numbers, for any real number x (except 0), there is exactly one number on the number line the same distance from 0 as x but on the opposite side of 0. For example, Figure 10 shows that the numbers 3 and -3 are each the same distance from 0 but are on opposite sides of 0. The numbers 3 and -3 are called **additive inverses,** or **opposites,** of each other.

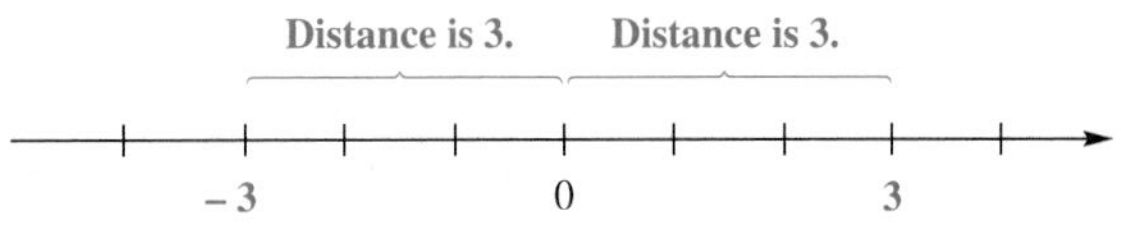

Figure 10

Additive Inverse

The **additive inverse** of a number x is the number that is the same distance from 0 on the number line as x, but on the opposite side of 0.

The additive inverse of the number 0 is 0 itself. In fact, 0 is the only real number that is its own additive inverse. Other additive inverses occur in pairs. For example, 4 and -4, and 5 and -5, are additive inverses of each other. Several pairs of additive inverses are shown in Figure 11.

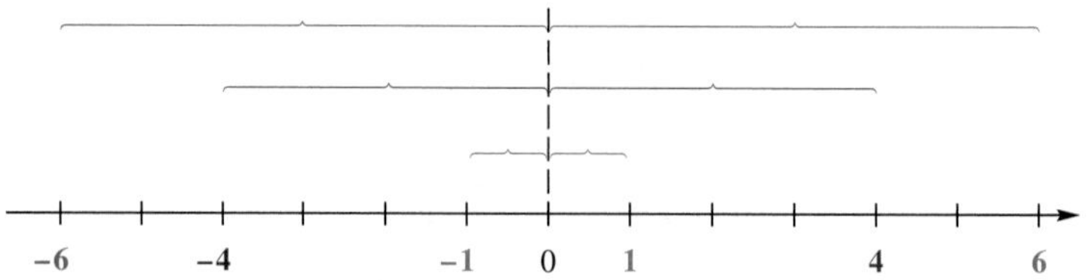

Figure 11

The additive inverse of a number can be indicated by writing the symbol $-$ in front of the number. With this symbol, the additive inverse of 7 is written -7. The additive inverse of -4 is written $-(-4)$, and can also be read "the opposite of -4" or "the negative of -4." Figure 11 suggests that 4 is an additive inverse of -4. Since a number can have only one additive inverse, the symbols 4 and $-(-4)$ must represent the same number, which means that

$$-(-4) = 4.$$

This idea can be generalized as follows.

Double Negative Rule

For any real number x,

$$-(-x) = x.$$

The following chart shows several numbers and their additive inverses.

Number	*Additive Inverse*
-4	$-(-4)$ or 4
0	0
19	-19
$-\frac{2}{3}$	$\frac{2}{3}$

NOTE The chart suggests that the additive inverse of a number is found by changing the sign of the number.

An important property of additive inverses will be studied later in this chapter: $a + (-a) = (-a) + a = 0$ for all real numbers a.

As mentioned above, additive inverses are numbers that are the same distance from 0 on the number line. See Figure 11. This idea can also be expressed by saying that a number and its additive inverse have the same absolute value. The **absolute value** of a real number can be defined as the distance between 0 and the number on the number line. The symbol for the absolute value of the number x is $|x|$, read "the absolute value of x." For example, the distance between 2 and 0 on the number line is 2 units, so that

$$|2| = 2.$$

Because the distance between -2 and 0 on the number line is also 2 units,

$$|-2| = 2.$$

Since distance is a physical measurement, which is never negative, the absolute value of a number is never negative. For example, $|12| = 12$ and $|-12| = 12$, since both 12 and -12 lie at a distance of 12 units from 0 on the number line. Also, since 0 is a distance of 0 units from 0, $|0| = 0$.

In symbols, the absolute value of x is defined as follows.

Formal Definition of Absolute Value

$$|x| = \begin{cases} x & \text{if } x \geq 0 \\ -x & \text{if } x < 0 \end{cases}$$

TEACHING TIP The definition of absolute value usually requires several numerical examples before it can be understood.

By this definition, if x is a positive number or 0, then its absolute value is x itself. For example, since 8 is a positive number, $|8| = 8$. However, if x is a negative number, then its absolute value is the additive inverse of x. This means that if $x = -9$, then $|-9| = -(-9) = 9$, since the additive inverse of -9 is 9.

CAUTION The formal definition of absolute value can be confusing if it is not read carefully. The "$-x$" in the second part of the definition *does not* represent a negative number. Since x is negative in the second part, $-x$ represents the opposite of a negative number, that is, a positive number. *The absolute value of a number is never negative.*

CHALKBOARD EXAMPLE

Simplify by removing the absolute value symbols.

$$-|32 - 2|$$

Answer: -30

EXAMPLE 3 Finding Absolute Value

Simplify by finding the absolute value.

(a) $|5| = 5$

(b) $|-5| = -(-5) = 5$

(c) $-|5| = -(5) = -5$

(d) $-|-14| = -(14) = -14$

(e) $|8 - 2| = |6| = 6$

(f) $-|8 - 2| = -|6| = -6$

Part (e) of Example 3 shows that absolute value bars are also grouping symbols. You must perform any operations that appear inside absolute value symbols before finding the absolute value.

OBJECTIVE 5 **Interpret real number meanings from a table of data.** A table of data provides a concise way of relating information. The final example gives an illustration of this.

EXAMPLE 4 Interpreting Data

The table shows the annual percent change from the previous year in the Consumer Price Index for 1986 and 1987.

CHALKBOARD EXAMPLE

In the table in Example 4, what is the largest percent change (positive or negative), and where is it found?

Answer: The largest percent change (positive or negative) is −21.9, found in gasoline in 1986.

	Percent Change	
Category	*1986*	*1987*
Food	3.2	4.1
Shelter	5.5	4.7
Rent, residential	5.8	4.1
Fuel and other utilities	−2.3	−1.1
Apparel and upkeep	.9	4.4
Private transportation	−4.7	−3.0
New cars	4.2	3.6
Gasoline	−21.9	−4.0
Public transportation	5.9	3.5
Medical care	7.5	6.6
Entertainment	3.4	3.3
Commodities	−.9	3.2

Source: Bureau of Labor Statistics, U.S. Dept. of Labor.

(a) What category in which year represents the greatest drop?

To find the greatest drop, we must find the negative entry with the largest absolute value. The entry for gasoline in 1986 is −21.9, so this is the category and year with the greatest drop.

(b) Which percent change is represented by a larger drop: 1986 fuel and other utilities or 1987 gasoline?

The two entries are −2.3 and −4.0, respectively. The larger drop is −4.0, so 1987 gasoline is the answer.

(c) What is the smallest percent change (positive or negative) in these two years?

The smallest percent change (positive or negative) is .9 and −.9, both found in 1986.

1.4 EXERCISES

1. 4 **2.** One example is 3.85. There are others. **3.** 0 **4.** One example is 5. There are others. **5.** One example is $\sqrt{12}$. There are others. **6.** 0 **7.** true **8.** false

In Exercises 1–6, give a number that satisfies the given condition.

1. An integer between 3.5 and 4.5

2. A rational number between 3.8 and 3.9

3. A whole number that is not positive and is less than 1

4. A whole number greater than 4.5

5. An irrational number that is between $\sqrt{11}$ and $\sqrt{13}$

6. A real number that is neither negative nor positive

In Exercises 7–10, decide whether each statement is true or false.

7. Every natural number is positive.

8. Every whole number is positive.

9. true **10.** true **11.** (a) 3, 7 (b) 0, 3, 7 (c) $-9, 0, 3, 7$ (d) $-9, -1\frac{1}{4}, -\frac{3}{5}, 0, 3, 5.9, 7$ (e) $-\sqrt{7}, \sqrt{5}$ (f) All are real numbers. **12.** (a) 3 (b) 0, 3 (c) $-5, -1, 0, 3$ (d) $-5.3, -5, -1, -\frac{1}{9}, 0, 1.2, 1.8, 3$ (e) $-\sqrt{3}, \sqrt{11}$ (f) All are real numbers. **13.** The *natural numbers* are the numbers with which we count. An example is 1. The *whole numbers* are the natural numbers with 0 also included. An example is 0. The *integers* are the whole numbers and their negatives. An example is -1. The *rational numbers* are the numbers that can be represented by a quotient of integers, such as $\frac{1}{2}$. The *irrational numbers,* such as $\sqrt{2}$, cannot be represented as a quotient of integers. The *real numbers* include all positive numbers, negative numbers, and zero. All the numbers listed are real. **14.** The decimal representation of a rational number will either terminate or repeat. **15.** 93,000 **16.** -1550 **17.** $-30°$ **18.** 5436 **19.** $-31{,}532$ **20.** 159,376 **21.** -8 **22.** 20; 25°; $-3°$

23. 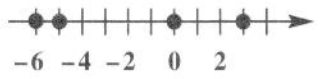

24. 

25. 

26. 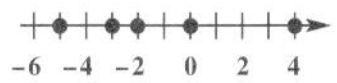

27. 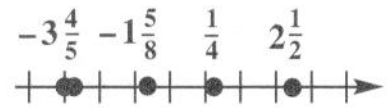

28. 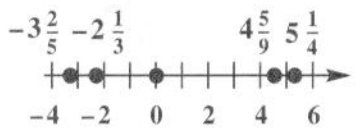

29. (a) A (b) A (c) B (d) B

9. Every integer is a rational number.

10. Every rational number is a real number.

For Exercises 11 and 12, see Example 1.

11. List all numbers from the set

$$\left\{-9, -\sqrt{7}, -1\frac{1}{4}, -\frac{3}{5}, 0, \sqrt{5}, 3, 5.9, 7\right\}$$

that are

(a) natural numbers; **(b)** whole numbers; **(c)** integers;
(d) rational numbers; **(e)** irrational numbers; **(f)** real numbers.

12. List all numbers from the set

$$\left\{-5.3, -5, -\sqrt{3}, -1, -\frac{1}{9}, 0, 1.2, 1.8, 3, \sqrt{11}\right\}$$

that are

(a) natural numbers; **(b)** whole numbers; **(c)** integers;
(d) rational numbers; **(e)** irrational numbers; **(f)** real numbers.

13. Explain in your own words the different sets of numbers introduced in this section, and give an example of each kind.

14. What two possible situations exist for the decimal representation of a rational number?

Use an integer to express each number representing a change in the following applications.

15. In February 1998, the number of housing starts in the United States increased from the previous month by 93,000 units. (*Source: The Wall Street Journal.*)

16. The Wolfsburg, Germany, Volkswagen plant turns out 1550 fewer cars per day than it did in 1991. (*Source:* Paul Klebnikov, "Bringing Back the Beetle," *Forbes,* April 7, 1997.)

17. The boiling point of chlorine is approximately 30° below 0° Fahrenheit.

18. The height of Mt. Arenal, an active volcano in Costa Rica, is 5436 feet above sea level. (*Source: The Universal Almanac,* 1997, John W. Wright, General Editor.)

19. Between 1980 and 1990, the population of the District of Columbia decreased by 31,532. (*Source:* U.S. Bureau of the Census.)

20. In 1994, the country of Taiwan produced 159,376 more passenger cars than commercial vehicles. (*Source:* American Automobile Manufacturers Association.)

21. The city of New Orleans lies 8 feet below sea level. (*Source:* U.S. Geological Survey, *Elevations and Distances in the United States,* 1990.)

22. When the wind speed is 20 miles per hour and the actual temperature is 25° Fahrenheit, the wind-chill factor is 3° below 0° Fahrenheit. (Give three responses.)

Graph each group of numbers on a number line. See Figure 6.

23. $0, 3, -5, -6$

24. $2, 6, -2, -1$

25. $-2, -6, -4, 3, 4$

26. $-5, -3, -2, 0, 4$

27. $\frac{1}{4}, 2\frac{1}{2}, -3\frac{4}{5}, -4, -1\frac{5}{8}$

28. $5\frac{1}{4}, 4\frac{5}{9}, -2\frac{1}{3}, 0, -3\frac{2}{5}$

29. Match each expression in Column I with its value in Column II. Some choices in Column II may not be used.

I	II
(a) $\lvert -7 \rvert$	**A.** 7
(b) $-(-7)$	**B.** -7
(c) $-\lvert -7 \rvert$	**C.** neither A nor B
(d) $-\lvert -(-7) \rvert$	**D.** both A and B

30. 2, 2, 2, −2 **31.** (a) 2 (b) 2 **32.** (a) 8 (b) 8 **33.** (a) −6 (b) 6 **34.** (a) −11 (b) 11 **35.** (a) −3 (b) 3 **36.** (a) −5 (b) 5 **37.** (a) 0 (b) 0 **38.** (a) 0 (b) 0 **39.** $a - b$ **40.** 0 **41.** −12 **42.** −14 **43.** −8 **44.** −16 **45.** 3 **46.** $|-2|$ or 2 **47.** $|-3|$ or 3 **48.** $|-8|$ or 8 **49.** $-|-6|$ or −6 **50.** $-|-3|$ or −3 **51.** $|5-3|$ or 2 **52.** $|7-2|$ or 5 **53.** true **54.** false **55.** true **56.** true **57.** true **58.** true **59.** false **60.** false **61.** true **62.** false **63.** false **64.** false **65.** Softwood plywood from 1995 to 1996 represents the greatest drop. **66.** The change in bright nails from 1994 to 1995 is greater. **67.** true **68.** Gypsum products from 1994 to 1995 represents the greatest increase.
Answers will vary in Exercises 69–78.
69. true: $a = 0$ or $b = 0$ or both $a = 0$ and $b = 0$; false: Choose any values for a and b so that neither a nor b is zero.

30. Fill in the blanks with the correct values: The opposite of −2 is ______, while the absolute value of −2 is ______. The additive inverse of −2 is ______, while the additive inverse of the absolute value of −2 is ______.

*Find **(a)** the opposite (or additive inverse) of each number and **(b)** the absolute value of each number.*

31. −2 **32.** −8 **33.** 6 **34.** 11
35. 7 − 4 **36.** 8 − 3 **37.** 7 − 7 **38.** 3 − 3

39. Look at Exercises 35 and 36 and use the results to complete the following: If $a - b > 0$, then the absolute value of $a - b$ in terms of a and b is ______.

40. Look at Exercises 37 and 38 and use the results to complete the following: If $a - b = 0$, then the absolute value of $a - b$ is ______.

Select the smaller of the two given numbers. See Examples 2 and 3.

41. −12, −4 **42.** −9, −14 **43.** −8, −1
44. −15, −16 **45.** $3, |-4|$ **46.** $5, |-2|$
47. $|-3|, |-4|$ **48.** $|-8|, |-9|$ **49.** $-|-6|, -|-4|$
50. $-|-2|, -|-3|$ **51.** $|5-3|, |6-2|$ **52.** $|7-2|, |8-1|$

Decide whether each statement is true or false. See Examples 2 and 3.

53. $6 > -(-2)$ **54.** $-8 > -(-2)$ **55.** $-4 \le -(-5)$
56. $-6 \le -(-3)$ **57.** $|-6| < |-9|$ **58.** $|-12| < |-20|$
59. $-|8| > |-9|$ **60.** $-|12| > |-15|$ **61.** $-|-5| \ge -|-9|$
62. $-|-12| \le -|-15|$ **63.** $|6-5| \ge |6-2|$ **64.** $|13-8| \le |7-4|$

The table shows the percent change in the Producer Price Index for selected commodities from 1994 to 1995 and from 1995 to 1996. Use the table to answer the questions in Exercises 65–68. See Example 4.

Commodity	Change from 1994 to 1995	Change from 1995 to 1996*
Interior solvent-based paint	16.4	11.1
Plastic construction products	10.9	−3.1
Softwood plywood	11.3	−14.1
Bright nails	1.8	−.3
Gypsum products	18.4	−.3

*1996 data is preliminary.
Source: U.S. Bureau of Labor Statistics.

65. Which category in what year represents the greatest drop?

66. Which of these is the greater number: the change in plastic construction products from 1995 to 1996 or the change in bright nails from 1994 to 1995?

67. True or false? The absolute value of the data in the first column for softwood plywood is less than the absolute value of the data in the second column for the same commodity.

68. Which category represents the greatest increase?

70. true: for all values; false: for no values 71. true: Choose a to be the opposite of b ($a = -b$); false: $a \neq -b$ 72. true: when $a + b \leq 0$; false: when $a + b > 0$
73. $\frac{1}{2}, \frac{5}{8}, 1\frac{3}{4}$ 74. $-1, -\frac{3}{4}, -5$ 75. $-3\frac{1}{2}, -\frac{2}{3}, \frac{3}{7}$
76. $\frac{1}{2}, -\frac{2}{3}, \frac{2}{7}$ 77. $\sqrt{5}, \pi, -\sqrt{3}$
78. $\frac{2}{3}, \frac{5}{6}, \frac{5}{2}$ 79. This is not true. The absolute value of 0 is 0, and 0 is not positive. A more accurate way of describing absolute value is to say that *absolute value is never negative,* or *absolute value is always nonnegative.* 80. true

▲ *For each statement give a pair of values for a and b that make it true, and then give a pair of values that make it false.*

69. $|a + b| = |a - b|$

70. $|a - b| = |b - a|$

71. $|a + b| = -|a + b|$

72. $|-(a + b)| = -(a + b)$

▲ *Give three numbers between −6 and 6 that satisfy each given condition.*

73. Positive real numbers but not integers

74. Real numbers but not positive numbers

75. Real numbers but not whole numbers

76. Rational numbers but not integers

77. Real numbers but not rational numbers

78. Rational numbers but not negative numbers

79. Students often say "Absolute value is always positive." Is this true? If not, explain why.

80. True or false: If a is negative, $|a| = -a$.

1.5 Addition and Subtraction of Real Numbers

OBJECTIVES

1. Add two numbers with the same sign.
2. Add positive and negative numbers.
3. Use the definition of subtraction.
4. Use the order of operations with real numbers.
5. Interpret words and phrases involving addition and subtraction.
6. Use signed numbers to interpret data.

FOR EXTRA HELP

SSG Sec. 1.5
SSM Sec. 1.5
Pass the Test Software
InterAct Math Tutorial Software
Video 2

In this section and the next section, we extend the rules for operations with positive numbers to the negative numbers, beginning with addition and subtraction.

OBJECTIVE 1 **Add two numbers with the same sign.** A number line can be used to explain addition of real numbers.

EXAMPLE 1 Adding Positive Numbers on a Number Line

Use a number line to find the sum $2 + 3$.

Add the positive numbers 2 and 3 on the number line by starting at 0 and drawing an arrow two units to the *right,* as shown in Figure 12. This arrow represents the number 2 in the sum $2 + 3$. Then, from the right end of this arrow draw another arrow three units to the right. The number below the end of this second arrow is 5, so $2 + 3 = 5$.

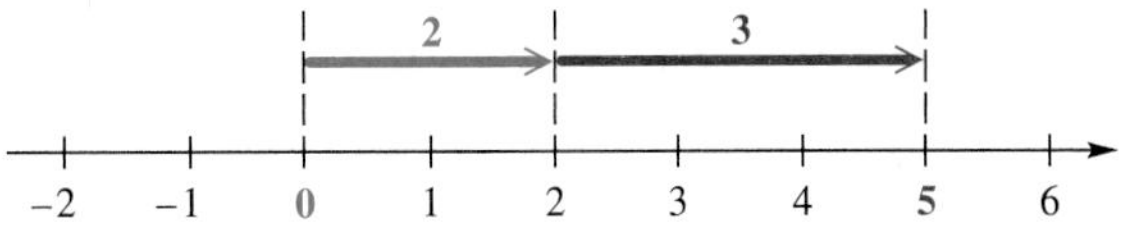

Figure 12

EXAMPLE 2 Adding Negative Numbers on a Number Line

Use a number line to find the sum $-2 + (-4)$. (Parentheses are placed around the -4 to avoid the confusing use of + and − next to each other.)

Add the negative numbers -2 and -4 on the number line by starting at 0 and drawing an arrow two units to the *left,* as shown in Figure 13. The arrow is drawn to the left to represent the addition of a *negative* number. From the left end of this first arrow, draw a second arrow four units to the left. The number below the end of this second arrow is -6, so $-2 + (-4) = -6$.

For Example 1

CHALKBOARD EXAMPLE

Use a number line to find the sum.

$$5 + 3$$

Answer: 8

INSTRUCTOR'S RESOURCES

PTB Sec. 1.5 **ISM** Sec. 1.5
AB Sec. 1.5

TEST GENERATOR

WORLD WIDE WEB
www.LialAlgebra.com

For Example 2

CHALKBOARD EXAMPLE

Use a number line to find the sum.

$$-2 + (-5)$$

Answer: -7

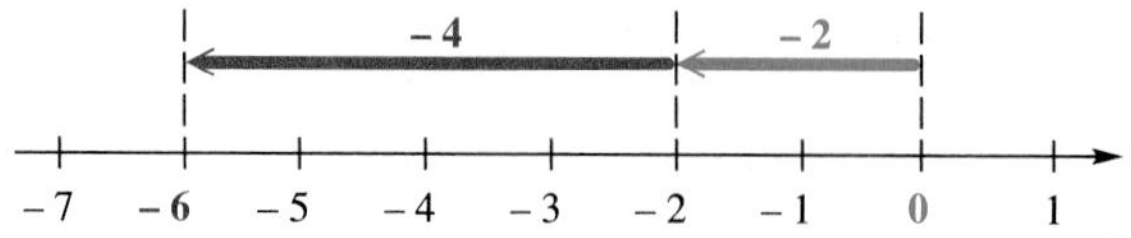

Figure 13

TEACHING TIP To avoid confusion, two operation symbols should not be written successively without using a parenthesis.

In Example 2, the sum of the two negative numbers -2 and -4 is a negative number whose distance from 0 is the sum of the distance of -2 from 0 and the distance of -4 from 0. That is, *the sum of two negative numbers is the negative of the sum of their absolute values.*

$$-2 + (-4) = -(|-2| + |-4|) = -(2 + 4) = -6$$

Adding Numbers with the Same Signs

To add two numbers with the *same* signs, add the absolute values of the numbers. The sum has the same sign as the numbers being added.

CHALKBOARD EXAMPLE

Find the sum.

$$-15 + (-4)$$

Answer: -19

EXAMPLE 3 Adding Two Negative Numbers

Find the sums.

(a) $-2 + (-9) = -(|-2| + |-9|) = -(2 + 9) = -11$

(b) $-8 + (-12) = -20$

(c) $-15 + (-3) = -18$

OBJECTIVE 2 Add positive and negative numbers. We can use a number line to explain the sum of a positive number and a negative number.

CHALKBOARD EXAMPLE

Use a number line to find the sum.

$$6 + (-3)$$

Answer: 3

EXAMPLE 4 Adding Numbers with Different Signs

Use a number line to find the sum $-2 + 5$.

Find the sum $-2 + 5$ on the number line by starting at 0 and drawing an arrow two units to the left. From the left end of this arrow, draw a second arrow five units to the right, as shown in Figure 14. The number below the end of the second arrow is 3, so $-2 + 5 = 3$.

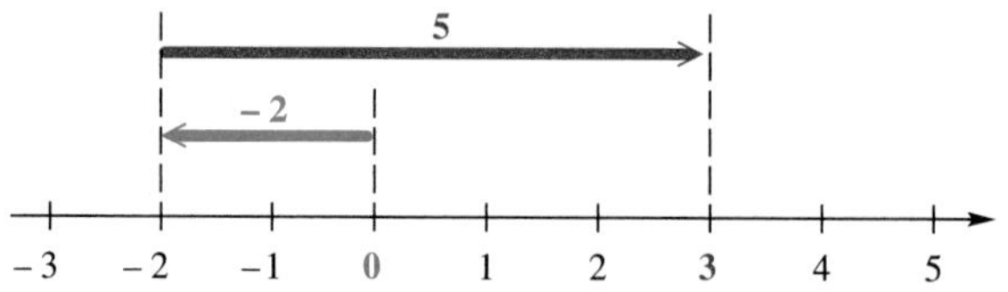

Figure 14

Adding Numbers with Different Signs

To add two numbers with *different* signs, subtract the smaller absolute value from the larger absolute value. The answer has the sign of the number with the larger absolute value.

For example, to add -12 and 5, find their absolute values: $|-12| = 12$ and $|5| = 5$. Then find the difference between these absolute values: $12 - 5 = 7$. Since $|-12| > |5|$, the sum will be negative, so that the final answer is $-12 + 5 = -7$.

While a number line is useful in showing the rules for addition, it is important to be able to do the problems quickly, "in your head."

CHALKBOARD EXAMPLE

Check the answer.

$$\frac{3}{4} + \left(-\frac{11}{8}\right) = -\frac{5}{8}$$

Answer: correct

EXAMPLE 5 Adding Mentally

Check each answer, trying to work the addition mentally. If you get stuck, use a number line.

(a) $7 + (-4) = 3$

(b) $-8 + 12 = 4$

(c) $-\frac{1}{2} + \frac{1}{8} = -\frac{4}{8} + \frac{1}{8} = -\frac{3}{8}$ Remember to get a common denominator first.

(d) $\frac{5}{6} + \left(-\frac{4}{3}\right) = -\frac{1}{2}$

(e) $-4.6 + 8.1 = 3.5$

(f) $-16 + 16 = 0$

(g) $42 + (-42) = 0$

TEACHING TIP Have students read problems to ensure understanding of symbols. Example 5(a), for instance, should be read "seven plus *negative* four equals three" (and not "seven plus minus four equals three").

Parts (f) and (g) in Example 5 suggest that the sum of a number and its additive inverse is 0. This is always true, and this property is discussed further in Section 1.7.

The rules for adding signed numbers are summarized as follows.

Adding Signed Numbers

Like signs Add the absolute values of the numbers. The sum has the same sign as the given numbers.

Unlike signs Find the difference between the larger absolute value and the smaller. The sum has the sign of the number with the larger absolute value.

OBJECTIVE 3 Use the definition of subtraction. We can illustrate subtraction of 4 from 7, written $7 - 4$, using a number line. As seen in Figure 15, we begin at 0 and draw an arrow seven units to the right. From the right end of this arrow, draw an arrow four units to the left. The number at the end of the second arrow shows that $7 - 4 = 3$.

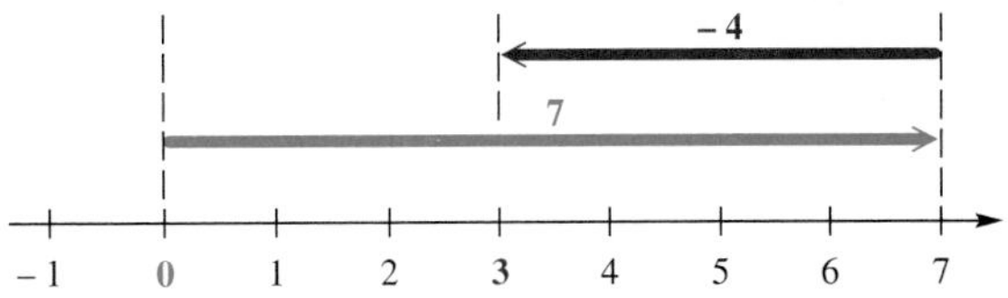

Figure 15

The procedure used to find $7 - 4$ is exactly the same procedure that would be used to find $7 + (-4)$, so that

$$7 - 4 = 7 + (-4).$$

This suggests that *subtraction* of a positive number from a larger positive number is the same as *adding* the additive inverse of the smaller number to the larger. This result leads to the definition of subtraction for all real numbers.

Definition of Subtraction

For any real numbers x and y,

$$x - y = x + (-y).$$

That is, to *subtract* y from x, *add the additive inverse* (or opposite) of y to x. This definition gives the following procedure for subtracting signed numbers.

Subtracting Signed Numbers

Step 1 Change the subtraction symbol to the addition symbol.

Step 2 Change the sign of the number being subtracted.

Step 3 Add.

CHALKBOARD EXAMPLE

Subtract $-8 - (-12)$.

Answer: 4

EXAMPLE 6 Using the Definition of Subtraction

Subtract.

No change ↓; Change − to +. ↓; Additive inverse of 3 ↓

(a) $12 - 3 = 12 + (-3) = 9$

(b) $5 - 7 = 5 + (-7) = -2$

No change ↓; Change − to +. ↓; Additive inverse of −5 ↓

(c) $-3 - (-5) = -3 + (5) = 2$

(d) $-6 - 9 = -6 + (-9) = -15$

(e) $8 - (-4) = 8 + (4) = 12$

TEACHING TIP Read Example 6(c) as "negative three minus negative five."

OBJECTIVE 4 Use the order of operations with real numbers. As before, with problems that have grouping symbols, first do any operations inside the parentheses and brackets. Work from the inside out.

CHALKBOARD EXAMPLE

Work the problem.

$$6 + [(-1 - 4) - 2]$$

Answer: -1

EXAMPLE 7 Adding and Subtracting with Grouping Symbols

Perform the indicated operations.

(a)

$-6 - [2 - (8 + 3)] = -6 - [2 - 11]$	Add.
$= -6 - [2 + (-11)]$	Use the definition of subtraction.
$= -6 - (-9)$	Add.
$= -6 + (9)$	Use the definition of subtraction.
$= 3$	Add.

(b)

$$\begin{aligned} 5 + [(-3 - 2) - (4 - 1)] &= 5 + [(-3 + (-2)) - 3] \\ &= 5 + [(-5) - 3] \\ &= 5 + [(-5) + (-3)] \\ &= 5 + (-8) \\ &= -3 \end{aligned}$$

(c) $\frac{2}{3} - \left[\frac{1}{12} - \left(-\frac{1}{4}\right)\right] = \frac{8}{12} - \left[\frac{1}{12} - \left(-\frac{3}{12}\right)\right]$ Get a common denominator.

$= \frac{8}{12} - \left[\frac{1}{12} + \frac{3}{12}\right]$ Use the definition of subtraction.

$= \frac{8}{12} - \frac{4}{12}$ Add.

$= \frac{4}{12}$ Subtract.

$= \frac{1}{3}$ Lowest terms

(d) $|4 - 7| + 2|6 - 3| = |-3| + 2|3|$ Work within absolute value symbols.

$= 3 + 2 \cdot 3$ Evaluate the absolute values.

$= 3 + 6$ Multiply.

$= 9$ Add.

OBJECTIVE 5 Interpret words and phrases involving addition and subtraction. We begin by working with addition words and phrases.

PROBLEM SOLVING

As we mentioned earlier, problem solving often requires translating words and phrases into symbols. The word *sum* is one of the words that indicates addition. The chart below lists some of the words and phrases that also signify addition.

Word or Phrase	*Example*	*Numerical Expression and Simplification*
Sum of	The *sum of* −3 and 4	$-3 + 4 = 1$
Added to	5 *added to* −8	$-8 + 5 = -3$
More than	12 *more than* −5	$-5 + 12 = 7$
Increased by	−6 *increased by* 13	$-6 + 13 = 7$
Plus	3 *plus* 14	$3 + 14 = 17$

CHALKBOARD EXAMPLE

Write a numerical expression for the phrase and simplify the expression.

7 increased by the sum of 8 and −3

Answer: $7 + [8 + (-3)] = 12$

EXAMPLE 8 Interpreting Words and Phrases Involving Addition

Write a numerical expression for each phrase, and simplify the expression.

(a) The sum of −8 and 4 and 6

$$-8 + 4 + 6 = -4 + 6 = 2$$

Add in order from left to right.

(b) 3 more than −5, increased by 12

$$-5 + 3 + 12 = -2 + 12 = 10$$

We now look at how we interpret words and phrases that involve subtraction.

PROBLEM SOLVING

In order to solve problems that involve subtraction, we must be able to interpret key words and phrases that indicate subtraction. *Difference* is one of them. Some of these are given in the chart below.

Word or Phrase	*Example*	*Numerical Expression and Simplification*
Difference between	The *difference between* −3 and −8	$-3 - (-8) = -3 + 8 = 5$
Subtracted from	12 *subtracted from* 18	$18 - 12 = 6$
Less	6 *less* 5	$6 - 5 = 1$
Less than	6 *less than* 5	$5 - 6 = 5 + (-6) = -1$
Decreased by	9 *decreased by* −4	$9 - (-4) = 9 + 4 = 13$
Minus	8 *minus* 5	$8 - 5 = 3$

CAUTION When you are subtracting two numbers, it is important that you write them in the correct order, because, in general, $a - b \neq b - a$. For example, $5 - 3 \neq 3 - 5$. For this reason, *think carefully before interpreting an expression involving subtraction.* (This difficulty did not arise for addition.)

CHALKBOARD EXAMPLE

Write a numerical expression for the phrase and simplify the expression.

−2 subtracted from the sum of 4 and −4

Answer: $[4 + (-4)] - (-2) = 2$

EXAMPLE 9 Interpreting Words and Phrases Involving Subtraction

Write a numerical expression for each phrase, and simplify the expression.

(a) The **difference between** −8 and 5

It is conventional to write the numbers in the order they are given when "difference between" is used.

$$-8 - 5 = -8 + (-5) = -13$$

(b) 4 **subtracted from** the sum of 8 and −3

Here addition is also used, as indicated by the word *sum.* First, add 8 and −3. Next, subtract 4 *from* this sum.

$$[8 + (-3)] - 4 = 5 - 4 = 1$$

(c) 4 **less than** −6

Be careful with order here. 4 must be taken *from* −6.

$$-6 - 4 = -6 + (-4) = -10$$

Notice that "4 less than −6" differs from "4 *is less than* −6." The second of these is symbolized as $4 < -6$ (which is a false statement).

(d) 8, **decreased by** 5 **less than** 12

First, write "5 less than 12" as $12 - 5$. Next, subtract $12 - 5$ from 8.

$$8 - (12 - 5) = 8 - 7 = 1$$

OBJECTIVE 6 **Use signed numbers to interpret data.** The Producer Price Index is the oldest continuous statistical series published by the Bureau of Labor Statistics. It measures the average changes in prices received by producers of all commodities produced in the United States. The next example shows how signed numbers can be used to interpret such data.

CHALKBOARD EXAMPLE

Refer to Figure 16 and use a signed number to represent the change in the PPI from 1989 to 1990.

Answer: .1

EXAMPLE 10 Using a Signed Number in Data Interpretation

The bar graph in Figure 16 gives the Producer Price Index (PPI) for construction materials between 1985 and 1990.

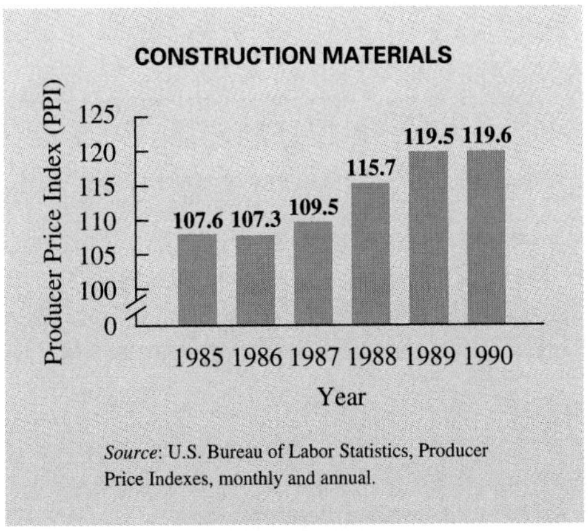

Figure 16

(a) Use a signed number to represent the change in the PPI from 1987 to 1988.

To find this change, we start with the index number from 1988 and subtract from it the index number from 1987.

$$\underbrace{115.7}_{\text{The 1988 index}} - \underbrace{109.5}_{\text{The 1987 index}} = \underbrace{6.2}_{\text{A positive number indicates an increase.}}$$

(b) Use a signed number to represent the change in the PPI from 1985 to 1986.

Use the same procedure as in part (a).

$$\underbrace{107.3}_{\text{The 1986 index}} - \underbrace{107.6}_{\text{The 1985 index}} = 107.3 + (-107.6) = \underbrace{-.3}_{\text{A negative number indicates a decrease.}}$$

CHALKBOARD EXAMPLE

The highest elevation in Argentina is Mt. Aconcagua, which is 6960 meters above sea level. The lowest point in Argentina is the Valdes Peninsula, 40 meters below sea level. Find the difference between the highest and lowest elevations.

Answer: 7000 meters

EXAMPLE 11 Solving a Problem Involving Subtraction

The record high temperature in the United States was 134° Fahrenheit, recorded at Death Valley, California, in 1913. The record low was −80°F, at Prospect Creek, Alaska, in 1971. See Figure 17. What is the difference between these highest and lowest temperatures? (*Source: The World Almanac and Book of Facts,* 1998.)

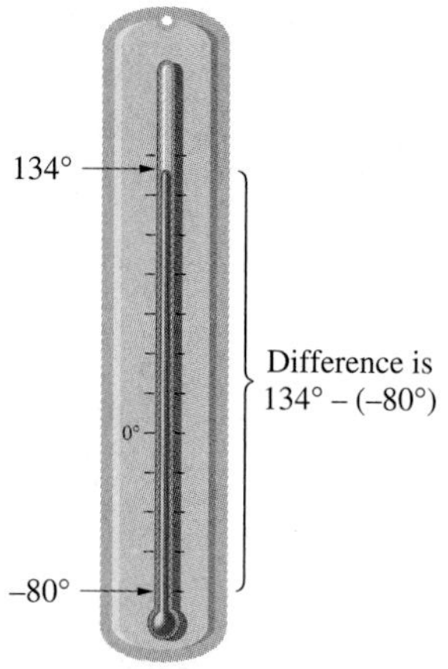

Figure 17

We must subtract the lowest temperature from the highest temperature.

$$134 - (-80) = 134 + 80 \quad \text{Use the definition of subtraction.}$$
$$= 214 \quad \text{Add.}$$

The difference between the two temperatures is 214°F.

1.5 EXERCISES

Fill in each blank with the correct response.

1. The sum of two negative numbers will always be a __________ number. (positive/negative)

2. The sum of a number and its opposite will always be __________.

3. To simplify the expression $8 + [-2 + (-3 + 5)]$, I should begin by adding ______ and ______, according to the rule for order of operations.

4. If I am adding a positive number and a negative number, and the negative number has the larger absolute value, the sum will be a __________ number. (positive/negative)

5. Explain in words how to add signed numbers. Consider the various cases and give examples.

6. Explain in words how to subtract signed numbers.

Find each sum or difference. See Examples 1–7.

7. $6 + (-4)$

8. $12 + (-9)$

9. $7 + (-10)$

10. $4 + (-8)$

11. $-7 + (-3)$

12. $-11 + (-4)$

13. $-10 + (-3)$

14. $-16 + (-7)$

15. $-12.4 + (-3.5)$

16. $-21.3 + (-2.5)$

17. $-8 + 7$

18. $-12 + 10$

19. $5 + [14 + (-6)]$

20. $7 + [3 + (-14)]$

21. $4 - 7$

22. $8 - 13$

1. negative **2.** zero (0) **3.** -3; 5 **4.** negative **5.** To add two numbers with the same sign, add their absolute values and keep the same sign for the sum. For example, $3 + 4 = 7$ and $-3 + (-4) = -7$. To add two numbers with different signs, subtract the smaller absolute value from the larger absolute value, and use the sign of the number with the larger absolute value. For example, $6 + (-4) = 2$ and $(-6) + 4 = -2$. **6.** To subtract $a - b$, write as $a + (-b)$ and follow the rules for addition. **7.** 2 **8.** 3 **9.** -3 **10.** -4 **11.** -10 **12.** -15 **13.** -13 **14.** -23 **15.** -15.9 **16.** -23.8 **17.** -1 **18.** -2 **19.** 13 **20.** -4 **21.** -3 **22.** -5

23. −4 24. −5 25. −10 26. −17 27. −16 28. −29 29. 11 30. 15 31. 19 32. 16 33. −4 34. −2 35. 5 36. 13 37. 0 38. 3 39. $\frac{3}{4}$ 40. $\frac{5}{3}$ 41. −8 42. −7 43. $\frac{15}{8}$ 44. $\frac{43}{40}$ 45. −6.3 46. −14.2 47. −24 48. 8 49. −16 50. −18 51. no 52. They must be nonzero opposites. 53. Answers will vary. One example is −8 − (−2) = −6. 54. Answers will vary. One example is −2 − (−8) = 6. 55. −1 56. −3 57. $\frac{17}{9}$ or $1\frac{8}{9}$ 58. $\frac{19}{10}$ or $1\frac{9}{10}$ 59. −5 + 12 + 6; 13 60. −3 + 5 + (−12); −10 61. [−19 + (−4)] + 14; −9 62. (−18 + 11) + (−2); −9 63. [−4 + (−10)] + 12; −2 64. [−7 + (−13)] + 14; −6 65. [8 + (−18)] + 4; −6 66. [−4 + (−6)] + 10; 0 67. 4 − (−8); 12 68. 7 − (−14); 21

23. $6 - 10$

24. $9 - 14$

25. $-7 - 3$

26. $-12 - 5$

27. $-10 - 6$

28. $-13 - 16$

29. $7 - (-4)$

30. $9 - (-6)$

31. $6 - (-13)$

32. $13 - (-3)$

33. $-7 - (-3)$

34. $-8 - (-6)$

35. $3 - (4 - 6)$

36. $6 - (7 - 14)$

37. $-3 - (6 - 9)$

38. $-4 - (5 - 12)$

39. $\frac{1}{2} - \left(-\frac{1}{4}\right)$

40. $\frac{1}{3} - \left(-\frac{4}{3}\right)$

41. $-8 + [3 + (-1) + (-2)]$

42. $-7 + [5 + (-8) + 3]$

43. $\frac{5}{8} - \left(-\frac{1}{2} - \frac{3}{4}\right)$

44. $\frac{9}{10} - \left(\frac{1}{8} - \frac{3}{10}\right)$

45. $[(-3.1) - 4.5] - (.8 - 2.1)$

46. $[(-7.8) - 9.3] - (.6 - 3.5)$

47. $[-5 + (-7)] + [-4 + (-9)] + [13 + (-12)]$

48. $[-3 + (-11)] + [13 + (-3)] + [19 + (-7)]$

49. $-4 + [(-6 - 9) - (-7 + 4)]$

50. $-8 + [(-3 - 10) - (-4 + 1)]$

51. Is it possible to add a negative number to another negative number and get a positive number? If so, give an example.

52. Under what conditions will the sum of a positive number and a negative number be a number which is neither negative nor positive?

53. Make up a subtraction problem so that the difference between two negative numbers is a negative number.

54. Make up a subtraction problem so that the difference between two negative numbers is a positive number.

Simplify each expression involving absolute value. Remember that absolute value bars serve as grouping symbols. See Example 7(d).

55. $|3 - 8| - |-2 + 8|$

56. $|-2 - 7| - |9 - (-3)|$

57. $\left|\frac{2}{3} - \frac{7}{3}\right| + 2\left|-\frac{4}{9} + \frac{5}{9}\right|$

58. $\left|-\frac{1}{5} + \frac{3}{5}\right| + 3\left|\frac{3}{10} - \frac{8}{10}\right|$

Write a numerical expression for each phrase and simplify. See Example 8.

59. The sum of −5 and 12 and 6

60. The sum of −3 and 5 and −12

61. 14 added to the sum of −19 and −4

62. −2 added to the sum of −18 and 11

63. The sum of −4 and −10, increased by 12

64. The sum of −7 and −13, increased by 14

65. 4 more than the sum of 8 and −18

66. 10 more than the sum of −4 and −6

Write a numerical expression for each phrase and simplify. See Example 9.

67. The difference between 4 and −8

68. The difference between 7 and −14

69. $-2 - 8; -10$
70. $-13 - 9; -22$
71. $[9 + (-4)] - 7; -2$
72. $[12 + (-7)] - 14; -9$
73. $[8 - (-5)] - 12; 1$
74. $[9 - (-2)] - 19; -8$
75. $13.2 billion 76. $8.6 billion
77. −$26.0 billion
78. −$11.3 billion
79. 50,395 feet 80. 37,486 feet
81. 1345 feet 82. 8274 feet

69. 8 less than -2

70. 9 less than -13

71. The sum of 9 and -4, decreased by 7

72. The sum of 12 and -7, decreased by 14

73. 12 less than the difference between 8 and -5

74. 19 less than the difference between 9 and -2

The bar graph shows the federal budget outlays for national defense for the years 1985 through 1994. Use a signed number to represent the change in outlay for each of the following time periods. See Example 10.

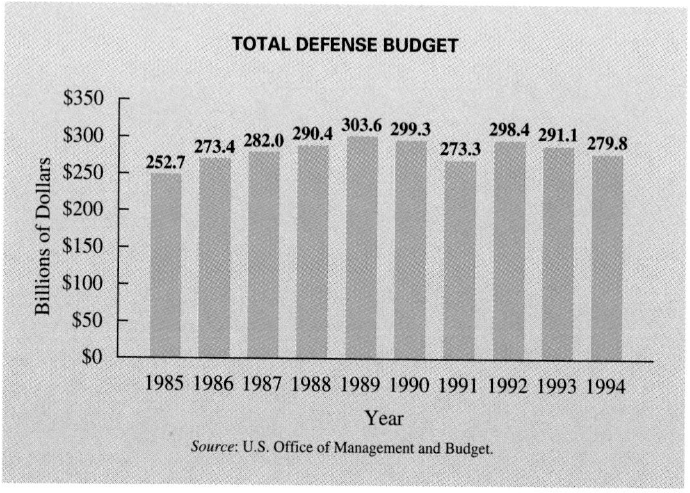

75. 1988 to 1989

76. 1986 to 1987

77. 1990 to 1991

78. 1993 to 1994

The two charts show the heights of some selected mountains and the depths of some selected trenches. Use the information given to find the answers in Exercises 79–82.

Mountain	*Height (in feet)*
Foraker	17,400
Wilson	14,246
Pikes Peak	14,110

Trench	*Depth (in feet, as a negative number)*
Philippine	−32,995
Cayman	−24,721
Java	−23,376

Source: The World Almanac and Book of Facts, 1998.

79. What is the difference between the height of Mt. Foraker and the depth of the Philippine Trench?

80. What is the difference between the height of Pikes Peak and the depth of the Java Trench?

81. How much deeper is the Cayman Trench than the Java Trench?

82. How much deeper is the Philippine Trench than the Cayman Trench?

83. (a) −10 (b) 5 (c) −12 (d) −31 84. (a) −10 (b) −6 (c) 21 (d) −48 85. 45°F 86. 112°F

Use this idea when working Exercises 83 and 84: If a player steals 65 bases in 1990 and 58 in 1991, the change in stolen bases is represented by $58 - 65 = -7$. *His number went* down *by* 7.

83. The bar graph shows the number of bases stolen by baseball player Rickey Henderson. Use a signed number to represent the change in the number of stolen bases from one year to the next. See Example 10. (*Source:* David S. Neft and Richard M. Cohen, *The Sports Encyclopedia: Baseball 1997.*)

(a) From 1991 to 1992 **(b)** From 1992 to 1993
(c) From 1989 to 1990 **(d)** From 1993 to 1994

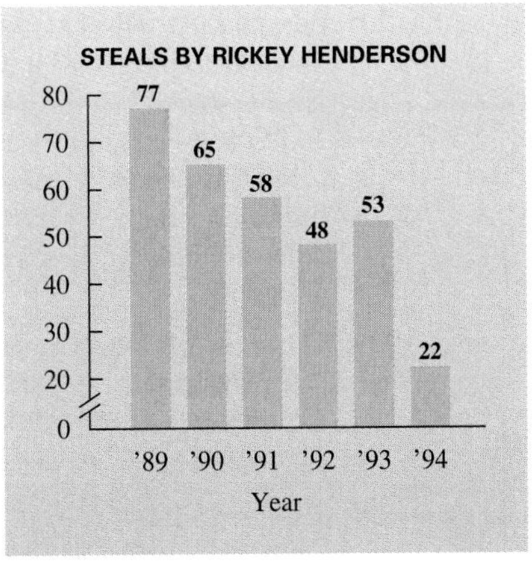

84. The bar graph shows the number of bases stolen by baseball player Kenny Lofton. Use a signed number to represent the change in the number of stolen bases from one year to the next. See Example 10. (*Source:* David S. Neft and Richard M. Cohen, *The Sports Encyclopedia: Baseball 1997.*)

(a) From 1993 to 1994 **(b)** From 1994 to 1995
(c) From 1995 to 1996 **(d)** From 1996 to 1997

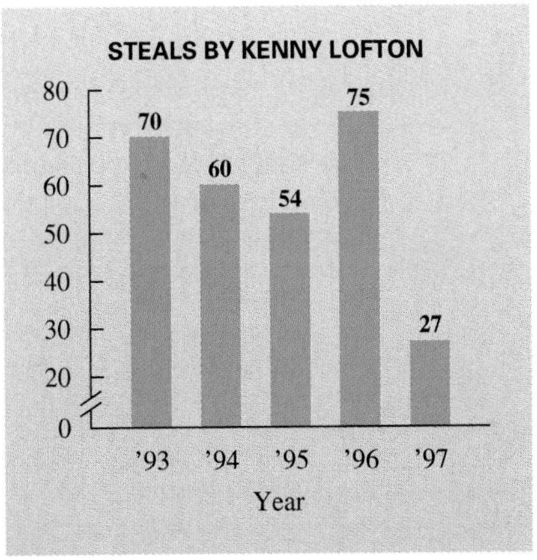

Solve each problem. (Refer to Figure 17 for Exercises 85–87.) See Example 11.

85. On January 23, 1943, the temperature rose 49°F in two minutes in Spearfish, South Dakota. If the starting temperature was −4°F, what was the temperature two minutes later? (*Source: The Guinness Book of World Records.*)

86. The lowest temperature ever recorded in Little Rock, Arkansas, was −5°F. The highest temperature ever recorded there was 117°F more than the lowest. What was this highest temperature? (*Source: The World Almanac and Book of Facts,* 1998.)

87. $-41°F$ **88.** 27 feet
89. 14,776 feet **90.** 543 feet
91. 365 pounds **92.** 37 yards
93. $323.83 **94.** 26.90 inches

87. The coldest temperature recorded in Chicago, Illinois, was $-27°F$ in 1985. The record low in Huron, South Dakota, was set in 1994 and was 14°F lower than $-27°F$. What was the record low in Huron? (*Source: The World Almanac and Book of Facts,* 1998.)

88. No one knows just why humpback whales love to heave their 45-ton bodies out of the water, but leap they do. (This activity is called *breaching.*) Mark and Debbie, two researchers based on the island of Maui, noticed that one of their favorite whales, "Pineapple," leaped 15 feet above the surface of the ocean while her mate cruised 12 feet below the surface. What is the difference between these two distances?

89. The top of Mt. Whitney, visible from Death Valley, has an altitude of 14,494 feet above sea level. The bottom of Death Valley is 282 feet below sea level. Using zero as sea level, find the difference between these two elevations. (*Source: The World Almanac and Book of Facts,* 1998.)

90. The highest point in Louisiana is Driskill Mountain, at an altitude of 535 feet. The lowest point is at Spanish Fort, 8 feet below sea level. Using zero as sea level, find the difference between these two elevations. (*Source: The World Almanac and Book of Facts,* 1998.)

91. A female polar bear weighed 660 pounds when she entered her winter den. She lost 45 pounds during each of the first two months of hibernation, and another 205 pounds before leaving the den with her two cubs in March. How much did she weigh when she left the den?

92. On three consecutive passes, Troy Aikman of the Dallas Cowboys passed for a gain of 6 yards, was sacked for a loss of 12 yards, and passed for a gain of 43 yards. What positive or negative number represents the total net yardage for the plays?

93. Kim Falgout owes $870.00 on her Master Card account. She returns two items costing $35.90 and $150.00 and receives credits for these on the account. Next, she makes a purchase of $82.50, and then two more purchases of $10.00 each. She makes a payment of $500.00. She then incurs a finance charge of $37.23. How much does she still owe?

94. A welder working with stainless steel must use precise measurements. Suppose a welder attaches two pieces of steel that are each 3.60 inches in length, and then attaches an additional three pieces that are each 9.10 inches long. She finally cuts off a piece that is 7.60 inches long. Find the length of the welded piece of steel.

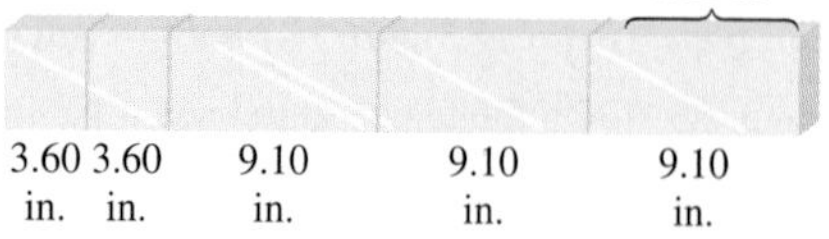

1.6 Multiplication and Division of Real Numbers

OBJECTIVES

1. Find the product of a positive number and a negative number.
2. Find the product of two negative numbers.
3. Identify factors of integers.
4. Use the reciprocal of a number to apply the definition of division.
5. Use the order of operations when multiplying and dividing signed numbers.
6. Evaluate expressions involving variables.
7. Interpret words and phrases involving multiplication and division.
8. Translate simple sentences into equations.

FOR EXTRA HELP

SSG Sec. 1.6
SSM Sec. 1.6
Pass the Test Software
InterAct Math Tutorial Software
Video 2

In this section we learn how to multiply with positive and negative numbers. We already know how to multiply positive numbers and that the product of two positive numbers is positive. We also know that the product of 0 and any positive number is 0, so we extend that property to all real numbers.

Multiplication by Zero

For any real number x, $x \cdot 0 = 0$.

OBJECTIVE 1 **Find the product of a positive number and a negative number.** In order to define the product of a positive and a negative number so that the result is consistent with the multiplication of two positive numbers, look at the following pattern.

$$\begin{aligned} 3 \cdot 5 &= 15 \\ 3 \cdot 4 &= 12 \\ 3 \cdot 3 &= 9 \\ 3 \cdot 2 &= 6 \\ 3 \cdot 1 &= 3 \\ 3 \cdot 0 &= 0 \\ 3 \cdot (-1) &= ? \end{aligned}$$

Numbers decrease by 3.

What should $3(-1)$ equal? The product $3(-1)$ represents the sum

$$-1 + (-1) + (-1) = -3,$$

so the product should be -3. Also,

$$3(-2) = -2 + (-2) + (-2) = -6$$

and

$$3(-3) = -3 + (-3) + (-3) = -9.$$

These results maintain the pattern in the list, which suggests the following rule.

Multiplying Numbers with Different Signs

For any positive real numbers x and y,

$$x(-y) = -(xy) \quad \text{and} \quad (-x)y = -(xy).$$

That is, the product of two numbers with opposite signs is negative.

CHALKBOARD EXAMPLE

Find the product.

$3(-5)$

Answer: -15

EXAMPLE 1 **Multiplying a Positive Number and a Negative Number**

Find the products using the multiplication rule given above.

(a) $8(-5) = -(8 \cdot 5) = -40$

(b) $5(-4) = -(5 \cdot 4) = -20$

(c) $(-7)(2) = -(7 \cdot 2) = -14$

(d) $(-9)(3) = -(9 \cdot 3) = -27$

INSTRUCTOR'S RESOURCES

PTB Sec. 1.6 **ISM** Sec. 1.6
AB Sec. 1.6

TEST GENERATOR

WORLD WIDE WEB
www.LialAlgebra.com

OBJECTIVE 2 Find the product of two negative numbers. The product of two positive numbers is positive, and the product of a positive and a negative number is negative. What about the product of two negative numbers? Look at another pattern.

$$\begin{aligned}(-5)(4) &= -20\\(-5)(3) &= -15\\(-5)(2) &= -10\\(-5)(1) &= -5\\(-5)(0) &= 0\\(-5)(-1) &= ?\end{aligned}$$

Numbers increase by 5.

The numbers on the left of the equals sign (in color) decrease by 1 for each step down the list. The products on the right increase by 5 for each step down the list. To maintain this pattern, $(-5)(-1)$ should be 5 more than $(-5)(0)$, or 5 more than 0, so

$$(-5)(-1) = 5.$$

The pattern continues with

$$\begin{aligned}(-5)(-2) &= 10\\(-5)(-3) &= 15\\(-5)(-4) &= 20\\(-5)(-5) &= 25,\end{aligned}$$

and so on. This pattern suggests the next rule.

Multiplying Two Negative Numbers

For any positive real numbers x and y,

$$(-x)(-y) = xy.$$

The product of two negative numbers is positive.

CHALKBOARD EXAMPLE

Find the product.

$(-8)(-5)$

Answer: 40

EXAMPLE 2 Multiplying Two Negative Numbers

Find the products using the multiplication rule given above.

(a) $(-9)(-2) = 9 \cdot 2 = 18$

(b) $(-6)(-12) = 6 \cdot 12 = 72$

(c) $(-8)(-1) = 8 \cdot 1 = 8$

(d) $(-15)(-2) = 15 \cdot 2 = 30$

A summary of multiplying signed numbers is given here.

Multiplying Signed Numbers

The product of two numbers having the *same* sign is *positive,* and the product of two numbers having *different* signs is *negative.*

OBJECTIVE 3 Identify factors of integers. In Section 1.1 the definition of a *factor* was given for whole numbers. (For example, since $9 \cdot 5 = 45$, both 9 and 5 are factors of 45.) The definition can now be extended to integers.

If the product of two integers is a third integer, then each of the two integers is a *factor* of the third. For example, $(-3)(-4) = 12$, so -3 and -4 are both factors of 12. The factors of 12 are $-12, -6, -4, -3, -2, -1, 1, 2, 3, 4, 6$, and 12.

The following chart shows several integers and the factors of those integers.

Integer	*Factors*
18	−18, −9, −6, −3, −2, −1, 1, 2, 3, 6, 9, 18
20	−20, −10, −5, −4, −2, −1, 1, 2, 4, 5, 10, 20
15	−15, −5, −3, −1, 1, 3, 5, 15
7	−7, −1, 1, 7
1	−1, 1

OBJECTIVE 4 Use the reciprocal of a number to apply the definition of division. In Section 1.5 we saw that the difference between two numbers is found by adding the additive inverse of the second number to the first. Similarly, the *quotient* of two numbers is found by *multiplying* by the *reciprocal,* or *multiplicative inverse.* By definition, since

$$8 \cdot \frac{1}{8} = \frac{8}{8} = 1 \qquad \text{and} \qquad \frac{5}{4} \cdot \frac{4}{5} = \frac{20}{20} = 1,$$

the reciprocal or multiplicative inverse of 8 is $\frac{1}{8}$, and of $\frac{5}{4}$ is $\frac{4}{5}$.

Reciprocal or Multiplicative Inverse

Pairs of numbers whose product is 1 are **reciprocals, or multiplicative inverses,** of each other.

The following chart shows several numbers and their multiplicative inverses.

Number	*Multiplicative Inverse (Reciprocal)*
4	$\frac{1}{4}$
$.3 = \frac{3}{10}$	$\frac{10}{3}$
-5	$\frac{1}{-5}$ or $-\frac{1}{5}$
$-\frac{5}{8}$	$-\frac{8}{5}$
0	None
1	1
−1	−1

Why is there no multiplicative inverse for the number 0? Suppose that k is to be the multiplicative inverse of 0. Then $k \cdot 0$ should equal 1. But $k \cdot 0 = 0$ for any number k. Since there is no value of k that is a solution of the equation $k \cdot 0 = 1$, the following statement can be made.

0 has no multiplicative inverse.

By definition, the quotient of x and y is the product of x and the multiplicative inverse of y.

Definition of Division

For any real numbers x and y, with $y \neq 0$, $\quad \frac{x}{y} = x \cdot \frac{1}{y}.$

TEACHING TIP It may be mentioned at this point that in an expression such as $1/(x-2)$, x cannot have the value of 2, because then the denominator would equal zero and the fraction would be undefined. Show also that a calculator returns an error message when $\frac{1}{0}$ is entered.

The definition of division indicates that y, the number to divide by, cannot be 0. The reason is that 0 has no multiplicative inverse, so that $\frac{1}{0}$ is not a number. **Because 0 has no multiplicative inverse, *division by 0 is undefined.* If a division problem turns out to involve division by 0, write "undefined."**

NOTE While division by zero $\left(\frac{a}{0}\right)$ is undefined, we may divide 0 by any nonzero number. In fact, if $a \neq 0$,

$$\frac{0}{a} = 0.$$

Since division is defined in terms of multiplication, all the rules for multiplying signed numbers also apply to dividing them.

CHALKBOARD EXAMPLE

Find the quotient using the definition of division.

$$\frac{-36}{6}$$

Answer: -6

EXAMPLE 3 Using the Definition of Division

Find the quotients using the definition of division.

(a) $\frac{12}{3} = 12 \cdot \frac{1}{3} = 4$

(b) $\frac{-10}{2} = -10 \cdot \frac{1}{2} = -5$

(c) $\frac{-14}{-7} = -14\left(\frac{1}{-7}\right) = 2$

(d) $-\frac{2}{3} \div \left(-\frac{4}{5}\right) = -\frac{2}{3} \cdot \left(-\frac{5}{4}\right) = \frac{5}{6}$

(e) $\frac{0}{13} = 0\left(\frac{1}{13}\right) = 0 \qquad \frac{0}{a} = 0 \quad (a \neq 0)$

(f) $\frac{-10}{0}$ Undefined

When dividing fractions, multiplying by the reciprocal works well. However, using the definition of division directly with integers is awkward. It is easier to divide in the usual way, then determine the sign of the answer. The following rule for division can be used instead of multiplying by the reciprocal.

Dividing Signed Numbers

The quotient of two numbers having the same sign is positive; the quotient of two numbers having *different* signs is *negative.*

Note that these are the same as the rules for multiplication.

CHALKBOARD EXAMPLE

Find the quotient.

$$\frac{-16}{-2}$$

Answer: 8

EXAMPLE 4 Dividing Signed Numbers

Find the quotients.

(a) $\dfrac{8}{-2} = -4$ **(b)** $\dfrac{-4.5}{-.09} = 50$

(c) $-\dfrac{1}{8} \div \left(-\dfrac{3}{4}\right) = -\dfrac{1}{8} \cdot \left(-\dfrac{4}{3}\right) = \dfrac{1}{6}$

From the definitions of multiplication and division of real numbers,

$$\frac{-40}{8} = -40 \cdot \frac{1}{8} = -5, \text{ and } \frac{40}{-8} = 40\left(\frac{1}{-8}\right) = -5, \text{ so that}$$

$$\frac{-40}{8} = \frac{40}{-8}.$$

Based on this example, the quotient of a positive and a negative number can be expressed in any of the following three forms.

For any positive real numbers x and y, $\dfrac{-x}{y} = \dfrac{x}{-y} = -\dfrac{x}{y}$.

The quotient of two negative numbers can be expressed as a quotient of two positive numbers.

For any positive real numbers x and y, $\dfrac{-x}{-y} = \dfrac{x}{y}$.

OBJECTIVE 5 Use the order of operations when multiplying and dividing signed numbers.

CHALKBOARD EXAMPLE

Perform the indicated operations.

$$\frac{-6(-8) + (-3)9}{(-2)[4 - (-3)]}$$

Answer: $-\dfrac{3}{2}$

EXAMPLE 5 Using the Order of Operations

Perform the indicated operations.

(a) $(-9)(2) - (-3)(2)$

First find all products, working from left to right.

$$(-9)(2) - (-3)(2) = \mathbf{-18 - (-6)}$$

Now perform the subtraction.

$$-18 - (-6) = -18 + 6 = -12$$

(b) $(-6)(-2) - (3)(-4) = 12 - (-12) = 12 + 12 = 24$

(c) $-5(-2 - 3) = -5(-5) = 25$

(d) $\dfrac{5(-2) - (3)(4)}{2(1 - 6)}$

Simplify the numerator and denominator separately. Then write in lowest terms.

$$\frac{5(-2) - (3)(4)}{2(1 - 6)} = \frac{-10 - 12}{2(-5)} = \frac{-22}{-10} = \frac{11}{5}$$

We now summarize the rules for operations with signed numbers.

Operations with Signed Numbers

Addition

Like signs Add the absolute values of the numbers. The sum has the same sign as the numbers.

Unlike signs Subtract the number with the smaller absolute value from the one with the larger. Give the sum the sign of the number having the larger absolute value.

Subtraction

Add the additive inverse, or opposite, of the second number.

Multiplication and Division

Like signs The product or quotient of two numbers with like signs is positive.

Unlike signs The product or quotient of two numbers with unlike signs is negative.

Division by 0 is undefined.

0 divided by a nonzero number equals 0.

TEACHING TIP Stress the difference between $\frac{0}{3}$ and $\frac{3}{0}$.

OBJECTIVE 6 **Evaluate expressions involving variables.** The next examples show numbers substituted for variables where the rules for multiplying and dividing signed numbers must be used.

CHALKBOARD EXAMPLE

Evaluate $2x^2 - 4y^2$ if $x = -2$ and $y = -3$.

Answer: -28

EXAMPLE 6 **Evaluating an Expression for Numerical Values**

Evaluate each expression, given that $x = -1$, $y = -2$, and $m = -3$.

(a) $(3x + 4y)(-2m)$

First substitute the given values for the variables. Then use the order of operations to find the value of the expression.

$$\begin{aligned} (3x + 4y)(-2m) &= [3(-1) + 4(-2)][-2(-3)] && \text{Put parentheses around the number for each variable.} \\ &= [-3 + (-8)][6] && \text{Find the products.} \\ &= (-11)(6) && \text{Add inside the parentheses.} \\ &= -66 && \text{Multiply.} \end{aligned}$$

TEACHING TIP At this point it may be a good idea to review the order of operations.

(b) $2x^2 - 3y^2$

Use parentheses as shown.

$$\begin{aligned} 2(-1)^2 - 3(-2)^2 &= 2(1) - 3(4) && \text{Apply the exponents.} \\ &= 2 - 12 && \text{Multiply.} \\ &= -10 && \text{Subtract.} \end{aligned}$$

(c) $\dfrac{4y^2 + x}{m}$

$$\frac{4(-2)^2+(-1)}{-3} = \frac{4(4) + (-1)}{-3} \quad \text{Apply the exponent.}$$

$$= \frac{16 + (-1)}{-3} \quad \text{Multiply.}$$

$$= \frac{15}{-3} \quad \text{Add.}$$

$$= -5 \quad \text{Divide.}$$

Notice how the fraction bar was used as a grouping symbol.

OBJECTIVE 7 **Interpret words and phrases involving multiplication and division.** Just as there are words and phrases that indicate addition and subtraction, certain ones also indicate multiplication and division.

PROBLEM SOLVING

The word *product* refers to multiplication. The chart below gives other key words and phrases that indicate multiplication.

Word or Phrase	*Example*	*Numerical Expression and Simplification*
Product of	The **product of** −5 and −2	$(-5)(-2) = 10$
Times	13 ***times*** −4	$13(-4) = -52$
Twice (meaning "2 times")	***Twice*** 6	$2(6) = 12$
Of (used with fractions)	$\frac{1}{2}$ ***of*** 10	$\frac{1}{2}(10) = 5$
Percent of	12% ***of*** −16	$.12(-16) = -1.92$

CHALKBOARD EXAMPLE

Write a numerical expression for the phrase and simplify.

Three-fifths of the sum of 2 and −7

Answer: $\frac{3}{5}[2 + (-7)] = -3$

EXAMPLE 7 **Interpreting Words and Phrases Involving Multiplication**

Write a numerical expression for each phrase and simplify. Use the order of operations.

(a) The **product of** 12 and the sum of 3 and −6
Here 12 is multiplied by "the sum of 3 and −6."

$$12[3 + (-6)] = 12(-3) = -36$$

(b) **Twice** the difference between 8 and −4

$$2[8 - (-4)] = 2[8 + 4] = 2(12) = 24$$

(c) Two-thirds **of** the sum of −5 and −3

$$\frac{2}{3}[-5 + (-3)] = \frac{2}{3}[-8] = -\frac{16}{3}$$

(d) 15% **of** the difference between 14 and −2
Remember that 15% = .15.

$$.15[14 - (-2)] = .15(14 + 2) = .15(16) = 2.4$$

PROBLEM SOLVING

The word *quotient* refers to the answer in a division problem. In algebra, quotients are usually represented with a fraction bar; the symbol ÷ is seldom used. When translating applied problems involving division, use a fraction bar. The chart gives some key phrases associated with division.

Phrase	*Example*	*Numerical Expression and Simplification*
Quotient of	The *quotient of* −24 and 3	$\frac{-24}{3} = -8$
Divided by	−16 *divided by* −4	$\frac{-16}{-4} = 4$
Ratio of	The *ratio of* 2 to 3	$\frac{2}{3}$

It is customary to write the first number named as the numerator and the second as the denominator when interpreting a phrase involving division, as shown in the next example.

CHALKBOARD EXAMPLE

Write a numerical expression for the phrase and simplify the expression.

The product of −9 and 2, divided by the difference between 5 and −1

Answer: $\frac{(-9)(2)}{5-(-1)} = -3$

EXAMPLE 8 Interpreting Words and Phrases Involving Division

Write a numerical expression for each phrase, and simplify the expression.

(a) The **quotient of** 14 and the sum of −9 and 2

"Quotient" indicates division. The number 14 is the numerator and "the sum of −9 and 2" is the denominator.

$$\frac{14}{-9+2} = \frac{14}{-7} = -2$$

(b) The product of 5 and −6, **divided by** the difference between −7 and 8

The numerator of the fraction representing the division is obtained by multiplying 5 and −6. The denominator is found by subtracting −7 and 8.

$$\frac{5(-6)}{-7-8} = \frac{-30}{-15} = 2$$

OBJECTIVE 8 Translate simple sentences into equations. In this section and the previous one, important words and phrases involving the four operations of arithmetic have been introduced. We can use these words and phrases to interpret sentences that translate into equations.

CHALKBOARD EXAMPLE

Write the following in symbols, using x as the variable, and find the solution.

The quotient of a number and −2 is 6.

Answer: $\frac{x}{-2} = 6$; −12

EXAMPLE 9 Translating Words Into an Equation

Write the following in symbols, using x as the variable, and guess or use trial and error to find the solution. All solutions come from the list of integers between −12 and 12, inclusive.

(a) Three **times** a number **is** −18.

The word *times* indicates multiplication, and the word *is* translates as the equals sign (=).

$$3x = -18$$

Since the integer between -12 and 12, inclusive, that makes this statement true is -6, the solution of the equation is -6.

(b) The sum of a number and 9 **is** 12.

$$x + 9 = 12$$

Since $3 + 9 = 12$, the solution of this equation is 3.

(c) The difference between a number and 5 **is** 0.

$$x - 5 = 0$$

Since $5 - 5 = 0$, the solution of this equation is 5.

(d) The quotient of 24 and a number **is** -2.

$$\frac{24}{x} = -2$$

Here, x must be a negative number, since the numerator is positive and the quotient is negative. Since $\frac{24}{-12} = -2$, the solution is -12.

CAUTION

It is important to recognize the distinction between the types of problems found in Examples 7 and 8 and Example 9. In Examples 7 and 8, the phrases translate as *expressions,* while in Example 9, the sentences translate as *equations.* Remember that an equation is a sentence with an = sign, while an expression is a phrase.

$$\frac{5(-6)}{-7-8} \quad \text{is an } \textbf{expression,}$$

$$3x = -18 \quad \text{is an } \textbf{equation.}$$

1.6 EXERCISES

1. greater than 0 2. less than 0 3. greater than 0 4. less than 0 5. less than 0 6. less than 0 7. greater than 0 8. less than 0 9. equal to 0 10. less than 0 11. 20 12. 24 13. -28 14. -40 15. 80 16. 240

Fill in each blank with one of the following: greater than 0, less than 0, equal to 0.

1. A positive number is ____________.

2. A negative number is ____________.

3. The product or the quotient of two numbers with the same sign is ____________.

4. The product or the quotient of two numbers with different signs is ____________.

5. If three negative numbers are multiplied together, the product is ____________.

6. If two negative numbers are multiplied and then their product is divided by a negative number, the result is ____________.

7. If a negative number is squared and the result is added to a positive number, the final answer is ____________.

8. The reciprocal of a negative number is ____________.

9. If three positive numbers, five negative numbers, and zero are multiplied, the product is ____________.

10. The fifth power of a negative number is ____________.

Find each product. See Examples 1 and 2.

11. $(-4)(-5)$ **12.** $(-4)(-6)$ **13.** $(-7)(4)$

14. $(-8)(5)$ **15.** $(-4)(-20)$ **16.** $(-8)(-30)$

17. 0 **18.** 0 **19.** $\frac{5}{6}$ **20.** $\frac{3}{10}$ **21.** $\frac{3}{2}$ **22.** 4 **23.** $-32, -16, -8, -4, -2, -1, 1, 2, 4, 8, 16, 32$ **24.** $-36, -18, -12, -9, -6, -4, -3, -2, -1, 1, 2, 3, 4, 6, 9, 12, 18, 36$ **25.** $-40, -20, -10, -8, -5, -4, -2, -1, 1, 2, 4, 5, 8, 10, 20, 40$ **26.** $-50, -25, -10, -5, -2, -1, 1, 2, 5, 10, 25, 50$ **27.** $-31, -1, 1, 31$ **28.** $-17, -1, 1, 17$ **29.** -3 **30.** -3 **31.** -2 **32.** -7 **33.** 16 **34.** 13 **35.** 0 **36.** 0 **37.** 25.63 **38.** 36.98 **39.** $\frac{3}{2}$ **40.** $\frac{3}{10}$ **41.** To multiply two signed numbers, multiply their absolute values. If the signs of the numbers are the same, the product is positive. If the signs of the numbers are different, the product is negative. For example, $3 \cdot 5 = 15$, $(-3)(-5) = 15$, and $(-3)(5) = -15$. **42.** Similar rules are used for division. For example, $\frac{15}{3} = 5$, $\frac{-15}{-3} = 5$, $\frac{-15}{3} = -5$, and $\frac{15}{-3} = -5$. **43.** -11 **44.** -2 **45.** -2 **46.** 7 **47.** 35 **48.** 15 **49.** 6 **50.** 8 **51.** -18 **52.** -105 **53.** 67 **54.** 36 **55.** -8 **56.** -21 **57.** 3 **58.** 5 **59.** 7 **60.** 24 **61.** 4 **62.** 4 **63.** -3 **64.** -3 **65.** First, substitute -3 for x and 4 for y to get $3(-3) + 2(4)$. Then perform the multiplications to get $-9 + 8$. Finally, add to get -1. **66.** negative **67.** 47 **68.** 68 **69.** 72 **70.** -228 **71.** $-\frac{78}{25}$ **72.** 1 **73.** 0 **74.** 0 **75.** -23 **76.** -6 **77.** 2 **78.** 0

17. $(-8)(0)$ **18.** $(-12)(0)$ **19.** $\left(-\frac{3}{8}\right)\left(-\frac{20}{9}\right)$

20. $\left(-\frac{5}{4}\right)\left(-\frac{6}{25}\right)$ **21.** $(-6)\left(-\frac{1}{4}\right)$ **22.** $(-8)\left(-\frac{1}{2}\right)$

Find all integer factors of each given number.

23. 32 **24.** 36 **25.** 40 **26.** 50 **27.** 31 **28.** 17

Find each quotient. See Examples 3 and 4.

29. $\frac{-15}{5}$ **30.** $\frac{-18}{6}$ **31.** $\frac{20}{-10}$ **32.** $\frac{28}{-4}$

33. $\frac{-160}{-10}$ **34.** $\frac{-260}{-20}$ **35.** $\frac{0}{-3}$ **36.** $\frac{0}{-6}$

37. $\frac{-10.252}{-.4}$ **38.** $\frac{-29.584}{-.8}$ **39.** $\left(-\frac{3}{4}\right) \div \left(-\frac{1}{2}\right)$ **40.** $\left(-\frac{3}{16}\right) \div \left(-\frac{5}{8}\right)$

41. Explain in words how to multiply signed numbers. Consider the various cases and give examples.

42. Explain in words how to divide signed numbers. Give examples.

Perform the indicated operations. See Examples 5(a), (b), and (c).

43. $7 - 3 \cdot 6$ **44.** $8 - 2 \cdot 5$

45. $-10 - (-4)(2)$ **46.** $-11 - (-3)(6)$

47. $-7(3 - 8)$ **48.** $-5(4 - 7)$

49. $(12 - 14)(1 - 4)$ **50.** $(8 - 9)(4 - 12)$

51. $(7 - 10)(10 - 4)$ **52.** $(5 - 12)(19 - 4)$

53. $(-2 - 8)(-6) + 7$ **54.** $(-9 - 4)(-2) + 10$

55. $3(-5) + |3 - 10|$ **56.** $4(-8) + |4 - 15|$

Perform each indicated operation. See Example 5(d).

57. $\frac{-5(-6)}{9 - (-1)}$ **58.** $\frac{-12(-5)}{7 - (-5)}$ **59.** $\frac{-21(3)}{-3 - 6}$

60. $\frac{-40(3)}{-2 - 3}$ **61.** $\frac{-10(2) + 6(2)}{-3 - (-1)}$ **62.** $\frac{8(-1) + 6(-2)}{-6 - (-1)}$

63. $\frac{-27(-2) - (-12)(-2)}{-2(3) - 2(2)}$ **64.** $\frac{-13(-4) - (-8)(-2)}{(-10)(2) - 4(-2)}$

65. Explain the method you would use to evaluate $3x + 2y$ if $x = -3$ and $y = 4$.

66. If x and y are both replaced by negative numbers, is the value of $4x + 8y$ positive or negative?

Evaluate each expression if $x = 6$, $y = -4$, and $a = 3$. See Example 6.

67. $5x - 2y + 3a$ **68.** $6x - 5y + 4a$ **69.** $(2x + y)(3a)$

70. $(5x - 2y)(-2a)$ **71.** $\left(\frac{1}{3}x - \frac{4}{5}y\right)\left(-\frac{1}{5}a\right)$ **72.** $\left(\frac{5}{6}x + \frac{3}{2}y\right)\left(-\frac{1}{3}a\right)$

73. $(-5 + x)(-3 + y)(3 - a)$ **74.** $(6 - x)(5 + y)(3 + a)$ **75.** $-2y^2 + 3a$

76. $5x - 4a^2$ **77.** $\frac{2y^2 - x}{a + 10}$ **78.** $\frac{xy + 8a}{x - y}$

79. $9 + (-9)(2); -9$
80. $-12 + 4(-7); -40$
81. $-4 - 2(-1)(6); 8$
82. $-1 - 2(-8)(2); 31$
83. $(1.5)(-3.2) - 9; -13.8$
84. $(4.2)(-8.5) - 3; -38.7$
85. $12[9 - (-8)]; 204$
86. $-3[3 - (-7)]; -30$
87. $\frac{-12}{-5 + (-1)}; 2$
88. $\frac{-20}{-8 + (-2)}; 2$
89. $\frac{15 + (-3)}{4(-3)}; -1$
90. $\frac{-18 + (-6)}{2(-4)}; 3$
91. $\frac{-\left(\frac{1}{2}\right)\left(\frac{1}{3}\right)}{-\frac{2}{3}}; \frac{9}{16}$
92. $\frac{\left(-\frac{2}{3}\right)\left(-\frac{3}{4}\right)}{\frac{1}{7}}; \frac{14}{15}$
93. $6x = -42; -7$
94. $4x = -36; -9$
95. $\frac{x}{3} = -3; -9$
96. $\frac{x}{4} = -1; -4$
97. $x - 6 = 4; 10$
98. $x - 7 = 2; 9$
99. $x + 5 = -5; -10$
100. $x + 6 = -3; -9$ 101. $8\frac{2}{5}$
102. $3\frac{1}{5}$ 103. 2 104. -7

Write a numerical expression for each phrase and simplify. See Examples 7 and 8.

79. The product of -9 and 2, added to 9

80. The product of 4 and -7, added to -12

81. Twice the product of -1 and 6, subtracted from -4

82. Twice the product of -8 and 2, subtracted from -1

83. Nine subtracted from the product of 1.5 and -3.2

84. Three subtracted from the product of 4.2 and -8.5

85. The product of 12 and the difference between 9 and -8

86. The product of -3 and the difference between 3 and -7

87. The quotient of -12 and the sum of -5 and -1

88. The quotient of -20 and the sum of -8 and -2

89. The sum of 15 and -3, divided by the product of 4 and -3

90. The sum of -18 and -6, divided by the product of 2 and -4

91. The product of $-\frac{1}{2}$ and $\frac{3}{4}$, divided by $-\frac{2}{3}$

92. The product of $-\frac{2}{3}$ and $-\frac{1}{5}$, divided by $\frac{1}{7}$

Write each statement in symbols, using x as the variable, and find the solution by guessing or by using trial and error. All solutions come from the set of integers between -12 and 12, inclusive. See Example 9.

93. Six times a number is -42.

94. Four times a number is -36.

95. The quotient of a number and 3 is -3.

96. The quotient of a number and 4 is -1.

97. 6 less than a number is 4.

98. 7 less than a number is 2.

99. When 5 is added to a number, the result is -5.

100. When 6 is added to a number, the result is -3.

To find the average of a group of numbers, we add the numbers and then divide the sum by the number of terms added. For example, to find the average of 14, 8, 3, 9, and 1, we add them and then divide by 5:

$$\frac{14 + 8 + 3 + 9 + 1}{5} = \frac{35}{5} = 7.$$

The average of these numbers is 7.

Exercises 101–106 involve finding the average of a group of numbers.

101. Find the average of 23, 18, 13, -4, and -8.

102. Find the average of 18, 12, 0, -4, and -10.

103. What is the average of all integers between -10 and 14, inclusive of both?

104. What is the average of all even integers between -18 and 4, inclusive of both?

105. $12.60 (to the nearest cent) **106.** 16,525 (in thousands, rounded) **107.** 0
108. The average will be positive if the sum of all the numbers is positive, and it will be negative if the sum of all the numbers is negative.
109. (a) 6 is divisible by 2. (b) 9 is not divisible by 2.
110. (a) $4 + 7 + 9 + 9 + 2 + 3 + 2 = 36$ is divisible by 3. (b) $2 + 4 + 4 + 3 + 8 + 7 + 1 = 29$ is not divisible by 3.
111. (a) 64 is divisible by 4. (b) 35 is not divisible by 4.
112. (a) 5 is divisible by 5. (b) 3 is not divisible by 5.

105. The table shows average hourly earnings of various fields in private industry for 1995. Find the average of these amounts.

Private Industry Group	*1995 Hourly Earnings*
Mining	$15.30
Construction	15.08
Manufacturing	12.37
Transportation, public utilities	14.23
Wholesale trade	12.43
Retail trade	7.69
Finance, insurance, real estate	12.33
Service	11.39

Source: U.S. Bureau of Labor Statistics.

106. During the 1990s, the number of labor union or employee association members has remained fairly constant. The chart shows the number of members, in thousands, for the years 1990–1996.

Year	*Number (in thousands)*
1990	16,740
1991	16,568
1992	16,390
1993	16,598
1994	16,748
1995	16,360
1996	16,269

Source: U.S. Bureau of Labor Statistics.

What is the average number, in thousands, for this seven-year period?

107. If the average of a group of numbers is 0, what is the sum of all the numbers?

108. Suppose there is a group of numbers with some positive and some negative. Under what conditions will the average be a positive number? Under what conditions will the average be negative?

The operation of division is used in divisibility tests. *A divisibility test allows us to determine whether a given number is divisible (without remainder) by another number. For example, a number is divisible by 2 if its last digit is divisible by 2, and not otherwise.*

109. Tell why **(a)** 3,473,986 is divisible by 2 and **(b)** 4,336,879 is not divisible by 2.

110. An integer is divisible by 3 if the sum of its digits is divisible by 3, and not otherwise. Show that **(a)** 4,799,232 is divisible by 3 and **(b)** 2,443,871 is not divisible by 3.

113. (a) 2 is divisible by 2 and $1 + 5 + 2 + 4 + 8 + 2 + 2 = 24$ is divisible by 3. (b) While 0 is divisible by 2, $2 + 8 + 7 + 3 + 5 + 9 + 0 = 34$ is not divisible by 3.
114. (a) 296 is divisible by 8. (b) 623 is not divisible by 8.
115. (a) $4 + 1 + 1 + 4 + 1 + 0 + 7 = 18$ is divisible by 9. (b) $2 + 2 + 8 + 7 + 3 + 2 + 1 = 25$ is not divisible by 9.
116. (a) $4 + 2 + 5 + 3 + 5 + 2 + 0 = 21$ is divisible by 3 and 20 is divisible by 4. (b) $4 + 2 + 4 + 9 + 4 + 7 + 4 = 34$ is not divisible by 3, and this is sufficient to show that the number is not divisible by 12.

111. An integer is divisible by 4 if its last two digits form a number divisible by 4, and not otherwise. Show that **(a)** 6,221,464 is divisible by 4 and **(b)** 2,876,335 is not divisible by 4.

112. An integer is divisible by 5 if its last digit is divisible by 5, and not otherwise. Show that **(a)** 3,774,595 is divisible by 5 and **(b)** 9,332,123 is not divisible by 5.

113. An integer is divisible by 6 if it is divisible by both 2 and 3, and not otherwise. Show that **(a)** 1,524,822 is divisible by 6 and **(b)** 2,873,590 is not divisible by 6.

114. An integer is divisible by 8 if its last three digits form a number divisible by 8, and not otherwise. Show that **(a)** 2,923,296 is divisible by 8 and **(b)** 7,291,623 is not divisible by 8.

115. An integer is divisible by 9 if the sum of its digits is divisible by 9, and not otherwise. Show that **(a)** 4,114,107 is divisible by 9 and **(b)** 2,287,321 is not divisible by 9.

116. An integer is divisible by 12 if it is divisible by both 3 and 4, and not otherwise. Show that **(a)** 4,253,520 is divisible by 12 and **(b)** 4,249,474 is not divisible by 12.

1.7 Properties of Real Numbers

OBJECTIVES

1. Use the commutative properties.
2. Use the associative properties.
3. Use the identity properties.
4. Use the inverse properties.
5. Use the distributive property.

FOR EXTRA HELP

SSG Sec. 1.7
SSM Sec. 1.7
Pass the Test Software
InterAct Math Tutorial Software
Video 2

TEACHING TIP Just as sweetness is one of the properties of sugar, the numbers we work with also have properties. This section covers some of these "number properties."

CHALKBOARD EXAMPLE

Use a commutative property to complete the statement.

$$5x = x \cdot \_\_$$

Answer: 5

If you were asked to find the sum $3 + 89 + 97$, you might mentally add $3 + 97$ to get 100, and then add $100 + 89$ to get 189. While the rule for order of operations says to add from left to right, we may change the order of the terms and group them in any way we choose without affecting the sum. These are examples of shortcuts that we use in everyday mathematics. These shortcuts are justified by the basic properties of addition and multiplication, discussed in this section. In these properties, *a, b,* and *c* represent real numbers.

OBJECTIVE 1 Use the commutative properties. The word *commute* means to go back and forth. Many people commute to work or to school. If you travel from home to work and follow the same route from work to home, you travel the same distance each time. The **commutative properties** say that if two numbers are added or multiplied in any order, the result is the same.

Commutative Properties

$$a + b = b + a$$
$$ab = ba$$

EXAMPLE 1 Using the Commutative Properties

Use a commutative property to complete each statement.

(a) $-8 + 5 = 5 + \_\_\_\_$

By the commutative property for addition, the missing number is -8, since $-8 + 5 = 5 + (-8)$.

(b) $(-2)(7) = \_\_\_\_(-2)$

By the commutative property for multiplication, the missing number is 7, since $(-2)(7) = (7)(-2)$.

INSTRUCTOR'S RESOURCES

PTB Sec. 1.7 **ISM** Sec. 1.7
AB Sec. 1.7
TEST GENERATOR
WORLD WIDE WEB www.LialAlgebra.com

OBJECTIVE 2 **Use the associative properties.** When we *associate* one object with another, we tend to think of those objects as being grouped together. The **associative properties** say that when we add or multiply three numbers, we can group the first two together or the last two together and get the same answer.

Associative Properties

$$(a + b) + c = a + (b + c)$$
$$(ab)c = a(bc)$$

CHALKBOARD EXAMPLE

Use an associative property to complete the statement.

$10 \cdot [(-8) \cdot (-3)] = \_\_$

Answer: $[10 \cdot (-8)] \cdot (-3)$

EXAMPLE 2 Using the Associative Properties

Use an associative property to complete each statement.

(a) $8 + (-1 + 4) = (8 + \_\_\_\_) + 4$
The missing number is -1.

(b) $[2 \cdot (-7)] \cdot 6 = 2 \cdot \_\_\_\_$
The completed expression on the right should be $2 \cdot [(-7) \cdot 6]$.

TEACHING TIP With the commutative properties, the number sequence changes on opposite sides of the equals sign, but with the associative properties, the number sequence on each side of the equals sign is the same.

By the associative property of addition, the sum of three numbers will be the same no matter how the numbers are "associated" in groups. For this reason, parentheses can be left out in many addition problems. For example, both

$$(-1 + 2) + 3 \qquad \text{and} \qquad -1 + (2 + 3)$$

can be written as

$$-1 + 2 + 3.$$

In the same way, parentheses also can be left out of many multiplication problems.

CHALKBOARD EXAMPLE

Is $(2 \cdot 4)6 = (4 \cdot 2)6$ an example of the associative property or of the commutative property?

Answer: commutative

EXAMPLE 3 Distinguishing between the Associative and Commutative Properties

(a) Is $(2 + 4) + 5 = 2 + (4 + 5)$ an example of the associative property or the commutative property?

The order of the three numbers is the same on both sides of the equals sign. The only change is in the grouping, or association, of the numbers. Therefore, this is an example of the associative property.

(b) Is $6(3 \cdot 10) = 6(10 \cdot 3)$ an example of the associative property or the commutative property?

The same numbers, 3 and 10, are grouped on each side. On the left, the 3 appears first, but on the right, the 10 appears first. Since the only change involves the order of the numbers, this statement is an example of the commutative property.

(c) Is $(8 + 1) + 7 = 8 + (7 + 1)$ an example of the associative property or the commutative property?

In the statement, both the order and the grouping are changed. On the left the order of the three numbers is 8, 1, and 7. On the right it is 8, 7, and 1. On the left the 8 and 1 are grouped, and on the right the 7 and 1 are grouped. Therefore, both the associative and the commutative properties are used.

CHALKBOARD EXAMPLE

Find the sum.

$43 + 26 + 17 + 24 + 6$

Answer: 116

EXAMPLE 4 Using the Commutative and Associative Properties

Find the sum: $23 + 41 + 2 + 9 + 25$.

The commutative and associative properties make it possible to choose pairs of numbers whose sums are easy to add.

$$\begin{aligned} 23 + 41 + 2 + 9 + 25 &= (41 + 9) + (23 + 2) + 25 \\ &= 50 + 25 + 25 \\ &= 100 \end{aligned}$$

OBJECTIVE 3 Use the identity properties. If a child wears a costume on Halloween, the child's appearance is changed, but his or her *identity* is unchanged. The identity of a real number is left unchanged when identity properties are applied. The **identity properties** say that the sum of 0 and any number equals that number, and the product of 1 and any number equals that number.

Identity Properties

$$a + 0 = a \quad \text{and} \quad 0 + a = a$$
$$a \cdot 1 = a \quad \text{and} \quad 1 \cdot a = a$$

The number 0 leaves the identity, or value, of any real number unchanged by addition. For this reason, 0 is called the **identity element for addition.** Since multiplication by 1 leaves any real number unchanged, 1 is the **identity element for multiplication.**

CHALKBOARD EXAMPLE

$5 + 0 = 5$ and $1 \cdot \frac{1}{3} = \frac{1}{3}$ are examples of the identity properties.

EXAMPLE 5 Using the Identity Properties

These statements are examples of the identity properties.

(a) $-3 + 0 = -3$

(b) $1 \cdot \frac{1}{2} = \frac{1}{2}$

We use the identity property for multiplication to write fractions in lowest terms and to get common denominators.

CHALKBOARD EXAMPLE

Use the identity property to simplify.

$$\frac{3}{8} - \frac{5}{24}$$

Answer: $\frac{1}{6}$

EXAMPLE 6 Using the Identity Property to Simplify Expressions

Simplify the following expressions.

(a) $\frac{49}{35}$

$$\begin{aligned} \frac{49}{35} &= \frac{7 \cdot 7}{5 \cdot 7} && \text{Factor.} \\ &= \frac{7}{5} \cdot \frac{7}{7} && \text{Write as a product.} \\ &= \frac{7}{5} \cdot 1 && \text{Divide.} \\ &= \frac{7}{5} && \text{Identity property} \end{aligned}$$

(b) $\frac{3}{4} + \frac{5}{24}$

$$\begin{aligned} \frac{3}{4} + \frac{5}{24} &= \frac{3}{4} \cdot 1 + \frac{5}{24} && \text{Identity property} \\ &= \frac{3}{4} \cdot \frac{6}{6} + \frac{5}{24} && \text{Get a common denominator.} \\ &= \frac{18}{24} + \frac{5}{24} && \text{Multiply.} \\ &= \frac{23}{24} && \text{Add.} \end{aligned}$$

TEACHING TIP The identity and inverse properties are often confused. Cover them with several examples.

OBJECTIVE 4 **Use the inverse properties.** Each day before you go to work or school, you probably put on your shoes before you leave. Before you go to sleep at night, you probably take them off, and this leads to the same situation that existed before you put them on. These operations from everyday life are examples of inverse operations. The **inverse properties** of addition and multiplication lead to the additive and multiplicative identities, respectively. Recall that $-a$ is the **additive inverse** of a and $\frac{1}{a}$ is the **multiplicative inverse,** or **reciprocal,** of the nonzero number a. The sum of the numbers a and $-a$ is 0, and the product of the nonzero numbers a and $\frac{1}{a}$ is 1.

Inverse Properties

$$a + (-a) = 0 \quad \text{and} \quad -a + a = 0$$

$$a \cdot \frac{1}{a} = 1 \quad \text{and} \quad \frac{1}{a} \cdot a = 1 \quad (a \neq 0)$$

CHALKBOARD EXAMPLE

Complete the statement so that it is an example of an inverse property.

$$-\frac{1}{9} \cdot \_\_\_ = 1$$

Answer: -9

EXAMPLE 7 **Using the Inverse Properties**

The following statements are examples of the inverse properties.

(a) $\frac{2}{3} \cdot \frac{3}{2} = 1$ **(b)** $(-5)\left(-\frac{1}{5}\right) = 1$

(c) $-\frac{1}{2} + \frac{1}{2} = 0$ **(d)** $4 + (-4) = 0$

In the next example, we show how the various properties are used to simplify an expression.

CHALKBOARD EXAMPLE

Use the additive inverse property to simplify the expression.

$$\frac{1}{2} + 3y + \left(-\frac{1}{2}\right)$$

Answer: $3y$

EXAMPLE 8 **Using Properties to Simplify an Expression**

Simplify $-2x + 10 + 2x$, using the properties discussed in this section.

$$\begin{aligned} -2x + 10 + 2x &= (-2x + 10) + 2x && \text{Order of operations} \\ &= [10 + (-2x)] + 2x && \text{Commutative property} \\ &= 10 + [(-2x) + 2x] && \text{Associative property} \\ &= 10 + 0 && \text{Inverse property} \\ &= 10 && \text{Identity property} \end{aligned}$$

Note that for *any* value of x, $-2x$ and $2x$ are additive inverses; this is why we can use the inverse property in this simplification.

NOTE The detailed procedure shown in Example 8 is seldom, if ever, used in practice. We include the example to show how the properties of this section apply, even though steps may be skipped when actually doing the simplification.

OBJECTIVE 5 **Use the distributive property.** The everyday meaning of the word *distribute* is "to give out from one to several." An important property of real number operations involves this idea.

Look at the following statements.

$$2(5 + 8) = 2(13) = 26$$
$$2(5) + 2(8) = 10 + 16 = 26$$

Since both expressions equal 26,

$$2(5 + 8) = 2(5) + 2(8).$$

This result is an example of the **distributive property,** the only property involving *both* addition and multiplication. With this property, a product can be changed to a sum or difference.

The distributive property says that multiplying a number a by a sum of numbers $b + c$ gives the same result as multiplying a by b and a by c and then adding the two products.

Distributive Property

$$a(b + c) = ab + ac \quad \text{and} \quad (b + c)a = ba + ca$$

As the arrows show, the a outside the parentheses is "distributed" over the b and c inside. Another form of the distributive property is valid for subtraction.

$$a(b - c) = ab - ac \quad \text{and} \quad (b - c)a = ba - ca$$

The distributive property also can be extended to more than two numbers.

$$a(b + c + d) = ab + ac + ad$$

NOTE The distributive property can be used "in reverse." For example, we can write

$$ac + bc = (a + b)c.$$

CHALKBOARD EXAMPLE

Use the distributive property to rewrite the expression.

$$3a - 3b$$

Answer: $3(a - b)$

EXAMPLE 9 **Using the Distributive Property**

Use the distributive property to rewrite each expression.

(a) $5(9 + 6) = 5 \cdot 9 + 5 \cdot 6$ Distributive property

$= 45 + 30$ Multiply.

$= 75$ Add.

(b) $4(x + 5 + y) = 4x + 4 \cdot 5 + 4y$ Distributive property

$= 4x + 20 + 4y$ Multiply.

(c) $-2(x + 3) = -2x + (-2)(3)$ Distributive property

$= -2x - 6$ Multiply.

(d) $3(k - 9) = 3k - 3 \cdot 9$ Distributive property

$= 3k - 27$ Multiply.

(e) $8(3r + 11t + 5z) = 8(3r) + 8(11t) + 8(5z)$ Distributive property

$= (8 \cdot 3)r + (8 \cdot 11)t + (8 \cdot 5)z$ Associative property

$= 24r + 88t + 40z$ Multiply.

(f) $6 \cdot 8 + 6 \cdot 2 = 6(8 + 2)$ Distributive property

$= 6(10) = 60$ Add, then multiply.

(g) $4x - 4m = 4(x - m)$ Distributive property

(h) $6x - 12 = 6 \cdot x - 6 \cdot 2 = 6(x - 2)$ Distributive property

The symbol $-a$ may be interpreted as $-1 \cdot a$. Similarly, when a negative sign precedes an expression within parentheses, it may also be interpreted as a factor of -1. The distributive property is used to remove parentheses from expressions such as $-(2y + 3)$. We do this by first writing $-(2y + 3)$ as $-1 \cdot (2y + 3)$.

$$-(2y + 3) = -1 \cdot (2y + 3)$$

$$= -1 \cdot (2y) + (-1) \cdot (3) \quad \text{Distributive property}$$

$$= -2y - 3 \quad \text{Multiply.}$$

CHALKBOARD EXAMPLE

Write without parentheses.

$-(-5y + 8)$

Answer: $5y - 8$

EXAMPLE 10 **Using the Distributive Property to Remove Parentheses**

Write without parentheses.

(a) $-(7r - 8) = -1(7r) + (-1)(-8)$ Distributive property

$= -7r + 8$ Multiply.

(b) $-(-9w + 2) = 9w - 2$

The properties discussed here are the basic properties that justify how we do algebra. You should know them by name because we will be referring to them frequently. Here is a summary of these properties.

TEACHING TIP Ask students to tell in their own words what the properties say about numbers. For example, the identity property for addition can be stated as follows: "If zero is added to any number, the result is the number that you started with."

Properties of Addition and Multiplication

For any real numbers a, b, and c, the following properties hold.

Commutative Properties $a + b = b + a$ $ab = ba$

Associative Properties $(a + b) + c = a + (b + c)$

$(ab)c = a(bc)$

Identity Properties There is a real number 0 such that

$a + 0 = a$ and $0 + a = a$.

There is a real number 1 such that

$a \cdot 1 = a$ and $1 \cdot a = a$.

Properties of Addition and Multiplication (continued)	
Inverse Properties	For each real number a, there is a single real number $-a$ such that $a + (-a) = 0$ and $(-a) + a = 0$. For each nonzero real number a, there is a single real number $\frac{1}{a}$ such that $a \cdot \frac{1}{a} = 1$ and $\frac{1}{a} \cdot a = 1$.
Distributive Property	$a(b + c) = ab + ac$ $\quad (b + c)a = ba + ca$

1.7 EXERCISES

1. B 2. F 3. C 4. I
5. B 6. D, F 7. B 8. A
9. G 10. H
11. commutative property
12. commutative property
13. associative property
14. associative property
15. associative property
16. associative property
17. inverse property
18. inverse property
19. inverse property
20. inverse property
21. identity property
22. identity property
23. commutative property
24. commutative property
25. distributive property
26. distributive property
27. identity property
28. identity property
29. distributive property
30. distributive property

Match each item in Column I with the correct choice(s) from Column II. Choices may be used once, more than once, or not at all.

I	II
1. Identity element for addition	**A.** $(5 \cdot 4) \cdot 3 = 5 \cdot (4 \cdot 3)$
2. Identity element for multiplication	**B.** 0
3. Additive inverse of a	**C.** $-a$
4. Multiplicative inverse, or reciprocal, of the nonzero number a	**D.** -1
5. The number that is its own additive inverse	**E.** $5 \cdot 4 \cdot 3 = 60$
6. The two numbers that are their own multiplicative inverses	**F.** 1
7. The only number that has no multiplicative inverse	**G.** $(5 \cdot 4) \cdot 3 = 3 \cdot (5 \cdot 4)$
8. An example of the associative property	**H.** $5(4 + 3) = 5 \cdot 4 + 5 \cdot 3$
9. An example of the commutative property	**I.** $\frac{1}{a}$
10. An example of the distributive property	

Decide whether each statement is an example of the commutative, associative, identity, inverse, or distributive property. See Examples 1, 2, 3, 5, 6, 7, and 9.

11. $7 + 18 = 18 + 7$

12. $13 + 12 = 12 + 13$

13. $5(13 \cdot 7) = (5 \cdot 13) \cdot 7$

14. $-4(2 \cdot 6) = (-4 \cdot 2) \cdot 6$

15. $-6 + (12 + 7) = (-6 + 12) + 7$

16. $(-8 + 13) + 2 = -8 + (13 + 2)$

17. $-6 + 6 = 0$

18. $12 + (-12) = 0$

19. $\left(\frac{2}{3}\right)\left(\frac{3}{2}\right) = 1$

20. $\left(\frac{5}{8}\right)\left(\frac{8}{5}\right) = 1$

21. $2.34 + 0 = 2.34$

22. $-8.456 + 0 = -8.456$

23. $(4 + 17) + 3 = 3 + (4 + 17)$

24. $(-8 + 4) + (-12) = -12 + (-8 + 4)$

25. $6(x + y) = 6x + 6y$

26. $14(t + s) = 14t + 14s$

27. $-\frac{5}{9} = -\frac{5}{9} \cdot \frac{3}{3} = -\frac{15}{27}$

28. $\frac{13}{12} = \frac{13}{12} \cdot \frac{7}{7} = \frac{91}{84}$

29. $5(2x) + 5(3y) = 5(2x + 3y)$

30. $3(5t) - 3(7r) = 3(5t - 7r)$

31. identity property
32. No. For example, $2 + (3 \times 4) \neq (2 + 3) \times (2 + 4)$.
33. The commutative properties allow us to change the order of two or more terms or factors, while the associative properties allow us to change the grouping of three or more terms or factors. The identity properties allow us to perform an operation so that the result is the number we started with. The inverse properties allow us to perform an operation that gives an identity element as a result.
34. The distributive property of multiplication with respect to addition says that a factor can be "distributed" to each term in a sum. For example, $3(x + y) = 3x + 3y$, and $-4(x + 2y + 3z) = -4x - 8y - 12z$.
35. $7 + r$ **36.** $9 + t$ **37.** s
38. w
39. $-6x + (-6) \cdot 7$; $-6x - 42$
40. $-5y + (-5) \cdot 2$; $-5y - 10$
41. $w + [5 + (-3)]$; $w + 2$
42. $b + [8 + (-10)]$; $b - 2$
43. 11 **44.** 13 **45.** 0
46. 0 **47.** $-.38$ **48.** $-.73$
49. 1 **50.** 1
51. Subtraction is not associative.
52. Division is not associative.
53. The expression following the first equals sign should be $-3(4) - 3(-6)$. The student forgot that 6 should be preceded by a $-$ sign. The correct work is $-3(4 - 6) = -3(4) - 3(-6) = -12 + 18 = 6$.
54. We must multiply $\frac{3}{4}$ by 1 in the form of a fraction, $\frac{3}{3}$: $\frac{3}{4} \cdot \frac{3}{3} = \frac{9}{12}$.
55. $(5 + 1)x$; $6x$
56. $(6 + 1)q$; $7q$ **57.** $4t + 12$
58. $5w + 20$ **59.** $-8r - 24$
60. $-11x - 44$ **61.** $-5y + 20$
62. $-9g + 36$ **63.** $-16y - 20z$
64. $-4b - 8a$ **65.** $8(z + w)$
66. $4(s + r)$ **67.** $7(2v + 5r)$
68. $13(5w + 4p)$
69. $24r + 32s - 40y$

31. The following conversation actually took place between one of the authors of this book and his son, Jack, when Jack was four years old:

DADDY:	"Jack, what is $3 + 0$?"
JACK:	"3."
DADDY:	"Jack, what is $4 + 0$?"
JACK:	"4. And Daddy, *string* plus zero equals *string*!"

What property of addition did Jack recognize?

32. The distributive property holds for multiplication with respect to addition. Is there a distributive property for addition with respect to multiplication? If not, give an example to show why.

33. Write a paragraph explaining in your own words the following properties of addition and multiplication: commutative, associative, identity, inverse.

34. Write a paragraph explaining in your own words the distributive property of multiplication with respect to addition. Give examples.

Use the indicated property to write a new expression that is equal to the given expression. Then simplify the new expression if possible. See Examples 1, 2, 5, 7, and 9.

35. $r + 7$; commutative
36. $t + 9$; commutative
37. $s + 0$; identity
38. $w + 0$; identity
39. $-6(x + 7)$; distributive
40. $-5(y + 2)$; distributive
41. $(w + 5) + (-3)$; associative
42. $(b + 8) + (-10)$; associative

Use the properties of this section to simplify each expression. See Examples 7 and 8.

43. $6t + 8 - 6t + 3$
44. $9r + 12 - 9r + 1$
45. $\frac{2}{3}x - 11 + 11 - \frac{2}{3}x$
46. $\frac{1}{5}y + 4 - 4 - \frac{1}{5}y$
47. $\left(\frac{9}{7}\right)(-.38)\left(\frac{7}{9}\right)$
48. $\left(\frac{4}{5}\right)(-.73)\left(\frac{5}{4}\right)$
49. $t + (-t) + \frac{1}{2}(2)$
50. $w + (-w) + \frac{1}{4}(4)$

51. Evaluate $25 - (6 - 2)$ and evaluate $(25 - 6) - 2$. Do you think subtraction is associative?

52. Evaluate $180 \div (15 \div 3)$ and evaluate $(180 \div 15) \div 3$. Do you think division is associative?

53. Suppose that a student shows you the following work.

$$-3(4 - 6) = -3(4) - 3(6) = -12 - 18 = -30$$

The student has made a very common error. Explain the student's mistake, and work the problem correctly.

54. Explain how the procedure of changing $\frac{3}{4}$ to $\frac{9}{12}$ requires the use of the multiplicative identity element, 1.

Use the distributive property to rewrite each expression. Simplify if possible. See Example 9.

55. $5x + x$
56. $6q + q$
57. $4(t + 3)$
58. $5(w + 4)$
59. $-8(r + 3)$
60. $-11(x + 4)$
61. $-5(y - 4)$
62. $-9(g - 4)$
63. $-\frac{4}{3}(12y + 15z)$
64. $-\frac{2}{5}(10b + 20a)$
65. $8 \cdot z + 8 \cdot w$
66. $4 \cdot s + 4 \cdot r$
67. $7(2v) + 7(5r)$
68. $13(5w) + 13(4p)$
69. $8(3r + 4s - 5y)$

70. $10u - 6v + 14w$
71. $(1 + 1 + 1)q$; $3q$
72. $(1 + 1 + 1 + 1)m$; $4m$
73. $(-5 + 1)x$; $-4x$
74. $(-9 + 1)p$; $-8p$
75. $-4t - 3m$ **76.** $-9x - 12y$
77. $5c + 4d$ **78.** $13x + 15y$
79. $3q - 5r + 8s$
80. $4z - 5w + 9y$
81. for example, "putting on your socks" and "putting on your shoes"
82. for example, "defective merchandise counter"

70. $2(5u - 3v + 7w)$ **71.** $q + q + q$ **72.** $m + m + m + m$
73. $-5x + x$ **74.** $-9p + p$

Use the distributive property to write each expression without parentheses. See Example 10.

75. $-(4t + 3m)$ **76.** $-(9x + 12y)$ **77.** $-(-5c - 4d)$
78. $-(-13x - 15y)$ **79.** $-(-3q + 5r - 8s)$ **80.** $-(-4z + 5w - 9y)$

81. The operations of "getting out of bed" and "taking a shower" are not commutative. Give an example of another pair of everyday operations that are not commutative.

82. The phrase "dog biting man" has two different meanings, depending on how the words are associated:

(dog biting) man dog (biting man)

Give another example of a three-word phrase that has different meanings depending on how the words are associated.

*Some exercise sets will include groups of exercises designated *Relating Concepts*. The goal of these groups is to relate topics currently being studied to ones studied earlier, to real-life situations, or to universally accepted truths. In general, the exercises in these groups should be worked in numerical order without skipping any. Answers to *all* of these exercises are given in the answer section.

These problems can be used as collaborative exercises with small groups of students.

RELATING CONCEPTS* (EXERCISES 83–86)

In Section 1.6 we used a pattern to see that the product of two negative numbers is a positive number. In the group of exercises that follows, we show another justification for determining the sign of the product of two negative numbers.

Work Exercises 83–86 in order.

83. Evaluate the expression $-3[5 + (-5)]$ by using the rules for order of operations.

84. Write the expression in Exercise 83 using the distributive property. Do not simplify the products.

85. The product -3×5 should be one of the terms you wrote when answering Exercise 84. Based on the results in Section 1.6, what is this product?

86. In Exercise 83, you should have obtained 0 as an answer. Now, consider the following, using the results of Exercises 83 and 85.

$$-3[5 + (-5)] = -3(5) + (-3)(-5)$$
$$0 = -15 + ?$$

The question mark represents the product $(-3)(-5)$. When added to -15, it must give a sum of 0. Therefore, how must we interpret $(-3)(-5)$?

Did you make the connection that a rule can be obtained in more than one way, with consistent results from each method?

83. 0 **84.** $-3(5) + (-3)(-5)$
85. -15 **86.** We must interpret $(-3)(-5)$ as 15, since it is the additive inverse of -15.

1.8 Simplifying Expressions

OBJECTIVES

1. Simplify expressions.
2. Identify terms and numerical coefficients.
3. Identify like terms.

OBJECTIVE 1 Simplify expressions. In this section we show how to simplify expressions using the properties of addition and multiplication introduced in the previous section.

EXAMPLE 1 Simplifying Expressions

Simplify the following expressions.

(a) $4x + 8 + 9$
Since $8 + 9 = 17$,

$$4x + 8 + 9 = 4x + 17.$$

INSTRUCTOR'S RESOURCES

PTB Sec. 1.8 **ISM** Sec. 1.8
AB Sec. 1.8

TEST GENERATOR

WORLD WIDE WEB
www.LialAlgebra.com

4 Combine like terms.

5 Simplify expressions from word phrases.

FOR EXTRA HELP

SSG Sec. 1.8
SSM Sec. 1.8

Pass the Test Software

InterAct Math Tutorial Software

Video 2

CHALKBOARD EXAMPLE

Simplify the expression.

$-(7 - 6k) + 9$

Answer: $2 + 6k$

(b) $4(3m - 2n)$

Use the distributive property first.

$$4(3m - 2n) = 4(3m) - 4(2n) \quad \text{Arrows denote distributive property.}$$
$$= (4 \cdot 3)m - (4 \cdot 2)n \quad \text{Associative property}$$
$$= 12m - 8n$$

(c) $6 + 3(4k + 5) = 6 + 3(4k) + 3(5)$ Distributive property

$= 6 + (3 \cdot 4)k + 3(5)$ Associative property

$= 6 + 12k + 15$

$= 6 + 15 + 12k$ Commutative property

$= 21 + 12k$

(d) $5 - (2y - 8) = 5 - 1 \cdot (2y - 8)$ Replace $-$ with -1.

$= 5 - 2y + 8$ Distributive property

$= 5 + 8 - 2y$ Commutative property

$= 13 - 2y$

NOTE In Example 1, parts (c) and (d), a different use of the commutative property would have resulted in answers of $12k + 21$ and $-2y + 13$. These answers also would be acceptable.

The steps using the commutative and associative properties will not be shown in the rest of the examples, but you should be aware that they are usually involved.

TEACHING TIP Distinguish between the three-term expression $a + b + c$ and the one-term expression abc. Also, ask students to identify the numerical coefficient of $\frac{x}{10}$, and explain why it is $\frac{1}{10}$, not 10.

OBJECTIVE 2 Identify terms and numerical coefficients. A **term** is a number, a variable, or a product or quotient of numbers and variables raised to powers.* Examples of terms include

$$-9x^2, \quad 15y, \quad -3, \quad 8m^2n, \quad \frac{2}{p}, \quad \text{and} \quad k.$$

The **numerical coefficient** of the term $9m$ is 9, the numerical coefficient of $-15x^3y^2$ is -15, the numerical coefficient of x is 1, and the numerical coefficient of 8 is 8. In the expression $\frac{x}{3}$, the numerical coefficient of x is $\frac{1}{3}$. Do you see why?

CAUTION It is important to be able to distinguish between *terms* and *factors.* For example, in the expression $8x^3 + 12x^2$, there are two *terms,* $8x^3$ and $12x^2$. On the other hand, in the one-term expression $(8x^3)(12x^2)$, $8x^3$ and $12x^2$ are *factors.*

Several examples of terms and their numerical coefficients follow.

*Another name for certain terms, **monomial,** is introduced in Chapter 4.

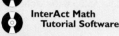

Term	Numerical Coefficient
$-7y$	-7
$8p$	8
$34r^3$	34
$-26x^5yz^4$	-26
$-k$	-1
$\frac{x}{7}$	$\frac{1}{7}$

OBJECTIVE 3 **Identify like terms.** Terms with exactly the same variables that have the same exponents are **like terms.** For example, $9m$ and $4m$ have the same variable and are like terms. Also, $6x^3$ and $-5x^3$ are like terms. The terms $-4y^3$ and $4y^2$ have different exponents and are **unlike terms.**

Here are some additional examples.

$5x$ and $-12x$	$3x^2y$ and $5x^2y$	Like terms
$4xy^2$ and $5xy$	$-7w^3z^3$ and $2xz^3$	Unlike terms

OBJECTIVE 4 **Combine like terms.** Recall the distributive property:

$$x(y + z) = xy + xz.$$

As seen in the previous section, this statement can also be written "backward" as

$$xy + xz = x(y + z).$$

This form of the distributive property may be used to find the sum or difference of like terms. For example,

$$3x + 5x = (3 + 5)x = 8x.$$

This process is called **combining like terms.**

TEACHING TIP In arithmetic, addition of fractions with the same denominators is an example of combining like terms.

NOTE Remember that *only like terms may be combined.* For example, $5x^2 + 2x \neq 7x^3$.

CHALKBOARD EXAMPLE

Combine terms.

$$5z + 9z - 4z$$

Answer: $10z$

EXAMPLE 2 **Combining Like Terms**

Combine like terms in the following expressions.

(a) $9m + 5m$

Use the distributive property as given above.

$$9m + 5m = (9 + 5)m = 14m$$

(b) $6r + 3r + 2r = (6 + 3 + 2)r = 11r$ Distributive property

(c) $4x + x = 4x + 1x = (4 + 1)x = 5x$ (Note: $x = 1x$.)

(d) $16y^2 - 9y^2 = (16 - 9)y^2 = 7y^2$

(e) $32y + 10y^2$ cannot be combined because $32y$ and $10y^2$ are unlike terms. The distributive property cannot be used here to combine coefficients.

When an expression involves parentheses, the distributive property is used both "forward" and "backward" to combine like terms, as shown in the following example.

CHALKBOARD EXAMPLE

Combine like terms.

$-(3 + 5k) + 7k$

Answer: $-3 + 2k$

EXAMPLE 3 Simplifying Expressions Involving Like Terms

Combine like terms in the following expressions.

(a) $14y + 2(6 + 3y) = 14y + 2(6) + 2(3y)$ Distributive property

$= 14y + 12 + 6y$ Multiply.

$= 20y + 12$ Combine like terms.

(b) $9k - 6 - 3(2 - 5k) = 9k - 6 - 3(2) - 3(-5k)$ Distributive property

$= 9k - 6 - 6 + 15k$ Multiply.

$= 24k - 12$ Combine like terms.

(c) $-(2 - r) + 10r = -1(2 - r) + 10r$ Replace $-$ with -1.

$= -1(2) - 1(-r) + 10r$ Distributive property

$= -2 + 1r + 10r$ Multiply.

$= -2 + 11r$ Combine like terms.

(d) $5(2a - 6) - 3(4a - 9) = 10a - 30 - 12a + 27$ Distributive property

$= -2a - 3$ Combine like terms.

Example 3(d) shows that the commutative property can be used with subtraction by treating the subtracted terms as the addition of their additive inverses.

TEACHING TIP Constants are like terms and may be combined.

NOTE Examples 2 and 3 suggest that like terms may be combined by adding or subtracting the coefficients of the terms and keeping the same variable factors.

OBJECTIVE 5 Simplify expressions from word phrases. Earlier we saw how to translate words, phrases, and statements into expressions and equations. Now we can simplify translated expressions by combining like terms.

CHALKBOARD EXAMPLE

Write the following phrase as a mathematical expression and simplify by combining terms.

Three times a number, subtracted from the sum of the number and 8

Answer: $(x + 8) - 3x$; $-2x + 8$

EXAMPLE 4 Converting Words to a Mathematical Expression

Convert to a mathematical expression, and simplify: The sum of 9, five times a number, four times the number, and six times the number.

The word "sum" indicates that the terms should be added. Use x to represent the number. Then the phrase translates as follows.

$9 + 5x + 4x + 6x$ Write as a mathematical expression.

$= 9 + 15x$ Combine like terms.

CAUTION In Example 4, we are dealing with an expression to be simplified, *not* an equation to be solved.

1. false **2.** false **3.** true
4. false **5.** (c) **6.** (c)
7. (a) **8.** (b) **9.** $4r + 11$
10. $7t + 14$ **11.** $5 + 2x - 6y$
12. $8 + 3s - 18t$ **13.** $-7 + 3p$
14. $-17 + 14r$ **15.** -12
16. -23 **17.** 5 **18.** -3
19. 1 **20.** 1 **21.** -1
22. -1 **23.** 74 **24.** 98
25. Answers will vary. For example, $-3x$ and $4x$.
26. Answers will vary. For example, x^2 and x^3.
27. like **28.** like **29.** unlike
30. unlike **31.** like **32.** like
33. unlike **34.** unlike
35. Apples and oranges are examples of unlike fruits, just like x and y are unlike terms. We cannot add x and y to get an expression any simpler than $x + y$; we cannot add, for example, 2 apples and 3 oranges to obtain 5 fruits that are all alike.
36. We use the distributive property to write $6t + 5t = (6 + 5)t$. Then we replace $6 + 5$ with 11 to get $11t$.
37. $9k - 5$ **38.** $4x + 4$
39. $-\frac{1}{3}t - \frac{28}{3}$ **40.** $\frac{49}{6}x - 9$
41. $-4.1r + 5.6$
42. $-4.7b + 6.3$ **43.** $-2y^2 + 3y^3$
44. $21x^4 + 7x^6$ **45.** $-19p + 16$
46. $-x + 21$ **47.** $-4y + 22$
48. $-5t + 61$ **49.** $-16y + 63$
50. $10t - 44$

1.8 EXERCISES

Decide whether each statement is true or false.

1. $6t + 5t^2 = 11t^3$

2. $9xy^2 - 3x^2y = 6xy$

3. $8r^2 + 3r - 12r^2 + 4r = -4r^2 + 7r$

4. $4 + 3t^3 = 7t^3$

In Exercises 5–8, choose the letter of the correct response.

5. Which one of the following is true for all real numbers x?
(a) $6 + 2x = 8x$ **(b)** $6 - 2x = 4x$
(c) $6x - 2x = 4x$ **(d)** $3 + 8(4x - 6) = 11(4x - 6)$

6. Which one of the following is an example of a pair of like terms?
(a) $6t, 6w$ **(b)** $-8x^2y, 9xy^2$ **(c)** $5ry, 6yr$ **(d)** $-5x^2, 2x^3$

7. Which one of the following is an example of a term with numerical coefficient 5?
(a) $5x^3y^7$ **(b)** x^5 **(c)** $\frac{x}{5}$ **(d)** 5^2xy^3

8. Which one of the following is a correct translation for "six times a number, subtracted from the product of eleven and the number" (if x represents the number)?
(a) $6x - 11x$ **(b)** $11x - 6x$ **(c)** $(11 + x) - 6x$ **(d)** $6x - (11 + x)$

Simplify each expression. See Example 1.

9. $4r + 19 - 8$

10. $7t + 18 - 4$

11. $5 + 2(x - 3y)$

12. $8 + 3(s - 6t)$

13. $-2 - (5 - 3p)$

14. $-10 - (7 - 14r)$

Give the numerical coefficient of each term.

15. $-12k$ **16.** $-23y$ **17.** $5m^2$ **18.** $-3n^6$ **19.** xw

20. pq **21.** $-x$ **22.** $-t$ **23.** 74 **24.** 98

25. Give an example of a pair of like terms in the variable x, such that one of them has a negative numerical coefficient, one has a positive numerical coefficient, and their sum has a positive numerical coefficient.

26. Give an example of a pair of unlike terms such that each term has x as the only variable factor.

Identify each group of terms as like *or* unlike.

27. $8r, -13r$ **28.** $-7a, 12a$ **29.** $5z^4, 9z^3$ **30.** $8x^5, -10x^3$

31. $4, 9, -24$ **32.** $7, 17, -83$ **33.** x, y **34.** t, s

35. There is an old saying, "You can't add apples and oranges." Explain how this saying can be applied to the goal of Objective 4 in this section.

36. Explain how the distributive property is used in combining $6t + 5t$ to get $11t$.

Simplify each expression by combining like terms. See Examples 1–3.

37. $4k + 3 - 2k + 8 + 7k - 16$

38. $9x + 7 - 13x + 12 + 8x - 15$

39. $-\frac{4}{3} + 2t + \frac{1}{3}t - 8 - \frac{8}{3}t$

40. $-\frac{5}{6} + 8x + \frac{1}{6}x - 7 - \frac{7}{6}$

41. $-5.3r + 4.9 - 2r + .7 + 3.2r$

42. $2.7b + 5.8 - 3b + .5 - 4.4b$

43. $2y^2 - 7y^3 - 4y^2 + 10y^3$

44. $9x^4 - 7x^6 + 12x^4 + 14x^6$

45. $13p + 4(4 - 8p)$

46. $5x + 3(7 - 2x)$

47. $-4(y - 7) - 6$

48. $-5(t - 13) - 4$

49. $-5(5y - 9) + 3(3y + 6)$

50. $-3(2t + 4) + 8(2t - 4)$

51. $4k - 7$ **52.** $-48j + 10$
53. $-23.7y - 12.6$
54. $43.2t - 28.8$
55. $(x + 3) + 5x$; $6x + 3$
56. $(x + 6) + 6x$; $7x + 6$
57. $(13 + 6x) - (-7x)$; $13 + 13x$
58. $(14 + 8x) - 5x$; $14 + 3x$
59. $2(3x + 4) - (-4 + 6x)$; 12
60. $3(12 + 8x) - (6 + 9x)$; $30 + 15x$
61. Wording will vary. One example is "the difference between 9 times a number and the sum of the number and 2."
62. Wording will vary. One example is "the difference between twice the sum of three times a number and 5, and twice the sum of the number and 4."

51. $-4(-3k + 3) - (6k - 4) - 2k + 1$

52. $-5(8j + 2) - (5j - 3) - 3j + 17$

53. $-7.5(2y + 4) - 2.9(3y - 6)$

54. $8.4(6t - 6) + 2.4(9 - 3t)$

Convert each phrase into a mathematical expression. Use x as the variable. Combine like terms when possible. See Example 4.

55. Five times a number, added to the sum of the number and three

56. Six times a number, added to the sum of the number and six

57. A number multiplied by -7, subtracted from the sum of 13 and six times the number

58. A number multiplied by 5, subtracted from the sum of 14 and eight times the number

59. Six times a number added to -4, subtracted from twice the sum of three times the number and 4 (*Hint: Twice* means two times.)

60. Nine times a number added to 6, subtracted from triple the sum of 12 and 8 times the number (*Hint: Triple* means three times.)

61. Write the expression $9x - (x + 2)$ using words, as in Exercises 55–60.

62. Write the expression $2(3x + 5) - 2(x + 4)$ using words, as in Exercises 55–60.

63. 2, 3, 4, 5 **64.** 1
65. (a) 1, 2, 3, 4 (b) 3, 4, 5, 6 (c) 4, 5, 6, 7
66. The value of $x + b$ also increases by 1 unit.
67. (a) 2, 4, 6, 8 (b) 2, 5, 8, 11 (c) 2, 6, 10, 14 **68.** m
69. (a) 7, 9, 11, 13 (b) 5, 8, 11, 14 (c) 1, 5, 9, 13 In comparison, we see that while the values themselves are different, the number of units of increase is the same as the corresponding parts of Exercise 67.
70. m

RELATING CONCEPTS (EXERCISES 63–70)

Work Exercises 63–70 in order. *They will help prepare you for graphing later in the text.*

63. Evaluate the expression $x + 2$ for the values of x shown in the chart.

x	x + 2
0	
1	
2	
3	

64. Based on your results from Exercise 63, complete the following statement: For every increase of 1 unit for x, the value of $x + 2$ increases by _____ unit(s).

65. Repeat Exercise 63 for these expressions:
(a) $x + 1$ **(b)** $x + 3$ **(c)** $x + 4$

66. Based on your results from Exercises 63 and 65, make a conjecture (an educated guess) about what happens to the value of an expression of the form $x + b$ for any value of b, as x increases by 1 unit.

67. Repeat Exercise 63 for these expressions:
(a) $2x + 2$ **(b)** $3x + 2$ **(c)** $4x + 2$

68. Based on your results from Exercise 67, complete the following statement: For every increase of 1 unit for x, the value of $mx + 2$ increases by _____ units.

69. Repeat Exercise 63 and compare your results to those in Exercise 67 for these expressions:
(a) $2x + 7$ **(b)** $3x + 5$ **(c)** $4x + 1$

70. Based on your results from Exercises 63–69, complete the following statement: For every increase of 1 unit for x, the value of $mx + b$ increases by _____ units.

Did you make the connection that an increase of 1 for x yields an increase of m for $mx + b$?

Instructor's Note

Grouping: 3 students
Time: 20 minutes

This activity uses arithmetic skills covered in this chapter and leads into the geometry problems in the next chapter. The three featured floor plans all have the same area but different perimeters.

A. Answers will vary.
B. 1. All areas are about 400 square feet. 2. Plan one: 71 feet; Plan two: 80 feet; Plan three: 106 feet
C. 1. They're all the same.
2. Plan one; Plan three
3. Answers will vary. For example, there is less heat loss to the outside, since there are fewer feet of perimeter per square foot of area.
4. Answers will vary.

CHAPTER 1 GROUP ACTIVITY

Comparing Shapes of Houses

Objective: Use arithmetic skills to make comparisons.

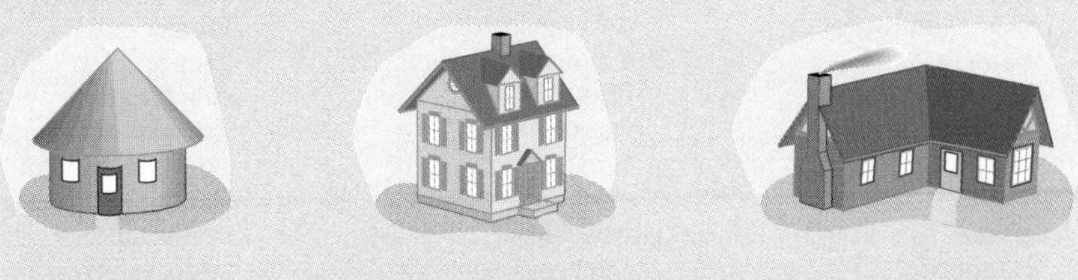

People throughout the world live in different shaped homes. As the population of the earth continues to grow, issues of housing and heating become increasingly important. This activity will explore perimeters and areas of some of these different shaped homes. Floor plans for three different homes are given below.

A. As a group, look at the dimensions of the given floor plans. Considering only the dimensions, which plan do you think has the greatest area?

B. Now have each student in your group pick one floor plan. For each plan, find the following. (Round all answers to the nearest whole number. In Plan one, let $\pi = 3.14$.)
1. The area of the plan
2. The perimeter or circumference (the distance around the outside) of the plan

C. Share your findings with the group and answer the following questions.
1. What did you determine about the areas of the three floor plans?
2. Which plan has the smallest perimeter? Which has the largest perimeter?
3. Why do you think houses with round floor plans might be more energy efficient?
4. What advantages do you think houses with square or rectangular floor plans have? Why do you think floor plans with these shapes are most common for homes today?

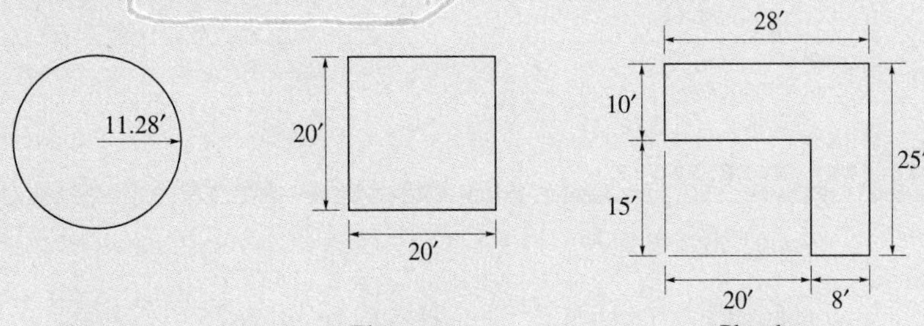

Plan one
$A = \pi r^2$
$C = 2\pi r$

Plan two
$A = s^2$

Plan three
$A = L \cdot W$

CHAPTER 1 SUMMARY

KEY TERMS

1.1 **natural numbers**
whole numbers
numerator
denominator
factor
product
factored
prime
composite
greatest common factor
lowest terms
reciprocal
quotient
sum
least common denominator (LCD)
mixed number
difference

1.2 **exponent (power)**
base
exponential expression
grouping symbols

1.3 **variable**
algebraic expression
equation
solution
set
element

1.4 **number line**
negative numbers
positive numbers
signed numbers
integers
graph
coordinate
rational numbers
set-builder notation
irrational numbers
real numbers
additive inverse (opposite)
absolute value

1.6 **multiplicative inverse (reciprocal)**

1.7 **commutative property**
associative property
identity property
identity element for addition
identity element for multiplication
inverse property
distributive property

1.8 **term**
numerical coefficient
like and unlike terms
combining like terms

NEW SYMBOLS

a^n	n factors of a
[ ]	square brackets (used as grouping symbols)
$=$	is equal to
$\neq$	is not equal to
$<$	is less than (read from left to right)
$>$	is greater than (read from left to right)
$\leq$	is less than or equal to (read from left to right)
$\geq$	is greater than or equal to (read from left to right)
{ }	set braces
$\{x \mid x$ **has a certain property**$\}$	set-builder notation
$-x$	the additive inverse, or opposite, of x
$\lvert x \rvert$	absolute value of x
$\frac{1}{x}$ **or** $1/x$	the multiplicative inverse, or reciprocal, of the nonzero number x
$a(b)$, $(a)(b)$, $a \cdot b$, **or** ab	a times b
$\frac{a}{b}$ **or** a/b	a divided by b

TEST YOUR WORD POWER

See how well you have learned the vocabulary in this chapter. Answers, with examples, are given at the bottom of the next page.

1. A **factor** is
(a) the answer in an addition problem
(b) the answer in a multiplication problem
(c) one of two or more numbers that are added to get another number
(d) one of two or more numbers that are multiplied to get another number.

2. A number is **prime** if
(a) it cannot be factored
(b) it has just one factor
(c) it has only itself and 1 as factors
(d) it has at least two different factors.

3. An **exponent** is
(a) a symbol that tells how many numbers are being multiplied
(b) a number raised to a power
(c) a number that tells how many times a factor is repeated
(d) one of two or more numbers that are multiplied.

4. A **variable** is
(a) a symbol used to represent an unknown number
(b) a value that makes an equation true
(c) a solution of an equation
(d) the answer in a division problem.

5. An **integer** is
(a) a positive or negative number
(b) a natural number, its opposite, or zero
(c) any number that can be graphed on a number line
(d) the quotient of two numbers.

6. A **coordinate** is
(a) the number that corresponds to a point on a number line
(b) the graph of a number
(c) any point on a number line
(d) the distance from 0 on a number line.

7. The **absolute value** of a number is
(a) the graph of the number
(b) the reciprocal of the number
(c) the opposite of the number
(d) the distance between 0 and the number on a number line.

8. A **term** is
(a) a numerical factor
(b) a number or a product or quotient of numbers and variables raised to powers
(c) one of several variables with the same exponents
(d) a sum of numbers and variables raised to powers.

9. A **numerical coefficient** is
(a) the numerical factor in a term
(b) the number of terms in an expression
(c) a variable raised to a power
(d) the variable factor in a term.

QUICK REVIEW

CONCEPTS / **EXAMPLES**

1.1 FRACTIONS

Operations with Fractions

Addition/Subtraction:

1. To add/subtract fractions with the same denominator, add/subtract the numerators and keep the same denominator.
2. To add/subtract fractions with different denominators, find the LCD and write each fraction with this LCD. Then follow the procedure above.

Multiplication: Multiply numerators and multiply denominators.

Division: Multiply the first fraction by the reciprocal of the second fraction.

Perform the operations.

$$\frac{2}{5}+\frac{7}{5}=\frac{2+7}{5}=\frac{9}{5}$$

$$\frac{2}{3}-\frac{1}{2}=\frac{4}{6}-\frac{3}{6} \quad \text{6 is the LCD.}$$

$$=\frac{4-3}{6}=\frac{1}{6}$$

$$\frac{4}{3}\cdot\frac{5}{6}=\frac{20}{18}=\frac{10}{9}$$

$$\frac{6}{5}\div\frac{1}{4}=\frac{6}{5}\cdot\frac{4}{1}=\frac{24}{5}$$

1.2 EXPONENTS, ORDER OF OPERATIONS, AND INEQUALITY

Order of Operations

Simplify within parentheses or above and below fraction bars first, in the following order.

1. Apply all exponents.
2. Do any multiplications or divisions from left to right.
3. Do any additions or subtractions from left to right. If no grouping symbols are present, start with Step 1.

Simplify $36 - 4(2^2 + 3)$.

$$36 - 4(2^2 + 3) = 36 - 4(4 + 3)$$
$$= 36 - 4(7)$$
$$= 36 - 28$$
$$= 8$$

Answers to Test Your Word Power

1. (d) *Example:* Since $2 \times 5 = 10$, the numbers 2 and 5 are factors of 10; other factors of 10 are $-10, -5, -2, -1, 1$, and 10. **2.** (c) *Examples:* 2, 3, 11, 41, 53 **3.** (c) *Example:* In 2^3, the number 3 is the exponent (or power), so 2 is a factor three times; $2^3 = 2 \cdot 2 \cdot 2 = 8$. **4.** (a) *Examples: a, b, c* **5.** (b) *Examples:* $-9, 0, 6$ **6.** (a) *Example:* The point graphed three units to the right of 0 on a number line has coordinate 3. **7.** (d) *Examples:* $|2| = 2$ and $|-2| = 2$ **8.** (b) *Examples:* $6, \frac{x}{2}, -4ab^2$ **9.** (a) *Examples:* The term 3 has numerical coefficient 3, $8z$ has numerical coefficient 8, and $-10x^4y$ has numerical coefficient -10.

1.3 VARIABLES, EXPRESSIONS, AND EQUATIONS

CONCEPTS	EXAMPLES
Evaluate an expression with a variable by substituting a given number for the variable.	Evaluate $2x + y^2$ if $x = 3$ and $y = -4$. $2x + y^2 = 2(3) + (-4)^2$ $= 6 + 16$ $= 22$
Values of a variable that make an equation true are solutions of the equation.	Is 2 a solution of $5x + 3 = 18$? $5(2) + 3 = 18$? $13 = 18$ False 2 is not a solution.

1.4 REAL NUMBERS AND THE NUMBER LINE

CONCEPTS	EXAMPLES
The Ordering of Real Numbers *a* is less than *b* if *a* is to the left of *b* on the number line.	Graph -2, 0, and 3. 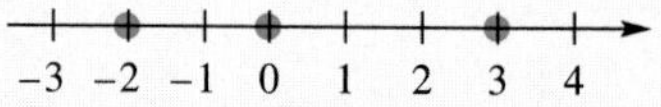 $-2 < 3$ $3 > 0$ $0 < 3$
The additive inverse of x is $-x$.	$-(5) = -5$ $-(-7) = 7$ $-0 = 0$
The absolute value of x, $\lvert x \rvert$, is the distance between x and 0 on the number line.	$\lvert 13 \rvert = 13$ $\lvert 0 \rvert = 0$ $\lvert -5 \rvert = 5$

1.5 ADDITION AND SUBTRACTION OF REAL NUMBERS

CONCEPTS	EXAMPLES
Rules for Addition To add two numbers with the same sign, add their absolute values. The sum has that same sign. To add two numbers with different signs, subtract their absolute values. The sum has the sign of the number with larger absolute value.	Add. $9 + 4 = 13$ $-8 + (-5) = -13$ $7 + (-12) = -5$ $-5 + 13 = 8$
Definition of Subtraction $x - y = x + (-y)$	Subtract. $5 - (-2) = 5 + 2 = 7$
Rules for Subtraction 1. Change the subtraction symbol to the addition symbol. 2. Change the sign of the number being subtracted. 3. Add, using the rules for addition.	$-3 - 4 = -3 + (-4) = -7$ $-2 - (-6) = -2 + 6 = 4$ $13 - (-8) = 13 + 8 = 21$

1.6 MULTIPLICATION AND DIVISION OF REAL NUMBERS

CONCEPTS	EXAMPLES
Rules for Multiplying and Dividing Signed Numbers The product (or quotient) of two numbers having the *same sign* is *positive;* the product (or quotient) of two numbers having *different signs* is *negative.*	Multiply or divide. $6 \cdot 5 = 30$ $(-7)(-8) = 56$ $\frac{20}{4} = 5$ $\frac{-24}{-6} = 4$ $(-6)(5) = -30$ $(6)(-5) = -30$ $\frac{-18}{9} = -2$ $\frac{49}{-7} = -7$

CONCEPTS	EXAMPLES
Definition of Division	
$\frac{x}{y} = x \cdot \frac{1}{y}, \quad y \neq 0$	$\frac{10}{2} = 10 \cdot \frac{1}{2} = 5$
Division by 0 is undefined.	$\frac{5}{0}$ is undefined.
0 divided by a nonzero number equals 0.	$\frac{0}{5} = 0$
1.7 PROPERTIES OF REAL NUMBERS	
Commutative	
$a + b = b + a$	$7 + (-1) = -1 + 7$
$ab = ba$	$5(-3) = (-3)5$
Associative	
$(a + b) + c = a + (b + c)$	$(3 + 4) + 8 = 3 + (4 + 8)$
$(ab)c = a(bc)$	$[(-2)(6)](4) = (-2)[(6)(4)]$
Identity	
$a + 0 = a \qquad 0 + a = a$	$-7 + 0 = -7 \qquad 0 + (-7) = -7$
$a \cdot 1 = a \qquad 1 \cdot a = a$	$9 \cdot 1 = 9 \qquad 1 \cdot 9 = 9$
Inverse	
$a + (-a) = 0 \qquad -a + a = 0$	$7 + (-7) = 0 \qquad -7 + 7 = 0$
$a \cdot \frac{1}{a} = 1 \qquad \frac{1}{a} \cdot a = 1 \quad (a \neq 0)$	$-2\left(-\frac{1}{2}\right) = 1 \qquad -\frac{1}{2}(-2) = 1$
Distributive	
$a(b + c) = ab + ac$	$5(4 + 2) = 5(4) + 5(2)$
$(b + c)a = ba + ca$	$(4 + 2)5 = 4(5) + 2(5)$
$a(b - c) = ab - ac$	$9(5 - 4) = 9(5) - 9(4)$
1.8 SIMPLIFYING EXPRESSIONS	
Only like terms may be combined.	Simplify $-3y^2 + 6y^2 + 14y^2 = 17y^2$.
	$4(3 + 2x) - 6(5 - x)$
	$= 12 + 8x - 30 + 6x$ Distributive property
	$= 14x - 18$

CHAPTER 1 REVIEW EXERCISES

1. $\frac{3}{4}$ 2. $\frac{59}{16}$

[1.1] *Perform each operation.**

1. $\frac{8}{5} \div \frac{32}{15}$

2. $\frac{3}{8} + 3\frac{1}{2} - \frac{3}{16}$

*For help with any of these exercises, refer to the section given in brackets.

3. $\frac{9}{40}$ **4.** 300 **5.** 625
6. $\frac{27}{125}$ **7.** .0000000032
8. .000000001 **9.** 27 **10.** 399
11. 39 **12.** 5 **13.** true
14. true **15.** false
16. $13 < 17$ **17.** $5 + 2 \neq 10$
18. (a) 1983, 1984, 1985, 1986, 1987, 1988, 1993, 1994, 1995
(b) 1987, 1988, 1989, 1990, 1991, 1992, 1994, 1995, 1996
(c) 67.4 billion

3. The pie chart illustrates how 800 people responded to a survey that asked "Do you believe that there was a conspiracy to assassinate John F. Kennedy?" What fractional part of the group did not have an opinion?

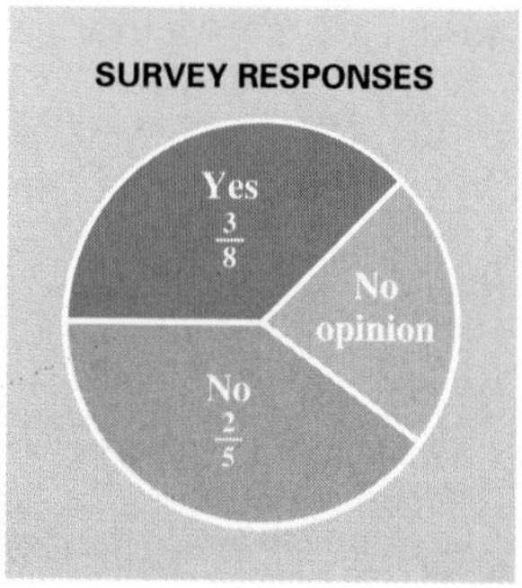

4. Based on the chart in Exercise 3, how many people responded "yes"?

[1.2] *Find the value of each exponential expression.*

5. 5^4 **6.** $\left(\frac{3}{5}\right)^3$ **7.** $(.02)^5$ **8.** $(.001)^3$

Find the value of each expression.

9. $8 \cdot 5 - 13$

10. $7[3 + 6(3^2)]$

11. $\dfrac{9(4^2 - 3)}{4 \cdot 5 - 17}$

12. $\dfrac{6(5 - 4) + 2(4 - 2)}{3^2 - (4 + 3)}$

Tell whether each statement is true or false.

13. $12 \cdot 3 - 6 \cdot 6 \leq 0$ **14.** $3[5(2) - 3] > 20$ **15.** $9 \leq 4^2 - 8$

Write each word statement in symbols.

16. Thirteen is less than seventeen.

17. Five plus two is not equal to ten.

18. Americans are literally bombarded by mail-order catalogs on a daily basis. The bar graph shows the estimated number of catalogs mailed to consumers and businesses during the years 1983–1996.

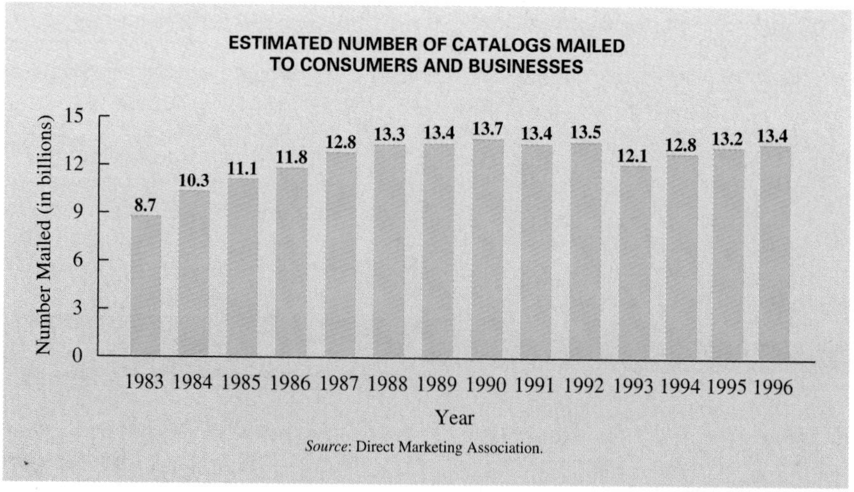

(a) In which years were *fewer than* 13.4 billion catalogs mailed?
(b) In which years were *at least* 12.8 billion mailed?
(c) How many *total* catalogs were mailed in the five years having the largest numbers of mailings?

19. 30 **20.** 60 **21.** 14
22. 13 **23.** $x + 6$ **24.** $8 - x$
25. $6x - 9$ **26.** $12 + \frac{3}{5}x$
27. yes **28.** no
29. $2x - 6 = 10$; 8
30. $4x = 8$; 2
31. 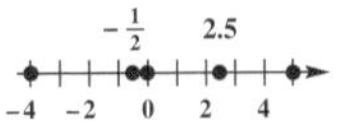

32. 

33. rational numbers, real numbers
34. irrational numbers, real numbers **35.** -10 **36.** -9
37. $-\frac{3}{4}$ **38.** $-|23|$ **39.** true
40. true **41.** true **42.** true
43. (a) 9 (b) 9 **44.** (a) 0 (b) 0 **45.** (a) -6 (b) 6
46. (a) $\frac{5}{7}$ (b) $\frac{5}{7}$ **47.** 12
48. -3 **49.** -19 **50.** -7
51. -6 **52.** -4 **53.** -17
54. $-\frac{29}{36}$ **55.** -21.8
56. -14 **57.** -10 **58.** -19
59. -11 **60.** -1 **61.** 7
62. $-\frac{43}{35}$ **63.** 10.31 **64.** -12
65. 2 **66.** -3

[1.3] *Find the numerical value of each expression if $x = 6$ and $y = 3$.*

19. $2x + 6y$ **20.** $4(3x - y)$ **21.** $\frac{x}{3} + 4y$ **22.** $\frac{x^2 + 3}{3y - x}$

Change each word phrase to an algebraic expression. Use x as the variable to represent the number.

23. Six added to a number
24. A number subtracted from eight
25. Nine subtracted from six times a number
26. Three-fifths of a number added to 12

Decide whether the given number is a solution of the given equation.

27. $5x + 3(x + 2) = 22$; 2
28. $\frac{t + 5}{3t} = 1$; 6

Change each word statement to an equation. Use x as the variable. Then find the solution from the set $\{0, 2, 4, 6, 8, 10\}$.

29. Six less than twice a number is 10.
30. The product of a number and 4 is 8.

[1.4] *Graph each group of numbers on a number line.*

31. $-4, -\frac{1}{2}, 0, 2.5, 5$
32. $-2, |-3|, -3, |-1|$

Classify each number, using the sets natural numbers, whole numbers, integers, rational numbers, irrational numbers, real numbers.

33. $\frac{4}{3}$
34. $\sqrt{6}$

Select the smaller number in each pair.

35. $-10, 5$ **36.** $-8, -9$ **37.** $-\frac{2}{3}, -\frac{3}{4}$ **38.** $0, -|23|$

Decide whether each statement is true or false.

39. $12 > -13$ **40.** $0 > -5$ **41.** $-9 < -7$ **42.** $-13 \geq -13$

*For the following, **(a)** find the opposite of the number and **(b)** find the absolute value of the number.*

43. -9 **44.** 0 **45.** 6 **46.** $-\frac{5}{7}$

Simplify each number by removing absolute value symbols.

47. $|-12|$ **48.** $-|3|$ **49.** $-|-19|$ **50.** $-|9 - 2|$

[1.5] *Perform the indicated operations.*

51. $-10 + 4$
52. $14 + (-18)$
53. $-8 + (-9)$
54. $\frac{4}{9} + \left(-\frac{5}{4}\right)$
55. $-13.5 + (-8.3)$
56. $(-10 + 7) + (-11)$
57. $[-6 + (-8) + 8] + [9 + (-13)]$
58. $(-4 + 7) + (-11 + 3) + (-15 + 1)$
59. $-7 - 4$
60. $-12 - (-11)$
61. $5 - (-2)$
62. $-\frac{3}{7} - \frac{4}{5}$
63. $2.56 - (-7.75)$
64. $(-10 - 4) - (-2)$
65. $(-3 + 4) - (-1)$
66. $-(-5 + 6) - 2$

67. $(-31 + 12) + 19; 0$
68. $[-4 + (-8)] + 13; 1$
69. $-4 - (-6); 2$
70. $[4 + (-8)] - 5; -9$
71. -2 72. -1 73. \$26.25
74. $-10°$F 75. $-\$29$
76. $-10°$ 77. It gained 4 yards.
78. 4,709,000 79. 36
80. -105 81. $\frac{1}{2}$ 82. 10.08
83. -20 84. -10 85. -24
86. -35 87. 4 88. -20
89. $-\frac{3}{4}$ 90. 11.3 91. -1
92. 2 93. 1 94. .5
95. -18 96. -18 97. 125
98. -423 99. $-4(5) - 9; -29$
100. $\frac{5}{6}[12 + (-6)]; 5$
101. $\frac{12}{8 + (-4)}; 3$
102. $\frac{-20(12)}{15 - (-15)}; -8$
103. $8x = -24; -3$
104. $\frac{x}{3} = -2; -6$

Write a numerical expression for each phrase, and simplify the expression.

67. 19 added to the sum of -31 and 12

68. 13 more than the sum of -4 and -8

69. The difference between -4 and -6

70. Five less than the sum of 4 and -8

Find the solution of the equation from the set $\{-3, -2, -1, 0, 1, 2, 3\}$ by guessing or by trial and error.

71. $x + (-2) = -4$

72. $12 + x = 11$

Solve each problem.

73. Like many people, Kareem Dunlap neglects to keep up his checkbook balance. When he finally balanced his account, he found the balance was $-\$23.75$, so he deposited \$50.00. What is his new balance?

74. The low temperature in Yellowknife, in the Canadian Northwest Territories, one January day was $-26°$F. It rose 16° that day. What was the high temperature?

75. Eric owed his brother \$28. He repaid \$13 but then borrowed another \$14. What positive or negative amount represents his present financial status?

76. If the temperature drops 7° below its previous level of $-3°$, what is the new temperature?

77. A football team gained 3 yards on the first play from scrimmage, lost 12 yards on the second play, and then gained 13 yards on the third play. How many yards did the team gain or lose altogether?

78. In 1985, the construction industry had 4,480,000 employees. By 1990, this number had increased by 759,000 employees, but it experienced a decrease of 530,000 employees by 1994. How many employees were there in 1994? (*Source:* U.S. Bureau of the Census.)

[1.6] *Perform the indicated operations.*

79. $(-12)(-3)$

80. $15(-7)$

81. $\left(-\frac{4}{3}\right)\left(-\frac{3}{8}\right)$

82. $(-4.8)(-2.1)$

83. $5(8 - 12)$

84. $(5 - 7)(8 - 3)$

85. $2(-6) - (-4)(-3)$

86. $3(-10) - 5$

87. $\frac{-36}{-9}$

88. $\frac{220}{-11}$

89. $-\frac{1}{2} \div \frac{2}{3}$

90. $-33.9 \div (-3)$

91. $\frac{-5(3) - 1}{8 - 4(-2)}$

92. $\frac{5(-2) - 3(4)}{-2[3 - (-2)] - 1}$

93. $\frac{10^2 - 5^2}{8^2 + 3^2 - (-2)}$

94. $\frac{(.6)^2 + (.8)^2}{(-1.2)^2 - (-.56)}$

Evaluate each expression if $x = -5$, $y = 4$, and $z = -3$.

95. $6x - 4z$

96. $5x + y - z$

97. $5x^2$

98. $z^2(3x - 8y)$

Write a numerical expression for each phrase, and simplify the expression.

99. Nine less than the product of -4 and 5

100. Five-sixths of the sum of 12 and -6

101. The quotient of 12 and the sum of 8 and -4

102. The product of -20 and 12, divided by the difference between 15 and -15

Write each sentence in symbols, using x as the variable, and find the solution by guessing or by trial and error. All solutions come from the list of integers between -12 and 12.

103. 8 times a number is -24.

104. The quotient of a number and 3 is -2.

105. (a) \$61,374,559
(b) \$2,245,068
106. \$2517.6 million
107. identity property
108. identity property
109. inverse property
110. inverse property
111. associative property
112. associative property
113. distributive property
114. commutative property
115. $(7 + 1)y$; $8y$
116. $-48 + 12t$
117. $3(2s + 5y)$
118. $4r - 5s$
119. $25 - (5 - 2) = 22$ and $(25 - 5) - 2 = 18$. Because different groupings lead to different results, we conclude that in general subtraction is not associative.
120. $180 \div (15 \div 5) = 60$ and $(180 \div 15) \div 5 = \frac{12}{5}$. Because different groupings lead to different results, we conclude that in general division is not associative.
121. $11m$
122. $16p^2$
123. $16p^2 + 2p$
124. $-4k + 12$
125. $-2m + 29$
126. $-5k - 1$

105. The payrolls and average salaries of 5 of the 28 major league baseball teams on opening day of 1998 are listed here.

Team	*Payroll*	*Average Salary*
Baltimore Orioles	\$68,988,134	\$2,555,116
New York Yankees	63,460,567	2,440,791
Cleveland Indians	59,583,500	2,127,982
Atlanta Braves	59,536,000	2,126,286
Texas Rangers	55,304,595	1,975,164

Source: The Associated Press.

(a) What is the average of the five payrolls? Round to the nearest dollar.
(b) What is the average of the five average salaries? Round to the nearest dollar.

106. The bar graph shows the 1993 sales in millions of dollars of four of the largest brands in the United States. What was the average of these sales?

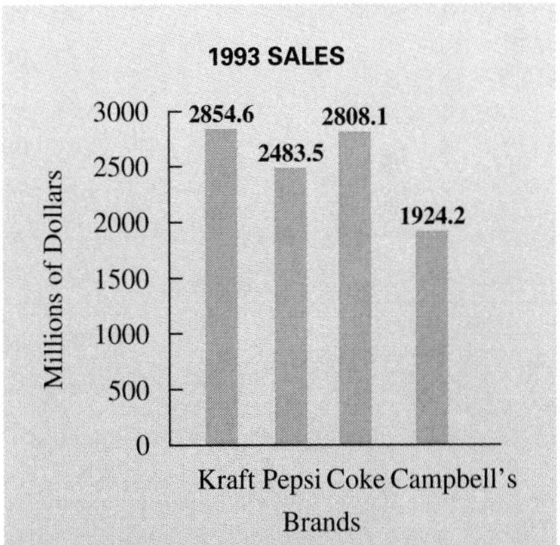

Source: Information Resources, Inc.

[1.7] *Decide whether each statement is an example of the commutative, associative, identity, inverse, or distributive property.*

107. $6 + 0 = 6$

108. $5 \cdot 1 = 5$

109. $-\frac{2}{3}\left(-\frac{3}{2}\right) = 1$

110. $17 + (-17) = 0$

111. $5 + (-9 + 2) = [5 + (-9)] + 2$

112. $w(xy) = (wx)y$

113. $3x + 3y = 3(x + y)$

114. $(1 + 2) + 3 = 3 + (1 + 2)$

Use the distributive property to rewrite each expression. Simplify if possible.

115. $7y + y$

116. $-12(4 - t)$

117. $3(2s) + 3(5y)$

118. $-(-4r + 5s)$

119. Evaluate $25 - (5 - 2)$ and $(25 - 5) - 2$. Use this example to explain why subtraction is not associative.

120. Evaluate $180 \div (15 \div 5)$ and $(180 \div 15) \div 5$. Use this example to explain why division is not associative.

[1.8] *Combine terms whenever possible.*

121. $2m + 9m$

122. $15p^2 - 7p^2 + 8p^2$

123. $5p^2 - 4p + 6p + 11p^2$

124. $-2(3k - 5) + 2(k + 1)$

125. $7(2m + 3) - 2(8m - 4)$

126. $-(2k + 8) - (3k - 7)$

127. 16 128. $\frac{25}{36}$ 129. $\frac{8}{3}$

130. $-\frac{1}{24}$ 131. 2 132. 77.6

133. $-1\frac{1}{2}$ 134. 11 135. $-\frac{28}{15}$

136. 24 137. $8x^2 - 21y^2$

138. $16t - 36$

139. When dividing 0 *by* a nonzero number, the quotient will be 0. However, dividing a number *by* 0 is undefined.

140. It is not correct because it does not consider the operation involved. Multiplying two negative numbers gives a positive number, but adding two negative numbers gives a negative number.

141. $5(x + 7)$; $5x + 35$

142. −13°F 143. −79

144. 22

MIXED REVIEW EXERCISES*

Perform the indicated operations.

127. $[(-2) + 7 - (-5)] + [-4 - (-10)]$

128. $\left(-\frac{5}{6}\right)^2$

129. $\frac{6(-4) + 2(-12)}{5(-3) + (-3)}$

130. $\frac{3}{8} - \frac{5}{12}$

131. $\frac{8^2 + 6^2}{7^2 + 1^2}$

132. $-16(-3.5) - 7.2(-3)$

133. $2\frac{5}{6} - 4\frac{1}{3}$

134. $-8 + [(-4 + 17) - (-3 - 3)]$

135. $-\frac{12}{5} \div \frac{9}{7}$

136. $(-8 - 3) - 5(2 - 9)$

137. $5x^2 - 12y^2 + 3x^2 - 9y^2$

138. $-4(2t + 1) - 8(-3t + 4)$

139. Write a sentence or two explaining the special considerations involving zero when dividing.

140. "Two negatives give a positive" is often heard from students. Is this correct? Use more precise language in explaining what this means.

141. Use x as the variable and write an expression for "the product of 5 and the sum of a number and 7." Then use the distributive property to rewrite the expression.

142. The highest temperature ever recorded in Albany, New York, was 99°F, while the lowest was 112° less than the highest. What was the lowest temperature ever recorded in Albany? (*Source: The World Almanac and Book of Facts,* 1998.)

The year-to-year percentage change in net corporate income for companies in the DJ-Global U.S. Index is given in the accompanying graph for the years 1994 through the second quarter of 1997.

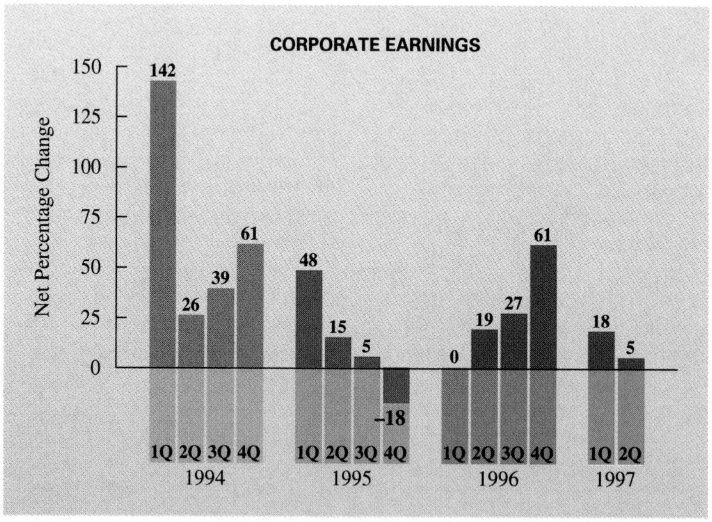

143. What signed number represents the change from the fourth quarter in 1994 to the fourth quarter in 1995?

144. What signed number represents the change from the third quarter in 1995 to the third quarter in 1996?

*The order of exercises in this final group does not correspond to the order in which topics occur in the chapter. This random ordering should help you prepare for the chapter test in yet another way.

CHAPTER 1 TEST

[1.1] 1. $\frac{7}{11}$ 2. $\frac{241}{120}$ 3. $\frac{19}{18}$
4. (a) 492 million (b) 861 million

[1.2] 5. true
[1.4] 6. (number line from −4 to 4)

7. rational numbers, real numbers
8. If -8 and -1 are both graphed on a number line, we see that the point for -8 is to the *left* of the point for -1. This indicates $-8 < -1$.
9. $\frac{-6}{2+(-8)}$; 1
[1.2] 10. (a) Japanese, Vietnamese, Filipino, Hawaiian, Other Asian and Pacific Islander (b) Asian Indian, Chinese, Korean

[1.1, 1.4–1.6] 11. 4 12. $-2\frac{5}{6}$

1. Write $\frac{63}{99}$ in lowest terms.

2. Add: $\frac{5}{8} + \frac{11}{12} + \frac{7}{15}$.

3. Divide: $\frac{19}{15} \div \frac{6}{5}$.

4. The pie chart indicates the market share of different means of intercity transportation, based on 1,230 million passengers carried.
(a) How many of these passengers used air travel?
(b) How many of these passengers did not use the bus?

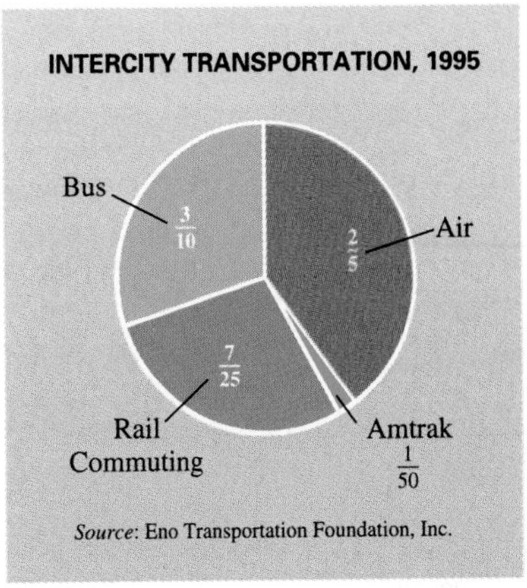

5. Decide whether $4[-20 + 7(-2)] \leq 135$ is true or false.

6. Graph the group of numbers $-1, -3, |-4|, |-1|$ on a number line.

7. To which of the following sets does $-\frac{2}{3}$ belong: natural numbers, whole numbers, integers, rational numbers, irrational numbers, real numbers?

8. Explain how a number line can be used to show that -8 is less than -1.

9. Write in symbols: The quotient of -6 and the sum of 2 and -8. Simplify the expression.

10. According to the bar graph shown here, which subgroups of Asian and Pacific Islander-owned businesses had revenues that were
(a) less than \$13 billion? **(b)** greater than or equal to \$16.2 billion?

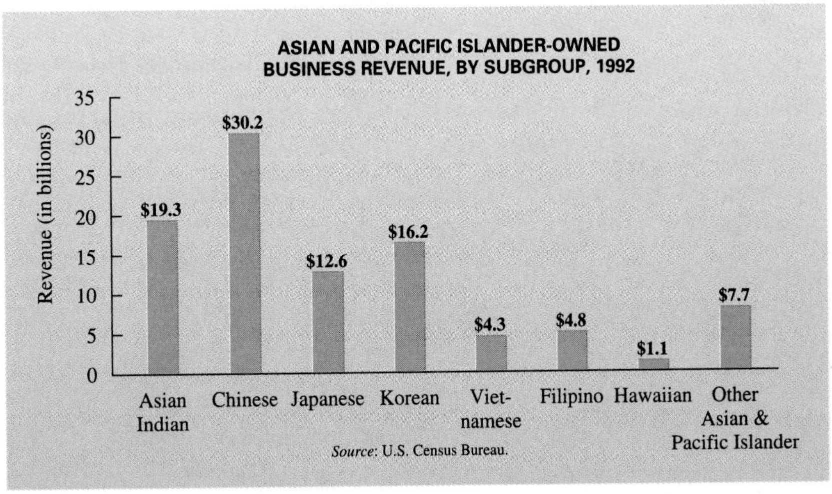

Perform the indicated operations.

11. $-2 - (5 - 17) + (-6)$

12. $-5\frac{1}{2} + 2\frac{2}{3}$

13. 2 **14.** 6 **15.** 108
16. 3 **17.** $\frac{30}{7}$
[1.3, 1.5, 1.6] **18.** 6 **19.** 4
[1.4–1.6] **20.** −70 **21.** 3
[1.5, 1.6] **22.** (a) −$3100
(b) $3600 (c) $6800 (d) $4200
23. 15 **24.** (a) −14 (b) 55
(c) 31

13. $-6 - [-7 + (2 - 3)]$

14. $4^2 + (-8) - (2^3 - 6)$

15. $(-5)(-12) + 4(-4) + (-8)^2$

16. $\dfrac{-7 - (-6 + 2)}{-5 - (-4)}$

17. $\dfrac{30(-1 - 2)}{-9[3 - (-2)] - 12(-2)}$

Find the solution for each equation from the set $\{-6, -4, -2, 0, 2, 4, 6\}$ *by guessing or by trial and error.*

18. $-x + 3 = -3$

19. $-3x = -12$

Evaluate each expression, given $x = -2$ *and* $y = 4$.

20. $3x - 4y^2$

21. $\dfrac{5x + 7y}{3(x + y)}$

Solve each problem.

22. The average sales prices of new single-family homes in the United States for the years 1990 through 1995 are shown in the chart. Determine the change from one year to the next by subtraction.

	Year	Average Sales Price	Change from Previous Year
	1990	$149,800	
	1991	$147,200	−$2600
(a)	1992	$144,100	________
(b)	1993	$147,700	________
(c)	1994	$154,500	________
(d)	1995	$158,700	________

Source: U.S. Bureau of the Census.

23. For a certain system of rating major league baseball relief pitchers, 3 points are awarded for a save, 3 points are awarded for a win, 2 points are subtracted for a loss, and 2 points are subtracted for a blown save. If Dennis Eckersley of the Boston Red Sox has 4 saves, 3 wins, 2 losses, and 1 blown save, how many points does he have?

24. The bar graph shows the number of corporate name changes that occurred during the years 1990–1996. These changes were mostly brought on by mergers and acquisitions. Use a signed number to represent each of the following.
(a) the change from 1990 to 1991
(b) the change from 1991 to 1996
(c) the change from 1992 to 1996

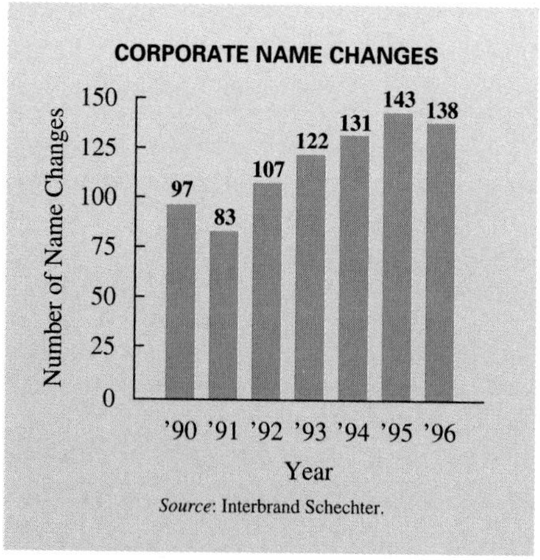

Source: Interbrand Schechter.

[1.7] 25. B 26. D 27. E
28. A 29. C
30. distributive property
31. (a) -18 (b) -18 (c) The distributive property assures us that the answers must be the same, because $a(b + c) = ab + ac$ for all a, b, c.
[1.8] 32. $21x$ 33. $15x - 3$

Match each property in Column I with the example of it in Column II.

I	II
25. Commutative	**A.** $3x + 0 = 3x$
26. Associative	**B.** $(5 + 2) + 8 = 8 + (5 + 2)$
27. Inverse	**C.** $-3(x + y) = -3x + (-3y)$
28. Identity	**D.** $-5 + (3 + 2) = (-5 + 3) + 2$
29. Distributive	**E.** $-\frac{5}{3}\left(-\frac{3}{5}\right) = 1$

30. What property is used to show that $3(x + 1) = 3x + 3$?

31. Consider the expression $-6[5 + (-2)]$.
(a) Evaluate it by first working within the brackets.
(b) Evaluate it by using the distributive property.
(c) Why must the answers in items (a) and (b) be the same?

Simplify by combining like terms.

32. $8x + 4x - 6x + x + 14x$

33. $5(2x - 1) - (x - 12) + 2(3x - 5)$

2 Linear Equations and Inequalities in One Variable

Sports

The first organized worldwide sporting event was the Summer Olympics I, held in 1896 in Athens, Greece. Three hundred eleven men from 13 nations gathered to compete. One hundred years later, the 1996 Summer Olympics, held in Atlanta, attracted 10,341 competitors—6562 men and 3779 women, from 197 nations. The results of an Olympic women's swimming event from 1968 to 1996 are given in the table. (Winning times are given in minutes and seconds to the nearest hundredth.)

200-Meter Freestyle

1968 Deborah Meyer, U.S.	2:10.50
1972 Shane Gould, Australia	2:03.56
1976 Kornelia Ender, East Germany	1:59.26
1980 Barbara Krause, East Germany	1:58.33
1984 Mary Wayte, U.S.	1:59.23
1988 Heike Friederich, East Germany	1:57.65
1992 Nicole Haislett, U.S.	1:57.90
1996 Claudia Poll, Costa Rica	1:58.16

Which year had the fastest winning time? In Section 2.6, we'll see how to determine the winning speed from this information.

In the United States, the first organized professional sport was baseball. The first baseball league was founded in 1876, and the first World Series playoff was held in 1882.* Both football and basketball worked their way from college campus to professional league status.

And, as you'll discover in this chapter, sports involves mathematics; the vast quantity of statistics kept by professional sports leagues fills numerous books.

*All data from *The Universal Almanac,* 1997, John W. Wright, General Editor.

2.1 The Addition and Multiplication Properties of Equality

OBJECTIVES

1. Identify linear equations.
2. Use the addition property of equality.
3. Use the multiplication property of equality.
4. Simplify equations, and then use the properties of equality.

FOR EXTRA HELP

SSG Sec. 2.1
SSM Sec. 2.1
Pass the Test Software
InterAct Math Tutorial Software
Video 3

Quite possibly the most important topic in the study of beginning algebra is the solution of equations. We will investigate many types of equations in this book, and the properties introduced in this section are essential in solving them.

OBJECTIVE 1 **Identify linear equations.** The simplest type of equation is a *linear equation.* Before we can solve a linear equation we must be able to recognize one.

Linear Equation

A **linear equation in one variable** can be written in the form

$$Ax + B = 0$$

for real numbers A and B, with $A \neq 0$.

For example,

$$4x + 9 = 0, \qquad 2x - 3 = 5, \qquad \text{and} \qquad x = 7$$

are linear equations; the last two can be written in the specified form using the properties to be developed in this section. However,

$$x^2 + 2x = 5, \qquad \frac{1}{x} = 6 \qquad \text{and} \qquad |2x + 6| = 0$$

are *not* linear equations.

Methods of solving linear equations are introduced in this section and in Section 2.2. As discussed in Section 1.3, a *solution* of an equation is a number that when substituted for the variable makes the equation true; that is, *satisfies* the equation. An equation is solved by finding its **solution set,** the set of all solutions. Equations that have exactly the same solution sets are **equivalent equations.** Linear equations are solved by using a series of steps to produce equivalent equations until an equation of the form

$$x = \text{a number}$$

is obtained.

OBJECTIVE 2 **Use the addition property of equality.** Consider the equation

$$x = 4.$$

Suppose we add 7 to both sides of the equation, getting

$$x + 7 = 11.$$

Now, adding $2x$ to both sides gives

$$3x + 7 = 2x + 11.$$

INSTRUCTOR'S RESOURCES

PTB Sec. 2.1 **ISM** Sec. 2.1
AB Sec. 2.1

TEST GENERATOR

WORLD WIDE WEB
www.LialAlgebra.com

Verify that $x = 4$ satisfies each of these three equations, so they are thus equivalent equations. To solve the third equation, $3x + 7 = 2x + 11$, all we have to do is reverse the steps shown earlier by adding the negative in each case. The **addition property of equality** justifies this idea.

Addition Property of Equality

If A, B, and C are mathematical expressions that represent real numbers, then the equations

$$A = B \quad \text{and} \quad A + C = B + C$$

have exactly the same solution. In words, the same expression may be added to both sides of an equation without changing the solution.

Note that, in the addition property, C represents a mathematical expression. This means that numbers or terms with variables, or even sums of terms that represent real numbers, can be added to both sides of an equation.

CHALKBOARD EXAMPLE

Solve $a + 2 = -3$.

Answer: $\{-5\}$

EXAMPLE 1 Using the Addition Property of Equality

Solve $x - 16 = 7$.

If the left side of this equation were just x, the solution would be found. Get x alone by using the addition property of equality and adding 16 on both sides.

$$x - 16 = 7$$
$$x - 16 + \mathbf{16} = 7 + \mathbf{16} \qquad \text{Add 16 on both sides.}$$
$$x = 23$$

Note that we combined the steps that change $x - 16 + 16$ to $x + 0$ and $x + 0$ to x. We will combine these steps from now on. Check by substituting 23 for x in the *original* equation.

$$x - 16 = 7 \qquad \text{Original equation}$$
$$23 - 16 = 7 \quad ? \qquad \text{Let } x = 23.$$
$$7 = 7 \qquad \text{True}$$

Since the check results in a true statement, $\{23\}$ is the solution set.

TEACHING TIP Equations can be thought of in terms of a balance. Thus, adding the same quantity to both sides does not affect the balance.

In this example, why was 16 added to both sides of the equation $x - 16 = 7$? The equation would be solved if it could be rewritten so that one side contained only the variable and the other side contained only a number. Since $x - 16 + 16 = x + 0 = x$, adding 16 on the left side simplifies that side to just x, the variable, as desired.

The addition property of equality says that the same expression may be *added* to both sides of an equation. As was shown in Chapter 1, subtraction is defined in terms of addition. Therefore, the addition property also permits *subtracting* the same expression from both sides of an equation.

CHALKBOARD EXAMPLE

Solve.

$9r + 4r + 6 - 2 = 9r + 4 + 3r$

Answer: {0}

EXAMPLE 2 Subtracting a Variable Expression to Solve an Equation

Solve the equation $3k - 12 + k + 2 = 5 + 3k + 2$.

Begin by combining like terms on each side of the equation to get

$$4k - 10 = 7 + 3k.$$

Next, get all terms that contain variables on the same side of the equation and all terms without variables on the other side. One way to start is to subtract $3k$ from both sides.

$4k - 10 - 3k = 7 + 3k - 3k$	Subtract $3k$ from both sides.
$k - 10 = 7$	Combine terms.
$k - 10 + 10 = 7 + 10$	Add 10 to both sides.
$k = 17$	

Check by substituting 17 for k in the original equation.

$3k - 12 + k + 2 = 5 + 3k + 2$		Original equation
$3(17) - 12 + 17 + 2 = 5 + 3(17) + 2$	?	Let $k = 17$.
$51 - 12 + 17 + 2 = 5 + 51 + 2$	?	Multiply.
$58 = 58$		True

The check results in a true statement, so the solution set is {17}.

NOTE Subtracting $3k$ in Example 2 is the same as adding $-3k$, since by the definition of subtraction, $a - b = a + (-b)$.

OBJECTIVE 3 Use the multiplication property of equality. The addition property of equality by itself is not enough to solve an equation like $3x + 2 = 17$.

$3x + 2 = 17$	
$3x + 2 - 2 = 17 - 2$	Subtract 2 from both sides.
$3x = 15$	

The variable x is not alone on one side of the equation; the equation has $3x$ instead. Another property is needed to change $3x = 15$ to $x =$ a number.

If $3x = 15$, then $3x$ and 15 both represent the same number. Multiplying both $3x$ and 15 by the same number will also result in an equality. The **multiplication property of equality** states that both sides of an equation can be multiplied by the same nonzero expression.

Multiplication Property of Equality

If A, B, and C are mathematical expressions that represent real numbers, then the equations

$$A = B \quad \text{and} \quad AC = BC$$

have exactly the same solution. (Assume that $C \neq 0$.) In words, both sides of an equation may be multiplied by the same nonzero expression without changing the solution.

This property can be used to solve $3x = 15$. The $3x$ on the left must be changed to $1x$, or x, instead of $3x$. To get x, multiply both sides of the equation by $\frac{1}{3}$. Use $\frac{1}{3}$ since it is the reciprocal of the coefficient of x. This works because $3 \cdot \frac{1}{3} = \frac{3}{3} = 1$.

$$3x = 15$$

$$\frac{1}{3}(3x) = \frac{1}{3} \cdot 15 \qquad \text{Multiply both sides by } \tfrac{1}{3}.$$

$$\left(\frac{1}{3} \cdot 3\right)x = \frac{1}{3} \cdot 15 \qquad \text{Associative property}$$

$$1x = 5 \qquad \text{Multiplicative inverse property}$$

$$x = 5 \qquad \text{Identity property}$$

The solution set of the equation is $\{5\}$. Check by substituting 5 for x in the original equation. From now on we shall combine the last two steps shown in this example.

Just as the addition property of equality permits subtracting the same number from both sides of an equation, the multiplication property of equality permits dividing both sides of an equation by the same nonzero number. For example, the equation $3x = 15$, solved above by multiplication, could also be solved by dividing both sides by 3, as follows.

$$3x = 15$$

$$\frac{3x}{3} = \frac{15}{3} \qquad \text{Divide by 3.}$$

$$x = 5 \qquad \text{Simplify.}$$

In practice, it is usually easier to multiply on each side if the coefficient of the variable is a fraction, and divide on each side if the coefficient is an integer. For example, to solve

$$-\frac{3}{4}x = 12$$

it is easier to multiply by $-\frac{4}{3}$ than to divide by $-\frac{3}{4}$. On the other hand, to solve

$$-5x = -20$$

it is easier to divide by -5 than to multiply by $-\frac{1}{5}$.

CHALKBOARD EXAMPLE

Solve $5.2p = 22.36$.

Answer: $\{4.3\}$

EXAMPLE 3 Dividing Each Side of an Equation by a Nonzero Number

Solve $2.1x = 6.09$.

Divide both sides of the equation by 2.1.

$$\frac{2.1x}{\mathbf{2.1}} = \frac{6.09}{\mathbf{2.1}}$$

$$x = 2.9 \qquad \text{Divide.}$$

You may wish to use a calculator to find the quotient $6.09 \div 2.1$. Check that the solution is 2.9.

$$2.1(\mathbf{2.9}) = 6.09 \quad ? \qquad \text{Let } x = 2.9.$$

$$6.09 = 6.09 \qquad \text{True}$$

The solution set is $\{2.9\}$.

In the next two examples, multiplication produces the solution more quickly than division.

CHALKBOARD EXAMPLE

Solve $\frac{a}{-2} = 8$.

Answer: $\{-16\}$

EXAMPLE 4 Using the Multiplication Property of Equality with a Fraction

Solve $\frac{a}{4} = 3$.

Replace $\frac{a}{4}$ by $\frac{1}{4}a$ since division by 4 is the same as multiplication by $\frac{1}{4}$. Get a alone by multiplying both sides by 4, the reciprocal of the coefficient of a.

$$\frac{a}{4} = 3$$

$$\frac{1}{4}a = 3 \qquad \text{Change } \tfrac{a}{4} \text{ to } \tfrac{1}{4}a.$$

$$\mathbf{4} \cdot \frac{1}{4}a = \mathbf{4} \cdot 3 \qquad \text{Multiply by 4.}$$

$$1a = 12 \qquad \text{Inverse property}$$

$$a = 12 \qquad \text{Identity property}$$

Check the answer.

$$\frac{a}{4} = 3 \qquad \text{Original equation}$$

$$\frac{\mathbf{12}}{4} = 3 \quad ? \qquad \text{Let } a = 12.$$

$$3 = 3 \qquad \text{True}$$

Since 12 satisfies the equation, the solution set is $\{12\}$.

CHALKBOARD EXAMPLE

Solve $-\frac{5}{6}t = -15$.

Answer: $\{18\}$

EXAMPLE 5 Using the Multiplication Property of Equality in Two Ways

Solve $\frac{3}{4}h = 6$.

We will show two ways to solve this equation.

Method 1

To get h alone, multiply both sides of the equation by $\frac{4}{3}$, the reciprocal of $\frac{3}{4}$.

$$\frac{3}{4}h = 6$$

$$\frac{4}{3}\left(\frac{3}{4}h\right) = \frac{4}{3} \cdot 6 \qquad \text{Multiply by } \tfrac{4}{3}.$$

$$1 \cdot h = \frac{4}{3} \cdot \frac{6}{1} \qquad \text{Inverse property}$$

$$h = 8 \qquad \text{Identity property}$$

Method 2

Begin by multiplying both sides of the equation by 4 to eliminate the denominator.

$$\frac{3}{4}h = 6$$

$$4\left(\frac{3}{4}h\right) = 4 \cdot 6 \qquad \text{Multiply by 4.}$$

$$3h = 24$$

$$\frac{3h}{3} = \frac{24}{3} \qquad \text{Divide by 3.}$$

$$h = 8$$

Using either method, the solution set is $\{8\}$. Check the answer by substitution in the original equation.

OBJECTIVE 4 **Simplify equations, and then use the properties of equality.** The next example uses the distributive property to first simplify an equation.

CHALKBOARD EXAMPLE

Solve.

$-3(m - 4) + 2(5 + 2m) = 29$

Answer: $\{7\}$

EXAMPLE 6 Simplifying an Equation before Solving

Solve $2(3 + 7x) - (1 + 15x) = 2$.

Use the distributive property to first simplify the equation.

$$2(3 + 7x) - 1(1 + 15x) = 2 \qquad \text{Replace the } - \text{ sign with } -1.$$

$$6 + 14x - 1 - 15x = 2 \qquad \text{Distributive property}$$

$$5 - x = 2 \qquad \text{Combine like terms.}$$

$$5 - x - 5 = 2 - 5 \qquad \text{Subtract 5.}$$

$$-x = -3$$

The variable is alone on the left, but its coefficient is -1, since $-x = -1 \cdot x$. When this occurs, simply multiply both sides by -1 (the reciprocal of -1).

$$-x = -3$$

$$-1 \cdot x = -3 \qquad -x = -1 \cdot x$$

$$-1(-1 \cdot x) = -1(-3) \qquad \text{Multiply by } -1.$$

$$x = 3$$

The solution set is $\{3\}$. Check by substituting into the original equation. (Incidentally, *dividing* by -1 would also allow us to solve an equation of the form $-x = a$.)

TEACHING TIP Many students have difficulty with signs when a subtraction sign appears in front of parentheses, as in Example 6. Be sure to warn them about such errors.

NOTE From the final steps in Example 6, we can see that the following is true.

If $-x = a$, then $x = -a$.

CONNECTIONS

The use of algebra to solve equations and applied problems is very old. The 3600-year-old Rhind Papyrus includes the following "word problem." "Aha, its whole, its seventh, it makes 19." This brief sentence describes the equation

$$x + \frac{x}{7} = 19.$$

The word *aha* was used as we would use the words "Let x equal . . . ." The solution of this equation is $16\frac{5}{8}$. The word *algebra* is from the work *Risab al-jabr m'al muquabalah,* written in the ninth century by Muhammed ibn Musa Al-Khowarizmi. The title means "the science of transposition and cancellation." These ideas are used in equation solving. From Latin versions of Khowarizmi's text, "al-jabr" became the broad term covering the art of equation solving.

2.1 EXERCISES

1. (a) and (c) **2.** (a) expression; $x + 15$ (b) expression; $y + 7$ (c) equation; $\{-1\}$ (d) equation; $\{-17\}$ **3.** The addition property of equality says that the same number (or expression) added to both sides of an equation results in an equivalent equation. Example: $-x$ can be added to both sides of $2x + 3 = x - 5$ to get the equivalent equation $x + 3 = -5$. The multiplication property of equality says that the same nonzero number (or expression) multiplied on both sides of the equation results in an equivalent equation. Example: Multiplying both sides of $7x = 4$ by $\frac{1}{7}$ gives the equivalent equation $x = \frac{4}{7}$. **4.** To find the solution to $-x = 5$, multiply by -1, or use the rule: If $-x = a$, then $x = -a$. **5.** $\{6\}$ **6.** $\{13\}$ **7.** $\{6.3\}$ **8.** $\{-20.6\}$ **9.** $\{-6\}$ **10.** $\{2\}$ **11.** $\{-2\}$ **12.** $\{5\}$ **13.** $\{4\}$ **14.** $\{-2\}$ **15.** $\{0\}$ **16.** $\{0\}$ **17.** $\{-2\}$ **18.** $\{7\}$ **19.** $\{-7\}$ **20.** $\{-4\}$ **21.** (a) and (b); A sample answer might be, "A linear equation in one variable is an equation that can be written using only one variable term with the variable to the first power."

1. Which of the pairs of equations are equivalent equations?
(a) $x + 2 = 6$ and $x = 4$ **(b)** $10 - x = 5$ and $x = -5$
(c) $x + 3 = 9$ and $x = 6$ **(d)** $4 + x = 8$ and $x = -4$

2. Decide whether each of the following is an expression or an equation. If it is an expression, simplify it. If it is an equation, solve it.
(a) $5x + 8 - 4x + 7$ **(b)** $-6y + 12 + 7y - 5$
(c) $5x + 8 - 4x = 7$ **(d)** $-6y + 12 + 7y = -5$

3. In your own words, state the addition and multiplication properties of equality and give examples.

4. State how you would find the solution of a linear equation if your next-to-last step reads "$-x = 5$."

Solve each equation using the addition property of equality. Check the solution. See Examples 1 and 2.

5. $x + 6 = 12$ **6.** $x - 4 = 9$ **7.** $x - 8.4 = -2.1$

8. $y + 15.5 = -5.1$ **9.** $\frac{2}{5}w - 6 = \frac{7}{5}w$ **10.** $-\frac{2}{7}z + 2 = \frac{5}{7}z$

11. $5.6x + 2 = 4.6x$ **12.** $9.1x - 5 = 8.1x$ **13.** $3p + 6 = 10 + 2p$

14. $8b - 4 = -6 + 7b$ **15.** $1.2y - 4 = .2y - 4$ **16.** $7.7r + 6 = 6.7r + 6$

17. $\frac{1}{2}x + 2 = -\frac{1}{2}x$ **18.** $\frac{1}{5}x - 7 = -\frac{4}{5}x$ **19.** $3x + 7 - 2x = 0$

20. $5x + 4 - 4x = 0$

21. Which of the following are not linear equations in one variable?
(a) $x^2 - 5x + 6 = 0$ **(b)** $x^3 = x$
(c) $3x - 4 = 0$ **(d)** $7x - 6x = 3 + 9x$

Define a linear equation in one variable in words.

22. Refer to the definition of linear equation in one variable given in this section. Why is the restriction $A \neq 0$ necessary?

22. If $A = 0$, the equation $Ax + B = C$ becomes $B = C$; with no variable, it is not a linear equation. **23.** $\{13\}$ **24.** $\{19\}$ **25.** $\{-4\}$ **26.** $\{11\}$ **27.** $\{0\}$ **28.** $\{0\}$ **29.** $\left\{\frac{7}{15}\right\}$ **30.** $\left\{\frac{103}{60}\right\}$ **31.** $\{7\}$ **32.** $\{-2\}$ **33.** $\{-4\}$ **34.** $\{21\}$ **35.** $\{13\}$ **36.** $\{-16\}$ **37.** $\{29\}$ **38.** $\{-11\}$ **39.** $\{18\}$ **40.** $\{-14\}$ **41.** If both sides of an equation are multiplied by 0, the resulting equation is $0 = 0$. This is true, but does not help to solve the equation. **42.** (c) **43.** $\{6\}$ **44.** $\{8\}$ **45.** $\{-5\}$ **46.** $\{-14\}$ **47.** $\left\{-\frac{18}{5}\right\}$ **48.** $\left\{-\frac{17}{2}\right\}$ **49.** $\{12\}$ **50.** $\{8\}$ **51.** $\{0\}$ **52.** $\{0\}$ **53.** $\{-48\}$ **54.** $\{-15\}$ **55.** $\{-12\}$ **56.** $\{-14\}$ **57.** $\left\{\frac{4}{7}\right\}$ **58.** $\left\{\frac{9}{5}\right\}$ **59.** $\{40\}$ **60.** $\{20\}$ **61.** $\{3\}$ **62.** $\{11\}$ **63.** $\{7\}$ **64.** $\{14\}$ **65.** $\{-35\}$ **66.** $\{-24\}$ **67.** $\left\{\frac{35}{2}\right\}$ **68.** $\left\{\frac{16}{3}\right\}$ **69.** $\left\{-\frac{27}{35}\right\}$ **70.** $\left\{-\frac{8}{15}\right\}$ **71.** $\{-12.2\}$ **72.** $\{8\}$ **73.** Answers will vary. For example, $\frac{3}{2}x = -6$. **74.** Answers will vary. For example, $100x = 17$. **75.** $3x = 2x + 17$; $\{17\}$ **76.** $7x - 6x = -9$; $\{-9\}$ **77.** $5x + 3x = 7x + 9$; $\{9\}$

Solve each equation. First simplify both sides of the equation as much as possible. Check each solution. See Examples 2 and 6.

23. $5t + 3 + 2t - 6t = 4 + 12$

24. $4x + 3x - 6 - 6x = 10 + 3$

25. $6x + 5 + 7x + 3 = 12x + 4$

26. $4x - 3 - 8x + 1 = -5x + 9$

27. $5.2q - 4.6 - 7.1q = -.9q - 4.6$

28. $-4.0x + 2.7 - 1.6x = -4.6x + 2.7$

29. $\frac{5}{7}x + \frac{1}{3} = \frac{2}{5} - \frac{2}{7}x + \frac{2}{5}$

30. $\frac{6}{7}s - \frac{3}{4} = \frac{4}{5} - \frac{1}{7}s + \frac{1}{6}$

31. $(5y + 6) - (3 + 4y) = 10$

32. $(8r - 3) - (7r + 1) = -6$

33. $2(p + 5) - (9 + p) = -3$

34. $4(k - 6) - (3k + 2) = -5$

35. $-6(2b + 1) + (13b - 7) = 0$

36. $-5(3w - 3) + (1 + 16w) = 0$

37. $10(-2x + 1) = -19(x + 1)$

38. $2(2 - 3r) = -5(r - 3)$

39. $-2(8p + 2) - 3(2 - 7p) = 2(4 + 2p)$

40. $-5(1 - 2z) + 4(3 - z) = 7(3 + z)$

41. In the statement of the multiplication property of equality in this section, there is a restriction that $C \neq 0$. What would happen if you should multiply both sides of an equation by 0?

42. Which one of the equations that follow does not require the use of the multiplication property of equality to solve it?

(a) $3x - 5x = 6$ **(b)** $-\frac{1}{4}x = 12$ **(c)** $5x - 4x = 7$ **(d)** $\frac{x}{3} = -2$

Solve each equation and check the solution. See Examples 1–6.

43. $5x = 30$

44. $7x = 56$

45. $3a = -15$

46. $5k = -70$

47. $10t = -36$

48. $4s = -34$

49. $-6x = -72$

50. $-8x = -64$

51. $2r = 0$

52. $5x = 0$

53. $\frac{1}{4}y = -12$

54. $\frac{1}{5}p = -3$

55. $-y = 12$

56. $-t = 14$

57. $-x = -\frac{4}{7}$

58. $-m = -\frac{9}{5}$

59. $.2t = 8$

60. $.9x = 18$

61. $4x + 3x = 21$

62. $9x + 2x = 121$

63. $5m + 6m - 2m = 63$

64. $11r - 5r + 6r = 168$

65. $\frac{x}{7} = -5$

66. $\frac{k}{8} = -3$

67. $-\frac{2}{7}p = -5$

68. $-\frac{3}{8}y = -2$

69. $-\frac{7}{9}c = \frac{3}{5}$

70. $-\frac{5}{6}d = \frac{4}{9}$

71. $-2.1m = 25.62$

72. $-3.9a = -31.2$

73. Write an equation that requires the use of the multiplication property of equality, where both sides must be multiplied by $\frac{2}{3}$, and the solution is a negative number.

74. Write an equation that requires the use of the multiplication property of equality, where both sides must be divided by 100, and the solution is not an integer.

Write an equation using the information given in the problem. Use x as the variable. Then solve the equation.

75. Three times a number is 17 more than twice the number. Find the number.

76. If six times a number is subtracted from seven times the number, the result is -9. Find the number.

77. If five times a number is added to three times the number, the result is the sum of seven times the number and 9. Find the number.

78. $4x = 6; \left\{\frac{3}{2}\right\}$ 79. $\frac{x}{-5} = 2;$ $\{-10\}$ 80. $\frac{2x}{5} = 4; \{10\}$

78. When a number is multiplied by 4, the result is 6. Find the number.

79. When a number is divided by -5, the result is 2. Find the number.

80. If twice a number is divided by 5, the result is 4. Find the number.

2.2 More on Solving Linear Equations

OBJECTIVES

1. Learn and use the four steps for solving a linear equation.
2. Solve equations with fractions or decimals as coefficients.
3. Recognize equations with no solutions or infinitely many solutions.
4. Write expressions for two related unknown quantities.

FOR EXTRA HELP

SSG Sec. 2.2
SSM Sec. 2.2
Pass the Test Software
InterAct Math Tutorial Software
Video 3

CHALKBOARD EXAMPLE

Solve.

$-3k - 5k - 6 + 11 = 2k - 5$

Answer: $\{1\}$

OBJECTIVE 1 **Learn and use the four steps for solving a linear equation.** To solve linear equations in general, follow these steps.

Solving a Linear Equation

Step 1 **Simplify each side separately.** Use the distributive property to clear parentheses and combine terms, as needed.

Step 2 **Isolate the variable terms on one side.** If necessary, use the addition property to get all variable terms on one side of the equation and all numbers on the other.

Step 3 **Isolate the variable.** Use the multiplication property, if necessary, to get the equation in the form x = a number.

Step 4 **Check.** Check the solution by substituting into the *original* equation.

EXAMPLE 1 **Using the Four Steps to Solve an Equation**

Solve $3r + 4 - 2r - 7 = 4r + 3$.

Step 1 $3r + 4 - 2r - 7 = 4r + 3$

$r - 3 = 4r + 3$ Combine like terms.

Step 2 $r - 3 - r = 4r + 3 - r$ Use the addition property of equality. Subtract r.

$-3 = 3r + 3$

$-3 - 3 = -3 + 3r + 3$ Add -3.

$-6 = 3r$

Step 3 $\frac{-6}{3} = \frac{3r}{3}$ Use the multiplication property of equality. Divide by 3.

$-2 = r$ or $r = -2$

Step 4 Substitute -2 for r in the original equation.

$3r + 4 - 2r - 7 = 4r + 3$

$3(-2) + 4 - 2(-2) - 7 = 4(-2) + 3$? Let $r = -2$.

$-6 + 4 + 4 - 7 = -8 + 3$? Multiply.

$-5 = -5$ True

The solution set of the equation is $\{-2\}$.

In Step 2 of Example 1, the terms were added and subtracted so that the variable term ended up on the right. Choosing differently would lead to the variable term being on the left side of the equation. Usually there is no advantage either way.

INSTRUCTOR'S RESOURCES

PTB Sec. 2.2 **ISM** Sec. 2.2 **AB** Sec. 2.2 | **TEST GENERATOR** | **WORLD WIDE WEB** www.LialAlgebra.com

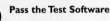

CHALKBOARD EXAMPLE

Solve.

$11 + 3(a + 1) = 5a + 16.$

Answer: $\{-1\}$

EXAMPLE 2 Using the Four Steps to Solve an Equation

Solve $4(k - 3) - k = k - 6$.

Step 1 Before combining like terms, use the distributive property to simplify $4(k - 3)$.

$$4(k - 3) - k = k - 6$$

$$4 \cdot k - 4 \cdot 3 - k = k - 6 \qquad \text{Distributive property}$$

$$4k - 12 - k = k - 6$$

$$3k - 12 = k - 6 \qquad \text{Combine like terms.}$$

Step 2

$$-k + 3k - 12 = -k + k - 6 \qquad \text{Add } -k.$$

$$2k - 12 = -6$$

$$2k - 12 + 12 = -6 + 12 \qquad \text{Add 12.}$$

$$2k = 6$$

Step 3

$$\frac{2k}{2} = \frac{6}{2} \qquad \text{Divide by 2.}$$

$$k = 3$$

Step 4 Check your answer by substituting 3 for k in the original equation. Remember to do the work inside the parentheses first.

$$4(k - 3) - k = k - 6$$

$$4(3 - 3) - 3 = 3 - 6 \quad ? \qquad \text{Let } k = 3.$$

$$4(0) - 3 = 3 - 6 \quad ?$$

$$0 - 3 = 3 - 6 \quad ?$$

$$-3 = -3 \qquad \text{True}$$

The solution set of the equation is $\{3\}$.

CHALKBOARD EXAMPLE

Solve $4y - (y + 7) = 9$.

Answer: $\left\{\frac{16}{3}\right\}$

EXAMPLE 3 Using the Four Steps to Solve an Equation

Solve $8a - (3 + 2a) = 3a + 1$.

Step 1 Simplify.

$$8a - (3 + 2a) = 3a + 1$$

$$8a - 3 - 2a = 3a + 1 \qquad \text{Distributive property}$$

$$6a - 3 = 3a + 1 \qquad \text{Combine terms.}$$

Step 2

$$-3a + 6a - 3 = -3a + 3a + 1 \qquad \text{Add } -3a.$$

$$3a - 3 = 1$$

$$3a - 3 + 3 = 1 + 3 \qquad \text{Add 3.}$$

$$3a = 4$$

Step 3

$$\frac{3a}{3} = \frac{4}{3} \qquad \text{Divide by 3.}$$

$$a = \frac{4}{3}$$

Step 4 Check the solution in the original equation.

$$8a - (3 + 2a) = 3a + 1$$

$$8\left(\frac{4}{3}\right) - \left[3 + 2\left(\frac{4}{3}\right)\right] = 3\left(\frac{4}{3}\right) + 1 \quad ? \qquad \text{Let } a = \tfrac{4}{3}.$$

$$\frac{32}{3} - \left[3 + \frac{8}{3}\right] = 4 + 1 \quad ?$$

$$\frac{32}{3} - \left[\frac{9}{3} + \frac{8}{3}\right] = 5 \quad ?$$

$$\frac{32}{3} - \frac{17}{3} = 5 \quad ?$$

$$5 = 5 \qquad \text{True}$$

The check shows that $\{\frac{4}{3}\}$ is the solution set.

Be very careful with signs when solving equations like the one in Example 3. When a subtraction sign appears immediately in front of a quantity in parentheses, such as in the expression

$$8 - (3 + 2a),$$

CAUTION remember that the $-$ sign acts like a factor of -1 and affects the sign of *every* term within the parentheses. Thus,

$$8 - (3 + 2a) = 8 + (-1)(3 + 2a) = 8 - 3 - 2a.$$

Change to — in *both* terms.

CHALKBOARD EXAMPLE

Solve.

$2 - 3(2 + 6z) = 4(z + 1) + 18$

Answer: $\left\{-\frac{13}{11}\right\}$

EXAMPLE 4 Using the Four Steps to Solve an Equation

Solve $4(8 - 3t) = 32 - 8(t + 2)$.

Step 1 Use the distributive property.

	$4(8 - 3t) = 32 - 8(t + 2)$	Given equation
	$32 - 12t = 32 - 8t - 16$	Distributive property
	$32 - 12t = 16 - 8t$	
Step 2	$32 - 12t + 12t = 16 - 8t + 12t$	Add $12t$.
	$32 = 16 + 4t$	
	$32 - 16 = 16 + 4t - 16$	Subtract 16.
	$16 = 4t$	
Step 3	$\frac{16}{4} = \frac{4t}{4}$	Divide by 4.
	$4 = t$ or $t = 4$	

Step 4 Check the solution.

$$\begin{aligned} 4(8-3t) &= 32-8(t+2) \\ 4(8-3\cdot 4) &= 32-8(4+2) \quad ? \quad \text{Let } t=4. \\ 4(8-12) &= 32-8(6) \quad ? \\ 4(-4) &= 32-48 \quad ? \\ -16 &= -16 \quad \text{True} \end{aligned}$$

Since 4 satisfies the equation, the solution set is $\{4\}$.

OBJECTIVE 2 Solve equations with fractions or decimals as coefficients. We can clear an equation of fractions by multiplying both sides by the least common denominator (LCD) of all denominators in the equation. It is a good idea to do this immediately after using the distributive property to clear parentheses. Most students make fewer errors working with integer coefficients.

CHALKBOARD EXAMPLE

Solve $\frac{1}{3}x - \frac{5}{12} = \frac{3}{4} + \frac{1}{2}x.$

Answer: $\{-7\}$

EXAMPLE 5 Solving an Equation with Fractions as Coefficients

Solve $\frac{2}{3}x - \frac{1}{2}x = -\frac{1}{6}x - 2.$

The LCD of all the fractions in the equation is 6. Start by multiplying both sides of the equation by 6.

$$\begin{aligned} \frac{2}{3}x - \frac{1}{2}x &= -\frac{1}{6}x - 2 \\ 6\left(\frac{2}{3}x - \frac{1}{2}x\right) &= 6\left(-\frac{1}{6}x - 2\right) \quad \text{Multiply by 6.} \\ 6\left(\frac{2}{3}x\right) + 6\left(-\frac{1}{2}x\right) &= 6\left(-\frac{1}{6}x\right) + 6(-2) \quad \text{Distributive property} \\ 4x - 3x &= -x - 12 \end{aligned}$$

Now use the four steps to solve this simpler equivalent equation.

Step 1 $x = -x - 12$ Combine like terms.

Step 2 $x + x = x - x - 12$ Add x.

$2x = -12$

Step 3 $\frac{2x}{2} = \frac{-12}{2}$ Divide by 2.

$x = -6$

Step 4 Check the answer.

$$\begin{aligned} \frac{2}{3}(-6) - \frac{1}{2}(-6) &= -\frac{1}{6}(-6) - 2 \quad ? \quad \text{Let } x=-6. \\ -4 + 3 &= 1 - 2 \quad ? \\ -1 &= -1 \quad \text{True} \end{aligned}$$

The solution set of the equation is $\{-6\}$.

CAUTION When clearing equations of fractions or decimals be sure to multiply *every* term on both sides of the equation by the least common denominator.

The multiplication property can also be used to clear an equation of decimals.

CHALKBOARD EXAMPLE

Solve.

$.5(2 - 3x) = 1.5 - .1(x + 7)$

Answer: $\left\{\frac{1}{7}\right\}$

EXAMPLE 6 Solving an Equation with Decimal Coefficients

Solve $.20t + .10(20 - t) = .18(20)$.

Step 1 Begin by clearing parentheses.

$$.20t + .10(20 - t) = .18(20)$$
$$.20t + .10(20) - .10t = .18(20) \qquad \text{Distributive property}$$

To clear decimals, multiply both sides of the equation by 100, since the decimals are all hundredths. A number can be multiplied by 100 by moving the decimal point two places to the right.

$$.20t + .10(20) - .10t = .18(20) \qquad \text{Multiply by 100.}$$
$$20t + 10(20) - 10t = 18(20)$$
$$20t + 200 - 10t = 360$$
$$10t + 200 = 360 \qquad \text{Combine like terms.}$$

Step 2

$$10t + 200 - \mathbf{200} = 360 - \mathbf{200} \qquad \text{Subtract 200.}$$
$$10t = 160$$

Step 3

$$\frac{10t}{\mathbf{10}} = \frac{160}{\mathbf{10}} \qquad \text{Divide by 10.}$$
$$t = 16$$

Step 4 Check that $\{16\}$ is the solution set by substituting into the original equation.

CAUTION To avoid errors when multiplying a term like $.10(20 - t)$ by 100, use the distributive property to clear parentheses first, then multiply each term by 100 to eliminate decimals.

OBJECTIVE 3 Recognize equations with no solutions or infinitely many solutions. Each equation that we have solved so far has had exactly one solution. As the next examples show, linear equations may also have no solutions or infinitely many solutions. (The four steps are not identified in these examples. See if you can identify them.)

CHALKBOARD EXAMPLE

Solve.

$3y - y + 10 = 2y - 4 + 14$

Answer: {all real numbers}

EXAMPLE 7 Solving an Equation That Has Infinitely Many Solutions

Solve $5x - 15 = 5(x - 3)$.

$$5x - 15 = 5(x - 3)$$
$$5x - 15 = 5x - 15 \qquad \text{Distributive property}$$
$$\mathbf{-5x} + 5x - 15 = \mathbf{-5x} + 5x - 15 \qquad \text{Add } -5x.$$
$$-15 = -15$$
$$\mathbf{15} - 15 = \mathbf{15} - 15 \qquad \text{Add 15.}$$
$$0 = 0$$

The variable has "disappeared." Since the last statement ($0 = 0$) is *true, any* real number is a solution. (We could have predicted this from the line in the solution that says $5x - 15 = 5x - 15$, which is certainly true for *any* value of x.) Indicate the solution set as {all real numbers}.

CAUTION When you are solving an equation like the one in Example 7, do not write $\{0\}$ as the solution set. While 0 is a solution, there are infinitely many other solutions.

CHALKBOARD EXAMPLE

Solve.

$$3z + 8 = 6(z - 1) - 3z$$

Answer: $\emptyset$

EXAMPLE 8 Solving an Equation That Has No Solution

Solve $2x + 3(x + 1) = 5x + 4$.

$2x + 3(x + 1) = 5x + 4$	
$2x + 3x + 3 = 5x + 4$	Distributive property
$5x + 3 = 5x + 4$	Combine terms.
$5x + 3 - 5x = 5x + 4 - 5x$	Subtract $5x$ from each side.
$3 = 4$	Combine terms.

Again, the variable has disappeared, but this time a *false* statement ($3 = 4$) results. When this happens, the equation has no solution. Its solution set is the **empty set,** or **null set,** symbolized $\emptyset$.

The chart below summarizes the solution sets of the three types of linear equations.

Final Equation	*Number of Solutions*	*Solution Set*
$x =$ a number	One	{a number}
A true statement with no variable, such as $0 = 0$	Infinite	{all real numbers}
A false statement with no variable, such as $3 = 4$	None	$\emptyset$

OBJECTIVE 4 Write expressions for two related unknown quantities. We continue our work with translating from words to symbols.

PROBLEM SOLVING

Often we are given a problem in which the sum of two quantities is a particular number, and we are asked to find the values of the two quantities. Example 9 shows how to express the unknown quantities in terms of a single variable.

CHALKBOARD EXAMPLE

One number is 5 more than twice another. If the first number is represented by x, write an expression for the second number.

Answer: $2x + 5$

EXAMPLE 9 Translating Phrases into an Algebraic Expression

Two numbers have a sum of 23. If one of the numbers is represented by k, find an expression for the other number.

First, suppose that the sum of two numbers is 23, and one of the numbers is 10. How would you find the other number? You would subtract 10 from 23 to get 13: $23 - 10 = 13$. So instead of using 10 as one of the numbers, use k as stated in the problem. The other number would be obtained in the same way. You must subtract k from 23. Therefore, an expression for the other number is $23 - k$.

NOTE The approach used in Example 9, first writing an expression with a trial number, can also be useful when translating applied problems to equations.

2.2 EXERCISES

1. *Step 1:* Clear parentheses and collect like terms, as needed. *Step 2:* Use the addition property to get all variable terms on one side of the equation and all numbers on the other. Then collect terms. *Step 3:* Use the multiplication property to get the equation in the form x = a number. *Step 4:* Check the solution. Examples will vary. 2. {2} 3. {4} 4. {5} 5. {−1} 6. $\left\{-\frac{5}{3}\right\}$ 7. $\left\{-\frac{4}{5}\right\}$ 8. $\left\{\frac{4}{3}\right\}$ 9. {−6} 10. {5} 11. $\emptyset$ 12. $\emptyset$ 13. {all real numbers} 14. {all real numbers} 15. {1} 16. {−6} 17. $\emptyset$ 18. $\emptyset$ 19. No, it is incorrect to divide by a variable. If $-3x$ is added to both sides, the equation becomes $4x = 0$, so $x = 0$ and {0} is the correct solution set. 20. (d) 21. Multiply both sides by the LCD of all fractions in the equation or by the power of 10 that makes all decimal numbers integers. 22. If $x = 2$, the denominator $x - 2$ becomes 0. Division by 0 is not defined. 23. {5} 24. {7} 25. {0} 26. {0} 27. $\left\{-\frac{7}{5}\right\}$ 28. $\left\{\frac{3}{25}\right\}$ 29. {120} 30. {60} 31. {6} 32. {4} 33. {15,000} 34. {5000}

1. In your own words, give the four steps used to solve a linear equation. Use an example to demonstrate the steps.

Solve each equation and check your solution. See Examples 1–4, 7, and 8.

2. $5k + 4 = 14$

3. $2m + 8 = 16$

4. $12h - 5 = 10h + 5$

5. $-4x - 1 = -2x + 1$

6. $-2p + 4 = -(5p + 1)$

7. $4x + 6 = -(x - 2)$

8. $3(4x + 2) = -6(x - 5)$

9. $5(2m + 3) = 3(4m + 9)$

10. $6(3w + 5) = 2(10w + 10)$

11. $6(4x - 1) = 12(2x + 3)$

12. $6(2x + 8) = 4(3x - 6)$

13. $3(2x - 4) = 6(x - 2)$

14. $3(6 - 4x) = 2(-6x + 9)$

15. $7r - 5r + 2 = 5r - r$

16. $9p - 4p + 6 = 7p - 3p$

17. $11x - 5(x + 3) - 6x = 0$

18. $6x - 4(x + 2) - 2x = 0$

19. After working correctly through several steps of a linear equation, a student obtains the equation $7x = 3x$. Then the student divides both sides by x to get $7 = 3$, and gives $\emptyset$ as the answer. Is this correct? If not, explain why.

20. Which one of the following linear equations does not have all real numbers as its solution?

(a) $5x = 4x + x$ (b) $2(x + 6) = 2x + 12$

(c) $\frac{1}{2}x = .5x$ (d) $3x = 2x$

21. Based on the discussion in this section, if an equation has decimals or common fractions as coefficients, what additional step will make the work easier?

22. Explain why an equation involving $\frac{1}{x - 2}$ cannot have 2 as a solution.

Solve each equation by clearing fractions or decimals. See Examples 5 and 6.

23. $\frac{3}{5}t - \frac{1}{10}t = t - \frac{5}{2}$

24. $-\frac{2}{7}r + 2r = \frac{1}{2}r + \frac{17}{2}$

25. $-\frac{1}{4}(x - 12) + \frac{1}{2}(x + 2) = x + 4$

26. $\frac{1}{9}(y + 18) + \frac{1}{3}(2y + 3) = y + 3$

27. $\frac{2}{3}k - \left(k + \frac{1}{4}\right) = \frac{1}{12}(k + 4)$

28. $-\frac{5}{6}q - \left(q - \frac{1}{2}\right) = \frac{1}{4}(q + 1)$

29. $.20(60) + .05x = .10(60 + x)$

30. $.30(30) + .15x = .20(30 + x)$

31. $1.00x + .05(12 - x) = .10(63)$

32. $.92x + .98(12 - x) = .96(12)$

33. $.06(10{,}000) + .08x = .072(10{,}000 + x)$

34. $.02(5000) + .03x = .025(5000 + x)$

*These problems can be used as collaborative exercises with small groups of students.

35. 800 **36.** Yes, you will get $(100 \cdot 2) \cdot 4 = 200 \cdot 4 = 800$. This is a result of the associative property. **37.** No, because $(100a)(100b) = 10{,}000ab \neq 100ab$. **38.** The distributive property involves the operation of *addition* as well. **39.** Yes, the associative property of multiplication is used here. **40.** no

RELATING CONCEPTS* (EXERCISES 35–40)

Work Exercises 35–40 in order.

35. Consider the term $100ab$. Evaluate it for $a = 2$ and $b = 4$.

36. Based on your study of Section 1.7, will you get the same answer for Exercise 35 if you evaluate $(100a)b$ for $a = 2$ and $b = 4$? Why or why not?

37. Is the term $(100a)(100b)$ equivalent to $100ab$? Why or why not?

38. If your answer to Exercise 37 is *no*, explain why the distributive property is not involved.

39. Consider the equation

$$.05(x + 2) + .10x = 2.00.$$

To simplify our work, we could begin by multiplying both sides by 100. When applied to the first term on the left, our expression would be

$$100 \cdot .05(x + 2).$$

Is this expression equivalent to $[100 \cdot .05](x + 2)$? (*Hint:* Compare to the expressions in Exercises 35 and 36, letting $a = .05$ and $b = x + 2$.)

40. Students often want to "distribute" the 100 to both .05 and $(x + 2)$ in the expression $100 \cdot .05(x + 2)$. Is this correct? (*Hint:* Compare your answer to those in Exercises 37 and 38.)

Did you make the connection that the distributive property is not used when the product of two quantities is multiplied by a third quantity?

41. {8} **42.** {2} **43.** {0} **44.** {0} **45.** {4} **46.** {−6} **47.** {20} **48.** {15} **49.** {all real numbers} **50.** {all real numbers} **51.** ∅ **52.** ∅ **53.** $11 - q$ **54.** $\frac{9}{k}$ **55.** $x + 7$ **56.** $65 - h$ **57.** $a + 12$; $a - 5$ **58.** $25r$ (cents) **59.** $\frac{t}{5}$ **60.** $3b + 2d$

Solve each equation and check the solution. See Examples 1–8.

41. $10(2x - 1) = 8(2x + 1) + 14$

42. $9(3k - 5) = 12(3k - 1) - 51$

43. $-(4y + 2) - (-3y - 5) = 3$

44. $-(6k - 5) - (-5k + 8) = -3$

45. $\frac{1}{2}(x + 2) + \frac{3}{4}(x + 4) = x + 5$

46. $\frac{1}{3}(x + 3) + \frac{1}{6}(x - 6) = x + 3$

47. $.10(x + 80) + .20x = 14$

48. $.30(x + 15) + .40(x + 25) = 25$

49. $4(x + 8) = 2(2x + 6) + 20$

50. $4(x + 3) = 2(2x + 8) - 4$

51. $9(v + 1) - 3v = 2(3v + 1) - 8$

52. $8(t - 3) + 4t = 6(2t + 1) - 10$

Write the answer to each problem as an algebraic expression. See Example 9.

53. Two numbers have a sum of 11. One of the numbers is q. Find the other number.

54. The product of two numbers is 9. One of the numbers is k. What is the other number?

55. A football player gained x yards rushing. On the next down he gained 7 yards. How many yards did he gain altogether?

56. A baseball player got 65 hits one season. He got h of the hits in one game. How many hits did he get in the rest of the games?

57. Mary is a years old. How old will she be in 12 years? How old was she 5 years ago?

58. Tom has r quarters. Find the value of the quarters in cents.

59. A bank teller has t dollars, all in five-dollar bills. How many five-dollar bills does the teller have?

60. A plane ticket costs b dollars for an adult and d dollars for a child. Find the total cost for 3 adults and 2 children.

2.3 An Introduction to Applications of Linear Equations

OBJECTIVES

1. Learn the six steps for solving applied problems.
2. Solve problems involving unknown numbers.
3. Solve problems involving sums of quantities.
4. Solve problems involving supplementary and complementary angles.
5. Solve problems involving consecutive integers.

FOR EXTRA HELP

SSG Sec. 2.3
SSM Sec. 2.3
Pass the Test Software
InterAct Math Tutorial Software
Video 3

■ Answers to Connections box: Polya's Step 1 corresponds to our Steps 1 and 2. Polya's Step 2 corresponds to our Step 3. Polya's Step 3 corresponds to our Steps 4 and 5. Polya's Step 4 corresponds to our Step 6. Trial and error or guessing and checking fit into Polya's Step 2, devising a plan.

CONNECTIONS

The purpose of algebra is to solve real problems. Since such problems are stated in words, not mathematical symbols, the first step in solving them is to translate the problem into one or more mathematical statements. This is the hardest step for most people. George Polya (1888–1985), a native of Budapest, Hungary, wrote the modern classic *How to Solve It.* In this book he proposed a four-step process for problem solving:

1. Understand the problem.
2. Devise a plan.
3. Carry out the plan.
4. Look back and check.

FOR DISCUSSION OR WRITING

Compare Polya's four-step process with the six steps given in this section. Identify which of the steps in our list match Polya's four steps. Trial and error is also a useful problem-solving tool. Where does this method fit into Polya's steps?

OBJECTIVE 1 Learn the six steps for solving applied problems. We now begin to look at how algebra is used to solve applied problems. It must be emphasized that many *meaningful* applications of mathematics require concepts that are beyond the level of this book. Some of the problems you will encounter will seem "contrived," and to some extent they are. But the skills you will develop in solving simple problems will help you in solving more realistic problems in chemistry, physics, biology, business, and other fields.

PROBLEM SOLVING

In earlier sections we learned how to translate words, phrases, and sentences into mathematical expressions and equations. Now we will use these translations to solve applied problems using algebra. While there is no specific method that enables you to solve all kinds of applied problems, the following six-step method is suggested.

Solving an Applied Problem

Step 1 **Decide what you are asked to find.** Read the problem carefully. Choose a variable to represent the number you are asked to find. *Write down* what the variable represents.

Step 2 **Write down any other pertinent information.** If there are other unknown quantities, express them using the variable. Draw figures or diagrams and use charts, if they apply.

Step 3 **Write an equation.** Translate the problem into an equation.

Step 4 **Solve the equation.** Use the properties to solve the equation.

Step 5 **Answer the question(s) posed.** Make sure you answer the question in the problem. In some cases, you need more than the solution of the equation.

Step 6 **Check.** Check your solution using the original words of the problem. Be sure your answer makes sense.

INSTRUCTOR'S RESOURCES

PTB Sec. 2.3 **ISM** Sec. 2.3
AB Sec. 2.3
TEST GENERATOR
WORLD WIDE WEB www.LialAlgebra.com

PROBLEM SOLVING

The third step is often the hardest. To translate the problem into an equation, write the given phrases as mathematical expressions. Since equal mathematical expressions are names for the same number, translate any words that mean *equal* or *same* as =. The = sign leads to an equation to be solved.

OBJECTIVE 2 Solve problems involving unknown numbers. Some of the simplest applied problems involve unknown numbers.

CHALKBOARD EXAMPLE

If 5 is added to the product of 9 and a number, the result is 19 less than the number. Find the number.

Answer: -3

EXAMPLE 1 Finding the Value of an Unknown Number

The product of 4, and a number decreased by 7, is 100. Find the number.

Step 1 Read the problem carefully. Decide what you are being told to find, and then choose a variable to represent the unknown quantity. In this problem, we are told to find a number, so we write

Let x = the number.

Step 2 There are no other unknown quantities to find.

Step 3 Translate the problem into an equation.

The product of 4,	and	a number	decreased by	7,	is	100.
↓		↓	↓	↓	↓	↓
$4 \cdot$	(	x	$-$	7)	=	100

Because of the commas in the given problem, writing the equation as $4x - 7 = 100$ is incorrect. The equation $4x - 7 = 100$ corresponds to the statement "The product of 4 and a number, decreased by 7, is 100."

Step 4 Solve the equation.

$$4(x - 7) = 100$$

$$4x - 28 = 100 \quad \text{Distributive property}$$

$$4x = 128 \quad \text{Add 28 to both sides.}$$

$$x = 32 \quad \text{Divide by 4.}$$

Step 5 The number is 32.

Step 6 Check the solution by using the original words of the problem. When 32 is decreased by 7, we get $32 - 7 = 25$. If 4 is multiplied by 25, we get 100, as the problem required. The answer, 32, is correct.

NOTE The commas in the statement of the problem in Example 1 are important in translating correctly.

OBJECTIVE 3 Solve problems involving sums of quantities. A common type of problem in elementary algebra involves finding two quantities when the sum of the quantities is known. In Example 9 of the previous section, we prepared for this type of problem by writing mathematical expressions for two related unknown quantities.

PROBLEM SOLVING

In general, to solve such problems, choose a variable to represent one of the unknowns and then represent the other quantity in terms of the same variable, using information obtained in the problem. Then write an equation based on the words of the problem. The next example illustrates these ideas.

CHALKBOARD EXAMPLE

On one day of their vacation, Annie drove three times as far as Jim. Altogether they drove 84 miles that day. Find the number of miles driven by each.

Answer: 21 miles for Jim; 63 miles for Annie

EXAMPLE 2 Finding the Numbers of Olympic Medals Won by the U.S.

In the 1996 Olympics, U.S. contestants won 12 more gold than silver medals. They won a total of 76 gold and silver medals. Find the number of each type of medal won. (*Source:* United States Olympic Committee.)

Step 1 Let $x =$ the number of silver medals.

Step 2 Let $x + 12 =$ the number of gold medals.

Step 3 Now write an equation.

The total	is	the number of silver	plus	the number of gold.
↓	↓	↓	↓	↓
76	=	x	+	$(x + 12)$

Step 4 Solve the equation.

$$76 = 2x + 12 \quad \text{Combine terms.}$$
$$76 - 12 = 2x + 12 - 12 \quad \text{Subtract 12.}$$
$$64 = 2x \quad \text{Combine terms.}$$
$$32 = x \quad \text{Divide by 2.}$$

Step 5 Because x represents the number of silver medals, the U.S. won 32 silver medals. Because $x + 12$ represents the number of gold medals, the U.S. won $32 + 12 = 44$ gold medals.

Step 6 Since there were 44 gold and 32 silver medals, the total number of medals was $44 + 32 = 76$. Because $44 - 32 = 12$, there were 12 more gold medals than silver medals. This information agrees with what is given in the problem, so the answers check.

TEACHING TIP It is appropriate at this point to mention how the nature of the applied problem restricts the set of possible solutions. For example, an answer such as -33 medals or $25\frac{1}{2}$ medals should be recognized as inappropriate.

NOTE

The problem in Example 2 could also have been solved by letting x represent the number of gold medals. Then $x - 12$ would represent the number of silver medals. The equation would then be

$$76 = x + (x - 12).$$

The solution of this equation is 44, which is the number of gold medals. The number of silver medals would then be $44 - 12 = 32$. The answers are the same, whichever approach is used.

CHALKBOARD EXAMPLE

On that same day, the owner of P.J.'s Coffeehouse found that the number of orders for croissants was $\frac{1}{6}$ the number of orders for muffins. If the total number for the two breakfast rolls was 56, how many orders were placed for croissants?

Answer: 8

EXAMPLE 3 Finding the Number of Orders for Tea

The owner of P. J.'s Coffeehouse found that on one day the number of orders for tea was $\frac{1}{3}$ the number of orders for coffee. If the total number of orders for the two drinks was 76, how many orders were placed for tea?

Step 1 Let x = the number of orders for coffee.

Step 2 Let $\frac{1}{3}x$ = the number of orders for tea.

Step 3 Use the fact that the total number of orders was 76 to write an equation.

The total	is	orders for coffee	plus	orders for tea.
↓	↓	↓	↓	↓
76	$=$	x	$+$	$\frac{1}{3}x$

Step 4 Now solve the equation.

$$76 = \frac{4}{3}x \qquad \text{Combine like terms.}$$

$$\frac{3}{4}(76) = \frac{3}{4}\left(\frac{4}{3}x\right) \qquad \text{Multiply by } \tfrac{3}{4}.$$

$$57 = x$$

Step 5 In this problem, *x does not represent the quantity that we are asked to find.* The number of orders placed for tea was $\frac{1}{3}x$. So $\frac{1}{3}(57) = 19$ is the number of orders for tea.

Step 6 The number of coffee orders (x) was 57 and the number of tea orders was 19; 19 is one-third of 57, and $19 + 57 = 76$. Since this agrees with the information given in the problem, the answer is correct.

PROBLEM SOLVING

In Example 3, it was easier to let the variable represent the quantity that was *not* asked for. This required an extra step in Step 5 to find the number of orders for tea. In some cases, this approach is easier than letting the variable represent the quantity that we are asked to find. Experience in solving problems will indicate when this approach is useful, and experience comes only from solving many problems!

CHALKBOARD EXAMPLE

At a meeting of the local computer user group, each member brought two nonmembers. If a total of 27 people attended, how many were members and how many were nonmembers?

Answer: 9 were members and 18 were nonmembers.

EXAMPLE 4 Analyzing a Gasoline/Oil Mixture

A lawn trimmer uses a mixture of gasoline and oil. For each ounce of oil the mixture contains 16 ounces of gasoline. If the tank holds 68 ounces of the mixture, how many ounces of oil and how many ounces of gasoline does it require when it is full?

Step 1 Let x = the number of ounces of oil required when full.

Step 2 Let $16x$ = the number of ounces of gasoline required when full.

	Amount of gasoline		Amount of oil		Total amount in tank
	↓		↓		↓
Step 3	$16x$	$+$	x	$=$	68

Step 4 $17x = 68$ Combine terms.

$x = 4$ Divide by 17.

Step 5 The trimmer requires 4 ounces of oil and $16(4) = 64$ ounces of gasoline when full.

Step 6 Since $4 + 64 = 68$, and 64 is 16 times 4, the answers check.

PROBLEM SOLVING

Sometimes it is necessary to find three unknown quantities in an applied problem. Frequently the three unknowns are compared in *pairs*. When this happens, it is usually easiest to let the variable represent the unknown found in both pairs. The next example illustrates this.

CHALKBOARD EXAMPLE

A piece of pipe is 50 inches long. It is cut into three pieces. The longest piece is 10 inches more than the middle-sized piece, and the shortest piece measures 5 inches less than the middle-sized piece. Find the lengths of the three pieces.

Answer: 25 inches, 15 inches, 10 inches

EXAMPLE 5 Dividing a Board into Pieces

The instructions for a woodworking project require three pieces of wood. The longest piece must be twice the length of the middle-sized piece, and the shortest piece must be 10 inches shorter than the middle-sized piece. Maria Gonzales has a board 70 inches long that she wishes to use. How long can each piece be?

Steps 1 and 2 Since the middle-sized piece appears in both pairs of comparisons, let x represent the length of the middle-sized piece. We have

x = the length of the middle-sized piece

$2x$ = the length of the longest piece

$x - 10$ = the length of the shortest piece.

A sketch is helpful here. See Figure 1.

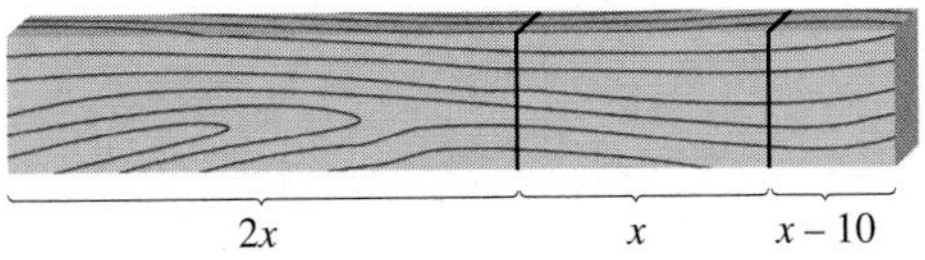

Figure 1

	Longest		Middle-sized		Shortest		Total length
	↓		↓		↓		↓
Step 3	$2x$	+	x	+	$(x - 10)$	=	70

Step 4

$4x - 10 = 70$ Combine terms.

$4x - 10 + \mathbf{10} = 70 + \mathbf{10}$ Add 10 to each side.

$4x = 80$ Combine terms.

$x = 20$ Divide by 4 on each side.

Step 5 The middle-sized piece is 20 inches long, the longest piece is $2(20) = 40$ inches long, and the shortest piece is $20 - 10 = 10$ inches long.

Step 6 Check to see that the sum of the lengths is 70 inches, and that all conditions of the problem are satisfied.

OBJECTIVE 4 Solve problems involving supplementary and complementary angles. The next example deals with concepts from geometry. An angle can be measured by a unit called the **degree** (°). Two angles whose sum is 90° are said to be **complementary,** or complements of each other. Two angles whose sum is 180° are said to be **supplementary,** or supplements of each other. See Figure 2. If x represents the degree measure of an angle, then

$90 - x$ represents the degree measure of its complement, and
$180 - x$ represents the degree measure of its supplement.

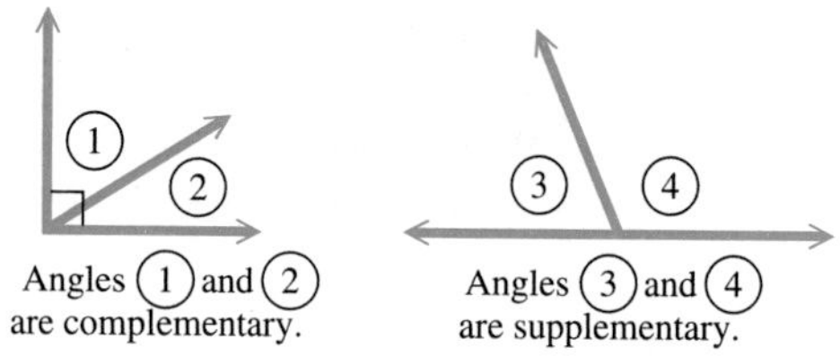

Figure 2

CHALKBOARD EXAMPLE

Find the measure of an angle such that the sum of the measures of its complement and its supplement is 174°.

Answer: 48°

EXAMPLE 6 Finding the Measure of an Angle

Find the measure of an angle whose supplement is 10° more than twice its complement.

Step 1 Let $x =$ the degree measure of the angle.

Step 2 Then $90 - x =$ the degree measure of its complement;
$180 - x =$ the degree measure of its supplement.

Step 3

Supplement	is	10	more than	twice	its complement.
↓	↓	↓	↓	↓	↓
$180 - x$	$=$	10	$+$	2	$\cdot\ (90 - x)$

Step 4 Solve the equation.

$$180 - x = 10 + 180 - 2x \quad \text{Distributive property}$$
$$180 - x = 190 - 2x \quad \text{Combine terms.}$$
$$180 - x + 2x = 190 - 2x + 2x \quad \text{Add } 2x.$$
$$180 + x = 190 \quad \text{Combine terms.}$$
$$180 + x - 180 = 190 - 180 \quad \text{Subtract } 180.$$
$$x = 10$$

Step 5 The measure of the angle is 10°.

Step 6 The complement of 10° is 80° and the supplement of 10° is 170°. 170° is equal to 10° more than twice 80° ($170 = 10 + 2(80)$ is true); therefore, the answer is correct.

OBJECTIVE 5 Solve problems involving consecutive integers. Two integers that differ by 1 are called **consecutive integers.** For example, 3 and 4, 6 and 7, and -2 and -1 are pairs of consecutive integers. In general, if x represents an integer, $x + 1$ represents the next larger consecutive integer.

Consecutive even integers, such as 8 and 10, differ by 2. Similarly, consecutive odd integers, such as 9 and 11, also differ by two. In general, if x represents an even inte-

TEACHING TIP Several numerical examples may be necessary before students understand the variable notation used in consecutive integer problems.

ger, $x + 2$ represents the next larger consecutive even integer. The same holds true for odd integers; that is, if x is an odd integer, $x + 2$ is the next larger odd integer.

CHALKBOARD EXAMPLE

If the sum of three consecutive integers is -45, what is the smallest integer?

Answer: -16

EXAMPLE 7 Finding Consecutive Integers

Two pages that face each other in this book have 569 as the sum of their page numbers. What are the page numbers?

Because the two pages face each other, they must have page numbers that are consecutive integers.

Step 1 Let x = the smaller page number.

Step 2 Then $x + 1$ = the larger page number.

Step 3 Because the sum of the page numbers is 569, the equation is

$$x + (x + 1) = 569.$$

Step 4 Solve the equation.

$$\begin{aligned} x + (x + 1) &= 569 \\ 2x + 1 &= 569 && \text{Combine like terms.} \\ 2x &= 568 && \text{Subtract 1.} \\ x &= 284 && \text{Divide by 2.} \end{aligned}$$

Step 5 The smaller page number is 284 and the larger page number is $284 + 1 = 285$.

Step 6 The sum of 284 and 285 is 569. Our answer is correct.

In the final example we do not number the steps. See if you can identify them.

CHALKBOARD EXAMPLE

Find two consecutive even integers such that six times the smaller added to the larger gives a sum of 86.

Answer: 12,14

EXAMPLE 8 Finding Consecutive Odd Integers

If the smaller of two consecutive odd integers is doubled, the result is 7 more than the larger of the two integers. Find the two integers.

Let x be the smaller integer. Since the two numbers are consecutive *odd* integers, then $x + 2$ is the larger. Now write an equation.

If the smaller is doubled	the result is	7	more than	the larger.
↓	↓	↓	↓	↓
$2x$	$=$	7	$+$	$x + 2$

Solve the equation.

$$\begin{aligned} 2x &= 7 + x + 2 \\ 2x &= 9 + x \\ 2x + (-x) &= 9 + x + (-x) \\ x &= 9 \end{aligned}$$

The first integer is 9 and the second is $9 + 2 = 11$. To check our answer we see that when 9 is doubled, we get 18, which is 7 more than the larger odd integer, 11. Our answer is correct.

2.3 EXERCISES

1. (c) 2. (d) 3. Your answer should include the following steps; *Step 1:* Read the problem carefully and assign a variable to the unknown; *Step 2:* Write down variable expressions for any other unknown quantities; *Step 3:* Write an equation; *Step 4:* Solve the equation; *Step 5:* Answer the question; *Step 6:* Check your solution. 4. Some examples are *is, are, was, were.*
5. 3 6. 5 7. −4 8. −3
9. 53 Republicans, 46 Democrats
10. 236 Republicans, 197 Democrats
11. 6562 men, 3779 women
12. Green Bay 35, Kansas City 10
13. Fernandez, 285; Irwin, 286

1. Which one of the following would not be a reasonable answer in an applied problem that requires finding the number of coins in a jar?

(a) 7 **(b)** 0 **(c)** $6\frac{2}{3}$ **(d)** 80

2. Which one of the following would not be a reasonable answer in an applied problem that requires finding someone's age (in years)?

(a) $5\frac{1}{2}$ **(b)** 7 **(c)** 12.25 **(d)** −4

3. Explain in your own words the six-step method for solving applied problems.

4. List some words that will translate as "=" in an applied problem.

Solve each problem. Use the six-step method. See Example 1.

5. If 1 is added to a number and this sum is doubled, the result is 5 more than the number. Find the number.

6. If 2 is subtracted from a number and this difference is tripled, the result is 4 more than the number. Find the number.

7. If 3 is added to twice a number and this sum is multiplied by 4, the result is the same as if the number is multiplied by 7 and 8 is added to the product. What is the number?

8. The sum of three times a number and 12 more than the number is the same as the difference between −6 and twice the number. What is the number?

Solve each problem. Use the six-step method. See Examples 2–5.

9. The U.S. Senate has 100 members. After the 1996 election, during the second session, there were 7 more Republicans than Democrats. One seat was vacant. How many Democrats and Republicans were there in the Senate? (*Source:* U.S. Congress, Joint Committee on Printing, *Congressional Directory,* biennial.)

10. The total number of Democrats and Republicans in the U.S. House of Representatives in 1996 was 433. There were 39 fewer Democrats than Republicans. How many members of each party were there? (*Source:* U.S. Congress, Joint Committee on Printing, *Congressional Directory,* biennial.)

11. There were 2783 more men than women competing in the 1996 Olympic Games in Atlanta. The total number of competitors was 10,341. How many men and how many women competed? (*Source: The Universal Almanac,* 1997, John W. Wright, General Editor.)

12. In the first Super Bowl, played in 1967, Green Bay and Kansas City scored a total of 45 points. Green Bay won by 25 points. What was the score of the first Super Bowl? (*Source: The Universal Almanac,* 1997, John W. Wright, General Editor.)

13. In the 1998 U.S. Senior Open, Hale Irwin finished with 1 stroke more than the winner, Vicente Fernandez. The sum of their scores was 571. Find their scores. (*Source:* Television coverage.)

14. **shorter piece: 15 inches; longer piece: 24 inches** **15.** **Airborne Express: 3; Federal Express: 9; United Parcel Service: 1**
16. **sorting mail: 1.5 hours; selling stamps: 3 hours; supervising: 2 hours**
17. **36 million miles** **18.** **2**
19. **Smoltz: $253\frac{2}{3}$; Maddux: 245; Wohlers: $77\frac{1}{3}$** **20.** **Gooden: $170\frac{2}{3}$; Cone: 72; Nelson: $74\frac{1}{3}$**
21. ***A* and *B*: 40°; *C*: 100°**

14. Nagaraj Nanjappa has a strip of paper 39 inches long. He wants to cut it into two pieces so that one piece will be 9 inches shorter than the other. How long should the two pieces be?

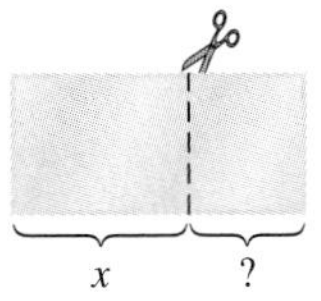

15. In one day Akilah Cadet received 13 packages. Federal Express delivered three times as many as Airborne Express, while United Parcel Service delivered 2 less than Airborne Express. How many packages did each service deliver to Akilah?

16. In her job at the post office, Janie Quintana works a 6.5-hour day. She sorts mail, sells stamps, and does supervisory work. One day she sold stamps twice as long as she sorted mail, and she supervised .5 hour longer than she sorted mail. How many hours did she spend at each task?

17. Venus is 31.2 million miles farther from the sun than Mercury, while Earth is 57 million miles farther from the sun than Mercury. If the total of the distances from these three planets to the sun is 196.2 million miles, how far away from the sun is Mercury? (All distances given here are *mean (average)* distances.) (*Source: The Universal Almanac,* 1997, John W. Wright, General Editor.)

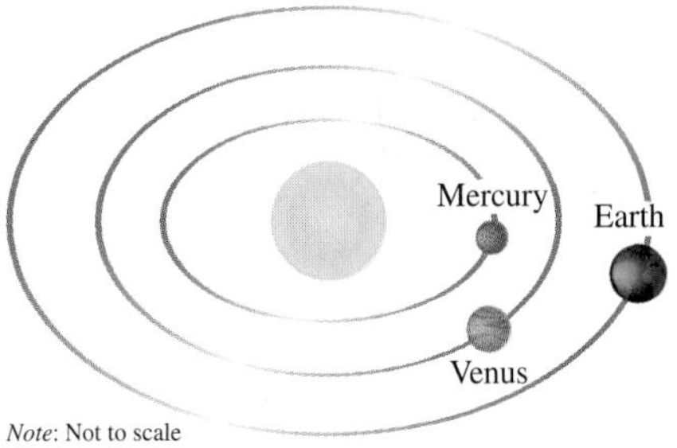

Note: Not to scale

18. Together, Saturn, Jupiter, and Mars have a total of 41 satellites (moons). Jupiter has 5 fewer satellites than Saturn, and Mars has 20 fewer satellites than Saturn. How many satellites does Mars have? (*Source: The Universal Almanac,* 1997, John W. Wright, General Editor.)

19. During the 1996 baseball season, Atlanta Braves pitchers John Smoltz, Greg Maddux, and Mark Wohlers pitched a total of 576 innings. Maddux pitched $8\frac{2}{3}$ fewer innings than Smoltz and $167\frac{2}{3}$ more innings than Wohlers. How many innings did each player pitch? (*Source:* Street and Smith's *Baseball Yearbook,* 1997.)

20. During their World Championship year of 1996, New York Yankees pitchers Dwight Gooden, David Cone, and Jeff Nelson pitched a total of 317 innings. Gooden pitched $96\frac{1}{3}$ innings more than Nelson, and Cone pitched $2\frac{1}{3}$ fewer innings than Nelson. How many innings did each player pitch? (*Source:* Street and Smith's *Baseball Yearbook,* 1997.)

21. The sum of the measures of the angles of any triangle is 180°. In triangle *ABC,* angles *A* and *B* have the same measure, while the measure of angle *C* is 60° larger than each of *A* and *B*. What are the measures of the three angles?

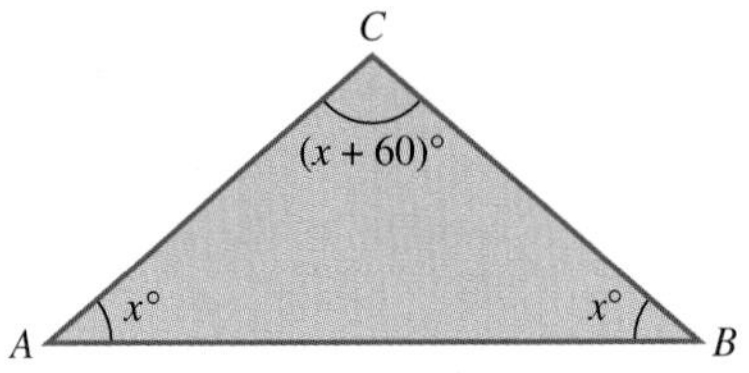

22. A: **154°; B and C: 13°** **23. exertion: 9443 calories; regulating body temperature: 1757 calories** **24. 1944 nickel: \$6; 1950 nickel: \$8** **25. 18 prescriptions**
26. 35 pounds
27. peanuts: $22\frac{1}{2}$ **ounces; cashews:** $4\frac{1}{2}$ **ounces**
28. active: 3.5 centigrams; inert: 332.5 centigrams
29. $k - m$ **30.** $\frac{r}{s}$ **31. no**
32. yes, 90°; yes, 45° **33.** $x - 1$
34. $x - 2, x, x + 2$

22. (See Exercise 21.) In triangle ABC, the measure of angle A is 141° more than the measure of angle B. The measure of angle B is the same as the measure of angle C. Find the measure of each angle.

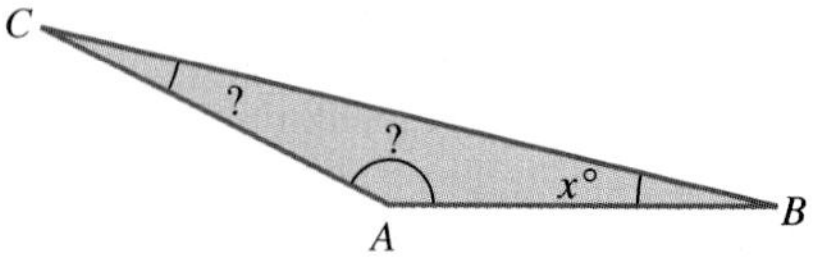

23. A husky running the Iditarod (a thousand-mile race between Anchorage and Nome, Alaska) burns $5\frac{3}{8}$ calories in exertion for every 1 calorie burned in thermoregulation in extreme cold. According to one scientific study, a husky in top condition burns an amazing total of 11,200 calories per day. How many calories are burned for exertion and how many are burned for regulation of body temperature?

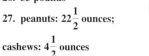

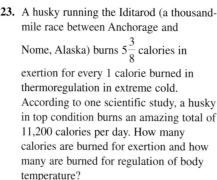

24. The 1997 edition of *A Guide Book of United States Coins* lists the value of a Mint State-65 (uncirculated) 1950 Jefferson nickel minted at Denver as $\frac{4}{3}$ the value of a similar condition 1944 nickel minted at Philadelphia. Together the total value of the two coins is \$14.00. What is the value of each coin?

25. A pharmacist found that at the end of the day she had $\frac{4}{3}$ as many prescriptions for antibiotics as she did for tranquilizers. She had 42 prescriptions altogether for these two types of drugs. How many did she have for tranquilizers?

26. In a mixture of concrete, there are 3 pounds of cement mix for every 1 pound of gravel. If the mixture contains a total of 140 pounds of these two ingredients, how many pounds of gravel are there?

27. A mixture of nuts contains only peanuts and cashews. For every ounce of cashews there are 5 ounces of peanuts. If the mixture contains a total of 27 ounces, how many ounces of each type of nut does the mixture contain?

28. An insecticide contains 95 centigrams of inert ingredient for every 1 centigram of active ingredient. If a quantity of the insecticide weighs 336 centigrams, how much of each type of ingredient does it contain?

Use the concepts of this section to answer each question.

29. If the sum of two numbers is k, and one of the numbers is m, how can you express the other number?

30. If the product of two numbers is r, and one of the numbers is s ($s \neq 0$), how can you express the other number?

31. Is there an angle whose supplement is equal to its complement? If so, what is the measure of the angle?

32. Is there an angle that is equal to its supplement? Is there an angle that is equal to its complement? If the answer is yes to either question, give the measure of the angle.

33. If x represents an integer, how can you express the next smaller consecutive integer in terms of x?

34. Express three consecutive even integers, all in terms of x, if x represents the middle of the three.

35. 80° **36.** 60° **37.** 26°
38. 39° **39.** 55° **40.** 50°
41. 68, 69 **42.** −179, −178
43. 10, 12 **44.** 13, 15
45. 101, 102 **46.** 18
47. 10, 11 **48.** 7, 8
49. 15, 17, 19 **50.** 20, 22, 24
51. 1993: \$2.78 billion; 1994: \$3.33 billion; 1995: \$3.53 billion

Solve each problem. See Example 6.

35. Find the measure of an angle whose supplement measures 10 times the measure of its complement.

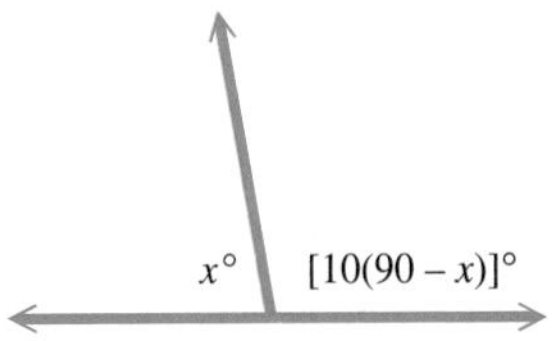

36. Find the measure of an angle whose supplement measures 4 times the measure of its complement.

37. Find the measure of an angle, if its supplement measures 38° less than three times its complement.

38. Find the measure of an angle, if its supplement measures 39° more than twice its complement.

39. Find the measure of an angle such that the sum of the measures of its complement and its supplement is 160°.

40. Find the measure of an angle such that the difference between the measures of its supplement and three times its complement is 10°.

Solve each problem. See Examples 7 and 8.

41. The sum of two consecutive integers is 137. Find the integers.

42. The sum of two consecutive integers is −357. Find the integers.

43. Find two consecutive even integers such that the smaller added to three times the larger gives a sum of 46.

44. Find two consecutive odd integers such that twice the larger is 17 more than the smaller.

45. Two pages that are back-to-back in this book have 203 as the sum of their page numbers. What are the page numbers?

46. If the sum of three consecutive even integers is 60, what is the smallest even integer?

47. When the smaller of two consecutive integers is added to three times the larger, the result is 43. Find the integers.

48. If five times the smaller of two consecutive integers is added to three times the larger, the result is 59. Find the integers.

49. If 6 is subtracted from the largest of three consecutive odd integers, with this result multiplied by 2, the answer is 23 less than the sum of the first and twice the second of the integers. Find the integers.

50. If the first and third of three consecutive even integers are added, the result is 22 less than three times the second integer. Find the integers.

Apply the ideas of this section to solve each problem based on the graph.

51. In 1994, the funding for Head Start programs increased by .55 billion dollars from the funding in 1993. In 1995, the increase was .20 billion dollars over the funding in 1994. For those three years the total funding was 9.64 billion dollars. How much was funded in each of these years? (*Source:* U.S. Department of Health and Human Services.)

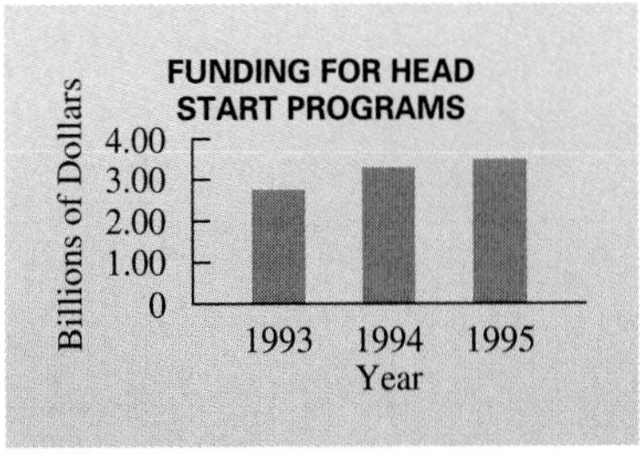

52. football: 2171; bicycling: 904; golf: 104

52. According to data provided by the National Safety Council, in 1992 the number of serious injuries per 100,000 participants in football, bicycling, and golf is illustrated in the graph. There were 800 more in bicycling than in golf, and there were 1267 more in football than in bicycling. Altogether there were 3179 serious injuries per 100,000 participants. How many such serious injuries were there in each sport?

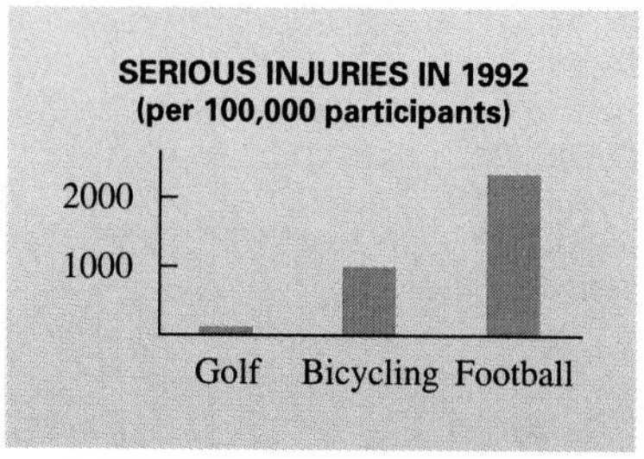

2.4 Formulas and Applications from Geometry

OBJECTIVES

1. Solve a formula for one variable given the values of the other variables.
2. Use a formula to solve a geometric application.
3. Solve problems about angle measures.
4. Solve a formula for a specified variable.

FOR EXTRA HELP

SSG Sec. 2.4
SSM Sec. 2.4
Pass the Test Software
InterAct Math Tutorial Software
Video 3

CHALKBOARD EXAMPLE

Find the value of the remaining variable.

$P = 2L + 2W;\quad P = 126,\ W = 25$

Answer: $L = 38$

Many applied problems can be solved with formulas. There are formulas for geometric figures such as squares and circles, for distance, for money earned on bank savings, and for converting English measurements to metric measurements, for example. For reference, the formulas used in this book are given on the inside covers.

OBJECTIVE 1 Solve a formula for one variable given the values of the other variables. Given the values of all but one of the variables in a formula, the value of the remaining variable can be found by using the methods introduced in this chapter for solving equations.

EXAMPLE 1 Using a Formula to Evaluate a Variable

Find the value of the remaining variable in each of the following.

(a) $A = LW;\quad A = 64,\ L = 10$

As shown in Figure 3, this formula gives the area of a rectangle with length L and width W. Substitute the given values into the formula and then solve for W.

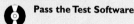

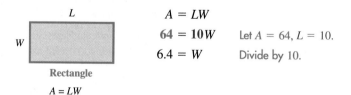

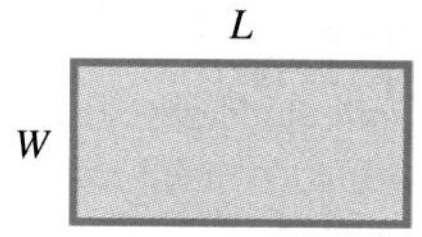

Rectangle

$A = LW$

Figure 3

$$A = LW$$
$$\mathbf{64} = \mathbf{10}W \quad \text{Let } A = 64,\ L = 10.$$
$$6.4 = W \quad \text{Divide by 10.}$$

Check that the width of the rectangle is 6.4.

(b) $A = \frac{1}{2}(b + B)h;\quad A = 210,\ B = 27,\ h = 10$

This formula gives the area of a trapezoid with parallel sides of lengths b and B and distance h between the parallel sides. See Figure 4. Again, begin by substituting the given values into the formula.

INSTRUCTOR'S RESOURCES

PTB Sec. 2.4 **ISM** Sec. 2.4 **AB** Sec. 2.4 | **TEST GENERATOR** | **WORLD WIDE WEB** www.LialAlgebra.com

TEACHING TIP Making a sketch when a geometric figure is mentioned in a problem often clarifies the problem.

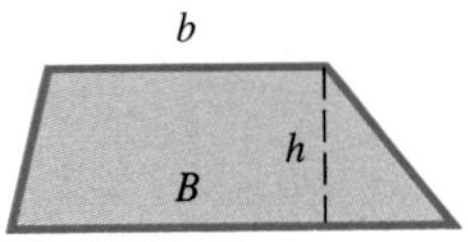

Trapezoid

$A = \frac{1}{2}(b + B)h$

Figure 4

TEACHING TIP After substitution of the given known variable values, the resulting equation is linear in one variable and is solved as in the previous sections.

$$A = \frac{1}{2}(b + B)h$$

$$\mathbf{210} = \frac{1}{2}(b + \mathbf{27})(\mathbf{10}) \qquad A = 210, B = 27, h = 10$$

Now solve for b.

$$210 = \frac{1}{2}(10)(b + 27) \qquad \text{Commutative property}$$

$$210 = 5(b + 27)$$

$$210 = 5b + 135 \qquad \text{Distributive property}$$

$$75 = 5b \qquad \text{Subtract 135.}$$

$$15 = b \qquad \text{Divide by 5.}$$

Check that the length of the shorter parallel side, b, is 15.

OBJECTIVE 2 Use a formula to solve a geometric application. For applications involving geometric figures, a drawing or diagram is useful in Step 2 of the six-step method.

Recall that the *perimeter* of a two-dimensional figure is the distance around the figure, or the sum of the lengths of its sides. Similarly, the **circumference** of a ball (or sphere) is the greatest distance around the ball. (See Figure 5.)

CHALKBOARD EXAMPLE

A rectangular lot is advertised for sale. The ad gives the length as 25 meters and the perimeter as 80 meters. A potential buyer wants to know the width of the lot. What is it?

Answer: 15 meters

EXAMPLE 2 Finding the Radius of a Basketball

A professional basketball has a circumference of at most 78 centimeters. See Figure 5. What is the radius of the circular cross section through the center of the ball?

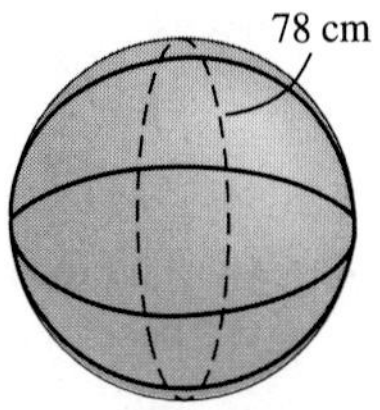

Figure 5

Step 1 We want to find the radius of the circular cross section shown, so let r = the radius in centimeters.

Step 2 See Figure 5.

Step 3 The formula for the circumference of a circle is

$$C = 2\pi r.$$

To find the radius r, we substitute 78 for C. We will use the approximate value 3.14 for π.

$$78 = 2(3.14)r \qquad C = 78,\ \pi \approx 3.14^*$$

Step 4 Solve the equation.

$$78 = 6.28r \qquad \text{Multiply.}$$
$$r = 12.42 \qquad \text{Divide by 6.28; round.}$$

Step 5 The radius of the indicated cross section is 12.42 centimeters.

Step 6 Check. If $r = 12.42$ and $\pi = 3.14$, the circumference is $2(3.14)(12.42)$ or about 78, as required.

The **area** of a geometric figure is a measure of the surface covered by the figure. Example 3 shows an application of area.

CHALKBOARD EXAMPLE

The area of a triangle is 120 square meters. The height is 24 meters. Find the length of the base of the triangle.

Answer: 10 meters

EXAMPLE 3 Finding the Height of a Triangular Sail

The area of a triangular sail of a sailboat is 126 square feet. The base of the sail is 12 feet. Find the height of the sail.

Step 1 Since we must find the height of the triangular sail,

let h = the height of the sail in feet.

Step 2 See Figure 6.

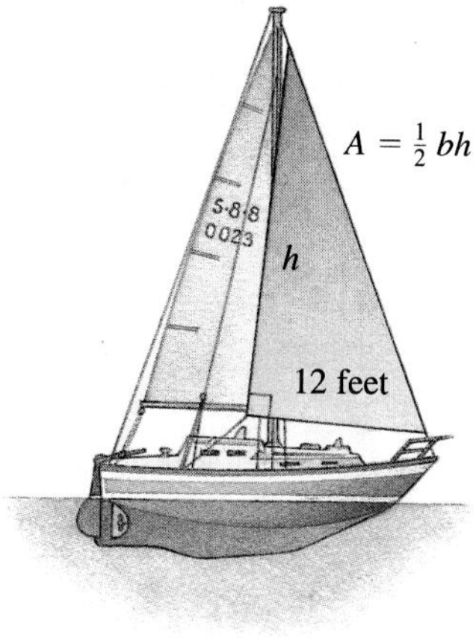

Figure 6

Step 3 The formula for the area of a triangle is $A = \frac{1}{2}bh$, where A is the area, b is the base, and h is the height. Using the information given in the problem, substitute 126 for A and 12 for b in the formula.

$$A = \frac{1}{2}bh$$

$$\mathbf{126} = \frac{1}{2}(\mathbf{12})h \qquad A = 126,\ b = 12$$

*The number π is *not exactly* equal to 3.14.

Step 4 Solve the equation.

$$126 = \frac{12}{2}h$$

$$126 = 6h$$

$$21 = h \qquad \text{Divide by 6.}$$

Step 5 The height of the sail is 21 feet.

Step 6 Check to see that the values $A = 126$, $b = 12$, and $h = 21$ satisfy the formula for the area of a triangle.

OBJECTIVE 3 **Solve problems about angle measures.** Angle measures are vital to the study of *geodesy,* the measurement of the earth's surface, and other fields. Methods of calculating angle measures are discussed in the next example.

Refer to Figure 7, which shows two intersecting lines forming angles that are numbered ①, ②, ③, and ④. Angles ① and ③ lie "opposite" each other. They are called **vertical angles.** Another pair of vertical angles are ② and ④. In geometry, it is shown that vertical angles have equal measures.

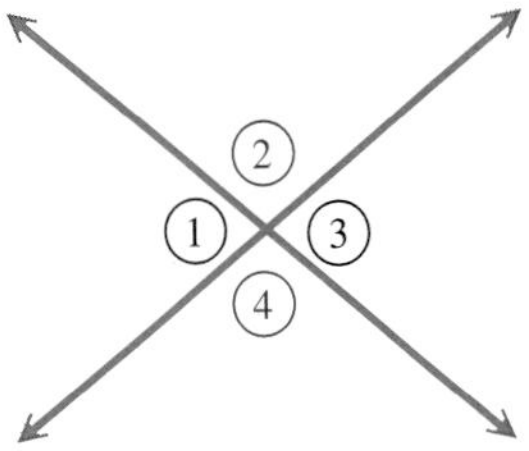

Figure 7

Now look at angles ① and ②. When their measures are added, we get 180°, the measure of a **straight angle,** so the angles are supplements. There are three other such pairs of angles: ② and ③, ③ and ④, ① and ④.

CHALKBOARD EXAMPLE

Find the measure of each marked angle in the figure.

Answer: 149°, 31°

EXAMPLE 4 Finding Angle Measures

Refer to the appropriate figures in each part.

(a) Find the measure of each marked angle in Figure 8.

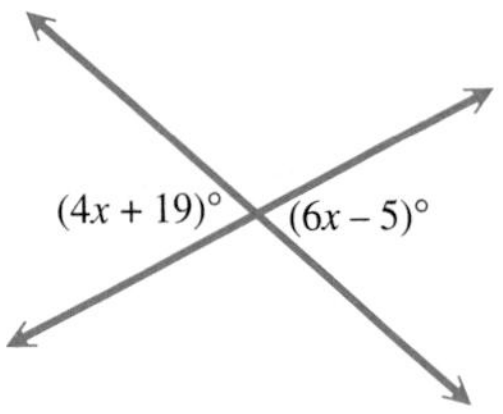

Figure 8

Since the marked angles are vertical angles, they have the same measures. Set $4x + 19$ equal to $6x - 5$ and solve.

$$\begin{aligned} 4x + 19 &= 6x - 5 & \\ -4x + 4x + 19 &= -4x + 6x - 5 & \text{Add } -4x. \\ 19 &= 2x - 5 & \\ 19 + 5 &= 2x - 5 + 5 & \text{Add 5.} \\ 24 &= 2x & \\ 12 &= x & \text{Divide by 2.} \end{aligned}$$

Now we must find the required angle measures. Since $x = 12$, one angle has measure $4(12) + 19 = 67$ degrees. The other has the same measure, since $6(12) - 5 = 67$ as well. Each angle measures 67°.

(b) Find the measure of each marked angle in Figure 9.

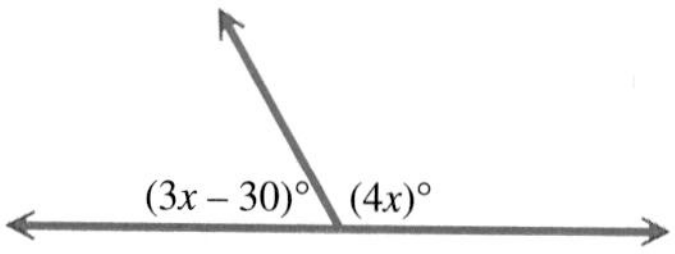

Figure 9

The measures of the marked angles must add to 180° since together they form a straight angle. The equation to solve is

$$\begin{aligned} (3x - 30) + 4x &= 180. & \\ 7x - 30 &= 180 & \text{Combine like terms.} \\ 7x - 30 + 30 &= 180 + 30 & \text{Add 30.} \\ 7x &= 210 & \\ x &= 30 & \text{Divide by 7.} \end{aligned}$$

To find the measures of the angles, substitute 30 for x in the two expressions.

$$\begin{aligned} 3x - 30 &= 3(30) - 30 = 90 - 30 = 60 \\ 4x &= 4(30) = 120 \end{aligned}$$

The two angle measures are 60° and 120°, which are supplementary. ■

CAUTION In Example 4, the answer is not the value of x. Remember to substitute the value of the variable into the expression for each angle you are asked to find.

TEACHING TIP Stress that solving for a specified variable uses exactly the same steps as solving any of the previous linear equations.

OBJECTIVE 4 Solve a formula for a specified variable. Sometimes it is necessary to solve a number of problems that use the same formula. For example, a surveying class might need to solve several problems that involve the formula for the area of a rectangle, $A = LW$. Suppose that in each problem the area (A) and the length (L) of a rectangle are given and the width (W) must be found. Rather than solving for W each time the formula is used, it would be simpler to *rewrite the formula* so that it is solved for W. This process is called **solving for a specified variable.** Solving a formula for a specified variable requires the same steps used earlier to solve equations with just one variable.

The formula for converting temperatures given in degrees Celsius to degrees Fahrenheit is

$$F = \frac{9}{5}C + 32.$$

The next example shows how to solve this formula for C.

CHALKBOARD EXAMPLE

Solve $A = p + prt$ for t.

Answer: $t = \dfrac{A - p}{pr}$

TEACHING TIP When solving an equation or formula for a specified variable, encourage students to circle or underline the variable they are solving for.

EXAMPLE 5 Solving for a Specified Variable

Solve $F = \frac{9}{5}C + 32$ for C.

To clear the fraction, multiply on both sides by 5.

$$F = \frac{9}{5}C + 32 \qquad \text{Solve for } C.$$

$$5F = 5\left(\frac{9}{5}C + 32\right) \qquad \text{Multiply by 5.}$$

$$5F = 9C + 160$$

Now, to get $9C$ alone on one side, subtract 160 from both sides. Then divide both sides by 9.

$$5F - 160 = 9C + 160 - 160 \qquad \text{Subtract 160.}$$

$$5F - 160 = 9C$$

$$\frac{5F - 160}{9} = C \qquad \text{Divide by 9.}$$

This result can be written in a different form using the distributive property to rewrite the numerator as $5(F - 32)$. Then we have

$$C = \frac{5(F - 32)}{9} = \frac{5}{9}(F - 32).$$

NOTE When solving a formula for a specified variable, treat that variable as if it were the *only* variable in the equation, and treat all others as if they were numbers. Use the method of solving equations described in Sections 2.1 and 2.2 to solve for the specified variable.

CHALKBOARD EXAMPLE

Solve $A = \frac{1}{2}(b + B)h$ for h.

Answer: $h = \dfrac{2A}{b + B}$

TEACHING TIP Stress that the distributive property does not apply in Example 6 when multiplying both sides by 2.

EXAMPLE 6 Solving for a Specified Variable

Solve $A = \frac{1}{2}(b + B)h$ for B.

This is the formula for the area of a trapezoid. To get B alone, begin by multiplying both sides by 2 to clear the fraction.

$$A = \frac{1}{2}(b + B)h$$

$$2A = 2 \cdot \frac{1}{2}(b + B)h \qquad \text{Multiply by 2. (Do not distribute the 2 on the right side.)}$$

$$2A = (b + B)h \qquad 2 \cdot \tfrac{1}{2} = \tfrac{2}{2} = 1$$

$$2A = bh + Bh \qquad \text{Distributive property}$$

Undo what was done to B by first subtracting bh on both sides. Then divide both sides by h.

$$2A - bh = Bh \qquad \text{Subtract } bh.$$

$$\frac{2A - bh}{h} = B \quad \text{or} \quad B = \frac{2A - bh}{h} \qquad \text{Divide by } h.$$

The result can be written in a different form as follows.

$$B = \frac{2A - bh}{h} = \frac{2A}{h} - \frac{bh}{h} = \frac{2A}{h} - b$$

Either form is correct.

2.4 EXERCISES

1. In your own words, explain what is meant by the *perimeter* of a geometric figure.
2. In your own words, explain what is meant by the *area* of a geometric figure.

Decide whether perimeter or area would be used to solve a problem concerning the measure of the quantity.

3. Carpeting for a bedroom
4. Sod for a lawn
5. Fencing for a yard
6. Baseboards for a living room
7. Tile for a bathroom
8. Fertilizer for a garden
9. Determining the cost for replacing a linoleum floor with a wood floor
10. Determining the cost for planting rye grass in a lawn for the winter

In the following exercises a formula is given along with the values of all but one of the variables in the formula. Find the value of the variable that is not given. See Example 1.

11. $P = 2L + 2W$ (perimeter of a rectangle); $L = 6, W = 4$
12. $P = 2L + 2W$; $L = 8, W = 5$
13. $P = 4s$ (perimeter of a square); $s = 6$
14. $P = 4s$; $s = 12$
15. $A = \frac{1}{2}bh$ (area of a triangle); $b = 10, h = 14$
16. $A = \frac{1}{2}bh$; $b = 8, h = 16$
17. $d = rt$ (distance formula); $d = 100, t = 2.5$
18. $d = rt$; $d = 252, r = 45$
19. $I = prt$ (simple interest); $p = 5000, r = .025, t = 7$
20. $I = prt$; $p = 7500, r = .035, t = 6$
21. $A = \frac{1}{2}h(b + B)$ (area of a trapezoid); $h = 7, b = 12, B = 14$
22. $A = \frac{1}{2}h(b + B)$; $h = 3, b = 19, B = 31$
23. $C = 2\pi r$ (circumference of a circle); $C = 8.164, \pi = 3.14$
24. $C = 2\pi r$; $C = 16.328, \pi = 3.14$
25. $A = \pi r^2$ (area of a circle); $r = 12, \pi = 3.14$

1. The perimeter of a geometric figure is the sum of the lengths of the sides. **2.** The area of a geometric figure is the number of square units enclosed by the figure. **3.** area **4.** area **5.** perimeter **6.** perimeter **7.** area **8.** area **9.** area **10.** area **11.** $P = 20$ **12.** $P = 26$ **13.** $P = 24$ **14.** $P = 48$ **15.** $A = 70$ **16.** $A = 64$ **17.** $r = 40$ **18.** $t = 5.6$ **19.** $I = 875$ **20.** $I = 1575$ **21.** $A = 91$ **22.** $A = 75$ **23.** $r = 1.3$ **24.** $r = 2.6$ **25.** $A = 452.16$

26. $A = 50.24$ 27. 4
28. $\frac{9}{5}C$ equals $\frac{9C}{5}$ because $\frac{9}{5}C = \frac{9}{5} \cdot \frac{C}{1} = \frac{9C}{5}$. 29. $V = 384$
30. $V = 150$ 31. $V = 48$
32. $V = 52$ 33. $V = 904.32$
34. $V = 7234.56$ 35. about 154,000 square feet 36. length: 36 inches; volume: 11,664 cubic inches

26. $A = \pi r^2$; $r = 4$, $\pi = 3.14$

27. If a formula contains exactly five variables, how many values would you need to be given in order to find the value of any one variable?

28. The formula for changing Celsius to Fahrenheit is given in Example 5 as $F = \frac{9}{5}C + 32$. Sometimes it is seen as $F = \frac{9C}{5} + 32$. These are both correct. Why is it true that $\frac{9}{5}C$ is equal to $\frac{9C}{5}$?

*The **volume** of a three-dimensional object is a measure of the space occupied by the object. For example, we would need to know the volume of a gasoline tank in order to know how many gallons of gasoline it would take to completely fill the tank. In each of the following exercises, a formula for the volume, V, of a three-dimensional object is given, along with values for the other variables. Solve for V. See Example 1.*

29. $V = LWH$ (volume of a rectangular-sided box); $L = 12$, $W = 8$, $H = 4$

30. $V = LWH$; $L = 10$, $W = 5$, $H = 3$

31. $V = \frac{1}{3}Bh$ (volume of a pyramid); $B = 36$, $h = 4$

32. $V = \frac{1}{3}Bh$; $B = 12$, $h = 13$

33. $V = \frac{4}{3}\pi r^3$ (volume of a sphere); $r = 6$, $\pi = 3.14$

34. $V = \frac{4}{3}\pi r^3$; $r = 12$, $\pi = 3.14$

Use a formula to write an equation for each application, and use the six-step problem-solving method of Section 2.3 to solve it. Formulas may be found on the inside covers of this book. See Examples 2 and 3. Use 3.14 for π, if applicable.

35. Recently, a prehistoric ceremonial site dating to about 3000 B.C. was discovered at Stanton Drew in southwestern England. The site, which is larger than Stonehenge, is a near perfect circle, consisting of nine concentric rings that probably held upright wooden posts. Around this timber temple is a wide, encircling ditch enclosing an area with a diameter of 443 feet. Find this enclosed area. (*Source: Archaeology,* vol. 51, no. 1, Jan./Feb. 1998.)

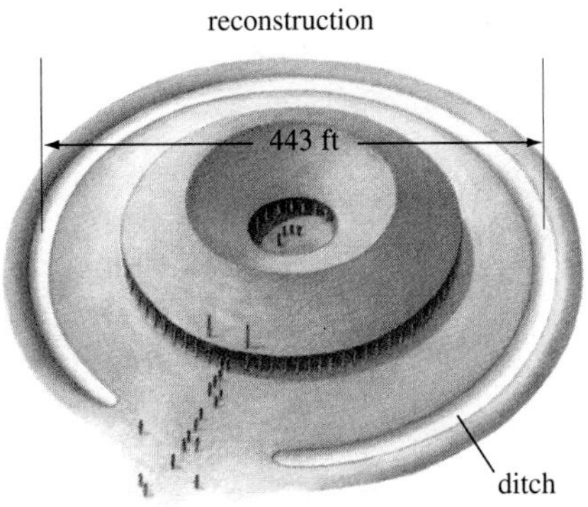

36. The U.S. Postal Service requires that any box sent through the mail have length plus girth (distance around) totaling no more than 108 inches. The maximum volume that meets this condition is contained by a box with a square end 18 inches on each side. What is the length of the box? What is the maximum volume?

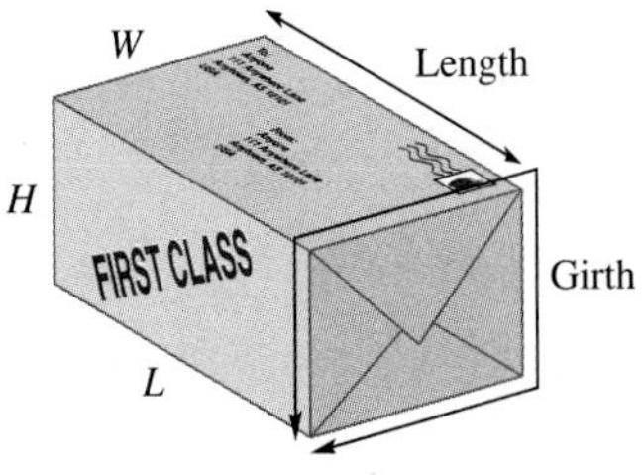

JOURNAL WRITING ▲ CHALLENGING SCIENTIFIC CALCULATOR GRAPHING CALCULATOR

37. 1978 feet **38.** 132.665 square feet **39.** perimeter: 172 inches; area: 1785 square inches **40.** perimeter: 13 inches; area: 10.5 square inches
41. $\frac{729}{32}$ or $22\frac{25}{32}$ cubic inches
42. 23,123,672 cubic feet
43. 23,800.10 square feet
44. 9036.24 square feet
45. 107°, 73° **46.** 48°, 132°
47. 75°, 75° **48.** 51°, 51°

37. The Skydome in Toronto, Canada, is the first stadium with a hard-shell, retractable roof. The steel dome is 630 feet in diameter. To the nearest foot, what is the circumference of this dome?

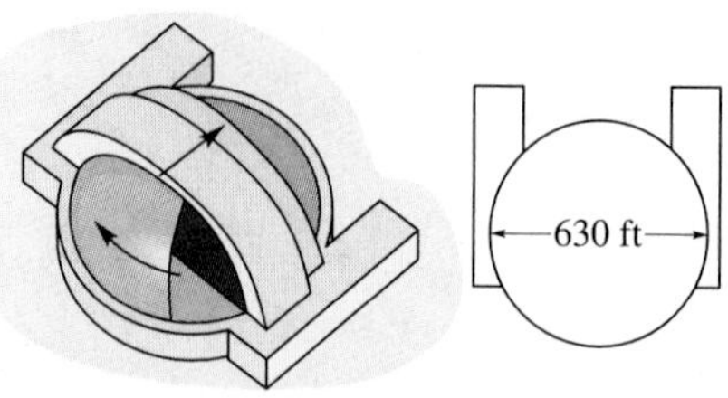

38. The largest drum ever constructed was played at the Royal Festival Hall in London in 1987. It had a diameter of 13 feet. What was the area of the circular face of the drum? (*Source: The Guinness Book of World Records.*)

39. The newspaper, *The Constellation,* printed in 1859 in New York City as part of the Fourth of July celebration, had length 51 inches and width 35 inches. What was the perimeter? What was the area? (*Source: The Guinness Book of World Records.*)

40. The *Daily Banner,* published in Roseberg, Oregon, in the nineteenth century, had page size 3 inches by 3.5 inches. What was the perimeter? What was the area? (*Source: The Guinness Book of World Records.*)

41. A color television set with a liquid crystal display was manufactured by Epson in 1985 and had dimensions 3 inches by $6\frac{3}{4}$ inches by $1\frac{1}{8}$ inches. What was the volume of this set? (*Source: The Guinness Book of World Records.*)

42. A sea lock at Zeebrugge, Belgium, measures 1640 feet by 187 feet by 75.4 feet. What is the volume of the lock? (It is a rectangular solid, so use the formula for the volume of such a figure.) (*Source: The Guinness Book of World Records.*)

43. The survey plat (a two-dimensional map plotted from a plane survey) in the figure shows two lots that form a trapezoid. The measures of the parallel sides are 115.80 feet and 171.00 feet. The height of the trapezoid is 165.97 feet. Find the combined area of the two lots. Round your answer to the nearest hundredth of a square foot.

44. Lot A in the figure is in the shape of a trapezoid. The parallel sides measure 26.84 feet and 82.05 feet. The height of the trapezoid is 165.97 feet. Find the area of Lot A. Round your answer to the nearest hundredth of a square foot.

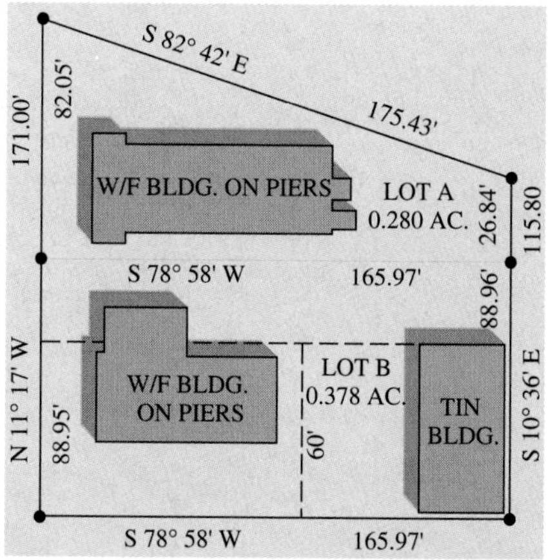

Property survey of lot in New Roads, Louisiana.

Find the measure of each marked angle. See Example 4.

45.

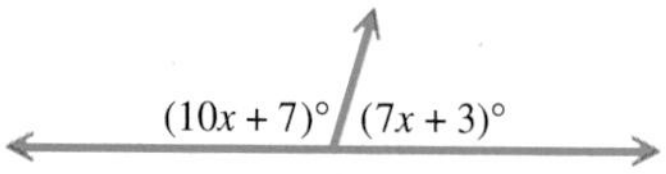

46.

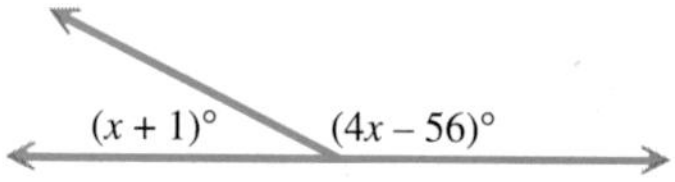

47.

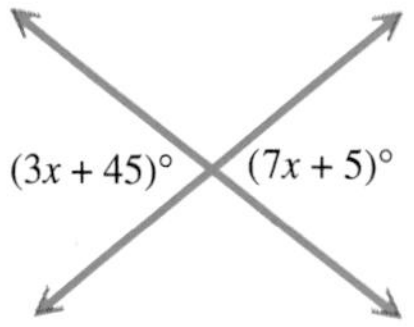

48.

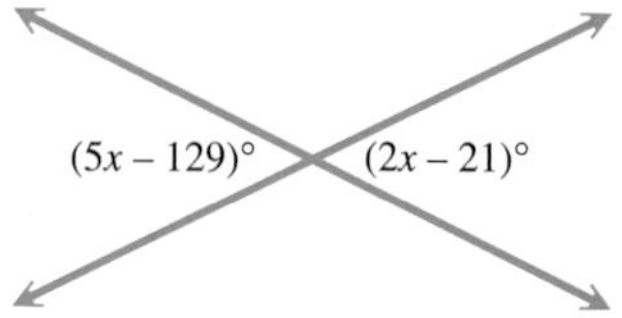

49. 139°, 139° 50. 105°, 105°

51. $r = \frac{d}{t}$ 52. $r = \frac{I}{pt}$

53. $p = \frac{I}{rt}$ 54. $H = \frac{V}{LW}$

55. $a = P - b - c$

56. $b = P - a - c$

57. $b = \frac{2A}{h}$ 58. $h = \frac{2A}{b}$

59. $r = \frac{A - p}{pt}$ 60. $W = \frac{P - 2L}{2}$

61. $h = \frac{V}{\pi r^2}$ 62. $h = \frac{3V}{\pi r^2}$

63. $m = \frac{y - b}{x}$ 64. $y = \frac{C - Ax}{B}$

65. (a) $P - 2L = 2W$ (b) $W = \frac{P - 2L}{2}$

66. (a) $\frac{P}{2} = L + W$ (b) $\frac{P}{2} - L = W$

67. (a) multiplication identity property (b) An expression divided by 1 is equal to itself. (c) rule for multiplication of fractions (d) rule for subtraction of fractions

68. $\frac{5T}{4} + 1$

49.

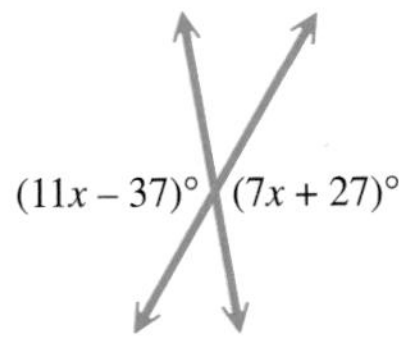

50.

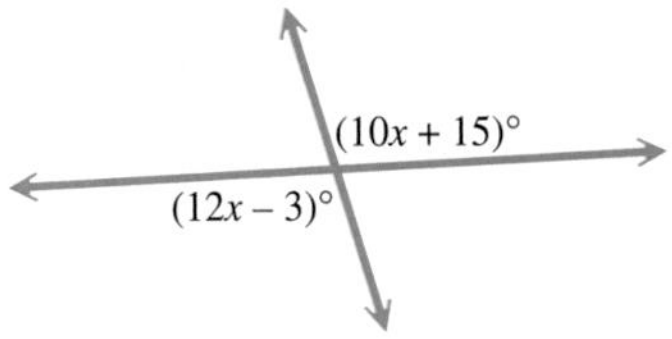

Solve each formula for the specified variable. See Examples 5 and 6.

51. $d = rt$ for r

52. $I = prt$ for r

53. $I = prt$ for p

54. $V = LWH$ for H

55. $P = a + b + c$ for a

56. $P = a + b + c$ for b

57. $A = \frac{1}{2}bh$ for b

58. $A = \frac{1}{2}bh$ for h

59. $A = p + prt$ for r

60. $P = 2L + 2W$ for W

61. $V = \pi r^2 h$ for h

62. $V = \frac{1}{3}\pi r^2 h$ for h

63. $y = mx + b$ for m

64. $Ax + By = C$ for y

RELATING CONCEPTS (EXERCISES 65–68)

Students in beginning algebra are often faced with the following problem: "I know I did the work correctly, but my answer doesn't look like the one given in the book. What's wrong?" In many cases, there are equally acceptable *equivalent* answers for a problem.

Work Exercises 65–68 in order, *to see how two seemingly different answers to a problem can both be correct.*

65. Solve the formula $P = 2L + 2W$ for W using the following steps.

(a) Subtract $2L$ from both sides.

(b) Divide both sides by 2 to get $\frac{P - 2L}{2} = W$.

66. Solve the formula $P = 2L + 2W$ for W using the following steps.

(a) Divide each term on both sides by 2.

(b) Subtract L from both sides to get $\frac{P}{2} - L = W$.

67. Compare the results in Exercises 65(b) and 66(b). They are both acceptable answers, but they are written in different forms. To show they are equivalent, follow the steps below and provide the justification for each step.

(a) $\frac{P}{2} - L = \frac{P}{2} - \frac{2}{2} \cdot L$

(b) $= \frac{P}{2} - \frac{2}{2} \cdot \frac{L}{1}$

(c) $= \frac{P}{2} - \frac{2L}{2}$

(d) $= \frac{P - 2L}{2}$

68. Suppose that the expression $\frac{5T + 4}{4}$ is obtained in a similar problem. What is another valid answer?

Did you make the connection that answers for a problem may be written in different, yet equivalent forms?

2.5 Ratios and Proportions

OBJECTIVES

1. Write ratios.
2. Decide whether proportions are true.
3. Solve proportions.
4. Solve applied problems using proportions.
5. Solve direct variation problems.

FOR EXTRA HELP

SSG Sec. 2.5
SSM Sec. 2.5
Pass the Test Software
InterAct Math Tutorial Software
Video 3

CHALKBOARD EXAMPLE

Write a ratio for the following phrase.

3 days to 2 weeks

Answer: $\frac{3}{14}$

OBJECTIVE 1 Write ratios. Ratios provide a way of comparing two numbers or quantities using division. A **ratio** is a quotient of two quantities with the same units.

The ratio of the number a to the number b is written

$$a \text{ to } b, \qquad a : b, \qquad \text{or} \qquad \frac{a}{b}.$$

This last way of writing a ratio is most common in algebra. Note that the order must be $\frac{a}{b}$, not $\frac{b}{a}$.

Percents are ratios where the second number is always 100. For example, 50% represents the ratio of 50 to 100, 27% represents the ratio of 27 to 100, and so on.

When ratios are used in comparing units of measure, the units should be the same. This is shown in Example 1.

EXAMPLE 1 Writing a Ratio

Write a ratio for each word phrase.

(a) The ratio of 5 hours to 3 hours
This ratio can be written as $\frac{5}{3}$.

(b) The ratio of 5 hours to 3 days
First convert 3 days to hours: 3 days $= 3 \cdot 24 = 72$ hours. The ratio of 5 hours to 3 days is thus

$$\frac{5 \text{ hours}}{72 \text{ hours}} \quad \text{or} \quad \frac{5}{72}.$$

NOTE Because the units of the two quantities in a ratio must be the same, ratios do not have units in their final forms.

OBJECTIVE 2 Decide whether proportions are true. A ratio is used to compare two numbers or amounts. A **proportion** is a statement that two ratios are equal. For example,

$$\frac{3}{4} = \frac{15}{20}$$

is a proportion that says that the ratios $\frac{3}{4}$ and $\frac{15}{20}$ are equal. In the proportion

$$\frac{a}{b} = \frac{c}{d},$$

a, b, c, and d are the **terms** of the proportion. Beginning with the proportion

$$\frac{a}{b} = \frac{c}{d}$$

and multiplying both sides by the common denominator, bd, gives

$$bd \cdot \frac{a}{b} = bd \cdot \frac{c}{d}$$

$$ad = bc.$$

INSTRUCTOR'S RESOURCES

PTB Sec. 2.5 **ISM** Sec. 2.5
AB Sec. 2.5

TEST GENERATOR

WORLD WIDE WEB
www.LialAlgebra.com

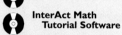

The same products *ad* and *bc* can be found by multiplying diagonally.

$$\frac{a}{b} = \frac{c}{d} \qquad \begin{matrix} bc \\ ad \end{matrix}$$

This is called **cross multiplication,** and *ad* and *bc* are called **cross products.**

Cross Products

If $\frac{a}{b} = \frac{c}{d}$, then the cross products *ad* and *bc* are equal.

Also, if $ad = bc$, then $\frac{a}{b} = \frac{c}{d}$ $(b, d \neq 0)$.

TEACHING TIP Some students erroneously try to "reduce" across the equals sign when working with proportions. Give examples to show why this cannot be done.

From the rule given above,

$$\text{if} \quad \frac{a}{b} = \frac{c}{d} \quad \text{then} \quad ad = bc.$$

However, if $\frac{a}{c} = \frac{b}{d}$, then $ad = cb$, or $ad = bc$. This means that the two proportions are equivalent and

$$\text{the proportion } \frac{a}{b} = \frac{c}{d} \text{ can also be written as } \frac{a}{c} = \frac{b}{d}.$$

Sometimes one form is more convenient to work with than the other.

CHALKBOARD EXAMPLE

Decide whether the proportion is true or false.

$$\frac{21}{15} = \frac{62}{45}$$

Answer: false

EXAMPLE 2 Deciding Whether a Proportion Is True

Decide whether the following proportions are true or false.

(a) $\frac{3}{4} = \frac{15}{20}$

Check to see whether the cross products are equal.

$$\frac{3}{4} = \frac{15}{20} \qquad \begin{matrix} \mathbf{4 \cdot 15 = 60} \\ \mathbf{3 \cdot 20 = 60} \end{matrix}$$

The cross products are equal, so the proportion is true.

(b) $\frac{6}{7} = \frac{30}{32}$

The cross products are $6 \cdot 32 = 192$ and $7 \cdot 30 = 210$. The cross products are different, so the proportion is false.

TEACHING TIP Caution students that finding cross products is *not* appropriate for solving

$$\frac{x}{3} + \frac{5}{4} = \frac{1}{2}$$

or for adding

$$\frac{2}{3} + \frac{1}{9}.$$

CAUTION

The cross product method cannot be used directly if there is more than one term on either side of the equals sign. For example, you cannot use the method directly to solve the equation

$$\frac{4}{x} + 3 = \frac{1}{9},$$

because there are two terms on the left side.

OBJECTIVE 3 Solve proportions. Four numbers are used in a proportion. If any three of these numbers are known, the fourth can be found.

CHALKBOARD EXAMPLE

Solve $\frac{a+6}{2a} = \frac{1}{5}$.

Answer: $\{-10\}$

EXAMPLE 3 Solving an Equation Using Cross Products

(a) Find x in the proportion

$$\frac{63}{x} = \frac{9}{5}.$$

The cross products must be equal, so

$$63 \cdot 5 = 9x$$
$$315 = 9x.$$

Divide both sides by 9 to get

$$35 = x.$$

The solution set is $\{35\}$.

(b) Solve the equation

$$\frac{m-2}{5} = \frac{m+1}{3}.$$

Find the cross products, and set them equal to each other.

$$3(m-2) = 5(m+1) \quad \text{Be sure to use parentheses.}$$
$$3m - 6 = 5m + 5 \quad \text{Distributive property}$$
$$3m = 5m + 11 \quad \text{Add 6.}$$
$$-2m = 11 \quad \text{Subtract } 5m.$$
$$m = -\frac{11}{2} \quad \text{Divide by } -2.$$

The solution set is $\{-\frac{11}{2}\}$.

TEACHING TIP Solving a proportion is a special case of solving equations with rational expressions, which will be discussed in detail in Section 6.6.

OBJECTIVE 4 Solve applied problems using proportions. Proportions occur in many practical applications.

CHALKBOARD EXAMPLE

If 7 shirts cost \$87.50, find the cost of 11 shirts.

Answer: \$137.50

EXAMPLE 4 Applying Proportions

A local store is offering 3 packs of toothpicks for \$.87. How much would it charge for 10 packs?

Let $x =$ the cost of 10 packs of toothpicks.

Set up a proportion. One ratio in the proportion can involve the number of packs, and the other can involve the costs. Make sure that the corresponding numbers appear in the numerator and the denominator.

$$\frac{\text{Cost of 3}}{\text{Cost of 10}} = \frac{3}{10}$$

$$\frac{.87}{x} = \frac{3}{10}$$

$$3x = .87(10) \quad \text{Cross products}$$

$$3x = 8.7$$

$$x = 2.90 \quad \text{Divide by 3.}$$

The 10 packs should cost \$2.90. As shown earlier, the proportion could also be written as $\frac{3}{.87} = \frac{10}{x}$, which would give the same cross products.

NOTE: Many people would solve the problem in Example 4 mentally as follows: Three packs cost \$.87, so one pack costs \$.87/3 = \$.29. Then ten packs will cost 10(\$.29) = \$2.90. If you do the problem this way, you are using proportions and probably not even realizing it!

An important application that uses proportions is *unit pricing*—deciding which size of an item offered in different sizes produces the best price per unit. For example, suppose you can buy 36 ounces of pancake syrup for \$3.89. To find the price per unit, set up the proportion

$$\frac{36 \text{ ounces}}{1 \text{ ounce}} = \frac{3.89}{x}$$

and solve for x.

$$36x = 3.89 \quad \text{Cross products}$$

$$x = \frac{3.89}{36} \quad \text{Divide by 36.}$$

$$x \approx .108 \quad \text{Use a calculator.}$$

Thus, the price for 1 ounce is \$.108, or about 11 cents. Notice that the unit price is the ratio of the cost for 36 ounces, \$3.89, to the number of ounces, 36, which means that the unit price for an item is found by dividing the cost by the number of units.

CHALKBOARD EXAMPLE

Find the best buy.

Extra crunchy peanut butter
12-ounce size: \$1.49
28-ounce size: \$1.99
40-ounce size: \$3.99

Answer: 28-ounce size

EXAMPLE 5 Determining Unit Price to Obtain the Best Buy

Besides the 36-ounce size discussed above, the local supermarket carries two other sizes of a popular brand of pancake syrup, priced as follows.

Size	*Price*
36-ounce	\$3.89
24-ounce	\$2.79
12-ounce	\$1.89

Which size is the best buy? That is, which size has the lowest unit price?

To find the best buy, divide the price by the number of units to get the price per ounce. Each result in the following table was found by using a calculator and rounding the answer to three decimal places.

Size	Unit Cost (dollars per ounce)
36-ounce	$\frac{\$3.89}{36} = \$.108$ ← The best buy
24-ounce	$\frac{\$2.79}{24} = \$.116$
12-ounce	$\frac{\$1.89}{12} = \$.158$

Since the 36-ounce size produces the lowest price per unit, it would be the best buy. (*Be careful:* Sometimes the largest container *does not* produce the lowest price per unit.)

OBJECTIVE 5 **Solve direct variation problems.** Suppose that gasoline costs \$1.50 per gallon. Then 1 gallon costs \$1.50, 2 gallons cost 2(\$1.50) = \$3.00, 3 gallons cost 3(\$1.50) = \$4.50, and so on. Using a proportion,

$$\frac{\text{Cost of 1 gallon}}{\text{Total cost}} = \frac{1 \text{ gallon}}{x \text{ gallons}}$$

$$x \cdot \text{Cost of 1 gallon} = 1 \cdot \text{Total cost.}$$

Thus, the total cost is obtained by multiplying the number of gallons by the price per gallon. In general, if k equals the price per gallon and x equals the number of gallons, then the total cost y is equal to kx. Notice that as number of gallons increases, total cost increases.

The preceding discussion is an example of variation. As in the gasoline example, two variables *vary directly* if one is a multiple of the other.

Direct Variation

y **varies directly** as x if there exists a number k such that

$$y = kx.$$

Another way to say this is that y **is directly proportional to** x.

CHALKBOARD EXAMPLE

The circumference of a circle varies directly as its radius. What is the circumference of a circle with a radius of 24 inches, if the circumference of a circle with a radius of 15 inches is 94.2 inches?

Answer: 150.72 inches

EXAMPLE 6 **Using Direct Variation to Find the Cost of Gasoline**

If 6 gallons of gasoline cost \$8.52, find the cost of 20 gallons.

Here, the total cost of the gasoline, y, varies directly as (or is directly proportional to) the number of gallons purchased, x. This means there is a number k such that $y = kx$. To find y when $x = 20$, use the fact that 6 gallons cost \$8.52.

$$y = kx$$

$$\mathbf{8.52} = k(\mathbf{6}) \quad \text{Let } x = 6,\ y = 8.52.$$

$$1.42 = k \quad \text{Divide by 6.}$$

Since $k = 1.42$,

$$y = 1.42x.$$

When $x = 20$,

$$y = 1.42(20) = 28.4,$$

so the cost to purchase 20 gallons of gasoline is \$28.40.

2.5 EXERCISES

Answers:

1. $\frac{5}{8}$ 2. $\frac{1}{3}$ 3. $\frac{1}{4}$ 4. $\frac{6}{5}$ 5. $\frac{2}{1}$ 6. $\frac{5}{6}$ 7. $\frac{3}{1}$ 8. $\frac{5}{8}$ 9. (d) 10. Answers will vary. Examples are 12 to 9, 40 to 30, and 20 to 15. 11. A ratio is a comparison, while a proportion is a statement that two ratios are equal. For example, $\frac{2}{3}$ is a ratio and $\frac{2}{3} = \frac{8}{12}$ is a proportion. 12. Cross products can only be used with proportions, that is, two fractions with an *equals sign* between them. 13. true 14. true 15. false 16. false 17. true 18. true 19. {35} 20. {7} 21. {27} 22. {16} 23. {−1} 24. $\left\{-\frac{31}{5}\right\}$ 25. 6.875 ounces 26. 4 gallons 27. \$670.48 28. 74.13 francs

Determine the ratio and write it in lowest terms. See Example 1.

1. 25 feet to 40 feet
2. 16 miles to 48 miles
3. 18 dollars to 72 dollars
4. 300 people to 250 people
5. 144 inches to 6 feet
6. 60 inches to 2 yards
7. 5 days to 40 hours
8. 75 minutes to 2 hours

9. Which one of the following ratios is not the same as the ratio 2 to 5?
(a) .4 **(b)** 4 to 10 **(c)** 20 to 50 **(d)** 5 to 2

10. Give three ratios that are equivalent to the ratio 4 to 3.

11. Explain the distinction between *ratio* and *proportion*. Give examples.

12. Suppose that someone told you to use cross products in order to multiply fractions. How would you explain to the person what is wrong with his or her thinking?

Decide whether each proportion is true or false. See Example 2.

13. $\frac{5}{35} = \frac{8}{56}$
14. $\frac{4}{12} = \frac{7}{21}$
15. $\frac{120}{82} = \frac{7}{10}$
16. $\frac{27}{160} = \frac{18}{110}$
17. $\frac{\frac{1}{2}}{5} = \frac{1}{10}$
18. $\frac{\frac{1}{3}}{6} = \frac{1}{18}$

Solve each equation. See Example 3.

19. $\frac{k}{4} = \frac{175}{20}$
20. $\frac{49}{56} = \frac{z}{8}$
21. $\frac{x}{6} = \frac{18}{4}$
22. $\frac{z}{80} = \frac{20}{100}$
23. $\frac{3y-2}{5} = \frac{6y-5}{11}$
24. $\frac{2p+7}{3} = \frac{p-1}{4}$

Solve each problem by setting up and solving a proportion. See Example 4.

25. A chain saw requires a mixture of 2-cycle engine oil and gasoline. According to the directions on a bottle of Oregon 2-cycle Engine Oil, for a 50 to 1 ratio requirement, approximately 2.5 fluid ounces of oil are required for 1 gallon of gasoline. For 2.75 gallons, how many fluid ounces of oil are required?

26. The directions on the bottle mentioned in Exercise 25 indicate that if the ratio requirement is 24 to 1, approximately 5.5 ounces of oil are required for 1 gallon of gasoline. If gasoline is to be mixed with 22 ounces of oil, how much gasoline is to be used?

27. In 1998, the average exchange rate between U.S. dollars and United Kingdom pounds was 1 pound to \$1.6762. Margaret went to London and exchanged her U.S. currency for U.K. pounds, and received 400 pounds. How much in U.S. money did Margaret exchange?

28. If 3 U.S. dollars can be exchanged for 4.5204 Swiss francs, how many Swiss francs can be obtained for \$49.20? (Round to the nearest hundredth.)

29. $9.30 30. $9.90
31. 4 feet 32. $25\frac{2}{3}$ inches
33. 12,500 fish 34. 23,300 fish
35. (a) $\frac{26}{100} = \frac{x}{350}$; $91 million
(b) $112 million; $11.2 million
(c) $119 million
36. (a) $\frac{22}{100} = \frac{x}{350}$; $77 million
(b) $234.5 million (c) $\frac{9}{10}$; $\frac{95}{350}$ or $\frac{19}{70}$ 37. 30 count
38. 2-ounce size

29. If 6 gallons of premium unleaded gasoline cost $3.72, how much would it cost to completely fill a 15-gallon tank?

30. If sales tax on a $16.00 compact disc is $1.32, how much would the sales tax be on a $120.00 compact disc player?

31. The distance between Kansas City, Missouri, and Denver is 600 miles. On a certain wall map, this is represented by a length of 2.4 feet. On the map, how many feet would there be between Memphis and Philadelphia, two cities that are actually 1000 miles apart?

32. The distance between Singapore and Tokyo is 3300 miles. On a certain wall map, this distance is represented by 11 inches. The actual distance between Mexico City and Cairo is 7700 miles. How far apart are they on the same map?

33. Biologists tagged 250 fish in Willow Lake on October 5. On a later date they found 7 tagged fish in a sample of 350. Estimate the total number of fish in Willow Lake to the nearest hundred.

34. On May 13 researchers at Argyle Lake tagged 420 fish. When they returned a few weeks later, their sample of 500 fish contained 9 that were tagged. Give an approximation of the fish population in Argyle Lake to the nearest hundred.

The Olympic Committee has come to rely more and more on television rights and major corporate sponsors to finance the games. The pie charts show the funding plans for the first Olympics in Athens and the 1996 Olympics in Atlanta. Use proportions and the figures to answer the questions in Exercises 35 and 36.

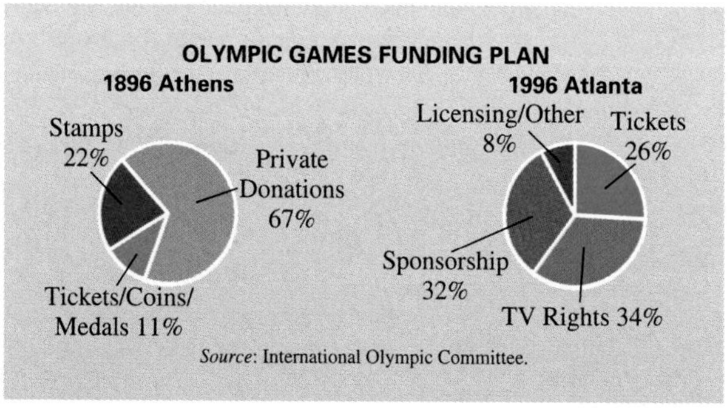

Source: International Olympic Committee.

35. In the 1996 Olympics, total revenue of $350 million was raised. There were 10 major sponsors.
(a) Write a proportion to find the amount of revenue provided by tickets. Solve it.
(b) What amount was provided by sponsors? Assuming the sponsors contributed equally, how much was provided per sponsor?
(c) What amount was raised by TV rights?

36. Suppose the amount of revenue raised in the 1896 Olympics was equivalent to the $350 million in 1996.
(a) Write a proportion for the amount of revenue provided by stamps and solve it.
(b) What amount (in dollars) would have been provided by private donations?
(c) In the 1988 Olympics, there were 9 major sponsors, and the total revenue was $95 million. What is the ratio of major sponsors in 1988 to those in 1996? What is the ratio of revenue in 1988 to revenue in 1996?

A supermarket was surveyed to find the prices charged for items in various sizes. Find the best buy (based on price per unit) for each particular item. See Example 5.

37. Trash bags
20 count: $3.09
30 count: $4.59

38. Black pepper
1-ounce size: $ .99
2-ounce size: $1.65
4-ounce size: $4.39

39. 31-ounce size **40.** 32-ounce size **41.** 32-ounce size **42.** 16-ounce size **43.** 4 **44.** 8 **45.** 1 **46.** 2

47. (a)

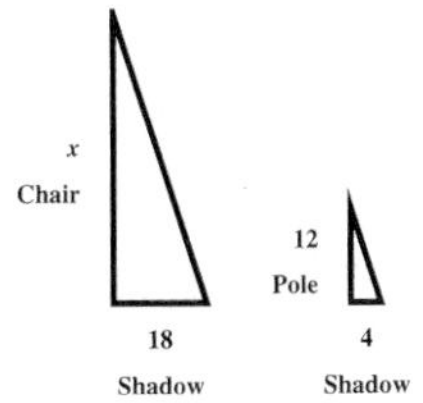

(b) 54 feet

48. (a)

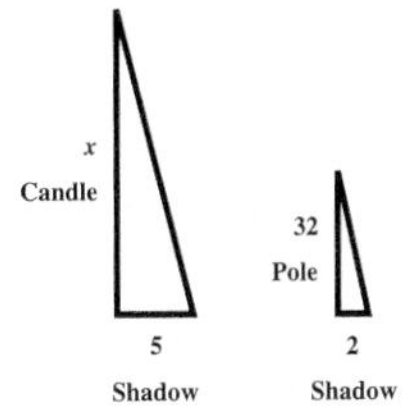

(b) 80 feet

39. Breakfast cereal
15-ounce size: $2.99
25-ounce size: $4.49
31-ounce size: $5.49

40. Cocoa mix
8-ounce size: $1.39
16-ounce size: $2.19
32-ounce size: $2.99

41. Tomato ketchup
14-ounce size: $ .89
32-ounce size: $1.19
64-ounce size: $2.95

42. Cut green beans
8-ounce size: $ .45
16-ounce size: $ .49
50-ounce size: $1.59

Two triangles are **similar** if they have the same shape (but not necessarily the same size). Similar triangles have sides that are proportional. The figure shows two similar triangles. Notice that the ratios of the corresponding sides are all equal to $\frac{3}{2}$:

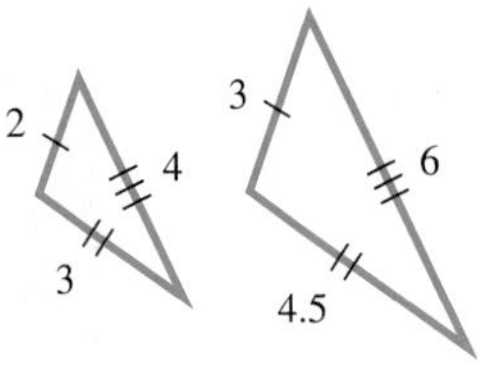

$$\frac{3}{2} = \frac{3}{2} \qquad \frac{4.5}{3} = \frac{3}{2} \qquad \frac{6}{4} = \frac{3}{2}.$$

If we know that two triangles are similar, we can set up a proportion to solve for the length of an unknown side.

▲ *Use a proportion to find the length x, given that the pair of triangles are similar.*

43.

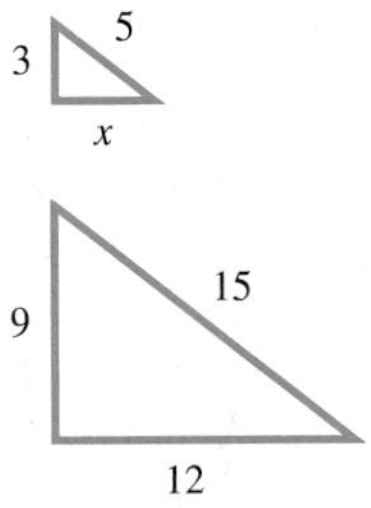

44.

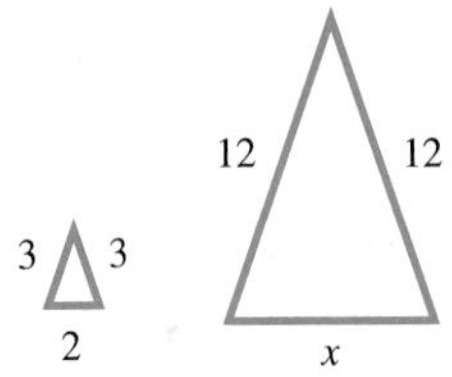

45.

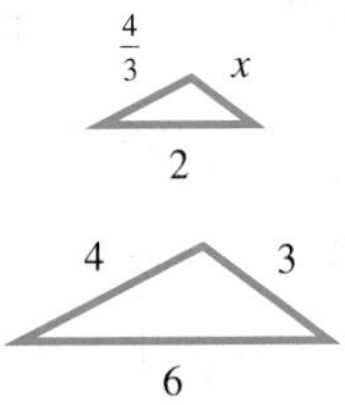

46.

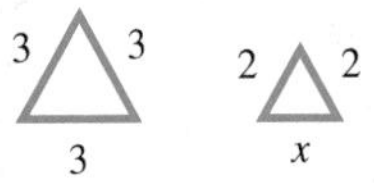

For the problems in Exercises 47 and 48, ***(a)*** *draw a sketch consisting of two right triangles, depicting the situation described, and* ***(b)*** *solve the problem. (*Source: The Guinness Book of World Records.*)*

47. An enlarged version of the chair used by George Washington at the Constitutional Convention casts a shadow 18 feet long at the same time a vertical pole 12 feet high casts a shadow 4 feet long. How tall is the chair?

48. One of the tallest candles ever constructed was exhibited at the 1897 Stockholm Exhibition. If it cast a shadow 5 feet long at the same time a vertical pole 32 feet high cast a shadow 2 feet long, how tall was the candle?

49. $242 **50.** $249 **51.** $255
52. $262 **53.** $4850
54. $15.17

The Consumer Price Index, issued by the U.S. Bureau of Labor Statistics, provides a means of determining the purchasing power of the U.S. dollar from one year to the next. Using the period from 1982 to 1984 as a measure of 100.0, the Consumer Price Index from 1990 to 1995 is shown here.

Year	*Consumer Price Index*
1990	130.7
1991	136.2
1992	140.3
1993	144.5
1994	148.2
1995	152.4

Source: Bureau of Labor Statistics.

To use the Consumer Price Index to predict a price in a particular year, we can set up a proportion and compare it with a known price in another year, as follows:

$$\frac{\text{Price in year } A}{\text{Index in year } A} = \frac{\text{Price in year } B}{\text{Index in year } B}.$$

Use the Consumer Price Index figures above to find the amount that would be charged for the use of the same amount of electricity that cost $225 in 1990. Give your answer to the nearest dollar.

49. in 1992 **50.** in 1993 **51.** in 1994 **52.** in 1995

53. The Consumer Price Index figures for shelter for the years 1981 and 1991 are 90.5 and 146.3. If shelter for a particular family cost $3000 in 1981, what would be the comparable cost in 1991? Give your answer to the nearest dollar.

54. Due to a volatile fuel oil market in the early 1980s, the price of fuel decreased during the first three quarters of the decade. The Consumer Price Index figures for 1982 and 1986 were 105.0 and 74.1. If it cost you $21.50 to fill your tank with fuel oil in 1982, how much would it have cost to fill the same tank in 1986? Give your answer to the nearest cent.

55. 30

56. (a) $5x = 12$ (b) $\frac{12}{5}$

57. (a) $5x = 12$ (b) $\frac{12}{5}$

58. They are the same. Solving by cross products yields the same solution as multiplying by the least common denominator.

RELATING CONCEPTS (EXERCISES 55–58)

In Section 2.2 we learned that to make the solution process easier, if an equation involves fractions, we can multiply both sides of the equation by the least common denominator of all the fractions in the equation. A proportion consists of two fractions equal to each other, so a proportion is a special case of this kind of equation.

Work Exercises 55–58 in order *to see how the process of solving by cross products is justified.*

55. In the equation $\frac{x}{6} = \frac{2}{5}$, what is the least common denominator of the two fractions?

56. Solve the equation in Exercise 55 as follows:
(a) Multiply both sides by the LCD. What is the equation that you obtain?
(b) Solve for x by dividing both sides by the coefficient of x. What is the solution?

57. Solve the equation in Exercise 55 as follows:
(a) Set the cross products equal. What is the equation you obtain?
(b) Repeat part (b) of Exercise 56.

58. Compare your results from Exercises 56(b) and 57(b). What do you notice?

Did you make the connection that solving a proportion by cross products is justified by the general method of solving an equation with fractions?

59. \$40.32
60. 15 square inches
61. $\frac{80}{3}$ or $26\frac{2}{3}$ inches
62. 20 centimeters
63. approximately 2593 miles

Solve each variation problem. See Example 6.

59. The interest on an investment varies directly as the rate of simple interest. If the interest is \$48 when the interest rate is 5%, find the interest when the rate is 4.2%.

60. For a given base, the area of a triangle varies directly as its height. Find the area of a triangle with a height of 6 inches, if the area is 10 square inches when the height is 4 inches.

61. The distance a spring stretches is directly proportional to the force applied. If a force of 30 pounds stretches a spring 16 inches, how far will a force of 50 pounds stretch the spring?

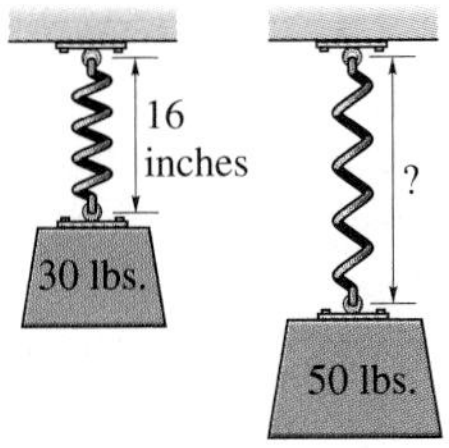

62. The perimeter of a square is directly proportional to the length of its side. What is the perimeter of a square with a 5-centimeter side, if a square with a 12-centimeter side has a perimeter of 48 centimeters?

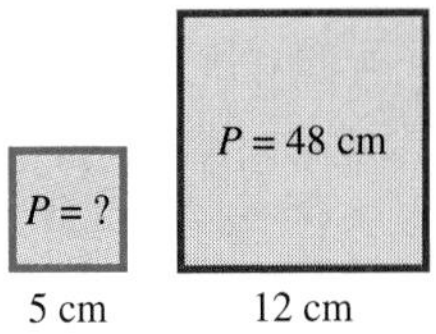

63. According to *The Guinness Book of World Records,* the longest recorded voyage in a paddleboat is 2226 miles in 103 days; the boat was propelled down the Mississippi River by the foot power of two boaters. Assuming a constant rate, how far would they have gone in 120 days? (*Hint:* Distance varies directly as time.)

2.6 More about Problem Solving

OBJECTIVES

1. Use percent in problems involving rates.
2. Solve problems involving mixtures.
3. Solve problems involving simple interest.
4. Solve problems involving denominations of money.
5. Solve problems involving distance, rate, and time.

FOR EXTRA HELP

SSG Sec. 2.6
SSM Sec. 2.6
Pass the Test Software
InterAct Math Tutorial Software
Video 4

OBJECTIVE 1 Use percent in problems involving rates. Recall that percent means "per hundred." Thus, percents are ratios where the second number is always 100. For example, 50% represents the ratio of 50 to 100 and 27% represents the ratio of 27 to 100.

PROBLEM SOLVING

Percents are often used in problems involving mixing different concentrations of a substance or different interest rates. In each case, to get the amount of pure substance or the interest, we multiply.

Mixture Problems	**Interest Problems (annual)**
base × rate (%) = percentage	principal × rate (%) = interest
$b \times r = p$	$p \times r = I$

In an equation, the percent always appears as a decimal. For example, 35% is written as .35, not 35.

INSTRUCTOR'S RESOURCES

PTB Sec. 2.6 **ISM** Sec. 2.6
AB Sec. 2.6
TEST GENERATOR
WORLD WIDE WEB
www.LialAlgebra.com

CHALKBOARD EXAMPLE

(a) What is the amount of pure acid in 40 liters of a 16% acid solution?
(b) Find the annual interest if \$5000 is invested at 4%.

Answer: (a) 6.4 liters
(b) \$200

EXAMPLE 1 Using Percent to Find a Percentage

(a) If a chemist has 40 liters of a 35% acid solution, then the amount of pure acid in the solution is

$$\underset{\text{Amount of solution}}{40} \times \underset{\text{Rate of concentration}}{.35} = \underset{\text{Amount of pure acid}}{14 \text{ liters.}}$$

(b) If \$1300 is invested for one year at 7% simple interest, the amount of interest earned in the year is

$$\underset{\text{Principal}}{\$1300} \times \underset{\text{Interest rate}}{.07} = \underset{\text{Interest earned}}{\$91.}$$

PROBLEM SOLVING

In the examples that follow, we use charts to organize the information in the problems. A chart enables us to more easily set up the equation for the problem, which is usually the most difficult step. The six steps described in Section 2.3 are used but are not always specifically numbered.

OBJECTIVE 2 Solve problems involving mixtures. In the next example, we use percent to solve a mixture problem.

CHALKBOARD EXAMPLE

A certain metal is 40% copper. How many kilograms of this metal must be mixed with 80 kilograms of a metal that is 70% copper to get a metal that is 50% copper?

Answer: 160 kilograms

EXAMPLE 2 Solving a Mixture Problem

A chemist needs to mix 20 liters of 40% acid solution with some 70% solution to get a mixture that is 50% acid. How many liters of the 70% solution should be used?

Step 1 Let x = the number of liters of 70% solution that are needed.

Step 2 Recall from part (a) of Example 1 that the amount of pure acid in this solution will be given by the product of the percent of strength and the number of liters of solution, or

$$\text{liters of pure acid in } x \text{ liters of 70\% solution} = .70x.$$

The amount of pure acid in the 20 liters of 40% solution is

$$\text{liters of pure acid in the 40\% solution} = .40(20) = 8.$$

The new solution will contain $20 + x$ liters of 50% solution. The amount of pure acid in this solution is

$$\text{liters of pure acid in the 50\% solution} = .50(20 + x).$$

The given information is summarized in the chart below.

Liters of Mixture	Rate (as a decimal)	Liters of Pure Acid
x	.70	$.70(x)$
20	.40	$.40(20)$
$20 + x$	.50	$.50(20 + x)$

TEACHING TIP Remind students that in a problem such as Example 2, the concentration of the final mixture must be between the concentrations of the two solutions making up the mixture.

Step 3 The number of liters of pure acid in the 70% solution added to the number of liters of pure acid in the 40% solution will equal the number of liters of pure acid in the final mixture, so the equation is

Pure acid in 70%	plus	pure acid in 40%	is	pure acid in 50%.
↓	↓	↓	↓	↓
$.70x$	$+$	$.40(20)$	$=$	$.50(20 + x)$.

Step 4 Clear parentheses, then multiply by 100 to clear decimals.

$$.70x + .40(20) = .50(20) + .50x \quad \text{Distributive property}$$
$$70x + 40(20) = 50(20) + 50x \quad \text{Multiply by 100.}$$

Solve for x.

$$70x + 800 = 1000 + 50x$$
$$20x + 800 = 1000 \quad \text{Subtract } 50x.$$
$$20x = 200 \quad \text{Subtract 800.}$$
$$x = 10 \quad \text{Divide by 20.}$$

Steps 5 and 6 Check this solution to see that the chemist needs to use 10 liters of 70% solution.

OBJECTIVE 3 Solve problems involving simple interest. The next example uses the formula for simple interest, $I = prt$. Remember that when $t = 1$, the formula becomes $I = pr$, as shown in the Problem-Solving box at the beginning of this section. Once again the idea of multiplying the total amount (principal) by the rate (rate of interest) gives the percentage (amount of interest).

CHALKBOARD EXAMPLE

With income earned by selling a patent, an engineer invests some money at 8%, and $3000 more than twice as much at 10%. The total annual income from the investments is $2540. Find the amount invested at 8%.

Answer: $8000

EXAMPLE 3 Solving a Simple Interest Problem

Elizabeth Thornton receives an inheritance. She plans to invest part of it at 9% and $2000 more than this amount in a less secure investment at 10%. To earn $1150 per year in interest, how much should she invest at each rate?

Let x = the amount invested at 9% (in dollars);

$x + 2000$ = the amount invested at 10% (in dollars).

Use a chart to arrange the information given in the problem.

Amount Invested in Dollars	*Rate of Interest*	*Interest for One Year*
x	.09	$.09x$
$x + 2000$	.10	$.10(x + 2000)$

We multiply amount by rate to get the interest earned. Since the total interest is to be $1150, the equation is

Interest at 9%	plus	interest at 10%	is	total interest.
↓	↓	↓	↓	↓
$.09x$	$+$	$.10(x + 2000)$	$=$	1150.

Clear parentheses; then clear decimals.

$$.09x + .10x + .10(2000) = 1150 \qquad \text{Distributive property}$$
$$9x + 10x + 10(2000) = 115{,}000 \qquad \text{Multiply by 100.}$$

Now solve for x.

$$9x + 10x + 20{,}000 = 115{,}000$$
$$19x + 20{,}000 = 115{,}000 \qquad \text{Combine terms.}$$
$$19x = 95{,}000 \qquad \text{Subtract 20,000.}$$
$$x = 5000 \qquad \text{Divide by 19.}$$

She should invest \$5000 at 9% and \$5000 + \$2000 = \$7000 at 10%.

NOTE Although decimals were cleared in Examples 2 and 3, the equations also can be solved without doing so.

OBJECTIVE 4 Solve problems involving denominations of money. If a cash drawer contains 37 quarters, the total value of the coins is

$$\underset{\text{Number of coins}}{37} \times \underset{\text{Denomination}}{\$.25} = \underset{\text{Total value}}{\$9.25.}$$

PROBLEM SOLVING

Problems that involve different denominations of money or items with different monetary values are very similar to mixture and interest problems. The basic relationship is the same.

Money Problems

Number × Value of one = Total value

A chart is helpful for these problems, too.

CHALKBOARD EXAMPLE

A man has \$2.55 in quarters and nickels. He has 9 more nickels than quarters. How many nickels and how many quarters does he have?

Answer: 16 nickels; 7 quarters

EXAMPLE 4 Solving a Problem about Money

A bank teller has 25 more five-dollar bills than ten-dollar bills. The total value of the money is \$200. How many of each denomination of bill does he have?

We must find the number of each denomination of bill that the teller has.

Let x = the number of ten-dollar bills;

$x + 25$ = the number of five-dollar bills.

Organize the given information in a chart.

Number of Bills	Denomination in Dollars	Dollar Value
x	10	$10x$
$x + 25$	5	$5(x + 25)$

TEACHING TIP The distinction between the number of bills and the *value* of that number of bills should be stressed.

Multiplying the number of bills by the denomination gives the monetary value. The value of the tens added to the value of the fives must be \$200:

Value of fives	plus	value of tens	is	\$200.
↓	↓	↓	↓	↓
$5(x + 25)$	$+$	$10x$	$=$	$200.$

Solve this equation.

$$5x + 125 + 10x = 200 \quad \text{Distributive property}$$
$$15x + 125 = 200 \quad \text{Combine terms.}$$
$$15x = 75 \quad \text{Subtract 125.}$$
$$x = 5 \quad \text{Divide by 15.}$$

Since x represents the number of tens, the teller has 5 tens and $5 + 25 = 30$ fives. Check that the value of this money is $5(\$10) + 30(\$5) = \$200$.

CHALKBOARD EXAMPLE

For a party, Guadalupe Saucedo bought some 15¢ candy and some 30¢ candy, paying \$15 in all. If there were 25 more pieces of 15¢ candy, how many pieces of the 30¢ kind did she buy?

Answer: 25 pieces

EXAMPLE 5 Solving a Problem about Ticket Prices

Lindsay Davenport, the winner of the 1998 U.S. Open women's tennis championship, plays World Team Tennis for the Sacramento Capitals. Tickets are \$35 for box seats and \$25 for all other seats. There are 5 times as many other seats as box seats. How many of each kind of ticket must the promoter sell for ticket sales revenue to reach \$16,000?

Choose a variable and use it to write the revenue from each kind of ticket.

Let x = the number of box seats;

$5x$ = the number of other seats.

Again, organize the information in a chart.

Number of Seats	*Ticket Price*	*Revenue*
x	35	$35x$
$5x$	25	$125x$

Total sales revenue is

Revenue from box seats	plus	revenue from other seats	is	total revenue.
↓	↓	↓	↓	↓
$35x$	$+$	$125x$	$=$	$16{,}000.$

To solve the equation, combine the terms on the left side of the equation, then divide.

$$160x = 16{,}000$$
$$x = 100 \quad \text{Divide by 160.}$$

Thus, there are 100 box seats and $5(100) = 500$ other seats. Check that this produces revenue of $100(\$35) + 500(\$25) = \$16{,}000$, as required.

OBJECTIVE 5 Solve problems involving distance, rate, and time. If an automobile travels at an average rate of 50 miles per hour for two hours, then it travels $50 \times 2 = 100$ miles. This is an example of the basic relationship between distance, rate, and time:

$$\text{distance} = \text{rate} \times \text{time},$$

TEACHING TIP This is a long section. For some classes, it may be desirable to spend two class meetings on this section, perhaps covering the distance, rate, and time problems at the second meeting.

given by the formula $d = rt$. By solving, in turn, for r and t in the formula, we obtain two other equivalent forms of the formula. The three forms are given below.

Distance, Rate, Time Relationship

$$d = rt \qquad r = \frac{d}{t} \qquad t = \frac{d}{r}$$

The following examples illustrate the uses of these formulas.

CHALKBOARD EXAMPLE

The record for men's freestyle swimming for 50 meters is 22.120 seconds, held by Tom Jager. What was Jager's average speed?

Answer: 2.260 meters per second (rounded)

EXAMPLE 6 Finding Distance, Rate, or Time

(a) The speed of sound is 1088 feet per second at sea level at 32°F. In 5 seconds under these conditions, sound travels

$$\underset{\text{Rate}}{\underset{\uparrow}{1088}} \times \underset{\text{Time}}{\underset{\uparrow}{5}} = \underset{\text{Distance}}{\underset{\uparrow}{5440 \text{ feet.}}}$$

Here, we found distance given rate and time, using $d = rt$.

(b) The winner of the first Indianapolis 500 race (in 1911) was Ray Harroun, driving a Marmon Wasp at an average speed of 74.59 miles per hour.* To complete the 500 miles, it took him

$$\frac{500}{74.59} = 6.70 \text{ hours} \quad \text{(rounded).}$$

(Distance → 500; Rate → 74.59; ← Time)

Here, we found time given rate and distance, using $t = \frac{d}{r}$. To convert .7 hour to minutes, multiply by 60: .7(60) = 42 minutes, so the race took Harroun 6 hours, 42 minutes to complete.

(c) In the 1996 Olympic Games in Atlanta, Claudia Poll of Costa Rica won the women's 200-meter freestyle swimming event in 1 minute, 58.16 seconds or 60 + 58.16 = 118.16 seconds.* Her rate was

$$\frac{200}{118.16} = 1.69 \text{ meters per second} \quad \text{(rounded).}$$

(Distance → 200; Time → 118.16; ← Rate)

Here, we found rate given distance and time, using $r = \frac{d}{t}$.

*Source: *The Universal Almanac,* 1997, John W. Wright, General Editor.

PROBLEM SOLVING

The next example shows how to solve a typical application of the formula $d = rt$. We continue to use the six-step method, but now we also include a sketch along with a chart in Step 2. The sketch shows what is happening in the problem and will help you set up the equation.

CHALKBOARD EXAMPLE

Two airplanes leave Boston at 12:00 noon and fly in opposite directions. If one flies at 410 mph and the other 120 mph faster, how long will it take them to be 3290 miles apart?

Answer: 3.5 hours

EXAMPLE 7 Solving a Motion Problem

Two cars leave Baton Rouge, Louisiana, at the same time and travel east on Interstate 10. One travels at a constant speed of 55 miles per hour and the other travels at a constant speed of 63 miles per hour. In how many hours will the distance between them be 24 miles?

Step 1 We are looking for time.

Let t = the number of hours until the distance between them is 24 miles.

Step 2 The sketch in Figure 10 shows what is happening in the problem.

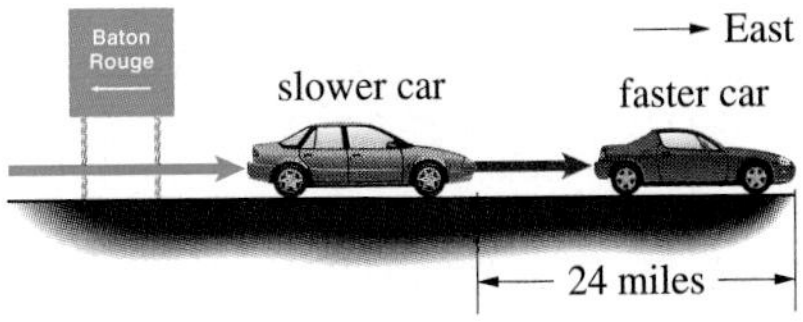

Figure 10

Now, construct a chart like the one below. Fill in the information given in the problem, and use t for the time traveled by each car. Multiply rate by time to get the expressions for distances traveled.

	Rate	Time	Distance
Faster car	63	t	$63t$
Slower car	55	t	$55t$

Difference is 24 miles.

Steps 3 and 4 Refer to Figure 10 and notice that the *difference* between the larger distance and the smaller distance is 24 miles. Now write the equation and solve it.

$$63t - 55t = 24$$

$$8t = 24 \quad \text{Combine terms.}$$

$$t = 3 \quad \text{Divide by 8.}$$

Steps 5 and 6 After 3 hours the faster car will have traveled $63 \times 3 = 189$ miles, and the slower car will have traveled $55 \times 3 = 165$ miles. Since $189 - 165 = 24$, the conditions of the problem are satisfied. It will take 3 hours for the distance between them to be 24 miles.

CHALKBOARD EXAMPLE

Two buses left the downtown terminal, traveling in opposite directions. One had an average speed of 10 miles per hour more than the other. Twelve minutes $\left(\frac{1}{5}\text{ hour}\right)$ later, they were 12 miles apart. What were their speeds?

Answer: 25 and 35 miles per hour

EXAMPLE 8 Solving a Motion Problem

Two planes leave Los Angeles at the same time. One heads south to San Diego; the other heads north to San Francisco. The San Francisco plane flies 50 miles per hour faster. In $\frac{1}{2}$ hour, the planes are 275 miles apart. What are their speeds?

Let r represent the speed of the slower plane. Then $r + 50$ will represent the speed of the faster plane. Fill in a chart.

	Rate	Time	Distance
Slower plane	r	$\frac{1}{2}$	$\frac{1}{2}r$
Faster plane	$r + 50$	$\frac{1}{2}$	$\frac{1}{2}(r + 50)$

Sum is 275 miles.

As Figure 11 shows, the planes are headed in *opposite* directions.

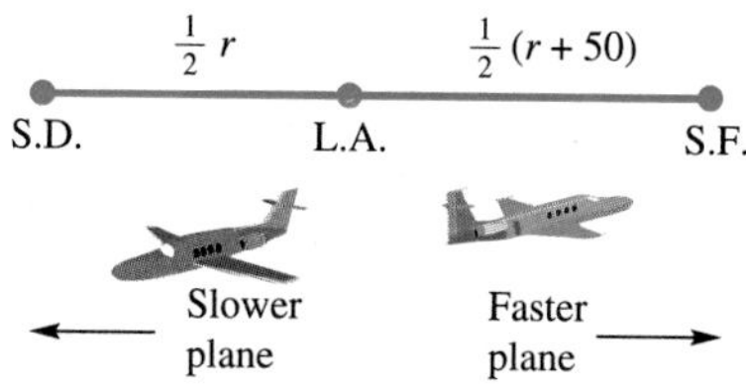

Figure 11

The *sum* of their distances equals 275 miles. We must solve the equation

$$\frac{1}{2}r + \frac{1}{2}(r + 50) = 275.$$

$$\begin{aligned} r + (r + 50) &= 550 && \text{Multiply by 2.} \\ 2r + 50 &= 550 && \text{Combine terms.} \\ 2r &= 500 && \text{Subtract 50.} \\ r &= 250 && \text{Divide by 2.} \end{aligned}$$

The slower plane (headed south) has a speed of 250 miles per hour. The speed of the faster plane is $250 + 50 = 300$ miles per hour. To check, verify that

$$\frac{1}{2}(250) + \frac{1}{2}(300) = 275 \text{ miles.}$$

NOTE

In motion problems like the ones in Examples 7 and 8, once you have filled in two pieces of information in each row of the chart, you can automatically fill in the third piece of information, using the appropriate formula relating distance, rate, and time. Set up the equation based on your sketch and the information in the chart.

Another way to solve the problems in this section is given in Chapter 8.

CAUTION Most difficulties in solving problems of the types discussed here occur because students do not take enough time to read the problem carefully and organize the information given in the problem. Organizing the information in a chart greatly increases the chances of solving the problem correctly.

2.6 EXERCISES

1. 35 milliliters **2.** 45 liters **3.** $350 **4.** $1500 **5.** $14.15 **6.** $17.50 **7.** 3.334 hours **8.** 3.240 hours **9.** 7.95 meters per second **10.** 9.54 meters per second **11.** $533 **12.** $10,694

Solve each problem. See Example 1 and the Problem-Solving box before Example 4.

1. How much pure acid is in 250 milliliters of a 14% acid solution?

2. How much pure alcohol is in 150 liters of a 30% alcohol solution?

3. If $10,000 is invested for one year at 3.5% simple interest, how much interest is earned?

4. If $25,000 is invested at 3% simple interest for 2 years, how much interest is earned?

5. What is the monetary amount of 283 nickels?

6. What is the monetary amount of 35 half-dollars?

In Exercises 7–8, find the time based on the information provided. Use a calculator and round your answer to the nearest thousandth. See Example 6.

	Event and Year	Participant	Distance	Rate
7.	Indianapolis 500, 1996	Buddy Lazier	500 miles	149.956 mph
8.	Daytona 500, 1996	Dale Jarrett	500 miles	154.308 mph

Source: Indianapolis Motor Speedway Hall of Fame and Museum.

In Exercises 9–10, find the rate based on the information provided. Use a calculator and round your answer to the nearest hundredth. See Example 6.

	Event and Year	Participant	Distance	Time
9.	Summer Olympics, 100-meter Hurdles, Women, 1996	Ludmila Engquist	100 meters	12.58 seconds
10.	Summer Olympics, 400-meter Relay, Women, 1996	United States	400 meters	41.95 seconds

Source: The Universal Almanac, 1997, John W. Wright, General Editor.

Use your knowledge of percent and the six-step method to solve these problems.

11. According to a Knight-Ridder Newspapers report, as of May 31, 1992, the nation's "consumer-debt burden" was 16.4%. This means that the average American had consumer debts, such as credit card bills, auto loans, and so on, totaling 16.4% of his or her take-home pay. Suppose that Paul Eldersveld has take-home pay of $3250 per month. What is 16.4% of his monthly take-home pay?

12. Several years ago, General Motors announced that it would raise prices on the following year's vehicles by an average of 1.6%. If a certain vehicle had an original price of $10,526 and this price was raised 1.6%, to the nearest dollar, what would the new price be?

13. 284% **14.** $11\frac{2}{3}\%$
15. (a) about $785 million (b) about $1726 million (c) about $471 million **16.** (a) about $26 million (b) about $104 million (c) about $78 million
17. 0 liters **18.** k liters
19. 160 gallons
20. 160 liters

13. The 1916 dime minted in Denver is quite rare. The 1979 edition of *A Guide Book of United States Coins* listed its value in extremely fine condition as $625. The 1997 value had increased to $2400. What was the percent increase in the value of this coin?

14. In 1963, the value of a 1903 Morgan dollar minted in New Orleans in uncirculated condition was $1500. Due to a discovery of a large hoard of these dollars late that year, the value plummeted. Its value as listed in the 1997 edition of *A Guide Book of United States Coins* was $175. What percent of its 1963 value was its 1997 value?

15. The pie chart shows the breakdown, by approximate percents, of age groups buying team sports equipment in 1995. According to the National Sporting Goods Association, about $15,691 million was spent that year for team equipment. How much was spent by each age group?
(a) 18–24
(b) 25–34
(c) 45–64

16. The pie chart shows the breakdown for the year 1995 of purchases of aerobic shoes. Approximately $372 million was spent on aerobic shoes that year. Determine the amount spent by each group.
(a) Under 14
(b) 35–44
(c) 45–64

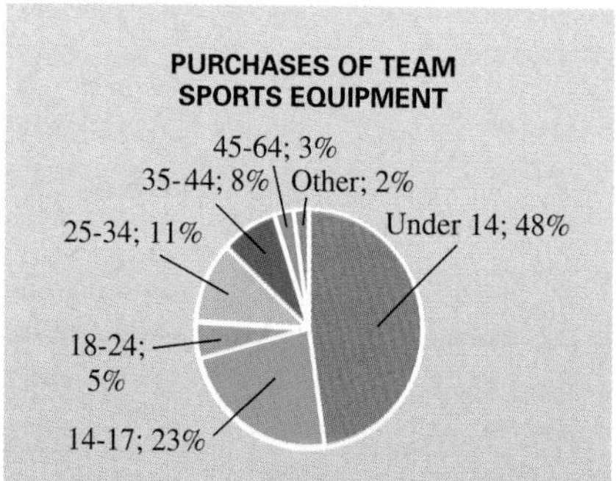

Source: National Sporting Goods Association.

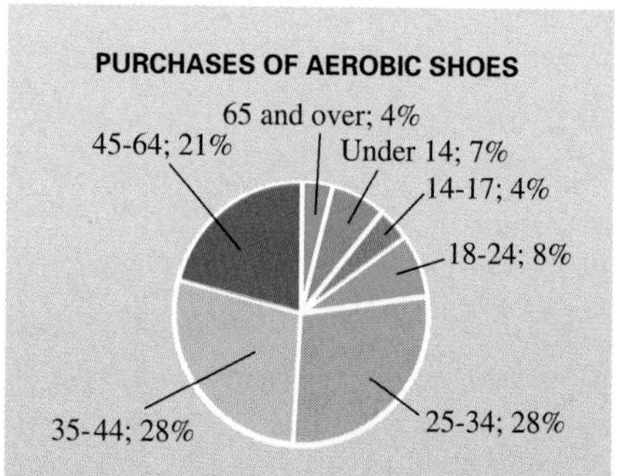

Source: National Sporting Goods Association.

17. Express the amount of alcohol in r liters of pure water.

18. Express the amount of alcohol in k liters of pure alcohol.

Work each mixture problem. Use the six-step method. See Example 2.

19. How many gallons of 50% antifreeze must be mixed with 80 gallons of 20% antifreeze to get a mixture that is 40% antifreeze?

Rate	*Gallons of Mixture*	*Gallons of Antifreeze*
.50	x	$.50x$
.20	80	$.20(80)$
.40	$x + 80$	$.40(x + 80)$

20. How many liters of 25% acid solution must be added to 80 liters of 40% solution to get a solution that is 30% acid?

Rate	*Liters of Mixture*	*Liters of Acid*
.25	x	$.25x$
.40	80	$.40(80)$
.30	$x + 80$	$.30(x + 80)$

21. $53\frac{1}{3}$ kilograms 22. $13\frac{1}{3}$ liters
23. 4 liters 24. $7\frac{1}{2}$ gallons
25. 25 milliliters 26. 20 milliliters 27. (d)
28. (a)
29. $5000 at 3%; $1000 at 5%

21. A certain metal is 20% tin. How many kilograms of this metal must be mixed with 80 kilograms of a metal that is 70% tin to get a metal that is 50% tin?

Rate	Kilograms of Metal	Kilograms of Pure Tin
.20	x	
.70		
.50		

22. A pharmacist has 20 liters of a 10% drug solution. How many liters of 5% solution must be added to get a mixture that is 8%?

Rate	Liters of Solution	Liters of Pure Drug
	20	20(.10)
.05		
.08		

23. How many liters of a 60% acid solution must be mixed with a 75% acid solution to get 20 liters of a 72% solution?

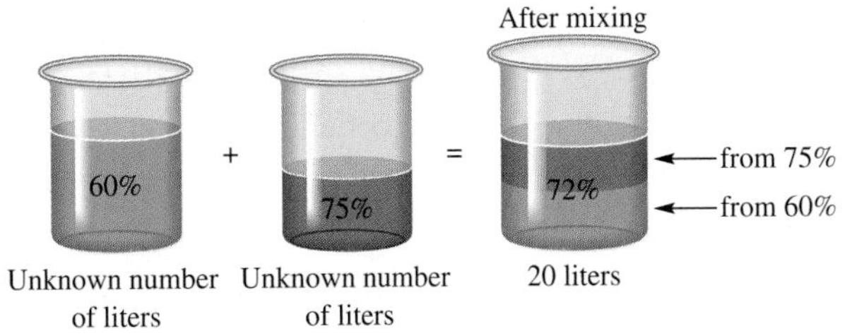

24. How many gallons of a 12% indicator solution must be mixed with a 20% indicator solution to get 10 gallons of a 14% solution?

25. Minoxidil is a drug that has recently proven to be effective in treating male pattern baldness. A pharmacist wishes to mix a solution that is 2% minoxidil. She has on hand 50 milliliters of a 1% solution, and she wishes to add some 4% solution to it to obtain the desired 2% solution. How much 4% solution should she add?

26. Water must be added to 20 milliliters of a 4% minoxidil solution to dilute it to a 2% solution. How many milliliters of water should be used?

Use the concepts of this section to answer each question.

27. Suppose that a chemist is mixing two acid solutions, one of 20% concentration and the other of 30% concentration. Which one of the following concentrations could *not* be obtained?
(a) 22% **(b)** 24%
(c) 28% **(d)** 32%

28. Suppose that pure alcohol is added to a 24% alcohol mixture. Which one of the following concentrations could *not* be obtained?
(a) 22% **(b)** 26%
(c) 28% **(d)** 30%

Work each investment problem using simple interest. See Example 3.

29. Li Nguyen invested some money at 3% and $4000 less than that amount at 5%. The two investments produced a total of $200 interest in one year. How much was invested at each rate?

30. **$12,000 at 2%; $15,000 at 3%**
31. **$40,000 at 3%; $110,000 at 4%**
32. **$2500 at 6%; $13,500 at 5%**
33. **25 fives** **34.** **10 nickels**
35. **84 fives; 42 tens**
36. **27 twenties; 15 fifties**
37. **20 pounds** **38.** **8 hamburgers; 4 bags of french fries**
39. **Yes. In Example 2, it would be reasonable to mix, say, $9\frac{1}{2}$ liters with 20 liters to get $29\frac{1}{2}$ liters of mixture. However, in Example 4, it is not possible to have a fractional part of a five- or ten-dollar bill.**

30. LaShondra Williams inherited some money from her uncle. She deposited part of the money in a savings account paying 2%, and $3000 more than that amount in a different account paying 3%. Her annual interest income was $690. How much did she deposit at each rate?

31. With income earned by selling the rights to his life story, an actor invests some of the money at 3% and $30,000 more than twice as much at 4%. The total annual interest earned from the investments is $5600. How much is invested at each rate?

32. An artist invests her earnings in two ways. Some goes into a tax-free bond paying 6%, and $6000 more than three times as much goes into mutual funds paying 5%. Her total annual interest income from the investments is $825. How much does she invest at each rate?

Work Exercises 33–38 involving different monetary rates. See Example 4.

33. A bank teller has some five-dollar bills and some twenty-dollar bills. The teller has 5 more twenties than fives. The total value of the money is $725. Find the number of five-dollar bills that the teller has.

Number of Bills	*Denomination*	*Value*
x	5	
$x + 5$	20	

34. A coin collector has $1.70 in dimes and nickels. She has 2 more dimes than nickels. How many nickels does she have?

Number of Coins	*Denomination*	*Value*
x	.05	$.05x$
	.10	

35. A cashier has a total of 126 bills, made up of fives and tens. The total value of the money is $840. How many of each kind does he have?

36. A convention manager finds that she has $1290, made up of twenties and fifties. She has a total of 42 bills. How many of each kind does she have?

37. A merchant wishes to mix candy worth $5 per pound with 40 pounds of candy worth $2 per pound to get a mixture that can be sold for $3 per pound. How many pounds of $5 candy should be used?

38. At Vern's Grill, hamburgers cost 90 cents each, and a bag of french fries costs 40 cents. How many hamburgers and how many bags of french fries can a customer buy with $8.80 if he wants twice as many hamburgers as bags of french fries?

Vern's Grill
Hamburgers 90¢ each
French fries 40¢ a bag
Soda pop 25¢ each
Shakes 90¢ each
Cherry pie 90¢ a slice
Ice cream 40¢ a scoop

39. Read Example 2. Can a problem of this type have a fraction as an answer? Now read Example 4. Can a problem of this type have a fraction as an answer? Explain.

40. (a) $.05x + .10(3400 - x) = 290$
(b) 1000 nickels; 2400 dimes
41. (a) $.05x + .10(3400 - x) = 290$
(b) \$1000 at 5%; \$2400 at 10%
42. They are the same.
43. No, you will get a different solution to the equation, but you *will* get the same answers to the problem.

RELATING CONCEPTS (EXERCISES 40–43)

Sometimes applied problems that may seem different actually apply the same concepts.

Work Exercises 40–43 in order, to see how this happens.

40. Consider the following problem: A hoard of coins consists of only nickels and dimes. There are 3400 coins, and the value of the money is \$290. How many of each denomination are there?
  (a) Write an equation you would use to solve the problem. Let x represent the number of nickels in the hoard.
  (b) Solve the problem.

41. Consider the following problem: An investor deposits \$3400 in two accounts. One account is a passbook savings account that pays 5% interest, and the other is a money market account that pays 10% interest. After one year, the total interest earned is \$290. How much did she invest at each rate?
  (a) Write an equation you would use to solve the problem. Let x represent the amount invested in the passbook savings account.
  (b) Solve the problem.

42. Compare the equations you wrote in Exercises 40(a) and 41(a). What do you notice about them?

43. If, in either of the problems in Exercises 40 and 41, you let x represent the other unknown quantity, will you get the same *solution* to the equation? Will you get the same *answers* to the problem?

Did you make the connection that applications that appear different may be solved in the same way?

44. The pie charts were constructed by taking each percent times 360° to determine the number of degrees that represents each wedge. **45.** 530 miles
46. 328 miles **47.** No, since the rate is given in miles per *hour,* the time must be in hours. So the distance is $45 \times \frac{1}{2} = \frac{45}{2} = 22\frac{1}{2}$ miles. **48.** (a)
49. 10 hours

44. How do you think the pie charts in Exercises 15 and 16 were constructed so that the sections represented the appropriate percentages accurately?

Solve each problem. See Example 6.

45. A driver averaged 53 miles per hour and took 10 hours to travel from Memphis to Chicago. What is the distance between Memphis and Chicago?

46. A small plane traveled from Warsaw to Rome, averaging 164 miles per hour. The trip took 2 hours. What is the distance from Warsaw to Rome?

47. Suppose that an automobile averages 45 miles per hour, and travels for 30 minutes. Is the distance traveled $45 \times 30 = 1350$ miles? If not, explain why not, and give the correct distance.

48. Which of the following choices is the best *estimate* for the average speed of a trip of 405 miles that lasted 8.2 hours?
  (a) 50 miles per hour **(b)** 30 miles per hour
  (c) 60 miles per hour **(d)** 40 miles per hour

Solve each problem. See Examples 7 and 8.

49. St. Louis and Portland are 2060 miles apart. A small plane leaves Portland, traveling toward St. Louis at an average speed of 90 miles per hour. Another plane leaves St. Louis at the same time, traveling toward Portland, averaging 116 miles per hour. How long will it take them to meet?

	r	t	d
Plane leaving Portland	90	t	$90t$
Plane leaving St. Louis	116	t	$116t$

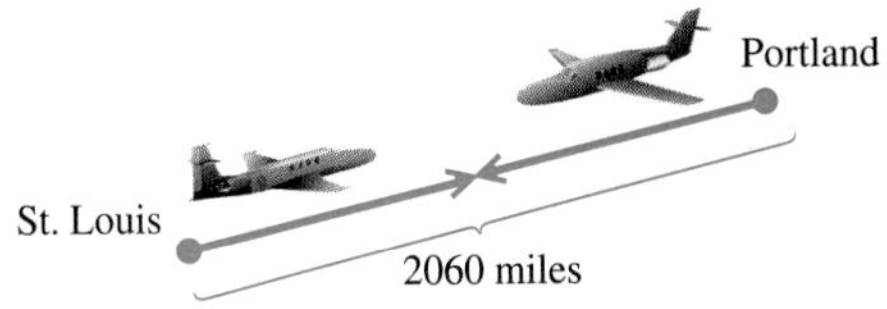

50. 5 hours 51. northbound: 60 miles per hour; southbound: 80 miles per hour 52. eastbound: 300 miles per hour; westbound: 450 miles per hour 53. 5 hours 54. 2.5 hours 55. single man: \$51.58; single woman: \$40.71 56. Yankees: \$115.25; Reds: \$79.31 57. Business/Government: 6,100,000; K–12: 5,000,000; HED: 900,000; Homes: 9,800,000

50. Atlanta and Cincinnati are 440 miles apart. John leaves Cincinnati, driving toward Atlanta at an average speed of 60 miles per hour. Pat leaves Atlanta at the same time, driving toward Cincinnati in her antique auto, averaging 28 miles per hour. How long will it take them to meet?

	r	t	d
John	60	t	$60t$
Pat	28	t	$28t$

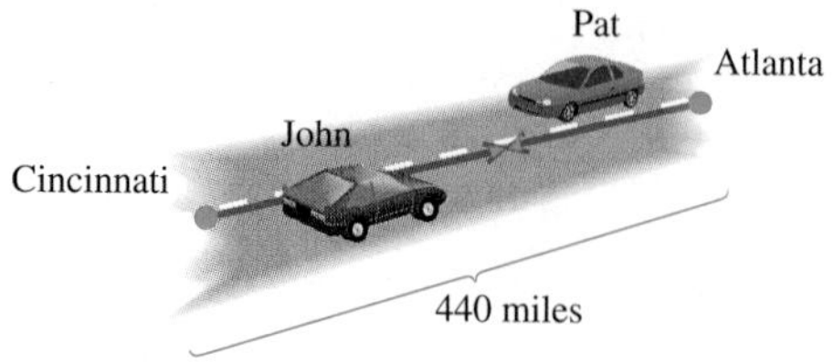

51. Two trains leave a city at the same time. One travels north and the other travels south at 20 miles per hour faster. In 2 hours the trains are 280 miles apart. Find their speeds.

	r	t	d
Northbound	x	2	
Southbound	$x + 20$	2	

52. Two planes leave an airport at the same time, one flying east, the other flying west. The eastbound plane travels 150 miles per hour slower. They are 2250 miles apart after 3 hours. Find the speed of each plane.

	r	t	d
Eastbound	$x - 150$	3	
Westbound	x	3	

53. At a given hour two steamboats leave a city in the same direction on a straight canal. One travels at 18 miles per hour and the other travels at 25 miles per hour. In how many hours will the boats be 35 miles apart?

54. From a point on a straight road, Lupe and Maria ride bicycles in the same direction. Lupe rides 10 miles per hour and Maria rides 12 miles per hour. In how many hours will they be 5 miles apart?

The remaining applications in this exercise set are not organized by type of problem. Use the problem-solving techniques described in this chapter to solve them.

55. According to a survey by Information Resources, Inc., single men spent an average of \$10.87 more than single women when buying frozen dinners during the 52 weeks that ended August 22, 1993. Together, a single man and a single woman would spend a total of \$92.29, according to the survey. What were the average expenditures for a single man and for a single woman?

56. Team Marketing Report, a sports-business newsletter, computes a Fan Cost Index for a trip to a major league baseball park. The index consists of the cost of four average-priced tickets, two small beers, four small sodas, four hot dogs, parking for one car, two game programs, and two twill baseball caps. For the 1994 season, the New York Yankees had the highest Fan Cost Index, while the Cincinnati Reds had the lowest. The Yankees' index was \$43.37 less than twice that of the Reds. What was the Fan Cost Index for each of these teams if the *total* of the two was \$194.56?

57. The pie chart shows the percents for the different categories of Macintosh computer units in use at the end of 1997. At that time there were approximately 21,782,000 units in use. Find the approximate number of units used in each category. (*Source:* Apple Computer, Inc., Cupertino, California.)

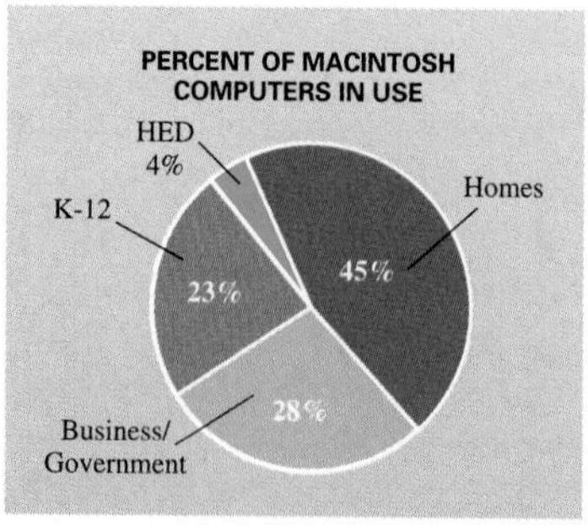

58. (a) .539 (b) .497
(c) .494
59. 120 miles per hour
60. \$8500 at 4%; \$3500 at 5%

58. At the start of play on September 22, 1997, the standings of the Central Division of the American League were as shown. "Winning percentage" is commonly expressed as a decimal rounded to the nearest thousandth. To find the winning percentage of a team, divide the number of wins by the total number of games played. Find the winning percentage of each of the following teams.
(a) Cleveland **(b)** Chicago
(c) Milwaukee

	Won	*Lost*
Cleveland	83	71
Chicago	77	78
Milwaukee	76	78
Kansas City	64	90
Minnesota	63	91

59. In an automobile race, a driver was 240 miles from the finish line after 3 hours. Another driver, who was in a later race, traveled at the same speed as the first driver. After 2.5 hours the second driver was 300 miles from the finish. Find the speed of each driver.

60. Fran Liberto sold a painting that she found at a garage sale to an art dealer. With the \$12,000 she got, she invested part at 4% in a certificate of deposit and the rest at 5% in a municipal bond. Her total annual income from the two investments was \$515. How much did she invest at each rate? (*Hint:* Let x represent the amount invested at 4%. Then $12{,}000 - x$ represents the amount invested at 5%.)

2.7 The Addition and Multiplication Properties of Inequality

OBJECTIVES

1. Graph intervals on a number line.
2. Use the addition property of inequality.
3. Use the multiplication property of inequality.
4. Solve linear inequalities.
5. Solve applied problems by using inequalities.
6. Solve three-part inequalities.

FOR EXTRA HELP

SSG Sec. 2.7
SSM Sec. 2.7
Pass the Test Software
InterAct Math Tutorial Software
Video 4

Solving inequalities is closely related to the methods of solving equations. In this section we introduce properties that are essential for solving inequalities.

Inequalities are algebraic expressions related by

$<$ "is less than,"
$\leq$ "is less than or equal to,"
$>$ "is greater than,"
$\geq$ "is greater than or equal to."

We solve an inequality by finding all real number solutions for it. For example, the solution set of $x \leq 2$ includes all real numbers that are less than or equal to 2, and not just the integers less than or equal to 2. For example, -2.5, -1.7, -1, $\frac{7}{4}$, $\frac{1}{2}$, $\sqrt{2}$, and 2 are all real numbers less than or equal to 2, and are therefore solutions of $x \leq 2$.

OBJECTIVE 1 Graph intervals on a number line. A good way to show the solution set of an inequality is by graphing. We graph all the real numbers satisfying $x \leq 2$ by placing a square bracket at 2 on a number line and drawing an arrow extending from the bracket to the left (to represent the fact that all numbers less than 2 are also part of the graph). The graph is shown in Figure 12.

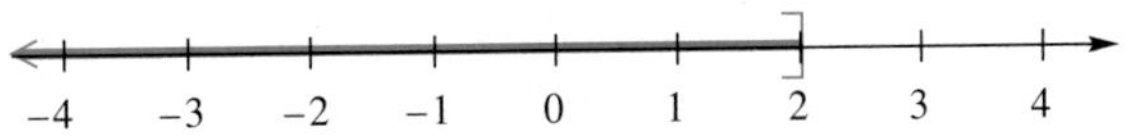

Figure 12

INSTRUCTOR'S RESOURCES

PTB Sec. 2.7 **ISM** Sec. 2.7
AB Sec. 2.7

TEST GENERATOR

WORLD WIDE WEB
www.LialAlgebra.com

TEACHING TIP Be sure to stress when to use a square bracket and when to use a parenthesis. Mention that the brackets and parentheses open in the direction of the shading.

The set of numbers less than or equal to 2 is an example of an **interval** on the number line. To write intervals, we use **interval notation.** For example, using this notation, the interval of all numbers less than or equal to 2 is written as $(-\infty, 2]$. The **negative infinity** symbol $-\infty$ does not indicate a number. It is used to show that the interval includes all real numbers less than 2. As on the number line, the square bracket indicates that 2 is part of the solution. A parenthesis is always used next to the infinity symbol. The set of real numbers is written in interval notation as $(-\infty, \infty)$.

CHALKBOARD EXAMPLE

Write each inequality in interval notation and graph the interval.

(a) $x \geq 3$

(b) $2 < x \leq 4$

Answer:

(a) $[3, \infty)$

(b) $(2, 4]$

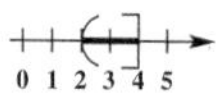

TEACHING TIP Some texts use open circles rather than parentheses and solid dots rather than square brackets. We have chosen to use parentheses and square brackets to indicate intervals throughout this text.

EXAMPLE 1 Graphing an Interval Written in Interval Notation on a Number Line

Write each inequality in interval notation and graph it.

(a) $x > -5$

The statement $x > -5$ says that x can represent any number greater than -5, but x cannot equal -5. The interval is written as $(-5, \infty)$. We show this on a graph by placing a parenthesis at -5 and drawing an arrow to the right, as in Figure 13. The parenthesis at -5 shows that -5 is not part of the graph.

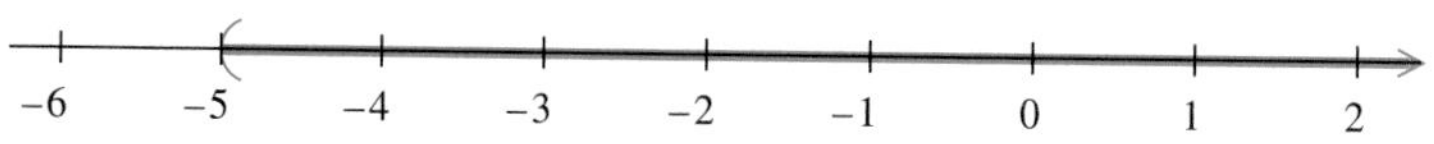

Figure 13

(b) $-1 \leq x < 3$

This statement is read "-1 is less than or equal to x *and* x is less than 3." Thus, we want the set of numbers that are *between* -1 and 3, with -1 included and 3 excluded. In interval notation, we write this as $[-1, 3)$, using a square bracket at -1 because it is part of the graph, and a parenthesis at 3, because 3 is not part of the graph. The graph is shown in Figure 14.

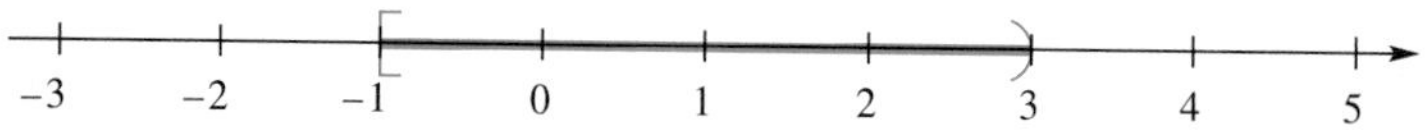

Figure 14

OBJECTIVE 2 Use the addition property of inequality. Inequalities such as $x + 4 \leq 9$ can be solved in much the same way as equations. Consider the inequality $2 < 5$. If 4 is added to both sides of this inequality, the result is

$$2 + 4 < 5 + 4$$
$$6 < 9,$$

a true sentence. Now subtract 8 from both sides:

$$2 - 8 < 5 - 8$$
$$-6 < -3.$$

The result is again a true sentence. These examples suggest the **addition property of inequality,** which states that the same real number can be added to both sides of an inequality without changing the solutions.

Addition Property of Inequality

For any algebraic expressions A, B, and C that represent real numbers, the inequalities

$$A < B \quad \text{and} \quad A + C < B + C$$

have exactly the same solutions. In words, the same expression may be added to both sides of an inequality without changing the solutions.

The addition property of inequality also works with $>$, $\leq$, or $\geq$. Just as with the addition property of equality, the same expression may also be *subtracted* from both sides of an inequality.

The following examples show how the addition property is used to solve inequalities. We will write solution sets in interval notation.

CHALKBOARD EXAMPLE

Solve the inequality, then graph.

$$-1 + 8r < 7r + 2$$

Answer: $(-\infty, 3)$

EXAMPLE 2 Using the Addition Property of Inequality

Solve the inequality $7 + 3k > 2k - 5$.

Use the addition property of inequality twice, once to get the terms containing k alone on one side of the inequality and a second time to get the integers together on the other side. (These steps can be done in either order.)

$7 + 3k > 2k - 5$	
$7 + 3k - \mathbf{2k} > 2k - 5 - \mathbf{2k}$	Subtract $2k$.
$7 + k > -5$	Combine terms.
$7 + k - \mathbf{7} > -5 - \mathbf{7}$	Subtract 7.
$k > -12$	

The solution set is $(-12, \infty)$. Its graph is shown in Figure 15.

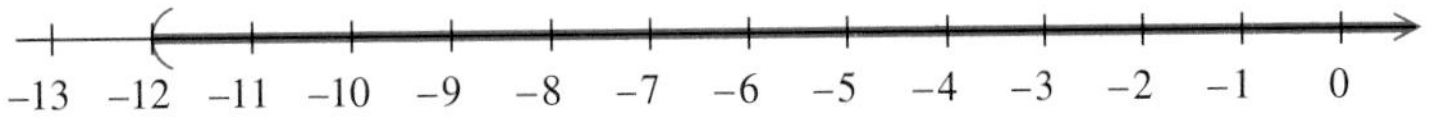

Figure 15

OBJECTIVE 3 Use the multiplication property of inequality. The addition property of inequality cannot be used to solve inequalities such as $4y \geq 28$. These inequalities require the multiplication property of inequality. To see how this property works, we look at some examples.

First, start with the inequality $3 < 7$ and multiply both sides by the positive number 2.

$3 < 7$	
$\mathbf{2}(3) < \mathbf{2}(7)$	Multiply both sides by 2.
$6 < 14$	True

Now multiply both sides of $3 < 7$ by the negative number -5.

$3 < 7$	
$\mathbf{-5}(3) < \mathbf{-5}(7)$	Multiply both sides by -5.
$-15 < -35$	False

To get a true statement when multiplying both sides by -5 requires reversing the direction of the inequality symbol.

$$3 < 7$$
$$-5(3) > -5(7) \quad \text{Multiply by } -5\text{; reverse the symbol.}$$
$$-15 > -35 \quad \text{True}$$

Take the inequality $-6 < 2$ as another example. Multiply both sides by the positive number 4.

$$-6 < 2$$
$$4(-6) < 4(2) \quad \text{Multiply by 4.}$$
$$-24 < 8 \quad \text{True}$$

Multiplying both sides of $-6 < 2$ by -5 and at the same time reversing the direction of the inequality symbol gives

$$-6 < 2$$
$$(-5)(-6) > (-5)(2) \quad \text{Multiply by } -5\text{; change} < \text{to} >.$$
$$30 > -10. \quad \text{True}$$

The two parts of the **multiplication property of inequality** are stated below.

Multiplication Property of Inequality

For any algebraic expressions A, B, and C that represent real numbers, with $C \neq 0$,

1. if C is *positive,* then the inequalities

$$A < B \quad \text{and} \quad AC < BC$$

have exactly the same solutions;

2. if C is *negative,* then the inequalities

$$A < B \quad \text{and} \quad AC > BC$$

have exactly the same solutions.

In other words, both sides of an inequality may be multiplied by the same expression representing a positive number without changing the solutions. If the expression represents a negative number, we must reverse the direction of the inequality symbol.

TEACHING TIP Students often forget to reverse the inequality symbol when multiplying or dividing by a negative number.

The multiplication property of inequality works with $>$, $\leq$, or $\geq$, as well. The multiplication property of inequality also permits *division* of both sides of an inequality by the same nonzero expression.

It is important to remember the differences in the multiplication property for positive and negative numbers.

TEACHING TIP Multiplying both sides of an inequality by a variable quantity should be avoided, since it is not known whether the variable quantity is positive or negative.

1. When both sides of an inequality are multiplied or divided by a positive number, the direction of the inequality symbol *does not change.* Adding or subtracting terms on both sides also does not change the symbol.
2. When both sides of an inequality are multiplied or divided by a negative number, the direction of the symbol *does change. Reverse the symbol of inequality only when you multiply or divide both sides by a negative number.*

CHALKBOARD EXAMPLE

Solve the inequality, then graph.

$$-2r > -12$$

Answer: $(-\infty, 6)$

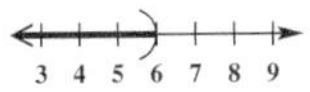

EXAMPLE 3 Using the Multiplication Property of Inequality

(a) Solve the inequality $3r < -18$.

Simplify this inequality by using the multiplication property of inequality and dividing both sides by 3. Since 3 is a positive number, the direction of the inequality symbol does not change.

$$3r < -18$$

$$\frac{3r}{3} < \frac{-18}{3} \qquad \text{Divide by 3.}$$

$$r < -6$$

The solution set is $(-\infty, -6)$. The graph is shown in Figure 16.

Figure 16

(b) Solve the inequality $-4t \geq 8$.

Here both sides of the inequality must be divided by -4, a negative number, which *does* change the direction of the inequality symbol.

$$-4t \geq 8$$

$$\frac{-4t}{-4} \leq \frac{8}{-4} \qquad \text{Divide by } -4\text{; symbol is reversed.}$$

$$t \leq -2$$

The solution set is $(-\infty, -2]$. The solutions are graphed in Figure 17.

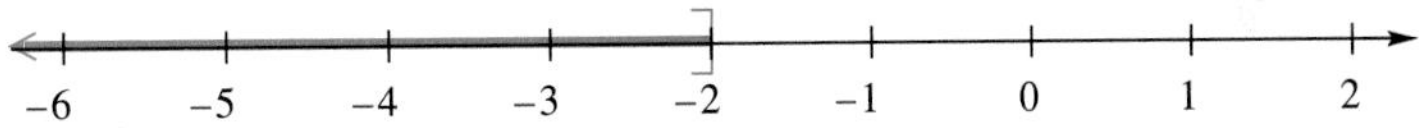

Figure 17

CAUTION Even though the number on the right side of the inequality in Example 3(a) is negative (-18), *do not reverse the direction of the inequality symbol.* Reverse the symbol only when multiplying or dividing by a negative number, as shown in Example 3(b).

OBJECTIVE 4 Solve linear inequalities. A **linear inequality** is an inequality that can be written in the form $ax + b < 0$, for real numbers a and b, with $a \neq 0$. ($<$ may be replaced with $>$, $\leq$, or $\geq$ in this definition.) To solve a linear inequality, follow these steps.

TEACHING TIP The steps for solving an inequality parallel those for solving an equation, with the additional requirement of reversal of the inequality symbol when multiplying or dividing by a negative number.

Solving a Linear Inequality

Step 1 **Simplify each side separately.** Use the properties to clear parentheses and combine terms.

Step 2 **Isolate the variable terms on one side.** Use the addition property of inequality to simplify the inequality to the form $ax < b$ or $ax > b$, where a and b are real numbers.

Step 3 **Isolate the variable.** Use the multiplication property of inequality to simplify the inequality to the form $x < d$ or $x > d$, where d is a real number.

Notice how these steps are used in the next example.

CHALKBOARD EXAMPLE

Solve and graph the solution set.

$$5r - r + 2 < 7r - 5$$

Answer: $\left(\frac{7}{3}, \infty\right)$

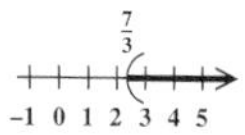

EXAMPLE 4 Solving a Linear Inequality

Solve the inequality $3z + 2 - 5 > -z + 7 + 2z$.

Step 1 Combine like terms and simplify.

$$3z + 2 - 5 > -z + 7 + 2z$$
$$3z - 3 > z + 7$$

Step 2 Use the addition property of inequality.

$$3z - 3 + \mathbf{3} > z + 7 + \mathbf{3} \qquad \text{Add 3.}$$
$$3z > z + 10$$
$$3z - z > z + 10 - z \qquad \text{Subtract } z.$$
$$2z > 10$$

Step 3 Use the multiplication property of inequality.

$$\frac{2z}{\mathbf{2}} > \frac{10}{\mathbf{2}} \qquad \text{Divide by 2.}$$
$$z > 5$$

Since 2 is positive, the direction of the inequality symbol was not changed in the third step. The solution set is $(5, \infty)$. Its graph is shown in Figure 18.

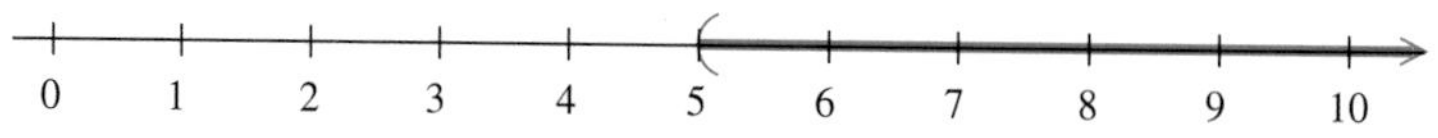

Figure 18

CHALKBOARD EXAMPLE

Solve and graph the solution set.

$$4(y - 1) - 3y > -15 - (2y + 1)$$

Answer: $(-4, \infty)$

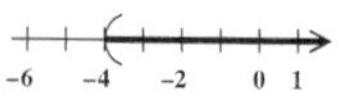

EXAMPLE 5 Solving a Linear Inequality

Solve $5(k - 3) - 7k \geq 4(k - 3) + 9$.

Step 1 Simplify and combine like terms.

$$5(k - 3) - 7k \geq 4(k - 3) + 9$$
$$5k - 15 - 7k \geq 4k - 12 + 9 \qquad \text{Distributive property}$$
$$-2k - 15 \geq 4k - 3 \qquad \text{Combine like terms.}$$

Step 2 Use the addition property.

$$-2k - 15 - \mathbf{4k} \geq 4k - 3 - \mathbf{4k} \qquad \text{Subtract } 4k.$$
$$-6k - 15 \geq -3$$
$$-6k - 15 + \mathbf{15} \geq -3 + \mathbf{15} \qquad \text{Add 15.}$$
$$-6k \geq 12$$

Step 3 Divide both sides by -6, a negative number. Change the direction of the inequality symbol.

$$\frac{-6k}{\mathbf{-6}} \leq \frac{12}{\mathbf{-6}} \qquad \text{Divide by } -6\text{; symbol is reversed.}$$
$$k \leq -2$$

The solution set is $(-\infty, -2]$. Its graph is shown in Figure 19.

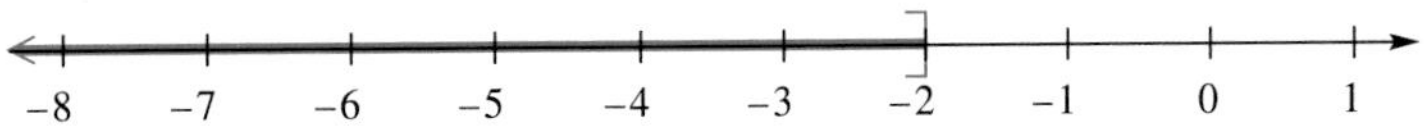

Figure 19

OBJECTIVE 5 **Solve applied problems by using inequalities.** Until now, the applied problems that we have studied have all led to equations.

PROBLEM SOLVING

Inequalities can be used to solve applied problems involving phrases that suggest inequality. The following chart gives some of the more common such phrases along with examples and translations.

Phrase	*Example*	*Inequality*
Is more than	A number ***is more than*** 4	$x > 4$
Is less than	A number ***is less than*** -12	$x < -12$
Is at least	A number ***is at least*** 6	$x \geq 6$
Is at most	A number ***is at most*** 8	$x \leq 8$

CAUTION Do not confuse statements like "5 is more than a number" with the phrase "5 more than a number." The first of these is expressed as "$5 > x$" while the second is expressed with addition, as "$x + 5$."

The next example shows an application of algebra that is important to anyone who has ever asked himself or herself "What score can I make on my next test and have a (particular grade) in this course?" It uses the idea of finding the average of a number of grades. In general, to find the average of n numbers, add the numbers, and divide by n.

CHALKBOARD EXAMPLE

Maggie has scores of 98, 86, and 88 on her first three tests in algebra. If she wants an average of at least 90 after her fourth test, what score must she make on her fourth test?

Answer: 88 or more

EXAMPLE 6 Finding an Average Test Score

Brent has test grades of 86, 88, and 78 on his first three tests in geometry. If he wants an average of at least 80 after his fourth test, what are the possible scores he can make on his fourth test?

Let $x =$ Brent's score on his fourth test. To find his average after 4 tests, add the test scores and divide by 4.

Average ↓ is at least ↓ 80. ↙

$$\frac{86 + 88 + 78 + x}{4} \geq 80$$

$$\frac{252 + x}{4} \geq 80 \qquad \text{Add the known scores.}$$

$$4\left(\frac{252 + x}{4}\right) \geq 4(80) \qquad \text{Multiply by 4.}$$

$$252 + x \geq 320$$

$$252 - 252 + x \geq 320 - 252 \qquad \text{Subtract 252.}$$

$$x \geq 68 \qquad \text{Combine terms.}$$

He must score 68 or more on the fourth test to have an average of *at least* 80.

CAUTION

Errors often occur when the phrases "at least" and "at most" appear in applied problems. Remember that

at least translates as **greater than or equal to**

and

at most translates as **less than or equal to.**

OBJECTIVE 6 Solve three-part inequalities. Inequalities that say that one number is *between* two other numbers are *three-part inequalities.* For example,

$$-3 < 5 < 7$$

says that 5 is between -3 and 7. Three-part inequalities can also be solved by using the addition and multiplication properties of inequality. The idea is to get the inequality in the form

$$\textbf{a number} < x < \textbf{another number,}$$

using "is less than." The solution set can then easily be graphed.

CHALKBOARD EXAMPLE

Solve $2 \leq 3p - 1 \leq 8$ and graph the solution set.

Answer: [1, 3]

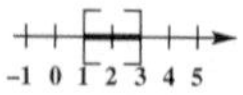

EXAMPLE 7 Solving Three-Part Inequalities

(a) Solve $4 \leq 3x - 5 < 6$ and graph the solutions.

This inequality is equivalent to the statement $4 \leq 3x - 5$ and $3x - 5 < 6$. Using the inequality properties, we could solve each of these simple inequalities separately, and then combine the solutions with "and." We get the same result if we simply work with all three parts at once. Working separately, the first step would be to add 5 on each side. Do this to all three parts.

$$4 \leq 3x - 5 < 6$$

$$4 + 5 \leq 3x - 5 + 5 < 6 + 5 \qquad \text{Add 5.}$$

$$9 \leq 3 < 11$$

Now divide each part by the positive number 3.

$$\frac{9}{3} \leq \frac{3x}{3} < \frac{11}{3} \qquad \text{Divide by 3.}$$

$$3 \leq x < \frac{11}{3}$$

The solution set is $[3, \frac{11}{3})$. Its graph is shown in Figure 20.

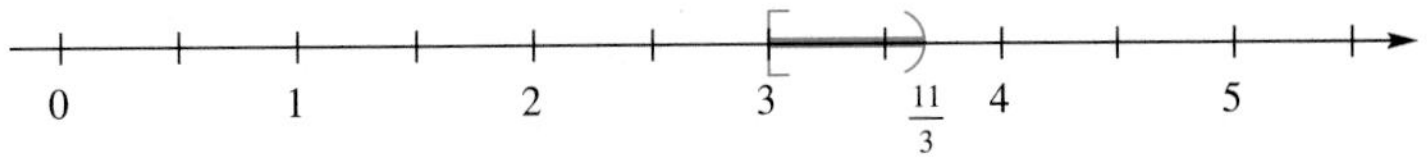

Figure 20

TEACHING TIP Be sure to work with all three parts of the inequality.

(b) Solve $-4 \leq \frac{2}{3}m - 1 < 8$ and graph the solutions.

Recall from Section 2.2 that fractions as coefficients in equations can be eliminated by multiplying both sides by the least common denominator of the fractions. The same is true for inequalities. One way to begin is to multiply all three parts by 3.

$$-4 \leq \frac{2}{3}m - 1 < 8$$

$$3(-4) \leq 3\left(\frac{2}{3}m - 1\right) < 3(8) \qquad \text{Multiply by 3.}$$

$$-12 \leq 2m - 3 < 24 \qquad \text{Distributive property}$$

Now add 3 to each part.

$$-12 + 3 \leq 2m - 3 + 3 < 24 + 3 \qquad \text{Add 3.}$$

$$-9 \leq 2m < 27$$

Finally, divide by 2 to get

$$-\frac{9}{2} \leq m < \frac{27}{2}.$$

The solution set is $\left[-\frac{9}{2}, \frac{27}{2}\right)$. Its graph is shown in Figure 21.

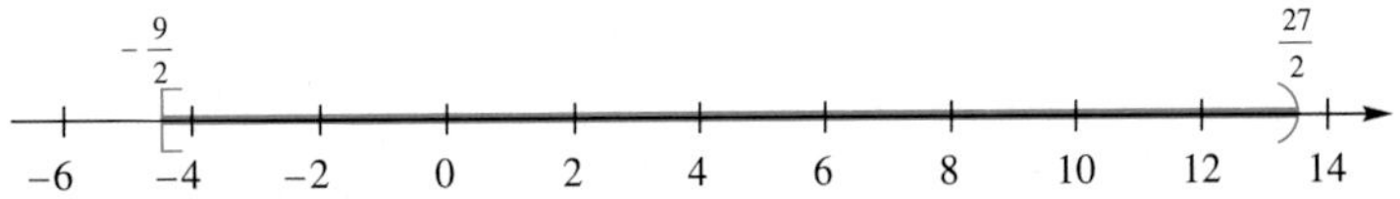

Figure 21

This inequality could also have been solved by first adding 1 to each part, and then multiplying each part by $\frac{3}{2}$.

■ The revenue is represented by $5x - 100$. The production cost is $125 + 4x$. The profit is represented by $R - C = (5x - 100) - (125 + 4x) = x - 225$. The solution of $x - 225 > 0$ is $x > 225$. In order to make a profit, more than 225 cassettes must be produced and sold.

CONNECTIONS

Many mathematical models involve inequalities rather than equations. This is often the case in economics. For example, a company that produces videocassettes has found that revenue from the sales of the cassettes is \$5 per cassette less sales costs of \$100. Production costs are \$125 plus \$4 per cassette. Profit (P) is given by revenue (R) less cost (C), so the company must find the production level x that makes

$$P = R - C > 0.$$

FOR DISCUSSION OR WRITING

Write an expression for revenue letting x represent the production level (number of cassettes to be produced). Write an expression for production costs using x. Write an expression for profit and solve the inequality shown above. Describe the solution in terms of the problem.

2.7 EXERCISES

1. Use a parenthesis if the symbol is $<$ or $>$. Use a square bracket if the symbol is $\le$ or $\ge$. **2.** The graph of $t \ge -7$ includes the endpoint, -7, while the graph of $t > -7$ does not. **3.** $x > -4$ **4.** $x \ge -4$ **5.** $x \le 4$ **6.** $x < 4$
7. $(-\infty, 4]$ 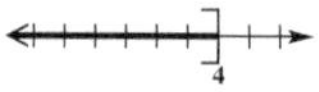
8. $(-\infty, -11]$ 
9. $(-\infty, -3)$ 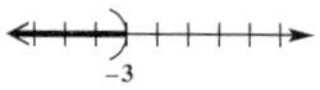
10. $(-\infty, 3)$ 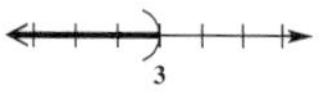
11. $(4, \infty)$ 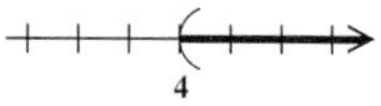
12. $(5, \infty)$ 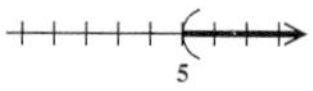
13. $[8, 10]$ 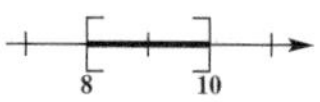
14. $[3, 5]$ 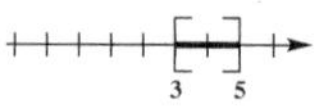
15. $(0, 10]$ 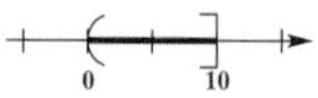
16. $[-3, 5)$ 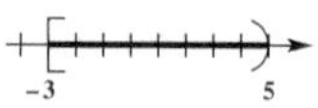
17. It would imply that $3 < -2$, a false statement. **18.** (a)
19. $z \ge 1$ 
20. $p \ge -8$ 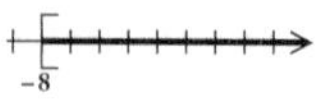

1. Explain how to determine whether to use a parenthesis or a square bracket at the endpoint when graphing an inequality on a number line.
2. How does the graph of $t \ge -7$ differ from the graph of $t > -7$?

Write an inequality involving the variable x that describes each set of numbers graphed. See Example 1.

3. 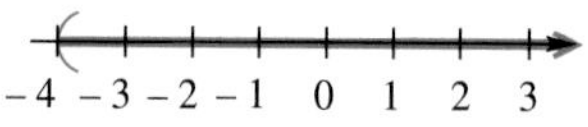
4. 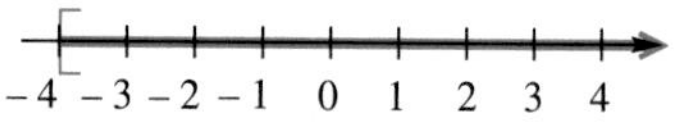
5. 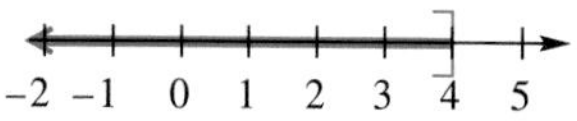
6. 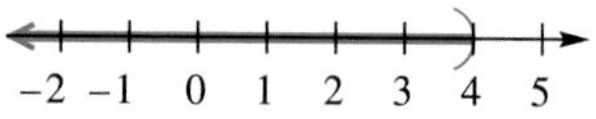

Write each inequality in interval notation and graph the interval on a number line. See Example 1.

7. $k \le 4$
8. $r \le -11$
9. $x < -3$
10. $y < 3$
11. $t > 4$
12. $m > 5$
13. $8 \le x \le 10$
14. $3 \le x \le 5$
15. $0 < y \le 10$
16. $-3 \le x < 5$
17. Why is it *wrong* to write $3 < x < -2$ to indicate that x is between -2 and 3?
18. If $p < q$ and $r < 0$, which one of the following statements is *false*?
    (a) $pr < qr$ **(b)** $pr > qr$ **(c)** $p + r < q + r$ **(d)** $p - r < q - r$

Solve each inequality and graph the solution set. See Example 2.

19. $z - 8 \ge -7$
20. $p - 3 \ge -11$
21. $2k + 3 \ge k + 8$
22. $3x + 7 \ge 2x + 11$
23. $3n + 5 < 2n - 6$
24. $5x - 2 < 4x - 5$
25. Under what conditions must the inequality symbol be reversed when solving an inequality?
26. Explain the steps you would use to solve the inequality $-5x > 20$.
27. Your friend tells you that when solving the inequality $6x < -42$ he reversed the direction of the inequality because of the presence of -42. How would you respond?
28. By what number must you *multiply* both sides of $.2x > 6$ to get just x on the left side?

JOURNAL WRITING ▲ CHALLENGING SCIENTIFIC CALCULATOR GRAPHING CALCULATOR

21. $k \geq 5$ 
22. $x \geq 4$ 
23. $n < -11$ 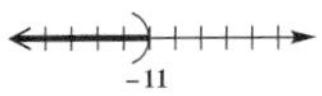
24. $x < -3$ 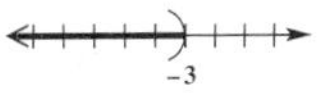
25. It must be reversed when multiplying or dividing by a negative number. **26.** Divide by -5 and reverse the inequality symbol to get $x < -4$. **27.** His method is incorrect. He divided *by* the positive number 6. The sign of the number divided *into* does not matter. **28.** 5
29. $(-\infty, 6)$ 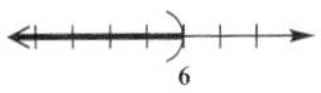
30. $(-\infty, 7)$ 
31. $[-10, \infty)$ 
32. $[-4, \infty)$ 
33. $(-\infty, -3)$ 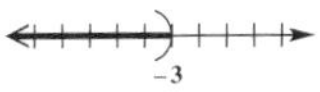
34. $(-\infty, -7)$ 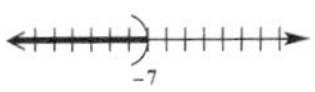
35. $(-\infty, 0]$ 
36. $(0, \infty)$ 
37. $(20, \infty)$ 
38. $(16, \infty)$ 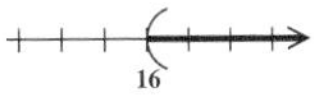
39. $[-3, \infty)$ 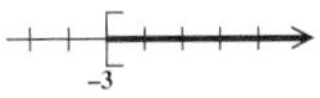
40. $(-\infty, 4]$ 
41. $[-5, \infty)$ 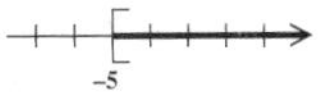
42. $(-\infty, 3)$ 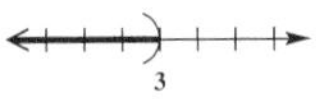
43. $(-\infty, 1)$ 
44. $[-1, \infty)$ 
45. $(-\infty, 0]$ 
46. $(0, \infty)$ 
47. $[4, \infty)$ 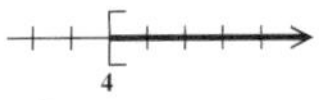
48. $(-\infty, -13)$ 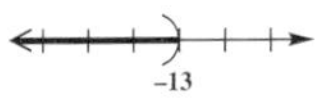
49. $(-\infty, 32)$ 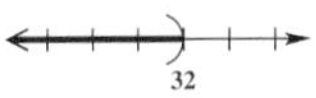

Solve each inequality. Write the solution set in interval notation and graph it. See Example 3.

29. $3x < 18$ **30.** $5x < 35$ **31.** $2y \geq -20$ **32.** $6m \geq -24$

33. $-8t > 24$ **34.** $-7x > 49$ **35.** $-x \geq 0$ **36.** $-k < 0$

37. $-\frac{3}{4}r < -15$ **38.** $-\frac{7}{8}t < -14$

39. $-.02x \leq .06$ **40.** $-.03v \geq -.12$

Solve each inequality. Write the solution set in interval notation and graph it. See Examples 4 and 5.

41. $5r + 1 \geq 3r - 9$ **42.** $6t + 3 < 3t + 12$

43. $6x + 3 + x < 2 + 4x + 4$ **44.** $-4w + 12 + 9w \geq w + 9 + w$

45. $-x + 4 + 7x \leq -2 + 3x + 6$ **46.** $14y - 6 + 7y > 4 + 10y - 10$

47. $5(x + 3) - 6x \leq 3(2x + 1) - 4x$ **48.** $2(x - 5) + 3x < 4(x - 6) + 1$

49. $\frac{2}{3}(p + 3) > \frac{5}{6}(p - 4)$ **50.** $\frac{7}{9}(y - 4) \leq \frac{4}{3}(y + 5)$

51. $4x - (6x + 1) \leq 8x + 2(x - 3)$ **52.** $2y - (4y + 3) > 6y + 3(y + 4)$

53. $5(2k + 3) - 2(k - 8) > 3(2k + 4) + k - 2$

54. $2(3z - 5) + 4(z + 6) \geq 2(3z + 2) + 3z - 15$

Write a three-part inequality involving the variable x that describes each set of numbers graphed. See Example 1(b).

55. 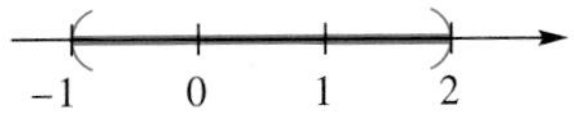

56. 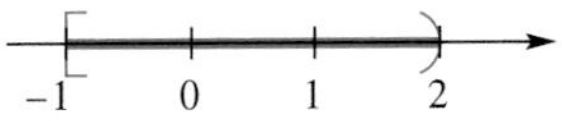

57. 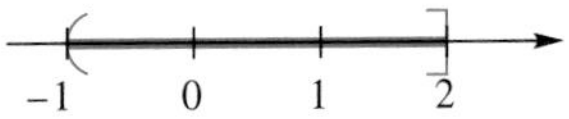

58. 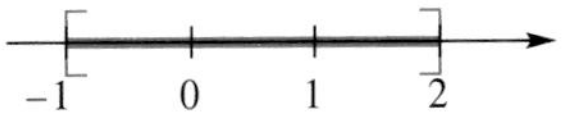

Solve each inequality. Write the solution set in interval notation and graph it. See Example 7.

59. $-5 \leq 2x - 3 \leq 9$ **60.** $-7 \leq 3x - 4 \leq 8$

61. $5 < 1 - 6m < 12$ **62.** $-1 \leq 1 - 5q \leq 16$

63. $10 < 7p + 3 < 24$ **64.** $-8 \leq 3r - 1 \leq -1$

65. $-12 \leq \frac{1}{2}z + 1 \leq 4$ **66.** $-6 \leq 3 + \frac{1}{3}a \leq 5$

67. $1 \leq 3 + \frac{2}{3}p \leq 7$ **68.** $2 < 6 + \frac{3}{4}y < 12$

69. $-7 \leq \frac{5}{4}r - 1 \leq -1$ **70.** $-12 \leq \frac{3}{7}a + 2 \leq -4$

RELATING CONCEPTS (EXERCISES 71–76)

The methods for solving linear equations and linear inequalities are quite similar. In Exercises 71–76, we show how the solutions of an inequality are closely connected to the solution of the corresponding equation.

Work these exercises in order.

71. Solve the equation $3x + 2 = 14$ and graph the solution set as a single point on the number line.

72. Solve the inequality $3x + 2 > 14$ and graph the solution set as an interval on the number line. How does this result compare to the one in Exercise 71?

(continued)

50. $\left[-\frac{88}{5}, \infty\right)$

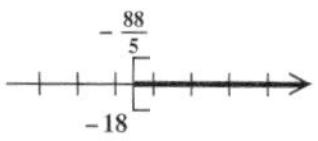

51. $\left[\frac{5}{12}, \infty\right)$

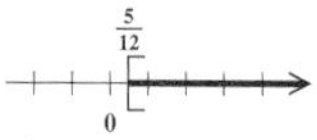

52. $\left(-\infty, -\frac{15}{11}\right)$

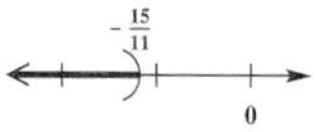

53. $(-21, \infty)$

54. $[-25, \infty)$

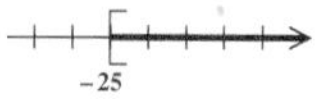

55. $-1 < x < 2$
56. $-1 \le x < 2$
57. $-1 < x \le 2$
58. $-1 \le x \le 2$

59. $[-1, 6]$

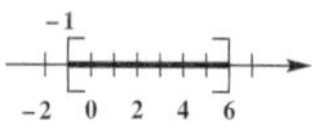

60. $[-1, 4]$

61. $\left(-\frac{11}{6}, -\frac{2}{3}\right)$

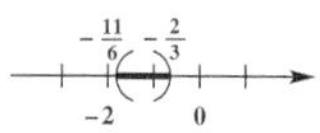

62. $\left[-3, \frac{2}{5}\right]$

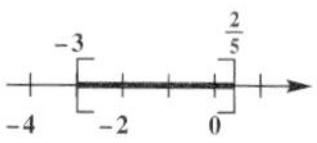

63. $(1, 3)$

64. $\left[-\frac{7}{3}, 0\right]$

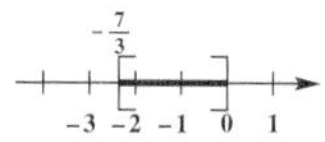

65. $[-26, 6]$

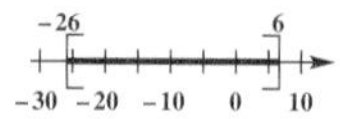

66. $[-27, 6]$

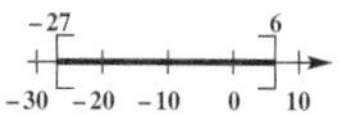

67. $[-3, 6]$

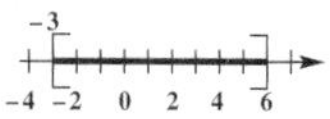

68. $\left(-\frac{16}{3}, 8\right)$

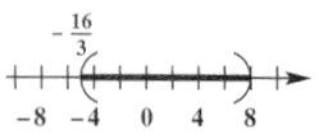

69. $\left[-\frac{24}{5}, 0\right]$

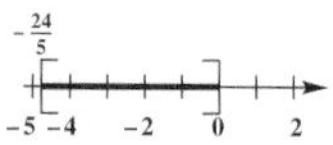

70. $\left[-\frac{98}{3}, -14\right]$

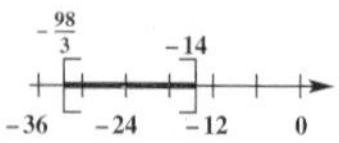

71. $\{4\}$

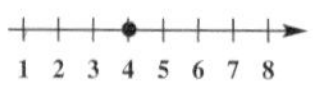

RELATING CONCEPTS (EXERCISES 71–76) (CONTINUED)

73. Solve the inequality $3x + 2 < 14$ and graph the solution set as an interval on the number line. How does this result compare to the one in Exercise 72?

74. If you were to graph all the solution sets from Exercises 71–73 on the same number line, what would the graph be? (This is called the *union* of all the solution sets.)

75. Based on your results from Exercises 71–74, if you were to graph the union of the solution sets of

$$-4x + 3 = -1, \qquad -4x + 3 > -1, \qquad \text{and} \qquad -4x + 3 < -1,$$

what do you think the graph would be?

76. Comment on the following statement: *Equality* is the boundary between *less than* and *greater than.*

Did you make the connection that the value that satisfies an equation separates the values that satisfy the corresponding less than and greater than inequalities?

Solve each problem by writing and solving an inequality. See Example 6.

77. Inkie Landry has grades of 76 and 81 on her first two algebra tests. If she wants an average of at least 80 after her third test, what possible scores can she make on her third test?

78. Mabimi Pampo has grades of 96 and 86 on his first two geometry tests. What possible scores can he make on his third test so that his average is at least 90?

79. The formula for converting Fahrenheit temperature to Celsius is

$$C = \frac{5}{9}(F - 32).$$

If the Celsius temperature on a certain summer day in Toledo is never more than 30°, how would you describe the corresponding Fahrenheit temperatures?

80. The formula for converting Celsius temperature to Fahrenheit is

$$F = \frac{9}{5}C + 32.$$

The Fahrenheit temperature of Key West, Florida, has never exceeded 95°. How would you describe this using Celsius temperature?

81. A product will break even or produce a profit if the revenue R from selling the product is at least equal to the cost C of producing it. Suppose that the cost C (in dollars) to produce x units of bicycle helmets is $C = 50x + 5000$, while the revenue R (in dollars) collected from the sale of x units is $R = 60x$. For what values of x does the product break even or produce a profit?

82. (See Exercise 81.) If the cost to produce x units of basketball cards is $C = 100x + 6000$ (in dollars), and the revenue collected from selling x units is $R = 500x$ (in dollars), for what values of x does the product break even or produce a profit?

83. For what values of x would the rectangle have perimeter of at least 400?

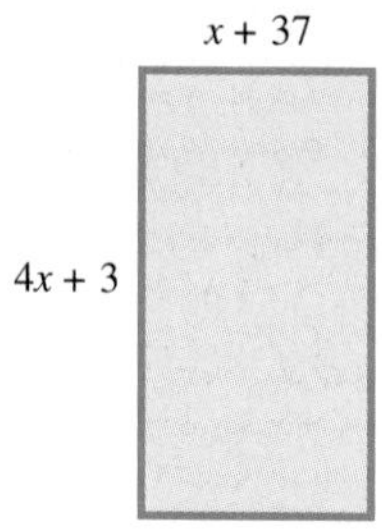

JOURNAL WRITING CHALLENGING SCIENTIFIC CALCULATOR GRAPHING CALCULATOR

72. $(4, \infty)$ The solutions are all the numbers to the right of 4.

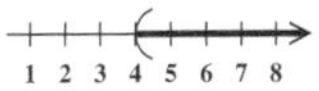

73. $(-\infty, 4)$ The solutions are all the numbers to the left of 4.

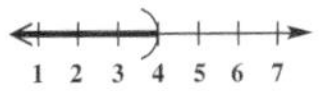

74. The graph would be all real numbers. **75.** The graph would be all real numbers. **76.** If a point on the number line satisfies an equation, the points on one side of that point will satisfy the corresponding less than inequality, and the points on the other side will satisfy the corresponding greater than inequality. **77.** 83 or greater **78.** 88 or greater **79.** It is never more than 86° Fahrenheit. **80.** It has never exceeded 35° Celsius. **81.** $x \geq 500$ **82.** $x \geq 15$ **83.** 32 or greater **84.** 14 or greater **85.** 15 minutes **86.** between 15 minutes and 18 minutes **87.** yes; yes **88.** no; no **89.** yes; no; no

90.

91.

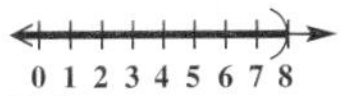

92. $(3, 8)$

93.

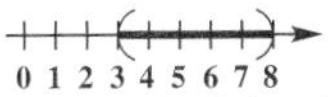

It is the graph of that interval $(3, 8)$.

94. $(-\infty, -4)$

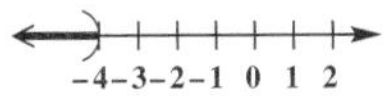

95. $\emptyset$

96.

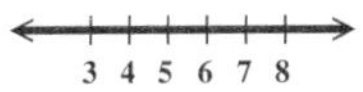

This graph includes both intervals graphed in Exercises 90 and 91. The result is all real numbers.

97. $(-\infty, 2)$

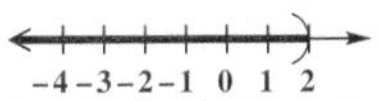

In Exercise 94 only the real numbers less than -4 were included, because both conditions had to be satisfied. Here all real numbers less than 2 satisfy at least one of the conditions.

98. $(-\infty, -3) \cup (-2, \infty)$

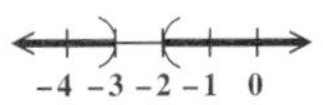

In Exercise 95, no real numbers satisfied both conditions. Here, the numbers in each interval satisfy at least one condition, so both intervals are in the solution set.

84. For what values of x would the triangle have perimeter of at least 72?

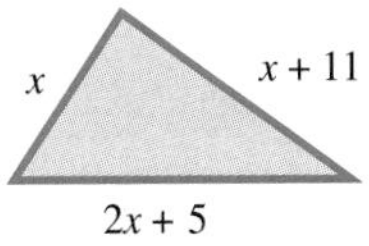

85. A long-distance phone call costs \$2.00 for the first three minutes plus \$.30 per minute for each minute or fractional part of a minute after the first three minutes. If x represents the number of minutes of the length of the call after the first three minutes, then $2 + .30x$ represents the cost of the call. If Jorge has \$5.60 to spend on a call, what is the maximum total time he can use the phone?

86. If the call described in Exercise 85 costs between \$5.60 and \$6.50, what are the possible total time lengths for the call?

RELATING CONCEPTS (EXERCISES 87–98)

The words *and* and *or* are very important in the context of inequalities. We use them to combine simple inequalities. For instance, the statement $-1 \leq x < 3$ means $-1 \leq x$ *and* $x < 3$, so that only values of x that satisfy *both* conditions are in the solution set, the interval $[-1, 3)$. On the other hand, the statement $x < -1$ *or* $x > 3$, that uses the connective *or,* will be true if $x < -1$ or if $x > 3$. Only one of the conditions must be true for the statement to be true. The two intervals that make this *or* statement true are combined, using the union symbol $\cup$, as $(-\infty, -1) \cup (3, \infty)$.

Work Exercises 87–98 in order.

87. Is the statement "Today is Tuesday *or* today is Wednesday" true if today is actually Wednesday? Is it true if it is actually Tuesday?

88. Is the statement "Today is Tuesday *and* today is Wednesday" true if today is actually Wednesday? Is it true if it is actually Tuesday?

89. Is the statement $x > 3$ and $x < 8$ true for $x = 5$? For $x = 10$? For $x = 3$?

90. Graph $x > 3$.

91. Graph $x < 8$.

92. Compare the graphs from Exercises 90 and 91. Give the values of x in interval notation that make *both* inequalities true.

93. Graph the solution set of $x > 3$ and $x < 8$. How does the graph compare to your answer to Exercise 92?

94. Give the solution set of $x < -4$ and $x < 2$ in interval notation and graph it.

95. What is the solution set of $3 - x > 6$ and $2x > -4$?

96. Graph the solution set of $x > 3$ or $x < 8$. Compare the graph with your answers to Exercises 90 and 91.

97. Give the solution set of $x < -4$ or $x < 2$ in interval notation and graph it. Compare with the answer to Exercise 94.

98. Give the solution set in interval notation and graph the solution set of $3 - x > 6$ or $2x > -4$. Compare with the answer to Exercise 95.

Did you make the connection that when *and* is used to connect two inequalities, both statements must be true for the combined statement to be true, but when *or* is used as the connective, the combined statement is true if either one or both statements are true?

Instructor's Note
Grouping: 4–5 students
Time: 20–30 minutes

You may want to lead your students through the conversions in the first row of the Olympic chart since this can cause difficulty.

CHAPTER 2 GROUP ACTIVITY

Are You a Race-Walker?

Objective: Use proportions to calculate walking speeds.

Materials: Students will need a large area for walking. Each group will need a stopwatch.

Race-walking at speeds exceeding 8 miles per hour is a high fitness, long-distance competitive sport. The table below contains the gold medal winners of the 1996 Olympic race-walking competition. Complete the table by applying the proportion given below to find the race-walker's steps per minute.

Use 10 km ≈ 6.21 miles. (Round all answers except those for steps per minute to the nearest thousandths. Round steps per minute to the nearest whole number.)

$$\frac{70 \text{ steps per minute}}{2 \text{ miles per hour}} = \frac{x \text{ steps per minute}}{y \text{ miles per hour}}$$

Event	*Gold Medal Winner*	*Country*	*Time in Hours: Minutes: Seconds*	*Time in Minutes*	*Time in Hours*	*y Miles per Hour*	*x Steps per Minute*
10 km Walk, Women	Yelena Nikolayeva	Russia	0:41:49	**[41.817]**	**[0.697]**	**[8.91]**	**[≈312]**
20 km Walk, Men	Jefferson Perez	Ecuador	1:20:07	**[80.117]**	**[1.335]**	**[9.303]**	**[≈326]**
50 km Walk, Men	Robert Korzeniowski	Poland	3:43:30	**[223.5]**	**[3.725]**	**[8.336]**	**[≈292]**

Source: 1999 World Almanac.

See answers in the first table. Answers will vary on the remainder of the activity.

A. Using a stopwatch, take turns counting how many steps each member of the group takes in one minute while walking at a normal pace. Record the results in the chart below. Then do it again at a fast pace. Record these results.

	Normal Pace		*Fast Pace*	
Name	x Steps per Minute	y Miles per Hour	x Steps per Minute	y Miles per Hour

B. Use the proportion above to convert the numbers from part A to miles per hour and complete the chart.

1. Find the average speed for the group at a normal pace and at a fast pace.
2. What is the minimum number of steps per minute you would have to take to be a race-walker?
3. At a fast pace did anyone in the group walk fast enough to be a race-walker? Explain how you decided.

CHAPTER 2 SUMMARY

KEY TERMS

2.1 linear equation in one variable
solution set
equivalent equations

2.2 empty (null) set

2.3 degree
complementary angles
supplementary angles
consecutive integers

2.4 perimeter
circumference
area
vertical angles
straight angle

2.5 ratio
proportion
cross products
vary directly as (is proportional to)

2.7 interval on the number line
interval notation
linear inequality
three-part inequality

NEW SYMBOLS

$\emptyset$	empty set
1°	one degree
a to b, $a{:}b$, or $\frac{a}{b}$	the ratio of a to b
(a, b)	interval notation for $a < x < b$
$[a, b]$	interval notation for $a \le x \le b$
∞	infinity
$-\infty$	negative infinity
$(-\infty, \infty)$	set of all real numbers

TEST YOUR WORD POWER

See how well you have learned the vocabulary in this chapter. Answers, with examples, are given at the bottom of the page.

1. A **solution set** is the set of numbers that
(a) make an expression undefined
(b) make an equation false
(c) make an equation true
(d) make an expression equal to 0.

2. The **empty set** is a set
(a) with 0 as its only element
(b) with an infinite number of elements
(c) with no elements
(d) of ideas.

3. Complementary angles are angles
(a) formed by two parallel lines
(b) whose sum is 90°
(c) whose sum is 180°
(d) formed by perpendicular lines.

4. Supplementary angles are angles
(a) formed by two parallel lines
(b) whose sum is 90°
(c) whose sum is 180°
(d) formed by perpendicular lines.

5. A **ratio**
(a) compares two quantities using a quotient
(b) says that two quotients are equal
(c) is a product of two quantities
(d) is a difference between two quantities.

6. A **proportion**
(a) compares two quantities using a quotient
(b) says that two quotients are equal
(c) is a product of two quantities
(d) is a difference between two quantities.

7. An **inequality** is
(a) a statement that two algebraic expressions are equal
(b) a point on a number line
(c) an equation with no solutions
(d) a statement with algebraic expressions related by $<$, $\le$, $>$, or $\ge$.

8. Interval notation is
(a) a portion of a number line
(b) a special notation for describing a point on a number line
(c) a way to use symbols to describe an interval on a number line
(d) a notation to describe unequal quantities.

Answers to Test Your Word Power

1. (c) *Example:* $\{8\}$ is the solution set of $2x + 5 = 21$. **2.** (c) *Example:* The empty set $\emptyset$ is the solution set of $5x + 3 = 5x + 4$. **3.** (b) *Example:* Angles with measures 35° and 55° are complementary angles. **4.** (c) *Example:* Angles with measures 112° and 68° are supplementary angles. **5.** (a) *Example:* $\frac{7 \text{ inches}}{12 \text{ inches}} = \frac{7}{12}$ **6.** (b) *Example:* $\frac{2}{8} = \frac{3}{12}$ **7.** (d) *Examples:* $x < 5$, $7 + 2y \ge 11$, $-5 < 2z - 1 \le 3$ **8.** (c) *Examples:* $(-\infty, 5]$, $(1, \infty)$, $[-3, 3)$

QUICK REVIEW

CONCEPTS	EXAMPLES
2.1 THE ADDITION AND MULTIPLICATION PROPERTIES OF EQUALITY	
The same expression may be added to (or subtracted from) each side of an equation without changing the solution.	Solve $x - 6 = 12$. $x - 6 + 6 = 12 + 6$ Add 6. $x = 18$ Combine terms. Solution set: $\{18\}$
Each side of an equation may be multiplied (or divided) by the same nonzero expression without changing the solution.	Solve $\frac{3}{4}x = -9$. $\frac{4}{3} \cdot \frac{3}{4}x = \frac{4}{3}(-9)$ Multiply by $\frac{4}{3}$. $x = -12$ Solution set: $\{-12\}$
2.2 MORE ON SOLVING LINEAR EQUATIONS	
Solving a Linear Equation	Solve the equation $2x + 3(x + 1) = 38$.
1. Clear parentheses and combine like terms to simplify each side.	$2x + 3x + 3 = 38$ Clear parentheses. $5x + 3 = 38$ Combine like terms.
2. Get the variable term on one side, a number on the other.	$5x + 3 - 3 = 38 - 3$ Subtract 3. $5x = 35$ Combine terms.
3. Get the equation into the form $x =$ a number.	$\frac{5x}{5} = \frac{35}{5}$ Divide by 5. $x = 7$
4. Check by substituting the result into the original equation.	$2x + 3(x + 1) = 38$ Check. $2(7) + 3(7 + 1) = 38$? Let $x = 7$. $14 + 24 = 38$? Multiply. $38 = 38$ True Solution set: $\{7\}$
2.3 AN INTRODUCTION TO APPLICATIONS OF LINEAR EQUATIONS	
Solving an Applied Problem Using the Six-Step Method	One number is 5 more than another. Their sum is 21. Find both numbers.
Step 1 Choose a variable to represent the unknown.	Let x be the smaller number.
Step 2 Determine expressions for any other unknown quantities, using the variable. Draw figures or diagrams and use charts if they apply.	Let $x + 5$ be the larger number.
Step 3 Write an equation.	$x + (x + 5) = 21$
Step 4 Solve the equation.	$2x + 5 = 21$ Combine terms. $2x + 5 - 5 = 21 - 5$ Subtract 5. $2x = 16$ Combine terms. $x = 8$ Divide by 2.

CONCEPTS	EXAMPLES
Step 5 Answer the question(s) asked in the problem.	The numbers are 8 and 13.
Step 6 Check your solution by using the original words of the problem. Be sure that the answer is appropriate and makes sense.	13 is 5 more than 8, and $8 + 13 = 21$. It checks.

2.4 FORMULAS AND APPLICATIONS FROM GEOMETRY

CONCEPTS	EXAMPLES
To find the values of one of the variables in a formula given values for the others, substitute the known values into the formula.	Find L if $A = LW$, given that $A = 24$ and $W = 3$. $24 = L \cdot 3$ $A = 24, W = 3$ $\frac{24}{3} = \frac{L \cdot 3}{3}$ Divide by 3. $8 = L$
To solve a formula for one of the variables, isolate that variable by treating the other variables as numbers and using the steps for solving equations.	Solve $A = \frac{1}{2}bh$ for b. $2A = 2\left(\frac{1}{2}bh\right)$ Multiply by 2. $2A = bh$ $\frac{2A}{h} = b$ Divide by h.

2.5 RATIOS AND PROPORTIONS

CONCEPTS	EXAMPLES
To write a ratio, express quantities in the same units.	Express as a ratio: 4 feet to 8 inches. 4 feet to 8 inches = 48 inches to 8 inches $= \frac{48}{8} = \frac{6}{1}$ or 6 to 1 or 6 : 1
To solve a proportion, use cross products.	Solve $\frac{x}{12} = \frac{35}{60}$. $60x = 12 \cdot 35$ Cross products $60x = 420$ Multiply. $\frac{60x}{60} = \frac{420}{60}$ Divide by 60. $x = 7$ Solution set: $\{7\}$
Solving Direct Variation Problems	At a constant speed (or rate) distance varies directly as time. Find the time to travel 85 miles at a constant speed if it takes 1.5 hours to go 67.5 miles at the same constant speed.
1. Write the variation equation using $y = kx$.	Let t represent the unknown time. The direct variation equation is $d = kt$. Substitute $d = 67.5$ and $t = 1.5$ to find k.
2. Find k by substituting the given values of x and y into the equation.	$67.5 = k(1.5)$ $45 = k$ Divide by 1.5.
3. Write the equation with the value of k from Step 2 and the given value of x or y. Solve for the remaining variable.	Now substitute $k = 45$ and $d = 85$ in the equation $d = kt$. $85 = 45t$ $1.89 = t$ Divide by 45; round up. It will take approximately 1.89 hours to travel 85 miles.

CONCEPTS	EXAMPLES

2.6 MORE ABOUT PROBLEM SOLVING

Problems involving applications of percent can be solved using charts.

A sum of money is invested at simple interest in two ways. Part is invested at 12%, and $20,000 less than that amount is invested at 10%. If the total interest for one year is $9000, find the amount invested at each rate.

Let x = amount invested at 12%;
$x - 20{,}000$ = amount invested at 10%.

Dollars Invested	Rate of Interest	Interest for One Year
x	.12	$.12x$
$x - 20{,}000$	.10	$.10(x - 20{,}000)$

$$.12x + .10(x - 20{,}000) = 9000$$
$$.12x + .10x + .10(-20{,}000) = 9000 \quad \text{Distributive property}$$
$$12x + 10x + 10(-20{,}000) = 900{,}000 \quad \text{Multiply by 100.}$$
$$12x + 10x - 200{,}000 = 900{,}000$$
$$22x - 200{,}000 = 900{,}000 \quad \text{Combine terms.}$$
$$22x = 1{,}100{,}000 \quad \text{Add 200,000.}$$
$$x = 50{,}000 \quad \text{Divide by 22.}$$

$50,000 is invested at 12% and $30,000 is invested at 10%.

The three forms of the formula relating distance, rate, and time, are $d = rt$, $r = \frac{d}{t}$, and $t = \frac{d}{r}$.

Two cars leave from the same point, traveling in opposite directions. One travels at 45 miles per hour and the other at 60 miles per hour. How long will it take them to be 210 miles apart?

Let t = time it takes for them to be 210 miles apart.

To solve a problem about distance, set up a sketch showing what is happening in the problem.

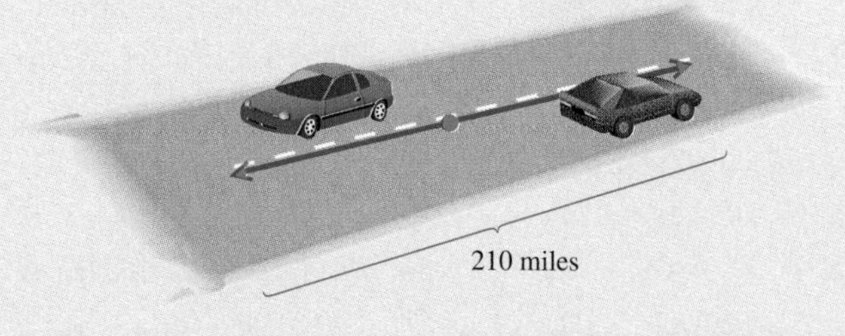

Make a chart using the information given in the problem, along with the unknown quantities.

The chart gives the information from the problem, with expressions for distance obtained by using $d = rt$.

	Rate	Time	Distance
One car	45	t	$45t$
Other car	60	t	$60t$

The sum of the distances, $45t$ and $60t$, must be 210 miles.

$$45t + 60t = 210$$
$$105t = 210 \quad \text{Combine like terms.}$$
$$t = 2 \quad \text{Divide by 2.}$$

It will take them 2 hours to be 210 miles apart.

CONCEPTS	EXAMPLES

2.7 THE ADDITION AND MULTIPLICATION PROPERTIES OF INEQUALITY

To solve an inequality:

1. Clear parentheses and combine like terms.
2. Add or subtract the same expression on each side to get the variable term on one side and a number on the other side.
3. Multiply or divide by the same expression on each side to get the form $x > a$ or $x < a$. (When multiplying or dividing by a negative expression, reverse the direction of the inequality symbol.)

Solve $3(1 - x) + 5 - 2x > 9 - 6$.

$3 - 3x + 5 - 2x > 9 - 6$	Clear parentheses.
$8 - 5x > 3$	Combine terms.
$8 - 5x - 8 > 3 - 8$	Subtract 8.
$-5x > -5$	Combine terms.
$\frac{-5x}{-5} < \frac{-5}{-5}$	Divide by -5; change $>$ to $<$.
$x < 1$	Lowest terms

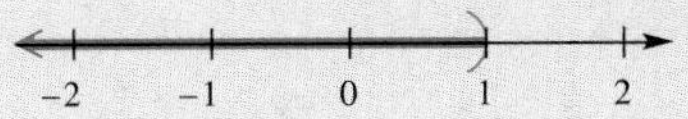

Solution set: $(-\infty, 1)$

To solve an inequality such as

$$4 < 2x + 6 < 8$$

work with all three expressions at the same time.

Solve $4 < 2x + 6 < 8$.

$4 - 6 < 2x + 6 - 6 < 8 - 6$	Subtract 6.
$-2 < 2x < 2$	Combine terms.
$\frac{-2}{2} < \frac{2x}{2} < \frac{2}{2}$	Divide by 2.
$-1 < x < 1$	

Solution set: $(-1, 1)$

CHAPTER 2 REVIEW EXERCISES

1. {6} **2.** {−12} **3.** {7} **4.** $\left\{\frac{2}{3}\right\}$ **5.** {11} **6.** {17} **7.** {5} **8.** {−4} **9.** {5} **10.** {−12} **11.** $\left\{\frac{64}{5}\right\}$ **12.** {4} **13.** {all real numbers} **14.** {−19} **15.** {all real numbers} **16.** {20} **17.** ∅ **18.** {all real numbers} **19.** 112 nations

[2.1–2.2] *Solve each equation.*

1. $m - 5 = 1$

2. $y + 8 = -4$

3. $3k + 1 = 2k + 8$

4. $5k = 4k + \frac{2}{3}$

5. $(4r - 2) - (3r + 1) = 8$

6. $3(2y - 5) = 2 + 5y$

7. $7k = 35$

8. $12r = -48$

9. $2p - 7p + 8p = 15$

10. $\frac{m}{12} = -1$

11. $\frac{5}{8}k = 8$

12. $12m + 11 = 59$

13. $3(2x + 6) - 5(x + 8) = x - 22$

14. $5x + 9 - (2x - 3) = 2x - 7$

15. $\frac{1}{2}r - \frac{r}{3} = \frac{r}{6}$

16. $.10(x + 80) + .20x = 14$

17. $3x - (-2x + 6) = 4(x - 4) + x$

18. $2(y - 3) - 4(y + 12) = -2(y + 27)$

[2.3] *Use the six-step method to solve each problem.*

19. Soccer is the world's most popular sport, with over 20 million participants. The number of participating nations in the 1990 World Cup was 6 less than in the 1986 World Cup. In 1994, 28 more nations participated than in 1986. How many nations participated in 1986 if there was a total of 358 participating nations in those three years? (*Source:* FIFA.)

JOURNAL WRITING ▲ CHALLENGING SCIENTIFIC CALCULATOR GRAPHING CALCULATOR

20. **Democrats: 75; Republicans: 45**
21. **Hawaii: 6425 square miles; Rhode Island: 1212 square miles**
22. **Seven Falls: 300 feet; Twin Falls: 120 feet** **23.** **80°** **24.** $h = 11$
25. $A = 28$ **26.** $r = 4.75$
27. $V = 904.32$ **28.** $L = \frac{A}{W}$
29. $h = \frac{2A}{b + B}$ **30.** **135°; 45°**
31. **100°; 100°**
32. **diameter: approximately 19.9 feet; radius: approximately 9.95 feet; area: approximately 311 square feet** **33.** **42.2°; 92.8°**
34. **Not enough information is given. We also need the value of** B.

20. In a recent year, the state of Florida had a total of 120 members in its House of Representatives, consisting of only Democrats and Republicans. There were 30 more Democrats than Republicans. How many representatives from each party were there?

21. The land area of Hawaii is 5213 square miles greater than the area of Rhode Island. Together, the areas total 7637 square miles. What is the area of each of the two states?

22. The height of Seven Falls in Colorado is $\frac{5}{2}$ the height of Twin Falls in Idaho. The sum of the heights is 420 feet. Find the height of each. (*Source: World Almanac and Book of Facts.*)

23. The supplement of an angle measures 10 times the measure of its complement. What is the measure of the angle?

[2.4] *A formula is given along with the values for all but one of the variables. Find the value of the variable that is not given.*

24. $A = \frac{1}{2}bh$; $A = 44$, $b = 8$

25. $A = \frac{1}{2}h(b + B)$; $b = 3$, $B = 4$, $h = 8$

26. $C = 2\pi r$; $C = 29.83$, $\pi = 3.14$

27. $V = \frac{4}{3}\pi r^3$; $r = 6$, $\pi = 3.14$

Solve the formula for each specified variable.

28. $A = LW$; for L

29. $A = \frac{1}{2}h(b + B)$; for h

Find the measure of each marked angle.

30.

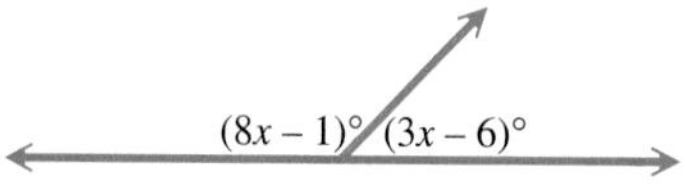

31.

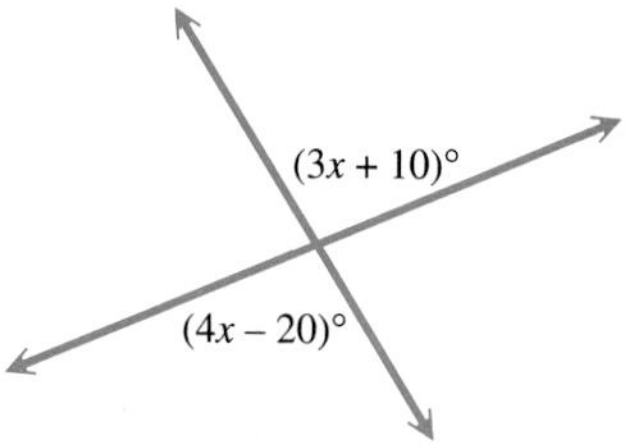

Solve each problem.

32. The Ziegfield Room in Reno, Nevada, has a circular turntable on which its showgirls dance. The circumference of the table is 62.5 feet. What is the diameter? What is the radius? What is the area? (Use $\pi = 3.14$.) (*Source: The Guinness Book of World Records,* 1997.)

33. A baseball diamond is a square with a side of 90 feet. The pitcher's rubber is located 60.5 feet from home plate as shown in the figure. Find the measures of the angles marked in the figure.

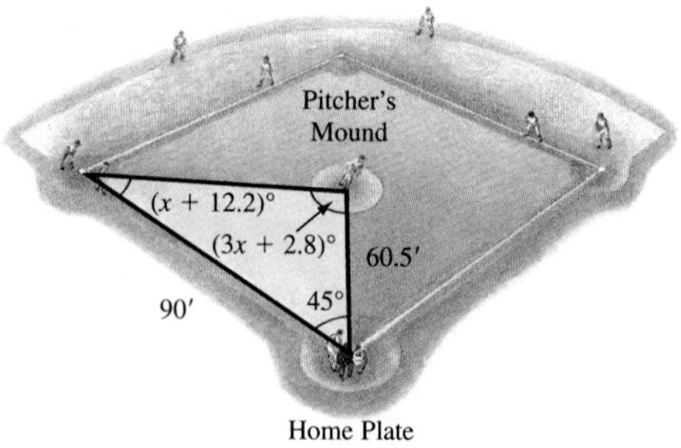

34. What is wrong with the following problem? "The formula for the area of a trapezoid is $A = \frac{1}{2}h(b + B)$. If $h = 12$ and $b = 14$, find the value of A."

JOURNAL WRITING ▲ CHALLENGING SCIENTIFIC CALCULATOR GRAPHING CALCULATOR

35. $\frac{3}{2}$ 36. $\frac{5}{14}$ 37. $\frac{3}{4}$
38. $\frac{1}{12}$ 39. $\left\{\frac{7}{2}\right\}$ 40. $\left\{-\frac{8}{3}\right\}$
41. $\left\{\frac{25}{19}\right\}$ 42. $6\frac{2}{3}$ pounds
43. .2 meter or 200 millimeters
44. \$3.06 45. 375 kilometers
46. 4 gold medals
47. 25.5-ounce size
48. approximately 55.2%
49. approximately \$15.4 million
50. 3.75 liters 51. \$5000 at 5%; \$5000 at 6% 52. 8.2 miles per hour

[2.5] *Give a ratio for each word phrase, writing fractions in lowest terms.*

35. 60 centimeters to 40 centimeters

36. 5 days to 2 weeks

37. 90 inches to 10 feet

38. 3 months to 3 years

Solve each equation.

39. $\frac{p}{21} = \frac{5}{30}$

40. $\frac{5 + x}{3} = \frac{2 - x}{6}$

41. $\frac{y}{5} = \frac{6y - 5}{11}$

Use proportions to solve each problem.

42. If 2 pounds of fertilizer will cover 150 square feet of lawn, how many pounds would be needed to cover 500 square feet?

43. In a camera, the ratio of the length of the image on the film to the object length equals the ratio of the distance from the lens to the image compared to the distance from the lens to the object. See the figure. How far from the lens should a child 1 meter tall be to fit on 35 mm film (35 mm by 35 mm), if the distance from the lens to the film is 7 mm?

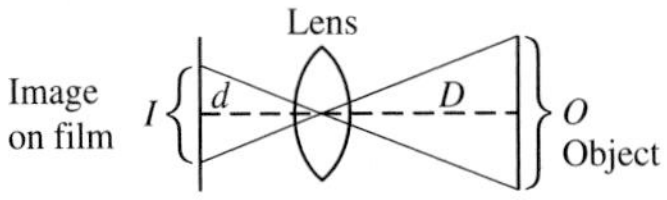

44. The tax on a \$24.00 item is \$2.04. How much tax would be paid on a \$36.00 item?

45. The distance between two cities on a road map is 32 centimeters. The two cities are actually 150 kilometers apart. The distance on the map between two other cities is 80 centimeters. How far apart are these cities?

46. In the 1996 Olympic Games held in Atlanta, Romania earned a total of 20 medals. Two of every ten medals were gold. How many gold medals were earned? (*Source: The Universal Almanac,* 1997, John W. Wright, General Editor.)

47. Which is the best buy for a popular breakfast cereal?
15-ounce size: \$2.69
20-ounce size: \$3.29
25.5-ounce size: \$3.49

[2.6] *Solve each problem.*

48. The San Francisco Giants baseball team has borrowed \$160 million of the \$290 million to build their new Pacific Bell Park, scheduled to open in 2002. Corporate sponsors will provide the rest of the cost. What percent of the cost of the new park is borrowed? (*Source:* Michael K. Ozanian, "Fields of Debt," *Forbes,* December 15, 1997, vol. 160, no. 13, p. 174.)

49. (Refer to Exercise 48.) It has been estimated that Bell Park's annual cash flow will be 1.07 times the cost of the interest on the loan of \$160 million. Suppose the money is borrowed at a 9% annual interest rate. How much will the annual cash flow amount to? (*Source:* Michael K. Ozanian, "Fields of Debt," *Forbes,* December 15, 1997, vol. 160, no. 13, p. 174.)

50. A nurse must mix 15 liters of a 10% solution of a drug with some 60% solution to get a 20% mixture. How many liters of the 60% solution will be needed?

51. Todd Cardella invested \$10,000 from which he earns an annual income of \$550 per year. He invested part of it at 5% annual interest and the remainder in bonds paying 6% interest. How much did he invest at each rate?

52. In 1846, the vessel *Yorkshire* traveled from Liverpool to New York, a distance of 3150 miles, in 384 hours. What was the *Yorkshire*'s average speed? Round your answer to the nearest tenth.

53. 13 hours **54.** $2\frac{1}{2}$ hours

55. $\frac{5}{6}$ hour or 50 minutes

56. 2.786 inches

57. 28 games

58. 80 newtons per square centimeter **59.** Answers will vary.

60. $[-4, \infty)$

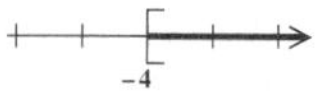

61. $(-\infty, 7)$ 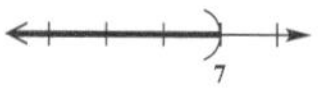

62. $[-5, 6)$ 

63. $[-3, \infty)$ 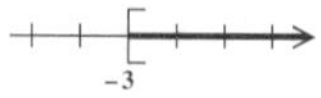

64. $(-\infty, 2)$ 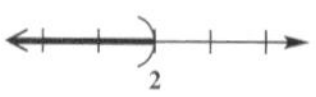

65. $[3, \infty)$ 

66. $[46, \infty)$ 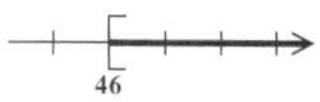

67. $(-\infty, -5)$ 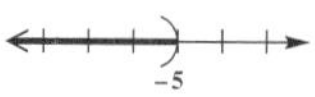

68. $(-\infty, -37)$ 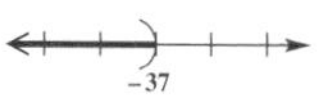

69. $\left[-2, \frac{3}{2}\right]$ 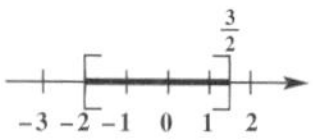

70. $\left(\frac{4}{3}, 5\right]$ 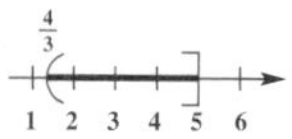

71. more than 2875

72. \$1.557 million

53. Sue Fredine drove from Louisville to Dallas, a distance of 819 miles, averaging 63 miles per hour. What was her driving time?

54. Two planes leave St. Louis at the same time. One flies north at 350 miles per hour and the other flies south at 420 miles per hour. In how many hours will they be 1925 miles apart?

55. Jim leaves his house on his bicycle and averages 5 miles per hour. His wife, Annie, leaves $\frac{1}{2}$ hour later, following the same path and averaging 8 miles per hour. How long will it take for Annie to catch up to Jim?

56. The circumference of a circle varies directly as its radius. A circle with a circumference of 5 inches has a radius of approximately .796 inches. Find the radius of a circle with a circumference of 17.5 inches.

57. Typically, the supply of an item varies directly as the price. When a computer game was priced at \$50, a merchant had a supply of 40 games. After the game went on sale for \$35, what was the supply?

58. The pressure exerted by a liquid at a given point is proportional to the depth of the point beneath the surface of the liquid. The pressure at 10 meters is $\frac{80}{3}$ newtons per square centimeter. What pressure is exerted at 30 meters?

59. Write your own applied problem, and use the six-step method to solve it.

[2.7] *Write each inequality in interval notation and graph it on a number line.*

60. $p \geq -4$ **61.** $x < 7$ **62.** $-5 \leq y < 6$

Solve each inequality and graph the solution set.

63. $y + 6 \geq 3$ **64.** $5t < 4t + 2$

65. $-6x \leq -18$ **66.** $8(k - 5) - (2 + 7k) \geq 4$

67. $4x - 3x > 10 - 4x + 7x$

68. $3(2w + 5) + 4(8 + 3w) < 5(3w + 2) + 2w$

69. $-3 \leq 2m + 1 \leq 4$ **70.** $9 < 3m + 5 \leq 20$

Solve each problem by writing an inequality.

71. In 1995 more than 6900 Americans were active water skiers. Of this group, the number of males was 1.4 times the number of females. Find the possible numbers of female water skiers. (*Source:* National Sporting Goods Association, Mount Prospect, Illinois.)

72. In 1995, the average National Football League quarterback's salary was \$1.307 million. In 1996, it rose to \$1.336 million. What salary in 1997 would make the average salary for these three years at least \$1.4 million? (*Source:* NFL Players Association.)

73. $\{7\}$ **74.** $r = \frac{I}{pt}$

75. $(-\infty, 2)$ **76.** $\{-9\}$

77. $\{70\}$ **78.** $\left\{\frac{13}{4}\right\}$

MIXED REVIEW EXERCISES*

Solve.

73. $\frac{y}{7} = \frac{y - 5}{2}$ **74.** $I = prt$ for r

75. $-2x > -4$ **76.** $2k - 5 = 4k + 13$

77. $.05x + .02x = 4.9$ **78.** $2 - 3(y - 5) = 4 + y$

*The order of the exercises in this final group does not correspond to the order in which topics occur in the chapter. This random ordering should help you in your preparation for the chapter test.

79. $\emptyset$ **80.** {all real numbers}
81. The first step is to distribute the negative sign over *both* terms in the parentheses, to get $3 - 8 - 4x$ on the left side of the equation. The student got $3 - 8 + 4x$ instead. The correct solution set is $\{-2\}$.
82. $2\frac{1}{2}$ cups **83.** Golden Gate Bridge: 4200 feet; Brooklyn Bridge: 1595 feet **84.** Zero makes the fractions undefined.
85. 64-ounce size **86.** 8 quarts
87. faster train: 80 miles per hour; slower train: 50 miles per hour
88. 9 centimeters or less
89. 8 centimeters
90. 50 meters or less
91. No. Only equations that have one fractional term on each side can be solved directly by cross multiplication. This equation has two terms on the left.

79. $9x - (7x + 2) = 3x + (2 - x)$

80. $\frac{1}{3}s + \frac{1}{2}s + 7 = \frac{5}{6}s + 5 + 2$

81. A student solved $3 - (8 + 4x) = 2x + 7$ and gave the answer as $\{6\}$. Verify that this answer is incorrect by checking it in the equation. Then explain the error and give the correct solution set. (*Hint:* The error involves the subtraction sign.)

82. A recipe for biscuit tortoni calls for $\frac{2}{3}$ cup of macaroon cookie crumbs. The recipe is for 8 servings. How many cups of macaroon cookie crumbs would be needed for 30 servings?

83. The Golden Gate Bridge in San Francisco is 2605 feet longer than the Brooklyn Bridge. Together, their spans total 5795 feet. How long is each bridge? (*Source: World Almanac and Book of Facts.*)

84. A student solves $\frac{3}{x} = \frac{5}{x}$ by cross multiplying and gets 0 as an answer. What is wrong with this answer?

85. Which is the best buy for apple juice?
32-ounce size: \$1.19
48-ounce size: \$1.79
64-ounce size: \$1.99

86. If 1 quart of oil must be mixed with 24 quarts of gasoline, how much oil would be needed for 192 quarts of gasoline?

87. Two trains are 390 miles apart. They start at the same time and travel toward one another, meeting 3 hours later. If the speed of one train is 30 miles per hour more than the speed of the other train, find the speed of each train.

88. One side of a triangle is 3 centimeters longer than the shortest side. The third side is twice as long as the shortest side. If the perimeter of the triangle cannot exceed 39 centimeters, find all possible lengths for the shortest side.

89. The shorter base of a trapezoid is 42 centimeters long, and the longer base is 48 centimeters long. The area of the trapezoid is 360 square centimeters. Find the height of the trapezoid.

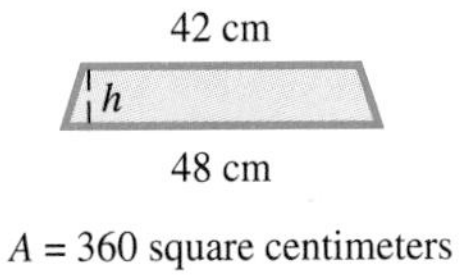

90. The perimeter of a square cannot be greater than 200 meters. Find the possible values for the length of a side.

91. Can $4 + \frac{5}{x} = \frac{2}{3}$ be solved directly by cross multiplication? Explain your answer.

CHAPTER 2 TEST

[2.1–2.2] **1.** $\{-6\}$ **2.** $\{21\}$ **3.** $\emptyset$ **4.** $\{30\}$ **5.** {all real numbers}

Solve each equation.

1. $5x + 9 = 7x + 21$

2. $-\frac{4}{7}x = -12$

3. $7 - (m - 4) = -3m + 2(m + 1)$

4. $.06(x + 20) + .08(x - 10) = 4.6$

5. $-8(2x + 4) = -4(4x + 8)$

[2.3] 6. Hawaii: 4021 square miles; Maui: 728 square miles; Kauai: 551 square miles
[2.4] 7. (a) $W = \frac{P - 2L}{2}$ or $W = \frac{P}{2} - L$ (b) 18 8. 75°, 75°
9. 50°
[2.5] 10. $\{-29\}$
11. 62.8 centimeters
12. 2300 miles
[2.6] 13. \$8000 at 3%; \$14,000 at 4.5% 14. 236% 15. 4 hours
[2.7] 16. $(-\infty, 4]$

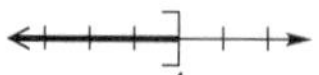

17. $(-2, 6]$

18. 2 is less than a number
19. East 132, West 120

Solve each problem.

6. The three largest islands in the Hawaiian island chain are Hawaii (the Big Island), Maui, and Kauai. Together, their areas total 5300 square miles. The island of Hawaii is 3293 square miles larger than the island of Maui, and Maui is 177 square miles larger than Kauai. What is the area of each island?

7. The formula for the perimeter of a rectangle is $P = 2L + 2W$.
(a) Solve for W.
(b) If $P = 116$ and $L = 40$, find the value of W.

Find the measure of each marked angle.

8.

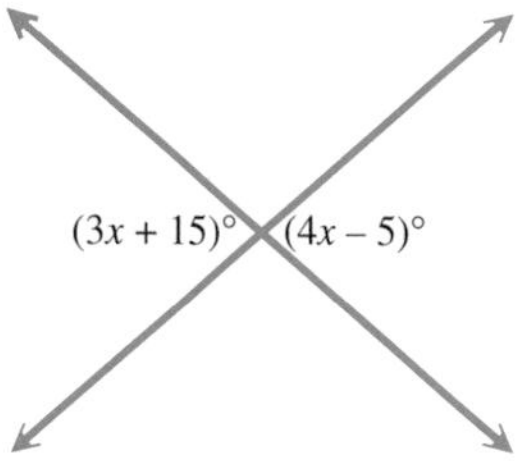

9. Find the measure of an angle if its supplement measures 10° more than three times its complement.

Solve the equation.

10. $\frac{y + 5}{3} = \frac{y - 3}{4}$

Solve each problem.

11. The circumference of a circle varies directly as the radius. A circle with a radius of 6 centimeters has a circumference of 37.68 centimeters. What is the circumference of a circle with a radius of 10 centimeters?

12. The distance between Milwaukee and Boston is 1050 miles. On a certain map this distance is represented by 21 inches. On the same map Seattle and Cincinnati are 46 inches apart. What is the actual distance between Seattle and Cincinnati?

13. Cathy Wacaser invested some money at 3% simple interest and \$6000 more than that amount at 4.5% simple interest. After one year her total interest from the two accounts was \$870. How much did she invest at each rate?

14. The Tampa Bay Lightning hockey team has debt of about \$177 million, with a franchise value of only \$75 million. What percent of the franchise value is the debt? (*Source:* Michael K. Ozanian, "Fields of Debt," *Forbes,* December 15, 1997, vol. 160, no. 13, p. 174(2).)

15. Two cars leave from the same point, traveling in opposite directions. One travels at a constant rate of 50 miles per hour while the other travels at a constant rate of 65 miles per hour. How long will it take for them to be 460 miles apart?

Solve each inequality and graph the solution set.

16. $-4x + 2(x - 3) \geq 4x - (3 + 5x) - 7$ **17.** $-10 < 3k - 4 \leq 14$

18. Which statement is translated as $2 < x$: 2 less than a number, or 2 is less than a number?

Exercises 19 and 20 refer to the 1997 All-Star basketball game held February 9 in Cleveland. (Source: *The ESPN 1998 Information Please Sports Almanac.*)

19. The East won the game by 12 points. The total of the two team scores was 252. What was the final score of the game?

20. 26 points

20. The high scorer of the game was Glenn Rice, playing for the Eastern Conference. He scored 7 more points than Latrell Sprewell, the high scorer on the Western Conference team. The total number of points scored by both players was 45. How many points did Rice score?

CUMULATIVE REVIEW EXERCISES CHAPTERS 1–2

[1.1] 1. $\frac{3}{4}$ 2. $\frac{37}{60}$ 3. $\frac{48}{5}$
[1.2] 4. $\frac{1}{2}x - 18$ 5. $\frac{6}{x + 12} = 2$
[1.4] 6. true [1.5–1.6] 7. 11
8. −8 9. 28
[1.3] 10. $-\frac{19}{3}$
[1.7] 11. distributive property
12. commutative property
[1.8] 13. $2k - 11$
[2.1-2.2] 14. $\{-1\}$ 15. $\{-1\}$
16. $\{-12\}$ [2.5] 17. $\{26\}$
[2.4] 18. $y = \frac{24 - 3x}{4}$
19. $n = \frac{A - P}{iP}$
[2.7] 20. $(-\infty, 1]$

21. $(-1, 2]$

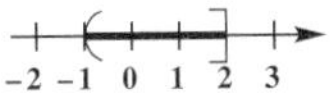

Beginning with this chapter, each chapter in the text will conclude with a set of cumulative review exercises designed to cover the major topics from the beginning of the course. This feature allows students to constantly review topics that have been introduced up to that point. Section references for these exercises are given with the answers at the back of the book.

1. Write $\frac{108}{144}$ in lowest terms.

Perform the indicated operations.

2. $\frac{5}{6} + \frac{1}{4} - \frac{7}{15}$

3. $\frac{9}{8} \cdot \frac{16}{3} \div \frac{5}{8}$

Translate from words to symbols. Use x as the variable, if necessary.

4. The difference between half a number and 18.

5. The quotient of 6 and 12 more than a number is 2.

6. True or false? $\frac{8(7) - 5(6 + 2)}{3 \cdot 5 + 1} \geq 1$

Perform the indicated operations.

7. $9 - (-4) + (-2)$

8. $\frac{-4(9)(-2)}{-3^2}$

9. $(-7 - 1)(-4) + (-4)$

10. Find the value of $\frac{3x^2 - y^3}{-4z}$ when $x = -2$, $y = -4$, and $z = 3$.

Name the property illustrated.

11. $7(k + m) = 7k + 7m$

12. $3 + (5 + 2) = 3 + (2 + 5)$

13. Simplify $-4(k + 2) + 3(2k - 1)$ by combining terms.

Solve each equation and check the solution.

14. $2r - 6 = 8r$

15. $4 - 5(a + 2) = 3(a + 1) - 1$

16. $\frac{2}{3}y + \frac{3}{4}y = -17$

17. $\frac{2x + 3}{5} = \frac{x - 4}{2}$

Solve each formula for the indicated variable.

18. $3x + 4y = 24$ for y

19. $A = P(1 + ni)$ for n

Solve each inequality. Graph the solution set.

20. $6(r - 1) + 2(3r - 5) \leq -4$

21. $-18 \leq -9z < 9$

[2.3] 22. 4 centimeters; 9 centimeters; 27 centimeters
[2.4] 23. 1029.92 feet
[2.5] 24. $\frac{25}{6}$ or $4\frac{1}{6}$ cups
[2.6] 25. 40 miles per hour; 60 miles per hour

Solve each problem.

22. For a woven hanging, Miguel Hidalgo needs three pieces of yarn, which he will cut from a 40-centimeter piece. The longest piece is to be 3 times as long as the middle-sized piece, and the shortest piece is to be 5 centimeters shorter than the middle-sized piece. What lengths should he cut?

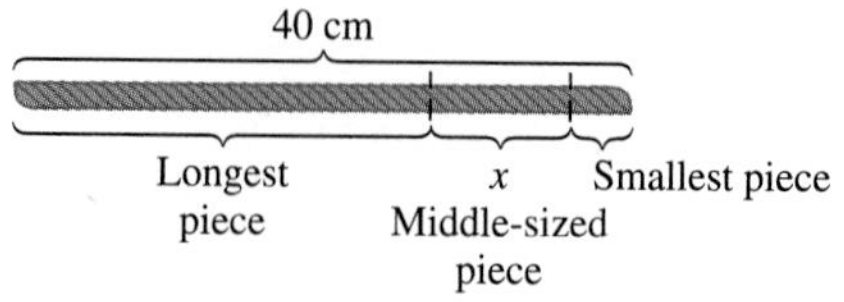

23. A radio telescope at the Max Planck Institute in Germany has a 328-foot diameter. What is the circumference of this telescope? (Use $\pi = 3.14$.) (*Source: The Guinness Book of World Records.*)

24. A cook wants to increase a recipe that serves 6 to make enough for 20 people. The recipe calls for $1\frac{1}{4}$ cups of grated cheese. How much cheese will be needed to serve 20?

25. Two cars are 400 miles apart. Both start at the same time and travel toward one another. They meet 4 hours later. If the speed of one car is 20 miles per hour faster than the other, what is the speed of each car?

Linear Equations in Two Variables

3

Calvin Coolidge, 30th president of the United States, once said "The chief business of America is business." Small businesses account for 99 percent of the 19 million nonfarm businesses in the United States today. Small businesses employ 55 percent of the private workforce, make 44 percent of all sales in America, and produce 38 percent of the nation's gross national product.*

The number of new consumer packaged-goods products introduced in the United States from 1991–1996 is illustrated in the line graph. The graph shows that, with the exception of 1994, this number has increased from year to year at a steady pace. What might account for the one-year decrease? We will return to this graph in the exercises for Section 3.1.

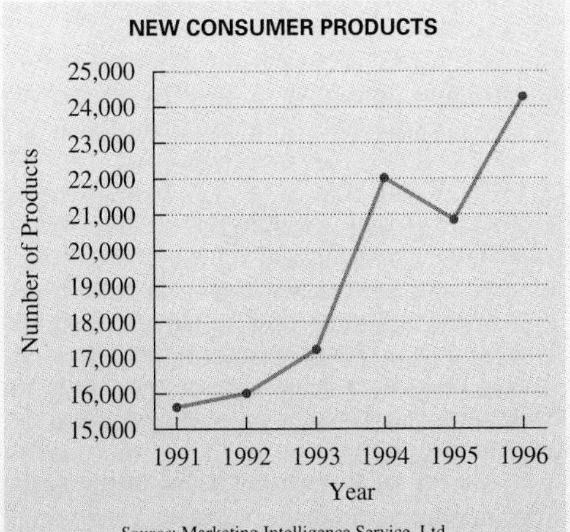

Source: Marketing Intelligence Service, Ltd.

Mathematics plays an important role in the business world. Preparing a business plan, pricing, borrowing money, determining market share, and tracking costs, revenue, and profit all require a good understanding of mathematics. Throughout this chapter many of the examples and exercises will further illustrate the use of mathematics in business management.

*Data from *The Universal Almanac,* 1997, John W. Wright, General Editor.

Business

INTERNET Visit our Web site at www.LialAlgebra.com

3.1 Linear Equations in Two Variables

OBJECTIVES

1. Interpret graphs.
2. Write a solution as an ordered pair.
3. Decide whether a given ordered pair is a solution of a given equation.
4. Complete ordered pairs for a given equation.
5. Plot ordered pairs.

FOR EXTRA HELP

SSG Sec. 3.1
SSM Sec. 3.1
Pass the Test Software
InterAct Math Tutorial Software
Video 5

CHALKBOARD EXAMPLE

Using Figure 1, estimate the amount of venture capital provided in 1994. The amount of capital in 1996 is double the amount for what year?

Answer: \$5.5 billion; 1993

Graphs are prevalent in our society. Pie charts (circle graphs) and bar graphs were introduced in Chapter 1, and we have seen many examples of them in the first two chapters of this book. It is important to be able to interpret graphs correctly.

OBJECTIVE 1 Interpret graphs. We begin with a bar graph where we must estimate the heights of the bars.

EXAMPLE 1 Interpreting Bar Graphs

Venture capital is money invested in new, often speculative, business enterprises. The amount of venture capital invested in companies has risen during the decade of the 1990s, as shown in the graph in Figure 1. Use the graph to determine the following.

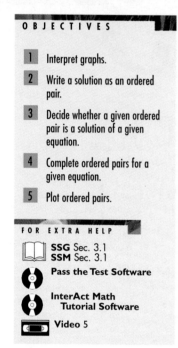

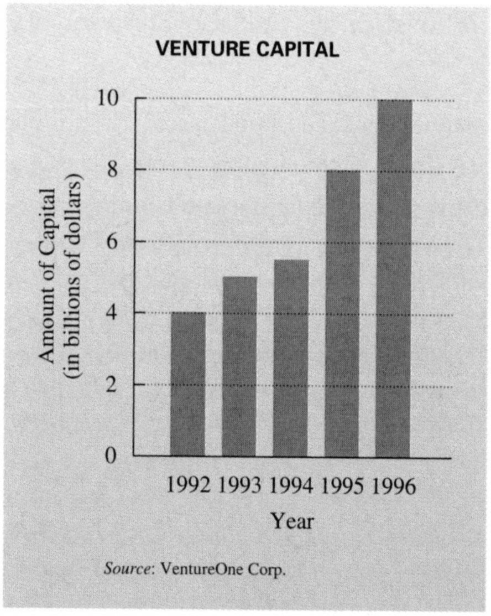

Figure 1

(a) What amount of venture capital was provided in 1992?

Move horizontally from the top of the bar for 1992 to the scale on the left to see that about \$4 billion was provided.

(b) In what year was that amount doubled?

Follow the line for \$8 billion across to the right. The bar for 1995 just touches that line, so the amount provided in 1992 was doubled in 1995.

INSTRUCTOR'S RESOURCES

PTB Sec. 3.1 ISM Sec. 3.1
AB Sec. 3.1
TEST GENERATOR
WORLD WIDE WEB
www.LialAlgebra.com

CHALKBOARD EXAMPLE

In Figure 2, in which year does the graph show the greatest increase in the number of deals? Estimate the number of deals in 1996 and compare it with the number in 1995.

Answer: From 1995 to 1996; 12,000 deals compared to about 9700 deals

EXAMPLE 2 Interpreting Line Graphs

The line graph in Figure 2 shows the total number of deals that involved venture capital in the 1990s. Use the graph to estimate the number of deals in 1993 and in 1995.

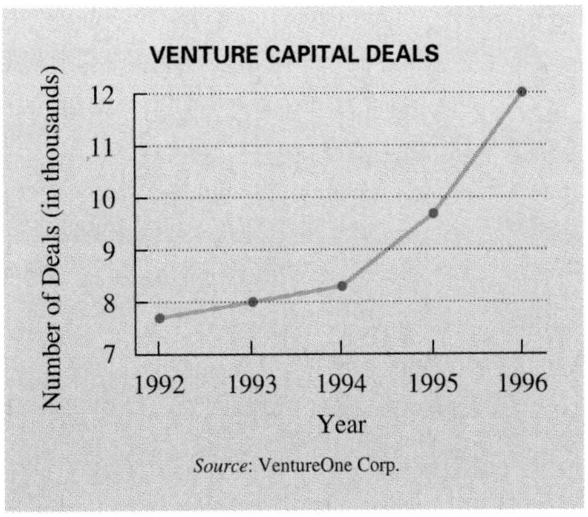

Figure 2

In 1993 the point lies on the line that corresponds to 8000, so 8000 venture capital deals were closed in 1993. The point for 1995 is about $\frac{2}{3}$, or .7, of the way between the lines for 9 and 10. Thus, we estimate 9700 deals were made in 1995.

We solved linear equations in one variable and explored their applications in Chapter 2. Now we want to extend those ideas to *linear equations in two variables.*

Linear Equation

A **linear equation in two variables** is an equation that can be put in the form

$$Ax + By = C,$$

where A, B, and C are real numbers and A and B are not both 0.

OBJECTIVE 2 Write a solution as an ordered pair. A solution of a linear equation in *two* variables requires *two* numbers, one for each variable. For example, the equation $y = 4x + 5$ is satisfied if x is replaced with 2 and y is replaced with 13, since

$$\mathbf{13} = 4(\mathbf{2}) + 5. \qquad \text{Let } x = 2;\ y = 13.$$

The pair of numbers $x = 2$ and $y = 13$ gives a solution of the equation $y = 4x + 5$. The phrase "$x = 2$ and $y = 13$" is abbreviated

$$\underbrace{(\overset{x\text{-value}}{2},\ \overset{y\text{-value}}{13})}_{\text{Ordered pair}}$$

with the x-value, 2, and the y-value, 13, given as a pair of numbers written inside parentheses. *The x-value is always given first.* A pair of numbers such as (2, 13) is called an

TEACHING TIP Ask students for other ordered pairs that satisfy the equation. Make the point that an infinite number of such pairs will satisfy a linear equation in two variables.

ordered pair. As the name indicates, the order in which the numbers are written is important. The ordered pairs (2, 13) and (13, 2) are not the same. The second pair indicates that $x = 13$ and $y = 2$. (Of course, letters other than x and y may be used in the equation with the numbers.)

OBJECTIVE 3 **Decide whether a given ordered pair is a solution of a given equation.** An ordered pair that is a solution of an equation is said to *satisfy* the equation.

CHALKBOARD EXAMPLE

Decide whether the ordered pair satisfies the given equation.

$(2, -5)$; $5x + 2y = 20$

Answer: no

EXAMPLE 3 Deciding Whether an Ordered Pair Satisfies an Equation

Decide whether the given ordered pair is a solution of the given equation.

(a) $(3, 2)$; $2x + 3y = 12$

To see whether (3, 2) is a solution of the equation $2x + 3y = 12$, we substitute 3 for x and 2 for y in the given equation.

$$2x + 3y = 12$$
$$2(3) + 3(2) = 12 \quad ? \qquad \text{Let } x = 3; \text{ let } y = 2.$$
$$6 + 6 = 12 \quad ?$$
$$12 = 12 \qquad \text{True}$$

This result is true, so (3, 2) satisfies $2x + 3y = 12$.

TEACHING TIP Students may tend to list ordered pairs backward if the equation is given, for example, as $x = 4y + 5$. Emphasize that the x-value of the ordered pair is always listed first.

(b) $(-2, -7)$; $m + 5n = 33$

$$(-2) + 5(-7) = 33 \quad ? \qquad \text{Let } m = -2; \text{ let } n = -7.$$
$$-2 + (-35) = 33 \quad ?$$
$$-37 = 33 \qquad \text{False}$$

This result is false, so $(-2, -7)$ is *not* a solution of $m + 5n = 33$.

OBJECTIVE 4 **Complete ordered pairs for a given equation.** Choosing a number for one variable in a linear equation makes it possible to find the value of the other variable, as shown in the next example.

CHALKBOARD EXAMPLE

Complete the given ordered pair for the equation $y = 2x - 9$.

$(\ \ , 7)$

Answer: (8, 7)

EXAMPLE 4 Completing an Ordered Pair

Complete the ordered pair $(7, \ \ )$ for the equation $y = 4x + 5$.

In this ordered pair, $x = 7$. (Remember that x always comes first.) To find the corresponding value of y, replace x with 7 in the equation $y = 4x + 5$.

$$y = 4(7) + 5 = 28 + 5 = 33$$

The ordered pair is (7, 33).

Ordered pairs often are displayed in a **table of values** as in the next example. The table may be written either vertically or horizontally.

CHALKBOARD EXAMPLE

Complete the table of values for the equation $2x - 3y = 12$. Then write the results as ordered pairs.

x	y
0	
	0
3	
	-3

Answer:

x	y
0	-4
6	0
3	-2
$\frac{3}{2}$	-3

$(0, -4)$, $(6, 0)$, $(3, -2)$, $\left(\frac{3}{2}, -3\right)$

EXAMPLE 5 Completing a Table of Values

Complete the given table of values for each equation. Then write the results as ordered pairs.

(a) $x - 2y = 8$

x	y
2	
10	
	0
	-2

To complete the first two ordered pairs, let $x = 2$ and $x = 10$, respectively.

If	$x = 2,$	If	$x = 10,$
then	$x - 2y = 8$	then	$x - 2y = 8$
becomes	$\mathbf{2} - 2y = 8$	becomes	$\mathbf{10} - 2y = 8$
	$-2y = 6$		$-2y = -2$
	$y = -3.$		$y = 1.$

Now complete the last two ordered pairs by letting $y = 0$ and $y = -2$, respectively.

If	$y = 0,$	If	$y = -2,$
then	$x - 2y = 8$	then	$x - 2y = 8$
becomes	$x - 2(\mathbf{0}) = 8$	becomes	$x - 2(\mathbf{-2}) = 8$
	$x - 0 = 8$		$x + 4 = 8$
	$x = 8.$		$x = 4.$

The completed table of values is as follows.

x	y
2	-3
10	1
8	0
4	-2

The corresponding ordered pairs are $(2, -3)$, $(10, 1)$, $(8, 0)$, and $(4, -2)$.

(b) $x = 5$

x	y
	-2
	6
	3

The given equation is $x = 5$. No matter which value of y might be chosen, the value of x is always the same, 5.

x	y
5	-2
5	6
5	3

The ordered pairs are $(5, -2)$, $(5, 6)$, and $(5, 3)$.

NOTE We can think of $x = 5$ in Example 5(b) as an equation in two variables by rewriting $x = 5$ as $x + 0y = 5$. This form of the equation shows that for any value of y, the value of x is 5. Similarly, $y = -2$ is the same as $0x + y = -2$.

Earlier in this book, we saw that linear equations in *one* variable had either one, zero, or an infinite number of real number solutions. Every linear equation in *two* variables has an infinite number of ordered pairs as solutions. Each choice of a number for one variable leads to a particular real number for the other variable.

To graph these solutions, represented as the ordered pairs (x, y), we need *two* number lines, one for each variable. These two number lines are drawn as shown in Figure 3. The horizontal number line is called the ***x*-axis.** The vertical line is called the ***y*-axis.** Together, the x-axis and y-axis form a **rectangular coordinate system.** It is also called the **Cartesian coordinate system,** in honor of René Descartes.

The coordinate system is divided into four regions, called **quadrants.** These quadrants are numbered counterclockwise, as shown in Figure 3. Points on the axes themselves are not in any quadrant. The point at which the x-axis and y-axis meet is called the **origin.** The origin, labeled 0 in Figure 3, is the point corresponding to (0, 0).

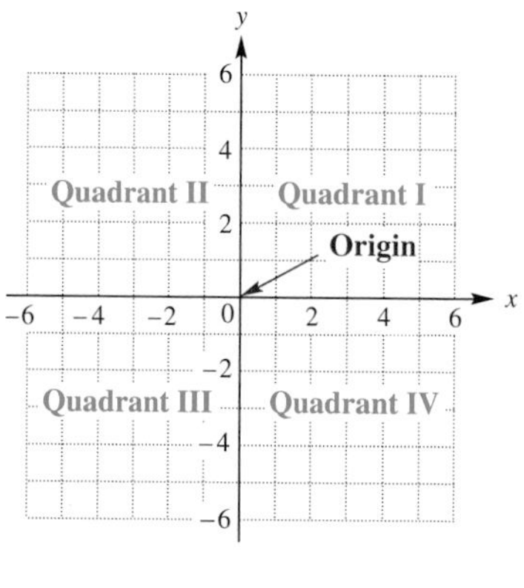

Figure 3

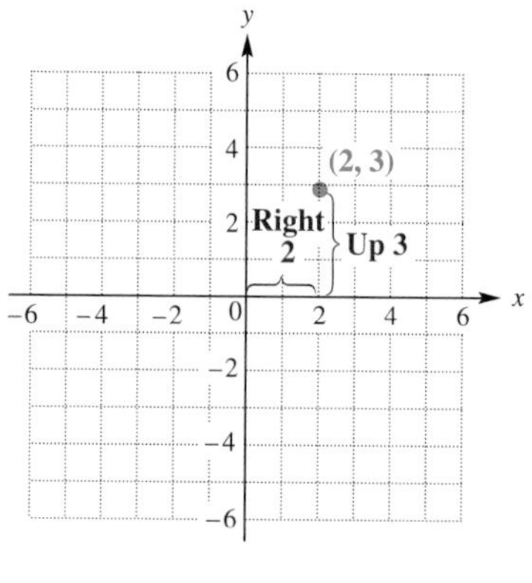

Figure 4

CONNECTIONS

The coordinate system that we use to plot points is credited to René Descartes (1596–1650), a French mathematician of the seventeenth century. It has been said that he developed the coordinate system while lying in bed, watching an insect move across the ceiling. He realized that he could locate the position of the insect at any given time by finding its distance from each of two perpendicular walls.

TEACHING TIP Relating the placing of a point in the coordinate plane to finding a town on a road map may help students visualize the process.

OBJECTIVE 5 **Plot ordered pairs.** By referring to the two axes, every point on the plane can be associated with an ordered pair. The numbers in the ordered pair are called the **coordinates** of the point. For example, locate the point associated with the ordered pair (2, 3) by starting at the origin. Since the x-coordinate is 2, go 2 units to the right along the x-axis. Then, since the y-coordinate is 3, turn and go up 3 units on a line parallel to the y-axis. This is called **plotting** the point (2, 3). (See Figure 4.) From now on we refer to the point with x-coordinate 2 and y-coordinate 3 as the point (2, 3).

NOTE When we graphed on a number line, one number corresponded to each point. On a plane, however, both numbers in the ordered pair are needed to locate a point. The ordered pair is a name for the point.

CHALKBOARD EXAMPLE

Plot the given ordered pairs on a coordinate system: (3, 5), (−2, 6), (−4, 0), (−5, −2), (5, −2), (0, −6).

Answer:

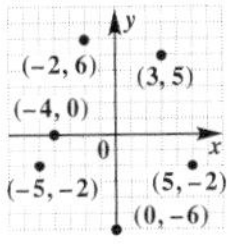

EXAMPLE 6 Plotting Ordered Pairs

Plot the given points on a coordinate system.

(a) (1, 5) **(b)** (−2, 3) **(c)** (−1, −4)

(d) (7, −2) **(e)** $\left(\frac{3}{2}, 2\right)$ **(f)** (5, 0)

Locate the point (−1, −4), for example, by first going 1 unit to the left along the x-axis. Then turn and go 4 units down, parallel to the y-axis. Plot the point $\left(\frac{3}{2}, 2\right)$, by going $\frac{3}{2}$ (or $1\frac{1}{2}$) units to the right along the x-axis. Then turn and go 2 units up, parallel to the y-axis. Figure 5 shows the graphs of the points in this example.

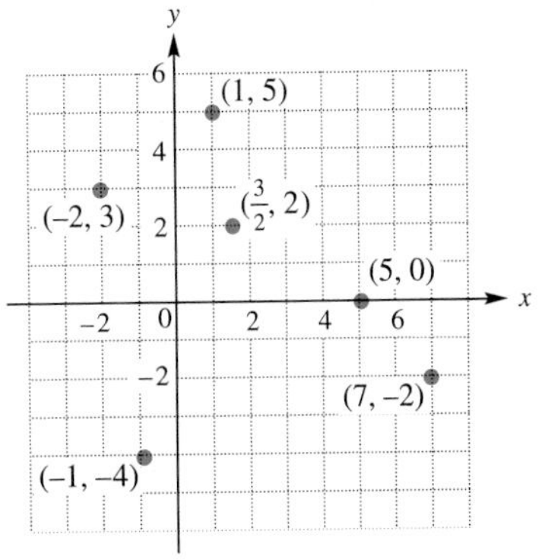

Figure 5

CHALKBOARD EXAMPLE

Complete the table of ordered pairs for the equation in Example 7.

x	1992	1993	1995
y			

Answer: 695, 721, 772

TEACHING TIP Remind students that there is no difference between horizontal and vertical tables of ordered pairs.

EXAMPLE 7 Completing Ordered Pairs to Estimate the Number of Business Incorporations

New business incorporations have increased steadily from 1990–1996. Their number can be closely approximated by the equation

$$y = 25.59x - 50{,}280,$$

where x is the year, and y is the number of incorporations in thousands.* This approximation is only valid for the years under examination, 1990–1996, since it was based on data for those years.

(a) Complete the table of ordered pairs for this linear equation. Round the y-values to the nearest thousand.

x	1990	1994	1996
y			

To find y when $x = 1990$, substitute into the equation.

$$y = 25.59(\mathbf{1990}) - 50{,}280 \qquad \text{Let } x = 1990.$$
$$= 644 \text{ (rounded)}$$

*Based on data from Dun & Bradstreet.

This means that, in 1990, there were 644,000 new incorporations. Find the y-values for 1994 and 1996 similarly. The completed table follows.

x	1990	1994	1996
y	644	746	798

(b) Interpret the ordered pair with $x = 1996$; with $x = 1994.5$.

The ordered pair (1996, 798) means that in 1996, 798,000 businesses were incorporated. We can interpret 1994.5 to mean halfway through 1994.

$$\begin{aligned} y &= 25.59(1994.5) - 50{,}280 && \text{Let } x = 1994.5. \\ &= 759 \end{aligned}$$

There were approximately 759,000 incorporations.

(c) Graph the ordered pairs found in part (a).

The ordered pairs are graphed in Figure 6. Notice how the axes are labeled. In this application, x represents the year, and y represents the number of corporations in thousands. Different scales are used on the two axes because the two sets of numbers differ so much in size. Here, each square represents one unit in the horizontal direction and 50 units in the vertical direction. Because the numbers in the first ordered pair are quite large, we show a break in both axes near the origin.

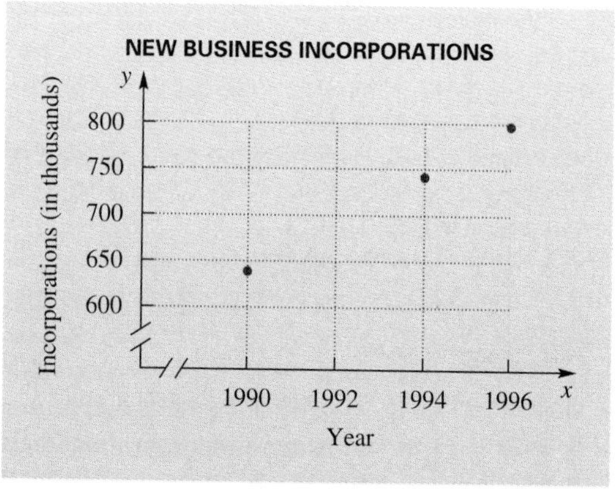

Figure 6

We can also think of ordered pairs as representing an input value x and an output value y. If we input x into the equation, the output is y. For instance, in Example 7, if we input the year into the equation

$$y = 25.59x - 50{,}280,$$

we get the number of incorporations (in thousands) in that year as the output. We encounter many examples of this type of relationship every day.

- The cost to fill the tank with gasoline depends on how many gallons are needed; the number of gallons is the input, and the cost is the output.
- The distance traveled depends on the traveling time; input a time, and the output is a distance.
- The growth of a plant depends on the amount of sun it gets; the input is the amount of sun, and the output is the growth.

This idea is illustrated in Figure 7 with an input-output "machine."

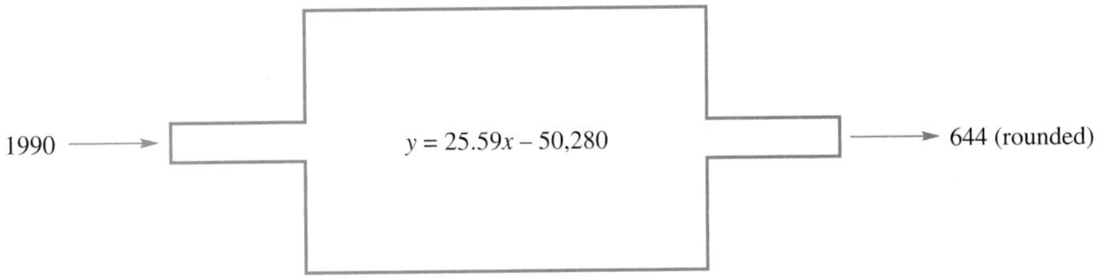

An input-output machine

Figure 7

3.1 EXERCISES

1. does; do not **2.** −4 **3.** II **4.** y **5.** 3 **6.** 6 **7.** between 1990 and 1991; $21,000 and between 1992 and 1993; $122,000 **8.** $52,000 **9.** 1993 **10.** 1992

Fill in each blank with the correct response.

1. The symbol (x, y) ____________ (does/does not) represent an ordered pair, while the symbols $[x, y]$ and $\{x, y\}$ ____________ (do/do not) represent ordered pairs.

2. The ordered pair (3, 2) is a solution of the equation $2x - 5y =$ _____.

3. The point whose graph has coordinates $(-4, 2)$ is in quadrant _____.

4. The point whose graph has coordinates (0, 5) lies along the _____-axis.

5. The ordered pair (4, _____) is a solution of the equation $y = 3$.

6. The ordered pair (_____, −2) is a solution of the equation $x = 6$.

Use the bar graph to respond to each statement or question. See Example 1.

7. Between which pairs of consecutive years did the winnings increase? How much was the increase in each case?

8. How much did winnings decline between 1991 and 1992?

9. Which year had the greatest winnings?

10. Which year had the least winnings?

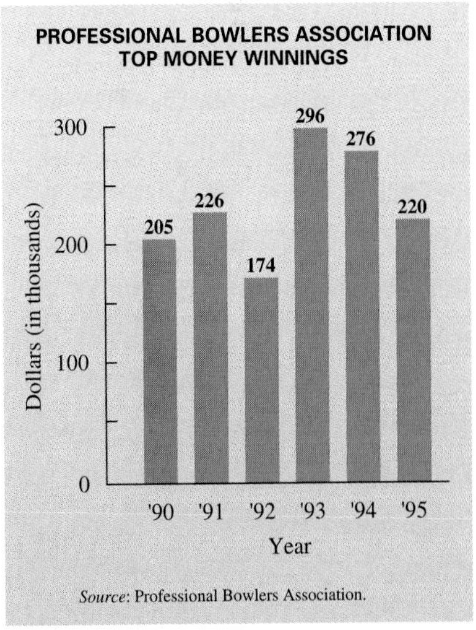

11. (d) **12.** 1993 **13.** inkjet **14.** The number of dot matrix printers shipped has decreased since 1992, while the number of the other two types increased. **15.** 1994–1995; 1993–1994 **16.** 1993: 17,200; 1994: 22,000; 1995: 21,000; 1996: 24,300 **17.** 4800; 3400; yes **18.** A linear equation in one variable can be written in the form $Ax + B = 0$, where $A \neq 0$. Examples are $2x + 5 = 0$, $3x + 6 = 2$, and $x = -5$. A linear equation in two variables can be written in the form $Ax + By = C$, where A and B cannot both equal 0. Examples are $2x + 3y = 8$, $3x = 5y$, and $x - y = 0$. **19.** yes **20.** yes **21.** yes **22.** yes **23.** no **24.** no

Use the line graph to respond to each statement or question. See Example 2.

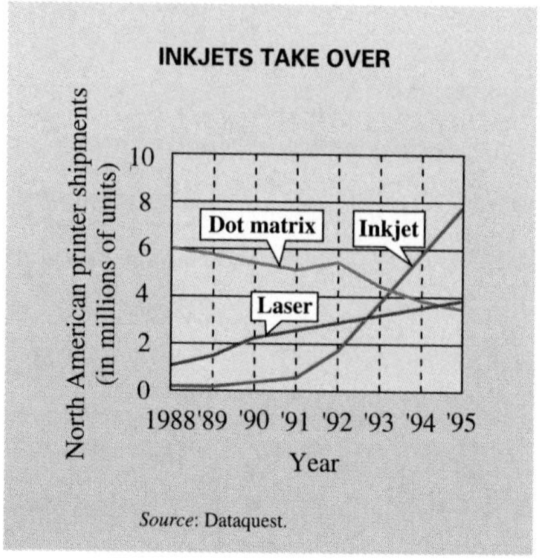

11. Which one of the following would be the best estimate for the total number of units of printer shipments in North America in 1989?

(a) 5.8 million **(b)** 1.5 million
(c) .2 million **(d)** 7.5 million

12. In what year did the number of inkjet printer shipments first exceed the number of laser printer shipments?

13. What type of printer has shown the most rapid increase in shipments since 1992?

14. Describe how the trend in shipments of dot matrix printers has differed from the shipments of inkjet and laser printers since 1992.

This line graph (first given in the chapter introduction) shows the number of new consumer packaged-goods products for the years 1991–1996. Use this graph for Exercises 15–17. See Example 2.

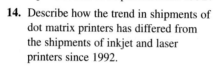

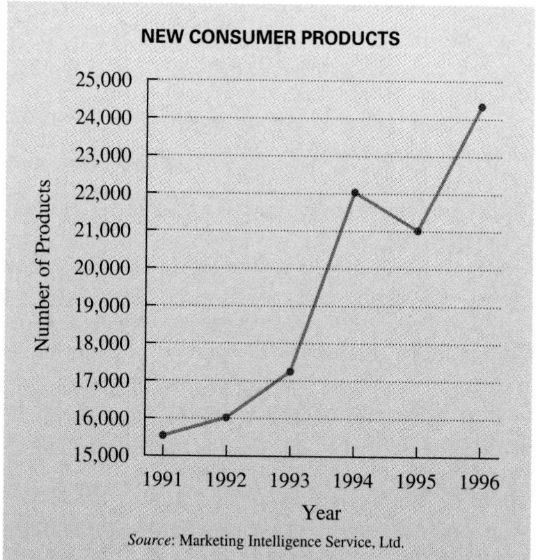

15. Between which two years did the number of new products decrease? Between which two years did the greatest increase occur?

16. Estimate the number of new products in 1993, 1994, 1995, and 1996.

17. Estimate the increase in new products from 1993 to 1994 and from 1995 to 1996. Does your answer confirm your answer to Exercise 15?

18. Define a linear equation in one variable and a linear equation in two variables, and give examples of each.

Decide whether the given ordered pair is a solution of the given equation. See Example 3.

19. $x + y = 9$; $(0, 9)$ **20.** $x + y = 8$; $(0, 8)$ **21.** $2p - q = 6$; $(4, 2)$

22. $2v + w = 5$; $(3, -1)$ **23.** $4x - 3y = 6$; $(2, 1)$ **24.** $5x - 3y = 15$; $(5, 2)$

25. yes 26. yes
27. yes 28. yes
29. no 30. no
31. Because there is an infinite number of real numbers that can replace either variable in a linear equation in two variables, there is an infinite number of solutions. A linear equation in one variable may have 0, 1, or an infinite number of solutions. 32. A: (2, 5); B: (−2, 2); C: (−5, 5); D: (−6, −2); E: (0, −1) 33. No, for two ordered pairs to be equal, the x-values must be equal, and the y-values must be equal. Here, $4 \neq -1$ and $-1 \neq 4$. 34. No, the ordered pair (3, 4) represents the point 3 units to the right of the origin and 4 units up from the x-axis. The ordered pair (4, 3) represents the point 4 units to the right of the origin and 3 units up from the x-axis. 35. 11
36. 7 37. $-\frac{7}{2}$ 38. −5
39. −4 40. −1 41. −5
42. −7 43. Substituting $\frac{1}{3}$ for x in $y = 6x + 2$ gives $y = 6\left(\frac{1}{3}\right) + 2 = 2 + 2 = 4$. Because $6\left(\frac{1}{7}\right) = \frac{6}{7}$, calculating y requires working with fractions.
44. b 45. 4; 6; −6
46. 8; 6; 3 47. 3; −5; −15
48. −9; 4; 9 49. −9; −9; −9
50. 12; 12; 12 51. −6; −6; −6
52. −10; −10; −10
53.–60.

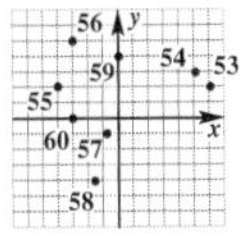

25. $y = 3x$; (2, 6) **26.** $x = -4y$; (−8, 2) **27.** $x = -6$; (−6, 5)

28. $y = 2$; (4, 2) **29.** $x + 4 = 0$; (−6, 2) **30.** $x - 6 = 0$; (4, 2)

31. Explain why there are an infinite number of solutions for a linear equation in two variables. How does this differ from the number of solutions for a linear equation in one variable?

32. Give the ordered pairs that correspond to the points labeled in the figure. (All coordinates are integers.)

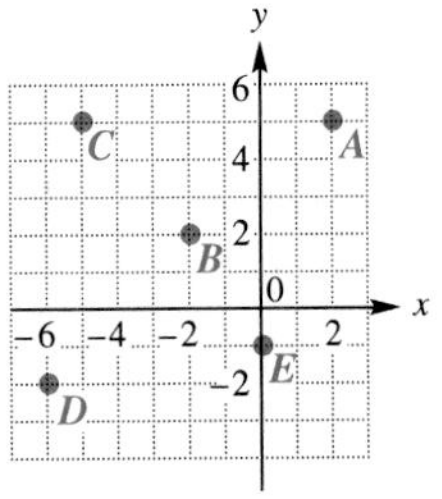

33. Do (4, −1) and (−1, 4) represent the same ordered pair? Explain why.

34. Do the ordered pairs (3, 4) and (4, 3) correspond to the same point on the plane? Explain why.

Complete each ordered pair for the equation $y = 2x + 7$. See Example 4.

35. (2,) **36.** (0,) **37.** (, 0) **38.** (, −3)

Complete the ordered pair for the equation $y = -4x - 4$. See Example 4.

39. (0,) **40.** (, 0) **41.** (, 16) **42.** (, 24)

43. Explain why it would be easier to find the corresponding y-value for $x = \frac{1}{3}$ than for $x = \frac{1}{7}$ in the equation $y = 6x + 2$.

44. For the equation $y = mx + b$, what is the y-value corresponding to $x = 0$ for *any* value of m?

Complete each table of values. See Example 5.

45. $2x + 3y = 12$

x	y
0	
	0
	8

46. $4x + 3y = 24$

x	y
0	
	0
	4

47. $3x - 5y = -15$

x	y
0	
	0
	−6

48. $4x - 9y = -36$

x	y
	0
0	
	8

49. $x = -9$

x	y
	6
	2
	−3

50. $x = 12$

x	y
	3
	8
	0

51. $y = -6$

x	y
8	
4	
−2	

52. $y = -10$

x	y
4	
0	
−4	

Plot each ordered pair in a rectangular coordinate system. See Example 6.

53. (6, 2) **54.** (5, 3) **55.** (−4, 2) **56.** (−3, 5)

57. $\left(-\frac{4}{5}, -1\right)$ **58.** $\left(-\frac{3}{2}, -4\right)$ **59.** (0, 4) **60.** (−3, 0)

Fill in each blank with the word positive *or the word* negative.

The point with coordinates (x, y) is in

61. quadrant III if x is ____________ and y is ____________.

62. quadrant II if x is ____________ and y is ____________.

63. quadrant IV if x is ____________ and y is ____________.

64. quadrant I if x is ____________ and y is ____________.

Complete each table of values and then plot the ordered pairs. See Examples 5 and 6.

65. $x - 2y = 6$

x	y
0	
	0
2	
	−1

66. $2x - y = 4$

x	y
0	
	0
1	
	−6

67. $3x - 4y = 12$

x	y
0	
	0
−4	
	−4

68. $2x - 5y = 10$

x	y
0	
	0
−5	
	−3

69. $y + 4 = 0$

x	y
0	
5	
−2	
−3	

70. $x - 5 = 0$

x	y
	1
	0
	6
	−4

Solve each problem. See Example 7.

71. Suppose that it costs \$5000 to start up a business selling snow cones. Furthermore, it costs \$.50 per cone in labor, ice, syrup, and overhead. Then the cost to make x snow cones is given by y dollars, where $y = .50x + 5000$. Express as an ordered pair each of the following.

(a) When 100 snow cones are made, the cost is \$5050. (*Hint:* What does x represent? What does y represent?)

(b) When the cost is \$6000, the number of snow cones made is 2000.

72. It costs a flat fee of \$20 plus \$5 per day to rent a pressure washer. Therefore, the cost to rent the pressure washer for x days is given by $y = 5x + 20$, where y is in dollars. Express as an ordered pair each of the following.

(a) When the washer is rented for 5 days, the cost is \$45.

(b) I paid \$50 when I returned the washer, so I must have rented it for 6 days.

73. In statistics, ordered pairs are used to decide whether two quantities are related in such a way that one can be predicted from the other. These ordered pairs are plotted on a graph, called a *scatter diagram.*

Some major league baseball fans are concerned by the increase in time to complete a game. Make a scatter diagram by plotting the following ordered pairs of years since 1980 and minutes beyond two hours to complete a game: (5, 40), (7, 48), (9, 46), (11, 49), (13, 48), (15, 57). As shown in the figure, the horizontal axis is used

LENGTHS OF BASEBALL GAMES

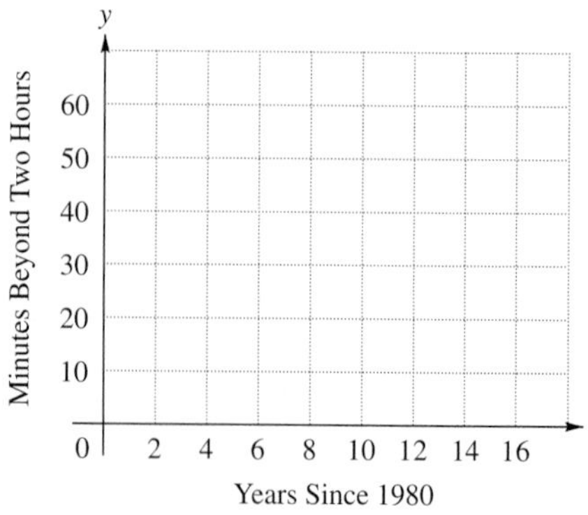

61. negative; negative
62. negative; positive
63. positive; negative
64. positive; positive
65. −3; 6; −2; 4

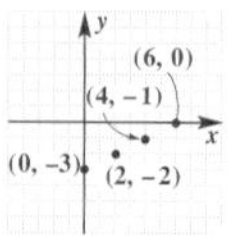

66. −4; 2; −2; −1

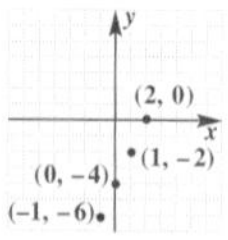

67. $-3; 4; -6; -\frac{4}{3}$

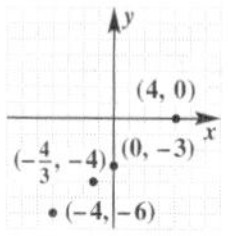

68. $-2; 5; -4; -\frac{5}{2}$

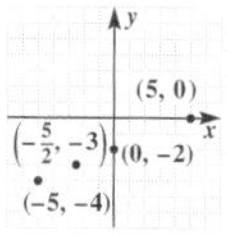

69. −4; −4; −4; −4

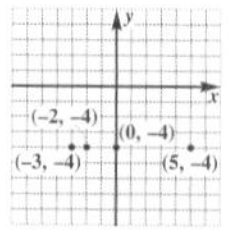

70. 5; 5; 5; 5

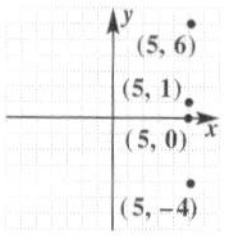

71. (a) (100, 5050) (b) (2000, 6000) **72.** (a) (5, 45) (b) (6, 50)

73. (a) Yes, there is an approximate linear relationship between the years since 1980 and the minutes beyond two hours to complete a game.

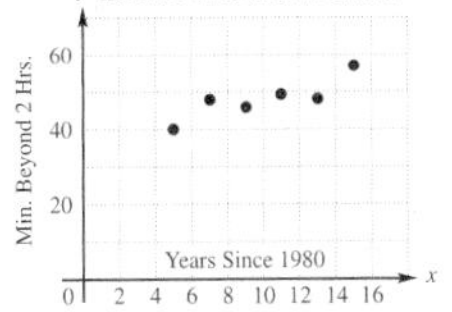

(b) Yes, a prediction could be made by determining the linear equation that describes the linear relationship. However, only input values for the years included in the given data, 1980–1996, should be used.
(c) input: years since 1980; output: minutes beyond two hours
74. (a) (20, 170), (40, 153), (60, 136), (80, 120) (b) input: ages between 20 and 80; output: upper limit of the target heart rate zone
75. (a) (30, 133), (40, 126), (60, 112), (70, 105) (b) input: age; output: lower limit of the target heart rate zone 76. between 140 and 170; between 126 and 153
77. No. To go beyond the given data at either end assumes that the graph continues in the same way. This may not be true.

to represent the years since 1980, and the vertical axis represents the minutes beyond two hours.

(a) Graph these points on a similar grid. Do the points lie in an approximately linear pattern?

(b) Could the number of years since 1980 be used to predict the length of a game?

(c) What is the input? What is the output?

(*Source: The New York Times,* May 30, 1995, p. B9.)

74. The maximum benefit for the heart from exercising occurs if the heart rate is in the target heart rate zone. The line graph shows the upper limit of the zone (in beats per minute) for various ages.

(a) Use the graph to estimate the ordered pairs in the form (x, y) that describe the upper limit of the target heart rate zone for ages 20, 40, 60, and 80.

(b) What is the input here? What is the output?

(*Source:* Robert V. Hockey, *Physical Fitness: The Pathway to Healthy Living,* Times Mirror/Mosby College Publishing, 1989, pp. 85–87.)

TARGET HEART RATE ZONE

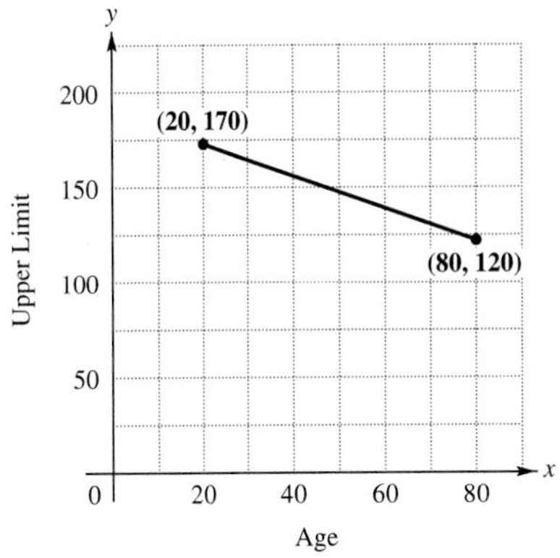

75. The lower limit of the target heart rate zone (see Exercise 74) is given in the accompanying graph of the equation $y = .7(220 - x)$, where the horizontal axis represents age and the vertical axis represents heart rate.

(a) Write ordered pairs (x, y) for ages 30, 40, 60, and 70 by estimating the y-values from the graph.

(b) What is the input here? What is the output?

TARGET HEART RATE ZONE

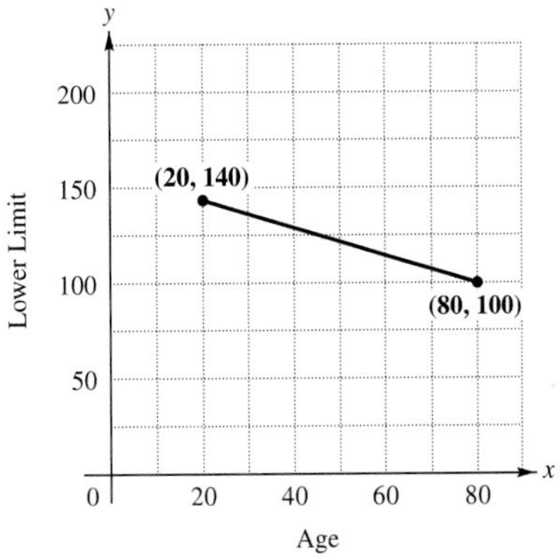

76. What is the heart rate zone for age 20? For age 40? (Refer to Exercises 74 and 75.)

77. Should the graph in Exercise 75 be used to estimate the lower limit of the target heart rate zone for ages below 20 or above 80? Why or why not?

78.
{(General Electric, 819), (IBM, 1867), (Hitachi, 963), (Canon, 1541), (Motorola, 1064)};
the companies are the inputs, the number of patents are the outputs.

78. The five corporations receiving the most patents for inventions in 1996 are listed below. Write a set of ordered pairs (company, number of patents) for these data. Give the inputs and outputs.

Corporation	*Number of Patents*
General Electric	819
IBM	1867
Hitachi	963
Canon	1541
Motorola	1064

Source: U.S. Patent and Trademark Office.

3.2 Graphing Linear Equations in Two Variables

OBJECTIVES

1. Graph linear equations.
2. Find intercepts.
3. Graph linear equations of the form $Ax + By = 0$.
4. Graph linear equations of the form $y = k$ or $x = k$.
5. Define a function.

FOR EXTRA HELP

SSG Sec. 3.2
SSM Sec. 3.2
Pass the Test Software
InterAct Math Tutorial Software
Video 5

In this section we will use a few ordered pairs that satisfy a linear equation to graph the equation.

OBJECTIVE 1 **Graph linear equations.** We know that infinitely many ordered pairs satisfy a linear equation. Some ordered pairs that are solutions of $x + 2y = 7$ are graphed in Figure 8.

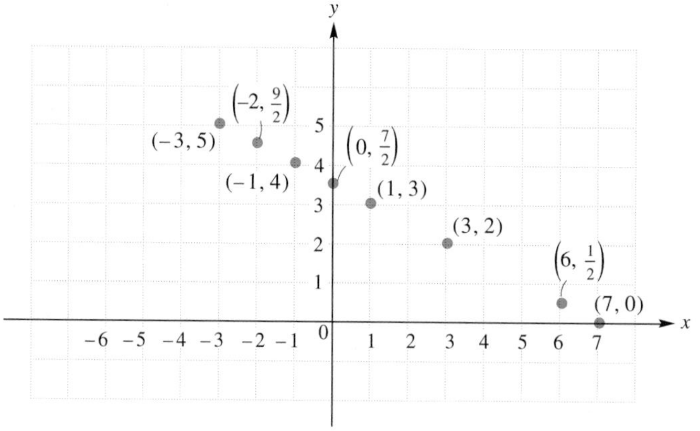

Figure 8

Notice that the points plotted in this figure all appear on a straight line, as shown in Figure 9. In fact, all ordered pairs satisfying the equation $x + 2y = 7$ correspond to points that lie on this same straight line, and the coordinates of any point on the line give a solution of the equation. This line is a "picture" of all the solutions of the equation $x + 2y = 7$. Only a portion of the line is shown here, but it extends indefinitely in both directions, as suggested by the arrowhead on each end of the line. The line is called the **graph** of the equation and the process of plotting the ordered pairs and drawing the line through the corresponding points is called **graphing.**

JOURNAL   WRITING   ▲ CHALLENGING   SCIENTIFIC CALCULATOR   GRAPHING CALCULATOR

INSTRUCTOR'S RESOURCES

PTB Sec. 3.2 **ISM** Sec. 3.2 **AB** Sec. 3.2 | **TEST GENERATOR** | **WORLD WIDE WEB** www.LialAlgebra.com

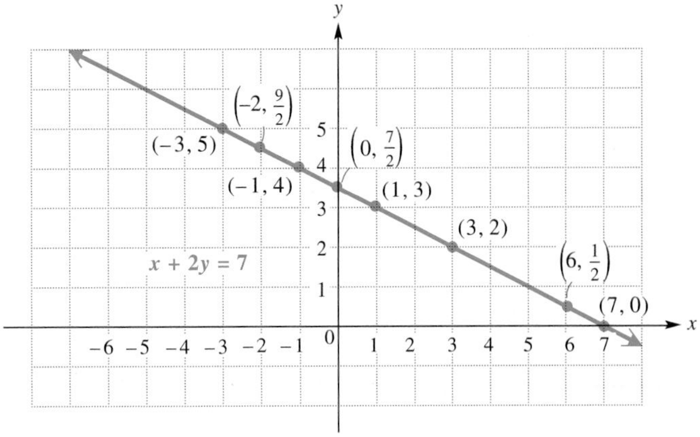

Figure 9

The preceding discussion can be generalized.

Graph of a Linear Equation

The graph of any linear equation in two variables is a straight line.

Notice that the word *line* appears in the name "*linear* equation."

Since two distinct points determine a line, we can graph a straight line by finding any two different points on the line. However, it is a good idea to plot a third point as a check.

CHALKBOARD EXAMPLE

Graph the linear equation $5x + 2y = -10$.

Answer:

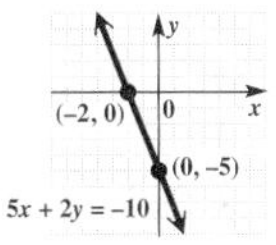

TEACHING TIP Emphasize that the table of values is just an alternate way of listing the ordered pairs (0, 3), (2, 0), (−2, 6).

EXAMPLE 1 Graphing a Linear Equation

Graph the linear equation $2y = -3x + 6$.

Although this equation is not in the form $Ax + By = C$, it *could* be put in that form, and so is a linear equation.

For most linear equations, two different points on the graph can be found by first letting $x = 0$, and then letting $y = 0$. Doing this gives the ordered pairs (0, 3) and (2, 0). We get a third ordered pair (as a check) by letting x or y equal some other number. For example, if $x = -2$, we find that $y = 6$, giving the ordered pair $(-2, 6)$. These three ordered pairs are shown in the table of values with Figure 10. Plot the corresponding points, then draw a line through them. This line, shown in Figure 10, is the graph of $2y = -3x + 6$.

x	y
0	3
2	0
−2	6

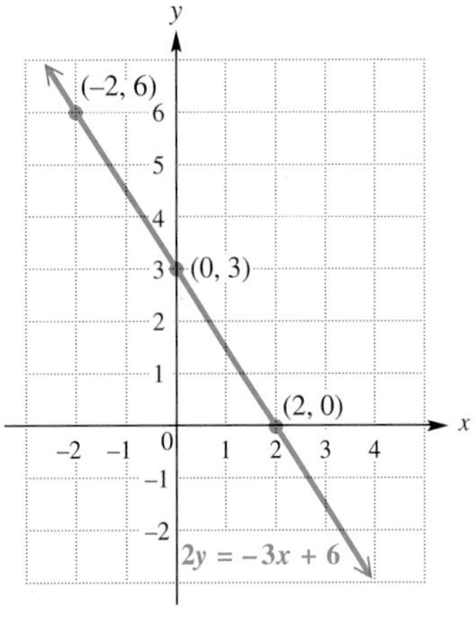

Figure 10

OBJECTIVE 2 **Find intercepts.** In Figure 10 the graph intersects (crosses) the y-axis at (0, 3) and the x-axis at (2, 0). For this reason (0, 3) is called the ***y*-intercept** and (2, 0) is called the ***x*-intercept** of the graph. The intercepts are particularly useful for graphing linear equations, as in Example 1.

Finding Intercepts

We find the x-intercept by letting $y = 0$ in the given equation and solving for x.

We find the y-intercept by letting $x = 0$ in the given equation and solving for y.

CHALKBOARD EXAMPLE

Find the intercepts for $5x + 2y = 10$. Draw the graph.

Answer: x-intercept (2, 0); y-intercept (0, 5)

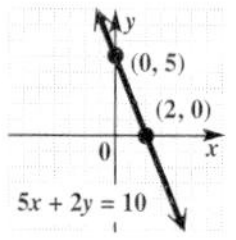

EXAMPLE 2 Finding Intercepts

Find the intercepts for the graph of $2x + y = 4$. Draw the graph.

Find the y-intercept by letting $x = 0$; find the x-intercept by letting $y = 0$.

$$\begin{aligned} 2x + y &= 4 \\ 2(0) + y &= 4 \\ 0 + y &= 4 \\ y &= 4 \end{aligned} \qquad \begin{aligned} 2x + y &= 4 \\ 2x + 0 &= 4 \\ 2x &= 4 \\ x &= 2 \end{aligned}$$

The y-intercept is (0, 4). The x-intercept is (2, 0). The graph with the two intercepts shown in color is given in Figure 11. We get a third point as a check. For example, choosing $x = 1$ gives $y = 2$. These three ordered pairs are shown in the table with Figure 11. Plot (0, 4), (2, 0), and (1, 2) and draw a line through them. This line, shown in Figure 11, is the graph of $2x + y = 4$.

x	y
0	4
2	0
1	2

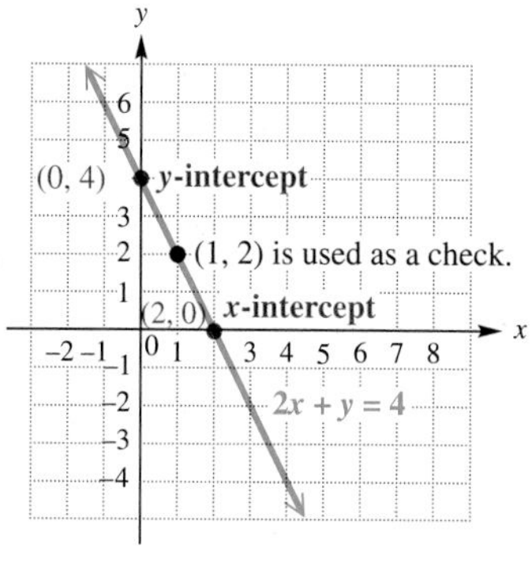

Figure 11

When choosing x- or y-values to find ordered pairs to plot, be careful to choose so that the resulting points are not too close together. For example, using $(-1, -1)$, (0, 0), and (1, 1) may result in an inaccurate line. It is better to choose points where the x-values differ by at least 2.

OBJECTIVE 3 **Graph linear equations of the form $Ax + By = 0$.** In earlier examples, the x- and y-intercepts were used to help draw the graphs. This is not always possible, as the following examples show. Example 3 shows what to do when the x- and y-intercepts are the same point.

CHALKBOARD EXAMPLE

Graph the linear equation
$4x - 2y = 0$.

Answer:

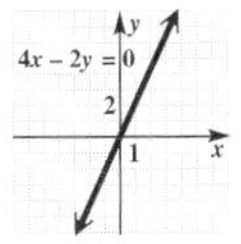

EXAMPLE 3 Graphing an Equation of the Form $Ax + By = 0$

Graph the linear equation $x - 3y = 0$.

If we let $x = 0$, then $y = 0$, giving the ordered pair (0, 0). Letting $y = 0$ also gives (0, 0). This is the same ordered pair, so choose two other values for x or y. Choosing 2 for y gives $x - 3 \cdot 2 = 0$, giving the ordered pair (6, 2). For a check point, we choose -6 for x getting -2 for y. This ordered pair, $(-6, -2)$, along with (0, 0) and (6, 2), was used to get the graph shown in Figure 12.

x	y
0	0
6	2
-6	-2

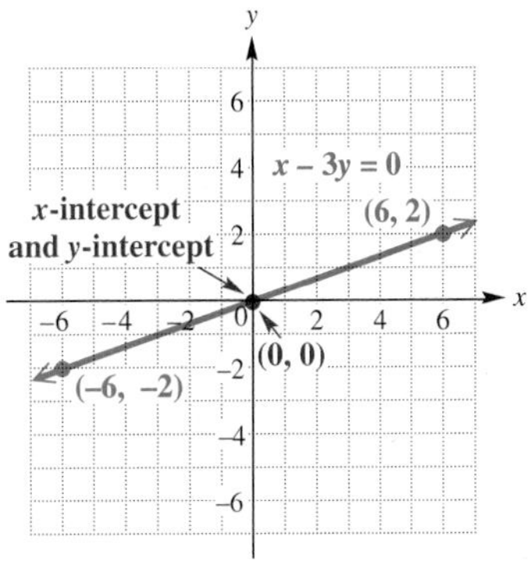

Figure 12

Example 3 can be generalized as follows.

Line through the Origin

If A and B are real numbers, the graph of a linear equation of the form

$$Ax + By = 0$$

goes through the origin (0, 0).

OBJECTIVE 4 Graph linear equations of the form $y = k$ or $x = k$. The equation $y = -4$ is a linear equation in which the coefficient of x is 0. (Write $y = -4$ as $0x + y = -4$ to see this.) Also, $x = 3$ is a linear equation in which the coefficient of y is 0. These equations lead to horizontal or vertical straight lines, as the next examples show.

CHALKBOARD EXAMPLE

Graph the linear equation
$y + 5 = 0$.

Answer:

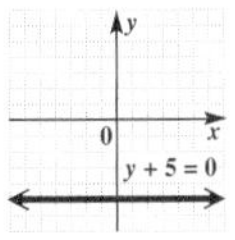

EXAMPLE 4 Graphing an Equation of the Form $y = k$

Graph the linear equation $y = -4$.

As the equation states, for any value of x, y is always equal to -4. To get ordered pairs that are solutions of this equation, we choose any numbers for x, always using -4 for y. Three ordered pairs that satisfy the equation are shown in the table of values with Figure 13. Drawing a line through these points gives the horizontal line shown in Figure 13.

x	y
−2	−4
0	−4
3	−4

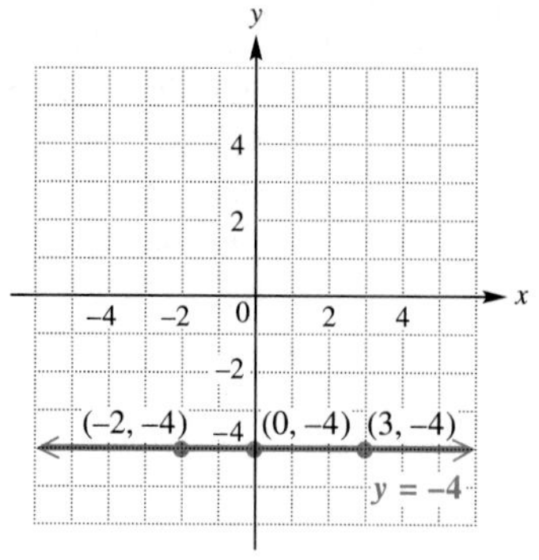

Figure 13

Horizontal Line

The graph of the linear equation $y = k$, where k is a real number, is the horizontal line going through the point $(0, k)$.

CHALKBOARD EXAMPLE

Graph the linear equation $x = 2$.

Answer:

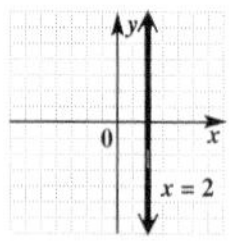

EXAMPLE 5 Graphing an Equation of the Form $x = k$

Graph the linear equation $x - 3 = 0$.

First add 3 to each side of the equation $x - 3 = 0$ to get $x = 3$. All the ordered pairs that are solutions of this equation have an x-value of 3. Any number can be used for y. We show three ordered pairs that satisfy the equation in the table of values with Figure 14. Drawing a line through these points gives the vertical line shown in Figure 14.

x	y
3	3
3	0
3	−2

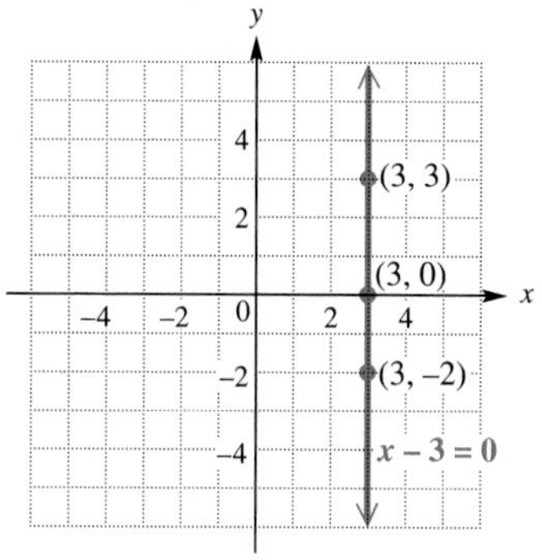

Figure 14

TEACHING TIP The equations of horizontal and vertical lines are often confused with each other. Mention $y = k$ is parallel to the x-axis and $x = k$ is parallel to the y-axis.

Vertical Line

The graph of the linear equation $x = k$, where k is a real number, is the vertical line going through the point $(k, 0)$.

In particular, notice that the horizontal line $y = 0$ is the x-axis and the vertical line $x = 0$ is the y-axis.

CONNECTIONS

Beginning in this chapter we include information on the basic features of graphing calculators. The most obvious feature is their ability to graph equations. We must solve the equation for y in order to input it into the calculator. Also, we must select an appropriate "window" for the graph. The window is determined by the minimum and maximum values of x and y. Graphing calculators have a standard window, often from $x = -10$ to $x = 10$ and from $y = -10$ to $y = 10$. We indicate this as $[-10, 10]$, $[-10, 10]$, with the x-interval shown first.

For example, to graph the equation $2x + y = 4$, discussed in Example 2, we first solve for y.

$$2x + y = 4$$
$$y = -2x + 4 \qquad \text{Subtract } 2x.$$

If we input this equation as $y = -2x + 4$ and choose the standard window, the calculator shows the following graph.

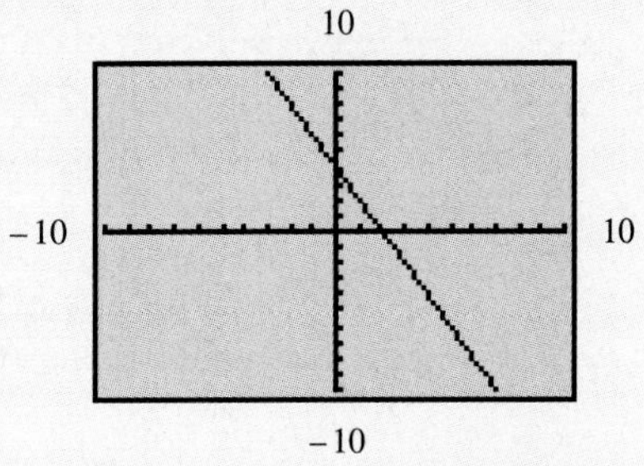

The x-value of the point on the graph where $y = 0$ (the x-intercept) gives the solution of the equation

$$y = 0,$$
or
$$-2x + 4 = 0. \qquad \text{Substitute } -2x + 4 \text{ for } y.$$

Since each tick mark on the x-axis represents 1, the graph shows that $(2, 0)$ is the x-intercept.

Working backwards, we use this idea to graphically find the solution of an equation in *one* variable as follows.

1. Rewrite the equation with 0 on one side.
2. Replace 0 with y and graph the equation in two variables.
3. Use the graph to find the x-value of the point where $y = 0$.

For example, we use this method to solve $6(3x + 5) = 2(10x + 10)$ as follows. First, rewrite the equation with 0 on one side.

$$6(3x + 5) = 2(10x + 10)$$
$$6(3x + 5) - 2(10x + 10) = 0 \qquad \text{Subtract } 2(10x + 10) \text{ on both sides.}$$

Second, replace 0 with y, and graph the equation in two variables.

$$6(3x + 5) - 2(10x + 10) = y$$

CONNECTIONS (CONTINUED)

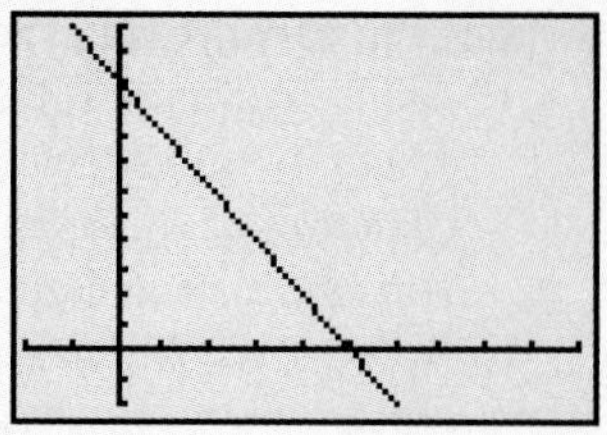

Third, use the graph to find the x-value of the point on the line where $y = 0$, the x-intercept. As shown in the figure, this occurs at $x = 5$. Verify this solution by substitution in the original equation.

FOR DISCUSSION OR WRITING

How would you rewrite these equations, with one side equal to 0, to input them into a graphing calculator for solution? (It is not necessary to clear parentheses or combine terms.)

1. $3x + 4 - 2x - 7 = 4x + 3$
2. $5x - 15 = 3(x - 2)$

1. $3x + 4 - 2x - 7 - 4x - 3 = 0$
2. $5x - 15 - 3(x - 2) = 0$

The different forms of straight-line equations and the methods of graphing them are given in the following summary.

Graphing Straight Lines

Equation	To Graph	Example
$y = k$	Draw a horizontal line, through $(0, k)$.	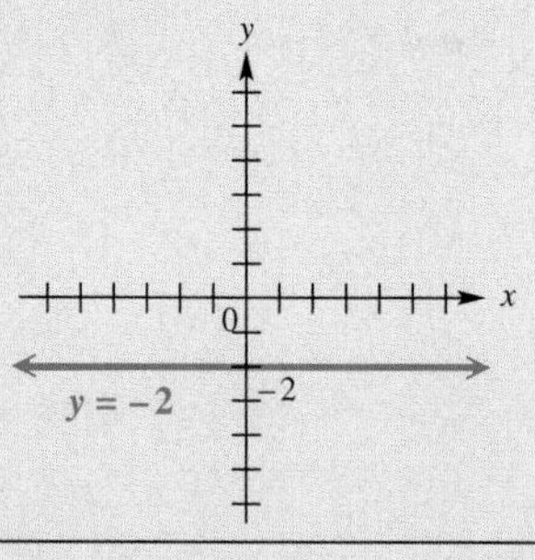
$x = k$	Draw a vertical line, through $(k, 0)$.	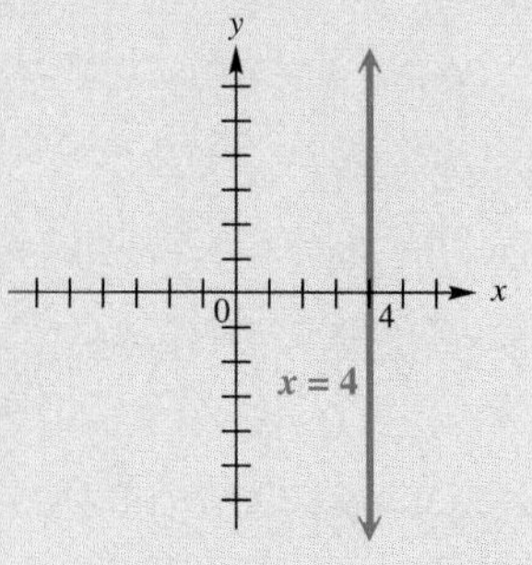

Graphing Straight Lines (continued)

Equation	*To Graph*	*Example*
$Ax + By = 0$	Graph goes through (0, 0). Get additional points that lie on the graph by choosing any value of x or y, except 0.	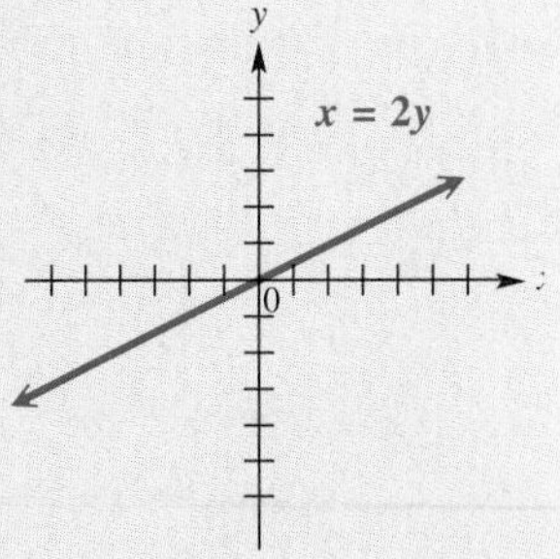
$Ax + By = C$ but not of the types above	Find any two points the line goes through. A good choice is to find the intercepts: let $x = 0$, and find the corresponding value of y; then let $y = 0$, and find x. As a check, get a third point by choosing a value of x or y that has not yet been used.	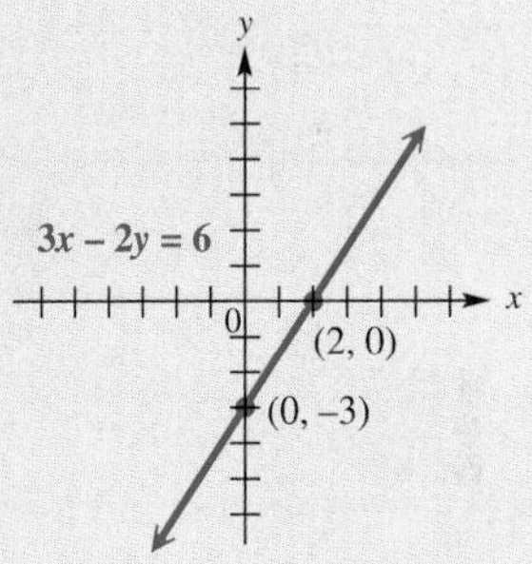

OBJECTIVE 5 Define a function. In the previous section, we saw several examples of relationships between two variables where each input value x produced an output value y. If each input x produces just *one* output y, we call the relationship a **function.** When a set of input values and a set of output values are related by a linear equation, $Ax + By = C$, each x-value corresponds to just one y-value, so linear equations define functions, unless $B = 0$. If $B = 0$, the equation becomes $Ax = C$ or $x = \frac{C}{A}$, where $\frac{C}{A}$ represents a real number. In this case, as we saw in Example 5, one input value x corresponds to an infinite number of output values y and the equation does not define a function.

CHALKBOARD EXAMPLE

Decide whether each set of ordered pairs defines a function.

(a) The ordered pairs that satisfy $2x - y = 10$

(b) The relationship between the price of an item and the total cost to purchase 5 of the items

(c) $\{(2, 0), (2, 1), (2, 2), (2, 3)\}$

Answer: **(a)** yes **(b)** yes **(c)** no

EXAMPLE 6 Deciding Whether a Set of Ordered Pairs Defines a Function

Which of the following sets defines a function?

(a) $\{(1, -2), (2, 3), (0, 4), (4, 7)\}$

The input 1 is paired with the output -2, the input 2 is paired with the output 3, and so on. Every input value corresponds to exactly one output value, so the set defines a function.

(b) $\{(2, -4), (1, -1), (0, 0), (1, 1), (2, 4)\}$

Here, the input value 2 corresponds to the output values -4 and 4, so this set does not define a function. (What two output values correspond to the input value 1?)

(c) The ordered pairs that satisfy the equation $2x + 3y = 12$

Because this is a linear equation with $B \neq 0$, it defines a function.

(d) The ordered pairs that satisfy the equation $5x = 10$

In the context of two variables, when the coefficient of y is zero, as here, any value can be used for y. The only restriction imposed by the equation is that x must equal

2. Thus, (2, 0), (2, 5), (2, −1), and any ordered pair with $x = 2$ satisfies this relationship. This is not a function, however, because $x = 2$ is paired with more than one real number.

(e) The ordered pairs that satisfy the equation $4y = 16$

Every ordered pair in this set has a y-value of 4. Since there is no restriction here on x, every real number is paired with $y = 4$. For example, (−1, 4), (2.5, 4), (100, 4), and so on belong to this set, which defines a function, because every x corresponds to exactly one y, namely 4.

(f) The relationship between number of gallons needed to fill the gas tank and cost for a fill-up

This defines a function with ordered pairs of the form (number of gallons, cost). For a specific number of gallons, there is exactly one cost.

(g) The relationship between time spent traveling at a constant speed and number of miles traveled

Here the ordered pairs have the form (time, number of miles). A specific amount of time will correspond to exactly one number of miles, so this, too, defines a function.

Notice that in those sets that define functions, *no x-value appears more than once* (with a different y-value). The sets that were not functions had the same x as the input in at least two ordered pairs.

3.2 EXERCISES

1. (c) 2. (e) 3. (d) 4. (a)
5. (b) 6. (c)
7. 5; 5; 3

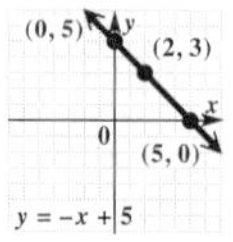

8. −2; 2; 3

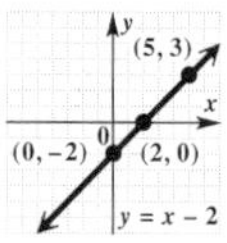

9. 1; 3; −1

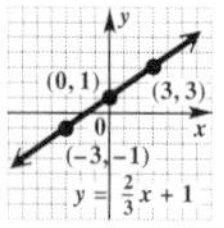

10. 2; −1; 5

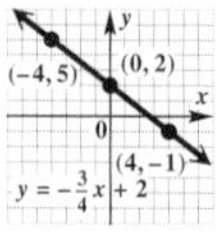

11. −6; −2; −5

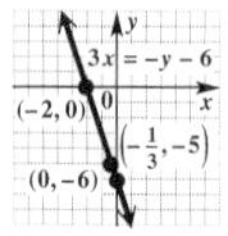

In Exercises 1–6, match the information about the graphs with the linear equations in (a)–(e).

(a) $x = 5$ **(b)** $y = -3$ **(c)** $2x - 5y = 8$ **(d)** $x + 4y = 0$ **(e)** $3x + y = -4$

1. The graph of the equation has x-intercept (4, 0).
2. The graph of the equation has y-intercept (0, −4).
3. The graph of the equation goes through the origin.
4. The graph of the equation is a vertical line.
5. The graph of the equation is a horizontal line.
6. The graph of the equation goes through (9, 2).

Complete the given ordered pairs using the given equation. Then graph each equation by plotting the points and drawing a line through them. See Examples 1 and 2.

7. $y = -x + 5$
(0,), (, 0), (2,)

8. $y = x - 2$
(0,), (, 0), (5,)

9. $y = \frac{2}{3}x + 1$
(0,), (3,), (−3,)

10. $y = -\frac{3}{4}x + 2$
(0,), (4,), (−4,)

11. $3x = -y - 6$
(0,), (, 0), $\left(-\frac{1}{3},\ \right)$

12. $x = 2y + 3$
(, 0), (0,), $\left(\ , \frac{1}{2}\right)$

Find the x-intercept and the y-intercept for the graph of each equation. See Example 2.

13. $2x - 3y = 24$ **14.** $-3x + 8y = 48$ **15.** $x + 6y = 0$ **16.** $3x - y = 0$

17. What is the equation of the x-axis?

18. What is the equation of the y-axis?

12. 3; $-\frac{3}{2}$; 4

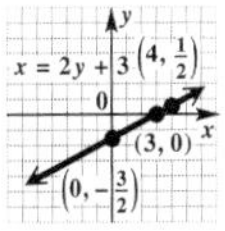

13. (12, 0); (0, −8)
14. (−16, 0); (0, 6)
15. (0, 0); (0, 0)
16. (0, 0); (0, 0)
17. $y = 0$ 18. $x = 0$
19. Choose a value other than 0 for either x or y. For example, if $x = -5, y = 4$.
20. Find two ordered pairs that satisfy the equation. Plot the corresponding points on a coordinate system. Draw a straight line through the two points. As a check, find a third ordered pair and verify that it lies on the line you drew.

21.
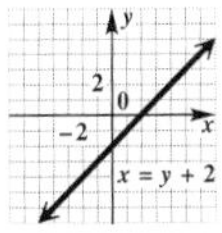

22.
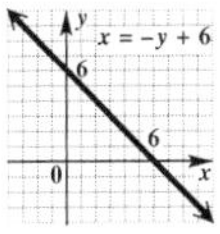

23.
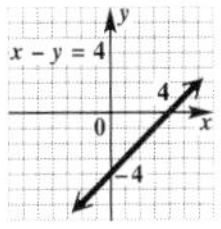

24.
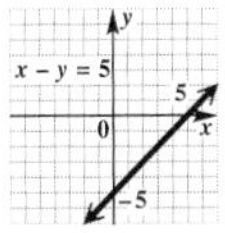

25.
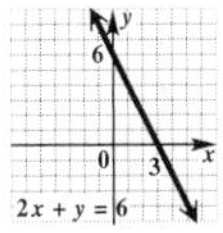

26.
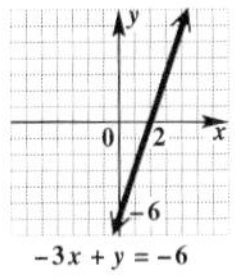

27.
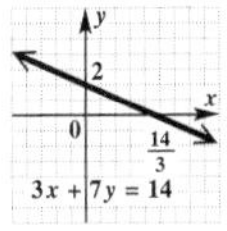

28.
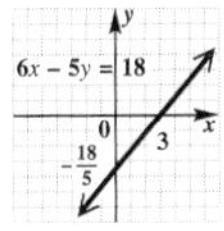

19. A student attempted to graph $4x + 5y = 0$ by finding intercepts. She first let $x = 0$ and found y; then she let $y = 0$ and found x. In both cases, the resulting point was (0, 0). She knew that she needed at least two different points to graph the line, but was unsure what to do next since finding intercepts gave her only one point. How would you explain to her what to do next?

20. Write a paragraph summarizing how to graph a linear equation in two variables.

Graph each linear equation. See Examples 1–5.

21. $x = y + 2$ **22.** $x = -y + 6$ **23.** $x - y = 4$ **24.** $x - y = 5$

25. $2x + y = 6$ **26.** $-3x + y = -6$ **27.** $3x + 7y = 14$ **28.** $6x - 5y = 18$

29. $y - 2x = 0$ **30.** $y + 3x = 0$ **31.** $y = -6x$ **32.** $y = 4x$

33. $y + 1 = 0$ **34.** $y - 3 = 0$ **35.** $x = -2$ **36.** $x = 4$

TECHNOLOGY INSIGHTS (EXERCISES 37–42)

In each exercise below, a calculator-generated graph of a linear equation in one variable with one side equal to 0 is shown. Accompanying the graph is the equation itself, where y is expressed in terms of x on the left side. Solve the equation using the methods of Section 2.2, and show that the solution you get is the same as the x-intercept (labeled "zero") on the calculator screen.

37. $8 - 2(3x - 4) - 2x = 0$

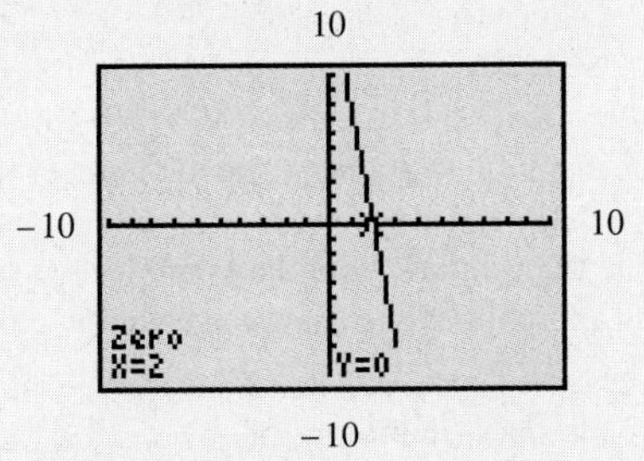

38. $5(2x - 1) - 4(2x + 1) - 7 = 0$

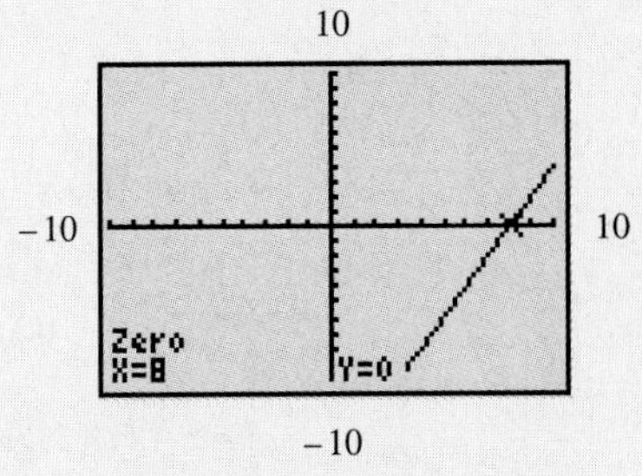

39. $.6x - .1x - x + 2.5 = 0$

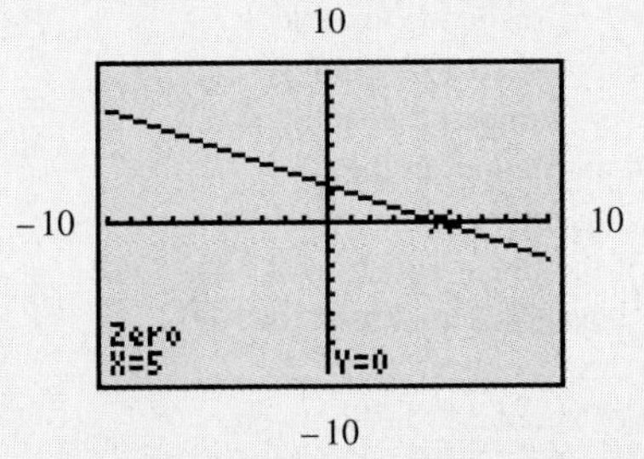

40. $-\frac{2}{7}x + 2x - \frac{1}{2}x - \frac{17}{2} = 0$

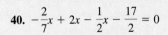

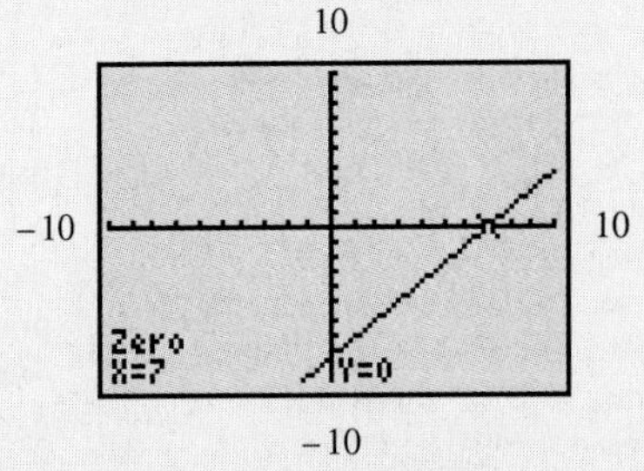

41. Use the results of Exercises 37–40 to explain how the x-intercept of the graph of an equation in two variables corresponds to the solution of an equation in one variable.

42. A horizontal line has no x-intercept. If you try to solve $5x - (3x + 2x) + 4 = 0$, you get no solution. What would the graph of $y = 5x - (3x + 2x) + 4$ look like on a graphing calculator?

JOURNAL WRITING ▲ CHALLENGING SCIENTIFIC CALCULATOR GRAPHING CALCULATOR

29.
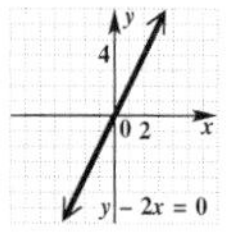

30.
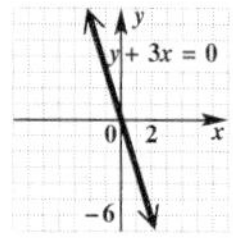

31.
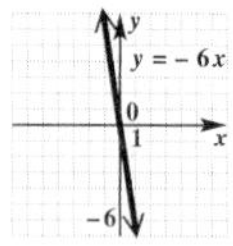

32.
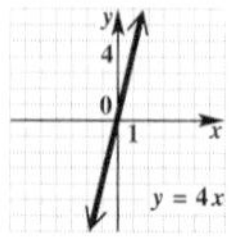

33.
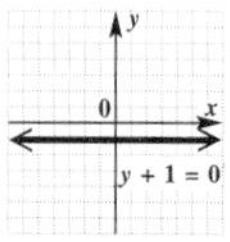

34.
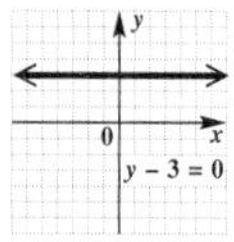

35.
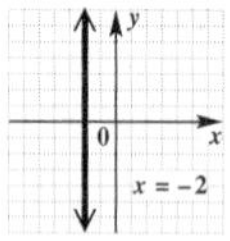

36.
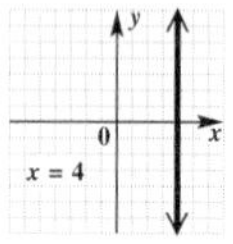

37. 2 **38.** 8 **39.** 5 **40.** 7 **41.** The y-value of each ordered pair gives the value of the expression in x for each x-value. The solution of the equation is the x-value that corresponds to $y = 0$, which is the x-intercept of the graph. **42.** It would be a horizontal line. **43.** (a) 1980: 52.2 thousand; 1985: 66.25 thousand; 1990: 80.3 thousand; 1994: 91.54 thousand (b) 1980: 54 thousand; 1985: 67 thousand; 1990: 77 thousand; 1994: 93 thousand (c) Yes, they are quite close. **44.** (a) 1988: \$44 billion; 1990: \$47.25 billion; 1992: \$50.5 billion; 1994: \$53.75 billion; 1996: \$57 billion (b) yes; Each year corresponds to exactly one sales amount.

Solve each problem.

43. The number of master's degrees earned in business management increased during the years 1971–1994 as shown in the figure. If $x = 0$ represents 1970, $x = 5$ represents 1975, $x = 10$ represents 1980, and so on, the number of master's degrees can be approximated by

$$y = 2.81x + 24.1,$$

where y is in thousands. (This is a *linear model* for the data.)

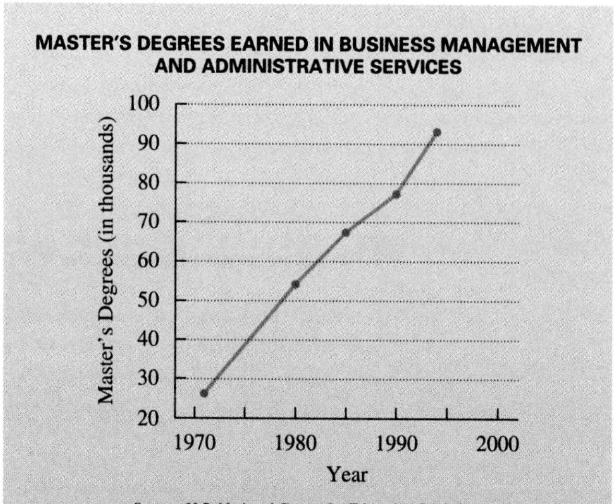

Source: U.S. National Center for Education Statistics, *Digest of Education Statistics*, annual.

(a) Use the equation to approximate the number of such degrees in the years 1980, 1985, 1990, and 1994.

(b) Estimate the y-values from the graph for the same years.

(c) Are the approximations using the equation close to the values you read from the graph?

44. Sporting goods sales (in billions of dollars) from 1988–1996 are approximated by the equation

$$y = 1.625x + 40.75,$$

where $x = 0$ corresponds to 1986, $x = 2$ corresponds to 1988, and so on. Sales for even-numbered years in that time period are plotted in the accompanying figure, which also shows the graph of the linear equation. As the graph indicates, the actual sales in 1992 (when the U.S. economy was depressed) of about \$47 billion were less than the sales approximated by the equation.

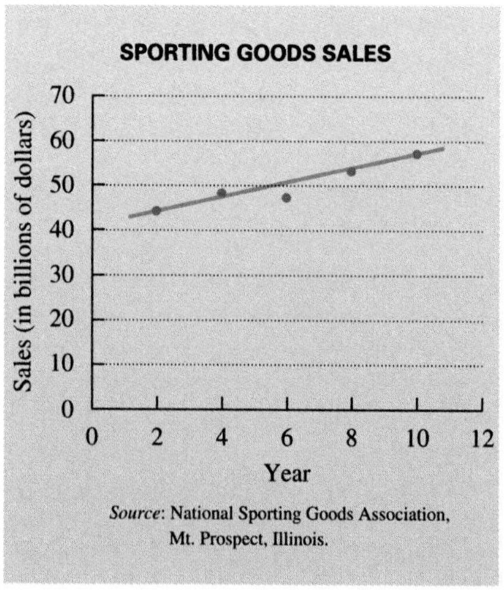

Source: National Sporting Goods Association, Mt. Prospect, Illinois.

(a) Use the *equation* to approximate the sales in each of the even-numbered years.

(b) Does this equation define a function? Why or why not?

45. (a) 163.2 centimeters
(b) 171 centimeters
(c) 151.5 centimeters
(d)

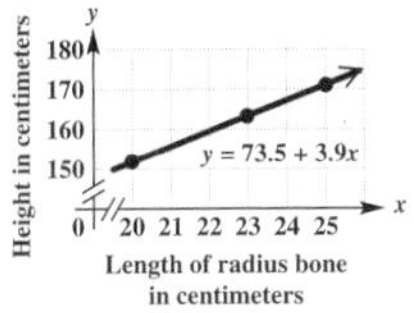

46. (a) 121 pounds (b) 132 pounds (c) 154 pounds
(d) 176 pounds
(e)

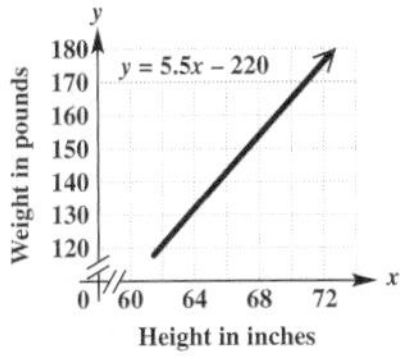

47. (a) \$1250 (b) \$1250
(c) \$1250 (d) \$6250

48. (a)

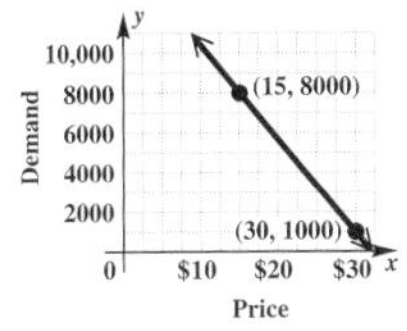

(b) about 10,300 (c) about \$23

45. The height y of a woman (in centimeters) is a function of the length of her radius bone x (from the wrist to the elbow) and is defined by $y = 73.5 + 3.9x$. Estimate the heights of women with radius bones of the following lengths.

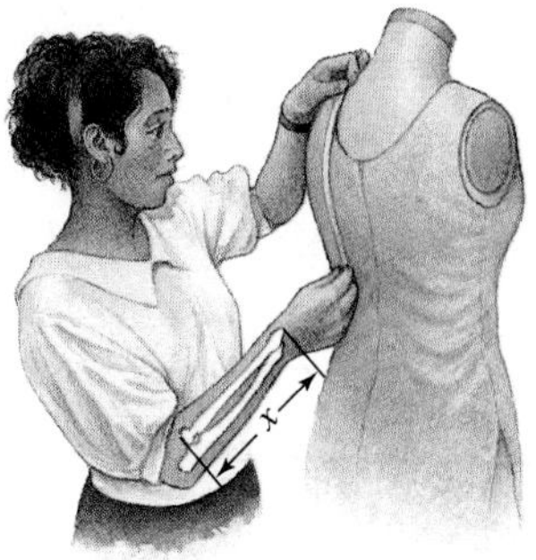

(a) 23 centimeters
(b) 25 centimeters
(c) 20 centimeters
(d) Graph $y = 73.5 + 3.9x$.

46. As a rough estimate, the weight of a man taller than about 60 inches is a linear function approximated by $y = 5.5x - 220$, where x is the height of the person in inches, and y is the weight in pounds. Estimate the weights of men whose heights are as follows.
(a) 62 inches **(b)** 64 inches
(c) 68 inches **(d)** 72 inches
(e) Graph $y = 5.5x - 220$.

47. The graph shows that the value of a certain automobile over its first five years is a function of the year. Use the graph to estimate the depreciation (loss in value) during the following years.

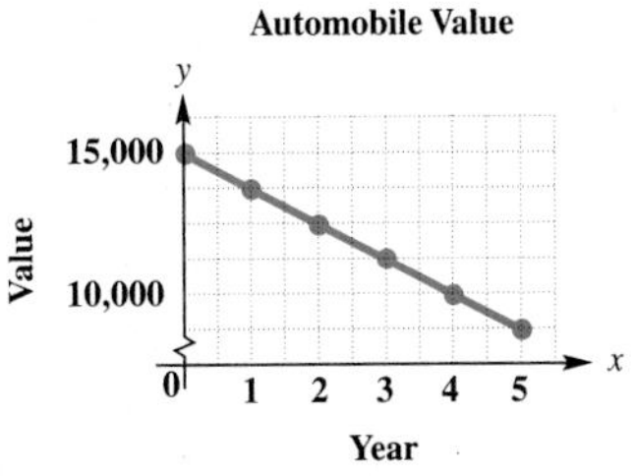

(a) First **(b)** Second **(c)** Fifth
(d) What is the total depreciation over the 5-year period?

48. The demand for an item is a function of its price. As price goes up, demand goes down. On the other hand, when price goes down, demand goes up. Suppose the demand for a certain Beanie Baby is 1000 when its price is \$30 and 8000 when it costs \$15.
(a) Let x be the price and y be the demand for the Beanie Baby. Graph the two given pairs of prices and demands.
(b) Assume the relationship is linear. Draw a line through the two points from part (a). From your graph estimate the demand if the price drops to \$10.
(c) Use the graph to estimate the price if the demand is 4000.

49. yes 50. yes 51. yes
52. yes 53. no 54. yes
55. yes 56. yes

Decide whether each set of ordered pairs defines a function. See Example 6.

49. $\{(0, 5), (2, 3), (4, 1), (6, -1), (8, -3)\}$

50. $\{(1, 3), (2, 3), (3, 3), (4, 3)\}$

51. The ordered pairs that satisfy $-x + 2y = 9$

52. The ordered pairs that satisfy $y = 4$

53. The ordered pairs that satisfy $x = 8$

54. The ordered pairs that satisfy $5x + y = 7$

55. The relationship in Exercise 43 between year and number of master's degrees

56. The relationship in Exercise 44 between year and sporting goods sales

3.3 The Slope of a Line

OBJECTIVES

1. Find the slope of a line given two points.
2. Find the slope from the equation of a line.
3. Use the slope to determine whether two lines are parallel, perpendicular, or neither.

FOR EXTRA HELP

SSG Sec. 3.3
SSM Sec. 3.3
Pass the Test Software
InterAct Math Tutorial Software
Video 5

TEACHING TIP Note the difference between x_2 and x^2. These are sometimes confused.

When two variables are related in such a way that the value of one depends on the value of the other, we can form ordered pairs of corresponding numbers. For example, if $x + y = 7$, then when $x = 2$, $y = 5$, and when $x = -3$, $y = 10$. We write these pairs as (2, 5) and (−3, 10), respectively, with the understanding that the first number in the ordered pair represents x and the second number represents y. To indicate two nonspecific ordered pairs that satisfy a particular equation relating x and y, we use *subscript notation.* We write the pairs as (x_1, y_1) and (x_2, y_2). (Read x_1 as "x-sub-one" and x_2 as "x-sub-two.")

We can graph a straight line if at least two different points on the line are known. A line also can be graphed by using just one point on the line if the "steepness" of the line is known.

OBJECTIVE 1 Find the slope of a line given two points. One way to measure the steepness of a line is to compare the vertical change in the line (the rise) to the horizontal change (the run) while moving along the line from one fixed point to another. This measure of steepness is called the *slope* of the line.

Figure 15 shows a line with the points (x_1, y_1) and (x_2, y_2). As we move along the line from the point (x_1, y_1) to the point (x_2, y_2), y changes by $y_2 - y_1$ units. This is the vertical change. Similarly, x changes by $x_2 - x_1$ units, the horizontal change. The ratio of the change in y to the change in x gives the slope of the line. We usually denote slope with the letter m. The slope of a line is defined as follows.

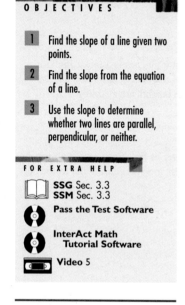

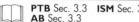

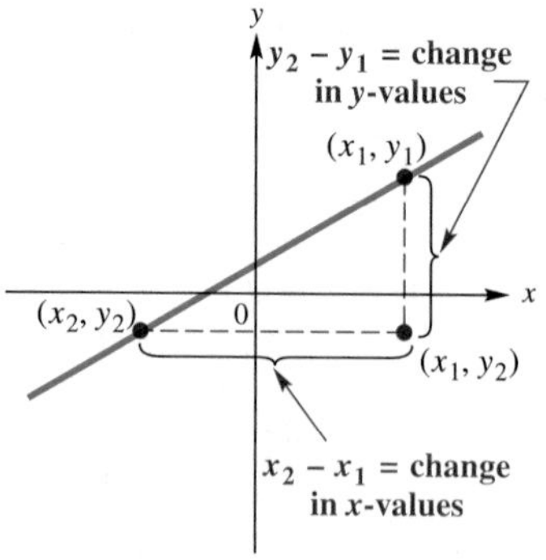

Figure 15

JOURNAL WRITING ▲ CHALLENGING SCIENTIFIC CALCULATOR GRAPHING CALCULATOR

INSTRUCTOR'S RESOURCES

PTB Sec. 3.3 **ISM** Sec. 3.3 **AB** Sec. 3.3 | **TEST GENERATOR** | **WORLD WIDE WEB** www.LialAlgebra.com

Slope Formula

The **slope** of the line through the points (x_1, y_1) and (x_2, y_2) is

$$m = \frac{\text{change in } y}{\text{change in } x} = \frac{y_2 - y_1}{x_2 - x_1} \quad \text{if } x_1 \neq x_2.$$

The slope of a line tells how fast y changes for each unit of change in x; that is, the slope gives the ratio of the change in y to the change in x. The change in y is called the **rise,** and the change in x is called the **run.**

Ski slopes, "grade" on a treadmill, and slope (or "fall") of a sewer pipe are some additional examples.

CONNECTIONS

The idea of slope is used in many everyday situations. For example, because $10\% = \frac{1}{10}$, a highway with a 10% grade (or slope) rises one meter for every 10 horizontal meters. The highway sign shown below is used to warn of a downgrade ahead that may be long or steep. Architects specify the pitch of a roof using slope; a $\frac{5}{12}$ roof means that the roof rises 5 feet for every 12 feet in the horizontal direction. The slope of a stairwell also indicates the ratio of the vertical rise to the horizontal run. The slope of the stairs in the figure is $\frac{8}{14}$.

Hill

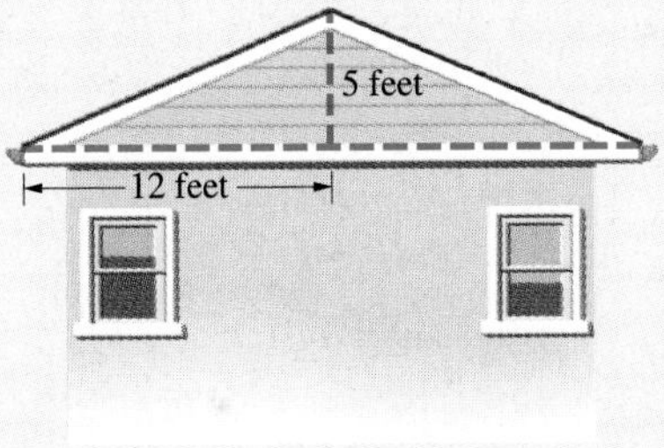

$\frac{5}{12}$ roof pitch

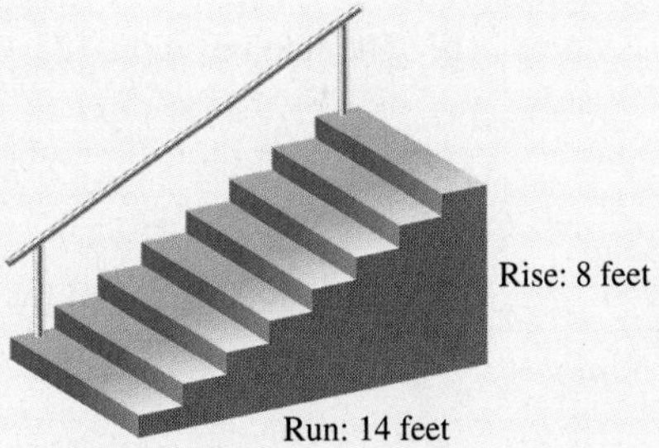

Slope of a stairwell

FOR DISCUSSION OR WRITING

Describe some other everyday examples of slope.

CHALKBOARD EXAMPLE

Find the slope of the line through (−3, 5) and (−4, 7).

Answer: −2

EXAMPLE 1 Finding the Slope of a Line

Find the slope of each of the following lines.

(a) The line through (−4, 7) and (1, −2)

Use the definition of slope. Let $(-4, 7) = (x_2, y_2)$ and $(1, -2) = (x_1, y_1)$. Then

$$\text{slope} = \frac{\text{change in } y}{\text{change in } x}$$

$$m = \frac{y_2 - y_1}{x_2 - x_1}$$

$$= \frac{7 - (-2)}{-4 - 1} = \frac{9}{-5} = -\frac{9}{5}.$$

As Figure 16 shows, this line has a slope of $-\frac{9}{5}$ (which can also be written as $\frac{-9}{5}$ or $\frac{9}{-5}$). One way of interpreting this is that the line drops vertically 9 units for a horizontal change of 5 units to the right.

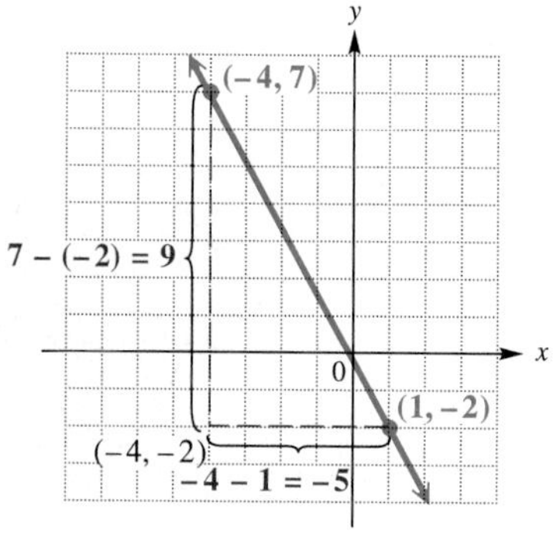

Figure 16

(b) The line through (12, −5) and (−9, −2)

$$m = \frac{-5 - (-2)}{12 - (-9)} = \frac{-3}{21} = -\frac{1}{7}$$

The same slope is found by subtracting in reverse order.

$$\frac{-2 - (-5)}{-9 - 12} = \frac{3}{-21} = -\frac{1}{7}$$

CAUTION It makes no difference which point is (x_1, y_1) or (x_2, y_2); however, it is important to be consistent. Start with the x- and y-values of one point (either one) and subtract the corresponding values of the other point.

TEACHING TIP Emphasize these points. Students should adopt the habit of checking their linear equation graphs to see if the slope of the graph agrees with the sign of the slope in the given equation.

In Example 1(a) the slope is negative and the corresponding line in Figure 16 falls from left to right. As Figure 17(a) shows, this is generally true of lines with negative slopes. Lines with positive slopes go up (rise) from left to right, as shown in Figure 17(b).

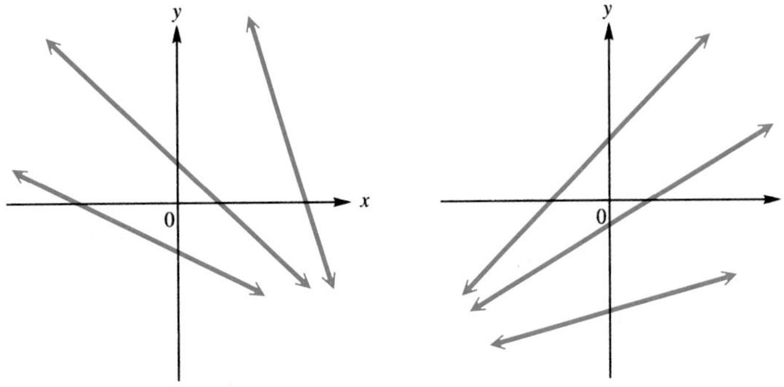

Lines with negative slopes
(a)

Lines with positive slopes
(b)

Figure 17

Positive and Negative Slopes

A line with positive slope rises from left to right.

A line with negative slope falls from left to right.

CHALKBOARD EXAMPLE

Find the slope of the line through (2, 5) and (−1, 5).

Answer: 0

EXAMPLE 2 Finding the Slope of a Horizontal Line

Find the slope of the line through (−8, 4) and (2, 4).

Use the definition of slope.

$$m = \frac{4 - 4}{-8 - 2} = \frac{0}{-10} = 0$$

As shown in Figure 18, the line through these two points is horizontal, with equation $y = 4$. *All horizontal lines have a slope of 0,* since the difference in y-values is always 0.

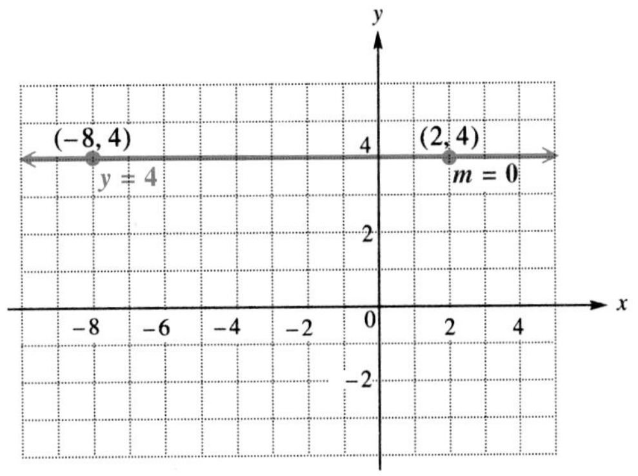

Figure 18

CHALKBOARD EXAMPLE

Find the slope of the line through (3, 1) and (3, −4).

Answer: undefined slope

TEACHING TIP Mention that this is why the formula for slope given at the beginning of this section had the restriction $x_1 \neq x_2$.

EXAMPLE 3 Finding the Slope of a Vertical Line

Find the slope of the line through (6, 2) and (6, −9).

$$m = \frac{2 - (-9)}{6 - 6} = \frac{11}{0} \quad \text{Undefined}$$

Since division by 0 is undefined, the slope is undefined. The graph in Figure 19 shows that the line through these two points is vertical, with equation $x = 6$. All points on a vertical line have the same x-value, so *the slope of any vertical line is undefined.*

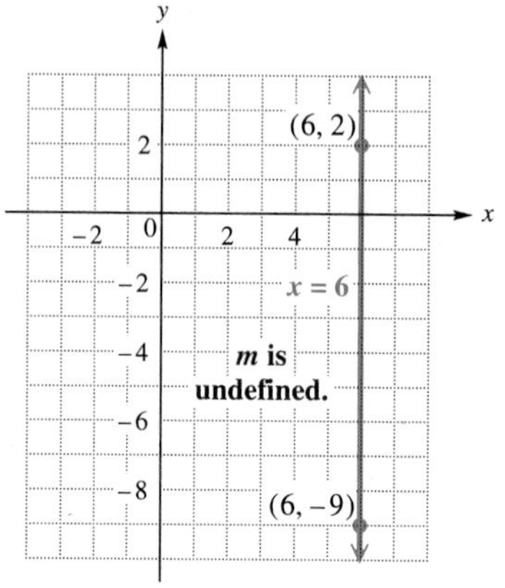

Figure 19

Slopes of Horizontal and Vertical Lines

Horizontal lines, with equations of the form $y = k$, **have slope 0.**

Vertical lines, with equations of the form $x = k$, **have undefined slope.**

OBJECTIVE 2 Find the slope from the equation of a line. The slope of a line also can be found directly from its equation. For example, the slope of the line

$$y = -3x + 5$$

can be found using any two points on the line. Get two points by choosing two different values of x, say −2 and 4, and finding the corresponding y-values.

If $x = -2$:	If $x = 4$:
$y = -3(-2) + 5$	$y = -3(4) + 5$
$y = 6 + 5$	$y = -12 + 5$
$y = 11.$	$y = -7.$

TEACHING TIP Use $2x + 3y = 12$ and show the slope is the same whether using two ordered pairs or solving for y to get the slope.

The ordered pairs are (−2, 11) and (4, −7). Now use the slope formula to find the slope.

$$m = \frac{11 - (-7)}{-2 - 4} = \frac{18}{-6} = -3$$

The slope, −3, is the same number as the coefficient of x in the equation $y = -3x + 5$. It can be shown that this always happens, *as long as the equation is solved for y.* This fact is used to find the slope of a line from its equation.

Finding the Slope of a Line from its Equation

Step 1 Solve the equation for y.

Step 2 The slope is given by the coefficient of x.

CHALKBOARD EXAMPLE

Find the slope of the line $3x + 2y = 9$.

Answer: $-\frac{3}{2}$

EXAMPLE 4 Finding Slope from an Equation

Find the slope of each of the following lines.

(a) $2x - 5y = 4$

Solve the equation for y.

$$2x - 5y = 4$$

$$-5y = -2x + 4 \qquad \text{Subtract } 2x \text{ from each side.}$$

$$y = \frac{2}{5}x - \frac{4}{5} \qquad \text{Divide each side by } -5.$$

The slope is given by the coefficient of x, so the slope is $m = \frac{2}{5}$.

(b) $8x + 4y = 1$

Solve the equation for y.

$$8x + 4y = 1$$

$$4y = -8x + 1 \qquad \text{Subtract } 8x \text{ from each side.}$$

$$y = -2x + \frac{1}{4} \qquad \text{Divide each side by } 4.$$

The slope of this line is given by the coefficient of x, -2.

OBJECTIVE 3 **Use the slope to determine whether two lines are parallel, perpendicular, or neither.** Two lines in a plane that never intersect are **parallel.** We use slopes to tell whether two lines are parallel. For example, Figure 20 shows the graph of $x + 2y = 4$ and the graph of $x + 2y = -6$. These lines appear to be parallel. Solve for y to find that both $x + 2y = 4$ and $x + 2y = -6$ have a slope of $-\frac{1}{2}$. Nonvertical parallel lines always have equal slopes.

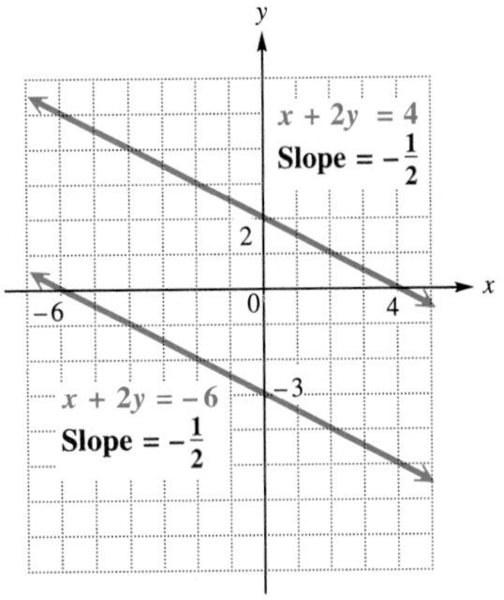

Figure 20

Figure 21 shows the graph of $x + 2y = 4$ and the graph of $2x - y = 6$. These lines appear to be **perpendicular** (meet at a 90° angle). Solving for y shows that the slope of $x + 2y = 4$ is $-\frac{1}{2}$, while the slope of $2x - y = 6$ is 2. The product of $-\frac{1}{2}$ and 2 is

$$-\frac{1}{2}(2) = -1.$$

This is true in general; the product of the slopes of two perpendicular lines, neither of which is vertical, is always -1. This means that the slopes of perpendicular lines are negative reciprocals: if one slope is the nonzero number a, the other is $-\frac{1}{a}$.

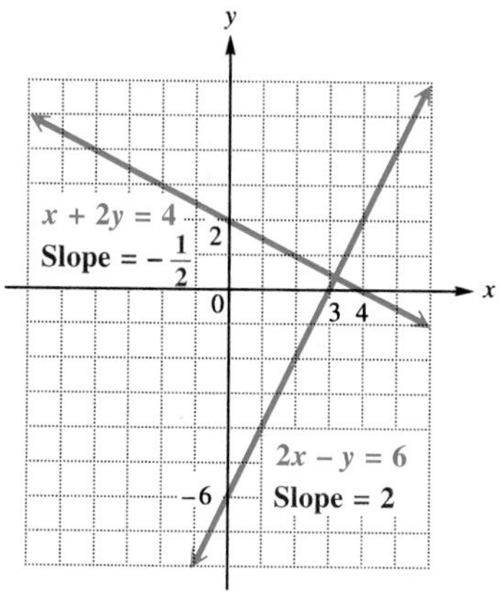

Figure 21

Parallel and Perpendicular Lines

Two nonvertical lines with the same slope are parallel; two perpendicular lines, neither of which is vertical, have slopes that are negative reciprocals of each other.

CHALKBOARD EXAMPLE

Decide whether the pair of lines is parallel, perpendicular, or neither.

$$3x - y = 4$$
$$x + 3y = 9$$

Answer: perpendicular

EXAMPLE 5 Deciding Whether Two Lines Are Parallel or Perpendicular

Decide whether the lines are *parallel, perpendicular,* or *neither.*

(a) $x + 2y = 7$

$-2x + y = 3$

Find the slope of each line by first solving each equation for y.

$$x + 2y = 7 \qquad\qquad -2x + y = 3$$

$$2y = -x + 7 \qquad\qquad y = 2x + 3$$

$$y = -\frac{1}{2}x + \frac{7}{2}$$

Slope: $-\frac{1}{2}$ Slope: 2

Since the slopes are not equal, the lines are not parallel. Check the product of the slopes: $-\frac{1}{2}(2) = -1$. The two lines are perpendicular because the product of their slopes is -1, indicating that the slopes are negative reciprocals.

TEACHING TIP Discuss the graphical meaning of *neither*: that is, they must intersect (but not at a right angle).

(b) $3x - y = 4$
$6x - 2y = 9$

Find the slopes. Both lines have a slope of 3, so the lines are parallel.

(c) $4x + 3y = 6$
$2x - y = 5$

Here the slopes are $-\frac{4}{3}$ and 2. These two straight lines are neither parallel nor perpendicular.

CONNECTIONS

Because the viewing window of a graphing calculator is a rectangle, the graphs of perpendicular lines will not appear perpendicular unless appropriate intervals are used for x and y. Graphing calculators usually have a key to select a "square" window automatically. In a square window, the x-interval is about 1.5 times the y-interval. The equations used in Figure 21 are graphed with the standard (nonsquare) window and then with a square window in the screens below.

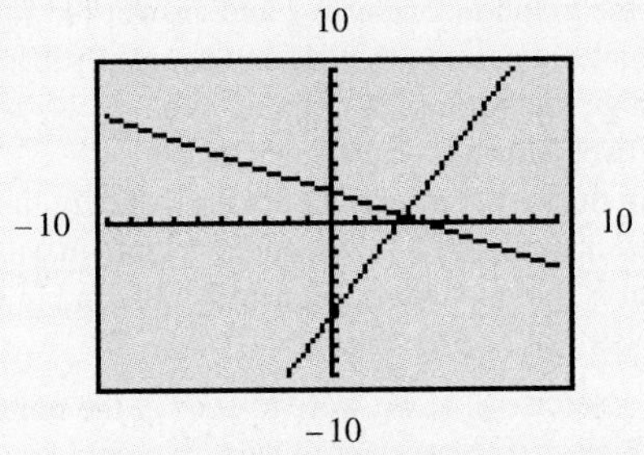

A standard (nonsquare) window
[−10, 10] by [−10, 10]
Lines do not appear perpendicular.

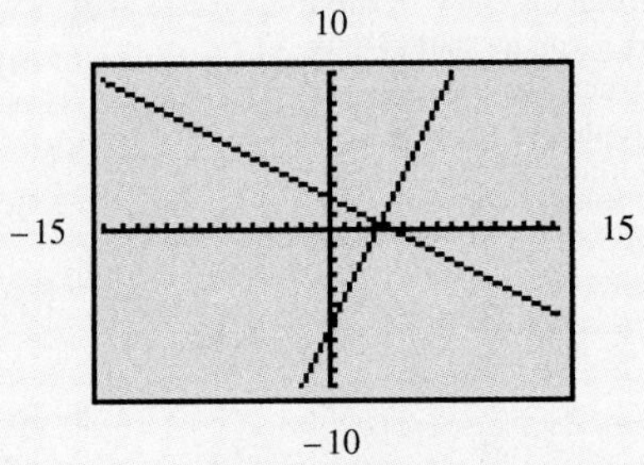

A square window
[−15, 15] by [−10, 10]
Lines appear perpendicular.

3.3 EXERCISES

1. Rise is the vertical change between two different points on a line. Run is the horizontal change between two different points on a line. 2. $\frac{3}{2}$ 3. 4

1. What is meant by "rise"? What is meant by "run"?

Use the coordinates of the indicated points to find the slope of each line. See Example 1.

2.

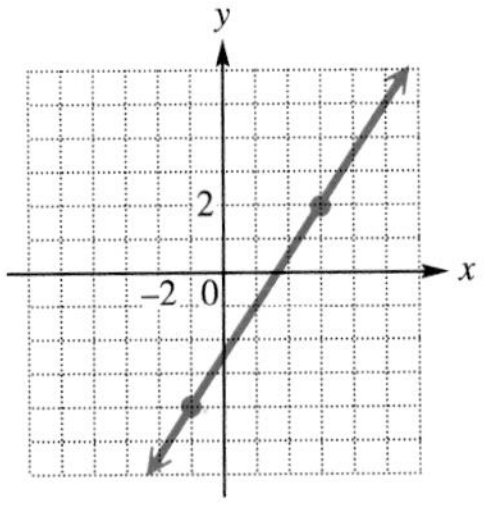

3.

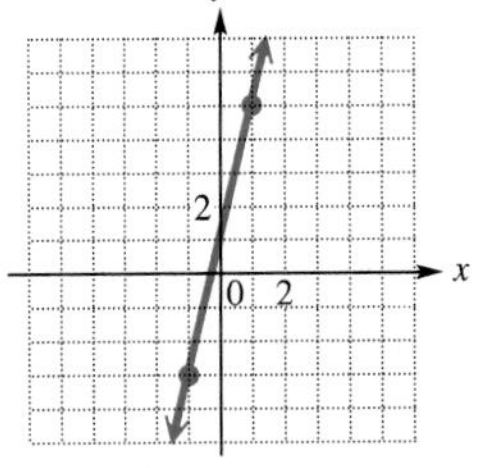

4. $-\frac{7}{4}$ 5. $-\frac{1}{2}$ 6. undefined
7. 0 8. (a) 6 (b) 4
(c) $\frac{6}{4}$ or $\frac{3}{2}$; slope of the line
9. Yes; it doesn't matter which point you start with. Both differences will be the negatives of the differences in Exercise 8, and the quotient will be the same.
In Exercises 10–13, sketches will vary.
10. The line must fall from left to right. 11. The line must rise from left to right. 12. The line must be vertical. 13. The line must be horizontal. 14. The slope of a line is the ratio (or quotient) of the rise, the change in y, and the run, the change in x.
15. Because he found the difference $3 - 5 = -2$ in the numerator, he should have subtracted in the same order in the denominator to get $-1 - 2 = -3$. The correct slope is $\frac{-2}{-3} = \frac{2}{3}$.
16. $\frac{7}{6}$ 17. $\frac{5}{4}$ 18. $-\frac{5}{8}$
19. $\frac{3}{2}$

4.

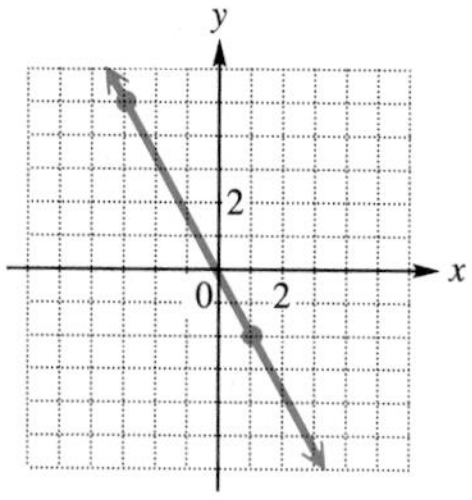

5.

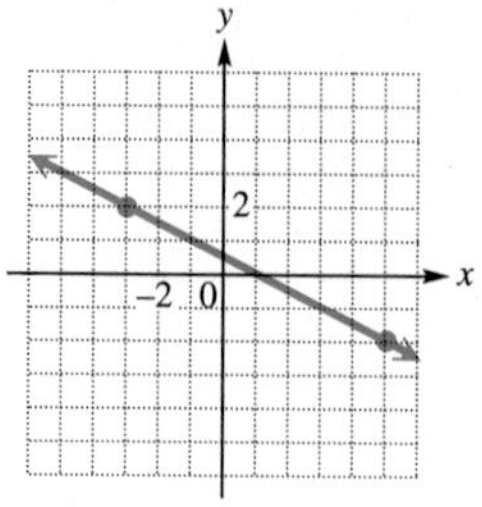

6.

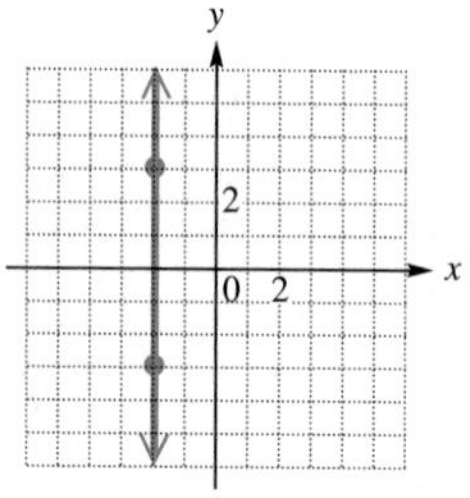

7.

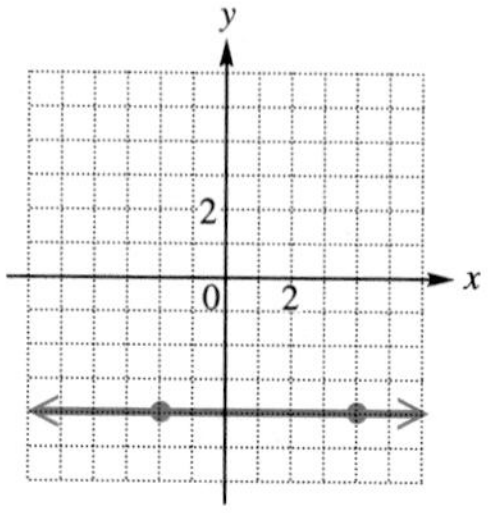

8. Look at the graph in Exercise 2 and answer the following.
(a) Start at the point $(-1, -4)$ and count vertically up to the horizontal line that goes through the other plotted point. What is this vertical change? (Remember: "up" means positive, "down" means negative.)
(b) From this new position, count horizontally to the other plotted point. What is this horizontal change? (Remember: "right" means positive, "left" means negative.)
(c) What is the quotient of the numbers found in parts (a) and (b)? What do we call this number?

9. Refer to Exercise 8. If we were to *start* at the point $(3, 2)$ and *end* at the point $(-1, -4)$, do you think that the answer to part (c) would be the same? Explain why or why not.

On a pair of axes similar to the one shown, sketch the graph of a straight line having the indicated slope.

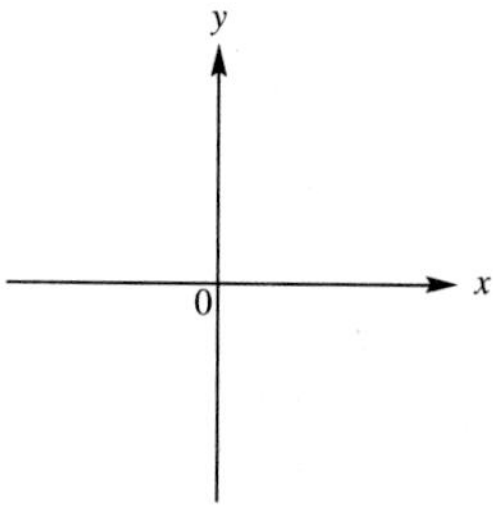

10. negative
11. positive
12. undefined
13. zero

14. Explain in your own words what is meant by *slope* of a line.

15. A student was asked to find the slope of the line through the points $(2, 5)$ and $(-1, 3)$. His answer, $-\frac{2}{3}$, was incorrect. He showed his work as

$$\frac{3-5}{2-(-1)} = \frac{-2}{3} = -\frac{2}{3}.$$

What was his error? Give the correct slope.

Find the slope of the line going through each pair of points. See Examples 1–3.

16. $(4, -1)$ and $(-2, -8)$ **17.** $(1, -2)$ and $(-3, -7)$
18. $(-8, 0)$ and $(0, -5)$ **19.** $(0, 3)$ and $(-2, 0)$

20. 0 21. 0
22. undefined 23. undefined
24. $\frac{1}{3}$ 25. $-\frac{1}{2}$
26. 2 27. 5 28. $-\frac{1}{2}$
29. $\frac{1}{4}$ 30. $\frac{3}{2}$ 31. $\frac{3}{2}$ 32. 0
33. undefined
34. (a) negative (b) negative
35. (a) negative (b) zero
36. (a) positive (b) zero
37. (a) positive (b) negative
38. (a) zero (b) positive
39. (a) zero (b) negative
40. They are the same; they have a product of −1. Vertical lines have undefined slope. 41. (a)
42. It is horizontal.
43. $-\frac{2}{5}$; $-\frac{2}{5}$; parallel
44. $\frac{4}{3}$; $\frac{4}{3}$; parallel
45. $\frac{8}{9}$; $-\frac{4}{3}$; neither
46. $\frac{5}{3}$; $\frac{3}{5}$; neither
47. $\frac{3}{2}$; $-\frac{2}{3}$; perpendicular
48. $\frac{3}{5}$; $-\frac{5}{3}$; perpendicular

20. $(6, -5)$ and $(-12, -5)$

21. $(4, 3)$ and $(-6, 3)$

22. $(-8, 6)$ and $(-8, -1)$

23. $(-12, 3)$ and $(-12, -7)$

▲ **24.** $(3.1, 2.6)$ and $(1.6, 2.1)$

▲ **25.** $\left(-\frac{7}{5}, \frac{3}{10}\right)$ and $\left(\frac{1}{5}, -\frac{1}{2}\right)$

Find the slope of each line. See Example 4.

26. $y = 2x - 3$ **27.** $y = 5x + 12$ **28.** $2y = -x + 4$ **29.** $4y = x + 1$

30. $-6x + 4y = 4$ **31.** $3x - 2y = 3$ **32.** $y = 4$ **33.** $x = 6$

The figure shows a line that has a positive slope (because it rises from left to right) and a positive y-value for the y-intercept (because it intersects the y-axis above the origin).

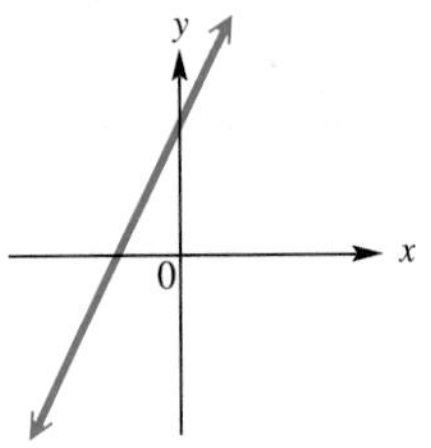

*For each line shown, decide whether **(a)** the slope is positive, negative, or zero and **(b)** the y-value of the y-intercept is positive, negative, or zero.*

34.

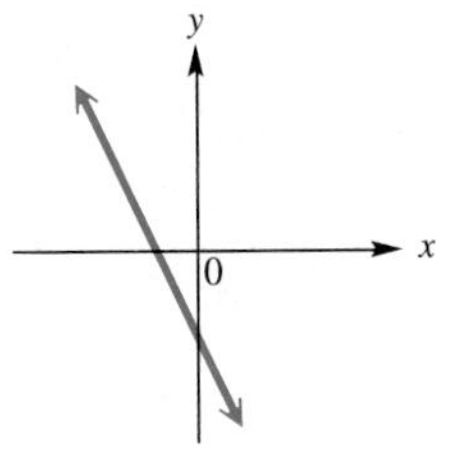

35.

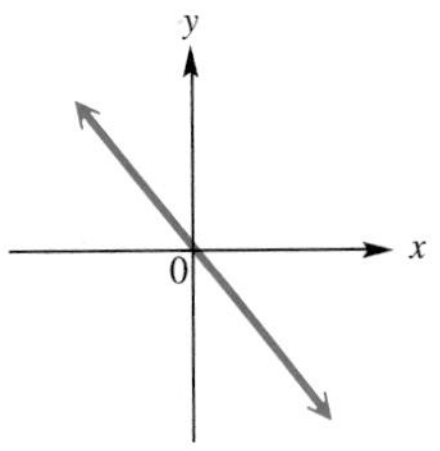

36.

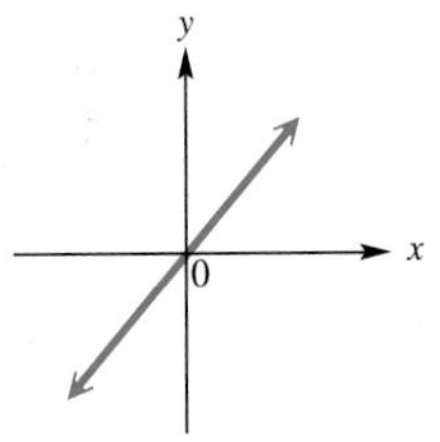

37.

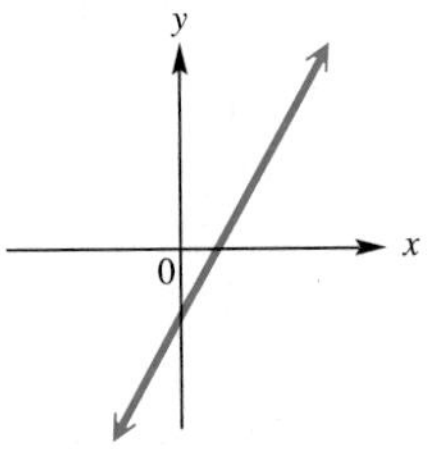

38.

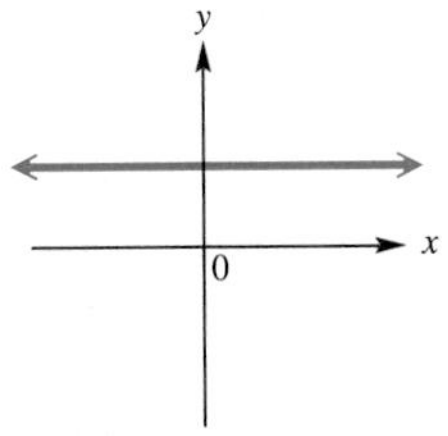

39.

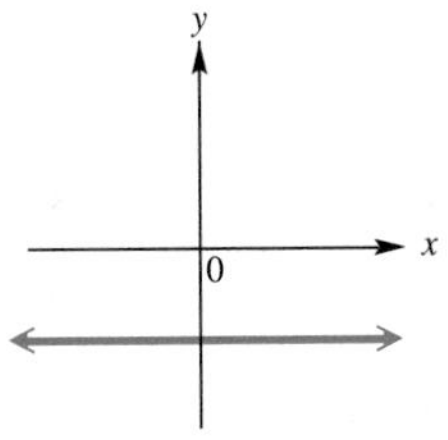

✎ **40.** If two nonvertical lines are parallel, what do we know about their slopes? If two lines are perpendicular and neither is parallel to an axis, what do we know about their slopes? Why must the lines be nonvertical?

41. If two lines are both vertical or both horizontal, which of the following are they?
(a) parallel **(b)** perpendicular **(c)** neither parallel nor perpendicular

42. If a line is vertical, what is true of any line that is perpendicular to it?

For each pair of equations, give the slopes of the lines and then determine whether the two lines are parallel, perpendicular, or neither parallel nor perpendicular. See Example 5.

43. $2x + 5y = 4$
$4x + 10y = 1$

44. $-4x + 3y = 4$
$-8x + 6y = 0$

45. $8x - 9y = 6$
$8x + 6y = -5$

46. $5x - 3y = -2$
$3x - 5y = -8$

47. $3x - 2y = 6$
$2x + 3y = 3$

48. $3x - 5y = -1$
$5x + 3y = 2$

49. $\frac{3}{10}$ 50. $\frac{8}{27}$

For Exercises 49 and 50, you may wish to refer to the Connections box preceding Example 1.

49. What is the slope (or pitch) of this roof?

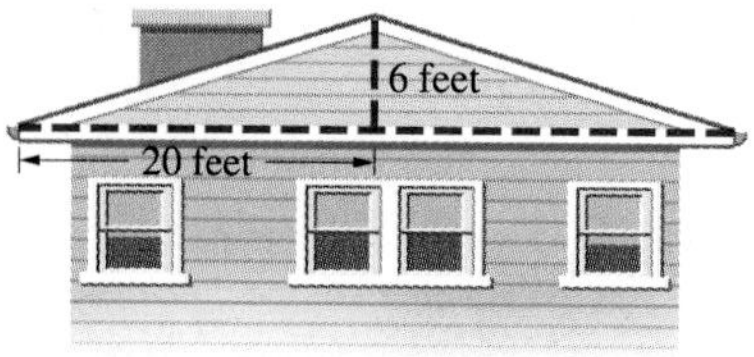

50. What is the slope (or grade) of this hill?

*These problems can be used as collaborative exercises with small groups of students.

51. 232,000

RELATING CONCEPTS* (EXERCISES 51–56)

Figure A depicts public school enrollment (in thousands) in grades 9–12 in the United States. Figure B gives the (average) number of public school students per computer.

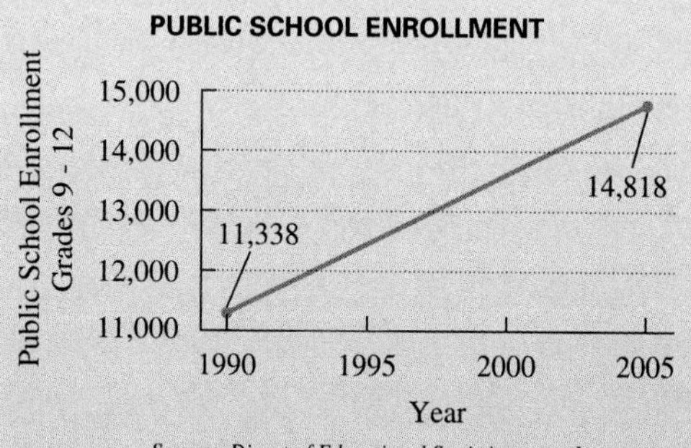

Source: Digest of Educational Statistics, annual, and *Projections of Educational Statistics*, annual.

Figure A

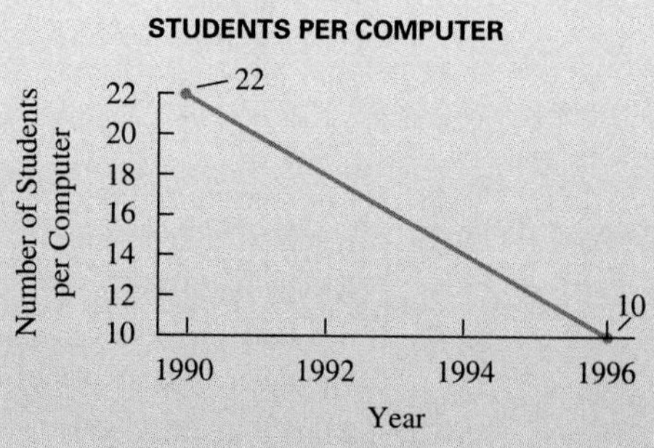

Source: Quality Education Data.

Figure B

Work Exercises 51–56 in order.

51. Use the ordered pairs (1990, 11,338) and (2005, 14,818) to find the slope of the line in Figure A.

52. positive; increased
53. 232,000 students 54. −2
55. negative; decreased
56. 2 students per computer

RELATING CONCEPTS (EXERCISES 51–56) (CONTINUED)

52. The slope of the line in Figure A is ______________ (positive/negative). This means that during the period represented, enrollment ______________ (increased/decreased).

53. The slope of a line represents its *rate of change.* Based on Figure A, what was the increase in students *per year* during the period shown?

54. Use the given ordered pairs to find the slope of the line in Figure B.

55. The slope of the line in Figure B is ______________ (positive/negative). This means that during the period represented, the number of students per computer ______________ (increased/decreased).

56. Based on Figure B, what was the decrease in students per computer *per year* during the period shown?

Did you make the connection between the sign of the slope of the line and the increase or decrease in the quantity represented by y?

57. (a) Possible ordered pairs are (1991, 37.8), (1992, 38.6), (1993, 39.4), (1994, 40.2), (1995, 41.0), (1996, 41.8). The input numbers are 1991, 1992, 1993, 1994, 1995, 1996. The output numbers (in thousands) are 37.8, 38.6, 39.4, 40.2, 41.0, 41.8. (b) Yes, they are all within .4 thousand (400).
58. (1992, 39)

Work each business-related problem.

57. The table gives the number of shopping centers in the years 1991–1996. The equation

$$y = .797x - 1549,$$

where x represents the year and y is the number of shopping centers in thousands, approximates these data quite well.

Number of Shopping Centers (in thousands)

Year	1991	1992	1993	1994	1995	1996
Number of Centers	38.0	39.0	39.6	40.4	41.2	42.1

Source: International Council of Shopping Centers.

(a) Give three ordered pairs for this *equation* and indicate the input and output numbers for each.

(b) Compare the y-values you find with the values given for the number of centers in the table. Are the approximations close?

58. The line graph shows the points for the ordered pairs from the table in Exercise 57. A straight line graph of the equation in Exercise 57 is also shown. Except for one point, the straight line is almost the same as the line graph. Which point is farthest from the straight line?

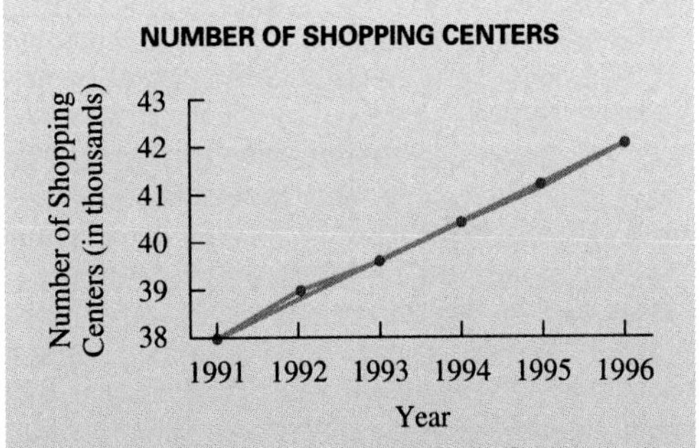

Source: International Council of Shopping Centers.

59. The change for each year is .1 billion (or 100,000,000) square feet, so the graph is a straight line.
60. The slope is .1 billion, which is the same as the yearly change in square footage.

59. The growth in retail square footage, in billions, is shown in the line graph. This graph looks like a straight line. If the change in square footage each year is the same, then it is a straight line. Find the change in square footage for the years shown in the graph. (*Hint:* To find the change in square footage from 1991 to 1992, subtract the y-value for 1991 from the y-value for 1992.) Is the graph a straight line?

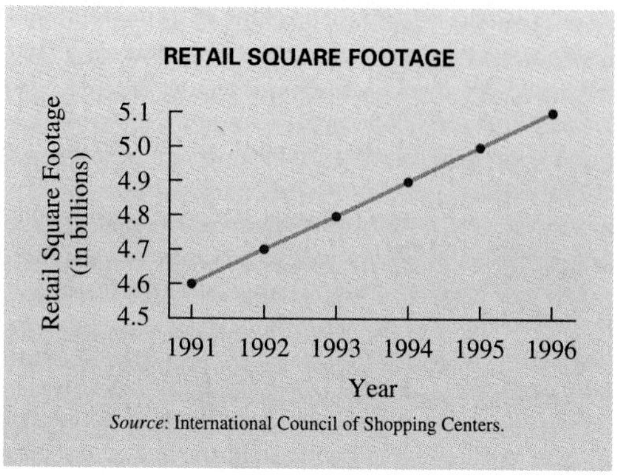

60. Find the slope of the line in Exercise 59 by using any two of the points shown on the line. How does the slope compare with the yearly change in square footage?

TECHNOLOGY INSIGHTS (EXERCISES 61–66)

61. 2 **62.** $\frac{2}{5}$

61. Two views of the same line are shown in the accompanying calculator screen, along with coordinates of two points displayed at the bottoms. What is the slope of this line?

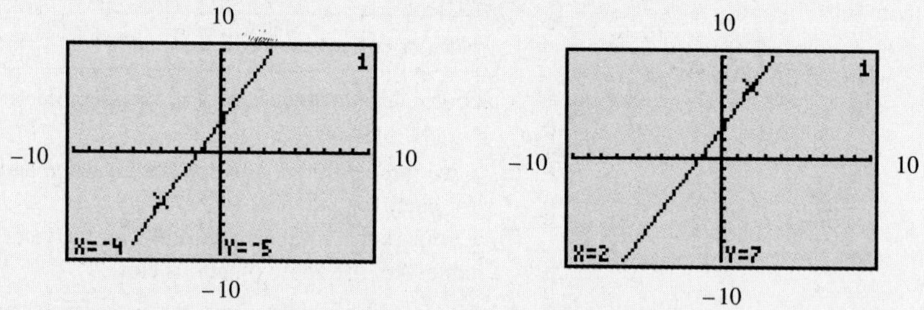

62. Repeat Exercise 61 for the line shown here.

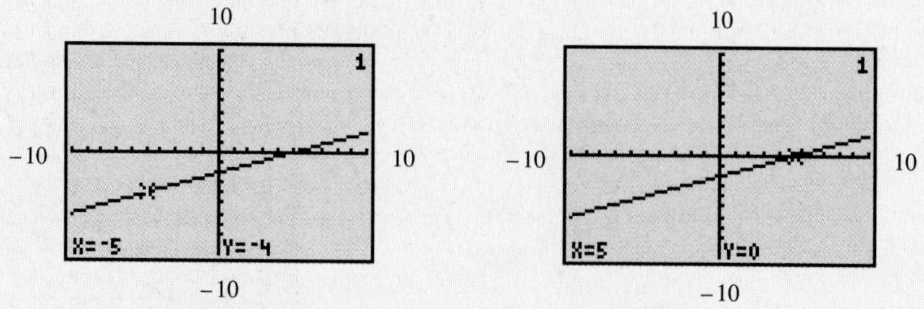

63. $\frac{2}{5}$ 64. $(-10, 0)$

65. $(0, 4)$ 66. (b)

TECHNOLOGY INSIGHTS (EXERCISES 61–66) (CONTINUED)

Some graphing calculators have the capability of displaying a table of points for a graph. The table shown here gives several points that lie on a line designated Y_1.

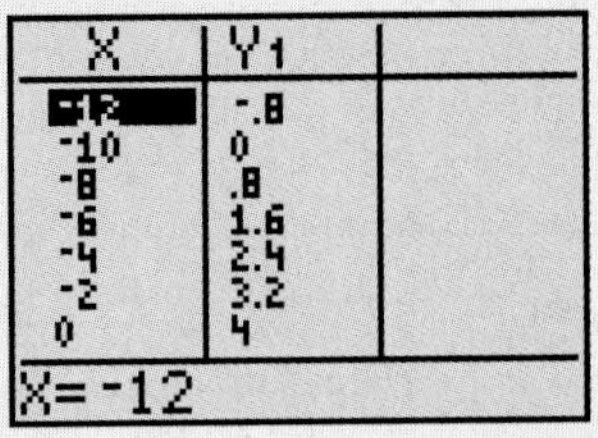

63. Use any pair of points displayed to find the slope of the line.
64. What is the x-intercept of the line?
65. What is the y-intercept of the line?
66. Which one of the two lines shown is the graph of Y_1?

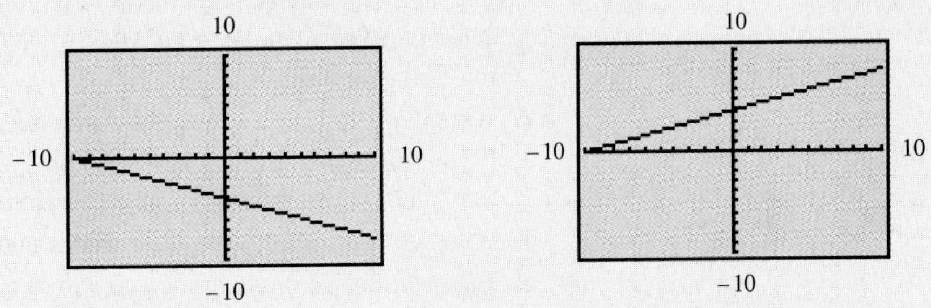

(a) (b)

CHAPTER 3 GROUP ACTIVITY

Instructor's Note
Grouping: 2 students
Time: 30 minutes

This activity can be done either on graph paper or using a graphing calculator. Another option is to assign each group a given price for the popcorn and then have each group report their equation to the class. Equations could then be graphed on a graphing calculator. In part B, answers will vary so groups may exchange papers and check each other's answers. (Students may need help determining graphing scale. This is a good place to apply the concepts of domain and range.)

Determining Business Profit

Objective: Use equations, tables, and graphs to make business decisions.

Graphs are used by businesses to analyze and make decisions. This activity will explore such a process.

The Parent Teacher Organization of a school decides to sell bags of popcorn as a fund-raiser. The parents want to estimate their profit. Costs include \$14 for popcorn (enough for 680 bags) and \$7 for bags to hold the popcorn.

A. Write a profit formula.

1. From the information above, determine the total costs. (Assume the organization is buying enough supplies for 680 bags of popcorn.)
2. If the bags sell for \$.25 each, write an expression for the total sales of n bags of popcorn.
3. Since Profit = Sales − Cost, write an equation that represents the profit P for n bags of popcorn sold.
4. Work in pairs. One person should use the profit equation from above to complete the table on the next page. The other person should decide on an appropriate scale and graph the profit equation.

(continued)

A. 1. \$21 2. $.25n$
3. $P = .25n - 21$
4.

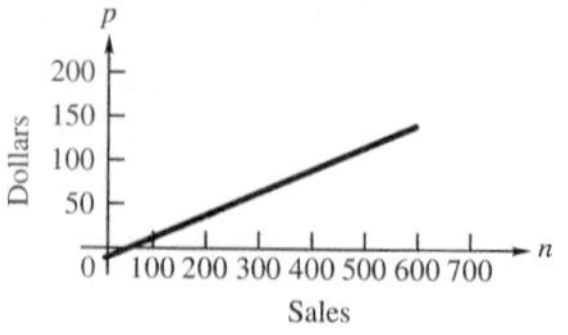

See also the answers within the table.
B. Answers will vary.
C. 1. A. 84; B. Answers will vary.
2. a loss 3. A. \$149;
B. Answers will vary.
4. Answers will vary and may include the idea that sales decrease as price increases.
5. Answers will vary.

n	P
0	[−21]
[84]	0
100	[4]

B. Choose a different price for a bag of popcorn (between \$.20 and \$.75).

1. Write the profit equation for this cost.
2. Switch roles, that is, if you drew the graph in part A, now make a table of values for this equation. Have your partner graph the profit equation on the same coordinate system you used in part A.

C. Compare your findings and answer the following questions.

1. What is the break-even point (that is, when profits are 0 or sales equal costs) for each equation?
2. What does it mean if you end up with a negative value for P?
3. If you sell all 680 bags, what will your profits be for the two different prices? Explain how you would estimate this from your graph.
4. What are the advantages and/or disadvantages of charging a higher price?
5. Together, decide on the price you would charge for a bag of popcorn. Explain why you chose this price.

CHAPTER 3 SUMMARY

KEY TERMS

3.1 **linear equation**, **ordered pair**, **table of values**, ***x*-axis**, ***y*-axis**, **rectangular (Cartesian) coordinate system**, **quadrants**, **origin**, **coordinates**, **plot**
3.2 **graph, graphing**, ***y*-intercept**, ***x*-intercept**, **function**
3.3 **subscript notation**, **slope**, **rise**, **run**, **parallel lines**, **perpendicular lines**

NEW SYMBOLS

(a, b)	an ordered pair	(x_1, y_1)	x-sub-one, y-sub-one	m	slope

TEST YOUR WORD POWER

See how well you have learned the vocabulary in this chapter. Answers, with examples, are given at the bottom of the next page.

1. An **ordered pair** is a pair of numbers written
(a) in numerical order between brackets
(b) between parentheses or brackets
(c) between parentheses in which order is important
(d) between parentheses in which order does not matter.

TEST YOUR WORD POWER (CONTINUED)

2. The **coordinates** of a point are
(a) the numbers in the corresponding ordered pair
(b) the solution of an equation
(c) the values of the x- and y-intercepts
(d) the graph of the point.

3. An **intercept** is
(a) the point where the x-axis and y-axis intersect
(b) a pair of numbers written in parentheses in which order matters
(c) one of the four regions determined by a rectangular coordinate system
(d) the point where a graph intersects the x-axis or the y-axis.

4. A **function** is
(a) the numbers in an ordered pair
(b) a set of ordered pairs in which each x-value corresponds to exactly one y-value
(c) a pair of numbers written between parentheses in which order matters
(d) the set of all ordered pairs that satisfy an equation.

5. The **slope** of a line is
(a) the measure of the run over the rise of the line
(b) the distance between two points on the line
(c) the ratio of the change in y to the change in x along the line
(d) the horizontal change compared to the vertical change of two points on the line.

6. Two lines in a plane are **parallel** if
(a) they represent the same line
(b) they never intersect
(c) they intersect at a 90° angle
(d) one has a positive slope and one has a negative slope.

7. Two lines in a plane are **perpendicular** if
(a) they represent the same line
(b) they never intersect
(c) they intersect at a 90° angle
(d) one has a positive slope and one has a negative slope.

QUICK REVIEW

CONCEPTS	EXAMPLES
3.1 LINEAR EQUATIONS IN TWO VARIABLES	
An ordered pair is a solution of an equation if it satisfies the equation.	Is $(2, -5)$ or $(0, -6)$ a solution of $4x - 3y = 18$? $4(2) - 3(-5) = 23 \neq 18$; $(2, -5)$ is not a solution. $4(0) - 3(-6) = 18$; $(0, -6)$ is a solution.
If a value of either variable in an equation is given, the other variable can be found by substitution.	Complete the ordered pair $(0,\quad)$ for $3x = y + 4$. $3(0) = y + 4$; $0 = y + 4$; $-4 = y$. The ordered pair is $(0, -4)$.
Plot the ordered pair $(-2, 4)$ by starting at the origin, going 2 units to the left, then going 4 units up.	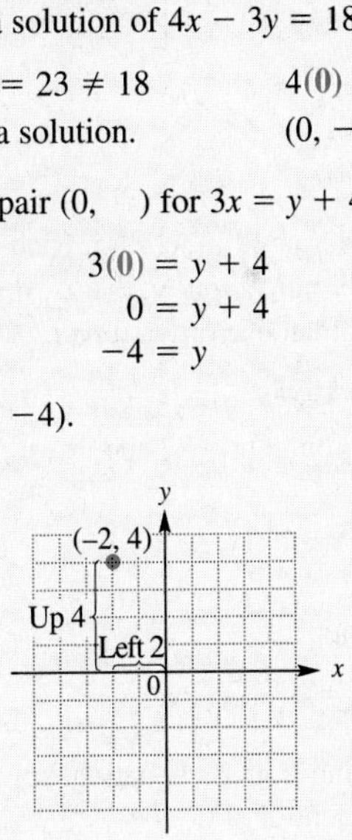

Answers to Test Your Word Power

1. (c) *Examples:* $(0, 3)$, $(3, 8)$, $(4, 0)$ **2.** (a) *Example:* The point associated with the ordered pair $(1, 2)$ has x-coordinate 1 and y-coordinate 2. **3.** (d) *Example:* The graph of the equation $4x - 3y = 12$ has x-intercept at $(3, 0)$ and y-intercept at $(0, -4)$. **4.** (b) *Example:* The set $\{(2, 0), (4, 3), (6, 6), (8, 9)\}$ defines a function. **5.** (c) *Example:* The line through $(3, 6)$ and $(5, 4)$ has slope $\frac{4-6}{5-3} = \frac{-2}{2} = -1$. **6.** (b) *Example:* See Figure 20 in Section 3.3. **7.** (c) *Example:* See Figure 21 in Section 3.3.

CONCEPTS	EXAMPLES
3.2 GRAPHING LINEAR EQUATIONS IN TWO VARIABLES	
The graph of $y = k$ is a horizontal line through $(0, k)$. The graph of $x = k$ is a vertical line through $(k, 0)$.	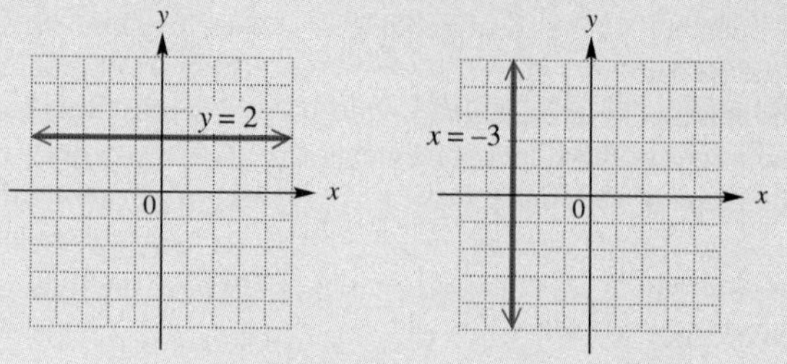
The graph of $Ax + By = 0$ goes through the origin. Find and plot another point that satisfies the equation. Then draw the line through the two points.	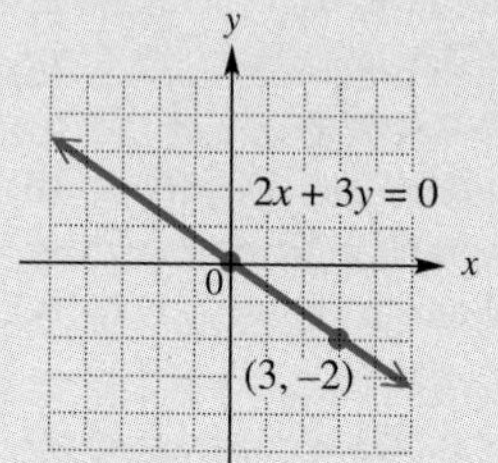
To graph a linear equation: **1.** Find at least two ordered pairs that satisfy the equation. **2.** Plot the corresponding points. **3.** Draw a straight line through the points.	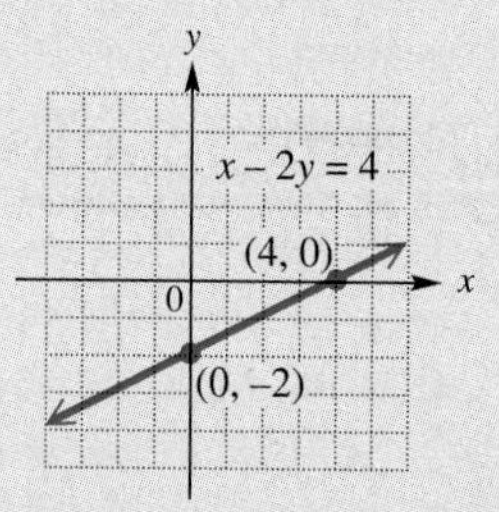
A set of ordered pairs defines a function if each input x corresponds to exactly one output y; that is, if no x-value occurs in more than one ordered pair.	The set $\{(4, 3), (8, 6), (12, 9)\}$ defines a function, but the set $\{(-1, 2), (0, 2), (-1, -2)\}$ does not because -1 is the x-value in two ordered pairs.
3.3 THE SLOPE OF A LINE	
The slope of the line through (x_1, y_1) and (x_2, y_2) is $m = \frac{y_2 - y_1}{x_2 - x_1}$ $(x_1 \neq x_2)$.	The line through $(-2, 3)$ and $(4, -5)$ has slope $m = \frac{-5 - 3}{4 - (-2)} = \frac{-8}{6} = -\frac{4}{3}$.
Horizontal lines have slope 0.	The line $y = -2$ has slope **0**.
Vertical lines have undefined slope.	The line $x = 4$ has **undefined slope.**
To find the slope of a line from its equation, solve for y. The slope is the coefficient of x.	Find the slope of $3x - 4y = 12$. $-4y = -3x + 12$ $y = \frac{3}{4}x - 3$ The slope is $\frac{3}{4}$.
Parallel lines have the same slope.	The lines $y = 3x - 1$ and $y = 3x + 4$ are parallel because both have slope 3.
The slopes of perpendicular lines are negative reciprocals (that is, their product is -1).	The lines $y = -3x - 1$ and $y = \frac{1}{3}x + 4$ are perpendicular because their slopes are -3 and $\frac{1}{3}$ and $-3\left(\frac{1}{3}\right) = -1$.

CHAPTER 3 REVIEW EXERCISES

[3.1] *Complete the given ordered pairs for each equation.*

1. $y = 3x + 2$; $(-1, \quad)$, $(0, \quad)$, $(\quad, 5)$

2. $4x + 3y = 6$; $(0, \quad)$, $(\quad, 0)$, $(-2, \quad)$

3. $x - 7 = 0$; $(\quad, -3)$, $(\quad, 0)$, $(\quad, 5)$

Decide whether the given ordered pair is a solution of the given equation.

4. $x + y = 7$; $(2, 5)$

5. $2x + y = 5$; $(-1, 3)$

6. $3x - y = 4$; $\left(\frac{1}{3}, -3\right)$

Name the quadrant in which each pair lies. Then plot each ordered pair in a rectangular coordinate system.

7. $(2, 3)$

8. $(-4, 2)$

9. $(3, 0)$

10. $(0, -6)$

11. If $xy > 0$, in what quadrant or quadrants must (x, y) lie?

12. On what axis does the point $(k, 0)$ lie for any real value of k?

[3.2] *Find the x- and y-intercepts for the line that is the graph of each equation, and graph the line.*

13. $y = 2x + 5$

14. $3x + 2y = 8$

15. $x + 2y = -4$

16. $x + y = 0$

Decide whether each set defines a function.

17. $\{(5, -3), (-4, -3), (-1, 8), (2, 8)\}$

18. The ordered pairs that satisfy the equation $3x - 9 = 0$

19. The following relationship between years and sponsorship spending: Sponsorship spending in North America (in billions of dollars) is closely approximated by $y = .425x - 1.225$, where x is the number of years since 1980. For example, in 1985, $x = 5$ and $y = .425(5) - 1.225 = .9$ billion dollars. (*Source: IEG Sponsorship Report.*)

20. Describe the input values and the output values for the relationship in Exercise 19.

[3.3] *Find the slope of each line.*

21. Through $(2, 3)$ and $(-4, 6)$

22. Through $(0, 6)$ and $(1, 6)$

23. Through $(2, 5)$ and $(2, 8)$

24. $y = 3x - 4$

25. $y = \frac{2}{3}x + 1$

26.

X	Y_1
0	1
1	2.5
2	4
3	5.5
4	7
5	8.5
6	10

X=0

---

1. -1; 2; 1 2. 2; $\frac{3}{2}$; $\frac{14}{3}$
3. 7; 7; 7 4. yes 5. no
6. yes 7. I 8. II 9. none
10. none

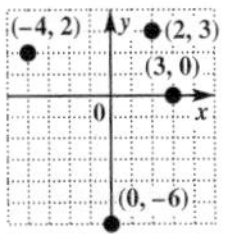

11. I or III 12. x-axis
13. $\left(-\frac{5}{2}, 0\right)$; $(0, 5)$

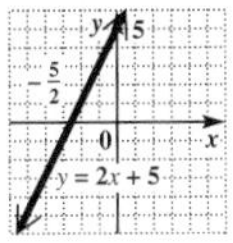

14. $\left(\frac{8}{3}, 0\right)$; $(0, 4)$

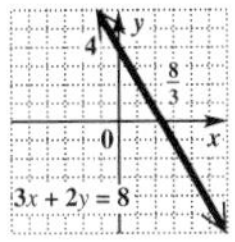

15. $(-4, 0)$; $(0, -2)$

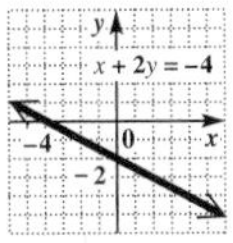

16. $(0, 0)$; $(0, 0)$

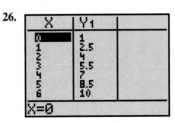

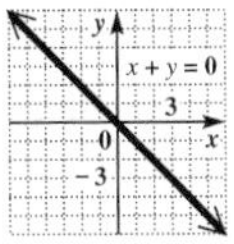

17. function 18. not a function 19. function
20. input values: years since 1980; output values: sponsorship spending in billions of dollars
21. $-\frac{1}{2}$ 22. 0 23. undefined
24. 3 25. $\frac{2}{3}$ 26. $\frac{3}{2}$

27. $-\frac{4}{7}$ **28.** 0 **29.** $\frac{1}{3}$
30. Two nonvertical perpendicular lines must have slopes with opposite signs, because their product is a negative number (-1), and thus one must be positive and the other one must be negative. **31.** parallel **32.** perpendicular **33.** neither

27.

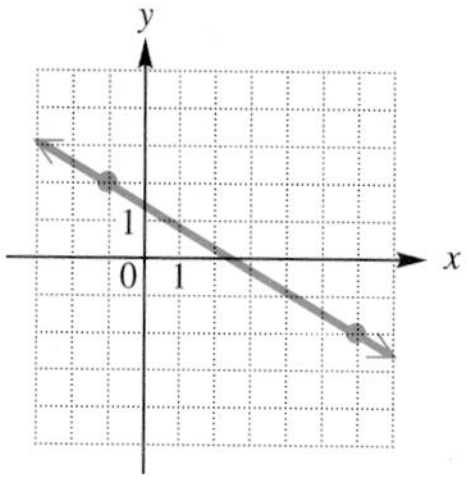

28. $y = 5$

29. A line perpendicular to the graph of $y = -3x + 3$

30. Explain why the signs of the slopes of perpendicular lines (neither of which is vertical) cannot be the same.

Decide whether the lines in each pair are parallel, perpendicular, or neither.

31. $3x + 2y = 6$
$6x + 4y = 8$

32. $x - 3y = 1$
$3x + y = 4$

33. $x - 2y = 8$
$x + 2y = 8$

34. 1992: 4.9%; 1993: 9.6%; 1994: 6.4%; $485 billion
35. 0; $\frac{8}{3}$; -9 **36.** $(-2, 0)$; none; undefined **37.** $\left(-\frac{3}{2}, 0\right)$; (0, 3); 2
38. $\left(\frac{4}{11}, 0\right)$; $\left(0, -\frac{4}{3}\right)$; $\frac{11}{3}$
39. $\frac{1}{3}$ **40.** 0 **41.** yes
42.

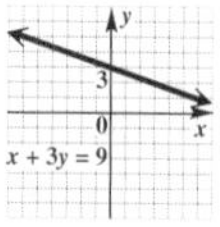

43.

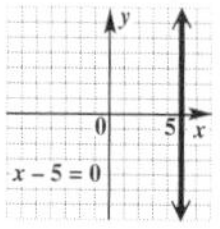

44.

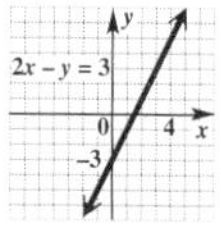

MIXED REVIEW EXERCISES

34. Use the bar graph to estimate the growth rates for motor vehicles and parts sales for the years 1992–1994. (The rate for 1992, for example, represents the percent increase in sales from 1991 to 1992.) Sales in 1993 were approximately $456 billion. Use the growth rate in 1994 to estimate sales in that year.

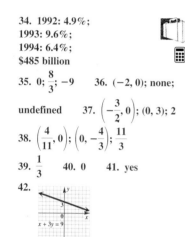

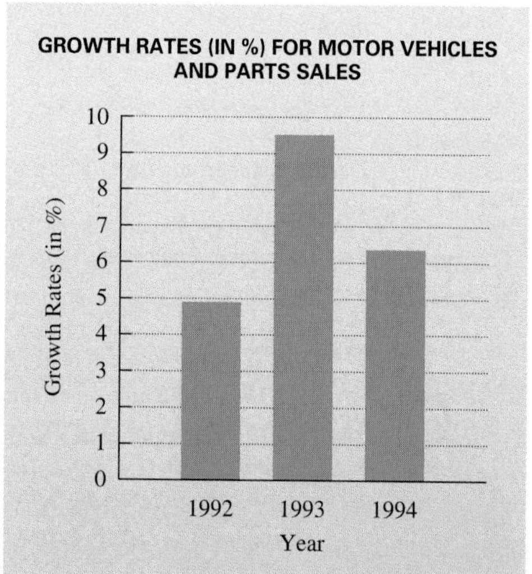

Source: U.S. Department of Commerce, *U.S. Industrial Outlook, 1994.*

35. Complete these ordered pairs for $x = 3y$: (0,), (8,), (, −3).

In Exercises 36–38, find the x- and y-intercepts and slope.

36. $x = -2$

37. $y = 2x + 3$

38. $11x - 3y = 4$

39. Find the slope of the line through $(4, -1)$ and $(-2, -3)$.

40. What is the slope of a line perpendicular to a line with undefined slope?

45. It is wise to plot three points before drawing the line, because if the three points do not lie in a line, then an error has been made and can be corrected.
46. Yes. The inputs are the five cities, and the outputs are the corresponding populations.

41. Is $\left(0, -\frac{16}{3}\right)$ a solution of $5x - 3y = 16$?

Graph each equation given in Exercises 42–44.

42. $x + 3y = 9$ **43.** $x - 5 = 0$ **44.** $2x - y = 3$

45. Two points determine a line. Explain why it is a good idea to plot three points before drawing the line.

46. The set of ordered pairs gives the five U.S. cities with the largest Hispanic populations in 1990. Data is in millions. (*Source:* U.S. Census Bureau.)

{(New York, 1.78), (Los Angeles, 1.39), (Chicago, .55), (San Antonio, .52), (Houston, .45)}

Does this set define a function? What are the inputs and the outputs?

CHAPTER 3 TEST

[3.1] 1. (b) 2. 16%
3. −6; −10; −15
4. −12; −12; −12
5. no [3.2] 6. To find the x-intercept, let $y = 0$, and to find the y-intercept, let $x = 0$.
7. (2, 0); (0, 6)

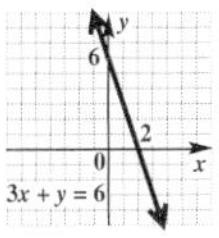

8. no x-intercept; (0, −3)

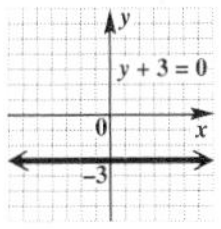

The graph shows the cost of 30 seconds of advertising time on the annual Super Bowl telecast. Use the graph to respond to Exercises 1 and 2.

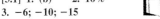

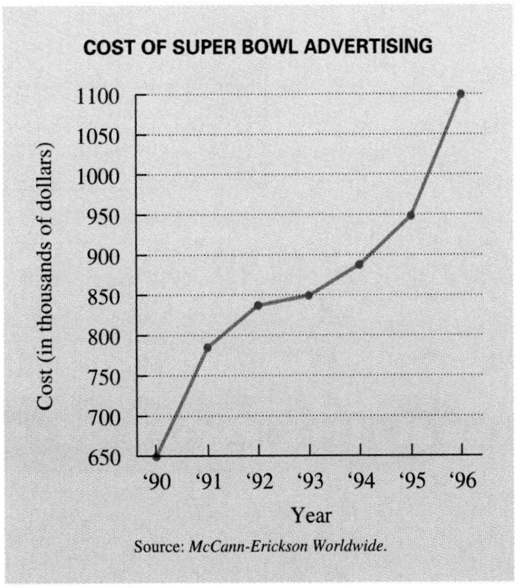

Source: *McCann-Erickson Worldwide.*

1. Which one of the following is the best estimate for the advertising cost in 1993?
(a) \$820,000 **(b)** \$848,000 **(c)** \$910,000

2. To the nearest percent, what was the percent increase in the advertising cost from 1995 to 1996?

Complete the ordered pairs for the given equation.

3. $3x + 5y = -30$; (0,), (, 0), (, 3)

4. $y + 12 = 0$; (0,), (−4,), $\left(\frac{5}{2}, \quad\right)$

5. Is (4, −1) a solution of $4x - 7y = 9$?

6. How do you find the x-intercept and the y-intercept for a linear equation in two variables?

Graph each linear equation. Give the x- and y-intercepts.

7. $3x + y = 6$ **8.** $y + 3 = 0$

9. Yes, because each input determines exactly one output. **10.** The inputs are the years from 1990 to 1996, and the outputs are the corresponding Super Bowl advertising costs in thousands of dollars.

[3.3] **11.** $-\frac{8}{3}$ **12.** -2 **13.** $\frac{5}{2}$

14. $\frac{1}{2}$ **15.** -4 **16.** $\frac{1}{4}$

9. Does the set of ordered pairs {(90, 650), (91, 780), (92, 830), (93, 850), (94, 880), (95, 950), (96, 1100)}, shown as points on the graph for Exercises 1 and 2, define a function? Explain why or why not.

10. Describe the inputs and outputs for the set of ordered pairs in Exercise 9.

Find the slope of each line in Exercises 11–16.

11. Through $(-4, 6)$ and $(-1, -2)$

12. $2x + y = 10$

13.

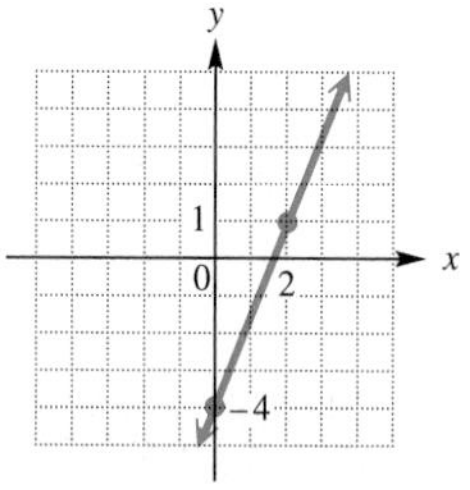

14. (These are two different views of the same line.)

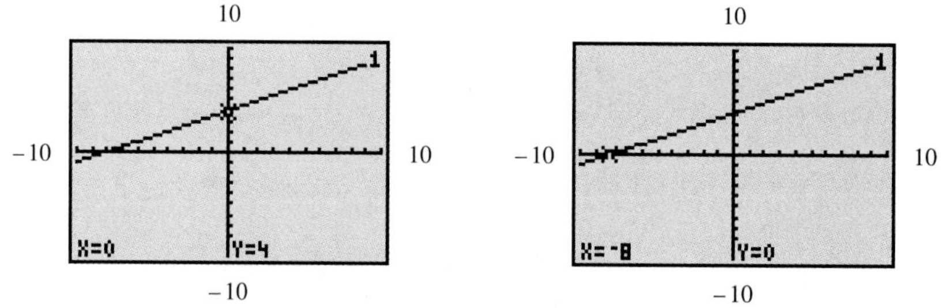

15. A line parallel to the graph of $y = -4x + 6$

16. A line perpendicular to the graph of $y = -4x + 6$

CUMULATIVE REVIEW EXERCISES CHAPTERS 1–3

[1.1] **1.** $\frac{551}{40}$ or $13\frac{31}{40}$ **2.** $\frac{6}{5}$

[1.5] **3.** 7 [1.6] **4.** $\frac{73}{18}$

[1.1] **5.** $\frac{5}{12}$ [1.4] **6.** true

[1.3] **7.** -134

[1.7] **8.** distributive property

[1.8] **9.** $-p + 2$

Perform the indicated operations.

1. $16\frac{7}{8} - 3\frac{1}{10}$

2. $\frac{3}{4} \div \frac{5}{8}$

3. $-11 + 20 + (-2)$

4. $\frac{(-3)^2 - (-4)(2^4)}{5 \cdot 2 - (-2)^3}$

5. Rafael and Elda Muñoz are painting a bedroom for their new baby. They painted $\frac{1}{4}$ of the room on Saturday and $\frac{1}{3}$ of the room on Sunday. How much of the room is still unpainted?

6. True or false? $\frac{4(3-9)}{2-6} \geq 6$

7. Find the value of $xz^3 - 5y^2$ when $x = -2$, $y = -4$, and $z = 3$.

8. What property does $3(-2 + x) = -6 + 3x$ illustrate?

9. Simplify $-4p - 6 + 3p + 8$ by combining terms.

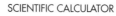

[2.4] 10. $h = \frac{3V}{\pi r^2}$
[2.2] 11. $\{-1\}$ 12. $\{2\}$
[2.2, 2.5] 13. $\{-13\}$
[2.7] 14. $(-\infty, 2]$
15. $\left(-\frac{8}{3}, -\frac{4}{3}\right)$
16. $300 or more
[2.4] 17. 6 miles
[2.6] 18. 10 liters
[3.1] 19.

x	y
12	89.45
28	81.95
36	78.20

20. $7000; $10,000

Solve.

10. $V = \frac{1}{3}\pi r^2 h$ for h

11. $6 - 3(1 + a) = 2(a + 5) - 2$

12. $-(m - 1) = 3 - 2m$

13. $\frac{y - 2}{3} = \frac{2y + 1}{5}$

14. $-5z \geq 4z - 18$

15. $2 < -6(z + 1) < 10$

Solve each problem.

16. Kimshana Lavoris earned $200 working part time during July, $375 during August, and $325 during September. If her average income for the four months from July through October must be at least $300, what possible amounts could she earn in October?

17. Mount Mayon in the Philippines is the most perfectly shaped conical volcano in the world. Its base is a perfect circle with a 39-mile circumference and it has a height of 8000 feet. (One mile is 5280 feet.) Find the radius of the circular base to the nearest mile. (*Hint:* This problem has some unneeded information.)

18. How much of a 20% chemical solution must be mixed with 30 liters of a 60% solution to get a 50% mixture?

19. The winning times in seconds for the women's 1000-meter speed skating event in the Winter Olympics for the years from 1960 to 1998 can be closely approximated by the equation

$$y = -.4685x + 95.07,$$

where x is the number of years since 1960. That is, $x = 5$ represents 1965, $x = 10$ represents 1970, and so on. Complete the table of ordered pairs for this linear equation. Round the y-values to the nearest hundredth of a second. (*Source: The Universal Almanac,* 1997, John W. Wright, General Editor.)

x	y
12	
28	
36	

20. Baby boomers are expected to inherit $10.4 trillion from their parents over the next 45 years, an average of $50,000 each. The pie chart shows how they plan to spend their inheritance. How much of the $50,000 is expected to go toward the purchase of a home? How much to retirement?

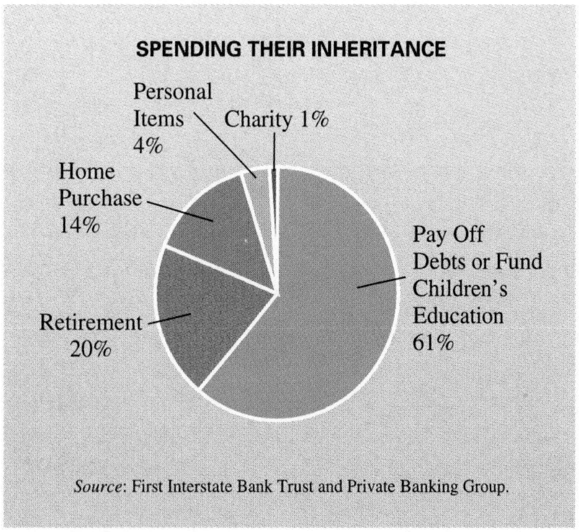

Consider the linear equation $3x + 2y = 12$. *Find the following.*

21. The x- and y-intercepts **22.** The graph **23.** The slope

24. Are the lines with equations $x + 5y = -6$ and $y = 5x - 8$ parallel, perpendicular, or neither?

25. California's exports to Pacific Rim nations, with the exception of China, fell during the first half of 1998, as shown in the bar graph. Is the set whose ordered pairs represent the data from the graph as (importer, % change) a function? Give the ordered pairs with the smallest change and the largest change. How is absolute value used in your answer?

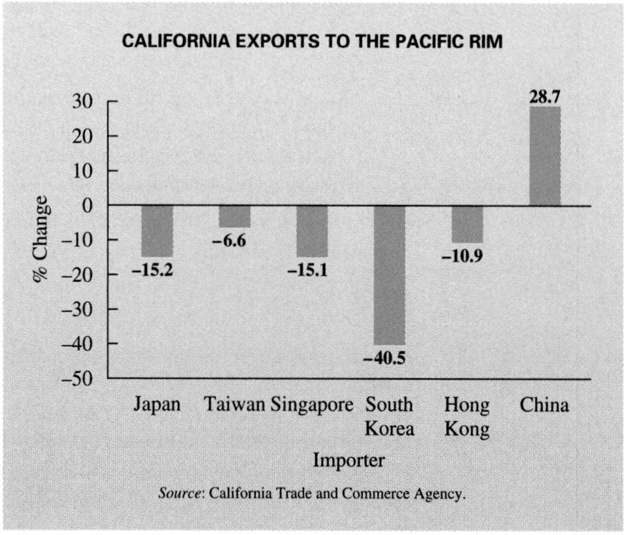

[3.2] 21. **x-intercept: (4, 0); y-intercept: (0, 6)**

22.

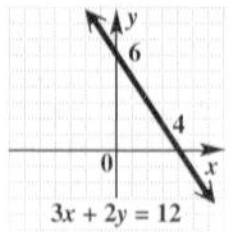

[3.3] 23. $-\frac{3}{2}$

24. perpendicular

25. Yes; (Taiwan, −6.6), (South Korea, −40.5); To determine the smallest and largest *change*, in the usual sense, we need to consider the absolute value of the changes.

Polynomials and Exponents

4

The number of passengers traveling by air has increased rapidly in the last decade. (See Exercise 46 in Section 4.1.) From 1985 to 1995, scheduled air carriers enjoyed an increase in net profits of $1,514,000,000. Surprisingly, in spite of more passengers traveling in planes that are consistently full, consumer complaints against U.S. airlines have generally decreased, as shown in the figure. There is one year, however, when all three graphs indicate an increase in each category from one year to the next, rather than a decrease. In which year did this occur?

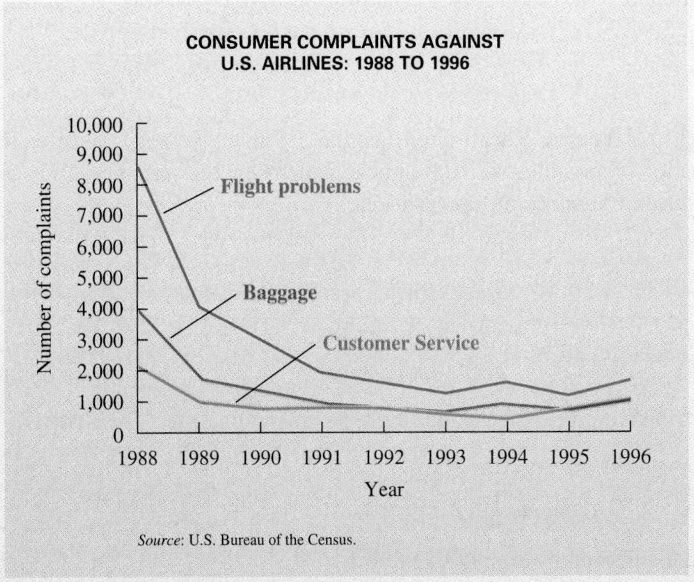

Source: U.S. Bureau of the Census.

The graphs in the figure can each be approximated by a *polynomial function.* Polynomials are one of the topics studied in this chapter. Throughout the chapter we will see other examples of polynomials that describe information about the aeronautics industry.

Aeronautics

Visit our Web site at www.LialAlgebra.com

4.1 Addition and Subtraction of Polynomials; Graphing Simple Polynomials

OBJECTIVES

1. Identify terms and coefficients.
2. Add like terms.
3. Know the vocabulary for polynomials.
4. Evaluate polynomials.
5. Add and subtract polynomials.
6. Graph equations defined by polynomials with degree 2.

FOR EXTRA HELP

SSG Sec. 4.1
SSM Sec. 4.1
Pass the Test Software
InterAct Math Tutorial Software
Video 6

OBJECTIVE 1 **Identify terms and coefficients.** In Chapter 1 we saw that in an expression such as

$$4x^3 + 6x^2 + 5x + 8,$$

the quantities $4x^3$, $6x^2$, $5x$, and 8 are called *terms.* As mentioned earlier, in the term $4x^3$, the number 4 is called the *numerical coefficient,* or simply the *coefficient,* of x^3. In the same way, 6 is the coefficient of x^2 in the term $6x^2$, 5 is the coefficient of x in the term $5x$, and 8 is the coefficient in the term 8. A constant term, like 8 in the expression above, can be thought of as $8x^0$, where

$$x^0 \text{ is defined to equal } 1.$$

We explain the reason for this definition later in this chapter.

CHALKBOARD EXAMPLE

Name the coefficient of each term in the expression $2x^3 - x$.

Answer: 2, −1

EXAMPLE 1 Identifying Coefficients

Name the (numerical) coefficient of each term in these expressions.

(a) $4x^3$

The coefficient is 4.

(b) $x - 6x^4$

The coefficient of x is 1 because $x = 1 \cdot x$. The coefficient of x^4 is -6 since $x - 6x^4$ can be written as the sum $x + (-6x^4)$.

(c) $5 - v^3$

The coefficient of the term 5 is 5 because $5 = 5v^0$. By writing $5 - v^3$ as a sum, $5 + (-v^3)$, or $5 + (-1v^3)$, the coefficient of v^3 can be identified as -1.

OBJECTIVE 2 **Add like terms.** Recall from Section 1.8 that *like terms* have exactly the same combination of variables with the same exponents on the variables. Only the coefficients may differ. Examples of like terms are

$$19m^5 \quad \text{and} \quad 14m^5,$$
$$6y^9, \quad -37y^9, \quad \text{and} \quad y^9,$$
$$3pq \quad \text{and} \quad -2pq,$$
$$2xy^2 \quad \text{and} \quad -xy^2.$$

Using the distributive property, we add like terms by adding their coefficients.

CHALKBOARD EXAMPLE

Add like terms.

$$r^2 + 3r + 5r^2$$

Answer: $6r^2 + 3r$

TEACHING TIP Note in Example 2(c) that m and m^2 are not like terms.

EXAMPLE 2 Adding Like Terms

Simplify each expression by adding like terms.

(a) $-4x^3 + 6x^3 = (-4 + 6)x^3 = 2x^3$ Distributive property

(b) $9x^6 - 14x^6 + x^6 = (9 - 14 + 1)x^6 = -4x^6$

(c) $12m^2 + 5m + 4m^2 = (12 + 4)m^2 + 5m = 16m^2 + 5m$

(d) $3x^2y + 4x^2y - x^2y = (3 + 4 - 1)x^2y = 6x^2y$

INSTRUCTOR'S RESOURCES

PTB Sec. 4.1 **ISM** Sec. 4.1
AB Sec. 4.1

TEST GENERATOR

WORLD WIDE WEB
www.LialAlgebra.com

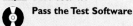

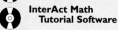

In Example 2(c), we cannot combine $16m^2$ and $5m$. These two terms are unlike because the exponents on the variables are different. *Unlike terms* have different variables or different exponents on the same variables.

OBJECTIVE 3 **Know the vocabulary for polynomials.** A **polynomial in x** is a term or the sum of a finite number of terms of the form ax^n, for any real number a and any whole number n. For example,

$$16x^8 - 7x^6 + 5x^4 - 3x^2 + 4$$

is a polynomial in x (the 4 can be written as $4x^0$). This polynomial is written in **descending powers** of the variable, since the exponents on x decrease from left to right. On the other hand,

$$2x^3 - x^2 + \frac{4}{x}$$

is not a polynomial in x, since a variable appears in a denominator. Of course, we could define *polynomial* using any variable and not just x, as in Example 2(c). In fact, polynomials may have terms with *more* than one variable, as in Example 2(d).

The **degree of a term** is the sum of the exponents on the variables. For example, $3x^4$ has degree 4, while $6x^{17}$ has degree 17. The term $5x$ has degree 1, -7 has degree 0 (since -7 can be written as $-7x^0$), and $2x^2y$ has degree $2 + 1 = 3$ (y has an exponent of 1.) The **degree of a polynomial** is the highest degree of any nonzero term of the polynomial. For example, $3x^4 - 5x^2 + 6$ is of degree 4, the polynomial $5x + 7$ is of degree 1, 3 (or $3x^0$) is of degree 0, and $x^2y + xy - 5xy^2$ is of degree 3.

Three types of polynomials are very common and are given special names. A polynomial with exactly three terms is called a **trinomial.** (*Tri-* means "three," as in *tri*angle.) Examples are

$$9m^3 - 4m^2 + 6, \qquad 19y^2 + 8y + 5, \qquad \text{and} \qquad -3m^5n^2 + 2n^3 - m^4.$$

A polynomial with exactly two terms is called a **binomial.** (*Bi-* means "two," as in *bi*cycle.) Examples are

$$-9x^4 + 9x^3, \qquad 8m^2 + 6m, \qquad \text{and} \qquad 3m^5n^2 - 9m^2n^4.$$

A polynomial with only one term is called a **monomial.** *(Mon(o)-* means "one," as in *mono*rail.) Examples are

$$9m, \qquad -6y^5, \qquad a^2b^2, \qquad \text{and} \qquad 6.$$

CHALKBOARD EXAMPLE

Simplify, give the degree, and tell whether the simplified polynomial is a monomial, binomial, trinomial, or none of these.

$$x^8 - x^7 + 2x^8$$

Answer: $3x^8 - x^7$; degree 8; binomial

EXAMPLE 3 **Classifying Polynomials**

For each polynomial, first simplify if possible by combining like terms. Then give the degree and tell whether it is a monomial, a binomial, a trinomial, or none of these.

(a) $2x^3 + 5$
The polynomial cannot be simplified. The degree is 3. The polynomial is a binomial.

(b) $4xy - 5xy + 2xy$
Add like terms to simplify: $4xy - 5xy + 2xy = xy$, which is a monomial of degree 2.

OBJECTIVE 4 **Evaluate polynomials.** A polynomial usually represents different numbers for different values of the variable, as shown in the next example.

CHALKBOARD EXAMPLE

Find the value of $2y^3 + 8y - 6$ when $y = -1$.

Answer: -16

EXAMPLE 4 Evaluating a Polynomial

Find the value of $3x^4 + 5x^3 - 4x - 4$ when $x = -2$ and when $x = 3$.

First, substitute -2 for x.

$$\begin{aligned}3x^4 + 5x^3 - 4x - 4 &= 3(-2)^4 + 5(-2)^3 - 4(-2) - 4\\ &= 3 \cdot 16 + 5 \cdot (-8) + 8 - 4\\ &= 48 - 40 + 8 - 4\\ &= 12\end{aligned}$$

Next, replace x with 3.

$$\begin{aligned}3x^4 + 5x^3 - 4x - 4 &= 3(3)^4 + 5(3)^3 - 4(3) - 4\\ &= 3 \cdot 81 + 5 \cdot 27 - 12 - 4\\ &= 362\end{aligned}$$

CAUTION Notice the use of parentheses around the numbers that are substituted for the variable in Example 4. This is particularly important when substituting a negative number for a variable that is raised to a power, so that the sign of the product is correct.

CONNECTIONS

In Section 3.2 we introduced the idea of a function: for every input x, there is one output y. Polynomials often provide a way of defining functions that approximate data collected over a period of time. For example, according to the U.S. National Aeronautics and Space Administration (NASA), the budget in millions of dollars for space station research for 1996–2001 can be approximated by the polynomial equation

$$y = -10.25x^2 - 126.04x + 5730.21,$$

where $x = 0$ represents 1996, $x = 1$ represents 1997, and so on, up to $x = 5$ representing 2001. The actual budget for 1998 was 5327 million dollars; an input of $x = 2$ (for 1998) gives approximately $y = 5437$. Considering the magnitude of the numbers, this is a very good approximation.

FOR DISCUSSION OR WRITING

Use the given polynomial equation to approximate the budget in other years between 1996 and 2001. Compare to the actual figures given here.

Year	*Budget (in millions of dollars)*
1996	5710
1997	5675
1998	5327
1999	5306
2000	5077
2001	4832

The polynomial gives the following: for 1996, 5730 million (greater than actual); for 1997, 5594 million (less than actual); for 1999, 5260 million (less than actual); for 2000, 5062 million (less than actual); for 2001, 4844 million (greater than actual).

OBJECTIVE 5 **Add and subtract polynomials.** Polynomials may be added, subtracted, multiplied, and divided.

Adding Polynomials

To add two polynomials, add like terms.

CHALKBOARD EXAMPLE

Add.

$$\begin{array}{r} 4x^3 - 3x^2 + 2x \\ \underline{6x^3 + 2x^2 - 3x} \end{array}$$

Answer: $10x^3 - x^2 - x$

EXAMPLE 5 **Adding Polynomials Vertically**

Add $6x^3 - 4x^2 + 3$ and $-2x^3 + 7x^2 - 5$.

Write like terms in columns.

$$\begin{array}{r} 6x^3 - 4x^2 + 3 \\ \underline{-2x^3 + 7x^2 - 5} \end{array}$$

Now add, column by column.

$$\begin{array}{rrr} 6x^3 & -4x^2 & 3 \\ \underline{-2x^3} & \underline{7x^2} & \underline{-\ 5} \\ 4x^3 & 3x^2 & -2 \end{array}$$

Add the three sums together.

$$4x^3 + 3x^2 + (-2) = 4x^3 + 3x^2 - 2$$

The polynomials in Example 5 also can be added horizontally, as shown in the next example.

CHALKBOARD EXAMPLE

Add.

$$(2x^4 - 6x^2 + 7) + (-3x^4 + 5x^2 + 2)$$

Answer: $-x^4 - x^2 + 9$

EXAMPLE 6 **Adding Polynomials Horizontally**

Add $6x^3 - 4x^2 + 3$ and $-2x^3 + 7x^2 - 5$.

Write the sum as

$$(6x^3 - 4x^2 + 3) + (-2x^3 + 7x^2 - 5).$$

Use the associative and commutative properties to rewrite this sum with the parentheses removed and with the subtractions changed to additions of inverses.

$$6x^3 + (-4x^2) + 3 + (-2x^3) + 7x^2 + (-5)$$

Place like terms together.

$$6x^3 + (-2x^3) + (-4x^2) + 7x^2 + 3 + (-5)$$

Combine like terms to get

$$4x^3 + 3x^2 + (-2), \qquad \text{or simply} \qquad 4x^3 + 3x^2 - 2,$$

the same answer found in Example 5.

Earlier, we defined the difference $x - y$ as $x + (-y)$. (We find the difference $x - y$ by adding x and the opposite of y.) For example,

$$7 - 2 = 7 + (-2) = 5 \qquad \text{and} \qquad -8 - (-2) = -8 + 2 = -6.$$

A similar method is used to subtract polynomials.

Subtracting Polynomials

To subtract two polynomials, change all the signs on the second polynomial and add the result to the first polynomial.

CHALKBOARD EXAMPLE

(a) Perform the subtraction $(7y^2 - 11y + 8) - (-3y^2 + 4y + 6)$.

(b) Subtract $(2y^3 - 7y^2 - 4y + 6)$ from $(14y^3 - 6y^2 + 2y - 5)$.

Answer:

(a) $10y^2 - 15y + 2$

(b) $12y^3 + y^2 + 6y - 11$

EXAMPLE 7 Subtracting Polynomials

(a) Perform the subtraction $(5x - 2) - (3x - 8)$.
By the definition of subtraction,

$$(5x - 2) - (3x - 8) = (5x - 2) + [-(3x - 8)].$$

As shown in Chapter 1, the distributive property gives

$$-(3x - 8) = -\mathbf{1}(3x - 8) = -3x + 8,$$

so

$$(5x - 2) - (3x - 8) = (5x - 2) + (-3x + 8) = 2x + 6.$$

(b) Subtract $6x^3 - 4x^2 + 2$ from $11x^3 + 2x^2 - 8$.
Write the problem.

$$(11x^3 + 2x^2 - 8) - (6x^3 - 4x^2 + 2)$$

Change all the signs in the second polynomial and add the two polynomials.

$$(11x^3 + 2x^2 - 8) + (-6x^3 + 4x^2 - 2) = 5x^3 + 6x^2 - 10$$

To check a subtraction problem, use the fact that if $a - b = c$, then $a = b + c$. For example, $6 - 2 = 4$, so we check by writing $6 = 2 + 4$, which is correct. Check the polynomial subtraction above by adding $6x^3 - 4x^2 + 2$ and $5x^3 + 6x^2 - 10$. Since the sum is $11x^3 + 2x^2 - 8$, the subtraction was performed correctly.

Subtraction also can be done in columns (vertically). We will use vertical subtraction in Section 4.6 when we study polynomial division.

CHALKBOARD EXAMPLE

Subtract.

$$\begin{array}{r} 14y^3 - 6y^2 + 2y \phantom{{}+6} \\ \underline{2y^3 - 7y^2 \phantom{{}+2y} + 6} \end{array}$$

Answer: $12y^3 + y^2 + 2y - 6$

EXAMPLE 8 Subtracting Polynomials Vertically

Use the method of subtracting by columns to find

$$(14y^3 - 6y^2 + 2y - 5) - (2y^3 - 7y^2 - 4y + 6).$$

Arrange like terms in columns.

$$\begin{array}{r} 14y^3 - 6y^2 + 2y - 5 \\ \underline{2y^3 - 7y^2 - 4y + 6} \end{array}$$

Change all signs in the second row, and then add.

$$\begin{array}{rl} 14y^3 - 6y^2 + 2y - 5 & \\ \underline{-2y^3 + 7y^2 + 4y - 6} & \text{Change all signs.} \\ 12y^3 + y^2 + 6y - 11 & \text{Add.} \end{array}$$

Either the horizontal or the vertical method may be used to add and subtract polynomials.

Polynomials in more than one variable are added and subtracted by combining like terms, just as with single variable polynomials.

CHALKBOARD EXAMPLE

Subtract.

$$(5m^3n + 3m^2n^2 - 4mn) - (7m^3n - m^2n^2 + 6mn)$$

Answer:
$-2m^3n + 4m^2n^2 - 10mn$

EXAMPLE 9 Adding and Subtracting Polynomials with More Than One Variable

Add or subtract as indicated.

(a) $(4a + 2ab - b) + (3a - ab + b)$

$$\begin{aligned}(4a + 2ab - b) + (3a - ab + b) &= 4a + 2ab - b + 3a - ab + b\\ &= 7a + ab\end{aligned}$$

(b) $(2x^2y + 3xy + y^2) - (3x^2y - xy - 2y^2)$

$$\begin{aligned}&(2x^2y + 3xy + y^2) - (3x^2y - xy - 2y^2)\\ &= 2x^2y + 3xy + y^2 - 3x^2y + xy + 2y^2\\ &= -x^2y + 4xy + 3y^2\end{aligned}$$

OBJECTIVE 6 Graph equations defined by polynomials with degree 2. In Chapter 3 we introduced graphs of straight lines. These graphs were defined by linear equations (which are actually polynomial equations of degree 1). By selective point-plotting, we can find the graphs of polynomial equations of degree 2.

CHALKBOARD EXAMPLE

Graph $y = 2x^2$.

Answer:

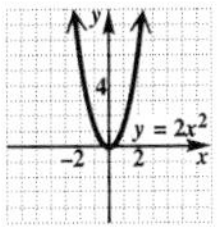

EXAMPLE 10 Graphing Equations Defined by Polynomials with Degree 2

(a) Graph $y = x^2$.

Select several values for x; then find the corresponding y-values. For example, selecting $x = 2$ gives

$$y = 2^2 = 4,$$

and so the point (2, 4) is on the graph of $y = x^2$. (Recall that in an ordered pair such as (2, 4), the x-value comes first and the y-value second.) We show some ordered pairs that satisfy $y = x^2$ in a table next to Figure 1. If the ordered pairs from the table are plotted on a coordinate system and a smooth curve drawn through them, the graph is as shown in Figure 1.

x	y
3	9
2	4
1	1
0	0
−1	1
−2	4
−3	9

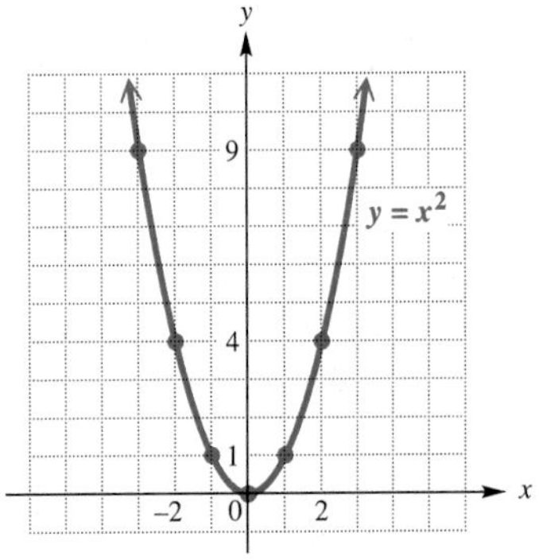

Figure 1

The graph of $y = x^2$ is the graph of a function, since each input x is related to just one output y. The curve in Figure 1 is called a **parabola.** The point (0, 0), the lowest point on this graph, is called the **vertex** of the parabola. The vertical line through the vertex (the y-axis here) is called the **axis** of the parabola. The axis of a parabola is a **line of symmetry** for the graph. If the graph is folded on this line, the two halves will match.

(b) Graph $y = -x^2 + 3$.

Once again plot points to obtain the graph. For example, if $x = -2$,

$$y = -(-2)^2 + 3 = -4 + 3 = -1.$$

This point and several others are shown in the table that accompanies the graph in Figure 2. The vertex of this parabola is (0, 3). This time the vertex is the *highest* point of the graph. The graph opens downward because x^2 has a negative coefficient.

x	y
-2	-1
-1	2
0	3
1	2
2	-1

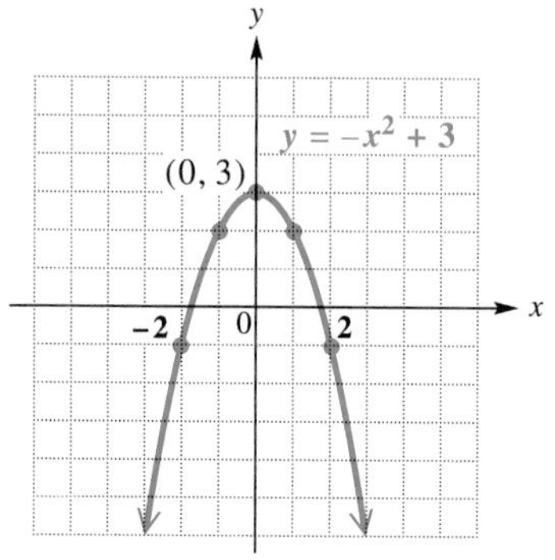

Figure 2

NOTE All polynomials of degree 2 have parabolas as their graphs. When graphing by plotting points, it is necessary to continue finding points until the vertex and points on either side of it are located. (In this section, all parabolas have their vertices on the x-axis or the y-axis.)

4.1 EXERCISES

1. 7; 5 2. two 3. 8
4. is not 5. 26 6. 4
7. 0 8. $5x^9$ 9. 1; 6
10. 1; -9 11. 1; 1
12. 1; 1 13. 2; -19, -1
14. 2; 2, -1 15. 3; 1, 8, 5
16. 3; 1, -2, -1

Fill in each blank with the correct response.

1. In the term $7x^5$, the coefficient is _____ and the exponent is _____.

2. The expression $5x^3 - 4x^2$ has __________ term(s). (how many?)

3. The degree of the term $-4x^8$ is _____.

4. The polynomial $4x^2 - y^2$ __________ an example of a trinomial. (is/is not)

5. When $x^2 + 10$ is evaluated for $x = 4$, the result is _____.

6. $5x$— $+ 3x^3 - 7x$ is a trinomial of degree 4.

7. $3xy + 2xy - 5xy =$ _____.

8. _____ is an example of a monomial with coefficient 5, in the variable x, having degree 9.

For each polynomial, determine the number of terms and name the coefficients of the terms. See Example 1.

9. $6x^4$ **10.** $-9y^5$ **11.** t^4 **12.** s^7

13. $-19r^2 - r$ **14.** $2y^3 - y$ **15.** $x + 8x^2 + 5x^3$ **16.** $v - 2v^3 - v^7$

17. $2m^5$ **18.** $-y^3$ **19.** $-r^5$ **20.** $-10y^2$ **21.** cannot be simplified **22.** cannot be simplified **23.** $-5x^5$ **24.** $7x^3$ **25.** $5p^9 + 4p^7$ **26.** $a^8 - 3a^2$ **27.** $-2y^2$ **28.** $-2r^5$ **29.** already simplified; 4; binomial **30.** already simplified; 3; binomial **31.** $11m^4 - 7m^3 - 3m^2$; 4; trinomial **32.** $-2p^5 + 4p^3 + 10p^2$; 5; trinomial **33.** x^4; 4; monomial **34.** r^6; 6; monomial **35.** 7; 0; monomial **36.** 9; 0; monomial **37.** (a) 36 (b) -12 **38.** (a) 19 (b) -2 **39.** (a) 14 (b) -19 **40.** (a) -5 (b) 1

In each polynomial add like terms whenever possible. Write the result in descending powers of the variable. See Example 2.

17. $-3m^5 + 5m^5$ **18.** $-4y^3 + 3y^3$ **19.** $2r^5 + (-3r^5)$

20. $-19y^2 + 9y^2$ **21.** $.2m^5 - .5m^2$ **22.** $-.9y + .9y^2$

23. $-3x^5 + 2x^5 - 4x^5$ **24.** $6x^3 - 8x^3 + 9x^3$ **25.** $-4p^7 + 8p^7 + 5p^9$

26. $-3a^8 + 4a^8 - 3a^2$ **27.** $-4y^2 + 3y^2 - 2y^2 + y^2$ **28.** $3r^5 - 8r^5 + r^5 + 2r^5$

For each polynomial first simplify, if possible, and write it in descending powers of the variable. Then give the degree of the resulting polynomial and tell whether it is a monomial, a binomial, a trinomial, or none of these. See Example 3.

29. $6x^4 - 9x$ **30.** $7t^3 - 3t$

31. $5m^4 - 3m^2 + 6m^4 - 7m^3$ **32.** $6p^5 + 4p^3 - 8p^5 + 10p^2$

33. $\frac{5}{3}x^4 - \frac{2}{3}x^4$ **34.** $\frac{4}{5}r^6 + \frac{1}{5}r^6$

35. $.8x^4 - .3x^4 - .5x^4 + 7$ **36.** $1.2t^3 - .9t^3 - .3t^3 + 9$

Find the value of each polynomial when **(a)** $x = 2$ *and when* **(b)** $x = -1$. *See Example 4.*

37. $2x^5 - 4x^4 + 5x^3 - x^2$ **38.** $2x^2 + 5x + 1$

39. $-3x^2 + 14x - 2$ **40.** $-2x^2 + 3$

41. -3 **42.** 14

TECHNOLOGY INSIGHTS (EXERCISES 41–42)

The graphing calculator screen shown here indicates that 1 has been stored into the memory location designated X, and then the polynomial $X^2 - 3X + 6$ has been evaluated. The result is 4. (This can be verified by hand using direct substitution, as shown in Example 4.)

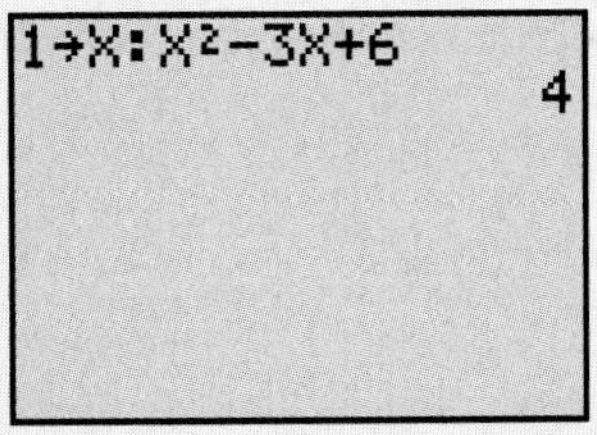

Predict the result the calculator will give for the following screens.

41.

42.

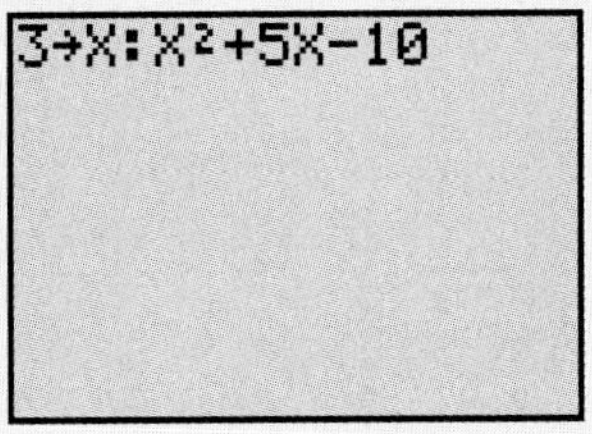

*These problems can be used as collaborative exercises with small groups of students.

43. 4; \$5.00   44. 6; \$27
45. 2.5; 130   46. 459.87 billion

RELATING CONCEPTS* (EXERCISES 43–46)

As explained earlier in this section, the polynomial equation

$$y = -10.25x^2 - 126.04x + 5730.21$$

gives a good approximation of NASA's budget for space station research, in millions of dollars, for 1996–2001, where $x = 0$ represents 1996, $x = 1$ represents 1997, and so on. If we evaluate the polynomial for a specific input value x, we will get one and only one output value y as a result. This idea is basic to the study of functions (first introduced in Section 3.2), one of the most important concepts in mathematics.

Work Exercises 43–46 in order.

43. If gasoline costs \$1.25 per gallon, then the monomial $1.25x$ gives the cost of x gallons. Evaluate this monomial for 4, and then use the result to fill in the blanks: If _____ gallons are purchased, the cost is _____.

44. If it costs \$15 to rent a chain saw plus \$2 per day, the binomial $2x + 15$ gives the cost to rent the chain saw for x days. Evaluate this polynomial for 6 and then use the result to fill in the blanks: If the saw is rented for _____ days, the cost is _____.

45. If an object is thrown upward under certain conditions its height in feet is given by the trinomial $-16x^2 + 60x + 80$, where x is in seconds. Evaluate this polynomial for 2.5 and then use the result to fill in the blanks: If _____ seconds have elapsed, the height of the object is _____ feet.

46. The polynomial $2.69x^2 + 4.75x + 452.43$ gives a good approximation for the number of revenue passenger miles, in billions, for the U.S. airline industry during the period from 1990 to 1995, where $x = 0$ represents 1990. Use this polynomial to approximate the number of revenue passenger miles in 1991. (*Hint:* Any power of 1 is equal to 1, so simply add the coefficients and the constant.) (*Source:* Air Transportation Association of America.)

Did you make the connection that a polynomial can be related to data, using a particular data value for input to get another value as output?

Add or subtract as indicated. See Examples 5 and 8.

47. $5m^2 + 3m$ 48. $10a^3 + a^2$
49. $4x^4 - 4x^2$ 50. $6y^5 - 6y^3$

47. Add.
$$\begin{array}{r} 3m^2 + 5m \\ \underline{2m^2 - 2m} \end{array}$$

48. Add.
$$\begin{array}{r} 4a^3 - 4a^2 \\ \underline{6a^3 + 5a^2} \end{array}$$

49. Subtract.
$$\begin{array}{r} 12x^4 - x^2 \\ \underline{8x^4 + 3x^2} \end{array}$$

50. Subtract.
$$\begin{array}{r} 13y^5 - y^3 \\ \underline{7y^5 + 5y^3} \end{array}$$

51. $\frac{7}{6}x^2 - \frac{2}{15}x + \frac{5}{6}$
52. $\frac{19}{21}y^2 - \frac{8}{15}y + \frac{53}{45}$
53. $6m^3 + m^2 + 12m - 14$
54. $4r^5 + 21r^4 - 4r^3$
55. $15m^3 - 13m^2 + 8m + 11$
56. $11a^4 - 4a^3 + 3a^2 - 2a + 7$
57. Answers will vary.
58. To add or subtract polynomials, we add or subtract the coefficients of the like terms. For example, to add $4x^2 + 3x - 7$ to $9x^2 + 8x - 3$, we add $4x^2$ and $9x^2$ to get $13x^2$, then add $3x$ and $8x$ to get $11x$, and add -7 and -3 to get -10. The final sum is $13x^2 + 11x - 10$.
59. $5m^2 - 14m + 6$
60. $-2x^2 - x + 1$
61. $4x^3 + 2x^2 + 5x$
62. $-b^6 + 5b^4 + b^2$
63. $-11y^4 + 8y^2 + y$
64. $-11t^5 + 9t^3 + 4t$
65. $a^4 - a^2 + 1$
66. $-7m^2 + 6m + 1$
67. $5m^2 + 8m - 10$
68. $3b^3 - b^2 - 4b - 2$
69. $-6x^2 - 12x + 12$
70. $-4t^2 - 12t + 27$
71. -10 72. 1 73. $4b - 5c$
74. $3t + 10s$ 75. $6x - xy - 7$
76. $2ab + 4a - 6b$
77. $-3x^2y - 15xy - 3xy^2$
78. $40t^3s^2 + 5t^2s^3 + 27ts^4$
79. $8x^2 + 8x + 6$
80. $10y^2 + 8y + 24$
81. (a) $23y + 5t$ (b) approximately 37.26°, 79.52°, and 63.22°
82. (a) $-5t^2s + 3ts^2 + 9ts$ (b) 67°, 83°, 30° 83. $-6x^2 + 6x - 7$
84. $-x^4 - 2x^3 - 3x^2 + x - 6$
85. $-7x - 1$
86. $-9t^3 + 13t - 1$

51. Add.

$$\begin{array}{r} \frac{2}{3}x^2 + \frac{1}{5}x + \frac{1}{6} \\ \frac{1}{2}x^2 - \frac{1}{3}x + \frac{2}{3} \\ \hline \end{array}$$

52. Add.

$$\begin{array}{r} \frac{4}{7}y^2 - \frac{1}{5}y + \frac{7}{9} \\ \frac{1}{3}y^2 - \frac{1}{3}y + \frac{2}{5} \\ \hline \end{array}$$

53. Add.

$$\begin{array}{r} 9m^3 - 5m^2 + 4m - 8 \\ -3m^3 + 6m^2 + 8m - 6 \\ \hline \end{array}$$

54. Add.

$$\begin{array}{r} 12r^5 + 11r^4 - 7r^3 - 2r^2 \\ -8r^5 + 10r^4 + 3r^3 + 2r^2 \\ \hline \end{array}$$

55. Subtract.

$$\begin{array}{r} 12m^3 - 8m^2 + 6m + 7 \\ -3m^3 + 5m^2 - 2m - 4 \\ \hline \end{array}$$

56. Subtract.

$$\begin{array}{r} 5a^4 - 3a^3 + 2a^2 - a + 6 \\ -6a^4 + a^3 - a^2 + a - 1 \\ \hline \end{array}$$

57. After reading Examples 5–8, explain whether you have a preference regarding horizontal or vertical addition and subtraction of polynomials.

58. Write a paragraph explaining how to add and subtract polynomials. Give an example using addition.

Perform the indicated operations. See Examples 6 and 7.

59. $(8m^2 - 7m) - (3m^2 + 7m - 6)$

60. $(x^2 + x) - (3x^2 + 2x - 1)$

61. $(16x^3 - x^2 + 3x) + (-12x^3 + 3x^2 + 2x)$

62. $(-2b^6 + 3b^4 - b^2) + (b^6 + 2b^4 + 2b^2)$

63. $(7y^4 + 3y^2 + 2y) - (18y^4 - 5y^2 + y)$

64. $(8t^5 + 3t^3 + 5t) - (19t^5 - 6t^3 + t)$

65. $(9a^4 - 3a^2 + 2) + (4a^4 - 4a^2 + 2) + (-12a^4 + 6a^2 - 3)$

66. $(4m^2 - 3m + 2) + (5m^2 + 13m - 4) - (16m^2 + 4m - 3)$

67. $[(8m^2 + 4m - 7) - (2m^2 - 5m + 2)] - (m^2 + m + 1)$

68. $[(9b^3 - 4b^2 + 3b + 2) - (-2b^3 - 3b^2 + b)] - (8b^3 + 6b + 4)$

69. $[(3x^2 - 2x + 7) - (4x^2 + 2x - 3)] - [(9x^2 + 4x - 6) + (-4x^2 + 4x + 4)]$

70. $[(6t^2 - 3t + 1) - (12t^2 + 2t - 6)] - [(4t^2 - 3t - 8) + (-6t^2 + 10t - 12)]$

71. Without actually performing the operations, determine mentally the coefficient of x^2 in the simplified form of $(-4x^2 + 2x - 3) - (-2x^2 + x - 1) + (-8x^2 + 3x - 4)$.

72. Without actually performing the operations, determine mentally the coefficient of x in the simplified form of $(-8x^2 - 3x + 2) - (4x^2 - 3x + 8) - (-2x^2 - x + 7)$.

Add or subtract as indicated. See Example 9.

73. $(6b + 3c) + (-2b - 8c)$

74. $(-5t + 13s) + (8t - 3s)$

75. $(4x + 2xy - 3) - (-2x + 3xy + 4)$

76. $(8ab + 2a - 3b) - (6ab - 2a + 3b)$

77. $(5x^2y - 2xy + 9xy^2) - (8x^2y + 13xy + 12xy^2)$

78. $(16t^3s^2 + 8t^2s^3 + 9ts^4) - (-24t^3s^2 + 3t^2s^3 - 18ts^4)$

For Exercises 79–82, use the formulas found on the inside covers.

▲ *Find the perimeter of each rectangle.*

79.

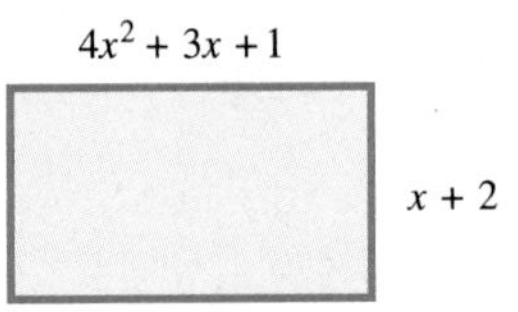

80.

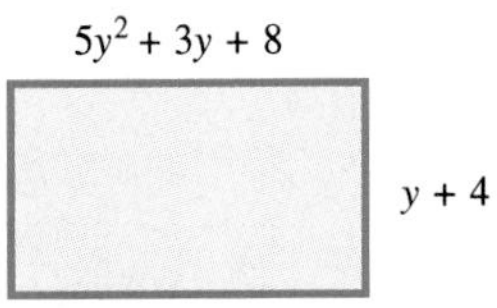

87. 0, −3, −4, −3, 0

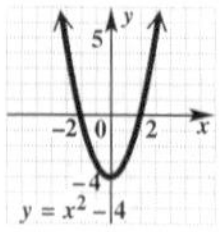

88. −2, −5, −6, −5, −2

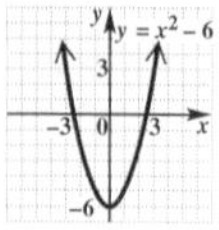

89. 7, 1, −1, 1, 7

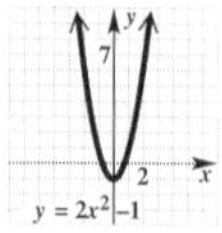

90. 10, 4, 2, 4, 10

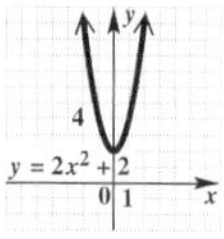

91. 0, 3, 4, 3, 0

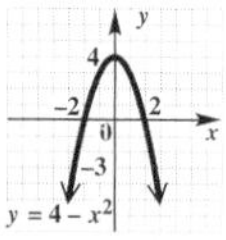

92. −2, 1, 2, 1, −2

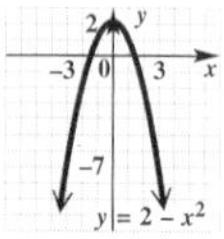

93. 4, 1, 0, 1, 4

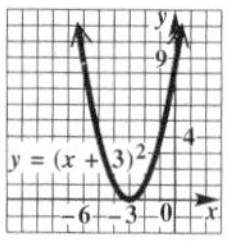

94. 4, 1, 0, 1, 4

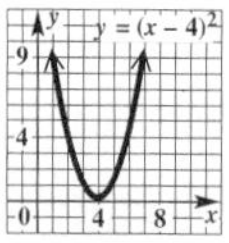

▲ *Find (**a**) a polynomial representing the perimeter of each triangle and (**b**) the measures of the angles of the triangle.*

81.

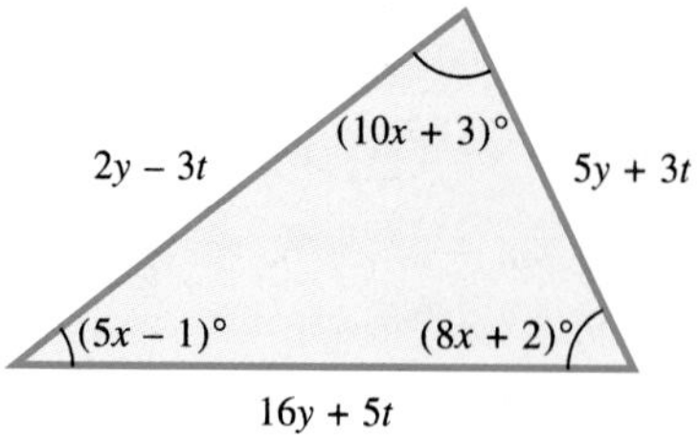

82.

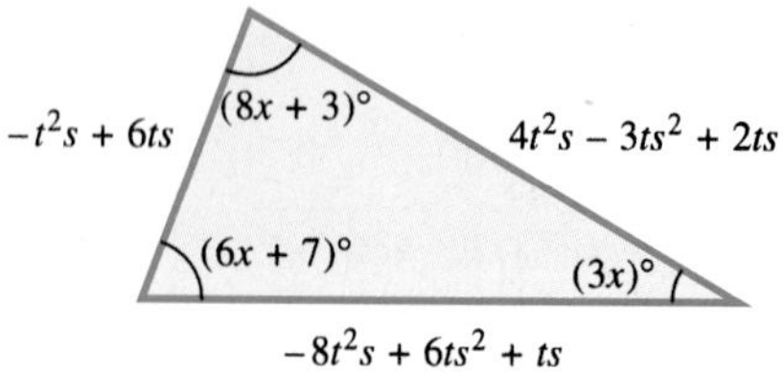

▲ *The concepts required to work Exercises 83–86 have been covered, but the usual wording of the problem has been changed. Perform the indicated operations.*

83. Subtract $9x^2 - 6x + 5$ from $3x^2 - 2$.

84. Find the difference when $9x^4 + 3x^2 + 5$ is subtracted from $8x^4 - 2x^3 + x - 1$.

85. Find the difference between the sum of $5x^2 + 2x - 3$ and $x^2 - 8x + 2$ and the sum of $7x^2 - 3x + 6$ and $-x^2 + 4x - 6$.

86. Subtract the sum of $9t^3 - 3t + 8$ and $t^2 - 8t + 4$ from the sum of $12t + 8$ and $t^2 - 10t + 3$.

Graph each of the following by completing the table of values. See Example 10.

87. $y = x^2 - 4$

x	y
−2	
−1	
0	
1	
2	

88. $y = x^2 - 6$

x	y
−2	
−1	
0	
1	
2	

89. $y = 2x^2 - 1$

x	y
−2	
−1	
0	
1	
2	

90. $y = 2x^2 + 2$

x	y
−2	
−1	
0	
1	
2	

91. $y = 4 - x^2$

x	y
−2	
−1	
0	
1	
2	

92. $y = 2 - x^2$

x	y
−2	
−1	
0	
1	
2	

93. $y = (x + 3)^2$

x	−5	−4	−3	−2	−1
y					

94. $y = (x - 4)^2$

x	2	3	4	5	6
y					

4.2 The Product Rule and Power Rules for Exponents

OBJECTIVES

1. Identify bases and exponents.
2. Use the product rule for exponents.

OBJECTIVE 1 Identify bases and exponents. Recall from Section 1.2 that in the expression 5^2, the number 5 is the *base* and 2 is the *exponent.* The expression 5^2 is called an *exponential expression.* Usually we do not write the exponent when it is 1; however, sometimes it is convenient to do so. In general, for any quantity a, $a^1 = a$.

JOURNAL WRITING ▲ CHALLENGING SCIENTIFIC CALCULATOR GRAPHING CALCULATOR

INSTRUCTOR'S RESOURCES

PTB Sec. 4.2 **ISM** Sec. 4.2 **AB** Sec. 4.2

TEST GENERATOR

WORLD WIDE WEB www.LialAlgebra.com

3 Use the rule $(a^m)^n = a^{mn}$.

4 Use the rule $(ab)^m = a^m b^m$.

5 Use the rule $\left(\frac{a}{b}\right)^m = \frac{a^m}{b^m}$.

FOR EXTRA HELP

SSG Sec. 4.2
SSM Sec. 4.2
Pass the Test Software
InterAct Math Tutorial Software
Video 6

CHALKBOARD EXAMPLE

Evaluate the exponential expression. Name the base and the exponent.

$$-2^5$$

Answer: -32; 2; 5

TEACHING TIP Cover several examples to stress these concepts, such as -2^3, $(-2)^3$, $-(-2)^3$, -2^4, $(-2)^4$, $-(-2)^4$.

EXAMPLE 1 Determining the Base and Exponent in an Exponential Expression

Evaluate each exponential expression. Name the base and the exponent.

	Base	Exponent
(a) $5^4 = 5 \cdot 5 \cdot 5 \cdot 5 = 625$	5	4
(b) $-5^4 = -1 \cdot 5^4 = -1 \cdot (5 \cdot 5 \cdot 5 \cdot 5) = -625$	5	4
(c) $(-5)^4 = (-5)(-5)(-5)(-5) = 625$	-5	4

CAUTION

Note the differences between parts (b) and (c) of Example 1. In -5^4 the lack of parentheses shows that the exponent 4 refers only to the base 5, and not -5; in $(-5)^4$ the parentheses show that the exponent 4 refers to the base -5. In summary, $-a^n$ and $(-a)^n$ are not necessarily the same.

Expression	Base	Exponent	Example
$-a^n$	a	n	$-3^2 = -(3 \cdot 3) = -9$
$(-a)^n$	$-a$	n	$(-3)^2 = (-3)(-3) = 9$

OBJECTIVE 2 Use the product rule for exponents. By the definition of exponents,

$$\begin{aligned} 2^4 \cdot 2^3 &= \underbrace{(2 \cdot 2 \cdot 2 \cdot 2)}_{\text{4 factors}}\underbrace{(2 \cdot 2 \cdot 2)}_{\text{3 factors}} \\ &= \underbrace{2 \cdot 2 \cdot 2 \cdot 2 \cdot 2 \cdot 2 \cdot 2}_{4+3=7 \text{ factors}} \\ &= 2^7. \end{aligned}$$

Also,

$$\begin{aligned} 6^2 \cdot 6^3 &= (6 \cdot 6)(6 \cdot 6 \cdot 6) \\ &= 6 \cdot 6 \cdot 6 \cdot 6 \cdot 6 \\ &= 6^5. \end{aligned}$$

Generalizing from these examples, $2^4 \cdot 2^3 = 2^{4+3} = 2^7$ and $6^2 \cdot 6^3 = 6^{2+3} = 6^5$, suggests the **product rule for exponents.**

Product Rule for Exponents

For any positive integers m and n, $a^m \cdot a^n = a^{m+n}$.
(Keep the same base; add the exponents.)
Example: $6^2 \cdot 6^5 = 6^{2+5} = 6^7$

CHALKBOARD EXAMPLE

Find the product.
(a) $(-7)^5 \cdot (-7)^3$
(b) $(-4p^5)(3p^8)$

Answer:
(a) $(-7)^8$
(b) $-12p^{13}$

EXAMPLE 2 Using the Product Rule

Use the product rule for exponents to find each result when possible.

(a) $6^3 \cdot 6^5 = 6^{3+5} = 6^8$ by the product rule.

(b) $(-4)^7(-4)^2 = (-4)^{7+2} = (-4)^9$

TEACHING TIP Be aware of the common student error:

$6^2 \cdot 6^3 = 36^5$ **Incorrect**

(c) $x^2 \cdot x = x^2 \cdot x^1 = x^{2+1} = x^3$

(d) $m^4m^3m^5 = m^{4+3+5} = m^{12}$

(e) $2^3 \cdot 3^2$

The product rule does not apply to the product $2^3 \cdot 3^2$, since the bases are different.

$$2^3 \cdot 3^2 = 8 \cdot 9 = 72$$

(f) $2^3 + 2^4$

The product rule does not apply to $2^3 + 2^4$, since this is a *sum,* not a *product.*

$$2^3 + 2^4 = 8 + 16 = 24$$

(g) $(2x^3)(3x^7)$

Since $2x^3$ means $2 \cdot x^3$ and $3x^7$ means $3 \cdot x^7$, we can use the associative and commutative properties to get

$$(2x^3)(3x^7) = (2 \cdot 3) \cdot (x^3 \cdot x^7) = 6x^{10}.$$

CAUTION Be sure that you understand the difference between *adding* and *multiplying* exponential expressions. For example,

$$8x^3 + 5x^3 = 13x^3, \quad \text{but} \quad (8x^3)(5x^3) = 8 \cdot 5x^{3+3} = 40x^6.$$

OBJECTIVE 3 **Use the rule $(a^m)^n = a^{mn}$.** We simplify an expression such as $(8^3)^2$ with the product rule for exponents.

$$(8^3)^2 = (8^3)(8^3) = 8^{3+3} = 8^6$$

The exponents in $(8^3)^2$ are multiplied to give the exponent in 8^6. As another example,

$$(5^2)^3 = 5^2 \cdot 5^2 \cdot 5^2 = 5^{2+2+2} = 5^6,$$

and $2 \cdot 3 = 6$. These examples suggest **power rule (a) for exponents.**

Power Rule (a) for Exponents

For any positive integers m and n, $(a^m)^n = a^{mn}$.
(Raise a power to a power by multiplying exponents.)
Example: $(3^2)^4 = 3^{2 \cdot 4} = 3^8$

CHALKBOARD EXAMPLE

Simplify $(6^2)^5$.

Answer: 6^{10}

EXAMPLE 3 Using Power Rule (a)

Use power rule (a) for exponents to simplify each expression.

(a) $(2^5)^3 = 2^{5 \cdot 3} = 2^{15}$

(b) $(5^7)^2 = 5^{7(2)} = 5^{14}$

(c) $(x^2)^5 = x^{2(5)} = x^{10}$

(d) $(n^3)^2 = n^{3(2)} = n^6$

OBJECTIVE 4 **Use the rule $(ab)^m = a^mb^m$.** The properties studied in Chapter 1 can be used to develop two more rules for exponents. Using the definition of an exponential expression and the commutative and associative properties, we can rewrite the expression $(4 \cdot 8)^3$ as follows.

TEACHING TIP The product rule and power rule (a) may be confused. Note the difference between $4^2 \cdot 4^3$ and $(4^2)^3$.

$$(4 \cdot 8)^3 = (4 \cdot 8)(4 \cdot 8)(4 \cdot 8) \quad \text{Definition of exponent}$$
$$= (4 \cdot 4 \cdot 4) \cdot (8 \cdot 8 \cdot 8) \quad \text{Commutative and associative properties}$$
$$= 4^3 \cdot 8^3 \quad \text{Definition of exponent}$$

This example suggests **power rule (b) for exponents.**

Power Rule (b) for Exponents

For any positive integer m, $(ab)^m = a^m b^m$.
(Raise a product to a power by raising each factor to the power.)
Example: $(2p)^5 = 2^5 p^5$

CHALKBOARD EXAMPLE
Simplify $(3a^2b^4)^5$.
Answer: $243a^{10}b^{20}$

EXAMPLE 4 Using Power Rule (b)

Use power rule (b) for exponents to simplify each expression.

(a) $(3xy)^2 = 3^2x^2y^2 = 9x^2y^2$

(b) $5(pq)^2 = 5(p^2q^2)$ Power rule (b)
$= 5p^2q^2$ Multiply.

(c) $3(2m^2p^3)^4 = 3[2^4(m^2)^4(p^3)^4]$ Power rule (b)
$= 3 \cdot 2^4m^8p^{12}$ Power rule (a)
$= 48m^8p^{12}$ Multiply.

(d) $(-5^6)^3 = (-1 \cdot 5^6)^3 = (-1)^3 \cdot (5^6)^3 = -1 \cdot 5^{18} = -5^{18}$

OBJECTIVE 5 Use the rule $\left(\frac{a}{b}\right)^m = \frac{a^m}{b^m}$. Since the quotient $\frac{a}{b}$ can be written as $a\left(\frac{1}{b}\right)$, we can use power rule (b), together with some of the properties of real numbers, to get **power rule (c) for exponents.**

Power Rule (c) for Exponents

For any positive integer m,

$$\left(\frac{a}{b}\right)^m = \frac{a^m}{b^m} \quad (b \neq 0).$$

(Raise a quotient to a power by raising both numerator and denominator to the power.)

Example: $\left(\frac{5}{3}\right)^2 = \frac{5^2}{3^2}$

CHALKBOARD EXAMPLE
Simplify $\left(\frac{3}{x}\right)^3$ $(x \neq 0)$.
Answer: $\frac{27}{x^3}$

EXAMPLE 5 Using Power Rule (c)

Use power rule (c) for exponents to simplify each expression.

(a) $\left(\frac{2}{3}\right)^5 = \frac{2^5}{3^5}$

(b) $\left(\frac{m}{n}\right)^3 = \frac{m^3}{n^3}$ $(n \neq 0)$

We list the rules for exponents discussed in this section below. These rules are basic to the study of algebra and should be *memorized.*

Rules for Exponents

For positive integers m and n:

		Examples
Product rule	$a^m \cdot a^n = a^{m+n}$	$6^2 \cdot 6^5 = 6^{2+5} = 6^7$
Power rules (a)	$(a^m)^n = a^{mn}$	$(3^2)^4 = 3^{2 \cdot 4} = 3^8$
(b)	$(ab)^m = a^m b^m$	$(2p)^5 = 2^5 p^5$
(c)	$\left(\frac{a}{b}\right)^m = \frac{a^m}{b^m} \quad (b \neq 0)$	$\left(\frac{5}{3}\right)^2 = \frac{5^2}{3^2}$

As shown in the next example, more than one rule may be needed to simplify an expression with exponents.

CHALKBOARD EXAMPLE

Simplify $\left(\frac{1}{5}\right)^4 (2x)^2$.

Answer: $\frac{4x^2}{625}$

TEACHING TIP Students may work Example 6(b) in an equally correct way:

$$\begin{aligned}(5x)^3(5x)^4 &= 5^3x^35^4x^4 \\ &= 5^3 \cdot 5^4x^3x^4 \\ &= 5^7x^7.\end{aligned}$$

EXAMPLE 6 Using Combinations of Rules

Use the rules for exponents to simplify each expression.

(a) $\left(\frac{2}{3}\right)^2 \cdot 2^3 = \frac{2^2}{3^2} \cdot \frac{2^3}{1}$ Power rule (c)

$= \frac{2^2 \cdot 2^3}{3^2 \cdot 1}$ Multiply the fractions.

$= \frac{2^5}{3^2}$ Product rule

(b) $(5x)^3(5x)^4 = (5x)^7$ Product rule

$= 5^7x^7$ Power rule (b)

(c) $(2x^2y^3)^4(3xy^2)^3 = 2^4(x^2)^4(y^3)^4 \cdot 3^3x^3(y^2)^3$ Power rule (b)

$= 2^4 \cdot 3^3x^8y^{12}x^3y^6$ Power rule (a)

$= 16 \cdot 27x^{11}y^{18}$ Product rule

$= 432x^{11}y^{18}$

(d) $(-x^3y)^2(-x^5y^4)^3$

Think of the negative sign in each factor as -1.

$(-1 \cdot x^3y)^2(-1 \cdot x^5y^4)^3$

$= (-1)^2(x^3)^2(y)^2 \cdot (-1)^3(x^5)^3(y^4)^3$ Power rule (b)

$= (-1)^2(x^6)(y^2)(-1)^3(x^{15})(y^{12})$ Power rule (a)

$= (-1)^5(x^{21})(y^{14})$ Product rule

$= -x^{21}y^{14}$

4.2 EXERCISES

Decide whether each statement is true or false.

1. $3^3 = 9$
2. $(-2)^4 = 2^4$
3. $(a^2)^3 = a^5$
4. $\left(\frac{1}{4}\right)^2 = \frac{1}{4^2}$

Write each expression using exponents.

5. $w \cdot w \cdot w \cdot w \cdot w \cdot w$
6. $t \cdot t \cdot t \cdot t \cdot t \cdot t \cdot t$
7. $\dfrac{1}{4 \cdot 4 \cdot 4 \cdot 4}$
8. $\dfrac{1}{3 \cdot 3 \cdot 3}$
9. $(-7x)(-7x)(-7x)(-7x)$
10. $(-8p)(-8p)$
11. $\left(\frac{1}{2}\right)\left(\frac{1}{2}\right)\left(\frac{1}{2}\right)\left(\frac{1}{2}\right)\left(\frac{1}{2}\right)\left(\frac{1}{2}\right)$
12. $\left(-\frac{1}{4}\right)\left(-\frac{1}{4}\right)\left(-\frac{1}{4}\right)\left(-\frac{1}{4}\right)\left(-\frac{1}{4}\right)$
13. Explain how the expressions $(-3)^4$ and -3^4 are different.
14. Explain how the expressions $(5x)^3$ and $5x^3$ are different.

Identify the base and the exponent for each exponential expression. In Exercises 15–18, also evaluate each expression. See Example 1.

15. 3^5
16. 2^7
17. $(-3)^5$
18. $(-2)^7$
19. $(-6x)^4$
20. $(-8x)^4$
21. $-6x^4$
22. $-8x^4$
23. Explain why the product rule does not apply to the expression $5^2 + 5^3$. Then evaluate the expression by finding the individual powers and adding the results.
24. Repeat Exercise 23 for the expression $(-4)^3 + (-4)^4$.

Use the product rule to simplify each expression. Write the answer in exponential form. See Example 2.

25. $5^2 \cdot 5^6$
26. $3^6 \cdot 3^7$
27. $4^2 \cdot 4^7 \cdot 4^3$
28. $5^3 \cdot 5^8 \cdot 5^2$
29. $(-7)^3(-7)^6$
30. $(-9)^8(-9)^5$
31. $t^3 \cdot t^8 \cdot t^{13}$
32. $n^5 \cdot n^6 \cdot n^9$
33. $(-8r^4)(7r^3)$
34. $(10a^7)(-4a^3)$
35. $(-6p^5)(-7p^5)$
36. $(-5w^8)(-9w^8)$
37. Explain why the product rule does not apply to the expression $3^2 \cdot 4^3$. Then evaluate the expression by finding the individual powers and multiplying the results.
38. Repeat Exercise 37 for the expression $(-3)^3 \cdot (-2)^5$.

1. false 2. true 3. false 4. true 5. w^6 6. t^7 7. $\frac{1}{4^4}$ 8. $\frac{1}{3^3}$ 9. $(-7x)^4$ 10. $(-8p)^2$ 11. $\left(\frac{1}{2}\right)^6$ 12. $\left(-\frac{1}{4}\right)^5$ 13. In $(-3)^4$, -3 is the base, while in -3^4, 3 is the base. $(-3)^4 = 81$, while $-3^4 = -81$. 14. In $(5x)^3$, $5x$ is the base, while in $5x^3$, x is the base. $(5x)^3 = 125x^3$, while $5x^3$ cannot be simplified further. 15. base: 3; exponent: 5; 243 16. base: 2; exponent: 7; 128 17. base: -3; exponent: 5; -243 18. base: -2; exponent: 7; -128 19. base: $-6x$; exponent: 4 20. base: $-8x$; exponent: 4 21. base: x; exponent: 4 22. base: x; exponent: 4 23. $5^2 + 5^3$ is a sum, not a product. $5^2 + 5^3 = 25 + 125 = 150$. 24. $(-4)^3 + (-4)^4$ is a sum, not a product. $(-4)^3 + (-4)^4 = -64 + 256 = 192$ 25. 5^8 26. 3^{13} 27. 4^{12} 28. 5^{13} 29. $(-7)^9$ 30. $(-9)^{13}$ 31. t^{24} 32. n^{20} 33. $-56r^7$ 34. $-40a^{10}$ 35. $42p^{10}$ 36. $45w^{16}$ 37. The product rule applies only when the bases are the same. $3^2 \cdot 4^3 = 9 \cdot 64 = 576$. 38. In the product $(-3)^3 \cdot (-2)^5$, the bases are not equal. $(-3)^3 \cdot (-2)^5 = -27 \cdot -32 = 864$.

39. -32; -32; 1
40. 64; -64; 0 **41.** 729; -729; 0
42. -243; -243; 1

The graphing calculator screen shown here on the left reinforces the fact that $(-3)^4$ and -3^4 are not equal, since the first is 81 and the second is -81. The screen on the right shows how a calculator displays whether a statement is true or false: If the calculator returns a 1, the statement is true; it returns a 0 for a false statement.

```
(-3)^4
              81
-3^4
             -81
```

```
(-3)^3=-3^3
               1
(-3)^4=-3^4
               0
```

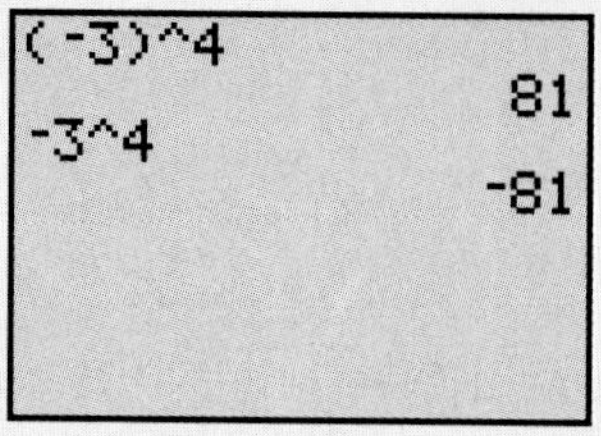

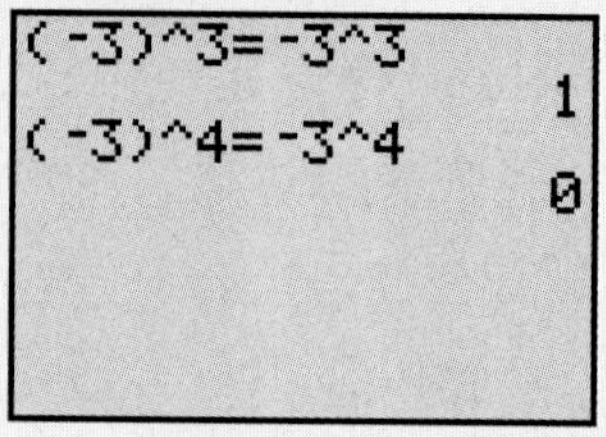

Evaluate each expression in the statement, and then tell whether the calculator would return a 1 or a 0.

39.
```
(-2)^5=-2^5
```

40.
```
(-2)^6=-2^6
```

41.
```
(-3)^6=-3^6
```

42.
```
(-3)^5=-3^5
```

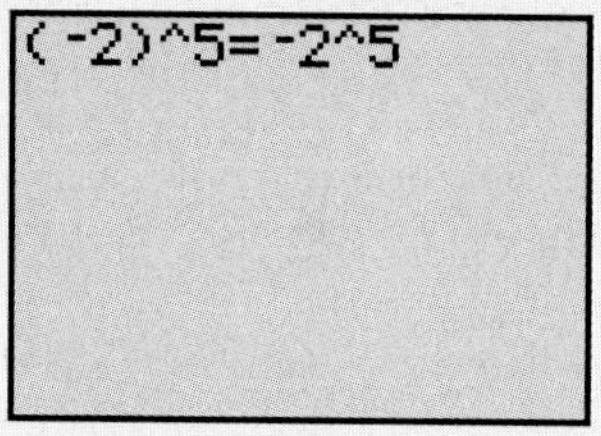

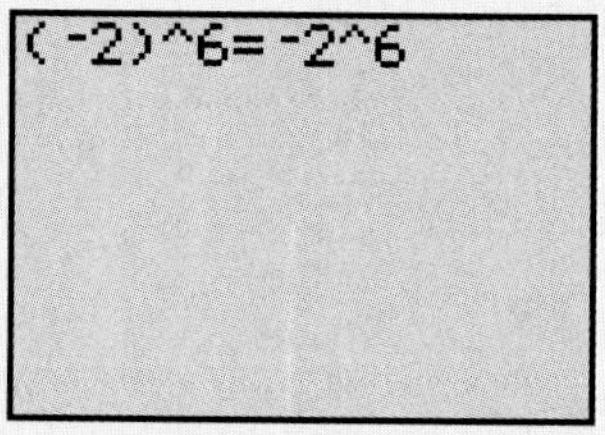

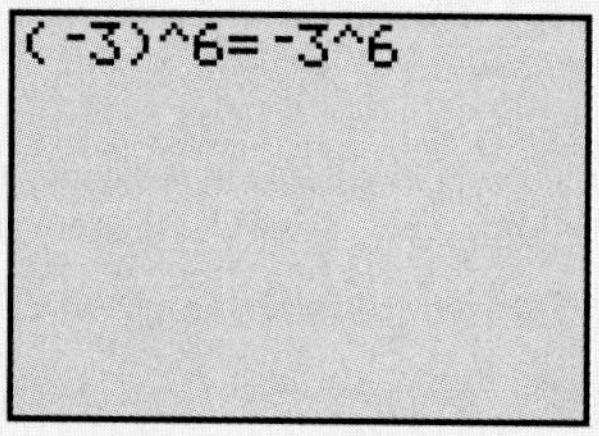

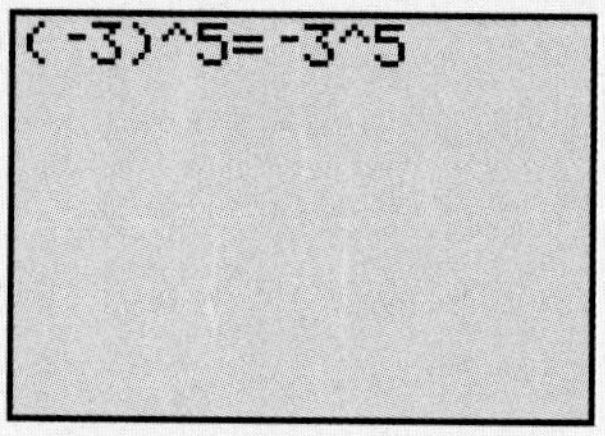

Use the power rules for exponents to simplify each expression. Write the answer in exponential form. See Examples 3–5.

43. 4^6 **44.** 8^{18} **45.** t^{20}
46. y^{30} **47.** $7^3 r^3$ **48.** $11^4 x^4$
49. $5^5 x^5 y^5$ **50.** $9^6 p^6 q^6$ **51.** 5^{12}
52. 9^{32} **53.** -8^{15} **54.** -7^{35}
55. $8 q^3 r^3$ **56.** $4 v^5 w^5$
57. $\dfrac{1}{2^3}$ **58.** $\dfrac{1}{3^5}$ **59.** $\dfrac{a^3}{b^3}$
60. $\dfrac{r^4}{t^4}$ **61.** $\dfrac{9^8}{5^8}$ **62.** $\dfrac{12^3}{7^3}$

43. $(4^3)^2$

44. $(8^3)^6$

45. $(t^4)^5$

46. $(y^6)^5$

47. $(7r)^3$

48. $(11x)^4$

49. $(5xy)^5$

50. $(9pq)^6$

51. $(-5^2)^6$

52. $(-9^4)^8$

53. $(-8^3)^5$

54. $(-7^5)^7$

55. $8(qr)^3$

56. $4(vw)^5$

57. $\left(\dfrac{1}{2}\right)^3$

58. $\left(\dfrac{1}{3}\right)^5$

59. $\left(\dfrac{a}{b}\right)^3 \quad (b \neq 0)$

60. $\left(\dfrac{r}{t}\right)^4 \quad (t \neq 0)$

61. $\left(\dfrac{9}{5}\right)^8$

62. $\left(\dfrac{12}{7}\right)^3$

63. $\frac{5^5}{2^5}$ 64. $\frac{3^{11}}{4^{11}}$ 65. $\frac{9^5}{8^3}$
66. $\frac{8^7}{5^4}$ 67. $2^{12}x^{12}$ 68. $6^{13}y^{13}$
69. -6^5p^5 70. 13^4q^4
71. $6^5x^{10}y^{15}$ 72. $5^7r^{35}t^{42}$
73. x^{21} 74. y^{35} 75. $2^2w^4x^{26}y^7$
76. $3^3x^{12}y^{11}z^{23}$ 77. $-r^{18}s^{17}$
78. $-t^{13}s^{39}$ 79. $\frac{5^3a^6b^{15}}{c^{18}}$
80. $\frac{6^4x^{12}y^{36}}{z^{20}}$ 81. This is incorrect. Using the product rule, it is simplified as follows: $(10^2)^3 = 10^{2 \cdot 3} = 10^6 = 1{,}000{,}000$.
82. The 4 is used as an *exponent* on 3. It is not multiplied by 3. The correct simplification is $3^4 \cdot x^8 \cdot y^{12} = 81x^8y^{12}$.
83. $12x^5$ 84. $\frac{3}{2}m^6$ 85. $6p^7$
86. $36\pi a^6$ 87. $125x^6$
88. $180x^6y^8$
89. $-a^4$, $-a^3$, $-(-a)^3$, $(-a)^4$; One way is to choose a positive number greater than 1 and substitute it for a in each expression. Then arrange the terms from smallest to largest.
90. Raising to an even power gives a positive number, while raising to an odd power gives a negative number.

Use a combination of the rules of exponents introduced in this section to simplify each expression. See Example 6.

63. $\left(\frac{5}{2}\right)^3 \cdot \left(\frac{5}{2}\right)^2$ **64.** $\left(\frac{3}{4}\right)^5 \cdot \left(\frac{3}{4}\right)^6$ **65.** $\left(\frac{9}{8}\right)^3 \cdot 9^2$

66. $\left(\frac{8}{5}\right)^4 \cdot 8^3$ **67.** $(2x)^9(2x)^3$ **68.** $(6y)^5(6y)^8$

69. $(-6p)^4(-6p)$ **70.** $(-13q)^3(-13q)$ **71.** $(6x^2y^3)^5$

72. $(5r^5t^6)^7$ **73.** $(x^2)^3(x^3)^5$ **74.** $(y^4)^5(y^3)^5$

75. $(2w^2x^3y)^2(x^4y)^5$ **76.** $(3x^4y^2z)^3(yz^4)^5$ **77.** $(-r^4s)^2(-r^2s^3)^5$

78. $(-ts^6)^4(-t^3s^5)^3$ **79.** $\left(\frac{5a^2b^5}{c^6}\right)^3 \quad (c \neq 0)$ **80.** $\left(\frac{6x^3y^9}{z^5}\right)^4 \quad (z \neq 0)$

81. A student tried to simplify $(10^2)^3$ as 1000^6. Is this correct? If not, how is it simplified using the product rule for exponents?

82. Explain why $(3x^2y^3)^4$ is *not* equivalent to $(3 \cdot 4)x^8y^{12}$.

Find the area of each figure. (Leave π in the answer for Exercise 86.) Use the formulas found on the inside covers.*

83.

84.

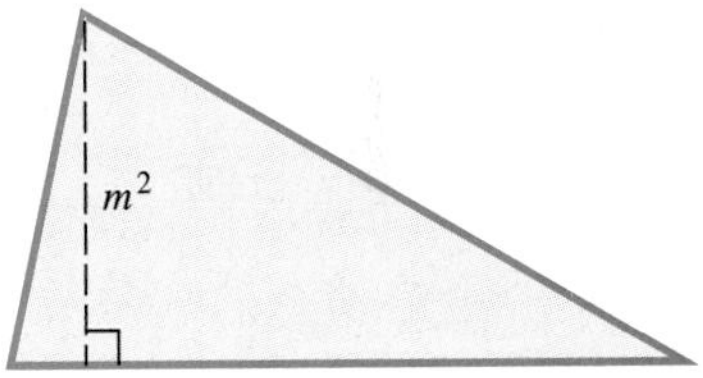

85.

86.

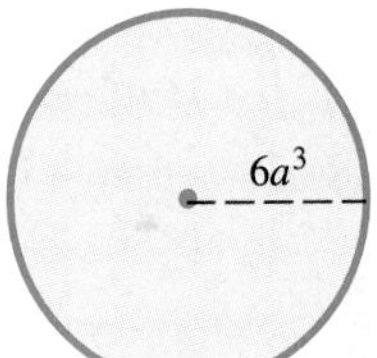

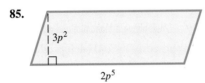

Find the volume of each figure. Use the formulas found on the inside covers.

87.

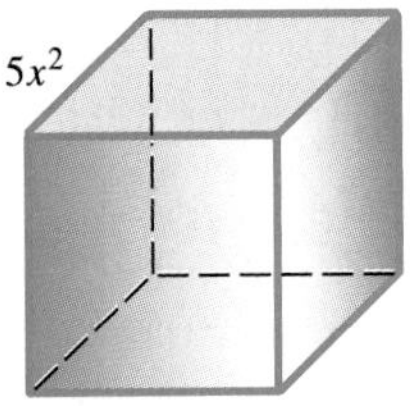

88.

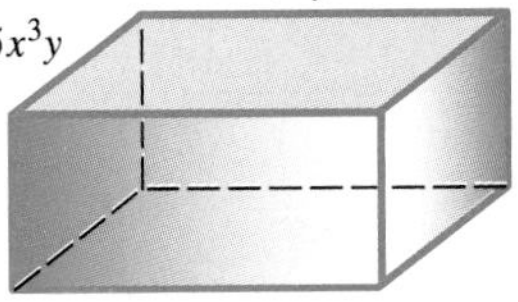

89. Assume a is a positive number greater than 1. Arrange the following terms in order from smallest to largest: $-(-a)^3$, $-a^3$, $(-a)^4$, $-a^4$. Explain how you decided on the order.

90. Describe a rule to tell whether an exponential expression with a negative base is positive or negative.

*The small square in the figures for Exercises 83–85 indicates a right angle (90°).

JOURNAL WRITING ▲ CHALLENGING SCIENTIFIC CALCULATOR GRAPHING CALCULATOR

91. \$304.16
92. \$449.95
93. \$1843.88
94. \$2207.63

In Chapter 2 we used the formula for simple interest, $I = prt$, which deals with interest paid only on the principal. With **compound interest,** interest is paid on the principal and the interest earned earlier. The formula for compound interest, which involves an exponential expression, is

$$A = P(1 + r)^n.$$

Here A is the amount accumulated from a principal of P dollars left untouched for n years with an annual interest rate r (expressed as a decimal).

Use this formula and a calculator to find A to the nearest cent in Exercises 91–94.

91. $P = \$250, r = .04, n = 5$

92. $P = \$400, r = .04, n = 3$

93. $P = \$1500, r = .035, n = 6$

94. $P = \$2000, r = .025, n = 4$

4.3 Multiplication of Polynomials

OBJECTIVES

1. Multiply a monomial and a polynomial.
2. Multiply two polynomials.
3. Multiply binomials by the FOIL method.

FOR EXTRA HELP

SSG Sec. 4.3
SSM Sec. 4.3
Pass the Test Software
InterAct Math Tutorial Software
Video 6

CHALKBOARD EXAMPLE

Find the product.

$$2x^4(3x^2 + 2x - 5)$$

Answer: $6x^6 + 4x^5 - 10x^4$

OBJECTIVE 1 **Multiply a monomial and a polynomial.** As shown in the previous section, the product of two monomials is found by using the rules for exponents and the commutative and associative properties. For example,

$$(-8m^6)(-9m^4) = -8(-9)(m^6)(m^4) = 72m^{6+4} = 72m^{10}.$$

CAUTION

Do not confuse *addition* of terms with *multiplication* of terms. For example,

$$7q^5 + 2q^5 = 9q^5, \quad \text{but} \quad (7q^5)(2q^5) = 7 \cdot 2q^{5+5} = 14q^{10}.$$

To find the product of a monomial and a polynomial with more than one term, we use the distributive property with the method just shown.

EXAMPLE 1 **Multiplying a Monomial and a Polynomial**

Use the distributive property to find each product.

(a) $4x^2(3x + 5)$

$$4x^2(3x + 5) = (4x^2)(3x) + (4x^2)(5) \quad \text{Distributive property}$$
$$= 12x^3 + 20x^2 \quad \text{Multiply monomials.}$$

(b) $-8m^3(4m^3 + 3m^2 + 2m - 1)$

$$-8m^3(4m^3 + 3m^2 + 2m - 1)$$
$$= (-8m^3)(4m^3) + (-8m^3)(3m^2) + (-8m^3)(2m) + (-8m^3)(-1) \quad \text{Distributive property}$$
$$= -32m^6 - 24m^5 - 16m^4 + 8m^3 \quad \text{Multiply monomials.}$$

OBJECTIVE 2 **Multiply two polynomials.** We can use the distributive property repeatedly to find the product of any two polynomials. For example, to find the product of the polynomials $x + 1$ and $x - 4$, think of $x - 4$ as a single quantity and use the distributive property as follows.

$$(x + 1)(x - 4) = x(x - 4) + 1(x - 4)$$

JOURNAL WRITING ▲ CHALLENGING SCIENTIFIC CALCULATOR GRAPHING CALCULATOR

INSTRUCTOR'S RESOURCES

PTB Sec. 4.3 ISM Sec. 4.3 AB Sec. 4.3

TEST GENERATOR

WORLD WIDE WEB www.LialAlgebra.com

Now use the distributive property twice to find $x(x - 4)$ and $1(x - 4)$.

$$\begin{aligned} x(x-4) + 1(x-4) &= x(x) + x(-4) + 1(x) + 1(-4) \\ &= x^2 - 4x + x - 4 \\ &= x^2 - 3x - 4 \end{aligned}$$

(We could have treated $x + 1$ as the single quantity instead to get $(x + 1)x + (x + 1)(-4)$ in the first step. Verify that this approach gives the same final result, $x^2 - 3x - 4$.)

We give a rule for multiplying any two polynomials below.

Multiplying Polynomials

To multiply two polynomials, multiply each term of the second polynomial by each term of the first polynomial and add the products.

CHALKBOARD EXAMPLE

Find the product.

$$(m^3 - 2m + 1) \times (2m^2 + 4m + 3)$$

Answer: $2m^5 + 4m^4 - m^3 - 6m^2 - 2m + 3$

EXAMPLE 2 Multiplying Two Polynomials

Multiply $(m^2 + 5)(4m^3 - 2m^2 + 4m)$.

Multiply each term of the second polynomial by each term of the first.

$$\begin{aligned} &(m^2 + 5)(4m^3 - 2m^2 + 4m) \\ &= m^2(4m^3) - m^2(2m^2) + m^2(4m) + 5(4m^3) - 5(2m^2) + 5(4m) \\ &= 4m^5 - 2m^4 + 4m^3 + 20m^3 - 10m^2 + 20m \end{aligned}$$

Now combine like terms.

$$= 4m^5 - 2m^4 + 24m^3 - 10m^2 + 20m$$

When at least one of the factors in a product of polynomials has three or more terms, the multiplication can be simplified by writing one polynomial above the other vertically.

CHALKBOARD EXAMPLE

Multiply.

$$\begin{array}{r} 3x^2 + 4x - 5 \\ x + 4 \\ \hline \end{array}$$

Answer:
$3x^3 + 16x^2 + 11x - 20$

EXAMPLE 3 Multiplying Vertically

Multiply $(x^3 + 2x^2 + 4x + 1)(3x + 5)$ using the vertical method.

Write the polynomials as follows.

$$\begin{array}{r} x^3 + 2x^2 + 4x + 1 \\ 3x + 5 \\ \hline \end{array}$$

It is not necessary to line up terms in columns, because any terms may be multiplied (not just like terms). Begin by multiplying each of the terms in the top row by 5.

Step 1

$$\begin{array}{rl} x^3 + 2x^2 + 4x + 1 & \\ 3x + 5 & \\ \hline 5x^3 + 10x^2 + 20x + 5 & \quad 5(x^3 + 2x^2 + 4x + 1) \end{array}$$

Notice how this process is similar to multiplication of whole numbers. Now multiply each term in the top row by $3x$. Be careful to place like terms in columns, since the final step will involve addition (as in multiplying two whole numbers).

Step 2

$$\begin{array}{rl} x^3 + 2x^2 + 4x + 1 & \\ 3x + 5 & \\ \hline 5x^3 + 10x^2 + 20x + 5 & \\ 3x^4 + 6x^3 + 12x^2 + 3x & \quad 3x(x^3 + 2x^2 + 4x + 1) \end{array}$$

TEACHING TIP Mention that polynomials are usually arranged in descending order before multiplying.

TEACHING TIP A good check for possible multiplication errors in the partial products is to note whether all terms in any vertical column are like terms.

Step 3 Add like terms.

$$\begin{array}{rrrrr} & x^3 + & 2x^2 + & 4x + & 1 \\ & & & 3x + & 5 \\ \hline & 5x^3 + & 10x^2 + & 20x + & 5 \\ 3x^4 + & 6x^3 + & 12x^2 + & 3x & \\ \hline 3x^4 + & 11x^3 + & 22x^2 + & 23x + & 5 \end{array}$$

The product is $3x^4 + 11x^3 + 22x^2 + 23x + 5$.

OBJECTIVE 3 Multiply binomials by the FOIL method. In algebra, many of the polynomials to be multiplied are both binomials (with just two terms). For these products a shortcut that eliminates the need to write out all the steps is used. To develop this shortcut, multiply $x + 3$ and $x + 5$ using the distributive property.

$$\begin{aligned} (x + 3)(x + 5) &= x(x + 5) + 3(x + 5) \\ &= x(x) + x(5) + 3(x) + 3(5) \\ &= x^2 + 5x + 3x + 15 \\ &= x^2 + 8x + 15 \end{aligned}$$

The first term in the second line, $x(x)$, is the product of the first terms of the two binomials.

$(x + 3)(x + 5)$ Multiply the first terms: $x(x)$.

The term $x(5)$ is the product of the first term of the first binomial and the last term of the second binomial. This is the **outer product.**

$(x + 3)(x + 5)$ Multiply the outer terms: $x(5)$.

The term $3(x)$ is the product of the last term of the first binomial and the first term of the second binomial. The product of these middle terms is the **inner product.**

$(x + 3)(x + 5)$ Multiply the inner terms: $3(x)$.

Finally, $3(5)$ is the product of the last terms of the two binomials.

$(x + 3)(x + 5)$ Multiply the last terms: $3(5)$.

The inner product and the outer product should be added mentally, so that the three terms of the answer can be written without extra steps as

$$(x + 3)(x + 5) = x^2 + 8x + 15.$$

A summary of these steps is given below. This procedure is sometimes called the **FOIL method,** which comes from the abbreviation for *First, Outer, Inner, Last.*

Multiplying Binomials by the FOIL Method

Step 1 **Multiply the first terms.** Multiply the two first terms of the binomials to get the first term of the answer.

Step 2 **Find the outer and inner products.** Find the outer product and the inner product and add them (mentally if possible) to get the middle term of the answer.

Step 3 **Multiply the last terms.** Multiply the two last terms of the binomials to get the last term of the answer.

CHALKBOARD EXAMPLE

Find the product by the FOIL method.

$(x - 2)(x - 6)$

Answer: $x^2 - 8x + 12$

EXAMPLE 4 Using the FOIL Method

Find the product $(x + 8)(x - 6)$ by the FOIL method.

Step 1 **F** Multiply the *first* terms: $x(x) = x^2$.

Step 2 **O** Find the product of the *outer* terms: $x(-6) = -6x$.

I Find the product of the *inner* terms: $8(x) = 8x$.

Add the outer and inner products mentally: $-6x + 8x = 2x$.

Step 3 **L** Multiply the *last* terms: $8(-6) = -48$.

The product $(x + 8)(x - 6)$ is the sum of the terms found in the three steps above, so

$$(x + 8)(x - 6) = x^2 - 6x + 8x - 48 = x^2 + 2x - 48.$$

As a shortcut, this product can be found in the following manner.

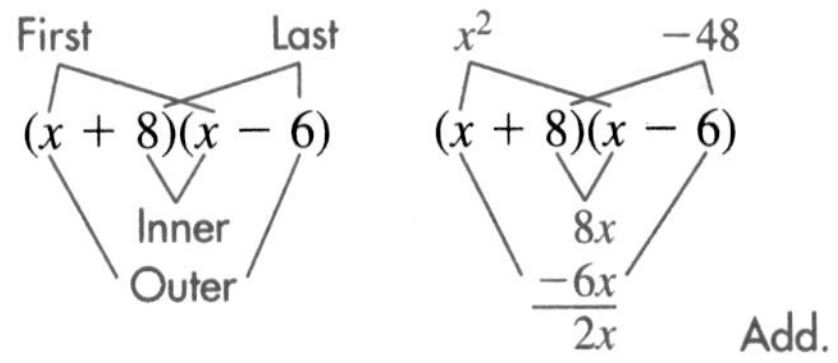

It is not possible to add the inner and outer products of the FOIL method if unlike terms result, as shown in the next example.

CHALKBOARD EXAMPLE

Multiply $(5x - 6)(2y + 3)$.

Answer:
$10xy + 15x - 12y - 18$

EXAMPLE 5 Using the FOIL Method

Multiply $(9x - 2)(3y + 1)$.

First	$(9x - 2)(3y + 1)$	$27xy$	
Outer	$(9x - 2)(3y + 1)$	$9x$	Unlike terms
Inner	$(9x - 2)(3y + 1)$	$-6y$	Unlike terms
Last	$(9x - 2)(3y + 1)$	-2	

F O I L

$$(9x - 2)(3y + 1) = 27xy + 9x - 6y - 2$$

CHALKBOARD EXAMPLE

Find the product.

$(-4y + x)(2y - 3x)$

Answer: $-8y^2 + 14xy - 3x^2$

EXAMPLE 6 Using the FOIL Method

Find the following products.

(a) $(2k + 5y)(k + 3y) = (2k)(k) + (2k)(3y) + (5y)(k) + (5y)(3y)$ (F O I L)

$= 2k^2 + 6ky + 5ky + 15y^2$

$= 2k^2 + 11ky + 15y^2$ Combine like terms.

(b) $(7p + 2q)(3p - q) = 21p^2 - pq - 2q^2$ FOIL

(c) $2x^2(x - 3)(3x + 4) = 2x^2(3x^2 - 5x - 12)$ FOIL

$= 6x^4 - 10x^3 - 24x^2$ Distributive property

4.3 EXERCISES

1. (a) B (b) D (c) A (d) C
2. (a) C (b) A (c) B (d) D
3. $40a^{14}$ 4. $15m^{10}$
5. $-6m^2 - 4m$ 6. $-30p - 15p^2$
7. $24p - 18p^2 + 36p^4$
8. $12x + 8x^2 + 20x^4$
9. $-16z^2 - 24z^3 - 24z^4$
10. $-21y - 35y^3 + 14y^4$
11. $14x^5y^3 + 21x^3y^2 - 28x^2y^2$
12. $-27x^3y^7 + 54x^2y^4 - 18x^2y^3$
13. $12x^3 + 26x^2 + 10x + 1$
14. $72y^3 - 70y^2 + 21y - 2$
15. $20m^4 - m^3 - 8m^2 - 17m - 15$
16. $3y^4 + 10y^3 - 7y^2 + 7y + 12$
17. $6x^6 - 3x^5 - 4x^4 + 4x^3 - 5x^2 + 8x - 3$
18. $2a^5 + a^4 - a^3 + a^2 - a + 3$
19. $5x^4 - 13x^3 + 20x^2 + 7x + 5$
20. $2m^4 - 7m^3 + 3m^2 + 17m - 15$
21. $n^2 + n - 6$
22. $r^2 + 2r - 48$
23. $x^2 - 36$ 24. $y^2 - 81$
25. $8r^2 - 10r - 3$
26. $10x^2 - 31x - 14$
27. $9x^2 - 4$ 28. $49x^2 - 9$
29. $9q^2 + 6q + 1$
30. $16w^2 + 56w + 49$
31. $3x^2 - 5xy - 2y^2$
32. $10p^2 - 13pm - 3m^2$
33. $-3t^2 - 14t + 24$
34. $-5x^2 + 19x - 18$
35. $6y^5 - 21y^4 - 45y^3$
36. $15t^6 + 40t^5 - 15t^4$

1. Match each product in Column I with the correct monomial in Column II.

I	II
(a) $(5x^3)(6x^5)$	**A.** $125x^{15}$
(b) $(-5x^5)(6x^3)$	**B.** $30x^8$
(c) $(5x^5)^3$	**C.** $-216x^9$
(d) $(-6x^3)^3$	**D.** $-30x^8$

2. Match each product in Column I with the correct polynomial in Column II.

I	II
(a) $(x - 5)(x + 3)$	**A.** $x^2 + 8x + 15$
(b) $(x + 5)(x + 3)$	**B.** $x^2 - 8x + 15$
(c) $(x - 5)(x - 3)$	**C.** $x^2 - 2x - 15$
(d) $(x + 5)(x - 3)$	**D.** $x^2 + 2x - 15$

Find each product. See Example 1.

3. $(-5a^9)(-8a^5)$ **4.** $(-3m^6)(-5m^4)$ **5.** $-2m(3m + 2)$

6. $-5p(6 + 3p)$ **7.** $3p(8 - 6p + 12p^3)$ **8.** $4x(3 + 2x + 5x^3)$

9. $-8z(2z + 3z^2 + 3z^3)$ **10.** $-7y(3 + 5y^2 - 2y^3)$

11. $7x^2y(2x^3y^2 + 3xy - 4y)$ **12.** $9xy^3(-3x^2y^4 + 6xy - 2x)$

Find each product. See Examples 2 and 3.

13. $(6x + 1)(2x^2 + 4x + 1)$ **14.** $(9y - 2)(8y^2 - 6y + 1)$

15. $(4m + 3)(5m^3 - 4m^2 + m - 5)$ **16.** $(y + 4)(3y^3 - 2y^2 + y + 3)$

17. $(2x - 1)(3x^5 - 2x^3 + x^2 - 2x + 3)$ **18.** $(2a + 3)(a^4 - a^3 + a^2 - a + 1)$

19. $(5x^2 + 2x + 1)(x^2 - 3x + 5)$ **20.** $(2m^2 + m - 3)(m^2 - 4m + 5)$

Find each product. Use the FOIL method. See Examples 4–6.

21. $(n - 2)(n + 3)$ **22.** $(r - 6)(r + 8)$ **23.** $(x + 6)(x - 6)$

24. $(y + 9)(y - 9)$ **25.** $(4r + 1)(2r - 3)$ **26.** $(5x + 2)(2x - 7)$

27. $(3x + 2)(3x - 2)$ **28.** $(7x + 3)(7x - 3)$ **29.** $(3q + 1)(3q + 1)$

30. $(4w + 7)(4w + 7)$ **31.** $(3x + y)(x - 2y)$ **32.** $(5p + m)(2p - 3m)$

33. $(-3t + 4)(t + 6)$ **34.** $(-5x + 9)(x - 2)$

35. $3y^3(2y + 3)(y - 5)$ **36.** $5t^4(t + 3)(3t - 1)$

37. $36x^2 + 24x + 4$
38. $3y^2 + 10y + 7$
39. $x^2 - 16$; $y^2 - 4$; $r^2 - 49$; In each case, the product is a binomial that is the difference of two squares.
40. $x^2 + 8x + 16$; $y^2 - 4y + 4$; $r^2 + 14r + 49$; In each case, the product is a trinomial. The first term is the square of the first term of each binomial factor. The middle term is twice the product of the two terms of each binomial. The last term is the square of the second term of each binomial.
41. $6p^2 - \frac{5}{2}pq - \frac{25}{12}q^2$
42. $-3x^2 + \frac{11}{4}xy - \frac{1}{2}y^2$
43. $2m^6 - 5m^3 - 12$
44. $4a^4 - 7a^2b^2 - 2b^4$
45. $2k^5 - 6k^3h^2 + k^2h^2 - 3h^4$
46. $4x^5 + 4x^3y - 5x^2y^4 - 5y^5$
47. $6p^8 + 15p^7 + 12p^6 + 36p^5 + 15p^4$
48. $5k^7 - 5k^6 + 20k^5 - 15k^4 + 15k^3 - 60k^2$
49. $-24x^8 - 28x^7 + 32x^6 + 20x^5$
50. $24x^8 - 12x^7 + 16x^6 - 16x^5 + 4x^4$
51. $14x + 49$
52. $2x^2 + 5x + 5$
53. $\pi x^2 - 9$
54. $(10 - \pi)x^2 + 17x + 3$
55. $30x + 60$

37. Find a polynomial that represents the area of this square.

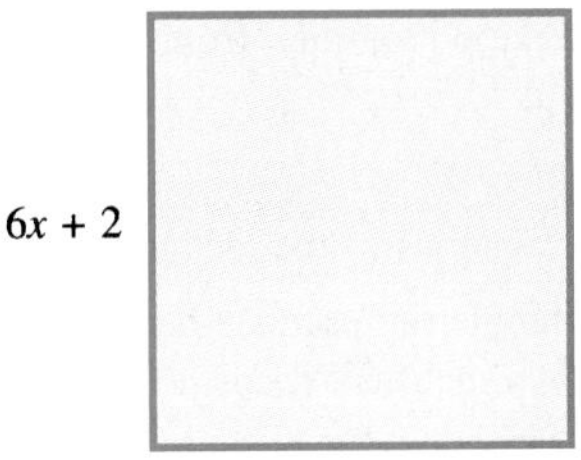

38. Find a polynomial that represents the area of this rectangle.

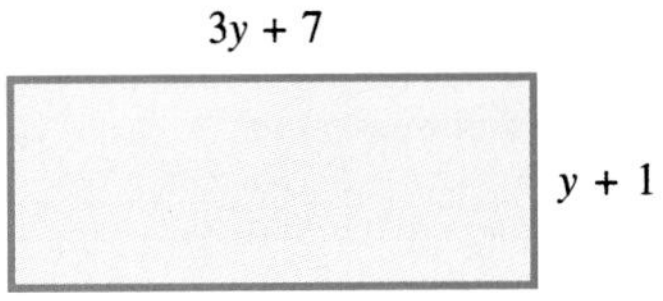

39. Perform the following multiplications:

$$(x + 4)(x - 4); \quad (y + 2)(y - 2); \quad (r + 7)(r - 7).$$

Observe your answers, and explain the pattern that can be found.

40. Repeat Exercise 39 for the following:

$$(x + 4)(x + 4); \quad (y - 2)(y - 2); \quad (r + 7)(r + 7).$$

▲ *Find each product.*

41. $\left(3p + \frac{5}{4}q\right)\left(2p - \frac{5}{3}q\right)$

42. $\left(-x + \frac{2}{3}y\right)\left(3x - \frac{3}{4}y\right)$

43. $(m^3 - 4)(2m^3 + 3)$

44. $(4a^2 + b^2)(a^2 - 2b^2)$

45. $(2k^3 + h^2)(k^2 - 3h^2)$

46. $(4x^3 - 5y^4)(x^2 + y)$

47. $3p^3(2p^2 + 5p)(p^3 + 2p + 1)$

48. $5k^2(k^2 - k + 4)(k^3 - 3)$

49. $-2x^5(3x^2 + 2x - 5)(4x + 2)$

50. $-4x^3(3x^4 + 2x^2 - x)(-2x + 1)$

▲ *Find a polynomial that represents the area of each shaded region. In Exercises 53 and 54 leave π in your answer. Use the formulas found on the inside covers.*

51.

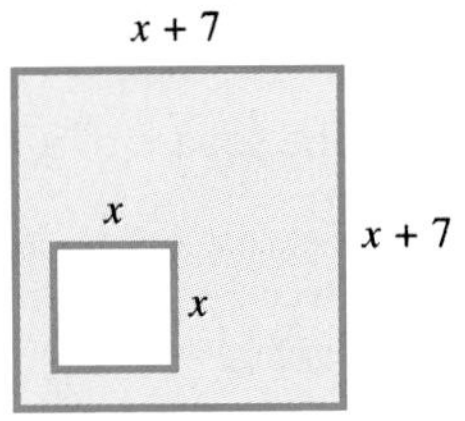

52.

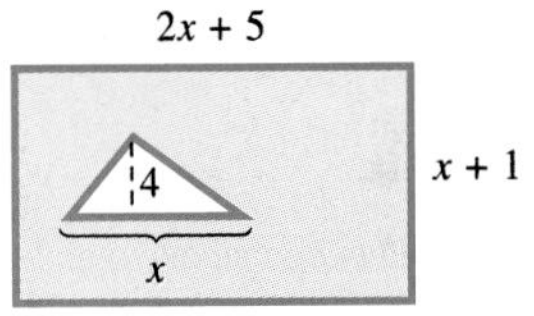

53.

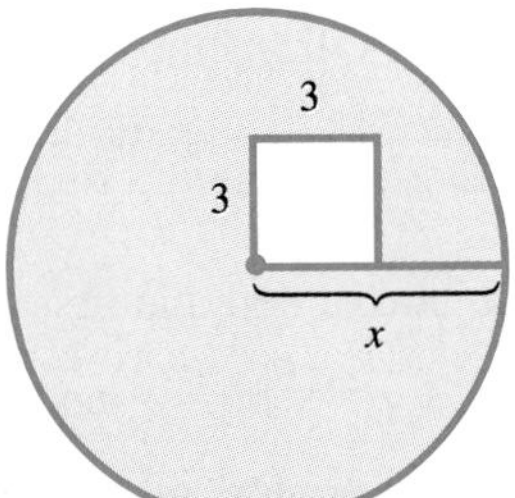

54.

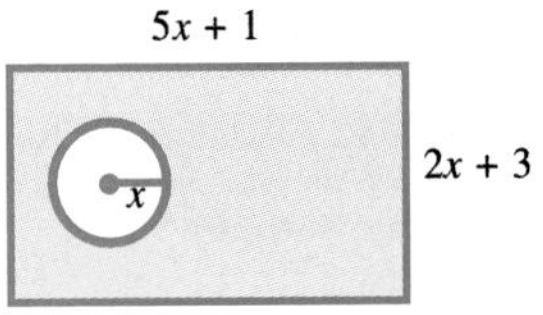

RELATING CONCEPTS (EXERCISES 55–62)

***Work Exercises 55–62 in order.** Refer to the figure as necessary.*

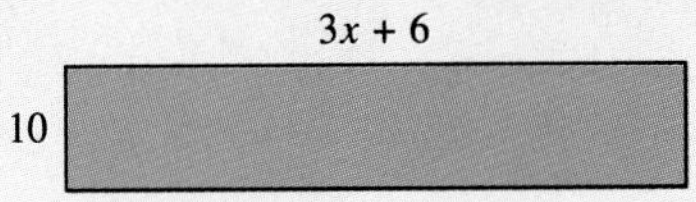

55. Find a polynomial that represents the area of the rectangle.

56. $30x + 60 = 600$ 57. 18
58. 10 yards by 60 yards
59. \$2100 60. 140 yards
61. \$1260 62. (a) $30kx + 60k$ (dollars) (b) $6rx + 32r$
63. Suppose that we want to multiply $(x + 3)(2x + 5)$. The letter F stands for *first*. We multiply the two first terms to get $2x^2$. The letter O represents *outer*. We next multiply the two outer terms to get $5x$. The letter I represents *inner*. The product of the two inner terms is $6x$. L stands for *last*. The product of the two last terms is 15. Very often the outer and inner products are like terms, as they are in this case. So we simplify $2x^2 + 5x + 6x + 15$ to get the final product, $2x^2 + 11x + 15$.
64. The trinomial has a *middle* term, and this must be taken into consideration when multiplying. For example, in $(x - 3)(2x^2 + 4x - 7)$, the term $4x$ is the middle term of the trinomial. The letter M does not appear in the word FOIL.

RELATING CONCEPTS (EXERCISES 55–62) (CONTINUED)

56. Suppose you know that the area of the rectangle is 600 square yards. Use this information and the polynomial from Exercise 55 to write an equation that allows you to solve for x.

57. Solve for x.

58. What are the dimensions of the rectangle (assume units are all in yards)?

59. Suppose the rectangle represents a lawn and it costs \$3.50 per square yard to lay sod on the lawn. How much will it cost to sod the entire lawn?

60. Use the result of Exercise 58 to find the perimeter of the lawn.

61. Again, suppose the rectangle represents a lawn and it costs \$9.00 per yard to fence the lawn. How much will it cost to fence the lawn?

62. **(a)** Suppose that it costs k dollars per square yard to sod the lawn. Determine a polynomial in the variables x and k that represents the cost to sod the entire lawn.

(b) Suppose that it costs r dollars per yard to fence the lawn. Determine a polynomial in the variables x and r that represents the cost to fence the lawn.

Did you make the connection that sodding requires knowing the area, while fencing requires knowing the perimeter?

63. Explain the FOIL method of multiplying two binomials. Give an example.

64. Why does the FOIL method not apply to the product of a binomial and a trinomial? Give an example.

4.4 Special Products

OBJECTIVES

1. Square binomials.
2. Find the product of the sum and difference of two terms.
3. Find higher powers of binomials.

FOR EXTRA HELP

SSG Sec. 4.4
SSM Sec. 4.4
Pass the Test Software
InterAct Math Tutorial Software
Video 6

CHALKBOARD EXAMPLE

Find $(t + u)^2$.

Answer: $t^2 + 2tu + u^2$

In this section, we develop patterns for certain binomial products that occur frequently.

OBJECTIVE 1 Square binomials. The square of a binomial can be found quickly by using the method shown in Example 1.

EXAMPLE 1 Squaring a Binomial

Find $(m + 3)^2$.

Squaring $m + 3$ by the FOIL method gives

$$(m + 3)(m + 3) = m^2 + 3m + 3m + 9 = m^2 + 6m + 9.$$

The result has the square of both the first and the last terms of the binomial:

$$m^2 = m^2 \quad \text{and} \quad 3^2 = 9.$$

The middle term is twice the product of the two terms of the binomial, since both the outer and inner products are $(m)(3)$ and

$$(m)(3) + (m)(3) = 2(m)(3) = 6m.$$

JOURNAL WRITING ▲ CHALLENGING SCIENTIFIC CALCULATOR GRAPHING CALCULATOR

INSTRUCTOR'S RESOURCES

PTB Sec. 4.4 ISM Sec. 4.4
AB Sec. 4.4

TEST GENERATOR

WORLD WIDE WEB
www.LialAlgebra.com

This example suggests the following rule.

Square of a Binomial

The square of a binomial is a trinomial consisting of the square of the first term, plus twice the product of the two terms, plus the square of the last term of the binomial. For x and y,

$$(x + y)^2 = x^2 + 2xy + y^2$$

and

$$(x - y)^2 = x^2 - 2xy + y^2.$$

CHALKBOARD EXAMPLE

Find $(5r - 6s)^2$.

Answer: $25r^2 - 60rs + 36s^2$

TEACHING TIP Remind students that in the square of a sum, all of the terms are positive. In the square of a difference, the middle term is negative.

EXAMPLE 2 Squaring Binomials

Use the rule to find each product.

(a) $(5z - 1)^2 = (5z)^2 - 2(5z)(1) + 1^2 = 25z^2 - 10z + 1$
Recall that $(5z)^2 = 5^2z^2 = 25z^2$.

(b) $(3b + 5r)^2 = (3b)^2 + 2(3b)(5r) + (5r)^2 = 9b^2 + 30br + 25r^2$

(c) $(2a - 9x)^2 = 4a^2 - 36ax + 81x^2$

(d) $\left(4m + \frac{1}{2}\right)^2 = (4m)^2 + 2(4m)\left(\frac{1}{2}\right) + \left(\frac{1}{2}\right)^2 = 16m^2 + 4m + \frac{1}{4}$

(e) $t(a + 2b)^2 = t(a^2 + 4ab + 4b^2)$ Square the binomial.
$= a^2t + 4abt + 4b^2t$ Distributive property

CAUTION A common error when squaring a binomial is forgetting the middle term of the product. In general, $(x + y)^2 \neq x^2 + y^2$ and $(x - y)^2 \neq x^2 - y^2$.

OBJECTIVE 2 Find the product of the sum and difference of two terms. Binomial products of the form $(x + y)(x - y)$ also occur frequently. In these products, one binomial is the sum of two terms, and the other is the difference of the same two terms. As an example, the product of $a + 2$ and $a - 2$ is

$$(a + 2)(a - 2) = a^2 - 2a + 2a - 4 = a^2 - 4.$$

Using the FOIL method, the product of $x + y$ and $x - y$ is the difference of two squares.

Product of the Sum and Difference of Two Terms

The product of the sum and difference of the two terms x and y is

$$(x + y)(x - y) = x^2 - y^2.$$

CAUTION The product $(x + y)(x - y)$ cannot be written as $(x + y)^2$ or as $(x - y)^2$, since one factor involves addition and the other involves subtraction.

CHALKBOARD EXAMPLE

Find the product.

$$(4 + y)(4 - y)$$

Answer: $16 - y^2$

EXAMPLE 3 Finding the Product of the Sum and Difference of Two Terms

Find each product.

(a) $(x + 4)(x - 4)$

Use the pattern for the sum and difference of two terms.

$$(x + 4)(x - 4) = x^2 - 4^2 = x^2 - 16$$

(b) $(3 - w)(3 + w)$

By the commutative property, this product is the same as $(3 + w)(3 - w)$.

$$(3 - w)(3 + w) = (3 + w)(3 - w) = 3^2 - w^2 = 9 - w^2$$

(c) $(a - b)(a + b) = a^2 - b^2$

CHALKBOARD EXAMPLE

Find the product.

$$\left(3r - \frac{1}{2}\right)\left(3r + \frac{1}{2}\right)$$

Answer: $9r^2 - \frac{1}{4}$

EXAMPLE 4 Finding the Product of the Sum and Difference of Two Terms

Find each product.

(a) $(5m + 3)(5m - 3)$

Use the rule for the product of the sum and difference of two terms.

$$(5m + 3)(5m - 3) = (5m)^2 - 3^2 = 25m^2 - 9$$

(b) $(4x + y)(4x - y) = (4x)^2 - y^2 = 16x^2 - y^2$

(c) $\left(z - \frac{1}{4}\right)\left(z + \frac{1}{4}\right) = z^2 - \frac{1}{16}$

(d) $r(x^2 + y)(x^2 - y) = r(x^4 - y^2)$

$$= rx^4 - ry^2$$

The product formulas of this section will be important later, particularly in Chapters 5 and 6. Therefore, it is important to memorize these formulas and practice using them.

OBJECTIVE 3 Find higher powers of binomials. The methods used in the previous section and this section can be combined to find higher powers of binomials.

CHALKBOARD EXAMPLE

Find the product.

$$(4x + 1)^3$$

Answer:
$64x^3 + 48x^2 + 12x + 1$

EXAMPLE 5 Finding Higher Powers of Binomials

Find each product.

(a) $(x + 5)^3$

$$\begin{aligned}(x + 5)^3 &= (x + 5)^2(x + 5) && a^3 = a^2 \cdot a\\ &= (x^2 + 10x + 25)(x + 5) && \text{Square the binomial.}\\ &= x^3 + 10x^2 + 25x + 5x^2 + 50x + 125 && \text{Multiply polynomials.}\\ &= x^3 + 15x^2 + 75x + 125 && \text{Combine like terms.}\end{aligned}$$

(b) $(2y - 3)^4$

$$\begin{aligned}(2y - 3)^4 &= (2y - 3)^2(2y - 3)^2 && a^4 = a^2 \cdot a^2\\ &= (4y^2 - 12y + 9)(4y^2 - 12y + 9) && \text{Square each binomial.}\\ &= 16y^4 - 48y^3 + 36y^2 - 48y^3 + 144y^2 && \text{Multiply polynomials.}\\ &\quad - 108y + 36y^2 - 108y + 81 &&\\ &= 16y^4 - 96y^3 + 216y^2 - 216y + 81 && \text{Combine like terms.}\end{aligned}$$

1. (a) $4x^2$ (b) $12x$ (c) 9 (d) $4x^2 + 12x + 9$ 2. To square a binomial, we square the first term, find twice the product of the two terms, and square the last term. For example, $(3x - 2)^2 = (3x)^2 + 2(3x)(-2) + (-2)^2 = 9x^2 - 12x + 4$. 3. $p^2 + 4p + 4$
4. $r^2 + 10r + 25$
5. $a^2 - 2ac + c^2$
6. $p^2 - 2py + y^2$
7. $16x^2 - 24x + 9$
8. $25y^2 + 20y + 4$
9. $64t^2 + 112st + 49s^2$
10. $49z^2 - 42zw + 9w^2$
11. $25x^2 + 4xy + \frac{4}{25}y^2$
12. $36m^2 - \frac{48}{5}mn + \frac{16}{25}n^2$
13. $4x^3 + 20x^2 + 25x$
14. $9t^3 - 6t^2 + t$
15. $-16r^2 + 16r - 4$
16. $-9y^2 + 48y - 64$
17. (a) $49x^2$ (b) 0 (c) $-9y^2$ (d) $49x^2 - 9y^2$. Because 0 is the identity element for addition, it is not necessary to write "+ 0."
18. To find the product of the sum and the difference of two terms, we find the difference of the squares of the two terms. For example, $(5x + 7y)(5x - 7y) = (5x)^2 - (7y)^2 = 25x^2 - 49y^2$.
19. $q^2 - 4$ 20. $x^2 - 64$
21. $4w^2 - 25$ 22. $9z^2 - 64$
23. $100x^2 - 9y^2$ 24. $169r^2 - 4z^2$
25. $4x^4 - 25$ 26. $81y^4 - 4$
27. $49x^2 - \frac{9}{49}$ 28. $81y^2 - \frac{4}{9}$
29. $9p^3 - 49p$ 30. $25q^3 - q$

31. $(a + b)^2$ 32. a^2

4.4 EXERCISES

1. Consider the square $(2x + 3)^2$.
   - **(a)** What is the square of the first term, $(2x)^2$?
   - **(b)** What is twice the product of the two terms, $2(2x)(3)$?
   - **(c)** What is the square of the last term, 3^2?
   - **(d)** Write the final product, which is a trinomial, using your results in parts (a)–(c).
2. Explain in your own words how to square a binomial. Give an example.

Find each square. See Examples 1 and 2.

3. $(p + 2)^2$ **4.** $(r + 5)^2$ **5.** $(a - c)^2$

6. $(p - y)^2$ **7.** $(4x - 3)^2$ **8.** $(5y + 2)^2$

9. $(8t + 7s)^2$ **10.** $(7z - 3w)^2$ **11.** $\left(5x + \frac{2}{5}y\right)^2$

12. $\left(6m - \frac{4}{5}n\right)^2$ **13.** $x(2x + 5)^2$ **14.** $t(3t - 1)^2$

15. $-(4r - 2)^2$ **16.** $-(3y - 8)^2$

17. Consider the product $(7x + 3y)(7x - 3y)$.
   - **(a)** What is the product of the first terms, $(7x)(7x)$?
   - **(b)** Multiply the outer terms, $(7x)(-3y)$. Then multiply the inner terms, $(3y)(7x)$. Add the results. What is this sum?
   - **(c)** What is the product of the last terms, $(3y)(-3y)$?
   - **(d)** Write the complete product using your answers in parts (a) and (c). Why is the sum found in part (b) omitted here?
18. Explain in your own words how to find the product of the sum and the difference of two terms. Give an example.

Find each product. See Examples 3 and 4.

19. $(q + 2)(q - 2)$ **20.** $(x + 8)(x - 8)$

21. $(2w + 5)(2w - 5)$ **22.** $(3z + 8)(3z - 8)$

23. $(10x + 3y)(10x - 3y)$ **24.** $(13r + 2z)(13r - 2z)$

25. $(2x^2 - 5)(2x^2 + 5)$ **26.** $(9y^2 - 2)(9y^2 + 2)$

27. $\left(7x + \frac{3}{7}\right)\left(7x - \frac{3}{7}\right)$ **28.** $\left(9y + \frac{2}{3}\right)\left(9y - \frac{2}{3}\right)$

29. $p(3p + 7)(3p - 7)$ **30.** $q(5q - 1)(5q + 1)$

RELATING CONCEPTS (EXERCISES 31–40)

Special products can be illustrated by using areas of rectangles. Use the figure, and ***work Exercises 31–36 in order*** *to justify the special product* $(a + b)^2 = a^2 + 2ab + b^2$.

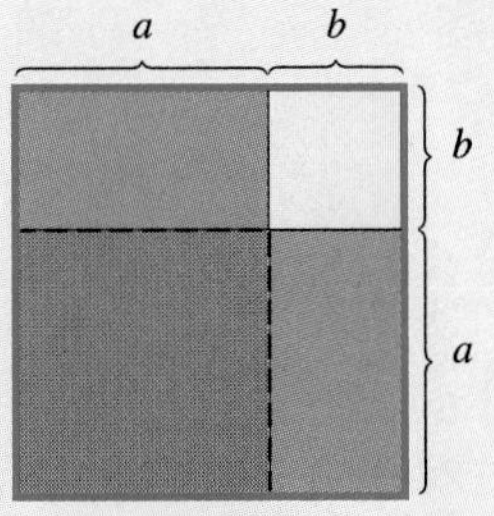

31. Express the area of the large square as the square of a binomial.
32. Give the monomial that represents the area of the red square.

33. $2ab$ 34. b^2
35. $a^2 + 2ab + b^2$
36. They both represent the area of the entire large square.
37. 1225
38. $30^2 + 2(30)(5) + 5^2$
39. 1225 40. They are equal.

RELATING CONCEPTS (EXERCISES 31–40) (CONTINUED)

33. Give the monomial that represents the sum of the areas of the blue rectangles.

34. Give the monomial that represents the area of the yellow square.

35. What is the sum of the monomials you obtained in Exercises 32–34?

36. Explain why the binomial square you found in Exercise 31 must equal the polynomial you found in Exercise 35.

To understand how the special product $(a + b)^2 = a^2 + 2ab + b^2$ can be applied to a purely numerical problem, ***work Exercises 37–40 in order.***

37. Evaluate 35^2 using either traditional paper-and-pencil methods or a calculator.

38. The number 35 can be written as $30 + 5$. Therefore, $35^2 = (30 + 5)^2$. Use the special product for squaring a binomial with $a = 30$ and $b = 5$ to write an expression for $(30 + 5)^2$. Do not simplify at this time.

39. Use the rules for order of operations to simplify the expression you found in Exercise 38.

40. How do the answers in Exercises 37 and 39 compare?

Did you make the connections among geometry, algebra, and arithmetic in these exercises?

The special product

$$(a + b)(a - b) = a^2 - b^2$$

can be used to perform some multiplication problems. For example,

$$51 \times 49 = (50 + 1)(50 - 1) = 50^2 - 1^2 = 2500 - 1 = 2499$$

$$102 \times 98 = (100 + 2)(100 - 2) = 100^2 - 2^2 = 10{,}000 - 4 = 9996.$$

Once these patterns are recognized, multiplications of this type can be done mentally.

Use this method to calculate each product mentally.

41. 101×99 **42.** 103×97

43. 201×199 **44.** 301×299

45. $20\frac{1}{2} \times 19\frac{1}{2}$ **46.** $30\frac{1}{3} \times 29\frac{2}{3}$

Find each product. See Example 5.

47. $(m - 5)^3$ **48.** $(p + 3)^3$ **49.** $(2a + 1)^3$

50. $(3m - 1)^3$ **51.** $(3r - 2t)^4$ **52.** $(2z + 5y)^4$

53. Explain how the expressions $x^2 + y^2$ and $(x + y)^2$ differ.

54. Does $a^3 + b^3$ equal $(a + b)^3$? Explain your answer.

41. 9999 42. 9991
43. 39,999 44. 89,999
45. $399\frac{3}{4}$ 46. $899\frac{8}{9}$
47. $m^3 - 15m^2 + 75m - 125$
48. $p^3 + 9p^2 + 27p + 27$
49. $8a^3 + 12a^2 + 6a + 1$
50. $27m^3 - 27m^2 + 9m - 1$
51. $81r^4 - 216r^3t + 216r^2t^2 - 96rt^3 + 16t^4$
52. $16z^4 + 160z^3y + 600z^2y^2 + 1000zy^3 + 625y^4$
53. $x^2 + y^2$ is the sum of squares, while $(x + y)^2$ is the square of a sum. $(x + y)^2 = x^2 + 2xy + y^2$, and thus there is another term, $2xy$.
54. In general, they are not equal, because $(a + b)^3 = a^3 + 3a^2b + 3ab^2 + b^3$.

55. $\frac{1}{2}m^2 - 2n^2$
56. $36p^2 + 12pq + q^2$
57. $9a^2 - 4$ 58. $8b^2 - \frac{1}{2}$
59. $\pi x^2 + 4\pi x + 4\pi$
60. $16x + 8$
61. $x^3 + 6x^2 + 12x + 8$
62. 512 cubic units

Determine a polynomial that represents the area of each figure. Use the formulas found on the inside covers.

55.

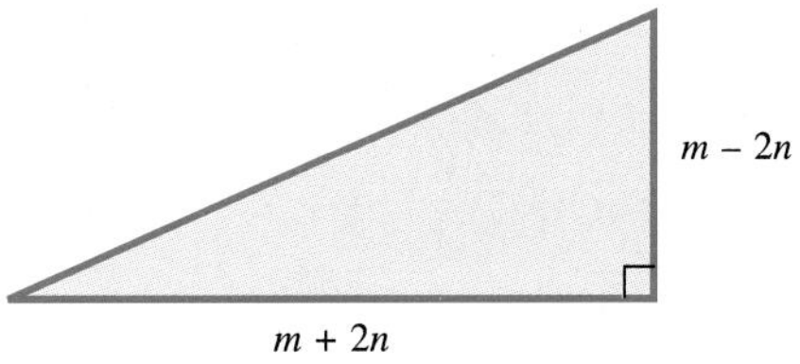

56.

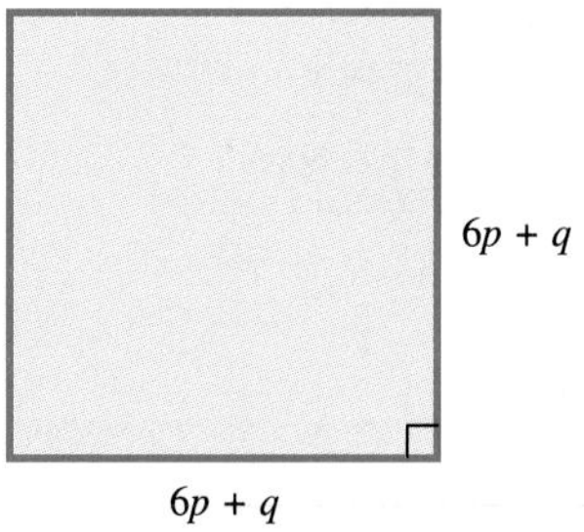

57.

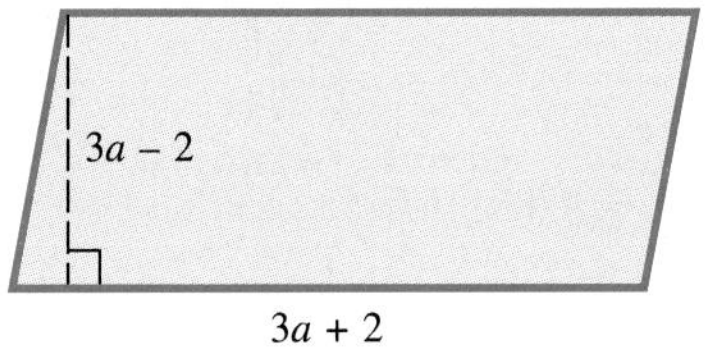

58.

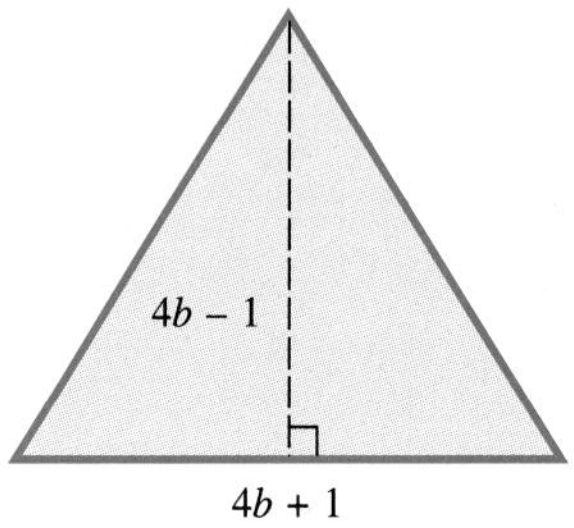

59.

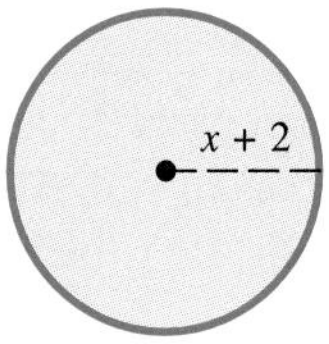

60.

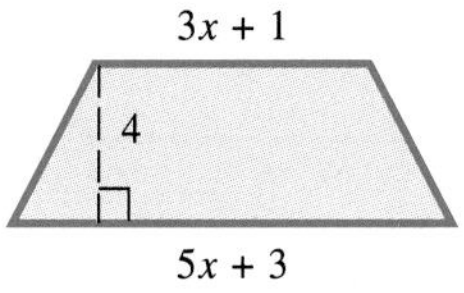

In Exercises 61 and 62, refer to the figure shown here.

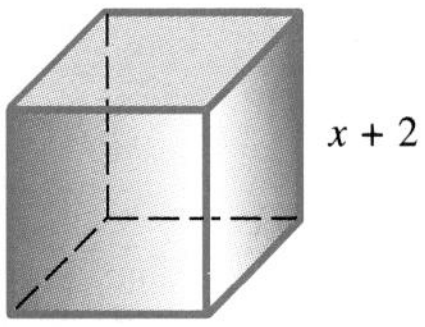

61. Find a polynomial that represents the volume of the cube.

62. If the value of x is 6, what is the volume of the cube?

4.5 Integer Exponents and the Quotient Rule

OBJECTIVES

1. Use zero as an exponent.
2. Use negative numbers as exponents.
3. Use the quotient rule for exponents.
4. Use combinations of rules.

In our work so far, all exponents have been positive integers. To develop meanings for exponents other than positive integers (such as 0 and negative integers), we want to define them in such a way that the same rules for exponents apply.

OBJECTIVE 1 Use zero as an exponent. Suppose we want to find a meaning for an expression such as

$$6^0,$$

where 0 is used as an exponent. If we were to multiply this by 6^2, for example, we would still want the product rule to be valid. Therefore, we would have

$$6^0 \cdot 6^2 = 6^{0+2} = 6^2.$$

JOURNAL WRITING ▲ CHALLENGING SCIENTIFIC CALCULATOR GRAPHING CALCULATOR

INSTRUCTOR'S RESOURCES

PTB Sec. 4.5 **ISM** Sec. 4.5 **AB** Sec. 4.5 | **TEST GENERATOR** | **WORLD WIDE WEB** www.LialAlgebra.com

FOR EXTRA HELP

SSG Sec. 4.5
SSM Sec. 4.5

Pass the Test Software

InterAct Math Tutorial Software

Video 7

So multiplying 6^2 by 6^0 should give 6^2. Since 6^0 is acting as if it were 1 here, we should define 6^0 to equal 1. This is the definition for 0 used as an exponent with any nonzero base.

Definition of Zero Exponent

For any nonzero real number a, $a^0 = 1$.

Example: $17^0 = 1$

CHALKBOARD EXAMPLE

Evaluate -7^0.

Answer: -1

EXAMPLE 1 Using Zero Exponents

Evaluate each exponential expression.

(a) $60^0 = 1$

(b) $(-60)^0 = 1$

(c) $-60^0 = -(1) = -1$

(d) $y^0 = 1$, if $y \neq 0$.

(e) $6y^0 = 6(1) = 6$, if $y \neq 0$.

(f) $(6y)^0 = 1$, if $y \neq 0$.

TEACHING TIP Cover several examples to clarify: 5^0, $(-5)^0$, -5^0, $-(-5)^0$, and so on.

CAUTION Notice the difference between parts (b) and (c) of Example 1. In Example 1(b), the base is -60 and the exponent is 0. Any nonzero base raised to a zero exponent is 1. But in Example 1(c), the base is 60. Then $60^0 = 1$, and $-60^0 = -1$.

OBJECTIVE 2 Use negative numbers as exponents. Now suppose that we want to give a meaning to

$$6^{-2}$$

so that the product rule is still valid. If we multiply 6^{-2} by 6^2, we get

$$6^{-2} \cdot 6^2 = 6^{-2+2} = 6^0 = 1.$$

The expression 6^{-2} is acting as if it were the reciprocal of 6^2 since their product is 1. The reciprocal of 6^2 may be written $\frac{1}{6^2}$, leading us to define 6^{-2} as $\frac{1}{6^2}$. This is a particular case of the definition of negative exponents.

Definition of a Negative Exponent

For any nonzero real number a and any integer n, $a^{-n} = \frac{1}{a^n}$.

Example: $3^{-2} = \frac{1}{3^2}$

By definition, a^{-n} and a^n are reciprocals, since

$$a^n \cdot a^{-n} = a^n \cdot \frac{1}{a^n} = 1.$$

The definition of a^{-n} also can be written as

$$a^{-n} = \frac{1}{a^n} = \left(\frac{1}{a}\right)^n.$$

For example,

$$6^{-3} = \left(\frac{1}{6}\right)^3 \quad \text{and} \quad \left(\frac{1}{3}\right)^{-2} = 3^2.$$

CHALKBOARD EXAMPLE

Simplify $2^{-1} + 5^{-1}$.

Answer: $\frac{7}{10}$

EXAMPLE 2 Using Negative Exponents

Simplify by using the definition of negative exponents.

(a) $3^{-2} = \frac{1}{3^2} = \frac{1}{9}$

(b) $5^{-3} = \frac{1}{5^3} = \frac{1}{125}$

(c) $\left(\frac{1}{2}\right)^{-3} = 2^3 = 8$

(d) $\left(\frac{2}{5}\right)^{-4} = \left(\frac{5}{2}\right)^4 = \frac{5^4}{2^4} = \frac{625}{16}$ $\qquad$ $\frac{2}{5}$ and $\frac{5}{2}$ are reciprocals.

(e) $4^{-1} - 2^{-1} = \frac{1}{4} - \frac{1}{2} = \frac{1}{4} - \frac{2}{4} = -\frac{1}{4}$

(f) $p^{-2} = \frac{1}{p^2}, \quad p \neq 0$

(g) $\frac{3}{3^{-1}} = \frac{3}{\frac{1}{3}} = 3 \div \frac{1}{3} = 3 \cdot 3 = 9$

(h) $\frac{2}{x^{-4}}, \quad x \neq 0$

$$\frac{2}{x^{-4}} = 2 \cdot \frac{1}{x^{-4}} = 2 \cdot x^4 \quad \text{or} \quad 2x^4$$

CAUTION

A negative exponent does not indicate a negative number; negative exponents lead to reciprocals.

Expression	*Example*	
a^{-n}	$3^{-2} = \frac{1}{3^2} = \frac{1}{9}$	Not negative
$-a^{-n}$	$-3^{-2} = -\frac{1}{3^2} = -\frac{1}{9}$	Negative

Examples 2(f) and 2(h) suggest that the definition of a negative exponent allows us to move factors in a fraction between the numerator and denominator if we also change the signs of the exponents. For example,

$$\frac{2^{-3}}{3^{-4}} = \frac{\frac{1}{2^3}}{\frac{1}{3^4}} \qquad \text{Definition of negative exponent}$$

$$= \frac{1}{2^3} \cdot \frac{3^4}{1} \qquad \text{Division by a fraction}$$

$$\frac{2^{-3}}{3^{-4}} = \frac{3^4}{2^3}. \qquad \text{Multiplication of fractions}$$

This fact is generalized below.

Changing from Negative to Positive Exponents

For any nonzero numbers a and b, and any integers m and n,

$$\frac{a^{-m}}{b^{-n}} = \frac{b^n}{a^m}.$$

Example: $\frac{3^{-5}}{2^{-4}} = \frac{2^4}{3^5}$

CHALKBOARD EXAMPLE

Write with only positive exponents. Assume all variables represent nonzero real numbers.

$$\frac{4h^{-5}}{m^{-2}k}$$

Answer: $\frac{4m^2}{h^5k}$

EXAMPLE 3 Changing from Negative to Positive Exponents

Write with only positive exponents. Assume all variables represent nonzero real numbers.

(a) $\frac{4^{-2}}{5^{-3}} = \frac{5^3}{4^2}$

(b) $\frac{m^{-5}}{p^{-1}} = \frac{p^1}{m^5} = \frac{p}{m^5}$

(c) $\frac{a^{-2}b}{3d^{-3}} = \frac{bd^3}{3a^2}$

(d) $x^3y^{-4} = \frac{x^3y^{-4}}{1} = \frac{x^3}{y^4}$

Be careful. We cannot change negative exponents to positive exponents using this rule if the exponents occur in a sum of terms. For example,

$$\frac{5^{-2} + 3^{-1}}{7 - 2^{-3}}$$

cannot be written with positive exponents using the rule given here. We would have to use the definition of a negative exponent to rewrite this expression with positive exponents, as

$$\frac{\frac{1}{5^2} + \frac{1}{3}}{7 - \frac{1}{2^3}}.$$

OBJECTIVE 3 Use the quotient rule for exponents. How do we simplify the quotient of two exponential expressions with the same base? We know that

$$\frac{6^5}{6^3} = \frac{6 \cdot 6 \cdot 6 \cdot 6 \cdot 6}{6 \cdot 6 \cdot 6} = 6^2.$$

Notice that $5 - 3 = 2$. Also,

$$\frac{6^2}{6^4} = \frac{6 \cdot 6}{6 \cdot 6 \cdot 6 \cdot 6} = \frac{1}{6^2} = 6^{-2}.$$

Here, $2 - 4 = -2$. These examples suggest the quotient rule for exponents.

Quotient Rule for Exponents

For any nonzero real number a and any integers m and n,

$$\frac{a^m}{a^n} = a^{m-n}.$$

(Keep the base; subtract the exponents.)

Example: $\frac{5^8}{5^4} = 5^{8-4} = 5^4$

CHALKBOARD EXAMPLE

Simplify. Write the answer with positive exponents.

$$\frac{8^4 \cdot m^9}{8^5 \cdot m^{10}}$$

Answer: $\frac{1}{8m}$

TEACHING TIP On some occasions, students may try to "reduce" $\frac{5^8}{5^6}$ as $\frac{1^8}{1^6}$. Stress that the expression does not say "5 divided by 5."

EXAMPLE 4 Using the Quotient Rule for Exponents

Simplify by using the quotient rule for exponents. Write answers with positive exponents.

(a) $\frac{5^8}{5^6} = 5^{8-6} = 5^2$

(b) $\frac{4^2}{4^9} = 4^{2-9} = 4^{-7} = \frac{1}{4^7}$

(c) $\frac{5^{-3}}{5^{-7}} = \frac{5^7}{5^3} = 5^{7-3} = 5^4$

(d) $\frac{q^5}{q^{-3}} = q^5q^3 = q^8, \quad q \neq 0$

(e) $\frac{3^2x^5}{3^4x^3} = \frac{3^2}{3^4} \cdot \frac{x^5}{x^3} = 3^{2-4} \cdot x^{5-3} = 3^{-2}x^2 = \frac{x^2}{3^2} = \frac{x^2}{9}, \quad x \neq 0$

(f) $\frac{(m+n)^{-2}}{(m+n)^{-4}} = \frac{(m+n)^4}{(m+n)^2} = (m+n)^{4-2} = (m+n)^2, \quad m \neq -n$

(g) $\frac{7x^{-3}y^2}{2^{-1}x^2y^{-5}} = \frac{7 \cdot 2^1y^2y^5}{x^2x^3} = \frac{14y^7}{x^5}$

Working as shown in Examples 4(c) and 4(d), first changing negative exponents to positive exponents, avoids the potential for errors.

We list all the definitions and rules for exponents given in this section and Section 4.2 in the following summary.

TEACHING TIP Stress that the product, quotient, and power rules are the same for positive or negative exponents.

Definitions and Rules for Exponents

If no denominators are zero, for any integers m and n:

			Examples
Product rule		$a^m \cdot a^n = a^{m+n}$	$7^4 \cdot 7^3 = 7^7$
Zero exponent		$a^0 = 1$	$(-3)^0 = 1$
Negative exponent		$a^{-n} = \frac{1}{a^n}$	$5^{-3} = \frac{1}{5^3}$
Quotient rule		$\frac{a^m}{a^n} = a^{m-n}$	$\frac{2^2}{2^5} = 2^{-3} = \frac{1}{2^3}$
Power rules	(a)	$(a^m)^n = a^{mn}$	$(4^2)^3 = 4^6$
	(b)	$(ab)^m = a^m b^m$	$(3k)^4 = 3^4 k^4$
	(c)	$\left(\frac{a}{b}\right)^m = \frac{a^m}{b^m}$	$\left(\frac{2}{3}\right)^{10} = \frac{2^{10}}{3^{10}}$
	(d)	$\frac{a^{-m}}{b^{-n}} = \frac{b^n}{a^m}$	$\frac{5^{-3}}{3^{-5}} = \frac{3^5}{5^3}$
	(e)	$\left(\frac{a}{b}\right)^{-m} = \left(\frac{b}{a}\right)^m$	$\left(\frac{4}{7}\right)^{-2} = \left(\frac{7}{4}\right)^2$

OBJECTIVE 4 Use combinations of rules. As shown in the next example, sometimes we may need to use more than one rule to simplify an expression.

CHALKBOARD EXAMPLE

Simplify. Assume all variables represent nonzero real numbers.

$$\frac{3^9 \cdot (x^2y)^{-2}}{3^3 \cdot x^{-4}y}$$

Answer: $\frac{3^6}{y^3}$

EXAMPLE 5 Using a Combination of Rules

Simplify each expression.

(a) $\frac{(4^2)^3}{4^5}$

Use power rule (a) and then the quotient rule.

$$\frac{(4^2)^3}{4^5} = \frac{4^6}{4^5} = 4^{6-5} = 4^1 = 4$$

(b) $(2x)^3(2x)^2$

Use the product rule first. Then use power rule (b).

$$(2x)^3(2x)^2 = (2x)^5 = 2^5x^5 \quad \text{or} \quad 32x^5$$

(c) $\left(\frac{2x^3}{5}\right)^{-4}$

Use power rules (a), (b), (c), and (e).

$$\left(\frac{2x^3}{5}\right)^{-4} = \left(\frac{5}{2x^3}\right)^4 \quad \text{Power rule (e)}$$

$$= \frac{5^4}{2^4(x^3)^4} \quad \text{Power rules (b) and (c)}$$

$$= \frac{5^4}{16x^{12}} \quad \text{Power rule (a)}$$

(d) $\left(\dfrac{3x^{-2}}{4^{-1}y^{3}}\right)^{-3} = \dfrac{3^{-3}x^{6}}{4^{3}y^{-9}}$ Power rules

$= \dfrac{x^{6}y^{9}}{3^{3}\cdot 4^{3}}$ Power rule (d)

$= \dfrac{x^{6}y^{9}}{27(64)}$ Definition of exponent

$= \dfrac{x^{6}y^{9}}{1728}$

NOTE Since the steps can be done in different orders, there are many equally good ways to simplify a problem like Example 5(d).

(All are in millions of dollars.) 3501.09; 4792.64; 5607.39; 6560.65; 7675.96

CONNECTIONS

In Exercises 91–94 of Section 4.2, we gave an example of an exponential expression that modeled the growth of money left at compound interest. Exponential expressions can also model the way in which quantities grow or decay. (These are examples of functions.) For instance, U.S. commercial space revenues for 1990–1995 can be modeled by the exponential equation

$$y = 3501.09(1.17)^{x},$$

where y is in millions of dollars, and $x = 0$ corresponds to 1990, $x = 1$ corresponds to 1991, and so on, up to $x = 5$ for 1995. Using a calculator, we can find that in 1991 the revenues were approximately $3501.09(1.17)^{1} = 4096.28$ million dollars. (This is actually a bit lower than the true amount of 4370 million dollars.) (*Source:* U.S. Department of Commerce, International Trade Administration, *U.S. Industrial Outlook 1994*; and unpublished data.)

FOR DISCUSSION OR WRITING

Use the model and a calculator to find the approximate revenues for 1990, 1992, 1993, 1994, and 1995.

4.5 EXERCISES

1. false **2.** true **3.** true **4.** true **5.** false **6.** true **7.** 1 **8.** 1 **9.** 1 **10.** 1 **11.** −1 **12.** −1 **13.** 0 **14.** 0 **15.** 0 **16.** 0

Decide whether each statement is true or false.

1. $(-2)^{-4} < 0$ **2.** $-2^{-4} < 0$ **3.** $(-5)^{0} > 0$

4. $-5^{0} < 0$ **5.** $1 - 12^{0} = 1$ **6.** $-(-3)^{0} = -1$

The given expression is either equal to 0, 1, *or* −1. *Decide which is correct. See Example 1.*

7. 9^{0} **8.** 5^{0} **9.** $(-4)^{0}$ **10.** $(-10)^{0}$

11. -9^{0} **12.** -5^{0} **13.** $(-2)^{0} - 2^{0}$ **14.** $(-8)^{0} - 8^{0}$

15. $\dfrac{0^{10}}{10^{0}}$ **16.** $\dfrac{0^{5}}{5^{0}}$

17. 2 **18.** 2 **19.** $\frac{1}{64}$
20. $\frac{1}{625}$ **21.** 16 **22.** 27
23. $\frac{49}{36}$ **24.** $\frac{27}{8}$ **25.** $\frac{1}{81}$
26. $-\frac{1}{64}$ **27.** $\frac{8}{15}$ **28.** $\frac{2}{3}$
29. 5^3 **30.** 11^3 **31.** $\frac{1}{9}$
32. $\frac{1}{7}$ **33.** 5^2 **34.** 6^3
35. x^{15} **36.** y^{10} **37.** 6^3
38. 5^2 **39.** $2r^4$ **40.** $3s^8$
41. $\frac{5^2}{4^3}$ **42.** $\frac{5^4}{6^2}$ **43.** $\frac{p^5}{q^8}$
44. $\frac{y^4}{x^8}$ **45.** r^9 **46.** a^{10}
47. $\frac{x^5}{6}$ **48.** $\frac{y^3}{3^2}$ **49.** $a + b$
50. $x + y$ **51.** $(x + 2y)^2$
52. $(p - 3q)^2$

Evaluate each expression. See Examples 1 and 2.

17. $7^0 + 9^0$ **18.** $8^0 + 6^0$ **19.** 4^{-3} **20.** 5^{-4}

21. $\left(\frac{1}{2}\right)^{-4}$ **22.** $\left(\frac{1}{3}\right)^{-3}$ **23.** $\left(\frac{6}{7}\right)^{-2}$ **24.** $\left(\frac{2}{3}\right)^{-3}$

25. $(-3)^{-4}$ **26.** $(-4)^{-3}$ **27.** $5^{-1} + 3^{-1}$ **28.** $6^{-1} + 2^{-1}$

Use the quotient rule to simplify each expression. Write the expression with positive exponents. Assume that all variables represent nonzero real numbers. See Examples 2–4.

29. $\frac{5^8}{5^5}$ **30.** $\frac{11^6}{11^3}$ **31.** $\frac{9^4}{9^5}$ **32.** $\frac{7^3}{7^4}$ **33.** $\frac{5}{5^{-1}}$

34. $\frac{6}{6^{-2}}$ **35.** $\frac{x^{12}}{x^{-3}}$ **36.** $\frac{y^4}{y^{-6}}$ **37.** $\frac{1}{6^{-3}}$ **38.** $\frac{1}{5^{-2}}$

39. $\frac{2}{r^{-4}}$ **40.** $\frac{3}{s^{-8}}$ **41.** $\frac{4^{-3}}{5^{-2}}$ **42.** $\frac{6^{-2}}{5^{-4}}$ **43.** p^5q^{-8}

44. $x^{-8}y^4$ **45.** $\frac{r^5}{r^{-4}}$ **46.** $\frac{a^6}{a^{-4}}$ **47.** $\frac{6^4x^8}{6^5x^3}$ **48.** $\frac{3^8y^5}{3^{10}y^2}$

49. $\frac{(a + b)^{-3}}{(a + b)^{-4}}$ **50.** $\frac{(x + y)^{-8}}{(x + y)^{-9}}$ **51.** $\frac{(x + 2y)^{-3}}{(x + 2y)^{-5}}$ **52.** $\frac{(p - 3q)^{-2}}{(p - 3q)^{-4}}$

53. 1 **54.** $\frac{5^2}{5^2}$ **55.** 5^0
56. $1 = 5^0$; This supports the definition for 0 as an exponent.

RELATING CONCEPTS (EXERCISES 53–56)

In Objective 1, we showed how 6^0 acts as 1 when it is applied to the product rule, thus motivating the definition for 0 as an exponent. We can also use the quotient rule to motivate this definition.

Work Exercises 53–56 in order.

53. Consider the expression $\frac{25}{25}$. What is its simplest form?

54. Because $25 = 5^2$, the expression $\frac{25}{25}$ can be written as the quotient of powers of 5. Write the expression in this way.

55. Apply the quotient rule for exponents to the expression you wrote in Exercise 54. Give the answer as a power of 5.

56. Your answers in Exercises 53 and 55 must be equal because they both represent $\frac{25}{25}$. Write this equality. What definition does this result support?

Did you make the connection that the definition of 0 as an exponent can be justified using division rules as well as multiplication rules?

57. 7^3 or 343 **58.** 5^4 or 625
59. $\frac{1}{x^2}$ **60.** $\frac{1}{y^5}$ **61.** $\frac{64x}{9}$
62. $\frac{625y}{8}$ **63.** $\frac{x^2z^4}{y^2}$ **64.** $\frac{p^{12}}{r^9q^3}$
65. $6x$ **66.** $10y$ **67.** $\frac{1}{m^{10}n^5}$
68. $\frac{m^{18}}{n^{13}}$

Use a combination of the rules for exponents to simplify each expression. Write answers with only positive exponents. Assume that all variables represent nonzero real numbers. See Example 5.

57. $\frac{(7^4)^3}{7^9}$ **58.** $\frac{(5^3)^2}{5^2}$ **59.** $x^{-3} \cdot x^5 \cdot x^{-4}$ **60.** $y^{-8} \cdot y^5 \cdot y^{-2}$

61. $\frac{(3x)^{-2}}{(4x)^{-3}}$ **62.** $\frac{(2y)^{-3}}{(5y)^{-4}}$ **63.** $\left(\frac{x^{-1}y}{z^2}\right)^{-2}$ **64.** $\left(\frac{p^{-4}q}{r^{-3}}\right)^{-3}$

65. $(6x)^4(6x)^{-3}$ **66.** $(10y)^9(10y)^{-8}$ **67.** $\frac{(m^7n)^{-2}}{m^{-4}n^3}$ **68.** $\frac{(m^8n^{-4})^2}{m^{-2}n^5}$

69. $\dfrac{1}{xyz}$ **70.** 1 **71.** x^3y^9
72. $\dfrac{z^{12}}{w^8}$ **73.** The student attempted to use the quotient rule with unequal bases. The correct way to simplify this expression is $\dfrac{16^3}{2^2} = \dfrac{(2^4)^3}{2^2} = \dfrac{2^{12}}{2^2} = 2^{10} = 1024.$
74. The student incorrectly assumed that the negative sign was part of the base. The correct way to simplify this expression is $-5^4 = -1(5^4) = -1 \cdot 625 = -625.$

69. $\dfrac{(x^{-1}y^2z)^{-2}}{(x^{-3}y^3z)^{-1}}$ **70.** $\dfrac{(a^{-2}b^{-3}c^{-4})^{-5}}{(a^2b^3c^4)^5}$ **71.** $\left(\dfrac{xy^{-2}}{x^2y}\right)^{-3}$ **72.** $\left(\dfrac{wz^{-5}}{w^{-3}z}\right)^{-2}$

73. Consider the following typical student **error**:

$$\frac{16^3}{2^2} = \left(\frac{16}{2}\right)^{3-2} = 8^1 = 8.$$

Explain what the student did incorrectly, and then give the correct answer.

74. Consider the following typical student **error**:

$$-5^4 = (-5)^4 = 625.$$

Explain what the student did incorrectly, and then give the correct answer.

4.6 Division of Polynomials

OBJECTIVES

1. Divide a polynomial by a monomial.
2. Divide a polynomial by a polynomial.

FOR EXTRA HELP

- **SSG** Sec. 4.6
- **SSM** Sec. 4.6
- **Pass the Test Software**
- **InterAct Math Tutorial Software**
- **Video** 7

OBJECTIVE 1 **Divide a polynomial by a monomial.** We add two fractions with a common denominator as follows.

$$\frac{a}{c} + \frac{b}{c} = \frac{a+b}{c}.$$

Looking at this statement in reverse gives us a rule for dividing a polynomial by a monomial.

Dividing a Polynomial by a Monomial

To divide a polynomial by a monomial, divide each term of the polynomial by the monomial:

$$\frac{a+b}{c} = \frac{a}{c} + \frac{b}{c} \quad (c \neq 0).$$

The parts of a division problem are named in the diagram.

$$\text{Dividend} \rightarrow \frac{12x^2 + 6x}{6x} = 2x + 1 \leftarrow \text{Quotient}$$
$$\text{Divisor} \rightarrow$$

CHALKBOARD EXAMPLE

Divide $12m^6 + 18m^5 + 30m^4$ by $6m^2$.

Answer: $2m^4 + 3m^3 + 5m^2$

EXAMPLE 1 **Dividing a Polynomial by a Monomial**

Divide $5m^5 - 10m^3$ by $5m^2$.

Use the rule above, with + replaced by −. Then use the quotient rule for exponents.

$$\frac{5m^5 - 10m^3}{5m^2} = \frac{5m^5}{5m^2} - \frac{10m^3}{5m^2} = m^3 - 2m$$

Recall from arithmetic that division problems can be checked by multiplication:

$$\frac{63}{7} = 9 \quad \text{because} \quad 7 \cdot 9 = 63.$$

To check the polynomial quotient, multiply $m^3 - 2m$ by $5m^2$. Because

$$5m^2(m^3 - 2m) = 5m^5 - 10m^3,$$

the quotient is correct.

INSTRUCTOR'S RESOURCES

PTB Sec. 4.6 **ISM** Sec. 4.6 **AB** Sec. 4.6 | **TEST GENERATOR** | **WORLD WIDE WEB** www.LialAlgebra.com

Since division by 0 is undefined, the quotient

$$\frac{5m^5 - 10m^3}{5m^2}$$

is undefined if $m = 0$. From now on we assume that no denominators are 0.

CHALKBOARD EXAMPLE

Divide $\dfrac{50m^4 - 30m^3 + 20m}{10m^3}$.

Answer: $5m - 3 + \dfrac{2}{m^2}$

EXAMPLE 2 Dividing a Polynomial by a Monomial

Divide $\dfrac{16a^5 - 12a^4 + 8a^2}{4a^3}$.

Divide each term of $16a^5 - 12a^4 + 8a^2$ by $4a^3$.

$$\frac{16a^5 - 12a^4 + 8a^2}{4a^3} = \frac{16a^5}{4a^3} - \frac{12a^4}{4a^3} + \frac{8a^2}{4a^3}$$

$$= 4a^2 - 3a + \frac{2}{a}$$

The result is not a polynomial because of the expression $\frac{2}{a}$, which has a variable in the denominator. While the sum, difference, and product of two polynomials are always polynomials, the quotient of two polynomials may not be.

Again, check by multiplying.

$$4a^3\left(4a^2 - 3a + \frac{2}{a}\right) = 4a^3(4a^2) - 4a^3(3a) + 4a^3\left(\frac{2}{a}\right)$$

$$= 16a^5 - 12a^4 + 8a^2$$

TEACHING TIP The most frequent error made here is with the third term of the quotient. Students often write $2a$ instead of $\dfrac{2}{a}$.

CHALKBOARD EXAMPLE

Divide

$\dfrac{-12p^5 - 8p^4 - 6p^3 + 5p^2}{-3p^3}$.

Answer: $4p^2 + \dfrac{8p}{3} + 2 - \dfrac{5}{3p}$

EXAMPLE 3 Dividing a Polynomial by a Monomial with a Negative Coefficient

Divide $-7x^3 + 12x^4 - 4x$ by $-4x$.

The polynomial should be written in descending powers before dividing. Write it as $12x^4 - 7x^3 - 4x$; then divide by $-4x$.

$$\frac{12x^4 - 7x^3 - 4x}{-4x} = \frac{12x^4}{-4x} - \frac{7x^3}{-4x} + \frac{-4x}{-4x}$$

$$= -3x^3 + \frac{7x^2}{4} + 1 = -3x^3 + \frac{7}{4}x^2 + 1$$

Check by multiplying.

CAUTION In Example 3, notice the quotient $\dfrac{-4x}{-4x} = 1$. It is a common error to leave this term out of the answer. Checking by multiplication will show that the answer $-3x^3 + \dfrac{7}{4}x^2$ is not correct.

OBJECTIVE 2 Divide a polynomial by a polynomial. We use a method of "long division" to divide a polynomial by a polynomial (other than a monomial). This method is similar to the method of long division used for two whole numbers. For comparison, the division of whole numbers is shown alongside the division of polynomials. Both polynomials must be written with descending powers before beginning the division process.

TEACHING TIP Students often make sign errors when performing the subtraction step in polynomial division. There are several strategies that can be used in the classroom to guard against such errors. One of the more popular ones is to have students circle the original signs in the bottom polynomial, and then write in the opposite sign. Then the addition is performed.

Step 1 Divide 27 into 6696.

$$27\overline{)6696}$$

Divide $2x + 3$ into $8x^3 - 4x^2 - 14x + 15$.

$$2x + 3\overline{)8x^3 - 4x^2 - 14x + 15}$$

Step 2 27 divides into 66 **2** times; $2 \cdot 27 =$ **54**.

$$\begin{array}{r} 2 \\ 27\overline{)6696} \\ \underline{54} \end{array}$$

$2x$ divides into $8x^3$ $\mathbf{4x^2}$ times; $4x^2(2x + 3) = \mathbf{8x^3 + 12x^2}$.

$$\begin{array}{r} 4x^2 \qquad\qquad\qquad\quad \\ 2x + 3\overline{)8x^3 - 4x^2 - 14x + 15} \\ \underline{\mathbf{8x^3 + 12x^2}} \qquad\qquad\quad \end{array}$$

Step 3 Subtract: $66 - 54 = 12$; then bring down the next digit.

$$\begin{array}{r} 2 \\ 27\overline{)6696} \\ \underline{54}\downarrow \\ 129 \end{array}$$

Subtract: $-4x^2 - 12x^2 = -16x^2$; then bring down the next term.

$$\begin{array}{r} 4x^2 \qquad\qquad\qquad\quad \\ 2x + 3\overline{)8x^3 - 4x^2 - 14x + 15} \\ \underline{8x^3 + 12x^2} \quad \downarrow \qquad\quad \\ -16x^2 - 14x \qquad\quad \end{array}$$

(To subtract two polynomials, change the sign of the second and then add.)

Step 4 27 divides into 129 **4** times; $4 \cdot 27 =$ **108**.

$$\begin{array}{r} 24 \\ 27\overline{)6696} \\ \underline{54} \\ 129 \\ \underline{\mathbf{108}} \end{array}$$

$2x$ divides into $-16x^2$ $\mathbf{-8x}$ times; $-8x(2x + 3) = \mathbf{-16x^2 - 24x}$.

$$\begin{array}{r} 4x^2 - 8x \qquad\quad \\ 2x + 3\overline{)8x^3 - 4x^2 - 14x + 15} \\ \underline{8x^3 + 12x^2} \qquad\qquad\quad \\ -16x^2 - 14x \qquad\quad \\ \underline{\mathbf{-16x^2 - 24x}} \qquad\quad \end{array}$$

Step 5 Subtract: $129 - 108 = 21$; then bring down the next digit.

$$\begin{array}{r} 24 \\ 27\overline{)6696} \\ \underline{54} \\ 129 \\ \underline{108}\downarrow \\ 216 \end{array}$$

Subtract: $-14x - (-24x) = 10x$; then bring down the next term.

$$\begin{array}{r} 4x^2 - 8x \qquad\quad \\ 2x + 3\overline{)8x^3 - 4x^2 - 14x + 15} \\ \underline{8x^3 + 12x^2} \qquad\qquad\quad \\ -16x^2 - 14x \qquad\quad \\ \underline{-16x^2 - 24x} \quad \downarrow \\ 10x + 15 \end{array}$$

TEACHING TIP Using the method suggested, Step 5 would look like this:

$$\begin{array}{r} 4x^2 - 8x \qquad\quad \\ 2x + 3\overline{)8x^3 - 4x^2 - 14x + 15} \\ - \quad - \qquad\qquad\qquad \\ \underline{\oplus 8x^3 \oplus 12x^2} \qquad\qquad\quad \\ -16x^2 - 14x \qquad\quad \\ + \qquad + \qquad\quad \\ \underline{\ominus 16x^2 \ominus 24x} \qquad\quad \\ 10x + 15 \end{array}$$

Step 6 27 divides into 216 **8** times; $8 \cdot 27 =$ **216**.

$$\begin{array}{r} 248 \\ 27\overline{)6696} \\ \underline{54} \\ 129 \\ \underline{108} \\ 216 \\ \underline{\mathbf{216}} \end{array}$$

6696 divided by 27 is 248. There is no remainder.

$2x$ divides into $10x$ **5** times; $5(2x + 3) =$ $\mathbf{10x + 15}$.

$$\begin{array}{r} 4x^2 - 8x + 5 \\ 2x + 3\overline{)8x^3 - 4x^2 - 14x + 15} \\ \underline{8x^3 + 12x^2} \\ -16x^2 - 14x \\ \underline{-16x^2 - 24x} \\ 10x + 15 \\ \underline{\mathbf{10x + 15}} \end{array}$$

$8x^3 - 4x^2 - 14x + 15$ divided by $2x + 3$ is $4x^2 - 8x + 5$. There is no remainder.

Step 7 Check by multiplying. $27 \cdot 248 = 6696$

Check by multiplying.

$$(2x + 3)(4x^2 - 8x + 5) = 8x^3 - 4x^2 - 14x + 15$$

Notice that at each step in the polynomial division process, the *first* term was divided into the *first* term.

CHALKBOARD EXAMPLE

Divide $x^2 + 4x + x^3 - 6$ by $x - 1$.

Answer: $x^2 + 2x + 6$

EXAMPLE 4 Dividing a Polynomial by a Polynomial

Divide $5x + 4x^3 - 8 - 4x^2$ by $2x - 1$.

Both polynomials must be written in descending powers of x. Rewrite the first polynomial as $4x^3 - 4x^2 + 5x - 8$. Then begin the division process.

$$\begin{array}{r} 2x^2 - x + 2 \\ 2x - 1\overline{)4x^3 - 4x^2 + 5x - 8} \\ \underline{4x^3 - 2x^2} \\ -2x^2 + 5x \\ \underline{-2x^2 + x} \\ 4x - 8 \\ \underline{4x - 2} \\ -6 \end{array}$$

Step 1 $2x$ divides into $4x^3$ $\mathbf{(2x^2)}$ times; $2x^2(2x - 1) = 4x^3 - 2x^2$.

Step 2 Subtract; bring down the next term.

Step 3 $2x$ divides into $-2x^2$ $\mathbf{(-x)}$ times; $-x(2x - 1) = -2x^2 + x$.

Step 4 Subtract; bring down the next term.

Step 5 $2x$ divides into $4x$ **2** times; $2(2x - 1) = 4x - 2$.

Step 6 Subtract. The remainder is -6.

In a division problem like this one, the division process stops when the degree of the remainder is less than the degree of the divisor. Thus, $2x - 1$ divides into $4x^3 - 4x^2 + 5x - 8$ with a quotient of $2x^2 - x + 2$ and a remainder of -6. Write the remainder as a fraction with $2x - 1$ as the denominator. The result is not a polynomial because of the remainder.

$$\frac{4x^3 - 4x^2 + 5x - 8}{2x - 1} = 2x^2 - x + 2 + \frac{-6}{2x - 1}$$

Step 7 Check by multiplying.

$$(2x-1)\left(2x^2-x+2+\frac{-6}{2x-1}\right)$$
$$=(2x-1)(2x^2)+(2x-1)(-x)+(2x-1)(2)+(2x-1)\left(\frac{-6}{2x-1}\right)$$
$$=4x^3-2x^2-2x^2+x+4x-2-6$$
$$=4x^3-4x^2+5x-8$$

CHALKBOARD EXAMPLE

Divide x^3-8 by $x-2$.

Answer: x^2+2x+4

EXAMPLE 5 Dividing into a Polynomial with Missing Terms

Divide x^3+2x-3 by $x-1$.

Here the polynomial x^3+2x-3 is missing the x^2 term. When terms are missing, use 0 as the coefficient for the missing terms; in this case, $0x^2$ is a "placeholder."

$$x^3+2x-3=x^3+0x^2+2x-3$$

Now divide.

$$\begin{array}{r} x^2+x+3\\ x-1\overline{)x^3+0x^2+2x-3}\\ \underline{x^3-x^2}\\ x^2+2x\\ \underline{x^2-x}\\ 3x-3\\ \underline{3x-3}\end{array}$$

TEACHING TIP Explain to students how placeholders here are used just like in whole numbers, where 0 distinguishes 302 from 32, for example.

The remainder is 0. The quotient is x^2+x+3. Check by multiplying.

$$(x^2+x+3)(x-1)=x^3+2x-3$$

CHALKBOARD EXAMPLE

Divide

$$\frac{2m^5+m^4+6m^3-3m^2-18}{m^2+3}.$$

Answer: $2m^3+m^2-6$

EXAMPLE 6 Dividing by a Polynomial with Missing Terms

Divide $x^4+2x^3+2x^2-x-1$ by x^2+1.

Since x^2+1 has a missing x term, write it as x^2+0x+1. Then proceed through the division process.

$$\begin{array}{r} x^2+2x+1\\ x^2+0x+1\overline{)x^4+2x^3+2x^2-x-1}\\ \underline{x^4+0x^3+x^2}\phantom{-x-1}\\ 2x^3+x^2-x\\ \underline{2x^3+0x^2+2x}\\ x^2-3x-1\\ \underline{x^2+0x+1}\\ -3x-2 \end{array} \quad \leftarrow\text{Remainder}$$

When the result of subtracting ($-3x-2$, in this case) is a polynomial of smaller degree than the divisor (x^2+0x+1), that polynomial is the remainder. Write the result as

$$x^2+2x+1+\frac{-3x-2}{x^2+1}.$$

1. -104 2. -104
3. They are both -104.
4. The answers should agree.

CONNECTIONS

In Section 4.1, we found the value of a polynomial in x for a given value of x by substituting that number for x. Surprisingly, we can accomplish the same thing by division. Suppose we want to find the value of $2x^3 - 4x^2 + 3x - 5$ for $x = -3$. Instead of substituting -3 for x in the polynomial, we divide the polynomial by $x - (-3) = x + 3$. The remainder will give the value of the polynomial for $x = -3$. In general, when a polynomial P is divided by $x - r$, the remainder is equal to P evaluated at $x = r$.

FOR DISCUSSION OR WRITING

1. Evaluate $2x^3 - 4x^2 + 3x - 5$ for $x = -3$.
2. Divide $2x^3 - 4x^2 + 3x - 5$ by $x + 3$. Give the remainder.
3. Compare the answers to Exercises 1 and 2. What do you notice?
4. Choose another polynomial and evaluate it both ways at some value of the variable. Do the answers agree?

1. $6x^2 + 8$; 2; $3x^2 + 4$ **2.** 0 **3.** $3x^2 + 4$; 2 (These may be reversed.); $6x^2 + 8$ **4.** is not **5.** The first is a polynomial divided by a monomial, covered in Objective 1. This section does not cover dividing a monomial by a polynomial of several terms. **6.** 2 **7.** $30x^3 - 10x + 5$ **8.** $60x^5 - 30x^2 + 40x$ **9.** $4m^3 - 2m^2 + 1$ **10.** $2t^3 - t + 1$ **11.** $4t^4 - 2t^2 + 2t$ **12.** $4r^3 - 2r^2 + 3r$ **13.** $a^4 - a + \frac{2}{a}$ **14.** $t^7 + t^6 + \frac{3}{t}$ **15.** $4x^3 - 3x^2 + 2x$ **16.** $8x^4 - 4x^3 + 10x^2$ **17.** $1 + 5x - 9x^2$ **18.** $1 - 6x^2 + 10x^3$ **19.** $\frac{12}{x} + 8 + 2x$ **20.** $\frac{3}{x} - 4 + 3x$ **21.** $\frac{4x^2}{3} + x + \frac{2}{3x}$ **22.** $\frac{5x^2}{3} - 2x + \frac{8}{3x}$

4.6 EXERCISES

Fill in each blank with the correct response.

1. In the statement $\frac{6x^2 + 8}{2} = 3x^2 + 4$, ______ is the dividend, ______ is the divisor, and ______ is the quotient.
2. The expression $\frac{3x + 12}{x}$ is undefined if $x =$ ______.
3. To check the division shown in Exercise 1, multiply ______ by ______ and show that the product is ______.
4. The expression $5x^2 - 3x + 6 + \frac{2}{x}$ ____________ a polynomial. (is/is not)
5. Explain why the division problem $\frac{16m^3 - 12m^2}{4m}$ can be performed using the methods of this section, while the division problem $\frac{4m}{16m^3 - 12m^2}$ cannot.
6. Suppose that a polynomial in the variable x has degree 5 and it is divided by a monomial in the variable x having degree 3. What is the degree of the quotient?

Perform each division. See Examples 1–3.

7. $\frac{60x^4 - 20x^2 + 10x}{2x}$
8. $\frac{120x^6 - 60x^3 + 80x^2}{2x}$
9. $\frac{20m^5 - 10m^4 + 5m^2}{5m^2}$
10. $\frac{12t^5 - 6t^3 + 6t^2}{6t^2}$
11. $\frac{8t^5 - 4t^3 + 4t^2}{2t}$
12. $\frac{8r^4 - 4r^3 + 6r^2}{2r}$
13. $\frac{4a^5 - 4a^2 + 8}{4a}$
14. $\frac{5t^8 + 5t^7 + 15}{5t}$

Divide each polynomial by $3x^2$. See Examples 1–3.

15. $12x^5 - 9x^4 + 6x^3$
16. $24x^6 - 12x^5 + 30x^4$
17. $3x^2 + 15x^3 - 27x^4$
18. $3x^2 - 18x^4 + 30x^5$
19. $36x + 24x^2 + 6x^3$
20. $9x - 12x^2 + 9x^3$
21. $4x^4 + 3x^3 + 2x$
22. $5x^4 - 6x^3 + 8x$

23. $9r^3 - 12r^2 + 2r + \frac{26}{3} - \frac{2}{3r}$

24. $4k^3 - 6k^2 - k + \frac{7}{2} - \frac{3}{2k}$

25. $-m^2 + 3m - \frac{4}{m}$

26. $-3r + 4 - \frac{5}{r^2}$

27. $4 - 3a + \frac{5}{a}$

28. $4y^3 - 2 + \frac{3}{y}$

29. $\frac{12}{x} - \frac{6}{x^2} + \frac{14}{x^3} - \frac{10}{x^4}$

30. $10x^3 - 7 + \frac{5}{x} - \frac{3}{x^2}$

31. $\frac{2}{3}x$ **would not be an acceptable form, because** $\frac{2}{3}x = \frac{2}{3} \cdot \frac{x}{1} = \frac{2x}{3}$**, which is *not* equivalent to** $\frac{2}{3x}$**.**

32. $4a^2 - 3a + 2a^{-1}$

33. $5x^3 + 4x^2 - 3x + 1$

34. $48m^2 + 96m + 24$

35. $-63m^4 - 21m^3 - 35m^2 + 14m$

36. **(a)** $n \geq m$ **(b)** $n - m$

Perform each division. See Examples 1–3.

23. $\dfrac{-27r^4 + 36r^3 - 6r^2 - 26r + 2}{-3r}$

24. $\dfrac{-8k^4 + 12k^3 + 2k^2 - 7k + 3}{-2k}$

25. $\dfrac{2m^5 - 6m^4 + 8m^2}{-2m^3}$

26. $\dfrac{6r^5 - 8r^4 + 10r^2}{-2r^4}$

27. $(20a^4 - 15a^5 + 25a^3) \div (5a^4)$

28. $(16y^5 - 8y^2 + 12y) \div (4y^2)$

29. $(120x^{11} - 60x^{10} + 140x^9 - 100x^8) \div (10x^{12})$

30. $(120x^{12} - 84x^9 + 60x^8 - 36x^7) \div (12x^9)$

✎ **31.** The quotient in Exercise 21 is $\frac{4x^2}{3} + x + \frac{2}{3x}$. Notice how the third term is written with x in the denominator. Would $\frac{2}{3}x$ be an acceptable form for this term? Why or why not?

32. Refer to the quotient given in Example 2 in this section. Write it as an equivalent expression using negative exponents as necessary.

▲ *Use the appropriate formula, found on the inside covers, to answer each question.*

33. The area of the rectangle is given by the polynomial $15x^3 + 12x^2 - 9x + 3$. What is the polynomial that expresses the length?

3

34. The area of the triangle is given by the polynomial $24m^3 + 48m^2 + 12m$. What is the polynomial that expresses the length of the base?

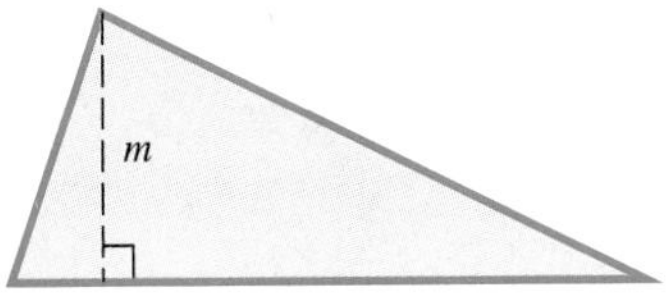

▲ **35.** The quotient of a certain polynomial and $-7m^2$ is $9m^2 + 3m + 5 - \frac{2}{m}$. Find the polynomial.

▲ **36.** Suppose that a polynomial of degree n is divided by a monomial of degree m to get a *polynomial* quotient.

(a) How do m and n compare in value?

(b) What is the expression that gives the degree of the quotient?

37. 1423

38. $(1 \times 10^3) + (4 \times 10^2) + (2 \times 10^1) + (3 \times 10^0)$

39. $x^3 + 4x^2 + 2x + 3$

40. **They are similar in that the coefficients of powers of ten are equal to the coefficients of the powers of** x**. They are different in that one is a constant while the other is a polynomial. They are equal if** $x = 10$ **(the base of our decimal system).**

RELATING CONCEPTS (EXERCISES 37–40)

Our system of numeration is called a decimal system. It is based on powers of ten. In a whole number such as 2846, each digit is understood to represent the number of powers of ten for its place value. The 2 represents two thousands (2×10^3), the 8 represents eight hundreds (8×10^2), the 4 represents four tens (4×10^1), and the 6 represents six ones (or units) (6×10^0). In expanded form we write

$$2846 = (2 \times 10^3) + (8 \times 10^2) + (4 \times 10^1) + (6 \times 10^0).$$

Keeping this information in mind, ***work Exercises 37–40 in order.***

37. Divide 2846 by 2, using paper-and-pencil methods: $2\overline{)2846}$.

38. Write your answer in Exercise 37 in expanded form.

39. Use the methods of this section to divide the polynomial $2x^3 + 8x^2 + 4x + 6$ by 2.

✎ **40.** Compare your answers in Exercises 38 and 39. How are they similar? How are they different? For what value of x does the answer in Exercise 39 equal the answer in Exercise 38?

Did you make the connection between division of whole numbers and division of polynomials?

41. $x + 2$ **42.** $m + 4$
43. $2y - 5$ **44.** $2y + 7$
45. $p - 4 + \frac{44}{p + 6}$
46. $x + 3 + \frac{-8}{x + 8}$
47. $r - 5$ **48.** $t + 7$
49. $6m - 1$ **50.** $6y + 7$
51. $2a - 14 + \frac{74}{2a + 3}$
52. $3w + 4 + \frac{18}{3w - 2}$
53. $4x^2 - 7x + 3$
54. $3t^2 - 5t + 6$
55. $4k^3 - k + 2$
56. $9r^3 - 2r + 6$
57. $5y^3 + 2y - 3$
58. $2r^2 + r - 3 + \frac{6}{r - 3}$
59. $3k^2 + 2k - 2 + \frac{6}{k - 2}$
60. $5z^2 - 11z + 32 + \frac{-62}{z + 2}$
61. $2p^3 - 6p^2 + 7p - 4 + \frac{14}{3p + 1}$
62. $3r^3 - r^2 - 2r + 5 + \frac{7}{2r - 3}$
63. $r^2 - 1 + \frac{4}{r^2 - 1}$
64. $t^2 + 3 + \frac{4}{t^2 + 1}$
65. $y^2 - y + 1$ **66.** $y^2 + y + 1$
67. $a^2 + 1$ **68.** $a^2 - 1$
69. $x^2 - 4x + 2 + \frac{9x - 4}{x^2 + 3}$
70. $3t^2 + 5t + 7 + \frac{12t + 37}{t^2 - 5}$
71. $x^3 + 3x^2 - x + 5$
72. $t^3 - 3t^2 - 1$
73. $\frac{3}{2}a - 10 + \frac{77}{2a + 6}$
74. $\frac{4}{3}x + 1 + \frac{-14}{3x + 6}$
75. The process stops when the degree of the remainder is less than the degree of the divisor.
76. The first term of the quotient should be $\frac{6x^3}{2x} = 3x^2$, not $4x^2$.
77. $x^2 + x - 3$ units
78. $2x^2 + 4x + 1$ units

Perform each division. See Example 4.

41. $\frac{x^2 - x - 6}{x - 3}$ **42.** $\frac{m^2 - 2m - 24}{m - 6}$ **43.** $\frac{2y^2 + 9y - 35}{y + 7}$

44. $\frac{2y^2 + 9y + 7}{y + 1}$ **45.** $\frac{p^2 + 2p + 20}{p + 6}$ **46.** $\frac{x^2 + 11x + 16}{x + 8}$

47. $(r^2 - 8r + 15) \div (r - 3)$ **48.** $(t^2 + 2t - 35) \div (t - 5)$

49. $\frac{12m^2 - 20m + 3}{2m - 3}$ **50.** $\frac{12y^2 + 20y + 7}{2y + 1}$ **51.** $\frac{4a^2 - 22a + 32}{2a + 3}$

52. $\frac{9w^2 + 6w + 10}{3w - 2}$ **53.** $\frac{8x^3 - 10x^2 - x + 3}{2x + 1}$ **54.** $\frac{12t^3 - 11t^2 + 9t + 18}{4t + 3}$

55. $\frac{8k^4 - 12k^3 - 2k^2 + 7k - 6}{2k - 3}$ **56.** $\frac{27r^4 - 36r^3 - 6r^2 + 26r - 24}{3r - 4}$

57. $\frac{5y^4 + 5y^3 + 2y^2 - y - 3}{y + 1}$ **58.** $\frac{2r^3 - 5r^2 - 6r + 15}{r - 3}$

59. $\frac{3k^3 - 4k^2 - 6k + 10}{k - 2}$ **60.** $\frac{5z^3 - z^2 + 10z + 2}{z + 2}$

61. $\frac{6p^4 - 16p^3 + 15p^2 - 5p + 10}{3p + 1}$ **62.** $\frac{6r^4 - 11r^3 - r^2 + 16r - 8}{2r - 3}$

Perform each division. See Examples 5 and 6.

63. $\frac{5 - 2r^2 + r^4}{r^2 - 1}$ **64.** $\frac{4t^2 + t^4 + 7}{t^2 + 1}$ **65.** $\frac{y^3 + 1}{y + 1}$

66. $\frac{y^3 - 1}{y - 1}$ **67.** $\frac{a^4 - 1}{a^2 - 1}$ **68.** $\frac{a^4 - 1}{a^2 + 1}$

69. $\frac{x^4 - 4x^3 + 5x^2 - 3x + 2}{x^2 + 3}$ **70.** $\frac{3t^4 + 5t^3 - 8t^2 - 13t + 2}{t^2 - 5}$

71. $\frac{2x^5 + 9x^4 + 8x^3 + 10x^2 + 14x + 5}{2x^2 + 3x + 1}$ **72.** $\frac{4t^5 - 11t^4 - 6t^3 + 5t^2 - t + 3}{4t^2 + t - 3}$

▲ **73.** $(3a^2 - 11a + 17) \div (2a + 6)$ ▲ **74.** $(4x^2 + 11x - 8) \div (3x + 6)$

75. Suppose that one of your classmates asks you the following question: "How do I know when to stop the division process in a problem like the one in Exercise 69?" How would you respond?

76. Suppose that someone asks you if the following division problem is correct:

$$(6x^3 + 4x^2 - 3x + 9) \div (2x - 3) = 4x^2 + 9x - 3.$$

Tell how, by looking only at the *first term* of the quotient, you immediately know that the problem has been worked incorrectly.

▲ *Find a polynomial that describes each quantity required. Use the formulas found on the inside covers.*

77. Give the length of the rectangle.

The area is $5x^3 + 7x^2 - 13x - 6$ square units.

78. Find the measure of the base of the parallelogram.

The area is $2x^3 + 2x^2 - 3x - 1$ square units.

79. $5x^2 - 11x + 14$ hours
80. $4x^4 - 5x^3 + 12x^2 - 15x + 1$ square yards

79. If the distance traveled is $5x^3 - 6x^2 + 3x + 14$ miles and the rate is $x + 1$ miles per hour, what is the time traveled?

80. If it costs $4x^5 + 3x^4 + 2x^3 + 9x^2 - 29x + 2$ dollars to fertilize a garden, and fertilizer costs $x + 2$ dollars per square yard, what is the area of the garden?

81. (a) is correct, (b) is incorrect.
82. (a) is incorrect, (b) is correct.
83. (a) is correct, (b) is incorrect.
84. Because any power of 1 is 1, to evaluate a polynomial for 1 we simply add the coefficients. If the divisor is $x - 1$, this method does not apply, since $1 - 1 = 0$ and division by 0 is undefined.

RELATING CONCEPTS (EXERCISES 81–84)

Students often would like to know whether the quotient obtained in a polynomial division problem is actually correct or whether an error was made. While the method described here is not 100% foolproof, it is quick and will at least give a fairly good idea as to the accuracy of the result. To illustrate, suppose that $4x^4 + 2x^3 - 14x^2 + 19x + 10$ is divided by $2x + 5$. Lakeisha and Stan obtain the following answers:

Lakeisha	*Stan*
$2x^3 - 4x^2 + 3x + 2$	$2x^3 - 4x^2 - 3x + 2.$

As a "quick check" we can evaluate Lakeisha's answer for $x = 1$ and Stan's answer for $x = 1$:

Lakeisha: When $x = 1$, her answer gives $2(1)^3 - 4(1)^2 + 3(1) + 2 = 3.$
Stan: When $x = 1$, his answer gives $2(1)^3 - 4(1)^2 - 3(1) + 2 = -3.$

Now, if the original quotient of the two polynomials is evaluated for $x = 1$, we get

$$\frac{4x^4 + 2x^3 - 14x^2 + 19x + 10}{2x + 5} = \frac{4(1)^4 + 2(1)^3 - 14(1)^2 + 19(1) + 10}{2(1) + 5} = \frac{21}{7} = 3.$$

Because Stan's answer, -3, is different from the quotient 3 just obtained, Stan can conclude his answer is incorrect. Lakeisha's answer, 3, agrees with the quotient just obtained, and while this does not *guarantee* that she is correct, at least she can feel better and go on to the next problem.

In Exercises 81–84, a division problem is given, along with two possible answers. One is correct and one is incorrect. Use the method just described to determine which one is correct and which one is not.

81. Problem $\dfrac{2x^2 + 3x - 14}{x - 2}$ Possible answers **(a)** $2x + 7$ **(b)** $2x - 7$

82. Problem $\dfrac{x^4 + 4x^3 - 5x^2 - 12x + 6}{x^2 - 3}$ Possible answers **(a)** $x^2 - 4x - 2$ **(b)** $x^2 + 4x - 2$

83. Problem $\dfrac{2y^3 + 17y^2 + 37y + 7}{2y + 7}$ Possible answers **(a)** $y^2 + 5y + 1$ **(b)** $y^2 - 5y + 1$

84. In the explanation preceding Exercise 81 we used 1 for the value of x to check our work. This is because a polynomial is easy to evaluate for 1. Why is this so? Why would we not be able to use 1 if the divisor is $x - 1$?

Did you make the connections between division of polynomials and evaluating polynomials?

4.7 An Application of Exponents: Scientific Notation

OBJECTIVES

1. Express numbers in scientific notation.
2. Convert numbers in scientific notation to numbers without exponents.
3. Use scientific notation in calculations.

FOR EXTRA HELP

SSG Sec. 4.7
SSM Sec. 4.7
Pass the Test Software
InterAct Math Tutorial Software
Video 7

CONNECTIONS

According to the Aerospace Industries Association of America, in 1996 aerospace industry profits were 7326 million dollars. Another way to express this number is 7,326,000,000. Isaac Asimov wrote in *Asimov on Numbers* that the radius of the nucleus of a carbon atom has been calculated as .0000000000038 centimeter. These examples use numbers from real life that are very large or very small, thus requiring many zeros when written in decimal form. Because of difficulties working with many zeros, these numbers are often expressed as powers of 10, as discussed in this section.

FOR DISCUSSION OR WRITING

1. Based on your experience with multiplication of whole numbers, how can you easily multiply a number by 100? By 10,000?
2. How can you easily divide a number by 100? By 10,000?
3. Discuss the role of the digit 0 in the following numbers: 32,000 and .00032.

Answers for Connections box:
1. Move the decimal point 2 places to the right and use zeros as placeholders as necessary; use the same procedure, but move 4 places to the right.
2. Move the decimal point 2 places to the left and use zeros as placeholders as necessary; use the same procedure, but move 4 places to the left.
3. In 32,000, zero is used as a placeholder for the hundreds, tens, and units place values. In .00032, zero is used as a placeholder for the tenths, hundredths, and thousandths place values.

OBJECTIVE 1 Express numbers in scientific notation. In **scientific notation,** a number is written in the form $a \times 10^n$, where n is an integer and $1 \leq |a| < 10$. When a number is multiplied by a power of 10, such as 10^1, $10^2 = 100$, $10^3 = 1000$, $10^{-1} = .1$, $10^{-2} = .01$, $10^{-3} = .001$, and so on, the net effect is to move the decimal point to the right if it is a positive power and to the left if it is a negative power. This is shown in the examples below. (In work with scientific notation, the times symbol, $\times$, is commonly used.)

$23.19 \times 10^1 = 23.19 \times 10 = 231.9$	Decimal point moves 1 place to the right.
$23.19 \times 10^2 = 23.19 \times 100 = 2319.$	Decimal point moves 2 places to the right.
$23.19 \times 10^3 = 23.19 \times 1000 = 23190.$	Decimal point moves 3 places to the right.
$23.19 \times 10^{-1} = 23.19 \times .1 = 2.319$	Decimal point moves 1 place to the left.
$23.19 \times 10^{-2} = 23.19 \times .01 = .2319$	Decimal point moves 2 places to the left.
$23.19 \times 10^{-3} = 23.19 \times .001 = .02319$	Decimal point moves 3 places to the left.

A number in scientific notation is always written with the decimal point after the first nonzero digit and then multiplied by the appropriate power of 10. For example, 35 is written 3.5×10^1, or 3.5×10; 56,200 is written 5.62×10^4, since

$$56{,}200 = 5.62 \times 10{,}000 = 5.62 \times 10^4.$$

To write a number in scientific notation, follow these steps.

Writing a Number in Scientific Notation

Step 1 Move the decimal point to the right of the first nonzero digit.

Step 2 Count the number of places you moved the decimal point.

Step 3 The number of places in Step 2 is the absolute value of the exponent on 10.

Step 4 The exponent on 10 is positive if you made the number smaller in Step 1. The exponent is negative if you made the number larger in Step 1. If the decimal point is not moved, the exponent is 0.

INSTRUCTOR'S RESOURCES

PTB Sec. 4.7 ISM Sec. 4.7
AB Sec. 4.7
TEST GENERATOR
WORLD WIDE WEB
www.LialAlgebra.com

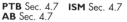

CHALKBOARD EXAMPLE

Write in scientific notation.

.0571

Answer: 5.71×10^{-2}

EXAMPLE 1 Using Scientific Notation

Write each number in scientific notation.

(a) 93,000,000

The number will be written in scientific notation as 9.3×10^n. To find the value of n, first compare 9.3 with the original number, 93,000,000. Is 9.3 larger or smaller? Here, 9.3 is *smaller* than 93,000,000. Therefore, we must multiply by a *positive* power of 10 so the product 9.3×10^n will equal the larger number.

Move the decimal point to follow the first nonzero digit (the 9). Count the number of places the decimal point was moved.

$$93{,}000{,}000 \quad \text{7 places}$$

Since the decimal point was moved 7 places, and since n is positive, $93{,}000{,}000 = 9.3 \times 10^7$.

(b) $63{,}200{,}000{,}000 = 6.3200000000 = 6.32 \times 10^{10}$

10 places

(c) .00462

Move the decimal point to the right of the first nonzero digit, and count the number of places the decimal point was moved.

$$.00462 \quad \text{3 places}$$

Since 4.62 is *larger* than .00462, the exponent must be *negative.*

$$.00462 = 4.62 \times 10^{-3}$$

(d) $.0000762 = 7.62 \times 10^{-5}$

NOTE Another way to choose the exponent when writing a number in scientific notation is this: If the original number is "large," like 93,000,000, use a *positive* exponent on 10, since positive is larger than negative. However, if the original number is "small," like .00462, use a *negative* exponent on 10, since negative is smaller than positive.

OBJECTIVE 2 Convert numbers in scientific notation to numbers without exponents. To convert a number written in scientific notation to a number without exponents, work in reverse. Multiplying by a positive power of 10 will make the number larger; multiplying by a negative power of 10 will make the number smaller.

CHALKBOARD EXAMPLE

Write without exponents.

8.7×10^5

Answer: 870,000

EXAMPLE 2 Writing Numbers without Exponents

Write each number without exponents.

(a) 6.2×10^3

Since the exponent is positive, make 6.2 larger by moving the decimal point 3 places to the right.

$$6.2 \times 10^3 = 6.200 = 6200.$$

(b) $4.283 \times 10^5 = 4.28300 = 428{,}300$ Move 5 places to the right.

(c) $7.04 \times 10^{-3} = .00704$ Move 3 places to the left.

As these examples show, the exponent tells the number of places and the direction that the decimal point is moved.

OBJECTIVE 3 Use scientific notation in calculations. The next example uses scientific notation with products and quotients.

CHALKBOARD EXAMPLE

Write the quotient without exponents.

$$\frac{4.8 \times 10^2}{2.4 \times 10^{-3}}$$

Answer: 200,000

EXAMPLE 3 Multiplying and Dividing with Scientific Notation

Write each product or quotient without exponents.

(a) $(6 \times 10^3)(5 \times 10^{-4})$

$(6 \times 10^3)(5 \times 10^{-4})$	$= (6 \times 5)(10^3 \times 10^{-4})$	Commutative and associative properties
	$= 30 \times 10^{-1}$	Product rule for exponents
	$= 3$	Write without exponents.

(b) $\dfrac{4 \times 10^{-5}}{2 \times 10^3} = \dfrac{4}{2} \times \dfrac{10^{-5}}{10^3} = 2 \times 10^{-8} = .00000002$

NOTE Multiplying or dividing numbers written in scientific notation may produce an answer in the form $a \times 10^0$. Since $10^0 = 1$, $a \times 10^0 = a$. For example,

$$(8 \times 10^{-4})(5 \times 10^4) = 40 \times 10^0 = 40.$$

The final example comes from Isaac Asimov's *Asimov on Numbers.*

CHALKBOARD EXAMPLE

Fill in the blank: "Divide (the radius of the Observable Universe) 1.23×10^{28} centimeters by the radius of the proton, which is 1.6×10^{-13} centimeters, and you have the answer: ______."

Answer: 7.7×10^{40}

EXAMPLE 4 Calculating Using Scientific Notation

Fill in the blank in this sentence: "We next suppose that the Greatest Possible Volume is packed perfectly tightly (leaving no empty spaces) with objects of the Smallest Possible Volume. If we divided 7.8×10^{84} by 1.7×10^{-40}, we find that the number of protons it takes to fill the Observable Universe is ______."

$\dfrac{7.8 \times 10^{84}}{1.7 \times 10^{-40}}$	$= \dfrac{7.8}{1.7} \times 10^{84-(-40)}$	Divide.
	$= \dfrac{7.8}{1.7} \times 10^{124}$	Subtract exponents.
	$\approx 4.6 \times 10^{124}$	Use a calculator.

4.7 EXERCISES

1. A 2. B 3. C 4. D
5. in scientific notation
6. in scientific notation
7. not in scientific notation; 5.6×10^6
8. not in scientific notation; 3.4×10^4
9. not in scientific notation; 8×10^1
10. not in scientific notation; 9×10^2
11. not in scientific notation; 4×10^{-3}
12. not in scientific notation; 7×10^{-4}

Match each number written in scientific notation in Column I with the correct choice from Column II.

I	II
1. 4.6×10^{-4}	**A.** .00046
2. 4.6×10^4	**B.** 46,000
3. 4.6×10^5	**C.** 460,000
4. 4.6×10^{-5}	**D.** .000046

Determine whether or not each number is written in scientific notation as defined in Objective 1. If it is not, write it as such.

5. 4.56×10^3 **6.** 7.34×10^5 **7.** 5,600,000 **8.** 34,000

9. $.8 \times 10^2$ **10.** $.9 \times 10^3$ **11.** .004 **12.** .0007

13. It is written as the product of a power of 10 and a number whose absolute value is between 1 and 10 (inclusive of 1). Some examples are 2.3×10^{-4} and 6.02×10^{23}. **14.** To multiply by a positive power of 10, move the decimal point to the right the same number of places as the exponent on 10, adding zeros as placeholders if necessary. To multiply by a negative power of 10, move the decimal point to the left. **15.** 5.876×10^9 **16.** 9.994×10^9 **17.** 8.235×10^4 **18.** 7.833×10^4 **19.** 7×10^{-6} **20.** 4×10^{-7} **21.** 2.03×10^{-3} **22.** 5.78×10^{-5} **23.** 750,000 **24.** 8,800,000 **25.** 5,677,000,000,000 **26.** 8,766,000,000 **27.** 6.21 **28.** 8.56 **29.** .00078 **30.** .000089 **31.** .000000005134 **32.** .0000000007123 **33.** 600,000,000,000 **34.** 120,000,000,000 **35.** 15,000,000 **36.** 1,600,000,000 **37.** 60,000 **38.** 80,000 **39.** .0003 **40.** .3 **41.** 40 **42.** 50 **43.** .000013 **44.** .00019

13. Explain in your own words what it means for a number to be written in scientific notation. Give examples.

14. Explain how to multiply a number by a positive power of ten. Then explain how to multiply a number by a negative power of ten.

Write each number in scientific notation. See Example 1.

15. 5,876,000,000 **16.** 9,994,000,000

17. 82,350 **18.** 78,330

19. .000007 **20.** .0000004

21. .00203 **22.** .0000578

Write each number without exponents. See Example 2.

23. 7.5×10^5 **24.** 8.8×10^6

25. 5.677×10^{12} **26.** 8.766×10^9

27. 6.21×10^0 **28.** 8.56×10^0

29. 7.8×10^{-4} **30.** 8.9×10^{-5}

31. 5.134×10^{-9} **32.** 7.123×10^{-10}

Use properties and rules for exponents to perform the indicated operations, and write each answer without exponents. See Example 3.

33. $(2 \times 10^8) \times (3 \times 10^3)$ **34.** $(4 \times 10^7) \times (3 \times 10^3)$

35. $(5 \times 10^4) \times (3 \times 10^2)$ **36.** $(8 \times 10^5) \times (2 \times 10^3)$

37. $(3 \times 10^{-4}) \times (2 \times 10^8)$ **38.** $(4 \times 10^{-3}) \times (2 \times 10^7)$

39. $\dfrac{9 \times 10^{-5}}{3 \times 10^{-1}}$ **40.** $\dfrac{12 \times 10^{-4}}{4 \times 10^{-3}}$

41. $\dfrac{8 \times 10^3}{2 \times 10^2}$ **42.** $\dfrac{5 \times 10^4}{1 \times 10^3}$

43. $\dfrac{2.6 \times 10^{-3}}{2 \times 10^2}$ **44.** $\dfrac{9.5 \times 10^{-1}}{5 \times 10^3}$

TECHNOLOGY INSIGHTS (EXERCISES 45–50)

Graphing calculators such as the TI-83 can display numbers in scientific notation (when in scientific mode), using the format shown in the screen on the left. For example, the calculator displays 5.4E3 to represent 5.4×10^3, the scientific notation form for 5400. 5.4E−4 means 5.4×10^{-4}. It will also perform operations with numbers entered in scientific notation, as shown in the screen on the right. Notice how the rules for exponents are applied.

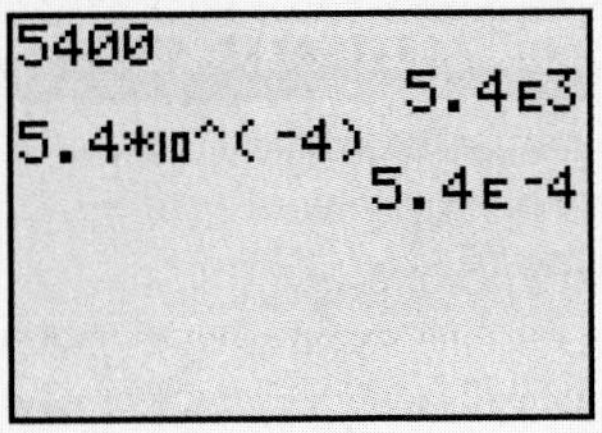

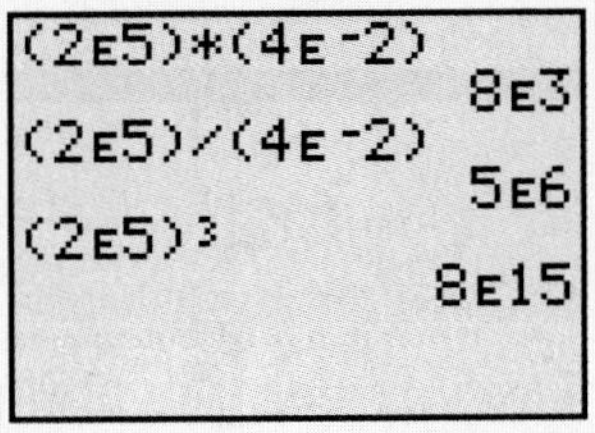

45. 4.7E−7 46. 2.1E−5
47. 2E7 48. 3E−7
49. 1E1 50. 1E3

TECHNOLOGY INSIGHTS (EXERCISES 45–50) (CONTINUED)

Predict the display the calculator would give for the expression shown in each screen.

45.
```
.00000047
```

46.
```
.000021
```

47.
```
(8E5)/(4E-2)
```

48.
```
(9E-4)/(3E3)
```

49.
```
(2E6)*(2E-3)/(4E
2)
```

50.
```
(5E-3)*(1E9)/(5E
3)
```

Each of the following statements comes from Astronomy! A Brief Edition *by James B. Kaler (Addison-Wesley, 1997). If the number in the statement is in scientific notation, write it in standard form without using exponents. If the number is in standard form, write it in scientific notation.*

51. 1×10^{10} 52. 4.7×10^6
53. 2,000,000,000
54. .0000000000001
55. $\$7.326 \times 10^9$
56. $\$4.6537 \times 10^{10}$

51. Multiplying this view over the whole sky yields a galaxy count of more than 10 *billion.* (page 496)

52. The circumference of the solar orbit is . . . about 4.7 million km . . . . (in reference to the orbit of Jupiter, page 395)

53. The solar luminosity requires that 2×10^9 kg of mass be converted into energy every second. (page 327)

54. At maximum, a cosmic ray particle—a mere atomic nucleus of only 10^{-13} cm across—can carry the energy of a professionally pitched baseball. (page 445)

Each of the following statements is based on data from the aerospace industry. Write each number in scientific notation.

55. In 1996, aerospace industry profits were \$7,326,000,000. (*Source:* Aerospace Industries Association of America.)

56. In 1995, the total receipts for international transportation transactions of the United States were \$46,537,000,000. (*Source:* U.S. Bureau of Economic Analysis.)

57. $\$3.0262 \times 10^{10}$
58. $\$1.664 \times 10^{10}$
59. about 15,300 seconds
60. about 4.24 hours
61. (a) 1995: 1.063×10^{11}; 1996: 1.124×10^{11}; 1997: 1.250×10^{11}
(b) 9.35×10^{9}
62. about 9.2×10^{8} acres

57. In 1994, the total value of U.S. aircraft shipments was \$30,262,000,000. (*Source:* U.S. Department of Commerce, International Trade Administration.)

58. In 1992, the total value of net orders for U.S. civil jet transport aircraft was \$16,640,000,000. (*Source:* Aerospace Industries Association of America.)

Solve each problem.

▲ **59.** The distance to Earth from the planet Pluto is 4.58×10^9 kilometers. In April 1983, Pioneer 10 transmitted radio signals from Pluto to Earth at the speed of light, 3.00×10^5 kilometers per second. How long (in seconds) did it take for the signals to reach Earth?

▲ **60.** In Exercise 59, how many hours did it take for the signals to reach Earth?

61. The accompanying graph depicts aerospace industry sales in current dollars. The figures at the tops of the bars represent billions of dollars.

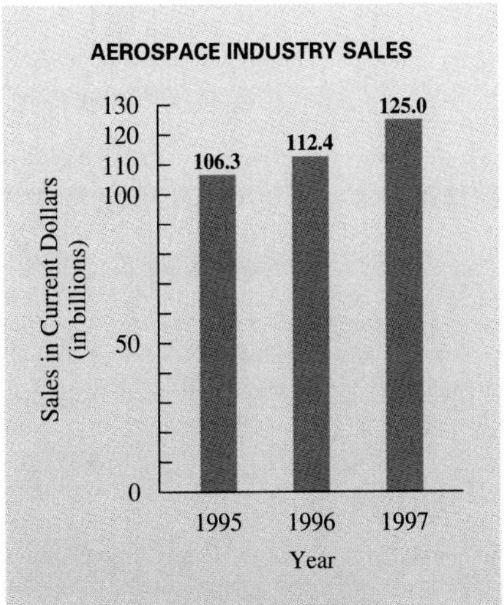

Source: U.S. Bureau of the Census, *Current Industrial Reports.*

(a) For each of the years, write the figure represented in scientific notation. The figure for 1996 is preliminary, and the figure for 1997 is an estimate.

(b) If a line segment is drawn between the tops of the bars for 1995 and 1997, what is its slope?

62. In a recent year, the state of Texas had about 1.3×10^6 farms with an average of 7.1×10^2 acres per farm. What was the total number of acres devoted to farmland in Texas that year? (*Source:* National Agricultural Statistics Service, U.S. Dept. of Agriculture.)

Instructor's Note
Grouping: 2 students
Time: 30 minutes

This activity allows students to explore how quadratic equations and graphs of parabolas can be used. This activity can be revisited when students learn to solve quadratic equations. Students may need help determining the scale for graphs.

A. See answers within the table.

B.

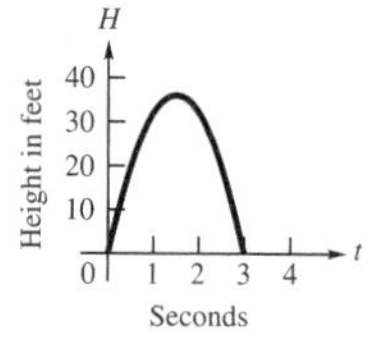

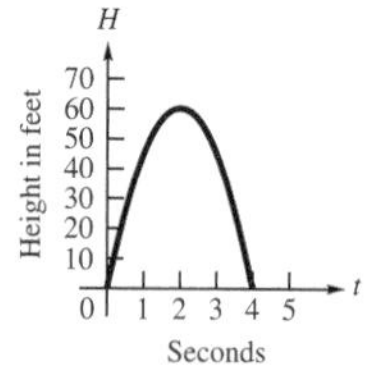

C. 1. first rocket: $t = 0$ and $t = 3$; second rocket: $t = 0$ and $t = 4$; $H = 0$ at takeoff and when the rocket hits the ground 2. first rocket: $H = 36$ feet; second rocket: $H = 64$ feet 3. first rocket: $t = 1.5$; second rocket: $t = 2$

CHAPTER 4 GROUP ACTIVITY

Measuring the Flight of a Rocket

Objective: Graph quadratic equations to solve an application problem.

A physics class observes the launch of two rockets that are slightly different. One has an initial velocity of 48 feet per second. The second one has an initial velocity of 64 feet per second. Follow the steps below to determine the maximum height each rocket will achieve and how long it will take it to reach that height.

This formula is used to find the height of a projectile when shot vertically into the air.

$$H = -16t^2 + Vt$$

H = height above the launch pad in feet, t = time in seconds, and V = initial velocity.

A. One student should complete the table for the first rocket. The other student should do the same for the second rocket. Use the formula for the height of a projectile and the values provided.

Initial Velocity = 48 Feet per Second

Time (seconds)	*Height (feet)*
0	[0]
.5	[20]
1	[32]
1.5	[36]
2	[32]
2.5	[20]
3	[0]

Initial Velocity = 64 Feet per Second

Time (seconds)	*Height (feet)*
0	[0]
.5	[28]
1	[48]
1.5	[60]
2	[64]
2.5	[60]
3	[48]
3.5	[28]
4	[0]

B. Graph the results. You must determine the scales for the two axes on the coordinate system. Clearly label each graph and its scale.

C. Answer the following questions for each rocket.

1. Find the time in seconds when the rocket is at height 0 feet. Why does this happen twice?
2. What is the maximum height the rocket reached?
3. At what time (in seconds) did the rocket reach its maximum height?

D. 1. second rocket 2. first rocket 3. Answers will vary. 4. no; only the second rocket at $t = 1$ and $t = 3$; Answers will vary. 5. Answers will vary.

D. Compare the data from both rockets and answer the following questions.

1. Which rocket had the greater maximum height? Explain how you decided.
2. Which rocket reached its maximum height first? Explain how you decided.
3. Consider the answers for Exercises D1 and D2 and discuss why this occurred.
4. Do both rockets reach a height of 48 feet? If yes, at what times? What observations about projectiles might this lead to?
5. What are some other possible uses for the formula of the height of a projectile?

CHAPTER 4 SUMMARY

KEY TERMS

4.1 polynomial	trinomial	vertex	inner product
descending powers	binomial	axis	FOIL
degree of a term	monomial	line of symmetry	**4.7** scientific notation
degree of a polynomial	parabola	**4.3** outer product	

NEW SYMBOLS

x^{-n} *x* to the negative *n* power

TEST YOUR WORD POWER

See how well you have learned the vocabulary in this chapter. Answers, with examples, are given at the bottom of the page.

1. A **polynomial** is an algebraic expression made up of
(a) a term or a finite product of terms with positive coefficients and exponents
(b) a term or a finite sum of terms with real coefficients and whole number exponents
(c) the product of two or more terms with positive exponents
(d) the sum of two or more terms with whole number coefficients and exponents.

2. The **degree of a term** is
(a) the number of variables in the term
(b) the product of the exponents on the variables
(c) the smallest exponent on the variables
(d) the sum of the exponents on the variables.

3. A **trinomial** is a polynomial with
(a) only one term
(b) exactly two terms
(c) exactly three terms
(d) more than three terms.

4. A **binomial** is a polynomial with
(a) only one term
(b) exactly two terms
(c) exactly three terms
(d) more than three terms.

5. A **monomial** is a polynomial with
(a) only one term
(b) exactly two terms
(c) exactly three terms
(d) more than three terms.

6. **FOIL** is a method for
(a) adding two binomials
(b) adding two trinomials
(c) multiplying two binomials
(d) multiplying two trinomials.

Answers to Test Your Word Power

1. (b) *Example:* $5x^3 + 2x^2 - 7$ **2.** (d) *Examples:* The term 6 has degree 0, $3x$ has degree 1, $-2x^8$ has degree 8, and $5x^2y^4$ has degree 6.
3. (c) *Example:* $2a^2 - 3ab + b^2$ **4.** (b) *Example:* $3t^3 + 5t$ **5.** (a) *Examples:* -5 and $4xy^5$
6. (c) *Example:* $(m + 4)(m - 3) = m(m) - 3m + 4m + 4(-3) = m^2 + m - 12$ (F, O, I, L)

QUICK REVIEW

CONCEPTS	EXAMPLES

4.1 ADDITION AND SUBTRACTION OF POLYNOMIALS; GRAPHING SIMPLE POLYNOMIALS

Addition
Add like terms.

Add.
$$\begin{array}{r} 2x^2 + 5x - 3 \\ \underline{5x^2 - 2x + 7} \\ 7x^2 + 3x + 4 \end{array}$$

Subtraction
Change the signs of the terms in the second polynomial and add to the first polynomial.

Subtract.
$$\begin{aligned} &(2x^2 + 5x - 3) - (5x^2 - 2x + 7) \\ &= (2x^2 + 5x - 3) + (-5x^2 + 2x - 7) \\ &= -3x^2 + 7x - 10 \end{aligned}$$

Graphing Simple Polynomials
To graph a simple polynomial equation such as $y = x^2 - 2$, plot points near the vertex. (In this chapter, all parabolas have a vertex on the x-axis or the y-axis.)

Graph $y = x^2 - 2$.

x	y
-2	2
-1	-1
0	-2
1	-1
2	2

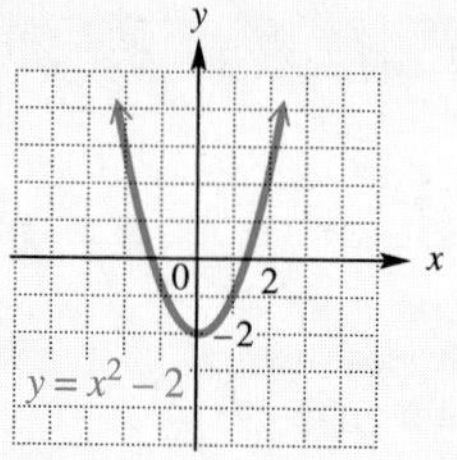

4.2 THE PRODUCT RULE AND POWER RULES FOR EXPONENTS

For any integers m and n with no denominators zero:

Perform the operations by using rules for exponents.

Product Rule

$a^m \cdot a^n = a^{m+n}$ — $2^4 \cdot 2^5 = 2^9$

Power Rules

(a) $(a^m)^n = a^{mn}$ — $(3^4)^2 = 3^8$

(b) $(ab)^m = a^m b^m$ — $(6a)^5 = 6^5 a^5$

(c) $\left(\dfrac{a}{b}\right)^m = \dfrac{a^m}{b^m}$ $(b \neq 0)$ — $\left(\dfrac{2}{3}\right)^4 = \dfrac{2^4}{3^4}$

4.3 MULTIPLICATION OF POLYNOMIALS

Multiplying Polynomials
Multiply each term of the first polynomial by each term of the second polynomial. Then add like terms.

Multiply.
$$\begin{array}{rrrrr} & 3x^3 & -\ 4x^2 & +\ 2x & -\ 7 \\ & & & 4x & +\ 3 \\ \hline & 9x^3 & -\ 12x^2 & +\ 6x & -\ 21 \\ 12x^4 & -\ 16x^3 & +\ 8x^2 & -\ 28x & \\ \hline 12x^4 & -\ 7x^3 & -\ 4x^2 & -\ 22x & -\ 21 \end{array}$$

FOIL Method for Multiplying Binomials

Multiply. $(2x + 3)(5x - 4)$

Step 1 Multiply the two first terms to get the first term of the answer.

$2x(5x) = 10x^2$

Step 2 Find the outer product and the inner product and mentally add them, when possible, to get the middle term of the answer.

$2x(-4) + 3(5x) = 7x$

Step 3 Multiply the two last terms to get the last term of the answer.

$3(-4) = -12$

The product of $(2x + 3)$ and $(5x - 4)$ is $10x^2 + 7x - 12$.

CONCEPTS	EXAMPLES

4.4 SPECIAL PRODUCTS

Square of a Binomial

$(a + b)^2 = a^2 + 2ab + b^2$

$(a - b)^2 = a^2 - 2ab + b^2$

Multiply.

$(3x + 1)^2 = 9x^2 + 6x + 1$

$(2m - 5n)^2 = 4m^2 - 20mn + 25n^2$

Product of the Sum and Difference of Two Terms

$(a + b)(a - b) = a^2 - b^2$

$(4a + 3)(4a - 3) = 16a^2 - 9$

4.5 INTEGER EXPONENTS AND THE QUOTIENT RULE

If $a \neq 0$, for integers m and n:

Simplify by using the rules for exponents.

Zero Exponent

$$a^0 = 1$$

$$15^0 = 1$$

Negative Exponent

$$a^{-n} = \frac{1}{a^n}$$

$$5^{-2} = \frac{1}{5^2} = \frac{1}{25}$$

Quotient Rule

$$\frac{a^m}{a^n} = a^{m-n}$$

$$\frac{4^8}{4^3} = 4^5$$

$$\frac{a^{-m}}{b^{-n}} = \frac{b^n}{a^m} \quad (b \neq 0)$$

$$\frac{4^{-2}}{3^{-5}} = \frac{3^5}{4^2}$$

$$\left(\frac{a}{b}\right)^{-m} = \left(\frac{b}{a}\right)^m \quad (b \neq 0)$$

$$\left(\frac{6}{5}\right)^{-3} = \left(\frac{5}{6}\right)^3$$

4.6 DIVISION OF POLYNOMIALS

Dividing a Polynomial by a Monomial
Divide each term of the polynomial by the monomial.

$$\frac{a + b}{c} = \frac{a}{c} + \frac{b}{c}$$

Divide.

$$\frac{4x^3 - 2x^2 + 6x - 8}{2x} = 2x^2 - x + 3 - \frac{4}{x}$$

Dividing a Polynomial by a Polynomial
Use "long division."

Divide.

$$\begin{array}{r} 2x - 5 \\ 3x + 4\overline{)6x^2 - 7x - 21} \\ \underline{6x^2 + 8x} \\ -15x - 21 \\ \underline{-15x - 20} \\ -1 \end{array} \quad \leftarrow \text{Remainder}$$

The final quotient is $2x - 5 + \frac{-1}{3x + 4}$.

4.7 AN APPLICATION OF EXPONENTS: SCIENTIFIC NOTATION

To write a number in scientific notation (as $a \times 10^n$), move the decimal point to follow the first nonzero digit. The number of places the decimal point is moved is the absolute value of n. If moving the decimal point makes the number smaller, n is positive. If it makes the number larger, n is negative. If the decimal point is not moved, n is 0.

Write in scientific notation.

$$247 = 2.47 \times 10^2$$

$$.0051 = 5.1 \times 10^{-3}$$

$$4.8 = 4.8 \times 10^0$$

Write without exponents.

$$3.25 \times 10^5 = 325{,}000$$

$$8.44 \times 10^{-6} = .00000844$$

CHAPTER 4 REVIEW EXERCISES

1. $22m^2$; degree 2; monomial **2.** $p^3 - p^2 + 4p + 2$; degree 3; none of these **3.** already in descending powers; degree 5; none of these **4.** $-8y^5 - 7y^4 + 9y$; degree 5; trinomial
5. $7r^4 - 4r^3 + 1$; degree 4; trinomial
6. $13x^3y^2 - 5xy^5 + 21x^2$
7. $a^3 + 4a^2$
8. $y^2 - 10y + 9$
9. $-13k^4 - 15k^2 + 18k$
10.

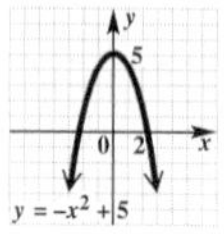

x	-2	-1	0	1	2
y	1	4	5	4	1

11.

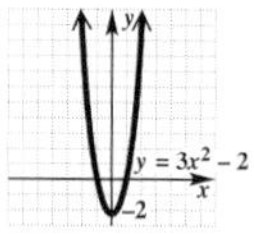

x	-2	-1	0	1	2
y	10	1	-2	1	10

12. The expression is a *sum* of powers of 7, not a *product.*
13. 4^{11} **14.** $(-5)^{11}$
15. $-72x^7$ **16.** $10x^{14}$
17. 19^5x^5 **18.** $(-4)^7y^7$
19. $5p^4t^4$ **20.** $\frac{7^6}{5^6}$
21. $6^2x^{16}y^4z^{16}$
22. $\frac{2^3m^9n^3}{p^6}$
23. $a^3 - 2a^2 - 7a + 2$
24. $6r^3 + 8r^2 - 17r + 6$
25. $5p^5 - 2p^4 - 3p^3 + 25p^2 + 15p$
26. $m^2 - 7m - 18$
27. $6k^2 - 9k - 6$
28. $2a^2 + 5ab - 3b^2$
29. $12k^2 - 32kq - 35q^2$
30. $s^3 - 3s^2 + 3s - 1$
31. $2x^2 + x - 6$
32. $25x^8 + 20x^6 + 4x^4$

[4.1] *Combine terms where possible in each polynomial. Write the answer in descending powers of the variable. Give the degree of the answer. Identify the polynomial as a monomial, binomial, trinomial, or none of these.*

1. $9m^2 + 11m^2 + 2m^2$

2. $-4p + p^3 - p^2 + 8p + 2$

3. $12a^5 - 9a^4 + 8a^3 + 2a^2 - a + 3$

4. $-7y^5 - 8y^4 - y^5 + y^4 + 9y$

5. $(12r^4 - 7r^3 + 2r^2) - (5r^4 - 3r^3 + 2r^2 - 1)$

6. Simplify $(5x^3y^2 - 3xy^5 + 12x^2) - (-9x^2 - 8x^3y^2 + 2xy^5)$.

Add or subtract as indicated.

7. Add.
$$\begin{array}{r} -2a^3 + 5a^2 \\ 3a^3 - a^2 \\ \hline \end{array}$$

8. Subtract.
$$\begin{array}{r} 6y^2 - 8y + 2 \\ 5y^2 + 2y - 7 \\ \hline \end{array}$$

9. Subtract.
$$\begin{array}{r} -12k^4 - 8k^2 + 7k \\ k^4 + 7k^2 - 11k \\ \hline \end{array}$$

Graph each equation by completing the table of values.

10. $y = -x^2 + 5$

x	-2	-1	0	1	2
y					

11. $y = 3x^2 - 2$

x	-2	-1	0	1	2
y					

[4.2] **12.** Why does the product rule for exponents not apply to the expression $7^2 + 7^4$?

Use the product rule, power rules, or both to simplify each expression. Write the answers in exponential form.

13. $4^3 \cdot 4^8$

14. $(-5)^6(-5)^5$

15. $(-8x^4)(9x^3)$

16. $(2x^2)(5x^3)(x^9)$

17. $(19x)^5$

18. $(-4y)^7$

19. $5(pt)^4$

20. $\left(\frac{7}{5}\right)^6$

21. $(6x^2z^4)^2(x^3yz^2)^4$

22. $\left(\frac{2m^3n}{p^2}\right)^3$

[4.3] *Find each product.*

23. $(a + 2)(a^2 - 4a + 1)$

24. $(3r - 2)(2r^2 + 4r - 3)$

25. $(5p^2 + 3p)(p^3 - p^2 + 5)$

26. $(m - 9)(m + 2)$

27. $(3k - 6)(2k + 1)$

28. $(a + 3b)(2a - b)$

29. $(6k + 5q)(2k - 7q)$

30. $(s - 1)^3$

31. Find a polynomial that represents the area of the rectangle shown.

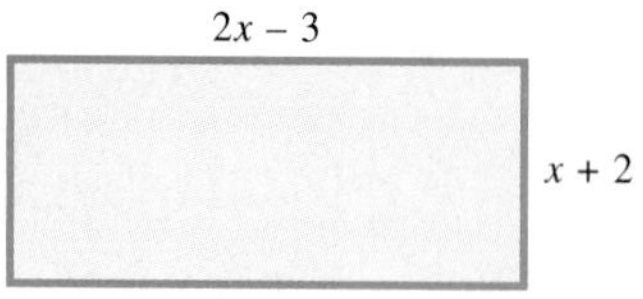

32. If the side of a square has a measure represented by $5x^4 + 2x^2$, what polynomial represents its area?

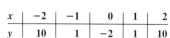

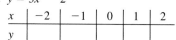

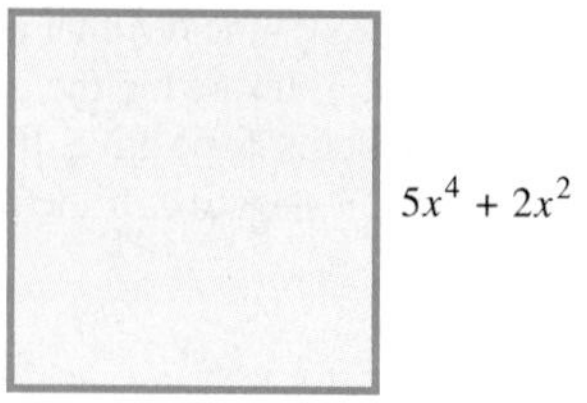

JOURNAL WRITING ▲ CHALLENGING SCIENTIFIC CALCULATOR GRAPHING CALCULATOR

33. $a^2 + 8a + 16$
34. $4r^2 + 20rt + 25t^2$
35. $36m^2 - 25$
36. $25a^2 - 36b^2$
37. $r^3 + 6r^2 + 12r + 8$
38. $25t^3 - 30t^2 + 9t$
39. (a) Answers will vary. For example, let $x = 1$ and $y = 2$. $(1 + 2)^2 \neq 1^2 + 2^2$, because $9 \neq 5$. (b) Answers will vary. For example, let $x = 1$ and $y = 2$. $(1 + 2)^3 \neq 1^3 + 2^3$, because $27 \neq 9$.
40. To find the third power of a binomial, such as $(a + b)^3$, first square the binomial and then multiply that result by the binomial:
$$(a + b)^3 = (a + b)^2(a + b) = (a^2 + 2ab + b^2)(a + b) = a^3 + 3a^2b + 3ab^2 + b^3.$$
41. In both cases, $x = 0$ and $y = 1$ lead to 1 on each side of the inequality. This would not be sufficient to show that *in general* the inequality is true. It would be necessary to choose other values of x and y.
42. $x^6 + 6x^4 + 12x^2 + 8$ cubic centimeters
43. $\frac{4}{3}\pi(x + 1)^3$ or $\frac{4}{3}\pi x^3 + 4\pi x^2 + 4\pi x + \frac{4}{3}\pi$ cubic inches
44. 2 **45.** 0 **46.** -1
47. $-\frac{1}{49}$ **48.** $\frac{64}{25}$ **49.** 5^8
50. $\frac{1}{81}$ **51.** $\frac{3}{4}$ **52.** $\frac{1}{36}$
53. x^2 **54.** y^7 **55.** $\frac{r^2}{6}$
56. $\frac{3^5}{p^3}$ **57.** $\frac{1}{a^3b^5}$ **58.** $72r^5$
59. $\frac{-5y^2}{3}$ **60.** $y^3 - 2y + 3$
61. $-2m^2n + mn + \frac{6n^3}{5}$
62. $2mn + 3m^4n^2 - 4n$

[4.4] *Find each product.*

33. $(a + 4)^2$ **34.** $(2r + 5t)^2$ **35.** $(6m - 5)(6m + 5)$

36. $(5a + 6b)(5a - 6b)$ **37.** $(r + 2)^3$ **38.** $t(5t - 3)^2$

39. Choose values for x and y to show that in general, the following hold true.

(a) $(x + y)^2 \neq x^2 + y^2$ (b) $(x + y)^3 \neq x^3 + y^3$

40. Write an explanation on how to raise a binomial to the third power. Give an example.

41. Refer to Exercise 39. Suppose that you happened to let $x = 0$ and $y = 1$. Would your results be sufficient to illustrate the truth, in general, of the inequalities shown? If not, what would you need to do as your next step in working the exercise?

42. What is the volume of a cube with one side having length $x^2 + 2$ centimeters?

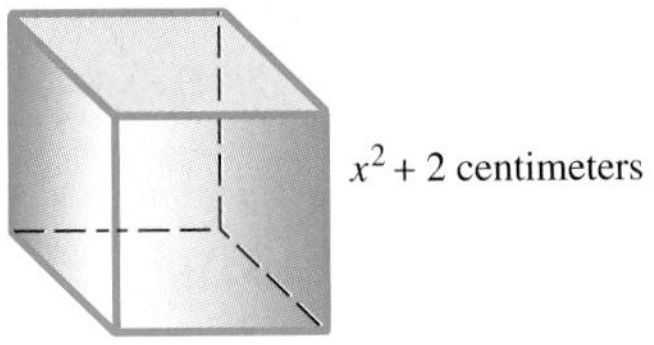

A cube

43. What is the volume of a sphere with radius $x + 1$ inches?

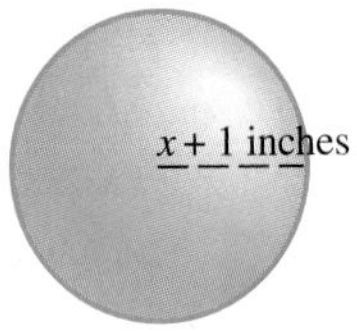

A sphere

[4.5] *Evaluate each expression.*

44. $6^0 + (-6)^0$ **45.** $(-23)^0 - (-23)^0$ **46.** -10^0

Write each expression using only positive exponents.

47. -7^{-2} **48.** $\left(\frac{5}{8}\right)^{-2}$ **49.** $(5^{-2})^{-4}$

50. $9^3 \cdot 9^{-5}$ **51.** $2^{-1} + 4^{-1}$ **52.** $\frac{6^{-5}}{6^{-3}}$

Simplify. Write each answer with only positive exponents. Assume that all variables represent nonzero real numbers.

53. $\frac{x^{-7}}{x^{-9}}$ **54.** $\frac{y^4 \cdot y^{-2}}{y^{-5}}$ **55.** $(6r^{-2})^{-1}$

56. $(3p)^4(3p^{-7})$ **57.** $\frac{ab^{-3}}{a^4b^2}$ **58.** $\frac{(6r^{-1})^2(2r^{-4})}{r^{-5}(r^2)^{-3}}$

[4.6] *Perform each division.*

59. $\frac{-15y^4}{9y^2}$ **60.** $\frac{6y^4 - 12y^2 + 18y}{6y}$

61. $(-10m^4n^2 + 5m^3n^2 + 6m^2n^4) \div (5m^2n)$

62. What polynomial, when multiplied by $6m^2n$, gives the product $12m^3n^2 + 18m^6n^3 - 24m^2n^2$?

63. This is not correct. The friend wrote the second term of the quotient as $-12x$ rather than $-2x$. Here is the correct method: $\frac{6x^2 - 12x}{6} = \frac{6x^2}{6} - \frac{12x}{6} = x^2 - 2x$.
64. $2r + 7$
65. $2a^2 + 3a - 1 + \frac{6}{5a - 3}$
66. $x^2 + 3x - 4$
67. $m^2 + 4m - 2$
68. 4.8×10^7
69. 2.8988×10^{10}
70. 8.24×10^{-8}
71. 24,000 **72.** 78,300,000
73. .000000897 **74.** 800
75. 4,000,000 **76.** .025
77. .000002 **78.** .0000000000016
79. 4.2×10^{42}
80. 7.47×10^2; 2×10^{11}

63. One of your friends in class simplified $\frac{6x^2 - 12x}{6}$ as $x^2 - 12x$. Is this correct? If not, what error did your friend make and how would you explain the correct method of performing the division?

Perform each division.

64. $\frac{2r^2 + 3r - 14}{r - 2}$

65. $\frac{10a^3 + 9a^2 - 14a + 9}{5a - 3}$

66. $\frac{x^4 + 3x^3 - 5x^2 - 3x + 4}{x^2 - 1}$

67. $\frac{m^4 + 4m^3 - 5m^2 - 12m + 6}{m^2 - 3}$

[4.7] *Write each number in scientific notation.*

68. 48,000,000 **69.** 28,988,000,000 **70.** .0000000824

Write each number without exponents.

71. 2.4×10^4 **72.** 7.83×10^7 **73.** 8.97×10^{-7}

Perform each indicated operation and write the answer without exponents.

74. $(2 \times 10^{-3}) \times (4 \times 10^5)$ **75.** $\frac{8 \times 10^4}{2 \times 10^{-2}}$ **76.** $\frac{12 \times 10^{-5} \times 5 \times 10^4}{4 \times 10^3 \times 6 \times 10^{-2}}$

The quote is taken from the source cited. Write each number in scientific notation in the quote without exponents.

77. The muon, a close relative of the electron produced by the bombardment of cosmic rays against the upper atmosphere, has a half-life of 2 millionths of a second (2×10^{-6}s). (Excerpt from *Conceptual Physics,* 6th edition, by Paul G. Hewitt. Copyright © by Paul G. Hewitt. Published by HarperCollins College Publishers.)

78. There are 13 red balls and 39 black balls in a box. Mix them up and draw 13 out one at a time without returning any ball . . . the probability that the 13 drawings each will produce a red ball is . . . 1.6×10^{-12}. (Warren Weaver, *Lady Luck,* New York: Doubleday & Company, Inc., 1963, pp. 298–299.)

The quote is taken from the source cited. Write each number in scientific notation.

79. An electron and a positron attract each other in two ways: the electromagnetic attraction of their opposite electric charges, and the gravitational attraction of their two masses. The electromagnetic attraction is

4,200,000,000,000,000,000,000,000,000,000,000,000,000,000,000

times as strong as the gravitational. (Isaac Asimov, *Isaac Asimov's Book of Facts,* New York: Bell Publishing Company, 1981, p. 106.)

80. A Boeing 747 would not do well in a dogfight; it is too big to maneuver with the lively agility required of an Air Force fighter. It performs best when it flies straight and level, and this is true of the worldwide airline industry as a whole. That industry features annual revenues that approach $200 billion. (T. A. Heppenheimer, *Turbulent Skies: The History of Commercial Aviation,* New York: John Wiley & Sons, 1995, p. 345.)

81. (a) 2.82×10^9
(b) 5.784×10^{11}
(c) 1.77×10^{10}
82. (a) 1×10^3 (b) 2×10^3
(c) 5×10^4 (d) 1×10^5

81. According to *The Wall Street Journal Almanac 1998,* the U.S. airline industry earned record profits in 1996 of \$2.82 billion and set records for both the number of passengers and the amount of cargo carried. Passenger traffic increased 7% to 578.4 billion revenue passenger miles, while cargo traffic rose 4.6% to 17.7 billion revenue ton miles.

Write each of the following numbers from the above paragraph using scientific notation.

(a) 2.82 billion **(b)** 578.4 billion **(c)** 17.7 billion

82. According to Campbell, Mitchell, and Reece in *Biology Concepts and Connections* (Benjamin Cummings, 1994, p. 230), "(t)he amount of DNA in a human cell is about 1000 times greater than the DNA in *E. coli.* Does this mean humans have 1000 times as many genes as the 2000 in *E. coli*? The answer is probably no; the human genome is thought to carry between 50,000 and 100,000 genes, which code for various proteins (as well as for tRNA and rRNA)."

Write each of the following numbers from the above quote using scientific notation.

(a) 1000 **(b)** 2000 **(c)** 50,000 **(d)** 100,000

83. (a) 2 (b) 92
(c) $x^3 + x^2 - 10x - 6$
(d) Both results are 94.
84. (a) 1 (b) 33
(c) $-x^3 - x^2 + 12x$
(d) Both results are -32.
85. (a) -6 (b) 12
(c) $x^4 - 2x^3 - 14x^2 + 30x + 9$
(d) Both results are -72.
86. (a) 3 (b) 183
(c) $x^2 + 4x + 1$ (d) Both results are 61.
87. Answers will vary.
88. We could not choose 3 because it would make the denominator equal to zero. It is the only invalid replacement for x.

RELATING CONCEPTS (EXERCISES 83–88)

In Exercises 83–88 we use the letters P and Q to denote the following polynomials:

$$P: \quad x - 3$$
$$Q: \quad x^3 + x^2 - 11x - 3.$$

Work these exercises in order.

83. (a) Evaluate P for $x = 5$.
(b) Evaluate Q for $x = 5$.
(c) Find the polynomial represented by $P + Q$.
(d) Evaluate the polynomial found in part (c) for $x = 5$ and verify that it is equal to the sum of the values you found in parts (a) and (b).

84. (a) Evaluate P for $x = 4$.
(b) Evaluate Q for $x = 4$.
(c) Find the polynomial represented by $P - Q$.
(d) Evaluate the polynomial found in part (c) for $x = 4$ and verify that it is equal to the difference between the values you found in parts (a) and (b).

85. (a) Evaluate P for $x = -3$.
(b) Evaluate Q for $x = -3$.
(c) Find the polynomial represented by $P \cdot Q$.
(d) Evaluate the polynomial found in part (c) for $x = -3$ and verify that it is equal to the product of the values you found in parts (a) and (b).

86. (a) Evaluate P for $x = 6$.
(b) Evaluate Q for $x = 6$.
(c) Find the polynomial represented by $\frac{Q}{P}$.
(d) Evaluate the polynomial found in part (c) for $x = 6$ and verify that it is equal to the quotient of the values you found in parts (a) and (b).

87. The concept illustrated in these exercises is this: If we evaluate P for a particular value of x and evaluate Q for the same value of x, and then perform an operation on P and Q, the resulting polynomial will give the appropriate value when evaluated for x. The values used in Exercises 83–86 were chosen arbitrarily. Repeat each exercise for a different value of x.

88. Refer to Exercise 86. Why could we *not* choose 3 as a replacement for x? Is there any other replacement for x that would not be valid?

Did you make the connection between operations with polynomials and operations with those same polynomials evaluated for particular values of the variables?

89. 2
90. $\frac{6^3r^6p^3}{5^3}$ 91. $144a^2 - 1$
92. $\frac{1}{16}$ 93. $\frac{1}{8^{12}}$
94. $p - 3 + \frac{5}{2p}$ 95. $\frac{2}{3m^3}$
96. $6k^3 - 21k - 6$ 97. r^{13}
98. $4r^2 + 20rs + 25s^2$
99. $-y^2 - 4y + 4$
100. $10r^2 + 21r - 10$
101. $y^2 + 5y + 1$
102. $\frac{5}{2} - \frac{4}{5xy} + \frac{3x}{2y^2}$
103. $10p^2 - 3p - 5$
104. $\frac{1}{5^{11}}$ 105. $49 - 28k + 4k^2$
106. $\frac{1}{x^4y^{12}}$

MIXED REVIEW EXERCISES

Perform the indicated operations. Write with positive exponents only. Assume all variables represent nonzero real numbers.

89. $5^0 + 7^0$

90. $\left(\frac{6r^2p}{5}\right)^3$

91. $(12a + 1)(12a - 1)$

92. 2^{-4}

93. $(8^{-3})^4$

94. $\frac{2p^3 - 6p^2 + 5p}{2p^2}$

95. $\frac{(2m^{-5})(3m^2)^{-1}}{m^{-2}(m^{-1})^2}$

96. $(3k - 6)(2k^2 + 4k + 1)$

97. $\frac{r^9 \cdot r^{-5}}{r^{-2} \cdot r^{-7}}$

98. $(2r + 5s)^2$

99. $(-5y^2 + 3y - 11) + (4y^2 - 7y + 15)$

100. $(2r + 5)(5r - 2)$

101. $\frac{2y^3 + 17y^2 + 37y + 7}{2y + 7}$

102. $(25x^2y^3 - 8xy^2 + 15x^3y) \div (10x^2y^3)$

103. $(6p^2 - p - 8) - (-4p^2 + 2p - 3)$

104. $\frac{5^8}{5^{19}}$

105. $(-7 + 2k)^2$

106. $\left(\frac{x}{y^{-3}}\right)^{-4}$

CHAPTER 4 TEST

[4.1] 1. $-7x^2 + 8x$; 2; binomial
2. $4n^4 + 13n^3 - 10n^2$; 4; trinomial
3. 4, −2, −4, −2, 4

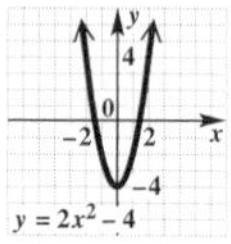

4. $-2y^2 - 9y + 17$
5. $-21a^3b^2 + 7ab^5 - 5a^2b^2$
6. $-12t^2 + 5t + 8$
[4.3] 7. $-27x^5 + 18x^4 - 6x^3 + 3x^2$
8. $t^2 - 5t - 24$
9. $8x^2 + 2xy - 3y^2$
[4.4] 10. $25x^2 - 20xy + 4y^2$
11. $100v^2 - 9w^2$

For each polynomial, combine terms when possible and write the polynomial in descending powers of the variable. Give the degree of the simplified polynomial. Decide whether the simplified polynomial is a monomial, binomial, trinomial, or none of these.

1. $5x^2 + 8x - 12x^2$

2. $13n^3 - n^2 + n^4 + 3n^4 - 9n^2$

3. Use the table to complete a set of ordered pairs that lie on the graph of $y = 2x^2 - 4$. Then graph the equation.

x	y
−2	
−1	
0	
1	
2	

Perform the indicated operations.

4. $(2y^2 - 8y + 8) + (-3y^2 + 2y + 3) - (y^2 + 3y - 6)$

5. $(-9a^3b^2 + 13ab^5 + 5a^2b^2) - (6ab^5 + 12a^3b^2 + 10a^2b^2)$

6. Subtract.

$$\begin{array}{r} 9t^3 - 4t^2 + 2t + 2 \\ \underline{9t^3 + 8t^2 - 3t - 6} \end{array}$$

7. $3x^2(-9x^3 + 6x^2 - 2x + 1)$

8. $(t - 8)(t + 3)$

9. $(4x + 3y)(2x - y)$

10. $(5x - 2y)^2$

11. $(10v + 3w)(10v - 3w)$

12. $2r^3 + r^2 - 16r + 15$
13. $9x^2 + 54x + 81$
[4.5] 14. $\frac{1}{625}$ 15. 2 16. $\frac{7}{12}$
[4.2] 17. $9x^3y^5$
[4.5] 18. 8^5 19. x^2y^6
20. Disagree, because $3^{-4} = \frac{1}{3^4} = \frac{1}{81}$, which is positive.
[4.6] 21. $4y^2 - 3y + 2 + \frac{5}{y}$
22. $-3xy^2 + 2x^3y^2 + 4y^2$
23. $3x^2 + 6x + 11 + \frac{26}{x-2}$
[4.7] 24. (a) 4.5×10^{10} (b) .0000036 (c) .00019
25. (a) 1.7×10^5; 1×10^3 (b) (more than) 1.7×10^8 pounds

12. $(2r - 3)(r^2 + 2r - 5)$

13. What polynomial expression represents the area of this square?

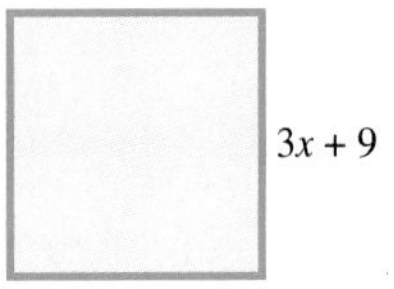

Evaluate each expression.

14. 5^{-4}

15. $(-3)^0 + 4^0$

16. $4^{-1} + 3^{-1}$

17. Use the rules for exponents to simplify $\frac{(3x^2y)^2(xy^3)^2}{(xy)^3}$. Assume x and y are nonzero.

Simplify, and write the answer using only positive exponents. Assume that variables represent nonzero numbers.

18. $\frac{8^{-1} \cdot 8^4}{8^{-2}}$

19. $\frac{(x^{-3})^{-2}(x^{-1}y)^2}{(xy^{-2})^2}$

20. Do you agree or disagree with the following statement?

3^{-4} represents a negative number.

Justify your answer.

Perform each division.

21. $\frac{8y^3 - 6y^2 + 4y + 10}{2y}$

22. $(-9x^2y^3 + 6x^4y^3 + 12xy^3) \div (3xy)$

23. $(3x^3 - x + 4) \div (x - 2)$

24. **(a)** Write 45,000,000,000 using scientific notation.
(b) Write 3.6×10^{-6} without using exponents.
(c) Write the quotient without using scientific notation: $\frac{9.5 \times 10^{-1}}{5 \times 10^3}$.

25. According to an article in *Aviation Week and Space Technology* (March 23, 1998), the MD-17 jet has the capacity to carry more than a 170,000-pound payload. Suppose we consider a group of 1000 such jets.
(a) Write the numbers 170,000 and 1000 using scientific notation.
(b) Multiply these two numbers to find the total payload that this group can carry. Express your answer using scientific notation.

CUMULATIVE REVIEW EXERCISES CHAPTERS 1–4

[1.1] 1. $\frac{7}{4}$ 2. 5 3. $\frac{19}{24}$
4. $-\frac{1}{20}$ 5. $31\frac{1}{4}$ cubic yards

Write each fraction in lowest terms.

1. $\frac{28}{16}$

2. $\frac{55}{11}$

Perform each operation.

3. $\frac{2}{3} + \frac{1}{8}$

4. $\frac{7}{4} - \frac{9}{5}$

5. A contractor installs toolsheds. Each requires $1\frac{1}{4}$ cubic yards of concrete. How much concrete would be needed for 25 sheds?

[1.5] 6. \$1836
[1.6] 7. 1, 3, 5, 9, 15, 45
[1.5] 8. positive
[1.6] 9. −8 10. 24 11. $\frac{1}{2}$
12. −4
[1.7] 13. associative property
14. distributive property
[1.5] 15. (a) −175 (b) 575
[1.8] 16. $-10x^2 + 21x - 29$
[2.1, 2.2] 17. $\left\{\frac{13}{4}\right\}$ **18.** $\emptyset$
[2.4] 19. $r = \frac{d}{t}$
[2.5] 20. {−5}
[2.1, 2.2] 21. {−12} 22. {20}
23. {all real numbers}
[2.3] 24. mouse: 160; elephant: 10
25. 4

6. A retailer has \$34,000 invested in her business. She finds that last year she earned 5.4% on this investment. How much did she earn?

7. List all positive integer factors of 45.

8. If $a < 0$ and $b > 0$, what is the sign of $b - a$?

Find the value of each expression if $x = -2$ and $y = 4$.

9. $\dfrac{4x - 2y}{x + y}$

10. $x^3 - 4xy$

Perform the indicated operations.

11. $\dfrac{(-13 + 15) - (3 + 2)}{6 - 12}$

12. $-7 - 3[2 + (5 - 8)]$

Decide which property justifies each statement.

13. $(9 + 2) + 3 = 9 + (2 + 3)$

14. $6(4 + 2) = 6(4) + 6(2)$

15. Use the bar graph to find a signed number that represents the change in the number of airline passenger deaths
(a) from 1994 to 1995
(b) from 1995 to 1996.

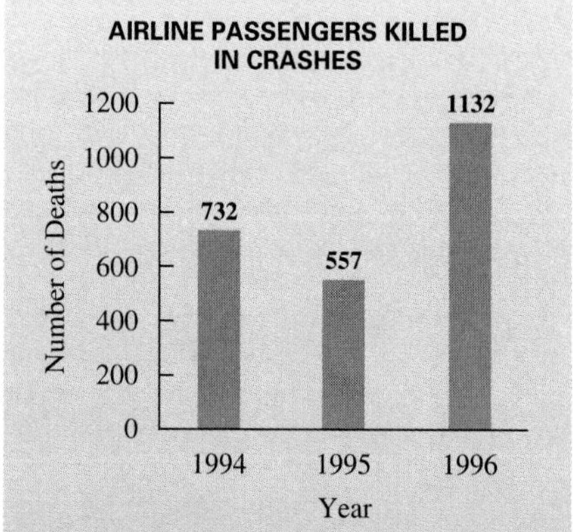

Source: International Civil Aviation Organization.

16. Simplify the expression $-3(2x^2 - 8x + 9) - (4x^2 + 3x + 2)$.

Solve each equation.

17. $2 - 3(t - 5) = 4 + t$

18. $2(5h + 1) = 10h + 4$

19. $d = rt$ for r

20. $\dfrac{x}{5} = \dfrac{x - 2}{7}$

21. $\dfrac{1}{3}p - \dfrac{1}{6}p = -2$

22. $.05x + .15(50 - x) = 5.50$

23. $4 - (3x + 12) = (2x - 9) - (5x - 1)$

Solve each problem.

24. A 1-ounce mouse takes about 16 times as many breaths as does a 3-ton elephant. If the two animals take a combined total of 170 breaths per minute, how many breaths does each take during that time period? (*Source:* Christopher McGowan, *Dinosaurs, Spitfires, and Sea Dragons,* Harvard University Press, 1991.)

25. If a number is subtracted from 8 and this difference is tripled, the result is 3 times the number. Find this number, and you will learn how many times a dolphin rests during a 24-hour period.

26. 1995 and 1996: 22; 1994: 24 [2.7] 27. $[10, \infty)$

28. $\left(-\infty, -\frac{14}{5}\right)$ 29. $[-4, 2)$

[2.4] 30. 11 feet and 22 feet

[3.1] 31.

x	0	-4	2	-2	4
y	2	0	3	1	4

[3.2] 32.

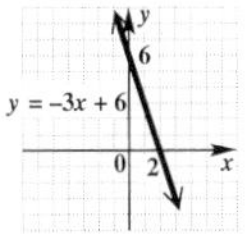

[4.1] 33.

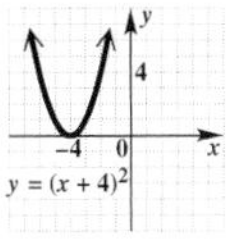

34. $11x^3 - 14x^2 - x + 14$

[4.3] 35. $18x^7 - 54x^6 + 60x^5$

36. $63x^2 + 57x + 12$

[4.4] 37. $25x^2 + 80x + 64$

[4.6] 38. $2x^2 - 3x + 1$

39. $y^2 - 2y + 6$

[4.5] 40. $\frac{5}{4}$ 41. 2

42. 1 43. $\frac{2b}{a^{10}}$

[4.7] 44. 3.45×10^4

45. .000000536

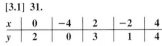

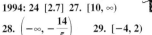

26. In 1995 and 1996, there were the same number of aircraft accidents involving passenger fatalities throughout worldwide scheduled air services. In 1994, there were 2 more than in each of the other years. How many accidents were there in each of these years if there was a total of 68 altogether? (*Source:* International Civil Aviation Organization.)

Solve each inequality.

27. $-8x \leq -80$

28. $-2(x + 4) > 3x + 6$

29. $-3 \leq 2x + 5 < 9$

Solve the problem.

30. One side of a triangle is twice as long as a second side. The third side of the triangle is 17 feet long. The perimeter of the triangle cannot be more than 50 feet. Find the longest possible values for the other two sides of the triangle.

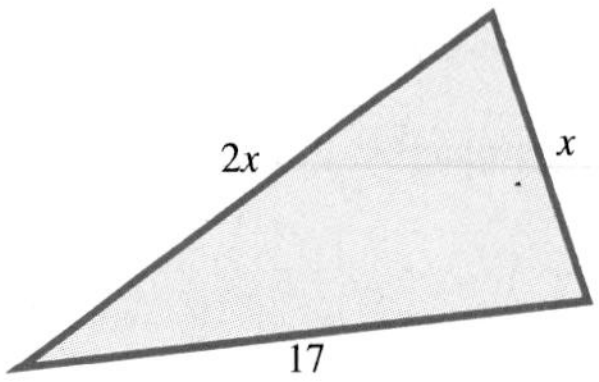

31. Complete the table of values for $-2x + 4y = 8$.

x	0		2		4
y		0		1	

32. Graph $y = -3x + 6$.

33. Graph $y = (x + 4)^2$, using the x-values $-6, -5, -4, -3$, and -2 to obtain a set of points.

Perform the indicated operations.

34. $(7x^3 - 12x^2 - 3x + 8) + (6x^2 + 4) - (-4x^3 + 8x^2 - 2x - 2)$

35. $6x^5(3x^2 - 9x + 10)$

36. $(7x + 4)(9x + 3)$

37. $(5x + 8)^2$

38. $\dfrac{14x^3 - 21x^2 + 7x}{7x}$

39. $\dfrac{y^3 - 3y^2 + 8y - 6}{y - 1}$

Evaluate each expression.

40. $4^{-1} + 3^0$

41. $2^{-4} \cdot 2^5$

42. $\dfrac{8^{-5} \cdot 8^7}{8^2}$

43. Write with positive exponents only: $\dfrac{(a^{-3}b^2)^2}{(2a^{-4}b^{-3})^{-1}}$.

44. Write in scientific notation: 34,500.

45. Write without exponents: 5.36×10^{-7}.

5 Factoring and Applications

Transportation

In the early days of the American frontier, it took four horses 75 days to haul one wagonload of goods a thousand miles. By the late 1800s, roads, canals, and railways connected "island" communities, greatly expediting the movement of people and goods. Whereas a stagecoach traveled 50 miles in a day, trains now covered 50 miles in an *hour*—over 700 miles per day. The vehicle that really changed twentieth-century American travel, however, was the automobile. Today emerging technologies are further transforming modern transportation systems. Just one example is the global network of computers moving millions of bits of data per second.

Much of this emerging technology depends on mathematics. Understanding mathematics makes it possible to model data using functions. In Chapter 4, we used polynomials of degree 2 to model data. Now, in this chapter, we introduce a method for solving polynomial equations of degree 2 called *quadratic equations,* and we give further examples of them as models. For instance, average fuel economy trends in miles per gallon for the automotive industry are closely approximated by the quadratic equation

$$y = -.04x^2 + .93x + 21,$$

where x represents the year. The years are coded so that $x = 0$ represents 1978, $x = 2$ represents 1980, $x = 4$ represents 1982, and so on. This equation was developed from the data in the table.

Fuel Economy

Year	1978	1980	1982	1984	1986	1988	1990	1992	1994	1996
Miles per Gallon	19.9	23.1	25.1	25.0	25.9	26.0	25.4	25.1	24.7	24.9

Source: National Highway Traffic Safety Administration.

Substituting 0 for x in the equation gives $y = 21$, the approximate average miles per gallon in 1978. For 1996, substituting $1996 - 1978 = 18$ for x in the equation gives $y = 24.78$ as the approximate average miles per gallon. Which of these is a closer approximation to the data in the table? We will return to this transportation model in Section 5.6, Example 4.

Visit our Web site at www.LialAlgebra.com

5.1 The Greatest Common Factor; Factoring by Grouping

OBJECTIVES

1. Find the greatest common factor of a list of terms.
2. Factor out the greatest common factor.
3. Factor by grouping.

FOR EXTRA HELP

SSG Sec. 5.1
SSM Sec. 5.1
Pass the Test Software
InterAct Math Tutorial Software
Video 8

Recall from Chapter 1 that to **factor** means to write a quantity as a product. That is, factoring is the opposite of multiplying. For example,

Multiplying	*Factoring*
$6 \cdot 2 = 12,$	$12 = 6 \cdot 2.$
↑ ↑ Factors ↑ Product	↑ Product ↑ ↑ Factors

Other factored forms of 12 are

$$(-6)(-2), \quad 3 \cdot 4, \quad (-3)(-4), \quad 12 \cdot 1, \quad \text{and} \quad (-12)(-1).$$

More than two factors may be used, so another factored form of 12 is $2 \cdot 2 \cdot 3$. The positive integer factors of 12 are

$$1, 2, 3, 4, 6, 12.$$

OBJECTIVE 1 Find the greatest common factor of a list of terms. An integer that is a factor of two or more integers is called a **common factor** of those integers. For example, 6 is a common factor of 18 and 24 since 6 is a factor of both 18 and 24. Other common factors of 18 and 24 are 1, 2, and 3. The **greatest common factor** of a list of integers is the largest common factor of those integers. Thus, 6 is the greatest common factor of 18 and 24, since it is the largest of the common factors of these numbers.

NOTE: Factors of a number are also divisors of the number. The greatest common factor is actually the same as the greatest common divisor.

CONNECTIONS

There are many rules for deciding what numbers divide into a given number. Here are some especially useful divisibility rules for small numbers. It is surprising how many people do not know them.

A Whole Number Divisible by:	*Must Have the Following Property:*
2	Ends in 0, 2, 4, 6, or 8
3	Sum of its digits is divisible by 3
4	Last two digits form a number divisible by 4
5	Ends in 0 or 5
6	Divisible by both 2 and 3
8	Last three digits form a number divisible by 8
9	Sum of its digits is divisible by 9
10	Ends in 0

Recall from Chapter 1 that a prime number has only itself and 1 as factors. In Section 1.1 we factored numbers into prime factors. This is the first step in finding the greatest

INSTRUCTOR'S RESOURCES

PTB Sec. 5.1 **ISM** Sec. 5.1 **AB** Sec. 5.1 | **TEST GENERATOR** | **WORLD WIDE WEB** www.LialAlgebra.com

common factor of a list of numbers. We find the greatest common factor (GCF) of a list of numbers as follows.

Finding the Greatest Common Factor (GCF)

Step 1 **Factor.** Write each number in prime factored form.

Step 2 **List common factors.** List each prime number that is a factor of every number in the list.

Step 3 **Choose smallest exponents.** Use as exponents on the common prime factors the *smallest* exponent from the prime factored forms. (If a prime does not appear in one of the prime factored forms, it cannot appear in the greatest common factor.)

Step 4 **Multiply.** Multiply the primes from Step 3. If there are no primes left after Step 3, the greatest common factor is 1.

TEACHING TIP The greatest common factor is often confused with the least common multiple (as in common denominator). Clearly distinguish between these two concepts.

CHALKBOARD EXAMPLE

Find the greatest common factor for the group of numbers.

12, 18, 26, 32

Answer: 2

EXAMPLE 1 Finding the Greatest Common Factor for Numbers

Find the greatest common factor for each list of numbers.

(a) 30, 45

First write each number in prime factored form.

$$30 = 2 \cdot 3 \cdot \mathbf{5}$$
$$45 = 3 \cdot 3 \cdot \mathbf{5}$$

Now, take each prime the *least* number of times it appears in all the factored forms. There is no 2 in the prime factored form of 45, so there will be no 2 in the greatest common factor. The least number of times 3 appears in all the factored forms is 1, and the least number of times 5 appears is also 1. From this, the GCF is

$$3^1 \cdot \mathbf{5^1} = 15.$$

(b) 72, 120, 432

Find the prime factored form of each number.

$$72 = 2 \cdot 2 \cdot 2 \cdot \mathbf{3} \cdot 3$$
$$120 = 2 \cdot 2 \cdot 2 \cdot \mathbf{3} \cdot 5$$
$$432 = 2 \cdot 2 \cdot 2 \cdot 2 \cdot \mathbf{3} \cdot 3 \cdot 3$$

The least number of times 2 appears in all the factored forms is 3, and the least number of times 3 appears is 1. There is no 5 in the prime factored form of either 72 or 432, so the GCF is

$$2^3 \cdot 3 = 24.$$

(c) 10, 11, 14

Write the prime factored form of each number.

$$10 = 2 \cdot 5$$
$$11 = 11$$
$$14 = 2 \cdot 7$$

There are no primes common to all three numbers, so the GCF is 1.

The greatest common factor can also be found for a list of variable terms. For example, the terms x^4, x^5, x^6, and x^7 have x^4 as the greatest common factor because each of these terms can be written with x^4 as a factor.

$$x^4 = 1 \cdot x^4, \qquad x^5 = x \cdot x^4, \qquad x^6 = x^2 \cdot x^4, \qquad x^7 = x^3 \cdot x^4$$

NOTE The exponent on a variable in the GCF is the *smallest* exponent that appears in the factors.

CHALKBOARD EXAMPLE

Find the greatest common factor for the list of terms.

$$-11r^9, \ -10r^{15}, \ -8r^{12}$$

Answer: r^9

EXAMPLE 2 Finding the Greatest Common Factor for Variable Terms

Find the greatest common factor for each list of terms.

(a) $21m^7, \ -18m^6, \ 45m^8, \ -24m^5$

$$\begin{aligned} 21m^7 &= 3 \cdot 7 \cdot m^7 \\ -18m^6 &= -1 \cdot 2 \cdot 3^2 \cdot m^6 \\ 45m^8 &= 3^2 \cdot 5 \cdot m^8 \\ -24m^5 &= -1 \cdot 2^3 \cdot 3 \cdot m^5 \end{aligned}$$

First, 3 is the greatest common factor of the coefficients 21, −18, 45, and −24. The smallest exponent on m is 5, so the GCF of the terms is $3m^5$.

(b) $x^4y^2, \ x^7y^5, \ x^3y^7, \ y^{15}$

$$\begin{aligned} x^4y^2 &= x^4 \cdot y^2 \\ x^7y^5 &= x^7 \cdot y^5 \\ x^3y^7 &= x^3 \cdot y^7 \\ y^{15} &= y^{15} \end{aligned}$$

There is no x in the last term, y^{15}, so x will not appear in the greatest common factor. There is a y in each term, however, and 2 is the smallest exponent on y. The GCF is y^2.

OBJECTIVE 2 Factor out the greatest common factor. We use the idea of a greatest common factor to write a polynomial (a sum) in factored form as a product. For example, the polynomial

$$3m + 12$$

has two terms, $3m$ and 12. The greatest common factor for these two terms is 3. We can write $3m + 12$ so that each term is a product with 3 as one factor.

$$3m + 12 = 3 \cdot m + 3 \cdot 4$$

Now use the distributive property.

$$3m + 12 = 3 \cdot m + 3 \cdot 4 = 3(m + 4)$$

The factored form of $3m + 12$ is $3(m + 4)$. This process is called **factoring out the greatest common factor.**

TEACHING TIP Factoring out the GCF and using the distributive property to multiply are inverse operations.

CAUTION The polynomial $3m + 12$ is *not* in factored form when written as

$$3 \cdot m + 3 \cdot 4.$$

The *terms* are factored, but the polynomial is not. The factored form of $3m + 12$ is the *product*

$$3(m + 4).$$

CHALKBOARD EXAMPLE

Factor out the greatest common factor.

$$8p^5q^2 + 16p^6q^3 - 12p^4q^7$$

Answer: $4p^4q^2(2p + 4p^2q - 3q^5)$

EXAMPLE 3 Factoring Out the Greatest Common Factor

Factor out the greatest common factor.

(a) $20m^5 + 10m^4 + 15m^3$

The GCF for the terms of this polynomial is $5m^3$.

$$\begin{aligned} 20m^5 + 10m^4 + 15m^3 &= (5m^3)(4m^2) + (5m^3)(2m) + (5m^3)3 \\ &= 5m^3(4m^2 + 2m + 3) \end{aligned}$$

Check by multiplying $5m^3$ and $4m^2 + 2m + 3$. You should get the original polynomial.

(b) $x^5 + x^3 = (x^3)x^2 + (x^3)1 = x^3(x^2 + 1)$

(c) $20m^7p^2 - 36m^3p^4 = 4m^3p^2(5m^4 - 9p^2)$

(d) $a(a + 3) + 4(a + 3)$

The binomial $a + 3$ is the greatest common factor here.

$$a(a + 3) + 4(a + 3) = (a + 3)(a + 4)$$

CAUTION Be sure to include the 1 in a problem like Example 3(b). Always check that the factored form can be multiplied out to give the original polynomial.

OBJECTIVE 3 Factor by grouping. Common factors are used in **factoring by grouping,** as explained in the next example.

CHALKBOARD EXAMPLE

Factor by grouping.

$$2mn - 8n + 3m - 12$$

Answer: $(m - 4)(2n + 3)$

TEACHING TIP You will need to provide several examples of factoring by grouping in order to help students understand this concept.

EXAMPLE 4 Factoring by Grouping

Factor by grouping.

(a) $2x + 6 + ax + 3a$

The first two terms have a common factor of 2, and the last two terms have a common factor of a.

$$2x + 6 + ax + 3a = 2(x + 3) + a(x + 3)$$

The expression is still not in factored form because it is the *sum* of two terms. Now, however, $x + 3$ is a common factor and can be factored out.

$$2x + 6 + ax + 3a = 2(x + 3) + a(x + 3) = (x + 3)(2 + a)$$

The final result is in factored form because it is a *product.* Note that the goal in factoring by grouping is to get a common factor, $x + 3$ here, so that the last step is possible.

Same

(b) $m^2 + 6m + 2m + 12 = m(m + 6) + 2(m + 6)$
$= (m + 6)(m + 2)$

(c) $6xy - 21x - 8y + 28 = 3x(2y - 7) - 4(2y - 7) = (2y - 7)(3x - 4)$

Must be same

Since the quantities in parentheses in the second step must be the same, it was necessary here to factor out -4 rather than 4.

CAUTION Use negative signs carefully when grouping, as in Example 4(c). Otherwise, sign errors may result.

Use these steps when factoring four terms by grouping.

Factoring by Grouping

Step 1 **Group terms.** Collect the terms into two groups so that each group has a common factor.

Step 2 **Factor within groups.** Factor out the greatest common factor from each group.

Step 3 **Factor the entire polynomial.** Factor a common binomial factor from the results of Step 2.

Step 4 **If necessary, rearrange terms.** If Step 2 does not result in a common binomial factor, try a different grouping.

CHALKBOARD EXAMPLE

Factor by grouping.

$6y^2 - 20w^2 + 15yw - 8yw$

Answer: $(2y + 5w)(3y - 4w)$

EXAMPLE 5 Rearranging Terms Before Factoring by Grouping

Factor by grouping.

(a) $10x^2 - 12y^2 + 15xy - 8xy$

Factoring out the common factor of 2 from the first two terms and the common factor of xy terms from the last two terms gives

$$10x^2 - 12y^2 + 15xy - 8xy = 2(5x^2 - 6y^2) + xy(15 - 8).$$

This did not lead to a common factor, so we try rearranging the terms. There is usually more than one way to do this. Let's try

$$10x^2 - 8xy - 12y^2 + 15xy,$$

grouping the first two terms and the last two terms as follows.

$$\begin{aligned} 10x^2 - 8xy - 12y^2 + 15xy &= 2x(5x - 4y) + 3y(-4y + 5x) \\ &= 2x(5x - 4y) + 3y(5x - 4y) \\ &= (5x - 4y)(2x + 3y) \end{aligned}$$

(b) $2xy + 12 - 3y - 8x$

We need to rearrange these terms to get two groups that each have a common factor. Trial and error suggests the following grouping.

$$2xy + 12 - 3y - 8x = (2xy - 3y) + (-8x + 12)$$
$$= y(2x - 3) - 4(2x - 3) \qquad \text{Factor each group.}$$
$$= (2x - 3)(y - 4) \qquad \text{Factor out the common binomial factor.}$$

5.1 EXERCISES

1. no; 6 2. no; p
3. multiplication 4. First, verify that you have factored completely. Then multiply the factors. The product should be the original polynomial. 5. Answers will vary. One example is 5, 10, 15.
6. Write each term in prime factored form. For the greatest common factor, choose the number or variable to the smallest exponent that appears in every term. The product of these numbers and variables is the greatest common factor. For example, the GCF of $6x^2$, $3x^3$, and $12x^4$ is $3 \cdot x^2$ or $3x^2$.
7. 8 8. 9 9. $10x^3$
10. $10z^4$ 11. $6m^3n^2$ 12. $5p^5r^3$
13. xy^2 14. a^3b 15. 6
16. 15 17. factored
18. factored 19. not factored
20. factored 21. yes; x^3y^2
22. no 23. 2 24. 2
25. x 26. x^2 27. $3m^2$
28. $2p^2$ 29. $2z^4$ 30. $3k^3$
31. $2mn^4$ 32. $3ab$
33. $-7x^3y^2$ 34. $-4m^2n$
35. $12(y - 2)$ 36. $18(p + 2)$
37. $10a(a - 2)$ 38. $15x^2(x - 2)$

1. Is 3 the greatest common factor of 18, 24, and 42? If not, what is?

2. Is pq the greatest common factor of pq^2, p^2, and p^2q^2? If not, what is?

3. Factoring is the opposite of what operation?

 4. How can you check your answer when you factor a polynomial?

5. Give an example of three numbers whose greatest common factor is 5.

6. Explain how to find the greatest common factor of a list of terms. Use examples.

Find the greatest common factor for each list of terms. See Examples 1 and 2.

7. $16y$, 24

8. $18w$, 27

9. $30x^3$, $40x^6$, $50x^7$

10. $60z^4$, $70z^8$, $90z^9$

11. $12m^3n^2$, $18m^5n^4$, $36m^8n^3$

12. $25p^5r^7$, $30p^7r^8$, $50p^5r^3$

13. $-x^4y^3$, $-xy^2$

14. $-a^4b^5$, $-a^3b$

15. $42ab^3$, $-36a$, $90b$, $-48ab$

16. $45c^3d$, $75c$, $90d$, $-105cd$

An expression is factored when it is written as a product, not a sum. Which of the following are not factored?

17. $2k^2(5k)$

18. $2k^2(5k + 1)$

19. $2k^2 + (5k + 1)$

20. $(2k^2 + 1)(5k + 1)$

21. Is $-xy$ a common factor of $-x^4y^3$ and $-xy^2$? If so, what is the other factor that when multiplied by $-xy$ gives $-x^4y^3$?

22. Is $-a^5b^2$ a common factor of $-a^4b^5$ and $-a^3b$?

Complete each factoring.

23. $12 = 6(\quad)$

24. $18 = 9(\quad)$

25. $3x^2 = 3x(\quad)$

26. $8x^3 = 8x(\quad)$

27. $9m^4 = 3m^2(\quad)$

28. $12p^5 = 6p^3(\quad)$

29. $-8z^9 = -4z^5(\quad)$

30. $-15k^{11} = -5k^8(\quad)$

31. $6m^4n^5 = 3m^3n(\quad)$

32. $27a^3b^2 = 9a^2b(\quad)$

33. $-14x^4y^3 = 2xy(\quad)$

34. $-16m^3n^3 = 4mn^2(\quad)$

Factor out the greatest common factor. See Example 3.

35. $12y - 24$

36. $18p + 36$

37. $10a^2 - 20a$

38. $15x^3 - 30x^2$

39. $5y^6(13y^4 + 7)$
40. $4a^3(25a^2 + 4)$
41. no common factor (except 1)
42. no common factor (except 1)
43. $8m^2n^2(n + 3)$
44. $19p^2y(1 - 2y^2)$
45. $13y^2(y^6 + 2y^2 - 3)$
46. $5x^3(x^2 + 5x - 4)$
47. $9qp^3(5q^3p^2 + 4p^3 + 9q)$
48. $5a^3z^2(25z^3 + 12az^2 - 17a^2)$
49. $a^3(a^2 + 2b^2 - 3a^2b^2 + 4ab^3)$
50. $x(x^5 + 5x^3y^3 - 6y^4 + 10y)$
51. $(x + 2)(c - d)$
52. $(5 - x)(r + t)$
53. $(m + 2n)(m + n)$
54. $(1 - 4p)(3p - 2q)$
55. not in factored form; $(7t + 4)(8 + x)$
56. not in factored form; $(5x - 1)(3r + 7)$
57. in factored form
58. in factored form
59. not in factored form; $(y + 4)(18x^2 + 7)$
60. not in factored form
61. The quantities in parentheses are not the same, so there is no common factor in the two terms, $12k^3(s - 3)$ and $7(s + 3)$.
62. Choose two groups of two terms with common factors. Factor each group. Look for a common factor of the two factored groups, and factor it out to get the final factored form. For example, to factor $x^2 + 5x + xy + 5y$, we could group the first two terms and the last two terms: $(x^2 + 5x) + (xy + 5y)$. Factor the groups: $x(x + 5) + y(x + 5)$. Now factor out the common binomial factor: $(x + 5)(x + y)$.
63. $(p + 4)(p + 3)$
64. $(m + 2)(m + 5)$
65. $(a - 2)(a + 5)$
66. $(y - 6)(y + 4)$
67. $(z + 2)(7z - a)$
68. $(m + 3p)(5m - 2p)$
69. $(3r + 2y)(6r - x)$
70. $(4s + 3y)(2s - t)$
71. $(a^2 + b^2)(3a + 2b)$
72. $(x^2 + y^2)(4x + 3y)$
73. $(1 - a)(1 - b)$
74. $(2 - x)(3 - y)$
75. $(4m - p^2)(4m^2 - p)$
76. $(5t - s^2)(2t^2 - s)$
77. $(5 - 2p)(m + 3)$

*These problems can be used as collaborative exercises with small groups of students.

39. $65y^{10} + 35y^6$ **40.** $100a^5 + 16a^3$

41. $11w^3 - 100$ **42.** $13z^5 - 80$

43. $8m^2n^3 + 24m^2n^2$ **44.** $19p^2y - 38p^2y^3$

45. $13y^8 + 26y^4 - 39y^2$ **46.** $5x^5 + 25x^4 - 20x^3$

47. $45q^4p^5 + 36qp^6 + 81q^2p^3$ **48.** $125a^3z^5 + 60a^4z^4 - 85a^5z^2$

49. $a^5 + 2a^3b^2 - 3a^5b^2 + 4a^4b^3$ **50.** $x^6 + 5x^4y^3 - 6xy^4 + 10xy$

51. $c(x + 2) - d(x + 2)$ **52.** $r(5 - x) + t(5 - x)$

53. $m(m + 2n) + n(m + 2n)$ **54.** $3p(1 - 4p) - 2q(1 - 4p)$

Students often have difficulty when factoring by grouping because they are not able to tell when the polynomial is indeed factored. For example,

$$5y(2x - 3) + 8t(2x - 3)$$

is not in factored form, because it is the *sum* of two terms, $5y(2x - 3)$ and $8t(2x - 3)$. However, because $2x - 3$ is a common factor of these two terms, the expression can now be factored as

$$(2x - 3)(5y + 8t).$$

The factored form is a *product* of two factors, $2x - 3$ and $5y + 8t$.

Determine whether each expression is in factored form or is not in factored form. If it is not in factored form, factor it if possible.

55. $8(7t + 4) + x(7t + 4)$ **56.** $3r(5x - 1) + 7(5x - 1)$

57. $(8 + x)(7t + 4)$ **58.** $(3r + 7)(5x - 1)$

59. $18x^2(y + 4) + 7(y + 4)$ **60.** $12k^3(s - 3) + 7(s + 3)$

61. Tell why it is not possible to factor the expression in Exercise 60.

62. Summarize the method of factoring a polynomial with four terms by grouping. Give an example.

Factor by grouping. See Examples 4 and 5.

63. $p^2 + 4p + 3p + 12$ **64.** $m^2 + 2m + 5m + 10$

65. $a^2 - 2a + 5a - 10$ **66.** $y^2 - 6y + 4y - 24$

67. $7z^2 + 14z - az - 2a$ **68.** $5m^2 + 15mp - 2mp - 6p^2$

69. $18r^2 + 12ry - 3xr - 2xy$ **70.** $8s^2 - 4st + 6sy - 3yt$

71. $3a^3 + 3ab^2 + 2a^2b + 2b^3$ **72.** $4x^3 + 3x^2y + 4xy^2 + 3y^3$

73. $1 - a + ab - b$ **74.** $6 - 3x - 2y + xy$

75. $16m^3 - 4m^2p^2 - 4mp + p^3$ **76.** $10t^3 - 2t^2s^2 - 5ts + s^3$

77. $5m + 15 - 2mp - 6p$ **78.** $y^2 - 3y - xy + 3x$

79. $18r^2 + 12ry - 3ry - 2y^2$ **80.** $3a^3 + 3ab^2 + 2a^2b + 2b^3$

81. $a^5 + 2a^5b - 3 - 6b$ **82.** $a^2b - 4a - ab^4 + 4b^3$

RELATING CONCEPTS* (EXERCISES 83–88)

In most cases, the choice of which pairs of terms to group when factoring by grouping can be made in several ways.

Work Exercises 83–88 in order *to see how this applies to the polynomial in Example 5(b).*

83. Start with the polynomial in Example 5(b), $2xy + 12 - 3y - 8x$, and rearrange the terms as follows: $2xy - 8x + (-3y) + 12$. What properties from Section 1.7 allow us to do this?

JOURNAL WRITING ▲ CHALLENGING SCIENTIFIC CALCULATOR GRAPHING CALCULATOR

78. $(y - x)(y - 3)$
79. $(6r - y)(3r + 2y)$
80. $(3a + 2b)(a^2 + b^2)$
81. $(a^5 - 3)(1 + 2b)$
82. $(a - b^3)(ab - 4)$
83. commutative property and associative property
84. $(2xy - 8x) + (-3y) + 12$; $2x$
85. $(2xy - 8x) + (-3y + 12)$; yes
86. $2x(y - 4) - 3(y - 4)$
87. No, because it is the difference between two terms, $2x(y - 4)$ and $3(y - 4)$.
88. $(2x - 3)(y - 4)$; yes

RELATING CONCEPTS (EXERCISES 83-88)

84. Group the first pair of terms in the rearranged polynomial. What is the greatest common factor of this pair?

85. Now group the second pair of terms in the rearranged polynomial. Is -3 a common factor of this second pair?

86. Factor the greatest common factor from the first pair and -3 from the second pair.

87. Is your result from Exercise 86 in factored form? If not, why?

88. If your answer to Exercise 87 is *no,* factor the polynomial. Is it the same result as the one shown in Example 5(b) in this section?

Did you make the connection that although there is more than one way to factor a polynomial by grouping, you should get the same answer?

89. (a) yes (b) When either one is multiplied out, the product is $1 - a + ab - b$.
90. No, the correct factored form is $18x^3y^2 + 9xy = 9xy(2x^2y + 1)$. If a polynomial has two terms, the product of the factors must have two terms, but $9xy(2x^2y) = 18x^3y^2$, which has just one term.

89. Refer to Exercise 73. The answer given in the back of the book is $(1 - a)(1 - b)$. A student factored this same polynomial and got the result $(a - 1)(b - 1)$.
(a) Is this student's answer correct?
(b) If your answer to part (a) is *yes,* explain why these two seemingly different answers are both acceptable.

90. A student factored $18x^3y^2 + 9xy$ as $9xy(2x^2y)$. Is this correct? If not, explain the error and factor correctly.

5.2 Factoring Trinomials

OBJECTIVES

1. Factor trinomials with a coefficient of 1 for the squared term.
2. Factor such polynomials after factoring out the greatest common factor.

FOR EXTRA HELP

SSG Sec. 5.2
SSM Sec. 5.2
Pass the Test Software
InterAct Math Tutorial Software
Video 8

Using FOIL, the product of the polynomials $k - 3$ and $k + 1$ is

$$(k - 3)(k + 1) = k^2 - 2k - 3.$$

Now suppose we are given the polynomial $k^2 - 2k - 3$ and want to rewrite it as the product $(k - 3)(k + 1)$. This product is called the **factored form** of $k^2 - 2k - 3$, and the process of finding the factored form is called **factoring.**

OBJECTIVE 1 Factor trinomials with a coefficient of 1 for the squared term. When factoring polynomials with integer coefficients, we use only integers for the numerical factors. For example, $x^2 + 5x + 6$ can be factored by finding integers a and b such that

$$x^2 + 5x + 6 = (x + a)(x + b).$$

To find these integers a and b, first multiply the two factors on the right-hand side of the equation:

$$(x + a)(x + b) = x^2 + ax + bx + ab.$$

By the distributive property,

$$x^2 + ax + bx + ab = x^2 + (a + b)x + ab.$$

JOURNAL WRITING ▲ CHALLENGING SCIENTIFIC CALCULATOR GRAPHING CALCULATOR

INSTRUCTOR'S RESOURCES

PTB Sec. 5.2 **ISM** Sec. 5.2
AB Sec. 5.2
TEST GENERATOR
WORLD WIDE WEB
www.LialAlgebra.com

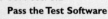

Comparing this result with $x^2 + 5x + 6$ shows that we must find integers a and b having a sum of 5 and a product of 6.

$$x^2 + 5x + \mathbf{6} = x^2 + (a + b)x + \mathbf{ab}$$

Sum of a and b is 5. Product of a and b is 6.

Since many pairs of integers have a sum of 5, it is best to begin by listing those pairs of integers whose product is 6. Both 5 and 6 are positive, so we consider only pairs in which both integers are positive.

Product	*Sum*	
$1 \cdot 6 = 6$	$1 + 6 = 7$	
$2 \cdot 3 = \mathbf{6}$	$2 + 3 = \mathbf{5}$	Sum is 5.

Both pairs have a product of 6, but only the pair 2 and 3 has a sum of 5. So 2 and 3 are the required integers, and

$$x^2 + 5x + 6 = (x + 2)(x + 3).$$

Check by multiplying the binomials using FOIL. Make sure that the sum of the outer and inner products produces the correct middle term.

$$(x + 2)(x + 3) = x^2 + 5x + 6$$

$$\begin{array}{r} 2x \\ 3x \\ \hline 5x \end{array} \quad \text{Add.}$$

This method of factoring can be used only for trinomials that have 1 as the coefficient of the squared term. Methods for factoring other trinomials are given in the next section.

CHALKBOARD EXAMPLE

Factor $y^2 + 12y + 20$.

Answer: $(y + 10)(y + 2)$

TEACHING TIP If your students have access to graphing calculators, you may wish to have them check their factoring by graphing the unfactored and factored forms of polynomials.

EXAMPLE 1 Factoring a Trinomial with All Terms Positive

Factor $m^2 + 9m + 14$.

Look for two integers whose product is 14 and whose sum is 9. List the pairs of integers whose product is 14. Then examine the sums. Again, only positive integers are needed because all signs are positive.

14, 1	$14 + 1 = 15$	
7, 2	$7 + 2 = \mathbf{9}$	Sum is 9.

From the list, 7 and 2 are the required integers, since $7 \cdot 2 = 14$ and $7 + 2 = 9$. Thus $m^2 + 9m + 14 = (m + 2)(m + 7)$.

NOTE: In Example 1, the answer also could have been written $(m + 7)(m + 2)$. Because of the commutative property of multiplication, the order of the factors does not matter.

The trinomials in the examples so far had all positive terms. If a trinomial has one or more negative terms, both positive and negative factors must be considered.

CHALKBOARD EXAMPLE

Factor $a^2 - 9a - 22$.

Answer: $(a - 11)(a + 2)$

EXAMPLE 2 Factoring a Trinomial with Two Negative Terms

Factor $p^2 - 2p - 15$.

Find two integers whose product is -15 and whose sum is -2. If these numbers do not come to mind right away, we can find them (if they exist) by listing all the pairs of integers whose product is -15. Because the last term, -15, is negative, we need pairs of integers with different signs.

$15, -1$	$15 + (-1) = 14$	
$5, -3$	$5 + (-3) = 2$	
$-15, 1$	$-15 + 1 = -14$	
$-5, 3$	$-5 + 3 = \mathbf{-2}$	Sum is -2.

The necessary integers are -5 and 3, and

$$p^2 - 2p - 15 = (p - 5)(p + 3).$$

CHALKBOARD EXAMPLE

Factor $m^2 - 2m + 5$.

Answer: prime

EXAMPLE 3 Deciding Whether a Polynomial Is Prime

(a) Factor $x^2 - 5x + 12$.

List all pairs of integers whose product is 12. Since the middle term is negative and the last term is positive, we need pairs with both numbers negative. Then examine the sums.

$-12, -1$	$-12 + (-1) = -13$
$-6, -2$	$-6 + (-2) = -8$
$-3, -4$	$-3 + (-4) = -7$

None of the pairs of integers has a sum of -5. Because of this, the trinomial $x^2 - 5x + 12$ *cannot be factored using only integer factors.* A polynomial that cannot be factored using only integer factors is called a **prime polynomial.**

(b) $k^2 - 8k + 11$

There is no pair of integers whose product is 11 and whose sum is -8, so $k^2 - 8k + 11$ is a prime polynomial.

TEACHING TIP Summarize the signs of the binomials when factoring a trinomial whose lead coefficient is positive:

1. If the last term of the trinomial is positive, both binomials will have the same "middle" sign as the second term.
2. If the last term of the trinomial is negative, the binomials will have one plus and one minus "middle" sign.

The procedure for factoring a trinomial of the form $x^2 + bx + c$ is summarized here.

Factoring $x^2 + bx + c$

Find two integers whose product is c and whose sum is b.

1. Both integers must be positive if b and c are positive.
2. Both integers must be negative if c is positive and b is negative.
3. One integer must be positive and one must be negative if c is negative.

CHALKBOARD EXAMPLE

Factor $r^2 - 6rs + 8s^2$.

Answer: $(r - 4s)(r - 2s)$

EXAMPLE 4 Factoring a Trinomial with Two Variables

Factor $z^2 - 2bz - 3b^2$.

To factor $z^2 - 2bz - 3b^2$, look for two expressions whose product is $-3b^2$ and whose sum is $-2b$. The expressions are $-3b$ and b, with

$$z^2 - 2bz - 3b^2 = (z - 3b)(z + b).$$

OBJECTIVE 2 Factor such polynomials after factoring out the greatest common factor. The trinomial in the next example does not have a coefficient of 1 for the squared term. (In fact, there is no squared term.) However, there may be a common factor.

CHALKBOARD EXAMPLE

Factor $3x^4 - 15x^3 + 18x^2$.

Answer: $3x^2(x - 3)(x - 2)$

EXAMPLE 5 Factoring a Trinomial with a Common Factor

Factor $4x^5 - 28x^4 + 40x^3$.

First, factor out the greatest common factor, $4x^3$.

$$4x^5 - 28x^4 + 40x^3 = 4x^3(x^2 - 7x + 10)$$

Now factor $x^2 - 7x + 10$. The integers -5 and -2 have a product of 10 and a sum of -7. The complete factored form is

$$4x^5 - 28x^4 + 40x^3 = 4x^3(x - 5)(x - 2).$$

CAUTION When factoring, always look for a common factor first. Remember to include the common factor as part of the answer. As a check, multiplying out the factored form should always give the original polynomial.

5.2 EXERCISES

1. a **and** b **must have different signs, one positive and one negative. 2.** a **and** b **must have the same sign, either both positive or both negative. 3. A prime polynomial is one that cannot be factored into polynomials with integer coefficients. 4. She forgot the common factor of** m**. Without it, the product of the factors does not equal the original polynomial. The correct answer is** $m(m + 1)(m + 2)$**. 5. 1 and 48, −1 and −48, 2 and 24, −2 and −24, 3 and 16, −3 and −16, 4 and 12, −4 and −12, 6 and 8, −6 and −8; the pair with a sum of −19 is −3 and −16. 6. 1 and 48, −1 and −48, 2 and 24, −2 and −24, 3 and 16, −3 and −16, 4 and 12, −4 and −12, 6 and 8, −6 and −8; the pair with a sum of 14 is 6 and 8. 7. 1 and −24, −1 and 24, 2 and −12, −2 and 12, 3 and −8, −3 and 8, 4 and −6, −4 and 6; the pair with a sum of −5 is 3 and −8. 8. 1 and −36, −1 and 36, 2 and −18, −2 and 18, 3 and −12, −3 and 12, 4 and −9, −4 and 9, 6 and −6; the pair with a sum of −16 is 2 and −18. 9. (c)**

1. In factoring a trinomial in x as $(x + a)(x + b)$, what must be true of a and b if the coefficient of the last term of the trinomial is negative?
2. In Exercise 1, what must be true of a and b if the coefficient of the last term is positive?
3. In your own words, explain the meaning of a *prime polynomial*.
4. A teacher asked a class to factor $m^3 + 3m^2 + 2m$. One student did not understand why her answer, $(m + 1)(m + 2)$ was incorrect. Explain her mistake, and give the correct answer.

In Exercises 5–8, list all pairs of integers with the given product. Then find the pair whose sum is given.

5. Product: 48 Sum: -19
6. Product: 48 Sum: 14
7. Product: -24 Sum: -5
8. Product: -36 Sum: -16
9. Which one of the following is the correct factored form of $x^2 - 12x + 32$?
   (a) $(x - 8)(x + 4)$ **(b)** $(x + 8)(x - 4)$
   (c) $(x - 8)(x - 4)$ **(d)** $(x + 8)(x + 4)$
10. Explain the steps you would use to factor $2x^3 + 8x^2 - 10x$.

Complete the following. See Examples 1–3.

11. $p^2 + 11p + 30 = (p + 5)(\quad)$
12. $x^2 + 10x + 21 = (x + 7)(\quad)$
13. $x^2 + 15x + 44 = (x + 4)(\quad)$
14. $r^2 + 15r + 56 = (r + 7)(\quad)$
15. $x^2 - 9x + 8 = (x - 1)(\quad)$
16. $t^2 - 14t + 24 = (t - 2)(\quad)$
17. $y^2 - 2y - 15 = (y + 3)(\quad)$
18. $t^2 - t - 42 = (t + 6)(\quad)$
19. $x^2 + 9x - 22 = (x - 2)(\quad)$
20. $x^2 + 6x - 27 = (x - 3)(\quad)$
21. $y^2 - 7y - 18 = (y + 2)(\quad)$
22. $y^2 - 2y - 24 = (y + 4)(\quad)$

RELATING CONCEPTS (EXERCISES 23-30)

To check your factoring of a trinomial, remember that when the factors are multiplied, the product must include *every* term of the original trinomial. In a trinomial such as $x^2 + x - 12$, students often factor so that the first and third terms are correct, but the middle term is incorrect, differing only in the sign.

***Work Exercises 23–30 in order** to see how this problem can be avoided.*

23. Add the pair of numbers: -3 and 4.

24. Add the pair of numbers: 3 and -4.

25. In Exercise 24, we added the *opposites* of the numbers in Exercise 23. How does the sum in Exercise 24 compare with the sum in Exercise 23?

26. Consider this factored form, given for $x^2 + x - 12$: $(x + 3)(x - 4)$. Multiply the two binomials. Is this the correct factored form? Why or why not?

27. Consider this factored form, given for $x^2 + x - 12$: $(x - 3)(x + 4)$. Multiply the two binomials. Is this the correct factored form? Why or why not?

28. In Exercises 26 and 27, one factored form is correct and one is not. For the one that is not, how does the sign of the middle term of the product compare with the signs of the second terms of the correct factored form?

29. Compare the two factored forms shown in Exercises 26 and 27, and use the result of Exercise 28 to complete the following statement: When I factor a trinomial into a product of binomials, and the middle term of the product is different only in sign, I should ______ in order to obtain the correct factored form.

30. Given that the factored form $(x + 5)(x - 3)$ is incorrect only in the sign of the middle term of the product for a particular trinomial, what would be the correct factored form?

Did you make the connection that if only the sign of the middle term is incorrect when you check the product of the binomial factors, then only the signs of the second terms in the binomials need to be changed?

Factor completely. If the polynomial cannot be factored, write prime. *See Examples 1–3.*

31. $y^2 + 9y + 8$
32. $a^2 + 9a + 20$
33. $b^2 + 8b + 15$
34. $x^2 + 6x + 8$
35. $m^2 + m - 20$
36. $p^2 + 4p - 5$
37. $y^2 - 8y + 15$
38. $y^2 - 6y + 8$
39. $x^2 + 4x + 5$
40. $t^2 + 11t + 12$
41. $t^2 - 8t + 16$
42. $s^2 - 10s + 25$
43. $r^2 - r - 30$
44. $q^2 - q - 42$
45. $n^2 - 12n - 35$
46. $x^2 - 4x + 12$

Factor completely. See Examples 4 and 5.

47. $r^2 + 3ra + 2a^2$
48. $x^2 + 5xa + 4a^2$
49. $t^2 - tz - 6z^2$
50. $a^2 - ab - 12b^2$
51. $x^2 + 4xy + 3y^2$
52. $p^2 + 9pq + 8q^2$
53. $v^2 - 11vw + 30w^2$
54. $v^2 - 11vx + 24x^2$
55. $4x^2 + 12x - 40$
56. $5y^2 - 5y - 30$
57. $2t^3 + 8t^2 + 6t$
58. $3t^3 + 27t^2 + 24t$
59. $2x^6 + 8x^5 - 42x^4$
60. $4y^5 + 12y^4 - 40y^3$
61. $m^3n - 10m^2n^2 + 24mn^3$
62. $y^3z + 3y^2z^2 - 54yz^3$

10. Factor out the greatest common factor, $2x$, to get $2x(x^2 + 4x - 5)$. Factor the trinomial, getting $2x(x - 1)(x + 5)$. **11.** $p + 6$ **12.** $x + 3$ **13.** $x + 11$ **14.** $r + 8$ **15.** $x - 8$ **16.** $t - 12$ **17.** $y - 5$ **18.** $t - 7$ **19.** $x + 11$ **20.** $x + 9$ **21.** $y - 9$ **22.** $y - 6$ **23.** 1 **24.** -1 **25.** The sums are opposites as well. **26.** The product is $x^2 - x - 12$. This is not correct because the middle term is incorrect. **27.** The product is $x^2 + x - 12$. This is correct because we obtain the exact trinomial we were given to factor. **28.** It is the opposite of what it should be. **29.** reverse the signs of the two second terms of the binomials **30.** $(x - 5)(x + 3)$ **31.** $(y + 8)(y + 1)$ **32.** $(a + 4)(a + 5)$ **33.** $(b + 3)(b + 5)$ **34.** $(x + 4)(x + 2)$ **35.** $(m + 5)(m - 4)$ **36.** $(p + 5)(p - 1)$ **37.** $(y - 5)(y - 3)$ **38.** $(y - 4)(y - 2)$ **39.** prime **40.** prime **41.** $(t - 4)^2$ or $(t - 4)(t - 4)$ **42.** $(s - 5)^2$ or $(s - 5)(s - 5)$ **43.** $(r - 6)(r + 5)$ **44.** $(q - 7)(q + 6)$ **45.** prime **46.** prime **47.** $(r + 2a)(r + a)$ **48.** $(x + 4a)(x + a)$ **49.** $(t + 2z)(t - 3z)$ **50.** $(a + 3b)(a - 4b)$ **51.** $(x + y)(x + 3y)$ **52.** $(p + q)(p + 8q)$ **53.** $(v - 5w)(v - 6w)$ **54.** $(v - 8x)(v - 3x)$ **55.** $4(x + 5)(x - 2)$ **56.** $5(y + 2)(y - 3)$ **57.** $2t(t + 1)(t + 3)$ **58.** $3t(t + 8)(t + 1)$ **59.** $2x^4(x - 3)(x + 7)$ **60.** $4y^3(y - 2)(y + 5)$ **61.** $mn(m - 6n)(m - 4n)$ **62.** $yz(y - 6z)(y + 9z)$ **63.** The factored form $(2x + 4)(x - 3)$ is incorrect because $2x + 4$ has a common factor of 2, which must be factored out for the trinomial to be completely factored.

JOURNAL WRITING ▲ CHALLENGING SCIENTIFIC CALCULATOR GRAPHING CALCULATOR

64. The factored form $(x - 3)(3x + 12)$ is incorrect because $3x + 12$ has a common factor of 3, which must be factored out.
65. $a^3(a + 4b)(a - b)$
66. $mn(m - 3n)(m + n)$
67. $yz(y + 3z)(y - 2z)$
68. $k^5(k - 5m)(k + 3m)$
69. $z^8(z - 7y)(z + 3y)$
70. $x^7(x + 8w)(x - 3w)$
71. $(a + b)(x + 4)(x - 3)$
72. $(x + y)(n^2 + n + 16)$
73. $(2p + q)(r - 9)(r - 3)$
74. $(3m - n)(k - 8)(k - 5)$
75. $a^2 + 13a + 36$
76. $y^2 - 4y - 21$

63. Use the FOIL method from Section 4.3 to show that $(2x + 4)(x - 3) = 2x^2 - 2x - 12$. If you are asked to completely factor $2x^2 - 2x - 12$, why would it be incorrect to give $(2x + 4)(x - 3)$ as your answer?

64. If you are asked to completely factor the polynomial $3x^2 + 9x - 12$, why would it be incorrect to give $(x - 1)(3x + 12)$ as your answer?

Use a combination of the factoring methods discussed in this section to factor each polynomial.

65. $a^5 + 3a^4b - 4a^3b^2$ **66.** $m^3n - 2m^2n^2 - 3mn^3$ **67.** $y^3z + y^2z^2 - 6yz^3$

68. $k^7 - 2k^6m - 15k^5m^2$ **69.** $z^{10} - 4z^9y - 21z^8y^2$ **70.** $x^9 + 5x^8w - 24x^7w^2$

▲ **71.** $(a + b)x^2 + (a + b)x - 12(a + b)$

▲ **72.** $(x + y)n^2 + (x + y)n + 16(x + y)$

▲ **73.** $(2p + q)r^2 - 12(2p + q)r + 27(2p + q)$

▲ **74.** $(3m - n)k^2 - 13(3m - n)k + 40(3m - n)$

▲ **75.** What polynomial can be factored as $(a + 9)(a + 4)$?

▲ **76.** What polynomial can be factored as $(y - 7)(y + 3)$?

5.3 More on Factoring Trinomials

OBJECTIVES

1. Factor trinomials by grouping when the coefficient of the squared term is not 1.
2. Factor trinomials using FOIL.

FOR EXTRA HELP

SSG Sec. 5.3
SSM Sec. 5.3
Pass the Test Software
InterAct Math Tutorial Software
Video 8

Trinomials such as $2x^2 + 7x + 6$, in which the coefficient of the squared term is *not* 1, can be factored with an extension of the method presented in the last section.

OBJECTIVE 1 Factor trinomials by grouping when the coefficient of the squared term is not 1. Recall that a trinomial such as $m^2 + 3m + 2$ is factored by finding two numbers whose product is 2 and whose sum is 3. To factor $2x^2 + 7x + 6$, we look for two integers whose product is $2 \cdot 6 = 12$ and whose sum is 7.

Sum is 7.

$$2x^2 + 7x + 6$$

Product is $2 \cdot 6 = 12$.

By considering the pairs of positive integers whose product is 12, the necessary integers are found to be 3 and 4. We use these integers to write the middle term, $7x$, as $7x = 3x + 4x$. With this, the trinomial $2x^2 + 7x + 6$ becomes

$$2x^2 + 7x + 6 = 2x^2 + \underbrace{3x + 4x}_{7x = 3x + 4x} + 6.$$

Factor the new polynomial by grouping as in Section 5.1.

$$\begin{aligned} 2x^2 + 3x + 4x + 6 &= x(2x + 3) + 2(2x + 3) \\ &= (2x + 3)(x + 2) \end{aligned}$$

The common factor of $2x + 3$ was factored out to get

$$2x^2 + 7x + 6 = (2x + 3)(x + 2).$$

Check by finding the product of $2x + 3$ and $x + 2$.

INSTRUCTOR'S RESOURCES

PTB Sec. 5.3 **ISM** Sec. 5.3
AB Sec. 5.3
TEST GENERATOR
WORLD WIDE WEB
www.LialAlgebra.com

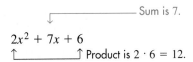

We could have written the middle term in the polynomial $2x^2 + 7x + 6$ as $7x = 4x + 3x$ to get

$$\begin{aligned} 2x^2 + 7x + 6 &= 2x^2 + 4x + 3x + 6 \\ &= 2x(x + 2) + 3(x + 2) \\ &= (x + 2)(2x + 3). \end{aligned}$$

Either result is correct.

CHALKBOARD EXAMPLE

Factor $3p^2 - 4p + 1$.

Answer: $(3p - 1)(p - 1)$

EXAMPLE 1 Factoring Trinomials by Grouping

Factor the trinomial.

(a) $6r^2 + r - 1$

We must find two integers with a product of $6(-1) = -6$ and a sum of 1.

Sum is 1.

$$6r^2 + r - 1 = 6r^2 + 1r - 1$$

Product is $6(-1) = -6$.

The integers are -2 and 3. We write the middle term, $+r$, as $-2r + 3r$, so that

$$6r^2 + r - 1 = 6r^2 - 2r + 3r - 1.$$

Factor by grouping on the right-hand side.

$$\begin{aligned} 6r^2 + r - 1 &= 6r^2 - 2r + 3r - 1 \\ &= 2r(3r - 1) + 1(3r - 1) && \text{The binomials must be the same.} \\ &= (3r - 1)(2r + 1) \end{aligned}$$

(b) $12z^2 - 5z - 2$

Look for two integers whose product is $12(-2) = -24$ and whose sum is -5. The required integers are 3 and -8, and

$$\begin{aligned} 12z^2 - 5z - 2 &= 12z^2 + 3z - 8z - 2 && -5z = 3z - 8z \\ &= 3z(4z + 1) - 2(4z + 1) && \text{Group terms and factor each group.} \\ &= (4z + 1)(3z - 2). && \text{Factor out } 4z + 1. \end{aligned}$$

(c) $10m^2 + mn - 3n^2$

Two integers whose product is $10(-3) = -30$ and whose sum is 1 are -5 and 6. Rewrite the trinomial with four terms.

$$\begin{aligned} 10m^2 + mn - 3n^2 &= 10m^2 - 5mn + 6mn - 3n^2 && mn = -5mn + 6mn \\ &= 5m(2m - n) + 3n(2m - n) && \text{Group terms and factor each group.} \\ &= (2m - n)(5m + 3n) && \text{Factor out the common factor.} \end{aligned}$$

OBJECTIVE 2 Factor trinomials using FOIL. The rest of this section shows an alternative method of factoring trinomials in which the coefficient of the squared term is not 1. This method uses trial and error.

To factor $2x^2 + 7x + 6$ (the trinomial factored at the beginning of this section) by trial and error, we must use FOIL backwards. We want to write $2x^2 + 7x + 6$ as the product of two binomials.

$$2x^2 + 7x + 6 = (\qquad)(\qquad)$$

The product of the two first terms of the binomials is $2x^2$. The possible factors of $2x^2$ are $2x$ and x or $-2x$ and $-x$. Since all terms of the trinomial are positive, we consider only positive factors. Thus, we have

$$2x^2 + 7x + 6 = (2x \qquad)(x \qquad).$$

The product of the two last terms, 6, can be factored as $1 \cdot 6$, $6 \cdot 1$, $2 \cdot 3$, or $3 \cdot 2$. Try each pair to find the pair that gives the correct middle term.

$(2x + 1)(x + 6)$ Incorrect

x
$12x$
$13x$ Add.

$(2x + 6)(x + 1)$ Incorrect

$6x$
$2x$
$8x$ Add.

Teaching Tip Understanding and using this concept will save lots of trial and check work.

Since $2x + 6 = 2(x + 3)$, the binomial $2x + 6$ has a common factor of 2, while $2x^2 + 7x + 6$ has no common factor other than 1. The product $(2x + 6)(x + 1)$ cannot be correct.

NOTE If the original polynomial has no common factor, then none of its binomial factors will either.

Now try the numbers 2 and 3 as factors of 6. Because of the common factor of 2 in $2x + 2$, $(2x + 2)(x + 3)$ will not work. Try $(2x + 3)(x + 2)$.

$(2x + 3)(x + 2) = 2x^2 + \mathbf{7x} + 6$ Correct

$3x$
$4x$
$7x$ Add.

Finally, we see that $2x^2 + 7x + 6$ factors as

$$2x^2 + 7x + 6 = (2x + 3)(x + 2).$$

Check by multiplying $2x + 3$ and $x + 2$.

CHALKBOARD EXAMPLE

Factor $6p^2 + 19p + 10$.

Answer: $(3p + 2)(2p + 5)$

EXAMPLE 2 Factoring a Trinomial with All Terms Positive Using FOIL

Factor $8p^2 + 14p + 5$.

The number 8 has several possible pairs of factors, but 5 has only 1 and 5 or -1 and -5. For this reason, it is easier to begin by considering the factors of 5. Ignore the negative factors since all coefficients in the trinomial are positive. If $8p^2 + 14p + 5$ can be factored, the factors will have the form

$$(\qquad + 5)(\qquad + 1).$$

The possible pairs of factors of $8p^2$ are $8p$ and p, or $4p$ and $2p$. Try various combinations, checking the middle term in each case.

$(8p + 5)(p + 1)$ Incorrect — $5p + 8p = 13p$ Add.

$(p + 5)(8p + 1)$ Incorrect — $40p + p = 41p$ Add.

$(4p + 5)(2p + 1)$ Correct — $10p + 4p = 14p$ Add.

Since the sum $14p$ is the correct middle term, the trinomial $8p^2 + 14p + 5$ factors as $(4p + 5)(2p + 1)$.

CHALKBOARD EXAMPLE

Factor $10m^2 - 23m + 12$.

Answer: $(2m - 3)(5m - 4)$

EXAMPLE 3 Factoring a Trinomial with a Negative Middle Term Using FOIL

Factor $6x^2 - 11x + 3$.

Since 3 has only 1 and 3 or -1 and -3 as factors, it is better here to begin by factoring 3. The last term of the trinomial $6x^2 - 11x + 3$ is positive and the middle term has a negative coefficient, so consider only negative factors. We need negative factors because the *product* of two negative factors is positive and their *sum* is negative, as required. Use -3 and -1 as factors of 3:

$$(\quad - 3)(\quad - 1).$$

The factors of $6x^2$ may be either $6x$ and x, or $2x$ and $3x$. Try $2x$ and $3x$.

$(2x - 3)(3x - 1)$ Correct — $-9x + (-2x) = -11x$ Add.

These factors give the correct middle term, so

$$6x^2 - 11x + 3 = (2x - 3)(3x - 1).$$

CHALKBOARD EXAMPLE

Factor $5p^2 + 13p - 6$.

Answer: $(5p - 2)(p + 3)$

EXAMPLE 4 Factoring a Trinomial with a Negative Last Term Using FOIL

Factor $8x^2 + 6x - 9$.

The integer 8 has several possible pairs of factors, as does -9. Since the last term is negative, one positive factor and one negative factor of -9 are needed. Since the coefficient of the middle term is small, it is wise to avoid large factors such as 8 or 9. Let us try 4 and 2 as factors of 8, and 3 and -3 as factors of -9, and check the middle term.

$(4x + 3)(2x - 3)$ Incorrect — $6x + (-12x) = -6x$ Add.

Now, try exchanging 3 and -3, since only the sign of the middle term is incorrect.

$$(4x - 3)(2x + 3) \quad \text{Correct}$$

$$\begin{array}{r} -6x \\ 12x \\ \hline 6x \end{array} \quad \text{Add.}$$

This time we got the correct middle term, so

$$8x^2 + 6x - 9 = (4x - 3)(2x + 3).$$

CHALKBOARD EXAMPLE

Factor $6m^2 + 11mn - 10n^2$.

Answer: $(3m - 2n)(2m + 5n)$

EXAMPLE 5 Factoring a Trinomial with Two Variables

Factor $12a^2 - ab - 20b^2$.

There are several pairs of factors of $12a^2$, including $12a$ and a, $6a$ and $2a$, and $3a$ and $4a$, just as there are many pairs of factors of $-20b^2$, including $-20b$ and b, $10b$ and $-2b$, $-10b$ and $2b$, $4b$ and $-5b$, and $-4b$ and $5b$. Once again, since the desired middle term is small, avoid the larger factors. Try the factors $6a$ and $2a$ and $4b$ and $-5b$.

$$(6a + 4b)(2a - 5b)$$

This cannot be correct, as mentioned before, since $6a + 4b$ has a common factor while the given trinomial has none. Try $3a$ and $4a$ with $4b$ and $-5b$.

$$(3a + 4b)(4a - 5b) = 12a^2 + ab - 20b^2 \quad \text{Incorrect}$$

Here the middle term has the wrong sign, so change the signs in the factors.

$$(3a - 4b)(4a + 5b) = 12a^2 - ab - 20b^2 \quad \text{Correct}$$

CHALKBOARD EXAMPLE

Factor $15y^3 + 55y^2 + 30y$.

Answer: $5y(3y + 2)(y + 3)$

EXAMPLE 6 Factoring a Trinomial with a Common Factor

Factor $28x^5 - 58x^4 - 30x^3$.

First factor out the greatest common factor, $2x^3$.

$$28x^5 - 58x^4 - 30x^3 = 2x^3(14x^2 - 29x - 15)$$

Now try to factor $14x^2 - 29x - 15$. Try $7x$ and $2x$ as factors of $14x^2$ and -3 and 5 as factors of -15.

$$(7x - 3)(2x + 5) = 14x^2 + 29x - 15 \quad \text{Incorrect}$$

The middle term differs only in sign, so change the signs in the two factors.

$$(7x + 3)(2x - 5) = 14x^2 - 29x - 15 \quad \text{Correct}$$

Finally, the factored form of $28x^5 - 58x^4 - 30x^3$ is

$$28x^5 - 58x^4 - 30x^3 = 2x^3(7x + 3)(2x - 5).$$

CAUTION Remember to include the common factor in the final result.

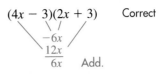

CHALKBOARD EXAMPLE

Factor $-4x^2 + 2x + 30$.

Answer: $-2(2x + 5)(x - 3)$

EXAMPLE 7 **Factoring a Trinomial with a Negative Common Factor**

Factor $-24a^3 - 42a^2 + 45a$.

The common factor could be $3a$ or $-3a$. If we factor out $-3a$, the first term of the trinomial factor will be positive, which makes it easier to factor.

$$-24a^3 - 42a^2 + 45a = -3a(8a^2 + 14a - 15) \quad \text{Factor out the greatest common factor.}$$
$$= -3a(4a - 3)(2a + 5) \quad \text{Use trial and error.}$$

5.3 EXERCISES

In Exercises 1–6 complete the steps in factoring the trinomial $2x^2 + x - 21$ by grouping.

1. Find the product of _____ and _____.

2. Find factors of _____ that have a sum of _____.

3. Write the middle term x as _____ + _____.

4. Factor the polynomial _____ by grouping as (_____) + (_____).

5. Factor out the common factor of _____.

6. The factored form is _____.

Fill in the blanks in Exercises 7–10.

7. $6a^2 + 7ab - 20b^2 = (3a$ _____)(_____ $+ 5b)$

8. $9m^2 - 3mn - 2n^2 = (3m$ _____)(_____ $- 2n)$

9. $3x^2 - 9x - 30 = 3(x^2$ _____) $= 3(x$ _____)(x _____)

10. $4z^3 - 10z^2 - 6z =$ _____($2z^2$ _____) $= 2z($_____ $- 3)(2z$ _____)

Decide which is the correct factored form of the given polynomial.

11. $4y^2 + 17y - 15$
(a) $(y + 5)(4y - 3)$
(b) $(2y - 5)(2y + 3)$

12. $12c^2 - 7c - 12$
(a) $(6c - 2)(2c + 6)$
(b) $(4c + 3)(3c - 4)$

13. $4k^2 + 13mk + 3m^2$
(a) $(4k + m)(k + 3m)$
(b) $(4k + 3m)(k + m)$

14. $2x^2 + 11x + 12$
(a) $(2x + 3)(x + 4)$
(b) $(2x + 4)(x + 3)$

15. For the polynomial $12x^2 + 7x - 12$, 2 is not a common factor. Explain why the binomial $2x - 6$, then, cannot be a factor of the polynomial.

16. Explain how the signs of the last terms of the two binomial factors of a trinomial are determined.

Factor completely. Use either method described in this section. See Examples 1–7.

17. $3a^2 + 10a + 7$

18. $7r^2 + 8r + 1$

19. $4r^2 + r - 3$

20. $4r^2 + 3r - 10$

21. $15m^2 + m - 2$

22. $6x^2 + x - 1$

23. $8m^2 - 10m - 3$

24. $12s^2 + 11s - 5$

25. $20x^2 + 11x - 3$

26. $20x^2 - 28x - 3$

27. $21m^2 + 13m + 2$

28. $38x^2 + 23x + 2$

29. $20y^2 + 39y - 11$

30. $10x^2 + 11x - 6$

31. $6b^2 + 7b + 2$

32. $6w^2 + 19w + 10$

1. 2; −21 **2.** −42; 1
3. $-6x$; $7x$ or $7x$; $-6x$
4. $2x^2 - 6x + 7x - 21$; $2x^2 - 6x$; $7x - 21$ **5.** $x - 3$
6. $(x - 3)(2x + 7)$ **7.** $-4b$; $2a$
8. $+ n$; $3m$
9. $-3x - 10$; $- 5$; $+2$ or $-3x - 10$; $+2$; $- 5$
10. $2z$; $-5z - 3$; z; $+ 1$
11. (a) **12.** (b) **13.** (a)
14. (a) **15.** The binomial $2x - 6$ cannot be a factor, because it has a common factor of 2, but the polynomial does not.
16. If the first term of the trinomial is positive and the last term is negative, the signs of the last two terms of the factors must be different. The signs are chosen so the middle term has the correct sign. If both the first and last terms of the trinomial are positive, then the signs of the last two terms of the factors must be the same—the sign on the middle term of the polynomial.
The order of the factors is irrelevant in Exercises 17–64.
17. $(3a + 7)(a + 1)$
18. $(7r + 1)(r + 1)$
19. $(4r - 3)(r + 1)$
20. $(4r - 5)(r + 2)$
21. $(3m - 1)(5m + 2)$
22. $(2x + 1)(3x - 1)$
23. $(4m + 1)(2m - 3)$
24. $(3s - 1)(4s + 5)$
25. $(4x + 3)(5x - 1)$
26. $(10x + 1)(2x - 3)$
27. $(3m + 1)(7m + 2)$
28. $(2x + 1)(19x + 2)$
29. $(4y - 1)(5y + 11)$
30. $(2x + 3)(5x - 2)$
31. $(2b + 1)(3b + 2)$
32. $(2w + 5)(3w + 2)$

33. $3(4x - 1)(2x - 3)$
34. $2(8b + 1)(3b - 5)$
35. $q(5m + 2)(8m - 3)$
36. $b(3a + 2)(5a + 4)$
37. $2m(m - 4)(m + 5)$
38. $3x(x - 2)(x + 6)$
39. $3n^2(5n - 3)(n - 2)$
40. $2a^2(3a + 2)(4a - 1)$
41. $3x^3(2x + 5)(3x - 5)$
42. $4z^3(8z + 3)(z - 1)$
43. $y^2(5x - 4)(3x + 1)$
44. $b^3(7a - 3)(2a + 3)$
45. $(3p + 4q)(4p - 3q)$
46. $(3m + 2n)(2m - 3n)$
47. $(5a + 2b)(5a + 3b)$
48. $(6x + y)(x - y)$
49. $(3a - 5b)(2a + b)$
50. $(5g + h)(5g - 2h)$
51. $m^4n(3m + 2n)(2m + n)$
52. $kq^4(6k + q)(2k - q)$
53. $(5 - x)(1 - x)$
54. $(7 + x)(1 + x)$
55. $(4 + 3x)(4 + x)$
56. $(2 + 7x)(9 + x)$
57. $-5x(2x + 7)(x - 4)$
58. $-6k(3k + 11)(k - 1)$
59. $-1(x + 7)(x - 3)$
60. $-1(x + 8)(x - 9)$
61. $-1(3x + 4)(x - 1)$
62. $-1(5x + 8)(x - 2)$
63. $-1(a + 2b)(2a + b)$
64. $-1(p - 4q)(3p - q)$
65. Yes, $(x + 7)(3 - x)$ is equivalent to $-1(x + 7)(x - 3)$, because $-1(x - 3) = -x + 3 = 3 - x$.
66. No. Multiplying out the first factored form gives $-x^2 + x + 72$, while the product of the factors of the second factored form is $x^2 - x - 72$, which is not an equivalent polynomial.
67. $(m + 1)^3(5q - 2)(5q + 1)$
68. $(y - 3)^2(6x + 1)(3x - 4)$
69. $(r + 3)^3(5x + 2y)(3x - 8y)$
70. $(k + 9)^7(2t + 5s)^2$
71. $-4, 4$
72. $-5, -1, 1, 5$
73. $-11, -7, 7, 11$
74. $-10, -6, 6, 10$

33. $24x^2 - 42x + 9$

34. $48b^2 - 74b - 10$

35. $40m^2q + mq - 6q$

36. $15a^2b + 22ab + 8b$

37. $2m^3 + 2m^2 - 40m$

38. $3x^3 + 12x^2 - 36x$

39. $15n^4 - 39n^3 + 18n^2$

40. $24a^4 + 10a^3 - 4a^2$

41. $18x^5 + 15x^4 - 75x^3$

42. $32z^5 - 20z^4 - 12z^3$

43. $15x^2y^2 - 7xy^2 - 4y^2$

44. $14a^2b^3 + 15ab^3 - 9b^3$

45. $12p^2 + 7pq - 12q^2$

46. $6m^2 - 5mn - 6n^2$

47. $25a^2 + 25ab + 6b^2$

48. $6x^2 - 5xy - y^2$

49. $6a^2 - 7ab - 5b^2$

50. $25g^2 - 5gh - 2h^2$

51. $6m^6n + 7m^5n^2 + 2m^4n^3$

52. $12k^3q^4 - 4k^2q^5 - kq^6$

53. $5 - 6x + x^2$

54. $7 + 8x + x^2$

55. $16 + 16x + 3x^2$

56. $18 + 65x + 7x^2$

57. $-10x^3 + 5x^2 + 140x$

58. $-18k^3 - 48k^2 + 66k$

If a trinomial has a negative coefficient for the squared term, such as $-2x^2 + 11x - 12$, it is usually easier to factor by first factoring out the common factor -1:

$$-2x^2 + 11x - 12 = -1(2x^2 - 11x + 12)$$
$$= -1(2x - 3)(x - 4).$$

Use this method to factor each trinomial. See Example 7.

59. $-x^2 - 4x + 21$

60. $-x^2 + x + 72$

61. $-3x^2 - x + 4$

62. $-5x^2 + 2x + 16$

63. $-2a^2 - 5ab - 2b^2$

64. $-3p^2 + 13pq - 4q^2$

65. The answer given in the back of the book for Exercise 59 is $-1(x + 7)(x - 3)$. Is $(x + 7)(3 - x)$ also a correct answer? Explain why or why not.

66. One answer for Exercise 60 is $-1(x + 8)(x - 9)$. Is $(-x - 8)(-x + 9)$ also a correct answer? Explain.

▲ *Factor each polynomial. Remember to factor out the greatest common factor as the first step.*

67. $25q^2(m + 1)^3 - 5q(m + 1)^3 - 2(m + 1)^3$

68. $18x^2(y - 3)^2 - 21x(y - 3)^2 - 4(y - 3)^2$

69. $15x^2(r + 3)^3 - 34xy(r + 3)^3 - 16y^2(r + 3)^3$

70. $4t^2(k + 9)^7 + 20ts(k + 9)^7 + 25s^2(k + 9)^7$

▲ *Find all integers k so that the trinomial can be factored using the methods of this section.* (Hint: *Try all possible factored forms with the given first and last terms. The coefficient of x will give the values of k.)*

71. $5x^2 + kx - 1$

72. $2c^2 + kc - 3$

73. $2m^2 + km + 5$

74. $3y^2 + ky + 3$

75. $5 \cdot 7$ 76. $(-5)(-7)$
77. The product of $3x - 4$ and $2x - 1$ is indeed $6x^2 - 11x + 4$.
78. The product of $4 - 3x$ and $1 - 2x$ is indeed $6x^2 - 11x + 4$.
79. The factors in Exercise 78 are the opposites of the factors in Exercise 77.
80. $(3 - 7t)(5 - 2t)$
81. $-a$; $-b$ 82. $-P$; $-Q$

RELATING CONCEPTS (EXERCISES 75–82)

One of the most common questions that beginning algebra students ask is this: "If my answer doesn't look exactly like the one given in the back of the book, is it necessarily incorrect?" Often there are several different equivalent forms of an answer that are all correct.

***Work Exercises 75–82 in order** to see how and why this is possible for factoring problems.*

75. Factor the integer 35 as the product of two prime numbers.

76. Factor the integer 35 as the product of the negatives of two prime numbers.

77. Verify the following factored form: $6x^2 - 11x + 4 = (3x - 4)(2x - 1)$.

78. Verify the following factored form: $6x^2 - 11x + 4 = (4 - 3x)(1 - 2x)$.

79. Compare the two valid factored forms in Exercises 77 and 78. How do the factors in each case compare?

80. Suppose you know that the correct factored form of a particular trinomial is $(7t - 3)(2t - 5)$. Based on your observations in Exercises 77–79, what is another valid factored form?

81. Look at your results in Exercises 75 and 76, and fill in the blanks: If an integer factors as the product of a and b, then it also factors as the product of _____ and _____.

82. Look at your results in Exercises 77 and 78 and fill in the blanks: If a trinomial factors as the product of the binomials P and Q, then it also factors as the product of the binomials _____ and _____.

Did you make the connection that the product of the negatives of two factors is the same as the product of the two factors?

5.4 Special Factoring Rules

OBJECTIVES

1. Factor the difference of two squares.
2. Factor a perfect square trinomial.
3. Factor the difference of two cubes.
4. Factor the sum of two cubes.

FOR EXTRA HELP

SSG Sec. 5.4
SSM Sec. 5.4
Pass the Test Software
InterAct Math Tutorial Software
Video 9

By reversing the rules for multiplying binomials that we learned in the last chapter, we get rules for factoring polynomials in certain forms.

OBJECTIVE 1 **Factor the difference of two squares.** Recall from the last chapter that

$$(x + y)(x - y) = x^2 - y^2.$$

Based on this product, we factor a **difference of two squares** as follows.

Difference of Two Squares

$$x^2 - y^2 = (x + y)(x - y)^*$$

*A pair of expressions like $x + y$ and $x - y$ are called *conjugates*.

JOURNAL WRITING ▲ CHALLENGING SCIENTIFIC CALCULATOR GRAPHING CALCULATOR

INSTRUCTOR'S RESOURCES

PTB Sec. 5.4 ISM Sec. 5.4
AB Sec. 5.4
TEST GENERATOR
WORLD WIDE WEB
www.LialAlgebra.com

CHALKBOARD EXAMPLE

Factor $p^2 - 100$.

Answer: $(p + 10)(p - 10)$

EXAMPLE 1 Factoring a Difference of Squares

Factor each difference of two squares.

(a) $x^2 - 49 = x^2 - 7^2 = (x + 7)(x - 7)$

(b) $y^2 - m^2 = (y + m)(y - m)$

(c) $z^2 - \frac{9}{16} = z^2 - \left(\frac{3}{4}\right)^2 = \left(z + \frac{3}{4}\right)\left(z - \frac{3}{4}\right)$

(d) $x^2 - 8$

Because 8 is not the square of an integer, this is a prime polynomial.

(e) $p^2 + 16$

The polynomial is not the *difference* of two squares. Using FOIL,

$$(p + 4)(p - 4) = p^2 - 16$$
$$(p - 4)(p - 4) = p^2 - 8p + 16$$

and

$$(p + 4)(p + 4) = p^2 + 8p + 16$$

so $p^2 + 16$ is a prime polynomial.

TEACHING TIP There are several general forms that students will need to recognize in this section. Covering the meaning of *difference of squares, sum of squares, difference of cubes,* and *sum of cubes* with several examples of each type will help students recognize these forms.

CAUTION As Example 1(e) suggests, after any common factor is removed, the sum of two squares cannot be factored.

CHALKBOARD EXAMPLE

Factor completely $50r^2 - 32$.

Answer: $2(5r + 4)(5r - 4)$

EXAMPLE 2 Factoring More Complex Differences of Squares

Factor completely.

(a) $9a^2 - 4b^2$

This is a difference of two squares because

$$9a^2 - 4b^2 = (3a)^2 - (2b)^2,$$

so $9a^2 - 4b^2 = (3a + 2b)(3a - 2b)$.

(b) $81y^2 - 36$

First factor out the common factor of 9.

$$81y^2 - 36 = 9(9y^2 - 4)$$
$$= 9(3y + 2)(3y - 2)$$

(c) $p^4 - 36 = (p^2)^2 - 6^2 = (p^2 + 6)(p^2 - 6)$

Neither $p^2 + 6$ nor $p^2 - 6$ can be factored further.

(d) $m^4 - 16 = (m^2)^2 - 4^2$

$= (m^2 + 4)(m^2 - 4)$ Difference of squares

$= (m^2 + 4)(m + 2)(m - 2)$ Difference of squares

CAUTION Remember to factor again when any of the *factors* is a difference of squares, as in Example 2(d).

OBJECTIVE 2 Factor a perfect square trinomial. The expressions 144, $4x^2$, and $81m^6$ are called *perfect squares,* since

$$144 = 12^2, \qquad 4x^2 = (2x)^2, \qquad \text{and} \qquad 81m^6 = (9m^3)^2.$$

A **perfect square trinomial** is a trinomial that is the square of a binomial. For example, $x^2 + 8x + 16$ is a perfect square trinomial since it is the square of the binomial $x + 4$:

$$x^2 + 8x + 16 = (x + 4)^2.$$

For a trinomial to be a perfect square, two of its terms must be perfect squares. For this reason, $16x^2 + 4x + 15$ cannot be a perfect square trinomial since only the term $16x^2$ is a perfect square.

On the other hand, even though two of the terms are perfect squares, the trinomial may not be a perfect square trinomial. For example, $x^2 + 6x + 36$ has two perfect square terms, but it is not a perfect square trinomial. (Try to find a binomial that can be squared to give $x^2 + 6x + 36$; it cannot be done.)

We can multiply to see that the square of a binomial gives the following perfect square trinomials.

Perfect Square Trinomial

$$x^2 + 2xy + y^2 = (x + y)^2$$
$$x^2 - 2xy + y^2 = (x - y)^2$$

The middle term of a perfect square trinomial is always twice the product of the two terms in the squared binomial. (This was shown in Section 4.4.) Use this to check any attempt to factor a trinomial that appears to be a perfect square.

While perfect square trinomials can be factored using the procedures of Sections 5.2 and 5.3, it is usually more efficient to recognize the pattern and factor accordingly. The following example illustrates this.

CHALKBOARD EXAMPLE

Factor $16a^2 + 56a + 49$.

Answer: $(4a + 7)^2$

EXAMPLE 3 Factoring a Perfect Square Trinomial

Factor the perfect square trinomial.

(a) $x^2 + 10x + 25$

The term x^2 is a perfect square and so is 25. Try to factor the trinomial as

$$x^2 + 10x + 25 = (x + 5)^2.$$

To check, take twice the product of the two terms in the squared binomial.

$$\text{Twice} \rightarrow 2 \cdot x \cdot 5 = 10x$$

First term of binomial ↑ ↑ Last term of binomial

Since $10x$ is the middle term of the trinomial, the trinomial is a perfect square and can be factored as $(x + 5)^2$.

(b) $x^2 - 22xz + 121z^2$

The first and last terms are perfect squares ($121 = 11^2$). Check to see whether the middle term of $x^2 - 22xz + 121z^2$ is twice the product of the first and last terms of the binomial $x - 11z$.

$$\text{Twice} \rightarrow 2 \cdot x \cdot 11z = 22xz$$

First term of binomial ↑ ↑ Last term of binomial

Since twice the product of the first and last terms of the binomial is the middle term, $x^2 - 22xz + 121z^2$ is a perfect square trinomial and

$$x^2 - 22xz + 121z^2 = (x - 11z)^2.$$

(c) $9m^2 - 24m + 16 = (3m)^2 - 2(3m)(4) + 4^2 = (3m - 4)^2$

Twice ↑ ↑ First term ↑ Last term

(d) $25y^2 + 20y + 16$

The first and last terms are perfect squares.

$$25y^2 = (5y)^2 \qquad \text{and} \qquad 16 = 4^2$$

Twice the product of the first and last terms of the binomial $5y + 4$ is

$$2 \cdot 5y \cdot 4 = 40y,$$

which is not the middle term of $25y^2 + 20y + 16$. This polynomial is not a perfect square. In fact, the polynomial cannot be factored even with the methods of Section 5.3; it is a prime polynomial.

NOTE The sign of the second term in the squared binomial is always the same as the sign of the middle term in the trinomial. Also, the first and last terms of a perfect square trinomial must be positive, since they are squares. For example, the polynomial $x^2 - 2x - 1$ cannot be a perfect square because the last term is negative.

OBJECTIVE 3 **Factor the difference of two cubes.** The difference of two squares was factored above; we can also factor the **difference of two cubes.** Use the following pattern.

Difference of Two Cubes

$$x^3 - y^3 = (x - y)(x^2 + xy + y^2)$$

This pattern *should be memorized.* Multiply on the right to see that the pattern gives the correct factors.

$$\begin{array}{rrrr}
 & x^2 & + xy & + y^2 \\
 & & x & - y \\
\hline
 & -x^2y & - xy^2 & - y^3 \\
x^3 & + x^2y & + xy^2 & \\
\hline
x^3 & & & - y^3
\end{array}$$

Notice the pattern of the terms in the factored form of $x^3 - y^3$.

- $x^3 - y^3 =$ (a binomial factor)(a trinomial factor)
- The binomial factor has the difference of the cube roots of the given terms.
- The terms in the trinomial factor are all positive.
- What you write in the binomial factor determines the trinomial factor:

First term squared + positive product of the terms + second term squared

$$x^3 - y^3 = (x - y)(x^2 + xy + y^2).$$

CAUTION

The polynomial $x^3 - y^3$ is not equivalent to $(x - y)^3$, because $(x - y)^3$ can also be written as

$$(x - y)^3 = (x - y)(x - y)(x - y)$$
$$= (x - y)(x^2 - 2xy + y^2)$$

but

$$x^3 - y^3 = (x - y)(x^2 + xy + y^2).$$

CHALKBOARD EXAMPLE

Factor $8k^3 - y^3$.

Answer:
$(2k - y)(4k^2 + 2ky + y^2)$

EXAMPLE 4 Factoring Differences of Cubes

Factor the following.

(a) $m^3 - 125$

Let $x = m$ and $y = 5$ in the pattern for the difference of two cubes.

$$x^3 - y^3 = (x - y)(x^2 + xy + y^2)$$
$$m^3 - 125 = m^3 - 5^3 = (m - 5)(m^2 + 5m + 5^2) \qquad \text{Let } x = m,\ y = 5.$$
$$= (m - 5)(m^2 + 5m + 25)$$

(b) $8p^3 - 27$

Since $8p^3 = (2p)^3$ and $27 = 3^3$, substitute into the rule using $2p$ for x and 3 for y.

$$8p^3 - 27 = (2p)^3 - 3^3$$
$$= (2p - 3)[(2p)^2 + (2p)3 + 3^2]$$
$$= (2p - 3)(4p^2 + 6p + 9)$$

(c) $4m^3 - 32 = 4(m^3 - 8)$

$$= 4(m^3 - 2^3)$$
$$= 4(m - 2)(m^2 + 2m + 4)$$

(d) $125t^3 - 216s^6 = (5t)^3 - (6s^2)^3$

$$= (5t - 6s^2)[(5t)^2 + (5t)(6s^2) + (6s^2)^2]$$
$$= (5t - 6s^2)(25t^2 + 30ts^2 + 36s^4)$$

CAUTION A common error in factoring the difference of two cubes, such as $x^3 - y^3 = (x - y)(x^2 + xy + y^2)$, is to try to factor $x^2 + xy + y^2$. It is easy to confuse this factor with a perfect square trinomial, $x^2 + 2xy + y^2$. Because there is no 2 in $x^2 + xy + y^2$, it is very unusual to be able to further factor an expression of the form $x^2 + xy + y^2$.

OBJECTIVE 4 **Factor the sum of two cubes.** A sum of two squares, such as $m^2 + 25$, cannot be factored using real numbers, but the **sum of two cubes** can be factored by the following pattern, *which should be memorized.*

Sum of Two Cubes

$$x^3 + y^3 = (x + y)(x^2 - xy + y^2)$$

Compare the pattern for the *sum* of two cubes with the pattern for the *difference* of two cubes. The only difference between them is the positive and negative signs.

$$x^3 - y^3 = (x - y)(x^2 + xy + y^2)$$

Positive; Same sign; Opposite sign

$$x^3 + y^3 = (x + y)(x^2 - xy + y^2)$$

Positive; Same sign; Opposite sign

Observing these relationships should help you to remember these patterns.

CHALKBOARD EXAMPLE

Factor $27m^3 + 125n^3$.

Answer: $(3m + 5n) \cdot (9m^2 - 15mn + 25n^2)$

EXAMPLE 5 **Factoring Sums of Cubes**

Factor.

(a) $k^3 + 27 = k^3 + 3^3$
$= (k + 3)(k^2 - 3k + 3^2)$
$= (k + 3)(k^2 - 3k + 9)$

(b) $8m^3 + 125 = (2m)^3 + 5^3$
$= (2m + 5)[(2m)^2 - (2m)(5) + 5^2]$
$= (2m + 5)(4m^2 - 10m + 25)$

(c) $1000a^6 + 27b^3 = (10a^2)^3 + (3b)^3$
$= (10a^2 + 3b)[(10a^2)^2 - (10a^2)(3b) + (3b)^2]$
$= (10a^2 + 3b)(100a^4 - 30a^2b + 9b^2)$

The methods of factoring discussed in this section are summarized here. All these rules should be memorized.

TEACHING TIP Throughout the exercises encourage students to use the correct phrase to describe each expression (e.g., "the sum of cubes").

Special Factorizations

Difference of two squares	$x^2 - y^2 = (x + y)(x - y)$
Perfect square trinomials	$x^2 + 2xy + y^2 = (x + y)^2$
	$x^2 - 2xy + y^2 = (x - y)^2$
Difference of two cubes	$x^3 - y^3 = (x - y)(x^2 + xy + y^2)$
Sum of two cubes	$x^3 + y^3 = (x + y)(x^2 - xy + y^2)$

Remember the *sum* of two *squares* can be factored only if the terms have a common factor.

5.4 EXERCISES

1. To help you factor the difference of squares, complete the following list of squares.

$1^2 =$ ____	$2^2 =$ ____	$3^2 =$ ____	$4^2 =$ ____	$5^2 =$ ____
$6^2 =$ ____	$7^2 =$ ____	$8^2 =$ ____	$9^2 =$ ____	$10^2 =$ ____
$11^2 =$ ____	$12^2 =$ ____	$13^2 =$ ____	$14^2 =$ ____	$15^2 =$ ____
$16^2 =$ ____	$17^2 =$ ____	$18^2 =$ ____	$19^2 =$ ____	$20^2 =$ ____

2. The following powers of x are all perfect squares: $x^2, x^4, x^6, x^8, x^{10}$. Based on this observation, we may make a conjecture (an educated guess) that if the power of a variable is divisible by ____ (with 0 remainder), then we have a perfect square.

3. To help you factor the sum or difference of cubes, complete the following list of cubes.

$1^3 =$ ____	$2^3 =$ ____	$3^3 =$ ____	$4^3 =$ ____	$5^3 =$ ____
$6^3 =$ ____	$7^3 =$ ____	$8^3 =$ ____	$9^3 =$ ____	$10^3 =$ ____

4. The following powers of x are all perfect cubes: $x^3, x^6, x^9, x^{12}, x^{15}$. Based on this observation, we may make a conjecture that if the power of a variable is divisible by ____ (with 0 remainder), then we have a perfect cube.

5. Identify the monomial as a perfect square, a perfect cube, both of these, or neither of these.
(a) $64x^6y^{12}$ **(b)** $125t^6$ **(c)** $49x^{12}$ **(d)** $81r^{10}$

6. What must be true for x^n to be both a perfect square and a perfect cube?

Factor each binomial completely. Use your answers in Exercises 1 and 2 as necessary. See Examples 1 and 2.

7. $y^2 - 25$ **8.** $t^2 - 16$ **9.** $9r^2 - 4$ **10.** $4x^2 - 9$

11. $36m^2 - \frac{16}{25}$ **12.** $100b^2 - \frac{4}{49}$ **13.** $36x^2 - 16$ **14.** $32a^2 - 8$

15. $196p^2 - 225$ **16.** $361q^2 - 400$ **17.** $16r^2 - 25a^2$ **18.** $49m^2 - 100p^2$

19. $100x^2 + 49$ **20.** $81w^2 + 16$ **21.** $p^4 - 49$ **22.** $r^4 - 25$

23. $x^4 - 1$ **24.** $y^4 - 16$ **25.** $p^4 - 256$ **26.** $16k^4 - 1$

Answers

1. 1; 4; 9; 16; 25; 36; 49; 64; 81; 100; 121; 144; 169; 196; 225; 256; 289; 324; 361; 400 2. 2
3. 1; 8; 27; 64; 125; 216; 343; 512; 729; 1000 4. 3 5. (a) both of these (b) a perfect cube (c) a perfect square (d) a perfect square 6. The exponent n must be divisible by 6 (with 0 remainder).
7. $(y + 5)(y - 5)$
8. $(t + 4)(t - 4)$
9. $(3r + 2)(3r - 2)$
10. $(2x + 3)(2x - 3)$
11. $\left(6m + \frac{4}{5}\right)\left(6m - \frac{4}{5}\right)$
12. $\left(10b + \frac{2}{7}\right)\left(10b - \frac{2}{7}\right)$
13. $4(3x + 2)(3x - 2)$
14. $8(2a + 1)(2a - 1)$
15. $(14p + 15)(14p - 15)$
16. $(19q + 20)(19q - 20)$
17. $(4r + 5a)(4r - 5a)$
18. $(7m + 10p)(7m - 10p)$
19. prime 20. prime
21. $(p^2 + 7)(p^2 - 7)$
22. $(r^2 + 5)(r^2 - 5)$
23. $(x^2 + 1)(x + 1)(x - 1)$
24. $(y^2 + 4)(y + 2)(y - 2)$
25. $(p^2 + 16)(p + 4)(p - 4)$
26. $(4k^2 + 1)(2k + 1)(2k - 1)$

27. When a student was directed to factor $x^4 - 81$ completely, his teacher did not give him full credit for the answer $(x^2 + 9)(x^2 - 9)$. The student argued that since his answer does indeed give $x^4 - 81$ when multiplied out, he should be given full credit. Was the teacher justified in her grading of this item? Why or why not?

28. The binomial $4x^2 + 16$ is a sum of two squares that *can* be factored. How is this binomial factored? When can the sum of two squares be factored?

Factor each trinomial completely. It may be necessary to factor out the greatest common factor first. See Example 3.

29. $w^2 + 2w + 1$
30. $p^2 + 4p + 4$
31. $x^2 - 8x + 16$
32. $x^2 - 10x + 25$
33. $t^2 + t + \frac{1}{4}$
34. $m^2 + \frac{2}{3}m + \frac{1}{9}$
35. $x^2 - 1.0x + .25$
36. $y^2 - 1.4y + .49$
37. $2x^2 + 24x + 72$
38. $3y^2 - 48y + 192$
39. $16x^2 - 40x + 25$
40. $36y^2 - 60y + 25$
41. $49x^2 - 28xy + 4y^2$
42. $4z^2 - 12zw + 9w^2$
43. $64x^2 + 48xy + 9y^2$
44. $9t^2 + 24tr + 16r^2$
45. $-50h^2 + 40hy - 8y^2$
46. $-18x^2 - 48xy - 32y^2$

Factor each binomial completely. Use your answers in Exercises 3 and 4 as necessary. See Examples 4 and 5.

47. $a^3 + 1$
48. $m^3 + 8$
49. $a^3 - 1$
50. $m^3 - 8$
51. $p^3 + q^3$
52. $x^3 + z^3$
53. $27x^3 - 1$
54. $64y^3 - 27$
55. $8p^3 + 729q^3$
56. $64x^3 + 125y^3$
57. $y^3 - 8x^3$
58. $w^3 - 216z^3$
59. $27a^3 - 64b^3$
60. $125m^3 - 8p^3$
61. $125t^3 + 8s^3$
62. $27r^3 + 1000s^3$

63. State and give the name of each of the five special factoring rules. For each rule, give the numbers of three exercises from this exercise set that are examples.

RELATING CONCEPTS (EXERCISES 64–67)

We have seen that multiplication and factoring are reverse processes. We know that multiplication and division are also related: To check a division problem, we multiply the quotient by the divisor to get the dividend.

To see how factoring and division are related, ***work Exercises 64–67 in order.***

64. Factor $10x^2 + 11x - 6$.
65. Use long division to divide $10x^2 + 11x - 6$ by $2x + 3$.
66. Could we have predicted the result in Exercise 65 from the result in Exercise 64? Explain.
67. Divide $x^3 - 1$ by $x - 1$. Use your answer to factor $x^3 - 1$.

Did you make the connection that the quotient in a division problem is a factor of the dividend?

▲ *Extend the methods of factoring presented so far in this chapter to factor each polynomial completely.*

68. $(m + n)^2 - (m - n)^2$
69. $(a - b)^3 - (a + b)^3$
70. $m^2 - p^2 + 2m + 2p$
71. $3r - 3k + 3r^2 - 3k^2$

Answers

27. The teacher was justified, because it was not factored completely; $x^2 - 9$ can be factored as $(x + 3)(x - 3)$. The complete factored form is $(x^2 + 9)(x + 3)(x - 3)$. **28.** $4x^2 + 16$ is factored as $4(x^2 + 4)$. The sum of two squares can be factored if the terms have a common factor. **29.** $(w + 1)^2$ **30.** $(p + 2)^2$ **31.** $(x - 4)^2$ **32.** $(x - 5)^2$ **33.** $\left(t + \frac{1}{2}\right)^2$ **34.** $\left(m + \frac{1}{3}\right)^2$ **35.** $(x - .5)^2$ **36.** $(y - .7)^2$ **37.** $2(x + 6)^2$ **38.** $3(y - 8)^2$ **39.** $(4x - 5)^2$ **40.** $(6y - 5)^2$ **41.** $(7x - 2y)^2$ **42.** $(2z - 3w)^2$ **43.** $(8x + 3y)^2$ **44.** $(3t + 4r)^2$ **45.** $-2(5h - 2y)^2$ **46.** $-2(3x + 4y)^2$ **47.** $(a + 1)(a^2 - a + 1)$ **48.** $(m + 2)(m^2 - 2m + 4)$ **49.** $(a - 1)(a^2 + a + 1)$ **50.** $(m - 2)(m^2 + 2m + 4)$ **51.** $(p + q)(p^2 - pq + q^2)$ **52.** $(x + z)(x^2 - xz + z^2)$ **53.** $(3x - 1)(9x^2 + 3x + 1)$ **54.** $(4y - 3)(16y^2 + 12y + 9)$ **55.** $(2p + 9q)(4p^2 - 18pq + 81q^2)$ **56.** $(4x + 5y)(16x^2 - 20xy + 25y^2)$ **57.** $(y - 2x)(y^2 + 2yx + 4x^2)$ **58.** $(w - 6z)(w^2 + 6wz + 36z^2)$ **59.** $(3a - 4b)(9a^2 + 12ab + 16b^2)$ **60.** $(5m - 2p)(25m^2 + 10mp + 4p^2)$ **61.** $(5t + 2s)(25t^2 - 10ts + 4s^2)$ **62.** $(3r + 10s)(9r^2 - 30rs + 100s^2)$ **63.** $x^2 - y^2 = (x + y)(x - y)$ Difference of Two Squares; Examples are Exercises 7–18 and 21–26.

$x^2 + 2xy + y^2 = (x + y)^2$ Perfect Square Trinomial; Examples are Exercises 29, 30, 33, 34, 37, 43, 44.

$x^2 - 2xy + y^2 = (x - y)^2$ Perfect Square Trinomial; Examples are Exercises 31, 32, 35, 36, 38–42, 45, 46.

$x^3 - y^3 = (x - y)(x^2 + xy + y^2)$ Difference of Two Cubes; Examples are Exercises 49, 50, 53, 54, 57–60.

$x^3 + y^3 = (x + y)(x^2 - xy + y^2)$ Sum of Two Cubes; Examples are Exercises 47, 48, 51, 52, 55, 56, 61, 62.

64. $(5x - 2)(2x + 3)$ **65.** $5x - 2$ **66.** Yes. If $10x^2 + 11x - 6$ factors as $(5x - 2)(2x + 3)$, then when $10x^2 + 11x - 6$ is divided by $2x + 3$, the quotient should be $5x - 2$.

JOURNAL   WRITING   ▲ CHALLENGING   SCIENTIFIC CALCULATOR   GRAPHING CALCULATOR

67. $x^2 + x + 1$; $(x - 1)(x^2 + x + 1)$
68. $4mn$ **69.** $-2b(3a^2 + b^2)$
70. $(m - p + 2)(m + p)$
71. $3(r - k)(1 + r + k)$
72. $(x + 1)^2 - 2^2$ or $(x + 1)^2 - 4$
73. $[(x + 1) - 2][(x + 1) + 2]$
74. $(x - 1)(x + 3)$
75. $x^2 + 2x - 3$
76. $(x - 1)(x + 3)$
77. The results are the same. (Preference is a matter of individual choice.)

RELATING CONCEPTS (EXERCISES 72–77)

It is possible to use two different methods of factoring a particular polynomial and get the same answer. To see this, ***work Exercises 72–77 in order,*** *referring to the polynomial*

$$(x^2 + 2x + 1) - 4.$$

72. Notice that the first three terms of the polynomial are grouped. Factor that trinomial, which is a perfect square trinomial, to obtain a difference of squares.

73. Factor the difference of squares obtained in Exercise 72.

74. Combine the constant terms within the factors from Exercise 73 to obtain the simplest forms of the factors.

75. Now start over and write the original polynomial $(x^2 + 2x + 1) - 4$ as a trinomial, by combining the constant terms.

76. Factor the trinomial obtained in Exercise 75 using the methods of Sections 5.2 and 5.3.

77. How do your results compare in Exercises 74 and 76? Which method do you prefer?

Did you make the connection that factoring a polynomial by two different methods should give the same answer?

78. No, it is not a perfect square because the middle term should be $30y$, not $14y$.
79. 10 **80.** 9 **81.** 9
82. 60

78. In the polynomial $9y^2 + 14y + 25$, the first and last terms are perfect squares. Can the polynomial be factored? If it can, factor it. If it cannot, explain why it is not a perfect square trinomial.

▲ *Find the value of the indicated variable.* (Hint: *Perform the squaring on the right side, then solve the equation using the steps given in Chapter 2.*)

79. Find a value of b so that $x^2 + bx + 25 = (x + 5)^2$.

80. For what value of c is $4m^2 - 12m + c = (2m - 3)^2$?

81. Find a so that $ay^2 - 12y + 4 = (3y - 2)^2$.

82. Find b so that $100a^2 + ba + 9 = (10a + 3)^2$.

SUMMARY Exercises on Factoring

These mixed exercises are included to give you practice in selecting an appropriate method for factoring a particular polynomial. As you factor a polynomial, ask yourself these questions to decide on a suitable factoring technique.

Factoring a Polynomial

Step 1 Is there a common factor?

Step 2 How many terms are in the polynomial?
Four terms Can the polynomial be factored by grouping?
Three terms Is it a perfect square trinomial? If the trinomial is not a perfect square, check to see whether the coefficient of the squared term is 1. If so, use the method of Section 5.2. If the coefficient of the squared term is not 1, use the general factoring methods of Section 5.3.
Two terms Check to see whether it is either the difference of two squares or the sum or difference of two cubes.

Step 3 Can any factors be factored further?

SUMMARY EXERCISES

Factor each polynomial completely.

1. $a^2 - 4a - 12$

2. $a^2 + 17a + 72$

3. $6y^2 - 6y - 12$

4. $7y^6 + 14y^5 - 168y^4$

5. $6a + 12b + 18c$

6. $m^2 - 3mn - 4n^2$

7. $p^2 - 17p + 66$

8. $z^2 - 6z + 7z - 42$

9. $10z^2 - 7z - 6$

10. $2m^2 - 10m - 48$

11. $m^2 - n^2 + 5m - 5n$

12. $15y + 5$

13. $8a^5 - 8a^4 - 48a^3$

14. $8k^2 - 10k - 3$

15. $z^2 - 3za - 10a^2$

16. $50z^2 - 100$

17. $x^2 - 4x - 5x + 20$

18. $100n^2r^2 + 30nr^3 - 50n^2r$

19. $6n^2 - 19n + 10$

20. $9y^2 + 12y - 5$

21. $16x + 20$

22. $m^2 + 2m - 15$

23. $6y^2 - 5y - 4$

24. $m^2 - 81$

25. $6z^2 + 31z + 5$

26. $5z^2 + 24z - 5 + 3z + 15$

27. $4k^2 - 12k + 9$

28. $8p^2 + 23p - 3$

29. $54m^2 - 24z^2$

30. $8m^2 - 2m - 3$

31. $3k^2 + 4k - 4$

32. $45a^3b^5 - 60a^4b^2 + 75a^6b^4$

33. $14k^3 + 7k^2 - 70k$

34. $5 + r - 5s - rs$

35. $y^4 - 16$

36. $20y^5 - 30y^4$

37. $8m - 16m^2$

38. $k^2 - 16$

39. $z^3 - 8$

40. $y^2 - y - 56$

41. $k^2 + 9$

42. $27p^{10} - 45p^9 - 252p^8$

43. $32m^9 + 16m^5 + 24m^3$

44. $8m^3 + 125$

45. $16r^2 + 24rm + 9m^2$

46. $z^2 - 12z + 36$

47. $15h^2 + 11hg - 14g^2$

48. $5z^3 - 45z^2 + 70z$

49. $k^2 - 11k + 30$

50. $64p^2 - 100m^2$

51. $3k^3 - 12k^2 - 15k$

52. $y^2 - 4yk - 12k^2$

53. $1000p^3 + 27$

54. $64r^3 - 343$

55. $6 + 3m + 2p + mp$

56. $2m^2 + 7mn - 15n^2$

57. $16z^2 - 8z + 1$

58. $125m^4 - 400m^3n + 195m^2n^2$

59. $108m^2 - 36m + 3$

60. $100a^2 - 81y^2$

61. $64m^2 - 40mn + 25n^2$

62. $4y^2 - 25$

63. $32z^3 + 56z^2 - 16z$

64. $10m^2 + 25m - 60$

65. $20 + 5m + 12n + 3mn$

66. $4 - 2q - 6p + 3pq$

67. $6a^2 + 10a - 4$

68. $36y^6 - 42y^5 - 120y^4$

69. $a^3 - b^3 + 2a - 2b$

70. $16k^2 - 48k + 36$

71. $64m^2 - 80mn + 25n^2$

72. $72y^3z^2 + 12y^2 - 24y^4z^2$

73. $8k^2 - 2kh - 3h^2$

74. $2a^2 - 7a - 30$

75. $(m + 1)^3 + 1$

76. $8a^3 - 27$

77. $10y^2 - 7yz - 6z^2$

78. $m^2 - 4m + 4$

79. $8a^2 + 23ab - 3b^2$

80. $a^4 - 625$

1. $(a - 6)(a + 2)$
2. $(a + 8)(a + 9)$
3. $6(y - 2)(y + 1)$
4. $7y^4(y + 6)(y - 4)$
5. $6(a + 2b + 3c)$
6. $(m - 4n)(m + n)$
7. $(p - 11)(p - 6)$
8. $(z + 7)(z - 6)$
9. $(5z - 6)(2z + 1)$
10. $2(m - 8)(m + 3)$
11. $(m + n + 5)(m - n)$
12. $5(3y + 1)$
13. $8a^3(a - 3)(a + 2)$
14. $(4k + 1)(2k - 3)$
15. $(z - 5a)(z + 2a)$
16. $50(z^2 - 2)$
17. $(x - 5)(x - 4)$
18. $10nr(10nr + 3r^2 - 5n)$
19. $(3n - 2)(2n - 5)$
20. $(3y - 1)(3y + 5)$
21. $4(4x + 5)$
22. $(m + 5)(m - 3)$
23. $(3y - 4)(2y + 1)$
24. $(m + 9)(m - 9)$
25. $(6z + 1)(z + 5)$
26. $(5z + 2)(z + 5)$
27. $(2k - 3)^2$
28. $(8p - 1)(p + 3)$
29. $6(3m + 2z)(3m - 2z)$
30. $(4m - 3)(2m + 1)$
31. $(3k - 2)(k + 2)$
32. $15a^3b^2(3b^3 - 4a + 5a^3b^2)$
33. $7k(2k + 5)(k - 2)$
34. $(5 + r)(1 - s)$
35. $(y^2 + 4)(y + 2)(y - 2)$
36. $10y^4(2y - 3)$
37. $8m(1 - 2m)$
38. $(k + 4)(k - 4)$
39. $(z - 2)(z^2 + 2z + 4)$
40. $(y - 8)(y + 7)$ **41.** prime
42. $9p^8(3p + 7)(p - 4)$
43. $8m^3(4m^6 + 2m^2 + 3)$
44. $(2m + 5)(4m^2 - 10m + 25)$
45. $(4r + 3m)^2$ **46.** $(z - 6)^2$
47. $(5h + 7g)(3h - 2g)$
48. $5z(z - 7)(z - 2)$
49. $(k - 5)(k - 6)$
50. $4(4p - 5m)(4p + 5m)$
51. $3k(k - 5)(k + 1)$
52. $(y - 6k)(y + 2k)$
53. $(10p + 3)(100p^2 - 30p + 9)$
54. $(4r - 7)(16r^2 + 28r + 49)$
55. $(2 + m)(3 + p)$
56. $(2m - 3n)(m + 5n)$
57. $(4z - 1)^2$
58. $5m^2(5m - 3n)(5m - 13n)$
59. $3(6m - 1)^2$

60. $(10a + 9y)(10a - 9y)$
61. prime 62. $(2y + 5)(2y - 5)$
63. $8z(4z - 1)(z + 2)$
64. $5(2m - 3)(m + 4)$
65. $(4 + m)(5 + 3n)$
66. $(2 - q)(2 - 3p)$
67. $2(3a - 1)(a + 2)$
68. $6y^4(3y + 4)(2y - 5)$
69. $(a - b)(a^2 + ab + b^2 + 2)$
70. $4(2k - 3)^2$ 71. $(8m - 5n)^2$
72. $12y^2(6yz^2 + 1 - 2y^2z^2)$
73. $(4k - 3h)(2k + h)$
74. $(2a + 5)(a - 6)$
75. $(m + 2)(m^2 + m + 1)$
76. $(2a - 3)(4a^2 + 6a + 9)$
77. $(5y - 6z)(2y + z)$
78. $(m - 2)^2$
79. $(8a - b)(a + 3b)$
80. $(a^2 + 25)(a + 5)(a - 5)$
81. $(x^3 - 1)(x^3 + 1)$
82. $(x - 1)(x^2 + x + 1)(x + 1) \cdot (x^2 - x + 1)$
83. $(x^2 - 1)(x^4 + x^2 + 1)$
84. $(x - 1)(x + 1)(x^4 + x^2 + 1)$
85. The result in Exercise 82 is completely factored.
86. Show that $x^4 + x^2 + 1 = (x^2 + x + 1)(x^2 - x + 1)$.
87. difference of squares
88. $(x - 3)(x^2 + 3x + 9)(x + 3) \cdot (x^2 - 3x + 9)$

RELATING CONCEPTS (EXERCISES 81–88)

A binomial may be *both* a difference of squares *and* a difference of cubes. One example of such a binomial is $x^6 - 1$. Using the techniques of Section 5.4, one factoring method will give the complete factored form, while the other will not.

Work Exercises 81–88 in order *to determine the method to use if you have to make such a decision.*

81. Factor $x^6 - 1$ as the difference of two squares.

82. The factored form obtained in Exercise 81 consists of a difference of cubes multiplied by a sum of cubes. Factor each binomial further.

83. Now start over and factor $x^6 - 1$ as the difference of two cubes.

84. The factored form obtained in Exercise 83 consists of a binomial which is a difference of squares and a trinomial. Factor the binomial further.

85. Compare your results in Exercises 82 and 84. Which one of these is the completely factored form?

86. Verify that the trinomial in the factored form in Exercise 84 is the product of the two trinomials in the factored form in Exercise 82.

87. Use the results of Exercises 81–86 to complete the following statement: In general, if I must choose between factoring first using the method for difference of squares or the method for difference of cubes, I should choose the ____________ method to eventually obtain the complete factored form.

88. Find the *complete* factored form of $x^6 - 729$ using the knowledge you have gained in Exercises 81–87.

Did you make the connection that when it is possible to factor a polynomial as the difference of squares or the difference of cubes, only one method gives a complete factored form?

5.5 Solving Quadratic Equations by Factoring

OBJECTIVES

1 Solve quadratic equations by factoring.

2 Solve other equations by factoring.

FOR EXTRA HELP

SSG Sec. 5.5
SSM Sec. 5.5
Pass the Test Software
InterAct Math Tutorial Software
Video 9

TEACHING TIP Review the distinction between expressions and equations.

In this section we introduce *quadratic equations,* equations that contain a squared term and no terms of higher degree.

Quadratic Equation

An equation that can be written in the form

$$ax^2 + bx + c = 0,$$

where a, b, and c are real numbers, with $a \neq 0$, is a **quadratic equation.**

The parabolas graphed in Section 4.1 are graphs of quadratic equations with $b = 0$, or with $b = 0$ and $c = 0$. The form $ax^2 + bx + c = 0$ is the **standard form** of a quadratic equation. For example,

$$x^2 + 5x + 6 = 0, \qquad 2x^2 - 5x = 3, \qquad \text{and} \qquad y^2 = 4$$

are all quadratic equations but only $x^2 + 5x + 6 = 0$ is in standard form.

OBJECTIVE 1 Solve quadratic equations by factoring. We use the **zero-factor property** to solve a quadratic equation by factoring.

JOURNAL WRITING ▲ CHALLENGING SCIENTIFIC CALCULATOR GRAPHING CALCULATOR

INSTRUCTOR'S RESOURCES

PTB Sec. 5.5 ISM Sec. 5.5
AB Sec. 5.5
TEST GENERATOR
WORLD WIDE WEB
www.LialAlgebra.com

Zero-Factor Property

If a and b are real numbers and **if $ab = 0$, then $a = 0$ or $b = 0$.**

In other words, if the product of two numbers is 0, then at least one of the numbers must be 0. This means that one number *must* be 0, but both *may be* 0.

CHALKBOARD EXAMPLE

Solve $(5x + 7)(2x + 3) = 0$.

Answer: $\left\{-\frac{7}{5}, -\frac{3}{2}\right\}$

EXAMPLE 1 Using the Zero-Factor Property

Solve the equation $(x + 3)(2x - 1) = 0$.

The product $(x + 3)(2x - 1)$ is equal to 0. By the zero-factor property, the only way that the product of these two factors can be 0 is if at least one of the factors is 0. Therefore, either $x + 3 = 0$ or $2x - 1 = 0$. Solve each of these two linear equations as in Chapter 2.

$$x + 3 = 0 \quad \text{or} \quad 2x - 1 = 0$$

$$x = -3 \quad \text{or} \quad 2x = 1 \qquad \text{Add 1 to both sides.}$$

$$x = \frac{1}{2} \qquad \text{Divide by 2.}$$

Since both of these equations have a solution, the equation $(x + 3)(2x - 1) = 0$ has two solutions, -3 and $\frac{1}{2}$. Check these answers by substituting -3 for x in the original equation. Then start over and substitute $\frac{1}{2}$ for x.

If $x = -3$, then

$$(-3 + 3)[2(-3) - 1] = 0 \quad ?$$

$$0(-7) = 0 \quad ?$$

$$0 = 0. \qquad \text{True}$$

If $x = \frac{1}{2}$, then

$$\left(\frac{1}{2} + 3\right)\left(2 \cdot \frac{1}{2} - 1\right) = 0 \quad ?$$

$$\frac{7}{2}(1 - 1) = 0 \quad ?$$

$$\frac{7}{2} \cdot 0 = 0 \quad ?$$

$$0 = 0. \qquad \text{True}$$

Both -3 and $\frac{1}{2}$ produce true statements, so the solution set is $\left\{-3, \frac{1}{2}\right\}$.

NOTE The word "or" as used in Example 1 means "one or the other or both."

In Example 1 the equation to be solved was presented with the polynomial in factored form. If the polynomial in an equation is not already factored, first make sure that the equation is in standard form. Then factor the polynomial.

CHALKBOARD EXAMPLE

Solve $r^2 = 8 - 2r$.

Answer: $\{-4, 2\}$

EXAMPLE 2 Solving a Quadratic Equation Not in Standard Form

Solve the equation $x^2 - 5x = -6$.

First, rewrite the equation with all terms on one side by adding 6 to both sides.

$$x^2 - 5x = -6$$

$$x^2 - 5x + 6 = 0 \qquad \text{Add 6.}$$

Now factor $x^2 - 5x + 6$. Find two numbers whose product is 6 and whose sum is -5. These two numbers are -2 and -3, so the equation becomes

$$(x - 2)(x - 3) = 0. \quad \text{Factor.}$$
$$x - 2 = 0 \quad \text{or} \quad x - 3 = 0 \quad \text{Zero-factor property}$$
$$x = 2 \quad \text{or} \quad x = 3 \quad \text{Solve each equation.}$$

Check both solutions by substituting first 2 and then 3 for x in the original equation. The solution set is $\{2, 3\}$.

In summary, follow these steps to solve quadratic equations by factoring.

Solving a Quadratic Equation by Factoring

Step 1 **Write in standard form.** Write the equation in standard form: all terms on one side of the equals sign, with 0 on the other side.

Step 2 **Factor.** Factor completely.

Step 3 **Use the zero-factor property.** Set each factor with a variable equal to 0, and solve the resulting equations.

Step 4 **Check.** Check each solution in the original equation.

TEACHING TIP Distinguish between solving linear and quadratic equations.

NOTE Not all quadratic equations can be solved by factoring. A more general method for solving such equations is given in Chapter 10.

CHALKBOARD EXAMPLE

Solve $3m^2 - 9m = 30$.

Answer: $\{5, -2\}$

EXAMPLE 3 Solving a Quadratic Equation with a Common Factor

Solve $4p^2 + 40 = 26p$.

Subtract $26p$ from each side and write in descending powers to get

$$4p^2 - 26p + 40 = 0.$$
$$2(2p^2 - 13p + 20) = 0 \quad \text{Factor out 2.}$$
$$2(2p - 5)(p - 4) = 0 \quad \text{Factor the trinomial.}$$
$$2 = 0 \quad \text{or} \quad 2p - 5 = 0 \quad \text{or} \quad p - 4 = 0 \quad \text{Zero-factor property}$$

The equation $2 = 0$ has no solution. Solve the equation in the middle by first adding 5 on both sides of the equation. Then divide both sides by 2. Solve the equation on the right by adding 4 to both sides.

$$2p - 5 = 0 \quad \text{or} \quad p - 4 = 0$$
$$2p = 5 \quad \text{or} \quad p = 4$$
$$p = \frac{5}{2}$$

The solutions of $4p^2 + 40 = 26p$ are $\frac{5}{2}$ and 4; check them by substituting in the original equation to see that the solution set is $\left\{\frac{5}{2}, 4\right\}$.

CAUTION A common error is to include 2 as a solution in Example 3. Only *variable* factors lead to solutions.

CHALKBOARD EXAMPLE

Solve $49m^2 - 9 = 0$.

Answer: $\left\{\frac{3}{7}, -\frac{3}{7}\right\}$

EXAMPLE 4 Solving Special Quadratic Equations

Solve the equations.

(a) $16m^2 - 25 = 0$

Factor the left-hand side of the equation as the difference of two squares.

$$(4m + 5)(4m - 5) = 0$$

$$4m + 5 = 0 \quad \text{or} \quad 4m - 5 = 0 \qquad \text{Zero-factor property}$$

$$4m = -5 \quad \text{or} \quad 4m = 5 \qquad \text{Solve each equation.}$$

$$m = -\frac{5}{4} \quad \text{or} \quad m = \frac{5}{4}$$

Check the two solutions, $-\frac{5}{4}$ and $\frac{5}{4}$, in the original equation. The solution set is $\left\{-\frac{5}{4}, \frac{5}{4}\right\}$.

(b) $y^2 = 2y$

First write the equation in standard form.

$$y^2 - 2y = 0 \qquad \text{Standard form}$$

$$y(y - 2) = 0 \qquad \text{Factor.}$$

$$y = 0 \quad \text{or} \quad y - 2 = 0 \qquad \text{Zero-factor property}$$

$$y = 2 \qquad \text{Solve.}$$

The solution set is $\{0, 2\}$.

(c) $k(2k + 1) = 3$

$$k(2k + 1) = 3$$

$$2k^2 + k = 3 \qquad \text{Distributive property}$$

$$2k^2 + k - 3 = 0 \qquad \text{Subtract 3.}$$

$$(k - 1)(2k + 3) = 0 \qquad \text{Factor.}$$

$$k - 1 = 0 \quad \text{or} \quad 2k + 3 = 0 \qquad \text{Zero-factor property}$$

$$k = 1 \quad \text{or} \quad 2k = -3$$

$$k = -\frac{3}{2}$$

The solution set is $\left\{1, -\frac{3}{2}\right\}$.

CAUTION

In Example 4(b) it is tempting to begin by dividing both sides of the equation by y to get $y = 2$. Note, however, that the other solution, 0, is not found by this method.

In Example 4(c) we could not use the zero-factor property to solve the equation in its given form because of the 3 on the right. Remember that the zero-factor property applies only to a product that equals 0.

OBJECTIVE 2 Solve other equations by factoring. We can also use the zero-factor property to solve equations that involve more than two factors with variables, as shown in Example 5. (These equations are *not* quadratic equations. Why not?)

CHALKBOARD EXAMPLE

Solve.

$(2x + 1)(2x^2 + 7x - 15) = 0$

Answer: $\left\{-\frac{1}{2}, \frac{3}{2}, -5\right\}$

EXAMPLE 5 Solving Equations with More Than Two Variable Factors

Solve the equations.

(a) $6z^3 - 6z = 0$

$$6z^3 - 6z = 0$$

$$6z(z^2 - 1) = 0 \quad \text{Factor out } 6z.$$

$$6z(z + 1)(z - 1) = 0 \quad \text{Factor } z^2 - 1.$$

By an extension of the zero-factor property this product can equal 0 only if at least one of the factors is 0. Write and solve three equations, one for each factor with a variable.

$$6z = 0 \quad \text{or} \quad z + 1 = 0 \quad \text{or} \quad z - 1 = 0$$

$$z = 0 \quad \text{or} \quad z = -1 \quad \text{or} \quad z = 1$$

Check by substituting, in turn, 0, -1, and 1 in the original equation. The solution set is $\{-1, 0, 1\}$.

(b) $(3x - 1)(x^2 - 9x + 20) = 0$

$$(3x - 1)(x^2 - 9x + 20) = 0$$

$$(3x - 1)(x - 5)(x - 4) = 0 \quad \text{Factor } x^2 - 9x + 20.$$

$$3x - 1 = 0 \quad \text{or} \quad x - 5 = 0 \quad \text{or} \quad x - 4 = 0 \quad \text{Zero-factor property}$$

$$x = \frac{1}{3} \quad \text{or} \quad x = 5 \quad \text{or} \quad x = 4$$

The solutions of the original equation are $\frac{1}{3}$, 4, and 5. Check each solution to verify that the solution set is $\{\frac{1}{3}, 4, 5\}$.

CAUTION In Example 5(b), it would be unproductive to begin by multiplying the two factors together. Keep in mind that the zero-factor property requires the product of two or more factors; this product must equal 0. Always consider first whether an equation is given in an appropriate form for the zero-factor property.

CONNECTIONS

Galileo Galilei (1564–1642) developed theories to explain physical phenomena and set up experiments to test his ideas. According to legend, Galileo dropped objects of different weights from the tower of Pisa to disprove the Aristotelian view that heavier objects fall faster than lighter objects. He developed the formula for freely falling objects described by $d = 16t^2$, where d is the distance in feet that an object falls (disregarding air resistance) in t seconds, regardless of weight.*

*From *Mathematical Ideas,* 8th edition, by Charles D. Miller, Vern E. Heeren, and E. John Hornsby, Jr., Addison-Wesley, 1997, page 442.

CONNECTIONS (CONTINUED)

The equation $d = 16t^2$ is a quadratic equation in the variable t. To find the number of seconds it would take an object to fall 256 feet, for example, substitute 256 for d and solve for t. On the other hand, to find how far the object would fall in 2 seconds, substitute 2 for t and calculate d. This idea of distance traveled depending on the time elapsed is an example of an important mathematical concept, the *function*. Functions were introduced in Chapter 3 and are discussed in more detail in Chapter 7.

1. 4 seconds 2. 64 feet

FOR DISCUSSION OR WRITING

1. How long would it take an object to fall 256 feet?
2. How far would an object fall in 2 seconds?

5.5 EXERCISES

1. $ax^2 + bx + c$ 2. standard 3. factoring 4. Set each factor equal to 0 and solve the linear equations. 5. $\left\{-\frac{8}{3}, -7\right\}$ 6. $\left\{-\frac{9}{5}, -6\right\}$ 7. $\{0, -4\}$ 8. $\{0, 10\}$ 9. To solve $2x(3x - 4) = 0$, set each *variable* factor equal to 0 to get $x = 0$ or $3x - 4 = 0$. Then solve the second equation to get $x = \frac{4}{3}$. 10. $\left\{4, -3, \frac{7}{2}\right\}$ 11. Because $(x - 9)^2 = (x - 9)(x - 9) = 0$ leads to two solutions of 9, we call 9 a double solution. 12. A quadratic equation is an equation in which the highest-degree term has degree 2. Some examples are $2x^2 + 3x = 5$, $x^2 - 4 = 0$, and $3x^2 + 4x = 0$. 13. $\{-2, -1\}$ 14. $\{-7, -1\}$ 15. $\{1, 2\}$ 16. $\{1, 3\}$ 17. $\{-8, 3\}$ 18. $\{-3, 5\}$ 19. $\{-1, 3\}$ 20. $\{-1, 4\}$ 21. $\{-2, -1\}$ 22. $\{-1, 3\}$ 23. $\{-4\}$ 24. $\{3\}$ 25. $\left\{-2, \frac{1}{3}\right\}$ 26. $\left\{-\frac{1}{2}, \frac{2}{3}\right\}$ 27. $\left\{-\frac{4}{3}, \frac{1}{2}\right\}$ 28. $\left\{-\frac{1}{2}, \frac{4}{3}\right\}$ 29. $\left\{-\frac{2}{3}\right\}$ 30. $\left\{-\frac{5}{6}\right\}$ 31. $\{-3, 3\}$ 32. $\{-10, 10\}$ 33. $\left\{-\frac{7}{4}, \frac{7}{4}\right\}$ 34. $\left\{-\frac{3}{2}, \frac{3}{2}\right\}$ 35. $\{-11, 11\}$ 36. $\{-20, 20\}$

Fill in the blanks.

1. A quadratic equation is an equation that can be put in the form _____ = 0.
2. The form $ax^2 + bx + c = 0$ is called _____ form.
3. If a quadratic equation is in standard form, to solve the equation we should begin by _____ the polynomial.
4. Once the polynomial in a quadratic equation is factored, what is the next step to solve the equation?

Solve each equation and check your solutions. See Example 1.

5. $(3k + 8)(k + 7) = 0$
6. $(5r + 9)(r + 6) = 0$
7. $t(t + 4) = 0$
8. $x(x - 10) = 0$
9. Students often become confused as to how to handle a constant, such as 2 in the equation $2x(3x - 4) = 0$. How would you explain to someone how to solve this equation and how to handle the constant 2?
10. As shown in Example 5, the zero-factor property can be extended to more than two factors. For example, to solve $(x - 4)(x + 3)(2x - 7) = 0$, we would set each factor equal to 0 and solve three equations. Find the solution set of this equation.
11. Why do you think that 9 is called a *double solution* of the equation $(x - 9)^2 = 0$?
12. Define *quadratic equation* in your own words, without using an algebraic expression. Then give examples.

Solve each equation and check your solutions. See Examples 2–4.

13. $y^2 + 3y + 2 = 0$
14. $p^2 + 8p + 7 = 0$
15. $y^2 - 3y + 2 = 0$
16. $r^2 - 4r + 3 = 0$
17. $x^2 = 24 - 5x$
18. $t^2 = 2t + 15$
19. $5x^2 = 15 + 10x$
20. $2m^2 = 8 + 6m$
21. $z^2 = -2 - 3z$
22. $p^2 = 2p + 3$
23. $m^2 + 8m + 16 = 0$
24. $b^2 - 6b + 9 = 0$
25. $3x^2 + 5x - 2 = 0$
26. $6r^2 - r - 2 = 0$
27. $6p^2 = 4 - 5p$
28. $6x^2 = 4 + 5x$
29. $9s^2 + 12s = -4$
30. $36x^2 + 60x = -25$
31. $2y^2 - 18 = 0$
32. $3m^2 - 300 = 0$
33. $16k^2 - 49 = 0$
34. $4w^2 - 9 = 0$
35. $n^2 = 121$
36. $x^2 = 400$

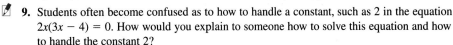

37. Another solution is -11.
38. Dividing by a variable is not allowed; 0 is also a solution.
39. $\{0, 7\}$ **40.** $\{0, 9\}$
41. $\left\{0, \frac{1}{2}\right\}$ **42.** $\left\{-\frac{1}{2}, 0\right\}$
43. $\{2, 5\}$ **44.** $\{2, 3\}$
45. $\left\{-4, \frac{1}{2}\right\}$ **46.** $\left\{-3, \frac{3}{2}\right\}$
47. $\left\{-12, \frac{11}{2}\right\}$ **48.** $\left\{\frac{2}{3}, 6\right\}$
49. $\{-1, 3\}$ **50.** $\{-3, -1\}$
51. $\left\{-\frac{5}{2}, \frac{1}{3}, 5\right\}$
52. $\left\{-\frac{4}{3}, -\frac{2}{3}, \frac{1}{2}\right\}$
53. $\left\{-\frac{7}{2}, -3, 1\right\}$
54. $\left\{-\frac{3}{2}, -1, \frac{4}{3}\right\}$
55. $\left\{-\frac{7}{3}, 0, \frac{7}{3}\right\}$
56. $\left\{-\frac{3}{4}, 0, \frac{3}{4}\right\}$
57. $\{-2, 0, 4\}$ **58.** $\{-2, 0, 3\}$
59. $\{-5, 0, 4\}$ **60.** $\{0, 2, 4\}$
61. $\{-3, 0, 5\}$ **62.** $\{-1, 0, 3\}$
63. $\left\{-\frac{4}{3}, -1, \frac{1}{2}\right\}$
64. $\left\{-\frac{3}{2}, -\frac{1}{2}, \frac{4}{3}\right\}$
65. $\left\{-\frac{2}{3}, 4\right\}$ **66.** $\left\{0, \frac{3}{4}, \frac{3}{2}\right\}$
67. To use the zero-factor property, one side of the equation must be 0. Before solving this equation, multiply the factors on the left side, and subtract 1 from both sides so the right side is 0.
68. Because both sides were divided by a variable, the solution $x = 0$ was lost. The correct procedure is to subtract $4x$ on both sides, so the right side is 0. Then factor on the left side and complete the solution.
69. $\{-.5, .1\}$

37. What is wrong with this reasoning for solving the equation in Exercise 35? "To solve $n^2 = 121$, I must find a number that, when multiplied by itself, gives 121. Because $11^2 = 121$, the solution of the equation is 11."

38. What is wrong with this reasoning in solving $x^2 = 7x$? "To solve $x^2 = 7x$, first divide both sides by x to get $x = 7$. Therefore, the solution is 7."

Solve each equation and check your solutions. See Examples 4 and 5.

39. $x^2 = 7x$ **40.** $t^2 = 9t$ **41.** $6r^2 = 3r$

42. $10y^2 = -5y$ **43.** $g(g - 7) = -10$ **44.** $r(r - 5) = -6$

45. $3z(2z + 7) = 12$ **46.** $4b(2b + 3) = 36$

47. $2(y^2 - 66) = -13y$ **48.** $3(t^2 + 4) = 20t$

49. $3x(x + 1) = (2x + 3)(x + 1)$ **50.** $2k(k + 3) = (3k + 1)(k + 3)$

51. $(2r + 5)(3r^2 - 16r + 5) = 0$ **52.** $(3m + 4)(6m^2 + m - 2) = 0$

53. $(2x + 7)(x^2 + 2x - 3) = 0$ **54.** $(x + 1)(6x^2 + x - 12) = 0$

55. $9y^3 - 49y = 0$ **56.** $16r^3 - 9r = 0$

▲ **57.** $r^3 - 2r^2 - 8r = 0$ ▲ **58.** $x^3 - x^2 - 6x = 0$

▲ **59.** $a^3 + a^2 - 20a = 0$ ▲ **60.** $y^3 - 6y^2 + 8y = 0$

▲ **61.** $r^4 = 2r^3 + 15r^2$ ▲ **62.** $x^3 = 3x + 2x^2$

▲ **63.** $6p^2(p + 1) = 4(p + 1) - 5p(p + 1)$ ▲ **64.** $6x^2(2x + 3) - 5x(2x + 3) = 4(2x + 3)$

▲ **65.** $(k + 3)^2 - (2k - 1)^2 = 0$ ▲ **66.** $(4y - 3)^3 - 9(4y - 3) = 0$

67. Verify that the solutions of $(x - 3)(x + 2) = 1$ are not found from the two equations

$$x - 3 = 1 \quad \text{and} \quad x + 2 = 1.$$

Explain why.

68. What is wrong with the following solution?

$$4x^2 = 4x$$
$$x = 1 \qquad \text{Divide both sides by } 4x.$$

Explain how to solve the equation correctly.

TECHNOLOGY INSIGHTS (EXERCISES 69–72)

In Section 3.2 we showed how an equation in one variable can be solved by getting 0 on one side, then replacing 0 with y to get a corresponding equation in two variables. The x-values of the x-intercepts of the graph of the two-variable equation then give the solutions of the original equation.

Use the calculator screens to determine the solution set of each quadratic equation. Verify your answers by substitution.

69. $x^2 + .4x - .05 = 0$

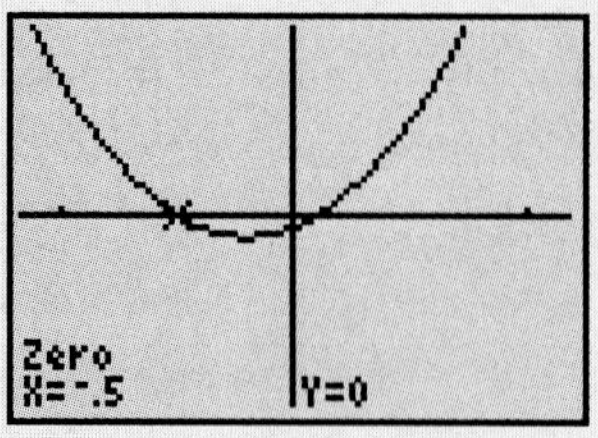

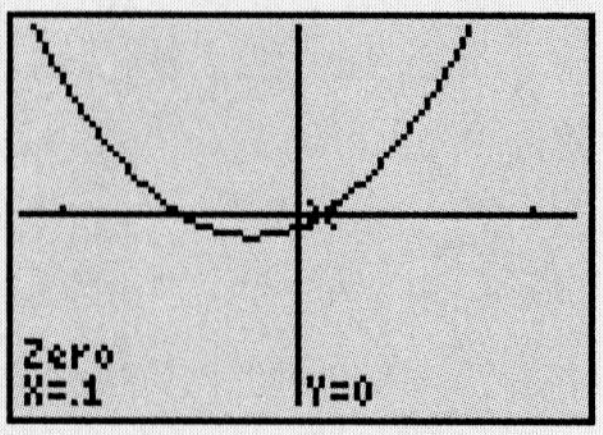

70. $\{1.5, 2.1\}$
71. $\{-1.1, -2.5\}$ 72. $\{-2.75, 3\}$

TECHNOLOGY INSIGHTS (EXERCISES 69–72) (CONTINUED)

70. $2x^2 - 7.2x + 6.3 = 0$

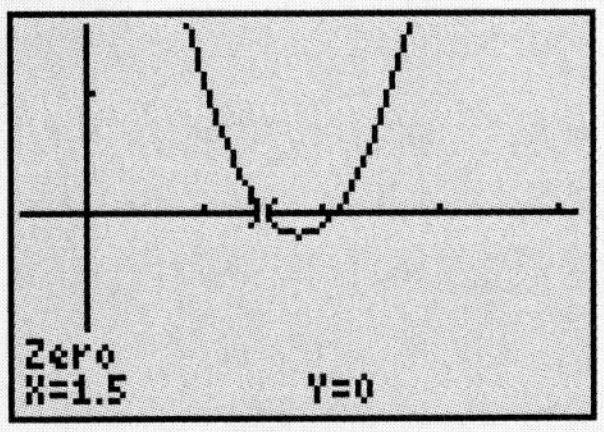

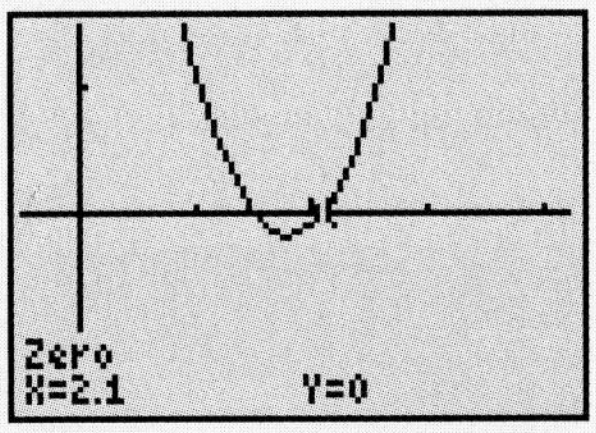

71. $2x^2 + 7.2x + 5.5 = 0$

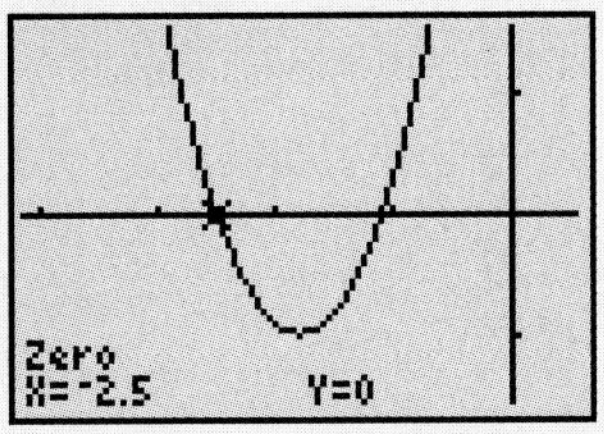

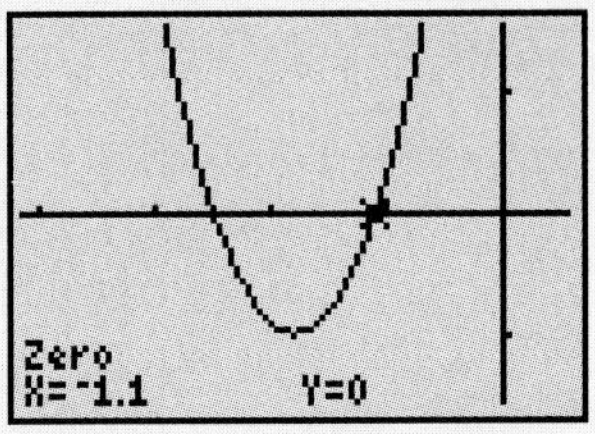

72. $4x^2 - x - 33 = 0$

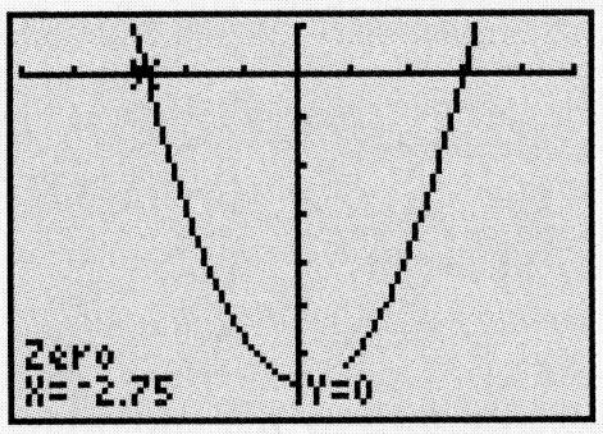

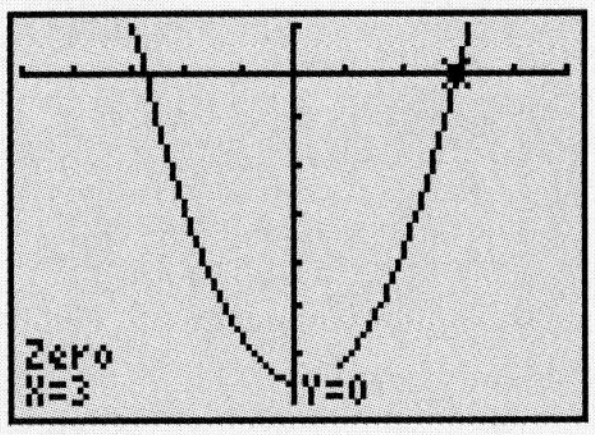

5.6 Applications of Quadratic Equations

OBJECTIVES

1. Solve problems about geometric figures.
2. Solve problems using the Pythagorean formula.
3. Solve problems using quadratic models.

FOR EXTRA HELP

SSG Sec. 5.6
SSM Sec. 5.6
Pass the Test Software
InterAct Math Tutorial Software
Video 9

We can now use factoring to solve quadratic equations that arise from applied problems.

OBJECTIVE 1 Solve problems about geometric figures. Most problems in this section require one of the formulas given on the inside covers. We still follow the six-step problem-solving method from Section 2.3 and continue the work with formulas and geometric problems begun in Section 2.4.

EXAMPLE 1 Solving an Area Problem

The Goldsteins are planning to add a rectangular porch to their house. The width of the porch will be 4 feet less than its length, and they want it to have an area of 96 square feet. Find the length and width of the porch.

Step 1 Let x = the length of the porch;

Step 2 $x - 4$ = the width (the width is 4 less than the length). See Figure 1.

INSTRUCTOR'S RESOURCES

PTB Sec. 5.6 **ISM** Sec. 5.6
AB Sec. 5.6

TEST GENERATOR

WORLD WIDE WEB
www.LialAlgebra.com

CHALKBOARD EXAMPLE

The length of a hall is five times the width. The area of the floor is 45 square meters. Find the length and width of the hall.

Answer: 15 meters, 3 meters

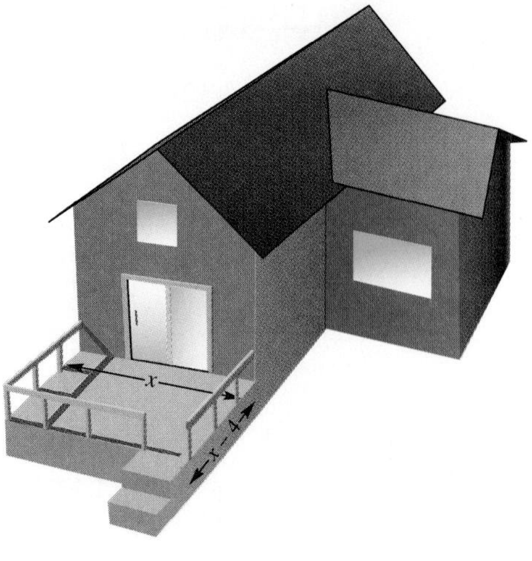

Figure 1

Step 3 The area of a rectangle is given by the formula

$$\text{area} = \text{length} \times \text{width.}$$

$$A = LW$$

$$96 = x(x - 4) \qquad \text{Let } A = 96,\ L = x,\ W = x - 4.$$

Step 4

$$96 = x^2 - 4x \qquad \text{Distributive property}$$

$$0 = x^2 - 4x - 96 \qquad \text{Subtract 96 from both sides.}$$

$$0 = (x - 12)(x + 8) \qquad \text{Factor.}$$

$$x - 12 = 0 \quad \text{or} \quad x + 8 = 0 \qquad \text{Zero-factor property}$$

$$x = 12 \quad \text{or} \quad x = -8 \qquad \text{Solve.}$$

Step 5 The solutions of the equations are 12 and −8. Since a rectangle cannot have a negative length, discard the solution −8. Then 12 feet is the length of the porch and $12 - 4 = 8$ feet is the width.

Step 6 As a check, note that the width is 4 less than the length and the area is $8 \cdot 12 = 96$ square feet, as required.

CAUTION When solving an applied problem, remember Step 6. Always check solutions against physical facts.

The next application involves *perimeter,* the distance around a figure, as well as area.

CHALKBOARD EXAMPLE

The width of a rectangular yard is 4 meters less than the length. The area is numerically 92 more than the perimeter. Find the length and width of the yard.

Answer: 14 meters, 10 meters

EXAMPLE 2 Solving an Area and Perimeter Problem

The length of a rectangular rug is 4 feet more than the width. The area of the rug is numerically 1 more than the perimeter. See Figure 2. Find the length and width of the rug.

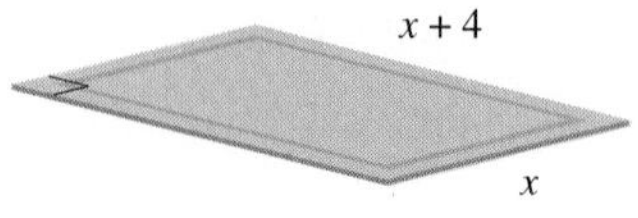

Figure 2

Let $x =$ the width of the rug.

Then $x + 4 =$ the length of the rug.

The area is the product of the length and width, so

$$A = LW.$$

Substituting $x + 4$ for the length and x for the width gives

$$A = (x + 4)x.$$

Now substitute into the formula for perimeter.

$$P = 2L + 2W$$
$$P = 2(x + 4) + 2x$$

According to the information given in the problem, the area is numerically 1 more than the perimeter. Write this as an equation.

The area	is	1	more than	the perimeter.
↓	↓	↓	↓	↓
$(x + 4)x$	$=$	1	$+$	$2(x + 4) + 2x$

Simplify and solve this equation.

$x^2 + 4x = 1 + 2x + 8 + 2x$	Distributive property
$x^2 + 4x = 9 + 4x$	Combine terms.
$x^2 = 9$	Subtract $4x$ from both sides.
$x^2 - 9 = 0$	Subtract 9 from both sides.
$(x + 3)(x - 3) = 0$	Factor.
$x + 3 = 0$ or $x - 3 = 0$	Zero-factor property
$x = -3$ or $x = 3$	

A rectangle cannot have a negative width, so ignore -3. The only valid solution is 3, so the width is 3 feet and the length is $3 + 4 = 7$ feet. Check to see that the area is numerically 1 more than the perimeter. The rug is 3 feet wide and 7 feet long.

OBJECTIVE 2 **Solve problems using the Pythagorean formula.** The next example requires the **Pythagorean formula** from geometry.

Pythagorean Formula

If a right triangle (a triangle with a 90° angle) has longest side of length c and two other sides of lengths a and b, then

$$a^2 + b^2 = c^2.$$

(See the figure.) The longest side, the **hypotenuse,** is opposite the right angle. The two shorter sides are the **legs** of the triangle.

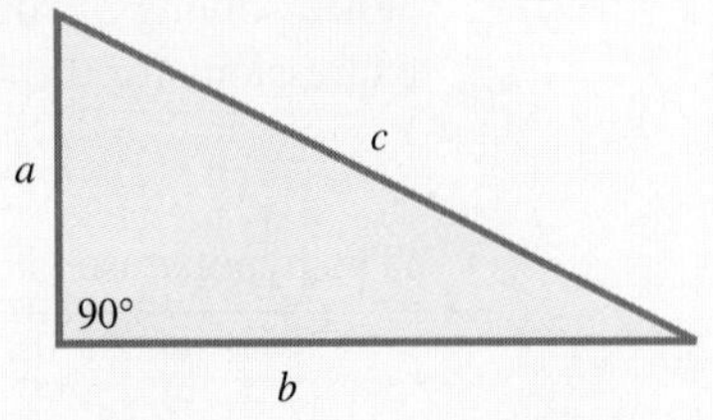

CHALKBOARD EXAMPLE

The hypotenuse of a right triangle is 3 inches longer than the longer leg. The shorter leg is 3 inches shorter than the longer leg. Find the lengths of the sides of the triangle.

Answer: 9 inches, 12 inches, 15 inches

EXAMPLE 3 Using the Pythagorean Formula

Ed and Mark leave their office with Ed traveling north and Mark traveling east. When Mark is 1 mile farther than Ed from the office, the distance between them is 2 miles more than Ed's distance from the office. Find their distances from the office and the distance between them.

Let x represent Ed's distance from the office.

Then $x + 1$ represents Mark's distance from the office,

and $x + 2$ represents the distance between them.

Label a right triangle as in Figure 3.

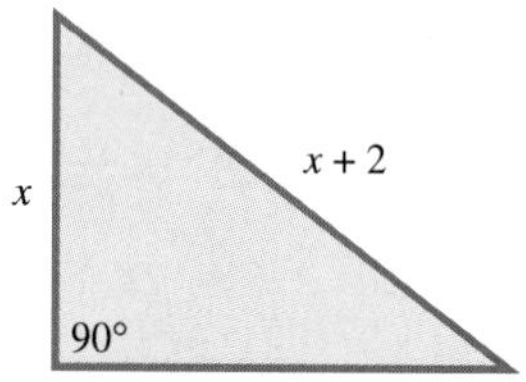

Figure 3

Substitute into the Pythagorean formula,

$$a^2 + b^2 = c^2$$
$$x^2 + (x + 1)^2 = (x + 2)^2.$$

Since $(x + 1)^2 = x^2 + 2x + 1$, and since $(x + 2)^2 = x^2 + 4x + 4$, the equation becomes

$$x^2 + x^2 + 2x + 1 = x^2 + 4x + 4.$$

$$x^2 - 2x - 3 = 0 \qquad \text{Standard form}$$

$$(x - 3)(x + 1) = 0 \qquad \text{Factor.}$$

$$x - 3 = 0 \quad \text{or} \quad x + 1 = 0 \qquad \text{Zero-factor property}$$

$$x = 3 \quad \text{or} \quad x = -1 \qquad \text{Solve.}$$

Since -1 cannot be the length of a side of a triangle, 3 is the only possible answer. Therefore, Ed is 3 miles north of the office, Mark is $3 + 1 = 4$ miles east of the office, and they are $3 + 2 = 5$ miles apart. Check to see that $3^2 + 4^2 = 5^2$ is true.

TEACHING TIP Stress that the Pythagorean formula cannot be applied to nonright triangles.

CAUTION

When solving a problem involving the Pythagorean formula, be sure that the expressions for the sides are properly placed.

$$\textbf{leg}^2 + \textbf{leg}^2 = \textbf{hypotenuse}^2$$

(The hypotenuse is opposite the right angle.)

The diagonal of the floor should be $\sqrt{208} \approx 14.4$ ft, or about 14 feet, 5 inches. The carpenter is off by 3 inches and must correct the error to avoid major construction problems.

CONNECTIONS

When a carpenter builds a floor for a rectangular room, it is essential that the corners of the floor are at right angles; otherwise, problems will occur when the walls are constructed, when flooring is laid, and so on. To check that the floor is "squared off," the carpenter can use the *converse* of the Pythagorean formula: If $a^2 + b^2 = c^2$, then the angle opposite side c is a right angle.

FOR DISCUSSION OR WRITING

Suppose a carpenter is building an 8-foot by 12-foot room. After the floor is built, the carpenter finds that the length of the diagonal of the floor is 14 feet, 8 inches. Is the floor "squared off" properly? If not, what should the diagonal measure?

OBJECTIVE 3 **Solve problems using quadratic models.** In the next example, we use the quadratic model for fuel economy from the beginning of this chapter.

CHALKBOARD EXAMPLE

Use the table in Example 4 to determine in what two years the number of miles per gallon was the same. Compare the corresponding values using the equation.

Answer: 1982 and 1992; The equation gives 24.08 for 1982 and 26.18 for 1992. The equation value is 1.02 less than the table value for 1982 and 1.08 more than the table value for 1992.

EXAMPLE 4 **Using a Quadratic Equation That Models Gasoline Mileage**

Earlier we gave the quadratic expression $-.04x^2 + .93x + 21$, that models the automotive industry fuel economy trend. If we set the expression equal to y, then y is the average miles per gallon for the industry in year x. Recall, x is coded so that $x = 0$ represents 1978, $x = 2$ represents 1980, and so on. The equation

$$y = -.04x^2 + .93x + 21$$

was developed from the following data.

Fuel Economy

Year	*Miles per Gallon*
1978	19.9
1980	23.1
1982	25.1
1984	25.0
1986	25.9
1988	26.0
1990	25.4
1992	25.1
1994	24.7
1996	24.9

Source: National Highway Traffic Safety Administration.

(a) From the table, in what year did the number of miles per gallon appear to peak?

Look down the second column for the largest number, 26.0 miles per gallon. Reading across to the first column, we see that this mileage corresponds to 1988.

(b) Use the equation to find the miles per gallon in 1992.

The miles per gallon in year x is given by y. In 1992, $x = 1992 - 1978 = 14$. Substitute 14 for x in the equation $y = -.04x^2 + .93x + 21$.

$$y = -.04(14)^2 + .93(14) + 21 = 26.18 \qquad \text{Use a calculator.}$$

From the table, the actual data for 1992 is 25.1, so this approximation is a little high.

(c) Repeat part (b) for 1986.

For 1986, $x = 1986 - 1978 = 8$, and

$$y = -.04(8)^2 + .93(8) + 21 = 25.88. \qquad \text{Let } x = 8.$$

How does this approximation compare to the actual data in the table?

(d) Which of the results in parts (b) and (c) is the better approximation of the data in the table?

The approximation for 1986 is much closer than the approximation for 1992.

5.6 EXERCISES

1. a variable; the unknown
2. expression; sketch (or figure)
3. an equation **4.** equation; question(s); Check
5. (a) $80 = (x + 8)(x - 8)$ (b) 12 (c) length: 20 units; width: 4 units
6. (a) $45 = (2x + 1)(x + 1)$ (b) 4 (c) base: 9 units; height: 5 units
7. (a) $60 = \frac{1}{2}(3x + 6)(x + 5)$ (b) 3 (c) base: 15 units; height: 8 units
8. (a) $192 = 4x(x + 2)$ (b) 6 (c) length: 8 units; width: 6 units

Complete these statements which review the six-step method for solving applied problems first introduced in Chapter 2.

1. Read the problem carefully, choose ______ to represent ______ and write it down.

2. Write down a mathematical ______ for any other unknown quantities. Draw a ______ if it would be helpful.

3. Translate the problem into ______.

4. Solve the ______ and answer the ______ in the problem. ______ your answer.

In Exercises 5–8, a figure and a corresponding geometric formula are given. Complete each problem using the problem-solving steps from Chapter 2.

(a) *Write an equation using the formula and the given information. (Step 3)*
(b) *Solve the equation, giving only the solution(s) that make sense in the problem. (Step 4)*
(c) *Use the solution(s) to find the indicated dimensions of the figure. (Step 5)*
(d) *Check your solution. (Step 6)*

5. Area of a rectangle: $A = LW$
The area of this rectangle is 80 square units. Find its length and its width.

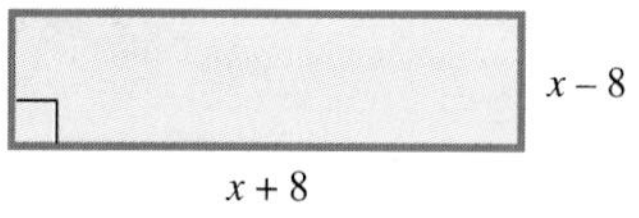

6. Area of a parallelogram: $A = bh$
The area of this parallelogram is 45 square units. Find its base and its height.

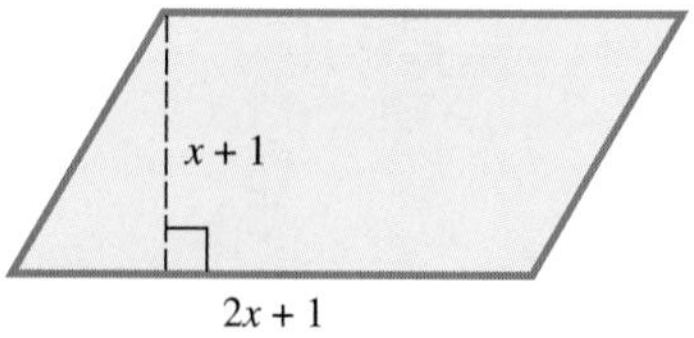

7. Area of a triangle: $A = \frac{1}{2}bh$
The area of this triangle is 60 square units. Find its base and its height.

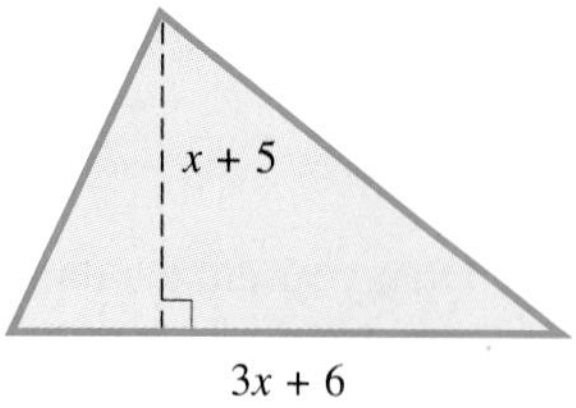

8. Volume of a rectangular Chinese box: $V = LWH$
The volume of this box is 192 cubic units. Find its length and its width.

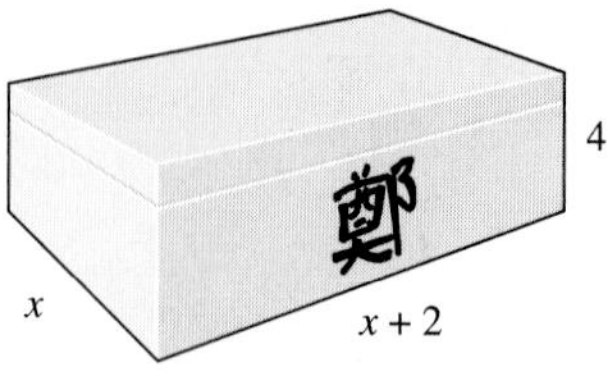

9. length: 7 inches; width: 4 inches **10.** length: 11 centimeters; width: 7 centimeters **11.** length: 11 inches; width: 8 inches **12.** length: 19 inches; width: 8 inches **13.** base: 12 inches; height: 5 inches **14.** 12 meters **15.** height: 13 inches; width: 10 inches **16.** length: 8 feet; width: 5 feet **17.** 12 centimeters **18.** 3 meters

Solve each problem. Check your answer to be sure that it is reasonable. Refer to the formulas found on the inside covers. See Examples 1 and 2.

9. The length of a VHS videocassette shell is 3 inches more than its width. The area of the rectangular top side of the shell is 28 square inches. Find the length and the width of the videocassette shell.

10. A plastic box that holds a standard audiocassette has a length 4 centimeters longer than its width. The area of the rectangular top of the box is 77 square centimeters. Find the length and the width of the box.

11. The dimensions of a certain IBM computer monitor screen are such that its length is 3 inches more than its width. If the length is increased by 1 inch while the width remains the same, the area is increased by 8 square inches. What are the dimensions of the screen?

12. The keyboard of the computer mentioned in Exercise 11 is 11 inches longer than it is wide. If both its length and width are increased by 2 inches, the area of the top of the keyboard is increased by 58 square inches. What are the length and the width of the keyboard?

13. The area of a triangle is 30 square inches. The base of the triangle measures 2 inches more than twice the height of the triangle. Find the measures of the base and the height.

14. A certain triangle has its base equal in measure to its height. The area of the triangle is 72 square meters. Find the equal base and height measure.

15. A ten-gallon aquarium holding African cichlids is 3 inches higher than it is wide. Its length is 21 inches, and its volume is 2730 cubic inches. What are the height and width of the aquarium?

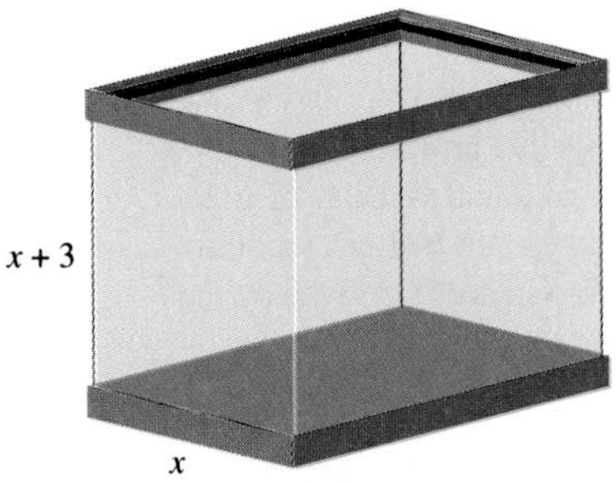

16. Nana Nantambu wishes to build a box to hold her tools. It is to be 2 feet high, and the width is to be 3 feet less than its length. If its volume is to be 80 cubic feet, find the length and the width of the box.

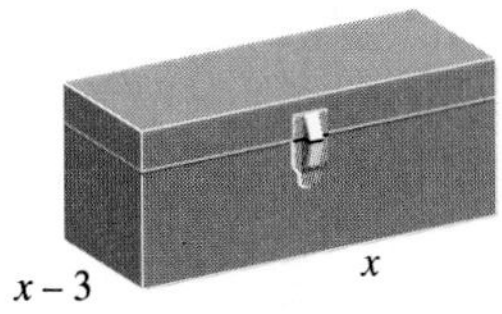

Use the Pythagorean formula to solve each problem. See Example 3.

17. The hypotenuse of a right triangle is 1 centimeter longer than the longer leg. The shorter leg is 7 centimeters shorter than the longer leg. Find the length of the longer leg of the triangle.

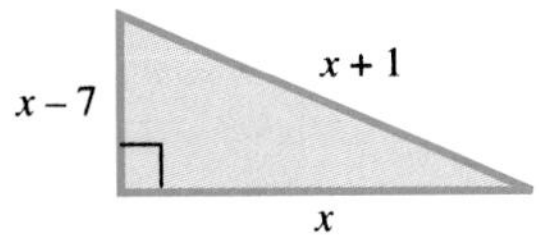

18. The longer leg of a right triangle is 1 meter longer than the shorter leg. The hypotenuse is 1 meter shorter than twice the shorter leg. Find the length of the shorter leg of the triangle.

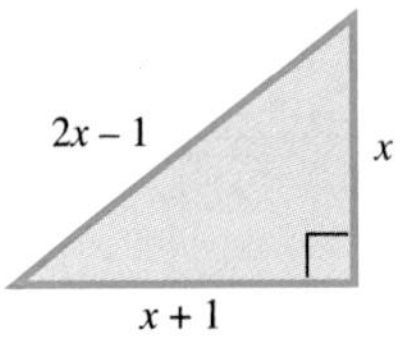

19. 8 feet **20.** 25 miles
21. 6 meters **22.** 8 feet
23. 1 second
24. $\frac{1}{2}$ second and $1\frac{1}{2}$ seconds
25. 3 seconds **26.** The negative solution, -1, does not make sense since t represents time.

19. A ladder is resting against a wall. The top of the ladder touches the wall at a height of 15 feet. Find the distance from the wall to the bottom of the ladder if the length of the ladder is one foot more than twice its distance from the wall.

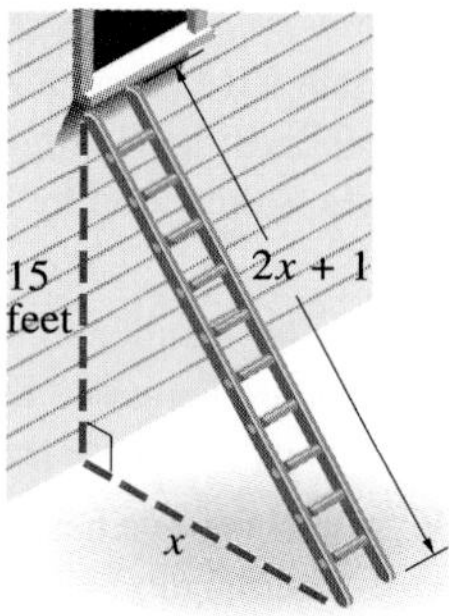

20. Two cars leave an intersection. One car travels north; the other travels east. When the car traveling north had gone 24 miles, the distance between the cars was four miles more than three times the distance traveled by the car heading east. Find the distance between the cars at that time.

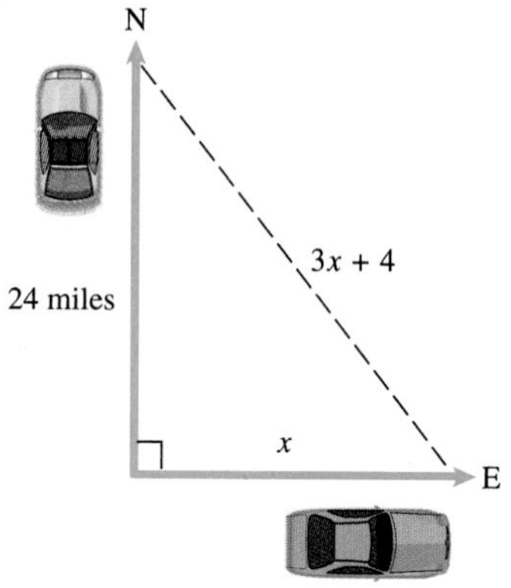

21. A garden has the shape of a right triangle with one leg 2 meters longer than the other. The hypotenuse is two meters less than twice the length of the shorter leg. Find the length of the shorter leg.

22. The hypotenuse of a right triangle is 1 foot more than twice the length of the shorter leg. The longer leg is 1 foot less than twice the length of the shorter leg. Find the length of the shorter leg.

If an object is propelled upward from a height of s feet at an initial velocity of v feet per second, then its height h after t seconds (disregarding air resistance*) is given by the equation

$$h = -16t^2 + vt + s,$$

where h is in feet. For example, if the object is propelled from a height of 48 feet with an initial velocity of 32 feet per second, its height h is given by the equation $h = -16t^2 + 32t + 48$.

Use this information in Exercises 23–26.

23. After how many seconds is the height 64 feet? (*Hint:* Let $h = 64$ and solve.)

24. After how many seconds is the height 60 feet?

25. After how many seconds does the object hit the ground? (*Hint:* When the object hits the ground, $h = 0$.)

26. The quadratic equation from Exercise 25 has two solutions, yet only one of them is appropriate for answering the question. Why is this so?

*From now on, in all problems of this sort involving an object propelled upward or dropped from a height, we give formulas that disregard air resistance.

27. 1 second and 3 seconds
28. 64 feet **29.** 4 seconds
30. The solution 0 represents the time at which it was propelled.
31. 9 centimeters
32. 16 square meters
33. 256 feet **34.** 1024 feet

If an object is propelled upward from ground level with an initial velocity of 64 *feet per second, its height h in feet t seconds later is given by the equation* $h = -16t^2 + 64t$. *Use this information in Exercises 27–30.*

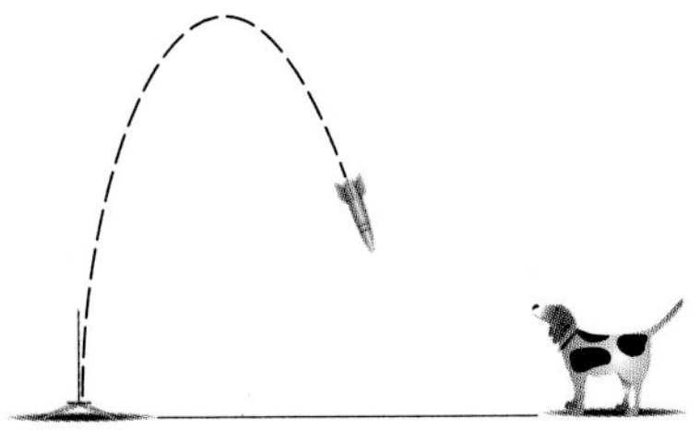

27. After how many seconds is the height 48 feet? (*Hint:* Let $h = 48$ and solve for t.)

28. It can be shown using concepts developed in other courses that the object reaches its maximum height 2 seconds after it is propelled. What is this maximum height?

29. After how many seconds does the object hit the ground? (*Hint:* Let $h = 0$.)

30. The quadratic equation from Exercise 29 has two solutions, yet only one of them is appropriate for answering the question. Why is this so?

Exercises 31 and 32 require the formula for the volume of a pyramid,

$$V = \frac{1}{3}Bh,$$

where B is the area of the base.

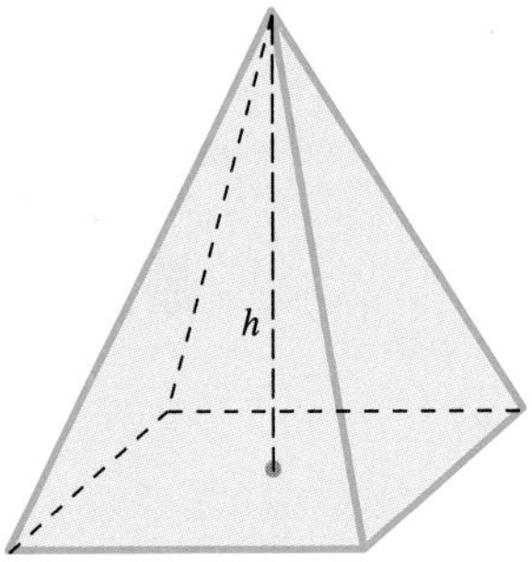

31. Suppose a pyramid has a rectangular base whose width is 3 centimeters less than the length. If the height is 8 centimeters and the volume is 144 cubic centimeters, find the length of the base.

32. The volume of a pyramid is 32 cubic meters. Suppose the numerical value of the height is 10 less than the numerical value of the area of the base. Find the area of the base.

If an object is dropped, the distance d in feet it falls in t seconds is given by

$$d = \frac{1}{2}gt^2,$$

where g is approximately 32 *feet per second*^{2}. *Find the distance an object would fall in the following times.*

33. 4 seconds

34. 8 seconds

35. 5.7 seconds
36. 7.1 seconds
37. 8.0 seconds
38. 8.5 seconds
39. −3, −2 or 4, 5
40. 0, 1 or 7, 8 41. 7, 9, 11
42. 5, 7, 9 43. −2, 0, 2 or 6, 8, 10 44. −2, 0, 2
45. (a) 1990 (b) 8.8 billion passengers; This approximation is .2 billion less than the value in the table. (c) 25 (d) 8.0 billion passengers; This is correct to the nearest one tenth billion. (e) 1995

How long would it take an object to fall from the top of the building described? Use a calculator as necessary, and round your answer to the nearest tenth of a second.

35. Navarre Building, New York City 512 feet

36. One Canada Square, London 800 feet

37. Central Plaza, Hong Kong 1028 feet

38. Las Vegas Stratosphere Tower, Las Vegas 1149 feet

Solve each problem. Refer to the discussion of consecutive integers and Examples 7 and 8 in Section 2.3.

39. The product of two consecutive integers is 11 more than their sum. Find the integers.

40. The product of two consecutive integers is 4 less than 4 times their sum. Find the integers.

41. Find three consecutive odd integers such that 3 times the sum of all three is 18 more than the product of the smaller two.

42. Find three consecutive odd integers such that the sum of all three is 42 less than the product of the larger two.

43. Find three consecutive even integers such that the sum of the squares of the smaller two is equal to the square of the largest.

44. Find three consecutive even integers such that the square of the sum of the smaller two is equal to twice the largest.

As we saw in Example 4, quadratic expressions are often used to approximate data. Exercises 45–47 further illustrate this kind of application.

45. The table shows the trend in transit ridership (in billions) from 1975 to 1995. The data is modeled fairly well by the quadratic expression $-.012x^2 + .381x + 5.96$. If y represents the number of passengers (in billions), then

$$y = -.012x^2 + .381x + 5.96.$$

Again, the years are coded, with $x = 5$ corresponding to 1975, $x = 10$ corresponding to 1980, and so on.

Transit Ridership Trend

Year	Number of Passengers (in billions)
1975	7.5
1980	8.5
1985	8.8
1990	9.0
1995	8.0

Source: American Public Transit Association.

(a) In what year shown in the table did the ridership peak?

(b) Use the equation to approximate the ridership y in 1990. (*Hint:* Let $x = 20$.) Does the equation give a good approximation of the value shown in the table?

(c) To what value does x correspond in 1995?

(d) Repeat part (b) for 1995.

(e) Which of the results in parts (b) and (d) is the better approximation of the actual data in the table?

46. (a) 8 (b) approximately 12,200 accidents **47.** (a) approximately 17,000 injuries (b) approximately .7 or 7 to 10; This indicates that in 1998, on the average, 1.4 people were injured per accident.
48. 192 cubic inches

46. Although motor vehicle accidents declined in New York City from 1990 to 1996, taxi accidents increased during that period, as indicated in the bar graph. The data in the graph is approximated by the quadratic equation

$$y = -143.3x^2 + 1823.3x + 6820.$$

Here, y represents the number of accidents in year x, where $x = 0$ corresponds to 1990, $x = 1$ corresponds to 1991, and so on.

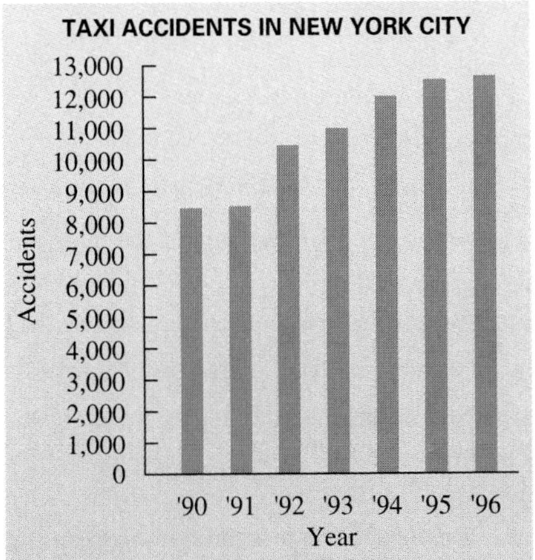

Source: The New York Times Metro, February 7, 1988.

(a) What value of x corresponds to 1998?

(b) According to the equation, if this trend continued, how many taxi accidents occurred in 1998?

47. (See Exercise 46.) The number of people injured in taxi accidents also increased over the years 1990 to 1996. The equation

$$y = -416.72x^2 + 4416.7x + 8500,$$

where y represents the number injured in year x, with $x = 0$ corresponding to 1990, $x = 1$ corresponding to 1991, and so on, approximates the data quite well.

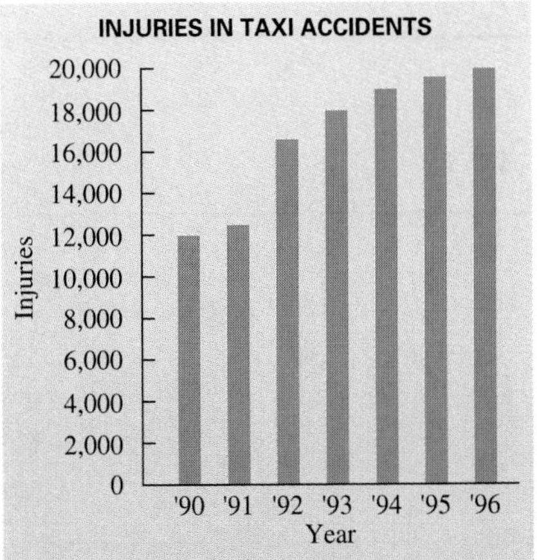

Source: The New York Times Metro, February 7, 1988.

(a) Use the equation to approximate the number injured in 1998.

(b) Find the ratio for 1998 (see Section 2.5) of the number of accidents (answer to Exercise 46(b)) to the number of people injured (answer to part (a) of this exercise). What does this tell us about the average number of people injured per accident?

48. A square piece of cardboard is to be formed into an open-topped box by cutting 3-inch squares from the corners and folding up the sides. See the figure. The volume V of the resulting box is given by the formula $V = 3(x - 6)^2$, where x is the original length of each side of the piece of cardboard. What is the volume of the box if the original length of each side of the piece of cardboard is 14 inches?

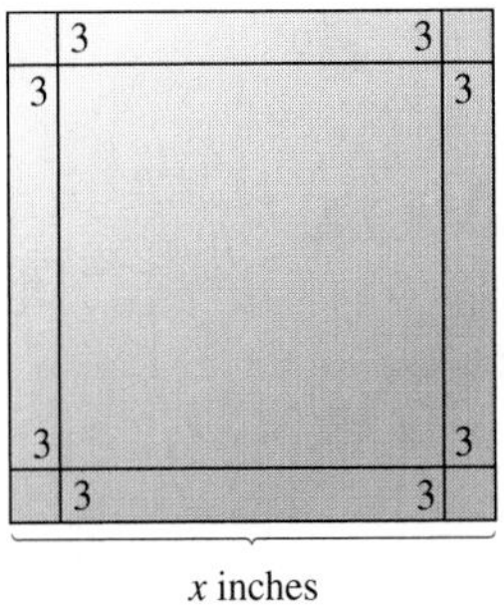

Original Piece of Cardboard

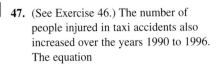

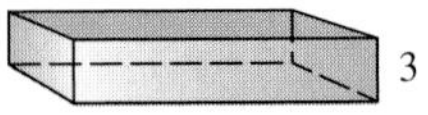

The Resulting Box

49. c^2 **50.** b^2 **51.** a^2
52. $a^2 + b^2 = c^2$; This is the Pythagorean formula.

RELATING CONCEPTS (EXERCISES 49–52)

One of the many known proofs of the Pythagorean formula is based on the figures shown.

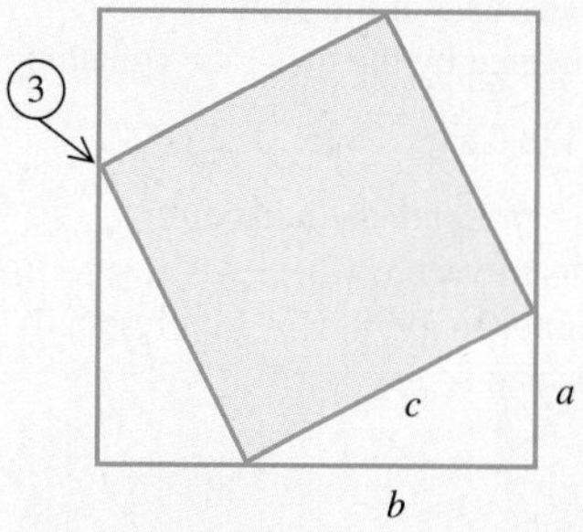

Figure A

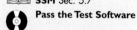

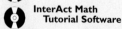

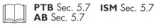

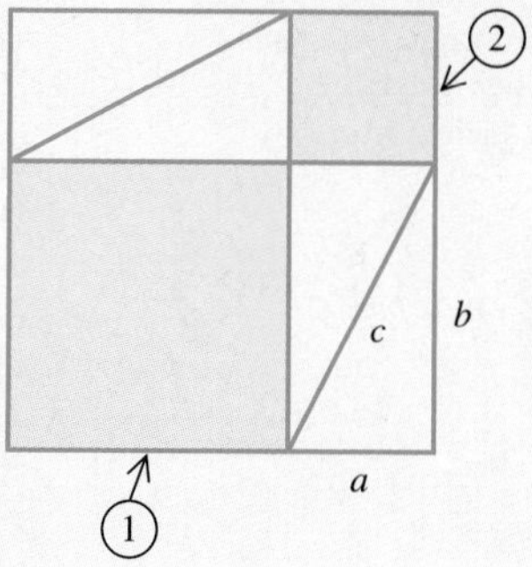

Figure B

Refer to the appropriate figure and ***answer Exercises 49–52 in order.***

49. What is an expression for the area of the dark square labeled ③ in Figure A?

50. The five regions in Figure A are equal in area to the six regions in Figure B. What is an expression for the area of the square labeled ① in Figure B?

51. What is an expression for the area of the square labeled ② in Figure B?

52. Represent this statement using algebraic expressions: The sum of the areas of the dark regions in Figure B is equal to the area of the dark region in Figure A. What does this equation represent?

Did you make the connection that the Pythagorean formula is the result of the fact that the dark areas in the two figures are equal?

5.7 Solving Quadratic Inequalities

OBJECTIVE

1 Solve quadratic inequalities and graph their solutions.

FOR EXTRA HELP

SSG Sec. 5.7
SSM Sec. 5.7
Pass the Test Software
InterAct Math Tutorial Software
Video 9

OBJECTIVE 1 **Solve quadratic inequalities and graph their solutions.** A **quadratic inequality** is an inequality that involves a second-degree polynomial. Examples of quadratic inequalities include

$$2x^2 + 3x - 5 < 0, \qquad x^2 \leq 4, \qquad \text{and} \qquad x^2 + 5x + 6 > 0.$$

Examples 1 and 2 show how to solve such inequalities.

EXAMPLE 1 **Solving a Quadratic Inequality Including Endpoints**

Solve $x^2 - 3x - 10 \leq 0$.

To begin, we find the solution of the corresponding quadratic equation,

$$x^2 - 3x - 10 = 0.$$

Factor to get

$$(x - 5)(x + 2) = 0$$

from which

$$x - 5 = 0 \quad \text{or} \quad x + 2 = 0$$
$$x = 5 \quad \text{or} \quad x = -2.$$

JOURNAL   WRITING   ▲ CHALLENGING   SCIENTIFIC CALCULATOR   GRAPHING CALCULATOR

INSTRUCTOR'S RESOURCES

PTB Sec. 5.7 **ISM** Sec. 5.7
AB Sec. 5.7

TEST GENERATOR

WORLD WIDE WEB
www.LialAlgebra.com

CHALKBOARD EXAMPLE

Solve and graph.

$$x^2 + 4x - 12 \le 0$$

Answer: $[-6, 2]$

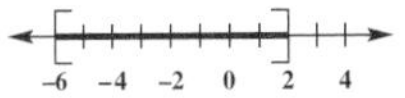

Since 5 and -2 are the only values that satisfy $x^2 - 3x - 10 = 0$, all other values of x will make $x^2 - 3x - 10$ either less than 0 (< 0) or greater than 0 (> 0). The values $x = 5$ and $x = -2$ determine three regions on the number line, as shown in Figure 4. Region A includes all numbers less than -2, Region B includes the numbers between -2 and 5, and Region C includes all numbers greater than 5.

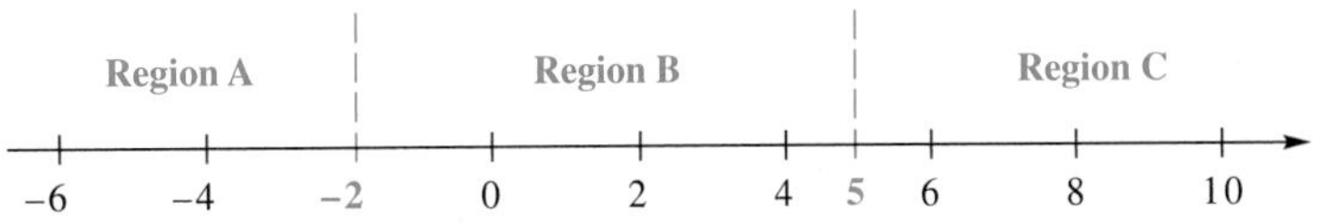

Figure 4

All values of x in a given region will cause $x^2 - 3x - 10$ to have the same sign (either positive or negative). Test one value of x from each region to see which regions satisfy $x^2 - 3x - 10 \le 0$. First, are the points in Region A part of the solution? As a trial value, choose any number less than -2, say -6.

$x^2 - 3x - 10 \le 0$		Original inequality
$(-6)^2 - 3(-6) - 10 \le 0$	?	Let $x = -6$.
$36 + 18 - 10 \le 0$	?	Simplify.
$44 \le 0$		False

Since $44 \le 0$ is false, the points in Region A do not belong to the solution.

What about Region B? Try the value $x = 0$.

$0^2 - 3(0) - 10 \le 0$	?	Let $x = 0$.
$-10 \le 0$		True

Since $-10 \le 0$ is true, the points in Region B do belong to the solution.

Try $x = 6$ to check Region C.

$6^2 - 3(6) - 10 \le 0$	?	Let $x = 6$.
$36 - 18 - 10 \le 0$	?	Simplify.
$8 \le 0$		False

Since $8 \le 0$ is false, the points in Region C do not belong to the solution.

The points in Region B are the only ones that satisfy $x^2 - 3x - 10 \le 0$. As shown in Figure 5, the solution includes the points in Region B together with the endpoints -2 and 5. The solution set is written as the interval $[-2, 5]$.

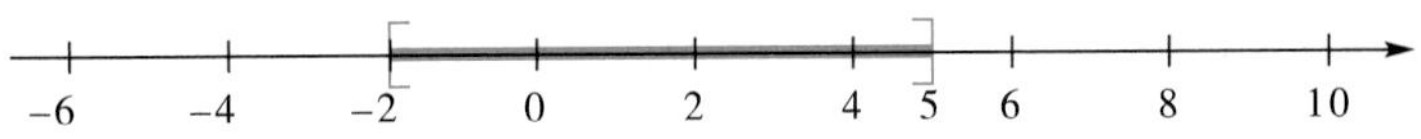

Figure 5

To summarize, we use the following steps to solve a quadratic inequality.

Solving a Quadratic Inequality

Step 1 **Write an equation.** Change the inequality to an equation.

Step 2 **Solve.** Use the zero-factor property to solve the equation from Step 1.

Step 3 **Determine the regions.** Use the solutions of the equation in Step 2 to determine regions on the number line.

Step 4 **Test each region.** Choose a number from each region. Substitute the number into the original inequality. If the number satisfies the inequality, all numbers in that region satisfy the inequality.

Step 5 **Write the solution set.** Write the solution set in interval notation.

CHALKBOARD EXAMPLE

Solve and graph.

$$x^2 - 3x - 4 > 0$$

Answer: $(-\infty, -1) \cup (4, \infty)$

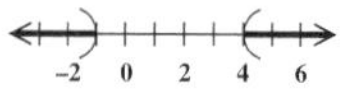

EXAMPLE 2 **Solving a Quadratic Inequality Excluding Endpoints**

Solve $-x^2 - 5x - 6 < 0$.

It will be easier to factor if we multiply both sides by -1. We must also remember to reverse the direction of the inequality symbol:

$$x^2 + 5x + 6 > 0.$$

Factoring the expression in the corresponding equation $x^2 + 5x + 6 = 0$, we get $(x + 2)(x + 3) = 0$. The solutions of the equation are -2 and -3. These points determine three regions on the number line. See Figure 6. This time, these points will not belong to the solution because only values of x that make $x^2 + 5x + 6$ *greater than* 0 are solutions.

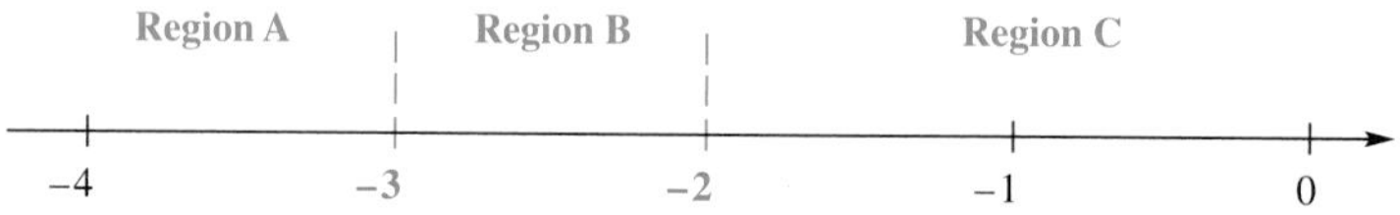

Figure 6

Do the points in Region A belong to the solution? Decide by selecting any number in Region A, such as -4. Does -4 satisfy the inequality?

$$-x^2 - 5x - 6 < 0 \qquad \text{Original inequality}$$
$$-(-4)^2 - 5(-4) - 6 < 0 \quad ? \qquad \text{Let } x = -4.$$
$$-16 + 20 - 6 < 0 \quad ? \qquad \text{Simplify.}$$
$$-2 < 0 \qquad \text{True}$$

Since $-2 < 0$ is true, all the points in Region A belong to the solution set of the inequality.

For Region B, choose a number between -3 and -2, say $-2\frac{1}{2}$, or $-\frac{5}{2}$.

$$-\left(-\frac{5}{2}\right)^2 - 5\left(-\frac{5}{2}\right) - 6 < 0 \quad ? \qquad \text{Let } x = -\frac{5}{2}.$$
$$-\frac{25}{4} + \frac{25}{2} - 6 < 0 \quad ? \qquad \text{Simplify.}$$
$$\frac{1}{4} < 0 \qquad \text{False}$$

Since $\frac{1}{4} < 0$ is false, no point in Region B belongs to the solution set.

For Region C, try the number 0.

$$-0^2 - 5(0) - 6 < 0 \quad ? \qquad \text{Let } x = 0.$$
$$-6 < 0 \qquad \text{True}$$

Since $-6 < 0$ is true, the points in Region C belong to the solution set.

The solutions include all values of x less than -3, together with all values of x greater than -2, as shown in the graph in Figure 7. Using inequalities, the solutions may be written as $x < -3$ *or* $x > -2$. Any number that satisfies *either* of these inequalities will be a solution, because *or* means "either one or the other or both." In interval notation, we write the solution set using the **union symbol** $\cup$ for *or*:

$$(-\infty, -3) \cup (-2, \infty).$$

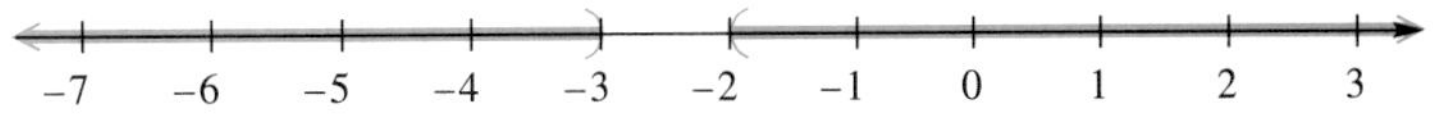

Figure 7

TEACHING TIP Students may try to write the inequality as $-2 < x < -3$. Note that this notation means "and." There is no number greater than -2 *and* also less than -3. Also, the inequality $-2 < x < -3$ implies that $-2 < -3$, which is false.

CAUTION There is no shortcut way to write the solution set $x < -3$ or $x > -2$.

CONNECTIONS

The number of items that must be sold for a company's revenue to equal its cost to produce those items is called the *break-even point.* If revenue, R, and cost, C, are given by expressions in x, where x is the number of items produced and sold, then the solution of the equation $R - C = 0$ gives the break-even point. For the company to make a profit, $R - C$ must be greater than 0, and the company would want to solve the inequality $R - C > 0$.

FOR DISCUSSION OR WRITING

Explain why $R - C > 0$ gives values of x that produce a profit. Suppose $R - C = x^2 - x - 12$. What values of x produce a profit? Why is only one part of the solution of the corresponding inequality valid here?

The profit is found by subtracting cost from revenue, represented by $R - C$. If $R - C = x^2 - x - 12$, the solution set of $R - C > 0$ is $x < -3$ or $x > 4$. Only $x > 4$ makes sense, since x must be positive in the context of the problem.

5.7 EXERCISES

1. Which of these inequalities is not a quadratic inequality?
   (a) $5x^2 - x + 7 < 0$ (b) $x^2 + 4 > 0$ (c) $3x + 5 \geq 0$
   (d) $(x - 1)(x + 2) \leq 0$

2. To solve a quadratic inequality, we use the ______ of the corresponding equation to determine ______ on the number line.

To prepare for using test values in the various regions, answer true or false depending on whether the given value of x satisfies or does not satisfy the inequality.

3. $(2x + 1)(x - 5) \geq 0$
   (a) $x = -\frac{1}{2}$ (b) $x = 5$ (c) $x = 0$ (d) $x = -6$

4. $(x - 6)(3x + 4) < 0$
   (a) $x = -3$ (b) $x = 6$ (c) $x = 0$ (d) $x = -\frac{4}{3}$

Answers:

1. (c) **2.** solutions; regions **3.** (a) true (b) true (c) false (d) true **4.** (a) false (b) false (c) true (d) false

5. $(-3, 3)$

6. $(-\infty, -2) \cup (2, \infty)$

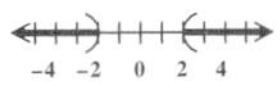

7. $(-\infty, -6] \cup [7, \infty)$

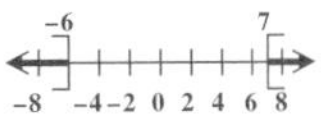

8. $[4, 5]$

9. $(-\infty, -3) \cup (-2, \infty)$

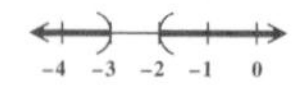

10. $(1, 2)$ 

11. $[-1, 5]$ 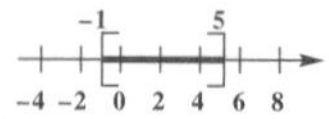

12. $\left[-\frac{1}{3}, 2\right]$ 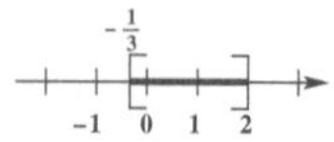

13. $\left(-1, \frac{2}{5}\right)$ 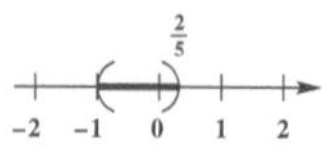

14. $(-\infty, -4) \cup \left(\frac{1}{2}, \infty\right)$

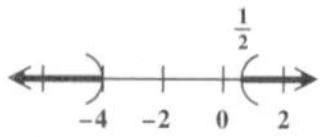

15. $\left(-\frac{1}{2}, \frac{4}{3}\right)$ 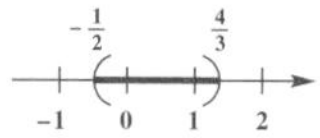

16. $\left(-\infty, -\frac{3}{2}\right) \cup \left(\frac{1}{3}, \infty\right)$

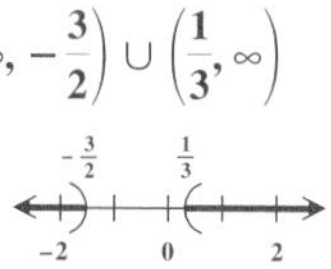

17. $(1, 6)$ 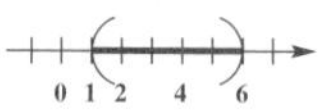

18. $\left[-\frac{3}{2}, 5\right]$ 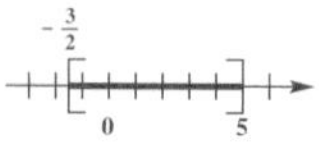

19. $\left(-\infty, -\frac{1}{2}\right) \cup \left(\frac{1}{3}, \infty\right)$

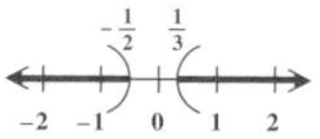

20. $\left[-\frac{1}{2}, \frac{2}{5}\right]$ 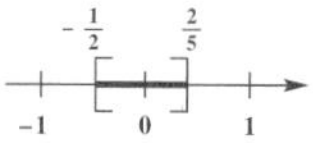

21. $\left(-\frac{2}{3}, -\frac{1}{4}\right)$ 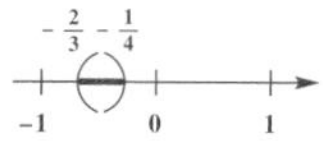

22. $(-4, 4)$ 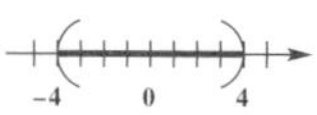

23. $(-\infty, -2) \cup (2, \infty)$

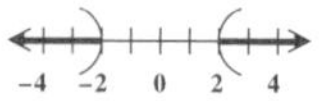

24. $(-\infty, -10] \cup [10, \infty)$

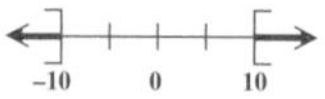

25. $(-\infty, -4) \cup (4, \infty)$

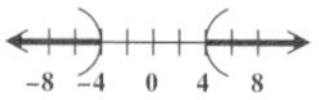

26. $(-\infty, -5] \cup [5, \infty)$

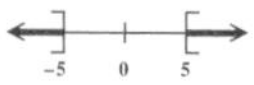

27. $[-1.5, .8]$

Solve each inequality and graph the solution set. See Examples 1 and 2.

5. $(a + 3)(a - 3) < 0$
6. $(b - 2)(b + 2) > 0$
7. $(a + 6)(a - 7) \geq 0$
8. $(z - 5)(z - 4) \leq 0$
9. $m^2 + 5m + 6 > 0$
10. $y^2 - 3y + 2 < 0$
11. $z^2 - 4z - 5 \leq 0$
12. $3p^2 - 5p - 2 \leq 0$
13. $5m^2 + 3m - 2 < 0$
14. $2k^2 + 7k - 4 > 0$
15. $6r^2 - 5r < 4$
16. $6r^2 + 7r > 3$
17. $q^2 - 7q < -6$
18. $2k^2 - 7k \leq 15$
19. $6m^2 + m - 1 > 0$
20. $30r^2 + 3r - 6 \leq 0$
21. $12p^2 + 11p + 2 < 0$
22. $a^2 - 16 < 0$
23. $9m^2 - 36 > 0$
24. $r^2 - 100 \geq 0$
25. $r^2 > 16$
26. $m^2 \geq 25$

TECHNOLOGY INSIGHTS (EXERCISES 27–30)

The calculator screens show the x-intercepts of the graph of the quadratic equation corresponding to each quadratic inequality. When the graph is above the x-axis, the values of y (which represents the quadratic expression) are positive. When the graph is below the x-axis, the values of y are negative. Therefore, the quadratic expression is positive for those x-values that correspond to positive y-values and negative for those x-values that correspond to negative y-values.

Use the graphs and the x-intercepts to determine the intervals that satisfy each inequality.

27. $x^2 + .7x - 1.2 \leq 0$

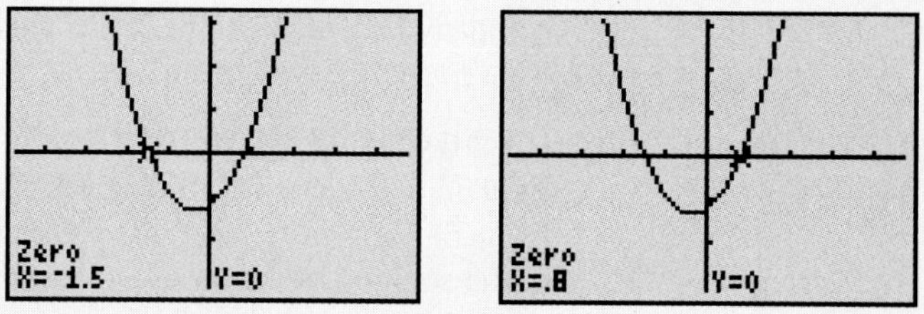

28. $3x^2 + 4x \geq 0$

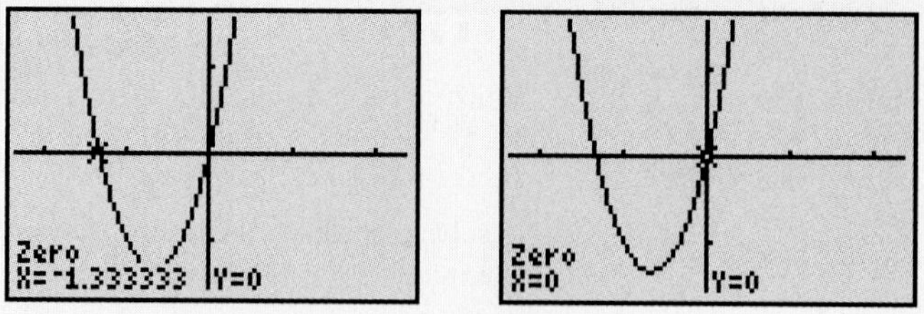

28. $(-\infty, -1.333333] \cup [0, \infty)$
29. $(-\infty, 1.5] \cup [3.5, \infty)$
30. $[-2.666667, .5]$
31. $\left[-2, \dfrac{1}{3}\right] \cup [4, \infty)$

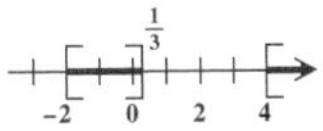

32. $(-\infty, -3] \cup \left[1, \dfrac{7}{2}\right]$

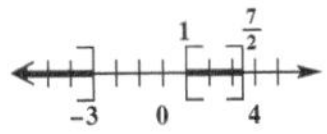

33. $(-\infty, -1) \cup (2, 4)$

34. $(-5, -2) \cup (3, \infty)$

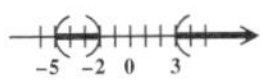

35. 2 seconds and 14 seconds **36.** between 2 seconds and 14 seconds **37.** between 0 and 2 seconds or between 14 and 16 seconds **38.** Because t represents time, only numbers greater than or equal to 0 can be used as solutions. Also, when $t = 16$, the height is 0, that is, ground level. If t is greater than 16, the height would be negative.

TECHNOLOGY INSIGHTS (EXERCISES 27–30) (CONTINUED)

29. $4x^2 - 20x + 21 \geq 0$

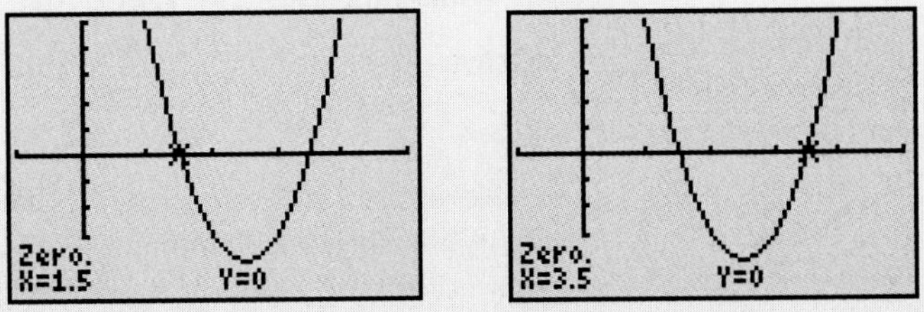

30. $6x^2 + 13x - 8 \leq 0$

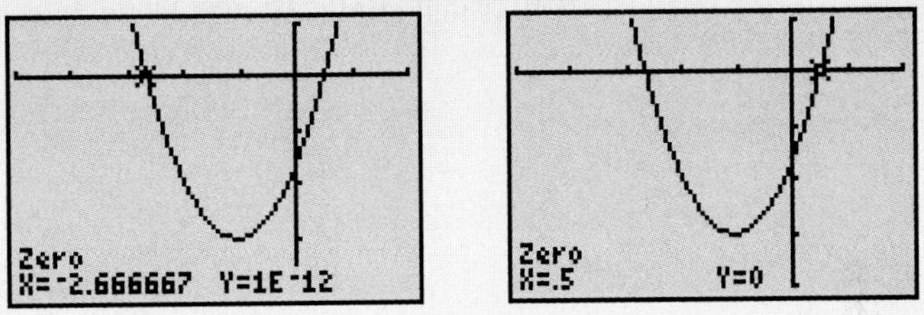

▲ *The given inequality is not quadratic, but it may be solved in a similar manner. Solve and graph each inequality.* (Hint: *Because these inequalities correspond to equations with three solutions, they determine four regions on a number line.)*

31. $(a + 2)(3a - 1)(a - 4) \geq 0$

32. $(2p - 7)(p - 1)(p + 3) \leq 0$

33. $(r - 2)(r^2 - 3r - 4) < 0$

34. $(m + 5)(m^2 - m - 6) > 0$

An object is propelled upward from ground level with an initial velocity of 256 feet per second. After t seconds, its height h is given by the equation $h = -16t^2 + 256t$, where h is in feet.

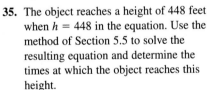

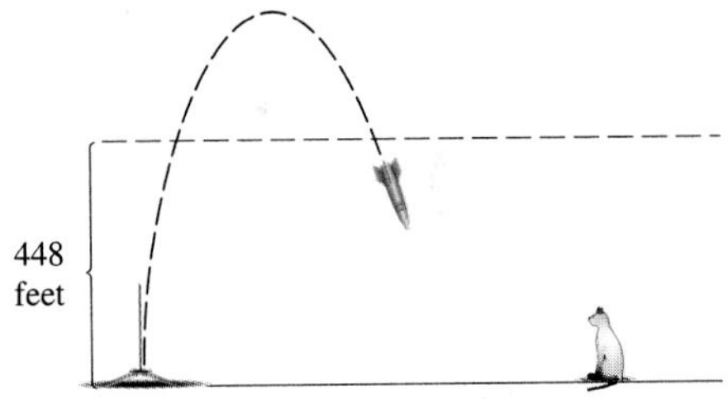

35. The object reaches a height of 448 feet when $h = 448$ in the equation. Use the method of Section 5.5 to solve the resulting equation and determine the times at which the object reaches this height.

36. The object is more than 448 feet above the ground when $h > 448$. This is given by the inequality $-16t^2 + 256t > 448$. Use the method of this section to find the time interval over which the object is more than 448 feet above the ground.

37. By solving $-16t^2 + 256t < 448$ and realizing that t must be greater than or equal to 0, we can determine the time intervals over which the object is less than 448 feet above the ground. Use the method of this section to find this time interval. (*Note:* We must also have $t \leq 16$.)

✎ **38.** Explain the restrictions on t described in Exercise 37.

Instructor's Note
Grouping: 3 students
Time: 20 minutes

This activity can be divided so each student solves the problems for a different acceleration rate (1, 2, or 3) or so that one student solves for time, one for speed, and one for distance. The first equation involves factoring using the difference of two squares where the second square is a decimal number found by first dividing $\frac{d}{a}$. Students will need help understanding how to do this. You may wish to work through the data for Car 1 with them. The other two equations involve only substitution. You could also do this activity after solving quadratic equations by taking the square root or using the quadratic formula (Chapter 10). Since this activity involves converting from seconds to hours, you may have students figure out how to do the conversions or you may want to tell them to divide seconds by 3600.

CHAPTER 5 GROUP ACTIVITY

Transportation for Tomorrow

Objective: Use quadratic equations to find vehicle speed and stopping distance given acceleration rate.

In 1998 major automotive manufacturers began to produce electric cars. In this activity the speed and stopping distances of three different electric vehicles with given acceleration rates will be examined. Acceleration is the rate of change of speed with respect to time and is given in feet per second per second. Complete the table below for Cars 1, 2, and 3 to find the time, speed, and distance for each vehicle. (Round to the nearest whole number.)

Car	*Acceleration Time*	*Speed*	*Stopping Distance*
1	**(18 seconds)**	**(60 miles per hour)**	**(228 feet)**
2	**(20 seconds)**	**(54 miles per hour)**	**(191 feet)**
3	**(17 seconds)**	**(64 miles per hour)**	**(255 feet)**

A. Car 1: The Toyota Rav4EV has an acceleration constant of $a = 4.89$ feet per second per second.*

1. Determine how long it would take for this vehicle to go $\frac{3}{10}$ mile (1584 feet) by solving the formula $d = at^2$, where d represents distance in feet, a is the acceleration constant, and t is time in seconds.
2. Now find the speed r in miles per hour the car has achieved using the distance formula $d = rt$. (Convert time to hours, distance to miles.)
3. Next find how many feet it will take the car to stop on a dry road at that speed (in miles per hour) using the following quadratic equation: $d = .045r^2 + 1.1r$.

B. Determine the same information for two other vehicles with the following acceleration constants.
Car 2: $a = 3.96$ feet per second per second
Car 3: $a = 5.48$ feet per second per second

**Source:* Ecomall.com.

CHAPTER 5 SUMMARY

KEY TERMS

5.1 factor
common factor
greatest common factor
factoring by grouping

5.2 factored form
factoring
prime polynomial

5.4 perfect square trinomial

5.5 quadratic equation
standard form

5.6 hypotenuse
legs

5.7 quadratic inequality

TEST YOUR WORD POWER

See how well you have learned the vocabulary in this chapter. Answers, with examples, are given at the bottom of the page.

1. Factoring is
- (a) a method of multiplying polynomials
- (b) the process of writing a polynomial as a product
- (c) the answer in a multiplication problem
- (d) a way to add the terms of a polynomial.

2. A polynomial is in **factored form** when
- (a) it is prime
- (b) it is written as a sum
- (c) the squared term has a coefficient of 1
- (d) it is written as a product.

3. A **perfect square trinomial** is a trinomial
- (a) that can be factored as the square of a binomial
- (b) that cannot be factored
- (c) that is multiplied by a binomial
- (d) where all terms are perfect squares.

4. A **quadratic equation** is a polynomial equation of
- (a) degree one
- (b) degree two
- (c) degree three
- (d) degree four.

5. A **hypotenuse** is
- (a) either of the two shorter sides of a triangle
- (b) the shortest side of a triangle
- (c) the side opposite the right angle in a triangle
- (d) the longest side in any triangle.

QUICK REVIEW

CONCEPTS	EXAMPLES
5.1 THE GREATEST COMMON FACTOR; FACTORING BY GROUPING	
Finding the Greatest Common Factor (GCF) 1. Include the largest numerical factor of every term. 2. Include each variable that is a factor of every term raised to the smallest exponent that appears in a term.	Find the greatest common factor of $4x^2y,\quad -6x^2y^3,\quad 2xy^2.$ $4x^2y = 2^2 \cdot x^2 \cdot y$ $-6x^2y^3 = -1 \cdot 2 \cdot 3 \cdot x^2 \cdot y^3$ $2xy^2 = 2 \cdot x \cdot y^2$ The greatest common factor is $2xy$.
Factoring by Grouping *Step 1* Group the terms so that each group has a common factor.	Factor $3x^2 + 5x - 24xy - 40y$. $(3x^2 + 5x) + (-24xy - 40y)$
Step 2 Factor out the greatest common factor in each group.	$x(3x + 5) - 8y(3x + 5)$
Step 3 Factor a common binomial factor from the result of Step 2.	$(3x + 5)(x - 8y)$
Step 4 If Step 3 cannot be performed, try a different grouping.	
5.2 FACTORING TRINOMIALS	
To factor $x^2 + bx + c$, find m and n such that $mn = c$ and $m + n = b$. $mn = c$ $\downarrow$ $x^2 + bx + c$ $\uparrow$ $m + n = b$ Then $x^2 + bx + c = (x + m)(x + n).$	Factor $x^2 + 6x + 8$. $mn = 8$ $\downarrow$ $x^2 + 6x + 8$ $\uparrow$ $m + n = 6$ $m = 2$ and $n = 4$ $x^2 + 6x + 8 = (x + 2)(x + 4)$

Answers to Test Your Word Power

1. (b) *Example:* $x^2 - 5x - 14 = (x - 7)(x + 2)$ **2.** (d) *Example:* The factored form of $x^2 - 5x - 14$ is $(x - 7)(x + 2)$. **3.** (a) *Example:* $a^2 + 2a + 1$ is a perfect square trinomial; its factored form is $(a + 1)^2$. **4.** (b) *Examples:* $y^2 - 3y + 2 = 0$, $x^2 - 9 = 0$, $2m^2 = 6m + 8$
5. (c) *Example:* In Figure 3 of Section 5.6, the hypotenuse is the side labeled $x + 2$.

CONCEPTS	EXAMPLES

5.3 MORE ON FACTORING TRINOMIALS

To factor $ax^2 + bx + c$:

By Grouping

Find m and n:

$$mn = ac$$

$$ax^2 + bx + c$$

$$m + n = b$$

Factor $3x^2 + 14x - 5$.

$$-15$$

$$mn = -15,\ m + n = 14$$

By Trial and Error

Use FOIL backwards.

By trial and error or by grouping.

$$3x^2 + 14x - 5 = (3x - 1)(x + 5)$$

5.4 SPECIAL FACTORING RULES

Difference of Squares

$x^2 - y^2 = (x + y)(x - y)$

Perfect Square Trinomial

$x^2 + 2xy + y^2 = (x + y)^2$

$x^2 - 2xy + y^2 = (x - y)^2$

Difference of Cubes

$x^3 - y^3 = (x - y)(x^2 + xy + y^2)$

Sum of Cubes

$x^3 + y^3 = (x + y)(x^2 - xy + y^2)$

Factor.

$$4x^2 - 9 = (2x + 3)(2x - 3)$$

$$9x^2 + 6x + 1 = (3x + 1)^2$$

$$4x^2 - 20x + 25 = (2x - 5)^2$$

$$m^3 - 8 = m^3 - 2^3 = (m - 2)(m^2 + 2m + 4)$$

$$27 + z^3 = 3^3 + z^3 = (3 + z)(9 - 3z + z^2)$$

5.5 SOLVING QUADRATIC EQUATIONS BY FACTORING

Zero-Factor Property

If a and b are real numbers and if $ab = 0$, then $a = 0$ or $b = 0$.

If $(x - 2)(x + 3) = 0$, then $x - 2 = 0$ or $x + 3 = 0$.

Solving a Quadratic Equation by Factoring

Step 1 Write in standard form.

Step 2 Factor.

Step 3 Use the zero-factor property.

Step 4 Check.

Solve $2x^2 = 7x + 15$.

$$2x^2 - 7x - 15 = 0$$

$$(2x + 3)(x - 5) = 0$$

$$2x + 3 = 0 \quad \text{or} \quad x - 5 = 0$$

$$2x = -3 \qquad\qquad x = 5$$

$$x = -\frac{3}{2}$$

Both solutions satisfy the original equation; the solution set is $\left\{-\frac{3}{2}, 5\right\}$.

5.6 APPLICATIONS OF QUADRATIC EQUATIONS

Pythagorean Formula

In a right triangle, the square of the hypotenuse equals the sum of the squares of the legs.

$$a^2 + b^2 = c^2$$

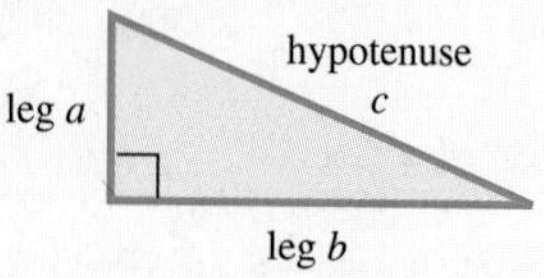

In a right triangle, one leg measures 2 feet longer than the other. The hypotenuse measures 4 feet longer than the shorter leg. Find the lengths of the three sides of the triangle.

Let $x =$ the length of the shorter leg. Then

$$x^2 + (x + 2)^2 = (x + 4)^2.$$

Verify that the solutions of this equation are -2 and 6. Discard -2 as a solution. Check that the sides are 6, $6 + 2 = 8$, and $6 + 4 = 10$ feet in length.

CONCEPTS	EXAMPLES

5.7 SOLVING QUADRATIC INEQUALITIES

Solving a Quadratic Inequality

Step 1 Change the inequality to an equation.

Step 2 Use the zero-factor property to solve the equation from Step 1.

Step 3 Use the solutions of the equation in Step 2 to determine regions on the number line.

Step 4 Choose a number from each region. Substitute the number into the original inequality. If the number satisfies the inequality, all numbers in that region satisfy the inequality.

Step 5 Write the solution set in interval notation.

Solve $2x^2 + 5x - 3 < 0$.

$$2x^2 + 5x - 3 = 0$$

$$(2x - 1)(x + 3) = 0$$

$$2x - 1 = 0 \quad \text{or} \quad x + 3 = 0$$

$$2x = 1 \quad \text{or} \quad x = -3$$

$$x = \frac{1}{2}$$

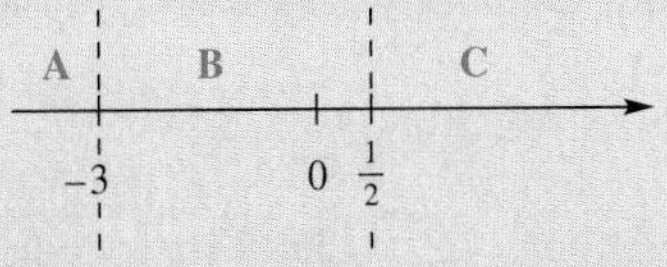

Choose -4 from A:

$$2(-4)^2 + 5(-4) - 3 < 0 \quad ?$$

$$9 < 0 \qquad \text{False}$$

Choose 0 from B:

$$2(0)^2 + 5(0) - 3 < 0 \quad ?$$

$$-3 < 0 \qquad \text{True}$$

Choose 1 from C:

$$2(1)^2 + 5(1) - 3 < 0 \quad ?$$

$$4 < 0 \qquad \text{False}$$

Only the numbers from Region B satisfy the inequality.

The solution set is written $\left(-3, \frac{1}{2}\right)$.

CHAPTER 5 REVIEW EXERCISES

1. $7(t + 2)$
2. $30z(2z^2 + 1)$
3. $(x - 4)(2y + 3)$
4. $(3y + 2)(2y + 3)$
5. $(x + 3)(x + 2)$
6. $(y - 5)(y - 8)$
7. $(q + 9)(q - 3)$
8. $(r - 8)(r + 7)$
9. $(r + 8s)(r - 12s)$
10. $(p + 12q)(p - 10q)$
11. $8p(p + 2)(p - 5)$
12. $3x^2(x + 2)(x + 8)$
13. $p^5(p - 2q)(p + q)$
14. $3r^3(r + 3s)(r - 5s)$
15. r and $6r$, $2r$ and $3r$
16. Factor out z.

[5.1] *Factor out the greatest common factor or factor by grouping.*

1. $7t + 14$

2. $60z^3 + 30z$

3. $2xy - 8y + 3x - 12$

4. $6y^2 + 9y + 4y + 6$

[5.2] *Factor completely.*

5. $x^2 + 5x + 6$

6. $y^2 - 13y + 40$

7. $q^2 + 6q - 27$

8. $r^2 - r - 56$

9. $r^2 - 4rs - 96s^2$

10. $p^2 + 2pq - 120q^2$

11. $8p^3 - 24p^2 - 80p$

12. $3x^4 + 30x^3 + 48x^2$

13. $p^7 - p^6q - 2p^5q^2$

14. $3r^5 - 6r^4s - 45r^3s^2$

[5.3]

15. To begin factoring $6r^2 - 5r - 6$, what are the possible first terms of the two binomial factors if we consider only positive integer coefficients?

16. What is the first step you would use to factor $2z^3 + 9z^2 - 5z$?

17. $(2k - 1)(k - 2)$
18. $(3r - 1)(r + 4)$
19. $(3r + 2)(2r - 3)$
20. $(5z + 1)(2z - 1)$
21. $(v + 3)(8v - 7)$
22. $4x^3(3x - 1)(2x - 1)$
23. $-3(x + 2)(2x - 5)$
24. $rs\,(5r + 6s)(2r + s)$
25. (b) 26. (d)
27. $(n + 7)(n - 7)$
28. $(5b + 11)(5b - 11)$
29. $(7y + 5w)(7y - 5w)$
30. $36(2p + q)(2p - q)$
31. prime 32. $(r - 6)^2$
33. $(3t - 7)^2$
34. $(m + 10)(m^2 - 10m + 100)$
35. $(5k + 4x)(25k^2 - 20kx + 16x^2)$
36. $(7x - 4)(49x^2 + 28x + 16)$
37. $\{-3, -1\}$ 38. $\{1, 4\}$
39. $\{3, 5\}$
40. $\left\{-\frac{4}{3}, 5\right\}$ 41. $\left\{-\frac{8}{9}, \frac{8}{9}\right\}$
42. $\{0, 8\}$ 43. $\{-1, 6\}$
44. $\{7\}$ 45. $\{6\}$
46. $\left\{-\frac{2}{5}, -2, -1\right\}$
47. length: 10 meters; width: 4 meters
48. length: 6 meters; width: 2 meters
49. length: 6 meters; height: 5 meters
50. 3 feet, 4 feet, 5 feet
51. 112 feet 52. 192 feet
53. 256 feet
54. after 8 seconds

Factor completely.

17. $2k^2 - 5k + 2$

18. $3r^2 + 11r - 4$

19. $6r^2 - 5r - 6$

20. $10z^2 - 3z - 1$

21. $8v^2 + 17v - 21$

22. $24x^5 - 20x^4 + 4x^3$

23. $-6x^2 + 3x + 30$

24. $10r^3s + 17r^2s^2 + 6rs^3$

[5.4]

25. Which one of the following is the difference of two squares?
(a) $32x^2 - 1$ **(b)** $4x^2y^2 - 25z^2$ **(c)** $x^2 + 36$ **(d)** $25y^3 - 1$

26. Which one of the following is a perfect square trinomial?
(a) $x^2 + x + 1$ **(b)** $y^2 - 4y + 9$ **(c)** $4x^2 + 10x + 25$ **(d)** $x^2 - 20x + 100$

Factor completely.

27. $n^2 - 49$

28. $25b^2 - 121$

29. $49y^2 - 25w^2$

30. $144p^2 - 36q^2$

31. $x^2 + 100$

32. $r^2 - 12r + 36$

33. $9t^2 - 42t + 49$

34. $m^3 + 1000$

35. $125k^3 + 64x^3$

36. $343x^3 - 64$

[5.5] *Solve each equation and check your solutions.*

37. $z^2 + 4z + 3 = 0$

38. $2m^2 - 10m + 8 = 0$

39. $x(x - 8) = -15$

40. $3z^2 - 11z - 20 = 0$

41. $81t^2 - 64 = 0$

42. $y^2 = 8y$

43. $3n(n - 5) = 18$

44. $t^2 - 14t + 49 = 0$

45. $t^2 = 12(t - 3)$

46. $(5z + 2)(z^2 + 3z + 2) = 0$

[5.6] *Solve each problem.*

47. The length of a rectangle is 6 meters more than the width. The area is 40 square meters. Find the length and width of the rectangle.

48. The length of a rectangle is three times the width. If the width were increased by 3 meters while the length remained the same, the new rectangle would have an area of 30 square meters. Find the length and width of the original rectangle.

49. The volume of a box is to be 120 cubic meters. The width of the box is to be 4 meters, and the height 1 meter less than the length. Find the length and height of the box.

50. The sides of a right triangle have lengths (in feet) that are consecutive integers. What are the lengths of the sides?

If an object is propelled straight up from ground level with an initial velocity of 128 feet per second, its height h in feet after t seconds is $h = 128t - 16t^2$.

Find the height of the object after the following periods of time.

51. 1 second

52. 2 seconds

53. 4 seconds

54. When does the object described above return to the ground?

55. 2 inches 56. 2 centimeters
57. (a) \$17.4 million (b) −\$48.6 million (c) The equation is based on the data for 1994 to 1996. To use it to predict much beyond 1996 may lead to incorrect results, since it is possible that the conditions the equation was based on will change.
58. $(-\infty, -5) \cup (3, \infty)$
59. $(-\infty, -4] \cup \left[\frac{1}{2}, \infty\right)$
60. $[2, 3]$
61. $(-\infty, -4] \cup \left[\frac{3}{2}, \infty\right)$

55. A 9-inch by 12-inch picture is to be placed on a cardboard mat so that there is an equal border around the picture. The area of the finished mat and picture is to be 208 square inches. How wide will the border be?

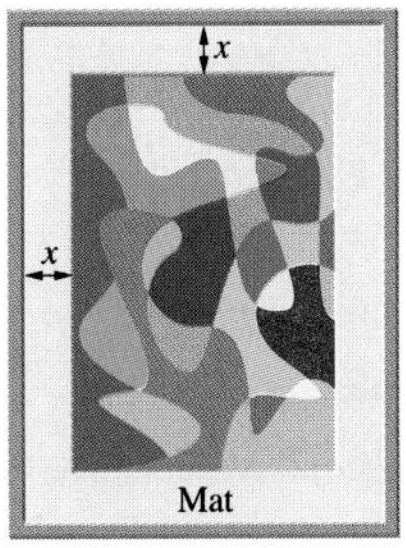

56. A box is made from a 12-centimeter by 10-centimeter piece of cardboard by cutting equal-sized squares from each corner and folding up the sides. The area of the bottom of the box is to be 48 square centimeters. Find the length of a side of the cutout squares.

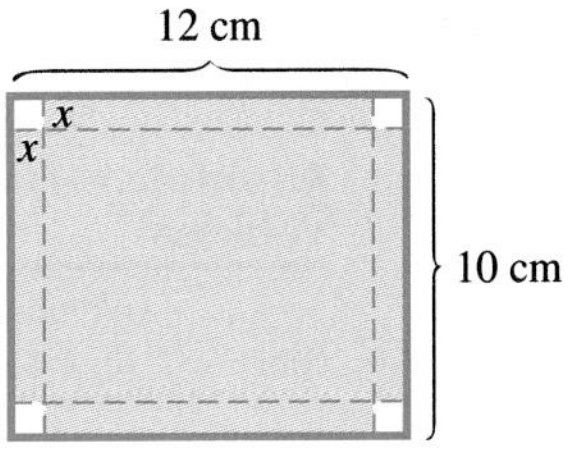

57. In 1994, Greyhound Lines, Inc. was near its second bankruptcy in three years. Since then, a new CEO, Craig R. Lentzsch, appears to have turned the company around. The bar graph and table of values show the operating income for 1994, when Lentzsch took over, 1995, and 1996. (*Source:* Company Reports, Rothschild Inc.)

Year	*Operating Income (in millions of dollars)*
1994	−65
1995	9.4
1996	37

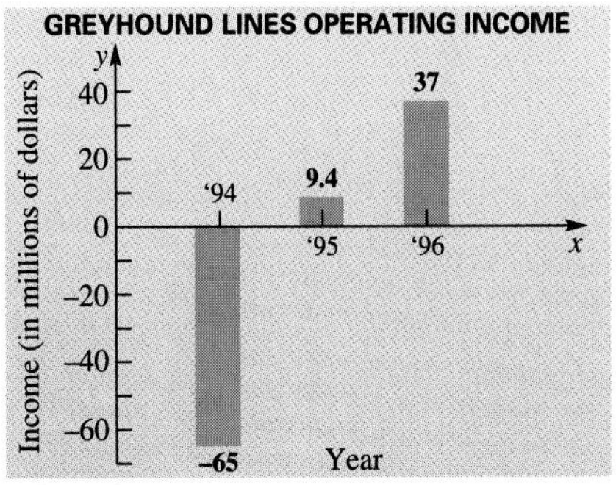

Using the data, we constructed the quadratic equation

$$y = -23.4x^2 + 285x - 831,$$

which gives the operating income y (in millions of dollars) in year x. We use $x = 4$ to represent 1994, $x = 5$ to represent 1995, and so on.

(a) Use the equation to predict the operating income in 1997.

(b) Use the equation to predict the operating income in 1998.

(c) Comment on the validity of the answers for parts (a) and (b).

[5.7] *Solve each inequality.*

58. $(q + 5)(q - 3) > 0$

59. $(2r - 1)(r + 4) \geq 0$

60. $m^2 - 5m + 6 \leq 0$

61. $2x^2 + 5x - 12 \geq 0$

62. $\left(-4, \frac{3}{2}\right)$
63. $(-\infty, -5) \cup (7, \infty)$

62. $2p^2 + 5p - 12 < 0$

63. Suppose you know that the solution set of a quadratic inequality involving the $<$ symbol is $(-5, 7)$. If the symbol is changed to $>$, what is the solution set of the new inequality?

64. $(z - x)(z - 10x)$
65. $(3k + 5)(k + 2)$
66. $(3m + 4p)(5m - 4p)$
67. $(y^2 + 25)(y + 5)(y - 5)$
68. $3m(2m + 3)(m - 5)$
69. $8abc(3b^2c - 7ac^2 + 9ab)$
70. prime
71. $6xyz(2xz^2 + 2y - 5x^2yz^3)$
72. $2a^3(a + 2)(a - 6)$
73. $(2r + 3q)(6r - 5q)$
74. $(10a + 3)(100a^2 - 30a + 9)$
75. $(7t + 4)^2$ 76. $\{0, 7\}$
77. $\{-5, 2\}$
78. $\left[-\frac{3}{4}, 1\right]$
79. (a) 417,000 vehicles (b) The estimate may be unreliable because the conditions that prevailed in the years 1995–1997 may have changed, causing either a greater increase or a greater decrease in the numbers of alternative-fueled vehicles.
80. $-6, -4$ or $6, 8$
81. width: 10 meters; length: 17 meters

MIXED REVIEW EXERCISES

Factor completely.

64. $z^2 - 11zx + 10x^2$

65. $3k^2 + 11k + 10$

66. $15m^2 + 20mp - 12mp - 16p^2$

67. $y^4 - 625$

68. $6m^3 - 21m^2 - 45m$

69. $24ab^3c^2 - 56a^2bc^3 + 72a^2b^2c$

70. $25a^2 + 15ab + 9b^2$

71. $12x^2yz^3 + 12xy^2z - 30x^3y^2z^4$

72. $2a^5 - 8a^4 - 24a^3$

73. $12r^2 + 18rq - 10rq - 15q^2$

74. $1000a^3 + 27$

75. $49t^2 + 56t + 16$

Solve.

76. $t(t - 7) = 0$

77. $x(x + 3) = 10$

78. $4x^2 - x - 3 \leq 0$

79. The numbers of alternative-fueled vehicles, in thousands, in use for the years 1995–1997 are given in the table.

Alternative-Fueled Vehicles

Year	Number (in thousands)
1995	333
1996	357
1997	386

Source: Energy Information Administration, Alternatives to Traditional Fuels, 1993.

Using statistical methods, we constructed the quadratic equation

$$y = 2.5x^2 - 453.5x + 20{,}850$$

to model the number of vehicles y in year x. Here we used $x = 95$ for 1995, $x = 96$ for 1996, and so on. Because only three years of data were used to determine the model, we must be particularly careful about using it to estimate for years before 1995 or after 1997.

(a) What prediction for 1998 is given by the equation?

(b) Why might the prediction for 1998 be unreliable?

80. The sum of two consecutive even integers is 34 less than their product. Find the integers.

81. The floor plan for a house is a rectangle with length 7 meters more than its width. The area is 170 square meters. Find the width and length of the house.

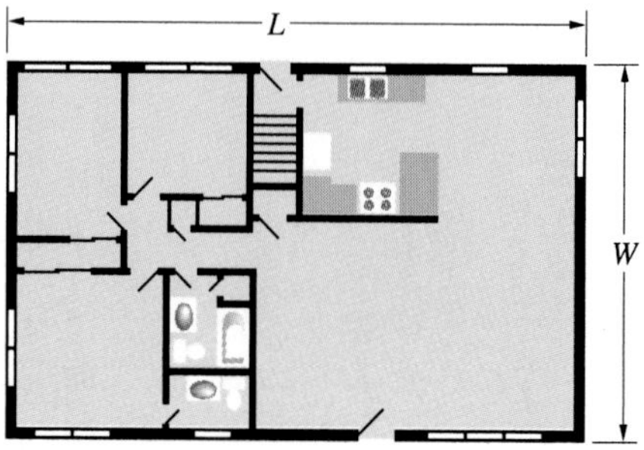

82. 6 meters
83. 34 miles **84.** 8 feet
85. 25 miles **86.** The polynomial is not complete factored because the factor $2x + 8$ has a common factor of 2. The complete factored form is $2(x + 4)(3x - 4)$.

82. The triangular sail of a schooner has an area of 30 square meters. The height of the sail is 4 meters more than the base. Find the base of the sail.

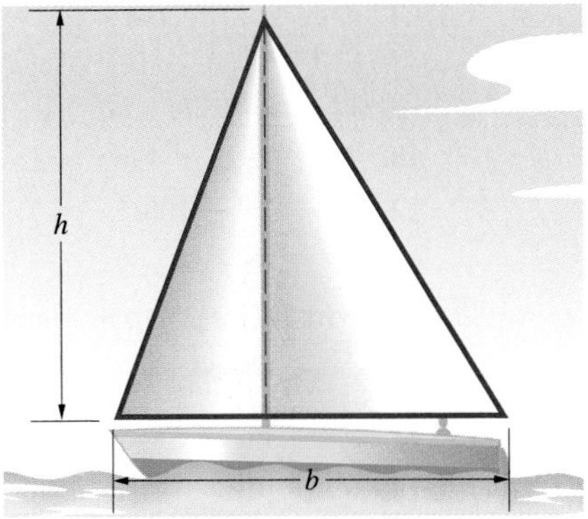

83. Two cars left an intersection at the same time. One traveled north. The other traveled 14 miles farther, but to the east. How far apart were they then, if the distance between them was 4 miles more than the distance traveled east?

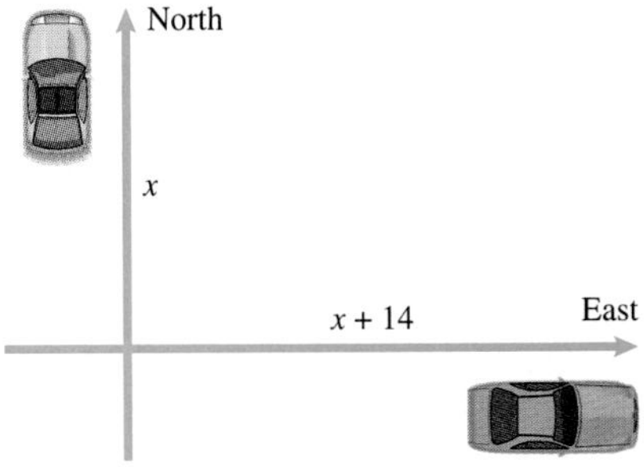

84. A ladder is leaning against a building. The distance from the bottom of the ladder to the building is 4 feet less than the length of the ladder. How high up the side of the building is the top of the ladder if that distance is 2 feet less than the length of the ladder?

85. A bicyclist heading east and a motorist traveling south left an intersection at the same time. When the motorist had gone 17 miles farther than the bicyclist, the distance between them was 1 mile more than the distance traveled by the motorist. How far apart were they then? (*Hint:* Draw a sketch.)

86. Although $(2x + 8)(3x - 4) = 6x^2 + 16x - 32$ is a true statement, the polynomial is not factored completely. Explain why and give the completely factored form.

CHAPTER 5 TEST

[5.1–5.4] **1.** (d)

1. Which one of the following is the correct, completely factored form of $2x^2 - 2x - 24$?
(a) $(2x + 6)(x - 4)$ **(b)** $(x + 3)(2x - 8)$
(c) $2(x + 4)(x - 3)$ **(d)** $2(x + 3)(x - 4)$

2. $m^2n(2mn + 3m - 5n)$
3. $(x + 3)(x - 8)$
4. $(2x + 3)(x - 1)$
5. $(5z - 1)(2z - 3)$
6. prime 7. prime
8. $(2 - a)(6 + b)$
9. $(3y + 8)(3y - 8)$
10. $(2x - 7y)^2$
11. $-2(x + 1)^2$
12. $3t^2(2t + 9)(t - 4)$
13. $(r - 5)(r^2 + 5r + 25)$
14. $8(k + 2)(k^2 - 2k + 4)$
15. The product $(p + 3)(p + 3) = p^2 + 6p + 9$, which does not equal $p^2 + 9$. The binomial $p^2 + 9$ is a prime polynomial.

[5.5] 16. $\left\{6, \frac{1}{2}\right\}$
17. $\left\{-\frac{2}{5}, \frac{2}{5}\right\}$ 18. $\{10\}$
19. $\{-3, 0, 3\}$

[5.6] 20. $1\frac{1}{2}$ seconds and $4\frac{1}{2}$ seconds
21. (a) $3x - 7$ (b) $2x - 1$ (c) 17 feet
22. July 1, 1995: 341,000; July 1, 1996: 368,000

[5.7] 23. $\left(-\frac{5}{2}, \frac{1}{3}\right)$

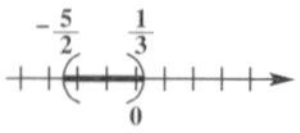

24. $(-\infty, -4] \cup [6, \infty)$

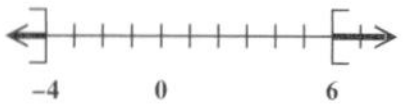

[5.5] 25. Another solution is $-\frac{2}{3}$.

Factor each polynomial completely. If it cannot be factored, write prime.

2. $2m^3n^2 + 3m^3n - 5m^2n^2$
3. $x^2 - 5x - 24$
4. $2x^2 + x - 3$
5. $10z^2 - 17z + 3$
6. $t^2 + 2t + 3$
7. $x^2 + 36$
8. $12 - 6a + 2b - ab$
9. $9y^2 - 64$
10. $4x^2 - 28xy + 49y^2$
11. $-2x^2 - 4x - 2$
12. $6t^4 + 3t^3 - 108t^2$
13. $r^3 - 125$
14. $8k^3 + 64$
15. Why is $(p + 3)(p + 3)$ not the correct factored form of $p^2 + 9$?

Solve each equation.

16. $2r^2 - 13r + 6 = 0$
17. $25x^2 - 4 = 0$
18. $x(x - 20) = -100$
19. $t^3 = 9t$

Solve each problem.

20. If an object is propelled from ground level at an initial velocity of 96 feet per second, after t seconds its height h in feet is given by the formula $h = -16t^2 + 96t$. After how many seconds will its height h be 108 feet?

21. A carpenter needs to cut a brace to support a wall stud. See the figure. The brace should be 7 feet less than three times the length of the stud. The brace will be fastened on the floor 1 foot less than twice the length of the stud away from the stud.

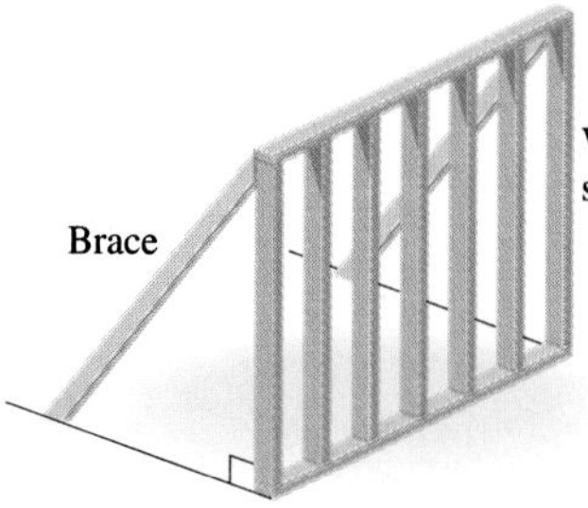

(a) Let x represent the length of the stud. Write an expression for the length of the brace.

(b) Write an expression for the distance from the wall to where the brace is fastened.

(c) How long should the brace be? (*Hint:* Use the Pythagorean formula.)

22. Assume the data given in Exercise 79 of the Chapter 5 Review Exercises represent the number of alternative-fueled vehicles at the beginning of each of the years shown. Then 95.5 would represent the number of such vehicles halfway through 1995 (July 1). Use the equation that models the data,

$$y = 2.5x^2 - 453.5x + 20{,}850,$$

to approximate the numbers of alternative-fueled vehicles on July 1 of 1995 and 1996. (Round your answers to the nearest whole number.)

Solve each inequality and graph the solutions.

23. $(3x - 1)(2x + 5) < 0$
24. $x^2 - 2x - 24 \geq 0$
25. Why isn't "$x = \frac{2}{3}$" the correct response to "Solve the equation $x^2 = \frac{4}{9}$"?

CUMULATIVE REVIEW EXERCISES CHAPTERS 1–5

[2.2] 1. $\{0\}$ **2.** $\{.05\}$ **3.** $\{6\}$
[2.4] 4. $P = \dfrac{A}{1 + rt}$
[2.3] 5. $110°$ and $70°$
6. exports: \$741 million; imports: \$426 million
[2.5] 7. 35 pounds
[3.1] 8. (a) negative; positive **(b)** negative; negative
9. (2, 27,000), (5, 63,000)
[3.2] 10. x-intercept: $\left(-\frac{1}{4}, 0\right)$; y-intercept: (0, 3)
11.

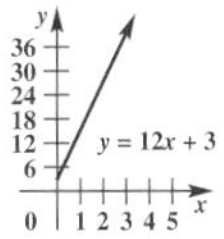

[3.3] 12. 209.2; A slope of 209.2 means that the number of radio stations increased on the average by about 209 per year.
[4.2, 4.5] 13. 4
14. $\frac{16}{9}$ **15.** 256 **16.** $\frac{1}{p^2}$

Solve each equation.

1. $3x + 2(x - 4) = 4(x - 2)$

2. $.3x + .9x = .06$

3. $\frac{2}{3}y - \frac{1}{2}(y - 4) = 3$

4. Solve for P: $A = P + Prt$.

Solve each problem.

5. Find the measures of the marked angles.

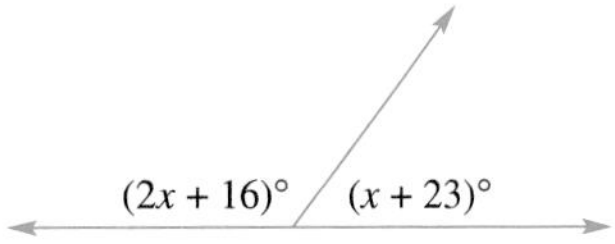

6. In a recent year, the United States exported to the Bahamas \$315 million more in goods than it imported. Together, the two amounts totaled \$1167 million. How much were the exports and how much were the imports?

7. In a mixture of concrete, there are 3 pounds of cement mix for every 1 pound of gravel. If the mixture contains a total of 140 pounds of these two ingredients, how many pounds of gravel are there?

8. Fill in each blank with *positive* or *negative.* The point with coordinates (a, b) is in
(a) quadrant II if a is _____ and b is _____.
(b) quadrant III if a is _____ and b is _____.

9. The sales of a small company are a function of the number of years x it has been in business. Sales (in thousands) are given by $y = 12x + 3$. Write ordered pairs for this function for the second and fifth years it has been in business.

10. What are the x- and y-intercepts of the graph of the sales equation $y = 12x + 3$?

11. Graph the linear equation $y = 12x + 3$ for $x \geq 0$.

12. The points on the graph show the number of U.S. radio stations in the years 1990–1995, along with the graph of a linear equation that models the data. Use the ordered pairs shown on the graph to find the slope of the line. Interpret the slope.

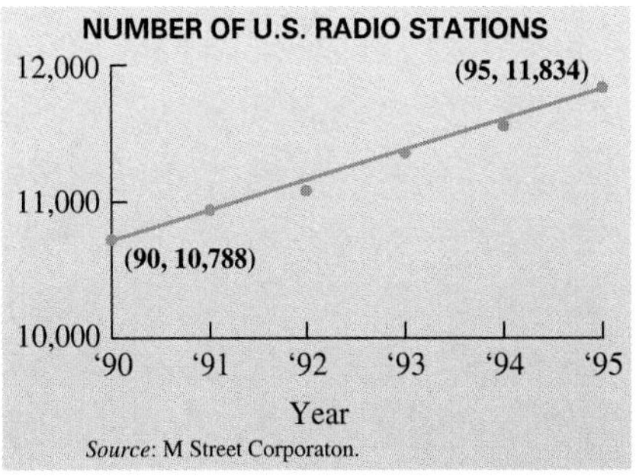

Evaluate each expression.

13. $2^{-3} \cdot 2^5$

14. $\left(\frac{3}{4}\right)^{-2}$

15. $\left(\frac{4^{-3} \cdot 4^4}{4^5}\right)^{-1}$

16. Simplify $\dfrac{(p^2)^3 p^{-4}}{(p^{-3})^{-1} p}$ and write the answer using only positive exponents. Assume $p \neq 0$.

[4.1] 17. $-4k^2 - 4k + 8$
[4.3] 18. $6m^8 - 15m^6 + 3m^4$
19. $3y^3 + 8y^2 + 12y - 5$
[4.4] 20. $4p^2 - 9q^2$
[4.6] 21. $4x^3 + 6x^2 - 3x + 10$
22. $6p^2 + 7p + 1 + \frac{7}{2p - 2}$
[5.2–5.3] 23. $(2a - 1)(a + 4)$
24. $(2m + 3)(5m + 2)$
25. $(5x + 3y)(3x - 2y)$
[5.4] 26. $(3x + 1)^2$
27. $-2(4t + 7z)^2$
28. $(5r + 9t)(5r - 9t)$
[5.1] 29. $25(4x^2 + 1)$
[5.3] 30. $2pq(3p + 1)(p + 1)$
[5.1] 31. $(2x + y)(a - b)$
[5.5] 32. $\left\{\frac{3}{2}, -2, 6\right\}$
33. $\left\{-\frac{2}{3}, \frac{1}{2}\right\}$
[5.7] 34. $(-\infty, -2] \cup \left[\frac{3}{2}, \infty\right)$

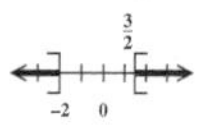

[5.6] 35. $-4, -2$ or $8, 10$
36. 5 meters, 12 meters, 13 meters

Perform the indicated operations.

17. $(2k^2 + 4k) - (5k^2 - 2) - (k^2 + 8k - 6)$

18. $3m^3(2m^5 - 5m^3 + m)$

19. $(y^2 + 3y + 5)(3y - 1)$

20. $(2p + 3q)(2p - 3q)$

21. $\dfrac{8x^4 + 12x^3 - 6x^2 + 20x}{2x}$

22. $(12p^3 + 2p^2 - 12p + 5) \div (2p - 2)$

Factor completely.

23. $2a^2 + 7a - 4$

24. $10m^2 + 19m + 6$

25. $15x^2 - xy - 6y^2$

26. $9x^2 + 6x + 1$

27. $-32t^2 - 112tz - 98z^2$

28. $25r^2 - 81t^2$

29. $100x^2 + 25$

30. $2pq + 6p^3q + 8p^2q$

31. $2ax - 2bx + ay - by$

Solve each equation.

32. $(2p - 3)(p + 2)(p - 6) = 0$

33. $6m^2 + m - 2 = 0$

34. Solve the inequality and graph the solution set: $2x^2 + x - 6 \geq 0$.

Solve each problem.

35. The difference between the squares of two consecutive even integers is 28 less than the square of the smaller integer. Find the two integers.

36. The length of the hypotenuse of a right triangle is twice the length of the shorter leg, plus 3 meters. The longer leg is 7 meters longer than the shorter leg. Find the lengths of the sides.

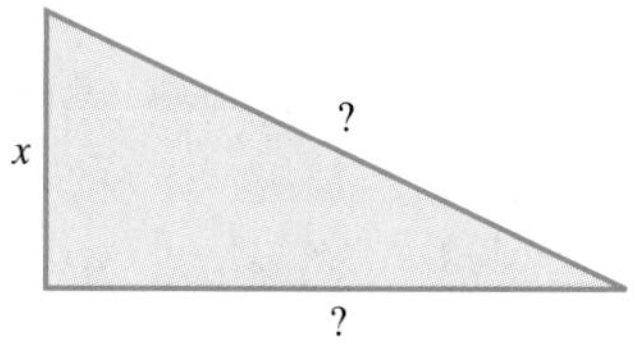

Rational Expressions 6

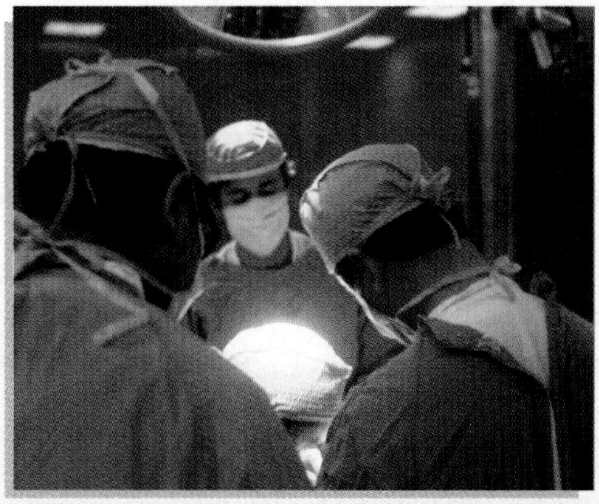

Health Care

One of the reasons that people are living longer today than in the past is the increased efficiency of health care. Organ transplants allow at-risk patients to survive longer than they would have a decade ago. Heart, liver, kidney, lung, pancreas, and cornea transplants are now available. The table shows the number of heart transplants in the United States for each year from 1990 through 1994.

Year	1990	1991	1992	1993	1994
Number of Heart Transplants	2108	2125	2171	2297	2340

Source: U.S. Department of Health and Human Services, Public Health Service, Division of Organ Transplantation, and United Network for Organ Sharing.

In 1995, the number of heart transplants was still on the rise, and compared to ten years before, in 1985, the increase was drastic. In those two years, the total number of transplants was 3080, and the ratio of the number in 1995 to the number in 1985 was approximately 33 to 10. How many transplants were performed in each of these two years? To answer this question, write an equation. If we let x represent the number of transplants in 1995, then $3080 - x$ represents the number of transplants in 1985. Writing this information as the ratio

$$\frac{\text{Number of transplants in 1995}}{\text{Number of transplants in 1985}},$$

we have

$$\frac{x}{3080 - x} = \frac{33}{10}.$$

The fractions on each side of this equation are *rational expressions,* the topic of this chapter. The equation is an example of a *rational equation.* In Section 6.6, we show how to solve this type of equation. You are asked to solve this problem in Exercise 71 of Section 6.7.

6.1 The Fundamental Property of Rational Expressions

OBJECTIVES

1. Find the values for which a rational expression is undefined.
2. Find the numerical value of a rational expression.
3. Write rational expressions in lowest terms.
4. Recognize equivalent forms of rational expressions.

FOR EXTRA HELP

SSG Sec. 6.1
SSM Sec. 6.1
Pass the Test Software
InterAct Math Tutorial Software
Video 10

TEACHING TIP Review the arithmetic operations with common fractions. Note for students that these same operations will be applied to algebraic fractions in this chapter.

The quotient of two integers (with divisor not zero) is called a rational number. In the same way, the quotient of two polynomials with divisor not equal to zero is called a *rational expression.* The techniques of factoring, studied in Chapter 5, are essential in working with rational expressions.

Rational Expression

A **rational expression** is an expression of the form

$$\frac{P}{Q}$$

where P and Q are polynomials, with $Q \neq 0$.

Examples of rational expressions include $\frac{-6x}{x^3+8}$, $\frac{9x}{y+3}$, and $\frac{2m^3}{9}$.

OBJECTIVE 1 Find the values for which a rational expression is undefined. A fraction with denominator 0 is *not* a rational expression, since division by 0 is not possible. For this reason, be careful when substituting a number for a variable in the denominator of a rational expression. For example, in

$$\frac{8x^2}{x-3}$$

x can take on any value except 3. When $x = 3$, the denominator becomes $3 - 3 = 0$, making the expression undefined.

NOTE The numerator of a rational expression may be *any* number.

To determine the values for which a rational expression is undefined, use the following procedure.

Determining When a Rational Expression Is Undefined

Step 1 Set the denominator of the rational expression equal to 0.

Step 2 Solve this equation.

Step 3 The solutions of the equation are the values that make the rational expression undefined.

This procedure is illustrated in Example 1.

CHALKBOARD EXAMPLE

Find all values for which the rational expression is undefined.

$$\frac{3r}{r^2+6r+8}$$

Answer: $-4, -2$

EXAMPLE 1 Finding Values That Make Rational Expressions Undefined

Find any values for which the following rational expressions are undefined.

(a) $\frac{p+5}{3p+2}$

Remember that the *numerator* may be any number; we must find any value of p that makes the *denominator* equal to 0 since division by 0 is undefined.

INSTRUCTOR'S RESOURCES

PTB Sec. 6.1 **ISM** Sec. 6.1 **AB** Sec. 6.1

TEST GENERATOR

WORLD WIDE WEB www.LialAlgebra.com

Step 1 Set the denominator equal to 0.

$$3p + 2 = 0$$

Step 2 Solve this equation.

$$3p = -2$$

$$p = -\frac{2}{3}$$

Step 3 Since $p = -\frac{2}{3}$ will make the denominator 0, the given expression is undefined for $-\frac{2}{3}$.

TEACHING TIP Students will occasionally list $m = 0$ as a value that makes the expression undefined. Remind them that a zero numerator is allowed, since the fraction will then be equal to zero.

(b) $\dfrac{9m^2}{m^2 - 5m + 6}$

Find the numbers that make the denominator 0 by solving the equation

$$m^2 - 5m + 6 = 0.$$

$$(m - 2)(m - 3) = 0 \qquad \text{Factor.}$$

$$m - 2 = 0 \quad \text{or} \quad m - 3 = 0 \qquad \text{Zero-factor property}$$

$$m = 2 \quad \text{or} \quad m = 3 \qquad \text{Solve.}$$

The original expression is undefined for 2 and for 3.

(c) $\dfrac{2r}{r^2 + 1}$

This denominator cannot equal 0 for any value of r, since r^2 is always greater than or equal to 0 and adding 1 makes the sum greater than 0. Thus, there are no values for which this rational expression is undefined.

OBJECTIVE 2 Find the numerical value of a rational expression.

CHALKBOARD EXAMPLE

Find the value of the rational expression when $x = 3$.

$$\frac{x}{2x + 1}$$

Answer: $\frac{3}{7}$

EXAMPLE 2 Evaluating a Rational Expression

Find the numerical value of $\dfrac{3x + 6}{2x - 4}$ for the given values of x.

(a) $x = 1$

Find the value of the rational expression by substituting 1 for x.

$$\frac{3x + 6}{2x - 4} = \frac{3(1) + 6}{2(1) - 4} \qquad \text{Let } x = 1.$$

$$= \frac{9}{-2}$$

$$= -\frac{9}{2}$$

(b) $x = 2$

Substituting 2 for x makes the denominator 0, so the rational expression is undefined when $x = 2$.

OBJECTIVE 3 Write rational expressions in lowest terms. A fraction such as $\frac{2}{3}$ is said to be in lowest terms. How can "lowest terms" be defined? We use the idea of greatest common factor to give this definition, which applies to all rational expressions.

Lowest Terms

A rational expression $\frac{P}{Q}$ $(Q \neq 0)$ is in **lowest terms** if the greatest common factor of its numerator and denominator is 1.

The properties of rational numbers also apply to rational expressions. We use the fundamental property of rational expressions to write a rational expression in lowest terms.

Fundamental Property of Rational Expressions

If $\frac{P}{Q}$ is a rational expression and if K represents any polynomial, where $K \neq 0$, then

$$\frac{PK}{QK} = \frac{P}{Q}.$$

This property is based on the identity property of multiplication, since

$$\frac{PK}{QK} = \frac{P}{Q} \cdot \frac{K}{K} = \frac{P}{Q} \cdot 1 = \frac{P}{Q}.$$

The fundamental property suggests the method used in the next example to write a fraction in lowest terms. We show the procedure with both a common fraction and a rational expression involving variables. Notice the similarity.

CHALKBOARD EXAMPLE

Write in lowest terms.

$$\frac{6p^3}{2p^2}$$

Answer: $3p$

EXAMPLE 3 Writing in Lowest Terms

Write each expression in lowest terms.

(a) $\frac{30}{72}$ **(b)** $\frac{14k^2}{2k^3}$

Begin by factoring.

$$\frac{30}{72} = \frac{2 \cdot 3 \cdot 5}{2 \cdot 2 \cdot 2 \cdot 3 \cdot 3} \qquad \frac{14k^2}{2k^3} = \frac{2 \cdot 7 \cdot k \cdot k}{2 \cdot k \cdot k \cdot k}$$

Group any factors common to the numerator and denominator.

$$\frac{30}{72} = \frac{5(2 \cdot 3)}{2 \cdot 2 \cdot 3(2 \cdot 3)} \qquad \frac{14k^2}{2k^3} = \frac{7(2 \cdot k \cdot k)}{k(2 \cdot k \cdot k)}$$

Use the fundamental property.

$$\frac{30}{72} = \frac{5}{2 \cdot 2 \cdot 3} = \frac{5}{12} \qquad \frac{14k^2}{2k^3} = \frac{7}{k}$$

CHALKBOARD EXAMPLE

Write $\frac{a^2 - b^2}{a^2 + 2ab + b^2}$ in lowest terms.

Answer: $\frac{a - b}{a + b}$

TEACHING TIP Students should be constantly reminded that rational expressions cannot be reduced to lowest terms until after the numerator and denominator have been factored.

EXAMPLE 4 Writing in Lowest Terms

Write each rational expression in lowest terms.

(a) $\frac{3x - 12}{5x - 20}$

Begin by factoring both numerator and denominator. Then use the fundamental property.

$$\frac{3x - 12}{5x - 20} = \frac{3(x - 4)}{5(x - 4)} = \frac{3}{5}$$

The rational expression $\frac{3x - 12}{5x - 20}$ equals $\frac{3}{5}$ for all values of x, where $x \neq 4$ (since the denominator of the original rational expression is 0 when x is 4).

(b) $\frac{m^2 + 2m - 8}{2m^2 - m - 6}$

$$\frac{m^2 + 2m - 8}{2m^2 - m - 6} = \frac{(m + 4)(m - 2)}{(2m + 3)(m - 2)} \quad \text{Factor.}$$

$$= \frac{m + 4}{2m + 3} \quad \text{Fundamental property}$$

Thus, $\frac{m^2 + 2m - 8}{2m^2 - m - 6} = \frac{m + 4}{2m + 3}$ for $m \neq -\frac{3}{2}$ and $m \neq 2$, since the denominator of the original expression is 0 for these values of m.

From now on, we will write statements of equality of rational expressions with the understanding that they apply only to those real numbers that make neither denominator equal to 0.

CAUTION

One of the most common errors in algebra occurs when students attempt to write rational expressions in lowest terms *before factoring.* The fundamental property is applied only *after* the numerator and denominator are expressed in factored form. For example, although x appears in both the numerator and denominator in Example 4(a), and 12 and 20 have a common factor of 4, the fundamental property cannot be used before factoring because $3x$, $5x$, 12, and 20 are *terms,* not *factors.* Terms are *added* or *subtracted;* factors are *multiplied* or *divided.* For example,

$$\frac{6 + 2}{3 + 2} = \frac{8}{5}, \quad \textbf{not} \quad \frac{6}{3} + \frac{2}{2} = 2 + 1 = 3.$$

Also,

$$\frac{2x + 3}{4x + 6} = \frac{2x + 3}{2(2x + 3)} = \frac{1}{2}, \quad \textbf{but} \quad \frac{x^2 + 6}{x + 3} \neq x + 2.$$

CHALKBOARD EXAMPLE

Write $\frac{z^2 - 5}{5 - z^2}$ in lowest terms.

Answer: -1

EXAMPLE 5 Writing in Lowest Terms (Factors Are Opposites)

Write $\frac{x - y}{y - x}$ in lowest terms.

At first glance, there does not seem to be any way in which $x - y$ and $y - x$ can be factored to get a common factor. This is not the case, however, since the numerator can be factored as

$$x - y = -1(-x + y) = -1(y - x).$$

Now use the fundamental property to simplify.

$$\frac{x - y}{y - x} = \frac{-1(y - x)}{1(y - x)} = \frac{-1}{1} = -1$$

Either the numerator or the denominator could have been factored in the first step.

In Example 5, the binomials in the numerator and denominator are opposites. A general rule for this situation follows.

A fraction in which the numerator and denominator show subtraction of the same terms but in opposite order is equal to -1. More generally, if a factor of the numerator is the opposite of a factor in the denominator, the quotient of those two factors is -1.

CHALKBOARD EXAMPLE

Write $\frac{9 - k}{9 + k}$ in lowest terms.

Answer: cannot be written in simpler form

EXAMPLE 6 Writing in Lowest Terms (Factors Are Opposites)

Write each rational expression in lowest terms.

(a) $\frac{2 - m}{m - 2}$

Since $2 - m$ and $m - 2$ (or $-2 + m$) are opposites, we use the rule above.

$$\frac{2 - m}{m - 2} = -1$$

(b) $\frac{4x^2 - 9}{6 - 4x}$

Factor the numerator and denominator and use the rule above.

$$\frac{4x^2 - 9}{6 - 4x} = \frac{(2x + 3)(2x - 3)}{2(3 - 2x)}$$

$$= \frac{2x + 3}{2}(-1)$$

$$= -\frac{2x + 3}{2}$$

(c) $\frac{3 + r}{3 - r}$

The quantity $3 - r$ *is not* the opposite of $3 + r$. This rational expression cannot be simplified.

TEACHING TIP Go over several examples to make sure this reduction is understood:

$$\frac{x + 3}{-x - 3}, \frac{-x + 3}{-3 + x}, \text{ and } \frac{-3 - x}{x + 3}$$

are all equal to -1. As seen in the next objective, there are often several *equivalent* ways of writing a rational expression.

OBJECTIVE 4 Recognize equivalent forms of rational expressions. When working with rational expressions, it is important to be able to recognize equivalent forms of an expression. For example, the common fraction $-\frac{5}{6}$ can also be written as $\frac{-5}{6}$ and as $\frac{5}{-6}$. Look again at Example 6(b). The form of the answer given there is only one of several acceptable forms. The $-$ sign representing the -1 factor is in front of the fraction, on the same line as the fraction bar. The -1 factor may be placed in front of the fraction, in the numerator, or in the denominator. Some other acceptable forms of the answer are

$$\frac{-(2x+3)}{2}, \quad \frac{-2x-3}{2}, \quad \text{and} \quad \frac{2x+3}{-2}.$$

However, $\frac{-2x+3}{2}$ is *not* an acceptable form, because the sign preceding 3 in the numerator should be $-$ rather than $+$. (The form $\frac{2x+3}{-2}$, listed above, is seldom used.)

CHALKBOARD EXAMPLE

Write four equivalent forms of the rational expression $-\frac{2x-6}{x+3}$.

Answer: $\frac{-(2x-6)}{x+3}$, $\frac{-2x+6}{x+3}$, $\frac{2x-6}{-(x+3)}$, $\frac{2x-6}{-x-3}$

EXAMPLE 7 Writing Equivalent Forms of a Rational Expression

Write four equivalent forms of the rational expression $-\frac{3x+2}{x-6}$.

If we let the negative sign preceding the fraction apply to the numerator, we have the equivalent form

$$\frac{-(3x+2)}{x-6}.$$

By distributing the negative sign in this expression, we have another equivalent form,

$$\frac{-3x-2}{x-6}.$$

If we let the negative sign apply to the denominator of the fraction, we get

$$\frac{3x+2}{-(x-6)}$$

or, distributing once again,

$$\frac{3x+2}{-x+6}.$$

CAUTION In Example 7, it would be incorrect to distribute the negative sign to *both* the numerator *and* the denominator. This would lead to the *opposite* of the original expression.

■ **1.** $3x^2 + 11x + 8$ cannot be factored, so this quotient cannot be reduced. By long division the quotient is

$$3x + 5 + \frac{-2}{x + 2}.$$

2. The numerator factors as $(x - 2)(x^2 + 2x + 4)$, so by reducing, the quotient is $x - 2$. Long division gives the same quotient.

CONNECTIONS

In Chapter 4 we used long division to find the quotient of two polynomials. For example, we found $(2x^2 + 5x - 12) \div (2x - 3)$ as follows:

$$
\begin{array}{r}
x + 4 \\
2x - 3\overline{)2x^2 + 5x - 12} \\
\underline{2x^2 - 3x} \\
8x - 12 \\
\underline{8x - 12} \\
0
\end{array}
$$

The quotient is $x + 4$. We also get the same quotient by expressing the division problem as a rational expression (fraction) and writing this rational expression in lowest terms.

$$\frac{2x^2 + 5x - 12}{2x - 3} = \frac{(2x - 3)(x + 4)}{2x - 3} = x + 4$$

FOR DISCUSSION OR WRITING
What kind of division problem has a quotient that cannot be found by reducing a fraction to lowest terms? Try using rational expressions to solve the following division problems. Then use long division to compare.

1. $(3x^2 + 11x + 8) \div (x + 2)$

2. $(x^3 - 8) \div (x^2 + 2x + 4)$

6.1 EXERCISES

1. (a) $3; -5$ (b) $q; -1$
2. (a) is not (b) are
3. A rational expression is a quotient of two polynomials, such as $\frac{x^2 + 3x - 6}{x + 4}$. One can think of this as an algebraic fraction.
4. The rational expression $\frac{a^2 - b^2}{a + b}$ is not in lowest terms. To simplify it, follow these steps:
$$\frac{a^2 - b^2}{a + b} = \frac{(a + b)(a - b)}{a + b} = a - b.$$
5. 0 6. 0 7. $-\frac{5}{3}$ 8. $-\frac{4}{3}$
9. $-3, 2$ 10. $1, 4$
11. never undefined
12. never undefined

1. Fill in the blanks with the correct responses.

(a) The rational expression $\frac{x + 5}{x - 3}$ is undefined when $x =$ _____ , and is equal to 0 when $x =$ _____ .

(b) The rational expression $\frac{p - q}{q - p}$ is undefined when $p =$ _____ , and in all other cases its simplified form is _____ .

2. Make the correct choice for the blank.

(a) $\frac{4 - r^2}{4 + r^2}$ _____ equal to -1.
 (is/is not)

(b) $\frac{5 + 2x}{3 - x}$ and $\frac{-5 - 2x}{x - 3}$ _____ equivalent rational expressions.
 (are/are not)

3. Define *rational expression* in your own words, and give an example.

4. Give an example of a rational expression that is not in lowest terms, and then show the steps required to write it in lowest terms.

Find any values for which each rational expression is undefined. See Example 1.

5. $\frac{12}{5y}$

6. $\frac{-7}{3z}$

7. $\frac{4x^2}{3x + 5}$

8. $\frac{2x^3}{3x + 4}$

9. $\frac{5m + 2}{m^2 + m - 6}$

10. $\frac{2r - 5}{r^2 - 5r + 4}$

11. $\frac{3x - 1}{x^2 + 2}$

12. $\frac{4q + 2}{q^2 + 9}$

13. (a) 1 (b) $\frac{17}{12}$
14. (a) $\frac{7}{10}$ (b) $\frac{8}{15}$
15. (a) 0 (b) $-\frac{10}{3}$
16. (a) $\frac{3}{2}$ (b) $-\frac{7}{3}$
17. (a) $\frac{9}{5}$ (b) undefined
18. (a) $-\frac{64}{15}$ (b) undefined
19. (a) $\frac{2}{7}$ (b) $\frac{13}{3}$
20. (a) undefined (b) $\frac{8}{25}$
21. Any number divided by itself is 1, *provided the number is not 0.* This expression is equal to $\frac{1}{x+2}$ for all values of x except -2 and 2.
22. No, the rational expression $\frac{2x-2}{x-2}$ can be written as $\frac{2(x-1)}{x-2}$, and does not simplify further. The fundamental property is applied to factors, *not* terms.
23. $3r^2$
24. $9p$ 25. $\frac{2}{5}$ 26. $\frac{5}{3}$
27. $\frac{x-1}{x+1}$ 28. $\frac{t-3}{t-1}$
29. $\frac{7}{5}$ 30. 2 31. $m-n$
32. $a+b$ 33. $\frac{3(2m+1)}{4}$
34. $\frac{5(2p+3)}{3}$ 35. $\frac{3m}{5}$
36. $3t$ 37. $\frac{3r-2s}{3}$
38. $\frac{4x+3y}{3}$ 39. $\frac{z-3}{z+5}$
40. $\frac{k+4}{k+5}$ 41. $k-3$
42. $t-5$ 43. $\frac{x+1}{x-1}$
44. $\frac{x+2}{x-2}$ 45. -1 46. -1
47. $-(m+1)$ 48. $-(a+b)$
49. -1 50. -1
51. already in lowest terms
52. already in lowest terms

*Find the numerical value of each rational expression when **(a)** $x = 2$ and **(b)** $x = -3$. See Example 2.*

13. $\frac{5x-2}{4x}$
14. $\frac{3x+1}{5x}$
15. $\frac{2x^2-4x}{3x}$
16. $\frac{4x^2-1}{5x}$
17. $\frac{(-3x)^2}{4x+12}$
18. $\frac{(-2x)^3}{3x+9}$
19. $\frac{5x+2}{2x^2+11x+12}$
20. $\frac{7-3x}{3x^2-7x+2}$

21. If 2 is substituted for x in the rational expression $\frac{x-2}{x^2-4}$, the result is $\frac{0}{0}$. An often-heard statement is "Any number divided by itself is 1." Does this mean that this expression is equal to 1 for $x = 2$? If not, explain.

22. For $x \neq 2$, the rational expression $\frac{2(x-2)}{x-2}$ is equal to 2. Can the same be said for $\frac{2x-2}{x-2}$? Explain.

Write each rational expression in lowest terms. See Examples 3 and 4.

23. $\frac{18r^3}{6r}$
24. $\frac{27p^2}{3p}$
25. $\frac{4(y-2)}{10(y-2)}$
26. $\frac{15(m-1)}{9(m-1)}$
27. $\frac{(x+1)(x-1)}{(x+1)^2}$
28. $\frac{(t+5)(t-3)}{(t-1)(t+5)}$
29. $\frac{7m+14}{5m+10}$
30. $\frac{8z-24}{4z-12}$
31. $\frac{m^2-n^2}{m+n}$
32. $\frac{a^2-b^2}{a-b}$
33. $\frac{12m^2-3}{8m-4}$
34. $\frac{20p^2-45}{6p-9}$
35. $\frac{3m^2-3m}{5m-5}$
36. $\frac{6t^2-6t}{2t-2}$
37. $\frac{9r^2-4s^2}{9r+6s}$
38. $\frac{16x^2-9y^2}{12x-9y}$
39. $\frac{zw+4z-3w-12}{zw+4z+5w+20}$
40. $\frac{km+4k+4m+16}{km+4k+5m+20}$
41. $\frac{5k^2-13k-6}{5k+2}$
42. $\frac{7t^2-31t-20}{7t+4}$
43. $\frac{2x^2-3x-5}{2x^2-7x+5}$
44. $\frac{3x^2+8x+4}{3x^2-4x-4}$

Write each expression in lowest terms. See Examples 5 and 6.

45. $\frac{6-t}{t-6}$
46. $\frac{2-k}{k-2}$
47. $\frac{m^2-1}{1-m}$
48. $\frac{a^2-b^2}{b-a}$
49. $\frac{q^2-4q}{4q-q^2}$
50. $\frac{z^2-5z}{5z-z^2}$
51. $\frac{p+6}{p-6}$
52. $\frac{5-x}{5+x}$

53. $x^2 + 3$ 54. $x^2 + 7x + 12$
Answers may vary in Exercises 55–60.
55. $\frac{-(x+4)}{x-3}, \frac{-x-4}{x-3}, \frac{x+4}{-(x-3)}, \frac{x+4}{-x+3}$
56. $\frac{-(x+6)}{x-1}, \frac{-x-6}{x-1}, \frac{x+6}{-(x-1)}, \frac{x+6}{-x+1}$
57. $\frac{-(2x-3)}{x+3}, \frac{-2x+3}{x+3}, \frac{2x-3}{-(x+3)}, \frac{2x-3}{-x-3}$
58. $\frac{-(5x-6)}{x+4}, \frac{-5x+6}{x+4}, \frac{5x-6}{-(x+4)}, \frac{5x-6}{-x-4}$
59. $-\frac{3x-1}{5x-6}, \frac{-(3x-1)}{5x-6}, -\frac{-3x+1}{-5x+6}, \frac{3x-1}{-5x+6}$
60. $-\frac{2x+9}{3x+1}, \frac{-(2x+9)}{3x+1}, -\frac{-2x-9}{-3x-1}, \frac{2x+9}{-3x-1}$
61. $\frac{m+n}{2}$ 62. $\frac{x^2+1}{x}$
63. $-\frac{b^2+ba+a^2}{a+b}$
64. $\frac{k^2-2k+4}{k-2}$ 65. $\frac{z+3}{z}$
66. $\frac{1-2r}{2}$ 67. $x + 3$
68. $x + 4$ 69. $x + 5$
70. $x + 6$

▲ **53.** The area of the rectangle is represented by $x^4 + 10x^2 + 21$. What is the width? $\left(\textit{Hint: Use } W = \frac{A}{L}.\right)$

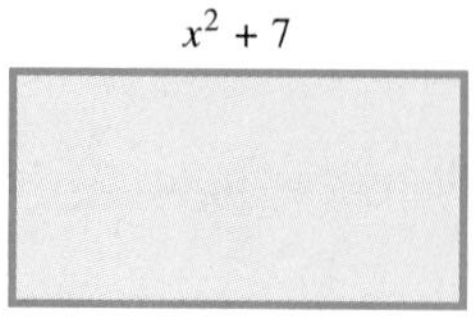

▲ **54.** The volume of the box is represented by

$$(x^2 + 8x + 15)(x + 4).$$

Find the polynomial that represents the area of the bottom of the box.

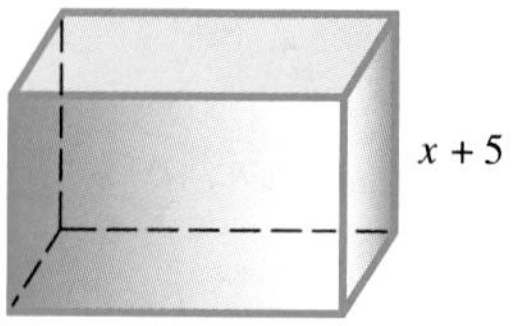

Write four equivalent expressions for each of the following. See Example 7.

55. $-\frac{x+4}{x-3}$ **56.** $-\frac{x+6}{x-1}$ **57.** $-\frac{2x-3}{x+3}$

58. $-\frac{5x-6}{x+4}$ **59.** $\frac{-3x+1}{5x-6}$ **60.** $\frac{-2x-9}{3x+1}$

▲ *Write each expression in lowest terms.*

61. $\frac{m^2 - n^2 - 4m - 4n}{2m - 2n - 8}$ **62.** $\frac{x^2y + y + x^2z + z}{xy + xz}$ **63.** $\frac{b^3 - a^3}{a^2 - b^2}$

64. $\frac{k^3 + 8}{k^2 - 4}$ **65.** $\frac{z^3 + 27}{z^3 - 3z^2 + 9z}$ **66.** $\frac{1 - 8r^3}{8r^2 + 4r + 2}$

TECHNOLOGY INSIGHTS (EXERCISES 67–72)

▲ In the tables shown in Exercises 67–72, the expressions for Y_1 and Y_2 are equivalent except for the value for which Y_1 shows an error. In the table at the right, $Y_1 = \frac{x^2 - 4}{x + 2}$. In each case, Y_1 is defined by a rational expression. Predict the denominator of Y_1.

X	Y1	Y2
-5	-7	-7
-4	-6	-6
-3	-5	-5
-2	ERROR	-4
-1	-3	-3
0	-2	-2
1	-1	-1

Y2=X-2

67.

X	Y1	Y2
-6	-9	-9
-5	-8	-8
-4	-7	-7
-3	ERROR	-6
-2	-5	-5
-1	-4	-4
0	-3	-3

Y2=X-3

68.

X	Y1	Y2
-7	-11	-11
-6	-10	-10
-5	-9	-9
-4	ERROR	-8
-3	-7	-7
-2	-6	-6
-1	-5	-5

Y2=X-4

69.

X	Y1	Y2
-8	13	13
-7	12	12
-6	11	11
-5	ERROR	10
-4	9	9
-3	8	8
-2	7	7

Y2=5-X

70.

X	Y1	Y2
-9	15	15
-8	14	14
-7	13	13
-6	ERROR	12
-5	11	11
-4	10	10
-3	9	9

Y2=6-X

71. $x - 3$ 72. $x - 4$

TECHNOLOGY INSIGHTS (EXERCISES 67–72) (CONTINUED)

71.

X	Y1	Y2
0	9	9
1	13	13
2	19	19
3	ERROR	27
4	37	37
5	49	49
6	63	63

Y2 = X²+3X+9

72.

X	Y1	Y2
1	21	21
2	28	28
3	37	37
4	ERROR	48
5	61	61
6	76	76
7	93	93

Y2 = X²+4X+16

6.2 Multiplication and Division of Rational Expressions

OBJECTIVES

1. Multiply rational expressions.
2. Divide rational expressions.

FOR EXTRA HELP

SSG Sec. 6.2
SSM Sec. 6.2
Pass the Test Software
InterAct Math Tutorial Software
Video 10

OBJECTIVE 1 Multiply rational expressions. The product of two fractions is found by multiplying the numerators and multiplying the denominators. Rational expressions are multiplied in the same way.

Multiplying Rational Expressions

The product of the rational expressions $\frac{P}{Q}$ and $\frac{R}{S}$ is

$$\frac{P}{Q} \cdot \frac{R}{S} = \frac{PR}{QS}.$$

The next example shows the multiplication of two common fractions and the multiplication of two rational expressions with variables so that you can compare the steps.

CHALKBOARD EXAMPLE

Multiply $\frac{8p^2q}{3} \cdot \frac{9}{pq^2}$.

Answer: $\frac{24p}{q}$

EXAMPLE 1 Multiplying Rational Expressions

Multiply. Write answers in lowest terms.

(a) $\frac{3}{10} \cdot \frac{5}{9}$ **(b)** $\frac{6}{x} \cdot \frac{x^2}{12}$

Find the product of the numerators and the product of the denominators.

$$\frac{3}{10} \cdot \frac{5}{9} = \frac{3 \cdot 5}{10 \cdot 9} \qquad \Big| \qquad \frac{6}{x} \cdot \frac{x^2}{12} = \frac{6 \cdot x^2}{x \cdot 12}$$

Use the fundamental property to write each product in lowest terms.

$$\frac{3}{10} \cdot \frac{5}{9} = \frac{1 \cdot 3 \cdot 5}{2 \cdot 5 \cdot 3 \cdot 3} = \frac{1}{6} \qquad \Big| \qquad \frac{6}{x} \cdot \frac{x^2}{12} = \frac{6 \cdot x \cdot x}{2 \cdot 6 \cdot x} = \frac{x}{2}$$

Notice in the second step above that the products were left in factored form since common factors must be identified to write the product in lowest terms.

INSTRUCTOR'S RESOURCES

PTB Sec. 6.2 ISM Sec. 6.2 AB Sec. 6.2 | TEST GENERATOR | WORLD WIDE WEB www.LialAlgebra.com

It is also possible to divide out common factors in the numerator and denominator *before* multiplying the rational expressions. For example,

NOTE

$$\frac{6}{5} \cdot \frac{35}{22} = \frac{2 \cdot 3}{5} \cdot \frac{5 \cdot 7}{2 \cdot 11}$$ Identify common factors.

$$= \frac{3 \cdot 7}{11}$$ Lowest terms

$$= \frac{21}{11}.$$ Multiply in numerator.

CHALKBOARD EXAMPLE

Multiply. Write the product in lowest terms.

$$\frac{3(p-q)}{p} \cdot \frac{q}{2(p-q)}$$

Answer: $\frac{3q}{2p}$

TEACHING TIP Note that the fractions are undefined if $x = 0$ or if x and y are additive inverses.

EXAMPLE 2 Multiplying Rational Expressions

Multiply. Write the product in lowest terms.

$$\frac{x+y}{2x} \cdot \frac{x^2}{(x+y)^2}$$

Use the definition of multiplication.

$$\frac{x+y}{2x} \cdot \frac{x^2}{(x+y)^2} = \frac{(x+y)x^2}{2x(x+y)^2}$$ Multiply numerators; multiply denominators.

$$= \frac{(x+y)x \cdot x}{2x(x+y)(x+y)}$$ Factor; identify common factors.

$$= \frac{x}{2(x+y)} \cdot \frac{x(x+y)}{x(x+y)}$$ Definition of multiplication

$$= \frac{x}{2(x+y)}$$ Lowest terms

Notice the factor $\frac{x(x+y)}{x(x+y)}$ in the third line. Since it equals 1, the final product is $\frac{x}{2(x+y)}$.

CHALKBOARD EXAMPLE

Multiply. Write the product in lowest terms.

$$\frac{x^2+7x+10}{3x+6} \cdot \frac{6x-6}{x^2+2x-15}$$

Answer: $\frac{2(x-1)}{x-3}$

EXAMPLE 3 Multiplying Rational Expressions

Multiply. Write the product in lowest terms.

$$\frac{x^2+3x}{x^2-3x-4} \cdot \frac{x^2-5x+4}{x^2+2x-3}$$

First factor the numerators and denominators. Then use the fundamental property to write the product in lowest terms.

$$\frac{x^2+3x}{x^2-3x-4} \cdot \frac{x^2-5x+4}{x^2+2x-3}$$

$$= \frac{x(x+3)}{(x-4)(x+1)} \cdot \frac{(x-4)(x-1)}{(x+3)(x-1)}$$ Factor.

$$= \frac{x(x+3)(x-4)(x-1)}{(x-4)(x+1)(x+3)(x-1)}$$ Multiply numerators; multiply denominators.

$$= \frac{x}{x+1}$$ Lowest terms

The quotients $\frac{x+3}{x+3}$, $\frac{x-4}{x-4}$, and $\frac{x-1}{x-1}$ are all equal to 1, justifying the final product $\frac{x}{x+1}$.

OBJECTIVE 2 Divide rational expressions. To develop a method for dividing rational numbers and rational expressions, consider the following problem. Suppose that you have $\frac{7}{8}$ gallon of milk and you wish to find how many quarts you have. Since a quart is $\frac{1}{4}$ gallon, you must ask yourself, "How many $\frac{1}{4}$s are there in $\frac{7}{8}$?" This would be interpreted as

$$\frac{7}{8} \div \frac{1}{4} \quad \text{or} \quad \frac{\frac{7}{8}}{\frac{1}{4}}$$

since the fraction bar means division.

The fundamental property of rational expressions discussed earlier can be applied to rational number values of P, Q, and K. With $P = \frac{7}{8}$, $Q = \frac{1}{4}$, and $K = 4$,

$$\frac{P}{Q} = \frac{P \cdot K}{Q \cdot K} = \frac{\frac{7}{8} \cdot 4}{\frac{1}{4} \cdot 4} = \frac{\frac{7}{8} \cdot 4}{1} = \frac{7}{8} \cdot \frac{4}{1}.$$

So, to divide $\frac{7}{8}$ by $\frac{1}{4}$, we must multiply $\frac{7}{8}$ by the reciprocal of $\frac{1}{4}$, namely 4. Since $\left(\frac{7}{8}\right)(4) = \frac{7}{2}$, there are $\frac{7}{2}$ or $3\frac{1}{2}$ quarts in $\frac{7}{8}$ gallon.

The discussion above illustrates the rule for dividing common fractions. To divide $\frac{a}{b}$ by $\frac{c}{d}$, multiply $\frac{a}{b}$ by the reciprocal of $\frac{c}{d}$. Division of rational expressions is defined in the same way.

Dividing Rational Expressions

If $\frac{P}{Q}$ and $\frac{R}{S}$ are any two rational expressions, with $\frac{R}{S} \neq 0$, then

$$\frac{P}{Q} \div \frac{R}{S} = \frac{P}{Q} \cdot \frac{S}{R} = \frac{PS}{QR}.$$

The next example shows the division of two common fractions and the division of two rational expressions involving variables.

CHALKBOARD EXAMPLE

Divide. Write the quotient in lowest terms.

$$\frac{9p^2}{3p+4} \div \frac{6p^3}{3p+4}$$

Answer: $\frac{3}{2p}$

EXAMPLE 4 Dividing Rational Expressions

Divide. Write answers in lowest terms.

(a) $\frac{5}{8} \div \frac{7}{16}$

(b) $\frac{y}{y+3} \div \frac{4y}{y+5}$

Multiply the first expression and the reciprocal of the second.

$$\frac{5}{8} \div \frac{7}{16} = \frac{5}{8} \cdot \frac{16}{7} \qquad \text{Reciprocal of } \tfrac{7}{16}$$

$$= \frac{5 \cdot 16}{8 \cdot 7}$$

$$= \frac{5 \cdot 8 \cdot 2}{8 \cdot 7}$$

$$= \frac{5 \cdot 2}{7}$$

$$= \frac{10}{7}$$

$$\frac{y}{y+3} \div \frac{4y}{y+5}$$

$$= \frac{y}{y+3} \cdot \frac{y+5}{4y} \qquad \text{Reciprocal of } \tfrac{4y}{y+5}$$

$$= \frac{y(y+5)}{(y+3)(4y)}$$

$$= \frac{y+5}{4(y+3)} \qquad \text{Fundamental property}$$

TEACHING TIP In Example 4, $y \neq 0$ (in addition to $y + 3 \neq 0$ and $y + 5 \neq 0$), because the reciprocal of 0 is undefined.

CHALKBOARD EXAMPLE

Divide. Write the quotient in lowest terms.

$$\frac{5a^2b}{2} \div \frac{10ab^2}{8}$$

Answer: $\frac{2a}{b}$

EXAMPLE 5 Dividing Rational Expressions

Divide. Write the quotient in lowest terms.

$$\frac{(3m)^2}{(2p)^3} \div \frac{6m^3}{16p^2}$$

Use the properties of exponents as necessary.

$$\frac{(3m)^2}{(2p)^3} \div \frac{6m^3}{16p^2} = \frac{(3m)^2}{(2p)^3} \cdot \frac{16p^2}{6m^3} \qquad \text{Multiply by reciprocal.}$$

$$= \frac{(3m)(3m)}{(2p)(2p)(2p)} \cdot \frac{16p^2}{6m^3} \qquad \text{Meaning of exponent}$$

$$= \frac{9 \cdot 16m^2p^2}{8 \cdot 6p^3m^3} \qquad \text{Multiply numerators. Multiply denominators.}$$

$$= \frac{3}{mp} \qquad \text{Lowest terms}$$

CHALKBOARD EXAMPLE

Divide. Write the quotient in lowest terms.

$$\frac{4x(x+3)}{2x+1} \div \frac{-x^2(x+3)}{4x^2-1}$$

Answer: $-\frac{4(2x-1)}{x}$

EXAMPLE 6 Dividing Rational Expressions

Divide. Write the quotient in lowest terms.

$$\frac{x^2-4}{(x+3)(x-2)} \div \frac{(x+2)(x+3)}{2x}$$

$$= \frac{x^2-4}{(x+3)(x-2)} \cdot \frac{2x}{(x+2)(x+3)} \qquad \text{Use the definition of division.}$$

$$= \frac{(x+2)(x-2)}{(x+3)(x-2)} \cdot \frac{2x}{(x+2)(x+3)} \qquad \text{Be sure numerators and denominators are factored.}$$

$$= \frac{(x+2)(x-2)(2x)}{(x+3)(x-2)(x+2)(x+3)} \qquad \text{Multiply numerators and multiply denominators.}$$

$$= \frac{2x}{(x+3)^2} \qquad \text{Use the fundamental property to write in lowest terms.}$$

In Example 6, only the numerator had to be factored. Remember that *all* numerators and denominators must be factored before the fundamental property can be applied.

CHALKBOARD EXAMPLE

Divide. Write the quotient in lowest terms.

$$\frac{ab - a^2}{a^2 - 1} \div \frac{a - b}{a^2 + 2a + 1}$$

Answer: $-\frac{a(a + 1)}{a - 1}$

EXAMPLE 7 Dividing Rational Expressions (Factors Are Opposites)

Divide. Write the quotient in lowest terms.

$$\frac{m^2 - 4}{m^2 - 1} \div \frac{2m^2 + 4m}{1 - m}$$

$$= \frac{m^2 - 4}{m^2 - 1} \cdot \frac{1 - m}{2m^2 + 4m}$$ Use the definition of division.

$$= \frac{(m + 2)(m - 2)}{(m + 1)(m - 1)} \cdot \frac{1 - m}{2m(m + 2)}$$ Factor; $1 - m$ and $m - 1$ differ only in sign.

$$= \frac{-1(m - 2)}{2m(m + 1)}$$ From Section 6.1, $\frac{1 - m}{m - 1} = -1$.

$$= \frac{2 - m}{2m(m + 1)}$$ Use the distributive property in the numerator.

TEACHING TIP Students often forget to write the -1 quotient when dividing out factors that are opposites of each other.

In summary, follow these steps to multiply and divide rational expressions.

Multiplying or Dividing Rational Expressions

Step 1 **Note the operation.** If the operation is division, use the definition of division to rewrite as multiplication.

Step 2 **Factor.** Factor all numerators and denominators completely.

Step 3 **Multiply.** Multiply numerators and multiply denominators.

Step 4 **Write in lowest terms.** Use the fundamental property to write the answer in lowest terms.

6.2 EXERCISES

1. (a) B (b) D (c) C (d) A

1. Match each multiplication problem in Column I with the correct product in Column II.

I	II
(a) $\frac{5x^3}{10x^4} \cdot \frac{10x^7}{2x}$	A. $\frac{2}{5x^5}$
(b) $\frac{10x^4}{5x^3} \cdot \frac{10x^7}{2x}$	B. $\frac{5x^5}{2}$
(c) $\frac{5x^3}{10x^4} \cdot \frac{2x}{10x^7}$	C. $\frac{1}{10x^7}$
(d) $\frac{10x^4}{5x^3} \cdot \frac{2x}{10x^7}$	D. $10x^7$

2. (a) D (b) C (c) A (d) B
3. $\frac{3a}{2}$ 4. 8 5. $-\frac{4x^4}{3}$
6. $-\frac{9m^6}{7}$ 7. $\frac{2}{c+d}$ 8. $\frac{2}{y-2}$
9. 5 10. $\frac{14q^2}{9}$ 11. $-\frac{3}{2t^4}$
12. $-\frac{27a^2}{2}$ 13. $\frac{1}{4}$ 14. $\frac{12}{5}$

2. Match each division problem in Column I with the correct quotient in Column II.

I	II
(a) $\frac{5x^3}{10x^4} \div \frac{10x^7}{2x}$	**A.** $\frac{5x^5}{2}$
(b) $\frac{10x^4}{5x^3} \div \frac{10x^7}{2x}$	**B.** $10x^7$
(c) $\frac{5x^3}{10x^4} \div \frac{2x}{10x^7}$	**C.** $\frac{2}{5x^5}$
(d) $\frac{10x^4}{5x^3} \div \frac{2x}{10x^7}$	**D.** $\frac{1}{10x^7}$

Multiply. Write each answer in lowest terms. See Examples 1 and 2.

3. $\frac{15a^2}{14} \cdot \frac{7}{5a}$ **4.** $\frac{27k^3}{9k} \cdot \frac{24}{9k^2}$ **5.** $\frac{12x^4}{18x^3} \cdot \frac{-8x^5}{4x^2}$

6. $\frac{12m^5}{-2m^2} \cdot \frac{6m^6}{28m^3}$ **7.** $\frac{2(c+d)}{3} \cdot \frac{18}{6(c+d)^2}$ **8.** $\frac{4(y-2)}{x} \cdot \frac{3x}{6(y-2)^2}$

Divide. Write each answer in lowest terms. See Examples 4 and 5.

9. $\frac{9z^4}{3z^5} \div \frac{3z^2}{5z^3}$ **10.** $\frac{35q^8}{9q^5} \div \frac{25q^6}{10q^5}$ **11.** $\frac{4t^4}{2t^5} \div \frac{(2t)^3}{-6}$

12. $\frac{-12a^6}{3a^2} \div \frac{(2a)^3}{27a}$ **13.** $\frac{3}{2y-6} \div \frac{6}{y-3}$ **14.** $\frac{4m+16}{10} \div \frac{3m+12}{18}$

*These problems can be used as collaborative exercises with small groups of students.

15. −4 causes the denominator in the first fraction to equal 0.
16. −5 causes the denominator in the second fraction to equal 0.
17. −7 causes the numerator in the divisor to equal 0, meaning that we would be dividing by 0. This is undefined.
18. We *are* allowed to divide 0 by a nonzero number.

RELATING CONCEPTS* (EXERCISES 15–18)

We know that division by 0 is undefined. For example, $5 \div 0$ cannot be a real number because there is no real number that when multiplied by 0 gives 5 as a product. When dividing rational expressions, we know that no denominator can be 0 in either of the expressions in the problem, and that furthermore, the numerator of the divisor cannot be 0.

Work Exercises 15–18 in order, *referring to the division problem*

$$\frac{x-6}{x+4} \div \frac{x+7}{x+5}.$$

15. Why must we have the restriction $x \neq -4$?

16. Why must we have the restriction $x \neq -5$?

17. Why must we have the restriction $x \neq -7$?

18. Why is 6 allowed as a replacement for x even though 6 causes the numerator in the first fraction to be 0?

Did you make the connection that division by 0 is not allowed in both the individual fractions and the problem as a whole?

19. Suppose I want to multiply $\frac{a^2-1}{6} \cdot \frac{9}{2a+2}$. I start by factoring where possible: $\frac{(a+1)(a-1)}{6} \cdot \frac{9}{2(a+1)}$. Next, I divide out common factors in the numerator and denominator to get $\frac{a-1}{2} \cdot \frac{3}{2}$. Finally, I multiply numerator times numerator and denominator times denominator to get the final product, $\frac{3(a-1)}{4}$.

19. Explain in your own words how to multiply rational expressions. Illustrate with an example.

20. Explain in your own words how to divide rational expressions. Illustrate with an example.

20. Suppose I want to divide $\frac{a^2-1}{6} \div \frac{2a+2}{9}$. I begin by changing $\div$ to $\cdot$ and replacing the divisor with its reciprocal to get $\frac{a^2-1}{6} \cdot \frac{9}{2a+2}$. Then I proceed as I did in the multiplication problem in the answer to Exercise 19.

21. $\frac{10}{9}$ 22. $\frac{2}{3}$ 23. $-\frac{3}{4}$

24. $-\frac{1}{4}$ 25. $-\frac{9}{2}$ 26. -1

27. $\frac{p+4}{p+2}$ 28. $\frac{z-2}{z+3}$

29. $\frac{(k-1)^2}{(k+1)(2k-1)}$ 30. $\frac{m-1}{2m-3}$

31. $\frac{4k-1}{3k-2}$ 32. $\frac{m-3}{2m-3}$

33. $\frac{m+4p}{m+p}$ 34. $\frac{r+6s}{r+s}$

35. $\frac{m+6}{m+3}$ 36. $\frac{z+4}{z-4}$

37. $\frac{y+3}{y+4}$ 38. $\frac{r-1}{r+6}$

39. $\frac{m}{m+5}$ 40. $\frac{m+4p}{m+p}$

41. $\frac{r+6s}{r+s}$ 42. $(x+1)(x+2)$

43. $\frac{(q-3)^2(q+2)^2}{q+1}$

44. $(x+4)(x+3)$

45. $\frac{x+10}{10}$ 46. $\frac{8}{m-8}$

47. $\frac{3-a-b}{2a-b}$

48. $(2r-t)(2r+t)$

49. $-\frac{(x+y)^2(x^2-xy+y^2)}{3y(y-x)(x-y)}$ or $\frac{(x+y)^2(x^2-xy+y^2)}{3y(x-y)^2}$

50. $\frac{-(b-2a)(a^2+b^3)}{(2a+b)^2}$ or $\frac{(2a-b)(a^2+b^3)}{(2a+b)^2}$

51. $\frac{5xy^2}{4q}$ 52. y^2-9

Multiply or divide. Write each answer in lowest terms. See Examples 3, 6, and 7.

21. $\frac{5x-15}{3x+9} \cdot \frac{4x+12}{6x-18}$

22. $\frac{8r+16}{24r-24} \cdot \frac{6r-6}{3r+6}$

23. $\frac{2-t}{8} \div \frac{t-2}{6}$

24. $\frac{4}{m-2} \div \frac{16}{2-m}$

25. $\frac{27-3z}{4} \cdot \frac{12}{2z-18}$

26. $\frac{5-x}{5+x} \cdot \frac{x+5}{x-5}$

27. $\frac{p^2+4p-5}{p^2+7p+10} \div \frac{p-1}{p+4}$

28. $\frac{z^2-3z+2}{z^2+4z+3} \div \frac{z-1}{z+1}$

29. $\frac{2k^2-k-1}{2k^2+5k+3} \div \frac{4k^2-1}{2k^2+k-3}$

30. $\frac{2m^2-5m-12}{m^2+m-20} \div \frac{4m^2-9}{m^2+4m-5}$

31. $\frac{2k^2+3k-2}{6k^2-7k+2} \cdot \frac{4k^2-5k+1}{k^2+k-2}$

32. $\frac{2m^2-5m-12}{m^2-10m+24} \div \frac{4m^2-9}{m^2-9m+18}$

33. $\frac{m^2+2mp-3p^2}{m^2-3mp+2p^2} \div \frac{m^2+4mp+3p^2}{m^2+2mp-8p^2}$

34. $\frac{r^2+rs-12s^2}{r^2-rs-20s^2} \div \frac{r^2-2rs-3s^2}{r^2+rs-30s^2}$

35. $\frac{m^2+3m+2}{m^2+5m+4} \cdot \frac{m^2+10m+24}{m^2+5m+6}$

36. $\frac{z^2-z-6}{z^2-2z-8} \cdot \frac{z^2+7z+12}{z^2-9}$

37. $\frac{y^2+y-2}{y^2+3y-4} \div \frac{y+2}{y+3}$

38. $\frac{r^2+r-6}{r^2+4r-12} \div \frac{r+3}{r-1}$

39. $\frac{2m^2+7m+3}{m^2-9} \cdot \frac{m^2-3m}{2m^2+11m+5}$

40. $\frac{m^2+2mp-3p^2}{m^2-3mp+2p^2} \div \frac{m^2+4mp+3p^2}{m^2+2mp-8p^2}$

41. $\frac{r^2+rs-12s^2}{r^2-rs-20s^2} \div \frac{r^2-2rs-3s^2}{r^2+rs-30s^2}$

42. $\frac{(x+1)^3(x+4)}{x^2+5x+4} \div \frac{x^2+2x+1}{x^2+3x+2}$

43. $\frac{(q-3)^4(q+2)}{q^2+3q+2} \div \frac{q^2-6q+9}{q^2+4q+4}$

44. $\frac{(x+4)^3(x-3)}{x^2-9} \div \frac{x^2+8x+16}{x^2+6x+9}$

▲ *In working each exercise, remember how grouping symbols are used (Section 1.2), how to factor sums and differences of cubes (Section 5.4), and how to factor by grouping (Section 5.1).*

45. $\frac{x+5}{x+10} \div \left(\frac{x^2+10x+25}{x^2+10x} \cdot \frac{10x}{x^2+15x+50}\right)$

46. $\frac{m-8}{m-4} \div \left(\frac{m^2-12m+32}{8m} \cdot \frac{m^2-8m}{m^2-8m+16}\right)$

47. $\frac{3a-3b-a^2+b^2}{4a^2-4ab+b^2} \cdot \frac{4a^2-b^2}{2a^2-ab-b^2}$

48. $\frac{4r^2-t^2+10r-5t}{2r^2+rt+5r} \cdot \frac{4r^3+4r^2t+rt^2}{2r+t}$

49. $\frac{-x^3-y^3}{x^2-2xy+y^2} \div \frac{3y^2-3xy}{x^2-y^2}$

50. $\frac{b^3-8a^3}{4a^3+4a^2b+ab^2} \div \frac{4a^2+2ab+b^2}{-a^3-ab^3}$

51. If the rational expression $\frac{5x^2y^3}{2pq}$ represents the area of a rectangle and $\frac{2xy}{p}$ represents the length, what rational expression represents the width?

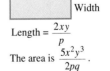

Width

Length = $\frac{2xy}{p}$

The area is $\frac{5x^2y^3}{2pq}$.

▲ **52.** If you are given the problem $\frac{4y+12}{2y-10} \div \frac{?}{y^2-y-20} = \frac{2(y+4)}{y-3}$, what must be the polynomial that is represented by the question mark?

6.3 The Least Common Denominator

OBJECTIVES

1 Find the least common denominator for a group of fractions.

2 Rewrite rational expressions with given denominators.

FOR EXTRA HELP

SSG Sec. 6.3
SSM Sec. 6.3

Pass the Test Software

InterAct Math
Tutorial Software

Video 10

OBJECTIVE 1 **Find the least common denominator for a group of fractions.** Just as with common fractions, adding or subtracting rational expressions (to be discussed in the next section) often requires finding a **least common denominator (LCD),** the least expression that all denominators divide into without a remainder. For example, the least common denominator for $\frac{2}{9}$ and $\frac{5}{12}$ is 36, since 36 is the smallest positive number that both 9 and 12 divide into evenly.

Least common denominators often can be found by inspection. For example, the LCD for $\frac{1}{6}$ and $\frac{2}{3m}$ is $6m$. In other cases, the LCD can be found by a procedure similar to that used in Chapter 5 for finding the greatest common factor.

> **Finding the Least Common Denominator**
>
> *Step 1* **Factor.** Factor each denominator into prime factors.
>
> *Step 2* **List the factors.** List each different denominator factor the *greatest* number of times it appears in any of the denominators.
>
> *Step 3* **Multiply.** Multiply the denominator factors from Step 2 to get the LCD.

When each denominator is factored into prime factors, every prime factor must divide evenly into the LCD.

In Example 1, the LCD is found both for numerical denominators and algebraic denominators.

TEACHING TIP When finding the LCD, remind students not to take each factor the *total* number of times it appears. Instead, use each factor the largest number of times it appears in any *single* denominator. For example, the LCD for $\frac{1}{x^3}$ and $\frac{1}{x^5}$ is x^5, not x^8.

┌─ **EXAMPLE 1** **Finding the Least Common Denominator**

Find the LCD for each pair of fractions.

(a) $\dfrac{1}{24}, \dfrac{7}{15}$ **(b)** $\dfrac{1}{8x}, \dfrac{3}{10x}$

Step 1 Write each denominator in factored form with numerical coefficients in prime factored form.

$$24 = 2 \cdot 2 \cdot 2 \cdot 3 \qquad\qquad 8x = 2 \cdot 2 \cdot 2 \cdot x$$
$$= 2^3 \cdot 3 \qquad\qquad\qquad = 2^3 \cdot x$$
$$15 = 3 \cdot 5 \qquad\qquad\qquad 10x = 2 \cdot 5 \cdot x$$

Step 2 We find the LCD by taking each different factor the *greatest* number of times it appears as a factor in any of the denominators.

The factor 2 appears three times in one product and not at all in the other, so the greatest number of times 2 appears is three. The greatest number of times both 3 and 5 appear is one.

Here 2 appears three times in one product and once in the other, so the greatest number of times 2 appears is three. The greatest number of times 5 appears is one, and the greatest number of times x appears in either product is one.

Step 3 $\text{LCD} = 2 \cdot 2 \cdot 2 \cdot 3 \cdot 5$
$= 2^3 \cdot 3 \cdot 5$
$= 120$

$\text{LCD} = 2 \cdot 2 \cdot 2 \cdot 5 \cdot x$
$= 2^3 \cdot 5 \cdot x$
$= 40x$

CHALKBOARD EXAMPLE

Find the LCD for the pair of fractions.

$$\frac{9}{8m^4} \text{ and } \frac{11}{12m^6}$$

Answer: $24m^6$

INSTRUCTOR'S RESOURCES

PTB Sec. 6.3 ISM Sec. 6.3
AB Sec. 6.3

TEST GENERATOR

WORLD WIDE WEB
www.LialAlgebra.com

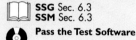

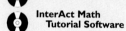

CHALKBOARD EXAMPLE

Find the LCD.

$$\frac{4}{16m^3n} \text{ and } \frac{5}{9m^5}$$

Answer: $144m^5n$

EXAMPLE 2 Finding the LCD

Find the LCD for $\frac{5}{6r^2}$ and $\frac{3}{4r^3}$.

Step 1 Factor each denominator.

$$6r^2 = 2 \cdot 3 \cdot r^2$$
$$4r^3 = 2 \cdot 2 \cdot r^3 = 2^2 \cdot r^3$$

Step 2 The greatest number of times 2 appears is two, the greatest number of times 3 appears is one, and the greatest number of times r appears is three; therefore,

Step 3 $\text{LCD} = 2^2 \cdot 3 \cdot r^3 = 12r^3.$

CHALKBOARD EXAMPLE

Find the LCD.

$$\frac{6}{x^2 - 4x} \text{ and } \frac{3x - 1}{x^2 - 16}$$

Answer: $x(x - 4)(x + 4)$

EXAMPLE 3 Finding the LCD

Find the LCD.

(a) $\frac{6}{5m}, \frac{4}{m^2 - 3m}$

Factor each denominator.

$$5m = 5 \cdot m$$
$$m^2 - 3m = m(m - 3)$$

Take each different factor the greatest number of times it appears as a factor.

$$\text{LCD} = 5 \cdot m \cdot (m - 3) = 5m(m - 3)$$

Since m is not a *factor* of $m - 3$, both factors, m and $m - 3$, must appear in the LCD.

TEACHING TIP Students have a tendency to list the least common denominator for m and $m - 3$ as only $m - 3$.

(b) $\frac{1}{r^2 - 4r - 5}, \frac{1}{r^2 - r - 20}, \frac{1}{r^2 - 10r + 25}$

Factor each denominator.

$$r^2 - 4r - 5 = (r - 5)(r + 1)$$
$$r^2 - r - 20 = (r - 5)(r + 4)$$
$$r^2 - 10r + 25 = (r - 5)^2$$

The LCD is $(r - 5)^2(r + 1)(r + 4)$.

(c) $\frac{1}{q - 5}, \frac{3}{5 - q}$

The expression $5 - q$ can be written as $-1(q - 5)$, since

$$-1(q - 5) = -q + 5 = 5 - q.$$

Because of this, either $q - 5$ or $5 - q$ can be used as the LCD.

OBJECTIVE 2 Rewrite rational expressions with given denominators. Once we find the LCD, we can use the fundamental property to write equivalent rational expressions with this LCD. The next example shows how to do this with both numerical and algebraic fractions.

CHALKBOARD EXAMPLE

Rewrite the rational expression with the indicated denominator.

$$\frac{7k}{5} = \frac{}{30}$$

Answer: $\frac{42k}{30}$

EXAMPLE 4 Writing a Fraction with a Given Denominator

Rewrite each expression with the indicated denominator.

(a) $\frac{3}{8} = \frac{}{40}$ **(b)** $\frac{9k}{25} = \frac{}{50k}$

For each example, first factor the denominator on the right. Then compare the denominator on the left with the one on the right to decide what factors are missing.

$$\frac{3}{8} = \frac{}{5 \cdot 8}$$

A factor of 5 is missing. Multiply by $\frac{5}{5}$ to get a denominator of 40.

$$\frac{3}{8} = \frac{3}{8} \cdot \frac{5}{5} = \frac{15}{40}$$

$\downarrow$ $\frac{5}{5} = 1$

$$\frac{9k}{25} = \frac{}{25 \cdot 2k}$$

Factors of 2 and k are missing. Get a denominator of $50k$ by multiplying by $\frac{2k}{2k}$.

$$\frac{9k}{25} = \frac{9k}{25} \cdot \frac{2k}{2k} = \frac{18k^2}{50k}$$

$\downarrow$ $\frac{2k}{2k} = 1$

Notice the use of the multiplicative identity property in each part of this example.

TEACHING TIP Multiplying by $\frac{5}{5}$ or $\frac{2k}{2k}$ changes only the form of the fraction, and not its value. Stress this application of the identity property of multiplication.

CHALKBOARD EXAMPLE

Rewrite the rational expression with the indicated denominator.

$$\frac{5k+1}{k^2+2k} = \frac{}{k(k+2)(k-1)}$$

Answer: $\frac{(5k+1)(k-1)}{k(k+2)(k-1)}$

EXAMPLE 5 Writing a Fraction with a Given Denominator

Rewrite the following rational expression with the indicated denominator.

$$\frac{12p}{p^2+8p} = \frac{}{p^3+4p^2-32p}$$

Factor $p^2 + 8p$ as $p(p+8)$. Compare with the denominator on the right which factors as $p(p+8)(p-4)$. The factor $p - 4$ is missing, so multiply $\frac{12p}{p(p+8)}$ by $\frac{p-4}{p-4}$.

$$\frac{12p}{p^2+8p} = \frac{12p}{p(p+8)} \cdot \frac{p-4}{p-4} \quad \text{Multiplicative identity property}$$

$$= \frac{12p(p-4)}{p(p+8)(p-4)} \quad \text{Multiplication of rational expressions}$$

$$= \frac{12p^2-48p}{p^3+4p^2-32p} \quad \text{Multiply the factors.}$$

NOTE In the next section we add and subtract rational expressions, which sometimes requires the steps illustrated in Examples 4 and 5. While it is beneficial to leave the denominator in factored form, we multiplied the factors in the denominator in Example 5 to give the answer in the same form as the original problem.

6.3 EXERCISES

1. (c) 2. (b) 3. (c) 4. (a)
5. 60 6. 50 7. 1800
8. 9000 9. x^5 10. m^8
11. $30p$ 12. $60k$ 13. $180y^4$
14. $50m^4$ 15. $15a^5b^3$
16. $9r^6s^8$ 17. $12p(p-2)$
18. $24k(k-2)$

Choose the correct response in Exercises 1–4.

1. Suppose that the greatest common factor of a and b is 1. Then the least common denominator for $\frac{1}{a}$ and $\frac{1}{b}$ is

(a) a. **(b)** b. **(c)** ab. **(d)** 1.

2. If a is a factor of b, then the least common denominator for $\frac{1}{a}$ and $\frac{1}{b}$ is

(a) a. **(b)** b. **(c)** ab. **(d)** 1.

3. The least common denominator for $\frac{11}{20}$ and $\frac{1}{2}$ is

(a) 40. **(b)** 2. **(c)** 20. **(d)** none of these.

4. Suppose that we wish to write the fraction $\frac{1}{(x-4)^2(y-3)}$ with denominator $(x-4)^3(y-3)^2$. We must multiply both the numerator and the denominator by
(a) $(x-4)(y-3)$. **(b)** $(x-4)^2$. **(c)** $x-4$. **(d)** $(x-4)^2(y-3)$.

Find the LCD for the fractions in each list. See Examples 1–3.

5. $\frac{-7}{15}, \frac{21}{20}$ **6.** $\frac{9}{10}, \frac{12}{25}$ **7.** $\frac{17}{100}, \frac{23}{120}, \frac{43}{180}$ **8.** $\frac{17}{250}, \frac{-21}{300}, \frac{127}{360}$

9. $\frac{9}{x^2}, \frac{8}{x^5}$ **10.** $\frac{12}{m^7}, \frac{13}{m^8}$ **11.** $\frac{-2}{5p}, \frac{15}{6p}$ **12.** $\frac{14}{15k}, \frac{9}{4k}$

13. $\frac{17}{15y^2}, \frac{55}{36y^4}$ **14.** $\frac{4}{25m^3}, \frac{-9}{10m^4}$ **15.** $\frac{13}{5a^2b^3}, \frac{29}{15a^5b}$ **16.** $\frac{-7}{3r^4s^5}, \frac{-22}{9r^6s^8}$

17. $\frac{7}{6p}, \frac{15}{4p-8}$ **18.** $\frac{7}{8k}, \frac{-23}{12k-24}$

19. $2^3 \cdot 3 \cdot 5$
20. $(t+4)^3(t-3)(t+8)$
21. The similarity is that 2 is replaced by $t+4$, 3 is replaced by $t-3$, and 5 is replaced by $t+8$.
22. The procedure used is the same. The only difference is that for algebraic fractions, the factors may contain variables, while in a common fraction, the factors are constants.

RELATING CONCEPTS (EXERCISES 19–22)

Suppose we want to find the LCD for the two common fractions

$$\frac{1}{24} \quad \text{and} \quad \frac{1}{20}.$$

In their prime factored forms, the denominators are

$$24 = 2^3 \cdot 3$$

and

$$20 = 2^2 \cdot 5.$$

Refer to this information as necessary and ***work Exercises 19–22 in order.***

19. What is the prime factored form of the LCD of the two fractions?

20. Suppose that two algebraic fractions have denominators $(t+4)^3(t-3)$ and $(t+4)^2(t+8)$. What is the factored form of the LCD of these?

21. What is the similarity between your answers in Exercises 19 and 20?

22. Comment on the following statement: The method for finding the LCD for two algebraic fractions is the same as the method for finding the LCD for two common fractions.

Did you make the connection between finding the LCD for common fractions and for algebraic fractions?

23. $18(r-2)$ 24. $30(p-6)$
25. $12p(p+5)^2$ 26. $5r(r+7)^2$
27. $8(y+2)(y+1)$
28. $9(m-2)(m-5)$
29. $m-3$ or $3-m$
30. $8-a$ or $a-8$
31. $p-q$ or $q-p$
32. $z-x$ or $x-z$
33. $a(a+6)(a-3)$
34. $y(y-5)(y+3)$
35. $(k+3)(k-5)(k+7)(k+8)$
36. $(z-2)(z-4)(z-5)(z+6)$
37. Yes, because $(2x-5)^2 = (5-2x)^2$. 38. Yes, because $(4t-3)(5t-6) = (3-4t)(6-5t)$.

Find the LCD for the fractions in each list. See Examples 1–3.

23. $\dfrac{37}{6r-12}, \dfrac{25}{9r-18}$

24. $\dfrac{-14}{5p-30}, \dfrac{5}{6p-36}$

25. $\dfrac{5}{12p+60}, \dfrac{17}{p^2+5p}, \dfrac{16}{p^2+10p+25}$

26. $\dfrac{13}{r^2+7r}, \dfrac{-3}{5r+35}, \dfrac{-7}{r^2+14r+49}$

27. $\dfrac{3}{8y+16}, \dfrac{22}{y^2+3y+2}$

28. $\dfrac{-2}{9m-18}, \dfrac{-9}{m^2-7m+10}$

29. $\dfrac{12}{m-3}, \dfrac{-4}{3-m}$

30. $\dfrac{-17}{8-a}, \dfrac{2}{a-8}$

31. $\dfrac{29}{p-q}, \dfrac{18}{q-p}$

32. $\dfrac{16}{z-x}, \dfrac{8}{x-z}$

33. $\dfrac{6}{a^2+6a}, \dfrac{-5}{a^2+3a-18}$

34. $\dfrac{8}{y^2-5y}, \dfrac{-2}{y^2-2y-15}$

35. $\dfrac{-5}{k^2+2k-35}, \dfrac{-8}{k^2+3k-40}, \dfrac{9}{k^2-2k-15}$

36. $\dfrac{19}{z^2+4z-12}, \dfrac{16}{z^2+z-30}, \dfrac{6}{z^2+2z-24}$

37. Suppose that $(2x-5)^2$ is the LCD for two fractions. Is $(5-2x)^2$ also acceptable as an LCD? Why?

38. Suppose that $(4t-3)(5t-6)$ is the LCD for two fractions. Is $(3-4t)(6-5t)$ also acceptable as an LCD? Why?

39. 7 40. 1 41. identity property of multiplication 42. 7 43. 1 44. identity property of multiplication

45. $\dfrac{60m^2k^3}{32k^4}$ 46. $\dfrac{15t^2y}{9y^2}$
47. $\dfrac{57z}{6z-18}$ 48. $\dfrac{6r}{15r-15}$
49. $\dfrac{-4a}{18a-36}$ 50. $\dfrac{-20y}{24y+72}$
51. $\dfrac{6(k+1)}{k(k-4)(k+1)}$
52. $\dfrac{15(m+8)}{m(m-9)(m+8)}$
53. $\dfrac{36r(r+1)}{(r-3)(r+2)(r+1)}$
54. $\dfrac{4m(m+2)}{(m-5)(m-3)(m+2)}$
55. $\dfrac{ab(a+2b)}{2a^3b+a^2b^2-ab^3}$
56. $\dfrac{2m(m-4)}{12m^3+14m^2-6m}$
57. $\dfrac{(t-r)(4r-t)}{t^3-r^3}$

RELATING CONCEPTS (EXERCISES 39–44)

Work Exercises 39–44 in order.

39. Suppose that you want to write $\frac{3}{4}$ as an equivalent fraction with denominator 28. By what number must you multiply both the numerator and the denominator?

40. If you write $\frac{3}{4}$ as an equivalent fraction with denominator 28, by what number are you actually multiplying the fraction?

41. What property of multiplication is being used when we write a common fraction as an equivalent one with a larger denominator? (See Section 1.7.)

42. Suppose that you want to write $\dfrac{2x+5}{x-4}$ as an equivalent fraction with denominator $7x-28$. By what number must you multiply both the numerator and the denominator?

43. If you write $\dfrac{2x+5}{x-4}$ as an equivalent fraction with denominator $7x-28$, by what number are you actually multiplying the fraction?

44. Repeat Exercise 41, changing "a common" to "an algebraic."

Did you make the connection between writing common fractions with larger denominators and writing algebraic fractions with more denominator factors?

Write each rational expression on the left with the indicated denominator. See Examples 4 and 5.

45. $\dfrac{15m^2}{8k} = \dfrac{}{32k^4}$

46. $\dfrac{5t^2}{3y} = \dfrac{}{9y^2}$

58. $\dfrac{(x-2)(3x-1)}{x^3-8}$

59. $\dfrac{2y(z-y)(y-z)}{y^4-z^3y}$ or $\dfrac{-2y(y-z)^2}{y^4-z^3y}$

60. $\dfrac{(2p+3q)(p^2-pq+q^2)}{(p+q)(p^3+q^3)}$

61. *Step 1:* Factor each denominator into prime factors. *Step 2:* List each different denominator factor the greatest number of times it appears in any of the denominators. *Step 3:* Multiply the factors in the list to get the LCD. For example, the least common denominator for $\dfrac{1}{(x+y)^3}$ and $\dfrac{-2}{(x+y)^2(p+q)}$ is $(x+y)^3(p+q)$.

62. Determine the "missing factors" by comparing the original denominator with the given denominator. Then multiply both the numerator and the denominator by the missing factors. For example, to write $\dfrac{2}{x+3}$ with denominator $(x+3)^2(x+1)$, the "missing factor" is $(x+3)(x+1)$, and the equivalent expression is $\dfrac{2(x+3)(x+1)}{(x+3)^2(x+1)}$.

47. $\dfrac{19z}{2z-6}=\dfrac{}{6z-18}$

48. $\dfrac{2r}{5r-5}=\dfrac{}{15r-15}$

49. $\dfrac{-2a}{9a-18}=\dfrac{}{18a-36}$

50. $\dfrac{-5y}{6y+18}=\dfrac{}{24y+72}$

51. $\dfrac{6}{k^2-4k}=\dfrac{}{k(k-4)(k+1)}$

52. $\dfrac{15}{m^2-9m}=\dfrac{}{m(m-9)(m+8)}$

53. $\dfrac{36r}{r^2-r-6}=\dfrac{}{(r-3)(r+2)(r+1)}$

54. $\dfrac{4m}{m^2-8m+15}=\dfrac{}{(m-5)(m-3)(m+2)}$

55. $\dfrac{a+2b}{2a^2+ab-b^2}=\dfrac{}{2a^3b+a^2b^2-ab^3}$

56. $\dfrac{m-4}{6m^2+7m-3}=\dfrac{}{12m^3+14m^2-6m}$

57. $\dfrac{4r-t}{r^2+rt+t^2}=\dfrac{}{t^3-r^3}$

58. $\dfrac{3x-1}{x^2+2x+4}=\dfrac{}{x^3-8}$

59. $\dfrac{2(z-y)}{y^2+yz+z^2}=\dfrac{}{y^4-z^3y}$

60. $\dfrac{2p+3q}{p^2+2pq+q^2}=\dfrac{}{(p+q)(p^3+q^3)}$

61. Write an explanation of how to find the least common denominator for a group of denominators. Give an example.

62. Write an explanation of how to write a rational expression as an equivalent rational expression with a given denominator. Give an example.

6.4 Addition and Subtraction of Rational Expressions

OBJECTIVES

1. Add rational expressions having the same denominator.
2. Add rational expressions having different denominators.
3. Subtract rational expressions.

FOR EXTRA HELP

SSG Sec. 6.4
SSM Sec. 6.4
Pass the Test Software
InterAct Math Tutorial Software
Video 10

To add and subtract rational expressions, we use our previous work on finding least common denominators and writing fractions with the LCD.

OBJECTIVE 1 **Add rational expressions having the same denominator.** We find the sum of two rational expressions with a procedure similar to the one used for adding two fractions.

Adding Rational Expressions

If $\frac{P}{Q}$ and $\frac{R}{Q}$ are rational expressions, then

$$\frac{P}{Q}+\frac{R}{Q}=\frac{P+R}{Q}.$$

Again, the first example shows how addition of rational expressions compares with that of rational numbers.

INSTRUCTOR'S RESOURCES

PTB Sec. 6.4 **ISM** Sec. 6.4
AB Sec. 6.4

TEST GENERATOR

WORLD WIDE WEB
www.LialAlgebra.com

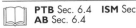

CHALKBOARD EXAMPLE

Add $\frac{x}{x+y} + \frac{1}{x+y}$.

Answer: $\frac{x+1}{x+y}$

EXAMPLE 1 Adding Rational Expressions with the Same Denominator

Add.

(a) $\frac{4}{7} + \frac{2}{7}$ **(b)** $\frac{3x}{x+1} + \frac{2x}{x+1}$

The denominators are the same, so the sum is found by adding the two numerators and keeping the same (common) denominator.

$$\frac{4}{7} + \frac{2}{7} = \frac{4+2}{7} = \frac{6}{7}$$

$$\frac{3x}{x+1} + \frac{2x}{x+1} = \frac{3x+2x}{x+1} = \frac{5x}{x+1}$$

OBJECTIVE 2 Add rational expressions having different denominators. Use the steps given below to add two rational expressions with different denominators. These are the same steps used to add fractions with different denominators.

Adding with Different Denominators

Step 1 **Find the LCD.** Find the least common denominator (LCD).

Step 2 **Rewrite fractions.** Rewrite each rational expression as an equivalent fraction with the LCD as the denominator.

Step 3 **Add.** Add the numerators to get the numerator of the sum. The LCD is the denominator of the sum.

Step 4 **Write in lowest terms.** Use the fundamental property to write the answer in lowest terms.

CHALKBOARD EXAMPLE

Add $\frac{m}{3n} + \frac{2}{7n}$.

Answer: $\frac{7m+6}{21n}$

EXAMPLE 2 Adding Rational Expressions with Different Denominators

Add.

(a) $\frac{1}{12} + \frac{7}{15}$ **(b)** $\frac{2}{3y} + \frac{1}{4y}$

Step 1 First find the LCD using the methods of the previous section.

$$\text{LCD} = 2^2 \cdot 3 \cdot 5 = 60 \qquad \text{LCD} = 2^2 \cdot 3 \cdot y = 12y$$

Step 2 Now rewrite each rational expression as a fraction with the LCD (either 60 or 12y) as the denominator.

$$\frac{1}{12} + \frac{7}{15} = \frac{1(5)}{12(5)} + \frac{7(4)}{15(4)} = \frac{5}{60} + \frac{28}{60}$$

$$\frac{2}{3y} + \frac{1}{4y} = \frac{2(4)}{3y(4)} + \frac{1(3)}{4y(3)} = \frac{8}{12y} + \frac{3}{12y}$$

TEACHING TIP Adding fractions with the same denominator is an example of adding like quantities. In the case of Example 2(a), five *sixtieths* plus twenty-eight *sixtieths* equals thirty-three *sixtieths*.

Step 3 Since the fractions now have common denominators, add the numerators.

Step 4 Write in lowest terms if necessary.

$$\frac{5}{60}+\frac{28}{60}=\frac{5+28}{60}=\frac{33}{60}=\frac{11}{20}$$

$$\frac{8}{12y}+\frac{3}{12y}=\frac{8+3}{12y}=\frac{11}{12y}$$

CHALKBOARD EXAMPLE

Add and write the sum in lowest terms.

$$\frac{-2}{p+1}+\frac{4p}{p^2-1}$$

Answer: $\frac{2}{p-1}$

EXAMPLE 3 Adding Rational Expressions

Add and write the sum in lowest terms.

$$\frac{2x}{x^2-1}+\frac{-1}{x+1}$$

Step 1 Since the denominators are different, find the LCD.

$$x^2-1=(x+1)(x-1)$$

$$x+1 \text{ is prime.}$$

The LCD is $(x+1)(x-1)$.

Step 2 Rewrite each rational expression as a fraction with common denominator $(x+1)(x-1)$.

$$\frac{2x}{x^2-1}+\frac{-1}{x+1}=\frac{2x}{(x+1)(x-1)}+\frac{-1(x-1)}{(x+1)(x-1)}$$ Multiply second fraction by $\frac{x-1}{x-1}$.

$$=\frac{2x}{(x+1)(x-1)}+\frac{-x+1}{(x+1)(x-1)}$$ Distributive property

Step 3

$$=\frac{2x-x+1}{(x+1)(x-1)}$$ Add numerators; keep the same denominator.

$$=\frac{x+1}{(x+1)(x-1)}$$ Combine like terms in the numerator.

Step 4

$$=\frac{1(x+1)}{(x+1)(x-1)}$$ Identity property for multiplication

$$=\frac{1}{x-1}$$ Fundamental property of rational expressions

CHALKBOARD EXAMPLE

Add and write the sum in lowest terms.

$$\frac{2k}{k^2-5k+4}+\frac{3}{k^2-1}$$

Answer: $\frac{2k^2+5k-12}{(k-4)(k-1)(k+1)}$

EXAMPLE 4 Adding Rational Expressions

Add and write the sum in lowest terms.

$$\frac{2x}{x^2+5x+6}+\frac{x+1}{x^2+2x-3}$$

Begin by factoring the denominators completely.

$$\frac{2x}{(x+2)(x+3)}+\frac{x+1}{(x+3)(x-1)}$$

TEACHING TIP If this fraction could be reduced to lower terms, the numerator would have a factor of $x + 2$, $x + 3$, or $x - 1$. Therefore, it is only necessary to check for possible factored forms of the numerator that would contain one of these binomials.

The LCD is $(x + 2)(x + 3)(x - 1)$. Use the fundamental property of rational expressions to rewrite each fraction with the LCD.

$$\frac{2x}{(x+2)(x+3)} + \frac{x+1}{(x+3)(x-1)}$$

$$= \frac{2x(x-1)}{(x+2)(x+3)(x-1)} + \frac{(x+1)(x+2)}{(x+3)(x-1)(x+2)}$$

$$= \frac{2x(x-1) + (x+1)(x+2)}{(x+2)(x+3)(x-1)} \qquad \text{Add numerators; keep the same denominator.}$$

$$= \frac{2x^2 - 2x + x^2 + 3x + 2}{(x+2)(x+3)(x-1)} \qquad \text{Distributive property}$$

$$= \frac{3x^2 + x + 2}{(x+2)(x+3)(x-1)} \qquad \text{Combine terms.}$$

Since $3x^2 + x + 2$ cannot be factored, the rational expression cannot be simplified further. It is usually best to leave the denominator in factored form since it is then easier to identify common factors in the numerator and denominator.

Rational expressions to be added or subtracted may have denominators that are opposites of each other. The next example illustrates this.

CHALKBOARD EXAMPLE

Add and write the sum in lowest terms.

$$\frac{m}{2m - 3n} + \frac{n}{3n - 2m}$$

Answer: $\frac{m - n}{2m - 3n}$ or $\frac{n - m}{3n - 2m}$

EXAMPLE 5 Adding Rational Expressions with Denominators That Are Opposites

Add and write the sum in lowest terms.

$$\frac{y}{y-2} + \frac{8}{2-y}$$

To get a common denominator of $y - 2$, multiply the second expression by -1 in both the numerator and the denominator.

$$\frac{y}{y-2} + \frac{8}{2-y} = \frac{y}{y-2} + \frac{8(-1)}{(2-y)(-1)} \qquad \text{Fundamental property}$$

$$= \frac{y}{y-2} + \frac{-8}{y-2} \qquad \text{Distributive property}$$

$$= \frac{y-8}{y-2} \qquad \text{Add numerators; keep the same denominator.}$$

If we had chosen to use $2 - y$ as the common denominator, the final answer would be in the form $\frac{8-y}{2-y}$, which is equivalent to $\frac{y-8}{y-2}$.

OBJECTIVE 3 Subtract rational expressions. To subtract rational expressions, use the following rule.

Subtracting Rational Expressions

If $\frac{P}{Q}$ and $\frac{R}{Q}$ are rational expressions, then

$$\frac{P}{Q} - \frac{R}{Q} = \frac{P - R}{Q}.$$

CHALKBOARD EXAMPLE

Subtract and write the difference in lowest terms.

$$\frac{5t}{t-1} - \frac{5}{t-1}$$

Answer: 5

EXAMPLE 6 Subtracting Rational Expressions

Subtract and write the difference in lowest terms.

$$\frac{2m}{m-1} - \frac{2}{m-1}$$

By the definition of subtraction,

$$\frac{2m}{m-1} - \frac{2}{m-1} = \frac{2m-2}{m-1} \qquad \text{Subtract numerators; keep the same denominator.}$$

$$= \frac{2(m-1)}{m-1} \qquad \text{Factor the numerator.}$$

$$= 2. \qquad \text{Write in lowest terms.}$$

CHALKBOARD EXAMPLE

Subtract and write the difference in lowest terms.

$$\frac{6}{a+2} - \frac{1}{a-3}$$

Answer: $\frac{5a-20}{(a+2)(a-3)}$

TEACHING TIP Stress the importance of being careful with signs. A common error would have the numerator of this fraction as $9x - 3x - 6$ rather than the correct one, $9x - 3x + 6$.

EXAMPLE 7 Subtracting Rational Expressions with Different Denominators

Subtract and write the difference in lowest terms.

$$\frac{9}{x-2} - \frac{3}{x} = \frac{9x}{x(x-2)} - \frac{3(x-2)}{x(x-2)} \qquad \text{The LCD is } x(x-2).$$

$$= \frac{9x - 3(x-2)}{x(x-2)} \qquad \text{Subtract numerators; keep the same denominator.}$$

$$= \frac{9x - 3x + 6}{x(x-2)} \qquad \text{Distributive property}$$

$$= \frac{6x+6}{x(x-2)} \qquad \text{Combine like terms in the numerator.}$$

NOTE We could factor the final numerator in Example 7 to get an answer in the form $\frac{6(x+1)}{x(x-2)}$; however, the fundamental property would not apply, since there are no common factors that would allow us to write the answer in lower terms. When a rational expression cannot be simplified further, we will leave numerators in polynomial form and denominators in factored form.

CHALKBOARD EXAMPLE

Subtract and write the difference in lowest terms.

$$\frac{4x}{x-1} - \frac{-3x-1}{1-x}$$

Answer: 1

EXAMPLE 8 Subtracting Rational Expressions with Denominators That Are Opposites

Subtract and write the difference in lowest terms.

$$\frac{3x}{x-5} - \frac{2x-25}{5-x}$$

The denominators are opposites, so either may be used as the common denominator. We choose $x - 5$.

$$\begin{aligned}
\frac{3x}{x-5} - \frac{2x-25}{5-x} &= \frac{3x}{x-5} - \frac{(2x-25)}{5-x} \cdot \frac{-1}{-1} && \text{Fundamental property} \\
&= \frac{3x}{x-5} - \frac{(-2x+25)}{x-5} && \text{Multiply.} \\
&= \frac{3x-(-2x+25)}{x-5} && \text{Subtract numerators.} \\
&= \frac{3x+2x-25}{x-5} && \text{Distributive property} \\
&= \frac{5x-25}{x-5} && \text{Combine terms.} \\
&= \frac{5(x-5)}{x-5} && \text{Factor.} \\
&= 5 && \text{Lowest terms}
\end{aligned}$$

CAUTION Sign errors often occur in subtraction problems like the ones in Examples 7 and 8. Use parentheses after the subtraction sign to avoid this common error. Remember that the numerator of the fraction being subtracted must be treated as a single quantity.

CHALKBOARD EXAMPLE

Subtract and write the difference in lowest terms.

$$\frac{3r}{r^2 - 5r} - \frac{4}{r^2 - 10r + 25}$$

Answer: $\frac{3r - 19}{(r-5)^2}$

EXAMPLE 9 Subtracting Rational Expressions

Subtract and write the difference in lowest terms.

$$\frac{6x}{x^2 - 2x + 1} - \frac{1}{x^2 - 1}$$

Begin by factoring.

$$\frac{6x}{x^2 - 2x + 1} - \frac{1}{x^2 - 1} = \frac{6x}{(x-1)(x-1)} - \frac{1}{(x-1)(x+1)}$$

From the factored denominators, we can now identify the common denominator, $(x - 1)(x - 1)(x + 1)$. Use the factor $x - 1$ twice, since it appears twice in the first denominator.

$$\begin{aligned}
&\frac{6x}{(x-1)(x-1)} - \frac{1}{(x-1)(x+1)} \\
&= \frac{6x(x+1)}{(x-1)(x-1)(x+1)} - \frac{1(x-1)}{(x-1)(x-1)(x+1)} && \text{Fundamental property} \\
&= \frac{6x(x+1) - 1(x-1)}{(x-1)(x-1)(x+1)} && \text{Subtract.} \\
&= \frac{6x^2 + 6x - x + 1}{(x-1)(x-1)(x+1)} && \text{Distributive property} \\
&= \frac{6x^2 + 5x + 1}{(x-1)(x-1)(x+1)} && \text{Combine like terms.}
\end{aligned}$$

The result may be written as $\frac{6x^2 + 5x + 1}{(x-1)^2(x+1)}$.

NOTE When adding and subtracting rational expressions, several different equivalent forms of the answer often exist. If your answer does not look exactly like the one given in the back of the book, check to see if you have written an equivalent form.

6.4 EXERCISES

RELATING CONCEPTS (EXERCISES 1–8)

Work Exercises 1–8 in order. *This will help you later in determining whether your answer is equivalent to the one given in the answer section if it doesn't match the form shown.*

Jill worked a problem involving the sum of two rational expressions correctly. Her answer was given as $\frac{5}{x-3}$. Jack worked the same problem and gave his answer as $\frac{-5}{3-x}$. We want to decide whether Jack's answer was also correct.

1. Evaluate the fraction $\frac{5}{7-3}$ using the rule for order of operations.

2. Multiply the fraction given in Exercise 1 by -1 in both the numerator and the denominator, using the distributive property as necessary. Leave it in unsimplified form.

3. In Exercise 2, what number did you actually multiply the *fraction* by?

4. Simplify the result you found in Exercise 2 and compare it to the one found in Exercise 1. How do they compare?

5. Now look at the answers given by Jill and Jack above. Based on what you learned in Exercises 1–4, determine whether Jack's answer was also correct. Why or why not?

6. Jason Jordan, a perceptive algebra student, made the following comment and was praised by his teacher for his insight: "I can see that if I change the sign of each term in a fraction, the result I get is equivalent to the fraction that I started with." Explain why Jason's observation is correct.

7. Jennifer Crum, another perceptive algebra student, followed Jason's comment with one of her own: "I can see that if I put a negative sign in front of a fraction and change the signs of all terms in either the numerator or the denominator, but not both, then the result I get is equivalent to the fraction that I started with." The teacher knew that some real education was happening in the classroom. Explain why Jennifer's observation is also correct.

8. Use the concepts of Exercises 1–7 to determine whether each rational expression is equivalent or not equivalent to $\frac{y-4}{3-y}$.

(a) $\frac{y-4}{y-3}$ **(b)** $\frac{-y+4}{-3+y}$ **(c)** $-\frac{y-4}{y-3}$

(d) $-\frac{y-4}{3-y}$ **(e)** $-\frac{4-y}{y-3}$ **(f)** $\frac{y+4}{3+y}$

Did you make the connection that rational expressions can be written in equivalent forms?

---

1. $\frac{5}{4}$ 2. $\frac{-5}{-7+3}$ 3. 1
4. $\frac{-5}{-4}=\frac{5}{4}$; The two answers are equal.
5. Jack's answer was correct. His answer can be obtained by multiplying Jill's correct answer by $\frac{-1}{-1}=1$, the identity element for multiplication.
6. Changing the sign of each term in the fraction is a way of multiplying by $\frac{-1}{-1}$ or 1, the identity element for multiplication.
7. Putting a negative sign in front of a fraction gives the opposite of the fraction. Changing the signs in either the numerator or the denominator also gives the opposite. Thus we have multiplied by $(-1)(-1)=1$, the identity element for multiplication.
8. (a) not equivalent (b) equivalent (c) equivalent (d) not equivalent (e) not equivalent (f) not equivalent
9. $\frac{11}{m}$ 10. $\frac{16}{p}$ 11. b
12. $-y$ 13. x 14. t
15. $y-6$ 16. $r-3$

Add or subtract. Write the answer in lowest terms. See Examples 1 and 6.

9. $\dfrac{4}{m}+\dfrac{7}{m}$ **10.** $\dfrac{5}{p}+\dfrac{11}{p}$ **11.** $\dfrac{a+b}{2}-\dfrac{a-b}{2}$

12. $\dfrac{x-y}{2}-\dfrac{x+y}{2}$ **13.** $\dfrac{x^2}{x+5}+\dfrac{5x}{x+5}$ **14.** $\dfrac{t^2}{t-3}+\dfrac{-3t}{t-3}$

15. $\dfrac{y^2-3y}{y+3}+\dfrac{-18}{y+3}$ **16.** $\dfrac{r^2-8r}{r-5}+\dfrac{15}{r-5}$

17. Explain with an example how to add or subtract rational expressions with the same denominators.

18. Explain with an example how to add or subtract rational expressions with different denominators.

Add or subtract. Write the answer in lowest terms. See Examples 2, 3, 4, and 7.

19. $\dfrac{z}{5}+\dfrac{1}{3}$ **20.** $\dfrac{p}{8}+\dfrac{3}{5}$ **21.** $\dfrac{5}{7}-\dfrac{r}{2}$

22. $\dfrac{10}{9}-\dfrac{z}{3}$ **23.** $-\dfrac{3}{4}-\dfrac{1}{2x}$ **24.** $-\dfrac{5}{8}-\dfrac{3}{2a}$

25. $\dfrac{x+1}{6}+\dfrac{3x+3}{9}$ **26.** $\dfrac{2x-6}{4}+\dfrac{x+5}{6}$ **27.** $\dfrac{x+3}{3x}+\dfrac{2x+2}{4x}$

28. $\dfrac{x+2}{5x}+\dfrac{6x+3}{3x}$ **29.** $\dfrac{7}{3p^2}-\dfrac{2}{p}$ **30.** $\dfrac{12}{5m^2}-\dfrac{5}{m}$

31. $\dfrac{x}{x-2}+\dfrac{4}{x+2}-\dfrac{8}{x^2-4}$ **32.** $\dfrac{2x}{x-1}+\dfrac{3}{x+1}-\dfrac{4}{x^2-1}$

33. $\dfrac{t}{t+2}+\dfrac{5-t}{t}-\dfrac{4}{t^2+2t}$ **34.** $\dfrac{2p}{p-3}+\dfrac{2+p}{p}-\dfrac{-6}{p^2-3p}$

35. What are the two possible LCDs that could be used for the sum

$$\frac{10}{m-2}+\frac{5}{2-m}?$$

36. If one form of the correct answer to a sum or difference of rational expressions is $\dfrac{4}{k-3}$, what would an alternate form of the answer be if the denominator is $3-k$?

Add or subtract. Write the answer in lowest terms. See Examples 5 and 8.

37. $\dfrac{4}{x-5}+\dfrac{6}{5-x}$ **38.** $\dfrac{10}{m-2}+\dfrac{5}{2-m}$ **39.** $\dfrac{-1}{1-y}+\dfrac{3-4y}{y-1}$

40. $\dfrac{-4}{p-3}-\dfrac{p+1}{3-p}$ **41.** $\dfrac{2}{x-y^2}+\dfrac{7}{y^2-x}$ **42.** $\dfrac{-8}{p-q^2}+\dfrac{3}{q^2-p}$

43. $\dfrac{x}{5x-3y}-\dfrac{y}{3y-5x}$ **44.** $\dfrac{t}{8t-9s}-\dfrac{s}{9s-8t}$ **45.** $\dfrac{3}{4p-5}+\dfrac{9}{5-4p}$

46. $\dfrac{8}{3-7y}-\dfrac{2}{7y-3}$

▲ *In these subtraction problems, the rational expression that follows the subtraction sign has a numerator with more than one term. Be very careful with signs and find each difference.*

47. $\dfrac{2m}{m-n}-\dfrac{5m+n}{2m-2n}$ **48.** $\dfrac{5p}{p-q}-\dfrac{3p+1}{4p-4q}$

49. $\dfrac{5}{x^2-9}-\dfrac{x+2}{x^2+4x+3}$ **50.** $\dfrac{1}{a^2-1}-\dfrac{a-1}{a^2+3a-4}$

---

17. To add or subtract rational expressions with the same denominators, combine the numerators and keep the same denominator. For example, $\dfrac{3x+2}{x-6}+\dfrac{-2x-8}{x-6}=\dfrac{x-6}{x-6}$. Then write in lowest terms. In this example, the sum simplifies to 1.
18. To add or subtract rational expressions with different denominators, find the LCD, write each fraction as an equivalent one with this LCD, and then proceed as in Exercise 17. For example, $\dfrac{3}{x}+\dfrac{2}{y}=\dfrac{3y}{xy}+\dfrac{2x}{xy}=\dfrac{3y+2x}{xy}$.
19. $\dfrac{3z+5}{15}$ **20.** $\dfrac{5p+24}{40}$
21. $\dfrac{10-7r}{14}$ **22.** $\dfrac{10-3z}{9}$
23. $\dfrac{-3x-2}{4x}$ **24.** $\dfrac{-5a-12}{8a}$
25. $\dfrac{x+1}{2}$ **26.** $\dfrac{2x-2}{3}$
27. $\dfrac{5x+9}{6x}$ **28.** $\dfrac{11x+7}{5x}$
29. $\dfrac{7-6p}{3p^2}$ **30.** $\dfrac{12-25m}{5m^2}$
31. $\dfrac{x+8}{x+2}$ **32.** $\dfrac{2x+7}{x+1}$
33. $\dfrac{3}{t}$ **34.** $\dfrac{3p-1}{p-3}$
35. $m-2$ or $2-m$
36. $\dfrac{-4}{3-k}$ or $-\dfrac{4}{3-k}$
37. $\dfrac{-2}{x-5}$ or $\dfrac{2}{5-x}$
38. $\dfrac{5}{m-2}$ or $\dfrac{-5}{2-m}$
39. -4 **40.** 1
41. $\dfrac{-5}{x-y^2}$ or $\dfrac{5}{y^2-x}$
42. $\dfrac{-11}{p-q^2}$ or $\dfrac{11}{q^2-p}$
43. $\dfrac{x+y}{5x-3y}$ or $\dfrac{-x-y}{3y-5x}$
44. $\dfrac{t+s}{8t-9s}$ or $\dfrac{-t-s}{9s-8t}$
45. $\dfrac{-6}{4p-5}$ or $\dfrac{6}{5-4p}$
46. $\dfrac{10}{3-7y}$ or $\dfrac{-10}{7y-3}$
47. $\dfrac{-(m+n)}{2(m-n)}$ **48.** $\dfrac{17p-1}{4(p-q)}$
49. $\dfrac{-x^2+6x+11}{(x+3)(x-3)(x+1)}$

50. $\dfrac{-a^2+a+5}{(a+1)(a-1)(a+4)}$
51. $\dfrac{-5q^2-13q+7}{(3q-2)(q+4)(2q-3)}$
52. $\dfrac{11y^2-y-11}{(2y-1)(y+3)(3y+2)}$
53. $\dfrac{9r+2}{r(r+2)(r-1)}$
54. $\dfrac{7k-9}{k(k+3)(k-1)}$
55. $\dfrac{2x^2+6xy+8y^2}{(x+y)(x+y)(x+3y)}$ or $\dfrac{2x^2+6xy+8y^2}{(x+y)^2(x+3y)}$
56. $\dfrac{2m^2-m+1}{(m+1)(m-1)(m+1)}$ or $\dfrac{2m^2-m+1}{(m+1)^2(m-1)}$
57. $\dfrac{15r^2+10ry-y^2}{(3r+2y)(6r-y)(6r+y)}$
58. $\dfrac{-4xz-7z^2}{(x-2z)(x+2z)(2x+5z)}$
59. $\dfrac{2k^2-10k+6}{(k-3)(k-1)^2}$
60. $\dfrac{-p^2-5p+29}{(p+2)^2(3p-1)}$
61. $\dfrac{7k^2+31k+92}{(k-4)(k+4)^2}$
62. $\dfrac{-3m^3-14m^2-66m+20}{(2m+5)(m+5)(m^2+5m+25)}$
63. (a) $\dfrac{9k^2+6k+26}{5(3k+1)}$ (b) $\dfrac{1}{4}$
64. $\dfrac{9p+8}{2p^2}$

51. $\dfrac{2q+1}{3q^2+10q-8}-\dfrac{3q+5}{2q^2+5q-12}$

52. $\dfrac{4y-1}{2y^2+5y-3}-\dfrac{y+3}{6y^2+y-2}$

▲ *Perform the indicated operations. See Examples 1–9.*

53. $\dfrac{4}{r^2-r}+\dfrac{6}{r^2+2r}-\dfrac{1}{r^2+r-2}$

54. $\dfrac{6}{k^2+3k}-\dfrac{1}{k^2-k}+\dfrac{2}{k^2+2k-3}$

55. $\dfrac{x+3y}{x^2+2xy+y^2}+\dfrac{x-y}{x^2+4xy+3y^2}$

56. $\dfrac{m}{m^2-1}+\dfrac{m-1}{m^2+2m+1}$

57. $\dfrac{r+y}{18r^2+12ry-3ry-2y^2}+\dfrac{3r-y}{36r^2-y^2}$

58. $\dfrac{2x-z}{2x^2-4xz+5xz-10z^2}-\dfrac{x+z}{x^2-4z^2}$

▲ *Perform the indicated operations. Remember the order of operations.*

59. $\left(\dfrac{-k}{2k^2-5k-3}+\dfrac{3k-2}{2k^2-k-1}\right)\dfrac{2k+1}{k-1}$

60. $\left(\dfrac{3p+1}{2p^2+p-6}-\dfrac{5p}{3p^2-p}\right)\dfrac{2p-3}{p+2}$

61. $\dfrac{k^2+4k+16}{k+4}\left(\dfrac{-5}{16-k^2}+\dfrac{2k+3}{k^3-64}\right)$

62. $\dfrac{m-5}{2m+5}\left(\dfrac{-3m}{m^2-25}-\dfrac{m+4}{125-m^3}\right)$

63. Refer to the rectangle in the figure.
(a) Find an expression that represents its perimeter. Give the simplified form.
(b) Find an expression that represents its area. Give the simplified form.

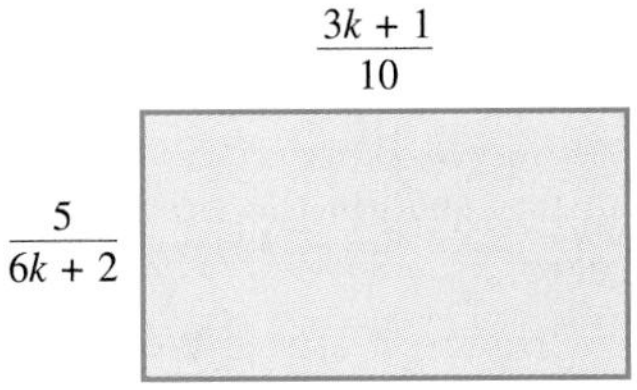

64. Refer to the triangle in the figure. Find an expression that represents its perimeter.

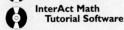

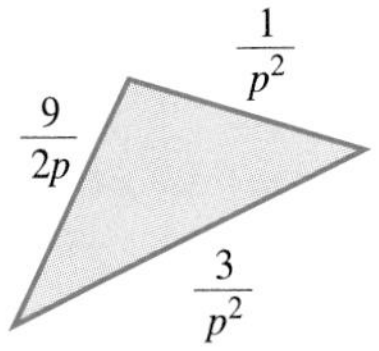

6.5 Complex Fractions

OBJECTIVES

1. Simplify a complex fraction by writing it as a division problem (Method 1).
2. Simplify a complex fraction by multiplying by the least common denominator (Method 2).

FOR EXTRA HELP

SSG Sec. 6.5
SSM Sec. 6.5
Pass the Test Software
InterAct Math Tutorial Software
Video 11

The quotient of two mixed numbers in arithmetic, such as $2\frac{1}{2}\div 3\frac{1}{4}$, can be written as a fraction:

$$2\frac{1}{2}\div 3\frac{1}{4}=\frac{2\frac{1}{2}}{3\frac{1}{4}}=\frac{2+\frac{1}{2}}{3+\frac{1}{4}}.$$

The last expression is the quotient of expressions that involve fractions. In algebra, some rational expressions also have fractions in the numerator, or denominator, or both.

Complex Fraction

A rational expression with fractions in the numerator, denominator, or both, is called a **complex fraction.**

INSTRUCTOR'S RESOURCES

PTB Sec. 6.5 **ISM** Sec. 6.5
AB Sec. 6.5
TEST GENERATOR
WORLD WIDE WEB
www.LialAlgebra.com

Examples of complex fractions include

$$\frac{2+\frac{1}{2}}{3+\frac{1}{4}}, \quad \frac{\frac{3x^2-5x}{6x^2}}{2x-\frac{1}{x}}, \quad \text{and} \quad \frac{3+x}{5-\frac{2}{x}}.$$

The parts of a complex fraction are named as follows.

$$\frac{\frac{2}{p}-\frac{1}{q}}{\frac{3}{p}+\frac{5}{q}}$$

← Numerator of complex fraction
← Main fraction bar
← Denominator of complex fraction

OBJECTIVE 1 Simplify a complex fraction by writing it as a division problem (Method 1). Since the main fraction bar represents division in a complex fraction, one method of simplifying a complex fraction involves division.

Method 1

To simplify a complex fraction:

Step 1 Write both the numerator and denominator as single fractions.

Step 2 Change the complex fraction to a division problem.

Step 3 Perform the indicated division.

TEACHING TIP Encourage students to learn both methods of simplifying complex fractions. Working the same problem side-by-side with each method may be helpful.

Once again, in this section the first example shows complex fractions from both arithmetic and algebra.

CHALKBOARD EXAMPLE

Simplify the complex fraction.

$$\frac{m+\frac{1}{2}}{\frac{6m+3}{4m}}$$

Answer: $\frac{2m}{3}$

EXAMPLE 1 Simplifying Complex Fractions by Method 1

Simplify each complex fraction.

(a) $\dfrac{\frac{2}{3}+\frac{5}{9}}{\frac{1}{4}+\frac{1}{12}}$ **(b)** $\dfrac{6+\frac{3}{x}}{\frac{x}{4}+\frac{1}{8}}$

Step 1 First, write each numerator as a single fraction.

$$\frac{2}{3}+\frac{5}{9}=\frac{2(3)}{3(3)}+\frac{5}{9}=\frac{6}{9}+\frac{5}{9}=\frac{11}{9}$$

$$6+\frac{3}{x}=\frac{6}{1}+\frac{3}{x}=\frac{6x}{x}+\frac{3}{x}=\frac{6x+3}{x}$$

Do the same thing with each denominator.

$$\frac{1}{4}+\frac{1}{12}=\frac{1(3)}{4(3)}+\frac{1}{12}=\frac{3}{12}+\frac{1}{12}=\frac{4}{12}$$

$$\frac{x}{4}+\frac{1}{8}=\frac{x(2)}{4(2)}+\frac{1}{8}=\frac{2x}{8}+\frac{1}{8}=\frac{2x+1}{8}$$

Step 2 The original complex fraction can now be written as follows.

$$\frac{\frac{11}{9}}{\frac{4}{12}} \qquad \Bigg| \qquad \frac{\frac{6x+3}{x}}{\frac{2x+1}{8}}$$

Step 3 Now use the rule for division and the fundamental property.

$$\frac{11}{9} \div \frac{4}{12} = \frac{11}{9} \cdot \frac{12}{4} = \frac{11 \cdot 3 \cdot 4}{3 \cdot 3 \cdot 4} = \frac{11}{3}$$

$$\frac{6x+3}{x} \div \frac{2x+1}{8} = \frac{6x+3}{x} \cdot \frac{8}{2x+1} = \frac{3(2x+1)}{x} \cdot \frac{8}{2x+1} = \frac{24}{x}$$

CHALKBOARD EXAMPLE

Simplify the complex fraction.

$$\frac{\frac{m^2n^3}{p}}{\frac{m^4n}{p^2}}$$

Answer: $\frac{n^2p}{m^2}$

EXAMPLE 2 Simplifying a Complex Fraction by Method 1

Simplify the complex fraction.

$$\frac{\frac{xp}{q^3}}{\frac{p^2}{qx^2}}$$

Here the numerator and denominator are already single fractions, so use the division rule and then the fundamental property.

$$\frac{xp}{q^3} \div \frac{p^2}{qx^2} = \frac{xp}{q^3} \cdot \frac{qx^2}{p^2} = \frac{x^3}{q^2p}$$

CHALKBOARD EXAMPLE

Simplify the complex fraction.

$$\frac{\frac{1}{a+1} + \frac{2}{b-2}}{\frac{2}{b-2} - \frac{1}{a+3}}$$

Answer: $\frac{(2a+b)(a+3)}{(a+1)(2a-b+8)}$

EXAMPLE 3 Simplifying a Complex Fraction by Method 1

Simplify the complex fraction.

$$\frac{\frac{3}{x+2} - 4}{\frac{2}{x+2} + 1} = \frac{\frac{3}{x+2} - \frac{4(x+2)}{x+2}}{\frac{2}{x+2} + \frac{1(x+2)}{x+2}}$$ Write both second terms with a denominator of $x + 2$.

$$= \frac{\frac{3 - 4(x+2)}{x+2}}{\frac{2 + 1(x+2)}{x+2}}$$ Subtract in the numerator. Add in the denominator.

$$= \frac{\frac{3 - 4x - 8}{x+2}}{\frac{2 + x + 2}{x+2}}$$ Distributive property

$$\frac{\frac{3-4x-8}{x+2}}{\frac{2+x+2}{x+2}} = \frac{\frac{-5-4x}{x+2}}{\frac{4+x}{x+2}} \quad \text{Combine terms.}$$

$$= \frac{-5-4x}{x+2} \cdot \frac{x+2}{4+x} \quad \text{Multiply by the reciprocal.}$$

$$= \frac{-5-4x}{4+x} \quad \text{Lowest terms}$$

OBJECTIVE 2 Simplify a complex fraction by multiplying by the least common denominator (Method 2). As an alternative method, a complex fraction may be simplified by a method that uses the fundamental property of rational expressions. Since any expression can be multiplied by a form of 1 to get an equivalent expression, we may multiply both the numerator and the denominator of a complex fraction by the same nonzero expression to get an equivalent complex fraction. If we choose the expression to be the LCD of all the fractions within the complex fraction, the complex fraction will be simplified. This is Method 2.

Method 2

To simplify a complex fraction:

Step 1 Find the LCD of all fractions within the complex fraction.

Step 2 Multiply both the numerator and the denominator of the complex fraction by this LCD using the distributive property as necessary. Write in lowest terms.

In the next example, Method 2 is used to simplify the complex fractions from Example 1.

CHALKBOARD EXAMPLE

Simplify the complex fraction.

$$\frac{2-\frac{6}{a}}{3+\frac{4}{a}}$$

Answer: $\frac{2a-6}{3a+4}$

EXAMPLE 4 Simplifying Complex Fractions by Method 2

Simplify each complex fraction.

(a) $\dfrac{\frac{2}{3}+\frac{5}{9}}{\frac{1}{4}+\frac{1}{12}}$ **(b)** $\dfrac{6+\frac{3}{x}}{\frac{x}{4}+\frac{1}{8}}$

Step 1 Find the LCD for all denominators in the complex fraction.

The LCD for 3, 9, 4, and 12 is 36. | The LCD for x, 4, and 8 is $8x$.

Step 2 Multiply numerator and denominator of the complex fraction by the LCD.

$$\frac{\frac{2}{3}+\frac{5}{9}}{\frac{1}{4}+\frac{1}{12}} = \frac{36\left(\frac{2}{3}+\frac{5}{9}\right)}{36\left(\frac{1}{4}+\frac{1}{12}\right)} \qquad \frac{6+\frac{3}{x}}{\frac{x}{4}+\frac{1}{8}} = \frac{8x\left(6+\frac{3}{x}\right)}{8x\left(\frac{x}{4}+\frac{1}{8}\right)}$$

TEACHING TIP Although we are finding the least common denominator, we do not use it as such. Multiplying the complex fraction by $\frac{36}{36}$ or by $\frac{8x}{8x}$ is again multiplying by one.

$$= \frac{36\left(\frac{2}{3}\right) + 36\left(\frac{5}{9}\right)}{36\left(\frac{1}{4}\right) + 36\left(\frac{1}{12}\right)}$$

$$= \frac{24 + 20}{9 + 3}$$

$$= \frac{44}{12} = \frac{4 \cdot 11}{4 \cdot 3}$$

$$= \frac{11}{3}$$

$$= \frac{8x(6) + 8x\left(\frac{3}{x}\right)}{8x\left(\frac{x}{4}\right) + 8x\left(\frac{1}{8}\right)} \quad \text{Distributive property}$$

$$= \frac{48x + 24}{2x^2 + x}$$

$$= \frac{24(2x + 1)}{x(2x + 1)} \quad \text{Factor.}$$

$$= \frac{24}{x} \quad \text{Lowest terms}$$

CHALKBOARD EXAMPLE

Simplify the complex fraction.

$$\frac{\frac{2}{a^2b} + \frac{3}{ab^2}}{\frac{4}{a^2b^2} - \frac{1}{ab}}$$

Answer: $\frac{2b + 3a}{4 - ab}$

EXAMPLE 5 Simplifying a Complex Fraction by Method 2

Simplify the complex fraction.

$$\frac{\frac{3}{5m} - \frac{2}{m^2}}{\frac{9}{2m} + \frac{3}{4m^2}}$$

The LCD for $5m$, m^2, $2m$, and $4m^2$ is $20m^2$. Multiply numerator and denominator by $20m^2$.

$$\frac{\frac{3}{5m} - \frac{2}{m^2}}{\frac{9}{2m} + \frac{3}{4m^2}} = \frac{20m^2\left(\frac{3}{5m} - \frac{2}{m^2}\right)}{20m^2\left(\frac{9}{2m} + \frac{3}{4m^2}\right)}$$

$$= \frac{20m^2\left(\frac{3}{5m}\right) - 20m^2\left(\frac{2}{m^2}\right)}{20m^2\left(\frac{9}{2m}\right) + 20m^2\left(\frac{3}{4m^2}\right)} \quad \text{Distributive property}$$

$$= \frac{12m - 40}{90m + 15}$$

TEACHING TIP Stress that the same answer is obtained regardless of the method used.

Either of the two methods shown in this section can be used to simplify a complex fraction. You may want to choose one method and stick with it to eliminate confusion. However, some students prefer to use Method 1 for problems like Example 2, which is the quotient of two fractions. They prefer Method 2 for problems like Examples 1, 3, 4, and 5, which have sums or differences in the numerators or denominators or both.

■ 1.413793103; 2

CONNECTIONS

Some numbers can be expressed as *continued fractions,* which are infinite complex fractions. For example, the irrational number $\sqrt{2}$ can be expressed as follows.

$$\sqrt{2} = 1 + \cfrac{1}{2 + \cfrac{1}{2 + \cfrac{1}{2 + \cfrac{1}{2 + \ddots}}}}$$

Better and better approximations of $\sqrt{2}$ can be found by using more and more terms of the fraction. We give the first three approximations here.

$$1 + \frac{1}{2} = 1.5$$

$$1 + \cfrac{1}{2 + \cfrac{1}{2}} = 1 + \cfrac{1}{\frac{5}{2}} = 1 + \frac{2}{5} = \frac{7}{5} = 1.4$$

$$1 + \cfrac{1}{2 + \cfrac{1}{2 + \cfrac{1}{2}}} = 1 + \cfrac{1}{2 + \cfrac{1}{\frac{5}{2}}} = 1 + \cfrac{1}{2 + \frac{2}{5}} = 1 + \cfrac{1}{\frac{12}{5}}$$

$$= 1 + \frac{5}{12} = \frac{17}{12} = 1.41\overline{6}$$

A calculator gives $\sqrt{2} \approx 1.414213562$ to nine decimal places.

FOR DISCUSSION OR WRITING

Give the next approximation of $\sqrt{2}$ and compare it to the calculator value shown above. How many places after the decimal agree?

6.5 EXERCISES

1. (a) 6; $\frac{1}{6}$ (b) 12; $\frac{3}{4}$

Note: In many problems involving complex fractions, several different equivalent forms of the answer exist. If your answer does not look exactly like the one given in the back of the book, check to see if your answer is an equivalent form.

1. Consider the complex fraction $\dfrac{\frac{1}{2} - \frac{1}{3}}{\frac{5}{6} - \frac{1}{12}}$. Answer each of the following, outlining Method 1 for simplifying this complex fraction.

(a) To combine the terms in the numerator, we must find the LCD of $\frac{1}{2}$ and $\frac{1}{3}$. What is this LCD? Determine the simplified form of the numerator of the complex fraction.

(b) To combine the terms in the denominator, we must find the LCD of $\frac{5}{6}$ and $\frac{1}{12}$. What is this LCD? Determine the simplified form of the denominator of the complex fraction.

(c) $\frac{1}{6} \div \frac{3}{4}$ (d) $\frac{2}{9}$

2. (a) 12 (b) $\frac{6-4}{10-1}$ (c) $\frac{2}{9}$

3. Choice (d) is correct, because every sign has been changed in the fraction. 4. Choice (a) is correct, because the given complex fraction must be positive, and choice (a) is the only positive number in the list. 5. Method 1 indicates to write the complex fraction as a division problem, and then perform the division. For example, to simplify $\frac{\frac{1}{2}}{\frac{2}{3}}$, we write $\frac{1}{2} \div \frac{2}{3}$. Then simplify as $\frac{1}{2} \cdot \frac{3}{2} = \frac{3}{4}$.

6. Method 2 indicates to multiply both the numerator and the denominator by the LCD of all the fractions within the complex fraction. For example, to simplify $\frac{\frac{1}{2}}{\frac{2}{3}}$, multiply by $\frac{6}{6}$ to get $\frac{3}{4}$.

7. -6 8. $-\frac{2}{3}$ 9. $\frac{1}{xy}$

10. p^2r 11. $\frac{2a^2b}{3}$ 12. $\frac{2r^3}{5t^4}$

13. $\frac{m(m+2)}{3(m-4)}$ 14. $\frac{3(q-5)}{q(q+5)}$

(c) Now use the results from parts (a) and (b) to write the complex fraction as a division problem using the symbol $\div$.

(d) Perform the operation from part (c) to obtain the final simplification.

2. Consider the same complex fraction given in Exercise 1, $\frac{\frac{1}{2}-\frac{1}{3}}{\frac{5}{6}-\frac{1}{12}}$. Answer each of the following, outlining Method 2 for simplifying this complex fraction.

(a) We must determine the LCD of all the fractions within the complex fraction. What is this LCD?

(b) Multiply every term in the complex fraction by the LCD found in part (a), but do not combine the terms in the numerator and the denominator yet.

(c) Combine the terms from part (b) to obtain the simplified form of the complex fraction.

3. Which one of the following complex fractions is equivalent to $\frac{3-\frac{1}{2}}{2-\frac{1}{4}}$? Answer this question without showing any work, and explain the reason for the equivalence.

(a) $\frac{3+\frac{1}{2}}{2+\frac{1}{4}}$ **(b)** $\frac{-3+\frac{1}{2}}{2-\frac{1}{4}}$ **(c)** $\frac{-3-\frac{1}{2}}{-2-\frac{1}{4}}$ **(d)** $\frac{-3+\frac{1}{2}}{-2+\frac{1}{4}}$

4. Only one of the choices below is equal to $\frac{\frac{1}{2}+\frac{1}{4}}{\frac{1}{3}+\frac{1}{12}}$. Which one is it? Answer this question without showing any work, and explain the reason for the equivalence.

(a) $\frac{9}{5}$ **(b)** $-\frac{9}{5}$ **(c)** $-\frac{5}{9}$ **(d)** -12

5. In your own words, describe Method 1 for simplifying complex fractions. Illustrate with the example $\frac{\frac{1}{2}}{\frac{2}{3}}$.

6. In your own words, describe Method 2 for simplifying complex fractions. Illustrate with the example $\frac{\frac{1}{2}}{\frac{2}{3}}$.

Simplify each complex fraction. Use either method. See Examples 1–5.

7. $\frac{-\frac{4}{3}}{\frac{2}{9}}$ **8.** $\frac{-\frac{5}{6}}{\frac{5}{4}}$ **9.** $\frac{\frac{x}{y^2}}{\frac{x^2}{y}}$ **10.** $\frac{\frac{p^4}{r}}{\frac{p^2}{r^2}}$

11. $\frac{\frac{4a^4b^3}{3a}}{\frac{2ab^4}{b^2}}$ **12.** $\frac{\frac{2r^4t^2}{3t}}{\frac{5r^2t^5}{3r}}$ **13.** $\frac{\frac{m+2}{3}}{\frac{m-4}{m}}$ **14.** $\frac{\frac{q-5}{q}}{\frac{q+5}{3}}$

15. $\dfrac{2}{x}$ 16. $\dfrac{8}{r}$ 17. $\dfrac{8}{x}$
18. $\dfrac{4}{m}$ 19. $\dfrac{a^2-5}{a^2+1}$
20. $\dfrac{q^2+1}{q^2+4}$ 21. $\dfrac{31}{50}$ 22. $\dfrac{49}{93}$
23. $\dfrac{y^2+x^2}{xy(y-x)}$ 24. $\dfrac{b+a}{ab}$
25. $\dfrac{40-12p}{85p}$ 26. $\dfrac{30-45m}{6+5m}$
27. $\dfrac{5y-2x}{3+4xy}$ 28. $\dfrac{p+2m^2}{4m^2p+mp}$
29. $\dfrac{a-2}{2a}$ 30. $\dfrac{m-3}{3m}$
31. $\dfrac{z-5}{4}$ 32. $\dfrac{a-1}{2}$
33. $\dfrac{-m}{m+2}$ 34. $\dfrac{x}{2-x}$
35. $\dfrac{3m(m-3)}{(m-1)(m-8)}$
36. $\dfrac{4(r-2)(r+2)}{(r-1)(5r+12)}$
37. division 38. identity property for multiplication

15. $\dfrac{\frac{2}{x}-3}{\frac{2-3x}{2}}$

16. $\dfrac{6+\frac{2}{r}}{\frac{3r+1}{4}}$

17. $\dfrac{\frac{1}{x}+x}{\frac{x^2+1}{8}}$

18. $\dfrac{\frac{3}{m}-m}{\frac{3-m^2}{4}}$

19. $\dfrac{a-\frac{5}{a}}{a+\frac{1}{a}}$

20. $\dfrac{q+\frac{1}{q}}{q+\frac{4}{q}}$

21. $\dfrac{\frac{5}{8}+\frac{2}{3}}{\frac{7}{3}-\frac{1}{4}}$

22. $\dfrac{\frac{6}{5}-\frac{1}{9}}{\frac{2}{5}+\frac{5}{3}}$

23. $\dfrac{\frac{1}{x^2}+\frac{1}{y^2}}{\frac{1}{x}-\frac{1}{y}}$

24. $\dfrac{\frac{1}{a^2}-\frac{1}{b^2}}{\frac{1}{a}-\frac{1}{b}}$

25. $\dfrac{\frac{2}{p^2}-\frac{3}{5p}}{\frac{4}{p}+\frac{1}{4p}}$

26. $\dfrac{\frac{2}{m^2}-\frac{3}{m}}{\frac{2}{5m^2}+\frac{1}{3m}}$

27. $\dfrac{\frac{5}{x^2y}-\frac{2}{xy^2}}{\frac{3}{x^2y^2}+\frac{4}{xy}}$

28. $\dfrac{\frac{1}{m^3p}+\frac{2}{mp^2}}{\frac{4}{mp}+\frac{1}{m^2p}}$

29. $\dfrac{\frac{1}{4}-\frac{1}{a^2}}{\frac{1}{2}+\frac{1}{a}}$

30. $\dfrac{\frac{1}{9}-\frac{1}{m^2}}{\frac{1}{3}+\frac{1}{m}}$

31. $\dfrac{\frac{1}{z+5}}{\frac{4}{z^2-25}}$

32. $\dfrac{\frac{1}{a+1}}{\frac{2}{a^2-1}}$

33. $\dfrac{\frac{1}{m+1}-1}{\frac{1}{m+1}+1}$

34. $\dfrac{\frac{2}{x-1}+2}{\frac{2}{x-1}-2}$

▲ 35. $\dfrac{\frac{1}{m-1}+\frac{2}{m+2}}{\frac{2}{m+2}-\frac{1}{m-3}}$

▲ 36. $\dfrac{\frac{5}{r+3}-\frac{1}{r-1}}{\frac{2}{r+2}+\frac{3}{r+3}}$

37. In a fraction, what operation does the fraction bar represent?

38. What property of real numbers justifies Method 2 of simplifying complex fractions?

39. $\dfrac{\frac{3}{8}+\frac{5}{6}}{2}$ 40. $\dfrac{29}{48}$ 41. $\dfrac{29}{48}$
42. Answers will vary.

RELATING CONCEPTS (EXERCISES 39–42)

In order to find the average of two numbers, we add them and divide by 2. Suppose that we wish to find the average of $\frac{3}{8}$ and $\frac{5}{6}$.

Work Exercises 39–42 in order, *to see how a complex fraction occurs in a problem like this.*

39. Write in symbols: the sum of $\frac{3}{8}$ and $\frac{5}{6}$, divided by 2. Your result should be a complex fraction.

40. Simplify the complex fraction from Exercise 39 using Method 1.

41. Simplify the complex fraction from Exercise 39 using Method 2.

✎ 42. Your answers in Exercises 40 and 41 should be the same. Which method did you prefer? Why?

Did you make the connection between finding the average of two fractions and simplifying a complex fraction?

43. $\frac{5}{3}$ 44. $\frac{65}{11}$ 45. $\frac{13}{2}$
46. $\frac{13}{5}$ 47. $\frac{19r}{15}$ 48. $\frac{q}{7}$

▲ *The expressions in Exercises 43–48 are called* continued fractions. *Simplify these continued fractions by starting at "the bottom" and working upward.*

43. $1 + \cfrac{1}{1 + \cfrac{1}{1 + 1}}$

44. $5 + \cfrac{5}{5 + \cfrac{5}{5 + 5}}$

45. $7 - \cfrac{3}{5 + \cfrac{2}{4 - 2}}$

46. $3 - \cfrac{2}{4 + \cfrac{2}{4 - 2}}$

47. $r + \cfrac{r}{4 - \cfrac{2}{6 + 2}}$

48. $\frac{2q}{7} - \cfrac{q}{6 + \cfrac{8}{4 + 4}}$

6.6 Solving Equations Involving Rational Expressions

OBJECTIVES

1. Distinguish between expressions with rational coefficients and equations with terms that are rational expressions.
2. Solve equations with rational expressions.
3. Solve a formula for a specified variable.

FOR EXTRA HELP

SSG Sec. 6.6
SSM Sec. 6.6
Pass the Test Software
InterAct Math Tutorial Software
Video 11

CHALKBOARD EXAMPLE

Identify the following as an expression or an equation. If it is an expression, simplify it. If it is an equation, solve it.

$$\frac{x}{3} + \frac{x}{4} = 10 + x$$

Answer: equation; {−24}

TEACHING TIP Spend enough time in class explaining how the least common denominator is used in different ways when adding and subtracting fractions as opposed to clearing fractions in an equation. Again, review the distinction between expressions and equations.

In Section 2.2 we solved equations with fractions as coefficients. By using the multiplication property of equality, we cleared the fractions by multiplying by the LCD. We continue this work here.

OBJECTIVE 1 Distinguish between expressions with rational coefficients and equations with terms that are rational expressions. Before solving equations with rational expressions, you must understand the difference between *sums* and *differences* of terms with rational coefficients, and *equations* with terms that are rational expressions. Sums and differences are *simplified,* while equations are *solved.*

EXAMPLE 1 Distinguishing between Expressions and Equations

Identify each of the following as an expression or an equation. If it is an expression, simplify it. If it is an equation, solve it.

(a) $\frac{3}{4}x - \frac{2}{3}x$

This is a difference of two terms, so it is an expression. (There is no equals sign.) Simplify by finding the LCD, writing each coefficient with this LCD, and combining like terms.

$$\frac{3}{4}x - \frac{2}{3}x = \frac{9}{12}x - \frac{8}{12}x \qquad \text{Get a common denominator.}$$

$$= \frac{1}{12}x \qquad \text{Combine like terms.}$$

(b) $\frac{3}{4}x - \frac{2}{3}x = \frac{1}{2}$

Because of the equals sign, this is an equation to be solved. We proceed as in Section 2.2, using the multiplication property of equality to clear fractions. The LCD is 12.

$$\frac{3}{4}x - \frac{2}{3}x = \frac{1}{2}$$

$$12\left(\frac{3}{4}x - \frac{2}{3}x\right) = 12\left(\frac{1}{2}\right) \qquad \text{Multiply by 12.}$$

$$12\left(\frac{3}{4}x\right) - 12\left(\frac{2}{3}x\right) = 12\left(\frac{1}{2}\right) \qquad \text{Distributive property}$$

INSTRUCTOR'S RESOURCES

PTB Sec. 6.6 ISM Sec. 6.6
AB Sec. 6.6
TEST GENERATOR
WORLD WIDE WEB
www.LialAlgebra.com

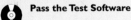

$$9x - 8x = 6 \qquad \text{Multiply.}$$
$$x = 6 \qquad \text{Combine like terms.}$$

The solution, 6, should be checked in the original equation. Because it leads to a true statement, the solution set is $\{6\}$.

The ideas of Example 1 can be summarized as follows.

When adding or subtracting, the LCD must be kept throughout the simplification. When solving an equation, the LCD is used to multiply both sides so that denominators are eliminated.

OBJECTIVE 2 Solve equations with rational expressions. The next few examples illustrate clearing an equation of fractions in order to solve it.

CHALKBOARD EXAMPLE

Solve and check your answer.

$$\frac{2m-3}{5} - \frac{m}{3} = -\frac{6}{5}$$

Answer: $\{-9\}$

TEACHING TIP Omitting the parentheses here is a common error.

EXAMPLE 2 Solving an Equation Involving Rational Expressions

Solve $\dfrac{p}{2} - \dfrac{p-1}{3} = 1$.

Multiply both sides by the LCD, 6.

$$6\left(\frac{p}{2} - \frac{p-1}{3}\right) = 6 \cdot 1$$
$$6\left(\frac{p}{2}\right) - 6\left(\frac{p-1}{3}\right) = 6 \qquad \text{Distributive property}$$
$$3p - 2(p-1) = 6$$

Be careful to put parentheses around $p - 1$; otherwise you may get an incorrect solution. Continue simplifying and solve the equation.

$$3p - 2p + 2 = 6 \qquad \text{Distributive property}$$
$$p + 2 = 6 \qquad \text{Combine like terms.}$$
$$p = 4 \qquad \text{Subtract 2.}$$

Check to see that $\{4\}$ is the solution set by replacing p with 4 in the original equation.

CAUTION The most common error in equations like the one found in Example 2 occurs when parentheses are not used for the numerator $p - 1$. If parentheses are not used, a sign error may result.

The equations in Examples 1(b) and 2 did not have variables in denominators. When solving equations that have a variable in a denominator, remember that the number 0 cannot be a denominator. Therefore, the solution cannot be a number that will make the denominator equal 0. Example 3 illustrates this.

CHALKBOARD EXAMPLE

Solve and check your answer.

$$1 - \frac{2}{x+1} = \frac{2x}{x+1}$$

Answer: When the equation is solved, -1 is found. However, since $x = -1$ leads to a 0 denominator in the original equation, the solution set is $\emptyset$.

EXAMPLE 3 Solving an Equation Involving Rational Expressions with No Solution

Solve $\dfrac{x}{x-2} = \dfrac{2}{x-2} + 2$.

Multiply both sides by the LCD, $x - 2$.

$$(x-2)\left(\frac{x}{x-2}\right) = (x-2)\left(\frac{2}{x-2}\right) + (x-2)(2)$$

$x = 2 + 2x - 4$	Distributive property
$x = -2 + 2x$	Combine terms.
$-x = -2$	Subtract $2x$.
$x = 2$	Multiply by -1.

The proposed solution is 2. If we substitute 2 into the original equation, we get

$$\frac{2}{2-2} = \frac{2}{2-2} + 2 \qquad ?$$

$$\frac{2}{0} = \frac{2}{0} + 2. \qquad ?$$

Notice that 2 makes both denominators equal 0. Because 0 cannot be the denominator of a fraction, the solution set is $\emptyset$.

While it is always a good idea to check solutions to guard against arithmetic and algebraic errors, it is *essential* to check proposed solutions when variables appear in denominators in the original equation. Some students like to determine which numbers cannot be solutions *before* solving the equation.

Solving Equations with Rational Expressions

Step 1 **Multiply by the LCD.** Multiply both sides of the equation by the least common denominator. (This clears the equation of fractions.)

Step 2 **Solve.** Solve the resulting equation.

Step 3 **Check.** Check each proposed solution by substituting it in the original equation. Reject any that cause a denominator to equal 0.

CHALKBOARD EXAMPLE

Solve and check your answer.

$$\frac{2}{p^2 - 2p} = \frac{3}{p^2 - p}$$

Answer: $\{4\}$

EXAMPLE 4 Solving an Equation Involving Rational Expressions

Solve $\dfrac{2}{x^2 - x} = \dfrac{1}{x^2 - 1}$.

Step 1 Begin by finding the LCD. Since $x^2 - x$ can be factored as $x(x-1)$, and $x^2 - 1$ can be factored as $(x+1)(x-1)$, the LCD is $x(x+1)(x-1)$.

$$\frac{2}{x(x-1)} = \frac{1}{(x+1)(x-1)} \qquad \text{Factor the denominators.}$$

Step 2 Notice that 0, -1, and 1 cannot be solutions of this equation. Multiply both sides of the equation by $x(x+1)(x-1)$.

$$x(x+1)(x-1)\frac{2}{x(x-1)} = x(x+1)(x-1)\frac{1}{(x+1)(x-1)}$$
$$2(x+1) = x$$
$$2x + 2 = x \qquad \text{Distributive property}$$
$$2 = -x \qquad \text{Subtract } 2x.$$
$$x = -2 \qquad \text{Multiply by } -1.$$

Step 3 The proposed solution is -2, which does not make any denominator equal 0. A check will verify that no arithmetic or algebraic errors have been made. Thus, the solution set is $\{-2\}$.

CHALKBOARD EXAMPLE

Solve $\frac{1}{x-2} + \frac{1}{5} = \frac{2}{5(x^2+4)}$.

Answer: $\{-4, -1\}$

EXAMPLE 5 Solving an Equation Involving Rational Expressions

Solve $\frac{1}{x-1} + \frac{1}{2} = \frac{2}{x^2-1}$.

Factor the denominator on the right.

$$\frac{1}{x-1} + \frac{1}{2} = \frac{2}{(x+1)(x-1)}$$

Notice that 1 and -1 cannot be solutions of this equation. Multiply both sides of the equation by the LCD, $2(x+1)(x-1)$.

$$2(x+1)(x-1)\left(\frac{1}{x-1} + \frac{1}{2}\right) = 2(x+1)(x-1)\frac{2}{(x+1)(x-1)}$$
$$2(x+1)(x-1)\frac{1}{x-1} + 2(x+1)(x-1)\frac{1}{2} = 2(x+1)(x-1)\frac{2}{(x+1)(x-1)}$$
$$2(x+1) + (x+1)(x-1) = 4$$
$$2x + 2 + x^2 - 1 = 4 \qquad \text{Distributive property}$$
$$x^2 + 2x + 1 = 4$$
$$x^2 + 2x - 3 = 0 \qquad \text{Get 0 on the right side.}$$

TEACHING TIP This section is the first time that most students have encountered equations where even though their algebra is correct, the apparent solution is not really a solution. Stress that the check is an essential part of the solution process.

Factoring gives

$$(x+3)(x-1) = 0.$$
$$x + 3 = 0 \quad \text{or} \quad x - 1 = 0 \qquad \text{Zero-factor property}$$
$$x = -3 \quad \text{or} \quad x = 1$$

-3 and 1 are proposed solutions. However, as noted above, 1 makes an original denominator equal 0, so 1 is not a solution. Substituting -3 for x gives a true statement, so $\{-3\}$ is the solution set.

CHALKBOARD EXAMPLE

Solve $\frac{6}{5a+10} - \frac{1}{a-5} = \frac{4}{a^2-3a-10}$.

Answer: $\{60\}$

EXAMPLE 6 Solving an Equation Involving Rational Expressions

Solve $\frac{1}{k^2+4k+3} + \frac{1}{2k+2} = \frac{3}{4k+12}$.

Factoring each denominator gives the equation

$$\frac{1}{(k+1)(k+3)} + \frac{1}{2(k+1)} = \frac{3}{4(k+3)}.$$

The LCD is $4(k+1)(k+3)$, indicating that -1 and -3 cannot be solutions of the equation. Multiply both sides by this LCD.

$$4(k+1)(k+3)\left(\frac{1}{(k+1)(k+3)} + \frac{1}{2(k+1)}\right) = 4(k+1)(k+3)\frac{3}{4(k+3)}$$

$$4(k+1)(k+3)\frac{1}{(k+1)(k+3)} + 2 \cdot 2(k+1)(k+3)\frac{1}{2(k+1)} = 4(k+1)(k+3)\frac{3}{4(k+3)}$$

$$4 + 2(k+3) = 3(k+1)$$

$$4 + 2k + 6 = 3k + 3 \qquad \text{Distributive property}$$

$$2k + 10 = 3k + 3$$

$$10 - 3 = 3k - 2k$$

$$7 = k$$

The proposed solution, 7, does not make an original denominator equal 0. A check shows that the algebra is correct, so $\{7\}$ is the solution set.

OBJECTIVE 3 **Solve a formula for a specified variable.** Solving a formula for a specified variable was discussed in Chapter 2. In the next example, this procedure is applied to a formula involving fractions.

CHALKBOARD EXAMPLE

Solve $a = \frac{v - w}{t}$ for v.

Answer: $v = at + w$

EXAMPLE 7 Solving for a Specified Variable

Solve the formula $\frac{1}{a} = \frac{1}{b} + \frac{1}{c}$ for c.

The LCD of all the fractions in the equation is *abc*, so multiply both sides by *abc*.

$$abc\left(\frac{1}{a}\right) = abc\left(\frac{1}{b} + \frac{1}{c}\right)$$

$$abc\left(\frac{1}{a}\right) = abc\left(\frac{1}{b}\right) + abc\left(\frac{1}{c}\right) \qquad \text{Distributive property}$$

$$bc = ac + ab$$

Since we are solving for c, get all terms with c on one side of the equation. Do this by subtracting ac from both sides.

$$bc - ac = ab \qquad \text{Subtract } ac.$$

Factor out the common factor c on the left.

$$c(b - a) = ab \qquad \text{Factor out } c.$$

Finally, divide both sides by the coefficient of c, which is $b - a$.

$$c = \frac{ab}{b - a}$$

TEACHING TIP Remind students that solving for a specified variable uses the same steps in the same sequence as solving an equation in one unknown. As in Chapter 2, encourage students to circle or underline the variable they are solving for.

CAUTION Students often have trouble in the step that involves factoring out the variable for which they are solving. In Example 7, we had to factor out c on the left side so that we could divide both sides by $b - a$.

When solving an equation for a specified variable, be sure that the specified variable appears alone on only one side of the equals sign in the final equation.

6.6 EXERCISES

Identify as an expression or an equation. If an expression appears, simplify it. If an equation appears, solve it. See Example 1.

1. $\frac{7}{8}x + \frac{1}{5}x$
2. $\frac{4}{7}x + \frac{3}{5}x$
3. $\frac{7}{8}x + \frac{1}{5}x = 1$
4. $\frac{4}{7}x + \frac{3}{5}x = 1$
5. $\frac{3}{5}y - \frac{7}{10}y$
6. $\frac{2}{3}y - \frac{7}{4}y$
7. $\frac{3}{5}y - \frac{7}{10}y = 1$
8. $\frac{2}{3}y - \frac{7}{4}y = -13$

When solving an equation with variables in denominators, we must determine the values that cause these denominators to equal 0, so we can reject these values if they appear as possible solutions. Find all values for which at least one denominator is equal to 0. Do not solve. See Examples 3–6.

9. $\frac{3}{x+2} - \frac{5}{x} = 1$
10. $\frac{7}{x} + \frac{9}{x-3} = 5$
11. $\frac{-1}{(x+3)(x-4)} = \frac{1}{2x+1}$
12. $\frac{8}{(x-8)(x+2)} = \frac{7}{3x-10}$
13. $\frac{4}{x^2+8x-9} + \frac{1}{x^2-4} = 0$
14. $\frac{-3}{x^2+9x-10} - \frac{12}{x^2-16} = 0$
15. Explain how the LCD is used in a different way when adding and subtracting rational expressions compared to solving equations with rational expressions.
16. If we multiply both sides of the equation $\frac{6}{x+5} = \frac{6}{x+5}$ by $x + 5$, we get $6 = 6$. Are all real numbers solutions of this equation? Explain.

Solve each equation and check. See Examples 1(b), 2, and 3.

17. $\frac{5}{m} - \frac{3}{m} = 8$
18. $\frac{4}{y} + \frac{1}{y} = 2$
19. $\frac{5}{y} + 4 = \frac{2}{y}$
20. $\frac{11}{q} = 3 - \frac{1}{q}$
21. $\frac{3x}{5} - 6 = x$
22. $\frac{5t}{4} + t = 9$
23. $\frac{4m}{7} + m = 11$
24. $a - \frac{3a}{2} = 1$
25. $\frac{z-1}{4} = \frac{z+3}{3}$
26. $\frac{r-5}{2} = \frac{r+2}{3}$
27. $\frac{3p+6}{8} = \frac{3p-3}{16}$
28. $\frac{2z+1}{5} = \frac{7z+5}{15}$
29. $\frac{2x+3}{x} = \frac{3}{2}$
30. $\frac{5-2y}{y} = \frac{1}{4}$
31. $\frac{k}{k-4} - 5 = \frac{4}{k-4}$
32. $\frac{-5}{a+5} = \frac{a}{a+5} + 2$
33. $\frac{q+2}{3} + \frac{q-5}{5} = \frac{7}{3}$
34. $\frac{t}{6} + \frac{4}{3} = \frac{t-2}{3}$
35. $\frac{x}{2} = \frac{5}{4} + \frac{x-1}{4}$
36. $\frac{8p}{5} = \frac{3p-4}{2} + \frac{5}{2}$

Answers

1. expression; $\frac{43}{40}x$
2. expression; $\frac{41}{35}x$
3. equation; $\left\{\frac{40}{43}\right\}$
4. equation; $\left\{\frac{35}{41}\right\}$
5. expression; $-\frac{1}{10}y$
6. expression; $-\frac{13}{12}y$
7. equation; $\{-10\}$
8. equation; $\{12\}$ 9. $-2, 0$
10. $0, 3$ 11. $-3, 4, -\frac{1}{2}$
12. $8, -2, \frac{10}{3}$ 13. $-9, 1, -2, 2$
14. $-10, 1, -4, 4$ 15. When adding and subtracting, the LCD is retained as part of the answer. When solving an equation, the LCD is used as the multiplier in applying the multiplication property of equality.
16. No. The value -5 causes both denominators to equal 0. The solution set contains all real numbers *except* -5.
17. $\left\{\frac{1}{4}\right\}$ 18. $\left\{\frac{5}{2}\right\}$ 19. $\left\{-\frac{3}{4}\right\}$
20. $\{4\}$ 21. $\{-15\}$ 22. $\{4\}$
23. $\{7\}$ 24. $\{-2\}$ 25. $\{-15\}$
26. $\{19\}$ 27. $\{-5\}$ 28. $\{-2\}$
29. $\{-6\}$ 30. $\left\{\frac{20}{9}\right\}$ 31. $\emptyset$
32. $\emptyset$ 33. $\{5\}$ 34. $\{12\}$
35. $\{4\}$ 36. $\{5\}$

JOURNAL   WRITING   ▲ CHALLENGING   SCIENTIFIC CALCULATOR   GRAPHING CALCULATOR

37. {1} 38. {−3} 39. {4}
40. {12} 41. {5} 42. {5}
43. {−2, 12} 44. {−4, 2}
45. ∅ 46. ∅ 47. {3}
48. {−5} 49. {3} 50. {−2}
51. $\left\{-\frac{1}{5}, 3\right\}$ 52. $\left\{-\frac{1}{3}, 3\right\}$
53. $\left\{-\frac{1}{2}, 5\right\}$ 54. $\left\{-\frac{6}{7}, 3\right\}$
55. {3} 56. {1}
57. $\left\{-\frac{1}{3}, 3\right\}$ 58. {4, 6}
59. $\left\{-6, \frac{1}{2}\right\}$ 60. {−4, 0}
61. {6} 62. {3}
63. Transform the equation so that the terms with *k* are on one side and the remaining term is on the other.
64. Factor out *k* on the left side.
65. $F = \frac{ma}{k}$ 66. $E = \frac{IR}{k}$
67. $a = \frac{kF}{m}$ 68. $R = \frac{kE}{I}$
69. $R = \frac{E - Ir}{I}$ or $R = \frac{E}{I} - r$
70. $r = \frac{E - IR}{I}$ or $r = \frac{E}{I} - R$
71. $A = \frac{h(B + b)}{2}$
72. $S = \frac{dn(a + L)}{2}$
73. $a = \frac{2S - ndL}{nd}$ or $a = \frac{2S}{nd} - L$
74. $B = \frac{2A - hb}{h}$ or $B = \frac{2A}{h} - b$
75. $y = \frac{xz}{x + z}$
76. $q = \frac{kp}{3p - k}$
77. $z = \frac{3y}{5 - 9xy}$ or $z = \frac{-3y}{9xy - 5}$
78. $a = \frac{bc}{c + b}$

Solve each equation and check. Be very careful with signs. See Example 2.

37. $\frac{a + 7}{8} - \frac{a - 2}{3} = \frac{4}{3}$

38. $\frac{x + 3}{7} - \frac{x + 2}{6} = \frac{1}{6}$

39. $\frac{p}{2} - \frac{p - 1}{4} = \frac{5}{4}$

40. $\frac{r}{6} - \frac{r - 2}{3} = -\frac{4}{3}$

41. $\frac{3x}{5} - \frac{x - 5}{7} = 3$

42. $\frac{8k}{5} - \frac{3k - 4}{2} = \frac{5}{2}$

Solve each equation and check your answer. See Examples 3–6.

43. $\frac{2}{m} = \frac{m}{5m + 12}$

44. $\frac{x}{4 - x} = \frac{2}{x}$

45. $\frac{-2}{z + 5} + \frac{3}{z - 5} = \frac{20}{z^2 - 25}$

46. $\frac{3}{r + 3} - \frac{2}{r - 3} = \frac{-12}{r^2 - 9}$

47. $\frac{3}{x - 1} + \frac{2}{4x - 4} = \frac{7}{4}$

48. $\frac{2}{p + 3} + \frac{3}{8} = \frac{5}{4p + 12}$

49. $\frac{y}{3y + 3} = \frac{2y - 3}{y + 1} - \frac{2y}{3y + 3}$

50. $\frac{2k + 3}{k + 1} - \frac{3k}{2k + 2} = \frac{-2k}{2k + 2}$

51. $\frac{5x}{14x + 3} = \frac{1}{x}$

52. $\frac{m}{8m + 3} = \frac{1}{3m}$

53. $\frac{2}{x - 1} - \frac{2}{3} = \frac{-1}{x + 1}$

54. $\frac{5}{p - 2} = 7 - \frac{10}{p + 2}$

55. $\frac{x}{2x + 2} = \frac{-2x}{4x + 4} + \frac{2x - 3}{x + 1}$

56. $\frac{5t + 1}{3t + 3} = \frac{5t - 5}{5t + 5} + \frac{3t - 1}{t + 1}$

57. $\frac{8x + 3}{x} = 3x$

58. $\frac{2}{y} = \frac{y}{5y - 12}$

▲ **59.** $\frac{3y}{y^2 + 5y + 6} = \frac{5y}{y^2 + 2y - 3} - \frac{2}{y^2 + y - 2}$

▲ **60.** $\frac{m}{m^2 + m - 2} + \frac{m}{m^2 - 1} = \frac{m}{m^2 + 3m + 2}$

▲ **61.** $\frac{x + 4}{x^2 - 3x + 2} - \frac{5}{x^2 - 4x + 3} = \frac{x - 4}{x^2 - 5x + 6}$

▲ **62.** $\frac{3}{r^2 + r - 2} - \frac{1}{r^2 - 1} = \frac{7}{2(r^2 + 3r + 2)}$

63. If you are solving a formula for the letter *k*, and your steps lead to the equation $kr - mr = km$, what would be your next step?

64. If you are solving a formula for the letter *k*, and your steps lead to the equation $kr - km = mr$, what would be your next step?

Solve each formula for the specified variable. See Example 7.

65. $m = \frac{kF}{a}$ for F

66. $I = \frac{kE}{R}$ for E

67. $m = \frac{kF}{a}$ for a

68. $I = \frac{kE}{R}$ for R

69. $I = \frac{E}{R + r}$ for R

70. $I = \frac{E}{R + r}$ for r

71. $h = \frac{2A}{B + b}$ for A

72. $d = \frac{2S}{n(a + L)}$ for S

73. $d = \frac{2S}{n(a + L)}$ for a

74. $h = \frac{2A}{B + b}$ for B

75. $\frac{1}{x} = \frac{1}{y} - \frac{1}{z}$ for y

76. $\frac{3}{k} = \frac{1}{p} + \frac{1}{q}$ for q

77. $9x + \frac{3}{z} = \frac{5}{y}$ for z

78. $\frac{1}{a} = \frac{1}{b} + \frac{1}{c}$ for a

79. The solution set is $\{-5\}$. The number 3 must be rejected.
80. The simplified form is $x + 5$. (a) It is the same as the actual solution. (b) It is the same as the rejected solution.
81. The solution set is $\{-3\}$. The number 1 must be rejected.
82. The simplified form is $\frac{x+3}{2(x+1)}$. (a) It is the same as the actual solution. (b) It is the same as the rejected solution.
83. Answers will vary.
84. If we transform so that 0 is on one side, then perform the operation(s) and *reduce to lowest terms* before solving, no rejected values will appear.

RELATING CONCEPTS (EXERCISES 79–84)

In Section 6.4 we saw how, after adding or subtracting two rational expressions, the result can often be simplified to lowest terms. In this section we learned that multiplying both sides of an equation by the LCD may lead to a solution that must be rejected because it causes a denominator to equal 0.

Work Exercises 79–84 in order, *to see a relationship between these two concepts.*

79. Solve the equation $\frac{x^2}{x-3} + \frac{2x-15}{x-3} = 0$ by multiplying both sides by $x - 3$. What is the solution? What is the number that must be rejected as a solution?

80. Combine the two rational expressions on the left side of the given equation in Exercise 79 and reduce to lowest terms.

(a) If you set the simplified form equal to 0 and solve, how does your solution compare to the actual solution in Exercise 79?

(b) If you set the common factor that you divided out equal to 0 and solve, how does your solution compare to the rejected solution in Exercise 79?

81. Consider the equation $\frac{1}{x-1} + \frac{1}{2} - \frac{2}{x^2-1} = 0$, which is equivalent to the equation solved in Example 5. According to Example 5, what is the solution? What is the number that must be rejected as a solution?

82. Combine the three rational expressions on the left side of the given equation in Exercise 81 and reduce to lowest terms. Repeat parts (a) and (b) of Exercise 80, comparing this time to the equation in Exercise 81.

83. Write a short paragraph summarizing what you have learned in working Exercises 79–82.

84. Devise a procedure whereby you could solve an equation involving rational expressions which would not lead to values that must be rejected.

Did you make the connection between rational expressions that can be simplified and rational equations with rejected solutions?

TECHNOLOGY INSIGHTS (EXERCISES 85–90)

The first line in the calculator screen on the left shows that 6 has been stored into memory location X. The expression $\frac{1}{X+4}$ is then calculated to be $\frac{1}{10}$, as expected. However, if the calculator is directed to evaluate $\frac{1}{X-6}$, an error message occurs, as shown in the screen on the right. The value 6 leads to division by 0, which is not allowed.

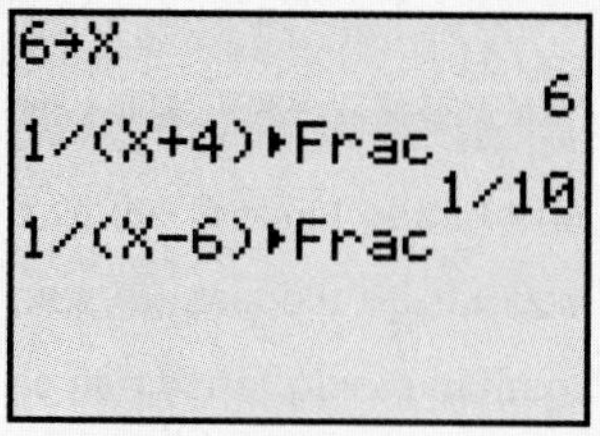

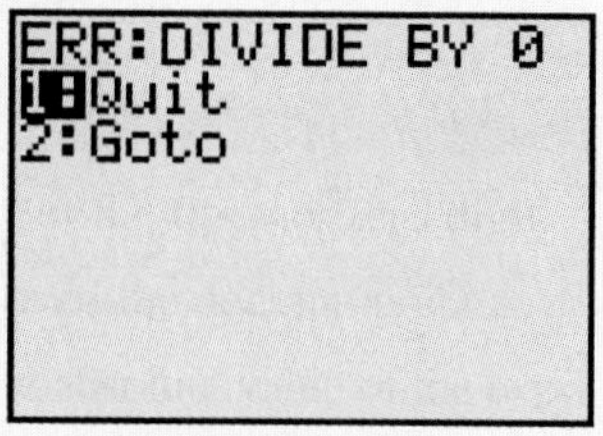

85. $-\frac{3}{10}$ 86. $-\frac{1}{4}$

87. An error message would occur.

88. An error message would occur.

89. An error message would occur.

90. An error message would occur.

TECHNOLOGY INSIGHTS (EXERCISES 85–90) (CONTINUED)

For the following screens, either evaluate the expression or tell whether an error message would occur for the expression entered.

85.
```
5→X
                5
-3/(X+5)▸Frac
```

86.
```
8→X
                8
-4/(X+8)▸Frac
```

87.
```
12→X
               12
-3/(12-X)
```

88.
```
16→X
               16
-8/(16-X)
```

89.
```
3→X
                3
1/((X+4)*(X-3))▸
Frac
```

90.
```
5→X
                5
1/((X-5)*(X-3))▸
Frac
```

SUMMARY Exercises on Operations and Equations with Rational Expressions

TEACHING TIP This section helps students to recognize the differences between equations and expressions. Ask students to identify each exercise as an equation or expression, and to describe what steps they would use to complete it.

We have performed the four operations of arithmetic with rational expressions and solved equations with rational expressions. The exercises in this summary include a mixed variety of problems of these types. To work them, recall the procedures explained in the earlier sections of this chapter. They are summarized here.

Multiplication of Rational Expressions	Multiply numerators and multiply denominators. Use the fundamental property to express in lowest terms.
Division of Rational Expressions	First, change the second fraction to its reciprocal; then multiply as described above.
Addition of Rational Expressions	Find the least common denominator (LCD) if necessary. Write all rational expressions with this LCD. Add numerators, and keep the same denominator. Express in lowest terms.

Subtraction of Rational Expressions	Find the LCD if necessary. Write all rational expressions with this LCD. Subtract numerators (use parentheses as required), and keep the same denominator. Express in lowest terms.
Solving Equations with Rational Expressions	Multiply both sides of the equation by the LCD of all the rational expressions in the equation. Solve, using methods described in earlier chapters. Be sure to check all proposed solutions and reject any that cause a denominator to equal 0.

A common error that was mentioned in Section 6.6 bears repeating here. Students often confuse *operations* on rational expressions with the *solution of equations* with rational expressions. For example, the four possible operations on the rational expressions $\frac{1}{x}$ and $\frac{1}{x-2}$ can be performed as follows.

Add:

$$\frac{1}{x}+\frac{1}{x-2}=\frac{x-2}{x(x-2)}+\frac{x}{x(x-2)}$$ Write with a common denominator.

$$=\frac{x-2+x}{x(x-2)}$$ Add numerators; keep the same denominator.

$$=\frac{2x-2}{x(x-2)}$$ Combine like terms.

Subtract:

$$\frac{1}{x}-\frac{1}{x-2}=\frac{x-2}{x(x-2)}-\frac{x}{x(x-2)}$$ Write with a common denominator.

$$=\frac{x-2-x}{x(x-2)}$$ Subtract numerators; keep the same denominator.

$$=\frac{-2}{x(x-2)}$$ Combine like terms.

Multiply:

$$\frac{1}{x}\cdot\frac{1}{x-2}=\frac{1}{x(x-2)}$$ Multiply numerators and multiply denominators.

Divide:

$$\frac{1}{x}\div\frac{1}{x-2}=\frac{1}{x}\cdot\frac{x-2}{1}=\frac{x-2}{x}$$ Change to multiplication by the reciprocal of the second fraction.

On the other hand, consider the *equation*

$$\frac{1}{x}+\frac{1}{x-2}=\frac{3}{4}.$$

Neither 0 nor 2 can be solutions of this equation, since each will cause a denominator to equal 0. We use the multiplication property of equality to multiply both sides by the LCD, $4x(x-2)$, giving an equation with no denominators.

$$4x(x-2)\frac{1}{x} + 4x(x-2)\frac{1}{x-2} = 4x(x-2)\frac{3}{4}$$

$4x - 8 + 4x = 3x^2 - 6x$	Distributive property
$0 = 3x^2 - 14x + 8$	Get 0 on one side.
$0 = (3x-2)(x-4)$	Factor.
$3x - 2 = 0 \quad \text{or} \quad x - 4 = 0$	Zero-factor property
$x = \frac{2}{3} \quad \text{or} \quad x = 4$	

Both $\frac{2}{3}$ and 4 are solutions since neither makes a denominator equal 0; the solution set is $\left\{\frac{2}{3}, 4\right\}$.

In conclusion, remember the following points when working exercises involving rational expressions.

Points to Remember When Working with Rational Expressions

1. The fundamental property is applied only after numerators and denominators have been *factored.*
2. When adding and subtracting rational expressions, the common denominator must be kept throughout the problem and in the final result.
3. Always look to see if the answer is in lowest terms; if it is not, use the fundamental property.
4. When solving equations with rational expressions, reject any proposed solution that causes an original denominator to equal 0.

SUMMARY EXERCISES

For each exercise, indicate "operation" if an operation is to be performed or "equation" if an equation is to be solved. Then perform the operation or solve the equation as the case may be.

1. $\frac{4}{p} + \frac{6}{p}$

2. $\frac{x^3y^2}{x^2y^4} \cdot \frac{y^5}{x^4}$

3. $\frac{1}{x^2+x-2} \div \frac{4x^2}{2x-2}$

4. $\frac{8}{m-5} = 2$

5. $\frac{2y^2+y-6}{2y^2-9y+9} \cdot \frac{y^2-2y-3}{y^2-1}$

6. $\frac{2}{k^2-4k} + \frac{3}{k^2-16}$

7. $\frac{x-4}{5} = \frac{x+3}{6}$

8. $\frac{3t^2-t}{6t^2+15t} \div \frac{6t^2+t-1}{2t^2-5t-25}$

9. $\frac{4}{p+2} + \frac{1}{3p+6}$

10. $\frac{1}{y} + \frac{1}{y-3} = -\frac{5}{4}$

11. $\frac{3}{t-1} + \frac{1}{t} = \frac{7}{2}$

12. $\frac{6}{y} - \frac{2}{3y}$

13. $\frac{5}{4z} - \frac{2}{3z}$

14. $\frac{k+2}{3} = \frac{2k-1}{5}$

1. operation; $\frac{10}{p}$ **2.** operation; $\frac{y^3}{x^3}$ **3.** operation; $\frac{1}{2x^2(x+2)}$ **4.** equation; $\{9\}$ **5.** operation; $\frac{y+2}{y-1}$ **6.** operation; $\frac{5k+8}{k(k-4)(k+4)}$ **7.** equation; $\{39\}$ **8.** operation; $\frac{t-5}{3(2t+1)}$ **9.** operation; $\frac{13}{3(p+2)}$ **10.** equation; $\left\{-1, \frac{12}{5}\right\}$ **11.** equation; $\left\{\frac{1}{7}, 2\right\}$ **12.** operation; $\frac{16}{3y}$

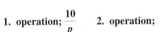

13. operation; $\frac{7}{12z}$
14. equation; {13}
15. operation; $\frac{3m+5}{(m+3)(m+2)(m+1)}$
16. operation; $\frac{k+3}{5(k-1)}$
17. equation; ∅ 18. equation; ∅
19. operation; $\frac{t+2}{2(2t+1)}$
20. equation; {−7}

15. $\frac{1}{m^2+5m+6}+\frac{2}{m^2+4m+3}$

16. $\frac{2k^2-3k}{20k^2-5k}\div\frac{2k^2-5k+3}{4k^2+11k-3}$

17. $\frac{2}{x+1}+\frac{5}{x-1}=\frac{10}{x^2-1}$

18. $\frac{3}{x+3}+\frac{4}{x+6}=\frac{9}{x^2+9x+18}$

19. $\frac{4t^2-t}{6t^2+10t}\div\frac{8t^2+2t-1}{3t^2+11t+10}$

20. $\frac{x}{x-2}+\frac{3}{x+2}=\frac{8}{x^2-4}$

6.7 Applications of Rational Expressions

OBJECTIVES

1. Solve problems about numbers and data.
2. Solve problems about distance.
3. Solve problems about work.
4. Solve problems about variation.

FOR EXTRA HELP

SSG Sec. 6.7
SSM Sec. 6.7
Pass the Test Software
InterAct Math Tutorial Software
Video 11

CHALKBOARD EXAMPLE

A certain number is added to the numerator and subtracted from the denominator of $\frac{5}{8}$. The new number equals the reciprocal of $\frac{5}{8}$. Find the number.

Answer: 3

TEACHING TIP Ask the student why it is important to check the solution in the words of the problem, and not in the equation that they solve for the problem. (The equation may be solved correctly, but set up incorrectly.)

Every time we learn to solve a new type of equation, we are able to apply our knowledge to solving new types of applications. In Section 6.6 we solved equations involving rational expressions; now we can solve applications that involve this type of equation. The six-step problem solving method of Chapter 2 still applies.

OBJECTIVE 1 Solve problems about numbers and data. We begin with an example about an unknown number.

EXAMPLE 1 Solving a Problem about an Unknown Number

If the same number is added to both the numerator and the denominator of the fraction $\frac{2}{5}$, the result is $\frac{2}{3}$. Find the number.

Step 1 Let x = the number added to the numerator and the denominator.

Step 2 Then

$$\frac{2+x}{5+x}$$

represents the result of adding the same number to both the numerator and denominator.

Step 3 Since this result is $\frac{2}{3}$, the equation is

$$\frac{2+x}{5+x}=\frac{2}{3}.$$

Step 4 Solve this equation by multiplying both sides by the LCD, $3(5+x)$.

$$3(5+x)\frac{2+x}{5+x}=3(5+x)\frac{2}{3}$$

$$3(2+x)=2(5+x)$$

$$6+3x=10+2x \qquad \text{Distributive property}$$

$$x=4 \qquad \text{Subtract } 2x\text{; subtract 6.}$$

Step 5 The number is 4.

Step 6 Check the solution in the words of the original problem. If 4 is added to both the numerator and denominator of $\frac{2}{5}$, the result is $\frac{6}{9}=\frac{2}{3}$, as required.

JOURNAL WRITING ▲ CHALLENGING SCIENTIFIC CALCULATOR GRAPHING CALCULATOR

INSTRUCTOR'S RESOURCES

PTB Sec. 6.7 **ISM** Sec. 6.7 **AB** Sec. 6.7
TEST GENERATOR
WORLD WIDE WEB www.LialAlgebra.com

CHALKBOARD EXAMPLE

In 1996, the ratio of deaths from accidents to deaths caused by diabetes was approximately 3 to 2. There were 32,000 more deaths due to accidents than to diabetes. How many deaths were caused by each of these?

Answer:
Accidents: 96,000
Diabetes: 64,000

EXAMPLE 2 Examining the Leading Causes of Death in the United States

In 1996, approximately 734,000 Americans died of heart disease, the leading cause of death in the United States. Cancer was not far behind. However, there was a large drop between cancer and the third leading cause, stroke. Approximately 380,000 more deaths were caused by cancer than stroke, and the ratio of cancer deaths to stroke deaths was 3.4 to 1. How many deaths were caused by cancer, and how many were caused by stroke?

Let x represent the number of deaths caused by stroke. Then $x + 380{,}000$ deaths were caused by cancer. Since the ratio of cancer deaths to stroke deaths was 3.4 to 1 (or simply 3.4), we can write the following equation.

$$\frac{x + 380{,}000}{x} = 3.4$$

Now solve the equation.

$$x + 380{,}000 = 3.4x \qquad \text{Multiply by } x.$$
$$380{,}000 = 2.4x \qquad \text{Subtract } x.$$
$$x \approx 158{,}000 \qquad \text{Divide by 2.4.}$$

Since x represents the number of stroke deaths, there were about 158,000 deaths due to stroke, and about $x + 380{,}000 = 538{,}000$ deaths due to cancer. A check shows that these numbers satisfy the conditions of the problem. (*Note:* These are only approximations based on actual data from the Centers for Disease Control and Prevention and the National Center for Health Statistics.)

OBJECTIVE 2 Solve problems about distance. Recall from Chapter 2 the following formulas relating distance, rate, and time.

CHALKBOARD EXAMPLE

(a) In 1994, Leroy Burrell of the United States set the world record in the 100-meter dash, 9.85 seconds. What was his average speed?
(b) The winner of the 2000-meter women's track event in 1994 was Sonia O'Sullivan of Ireland, whose average speed was 369 meters per minute. What was her winning time?

Answer:
(a) 10.15 meters per second
(b) 5.42 minutes

Distance, Rate, and Time Relationship

$$d = rt \qquad r = \frac{d}{t} \qquad t = \frac{d}{r}$$

You may wish to refer to Example 6 in Section 2.6, which illustrates the basic use of these formulas.

PROBLEM SOLVING

In Section 2.6 we solved applications involving distance, rate, and time. Recall the importance of setting up a chart to organize the information given in these problems.

CHALKBOARD EXAMPLE

A boat can go 20 miles against a current in the same time it can go 60 miles with the current. The current flows at 4 miles per hour. Find the speed of the boat with no current.

Answer: 8 miles per hour

EXAMPLE 3 Solving a Problem about Distance, Rate, and Time

The Big Muddy River has a current of 3 miles per hour. A motorboat takes the same amount of time to go 12 miles downstream as it takes to go 8 miles upstream. What is the speed of the boat in still water?

This problem requires the distance formula, $d = rt$. Let x = the speed of the boat in still water. Since the current pushes the boat when the boat is going downstream, the speed of the boat downstream will be the sum of the speed of the boat and the speed of the current, or $x + 3$ miles per hour. Also, the boat's speed going upstream is $x - 3$ miles per hour. See Figure 1.

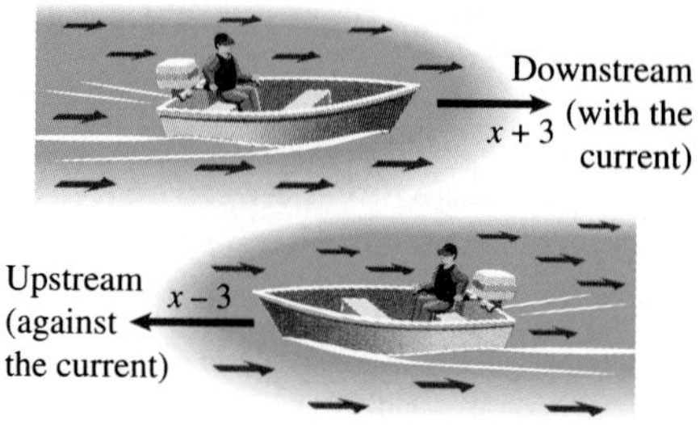

Figure 1

Summarize all of the information in a chart.

	d	r	t
Downstream	12	$x + 3$	
Upstream	8	$x - 3$	

Fill in the last column, representing time, by using the formula $t = \frac{d}{r}$.

For the time upstream,

$$\frac{d}{r} = \frac{8}{x - 3},$$

and for the time downstream,

$$\frac{d}{r} = \frac{12}{x + 3}.$$

Now complete the chart.

	d	r	t
Downstream	12	$x + 3$	$\frac{12}{x+3}$
Upstream	8	$x - 3$	$\frac{8}{x-3}$

Times are equal.

According to the original problem, the time upstream equals the time downstream. The two times from the chart must therefore be equal, giving the equation

$$\frac{12}{x + 3} = \frac{8}{x - 3}.$$

Solve this equation by multiplying both sides by the LCD, $(x + 3)(x - 3)$.

$$(x + 3)(x - 3)\frac{12}{x + 3} = (x + 3)(x - 3)\frac{8}{x - 3}$$

$$12(x - 3) = 8(x + 3)$$

$$12x - 36 = 8x + 24 \quad \text{Distributive property}$$

$$4x = 60 \quad \text{Subtract } 8x\text{; add } 36.$$

$$x = 15 \quad \text{Divide by } 4.$$

The speed of the boat in still water is 15 miles per hour. Check this solution by first finding the speed of the boat downstream, which is $15 + 3 = 18$ miles per hour. Traveling 12 miles would take

$$d = rt$$
$$12 = 18t$$
$$t = \frac{2}{3} \text{ hour.}$$

On the other hand, the speed of the boat upstream is $15 - 3 = 12$ miles per hour, and traveling 8 miles would take

$$d = rt$$
$$8 = 12t$$
$$t = \frac{2}{3} \text{ hour.}$$

The time upstream equals the time downstream, as required.

■ 54.5 miles per hour; 57.4 miles per hour; yes; More time is spent at the slower speed, and thus the average speed is less than the average of the two speeds.

CONNECTIONS

Sometimes what seems to be the obvious answer is not correct. Consider the following problem.

A car travels from *A* to *B* at 40 miles per hour and returns at 60 miles per hour. What is its average rate for the entire trip?

The correct answer is not 50 miles per hour! Remembering the distance-rate-time relationship and letting $x =$ the distance between *A* and *B*, we can simplify a complex fraction to find the correct answer.

$$\text{Average rate for entire trip} = \frac{\text{Total distance}}{\text{Total time}}$$
$$= \frac{x + x}{\frac{x}{40} + \frac{x}{60}}$$
$$= \frac{2x}{\frac{3x}{120} + \frac{2x}{120}}$$
$$= \frac{2x}{\frac{5x}{120}}$$
$$= 2x \cdot \frac{120}{5x}$$
$$= 48$$

The average rate for the entire trip is 48 miles per hour.

FOR DISCUSSION OR WRITING

Find the average rate for the entire trip if the speed from *A* to *B* is 50 miles per hour. Repeat for 55 miles per hour. Will the result always be less than $\frac{1}{2}$ the sum of the two speeds? What might account for this?

OBJECTIVE 3 **Solve problems about work.** Suppose that you can mow your lawn in 4 hours. After 1 hour, you will have mowed $\frac{1}{4}$ of the lawn. After 2 hours, you will have mowed $\frac{2}{4}$ or $\frac{1}{2}$ of the lawn, and so on. This idea is generalized as follows.

Rate of Work

If a job can be completed in t units of time, then the rate of work is

$$\frac{1}{t} \text{ job per unit of time.}$$

PROBLEM SOLVING

The relationship between problems involving work and problems involving distance is a very close one. Recall that the formula $d = rt$ says that distance traveled is equal to rate of travel multiplied by time traveled. Similarly, the fractional part of a job accomplished is equal to the rate of the work multiplied by the time worked. In the lawn-mowing example, after 3 hours, the fractional part of the job done is

$$\underbrace{\frac{1}{4}}_{\text{Rate of work}} \cdot \underbrace{3}_{\text{Time worked}} = \underbrace{\frac{3}{4}}_{\text{Fractional part of job done}}.$$

After 4 hours, $(\frac{1}{4})(4) = 1$ whole job has been done.

These ideas are used in solving problems about the length of time needed to do a job.

CHALKBOARD EXAMPLE

Al and Mario operate a small roofing company. Mario can roof an average house alone in 9 hours. Al can roof a house alone in 8 hours. How long will it take them to do the job if they work together?

Answer: $\frac{72}{17}$ or $4\frac{4}{17}$ hours

EXAMPLE 4 Solving a Problem about Work

With a riding lawn mower, Mateo, the groundskeeper in a large park, can cut the lawn in 8 hours. With a small mower, his assistant Chet needs 14 hours to cut the same lawn. If both Mateo and Chet work on the lawn, how long will it take to cut it?

Let x = the number of hours it will take for Mateo and Chet to mow the lawn, working together.

Certainly, x will be less than 8, since Mateo alone can mow the lawn in 8 hours. Begin by making a chart. Based on the previous discussion, Mateo's rate alone is $\frac{1}{8}$ job per hour, and Chet's rate is $\frac{1}{14}$ job per hour.

	Rate	*Time Working Together*	*Fractional Part of the Job Done When Working Together*
Mateo	$\frac{1}{8}$	x	$\frac{1}{8}x$
Chet	$\frac{1}{14}$	x	$\frac{1}{14}x$

Sum is 1 whole job.

Since together Mateo and Chet complete 1 whole job, we must add their individual fractional parts and set the sum equal to 1.

$$\begin{array}{ccccc} \text{Fractional part done by Mateo} & + & \text{Fractional part done by Chet} & = & \text{1 whole job} \\ \frac{1}{8}x & + & \frac{1}{14}x & = & 1 \end{array}$$

$$56\left(\frac{1}{8}x + \frac{1}{14}x\right) = 56(1) \qquad \text{Multiply by LCD, 56.}$$

$$56\left(\frac{1}{8}x\right) + 56\left(\frac{1}{14}x\right) = 56(1) \qquad \text{Distributive property}$$

$$7x + 4x = 56$$

$$11x = 56 \qquad \text{Combine like terms.}$$

$$x = \frac{56}{11} \qquad \text{Divide by 11.}$$

Working together, Mateo and Chet can mow the lawn in $\frac{56}{11}$ hours, or $5\frac{1}{11}$ hours. (Is this answer reasonable?)

NOTE

An alternative approach in work problems is to consider the part of the job that can be done in 1 hour. For instance, in Example 4 Mateo can do the entire job in 8 hours, and Chet can do it in 14. Thus, their work rates, as we saw in Example 4, are $\frac{1}{8}$ and $\frac{1}{14}$, respectively. Since it takes them x hours to complete the job when working together, in one hour they can mow $\frac{1}{x}$ of the lawn. The amount mowed by Mateo in one hour plus the amount mowed by Chet in one hour must equal the amount they can do together. This leads to the equation

$$\underset{\text{Amount by Mateo} \rightarrow}{} \frac{1}{8} + \overset{\text{Amount by Chet}}{\frac{1}{14}} = \frac{1}{x}. \quad \leftarrow \text{Amount together}$$

Compare this with the equation in Example 4. Multiplying both sides by $56x$ leads to

$$7x + 4x = 56,$$

the same equation found in the third step in the example. The same solution results.

OBJECTIVE 4 Solve problems about variation. In Section 2.5 we discussed direct variation. (See Example 6 in that section.) Recall that y *varies directly* as x if there exists a constant (a numerical value) k such that $y = kx$. This value is called the **constant of variation.** For example, the circumference C of a circle varies directly as its diameter d, because $C = \pi d$. Here, the constant of variation is π.

Some types of variation involve rational expressions. Suppose that a rectangle has area 48 square units and length 24 units. Then its width is 2 units. However, if the area stays the same (48 square units) and the length decreases to 12 units, the width increases to 4 units. See Figure 2.

As length decreases, width increases. This is an example of **inverse variation,** since $W = \dfrac{A}{L}$.

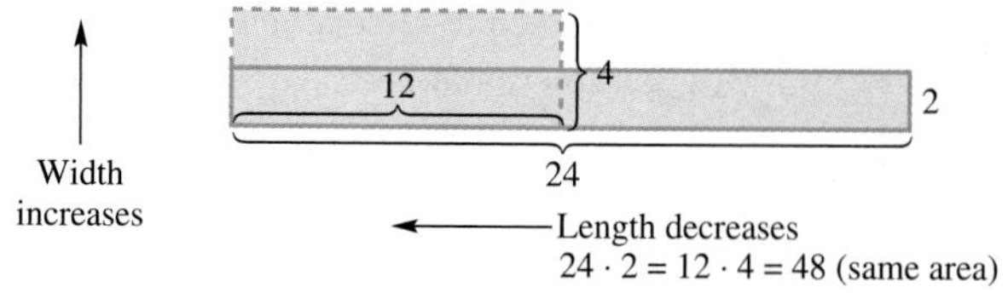

Figure 2

This and other types of variation are defined below, where k represents a constant.

Types of Variation

m varies inversely as p	$m = \dfrac{k}{p}$
y varies directly as the square of x	$y = kx^2$
r varies inversely as the square of s	$r = \dfrac{k}{s^2}$

CHALKBOARD EXAMPLE

If the cost of producing pairs of rubber gloves varies inversely as the number of pairs produced, and 5000 pairs can be produced for \$.50 per pair, how much will it cost per pair to produce 10,000 pairs?

Answer: \$.25

EXAMPLE 5 Using Inverse Variation

In the manufacturing of a certain medical syringe, the cost of producing the syringe varies inversely as the number produced. If 10,000 syringes are produced, the cost is \$2 per unit. Find the cost per unit to produce 25,000 syringes.

Let x = the number of syringes produced and
c = the cost per unit.

Since c varies inversely as x, there is a constant k such that

$$c = \frac{k}{x}.$$

Find k by replacing c with 2 and x with 10,000.

$$2 = \frac{k}{10{,}000}$$

$$20{,}000 = k \qquad \text{Multiply by 10,000.}$$

Since $c = \dfrac{k}{x}$,

$$c = \frac{20{,}000}{25{,}000} = .80. \qquad \text{Let } k = 20{,}000 \text{ and } x = 25{,}000.$$

The cost per unit to make 25,000 syringes is \$.80.

CHALKBOARD EXAMPLE

Suppose that y varies inversely as the square of x. If $y = 5$ when $x = 2$, find y when $x = 10$.

Answer: $\dfrac{1}{5}$

EXAMPLE 6 Using Inverse Variation

In electric current flow, it is found that the resistance offered by a fixed length of wire of a given material (measured in units called *ohms*) varies inversely as the square of the diameter of the wire. If a wire .01 centimeter in diameter has a resistance of .40 ohm, what is the resistance of a wire of the same length and material but .03 centimeter in diameter?

If R is the resistance and d the diameter, then

$$R = \frac{k}{d^2}$$

for some constant k. (The value of the constant k depends on the material and the temperature.) Since a wire .01 centimeter in diameter has a resistance of .40 ohm,

$$.40 = \frac{k}{(.01)^2} \qquad \text{Let } R = .40 \text{ and } d = .01.$$

$$k = .00004. \qquad \text{Multiply by } (.01)^2 = .0001.$$

Thus, for this particular length of wire,

$$R = \frac{.00004}{d^2}. \qquad \text{Let } k = .00004.$$

For the same length wire with $d = .03$,

$$R = \frac{.00004}{(.03)^2} = \frac{2}{45} \text{ ohm.}$$

6.7 EXERCISES

Use Steps 1–3 of the six-step method to set up the equation you would use to solve each problem. Do not actually solve the equation. See Example 1.

1. The numerator of the fraction $\frac{5}{6}$ is increased by an amount so that the value of the resulting fraction is $\frac{13}{3}$. By what amount was the numerator increased?

(a) Let x = ______. (Step 1)

(b) Write an expression for "the numerator of the fraction $\frac{5}{6}$ is increased by an amount." (Step 2)

(c) Set up an equation to solve the problem. (Step 3)

2. If the same number is added to the numerator and subtracted from the denominator of $\frac{23}{12}$, the resulting fraction is equal to $\frac{3}{2}$. What is the number?

(a) Let x = ______. (Step 1)

(b) Write an expression for "a number is added to the numerator of $\frac{23}{12}$." Then write an expression for "the same number is subtracted from the denominator of $\frac{23}{12}$." (Step 2)

(c) Set up an equation to solve the problem. (Step 3)

Use the six-step method to solve each problem. See Example 1.

3. In a certain fraction, the denominator is 6 more than the numerator. If 3 is added to both the numerator and the denominator, the resulting fraction is equivalent to $\frac{5}{7}$. What was the original fraction?

1. (a) an amount (b) $5 + x$ (c) $\frac{5+x}{6} = \frac{13}{3}$
2. (a) a number (b) $23 + x$; $12 - x$ (c) $\frac{23+x}{12-x} = \frac{3}{2}$
3. $\frac{12}{18}$

4. $\frac{12}{3}$ 5. $\frac{1386}{97}$ 6. 36
7. 1989: 33,579 (thousand); 1990: 34,203 (thousand)
8. 1984: 4440; 1985: 8200
9. female: 144,000; male: 576,000
10. United: 40; Kaiser: 12
11. 24.15 kilometers per hour
12. 39.25 seconds

4. The numerator of a certain fraction is 4 times the denominator. If 6 is added to both the numerator and the denominator, the resulting fraction is equivalent to 2. What was the original fraction?

5. A quantity, its $\frac{2}{3}$, its $\frac{1}{2}$, and its $\frac{1}{7}$, added together, become 33. What is the quantity? (From the *Rhind Mathematical Papyrus.*)

6. A quantity, its $\frac{3}{4}$, its $\frac{1}{2}$, and its $\frac{1}{3}$, added together, become 93. What is the quantity?

Solve each problem using the six-step method. See Example 2.

7. (All enrollments are in thousands.) The Medicare enrollment in 1990 was 624 more than in 1989. The ratio of these enrollments was $\frac{877}{861}$. What was the enrollment in each of these years? (*Source:* U.S. Health Care Financing Administration.)

8. In 1985, during the early years of the appearance of AIDS, there were about 3760 more cases reported than in 1984. The ratio of these numbers of cases was $\frac{205}{111}$. How many cases were reported in each of these years? (*Source:* Centers for Disease Control.)

9. In 1995, the number of female physicians in the United States was approximately $\frac{1}{4}$ the number of male physicians. The total number of physicians was approximately 720,000. How many female and how many male physicians were there? (*Source:* American Medical Association.)

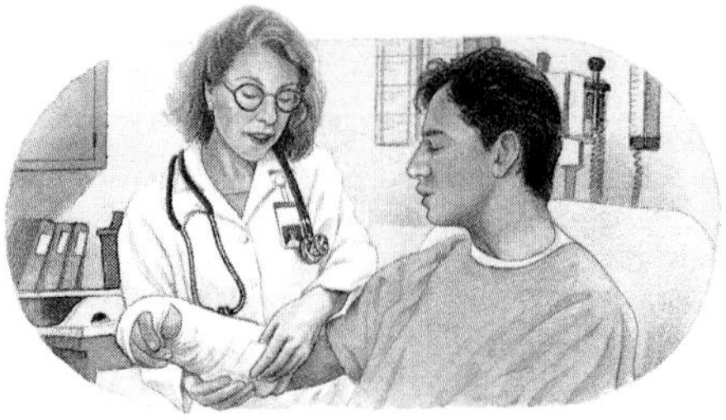

10. Two of the largest national managed care firms (as ranked by total HMO enrollments) in 1996 were United Health Care Corporation and Kaiser Foundation Health Plans, Inc. The number of plans provided by United was $\frac{10}{3}$ that of Kaiser. Together, the number of plans was 52. How many plans were provided by each company? (*Source:* InterStudy Publications.)

Solve each problem. See Example 6 in Section 2.6. (Source: Sports Illustrated 1998 Sports Almanac.)

11. In the 1994 Olympics, Vladimir Smimov of Kazakhstan won the 50-kilometer nordic skiing event in 2.07 hours. What was his average rate?

12. Bonnie Blair won the women's 500-meter speed skating event in the 1994 Olympics. Her average rate was 12.74 meters per second. What was her time?
to do 500m

13. 3.429 hours
14. 3.372 hours
15. 7.91 meters per second
16. 4.15 yards per second
17. $\frac{500}{x-10} = \frac{600}{x+10}$
18. $\frac{8}{4-x} = \frac{24}{4+x}$
19. $\frac{D}{R} = \frac{d}{r}$
20. into a headwind: $m - 5$ miles per hour; with a tailwind: $m + 5$ miles per hour
21. 8 miles per hour
22. 10 miles per hour

13. The winner of the 1997 Indianapolis 500 (mile) race was Arie Luyendyk, with an average rate of 145.827 miles per hour. What was his time in hours?

14. In 1997, Jeff Gordon drove his Chevrolet to victory in the Daytona 500 (mile) stock car race. His average rate was 148.295 miles per hour. What was his time in hours?

15. The NCAA Division I women's champion for the 400-meter dash in 1997 was LaTarsha Stroman of Louisiana State University. Her winning time was 50.60 seconds. What was her rate in meters per second?

16. The NCAA Division I women's champion for the 200-yard freestyle event in 1996–1997 was Martina Moracova of Southern Methodist University. Her time was 48.18 seconds. What was her rate in yards per second?

Set up the equation you would use to solve each problem. Do not actually solve. See Example 3.

17. Julio flew his airplane 500 miles against the wind in the same time it took him to fly it 600 miles with the wind. If the speed of the wind was 10 miles per hour, what was the average speed of his plane? (Let x = speed of the plane in still air.)

	d	r	t
Against the wind	500	$x - 10$	
With the wind	600	$x + 10$	

18. Luvenia can row 4 miles per hour in still water. It takes as long to row 8 miles upstream as 24 miles downstream. How fast is the current? (Let x = speed of the current.)

	d	r	t
Upstream	8	$4 - x$	
Downstream	24	$4 + x$	

Solve each problem. See Example 3.

19. Suppose Stephanie walks D miles at R miles per hour in the same time that Wally walks d miles at r miles per hour. Give an equation relating D, R, d, and r.

20. If a migrating hawk travels m miles per hour in still air, what is its rate when it flies into a steady headwind of 5 miles per hour? What is its rate with a tailwind of 5 miles per hour?

21. A boat can go 20 miles against a current in the same time that it can go 60 miles with the current. The current is 4 miles per hour. Find the speed of the boat in still water.

22. A plane flies 350 miles with the wind in the same time that it can fly 310 miles against the wind. The plane has a still-air speed of 165 miles per hour. Find the speed of the wind.

23. $18\frac{1}{2}$ **miles per hour**
24. **900 miles**
25. **N'Deti: 12.02 miles per hour; McDermott: 8.78 miles per hour**
26. **Kuscsik: 8.1 miles per hour; Pippig: 10.8 miles per hour**
27. $\frac{1}{10}$ **job per hour**
28. $\frac{2}{3}$ **of the job**
29. $\frac{1}{8}x + \frac{1}{6}x = 1$ **or** $\frac{1}{8} + \frac{1}{6} = \frac{1}{x}$
30. $\frac{1}{2}t + \frac{1}{3}t = 1$ **or** $\frac{1}{2} + \frac{1}{3} = \frac{1}{t}$
31. $4\frac{4}{17}$ **hours**
32. $2\frac{2}{9}$ **hours**
33. $5\frac{5}{11}$ **hours**
34. $4\frac{8}{19}$ **hours** **35.** **3 hours**
36. $5\frac{1}{3}$ **days**

23. The distance from Seattle, Washington, to Victoria, British Columbia, is about 148 miles by ferry. It takes about 4 hours less to travel by the same ferry from Victoria to Vancouver, British Columbia, a distance of about 74 miles. What is the average speed of the ferry?

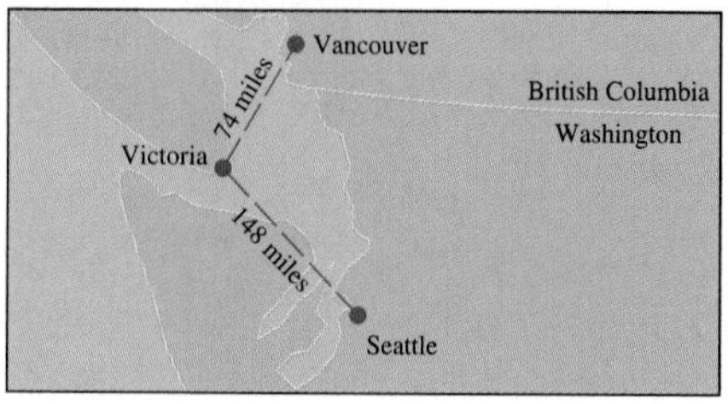

24. Sandi Goldstein flew from Dallas to Indianapolis at 180 miles per hour and then flew back at 150 miles per hour. The trip at the slower speed took 1 hour longer than the trip at the higher speed. Find the distance between the two cities.

25. Cosmas N'Deti of Kenya won the 1995 Boston Marathon (a 26-mile race) in (about) .8 of an hour less time than the winner of the first Boston Marathon in 1897, John H. McDermott of New York. Find each runner's speed, if McDermott's speed was (about) .73 times N'Deti's speed. (*Hint:* Don't round until the end of the calculations.)

26. Women first ran in the Boston Marathon in 1972, when Nina Kuscsik of New York won the race. In 1995, the winner was Uta Pippig of Germany, whose time was .8 of an hour less than Kuscsik's in 1972. If Pippig ran $\frac{4}{3}$ as fast as Kuscsik, find each runner's speed. (See Exercise 25 for distance.)

27. If it takes Elayn 10 hours to do a job, what is her rate?

28. If it takes Clay 12 hours to do a job, how much of the job does he do in 8 hours?

In Exercises 29 and 30, set up the equation you would use to solve each problem. Do not actually solve. See Example 4.

29. Working alone, Jorge can paint a room in 8 hours. Caterina can paint the same room working alone in 6 hours. How long will it take them if they work together? (Let x represent the time working together.)

30. Edwin Bedford can tune up his Chevy in 2 hours working alone. Beau can do the job in 3 hours working alone. How long would it take them if they worked together? (Let t represent the time working together.)

Solve each problem. See Example 4.

31. Geraldo and Luisa Hernandez operate a small laundry. Luisa, working alone, can clean a day's laundry in 9 hours. Geraldo can clean a day's laundry in 8 hours. How long would it take them if they work together?

32. Lea can groom the horses in her boarding stable in 5 hours, while Tran needs 4 hours. How long will it take them to groom the horses if they work together?

33. A pump can pump the water out of a flooded basement in 10 hours. A smaller pump takes 12 hours. How long would it take to pump the water from the basement using both pumps?

34. Doug Todd's copier can do a printing job in 7 hours. Scott's copier can do the same job in 12 hours. How long would it take to do the job using both copiers?

35. An experienced employee can enter tax data into a computer twice as fast as a new employee. Working together, it takes the employees 2 hours. How long would it take the experienced employee working alone?

36. One roofer can put a new roof on a house three times faster than another. Working together they can roof a house in 4 days. How long would it take the faster roofer working alone?

37. $2\frac{7}{10}$ hours 38. 36 hours

39. $9\frac{1}{11}$ minutes 40. $8\frac{1}{4}$ hours

41. direct 42. direct

43. direct 44. inverse

45. direct 46. inverse

47. inverse 48. direct 49. 9

50. 15 51. $\frac{16}{5}$ 52. $\frac{768}{49}$

53. $\frac{4}{9}$ 54. $\frac{64}{25}$ 55. increases

56. decreases

▲ **37.** One pipe can fill a swimming pool in 6 hours, and another pipe can do it in 9 hours. How long will it take the two pipes working together to fill the pool $\frac{3}{4}$ full?

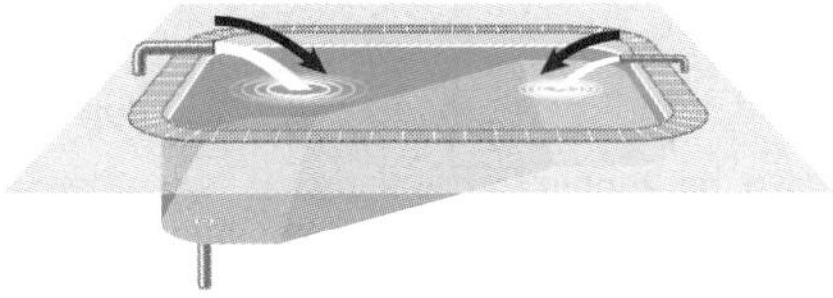

▲ **38.** An inlet pipe can fill a swimming pool in 9 hours, and an outlet pipe can empty the pool in 12 hours. Through an error, both pipes are left open. How long will it take to fill the pool?

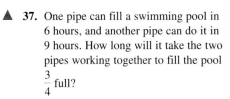

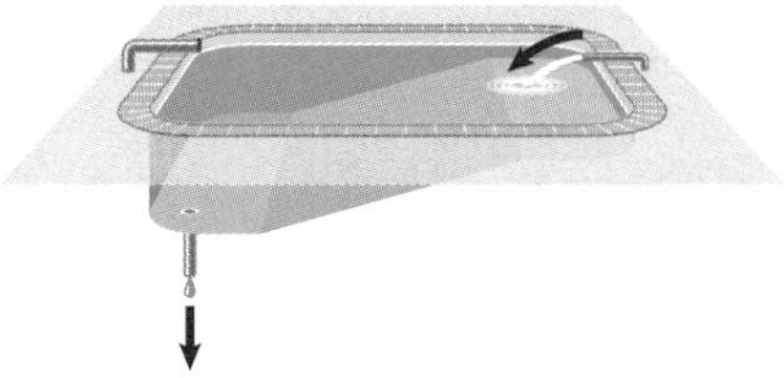

▲ **39.** A cold water faucet can fill a sink in 12 minutes, and a hot water faucet can fill it in 15. The drain can empty the sink in 25 minutes. If both faucets are on and the drain is open, how long will it take to fill the sink?

▲ **40.** Refer to Exercise 38. Assume the error was discovered after both pipes had been running for 3 hours, and the outlet pipe was then closed. How much more time would then be required to fill the pool? (*Hint:* How much of the job had been done when the error was discovered?)

*Use personal experience or intuition to determine whether the variation between the indicated quantities is direct or inverse.**

41. The number of different lottery tickets you buy and your probability of winning that lottery

42. The rate and the distance traveled by a pickup truck in 3 hours

43. The amount of pressure put on the accelerator of a car and the speed of the car

44. The number of days from now until December 25 and the magnitude of the frenzy of Christmas shopping

45. The surface area of a balloon and its diameter

46. Your age and the probability that you believe in Santa Claus

47. The number of days until the end of the baseball season and the number of home runs that Sammy Sosa has

48. The amount of gasoline you pump and the amount you will pay

Solve each problem involving variation. See Example 6 in Section 2.6 and Examples 5 and 6 in this section.

49. If x varies directly as y, and $x = 27$ when $y = 6$, find x when $y = 2$.

50. If z varies directly as x, and $z = 30$ when $x = 8$, find z when $x = 4$.

51. If m varies inversely as p^2, and $m = 20$ when $p = 2$, find m when p is 5.

52. If a varies inversely as b^2, and $a = 48$ when $b = 4$, find a when $b = 7$.

53. If p varies inversely as q^2, and $p = 4$ when $q = \frac{1}{2}$, find p when $q = \frac{3}{2}$.

54. If z varies inversely as x^2, and $z = 9$ when $x = \frac{2}{3}$, find z when $x = \frac{5}{4}$.

55. If the constant of variation is positive and y varies directly as x, then as x increases, y ________________.
(increases/decreases)

56. If the constant of variation is positive and y varies inversely as x, then as x increases, y ________________.
(increases/decreases)

*The authors thank Linda Kodama of Kapiolani Community College for suggesting the inclusion of exercises of this type.

57. $106\frac{2}{3}$ miles per hour
58. 8.68 pounds per square inch
59. 25 kilograms per hour
60. $12\frac{1}{2}$ amps
61. 20 pounds per square foot
62. $7\frac{1}{2}$ pounds per square inch
63. 15 feet 64. $42\frac{2}{3}$ inches
65. 144 feet
66. $14\frac{22}{27}$ foot-candles

57. Over a specified distance, speed varies inversely with time. If a Dodge Viper on a test track goes a certain distance in one-half minute at 160 miles per hour, what speed is needed to go the same distance in three-fourths of a minute?

58. The pressure exerted by water at a given point varies directly with the depth of the point beneath the surface of the water. Water exerts 4.34 pounds per square inch for every 10 feet traveled below the water's surface. What is the pressure exerted on a scuba diver at 20 feet?

59. In the inversion of raw sugar, the rate of change of the amount of raw sugar varies directly as the amount of raw sugar remaining. The rate is 200 kilograms per hour when there are 800 kilograms left. What is the rate of change per hour when only 100 kilograms are left?

60. The current in a simple electrical circuit varies inversely as the resistance. If the current is 20 amps when the resistance is 5 ohms, find the current when the resistance is 8 ohms.

61. If the temperature is constant, the pressure of a gas in a container varies inversely as the volume of the container. If the pressure is 10 pounds per square foot in a container with 3 cubic feet, what is the pressure in a container with 1.5 cubic feet?

62. If the volume is constant, the pressure of gas in a container varies directly as the temperature. If the pressure is 5 pounds per square inch at a temperature of 200 Kelvin, what is the pressure at a temperature of 300 Kelvin?

63. For a constant area, the length of a rectangle varies inversely as the width. The length of a rectangle is 27 feet when the width is 10 feet. Find the width of a rectangle with the same area if the length is 18 feet.

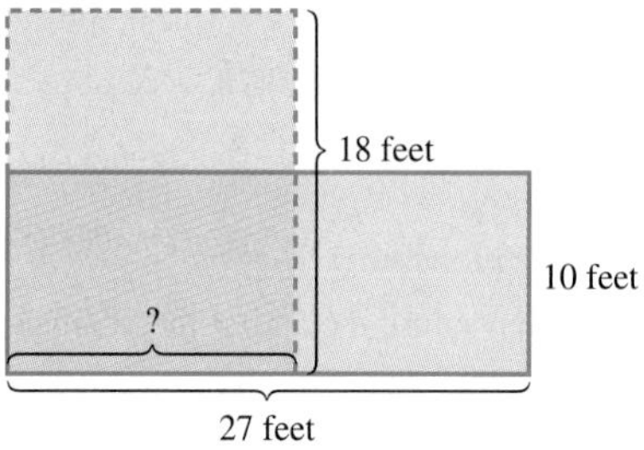

64. Hooke's law for an elastic spring states that the distance a spring stretches varies directly with the force applied. If a force of 75 pounds stretches a certain spring 16 inches, how much will a force of 200 pounds stretch the spring?

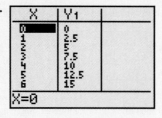

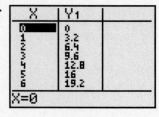

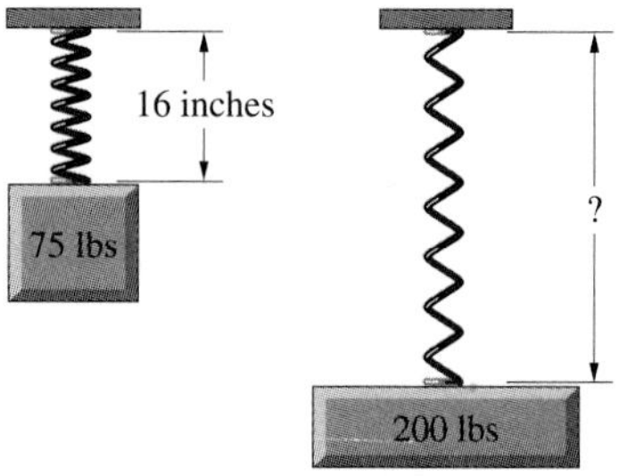

65. For a body falling freely from rest (disregarding air resistance), the distance the body falls varies directly as the square of the time. If an object is dropped from the top of a tower 400 feet high and hits the ground in 5 seconds, how far did it fall in the first 3 seconds?

66. The illumination produced by a light source varies inversely as the square of the distance from the source. If the illumination produced 4 feet from a light source is 75 foot-candles, find the illumination produced 9 feet from the same source.

67. direct 68. direct

TECHNOLOGY INSIGHTS (EXERCISES 67–70)

In each table, Y_1 either varies directly or varies inversely as x. Tell which type of variation is illustrated.

67.

X	Y1
0	0
1	2.5
2	5
3	7.5
4	10
5	12.5
6	15

X=0

68.

X	Y1
0	0
1	3.2
2	6.4
3	9.6
4	12.8
5	16
6	19.2

X=0

69. inverse **70.** inverse

TECHNOLOGY INSIGHTS (EXERCISES 67–70) (CONTINUED)

69.

X	Y1
1	24
2	12
3	8
4	6
5	4.8
6	4
7	3.4286

X=1

70.

X	Y1
1	48
2	24
3	16
4	12
5	9.6
6	8
7	6.8571

X=1

71. approximately 2364 in 1995 and 716 in 1985 (The actual numbers were 2361 and 719.) **72.** approximately 3923 in 1995 and 603 in 1985 (The actual numbers were 3924 and 602.) **73.** Answers will vary. Here is one possibility: I start with the fraction $\frac{11}{8}$. If I add -2 to both the numerator and the denominator, I get $\frac{9}{6}$, which simplifies to $\frac{3}{2}$. The problem is stated as follows: "If a number is added to both the numerator and the denominator of $\frac{11}{8}$, the resulting fraction is equal to $\frac{3}{2}$. What is the number?" To solve, let x = the number. The equation is $\frac{11+x}{8+x} = \frac{3}{2}$. This leads to $2(11 + x) = 3(8 + x)$, which leads to $22 + 2x = 24 + 3x$, or $-2 = x$. The number is -2. **74.** The equation would be erroneously written $\frac{8}{4+x} = \frac{24}{4-x}$. Solving for x gives $x = -2$. Because x represents the speed of the current, it cannot be negative. Therefore, the student should realize that there is a problem in the solution process.

*Refer to the discussion in the chapter opener pertaining to heart transplants. Use the strategy discussed there to solve each of these problems. (*Sources: *U.S. Department of Health and Human Services, Public Health Service, Division of Organ Transplantation, and United Network for Organ Sharing.)*

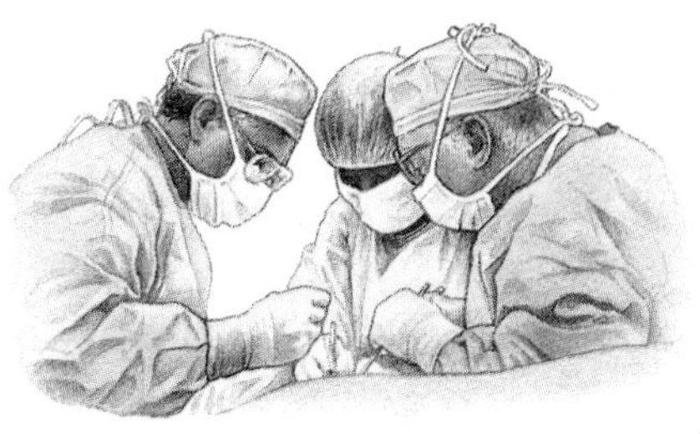

71. In 1995, the number of heart transplants continued to show the increase it had exhibited during the decade preceding it. In the years 1985 and 1995, the total number of heart transplants was 3080, and the ratio of the number in 1995 to the number in 1985 was approximately 33 to 10. How many heart transplants were performed in each of these two years?

72. In the years 1985 and 1995, the total number of liver transplants was 4526. The ratio of the number in 1995 to the number in 1985 was approximately 65 to 10. How many liver transplants were performed in each of these two years?

73. Students often wonder how teachers and textbook authors make up problems for them to solve. One way to do this is to start with the answer and work backwards. For example, suppose that we start with the fraction $\frac{7}{3}$ and add 3 to both the numerator and the denominator. We get $\frac{10}{6}$, which simplifies to $\frac{5}{3}$. Based on this observation, we can write the following problem.

> If a number is added to both the numerator and the denominator of $\frac{7}{3}$, the resulting fraction is equal to $\frac{5}{3}$. What is the number?

Because of how we constructed the problem, the answer must be 3.

Make up your own problem similar to this one, and then solve it using an equation involving rational expressions.

74. Refer to the chart in Exercise 18. Suppose that a student made the error of interchanging the positions of the expressions $4 - x$ and $4 + x$, but used the correct method of setting up the equation, applying the formula $t = \frac{d}{r}$. Solve the equation the student used, and explain how the student should immediately know that there is something wrong in the setup.

Instructor's Note
Grouping: 3 students
Time: 20 minutes

Groups can divide up the work in this activity by having one person calculate total fat grams for a sedentary person, another for an active person, and the third for a very active person.

CHAPTER 6 GROUP ACTIVITY

How Much Fat Do You Need?

Objective: Use ratios and proportions to calculate total fat grams needed in an individual's daily diet.

Increased risks for some leading causes of death such as heart disease, diabetes, high blood pressure, and some cancers are associated with diets high in fat. Studies show that diets low in fat are linked to the lowest rates of heart disease and some forms of cancer. How do you decide how much fat is necessary for each individual? This activity will show how to calculate total fat grams necessary to maintain or reach ideal body weight.

To calculate daily caloric need, follow these steps.

Step 1 Determine activity level factor.

A sedentary person's activity level factor = 12.

An active person's activity level factor = 15.

A very active person's activity level factor = 18.

Step 2 Determine daily caloric need.

Multiply the activity level from Step 1 by your weight in pounds. (You can use your actual weight or your ideal weight.)

A. As a group, use the above information and the following ratio to write a proportion that can be used to calculate maximum total fat grams needed daily to maintain a person's weight or to reach his/her ideal weight.

$$\frac{400 \text{ calories}}{11 \text{ total fat grams}}$$

B. Use the previous information to complete the chart below. Round answers to the nearest tenth.

Total Fat Grams Needed Daily

Person	*Weight*	*Sedentary*	*Active*	*Very Active*
A	134	[44.2 grams]	[55.3 grams]	[66.33 grams]
B	200	[66 grams]	[82.5 grams]	[99 grams]
C	153	[50.5 grams]	[63.1 grams]	[75.7 grams]

C. Now complete the chart for the members in your group. Complete the three activity levels. (Actual weight or ideal weight can be used.) [Answers will vary.]

Total Fat Grams Needed Daily

Person	*Weight*	*Sedentary*	*Active*	*Very Active*

CHAPTER 6 SUMMARY

KEY TERMS

6.1 **rational expression**
lowest terms

6.3 **least common denominator (LCD)**

6.5 **complex fraction**

6.7 **constant of variation**
inverse variation

TEST YOUR WORD POWER

See how well you have learned the vocabulary in this chapter. Answers, with examples, are given at the bottom of the page.

1. A **rational expression** is
   (a) an algebraic expression made up of a term or the sum of a finite number of terms with real coefficients and whole number exponents
   (b) a polynomial equation of degree two
   (c) an expression with one or more fractions in the numerator, denominator, or both
   (d) the quotient of two polynomials with denominator not zero.

2. A **complex fraction** is
   (a) an algebraic expression made up of a term or the sum of a finite number of terms with real coefficients and whole number exponents
   (b) a polynomial equation of degree two
   (c) a rational expression with one or more fractions in the numerator, denominator, or both
   (d) the quotient of two polynomials with denominator not zero.

QUICK REVIEW

CONCEPTS	EXAMPLES
6.1 THE FUNDAMENTAL PROPERTY OF RATIONAL EXPRESSIONS	
To find the values for which a rational expression is not defined, set the denominator equal to zero and solve the equation.	Find the values for which the expression $\frac{x-4}{x^2-16}$ is not defined. $x^2 - 16 = 0$; $(x-4)(x+4) = 0$; $x - 4 = 0$ or $x + 4 = 0$; $x = 4$ or $x = -4$. The rational expression is not defined for 4 or -4.
To write a rational expression in lowest terms, (1) factor, and (2) use the fundamental property to remove common factors from the numerator and denominator.	Write $\frac{x^2-1}{(x-1)^2}$ in lowest terms. $\frac{x^2-1}{(x-1)^2} = \frac{(x-1)(x+1)}{(x-1)(x-1)} = \frac{x+1}{x-1}$

(continued)

Answers to Test Your Word Power

1. (d) *Examples:* $-\frac{3}{4y}, \frac{5x^3}{x+2}, \frac{a+3}{a^2-4a-5}$ **2.** (c) *Examples:* $\frac{\frac{2}{3}}{\frac{4}{7}}, \frac{\frac{1}{x}-\frac{1}{y}}{x+\frac{1}{y}}, \frac{\frac{2}{a+1}}{a^2-1}$

CONCEPTS	EXAMPLES
There are often several different equivalent forms of a rational expression.	Give two equivalent forms of $-\frac{x-1}{x+2}$. Distribute the $-$ sign in the numerator to get $\frac{-x+1}{x+2}$; do so in the denominator to get $\frac{x-1}{-x-2}$. (There are other forms as well.)

6.2 MULTIPLICATION AND DIVISION OF RATIONAL EXPRESSIONS

CONCEPTS	EXAMPLES
Multiplication	Multiply. $\frac{3x+9}{x-5} \cdot \frac{x^2-3x-10}{x^2-9}$
1. Factor.	$= \frac{3(x+3)}{x-5} \cdot \frac{(x-5)(x+2)}{(x+3)(x-3)}$
2. Multiply numerators and multiply denominators.	$= \frac{3(x+3)(x-5)(x+2)}{(x-5)(x+3)(x-3)}$
3. Write in lowest terms.	$= \frac{3(x+2)}{x-3}$
Division	Divide. $\frac{2x+1}{x+5} \div \frac{6x^2-x-2}{x^2-25}$
1. Factor.	$= \frac{2x+1}{x+5} \div \frac{(2x+1)(3x-2)}{(x+5)(x-5)}$
2. Multiply the first rational expression by the reciprocal of the second.	$= \frac{2x+1}{x+5} \cdot \frac{(x+5)(x-5)}{(2x+1)(3x-2)}$
3. Write in lowest terms.	$= \frac{x-5}{3x-2}$

6.3 THE LEAST COMMON DENOMINATOR

CONCEPTS	EXAMPLES
Finding the LCD	Find the LCD for $\frac{3}{k^2-8k+16}$ and $\frac{1}{4k^2-16k}$.
1. Factor each denominator into prime factors.	$k^2-8k+16=(k-4)^2$ $4k^2-16k=4k(k-4)$
2. List each different factor the greatest number of times it appears.	$\text{LCD} = (k-4)^2 \cdot 4 \cdot k$
3. Multiply the factors from Step 2 to get the LCD.	$= 4k(k-4)^2$
Writing a Rational Expression with the LCD as Denominator	Find the numerator: $\frac{5}{2z^2-6z} = \frac{}{4z^3-12z^2}$.
1. Factor both denominators.	$\frac{5}{2z(z-3)} = \frac{}{4z^2(z-3)}$
2. Decide what factors the denominator must be multiplied by to equal the LCD.	$2z(z-3)$ must be multiplied by $2z$.
3. Multiply the rational expression by that factor divided by itself (multiply by 1).	$\frac{5}{2z(z-3)} \cdot \frac{2z}{2z} = \frac{10z}{4z^2(z-3)} = \frac{10z}{4z^3-12z^2}$

CONCEPTS	EXAMPLES
6.4 ADDITION AND SUBTRACTION OF RATIONAL EXPRESSIONS	
Adding Rational Expressions	Add. $\frac{2}{3m+6}+\frac{m}{m^2-4}$
1. Find the LCD.	$3m+6=3(m+2)$ $m^2-4=(m+2)(m-2)$ The LCD is $3(m+2)(m-2)$.
2. Rewrite each rational expression with the LCD as denominator.	$=\frac{2(m-2)}{3(m+2)(m-2)}+\frac{3m}{3(m+2)(m-2)}$
3. Add the numerators to get the numerator of the sum. The common denominator is the denominator of the sum.	$=\frac{2m-4+3m}{3(m+2)(m-2)}$
4. Write in lowest terms.	$=\frac{5m-4}{3(m+2)(m-2)}$
Subtracting Rational Expressions Follow the same steps as for addition, but subtract in Step 3.	Subtract. $\frac{6}{k+4}-\frac{2}{k}$ The LCD is $k(k+4)$. $\frac{6k}{(k+4)k}-\frac{2(k+4)}{k(k+4)}=\frac{6k-2(k+4)}{k(k+4)}$ $=\frac{6k-2k-8}{k(k+4)}=\frac{4k-8}{k(k+4)}$
6.5 COMPLEX FRACTIONS	
Simplifying Complex Fractions	Simplify.
Method 1 Simplify the numerator and denominator separately. Then divide the simplified numerator by the simplified denominator.	**(1)** $\frac{\frac{1}{a}-a}{1-a}=\frac{\frac{1}{a}-\frac{a^2}{a}}{1-a}=\frac{\frac{1-a^2}{a}}{1-a}$ $=\frac{1-a^2}{a}\cdot\frac{1}{1-a}$ $=\frac{(1-a)(1+a)}{a(1-a)}=\frac{1+a}{a}$
Method 2 Multiply numerator and denominator of the complex fraction by the LCD of all the denominators in the complex fraction. Write in lowest terms.	**(2)** $\frac{\frac{1}{a}-a}{1-a}=\frac{\frac{1}{a}-a}{1-a}\cdot\frac{a}{a}=\frac{\frac{a}{a}-a^2}{(1-a)a}$ $=\frac{1-a^2}{(1-a)a}=\frac{(1+a)(1-a)}{(1-a)a}$ $=\frac{1+a}{a}$
6.6 SOLVING EQUATIONS INVOLVING RATIONAL EXPRESSIONS	
Solving Equations with Rational Expressions	Solve $\frac{2}{x-1}+\frac{3}{4}=\frac{5}{x-1}$.
1. Find the LCD of all denominators in the equation.	The LCD is $4(x-1)$. Note that 1 cannot be a solution.
2. Multiply each side of the equation by the LCD.	$4(x-1)\left(\frac{2}{x-1}+\frac{3}{4}\right)=4(x-1)\left(\frac{5}{x-1}\right)$ $4(x-1)\left(\frac{2}{x-1}\right)+4(x-1)\left(\frac{3}{4}\right)=4(x-1)\left(\frac{5}{x-1}\right)$

(continued)

CONCEPTS	EXAMPLES
3. Solve the resulting equation, which should have no fractions.	$$8 + 3(x - 1) = 20$$ $$8 + 3x - 3 = 20$$ $$3x = 15$$ $$x = 5$$
4. Check each proposed solution.	The proposed solution, 5, checks. The solution set is $\{5\}$.

6.7 APPLICATIONS OF RATIONAL EXPRESSIONS

Solving Problems about Distance

Use the six-step method.

Step 1 State what the variable represents.

Step 2 Use a chart to identify distance, rate, and time. Solve $d = rt$ for the unknown quantity in the chart.

Step 3 From the wording in the problem, decide the relationship between the quantities. Use those expressions to write an equation.

Step 4 Solve the equation.

Steps 5 and 6 Answer the question; check the solution.

On a trip from Sacramento to Monterey, Marge traveled at an average speed of 60 miles per hour. The return trip, at an average speed of 64 miles per hour, took $\frac{1}{4}$ hour less. How far did she travel between the two cities?

Let x = the unknown distance.

	d	r	$t = \dfrac{d}{r}$
Going	x	60	$\dfrac{x}{60}$
Returning	x	64	$\dfrac{x}{64}$

Since the time for the return trip was $\frac{1}{4}$ hour less, the time going equals the time returning plus $\frac{1}{4}$.

$$\frac{x}{60} = \frac{x}{64} + \frac{1}{4}$$
$$16x = 15x + 240 \qquad \text{Multiply by 960.}$$
$$x = 240 \qquad \text{Subtract } 15x.$$

She traveled 240 miles.

The trip there took $\frac{240}{60} = 4$ hours while the return trip took $\frac{240}{64} = 3.75$ hours, which is $\frac{1}{4}$ hour less time. The solution checks.

Solving Problems about Work

Use the six-step method.

Step 1 State what the variable represents.

Step 2 Put the information from the problem in a chart. If a job is done in t units of time, the rate is $\frac{1}{t}$.

It takes the regular mail carrier 6 hours to cover her route. A substitute takes 8 hours to cover the same route. How long would it take them to cover the route together?

Let x = the number of hours it would take them working together.

The rate of the regular carrier is $\frac{1}{6}$ job per hour; the rate of the substitute is $\frac{1}{8}$ job per hour. Multiply rate times time to get the fractional part of the job done.

	Rate	Time	Part of the Job Done
Regular	$\dfrac{1}{6}$	x	$\dfrac{1}{6}x$
Substitute	$\dfrac{1}{8}$	x	$\dfrac{1}{8}x$

CONCEPTS	EXAMPLES

Step 3 Write the equation. The sum of the fractional parts should equal 1 (whole job).

The equation is

$$\frac{1}{6}x + \frac{1}{8}x = 1.$$

Step 4 Solve the equation.

The solution of the equation is $\frac{24}{7}$.

Steps 5 and 6 Answer the question; check the solution.

It would take them $\frac{24}{7}$ or $3\frac{3}{7}$ hours to cover the route together.

Solving Inverse Variation Problems
1. Write the variation equation $y = \frac{k}{x}$.

If a varies inversely as b, and $a = 4$ when $b = 4$, find a when $b = 6$. The equation for inverse variation is $a = \frac{k}{b}$.

2. Find k by substituting the given values of x and y into the equation.

Substitute $a = 4$ and $b = 4$.

$$4 = \frac{k}{4}$$

$$k = 16$$

3. Write the equation with the value of k from Step 2 and the given value of x or y. Solve for the remaining variable.

Let $k = 16$ and $b = 6$ in the variation equation.

$$a = \frac{16}{6} = \frac{8}{3}$$

CHAPTER 6 REVIEW EXERCISES

1. 3 **2.** 0 **3.** $-5, -\frac{2}{3}$

4. Set the denominator equal to 0 and solve the equation. Any solutions are values for which the rational expression is undefined.

5. (a) $\frac{11}{8}$ (b) $\frac{13}{22}$

6. (a) undefined (b) 1

7. $\frac{b}{3a}$ **8.** -1 **9.** $\frac{-(2x + 3)}{2}$

10. $\frac{2p + 5q}{5p + q}$

Answers may vary in Exercises 11 and 12.

11. $\frac{-(4x - 9)}{2x + 3}, \frac{-4x + 9}{2x + 3}, \frac{4x - 9}{-(2x + 3)},$

$\frac{4x - 9}{-2x - 3}$

12. $\frac{-8 + 3x}{-3 - 6x}, \frac{-(-8 + 3x)}{3 + 6x},$

$\frac{8 - 3x}{-(-3 - 6x)}, -\frac{-8 + 3x}{3 + 6x}$

13. $\frac{72}{p}$ **14.** 2 **15.** $\frac{5}{8}$

16. $\frac{r + 4}{3}$ **17.** $\frac{3a - 1}{a + 5}$

18. $\frac{y - 2}{y - 3}$

[6.1] *Find any values of the variable for which the rational expression is undefined.*

1. $\frac{4}{x - 3}$

2. $\frac{y + 3}{2y}$

3. $\frac{2k + 1}{3k^2 + 17k + 10}$

4. How would you determine the values of the variable for which a rational expression is undefined?

Find the numerical value of each rational expression when (a) $x = -2$ and (b) $x = 4$.

5. $\frac{4x - 3}{5x + 2}$

6. $\frac{3x}{x^2 - 4}$

Write each rational expression in lowest terms.

7. $\frac{5a^3b^3}{15a^4b^2}$

8. $\frac{m - 4}{4 - m}$

9. $\frac{4x^2 - 9}{6 - 4x}$

10. $\frac{4p^2 + 8pq - 5q^2}{10p^2 - 3pq - q^2}$

Write four equivalent expressions for each of the following.

11. $-\frac{4x - 9}{2x + 3}$

12. $\frac{8 - 3x}{3 + 6x}$

[6.2] *Find each product or quotient and write the answer in lowest terms.*

13. $\frac{18p^3}{6} \cdot \frac{24}{p^4}$

14. $\frac{8x^2}{12x^5} \cdot \frac{6x^4}{2x}$

15. $\frac{x - 3}{4} \cdot \frac{5}{2x - 6}$

16. $\frac{2r + 3}{r - 4} \cdot \frac{r^2 - 16}{6r + 9}$

17. $\frac{6a^2 + 7a - 3}{2a^2 - a - 6} \div \frac{a + 5}{a - 2}$

18. $\frac{y^2 - 6y + 8}{y^2 + 3y - 18} \div \frac{y - 4}{y + 6}$

📄 JOURNAL 📝 WRITING ▲ CHALLENGING 🖩 SCIENTIFIC CALCULATOR 🖥 GRAPHING CALCULATOR

19. $\dfrac{p+5}{p+1}$ **20.** $\dfrac{3z+1}{z+3}$

21. $108y^4$

22. $(x+3)(x+1)(x+4)$

23. $\dfrac{15a}{10a^4}$ **24.** $\dfrac{-54}{18-6x}$

25. $\dfrac{15y}{50-10y}$

26. $\dfrac{4b(b+2)}{(b+3)(b-1)(b+2)}$

27. $\dfrac{15}{x}$ **28.** $-\dfrac{2}{p}$ **29.** $\dfrac{4k-45}{k(k-5)}$

30. $\dfrac{28+11y}{y(7+y)}$ **31.** $\dfrac{-2-3m}{6}$

32. $\dfrac{3(16-x)}{4x^2}$

33. $\dfrac{7a+6b}{(a-2b)(a+2b)}$

34. $\dfrac{-k^2-6k+3}{3(k+3)(k-3)}$

35. $\dfrac{5z-16}{z(z+6)(z-2)}$

36. $\dfrac{-13p+33}{p(p-2)(p-3)}$

37. (a) $\dfrac{a}{b}$ (b) $\dfrac{a}{b}$ (c) Answers will vary.

38. $\dfrac{10}{13}$ **39.** $\dfrac{4(y-3)}{y+3}$

40. $\dfrac{(q-p)^2}{pq}$ **41.** $\dfrac{xw+1}{xw-1}$

42. $\dfrac{1-r-t}{1+r+t}$

43. It would cause the first and third denominators to equal 0.

44. $\{-16\}$ **45.** $\emptyset$

46. $\{3\}$

---

19. $\dfrac{2p^2+13p+20}{p^2+p-12} \cdot \dfrac{p^2+2p-15}{2p^2+7p+5}$

20. $\dfrac{3z^2+5z-2}{9z^2-1} \cdot \dfrac{9z^2+6z+1}{z^2+5z+6}$

[6.3] *Find the least common denominator for each list of fractions.*

21. $\dfrac{4}{9y}, \dfrac{7}{12y^2}, \dfrac{5}{27y^4}$

22. $\dfrac{3}{x^2+4x+3}, \dfrac{5}{x^2+5x+4}$

Rewrite each rational expression with the given denominator.

23. $\dfrac{3}{2a^3} = \dfrac{}{10a^4}$

24. $\dfrac{9}{x-3} = \dfrac{}{18-6x}$

25. $\dfrac{-3y}{2y-10} = \dfrac{}{50-10y}$

26. $\dfrac{4b}{b^2+2b-3} = \dfrac{}{(b+3)(b-1)(b+2)}$

[6.4] *Add or subtract and write each answer in lowest terms.*

27. $\dfrac{10}{x} + \dfrac{5}{x}$

28. $\dfrac{6}{3p} - \dfrac{12}{3p}$

29. $\dfrac{9}{k} - \dfrac{5}{k-5}$

30. $\dfrac{4}{y} + \dfrac{7}{7+y}$

31. $\dfrac{m}{3} - \dfrac{2+5m}{6}$

32. $\dfrac{12}{x^2} - \dfrac{3}{4x}$

33. $\dfrac{5}{a-2b} + \dfrac{2}{a+2b}$

34. $\dfrac{4}{k^2-9} - \dfrac{k+3}{3k-9}$

35. $\dfrac{8}{z^2+6z} - \dfrac{3}{z^2+4z-12}$

36. $\dfrac{11}{2p-p^2} - \dfrac{2}{p^2-5p+6}$

[6.5]

37. Simplify the complex fraction $\dfrac{\frac{a^4}{b^2}}{\frac{a^3}{b}}$ by

(a) Method 1 as described in Section 6.5.

(b) Method 2 as described in Section 6.5.

(c) Explain which method you prefer, and why.

Simplify each complex fraction.

38. $\dfrac{\frac{2}{3}-\frac{1}{6}}{\frac{1}{4}+\frac{2}{5}}$

39. $\dfrac{\frac{y-3}{y}}{\frac{y+3}{4y}}$

40. $\dfrac{\frac{1}{p}-\frac{1}{q}}{\frac{1}{q-p}}$

41. $\dfrac{x+\frac{1}{w}}{x-\frac{1}{w}}$

42. $\dfrac{\frac{1}{r+t}-1}{\frac{1}{r+t}+1}$

[6.6]

43. Before even beginning the solution process, how do you know that 2 cannot be a solution to the equation found in Exercise 46 below?

Solve each equation and check your solutions.

44. $\dfrac{4-z}{z} + \dfrac{3}{2} = \dfrac{-4}{z}$

45. $\dfrac{3y-1}{y-2} = \dfrac{5}{y-2} + 1$

46. $\dfrac{3}{m-2} + \dfrac{1}{m-1} = \dfrac{7}{m^2-3m+2}$

47. $t = \frac{Ry}{m}$ 48. $y = \frac{4x + 5}{3}$
49. $m = \frac{4 + p^2q}{3p^2}$ 50. $\frac{9}{5}$
51. $\frac{2}{6}$ 52. about 1.527 hours
53. $3\frac{1}{13}$ hours 54. 2 hours
55. inverse 56. approximately 33,230 males and 5275 females
57. 4 centimeters 58. 5 hours

Solve each formula for the specified variable.

47. $m = \frac{Ry}{t}$ for t **48.** $x = \frac{3y - 5}{4}$ for y **49.** $p^2 = \frac{4}{3m - q}$ for m

[6.7] *Solve each problem. Use the six-step method.*

50. In a certain fraction, the denominator is 4 less than the numerator. If 3 is added to both the numerator and the denominator, the resulting fraction is equal to $\frac{3}{2}$. Find the original fraction.

51. The denominator of a certain fraction is 3 times the numerator. If 2 is added to the numerator and subtracted from the denominator, the resulting fraction is equal to 1. Find the original fraction.

52. On August 18, 1996, Scott Sharp won the True Value 200-mile Indy race driving a Ford with an average speed of 130.934 miles per hour. What was his time? (*Source: Sports Illustrated 1998 Sports Almanac.*)

53. A man can plant his garden in 5 hours, working alone. His daughter can do the same job in 8 hours. How long would it take them if they worked together?

54. The head gardener can mow the lawns in the city park twice as fast as his assistant. Working together, they can complete the job in $1\frac{1}{3}$ hours. How long would it take the head gardener working alone?

55. The longer the term of your subscription to *Monitoring Times,* the less you will have to pay per year. Is this an example of direct or inverse variation?

56. In 1994, 38,505 American deaths were due to firearms. Of this total, the approximate ratio of male deaths to female deaths was 63 to 10. How many males and how many females died as a result of using firearms? (*Source:* National Center for Health Statistics, Monthly Vital Statistics Reports.)

57. If a parallelogram has a fixed area, the height varies inversely as the base. A parallelogram has a height of 8 centimeters and a base of 12 centimeters. Find the height if the base is changed to 24 centimeters.

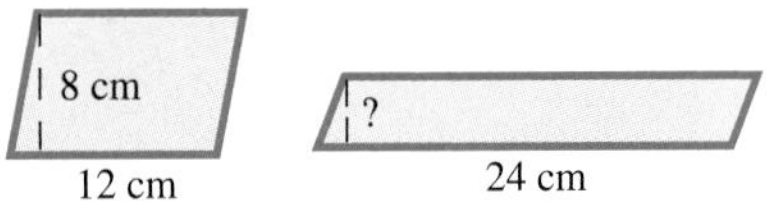

58. At a given hour, two steamboats leave a city in the same direction on a straight canal. One travels at 18 miles per hour, and the other travels at 25 miles per hour. In how many hours will the boats be 35 miles apart?

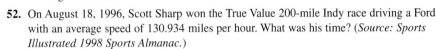

59. (a) −3 (b) −1 (c) −3, −1

RELATING CONCEPTS (EXERCISES 59–68)

In these exercises, we summarize the various concepts involving rational expressions we have covered.

Work Exercises 59–68 in order.

Let P, Q, and R be rational expressions defined as follows.

$$P = \frac{6}{x + 3} \qquad Q = \frac{5}{x + 1} \qquad R = \frac{4x}{x^2 + 4x + 3}$$

59. Find the value or values for which the expression is undefined.
(a) P **(b)** Q **(c)** R

60. $\frac{15}{2x}$

61. If $x = 0$, the divisor R is equal to 0, and division by 0 is undefined.

62. $(x + 3)(x + 1)$

63. $\frac{7}{x + 1}$

64. $\frac{11x + 21}{4x}$

65. $\emptyset$

66. We know that -3 is not allowed because P and R are undefined for $x = -3$.

67. Rate is equal to distance divided by time. Here, distance is 6 miles and time is $x + 3$ minutes, so rate $= \frac{6}{x + 3}$, which is the expression for P.

68. $\frac{6}{5}, \frac{5}{2}$

RELATING CONCEPTS (EXERCISES 59–68) (CONTINUED)

60. Find and express in lowest terms: $(P \cdot Q) \div R$.

61. Why is $(P \cdot Q) \div R$ not defined if $x = 0$?

62. Find the LCD for P, Q, and R.

63. Perform the operations and express in lowest terms: $P + Q - R$.

64. Simplify the complex fraction $\frac{P + Q}{R}$.

65. Solve the equation $P + Q = R$.

66. How does your answer to Exercise 59 help you work Exercise 65?

67. Suppose that a car travels 6 miles in $x + 3$ minutes. Explain why P represents the rate of the car (in miles per minute).

68. For what value or values of x is $R = \frac{40}{77}$?

Did you make the connections between the various operations with rational expressions?

69. $\frac{(5 + 2x - 2y)(x + y)}{(3x + 3y - 2)(x - y)}$

70. $\frac{m + 7}{(m - 1)(m + 1)}$

71. $8p^2$ **72.** $\frac{1}{6}$ **73.** 3

74. $\frac{z + 7}{(z + 1)(z - 1)^2}$

75. $r = \frac{3kz}{5k - z}$ or $r = \frac{-3kz}{z - 5k}$

76. $\{-4\}$

77. approximately 7443 for hearts and 3722 for livers (The actual numbers were 7467 and 3698.)

78. 150 kilometers per hour

79. $\frac{36}{5}$ **80. 6 days**

MIXED REVIEW EXERCISES

Perform the indicated operations.

69. $\dfrac{\frac{5}{x - y} + 2}{3 - \frac{2}{x + y}}$

70. $\frac{4}{m - 1} - \frac{3}{m + 1}$

71. $\frac{8p^5}{5} \div \frac{2p^3}{10}$

72. $\frac{r - 3}{8} \div \frac{3r - 9}{4}$

73. $\dfrac{\frac{5}{x} - 1}{\frac{5 - x}{3x}}$

74. $\frac{4}{z^2 - 2z + 1} - \frac{3}{z^2 - 1}$

Solve.

75. $\frac{1}{k} + \frac{3}{r} = \frac{5}{z}$ for r

76. $\frac{5 + m}{m} + \frac{3}{4} = \frac{-2}{m}$

77. In 1996, the total number of people waiting for heart and liver transplants was 11,165. The ratio of those waiting for heart transplants to those waiting for liver transplants was approximately 2 to 1. How many of each type of candidate were there? (*Sources:* U.S. Department of Health and Human Services, Public Health Service, Division of Organ Transplantation, and United Network for Organ Sharing.)

78. Emily Falzon flew her plane 400 kilometers with the wind in the same time it took her to go 200 kilometers against the wind. The speed of the wind is 50 kilometers per hour. Find the speed of the plane in still air.

79. If x varies directly as y, and $x = 12$ when $y = 5$, find x when $y = 3$.

80. When Mario and Luigi work together on a job, they can do it in $3\frac{3}{7}$ days. Mario can do the job working alone in 8 days. How long would it take Luigi working alone?

CHAPTER 6 TEST

[6.1] 1. $-2, 4$

2. (a) $\frac{11}{6}$ (b) undefined

3. (Answers may vary.) $\frac{-(6x-5)}{2x+3}$, $\frac{-6x+5}{2x+3}$, $\frac{6x-5}{-(2x+3)}$, $\frac{6x-5}{-2x-3}$ 4. $-3x^2y^3$

5. $\frac{3a+2}{a-1}$

[6.2] 6. $\frac{25}{27}$ 7. $\frac{3k-2}{3k+2}$

8. $\frac{a-1}{a+4}$

[6.3] 9. $150p^5$

10. $(2r+3)(r+2)(r-5)$

11. $\frac{240p^2}{64p^3}$ 12. $\frac{21}{42m-84}$

[6.4] 13. 2

14. $\frac{-14}{5(y+2)}$

15. $\frac{-x^2+x+1}{3-x}$ or $\frac{x^2-x-1}{x-3}$

16. $\frac{-m^2+7m+2}{(2m+1)(m-5)(m-1)}$

[6.5] 17. $\frac{2k}{3p}$ 18. $\frac{-2-x}{4+x}$

[6.6] 19. $-1, 4$ 20. $\left\{-\frac{1}{2}\right\}$

21. $D = \frac{dF-k}{F}$ or $D = \frac{k-dF}{-F}$

[6.7] 22. 3 miles per hour

1. Find any values for which $\frac{3x-1}{x^2-2x-8}$ is undefined.

2. Find the numerical value of $\frac{6r+1}{2r^2-3r-20}$ when

(a) $r = -2$ and **(b)** $r = 4$.

3. Write four rational expressions equivalent to $-\frac{6x-5}{2x+3}$.

Write each rational expression in lowest terms.

4. $\frac{-15x^6y^4}{5x^4y}$

5. $\frac{6a^2+a-2}{2a^2-3a+1}$

Multiply or divide. Write each answer in lowest terms.

6. $\frac{5(d-2)}{9} \div \frac{3(d-2)}{5}$

7. $\frac{6k^2-k-2}{8k^2+10k+3} \cdot \frac{4k^2+7k+3}{3k^2+5k+2}$

8. $\frac{4a^2+9a+2}{3a^2+11a+10} \div \frac{4a^2+17a+4}{3a^2+2a-5}$

Find the least common denominator for each list of fractions.

9. $\frac{-3}{10p^2}, \frac{21}{25p^3}$, and $\frac{-7}{30p^5}$

10. $\frac{r+1}{2r^2+7r+6}$ and $\frac{-2r+1}{2r^2-7r-15}$

Rewrite each rational expression with the given denominator.

11. $\frac{15}{4p} = \frac{}{64p^3}$

12. $\frac{3}{6m-12} = \frac{}{42m-84}$

Add or subtract. Write each answer in lowest terms.

13. $\frac{4x+2}{x+5} + \frac{-2x+8}{x+5}$

14. $\frac{-4}{y+2} + \frac{6}{5y+10}$

15. $\frac{x+1}{3-x} + \frac{x^2}{x-3}$

16. $\frac{3}{2m^2-9m-5} - \frac{m+1}{2m^2-m-1}$

Simplify each complex fraction.

17. $\dfrac{\frac{2p}{k^2}}{\frac{3p^2}{k^3}}$

18. $\dfrac{\frac{1}{x+3}-1}{1+\frac{1}{x+3}}$

19. What are the two numbers that would have to be rejected as possible solutions for the equation $\frac{2}{x+1} - \frac{3}{x-4} = 6$?

20. Solve the equation $\frac{2x}{x-3} + \frac{1}{x+3} = \frac{-6}{x^2-9}$. Be sure to check your solution(s).

21. Solve the formula $F = \frac{k}{d-D}$ for D.

Solve each problem.

22. A boat goes 7 miles per hour in still water. It takes as long to go 20 miles upstream as 50 miles downstream. Find the speed of the current.

23. $2\frac{2}{9}$ hours 24. 27 days

23. A man can paint a room in his house, working alone, in 5 hours. His wife can do the job in 4 hours. How long will it take them to paint the room if they work together?

24. Under certain conditions, the length of time that it takes for fruit to ripen during the growing season varies inversely as the average maximum temperature during the season. If it takes 25 days for fruit to ripen with an average maximum temperature of 80°, find the number of days it would take at 75°. Round your answer to the nearest whole number.

CUMULATIVE REVIEW EXERCISES CHAPTERS 1–6

[1.2, 1.5, 1.6] 1. 2
[2.2] 2. {17}
[2.4] 3. $b = \frac{2A}{h}$
[2.5] 4. $\left\{-\frac{2}{7}\right\}$
[2.7] 5. $[-8, \infty)$
6. $(4, \infty)$
[3.1] 7. (a) $(-3, 0)$ (b) $(0, -4)$
[3.2] 8.

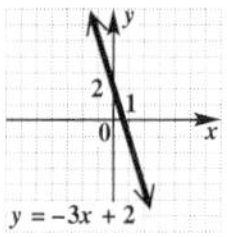

[4.1] 9.

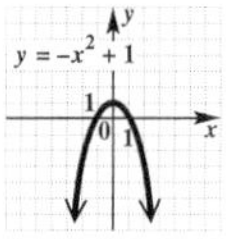

[4.2, 4.5] 10. $\frac{1}{2^4x^7}$ 11. $\frac{1}{m^6}$
12. $\frac{q}{4p^2}$
[4.1] 13. $k^2 + 2k + 1$
[4.2] 14. $72x^6y^7$
[4.4] 15. $4a^2 - 4ab + b^2$
[4.3] 16. $3y^3 + 8y^2 + 12y - 5$
[4.6] 17. $6p^2 + 7p + 1 + \frac{3}{p - 1}$
[4.7] 18. 1.4×10^5 seconds
[5.3] 19. $(4t + 3v)(2t + v)$
20. prime
[5.4] 21. $(4x^2 + 1)(2x + 1)(2x - 1)$
[5.5] 22. $\{-3, 5\}$
23. $\left\{5, -\frac{1}{2}, \frac{2}{3}\right\}$
[5.6] 24. -2 or -1
25. 6 meters

1. Use the order of operations to evaluate $3 + 4\left(\frac{1}{2} - \frac{3}{4}\right)$.

Solve.

2. $3(2y - 5) = 2 + 5y$ **3.** $A = \frac{1}{2}bh$ for b **4.** $\frac{2 + m}{2 - m} = \frac{3}{4}$

5. $5y \leq 6y + 8$ **6.** $5m - 9 > 2m + 3$

7. For the graph of $4x + 3y = -12$,
(a) what is the x-intercept and (b) what is the y-intercept?

Sketch each graph.

8. $y = -3x + 2$ **9.** $y = -x^2 + 1$

Simplify each expression. Write with only positive exponents.

10. $\frac{(2x^3)^{-1} \cdot x}{2^3x^5}$ **11.** $\frac{(m^{-2})^3m}{m^5m^{-4}}$ **12.** $\frac{2p^3q^4}{8p^5q^3}$

Perform the indicated operations.

13. $(2k^2 + 3k) - (k^2 + k - 1)$ **14.** $8x^2y^2(9x^4y^5)$

15. $(2a - b)^2$ **16.** $(y^2 + 3y + 5)(3y - 1)$

17. $\frac{12p^3 + 2p^2 - 12p + 4}{2p - 2}$

18. A computer can do one operation in 1.4×10^{-7} seconds. How long would it take for the computer to do one trillion (10^{12}) operations?

Factor completely.

19. $8t^2 + 10tv + 3v^2$ **20.** $8r^2 - 9rs + 12s^2$ **21.** $16x^4 - 1$

Solve each equation.

22. $r^2 = 2r + 15$

23. $(r - 5)(2r + 1)(3r - 2) = 0$

Solve each problem.

24. One number is 4 more than another. The product of the numbers is 2 less than the smaller number. Find the smaller number.

25. The length of a rectangle is 2 meters less than twice the width. The area is 60 square meters. Find the width of the rectangle.

26. 30 (The maximum percent is 30%.)
[6.1] 27. (a) 28. (d)
[6.4] 29. $\frac{4}{q}$
30. $\frac{3r + 28}{7r}$
31. $\frac{7}{15(q-4)}$
32. $\frac{-k-5}{k(k+1)(k-1)}$
[6.2] 33. $\frac{7(2z+1)}{24}$
[6.5] 34. $\frac{195}{29}$
[6.6] 35. 4, 0
36. $\left\{\frac{21}{2}\right\}$
37. $\{-2, 1\}$
[6.7] 38. 150 miles
39. $1\frac{1}{5}$ hours
40. 20 pounds

26. In 1996, the U.S. Department of Health and Human Services and the Department of Agriculture set guidelines for adult diets. A maximum percent of one's diet was suggested for calories from fat. It is a number such that when it is divided by 60, the quotient is the same as when it is divided into 15. Find this positive number to discover the maximum suggested percent from fat.

27. One of the following is equal to 1 for *all* real numbers. Which one is it?

(a) $\frac{k^2+2}{k^2+2}$ **(b)** $\frac{4-m}{4-m}$ **(c)** $\frac{2x+9}{2x+9}$ **(d)** $\frac{x^2-1}{x^2-1}$

28. Which one of the following rational expressions is *not* equivalent to $\frac{4-3x}{7}$?

(a) $-\frac{-4+3x}{7}$ **(b)** $-\frac{4-3x}{-7}$ **(c)** $\frac{-4+3x}{-7}$ **(d)** $\frac{-(3x+4)}{7}$

Perform each operation and write the answer in lowest terms.

29. $\frac{5}{q} - \frac{1}{q}$

30. $\frac{3}{7} + \frac{4}{r}$

31. $\frac{4}{5q-20} - \frac{1}{3q-12}$

32. $\frac{2}{k^2+k} - \frac{3}{k^2-k}$

33. $\frac{7z^2+49z+70}{16z^2+72z-40} \div \frac{3z+6}{4z^2-1}$

34. Simplify the complex fraction $\frac{\frac{4}{a}+\frac{5}{2a}}{\frac{7}{6a}-\frac{1}{5a}}$.

35. What values of x cannot possibly be solutions of the equation $\frac{1}{x-4} = \frac{3}{2x}$?

Solve each equation. Check your solutions.

36. $\frac{r+2}{5} = \frac{r-3}{3}$

37. $\frac{1}{x} = \frac{1}{x+1} + \frac{1}{2}$

Solve each problem.

38. On a business trip, Arlene traveled to her destination at an average speed of 60 miles per hour. Coming home, her average speed was 50 miles per hour, and the trip took $\frac{1}{2}$ hour longer. How far did she travel each way?

39. Juanita can weed the yard in 3 hours. Benito can weed the yard in 2 hours. How long would it take them if they worked together?

40. The force required to compress a spring varies directly as the change in the length of the spring. If a force of 12 pounds is required to compress a certain spring 3 inches, how much force is required to compress the spring 5 inches?

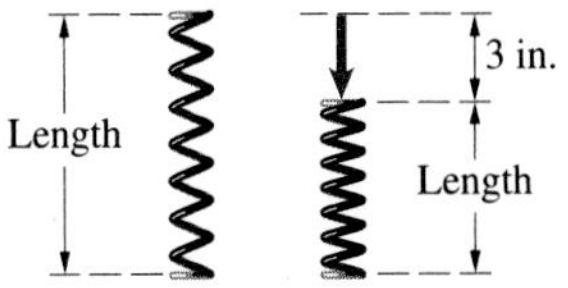

7 Equations of Lines, Inequalities, and Functions

Government

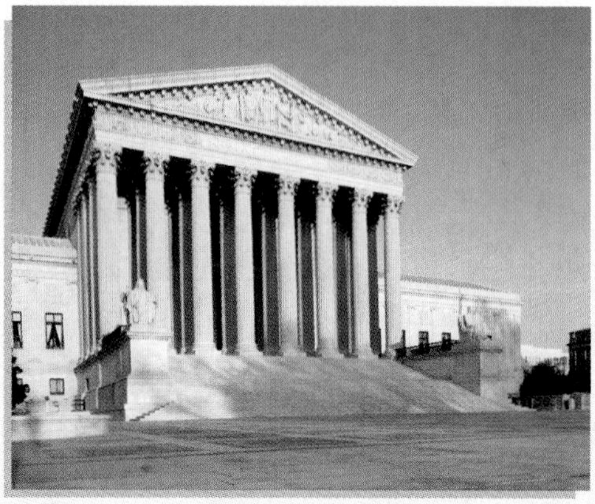

We began the study of linear equations and functions in Chapter 3. Now we continue these topics in this chapter. The Chapter 3 theme was business. One big business is government, our theme for this chapter. The federal government alone employed 2.9 million civilians in 1996 in addition to the armed forces. State and local government employees numbered almost 4 million in 1995, according to the U.S. Justice Department. Government branches collect and disperse vast amounts of mathematical data, much of it in the form of graphs. One example is shown here. From the graph, does it appear that the number of prisoners (in thousands) in state and federal institutions increased linearly over the years from 1988 to 1996? We will return to this example later, in Section 7.3.

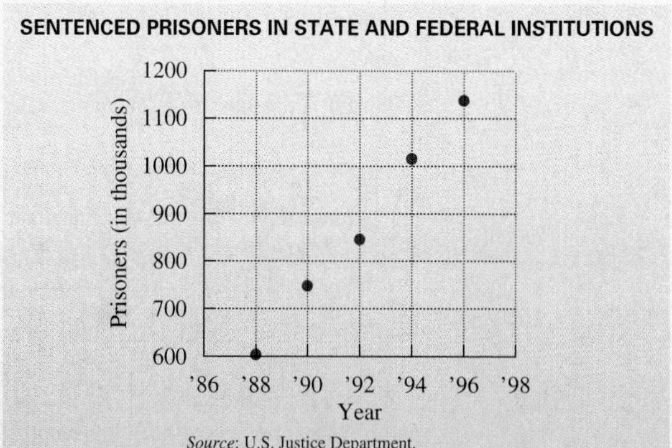

Source: U.S. Justice Department.

Throughout this chapter we provide many government-related applications of linear functions and their graphs. Other examples include changes in public school expenditures, storage of nuclear waste, and growth in the U.S. foreign-born population.

7.1 Equations of a Line

OBJECTIVES

1. Write an equation of a line given its slope and y-intercept.
2. Graph a line given its slope and a point on the line.
3. Write an equation of a line given its slope and any point on the line.
4. Write an equation of a line given two points on the line.
5. Find the equation of a line that fits a data set.

FOR EXTRA HELP

SSG Sec. 7.1
SSM Sec. 7.1
Pass the Test Software
InterAct Math Tutorial Software
Video 12

TEACHING TIP Before beginning this section, suggest that students look back at the Quick Review for Chapter 3.

CHALKBOARD EXAMPLE

Find an equation of the line with slope -1 and y-intercept $(0, 8)$.

Answer: $y = -x + 8$

Section 3.3 showed how to find the slope (steepness) of a line from the equation of the line by solving the equation for y. In that form, the slope is the coefficient of x. For example, the slope of the line with equation $y = 2x + 3$ is 2, the coefficient of x. What does the number 3 represent? If $x = 0$, the equation becomes

$$y = 2(0) + 3 = 0 + 3 = 3.$$

Since $y = 3$ corresponds to $x = 0$, $(0, 3)$ is the y-intercept of the graph of $y = 2x + 3$. An equation like $y = 2x + 3$ that is solved for y is said to be in **slope-intercept form** because both the slope and the y-intercept of the line can be read directly from the equation.

Slope-Intercept Form

The slope-intercept form of the equation of a line with slope m and y-intercept $(0, b)$ is

$$y = mx + b.$$

The slope-intercept form is the most useful form for a linear equation, because it is the one that describes a linear function.

OBJECTIVE 1 **Write an equation of a line given its slope and y-intercept.** Given the slope and y-intercept of a line, we can use the slope-intercept form to find an equation of the line.

EXAMPLE 1 **Finding an Equation of a Line**

Find an equation of the line with slope $\frac{2}{3}$ and y-intercept $(0, -1)$.

Here $m = \frac{2}{3}$ and $b = -1$, so the equation is

$$y = mx + b$$

$$y = \frac{2}{3}x - 1.$$

CONNECTIONS

Businesses must consider the amount of value lost, called **depreciation,** during each year of a machine's useful life. The simplest way to calculate depreciation is to assume that an item with a useful life of n years loses $\frac{1}{n}$ of its value each year. Historically, if the equipment had salvage value, the depreciation was calculated on the **net cost,** the difference between the purchase price and the salvage value. If P represents the purchase price and S the salvage value of an item with a useful life of n years, then the annual depreciation would be

$$D = \frac{1}{n}(P - S).$$

Because this is a linear equation, this is called **straight-line depreciation.**

However, from a practical viewpoint, it is often difficult to determine the salvage value when the equipment is new. Also, for some equipment, such as computers, there is no residual dollar value at the end of the useful life due to

INSTRUCTOR'S RESOURCES

PTB Sec. 7.1 **ISM** Sec. 7.1 **AB** Sec. 7.1

TEST GENERATOR

WORLD WIDE WEB www.LialAlgebra.com

CONNECTIONS (CONTINUED)

obsolescence. In actual practice now, it is customary to find straight-line depreciation by the simpler linear equation

$$D = \frac{1}{n}P,$$

which assumes no salvage value.

FOR DISCUSSION OR WRITING

1. Find the depreciation using both methods for a \$50,000 (new) asset with a useful life of 10 years and a salvage value of \$15,000. How much is "written off" in each case over the ten-year period?
2. The depreciation equation is given in slope-intercept form. What does the slope represent here? What does the y-value of the y-intercept represent?

Source: Joel E. Halle, CPA.

1. \$3500, \$5000; \$35,000, \$50,000 2. The loss in value each year; the depreciation when the item is brand new ($D = 0$)

OBJECTIVE 2 Graph a line given its slope and a point on the line. We can use the slope and y-intercept to graph a line. For example, to graph $y = \frac{2}{3}x - 1$, first locate the y-intercept, $(0, -1)$, on the y-axis. From the definition of slope and the fact that the slope of this line is $\frac{2}{3}$,

$$m = \frac{\text{difference in } y\text{-values}}{\text{difference in } x\text{-values}} = \frac{2}{3}.$$

Another point P on the graph of the line can be found by counting from the y-intercept 2 units up and then counting 3 units to the right. We then draw the line through point P and the y-intercept, as shown in Figure 1. This method can be extended to graph a line given its slope and any point on the line, not just the y-intercept.

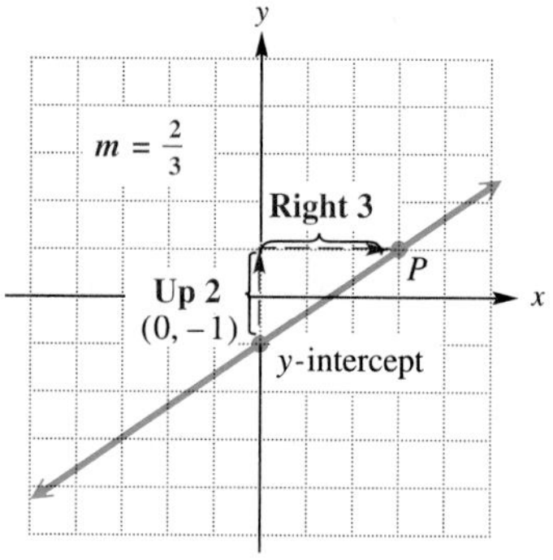

Figure 1

CHALKBOARD EXAMPLE

Graph the line through $(2, -3)$ with the slope $-\frac{1}{3}$.

Answer:

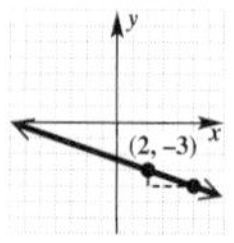

EXAMPLE 2 Graphing a Line Given a Point and the Slope

Graph the line through $(-2, 3)$ with slope -4.

First, locate the point $(-2, 3)$. Write the slope as

$$m = \frac{\text{difference in } y\text{-values}}{\text{difference in } x\text{-values}} = -4 = \frac{-4}{1}.$$

We locate another point on the line by counting 4 units down (because of the negative sign) and then 1 unit to the right. Finally, we draw the line through this new point P and the given point $(-2, 3)$. See Figure 2.

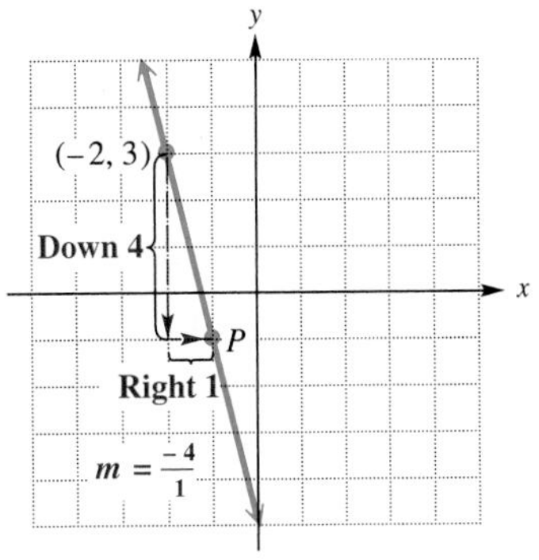

Figure 2

TEACHING TIP Emphasize that the count of the 4 units down, l unit to the right starts at the given point, not at the origin.

NOTE In Example 2, we could have written the slope as $\frac{4}{-1}$ instead. In this case, we would move 4 units up from $(-2, 3)$ and then 1 unit to the left (because of the negative sign). Verify that this produces the same line.

OBJECTIVE 3 **Write an equation of a line given its slope and any point on the line.** Let m represent the slope of the line and let (x_1, y_1) represent the given point on the line. Let (x, y) represent any other point on the line. Then by the definition of slope,

$$\frac{y - y_1}{x - x_1} = m$$

or

$$y - y_1 = m(x - x_1).$$

This result is the **point-slope form** of the equation of a line.

Point-Slope Form

The point-slope form of the equation of a line with slope m going through (x_1, y_1) is

$$y - y_1 = m(x - x_1).$$

This very important result should be memorized.

CHALKBOARD EXAMPLE

Find the equation of the line through (5, 2) with slope $-\frac{1}{3}$. Write the equation in slope-intercept form.

Answer: $y = -\frac{1}{3}x + \frac{11}{3}$

EXAMPLE 3 Using the Point-Slope Form to Write an Equation

Find an equation of each of the following lines. Write the equation in slope-intercept form.

(a) Through $(-2, 4)$, with slope -3

The given point is $(-2, 4)$ so $x_1 = -2$ and $y_1 = 4$. Also, $m = -3$. Substitute these values into the point-slope form.

$$\begin{aligned} y - y_1 &= m(x - x_1) && \\ y - 4 &= -3[x - (-2)] && \\ y - 4 &= -3(x + 2) && \\ y - 4 &= -3x - 6 && \text{Distributive property} \\ y &= -3x - 2 && \text{Add 4.} \end{aligned}$$

The last equation is in slope-intercept form.

(b) Through $(4, 2)$, with slope $\frac{3}{5}$

Use $x_1 = 4$, $y_1 = 2$, and $m = \frac{3}{5}$ in the point-slope form.

$$\begin{aligned} y - y_1 &= m(x - x_1) && \\ y - 2 &= \frac{3}{5}(x - 4) && \\ y - 2 &= \frac{3}{5}x - \frac{12}{5} && \text{Distributive property} \\ y &= \frac{3}{5}x - \frac{12}{5} + \frac{10}{5} && \text{Add } 2 = \tfrac{10}{5} \text{ to both sides.} \\ y &= \frac{3}{5}x - \frac{2}{5} && \text{Combine terms.} \end{aligned}$$

TEACHING TIP Whenever a problem asks for an equation of a line, the point-slope formula should immediately come to mind.

NOTE: We did not clear fractions after the substitution step in Example 3(b) because we want the equation in slope-intercept form—that is, solved for y. Graphing calculators require equations to be in this form, so it has become the most useful.

OBJECTIVE 4 Write an equation of a line given two points on the line. We can also use the point-slope form to find an equation of a line when two points on the line are known.

CHALKBOARD EXAMPLE

Find an equation of the line through the points $(2, 5)$ and $(-1, 6)$. Write the equation in slope-intercept form.

Answer: $y = -\frac{1}{3}x + \frac{17}{3}$

EXAMPLE 4 Finding the Equation of a Line Given Two Points

Find an equation of the line through the points $(-2, 5)$ and $(3, 4)$. Write the equation in slope-intercept form.

First, find the slope of the line, using the definition of slope.

$$\text{slope} = \frac{y_2 - y_1}{x_2 - x_1} = \frac{5 - 4}{-2 - 3} = \frac{1}{-5} = -\frac{1}{5}$$

Now we use either $(-2, 5)$ or $(3, 4)$ and the point-slope form. We choose $(3, 4)$.

$$\begin{aligned} y - y_1 &= m(x - x_1) && \\ y - 4 &= -\frac{1}{5}(x - 3) && \text{Let } y_1 = 4,\ m = -\tfrac{1}{5},\ x_1 = 3. \\ y - 4 &= -\frac{1}{5}x + \frac{3}{5} && \text{Distributive property} \end{aligned}$$

$$y = -\frac{1}{5}x + \frac{3}{5} + \frac{20}{5} \qquad \text{Add } 4 = \tfrac{20}{5} \text{ to both sides.}$$

$$y = -\frac{1}{5}x + \frac{23}{5} \qquad \text{Combine terms.}$$

The same result would be found by using $(-2, 5)$ for (x_1, y_1).

In Chapter 3, most linear equations were given in the form $Ax + By = C$. A linear equation can be written in this form in many different, equally correct, ways. To avoid confusion, we define the following *standard form of a linear equation.*

Standard Form

A linear equation is in standard form if it is written as

$$Ax + By = C,$$

where A, B, and C are integers and $A > 0$, $B \neq 0$.

CAUTION

The above definition of "standard form" is not the same in all texts. Also, for example, $3x + 4y = 12$, $6x + 8y = 24$, and $9x + 12y = 36$ all represent the same set of ordered pairs. Let us agree that $3x + 4y = 12$ is preferable to the other forms because the greatest common factor of 3, 4, and 12 is 1.

A summary of the types of linear equations is given here.

Linear Equations

$x = k$	**Vertical line** Slope is undefined. x-intercept is $(k, 0)$.
$y = k$	**Horizontal line** Slope is 0. y-intercept is $(0, k)$.
$y = mx + b$	**Slope-intercept form** Slope is m. y-intercept is $(0, b)$.
$y - y_1 = m(x - x_1)$	**Point-slope form** Slope is m. Line goes through (x_1, y_1).
$Ax + By = C$	**Standard form** Slope is $-\frac{A}{B}$. x-intercept is $\left(\frac{C}{A}, 0\right)$. y-intercept is $\left(0, \frac{C}{B}\right)$.

OBJECTIVE 5 **Find the equation of a line that fits a data set.** Given a set of data or a graph, we often want to find an equation that can be used to estimate or predict output values that are reasonable for additional input values. Example 5 discusses a procedure to accomplish this for data that fit a linear pattern—that is, whose graph consists of points lying close to a line.

CHALKBOARD EXAMPLE

The number of U.S. children (in thousands) educated at home for selected years is given in the table. Letting $x = 3$ represent 1993, and so on, use the first and last data points to write an equation in slope-intercept form that fits the data.

School Year	x	*Number of Children*
1993	3	588
1994	4	735
1995	5	800
1996	6	920
1997	7	1100

Source: National Home Education Research Institute, Salem, OR.

Answer: $y = 128x + 204$

EXAMPLE 5 Finding an Equation of a Line That Describes Data

The total expenditures (in billions of dollars) for public elementary and secondary education for each year in the first half of the nineties are given below. Plot the data and find an equation that approximates it.

School Year	x	*Expenditures*
1990	0	229
1991	1	241
1992	2	253
1993	3	265
1994	4	279
1995*	5	291

*Estimated
Source: U.S. Department of Education.

Letting y represent the total expenditures in year x, where $x = 0$ represents 1990, $x = 1$ represents 1991, and so on, we plot the data as shown in Figure 3.

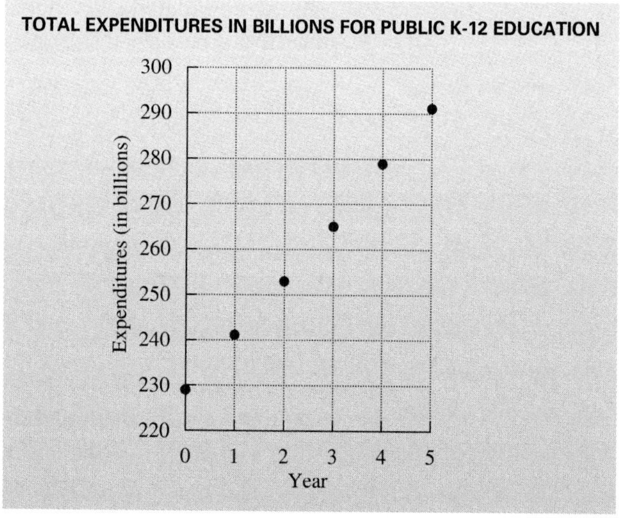

Figure 3

The points appear to lie approximately in a straight line. We can use two of the data pairs and the point-slope form of the equation of a line to get an equation that describes the relationship between the year and the expenditures. Suppose we choose from the table the ordered pairs (0, 229) and (4, 279). First, we need to find the slope of the line through these points.

$$m = \frac{y_2 - y_1}{x_2 - x_1} \qquad \text{Slope formula}$$

$$m = \frac{279 - 229}{4 - 0} \qquad \text{Let } (4, 279) = (x_2, y_2) \text{ and } (0, 229) = (x_1, y_1).$$

$$m = 12.5$$

Now, use the point-slope form to find the equation of the line.

$$y - y_1 = m(x - x_1) \quad \text{Point-slope form}$$
$$y - 229 = 12.5(x - 0) \quad \text{Substitute for } y_1, m, \text{ and } x_1.$$
$$y - 229 = 12.5x \quad \text{Distributive property}$$
$$y = 12.5x + 229 \quad \text{Add 229 to both sides.}$$

To see how well the equation $y = 12.5x + 229$ approximates the ordered pairs (x, y) in the table of data, let $x = 2$ (for 1992) and find y.

$$y = 12.5x + 229 \quad \text{Equation of the line}$$
$$y = 12.5(2) + 229 \quad \text{Substitute 2 for } x.$$
$$y = 254$$

The corresponding value in the table for 1992 is 253, so the equation appears to approximate the data quite well. Thus, the equation could be used with some caution to predict values for years that are not included in the table.

NOTE In Example 5, if we had chosen two different data points, we would get a slightly different equation.

Here is a summary of what is needed to find the equation of a line.

Finding the Equation of a Line

To find the equation of a line, you need

1. a point on the line, and
2. the slope of the line.

If two points are known, first find the slope, and then use the point-slope form.

7.1 EXERCISES

1. D 2. C 3. B 4. A
5. The slope m of a vertical line is undefined, so it is not possible to write the equation of the line in the form $y = mx + b$.

Match the correct equation in Column II with the description given in Column I.

I	II
1. slope $= -2$, through the point $(4, 1)$	**A.** $y = 4x$
2. slope $= -2$, y-intercept $(0, 1)$	**B.** $y = \frac{1}{4}x$
3. passing through the points $(0, 0)$ and $(4, 1)$	**C.** $y = -2x + 1$
4. passing through the points $(0, 0)$ and $(1, 4)$	**D.** $y - 1 = -2(x - 4)$

5. Explain why the equation of a vertical line cannot be written in the form $y = mx + b$.

6. (a) C (b) B (c) A (d) D
7. $y = 3x - 3$ 8. $y = 2x - 4$
9. $y = -x + 3$
10. $y = -\frac{1}{2}x + 2$
11. $y = 4x - 3$
12. $y = -5x + 6$
13. $y = 3$
14. $y = 3x$
15.
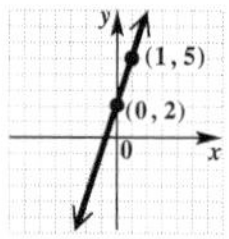

16.
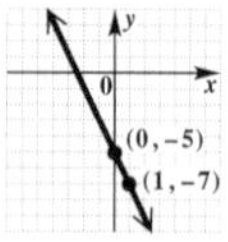

17.
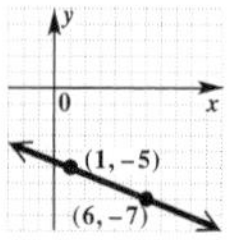

18.
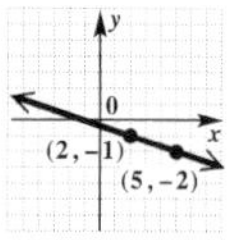

19.
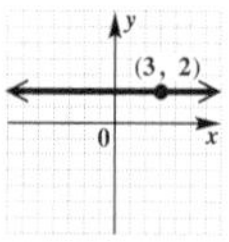

20.
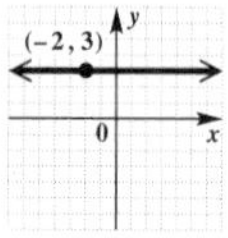

21.
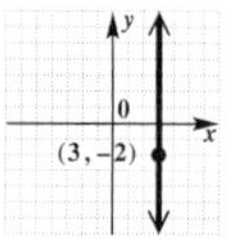

22.
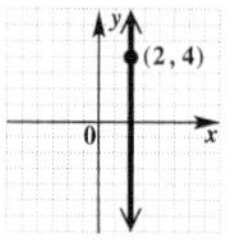

6. Match each equation with the graph that would most closely resemble its graph.

(a) $y = x + 3$
(b) $y = -x + 3$
(c) $y = x - 3$
(d) $y = -x - 3$

A.
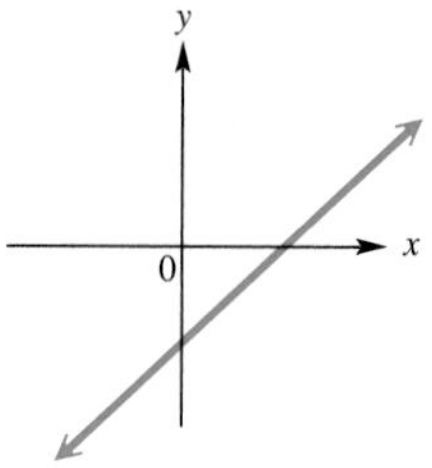

B.
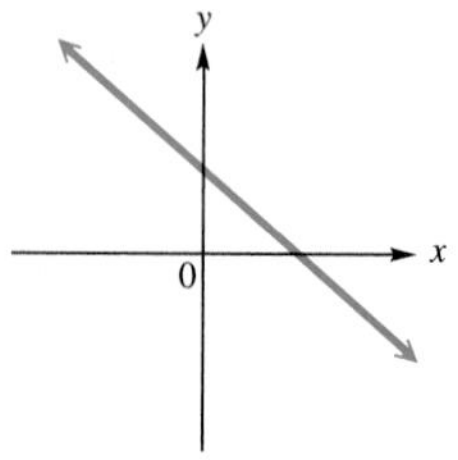

C.
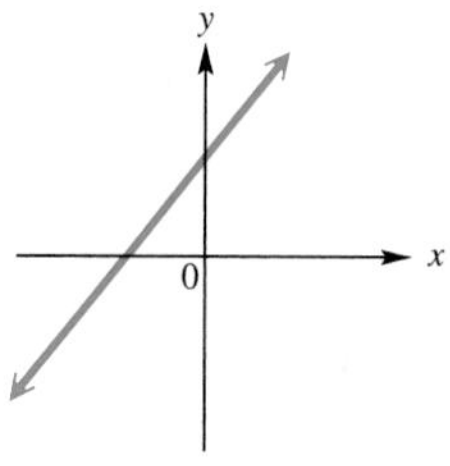

D.
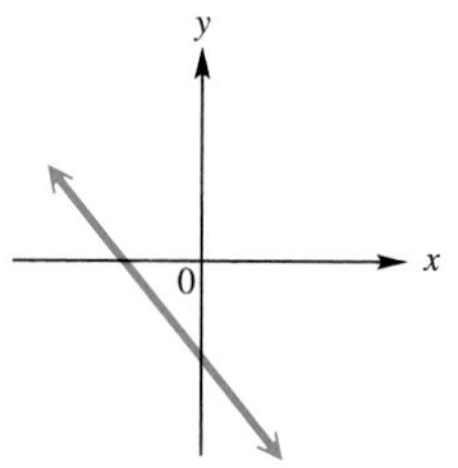

*Use the geometric interpretation of slope (*rise *divided by* run, *from Section 3.3) to find the slope of each line. Then, by identifying the y-intercept from the graph, write the slope-intercept form of the equation of the line.*

7.
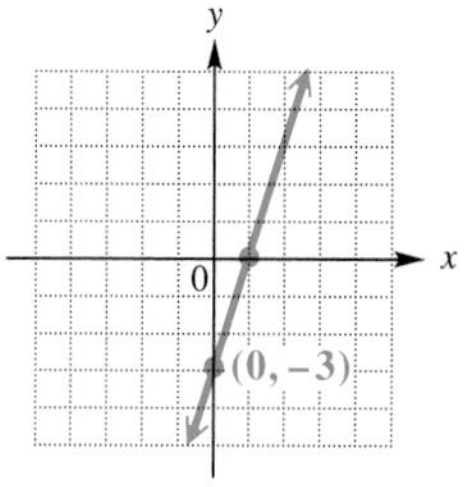

8.
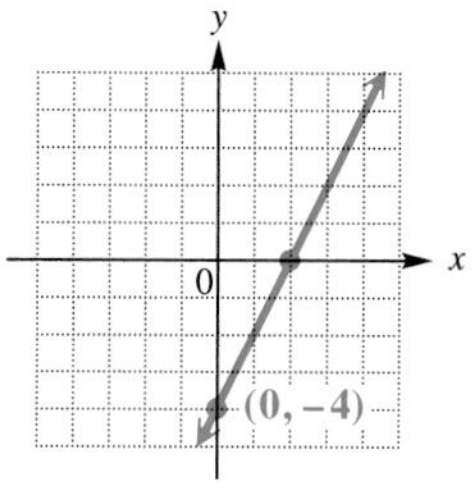

9.
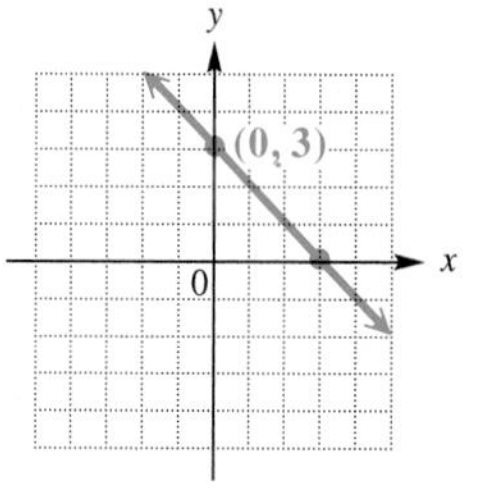

10.
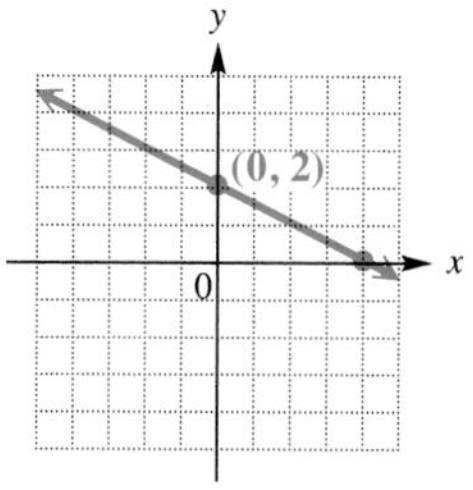

Write the equation of each line with the given slope and y-intercept. See Example 1.

11. $m = 4$, $(0, -3)$
12. $m = -5$, $(0, 6)$
13. $m = 0$, $(0, 3)$
14. $m = 3$, $(0, 0)$

Graph each line going through the given point and having the given slope. (In Exercises 19–22, recall the types of lines having slope 0 and undefined slope.) See Example 2.

15. $(0, 2)$, $m = 3$
16 $(0, -5)$, $m = -2$
17. $(1, -5)$, $m = -\frac{2}{5}$
18. $(2, -1)$, $m = -\frac{1}{3}$
19. $(3, 2)$, $m = 0$
20. $(-2, 3)$, $m = 0$
21. $(3, -2)$, undefined slope
22. $(2, 4)$, undefined slope

23. the y-axis
24. the x-axis
25. $y = 2x - 7$ **26.** $y = 3x + 1$
27. $y = \frac{2}{3}x + \frac{19}{3}$
28. $y = \frac{3}{4}x + 4$
29. $y = -\frac{4}{5}x + \frac{9}{5}$
30. $y = -\frac{7}{2}x + \frac{45}{2}$

23. What is the common name given to a vertical line whose x-intercept is the origin?

24. What is the common name given to a line with slope 0 whose y-intercept is the origin?

Write an equation for each line passing through the given point and having the given slope. Write the equation in slope-intercept form. See Example 3.

25. $(4, 1)$, $m = 2$
26. $(2, 7)$, $m = 3$
27. $(-2, 5)$, $m = \frac{2}{3}$
28. $(-4, 1)$, $m = \frac{3}{4}$
29. $(6, -3)$, $m = -\frac{4}{5}$
30. $(7, -2)$, $m = -\frac{7}{2}$

*These problems can be used as collaborative exercises with small groups of students.

31. $4 = -3(-2) + b$
32. $b = -2$ **33.** $y = -3x - 2$
34. The equations are the same.

RELATING CONCEPTS* (EXERCISES 31–34)

In the examples of this section, we used the point-slope form $y - y_1 = m(x - x_1)$ to find the equation of a line given a point on the line and the slope of the line. Another method uses the slope-intercept (or function) form, $y = mx + b$. To develop this method, we will use the same problem as in Example 3(a): Find an equation of the line through $(-2, 4)$ with slope -3.

Work Exercises 31–34 in order.

31. Substitute the given information for m, x, and y into the equation $y = mx + b$.

32. Use your answer from Exercise 31 to find the value of b.

33. Using the given slope and the value of b from Exercise 32, what is the equation of the line in the form $y = mx + b$?

34. Compare your answer from Exercise 33 with the equation found in Example 3(a). What do you notice?

Did you make the connection between using the two different forms of linear equations to write the equation of a line?

35. $y = x$ (There are other forms as well.) **36.** The slope-intercept form is $y = mx + b$, where m is the slope, and $(0, b)$ is the y-intercept. Thus, the slope and the y-intercept must be given (or found) to use this form. The point-slope form is $y - y_1 = m(x - x_1)$, where (x_1, y_1) is any point on the line, and m is the slope. To use this form, a point on the line and the slope must be given (or found). Examples will vary.
37. $y = x - 3$ **38.** $y = x + 6$
39. $y = -\frac{5}{7}x - \frac{54}{7}$
40. $y = -\frac{3}{5}x - \frac{11}{5}$
41. $y = -\frac{2}{3}x - 2$
42. $y = \frac{1}{2}x + 2$
43. $y = \frac{1}{3}x + \frac{4}{3}$
44. $y = -\frac{1}{3}x + \frac{22}{9}$

▲ 35. If a line passes through the origin and a second point whose x- and y-coordinates are equal, what is an equation of the line?

36. Describe in your own words the slope-intercept and point-slope forms of the equation of a line. Tell what information must be given to use each form to write an equation. Include examples.

Write an equation for the line passing through the given pair of points. Write each equation in slope-intercept form. See Example 4.

37. $(8, 5)$ and $(9, 6)$
38. $(4, 10)$ and $(6, 12)$
39. $(-1, -7)$ and $(-8, -2)$
40. $(-2, -1)$ and $(3, -4)$
41. $(0, -2)$ and $(-3, 0)$
42. $(-4, 0)$ and $(0, 2)$

▲ 43. $\left(\frac{1}{2}, \frac{3}{2}\right)$ and $\left(-\frac{1}{4}, \frac{5}{4}\right)$

▲ 44. $\left(-\frac{2}{3}, \frac{8}{3}\right)$ and $\left(\frac{1}{3}, \frac{7}{3}\right)$

45. The table on the next page gives the heavy metal nuclear waste (in thousands of metric tons) from spent reactor fuel now stored temporarily at reactor sites, awaiting permanent storage. The figures for the year 2000 and beyond are estimates of the U.S. Department of Energy. (*Source:* "Burial of Radioactive Nuclear Waste under the Seabed," *Scientific American,* January 1998, p. 62.)

45. (a) (5, 42), (15, 61), (25, 76)
(b) yes

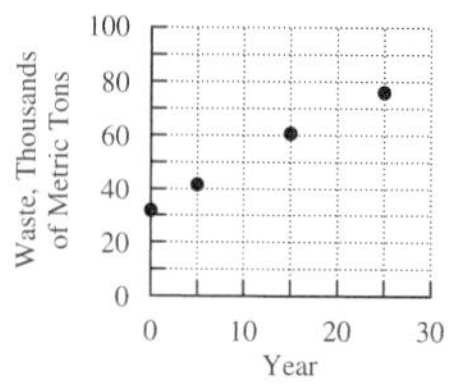

(c) $y = 1.76x + 32$ (d) $y = 49.6$, so there will be about 49,600 metric tons in 2005.
46. (a) $y = -.96x + 5.6$
(b) .8 million or 800,000 tons
(c) during 2001 (Since $.83333(12) = 10$, the landfill capacity will be 0 in October.) **47.** (a) \$400
(b) \$.25 (c) $y = .25x + 400$
(d) \$425 (e) 1500

Heavy Metal Nuclear Waste

Year x	Waste y
1995	32
2000	42
2010	61
2020	76

Let $x = 0$ represent 1995, $x = 5$ represent 2000 (since $2000 - 1995 = 5$), and so on. Refer to Example 5 and do the following.

(a) For 1995, the ordered pair is (0, 32). Write ordered pairs for the data for the other years given in the table.
(b) Plot the ordered pairs (x, y). Do the points lie approximately in a line?
(c) Use the ordered pairs (0, 32) and (25, 76) to find the equation of a line that approximates the other ordered pairs. Use the form $y = mx + b$.
(d) Use the equation from part (c) to estimate the amount of nuclear waste in 2005. (*Hint:* What is the value of x for 2005?)

46. The Waste Management and Recycling Division of Sacramento County is responsible for managing the disposal of solid waste, including the operation of a landfill. The graph shows the remaining capacity (in millions of tons) at the county landfill at several times during the past few years. These points appear to lie close to a line.

(a) Write an equation in slope-intercept form of the line shown through the points (0, 5.6) and (2.5, 3.2).
(b) Use the equation from part (a) to estimate the remaining capacity of the landfill in 2001.
(c) During what year will the capacity of the landfill be used up? (*Hint:* Find the point where $y = 0$—that is, the x-intercept.) Estimate the month when it will occur.

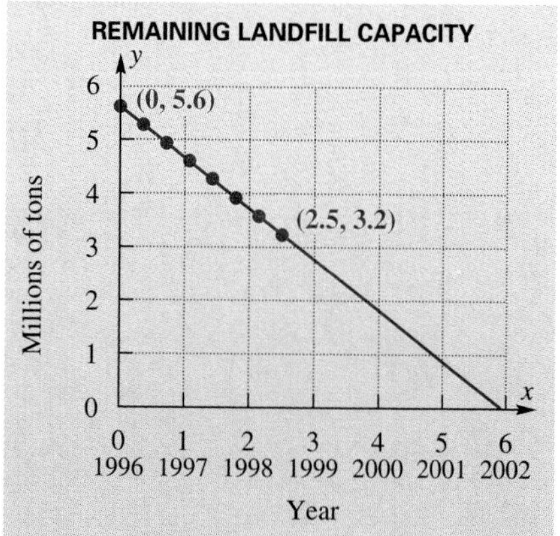

Source: Waste Management and Recycling Division, Sacramento County Public Works Agency, 1998 Report.

The cost to produce x items, in some cases, is expressed as $y = mx + b$. *The number b gives the* fixed cost *(that is, the cost that is the same no matter how many items are produced), and the number m is the* variable cost *(the cost to produce an additional item). Write the cost equation for each of the following, and answer the questions.*

47. It costs \$400 to start up a business selling ice cream cones. Each cone costs \$.25 to produce.
(a) What is the fixed cost?
(b) What is the variable cost?
(c) Write the cost equation.
(d) What will be the cost to produce 100 cones, based on the cost equation?
(e) How many cones will be produced if total cost is \$775?

48. (a) \$2000 (b) \$.02
(c) $y = .02x + 2000$
(d) \$2200 (e) 30,000
49. (a) (1, 822.3), (2, 774.9)
(b) $y = -47.4x + 869.7$
(c) The slope represents the change in sales from 1996 to 1997. The negative slope indicates that sales *decreased*.
50. (a) (1, 586.8), (2, 733.6)
(b) $y = 146.8x + 440$ (c) The slope represents the change in sales from 1996 to 1997. Here, the positive value shows that sales increased.

48. It costs \$2000 to purchase a copier and each copy costs \$.02 to make.
 (a) What is the fixed cost?
 (b) What is the variable cost?
 (c) Write the cost equation.
 (d) What will be the cost to produce 10,000 copies, based on the cost equation?
 (e) How many copies will be produced if total cost is \$2600?

▲ *The sales of a company for a given year can be written as an ordered pair in which the first number, x, gives the year and the second number, y, gives the sales for that year. If the sales increase at the same rate each year, a linear equation for sales (in millions of dollars) can be found. Merck & Co. sales for 1996 (year 1) and 1997 (year 2) are given for two different products.*

Antibiotics		**Vaccines/Biologicals**	
Year x	*Sales* y	*Year* x	*Sales* y
1	822.3	1	586.8
2	774.9	2	733.6

Source: Merck & Co., Inc. 1997 Annual Report.

49. **(a)** Write two ordered pairs in the form (year, sales) for antibiotics.
 (b) Write the sales equation in the form $y = mx + b$.
 (c) What does the slope, m, represent in this situation?

50. **(a)** Write two ordered pairs in the form (year, sales) for vaccines/biologicals.
 (b) Write the sales equation in the form $y = mx + b$.
 (c) What does the slope, m, represent in this situation?

51. $y = -3x + 6$
52. $y = \frac{3}{4}x + 1$

TECHNOLOGY INSIGHTS (EXERCISES 51–54)

▲ Two views of the same line are shown on a calculator screen.

Use the displays at the bottom of the screen to find an equation of the form $y = mx + b$ for each line.

51.

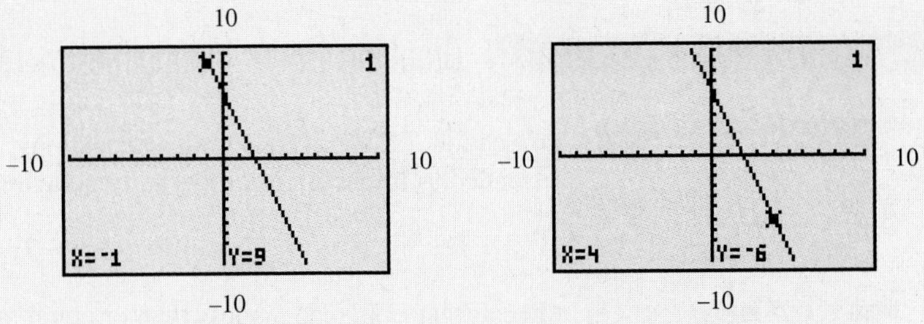

52.

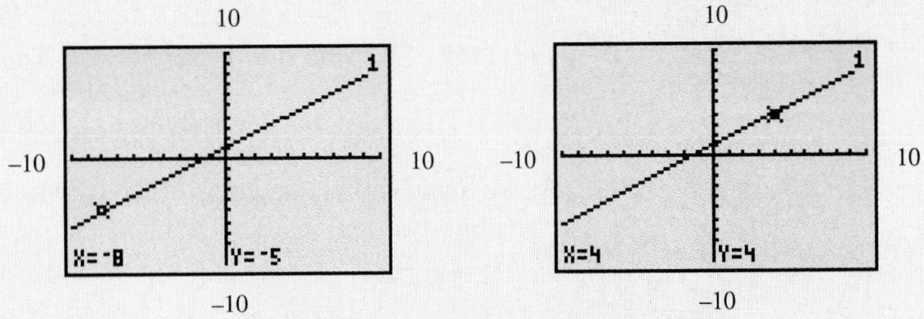

(continued)

53. $Y_1 = \frac{3}{4}x + 1$

54. $Y_1 = -4x - 2$

A table of points, generated by a graphing calculator, is shown for a line Y_1.

Use any two points to find the equation of each line in slope-intercept form.

53.

X	Y1
-2	-.5
-1	.25
0	1
1	1.75
2	2.5
3	3.25
4	4

X= -2

54.

X	Y1
-4	14
-3	10
-2	6
-1	2
0	-2
1	-6
2	-10

X= -4

55. (0, 32); (100, 212)

56. $\frac{9}{5}$ 57. $F - 32 = \frac{9}{5}(C - 0)$ or $F - 212 = \frac{9}{5}(C - 100)$

58. $F = \frac{9}{5}C + 32$

59. $C = \frac{5}{9}(F - 32)$ or $C = \frac{5}{9}F - \frac{160}{9}$ 60. 86°

61. 10° 62. −40°

RELATING CONCEPTS (EXERCISES 55–62)

If we think of ordered pairs of the form (C, F), then the two most common methods of measuring temperature, Celsius and Fahrenheit, can be related as follows: When $C = 0$, $F = 32$, and when $C = 100$, $F = 212$. ***Work Exercises 55–62 in order.***

55. Write two ordered pairs relating these two temperature scales.

56. Find the slope of the line joining the two points.

57. Use the point-slope form to find an equation of the line. (Your variables should be C and F rather than x and y.)

58. Write an equation for F in terms of C.

59. Use the equation from Exercise 58 to write an equation for C in terms of F.

60. Use the equation from Exercise 58 to find the Fahrenheit temperature when $C = 30$.

61. Use the equation from Exercise 59 to find the Celsius temperature when $F = 50$.

62. For what temperature is $F = C$? (*Hint:* Substitute one variable for the other in either equation from Exercise 58 or 59.)

Did you make the connection that Celsius and Fahrenheit temperatures are related by a linear equation?

7.2 Graphing Linear Inequalities in Two Variables

OBJECTIVES

1. Graph ≤ or ≥ linear inequalities.
2. Graph < or > linear inequalities.
3. Graph inequalities with a boundary through the origin.

In Section 3.2 we graphed linear equations, such as $2x + 3y = 6$. Now this work is extended to **linear inequalities in two variables,** such as

$$2x + 3y \le 6.$$

(Recall that ≤ is read "is less than or equal to.")

OBJECTIVE 1 Graph ≤ or ≥ linear inequalities. The inequality $2x + 3y \le 6$ means that

$$2x + 3y < 6 \quad \text{or} \quad 2x + 3y = 6.$$

As we found earlier, the graph of $2x + 3y = 6$ is a line. This **boundary line** divides the plane into two regions. The graph of the solutions of the inequality $2x + 3y < 6$ will

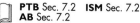

FOR EXTRA HELP

SSG Sec. 7.2
SSM Sec. 7.2
Pass the Test Software
InterAct Math Tutorial Software
Video 12

include only *one* of these regions. We find the required region by solving the given inequality for *y*.

$$2x + 3y \leq 6$$
$$3y \leq -2x + 6 \qquad \text{Subtract } 2x.$$
$$y \leq -\frac{2}{3}x + 2 \qquad \text{Divide by 3.}$$

From this last statement, ordered pairs in which *y is less than or equal to* $-\frac{2}{3}x + 2$ will be solutions to the inequality. The ordered pairs in which *y* is equal to $-\frac{2}{3}x + 2$ are on the boundary line, so the pairs in which *y is less than* $-\frac{2}{3}x + 2$ will be *below* that line. To indicate the solution, shade the region below the line, as in Figure 4. The shaded region, along with the line, is the desired graph.

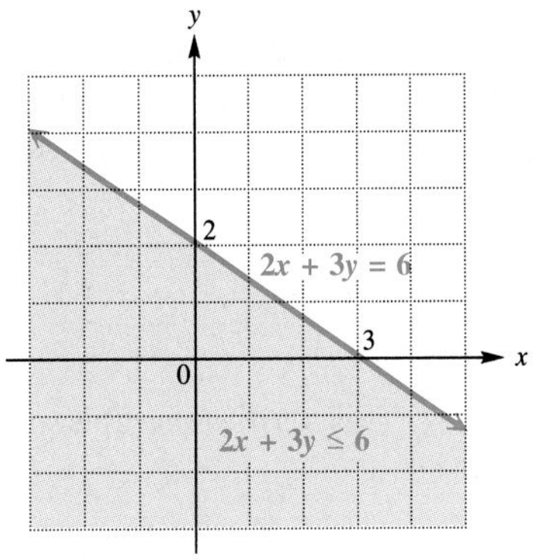

Figure 4

As an alternative method, a test point gives a quick way to find the correct region to shade. We choose any point *not* on the line. Because (0, 0) is easy to substitute into an inequality, it is often a good choice, and we use it here. Substitute 0 for *x* and 0 for *y* in the given inequality to see whether the resulting statement is true or false. In the example above,

$$2x + 3y \leq 6 \qquad \text{Original inequality}$$
$$2(0) + 3(0) \leq 6 \quad ? \qquad \text{Let } x = 0 \text{ and } y = 0.$$
$$0 + 0 \leq 6 \quad ?$$
$$0 \leq 6. \qquad \text{True}$$

Since the last statement is true, shade the region that includes the test point (0, 0). This agrees with the result shown in Figure 4.

CAUTION If the boundary line passes through the origin, do not use (0, 0) as the test point. In that case, the point (0, 0) lies on the line. Instead, choose any ordered pair that does not lie on the line.

OBJECTIVE 2 Graph < or > linear inequalities. Inequalities that do not include the equals sign are graphed in a similar way.

CHALKBOARD EXAMPLE

Graph the inequality $3x + 5y > 15$.

Answer:

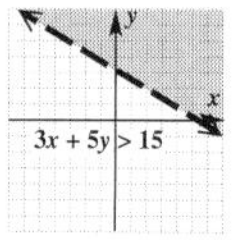

TEACHING TIP Emphasize that the coordinates of any point in the shaded region satisfy the inequality, while those on the boundary do not. Select ordered pairs inside and outside the shaded region and substitute them in the inequality to show this.

EXAMPLE 1 Graphing a Linear Inequality

Graph the inequality $x - y > 5$.

This inequality does not include the equals sign. Therefore, the points on the line $x - y = 5$ do *not* belong to the graph. However, the line still serves as a boundary for two regions, one of which satisfies the inequality. To graph the inequality, first graph the equation $x - y = 5$. We use a dashed line to show that the points on the line are *not* solutions of the inequality $x - y > 5$. Now we choose a test point to see which side of the line satisfies the inequality. We choose $(1, -2)$ this time.

$x - y > 5$		Original inequality
$1 - (-2) > 5$	?	Let $x = 1$ and $y = -2$.
$3 > 5$		False

Since $3 > 5$ is false, the graph of the inequality is *not* the region that contains $(1, -2)$. We shade the other region, as shown in Figure 5. This shaded region is the required graph. To check that the correct region is shaded, select a test point in the shaded region and substitute for x and y in the inequality $x - y > 5$. For example, we use $(4, -3)$ from the shaded region as follows.

$x - y > 5$		
$4 - (-3) > 5$	?	
$7 > 5$		True

This verifies that the correct region was shaded in Figure 5.

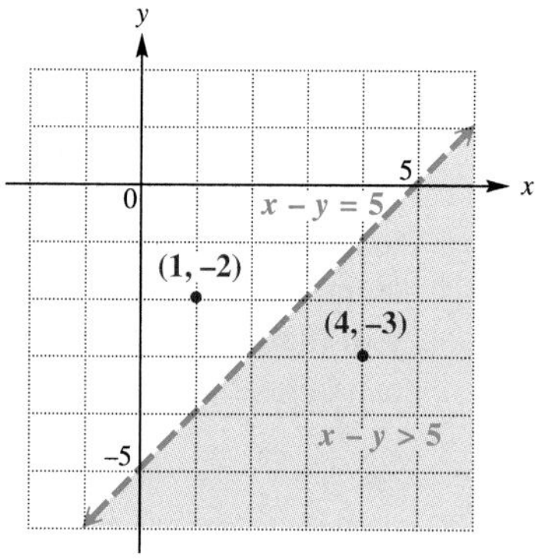

Figure 5

A summary of the steps used to graph a linear inequality in two variables follows.

Graphing a Linear Inequality

Step 1 **Graph the boundary.** Graph the line that is the boundary of the region. Use the methods of Section 3.2. Draw a solid line if the inequality has $\leq$ or $\geq$; draw a dashed line if the inequality has $<$ or $>$.

Step 2 **Shade the appropriate side.** Use any point not on the line as a test point. Substitute for x and y in the *inequality.* If a true statement results, shade the side containing the test point. If a false statement results, shade the other side.

CHALKBOARD EXAMPLE

Graph the inequality $y < 4$.

Answer:

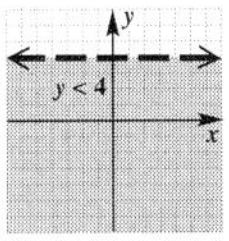

EXAMPLE 2 Graphing a Linear Inequality with a Vertical Boundary Line

Graph the inequality $x \leq 3$.

First, we graph $x = 3$, a vertical line going through the point (3, 0). We use a solid line (why?) and choose (0, 0) as a test point.

$x \leq 3$		Original inequality
$0 \leq 3$	?	Let $x = 0$.
$0 \leq 3$		True

Since $0 \leq 3$ is true, we shade the region containing (0, 0), as in Figure 6.

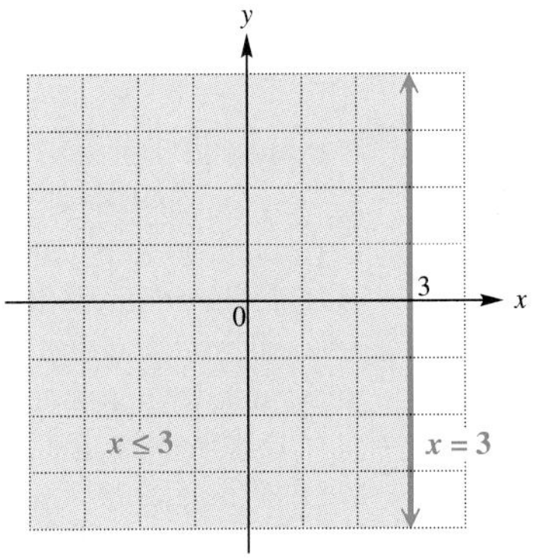

Figure 6

TEACHING TIP Emphasize that the coordinates of any point in the shaded region or on the line satisfy the inequality.

OBJECTIVE 3 **Graph inequalities with a boundary through the origin.** Remember that when we graph an inequality having a boundary line that goes through the origin, (0, 0) cannot be used as a test point.

CHALKBOARD EXAMPLE

Graph the inequality $x \geq -3y$.

Answer:

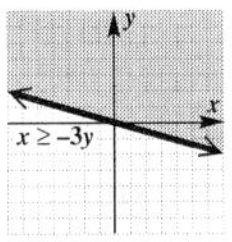

EXAMPLE 3 Graphing a Linear Inequality with a Boundary Line through the Origin

Graph the inequality $x \leq 2y$.

Begin by graphing $x = 2y$. Some ordered pairs that can be used to graph this line are (0, 0), (6, 3), and (4, 2). Draw a solid line. Since (0, 0) is on the line $x = 2y$, it cannot be used as a test point. We must choose a test point off the line. For example, choose (1, 3), which is not on the line.

$x \leq 2y$		Original inequality
$1 \leq 2(3)$	?	Let $x = 1$ and $y = 3$.
$1 \leq 6$		True

Since $1 \leq 6$ is true, we shade the side of the graph containing the test point (1, 3). (See Figure 7 on the next page.)

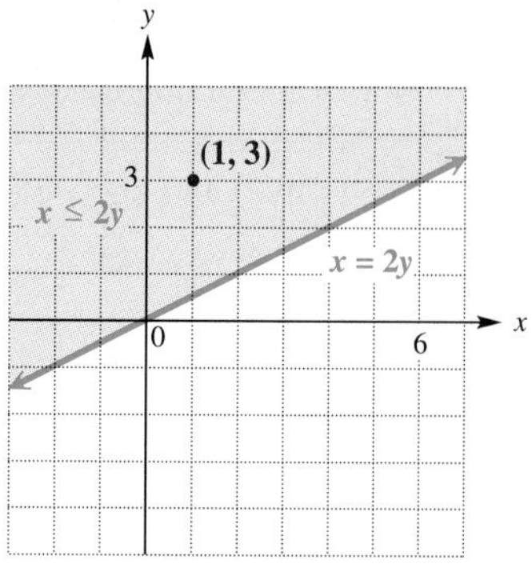

Figure 7

1. Answers will vary.
2. (a) $(5, \infty)$ (b) $(-\infty, 5)$ (c) $\left[3, \frac{11}{3}\right)$

CONNECTIONS

Graphing calculators have a feature that allows us to shade regions in the plane, so they can be used to graph a linear inequality in two variables. Since all calculators are different, consult the manual for directions for your calculator. The calculator will not draw the graph as a dashed line, so it is still necessary to understand what is and is not included in the solution set.

To solve the inequalities in one variable, $-2x + 4 > 0$ and $-2x + 4 < 0$, we use the graph of $y = -2x + 4$, shown below. For $y = -2x + 4 > 0$, we want the values of x such that $y > 0$, so that the line is *above* the x-axis. From the graph, we see that this is the case for $x < 2$. Thus, the solution set is $(-\infty, 2)$. Similarly, the solution set of $y = -2x + 4 < 0$ is $(2, \infty)$, because the line is *below* the x-axis when $x > 2$.

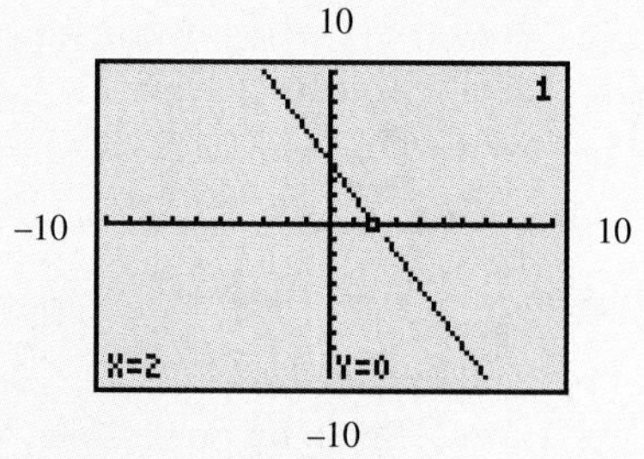

FOR DISCUSSION OR WRITING

1. Discuss the pros and cons of using a calculator to solve a linear inequality in one variable.
2. Use a graphing calculator to solve the following inequalities in one variable from Section 2.7.
   (a) $3x + 2 - 5 > -x + 7 + 2x$ (Example 4)
   (b) $3x + 2 - 5 < -x + 7 + 2x$ (Use the result from part (a).)
   (c) $4 \le 3x - 5 < 6$ (Example 7)

7.2 EXERCISES

1. $>$ 2. $<$ 3. $\geq$ 4. $\leq$
5. $\leq$ 6. $\geq$ 7. false
8. true 9. true 10. true

11.
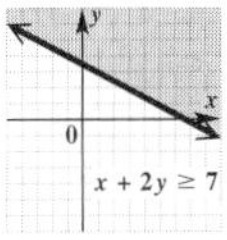

12.
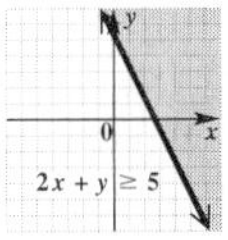

13.
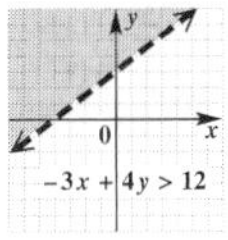

The following statements were taken from recent articles in newspapers or magazines. Each includes a phrase that can be symbolized with one of the inequality symbols $<$, $\leq$, $>$, or $\geq$. In Exercises 1–6, complete each sentence with the appropriate inequality symbol.

1. The landslides of November 14 and January 19 caused more than \$1 million in damage to the condominium complex. (*Source:* Bunky Bakutis, "Owners Thank Makaha Flood Heroes," *The Honolulu Advertiser,* January 28, 1997, B3.)

Damages from landslides were ______ \$1 million.

2. In 1989, the U.S. Labor Department's Wage and Hour Division had 1000 compliance officers. Now it has fewer than 800. (*Source:* Michael Doyle, "3-Nation Meeting to Assess Battle against Child Labor," *The Sacramento Bee,* February 24, 1997.)

The department has ______ 800 officers.

3. More than \$3 billion in student loans were at least 60 days past due in 1994. (*Source:* Gilbert Chan, "Lessons on Loans," *The Sacramento Bee,* February 13, 1997.)

Student loans were ______ 60 days past due.

4. The committees supervise housing projects with up to 100 families. (*Source:* "Doug Is It," *Fortune,* May 25, 1998, p. 82.)

The housing projects have ______ 100 families.

5. Internal combustion engines can reach fuel efficiencies as high as 40 percent at a constant speed. (*Source:* "The Case for Electric Vehicles," *Scientific American,* November 1996.)

Fuel efficiencies are ______ 40 percent.

6. A car must travel through the intersection at no less than 15 miles an hour before the driver receives a ticket. (*Source:* "The Case for Electric Vehicles," *Scientific American,* November 1996.)

The speed must be ______ 15 miles per hour.

Answer true or false to each of the following.

7. The point (4, 0) lies on the graph of $3x - 4y < 12$.

8. The point (4, 0) lies on the graph of $3x - 4y \leq 12$.

9. Both points (4, 1) and (0, 0) lie on the graph of $3x - 2y \geq 0$.

10. The graph of $y > x$ does not contain points in quadrant IV.

In Exercises 11–16, the straight line boundary has been drawn. Complete the graph by shading the correct region. See Examples 1–3.

11. $x + 2y \geq 7$

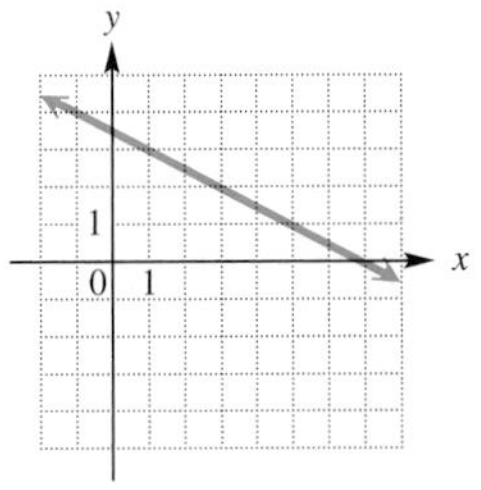

12. $2x + y \geq 5$

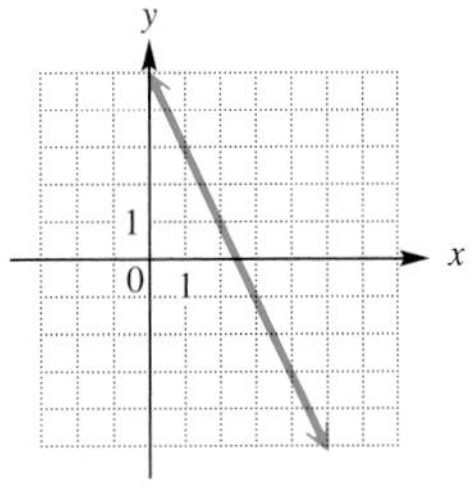

13. $-3x + 4y > 12$

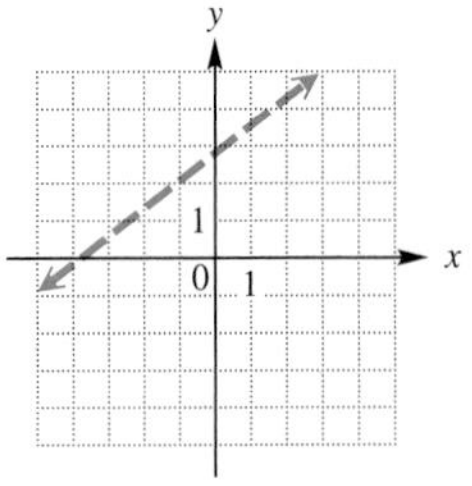

14. 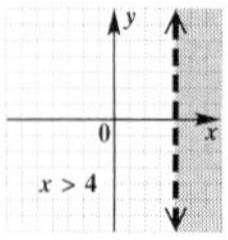

15. 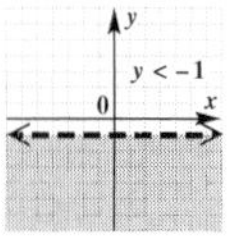

16. 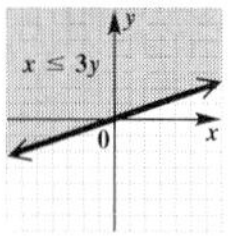

17. Graph the boundary line that satisfies the corresponding equation. Make the line dashed if the inequality is $<$ or $>$. Make it solid if the inequality is $\leq$ or $\geq$. Use a test point to decide whether the inequality is satisfied by points below or above the line. Shade the region determined by the test point. 18. If the boundary line goes through the origin, which is the point (0, 0), that point is on the line. Therefore, it will not indicate which side of the boundary contains points that satisfy the inequality.

19. 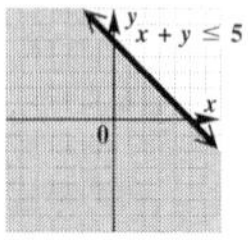

20. 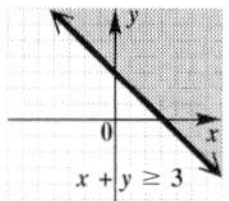

21. 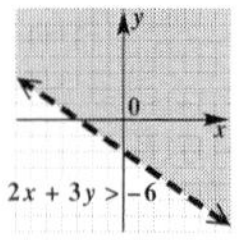

22. 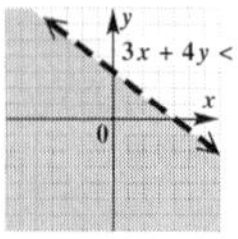

23. 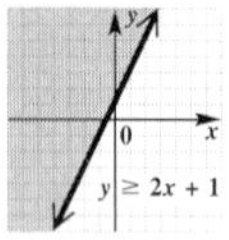

24. 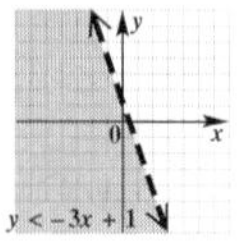

14. $x > 4$ **15.** $y < -1$ **16.** $x \leq 3y$

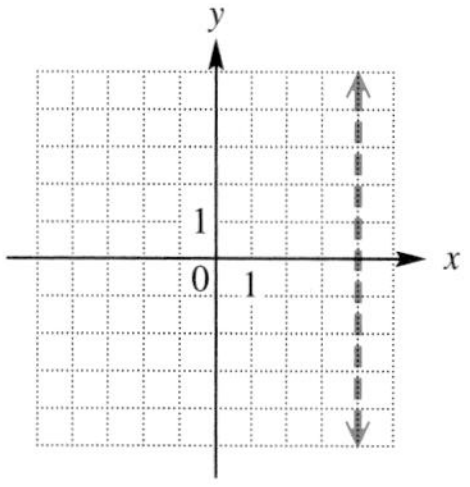

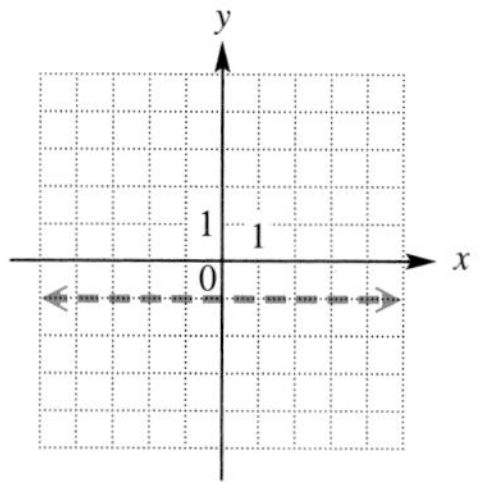

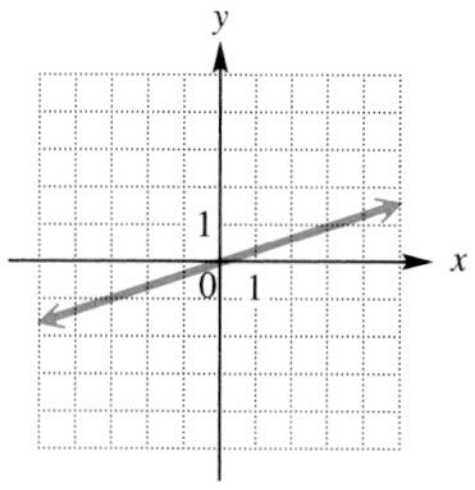

17. Explain in your own words how to graph a linear inequality in two variables.

18. Explain why the point (0, 0) is not an appropriate choice for a test point when graphing an inequality whose boundary goes through the origin.

Graph each linear inequality. See Examples 1–3.

19. $x + y \leq 5$	**20.** $x + y \geq 3$	**21.** $2x + 3y > -6$	**22.** $3x + 4y < 12$
23. $y \geq 2x + 1$	**24.** $y < -3x + 1$	**25.** $x \leq -2$	**26.** $x \geq 1$
27. $y < 5$	**28.** $y < -3$	**29.** $y \geq 4x$	**30.** $y \leq 2x$

31. Explain why the graph of $y > x$ cannot lie in quadrant IV.

32. Explain why the graph of $y < x$ cannot lie in quadrant II.

TECHNOLOGY INSIGHTS (EXERCISES 33–36)

A calculator was used to generate the shaded graphs in choices A–D.

A.

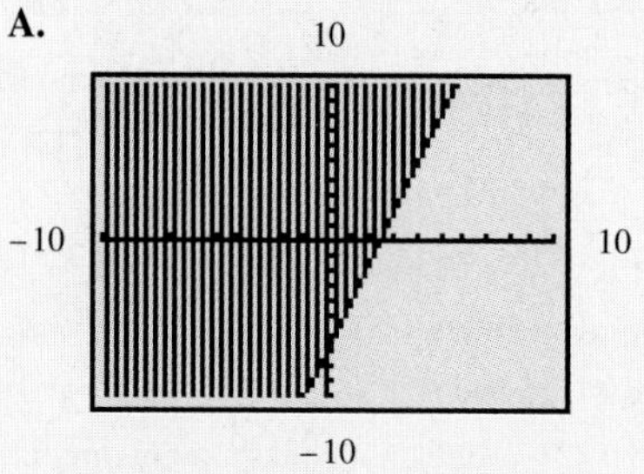

B.

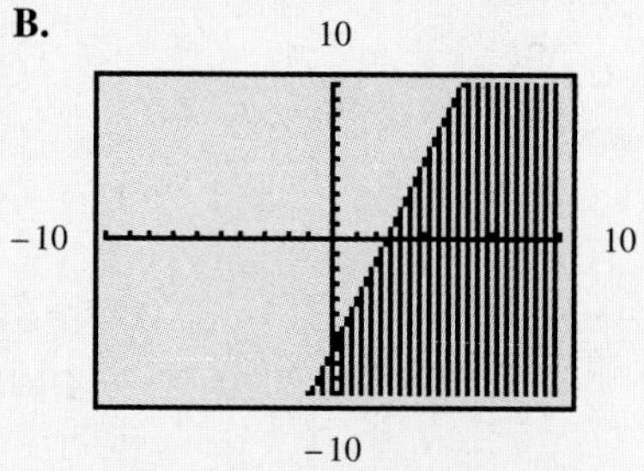

C.

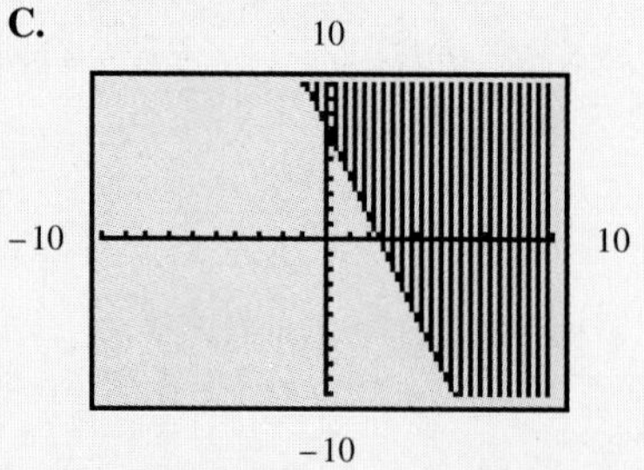

D.

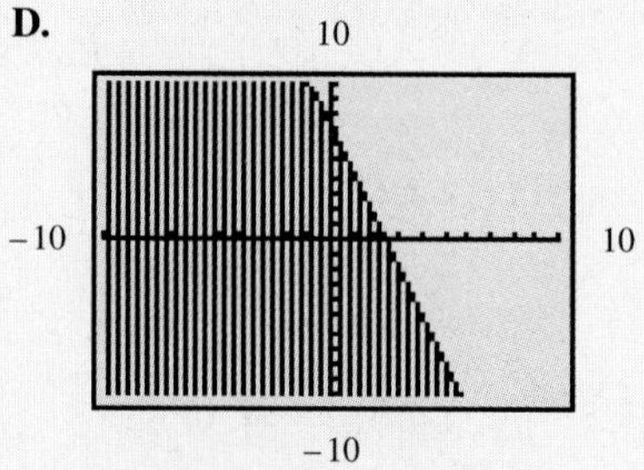

Match each inequality with the appropriate choice.

33. $y \geq 3x - 7$ **34.** $y \leq 3x - 7$ **35.** $y \geq -3x + 7$ **36.** $y \leq -3x + 7$

▲ *For the given information (a) graph the inequality. Here $x \geq 0$ and $y \geq 0$, so graph only the part of the inequality in quadrant I. (b) Give some ordered pairs that satisfy the inequality.*

37. A company will ship x units of merchandise to outlet I and y units of merchandise to outlet II. The company must ship a total of at least 500 units to these two outlets. This can be expressed by writing

$$x + y \geq 500.$$

38. A toy manufacturer makes stuffed bears and geese. It takes 20 minutes to sew a bear and 30 minutes to sew a goose. There is a total of 480 minutes of sewing time available to make x bears and y geese. These restrictions lead to the inequality

$$20x + 30y \leq 480.$$

39. The number of civilian federal employees (in thousands) is approximated by $y = -51.1x + 3153$, where x represents the year. Here, $x = 1$ corresponds to 1991, $x = 2$ corresponds to 1992, and so on. Thus, the total number of federal employees (including noncivilians) in each year is described by the inequality

$$y \geq -51.1x + 3153.$$

(*Hint:* Use tick marks of 1 to 10 on the x-axis and 2500 to 3500, at intervals of 200, on the y-axis.) (*Source:* U.S. Bureau of the Census.)

40. The equation $y = 57.1x + 1798$ approximates the number of free lunches served in public schools each year in millions. The year 1992 is represented by $x = 2$, 1993 is represented by $x = 3$, and so on. Free school breakfasts are also served in some (but not all) of the same schools. The number of children receiving both meals is described by the inequality

$$y \leq 57.1x + 1798.$$

(*Hint:* Use tick marks of 1 to 10 on the x-axis and 1800, 1900, 2000, 2100, 2200, 2300, and 2400 on the y-axis.) (*Source:* U.S. Department of Agriculture, Food and Consumer Service.)

TECHNOLOGY INSIGHTS (EXERCISES 41–44)

The calculator graph at the top of the next page is that of $y = -2(x + 1) + 4(x - 1)$. If we solve the equation

$$-2(x + 1) + 4(x - 1) = 0,$$

we obtain the solution 3. Because the graph of the line lies *above* the x-axis for values of x greater than 3, the solution set of

"y is greater than 0" means *above* the x-axis
↓

$$-2(x + 1) + 4(x - 1) > 0$$

is the set of numbers greater than 3. On the other hand, because the graph of the line lies *below* the x-axis for values of x less than 3, the solution set of

"y is less than 0" means *below* the x-axis
↓

$$-2(x + 1) + 4(x - 1) < 0$$

(continued)

25.
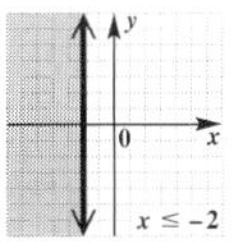

26.
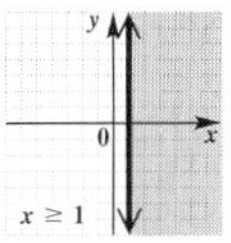

27.
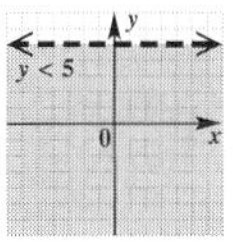

28.
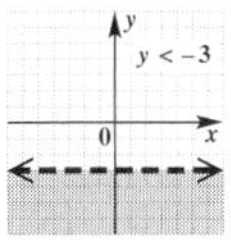

29.
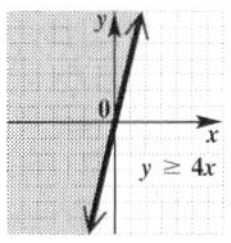

30.
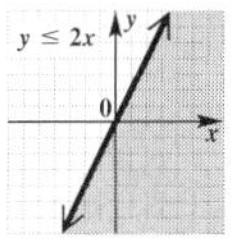

31. Every point in quadrant IV has a positive x-value and a negative y-value. Substituting into $y > x$ would imply that a negative number is greater than a positive number, which is always false. Thus, the graph of $y > x$ cannot lie in quadrant IV.
32. In quadrant II, every x-value is negative and every y-value is positive. The inequality $y < x$ is always false if y is positive and x is negative. Thus, the graph of $y < x$ cannot lie in quadrant II.
33. A **34.** B **35.** C
36. D
In part (b) of Exercises 37–40, other answers are possible.
37. (a)
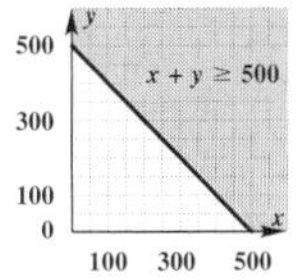

(b) (500, 0), (200, 400)

38. (a)

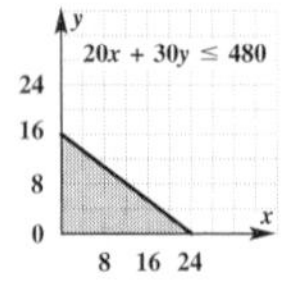

(b) (16, 1), (20, 2)

39. (a)

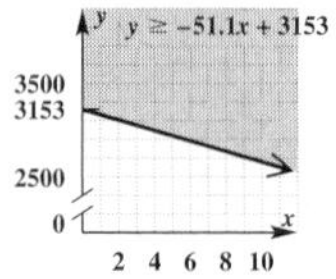

(b) (0, 3153), (2, 3050.8), (10, 2642)

40. (a)

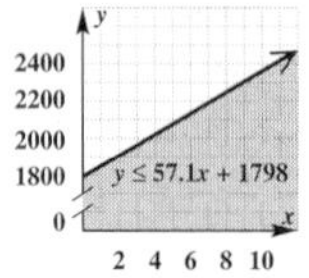

(b) (0, 1798), (5, 2083.5), (10, 2369)

41. (a) $\{-2\}$ (b) $(-2, \infty)$ (c) $(-\infty, -2)$

42. (a) $\{5\}$ (b) $(5, \infty)$ (c) $(-\infty, 5)$

43. (a) $\{-4\}$ (b) $(-\infty, -4)$ (c) $(-4, \infty)$

44. (a) $\{-6\}$ (b) $(-\infty, -6)$ (c) $(-6, \infty)$

TECHNOLOGY INSIGHTS (EXERCISES 41–44) (CONTINUED)

is the set of numbers less than 3.

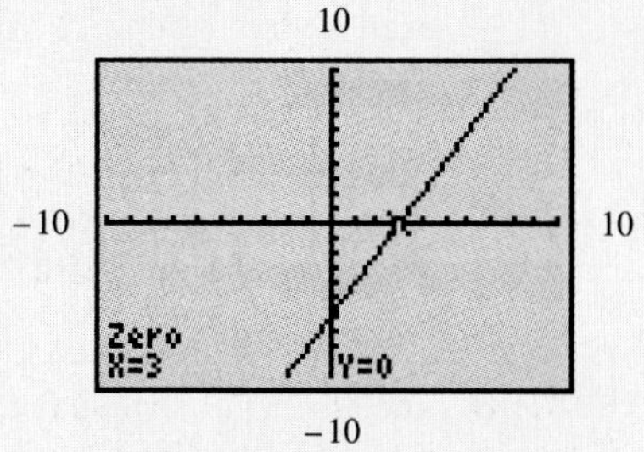

Use the graph to solve the equation and inequalities listed in parts (a)–(c) below the graph.

41. $y = -3x + 2(2x + 1)$

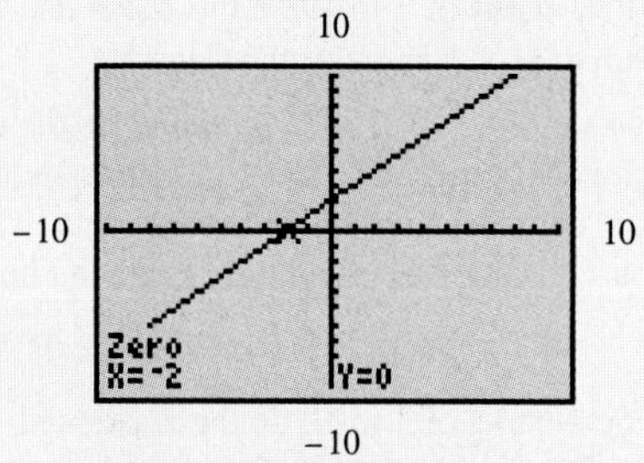

(a) $-3x + 2(2x + 1) = 0$
(b) $-3x + 2(2x + 1) > 0$
(c) $-3x + 2(2x + 1) < 0$

42. $y = -2(x + 1) + 4(x - 2)$

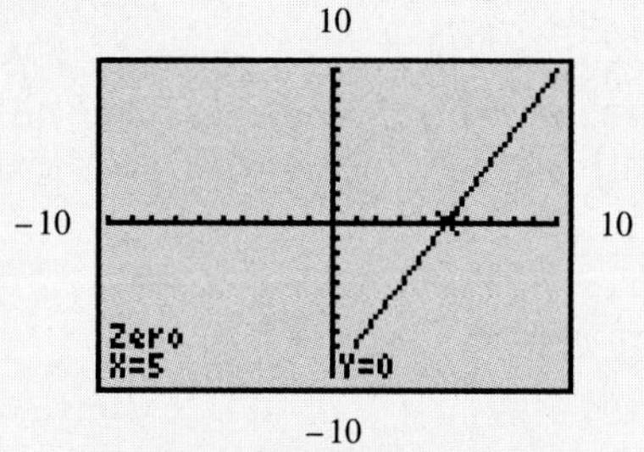

(a) $-2(x + 1) + 4(x - 2) = 0$
(b) $-2(x + 1) + 4(x - 2) > 0$
(c) $-2(x + 1) + 4(x - 2) < 0$

43. $y = 4x - (6x + 8)$

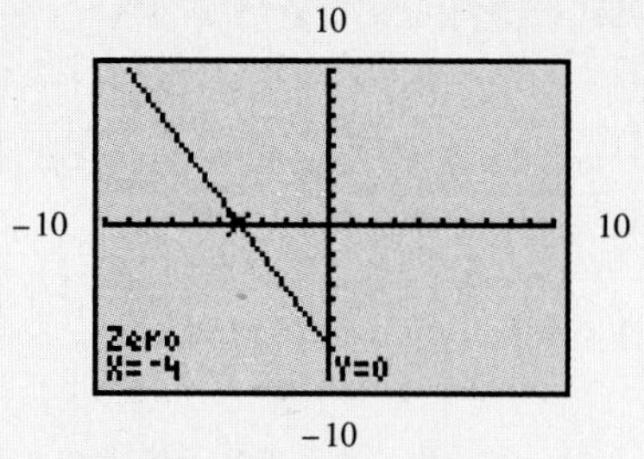

(a) $4x - (6x + 8) = 0$
(b) $4x - (6x + 8) > 0$
(c) $4x - (6x + 8) < 0$

44. $y = 6x - 3(3x + 6)$

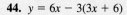

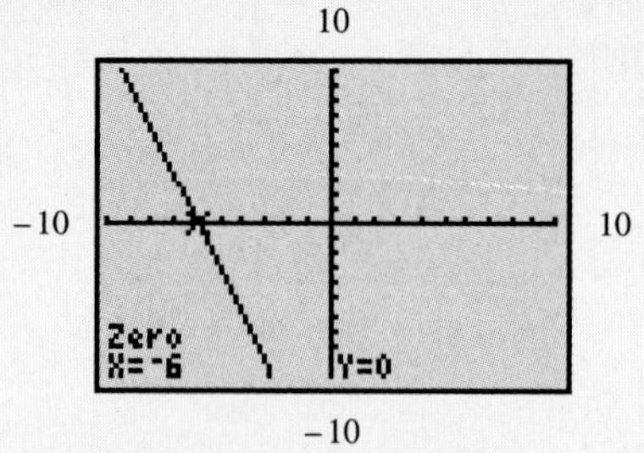

(a) $6x - 3(3x + 6) = 0$
(b) $6x - 3(3x + 6) > 0$
(c) $6x - 3(3x + 6) < 0$

7.3 Functions

OBJECTIVES

1. Understand the definition of a relation.
2. Understand the definition of a function.
3. Decide whether an equation defines a function.
4. Find domains and ranges.
5. Use $f(x)$ notation.
6. Apply the function concept in an application.

FOR EXTRA HELP

SSG Sec. 7.3
SSM Sec. 7.3
Pass the Test Software
InterAct Math Tutorial Software
Video 12

CHALKBOARD EXAMPLE

Use ordered pairs to define the relation.

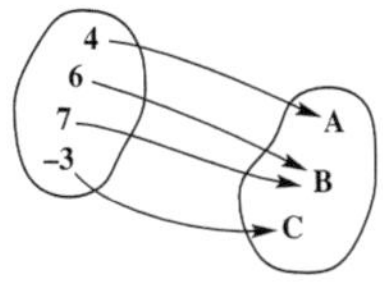

Answer:
{(4, A), (6, B), (7, B), (−3, C)}

In Section 3.1, the equation $y = 25.59x - 50{,}280$ was used to find the number of new business incorporations in thousands. Choosing values for x and using the equation to find the corresponding values of y led to a set of ordered pairs (x, y). In each ordered pair, y (the number of thousands of incorporations) was *related* to x (the number of years since 1960) by the equation.

OBJECTIVE 1 Understand the definition of a relation. In an ordered pair (x, y), x and y are called the **components** of the ordered pair. Any set of ordered pairs is called a **relation.** The set of all first components in the ordered pairs of a relation is the **domain** of the relation, and the set of all second components in the ordered pairs is the **range** of the relation.

EXAMPLE 1 Using Ordered Pairs to Define a Relation

(a) The relation {(0, 1), (2, 5), (3, 8), (4, 2)} has domain {0, 2, 3, 4} and range {1, 2, 5, 8}. The correspondence between the elements of the domain and the elements of the range is shown in Figure 8.

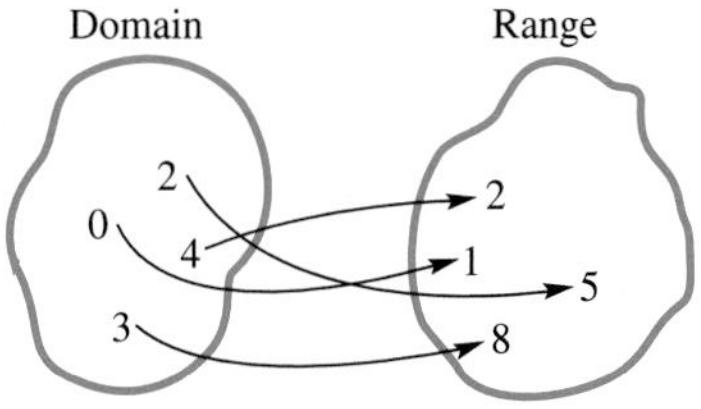

Figure 8

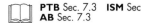

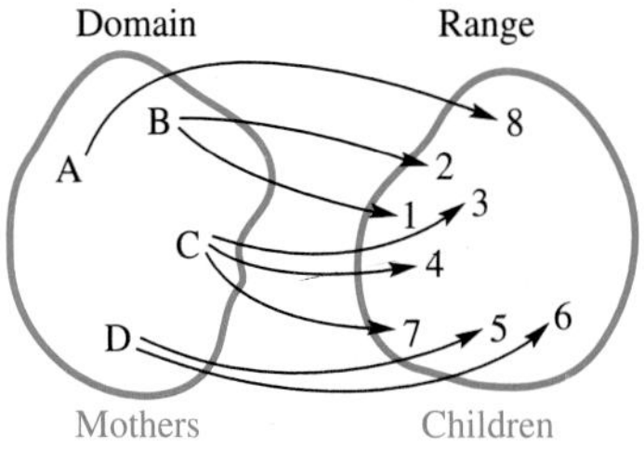

Figure 9

(b) Figure 9 shows a relation where the domain elements represent four mothers and the range elements represent eight children. The figure shows the correspondence between each mother and her children. The relation also could be written as the set of ordered pairs {(A, 8), (B, 1), (B, 2), (C, 3), (C, 4), (C, 7), (D, 5), (D, 6)}.

OBJECTIVE 2 Understand the definition of a function. In Section 3.2, we discussed a special type of relation, called a *function.*

Function

A **function** is a set of ordered pairs in which each first component corresponds to exactly one second component.

TEACHING TIP Note that we are dealing with a set in which the individual elements are ordered pairs.

In Chapter 3, we called the first component the *input* and the second component the *output,* because the second component was determined by the first. By this definition, the relation in Example 1(a) is a function. However, the relation in Example 1(b) is *not* a function, because at least one first component (mother) corresponds to more than one second component (child). Notice that if the ordered pairs in Example 1(b) are reversed, with the child as the first component and the mother as the second component, the result *is* a function.

TEACHING TIP It might be helpful to briefly review with students the material in Section 3.2, Objective 5, on functions.

INSTRUCTOR'S RESOURCES

PTB Sec. 7.3 **ISM** Sec. 7.3
AB Sec. 7.3
TEST GENERATOR
WORLD WIDE WEB
www.LialAlgebra.com

The simple relations given here were defined by listing the ordered pairs or by showing the correspondence with a figure. As we have seen, most useful functions have an infinite number of ordered pairs and are usually defined with an equation that tells how to get the second component given the first. We have been using equations with x and y as the variables, where x represents the first component (input) and y the second component (output) in the ordered pairs.

Here are some everyday examples of functions.

1. The **cost** y in dollars charged by an express mail company is a function of the **weight in pounds** x determined by the equation $y = 1.5(x - 1) + 9$.
2. In one state, the sales tax is 6% of the price of an item. The **tax** y on a particular item is a function of the **price** x, so that $y = .06x$.
3. The **distance** d traveled by a car moving at a constant speed of 45 miles per hour is a function of the **time** t. Thus, $d = 45t$.

OBJECTIVE 3 Decide whether an equation defines a function. Given a graph of an equation, the definition of a function can be used to decide whether the graph represents a function or not. By the definition of a function, each value of x must lead to exactly one value of y. In Figure 10 the indicated value of x leads to two values of y, so this graph is not the graph of a function. A vertical line can be drawn that intersects this graph in more than one point.

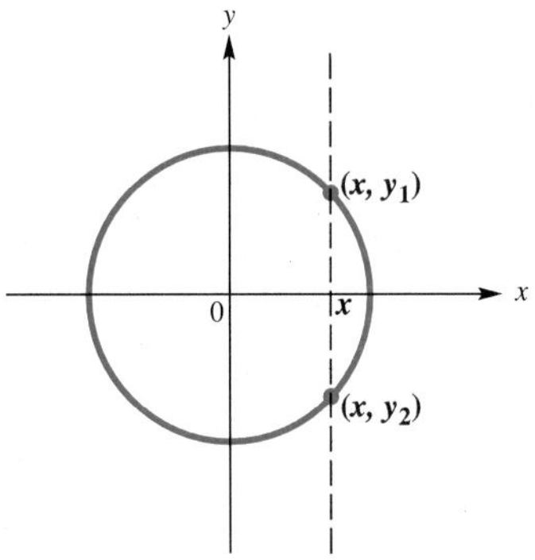

Figure 10

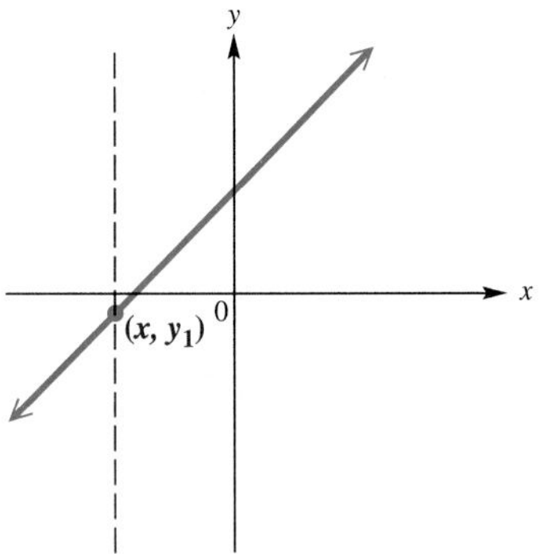

Figure 11

On the other hand, in Figure 11 any vertical line will intersect the graph in no more than one point. Because of this, the graph in Figure 11 is the graph of a function. This idea gives the **vertical line test** for a function.

Vertical Line Test

If a vertical line intersects a graph in more than one point, the graph is not the graph of a function.

As Figure 11 suggests, any nonvertical line is the graph of a function. For this reason, any linear equation of the form $y = mx + b$ defines a function. (Recall that a vertical line has undefined slope.)

CHALKBOARD EXAMPLE

Determine whether the relation is a function.

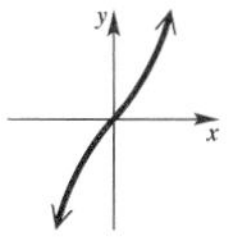

Answer: Yes, it is a function.

EXAMPLE 2 Deciding Whether a Relation Defines a Function

Decide whether or not the relations defined as follows are functions.

(a) $y = 2x - 9$

This linear equation is in the form $y = mx + b$. Since the graph of this equation is a line that is not vertical, the equation defines a function.

(b)

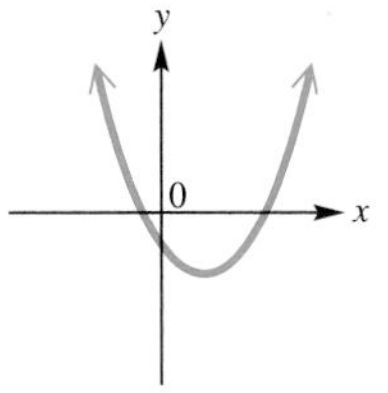

Use the vertical line test. Any vertical line will cross the graph just once, so this is the graph of a function.

(c)

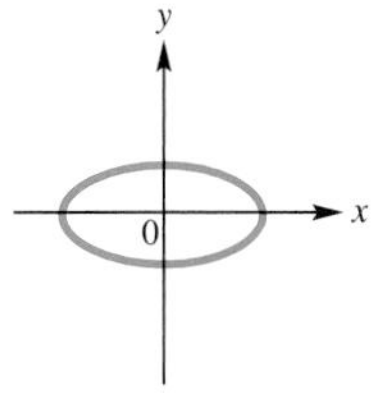

The vertical line test shows that this graph is not the graph of a function; a vertical line could cross the graph twice.

(d) $x = 4$

The graph of $x = 4$ is a vertical line, so the equation does not define a function.

(e) $x = y^2$

We need to decide whether any x-value corresponds to more than one y-value. Suppose $x = 9$. Then

$$x = y^2 \qquad \text{becomes} \qquad y^2 = 9.$$

The equation $y^2 = 9$ has *two* solutions: $y = 3$ and $y = -3$. Because the *one* x-value, 9, leads to *two* y-values, the equation $x = y^2$ does not define a function. The graph of the equation in Figure 12 supports our conclusion.

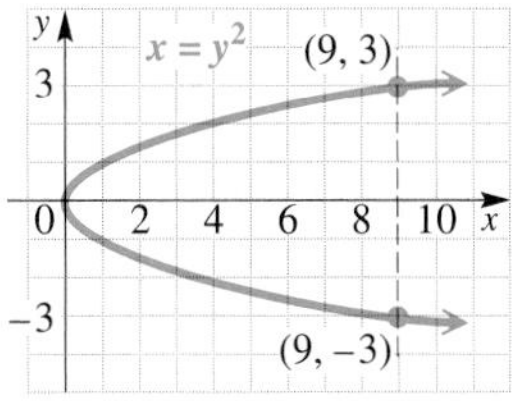

Figure 12

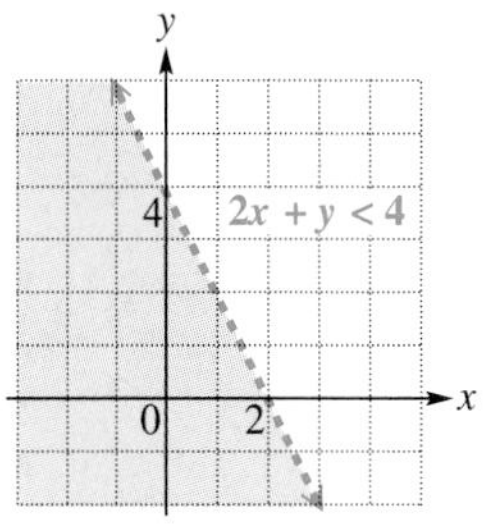

Figure 13

(f) $2x + y < 4$

Since this is an inequality, any choice of a value for x will lead to an *infinite* number of y-values. For example, if $x = 3$,

$$\begin{aligned} 2x + y &< 4 \\ 2(3) + y &< 4 \qquad \text{Let } x = 3. \\ 6 + y &< 4 \\ y &< -2. \end{aligned}$$

If $x = 3$, y can be *any* value less than -2. Since $x = 3$ leads to more than one y-value, the inequality $2x + y < 4$ does not define a function. Again, our conclusion is supported by the graph of the inequality in Figure 13.

Example 2 suggests some generalizations. Inequalities never define functions, since each x-value leads to an infinite number of y-values. An equation in which y is squared cannot define a function because most x-values will lead to two y-values. This is true for any *even* power of y, such as y^4, y^6, and so on. Similarly, an equation involving $|y|$ does not define a function, because some x-values lead to more than one y-value.

OBJECTIVE 4 **Find domains and ranges.** By the definitions of domain and range given for relations, the set of all numbers that can be used as replacements for x in a function is the domain of the function, and the set of all possible values of y is the range of the function.

CHALKBOARD EXAMPLE

Find the domain and range for the function $y = x^2 + 4$.

Answer: Domain: $(-\infty, \infty)$; Range: $[4, \infty)$

EXAMPLE 3 **Finding the Domain and Range of a Function**

Find the domain and range for the following functions.

(a) $y = 2x - 4$

Any number at all may be used for x, so the domain is the set of all real numbers. Also, any number may be used for y, so the range is also the set of all real numbers. As indicated in Figure 14, the graph of the equation is a straight line that extends infinitely in both directions, confirming that both the domain and range are $(-\infty, \infty)$.

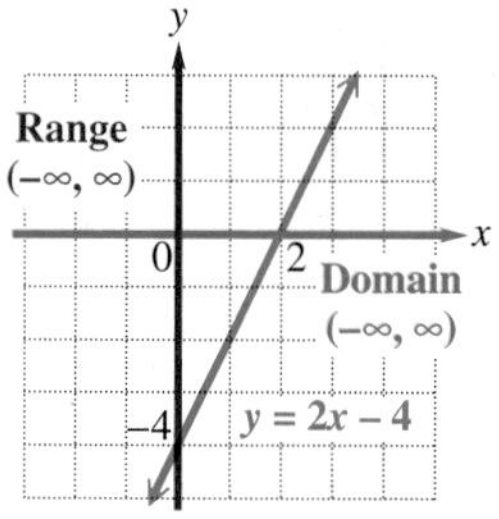

Figure 14

(b) $y = x^2$

Any number can be squared, so the domain is the set of all real numbers. However, since the square of a real number cannot be negative and since $y = x^2$, the values of y cannot be negative, making the range the set of all nonnegative numbers, or $[0, \infty)$, in interval notation. From Section 4.1, the graph of this equation is a parabola that opens upward. (See Figure 15.) While x can take any real number value, notice that y is always greater than or equal to 0.

x	y
0	0
1	1
−1	1
2	4
−2	4
3	9
−3	9

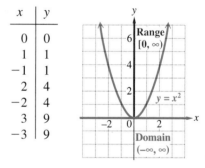

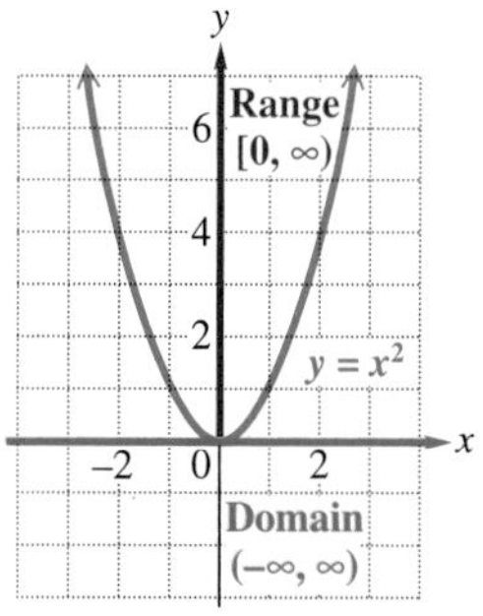

Figure 15

OBJECTIVE 5 **Use $f(x)$ notation.** The letters f, g, and h are commonly used to name functions. For example, the function $y = 3x + 5$ may be written

$$f(x) = 3x + 5,$$

where $f(x)$ is read "f of x." The notation $f(x)$ is another way of writing y in a function. For the function $f(x) = 3x + 5$, if $x = 7$ then

$$f(7) = 3 \cdot 7 + 5 \qquad \text{Let } x = 7.$$
$$= 21 + 5 = 26.$$

TEACHING TIP Stress that y and $f(x)$ can be used interchangeably in the equation that defines a function.

Read this result, $f(7) = 26$, as "f of 7 equals 26." The notation $f(7)$ means the value of y when x is 7. The statement $f(7) = 26$ says that the value of y is 26 when x is 7. It also indicates that the point (7, 26) lies on the graph of f.

To find $f(-3)$, replace x with -3.

$$f(-3) = 3(-3) + 5 \qquad \text{Let } x = -3.$$
$$f(-3) = -9 + 5 = -4$$

CAUTION The notation $f(x)$ does *not* mean f times x; $f(x)$ means the function value of x. It represents the y-value that corresponds to x.

Writing a Fraction in Lowest Terms

In the notation $f(x)$,
- f is the name of the function,
- x is the domain value,
- $f(x)$ is the range value y for the domain value x.

CHALKBOARD EXAMPLE

For the function $f(x) = 6x - 2$, find $f(-1)$.

Answer: -8

EXAMPLE 4 Using Function Notation

For the function with $f(x) = x^2 - 3$, find the following.

(a) $f(2)$
Replace x with 2.

$$f(x) = x^2 - 3$$
$$f(2) = 2^2 - 3 \qquad \text{Let } x = 2.$$
$$f(2) = 4 - 3 = 1$$

(b) $f(0) = 0^2 - 3 = 0 - 3 = -3$

(c) $f(-3) = (-3)^2 - 3 = 9 - 3 = 6$

(d) The ordered pair with $x = -3$
From part (c), when $x = -3$, $y = f(-3) = 6$, so the ordered pair is $(-3, 6)$.

TEACHING TIP Function notation will require several examples, as the notation can be confusing for students.

OBJECTIVE 6 **Apply the function concept in an application.** Because a function assigns each element in its domain to exactly one element in its range, the function concept is used in real data applications where two quantities are related. At the beginning of this chapter, a graph showed the number of prisoners (in thousands) in state and federal institutions. Our final example discusses this application.

CHALKBOARD EXAMPLE

The numbers of U.S. children (in thousands) educated at home for selected years are given in the table.

School Year	*Number of Children*
1993	588
1994	735
1995	800
1996	920
1997	1100

Source: National Home Education Research Institute, Salem, OR.

(continued on next page)

EXAMPLE 5 Applying the Function Concept to a Prison Population

The information shown in the graph at the beginning of this chapter is presented here in table form.

Sentenced Prisoners in State and Federal Institutions (in thousands)

Year	*Number of Prisoners*
1988	604
1990	740
1992	846
1994	1016
1996	1138

Source: U.S. Justice Department.

Use the table to respond to the following problems.

(a) Write a set of ordered pairs that defines a function f for this data.
(b) Give the domain and range of f.
(c) Find $f(1995)$.
(d) In what year did the number of children equal 920 thousand?

Answer:
(a) $f = \{(1993, 588), (1994, 735), (1995, 800), (1996, 920), (1997, 1100)\}$
(b) domain: $\{1993, 1994, 1995, 1996, 1997\}$; range: $\{588, 735, 800, 920, 1100\}$
(c) 800 thousand
(d) 1996

(a) Write a set of ordered pairs that define a function f.

If we choose the years as the domain elements and the number of prisoners as the range elements, the information in the table can be written as a set of five ordered pairs. In set notation, the function is

$$f = \{(1988, 604), (1990, 740), (1992, 846), (1994, 1016), (1996, 1138)\}.$$

(b) What is the domain of f? What is the range?

The domain is the set of years, or of x-values: $\{1988, 1990, 1992, 1994, 1996\}$. The range is the set of numbers of prisoners, or y-values: $\{604, 740, 846, 1016, 1138\}$.

(c) Find $f(1990)$ and $f(1996)$.

From the table or the set of ordered pairs in part (a),

$$f(1990) = 740 \qquad \text{and} \qquad f(1996) = 1138.$$

(d) For what x-value does $f(x)$ equal 846? 604?

Use the table or the ordered pairs given in part (a) to find $f(1992) = 846$ and $f(1988) = 604$.

(e) Suppose $f(1998) = 1286$. What does this tell us about the prison population?

It means that, in 1998, there were 1286 thousand (1,286,000) prisoners in state and federal prisons.

7.3 EXERCISES

1. 3; 3; (1, 3) **2.** 4; 4; (2, 4) **3.** 5; 5; (3, 5) **4.** $t + 2$; $t + 2$; $(t, t + 2)$ **5.** The graph consists of the four points (0, 2), (1, 3), (2, 4), and (3, 5). **6.** The graph is the line through any two of the points (0, 2), (1, 3), (2, 4), and (3, 5). **7.** not a function; domain: $\{-4, -2, 0\}$; range: $\{3, 1, 5, -8\}$ **8.** function; domain: $\{3, 1, 0, -1, -2\}$; range: $\{7, 4, -2, -1, 5\}$ **9.** function; domain: {A, B, C, D, E}; range: {2, 3, 6, 4} **10.** not a function; domain: {4, 2, 8, 6, 10}; range: $\{-1, -2, -3, -6, -8\}$

Complete the following table for the linear function defined by $f(x) = x + 2$.

	x	$x + 2$	$f(x)$	(x, y)
	0	2	2	(0, 2)
1.	1			
2.	2			
3.	3			
4.	t			

5. Describe the graph of function f in Exercises 1–4 if the domain is $\{0, 1, 2, 3\}$.

6. Describe the graph of function f in Exercises 1–4 if the domain is the set of real numbers $(-\infty, \infty)$.

Decide whether the relation is or is not a function. Give the domain and the range in Exercises 7–10. See Examples 1 and 2.

7. $\{(-4, 3), (-2, 1), (0, 5), (-2, -8)\}$

8. $\{(3, 7), (1, 4), (0, -2), (-1, -1), (-2, 5)\}$

9.

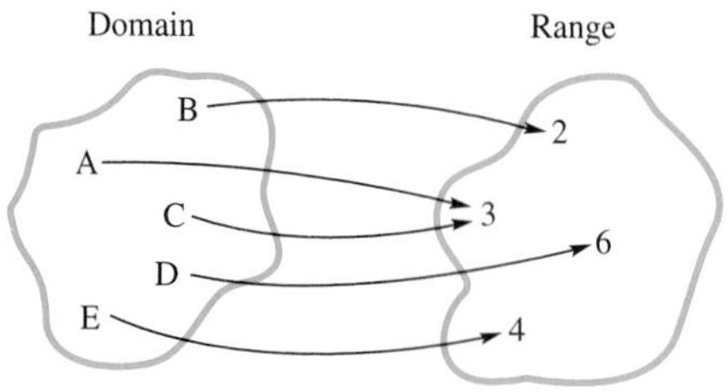

10.

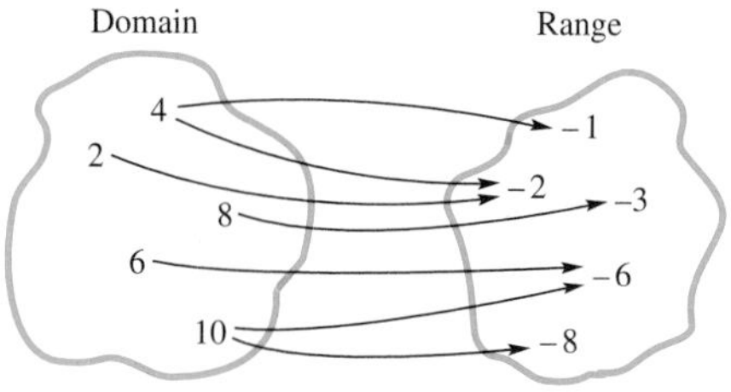

11. function 12. not a function
13. not a function 14. function
15. function 16. function
17. function 18. function
19. function 20. not a function
21. not a function
22. not a function
23. domain: $(-\infty, \infty)$; range: $[2, \infty)$
24. domain: $(-\infty, \infty)$; range: $[-3, \infty)$
25. domain: $(-\infty, \infty)$; range: $(-\infty, \infty)$
26. domain: $(-\infty, \infty)$; range: $(-\infty, \infty)$
27. domain: $[0, \infty)$; range: $[0, \infty)$
28. domain: $(-\infty, \infty)$; range: $[0, \infty)$

11.

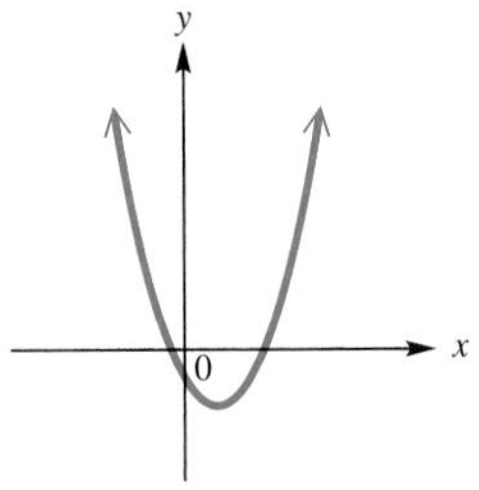

12.

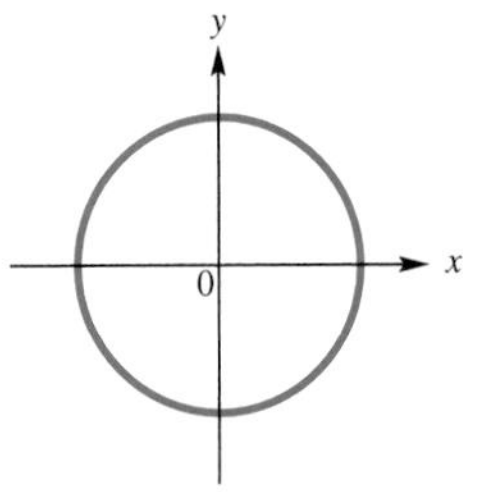

13.

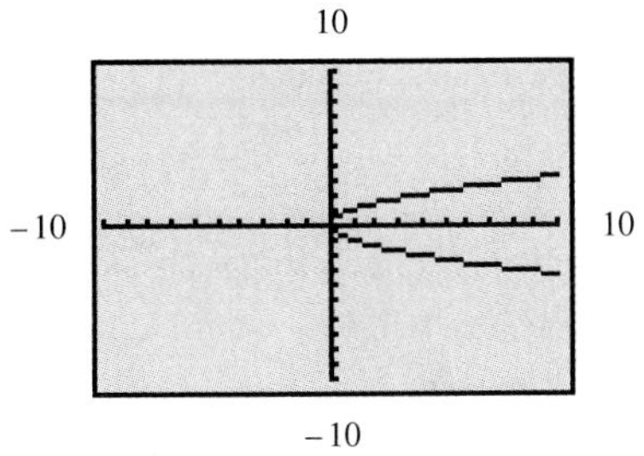

14.

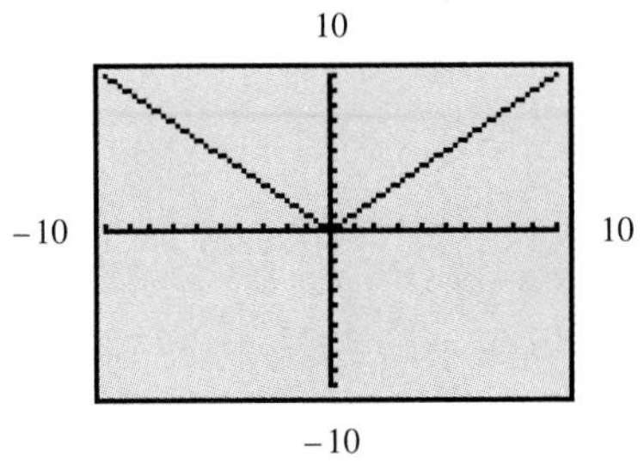

Decide whether each equation or inequality defines y as a function of x. (Remember that to be a function, every value of x must give one and only one value of y.) See Example 2.

15. $y = 5x + 3$ **16.** $y = -7x + 12$ **17.** $y = x^2$
18. $y = -3x^2$ **19.** $y = x^3$ **20.** $x = (y - 1)^2$
21. $x = y^4$ **22.** $y < 3x$

Find the domain and the range for the function. See Example 3.

23. $f(x) = x^2 + 2$ **24.** $f(x) = x^2 - 3$ **25.** $f(x) = 2x + 1$
26. $f(x) = -x + 3$ **27.** $f(x) = \sqrt{x}$ **28.** $f(x) = |x|$

29. (2, 4) 30. (−1, −4)
31. $\frac{8}{3}$ 32. $f(x) = \frac{8}{3}x - \frac{4}{3}$

RELATING CONCEPTS (EXERCISES 29–32)

A function such as $f(x) = 3x - 4$ can be graphed by replacing $f(x)$ with y and then using the methods described earlier. Let us assume that some function is written in the form $f(x) = mx + b$, for particular values of m and b.

Answer Exercises 29–32 in order.

29. If $f(2) = 4$, name the coordinates of one point on the line.

30. If $f(-1) = -4$, name the coordinates of another point on the line.

31. Use the results of Exercises 29 and 30 to find the slope of the line.

32. Use the slope-intercept form of the equation of a line to write the function in the form $f(x) = mx + b$.

Did you make the connection between the slope-intercept form of the equation of a line and the equation that defines a function?

33. (a) 11 (b) 3 (c) −9
34. (a) −1 (b) 5 (c) 14
35. (a) 4 (b) 2 (c) 14
36. (a) 10 (b) 0 (c) −30
37. (a) 2 (b) 0 (c) 3
38. (a) 9 (b) 7 (c) 4

*For each function f, find **(a)** $f(2)$, **(b)** $f(0)$, **(c)** $f(-3)$. See Example 4.*

33. $f(x) = 4x + 3$ **34.** $f(x) = -3x + 5$ **35.** $f(x) = x^2 - x + 2$
36. $f(x) = x^3 + x$ **37.** $f(x) = |x|$ **38.** $f(x) = |x + 7|$

39. (a) 4 (b) 2
40. (a) -8 (b) 64
41. {(1970, 9.6), (1980, 14.1), (1990, 19.8), (1997, 25.8)}; yes
42. domain: {1970, 1980, 1990, 1997}; range: {9.6, 14.1, 19.8, 25.8}
43. $g(1980) = 14.1$; $g(1990) = 19.8$
44. 1997
45. For the year 2000, the function predicts 29.3 million foreign-born residents in the United States.
46.

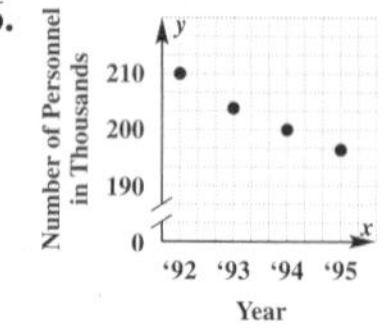

yes

A calculator can be thought of as a function machine. We input a number value (from the domain), and then by pressing the appropriate key, we obtain an output value (from the range). Use your calculator, follow the directions, and then answer the questions.

39. Suppose we enter the value 4 and then take the square root (that is, activate the square root *function,* using the key labeled $\sqrt{x}$).
(a) What is the domain value here?
(b) What is the range value obtained?

40. Suppose we enter the value -8 and then activate the squaring function, using the key labeled x^2.
(a) What is the domain value here?
(b) What range value is obtained?

The number of U.S. foreign-born residents has grown by more than 30 percent since 1990. The graph shows the number of immigrants (in millions) for selected years. Use the information in the graph for Exercises 41–45. See Example 5.

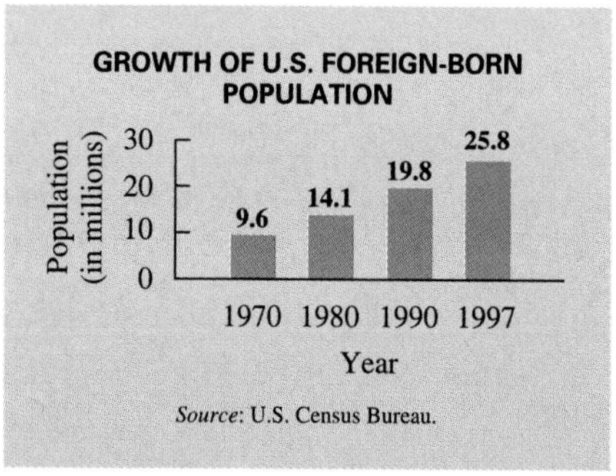

41. Write the information in the graph as a set of ordered pairs. Does this set define a function?

42. Name the function in Exercise 41 g. Give the domain and range of g.

43. Find $g(1980)$ and $g(1990)$.

44. For what value of x does $g(x) = 25.8$?

45. Suppose $g(2000) = 29.3$. What does this tell you in the context of the application?

The data give the numbers (in thousands) of U.S. active-duty female military personnel during the years 1992–1995. Use the information in the table for Exercises 46–50.

Active-Duty Female Military Personnel (in thousands)

Year	*Number*
1992	210
1993	204
1994	200
1995	196

Source: U.S. Department of Defense.

46. Plot the ordered pairs (year, number) from the table. Do the points suggest that a linear function would give a reasonable approximation of the data?

47. $y = -\frac{14}{3}x + 9506$
48. 1993: 205 thousand; 1994: 201 thousand
49. $y = -4x + 8176$
50. 1992: 208 thousand; 1995: 196 thousand; Both equations give good approximations. The equation from Exercise 47 gives answers that each vary by 1 thousand from the data. The equation from Exercise 49 gives one answer that varies by 2 thousand and one answer that is exact. 51. 4 52. 4 53. 1 54. 1 55. 1 56. (0, 1) 57. $y = x + 1$
58. (a) domain: {1, 2, 3, 4, . . .}; range: {11, 18, 25, 32, . . .} (b) The next line segment goes from $x = 2$ to $x = 3$ with $y = 4 + 3 \cdot 7 = 25$. These horizontal segments go up by 7 between consecutive domain values. Each segment has a parenthesis on the left and a bracket on the right. (c) No vertical line intersects more than one point on the graph, so this is the graph of a function. (d) the number of days the saw is rented; the cost in dollars to rent the saw 59. They are alike because a relation and a function are both sets of ordered pairs. They are different, however, because a relation is *any* set of ordered pairs. In a function, each first component must correspond to exactly one second component. 60. The domain of a function is the set of all possible replacements for x; the range is the set of all possible values of y.

47. Use the first and last data pairs in the table to write an equation relating the data. (*Hint:* See Section 7.1, Example 5.)

48. Use your equation from Exercise 47 to approximate the number of active-duty female military personnel in 1993 and 1994.

49. Use the second and third data pairs to write an equation relating the data.

50. Use your equation from Exercise 49 to approximate the number of active-duty female military personnel in 1992 and 1995. Which of the equations in Exercises 47 and 49 gives better approximations?

TECHNOLOGY INSIGHTS (EXERCISES 51–57)

The table was generated by a graphing calculator. The expression Y_1 *represents* $f(x)$.

X	Y1
-2	-1
-1	0
0	1
1	2
2	3
3	4
4	5

X= -2

51. When $x = 3$, $Y_1 =$ ______?

52. What is $f(3)$?

53. When $Y_1 = 2$, $x =$ ______?

54. If $f(x) = 2$, what is the value of x?

55. The points represented in the table all lie in a straight line. What is the slope of the line?

56. What is the y-intercept of the line?

57. Write the function in the form $y = mx + b$, for the appropriate values of m and b.

Not every function can be readily expressed as an equation. Here is an example of such a function from everyday life.

58. A chain-saw rental firm charges \$7 per day or fraction of a day to rent a saw, plus a fixed fee of \$4 for resharpening the blade. Let y represent the cost of renting a saw for x days. A portion of the graph is shown here.

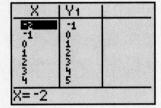

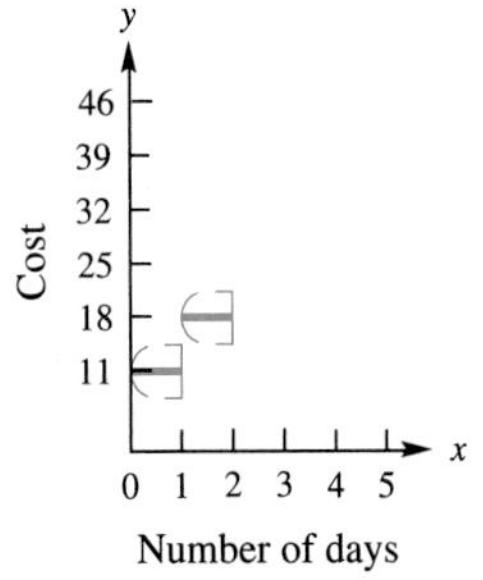

(a) Give the domain and range of this relation.
(b) Explain how the graph can be continued.
(c) Explain how the vertical line test applies here.
(d) What do the numbers in the domain represent here? What do the numbers in the range represent?

59. Define *relation* and *function.* Compare the two definitions. How are they alike? How are they different?

60. In your own words, explain the meaning of domain and range of a function.

Instructor's Note

Grouping: 3–4 students
Time: 45 minutes

Students may need help determining the appropriate scale when graphing relations and functions in this activity. Use the words domain and range when talking about scale. You may want to have students use a graphing calculator for parts of the activity. This activity can be split into two parts; have students do the second part (Exercise C) either the next day or as homework.

A. 1. One possible graph is shown here. **2. yes**

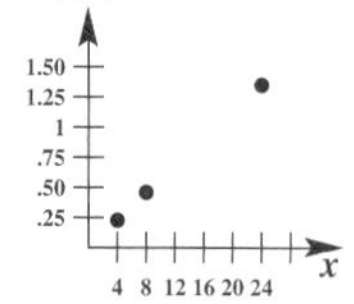

3. yes 4. domain: {4, 8, 24}; range: {.23, .46, 1.38}
5. $f(x) = .0575x$
B. 1. **2. yes**

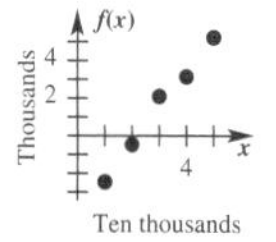

3. no 4. domain: {10,000, 20,000, 30,000, 40,000, 50,000}; range: {−2527, −358, 2078, 3578, 5078}
C. 1. $f(x) = .10x$
2.

x	$f(x)$
10,000	1000
20,000	2000
30,000	3000
40,000	4000
50,000	5000

CHAPTER 7 GROUP ACTIVITY

Taxes as Functions

Objective: Use data to determine if a relation is a function, determine domain and range, and graph the function.

Different forms of taxation fund governments. There are various types of taxes, including income, property, estate, and sales taxes. Taxes are determined in different ways. This activity looks at some of these taxes and asks you to determine if they can be modeled by linear functions. Take turns in your group graphing and answering the questions below.

A. The following table is for sales tax in Sierra County, NM.

x = Cost of Item	*y = Sales Tax*
$ 4.00	$ .23
$ 8.00	$ .46
$24.00	$1.38

1. Plot the points on a graph. (Determine the scale for the domain and range of your graph.)
2. Using the criteria for functions, determine if this is a function.
3. Is this a linear function?
4. What is the domain? What is the range?
5. Write the equation of the line that fits the data. Use $f(x)$ notation if applicable. (*Hint:* Start by finding the slope of the line.)

B. The table below is for income tax for a family of four in the U.S. in 1994.

x = Income	*y = Tax*
$10,000	−$2527
$20,000	−$358
$30,000	$2078
$40,000	$3578
$50,000	$5078

Source: 1995 Information Please Almanac.

Complete parts (1)–(4) above for this tax.

(continued)

3.
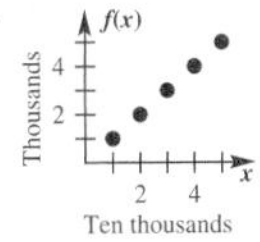

4. Taxes are higher for all incomes except \$50,000.

5. Answers will vary.

C. In the 1996 presidential election, one candidate proposed a flat income tax. For this activity, consider a flat tax rate of 10% of a given taxpayer's income.

1. Write the function that would represent this tax.
2. Using the incomes given in the table in B, determine the flat tax for each income. Make your own table.
3. Plot these points.
4. How does the flat tax compare to the 1994 taxes?
5. Who might oppose/support a flat tax?

CHAPTER 7 SUMMARY

KEY TERMS

7.2 **linear inequality in two variables**
boundary line

7.3 **components**
relation
domain
range
function

NEW SYMBOLS

$f(x)$ function f of x

TEST YOUR WORD POWER

See how well you have learned the vocabulary in this chapter. Answers, with examples, are given at the bottom of the page.

1. A **relation** is
   (a) any set of ordered pairs
   (b) a set of ordered pairs in which each first component corresponds to exactly one second component
   (c) two sets of ordered pairs that are related
   (d) a graph of ordered pairs.

2. The **domain** of a relation is
   (a) the set of all x- and y-values in the ordered pairs of the relation
   (b) the difference between the components in an ordered pair of the relation
   (c) the set of all first components in the ordered pairs of the relation
   (d) the set of all second components in the ordered pairs of the relation.

3. The **range** of a relation is
   (a) the set of all x- and y-values in the ordered pairs of the relation
   (b) the difference between the components in an ordered pair of the relation
   (c) the set of all first components in the ordered pairs of the relation
   (d) the set of all second components in the ordered pairs of the relation.

4. A **function** is
   (a) any set of ordered pairs
   (b) a set of ordered pairs in which each first component corresponds to exactly one second component
   (c) two sets of ordered pairs that are related
   (d) a graph of ordered pairs.

Answers to Test Your Word Power

1. (a) *Example:* $\{(0, 2), (2, 4), (3, 6), (-1, 3)\}$ **2.** (c) *Example:* The domain in the relation given above is the set of x-values, that is, $\{0, 2, 3, -1\}$. **3.** (d) *Example:* The range of the relation given above is the set of y-values, that is, $\{2, 4, 6, 3\}$. **4.** (b) *Example:* The relation given above is a function since each x-value corresponds to exactly one y-value.

QUICK REVIEW

CONCEPTS	EXAMPLES

7.1 EQUATIONS OF A LINE

Slope-Intercept Form
$y = mx + b$
m is the slope.
$(0, b)$ is the y-intercept.

Find an equation of the line with slope 2 and y-intercept $(0, -5)$.

The equation is $y = 2x - 5$.

Point-Slope Form
$y - y_1 = m(x - x_1)$
m is the slope.
(x_1, y_1) is a point on the line.

Find an equation of the line with slope $-\frac{1}{2}$ through $(-4, 5)$.

$$y - 5 = -\frac{1}{2}(x - (-4))$$
$$y - 5 = -\frac{1}{2}(x + 4)$$
$$y - 5 = -\frac{1}{2}x - 2$$
$$y = -\frac{1}{2}x + 3$$

Standard Form
$Ax + By = C$
A, B, and C are integers and $A > 0$, $B \neq 0$.

This equation is written in standard form as

$$x + 2y = 6,$$

with $A = 1$, $B = 2$, and $C = 6$.

7.2 GRAPHING LINEAR INEQUALITIES IN TWO VARIABLES

1. Graph the line that is the boundary of the region. Make it solid if the inequality is $\leq$ or $\geq$; make it dashed if the inequality is $<$ or $>$.
2. Use any point not on the line as a test point. Substitute for x and y in the inequality. If the result is true, shade the side of the line containing the test point; if the result is false, shade the other side.

Graph $2x + y \leq 5$.
Graph the line $2x + y = 5$.
Make it solid because of $\leq$.

Use $(1, 0)$ as a test point.

$2(1) + 0 \leq 5$?
$2 \leq 5$ True

Shade the side of the line containing $(1, 0)$.

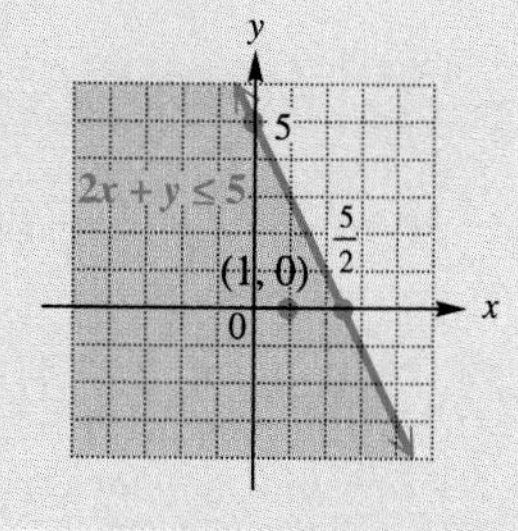

7.3 FUNCTIONS

Vertical Line Test
If a vertical line intersects a graph in more than one point, the graph is not the graph of a function.

The graph shown is not the graph of a function.

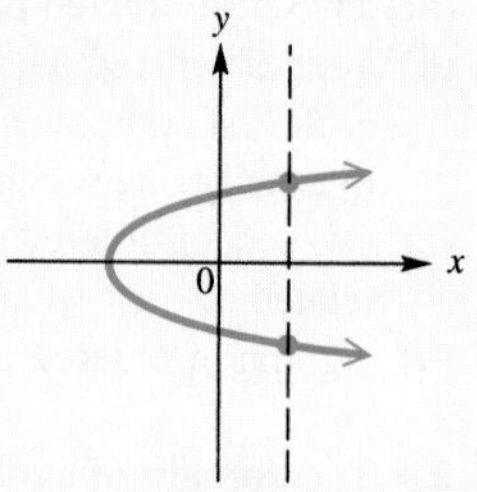

The **domain** of a function is the set of numbers that can replace x in the expression for the function. The **range** is the set of y-values that result as x is replaced by each number in the domain. To find $f(x)$ for a specific value of x, replace x by that value in the expression for the function.

Let $f(x) = (x - 1)^2$. Find
(a) the domain of f;
(b) the range of f;
(c) $f(-1)$.

(a) The domain is $(-\infty, \infty)$.
(b) The range is all y such that $y \geq 0$ or $[0, \infty)$, because y equals the square of $x - 1$, so y must be nonnegative.
(c) $f(-1) = (-1 - 1)^2 = (-2)^2 = 4$

CHAPTER 7 REVIEW EXERCISES

1. $y = -x + \frac{2}{3}$ 2. $y = -\frac{1}{2}x + 4$
3. $y = x - 7$ 4. $y = \frac{2}{3}x + \frac{14}{3}$
5. $y = -\frac{3}{4}x - \frac{1}{4}$
6. $y = -\frac{1}{4}x + \frac{3}{2}$ 7. $y = 1$
8. $x = \frac{1}{3}$; It is not possible to express this equation as $y = mx + b$.
9.

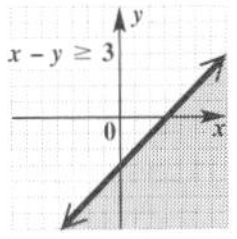

10.

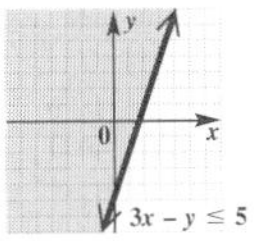

11.

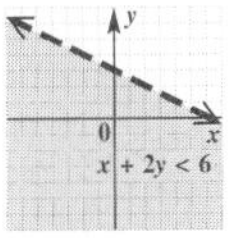

12.

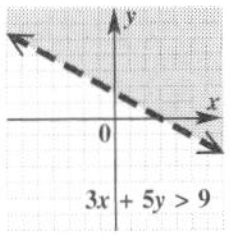

13.

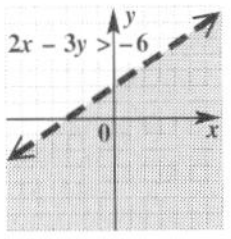

14.

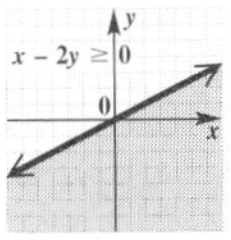

15. not a function; domain: $\{-2, 0, 2\}$; range: $\{4, 8, 5, 3\}$
16. function; domain: $\{8, 7, 6, 5, 4\}$; range: $\{3, 4, 5, 6, 7\}$
17. not a function
18. function 19. function
20. function 21. not a function
22. domain: $(-\infty, \infty)$; range: $(-\infty, \infty)$
23. domain: $(-\infty, \infty)$; range: $[1, \infty)$
24. domain: $(-\infty, \infty)$; range: $[0, \infty)$
25. (a) 8 (b) -1

[7.1] *Write an equation for each line in the form $y = mx + b$, if possible.*

1. $m = -1, b = \frac{2}{3}$
2. Through (2, 3) and (−4, 6)
3. Through (4, −3), $m = 1$
4. Through (−1, 4), $m = \frac{2}{3}$
5. Through (1, −1), $m = -\frac{3}{4}$
6. $m = -\frac{1}{4}, b = \frac{3}{2}$
7. Slope 0, through (−4, 1)
8. Through $\left(\frac{1}{3}, -\frac{5}{4}\right)$ with undefined slope

[7.2] *Complete the graph of each linear inequality by shading the correct region.*

9. $x - y \geq 3$

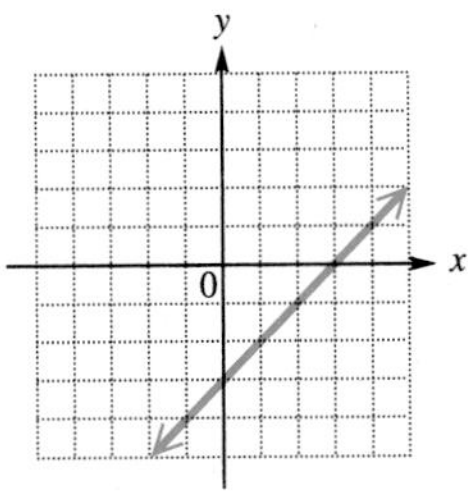

10. $3x - y \leq 5$

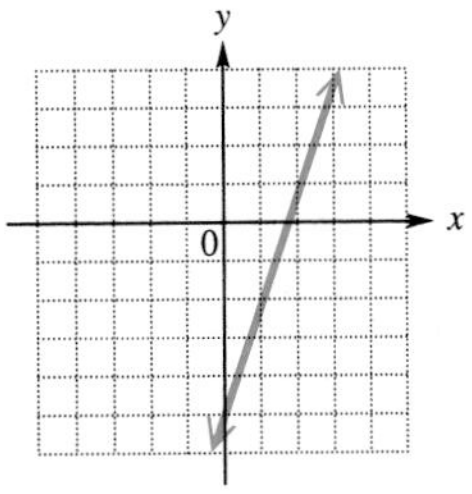

11. $x + 2y < 6$

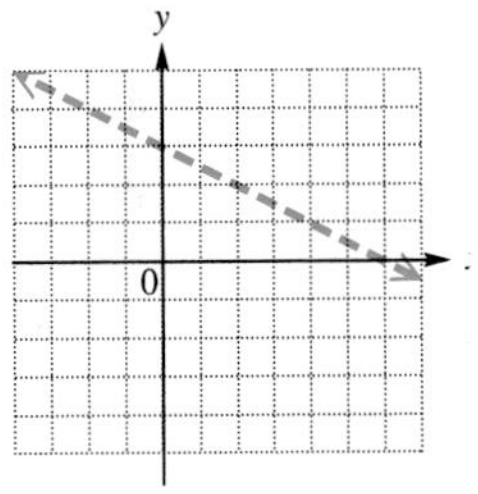

Graph each linear inequality.

12. $3x + 5y > 9$
13. $2x - 3y > -6$
14. $x - 2y \geq 0$

[7.3] *Decide whether each relation is or is not a function. In Exercises 15 and 16, give the domain and the range.*

15. $\{(-2, 4), (0, 8), (2, 5), (2, 3)\}$
16. $\{(8, 3), (7, 4), (6, 5), (5, 6), (4, 7)\}$

17.

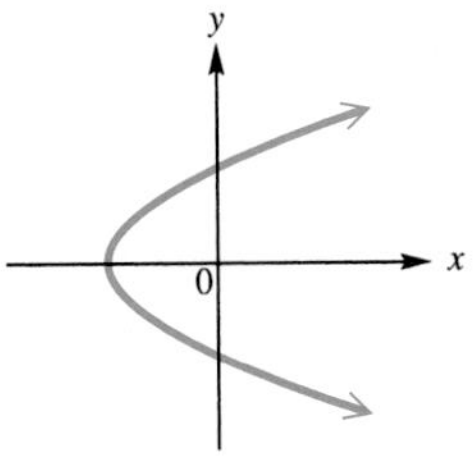

18.

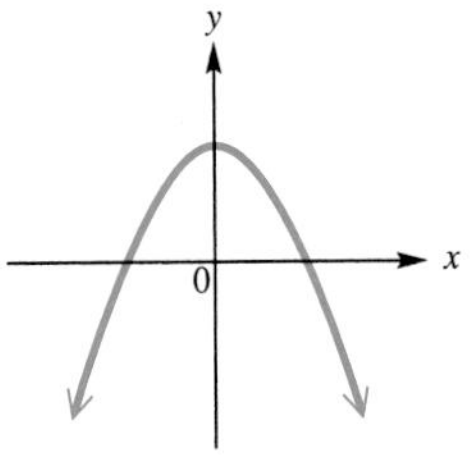

19. $2x + 3y = 12$
20. $y = x^2$
21. $x - 5y < 10$

Find the domain and range of the function.

22. $4x - 3y = 12$
23. $y = x^2 + 1$
24. $y = |x - 1|$

Find (a) $f(2)$ and (b) $f(-1)$.

25. $f(x) = 3x + 2$
26. $f(x) = 2x^2 - 1$
27. $f(x) = |x + 3|$

26. (a) 7 (b) 1
27. (a) 5 (b) 2
28. (a) 4.58; The equation predicts net profits of $4.58 billion for 1998.
(b) $m = .48$ (c) The slope of .48 indicates that net profits will increase by $.48 billion each year.
29. $y = -1$
30. $y = -\frac{1}{4}x - \frac{5}{4}$
31. $y = -3x + 30$
32. $y = -\frac{4}{7}x - \frac{23}{7}$
33.
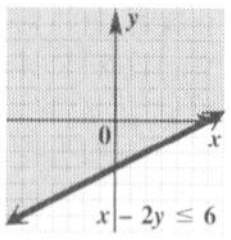

34.
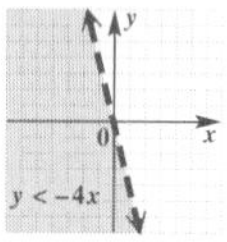

35.
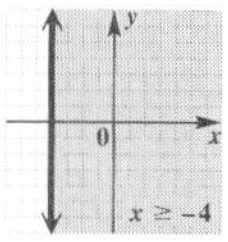

36. 1 37. Either a > inequality or a < inequality has a dashed boundary line to indicate that the points on the line are not included in the solution set.

28. The net profits (in billions of dollars) for Coca Cola have increased in recent years as shown in the graph.

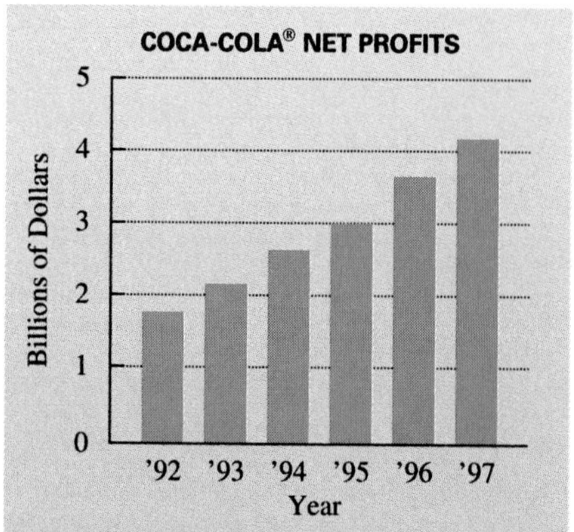

Source: Betsy Morris, "Doug Is It," *Fortune*, May 25, 1998, p. 78.

The tops of the bars appear to lie in a linear pattern. The linear equation $y = f(x) = .48x - 42.46$ gives a good approximation of the line through the centers of the tops of the bars.

(a) Use the equation to find $f(98)$ and interpret the result in terms of the application. (*Hint:* $x = 98$ represents the year 1998 and would be shown in the graph as '98.)

(b) Use the equation to find the slope of the line.

(c) Discuss the relationship between the slope of the line and the trend in net profits for Coca Cola.

MIXED REVIEW EXERCISES

Write an equation in slope-intercept form for each line described in Exercises 29–32.

29. Through $(6, -1)$ and $(5, -1)$

30. $m = -\frac{1}{4}$; $b = -\frac{5}{4}$

31. Through $(8, 6)$; $m = -3$

32. Through $(3, -5)$ and $(-4, -1)$

Graph each inequality.

33. $x - 2y \le 6$

34. $y < -4x$

35. $x \ge -4$

36. Find $f(2)$ if $f(x) = (3 - 2x)^2$.

37. What kind of inequality has a dashed line as its boundary? Explain why.

38.
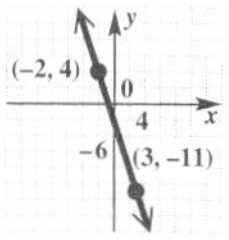

39. −3 40. $y = -3x - 2$
41. $\left(-\frac{2}{3}, 0\right)$, $(0, -2)$
42.
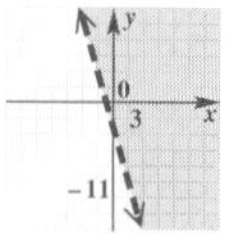

43. −32 44. domain: $(-\infty, \infty)$; range: $(-\infty, \infty)$

RELATING CONCEPTS (EXERCISES 38–44)

Use the concepts of this chapter and Chapter 3 in ***working Exercises 38–44 in order.***

38. Plot the points $(-2, 4)$ and $(3, -11)$ and draw the line joining them.

39. Find the slope of the line in Exercise 38.

40. Find the slope-intercept form of the equation of the line in Exercise 38.

41. What are the x- and y-intercepts of this line?

42. Suppose that $y = mx + b$ is the equation of the line. Graph the inequality $y > mx + b$.

43. Replace y with $f(x)$ in your answer to Exercise 40. Find $f(10)$.

44. Give the domain and the range of the function described in Exercise 43.

Did you make the connection between the equations of a line, the graphs of linear equations and inequalities, and functions?

45. (a) \$2.910 billion; 1996 (b) domain: {1994, 1995, 1996, 1997}; range: {2.677, 2.910, 3.297, 3.561} (c) $y = .3255x + 2.5845$; yes (d) \$3.2355 billion; Yes, the result is reasonably close. 46. (a) yes; yes (b) domain: {1994, 1995, 1996, 1997}; range: {2.552, 2.625, 2.695, 2.752} (c) $y = .0715x + 2.552$ (d) \$2.838 billion

45. The judicial branch of the U.S. government had expenditures (in billions of dollars) for the years 1994–1997 as shown in the graph. Use the graph to answer the following questions.

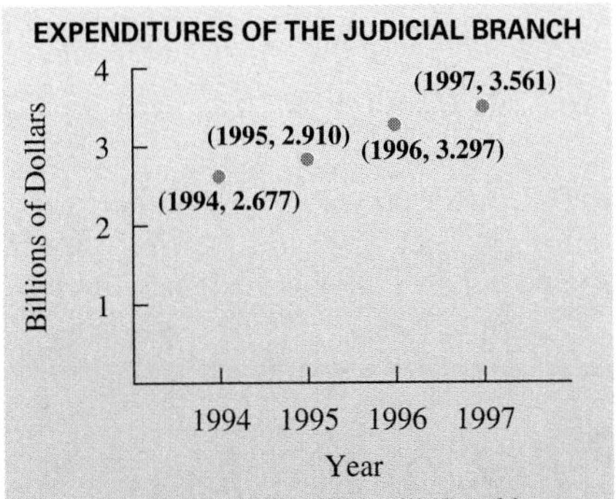

Source: Office of Management and Budget of the U.S. Government.

(a) Let function f be defined by the four ordered pairs in the graph. What is $f(1995)$? If $f(x) = 3.297$, what is x?

(b) What are the domain and range of f?

(c) The points in the graph indicate that a linear equation would approximate the data quite well. Let $x = 0$ represent 1994, $x = 1$ represent 1995, and so on. Write a linear equation for f using the points for 1995 and 1997. Would you expect to get a slightly different equation if you used a different pair of points?

(d) Use the equation from part (c) to determine $f(1996)$. Is the result close to the actual value given in the graph?

46. The graph shows the expenditures (in billions of dollars) of the legislative branch of the U.S. government. Use the graph to respond to the following.

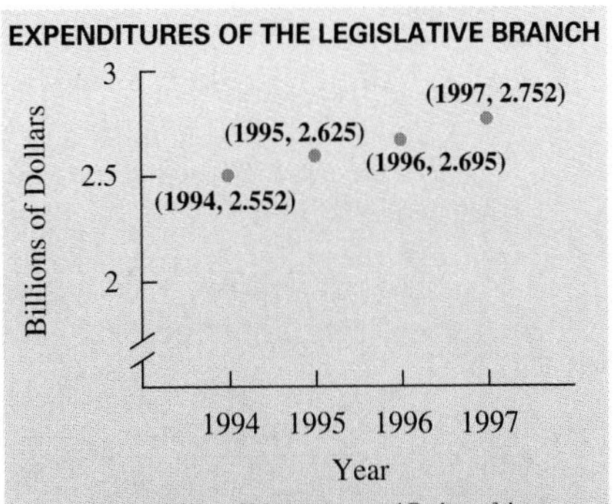

Source: Office of Management and Budget of the U.S. Government.

(a) Is this the graph of a function? Could it be approximated reasonably by a linear function?

(b) Give the domain and range.

(c) Write a linear equation that approximates the given data points, using the years 1994 and 1996. Let $x = 0$ represent 1994, $x = 1$ represent 1995, and so on.

(d) Use the equation from part (c) to predict legislative branch expenditures in 1998.

CHAPTER 7 TEST

[7.1] 1. $y = 2x + 6$
2. $y = \frac{5}{2}x - 4$
3. $y = \frac{1}{2}x + 4$
4. $2x + 3y = 15$
[7.2] 5.

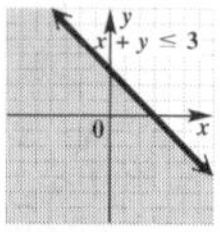

6.

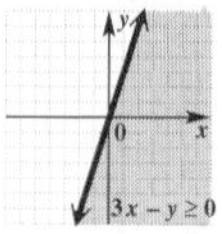

[7.3] 7. 20 **8.** not a function
9. function; domain: $(-\infty, \infty)$; range: $[2, \infty)$
10. function; domain: $(-\infty, \infty)$; range: $(-\infty, \infty)$
11. not a function
12. Every vertical line intersects the graph (a line) in only one point.

Write an equation for each line in slope-intercept form.

1. Through $(-1, 4)$ with slope 2

2.

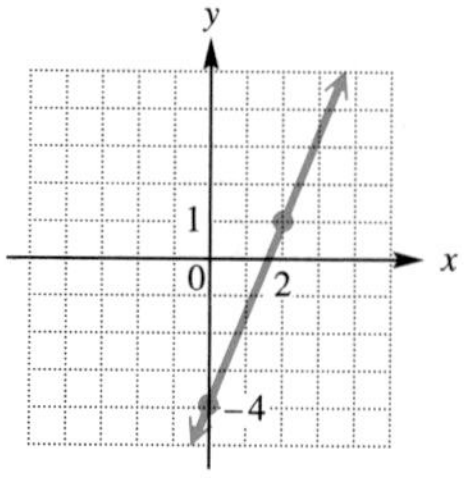

3.

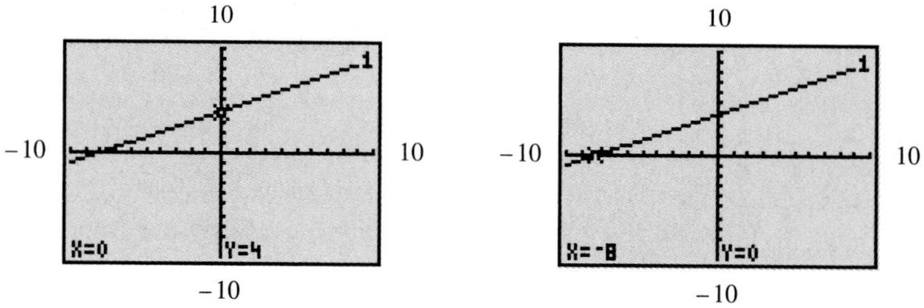

(These are two different views of the same line.)

4. Write the linear equation $y = -\frac{2}{3}x + 5$ in standard form.

Graph the linear inequality.

5. $x + y \leq 3$

6. $3x - y \geq 0$

7. If $f(x) = -4x + 8$, find $f(-3)$.

Decide whether each of the following represents a function. If it does, give the domain and the range.

8.

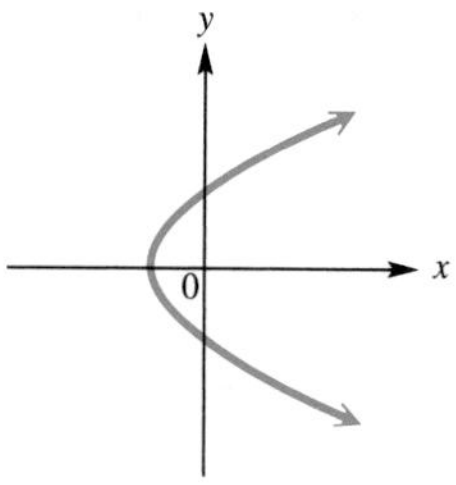

9. $y = x^2 + 2$

10. $y = 2x + 1$

11. $\{(0, 2), (0, -2), (1, 5), (2, 10)\}$

12. Why does the graph of $y = mx + b$, for real numbers m and b, satisfy the conditions for the graph of a function?

13. yes; $4459 billion or $4,459,000,000,000
14. $x = 1996$
15. 234.6 (billion per year)
16. Consumer expenditures are increasing by $234.6 billion each year.

The table gives government data on consumer spending (in billions of dollars) during the years 1990 through 1996.

Personal Consumption

Year	Expenditures (in billions of dollars)
1990	3839
1991	3975
1992	4220
1993	4459
1994	4717
1995	4958
1996	5208

Source: U.S. Commerce Department, Bureau of Economic Analysis.

13. Does this set of data define a function f? If so, what is $f(1993)$?

14. If the set of data defines a function f and $f(x) = 5208$, what is x?

15. The data points are closely approximated by the line $y = 234.6x - 463{,}161$. What is the slope of this line?

16. What does the slope of the line tell you about the annual increase in consumer expenditures?

CUMULATIVE REVIEW EXERCISES CHAPTERS 1–7

[2.2] 1. $\{-65\}$

[2.4] 2. $t = \frac{A - p}{pr}$

[5.5] 3. $\left\{-1, -\frac{1}{7}\right\}$

[6.6] 4. $\{3\}$ 5. $\{5\}$

[2.7] 6. $(-2.6, \infty)$

7. $(0, \infty)$

8. $(-\infty, -4]$

[4.1–4.5] 9. $\frac{1}{x^2y}$ 10. $\frac{y^7}{x^{13}z^2}$

11. $\frac{m^6}{8n^9}$ 12. $2x^2 - 4x + 38$

13. $15x^2 + 7xy - 2y^2$

14. $x^3 + 8y^3$

[4.6] 15. $m^2 - 2m + 3$

Solve each equation.

1. $-5(8 - 2z) + 4(7 - z) = 7(8 + z) - 3$

2. $A = p + prt$ for t

3. $7x^2 + 8x + 1 = 0$

4. $\frac{4}{x - 2} = 4$

5. $\frac{2}{x - 1} = \frac{5}{x - 1} - \frac{3}{4}$

Solve each inequality and graph the solution set.

6. $-2.5x < 6.5$

7. $4(x + 3) - 5x < 12$

8. $\frac{2}{3}y - \frac{1}{6}y \le -2$

Write with only positive exponents. Assume that all variables represent positive real numbers.

9. $(x^2y^{-3})(x^{-4}y^2)$

10. $\frac{x^{-6}y^3z^{-1}}{x^7y^{-4}z}$

11. $(2m^{-2}n^3)^{-3}$

Perform the indicated operations.

12. $2(3x^2 - 8x + 1) - 4(x^2 - 3x - 9)$

13. $(3x + 2y)(5x - y)$

14. $(x + 2y)(x^2 - 2xy + 4y^2)$

15. $\frac{m^3 - 3m^2 + 5m - 3}{m - 1}$

[5.1–5.4] 16. $(y + 6k)(y - 2k)$
17. $(3x^2 - 5y)(3x^2 + 5y)$
18. $5x^2(5x - 13y)(5x - 3y)$
19. $(f + 10)^2$ **20.** prime
[6.4] 21. 1 **22.** $\frac{6x + 22}{(x + 1)(x + 3)}$ or $\frac{2(3x + 11)}{(x + 1)(x + 3)}$
[6.2] 23. $\frac{4(x - 5)}{3(x + 5)}$ **24.** $\frac{x + 1}{x}$
25. $\frac{(x + 3)^2}{3x}$ **26.** $\frac{4xy^4}{z^2}$
[6.5] 27. $\frac{5}{8}$ **28.** 6
[3.3] 29. $-\frac{4}{3}$ **30.** 0
[7.1] 31. $y = -4x + 15$
32. $y = 4x$
[3.2] 33.

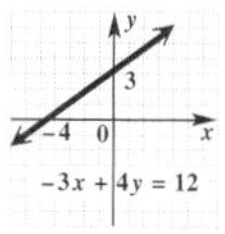

[7.2] 34.

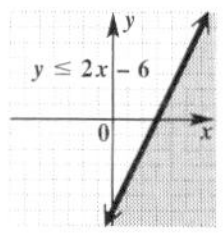

35.

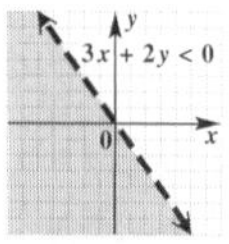

[2.3] 36. corporate income taxes: \$171.8 billion; individual income taxes: \$656.4 billion
[2.6] 37. 15°, 35°, 130°
[5.6] 38. 7 inches
[2.5] 39. 14
[6.7] 40. 1 hour

Factor each polynomial completely.

16. $y^2 + 4yk - 12k^2$

17. $9x^4 - 25y^2$

18. $125x^4 - 400x^3y + 195x^2y^2$

19. $f^2 + 20f + 100$

20. $100x^2 + 49$

Perform each indicated operation. Express the answer in lowest terms.

21. $\frac{3}{2x + 6} + \frac{2x + 3}{2x + 6}$

22. $\frac{8}{x + 1} - \frac{2}{x + 3}$

23. $\frac{x^2 - 25}{3x + 6} \cdot \frac{4x + 8}{x^2 + 10x + 25}$

24. $\frac{x^2 + 2x - 3}{x^2 - 5x + 4} \cdot \frac{x^2 - 3x - 4}{x^2 + 3x}$

25. $\frac{x^2 + 5x + 6}{3x} \div \frac{x^2 - 4}{x^2 + x - 6}$

26. $\frac{6x^4y^3z^2}{8xyz^4} \div \frac{3x^2}{16y^2}$

Simplify each complex fraction.

27. $\dfrac{\frac{2}{3} - \frac{1}{4}}{\frac{1}{2} + \frac{1}{6}}$

28. $\dfrac{\frac{12}{x + 6}}{\frac{4}{2x + 12}}$

Find the slope of each line described.

29. Through $(-4, 5)$ and $(2, -3)$

30. Horizontal, through $(4, 5)$

Give the equation of each line described in the form $y = mx + b$.

31. Through $(4, -1)$, $m = -4$

32. Through $(0, 0)$ and $(1, 4)$

Graph each equation or inequality.

33. $-3x + 4y = 12$

34. $y \leq 2x - 6$

35. $3x + 2y < 0$

Solve each problem.

36. In 1996, the U.S. government raised \$484.6 billion more from individual income taxes than from corporate income taxes. The total amount raised from these two sources was \$828.2 billion. How much was raised from each source? (*Source:* Office of Management and Budget.)

37. Find the measure of each angle of the triangle.

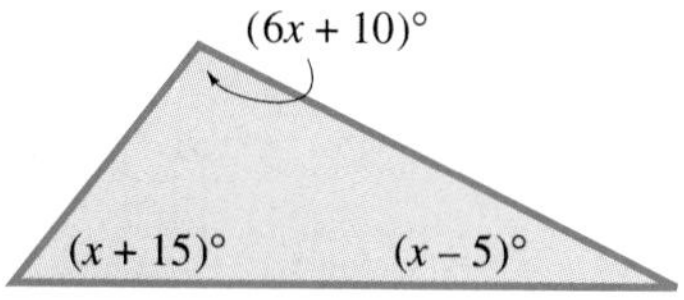

38. The length of the shorter leg of a right triangle is tripled and 4 inches is added to the result, giving the length of the hypotenuse. The longer leg is 10 inches longer than twice the shorter leg. Find the length of the shorter leg of the triangle.

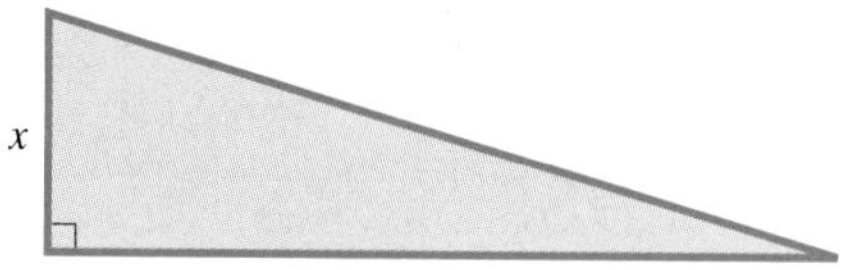

39. If x varies directly as y and $x = 4$ when $y = 12$, find x when $y = 42$.

40. If a man can mow his lawn in 3 hours and his wife can do the same job in 1.5 hours, how long will it take them to do the job together?

Linear Systems 8

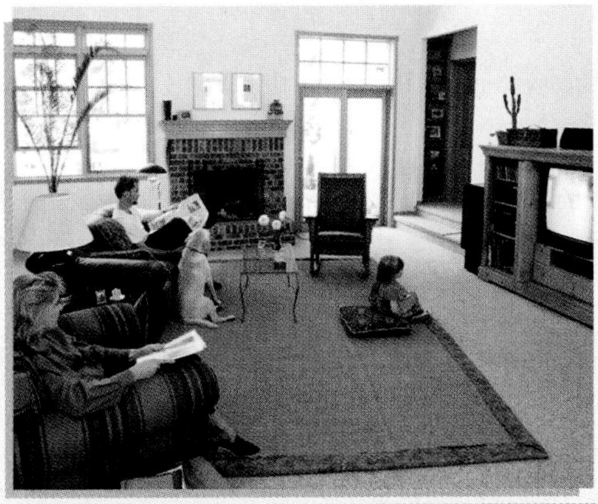

Consumers are spending more money than ever on entertainment. The average American household spent $1612 in 1995 on leisure pursuits including books, TV, movies, theater, and sporting events—almost as much as for health care, according to a survey by the U.S. Labor Department. This voracious appetite for entertainment can make even leisure time exhausting. Some accommodate this plethora of entertainment choices by "multitasking"; for example, watching TV or listening to the radio as they read, channel-surfing to watch several programs at once, or using the picture-in-picture feature available in many television models to follow two sporting events at the same time.

Entertainment

In recent years, movies and television have been the big winners in the entertainment marketplace. From 1991 to 1996, movie box-office gross increased from $4803.2 million to $5911.5 million, or about $1.1 billion, as shown in the graph.

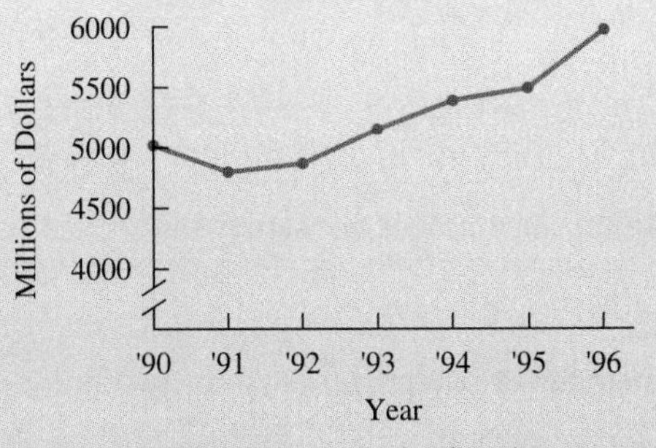

Source: Motion Picture Association of America.

Although the box-office gross dipped from 1990 to 1991, it has steadily increased since then. For what year does the graph show the greatest increase? Exercises 43–48 of Section 8.3 present more information on the increase in movie attendance. Linear equations and inequalities are also applied to other areas of the entertainment industry in this chapter.

8.1 Solving Systems of Linear Equations by Graphing

OBJECTIVES

1. Decide whether a given ordered pair is a solution of a system.
2. Solve linear systems by graphing.
3. Solve special systems by graphing.
4. Identify special systems without graphing.
5. Recognize how a graphing calculator is used to solve a linear system.

FOR EXTRA HELP

SSG Sec. 8.1
SSM Sec. 8.1
Pass the Test Software
InterAct Math Tutorial Software
Video 13

CHALKBOARD EXAMPLE

Decide whether the given ordered pair is a solution of the system.

$(4, -1)$

$5x + 6y = 14$

$2x + 5y = 3$

Answer: yes

A **system of linear equations** consists of two or more linear equations with the same variables. Examples of systems of linear equations include

$$\begin{aligned} 2x + 3y &= 4 \\ 3x - y &= -5 \end{aligned} \qquad \begin{aligned} x + 3y &= 1 \\ -y &= 4 - 2x \end{aligned} \qquad \text{and} \qquad \begin{aligned} x - y &= 1 \\ y &= 3. \end{aligned}$$

In the system on the right, think of $y = 3$ as an equation in two variables by writing it as $0x + y = 3$.

OBJECTIVE 1 Decide whether a given ordered pair is a solution of a system. A **solution of a system** of linear equations is an ordered pair (x, y) that makes both equations true at the same time.

EXAMPLE 1 Determining Whether an Ordered Pair Is a Solution

Is $(4, -3)$ a solution of the following systems?

(a) $x + 4y = -8$
$3x + 2y = 6$

To decide whether $(4, -3)$ is a solution of the system, substitute 4 for x and -3 for y in each equation.

$$\begin{aligned} x + 4y &= -8 \\ 4 + 4(-3) &= -8 \quad ? \\ 4 + (-12) &= -8 \quad ? \\ -8 &= -8 \quad \text{True} \end{aligned} \qquad \begin{aligned} 3x + 2y &= 6 \\ 3(4) + 2(-3) &= 6 \quad ? \\ 12 + (-6) &= 6 \quad ? \\ 6 &= 6 \quad \text{True} \end{aligned}$$

Since $(4, -3)$ satisfies both equations, it is a solution.

(b) $2x + 5y = -7$
$3x + 4y = 2$

Again, substitute 4 for x and -3 for y in both equations.

$$\begin{aligned} 2x + 5y &= -7 \\ 2(4) + 5(-3) &= -7 \quad ? \\ 8 + (-15) &= -7 \quad ? \\ -7 &= -7 \quad \text{True} \end{aligned} \qquad \begin{aligned} 3x + 4y &= 2 \\ 3(4) + 4(-3) &= 2 \quad ? \\ 12 + (-12) &= 2 \quad ? \\ 0 &= 2 \quad \text{False} \end{aligned}$$

Here $(4, -3)$ is not a solution since it does not satisfy the second equation.

We discuss several methods of solving a system of two linear equations with two variables in this chapter.

OBJECTIVE 2 Solve linear systems by graphing. The set of all ordered pairs that are solutions of a system is its **solution set.** One way to find the solution set of a system of two linear equations is to graph both equations on the same axes. The graph of each line shows points whose coordinates satisfy the equation of that line. The coordinates of any point where the lines intersect give a solution of the system. Since two *different* straight lines can intersect at no more than one point, there can never be more than one solution for such a system.

TEACHING TIP Explain that if $(4, -3)$ satisfies both equations, then the graphs of both equations must pass through that point.

INSTRUCTOR'S RESOURCES

PTB Sec. 8.1 **ISM** Sec. 8.1 **AB** Sec. 8.1

TEST GENERATOR

WORLD WIDE WEB www.LialAlgebra.com

CHALKBOARD EXAMPLE

Solve the system by graphing.

$$5x - 3y = 9$$
$$x + 2y = 7$$

Answer: $\{(3, 2)\}$

EXAMPLE 2 Solving a System by Graphing

Solve the following system of equations by graphing both equations on the same axes.

$$2x + 3y = 4$$
$$3x - y = -5$$

As shown in Chapter 3, we graph these two equations by plotting several points for each line. Some ordered pairs that satisfy each equation are shown below.

$2x + 3y = 4$

x	y
0	$\frac{4}{3}$
2	0
-2	$\frac{8}{3}$

$3x - y = -5$

x	y
0	5
$-\frac{5}{3}$	0
-2	-1

The lines in Figure 1 suggest that the graphs intersect at the point $(-1, 2)$. Check this by substituting -1 for x and 2 for y in both equations. Since $(-1, 2)$ satisfies both equations, the solution set of this system is $\{(-1, 2)\}$.

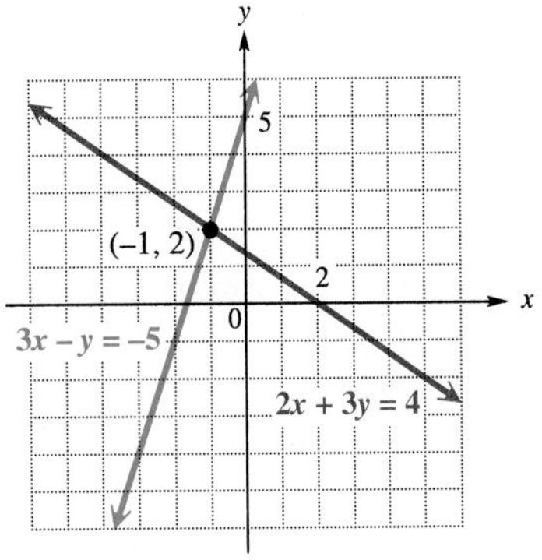

Figure 1

TEACHING TIP Students will learn quickly the drawbacks of the graphing method. A good oral exercise is to ask them what some of these drawbacks are.

CAUTION A difficulty with the graphing method is that it may not be possible to determine from the graph the exact coordinates of the point that represents the solution, particularly if the solution does not involve integers. For this reason, we give algebraic methods of solution later in this chapter. The graphing method does, however, show geometrically how solutions are found.

OBJECTIVE 3 Solve special systems by graphing. Sometimes the graphs of the two equations in a system either do not intersect at all or are the same line, as in the systems of Example 3.

CHALKBOARD EXAMPLE

Solve each system by graphing.

(a) $3x - y = 4$
$6x - 2y = 12$

(b) $2x - 5y = 8$
$4x - 10y = 16$

Answer:
(a) $\emptyset$
(b) infinite number of solutions

EXAMPLE 3 Solving Special Systems

Solve each system by graphing.

(a) $2x + y = 2$
$2x + y = 8$

The graphs of these lines are shown in Figure 2. The two lines are parallel and have no points in common. For such a system, there is no solution; we write the solution set as $\emptyset$.

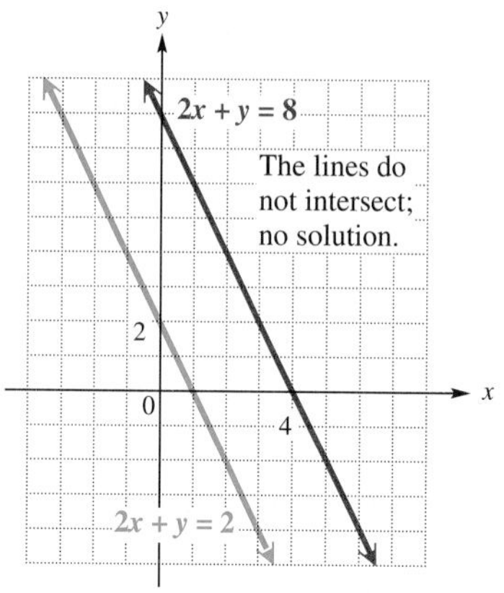

Figure 2

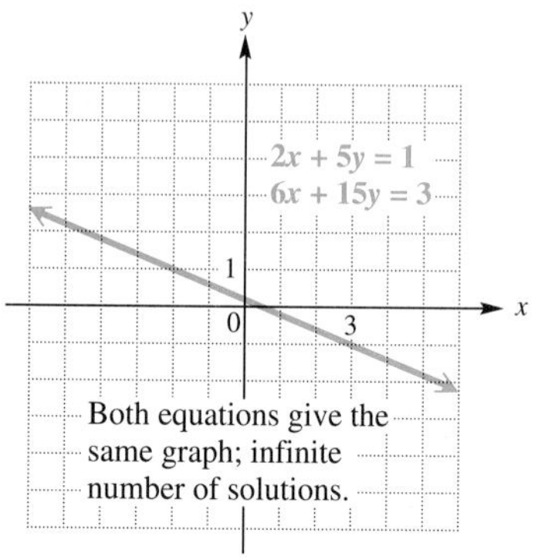

Figure 3

(b) $2x + 5y = 1$
$6x + 15y = 3$

The graphs of these two equations are the same line. See Figure 3. The second equation can be obtained by multiplying both sides of the first equation by 3. In this case, every point on the line is a solution of the system, and the solution set contains an infinite number of ordered pairs. We write "infinite number of solutions" to indicate this case.

The system in Example 2 has exactly one solution. A system with a solution is called a **consistent system.** A system with no solution, such as the one in Example 3(a), is called an **inconsistent system.** The equations in Example 2 are **independent equations,** equations that have different graphs. The equations of the system in Example 3(b) have the same graph. Because they are different forms of the same equation, these equations are called **dependent equations.** Examples 2 and 3 show the three cases that may occur when solving a system of equations with two variables.

Possible Types of Solutions

1. The graphs intersect at exactly one point, which gives the (single) ordered pair solution of the system. The **system is consistent** and the **equations are independent.**

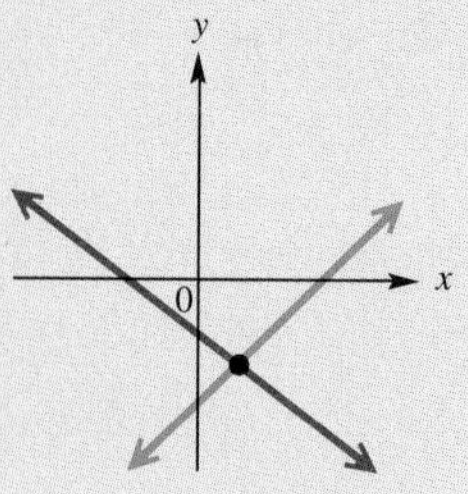

Possible Types of Solutions (continued)

2. The graphs are parallel lines, so there is no solution and the solution set is $\emptyset$. The **system is inconsistent.**

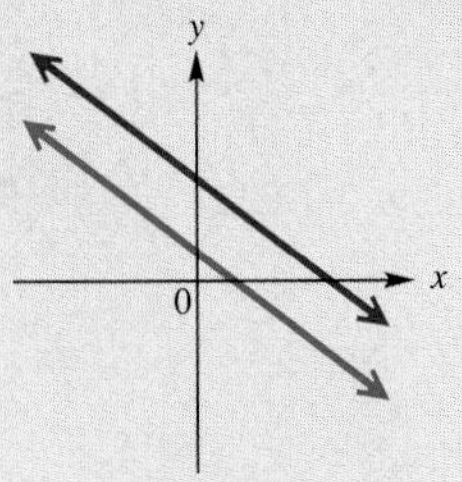

3. The graphs are the same line. There are an infinite number of solutions. The **equations are dependent.**

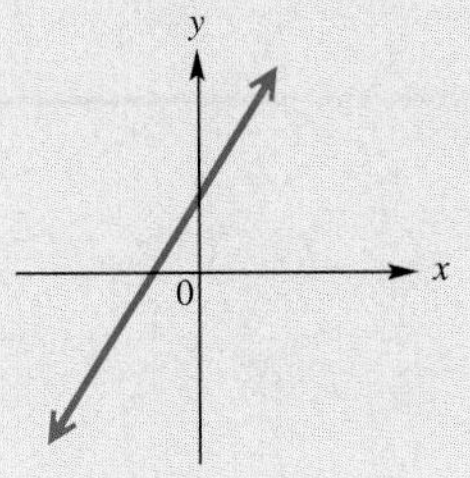

OBJECTIVE 4 Identify special systems without graphing. Example 3 showed that the graphs of an inconsistent system are parallel lines and the graphs of a system of dependent equations are the same line. We can recognize these special kinds of systems without graphing by using slopes.

CHALKBOARD EXAMPLE

Describe the system without graphing.

$$-x + 3y = 2$$
$$2x - 6y = -4$$

Answer: The equations represent the same line. The system has an infinite number of solutions.

EXAMPLE 4 Identifying the Three Cases Using Slopes

Describe each system without graphing.

(a) $3x + 2y = 6$
$-2y = 3x - 5$

Write each equation in slope-intercept form by solving for y.

$$3x + 2y = 6 \qquad\qquad -2y = 3x - 5$$
$$2y = -3x + 6 \qquad\qquad y = -\frac{3}{2}x + \frac{5}{2}$$
$$y = -\frac{3}{2}x + 3$$

Both equations have slope $-\frac{3}{2}$ but they have different y-intercepts, 3 and $\frac{5}{2}$. In Chapter 3 we found that lines with the same slope are parallel, so these equations have graphs that are parallel lines. The system has no solution.

(b) $2x - y = 4$
$x = \frac{y}{2} + 2$

Again, write the equations in slope-intercept form.

$$2x - y = 4 \qquad\qquad x = \frac{y}{2} + 2$$
$$-y = -2x + 4 \qquad\qquad \frac{y}{2} + 2 = x$$
$$y = 2x - 4 \qquad\qquad \frac{y}{2} = x - 2$$
$$y = 2x - 4$$

The equations are exactly the same; their graphs are the same line. The system has an infinite number of solutions.

(c) $x - 3y = 5$
$2x + y = 8$

In slope-intercept form, the equations are as follows.

$$x - 3y = 5 \qquad\qquad 2x + y = 8$$
$$-3y = -x + 5 \qquad\qquad y = -2x + 8$$
$$y = \frac{1}{3}x - \frac{5}{3}$$

The graphs of these equations are neither parallel lines nor the same line since the slopes are different. This system has exactly one solution.

OBJECTIVE 5 **Recognize how a graphing calculator is used to solve a linear system.** The next example shows the graphical method of solution of a system used by a graphing calculator.

CHALKBOARD EXAMPLE

The screen here shows how a graphing calculator illustrates the solution of the system from the Chalkboard Example for text Example 2,

$$5x - 3y = 9$$
$$x + 2y = 7.$$

What is the solution set of the system of equations?

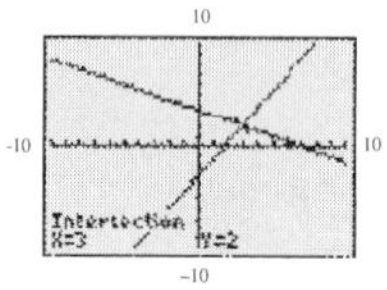

Answer: $\{(3, 2)\}$

EXAMPLE 5 Illustrating the Solution of a System with a Graphing Calculator

Figures 4(a) and (b) illustrate how a graphing calculator can represent the solution of the system in Example 1(a),

$$x + 4y = -8$$
$$3x + 2y = 6.$$

In order to enter the equations, first solve each one for y.

$$x + 4y = -8 \qquad\qquad 3x + 2y = 6$$
$$4y = -x - 8 \qquad\qquad 2y = -3x + 6$$
$$y = -\frac{1}{4}x - 2 \qquad\qquad y = -\frac{3}{2}x + 3$$

The calculator allows us to enter several equations to be graphed at the same time. Designate the first one Y_1 and the second one Y_2. Graph the two equations using a standard window, and then use the capability of the calculator to find the coordinates of the point of intersection of the graphs. The display at the bottom of Figure 4(b) indicates that the solution set is $\{(4, -3)\}$.

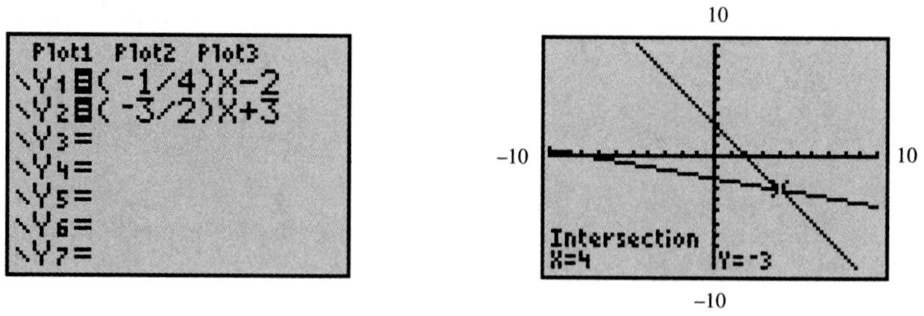

(a) (b)

Figure 4

1. (a) B (b) C (c) D (d) A
2. It is not a solution of the system because it is not a solution of the second equation, $2x + y = 4$.
3. yes **4.** yes **5.** no
6. no **7.** yes **8.** yes
9. yes **10.** yes **11.** no
12. One example is
$3x + y = 7$
$x - y = 9$.
To find the equations of such a system, just make up a typical left-hand side, like $3x + y$ or $x - y$. Then substitute 4 for x and -5 for y and evaluate to determine what the right-hand side should be.

8.1 EXERCISES

1. Each ordered pair in Column I is a solution of one of the systems graphed in Column II. Because of the location of the point of intersection, you should be able to determine the correct system for each solution. Match each system from Column II with its solution from Column I.

I

(a) (3, 4)

(b) (−2, 3)

(c) (−4, −1)

(d) (5, −2)

II

A.

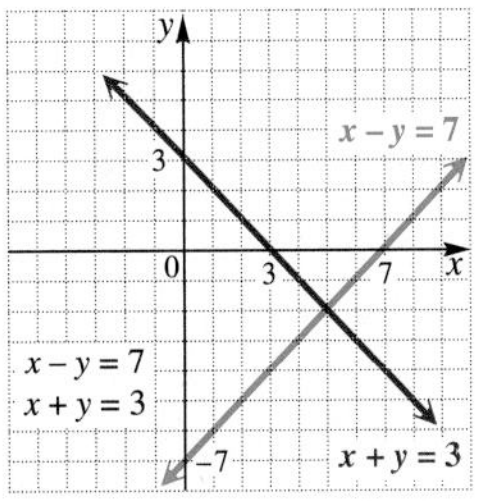

B.

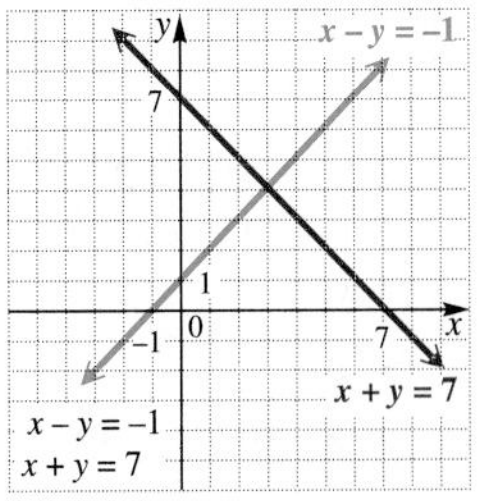

C.

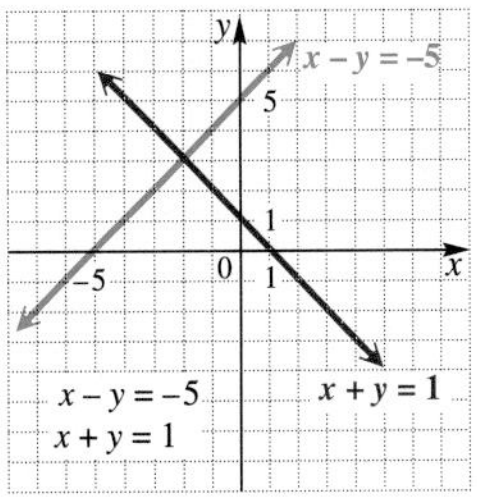

D.

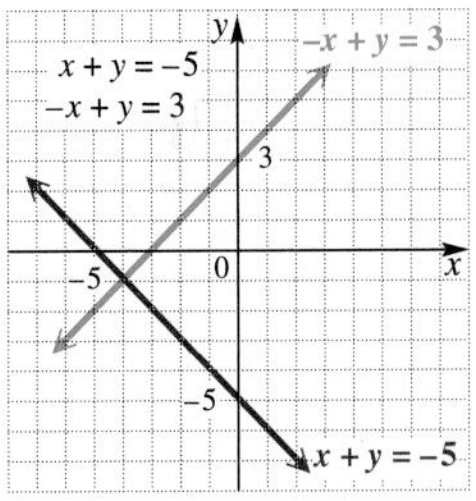

2. When a student was asked to determine whether the ordered pair $(1, -2)$ is a solution of the system

$$x + y = -1$$
$$2x + y = 4,$$

he answered "yes." His reasoning was that the ordered pair satisfies the equation $x + y = -1$: $1 + (-2) = -1$ is true. Why is the student's answer wrong?

Decide whether the given ordered pair is a solution of the given system. See Example 1.

3. (6, 2)
$3x + y = 20$
$2x + 3y = 18$

4. (3, 4)
$2x + y = 10$
$3x + 2y = 17$

5. (2, −3)
$x + y = -1$
$2x + 5y = 19$

6. (4, 3)
$x + 2y = 10$
$3x + 5y = 3$

7. (−1, −3)
$3x + 5y = -18$
$4x + 2y = -10$

8. (−9, −2)
$2x - 5y = -8$
$3x + 6y = -39$

9. (7, −2)
$4x = 26 - y$
$3x = 29 + 4y$

10. (9, 1)
$2x = 23 - 5y$
$3x = 24 + 3y$

11. (6, −8)
$-2y = x + 10$
$3y = 2x + 30$

12. Write an example of a system with solution (4, −5). Then explain how you found the equations in your system.

13. (a); The ordered pair solution must be in quadrant II, and $(-4, -4)$ is in quadrant III.
14. (d); The ordered pair solution must be on the y-axis, with $y < 0$.
15. $\{(4, 2)\}$

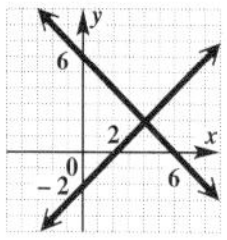

16. $\{(1, -2)\}$

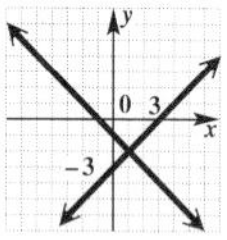

17. $\{(0, 4)\}$

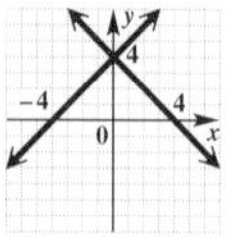

18. $\{(0, -5)\}$

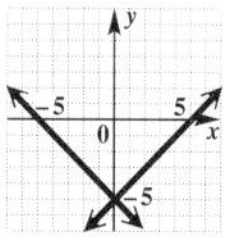

19. $\{(4, -1)\}$

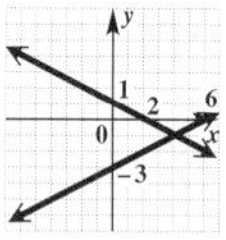

20. $\{(1, -2)\}$

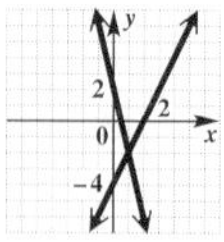

In Exercises 21–29, we do not show the graphs.
21. $\{(1, 3)\}$ 22. $\{(3, 2)\}$
23. $\{(0, 2)\}$
24. $\emptyset$ (inconsistent system)
25. $\emptyset$ (inconsistent system)
26. infinite number of solutions (dependent equations)
27. infinite number of solutions (dependent equations)
28. $\{(0, -3)\}$
29. $\{(4, -3)\}$
30. There is no way that the sum of two numbers can be both 2 and 4 at the same time.
31. Two nonparallel lines will intersect in only one point (if they are distinct) or infinitely many points (if they are the same). They cannot intersect in exactly two points.

13. Which one of the ordered pairs below could not possibly be a solution of the system graphed? Why is it the only valid choice?

(a) $(-4, -4)$ **(b)** $(-2, 2)$
(c) $(-4, 4)$ **(d)** $(-3, 3)$

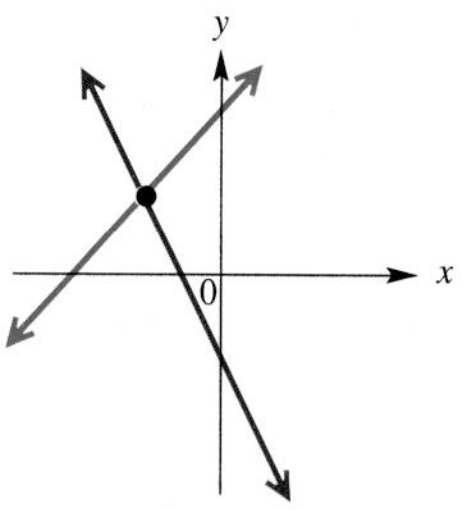

14. Which one of the ordered pairs below could possibly be a solution of the system graphed? Why is it the only valid choice?

(a) $(2, 0)$ **(b)** $(0, 2)$
(c) $(-2, 0)$ **(d)** $(0, -2)$

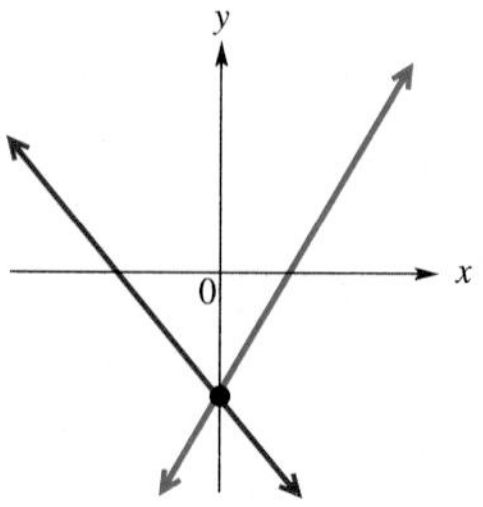

Solve each system of equations by graphing both equations on the same axes. If the system is inconsistent or the equations are dependent, say so. See Examples 2–4.

15. $x - y = 2$
$x + y = 6$

16. $x - y = 3$
$x + y = -1$

17. $x + y = 4$
$y - x = 4$

18. $x + y = -5$
$x - y = 5$

19. $x - 2y = 6$
$x + 2y = 2$

20. $2x - y = 4$
$4x + y = 2$

21. $3x - 2y = -3$
$-3x - y = -6$

22. $2x - y = 4$
$2x + 3y = 12$

23. $2x - 3y = -6$
$y = -3x + 2$

24. $x + 2y = 4$
$2x + 4y = 12$

25. $2x - y = 6$
$4x - 2y = 8$

26. $2x - y = 4$
$4x = 2y + 8$

27. $3x = 5 - y$
$6x + 2y = 10$

28. $-3x + y = -3$
$y = x - 3$

29. $3x - 4y = 24$
$y = -\frac{3}{2}x + 3$

30. How can you tell, without graphing, that the system

$$x + y = 2$$
$$x + y = 4$$

has no solution?

31. Explain why a system of two linear equations cannot have exactly two solutions.

32. Explain the three situations that may occur (regarding the number of solutions) when solving a system of two linear equations in two variables by graphing.

33. Is it possible for a system of *three* linear equations in two variables to have a single solution? If so, give an example.

Listening to recorded music has been an extremely popular form of entertainment for many years. During the early years, the most common format for recorded music was the vinyl phonograph record, available in various speeds (78, 45, and $33\frac{1}{3}$ revolutions per minute). The long-playing record (LP) and single (45) were the mainstays in the 1950s and remained so until the early 1980s. Another popular format among audiophiles was the reel-to-reel tape. The 1960s and 1970s saw the emergence of the now defunct four- and eight-track tape formats, and during those years the familiar audiocassette (or simply "cassette") was in its infancy. Music lovers today rely on recordings that are almost exclusively found in the cassette and compact disc (CD) formats.

32. If the lines have the same slope and different y-intercepts, they are parallel and the system has no solution. If the slopes are the same and the y-intercepts are the same, they coincide and the system has infinitely many solutions. If the slopes are unequal, they intersect in only one point and the system has only one solution. **33.** Yes, it is possible. For example, the system
$x + y = 5$
$x - y = -1$
$2x - y = 1$
has the single solution (2, 3).
34. 1991 **35.** about 350 million for each format **36.** (1987, 100 million) **37.** 200 million **38.** from 1983 to 1987 **39.** between 1984 and 1986, and between 1988 and 1990 **40.** positive **41.** (a) neither (b) intersecting lines (c) one solution **42.** (a) neither (b) intersecting lines (c) one solution **43.** (a) dependent (b) one line (c) infinite number of solutions **44.** (a) inconsistent (b) parallel lines (c) no solution **45.** (a) inconsistent (b) parallel lines (c) no solution **46.** (a) neither (b) intersecting lines (c) one solution

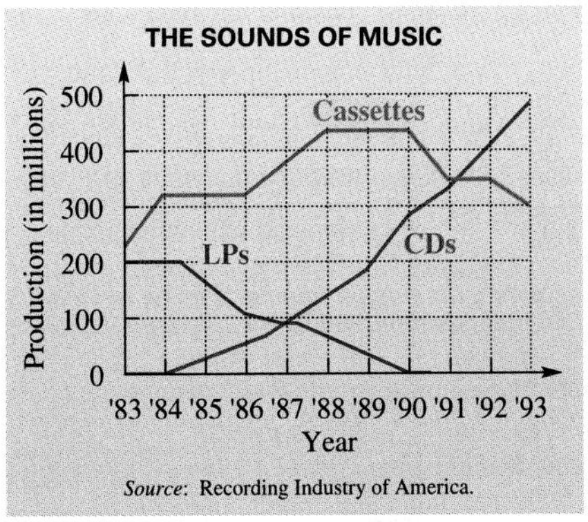

The graph above shows how the production of vinyl LPs, cassettes, and CDs changed over the years from 1983 to 1993. Use the graph to respond to Exercises 34–40.

34. In what year did cassette production and CD production reach equal levels?

35. (See Exercise 34.) What were those production levels?

36. Express as an ordered pair in the form (year, production level) the point of intersection of the graphs of LP production and CD production.

37. For 1987, what was the *total* production for LPs and CDs?

38. For which years in this period was the production of CDs less than the production of LPs?

39. Between which two nonconsecutive years was the production of cassettes approximately constant?

40. If a straight line were used to model the graph for CD production from 1984 to 1993, would its slope be positive, negative, or zero?

Without graphing, answer the following questions for each linear system. See Example 4.
(a) *Is the system inconsistent, are the equations dependent, or neither?*
(b) *Is the graph a pair of intersecting lines, a pair of parallel lines, or one line?*
(c) *Does the system have one solution, no solution, or an infinite number of solutions?*

41. $y - x = -5$
$x + y = 1$

42. $2x + y = 6$
$x - 3y = -4$

43. $x + 2y = 0$
$4y = -2x$

44. $y = 3x$
$y + 3 = 3x$

45. $5x + 4y = 7$
$10x + 8y = 4$

46. $2x + 3y = 12$
$2x - y = 4$

An application of mathematics in economics deals with **supply and demand.** Typically, as the price of an item increases, the demand for the item decreases, while the supply increases. (There are exceptions to this, however.) If supply and demand can be described by straight-line equations, the point at which the lines intersect determines the **equilibrium supply** and **equilibrium demand.** Suppose that an economist has studied the supply and demand for aluminum siding and has concluded that the price per unit, p, and the demand, x, are related by the demand equation

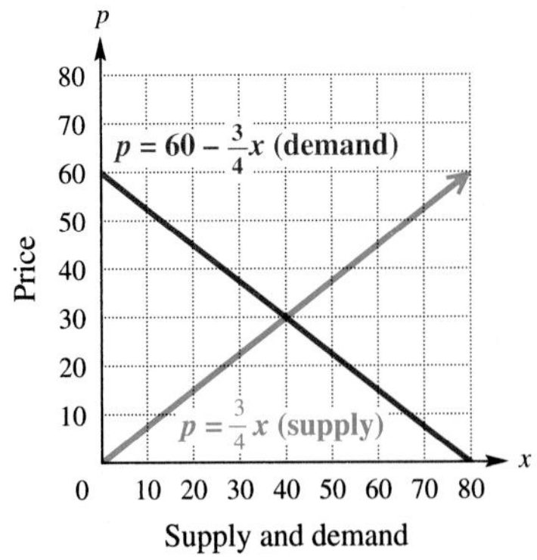

$p = 60 - \frac{3}{4}x$, while the supply is given by the equation $p = \frac{3}{4}x$. The graphs of these two equations are shown here.

47. 40 **48.** 30 **49.** (40, 30)
50. Supply exceeds demand.

Use the graph on the previous page to answer each question.

47. At what value of x does supply equal demand?

48. At what value of p does supply equal demand?

49. What are the coordinates of the point of intersection of the two lines?

50. When $x > 40$, does demand exceed supply or does supply exceed demand?

51. B **52.** D **53.** A
54. C

TECHNOLOGY INSIGHTS (EXERCISES 51–54)

Match the graphing calculator screens from choices A–D with the appropriate system in Exercises 51–54. See Example 5.

A. 10

−10 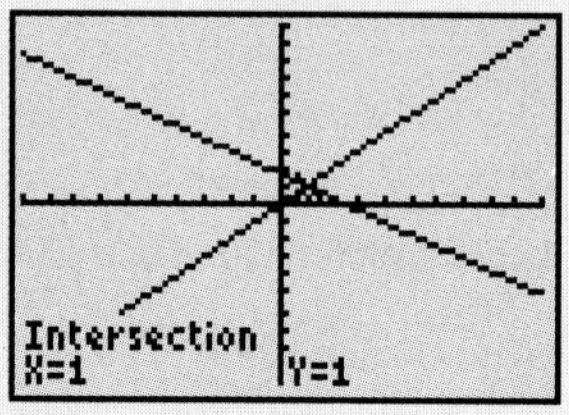 10

−10

B. 10

−10 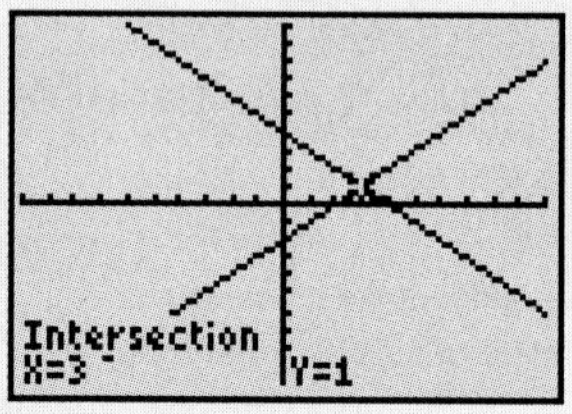 10

−10

C. 10

−10 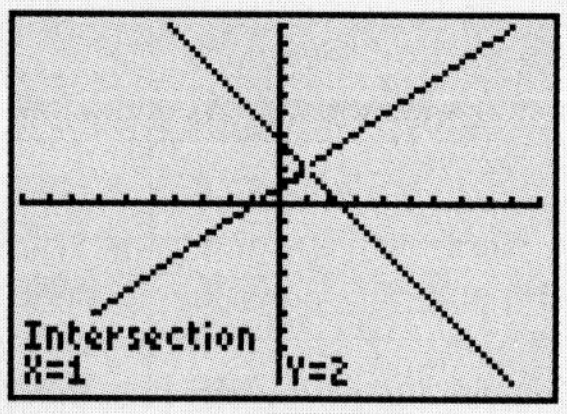 10

−10

D. 10

−10 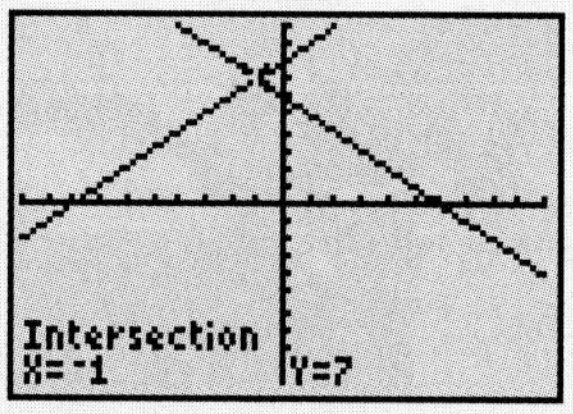 10

−10

51. $x + y = 4$
$x - y = 2$

52. $x + y = 6$
$x - y = -8$

53. $2x + 3y = 5$
$x - y = 0$

54. $3x + 2y = 7$
$-x + y = 1$

55. $\{(-1, 5)\}$ **56.** $\{(4, -3)\}$

Use a graphing calculator to solve each system. See Example 5.

55. $3x + y = 2$
$2x - y = -7$

56. $x + 2y = -2$
$2x - y = 11$

*These problems can be used as collaborative exercises with small groups of students.

57. $\{2\}$ **58.** 5

RELATING CONCEPTS* (EXERCISES 57–62)

In these exercises we relate solving linear equations in one variable and solving a linear system of two equations in two variables.

Work Exercises 57–62 in order.

57. Solve the linear equation $\frac{1}{2}x + 4 = 3x - 1$ using the methods described in Chapter 2.

58. Check your solution for the equation in Exercise 57 by substituting back into the original equation. What is the value that you get for both the left and right sides after you make this substitution?

59. {(2, 5)}

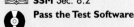

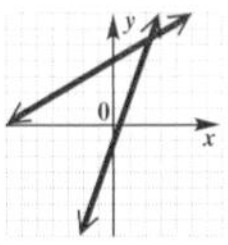

60. The x-coordinate, 2, is equal to the solution of the equation.
61. The y-coordinate, 5, is equal to the value we obtained on both sides when checking. 62. 5; 3; 5

RELATING CONCEPTS (EXERCISES 57–62) (CONTINUED)

59. Graph the linear system and find the solution set.

$$y = \frac{1}{2}x + 4$$
$$y = 3x - 1$$

60. How does the x-coordinate of the solution of the system in Exercise 59 compare to the solution of the linear equation in Exercise 57?

61. How does the y-coordinate of the solution of the system in Exercise 59 compare to the value you obtained in the check in Exercise 58?

62. Based on your observations in Exercises 57–61, fill in the blanks with the correct responses.

The solution of the linear equation $\frac{2}{3}x + 3 = -x + 8$ is 3. When we substitute 3 back into the equation, we get the value _____ on both sides, verifying that 3 is indeed the solution. Now if we graph the system

$$y = \frac{2}{3}x + 3$$
$$y = -x + 8,$$

we find that the solution set of the system is {(_____, _____)}.

Did you make the connection between solving equations in one variable and solving systems in two variables?

8.2 Solving Systems of Linear Equations by Substitution

OBJECTIVES

1. Solve linear systems by substitution.
2. Solve special systems.
3. Solve linear systems with fractions.

FOR EXTRA HELP

SSG Sec. 8.2
SSM Sec. 8.2
Pass the Test Software
InterAct Math Tutorial Software
Video 13

CHALKBOARD EXAMPLE

Solve the system.

$$2x + 7y = -12$$
$$x = -2y$$

Answer: {(8, −4)}

Graphing to solve a system of equations has a serious drawback. It is difficult to estimate a solution such as $\left(\frac{1}{3}, -\frac{5}{6}\right)$ accurately from a graph (unless a graphing calculator is used).

OBJECTIVE 1 Solve linear systems by substitution. An algebraic method involving substitution is particularly useful for solving systems where either equation is solved, or can be solved easily, for one of the variables.

EXAMPLE 1 Solving a System by Substitution

Solve the system

$$3x + 5y = 26$$
$$y = 2x.$$

The second equation is solved for y. Substituting $2x$ for y in the first equation gives

$$\begin{aligned} 3x + 5y &= 26 \\ 3x + 5(2x) &= 26 \qquad \text{Let } y = 2x. \\ 3x + 10x &= 26 \\ 13x &= 26 \\ x &= 2. \end{aligned}$$

Since $y = 2x$ and $x = 2$, $y = 2(2) = 4$. Check that the solution set of the given system is {(2, 4)} by substituting 2 for x and 4 for y in *both* equations.

INSTRUCTOR'S RESOURCES

PTB Sec. 8.2 **ISM** Sec. 8.2
AB Sec. 8.2

TEST GENERATOR

WORLD WIDE WEB
www.LialAlgebra.com

TEACHING TIP Mention that the substitution method is also essential in solving some nonlinear systems of equations, a topic covered in intermediate and college algebra courses.

TEACHING TIP Emphasize that the solution set found by the substitution method will be the same as the solution set found by graphing. The solution set is the same; only the method is different.

CHALKBOARD EXAMPLE

Use substitution to solve the system.

$$x + 1 = -4y$$
$$2x - 5y = 11$$

Answer: $\{(3, -1)\}$

To solve a system by substitution, follow these steps.

Solving Linear Systems by Substitution

Step 1 **Solve for a variable.** Solve one of the equations for either variable. (If one of the variables has coefficient 1 or -1, choose it, since the substitution method is usually easier this way.)

Step 2 **Substitute.** Substitute for that variable in the other equation. The result should be an equation with just one variable.

Step 3 **Solve.** Solve the equation from Step 2.

Step 4 **Substitute.** Substitute the result from Step 3 into the equation from Step 1 to find the value of the other variable.

Step 5 **Find the solution set.** Check the solution in both of the given equations. Then write the solution set.

EXAMPLE 2 Solving a System by Substitution

Use substitution to solve the system

$$2x = 8 - 3y$$
$$4x + 3y = 4.$$

Step 1 The substitution method requires that an equation be solved for one of the variables. Choose the first equation of the system and solve it for x.

$$2x = 8 - 3y$$
$$x = \frac{8 - 3y}{2} \quad \text{Divide both sides by 2.}$$

Step 2 Now we substitute this value for x in the second equation of the system.

$$4x + 3y = 4$$
$$4\left(\frac{8 - 3y}{2}\right) + 3y = 4 \quad \text{Let } x = \frac{8 - 3y}{2}.$$

Step 3 Solve this equation.

$$2(8 - 3y) + 3y = 4 \quad \text{Divide 4 by 2.}$$
$$16 - 6y + 3y = 4 \quad \text{Distributive property}$$
$$-3y = -12 \quad \text{Combine terms; subtract 16.}$$
$$y = 4 \quad \text{Divide by } -3.$$

Step 4 Find x by letting $y = 4$ in $x = \dfrac{8 - 3y}{2}$.

$$x = \frac{8 - 3 \cdot 4}{2} = \frac{8 - 12}{2} = \frac{-4}{2} = -2$$

Step 5 The solution set of the given system is $\{(-2, 4)\}$. Check the solution in both equations.

OBJECTIVE 2 Solve special systems. In the previous section we solved inconsistent systems with graphs that are parallel lines and systems of dependent equations with graphs that are the same line. We can also solve these special systems with the substitution method.

CHALKBOARD EXAMPLE

Use substitution to solve the system.

$$y = 8x + 4$$
$$16x - 2y = 8$$

Answer: $\emptyset$

EXAMPLE 3 Solving an Inconsistent System by Substitution

Use substitution to solve the system

$$x = 5 - 2y$$
$$2x + 4y = 6.$$

Substitute $5 - 2y$ for x in the second equation.

$$2x + 4y = 6$$
$$2(\mathbf{5 - 2y}) + 4y = 6 \qquad \text{Let } x = 5 - 2y.$$
$$10 - 4y + 4y = 6 \qquad \text{Distributive property}$$
$$10 = 6 \qquad \text{False}$$

This false result means that the system is inconsistent and its solution set is $\emptyset$. The equations of the system have graphs that are parallel lines. See Figure 5.

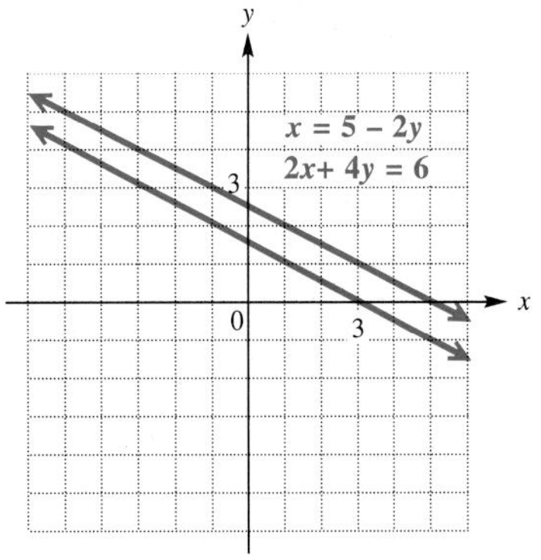

Figure 5

CAUTION It is a common error to give "false" as the answer to an inconsistent system. The correct response is $\emptyset$.

CHALKBOARD EXAMPLE

Solve the following system by the substitution method.

$$-x - 3y = 7$$
$$4x + 12y = -28$$

Answer: infinite number of solutions

EXAMPLE 4 Solving a System with Dependent Equations by Substitution

Solve the following system by the substitution method.

$$3x - y = 4 \qquad (1)$$
$$-9x + 3y = -12 \qquad (2)$$

Begin by solving the first equation for y to get $y = 3x - 4$. Substitute $3x - 4$ for y in equation (2) and solve the resulting equation.

$$-9x + 3(\mathbf{3x - 4}) = -12$$
$$-9x + 9x - 12 = -12 \qquad \text{Distributive property}$$
$$0 = 0 \qquad \text{Add 12; combine terms.}$$

This true result means that every solution of one equation is also a solution of the other, so the system has an infinite number of solutions: all the ordered pairs corresponding to points that lie on the common graph. A graph of the equations of this system is shown in Figure 6.

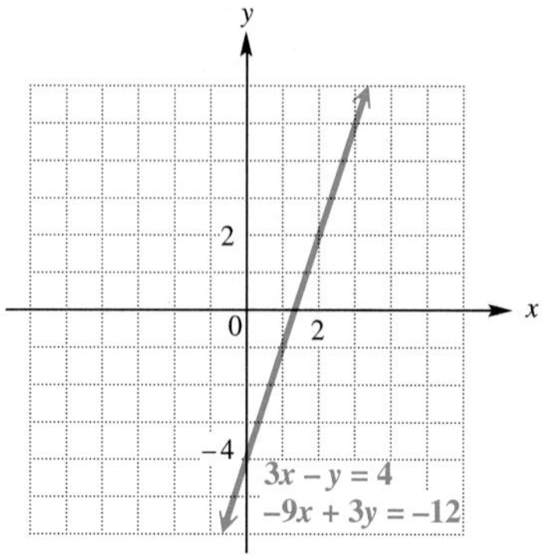

Figure 6

CAUTION It is a common error to give "true" as the answer to a dependent system. The correct response is "infinite number of solutions."

OBJECTIVE 3 Solve linear systems with fractions. When a system includes equations with fractions as coefficients, eliminate the fractions by multiplying both sides by a common denominator. Then solve the resulting system.

CHALKBOARD EXAMPLE

Solve the system.

$$\frac{2}{3}x + \frac{1}{2}y = 6$$
$$\frac{1}{2}x - \frac{3}{4}y = 0$$

Answer: {(6, 4)}

EXAMPLE 5 Solving by Substitution with Fractions as Coefficients

Solve the system

$$4x + \frac{1}{3}y = \frac{8}{3} \qquad (1)$$

$$\frac{1}{2}x + \frac{3}{4}y = -\frac{5}{2}. \qquad (2)$$

Begin by eliminating fractions. Clear equation (1) of fractions by multiplying both sides by 3.

$$3\left(4x + \frac{1}{3}y\right) = 3 \cdot \frac{8}{3} \qquad \text{Multiply by 3.}$$

$$3(4x) + 3\left(\frac{1}{3}y\right) = 3 \cdot \frac{8}{3} \qquad \text{Distributive property}$$

$$12x + y = 8 \qquad (3)$$

Now clear equation (2) of fractions by multiplying both sides by the common denominator 4.

$$4\left(\frac{1}{2}x + \frac{3}{4}y\right) = 4\left(-\frac{5}{2}\right) \qquad \text{Multiply by 4.}$$

$$4\left(\frac{1}{2}x\right) + 4\left(\frac{3}{4}y\right) = 4\left(-\frac{5}{2}\right) \qquad \text{Distributive property}$$

$$2x + 3y = -10 \qquad (4)$$

The given system of equations has been simplified to

$$12x + y = 8 \tag{3}$$
$$2x + 3y = -10. \tag{4}$$

Solve the system by the substitution method. Equation (3) can be solved for y by subtracting $12x$ on each side.

$$12x + y = 8 \tag{3}$$
$$y = -12x + 8$$

Now substitute the result for y in equation (4).

$$2x + 3(-12x + 8) = -10 \qquad \text{Let } y = -12x + 8.$$
$$2x - 36x + 24 = -10 \qquad \text{Distributive property}$$
$$-34x = -34 \qquad \text{Combine terms; subtract 24.}$$
$$x = 1 \qquad \text{Divide by } -34.$$

Substitute $x = 1$ in $y = -12x + 8$ to get $y = -4$. Check by substituting 1 for x and -4 for y in the original equations. The solution set is $\{(1, -4)\}$.

8.2 EXERCISES

1. No, it is not correct, because the solution set is $\{(3, 0)\}$. The y-value in the ordered pair must also be determined. 2. No. Obtaining $0 = 0$ means that the system has infinitely many solutions, not that $(0, 0)$ is a solution. 3. $\{(3, 9)\}$
4. $\{(2, -10)\}$ 5. $\{(7, 3)\}$
6. $\{(-2, 1)\}$ 7. $\{(0, 5)\}$
8. $\{(-3, 0)\}$ 9. $\{(-4, 8)\}$
10. $\{(3, -1)\}$ 11. $\{(3, -2)\}$
12. $\{(7, 1)\}$ 13. infinite number of solutions 14. infinite number of solutions 15. $\left\{ \left(\frac{1}{3}, -\frac{1}{2} \right) \right\}$
16. $\left\{ \left(\frac{5}{3}, \frac{1}{2} \right) \right\}$ 17. $\emptyset$
18. $\emptyset$ 19. infinite number of solutions 20. infinite number of solutions

 1. A student solves the system

$$5x - y = 15$$
$$7x + y = 21$$

and finds that $x = 3$, which is the correct value for x. The student gives the solution set as $\{3\}$. Is this correct? Explain.

 2. A student solves the system

$$x + y = 4$$
$$2x + 2y = 8$$

and obtains the equation $0 = 0$. The student gives the solution set as $\{(0, 0)\}$. Is this correct? Explain.

Solve each system by the substitution method. Check your solution. See Examples 1–4.

3. $x + y = 12$
$ y = 3x$

4. $x + 3y = -28$
$y = -5x$

5. $3x + 2y = 27$
$ x = y + 4$

6. $4x + 3y = -5$
$x = y - 3$

7. $3x + 5y = 25$
$ x - 2y = -10$

8. $5x + 2y = -15$
$2x - y = -6$

9. $3x + 4 = -y$
$2x + y = 0$

10. $2x - 5 = -y$
$x + 3y = 0$

11. $7x + 4y = 13$
$x + y = 1$

12. $3x - 2y = 19$
$x + y = 8$

13. $3x - y = 5$
$ y = 3x - 5$

14. $4x - y = -3$
$ y = 4x + 3$

15. $6x - 8y = 6$
$-3x + 2y = -2$

16. $3x + 2y = 6$
$-6x + 4y = -8$

17. $2x + 8y = 3$
$ x = 8 - 4y$

18. $2x + 10y = 3$
$x = 1 - 5y$

19. $12x - 16y = 8$
$3x = 4y + 2$

20. $6x + 9y = 6$
$2x = 2 - 3y$

 21. Professor Brandsma gave the following item on a test in introductory algebra:
Solve the following system by the substitution method.

$$3x - y = 13$$
$$2x + 5y = 20$$

21. The first student had less work to do, because the coefficient of y in the first equation is -1. The second student had to divide by 2, introducing fractions into the expression for x. 22. (a) A false statement, such as $0 = 3$, occurs. (b) A true statement, such as $0 = 0$, occurs. 23. $\{(2, -3)\}$ 24. $\{(18, -12)\}$ 25. $\{(3, 2)\}$ 26. $\{(3, 2)\}$ 27. $\{(-2, 1)\}$ 28. $\{(3, -1)\}$ 29. 1993 30. 1994–1995 31. To find the total cost, multiply the number of bicycles (x) by the cost per bicycle ($400), and add the fixed cost ($5000). Thus, $y_1 = 400x + 5000$ gives this total cost (in dollars). 32. $y_2 = 600x$ 33. $y_1 = 400x + 5000$, $y_2 = 600x$; solution set: $\{(25, 15{,}000)\}$ 34. 25; 15,000; 15,000 35. $\{(2, 4)\}$

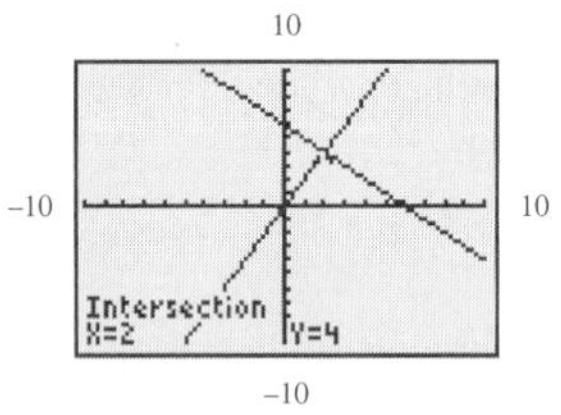

36. $\{(-1, -8)\}$

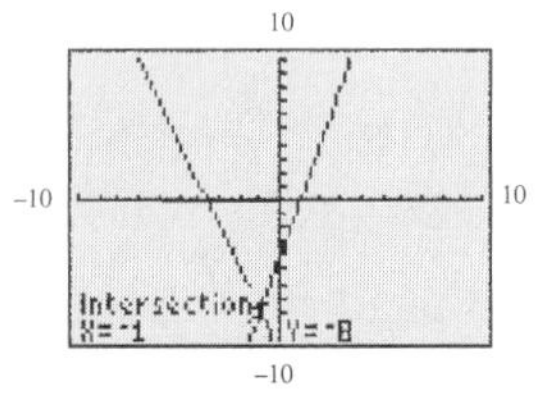

37. $\{(1, 5)\}$

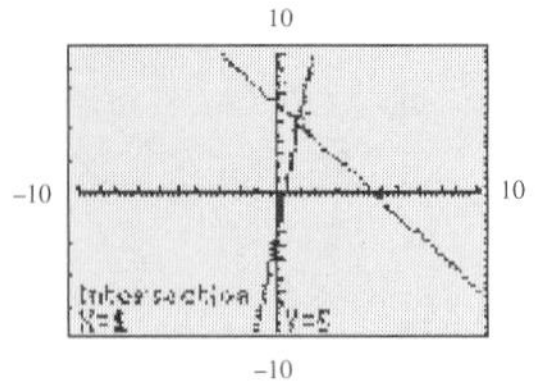

One student worked the problem by solving first for y in the first equation. Another student worked it by solving first for x in the second equation. Both students got the correct solution, (5, 2). Which student, do you think, had less work to do? Explain.

22. When you use the substitution method, how can you tell that a system has
(a) no solution?
(b) an infinite number of solutions?

In each system given, begin by clearing fractions and then solve by substitution. See Example 5.

23. $\frac{5}{3}x + 2y = \frac{1}{3} + y$
$2x - 3 + \frac{y}{3} = -2 + x$

24. $\frac{x}{6} + \frac{y}{6} = 1$
$-\frac{1}{2}x - \frac{1}{3}y = -5$

25. $\frac{x}{2} - \frac{y}{3} = \frac{5}{6}$
$\frac{x}{5} - \frac{y}{4} = \frac{1}{10}$

26. $\frac{x}{3} - \frac{3y}{4} = -\frac{1}{2}$
$\frac{2x}{3} + \frac{y}{2} = 3$

27. $\frac{x}{5} + 2y = \frac{8}{5}$
$\frac{3x}{5} + \frac{y}{2} = -\frac{7}{10}$

28. $\frac{x}{2} + \frac{y}{3} = \frac{7}{6}$
$\frac{x}{4} - \frac{3y}{2} = \frac{9}{4}$

Attending professional sporting events ranks among the most popular forms of entertainment. Use systems of equations to solve the sports entertainment problems in Exercises 29 and 30.

29. During the period from 1991 to 1996, average ticket prices rose in the National Football League from $25.21 to $35.74. If we let $x = 1$ represent 1991, $x = 2$ represent 1992, and so on, the linear equation $y = 2.1x + 22.8$ gives a good approximation for this average price, where y is in dollars. To determine the year in which the average ticket price was $28.68, solve the system

$$y = 2.1x + 22.8$$
$$y = 28.68.$$

The rounded x-value will then give us the year. Solve this system by substitution to determine the year. (*Hint:* After finding the value of x, you must then determine which year it represents.) (*Source:* Team Marketing Report, Chicago.)

30. During the period from the 1991–1992 season through the 1996–1997 season, the average price of a ticket to a National Basketball Association game rose from $23.24 to $34.08. If we let $x = 1$ represent the 1991–1992 season, $x = 2$ represent 1992–1993, and so on, the linear equation $y = 2.14x + 20.81$ gives a good approximation for this average price, where y is in dollars. To determine the period in which the average ticket price was $28.63, solve the system

$$y = 2.14x + 20.81$$
$$y = 28.63.$$

The rounded x-value will then give us the first year of the period. Solve this system by substitution to determine the period. See the hint in Exercise 29. (*Source:* Team Marketing Report, Chicago.)

38. $\{(2, -5)\}$

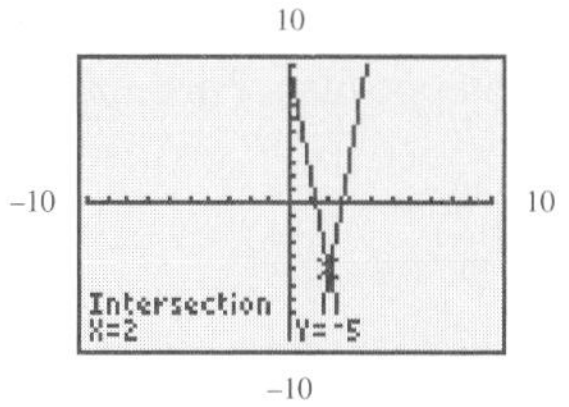

39. $\{(5, -3)\}$; **The equations to input are** $y_1 = \frac{5 - 4x}{5}$ **and** $y_2 = \frac{1 - 2x}{3}$.

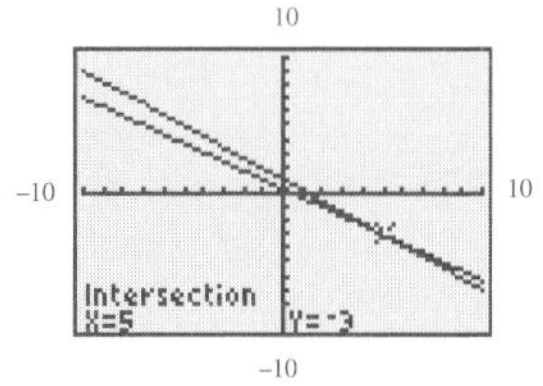

40. $\left\{\left(\frac{19}{3}, -5\right)\right\}$; **The equations to input are** $y_1 = \frac{13 - 6x}{5}$ **and** $y_2 = \frac{4 - 3x}{3}$.

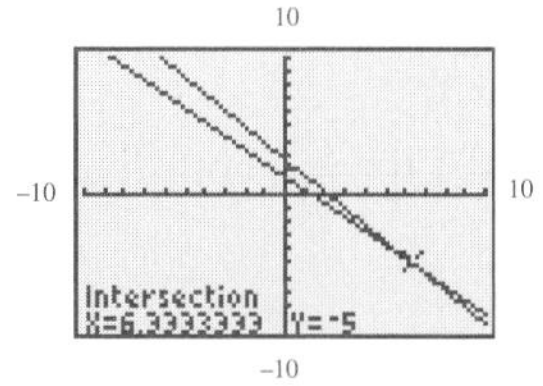

41. Adjust the viewing window so that it does appear.

42. The substitution method would require arithmetic with decimals. A graphing calculator would provide the solution without requiring the arithmetic.

RELATING CONCEPTS (EXERCISES 31–34)

A system of linear equations can be used to model the cost and the revenue of a business.

Work Exercises 31–34 in order.

31. Suppose that you start a business manufacturing and selling bicycles, and it costs you \$5000 to get started. You determine that each bicycle will cost \$400 to manufacture. Explain why the linear equation $y_1 = 400x + 5000$ gives your *total* cost to manufacture x bicycles (y_1 in dollars).

32. You decide to sell each bike for \$600. What expression in x represents the revenue you will take in if you sell x bikes? Write an equation using y_2 to express your revenue when you sell x bikes (y_2 in dollars).

33. Form a system from the two equations in Exercises 31 and 32 and then solve the system.

34. The value of x from Exercise 33 is the number of bikes it takes to *break even.* Fill in the blanks: When _____ bikes are sold, the break-even point is reached. At that point, you have spent _____ dollars and taken in _____ dollars.

Did you make the connections between x and the number of bikes, and y and the amount taken in?

Solve each system by substitution. Then graph both lines in the same viewing window of a graphing calculator and use the intersection feature to support your answer. See Example 5 in Section 8.1. (In Exercises 39 and 40 you will need to solve each equation for y first before graphing.)

35. $y = 6 - x$
$y = 2x$

36. $y = 4x - 4$
$y = -3x - 11$

37. $y = -\frac{4}{3}x + \frac{19}{3}$
$y = \frac{15}{2}x - \frac{5}{2}$

38. $y = -\frac{15}{2}x + 10$
$y = \frac{25}{3}x - \frac{65}{3}$

39. $4x + 5y = 5$
$2x + 3y = 1$

40. $6x + 5y = 13$
$3x + 3y = 4$

41. If the point of intersection does not appear on your screen when solving a linear system using a graphing calculator, how can you find the point of intersection?

42. Suppose that you were asked to solve the system

$$y = 1.73x + 5.28$$
$$y = -2.94x - 3.85.$$

Why would it probably be easier to solve this system using a graphing calculator than using the substitution method?

8.3 Solving Systems of Linear Equations by Elimination

OBJECTIVES

1. Solve linear systems by elimination.
2. Multiply when using the elimination method.

OBJECTIVE 1 Solve linear systems by elimination. An algebraic method that depends on the addition property of equality can also be used to solve systems. As mentioned earlier, adding the same quantity to each side of an equation results in equal sums.

$$\text{If} \quad A = B, \quad \text{then} \quad A + C = B + C.$$

This addition can be taken a step further. Adding *equal* quantities, rather than the *same* quantity, to both sides of an equation also results in equal sums.

$$\text{If} \quad A = B \quad \text{and} \quad C = D, \quad \text{then} \quad A + C = B + D.$$

INSTRUCTOR'S RESOURCES

PTB Sec. 8.3 **ISM** Sec. 8.3
AB Sec. 8.3

TEST GENERATOR

WORLD WIDE WEB
www.LialAlgebra.com

3 Use an alternative method to find the second value in a solution.

4 Use the elimination method to solve special systems.

FOR EXTRA HELP

SSG Sec. 8.3
SSM Sec. 8.3
Pass the Test Software
InterAct Math Tutorial Software
Video 13

CHALKBOARD EXAMPLE

Use the elimination method to solve the system.

$$3x - y = 7$$
$$2x + y = 3$$

Answer: $\{(2, -1)\}$

TEACHING TIP Remind students that the solution (4, 1) is the point of intersection of the two lines, and that it is the solution that would be obtained by using the graphing method.

Using the addition property to solve systems is called the **elimination method.** When using this method, the idea is to *eliminate* one of the variables. To do this, one of the variables in the two equations must have coefficients that are opposites. For most systems, this method is more efficient than graphing.

EXAMPLE 1 Using the Elimination Method

Use the elimination method to solve the system

$$x + y = 5$$
$$x - y = 3.$$

Each equation in this system is a statement of equality, so, as we discussed above, the sum of the right sides equals the sum of the left sides. Adding in this way gives

$$(x + y) + (x - y) = 5 + 3.$$

Combine terms and simplify to get

$$2x = 8$$
$$x = 4. \qquad \text{Divide by 2.}$$

Notice that y has been eliminated. The result, $x = 4$, gives the x-value of the solution of the given system. To find the y-value of the solution, substitute 4 for x in either of the two equations of the system. Choosing the first equation, $x + y = 5$, gives

$$x + y = 5$$
$$4 + y = 5 \qquad \text{Let } x = 4.$$
$$y = 1.$$

The solution, (4, 1), can be checked by substituting 4 for x and 1 for y in both equations of the given system.

$$x + y = 5 \qquad\qquad x - y = 3$$
$$4 + 1 = 5 \quad ? \qquad\qquad 4 - 1 = 3 \quad ?$$
$$5 = 5 \quad \text{True} \qquad\qquad 3 = 3 \quad \text{True}$$

Since both results are true, the solution set of the given system is $\{(4, 1)\}$.

CAUTION A system is not completely solved until values for *both* x and y are found. Do not stop after finding the value of only one variable. Remember to write the solution set as a set containing an ordered pair.

In general, we use the following steps to solve a linear system of equations by the elimination method.

Solving Linear Systems by Elimination

Step 1 **Write in standard form.** Write both equations of the system in standard form $Ax + By = C$.

Step 2 **Multiply.** Multiply one or both equations by appropriate numbers (if necessary) so that the coefficients of x (or y) are opposites of each other.

Step 3 **Add.** Add the two equations to get an equation with only one variable.

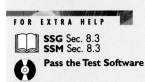

Solving Linear Systems by Elimination (continued)

Step 4 **Solve.** Solve the equation from Step 3.

Step 5 **Substitute.** Substitute the solution from Step 4 into either of the original equations to find the value of the remaining variable.

Step 6 **Find the solution set.** Check the solution in both of the original equations. Then write the solution set.

It does not matter which variable is eliminated first. Usually we choose the one that is more convenient to work with.

CHALKBOARD EXAMPLE

Solve the system.

$$x + 2 = -3y$$
$$3x - 4x + 7 = 2y + 7$$

Answer: $\{(4, -2)\}$

EXAMPLE 2 Using the Elimination Method

Solve the system.

$$y + 11 = 2x$$
$$4 + 5x + y = 2y + 30$$

Step 1 Rewrite both equations in the form $Ax + By = C$, getting the system

$$-2x + y = -11 \qquad \text{Subtract } 2x \text{ and } 11.$$
$$5x - y = 26. \qquad \text{Subtract 4 and } 2y.$$

Step 2 Because the coefficients of y are 1 and -1, adding will eliminate y. It is not necessary to multiply either equation by a number.

Step 3 Add the two equations. This time we use vertical addition.

$$\begin{array}{r} -2x + y = -11 \\ 5x - y = 26 \\ \hline 3x = 15 \end{array} \qquad \text{Add in columns.}$$

Step 4 Solve the equation.

$$3x = 15$$
$$x = 5 \qquad \text{Divide by 3.}$$

Step 5 Find the value of y by substituting 5 for x in either of the original equations. Choosing the first gives

$$y + 11 = 2x$$
$$y + 11 = 2(5) \qquad \text{Let } x = 5.$$
$$y = 10 - 11 \qquad \text{Subtract 11.}$$
$$y = -1.$$

Step 6 The solution set is $\{(5, -1)\}$. Check the solution by substitution into both of the original equations. Let $x = 5$ and $y = -1$.

$$y + 11 = 2x \qquad\qquad 4 + 5x + y = 2y + 30$$
$$(-1) + 11 = 2(5) \quad ? \qquad\qquad 4 + 5(5) + (-1) = 2(-1) + 30 \quad ?$$
$$10 = 10 \quad \text{True} \qquad\qquad 28 = 28 \quad \text{True}$$

Since $(5, -1)$ is a solution of *both* equations, our solution set is correct.

TEACHING TIP As explained in the text, some systems are easier to solve by elimination than by substitution. A good oral exercise is to ask for the situations in which substitution is a more reasonable choice, if a choice of method is given.

OBJECTIVE 2 Multiply when using the elimination method. In both examples above, a variable was eliminated by adding the equations. Sometimes we need to multiply both sides of one or both equations in a system by some number before adding will eliminate a variable.

CHALKBOARD EXAMPLE

Solve the system.

$$4x - 5y = -18$$
$$3x + 2y = -2$$

Answer: $\{(-2, 2)\}$

EXAMPLE 3 Multiplying Both Equations When Using the Elimination Method

Solve the system.

$$2x + 3y = -15 \quad \textbf{(1)}$$
$$5x + 2y = 1 \quad \textbf{(2)}$$

Adding the two equations gives $7x + 5y = -14$, which does not eliminate either variable. However, we can multiply each equation by a suitable number so that the coefficients of one of the two variables are opposites. For example, to eliminate x, multiply both sides of equation (1) by 5, and both sides of equation (2) by -2.

$$\begin{array}{rl} 10x + 15y = -75 & \text{Multiply equation (1) by 5.} \\ -10x - 4y = -2 & \text{Multiply equation (2) by } -2. \\ \hline 11y = -77 & \text{Add.} \\ y = -7 & \end{array}$$

Substituting -7 for y in either equation (1) or (2) gives $x = 3$. Check that the solution set of the system is $\{(3, -7)\}$.

OBJECTIVE 3 Use an alternative method to find the second value in a solution. Sometimes it is easier to find the value of the second variable in a solution by using the elimination method twice. The next example shows this approach.

CHALKBOARD EXAMPLE

Solve the system.

$$3y = 8 + 4x$$
$$6x = 9 - 2y$$

Answer: $\left\{\left(\frac{11}{26}, \frac{42}{13}\right)\right\}$

EXAMPLE 4 Finding the Second Value Using an Alternative Method

Solve the system.

$$4x = 9 - 3y \quad \textbf{(1)}$$
$$5x - 2y = 8 \quad \textbf{(2)}$$

Rearrange the terms in equation (1) so that like terms are aligned in columns. Add $3y$ to both sides to get the system

$$4x + 3y = 9 \quad \textbf{(3)}$$
$$5x - 2y = 8.$$

One way to proceed is to eliminate y by multiplying both sides of equation (3) by 2 and both sides of equation (2) by 3, and then adding.

$$\begin{array}{r} 8x + 6y = 18 \\ 15x - 6y = 24 \\ \hline 23x \qquad = 42 \\ x = \dfrac{42}{23} \end{array}$$

TEACHING TIP Show how to find the value of y in Example 4 by substituting $\frac{42}{23}$ for x in one of the original equations so that students can see the advantage of the method used in this example.

Substituting $\frac{42}{23}$ for x in one of the given equations would give y, but the arithmetic involved would be messy. Instead, solve for y by starting again with the original equations and eliminating x. Multiply both sides of equation (3) by 5 and both sides of equation (2) by -4, and then add.

$$\begin{aligned} 20x + 15y &= 45 \\ -20x + 8y &= -32 \\ \hline 23y &= 13 \\ y &= \frac{13}{23} \end{aligned}$$

Check that the solution set is $\left\{\left(\frac{42}{23}, \frac{13}{23}\right)\right\}$.

When the value of the first variable is a fraction, the method used in Example 4 helps avoid arithmetic errors. Of course, this method could be used to solve any system of equations.

OBJECTIVE 4 Use the elimination method to solve special systems. The next example shows the elimination method when a system is inconsistent or the equations of the system are dependent. To contrast the elimination method with the substitution method, in part (b) we use the same system solved in Example 4 of the previous section.

CHALKBOARD EXAMPLE

Solve the system by the elimination method.

$$\begin{aligned} 3x + y &= -7 \\ 6x + 2y &= 5 \end{aligned}$$

Answer: $\emptyset$

EXAMPLE 5 Using the Elimination Method for an Inconsistent System or Dependent Equations

Solve each system by the elimination method.

(a) $2x + 4y = 5$
$4x + 8y = -9$

Multiply both sides of $2x + 4y = 5$ by -2; then add to $4x + 8y = -9$.

$$\begin{aligned} -4x - 8y &= -10 \\ 4x + 8y &= -9 \\ \hline 0 &= -19 \quad \text{False} \end{aligned}$$

The false statement $0 = -19$ shows that the given system has solution set $\emptyset$.

(b) $3x - y = 4$
$-9x + 3y = -12$

Multiply both sides of the first equation by 3; then add the two equations.

$$\begin{aligned} 9x - 3y &= 12 \\ -9x + 3y &= -12 \\ \hline 0 &= 0 \quad \text{True} \end{aligned}$$

As before, this result indicates that every solution of one equation is also a solution of the other; there are an infinite number of solutions.

Summary of Situations That May Occur

One of three situations may occur when any of the methods described so far is used to solve a linear system of equations.

1. The result is a statement such as $x = 2$ or $y = -3$. The solution set will have exactly one ordered pair. The graphs of the equations of the system will intersect at exactly one point.

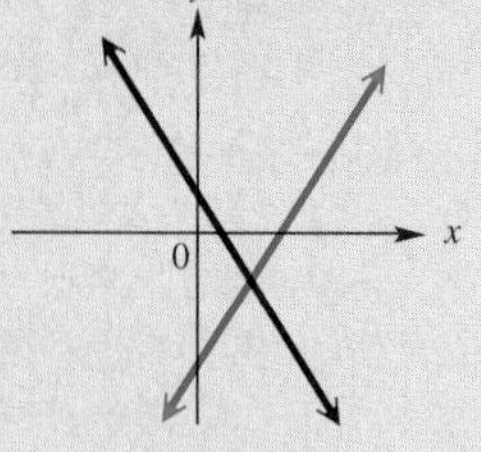

2. The final result is a false statement, such as $0 = 4$. In this case, the graphs are parallel lines and the solution set is $\emptyset$.

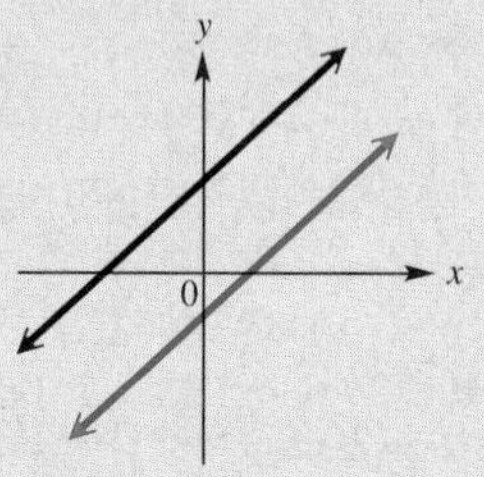

3. The final result is a true statement, such as $0 = 0$. The graphs of the equations of the system are the same line, and an infinite number of ordered pairs are solutions.

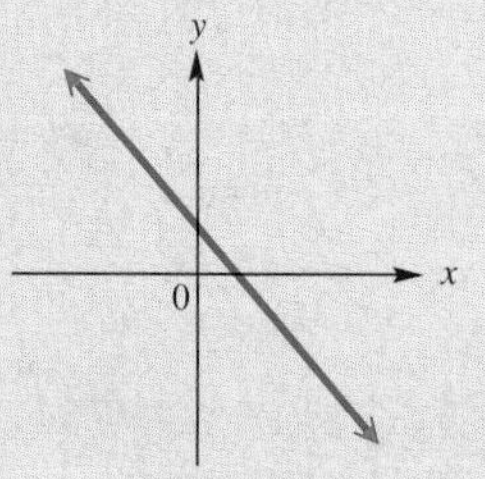

When no method of solution of a system is specified and a choice of substitution or elimination is allowed, use the following guidelines.

1. If one of the equations of the system is already solved for one of the variables, such as

$$\begin{aligned} 3x + 4y &= 9 \\ y &= 2x - 6 \end{aligned} \qquad \text{or} \qquad \begin{aligned} -5x + 3y &= 9 \\ x &= 3y - 7, \end{aligned}$$

the substitution method is the better choice.

2. If both equations are in standard $Ax + By = C$ form, such as

$$\begin{aligned} 4x - 11y &= 3 \\ -2x + 3y &= 4, \end{aligned}$$

and none of the variables has coefficient -1 or 1, the elimination method is the better choice.

3. If one or both of the equations are in standard form and the coefficient of one of the variables is -1 or 1, such as

$$\begin{aligned} 3x + y &= -2 \\ -5x + 2y &= 4 \end{aligned} \qquad \text{or} \qquad \begin{aligned} -x + 3y &= -4 \\ 3x - 2y &= 8, \end{aligned}$$

use the elimination method, or solve for the variable with coefficient -1 or 1 and then use the substitution method.

8.3 EXERCISES

1. true 2. false; Multiply by -3. 3. true 4. true 5. $\{(4, 6)\}$ 6. $\{(2, 8)\}$ 7. $\{(-1, -3)\}$ 8. $\{(-5, -5)\}$ 9. $\{(-2, 3)\}$ 10. $\{(2, 5)\}$ 11. $\left\{\left(-\frac{2}{3}, \frac{17}{2}\right)\right\}$ 12. $\left\{\left(-\frac{2}{5}, \frac{93}{10}\right)\right\}$ 13. $\{(3, -6)\}$ 14. $\{(5, -2)\}$ 15. $\{(7, 4)\}$ 16. $\{(2, 9)\}$ 17. $\{(0, 3)\}$ 18. $\{(0, 7)\}$ 19. $\{(3, 0)\}$ 20. $\{(-4, 0)\}$ 21. $\left\{\left(-\frac{32}{23}, -\frac{17}{23}\right)\right\}$ 22. $\left\{\left(-\frac{5}{7}, -\frac{2}{7}\right)\right\}$ 23. It would be easiest to solve for y in the first equation, to get $y = 5 - 2x$. Then substitute $5 - 2x$ for y in the second equation to get $5x + 3(5 - 2x) = 11$. Solve this equation for x. Substitute back into either of the original equations to find the value of y. Then check the solution in both equations of the original system. 24. One way to do this would be to multiply the first equation by 5 and the second equation by 3. Then add to eliminate y, and solve for x. Substitute back into either of the original equations to find the value of y. Check the solution in both equations of the original system. 25. $\{(-3, 4)\}$ 26. $\{(7, 0)\}$ 27. infinite number of solutions 28. infinite number of solutions 29. $\{(0, 6)\}$ 30. $\{(6, 5)\}$ 31. $\emptyset$ 32. $\emptyset$

Answer true or false for each of the following statements. If false, tell why.

1. The ordered pair (0, 0) *must* be a solution of a system of the form

$$\begin{aligned} Ax + By &= 0 \\ Cx + Dy &= 0. \end{aligned}$$

2. To eliminate the y-terms in the system

$$\begin{aligned} 2x + 12y &= 7 \\ 3x + 4y &= 1, \end{aligned}$$

we should multiply the bottom equation by 3 and then add.

3. The system

$$\begin{aligned} x + y &= 1 \\ x + y &= 2 \end{aligned}$$

has $\emptyset$ as its solution set.

4. The ordered pair $(4, -5)$ cannot be a solution of a system that contains the equation $5x - 4y = 0$.

*Solve each system by the elimination method. Check your answer. See Examples 1, 3, and 4.**

5. $x - y = -2$, $x + y = 10$
6. $x + y = 10$, $x - y = -6$
7. $2x + y = -5$, $x - y = 2$
8. $2x + y = -15$, $-x - y = 10$
9. $3x + 2y = 0$, $-3x - y = 3$
10. $5x - y = 5$, $-5x + 2y = 0$
11. $6x - 2y = -21$, $-6x + 8y = 72$
12. $6x - 2y = -21$, $6x + 8y = 72$
13. $2x - y = 12$, $3x + 2y = -3$
14. $x + y = 3$, $-3x + 2y = -19$
15. $3x + 3y = 33$, $5x - 2y = 27$
16. $4x - 3y = -19$, $3x + 2y = 24$
17. $5x + 4y = 12$, $3x + 5y = 15$
18. $2x + 3y = 21$, $5x - 2y = -14$
19. $5x - 4y = 15$, $-3x + 6y = -9$
20. $4x + 5y = -16$, $5x - 6y = -20$
21. $3x - 7y = 1$, $-5x + 4y = 4$
22. $-4x + 3y = 2$, $5x - 2y = -3$
23. Explain the steps you would use to solve the system $\begin{aligned} 2x + y &= 5 \\ 5x + 3y &= 11 \end{aligned}$ by substitution.
24. Explain the steps you would use to solve the system $\begin{aligned} 4x - 3y &= -7 \\ 6x + 5y &= 18 \end{aligned}$ by elimination.

Solve each system by elimination. See Examples 2 and 5.

25. $5x - 4y - 8x - 2 = 6x + 3y - 3$, $4x - y = -2y - 8$
26. $2x - 8y + 3y + 2 = 5y + 16$, $8x - 2y = 4x + 28$
27. $-2x + 3y = 12 + 2y$, $2x - 5y + 4 = -8 - 4y$
28. $2x + 5y = 7 + 4y - x$, $5x + 3y + 8 = 22 - x + y$
29. $7x - 9 + 2y - 8 = -3y + 4x + 13$, $4y - 8x = -8 + 9x + 32$
30. $y + 9 = 3x - 2y + 6$, $5 - 3x + 24 = -2x + 4y + 3$
31. $6x + 4y - 8 = x + 2y - 5$, $-3x + 4y - 1 = -8x + 2y + 8$
32. $6x - 4y + 3 = 9x - 12y + 4$, $5x + y + 2 = 8x - 7y + 6$

*The authors thank Mitchel Levy of Broward Community College for his suggestions for this group of exercises.

33. (a) $\{(1, 4)\}$ (b) $\{(1, 4)\}$ (c) Answers will vary.
34. (a) $\{(-5, 2)\}$ (b) $\{(-5, 2)\}$ (c) Answers will vary.
35. Yes, they should both get the same answer, since both procedures are mathematically valid.
36. Neither is correct. Each student found only one value (the first x, the second y). The solution set is $\{(12, -6)\}$.
37. $\{(0, 3)\}$
38. $\{(2, -3)\}$
39. $\{(24, -12)\}$
40. $\{(10, -12)\}$
41. $\{(3, 2)\}$
42. $\{(-4, 2)\}$

*In Exercises 33 and 34 **(a)** solve the system by the elimination method, **(b)** solve the system by the substitution method, and **(c)** tell which method you prefer for that particular system, and why.*

33. $4x - 3y = -8$
$x + 3y = 13$

34. $2x + 5y = 0$
$x = -3y + 1$

Exercises 35 and 36 refer to the system

$$\frac{1}{3}x - \frac{1}{2}y = 7$$
$$\frac{1}{6}x + \frac{1}{3}y = 0.$$

35. One student solved the system by multiplying both equations by 6 to clear fractions, and another student multiplied by 12. Assuming they do all other work correctly, should they both get the same answer? Explain.

36. One student solved the system and wrote as his answer $\{12\}$ while another solved it and wrote $\{-6\}$ as her answer. Who, if either, was correct? Why?

▲ *Solve each system by any method. First clear all fractions.*

37. $x + \frac{1}{3}y = y - 2$
$\frac{1}{4}x + y = x + y$

38. $\frac{5}{3}x + 2y = \frac{1}{3} + y$
$3x - 3 + \frac{y}{3} = -2 + 2x$

39. $\frac{x}{6} + \frac{y}{6} = 2$
$-\frac{1}{2}x - \frac{1}{3}y = -8$

40. $\frac{x}{2} - \frac{y}{3} = 9$
$\frac{x}{5} - \frac{y}{4} = 5$

41. $\frac{x}{3} - \frac{3y}{4} = -\frac{1}{2}$
$\frac{x}{6} + \frac{y}{8} = \frac{3}{4}$

42. $\frac{x}{5} + 2y = \frac{16}{5}$
$\frac{3x}{5} + \frac{y}{2} = -\frac{7}{5}$

RELATING CONCEPTS (EXERCISES 43–48)

Attending the movies is one of America's favorite forms of entertainment. The graph shows how attendance gradually increased from 1991 to 1996. In 1991, attendance was 1141 million, as represented by the point $P(1991, 1141)$. In 1996, attendance was 1339 million, as represented by the point $Q(1996, 1339)$. We can find an equation of line segment PQ using a system of equations, and then use the equation to approximate the attendance in any of the years between 1991 and 1996.

Work Exercises 43–48 in order.

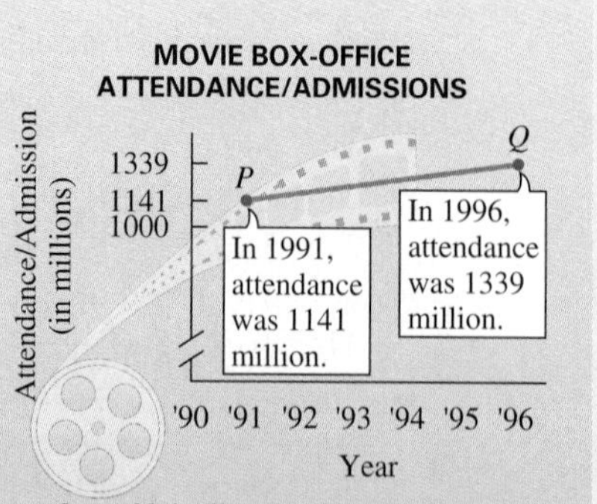

Source: Motion Picture Association of America.

43. $1141 = 1991a + b$
44. $1339 = 1996a + b$
45. $1991a + b = 1141$
$1996a + b = 1339$; solution set: $\{(39.6, -77{,}702.6)\}$
46. $y = 39.6x - 77{,}702.6$
47. 1220.2 (million); This is slightly less than the actual figure.
48. It is not realistic to expect the data to lie in a perfectly straight line; as a result, the quantity obtained from an equation determined in this way will probably be "off" a bit. One cannot put too much faith in models such as this one, because not all data points are linear in nature.

RELATING CONCEPTS (EXERCISES 43–48) (CONTINUED)

43. The line segment has an equation that can be written in the form $y = ax + b$. Using the coordinates of point P and letting $x = 1991$ and $y = 1141$, write an equation in the variables a and b.

44. Using the coordinates of point Q and letting $x = 1996$ and $y = 1339$, write a second equation in the variables a and b.

45. Write the system of equations formed from the two equations in Exercises 43 and 44, and solve the system using the elimination method.

46. What is the equation of the segment PQ?

47. Let $x = 1993$ in the equation of Exercise 46, and solve for y. How does the result compare with the actual figure of 1244 million?

48. The data for the years 1991 through 1996 do not lie in a perfectly straight line. Explain the pitfalls of relying too heavily on using the equation in Exercise 46 to predict attendance.

Did you make the connection that the equation of a line can be used to interpret and model real-life data?

8.4 Applications of Linear Systems

OBJECTIVES

1. Solve problems about unknown numbers.
2. Solve problems about quantities and their costs.
3. Solve problems about mixtures.
4. Solve problems about rate or speed.

FOR EXTRA HELP

SSG Sec. 8.4
SSM Sec. 8.4
Pass the Test Software
InterAct Math Tutorial Software
Video 13

Many applied problems are more easily translated into equations if two variables are used. With two variables, a system of two equations is needed to find the solution.

PROBLEM SOLVING

Recall from Chapter 2 the six-step method for solving applied problems. Those steps can be modified slightly to allow for two variables and two equations.

Solving Applied Problems with Two Variables

Step 1 **Choose the variables.** Choose a variable to represent each of two unknown values that must be found. Write down what each variable represents.

Step 2 **Draw figures or diagrams and use charts.** If figures, diagrams, or charts can help, prepare them.

Step 3 **Write two equations.** Translate the problem into a system of two equations using both variables.

Step 4 **Solve the system.** Solve the system of equations, using either the elimination or the substitution method.

Step 5 **Answer the question(s).** Answer the question or questions asked in the problem.

Step 6 **Check.** Check your answer by using the original words of the problem. Be sure that your answer makes sense.

INSTRUCTOR'S RESOURCES

PTB Sec. 8.4 ISM Sec. 8.4
AB Sec. 8.4

TEST GENERATOR

WORLD WIDE WEB
www.LialAlgebra.com

OBJECTIVE 1 Solve problems about unknown numbers.

CHALKBOARD EXAMPLE

The two top-selling Disney videos of 1996 were *Toy Story* and *Pocahontas*. Together they sold 38 million copies. *Pocahontas* sold 4 million fewer copies than *Toy Story*. How many copies of each title were sold?

Answer: *Toy Story*: 21 million; *Pocahontas*: 17 million

EXAMPLE 1 Solving a Problem about Two Unknown Numbers

Among the ten top-selling videos in 1996, only three were not Disney films. Two of those were *Independence Day* (Fox) and *Twister* (Warner). Together, these two titles sold 31.5 million copies. *Independence Day* sold 12.5 million more copies than *Twister*. How many copies of each title were sold? (*Source:* Paul Kagan Associates Inc.)

Step 1 Let x = the number of copies, in millions, of *Twister* sold;
y = the number of copies, in millions, of *Independence Day* sold.

Step 2 A figure, diagram, or chart will not help in this problem, so go on to Step 3.

Step 3 From the information in the problem, set up a system of equations.

$$x + y = 31.5$$ The total number sold was 31.5 million.

$$y = x + 12.5$$ The number of *Independence Day* titles sold was 12.5 million more than that of *Twister*.

Step 4 Solve the system from Step 3. Use the substitution method, since the second equation is already solved for y.

$$x + (x + 12.5) = 31.5$$ Substitute $x + 12.5$ for y in the first equation.

$$2x + 12.5 = 31.5$$ Combine like terms.

$$2x = 31.5 - 12.5$$ Subtract 12.5.

$$2x = 19$$

$$x = 9.5$$ Divide by 2.

Substitute 9.5 for x in either equation to find $y = 22$.

Step 5 *Twister* sold 9.5 million copies, and *Independence Day* sold 22 million copies.

Step 6 The sum of 9.5 and 22 is 31.5, and 22 is 12.5 more than 9.5. The solution satisfies the conditions of the problem.

TEACHING TIP The type of problem in Exercises 7 and 8 in the exercise set is an excellent choice to begin a presentation of applications of linear systems. Setting up the system for an application is usually the most difficult part for the student; however, this system is very easy to set up.

CAUTION If an applied problem asks for *two* values (as in Example 1), be sure to give both of them, not just one, in your answer.

OBJECTIVE 2 Solve problems about quantities and their costs. We can also use a linear system to solve a common type of applied problem involving two quantities and their costs.

CHALKBOARD EXAMPLE

The average movie ticket (to the nearest U.S. dollar) costs \$10 in Geneva and \$8 in Paris (*Source: Parade Magazine*, September 13, 1998, p. 25). If a group of 36 people from these two cities paid \$298 for tickets to see *As Good as It Gets*, how many people from each city were there?

Answer: Geneva: 5; Paris: 31

EXAMPLE 2 Solving a Problem about Quantities and Costs

The 1997 box office smash *Titanic* earned more in Europe than in the United States. This may be because average movie prices in Europe exceed those of the United States (*Source: Parade Magazine,* September 13, 1998, p. 25). For example, while the average movie ticket (to the nearest dollar) costs \$5 in the United States, it costs an equivalent of \$11 in London. Suppose that a group of 41 Americans and Londoners who paid these average prices spent a total of \$307 for tickets. How many from each country were in the group?

Step 1 Let x = the number of Americans in the group;
y = the number of Londoners in the group.

Step 2 Summarize the information given in the problem in a chart. The entries in the "Total Value" column were found by multiplying the number of tickets sold by the price per ticket.

	Number of Tickets	Price per Ticket (in dollars)	Total Value
American	x	5	$5x$
Londoner	y	11	$11y$
Total	41		307

TEACHING TIP Explain that one equation involves the total *number* of tickets, while the other involves the total *value* of the tickets.

Step 3 The total number of tickets was 41, so one equation is

$$x + y = 41.$$

Since the total value was $307, the final column leads to

$$5x + 11y = 307.$$

These two equations form a system.

$$x + y = 41 \quad (1)$$
$$5x + 11y = 307 \quad (2)$$

Step 4 Solve the system of equations using the elimination method. First, eliminate the x terms. Multiply both sides of equation (1) by -5 to get

$$-5x - 5y = -205.$$

Then add this result to equation (2).

$$\begin{aligned} -5x - 5y &= -205 \\ 5x + 11y &= 307 \\ \hline 6y &= 102 \\ y &= 17 \end{aligned}$$

Substitute 17 for y in equation (1) to get

$$\begin{aligned} x + y &= 41 \quad (1) \\ x + 17 &= 41 \\ x &= 24. \end{aligned}$$

Step 5 There were 24 Americans and 17 Londoners in the group.

Step 6 The sum of 24 and 17 is 41, so the number of movie-goers is correct. Since 24 Americans paid $5 each and 17 Londoners paid $11 each, the total of the admission prices is $5(24) + $11(17) = $307, which agrees with the total amount stated in the problem. ■

OBJECTIVE 3 Solve problems about mixtures. In Chapter 2 we solved mixture problems using one variable. Many mixture problems can also be solved using a system of two equations in two variables.

CHALKBOARD EXAMPLE

How many liters of 25% alcohol solution must be mixed with 12% solution to get 13 liters of 15% solution?

Answer: 3 liters of 25%; 10 liters of 12%

EXAMPLE 3 Solving a Mixture Problem Involving Percent

A pharmacist needs 100 liters of 50% alcohol solution. She has on hand 30% alcohol solution and 80% alcohol solution, which she can mix. How many liters of each will be required to make the 100 liters of 50% alcohol solution?

Step 1 Let x = the number of liters of 30% alcohol needed;
y = the number of liters of 80% alcohol needed.

Step 2 Summarize the information in a chart. Percents are represented in decimal form.

Rate	*Liters of Mixture*	*Liters of Pure Alcohol*
.30	x	$.30x$
.80	y	$.80y$
.50	100	.50(100)

Figure 7 gives an idea of what is actually happening in this problem.

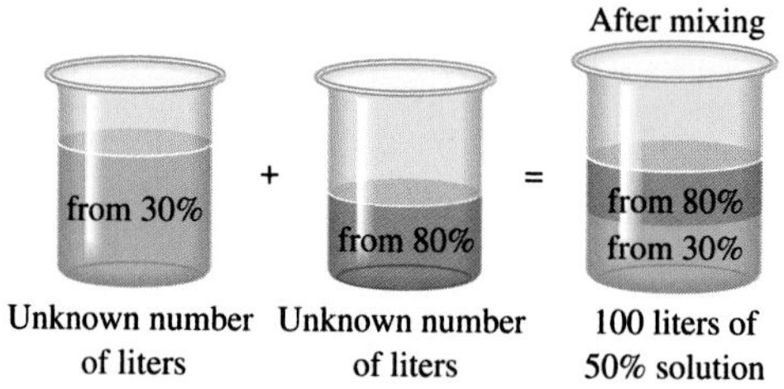

Figure 7

Step 3 We must write two equations. Since the total number of liters in the final mixture will be 100, the first equation is

$$x + y = 100.$$

To find the amount of pure alcohol in each mixture, multiply the number of liters by the concentration. The amount of pure alcohol in the 30% solution added to the amount of pure alcohol in the 80% solution will equal the amount of pure alcohol in the final 50% solution. This gives the second equation,

$$.30x + .80y = .50(100).$$

These two equations form a system.

$$x + y = 100$$
$$.30x + .80y = 50 \qquad .50(100) = 50$$

Step 4 Solve this system by the substitution method. Solving the first equation of the system for x gives $x = 100 - y$. Substitute $100 - y$ for x in the second equation to get

$$.30(100 - y) + .80y = 50 \qquad \text{Let } x = 100 - y.$$
$$30 - .30y + .80y = 50 \qquad \text{Distributive property}$$
$$.50y = 20 \qquad \text{Combine terms; subtract 30.}$$
$$y = 40. \qquad \text{Divide by .50.}$$

Then $x = 100 - y = 100 - 40 = 60.$

Step 5 The pharmacist should use 60 liters of the 30% solution and 40 liters of the 80% solution.

Step 6 Since $60 + 40 = 100$ and $.30(60) + .80(40) = 50$, this mixture will give the 100 liters of 50 percent solution, as required in the original problem.

NOTE The system in this problem could have been solved by the elimination method. Also, we could have cleared decimals by multiplying both sides of the second equation by 10.

OBJECTIVE 4 **Solve problems about rate or speed.** Problems that use the distance formula $d = rt$ were first introduced in Chapter 2. In some cases, these problems can be solved by using a system of two linear equations. Keep in mind that setting up a chart and drawing a sketch will help.

CHALKBOARD EXAMPLE

In one hour Ann can row 2 miles against the current or 10 miles with the current. Find the speed of the current and Ann's speed in still water.

Answer: 4 miles per hour; 6 miles per hour

EXAMPLE 4 **Solving a Problem about Distance, Rate, and Time**

Two executives in cities 400 miles apart drive to a business meeting at a location on the line between their cities. They meet after 4 hours. Find the speed of each car if one car travels 20 miles per hour faster than the other.

Step 1 Let x = the speed of the faster car;
y = the speed of the slower car.

Step 2 We use the formula that relates distance, rate, and time, $d = rt$. Since each car travels for 4 hours, the time, t, for each car is 4. This information is shown in the chart. The distance is found by using the formula $d = rt$ and the expressions already entered in the chart.

	r	t	d
Faster car	x	4	$4x$
Slower car	y	4	$4y$

Find d from $d = rt$.

Draw a sketch showing what is happening in the problem. See Figure 8.

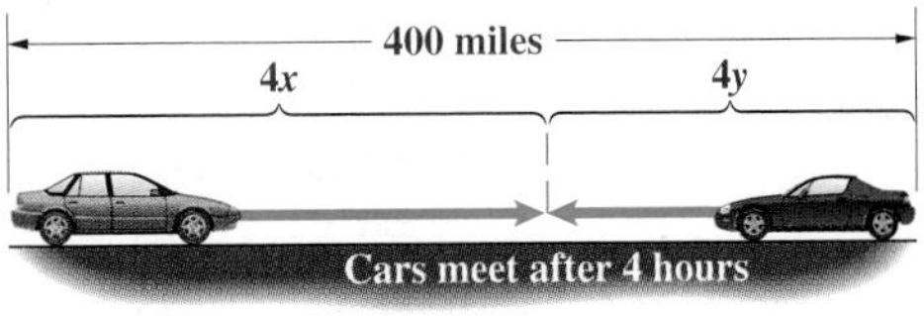

Figure 8

Step 3 As shown in the figure, since the total distance traveled by both cars is 400 miles, one equation is

$$4x + 4y = 400.$$

Because the faster car goes 20 miles per hour faster than the slower car, the second equation is

$$x = 20 + y.$$

Step 4 This system of equations,

$$4x + 4y = 400$$
$$x = 20 + y,$$

can be solved by substitution. Replace x with $20 + y$ in the first equation of the system and solve for y.

$$4(20 + y) + 4y = 400 \quad \text{Let } x = 20 + y.$$
$$80 + 4y + 4y = 400 \quad \text{Distributive property}$$
$$80 + 8y = 400 \quad \text{Combine like terms.}$$
$$8y = 320 \quad \text{Subtract 80.}$$
$$y = 40 \quad \text{Divide by 8.}$$

Step 5 Since $x = 20 + y$, and $y = 40$,

$$x = 20 + 40 = 60.$$

The speeds of the two cars are 40 miles per hour and 60 miles per hour.

Step 6 Check the answers. Since each car travels for 4 hours, the total distance traveled is

$$4(40) + 4(60) = 160 + 240 = 400$$

miles, as required. ■

The problems in Examples 2–4 also could be solved using only one variable. Many students find that the solution is simpler with two variables.

CAUTION Be careful! When you use two variables to solve a problem, you must write two equations.

8.4 EXERCISES

1. (d) 2. (a) 3. (b) 4. (c)

Choose the correct response in Exercises 1–5.

1. Which one of the following represents the monetary value of x 20-dollar bills?

(a) $\frac{x}{20}$ dollars **(b)** $\frac{20}{x}$ dollars **(c)** $20 + x$ dollars **(d)** $20x$ dollars

2. Which one of the following represents the cost of t pounds of candy that sells for \$1.95 per pound?

(a) $\$1.95t$ **(b)** $\frac{\$1.95}{t}$ **(c)** $\frac{t}{\$1.95}$ **(d)** $\$1.95 + t$

3. Suppose that x liters of a 40% acid solution are mixed with y liters of a 35% solution to obtain 100 liters of a 38% solution. One equation in a system for solving this problem is $x + y = 100$. Which one of the following is the other equation?

(a) $.35x + .40y = .38(100)$ **(b)** $.40x + .35y = .38(100)$
(c) $35x + 40y = 38$ **(d)** $40x + 35y = .38(100)$

4. What is the speed of a plane that travels at a rate of 560 miles per hour *into* a wind of r miles per hour?

(a) $560 + r$ mph **(b)** $\frac{560}{r}$ mph **(c)** $560 - r$ mph **(d)** $r - 560$ mph

5. (c) **6.** **Choose two variables to represent two unknown quantities. Write a system of two equations in these two variables, and then solve the system. Check to see that the two values satisfy all conditions of the problem.**
7. **the second number;** $x - y = 48$**; The two numbers are 73 and 25.**
8. **the second number;** $x - y = 11$**; The two numbers are 106 and 95.**
9. **Boyz II Men: 134; Bruce Springsteen & the E St. Band: 40**
10. **country: 2502; adult contemporary: 1521**
11. **Terminal Tower: 708 feet; Society Center: 948 feet**
12. **Labrador Retrievers: 149,505; Rottweilers: 89,867**

5. What is the speed of a plane that travels at a rate of 560 miles per hour *with* a wind of r miles per hour?

(a) $\frac{r}{560}$ mph **(b)** $560 - r$ mph **(c)** $560 + r$ mph **(d)** $r - 560$ mph

6. Using the six-step method for solving an applied problem with two variables, write a short paragraph describing the general procedure you will use to solve the problems that follow in this exercise set.

The problems in Exercises 7 and 8 are good warm-up problems for the ones that follow. In each case, refer to the six-step method, fill in the blanks for Steps 1 and 3, and then complete the solution by applying Steps 4–6.

7. The sum of two numbers is 98 and the difference between them is 48. Find the two numbers.

Step 1 Let x = the first number and let y = ______________.

Step 2 Not necessary

Step 3 First equation: $x + y = 98$
Second equation: ______________

8. The sum of two numbers is 201 and the difference between them is 11. Find the two numbers.

Step 1 Let x = the first number and let y = ______________.

Step 2 Not necessary

Step 3 First equation: $x + y = 201$
Second equation: ______________

Write a system of equations for each problem, and then solve the problem. Use the six-step method. See Example 1.

9. During 1995, two of the top-grossing concert tours were by Boyz II Men and Bruce Springsteen & the E St. Band. Together the two tours visited 174 cities. Boyz II Men visited 94 cities more than Bruce Springsteen. How many cities did each group visit? (*Source:* Pollstar.)

10. In 1997, the two top formats of U.S. commercial radio stations were country and adult contemporary. There were 981 fewer adult contemporary stations than country stations, and together they comprised a total of 4023 stations. How many stations of each format were there? (*Source:* M Street Corporation.)

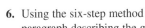

11. The Terminal Tower in Cleveland, Ohio, is 240 feet shorter than the Society Center, also in Cleveland. The total of the heights of the two buildings is 1656 feet. Find the heights of the buildings. (*Source: The World Almanac and Book of Facts,* 1998.)

12. Labrador Retrievers and Rottweilers were the two breeds of dogs that led the number of registrations by the American Kennel Club in 1996. Together they numbered 239,372. There were 59,638 more Labrador Retrievers registered than Rottweilers. How many of each breed were registered? (*Source:* American Kennel Club.)

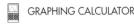

13. 46 ones; 28 tens
14. 20 fives; 49 tens
15. 2 copies of *Godzilla*; 5 Aerosmith compact discs
16. 1 copy of *Rudolph The Movie*; 4 *Touched by an Angel* soundtracks
17. $2500 at 4%; $5000 at 5%
18. $30,000 at 3%; $10,000 at 2%
19. Japan: $17.19; Switzerland: $13.15

Write a system of equations for each problem, and then solve the problem. Use the six-step method. See Example 2.

13. A motel clerk counts his $1 and $10 bills at the end of a day. He finds that he has a total of 74 bills having a combined monetary value of $326. Find the number of bills of each denomination that he has.

Denomination of Bill	Number of Bills	Total Value
$1	x	
$10	y	
Totals	74	$326

14. Latarsha Johnson is a bank teller. At the end of a day, she has a total of 69 $5 and $10 bills. The total value of the money is $590. How many of each denomination does she have?

Denomination of Bill	Number of Bills	Total Value
$5	x	$5x
$10	y	
Totals		

15. On November 1, 1998, K-Mart advertised videocassettes and compact discs in its Sunday newspaper insert. Cheryl Joslyn went shopping and bought each of her 7 nephews a gift, either a copy of the movie *Godzilla* or the latest Aerosmith compact disc. The movie cost $14.95 and the compact disc cost $16.88, and she spent a total of $114.30. How many movies and how many compact discs did she buy? (*Source: Times Picayune.*)

16. Terry McGinnis saw the K-Mart ad (see Exercise 15) and she too went shopping. She bought each of her 5 nieces a gift, either a copy of *Rudolph The Movie* or the soundtrack to *Touched by an Angel.* The movie cost $14.99 and the soundtrack cost $13.88, and she spent a total of $70.51. How many movies and how many soundtracks did she buy? (*Source: Times Picayune.*)

17. Derrick Washington has twice as much money invested at 5% simple annual interest as he does at 4%. If his yearly income from these two investments is $350, how much does he have invested at each rate?

18. Chuck Fitch invested his textbook royalty income in two accounts, one paying 3% annual simple interest and the other paying 2% interest. He earned a total of $1100 interest. If he invested three times as much in the 3% account as he did in the 2% account, how much did he invest at each rate?

19. Average movie ticket prices in the United States are, in general, lower than in other countries. It would cost $77.87 to buy three tickets in Japan plus two tickets in Switzerland. Three tickets in Switzerland plus two tickets in Japan would cost $73.83. How much does an average movie ticket cost in each of these countries? (*Source:* Business Traveler International.)

20. Germany: $8.93; France: $8.85
21. 80 liters of 40% solution; 40 liters of 70% solution
22. 24 liters of 90% solution; 96 liters of 75% solution
23. 30 pounds at $6 per pound; 60 pounds at $3 per pound
24. 30 pounds at $1.20 per pound; 15 pounds at $1.80 per pound
25. 60 barrels at $40 per barrel; 40 barrels at $60 per barrel
26. 25 bags at $70 per bag; 15 bags at $90 per bag

20. (See Exercise 19.) Four movie tickets in Germany plus three movie tickets in France would cost $62.27. Three tickets in Germany plus four tickets in France would cost $62.19. How much does an average movie ticket cost in each of these countries? (*Source:* Business Traveler International.)

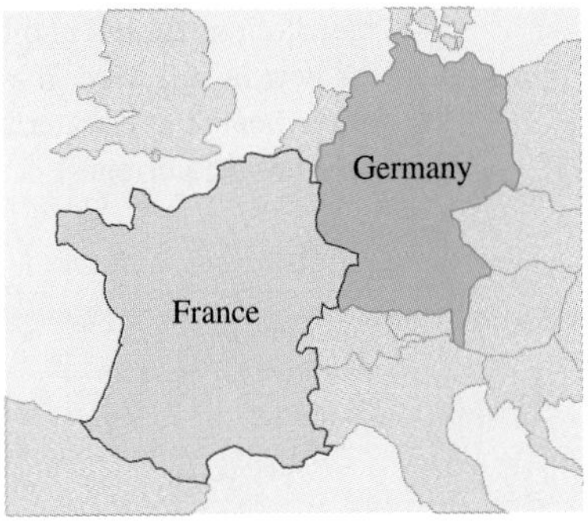

Write a system of equations for each problem, and then solve the problem. Use the six-step method. See Example 3.

21. A 40% dye solution is to be mixed with a 70% dye solution to get 120 liters of a 50% solution. How many liters of the 40% and 70% solutions will be needed?

Rate	Liters of Solution	Liters of Pure Dye
.40	x	
.70	y	
.50	120	

22. A 90% antifreeze solution is to be mixed with a 75% solution to make 120 liters of a 78% solution. How many liters of the 90% and 75% solutions will be used?

Rate	Liters of Solution	Liters of Pure Antifreeze
.90	x	
.75	y	
.78	120	

23. A merchant wishes to mix coffee worth $6 per pound with coffee worth $3 per pound to get 90 pounds of a mixture worth $4 per pound. How many pounds of the $6 and the $3 coffee will be needed?

Pounds	Dollars per Pound	Cost
x	6	
y		
90		

24. A grocer wishes to blend candy selling for $1.20 a pound with candy selling for $1.80 a pound to get a mixture that will be sold for $1.40 a pound. How many pounds of the $1.20 and the $1.80 candy should be used to get 45 pounds of the mixture?

Pounds	Dollars per Pound	Cost
x		
y	1.80	
45		

25. How many barrels of pickles worth $40 per barrel and pickles worth $60 per barrel must be mixed to obtain 100 barrels of a mixture worth $48 per barrel?

26. The owner of a nursery wants to mix some fertilizer worth $70 per bag with some worth $90 per bag to obtain 40 bags of mixture worth $77.50 per bag. How many bags of each type should she use?

27. boat: 10 miles per hour; current: 2 miles per hour 28. boat: 5 miles per hour; current: 3 miles per hour 29. plane: 470 miles per hour; wind: 30 miles per hour 30. plane: 160 miles per hour; wind: 40 miles per hour 31. car leaving Cincinnati: 55 miles per hour; car leaving Toledo: 70 miles per hour 32. 35 miles per hour and 65 miles per hour 33. Roberto: 3 miles per hour; Juana: 2.5 miles per hour 34. Abbot: 80 miles per hour; Costello: 40 miles per hour

Write a system of equations for each problem, and then solve the problem. Use the six-step method. See Example 4.

27. A boat takes 3 hours to go 24 miles upstream. It can go 36 miles downstream in the same time. Find the speed of the current and the speed of the boat in still water if x = the speed of the boat in still water and y = the speed of the current.

	d	r	t
Downstream	36	$x + y$	
Upstream	24	$x - y$	

28. It takes a boat $1\frac{1}{2}$ hours to go 12 miles downstream, and 6 hours to return. Find the speed of the boat in still water and the speed of the current. Let x = the speed of the boat in still water and y = the speed of the current.

	d	r	t
Downstream	12	$x + y$	$\frac{3}{2}$
Upstream			6

29. If a plane can travel 440 miles per hour into the wind and 500 miles per hour with the wind, find the speed of the wind and the speed of the plane in still air.

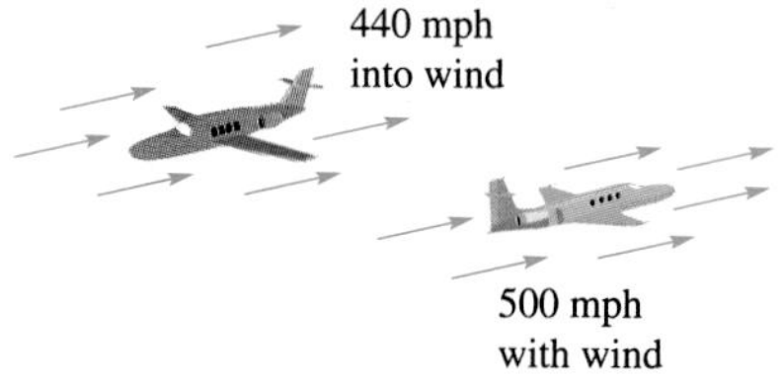

30. A small plane travels 200 miles per hour with the wind and 120 miles per hour against it. Find the speed of the wind and the speed of the plane in still air.

31. Toledo and Cincinnati are 200 miles apart. A car leaves Toledo traveling toward Cincinnati, and another car leaves Cincinnati at the same time, traveling toward Toledo. The car leaving Toledo averages 15 miles per hour more than the other, and they meet after 1 hour and 36 minutes. What are the rates of the cars?

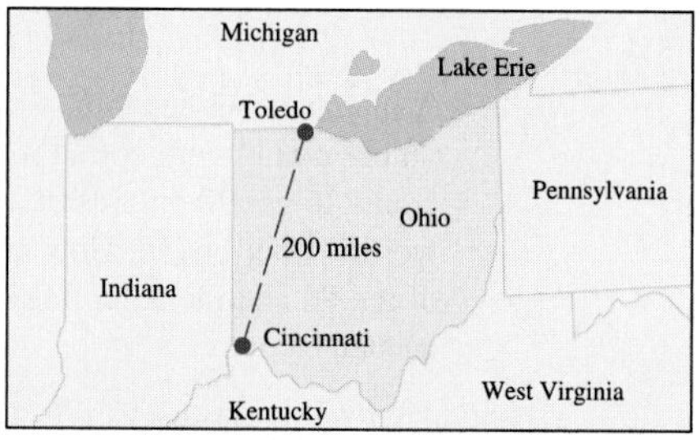

32. Kansas City and Denver are 600 miles apart. Two cars start from these cities, traveling toward each other. They meet after 6 hours. Find the rate of each car if one travels 30 miles per hour slower than the other.

▲ **33.** At the beginning of a walk for charity, Roberto and Juana are 5.5 miles apart. If they leave at the same time and walk in the same direction, Roberto overtakes Juana in 11 hours. If they walk toward each other, they meet in 1 hour. What are their speeds?

▲ **34.** Mr. Abbot left Farmersville in a plane at noon to travel to Exeter. Mr. Costello left Exeter in his automobile at 2 P.M. to travel to Farmersville. It is 400 miles from Exeter to Farmersville. If the sum of their speeds was 120 miles per hour, and if they crossed paths at 4 P.M., find the speed of each.

8.5 Solving Systems of Linear Inequalities

OBJECTIVES

1 Solve systems of linear inequalities by graphing.

2 Recognize how a graphing calculator is used to solve a system of linear inequalities.

FOR EXTRA HELP

SSG Sec. 8.5
SSM Sec. 8.5
Pass the Test Software
InterAct Math Tutorial Software
Video 13

Graphing the solution set of a linear inequality was discussed in Section 7.2. To graph the solution set of $x + 3y > 12$, for example, remember that you first graph the line $x + 3y = 12$ by finding a few ordered pairs that satisfy the equation. Because the points on this boundary line do not satisfy the inequality, use a dashed line. To decide which side of the line to shade, choose any test point not on the line, such as (0, 0). Then substitute 0 for x and 0 for y in the given inequality.

$$x + 3y > 12$$
$$0 + 3(0) > 12 \quad ? \quad \text{Let } x = 0,\ y = 0.$$
$$0 > 12 \quad \text{False}$$

Since the test point does not satisfy the inequality, shade the region on the side of the boundary line that does not include (0, 0), as in Figure 9.

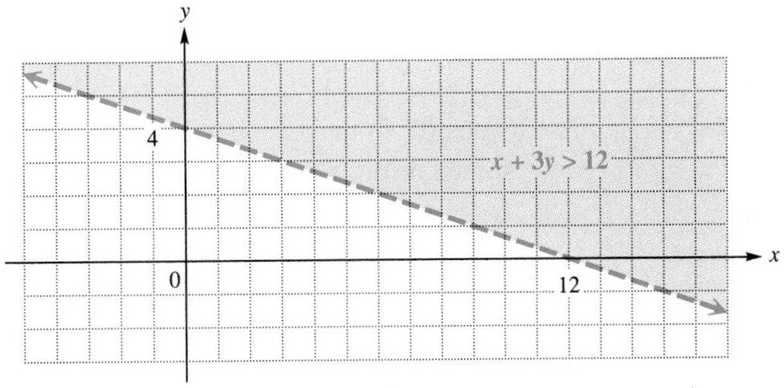

Figure 9

OBJECTIVE 1 **Solve systems of linear inequalities by graphing.** The same method is used to find the solution set of a system of two linear inequalities. A **system of linear inequalities** consists of two or more linear inequalities. The **solution set of a system of linear inequalities** includes all ordered pairs that make all inequalities of the system true at the same time.

To solve a system of linear inequalities, use the following steps.

Graphing a System of Linear Inequalities

Step 1 **Graph the inequalities.** Graph each inequality using the method of Section 7.2.

Step 2 **Choose the intersection.** Indicate the solution of the system by using dark shading on the intersection of the two graphs (the region where the two graphs overlap).

CHALKBOARD EXAMPLE

Graph the solution set of the system.

$$x - 2y \le 8$$
$$3x + y \ge 6$$

Answer:

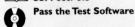

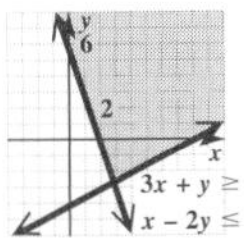

EXAMPLE 1 **Solving a System of Two Linear Inequalities**

Graph the solution set of the linear system.

$$3x + 2y \le 6$$
$$2x - 5y \ge 10$$

INSTRUCTOR'S RESOURCES

PTB Sec. 8.5 **ISM** Sec. 8.5
AB Sec. 8.5
TEST GENERATOR
WORLD WIDE WEB www.LialAlgebra.com

Begin by graphing $3x + 2y \leq 6$. To do this, we graph $3x + 2y = 6$ as a solid line. Since (0, 0) makes the inequality true, we shade the region containing (0, 0), as shown in Figure 10.

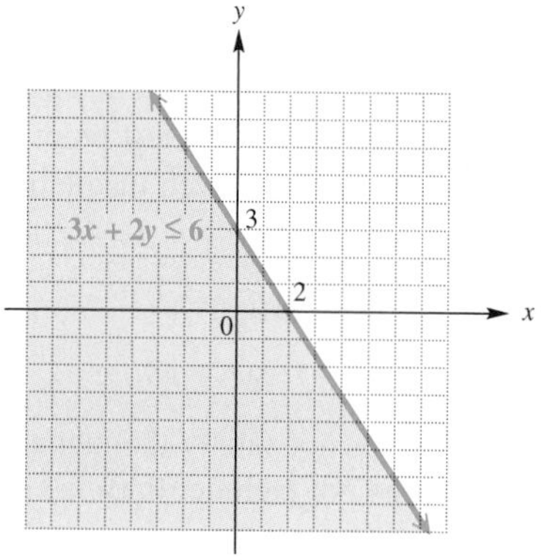

Figure 10

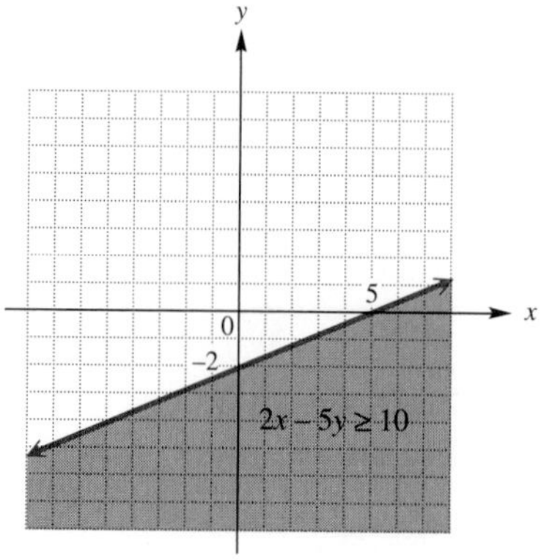

Figure 11

Now graph $2x - 5y \geq 10$. The solid line boundary is the graph of $2x - 5y = 10$. Since (0, 0) makes the inequality false, shade the region that does not contain (0, 0), as shown in Figure 11.

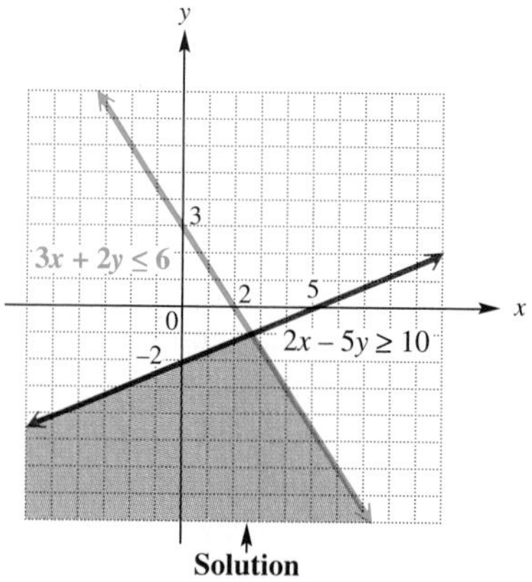

Figure 12

The solution set of the system is given by the intersection (overlap) of the regions of the graphs in Figures 10 and 11. The solution is the shaded region in Figure 12, and includes portions of the two boundary lines.

CHALKBOARD EXAMPLE

Graph the solution set of the system.

$$x + 2y < 0$$
$$3x - 4y < 12$$

Answer:

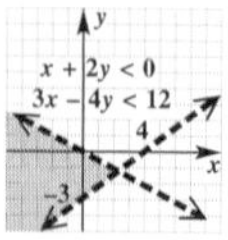

NOTE In practice, we usually do all the work on one set of axes at the same time. In the following examples, only one graph is shown. Be sure that the region of the solution set is clearly indicated.

EXAMPLE 2 Solving a System of Two Linear Inequalities

Graph the solution set of the system.

$$x - y > 5$$
$$2x + y < 2$$

Figure 13 shows the graphs of both $x - y > 5$ and $2x + y < 2$. Dashed lines show that the graphs of the inequalities do not include their boundary lines. The solution set of the system is the shaded region in the figure. The solution set does not include either boundary line.

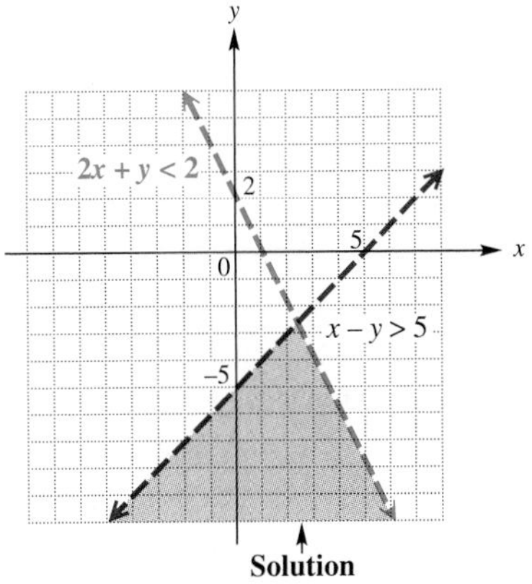

Figure 13

CHALKBOARD EXAMPLE

Graph the solution set of the system.

$$2x + 3y \geq -6$$
$$x \leq 3$$
$$y \leq 1$$

Answer:

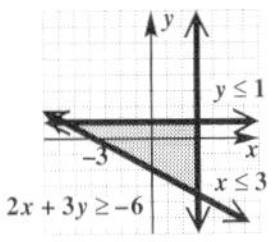

EXAMPLE 3 Solving a System of Three Linear Inequalities

Graph the solution set of the system.

$$4x - 3y \leq 8$$
$$x \geq 2$$
$$y \leq 4$$

Recall that $x = 2$ is a vertical line through the point (2, 0), and $y = 4$ is the horizontal line through (0, 4). The graph of the solution set is the shaded region in Figure 14.

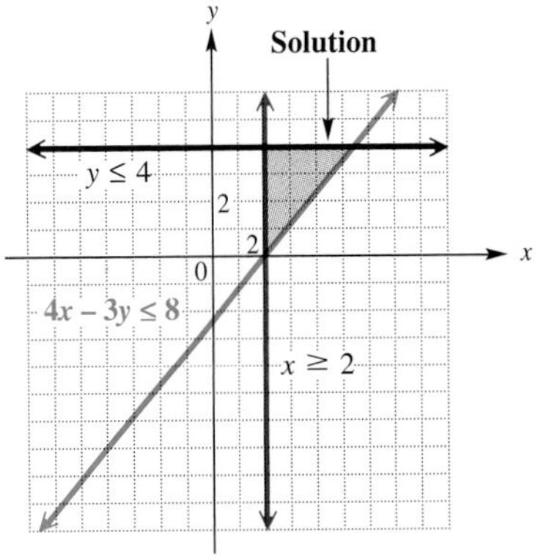

Figure 14

OBJECTIVE 2 **Recognize how a graphing calculator is used to solve a system of linear inequalities.** The solution set of a system of linear inequalities can be represented on a graphing calculator screen, as shown in Example 4.

CHALKBOARD EXAMPLE

The screen here shows how a graphing calculator illustrates the solution set of the system from the Chalkboard Example for text Example 2,

$$x + 2y < 0$$
$$3x - 4y < 12.$$

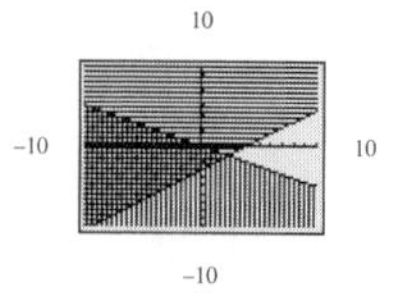

Describe the region that represents the solution set.

Answer: The cross-hatched region represents the solution set.

EXAMPLE 4 Illustrating the Solution Set of a System of Linear Inequalities with a Graphing Calculator

Figures 15(a) and (b) illustrate how a graphing calculator can represent the solution set of the system

$$y < 3x + 2$$
$$y > -2x - 5.$$

The first inequality can be graphed by directing the calculator to shade *below* the line $Y_1 = 3x + 2$ (because of the $<$ symbol). The second can be graphed by shading *above* the line $Y_2 = -2x - 5$ (because of the $>$ symbol). Figure 15(a) shows the appropriate screen for these directions on a TI-83 calculator. Graphing the inequalities in the standard viewing window gives the screen in Figure 15(b). The cross-hatched region is the intersection of the solution sets of the two individual inequalities and represents the solution set of the system.

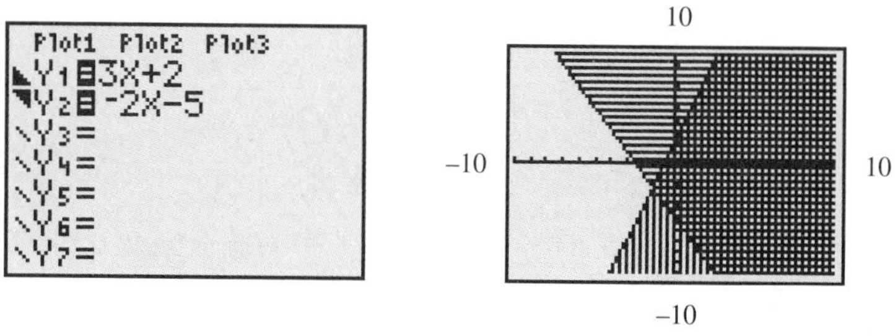

(a) (b)

Figure 15

If the inequalities in a system are not solved for y as in this example, it will be necessary to do so. Notice that it is not possible to determine whether the boundary lines are included or excluded in the solution set. For this system, neither one is included.

8.5 EXERCISES

1. C 2. A 3. B 4. D

5.

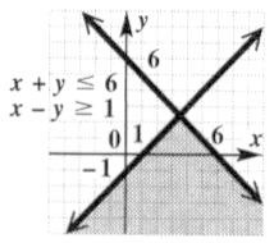

6.

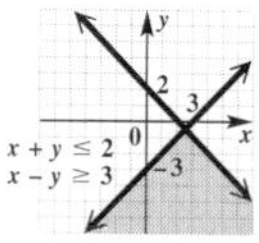

7.

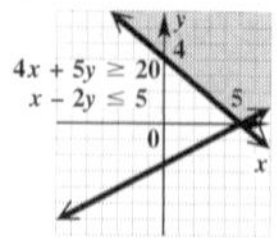

8.

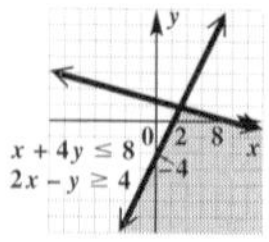

9.

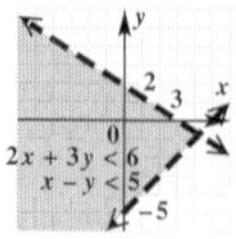

10.

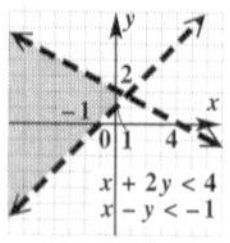

Match the system of inequalities with the correct graph from choices A–D.

1. $x \geq 5$
   $y \leq -3$
2. $x \leq 5$
   $y \geq -3$
3. $x > 5$
   $y < -3$
4. $x < 5$
   $y > -3$

A.

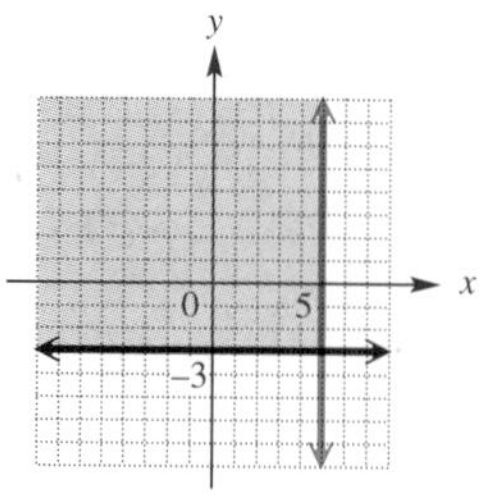

B.

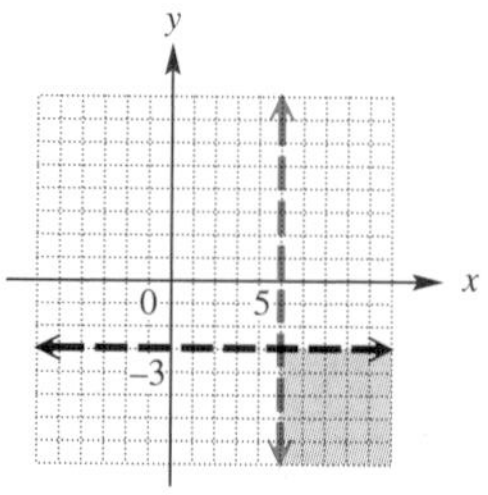

C.

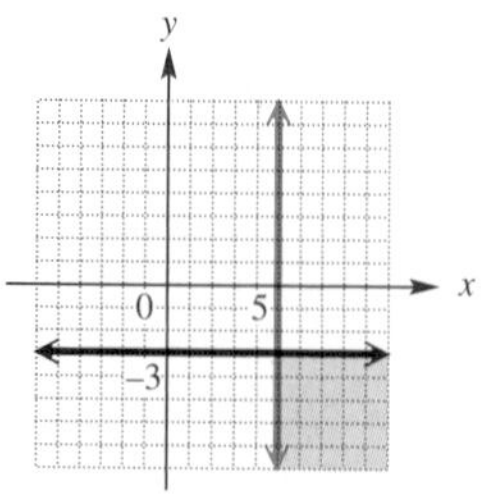

D.

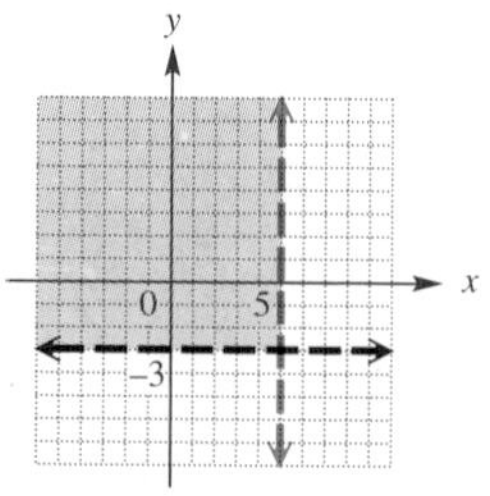

JOURNAL WRITING ▲ CHALLENGING SCIENTIFIC CALCULATOR GRAPHING CALCULATOR

11.

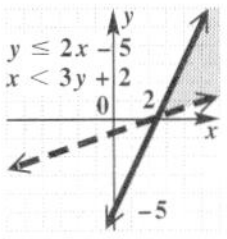

12.

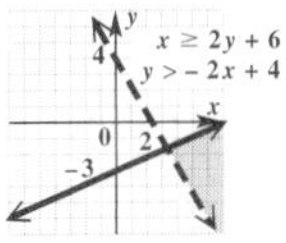

13.

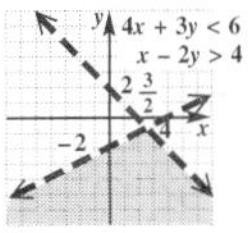

14.

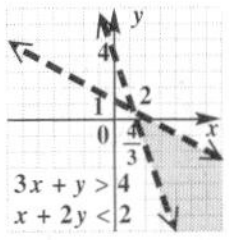

15.

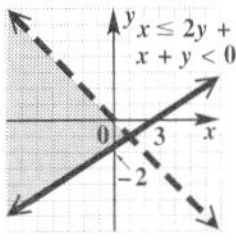

16.

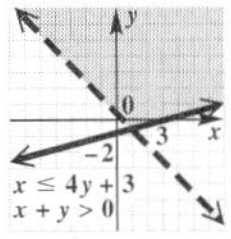

17.

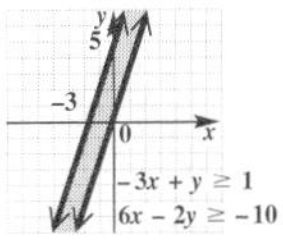

18.

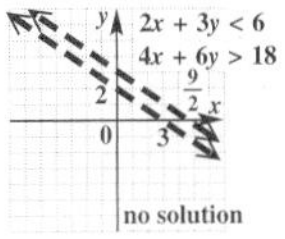

19.

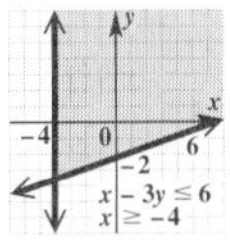

20. Graph each inequality separately, being careful to use a solid or dashed line as appropriate, and shade the correct region. Then choose the *intersection* of the two regions as the solution set of the system.

21.

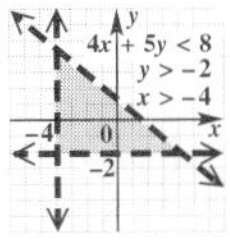

22.

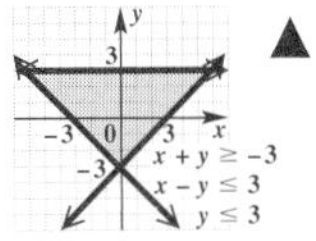

23.

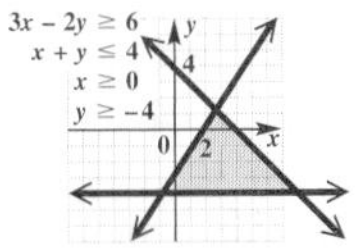

24. Substitute 0 for x and 0 for y in all the inequalities, and show that they all become true statements.

25. (4, 0), (4, 5), (9, 5)

26. (1, 2), (1, 4), (3, 2)

27. (−3, 3), (5, 3), (5, −5)

28. (−3, 4), (−3, −3), (4, 4)

Graph the solution set of each system of linear inequalities. See Examples 1 and 2.

5. $x + y \le 6$, $x - y \ge 1$

6. $x + y \le 2$, $x - y \ge 3$

7. $4x + 5y \ge 20$, $x - 2y \le 5$

8. $x + 4y \le 8$, $2x - y \ge 4$

9. $2x + 3y < 6$, $x - y < 5$

10. $x + 2y < 4$, $x - y < -1$

11. $y \le 2x - 5$, $x < 3y + 2$

12. $x \ge 2y + 6$, $y > -2x + 4$

13. $4x + 3y < 6$, $x - 2y > 4$

14. $3x + y > 4$, $x + 2y < 2$

15. $x \le 2y + 3$, $x + y < 0$

16. $x \le 4y + 3$, $x + y > 0$

17. $-3x + y \ge 1$, $6x - 2y \ge -10$

18. $2x + 3y < 6$, $4x + 6y > 18$

19. $x - 3y \le 6$, $x \ge -4$

20. Suppose that your friend was absent from class (because of a bad headache) on the day that systems of inequalities were covered. He asked you to write a short explanation of the procedure. Do this for him.

Graph the solution set of each system. See Example 3.

21. $4x + 5y < 8$, $y > -2$, $x > -4$

22. $x + y \ge -3$, $x - y \le 3$, $y \le 3$

23. $3x - 2y \ge 6$, $x + y \le 4$, $x \ge 0$, $y \ge -4$

24. How can we determine that the point (0, 0) is a solution of the system in Exercise 21 without actually graphing the system?

In a procedure called linear programming, studied in finite mathematics courses, it is necessary to find the "corner points" of a shaded region. For example, the system in Example 3 has corner points at (2, 0), (2, 4), and (5, 4). See Figure 14. Find all corner points for the following systems.

25. $y \ge x - 4$, $y \le 5$, $x \ge 4$

26. $y \le -x + 5$, $y \ge 2$, $x \ge 1$

27. $x + y \ge 0$, $y \le 3$, $x \le 5$

28. $x - y \le 0$, $x \ge -3$, $y \le 4$

JOURNAL WRITING CHALLENGING SCIENTIFIC CALCULATOR GRAPHING CALCULATOR

29. D 30. C 31. A
32. B

TECHNOLOGY INSIGHTS (EXERCISES 29–32)

Match each system of inequalities with its calculator-generated solution set. See Example 4.

A.

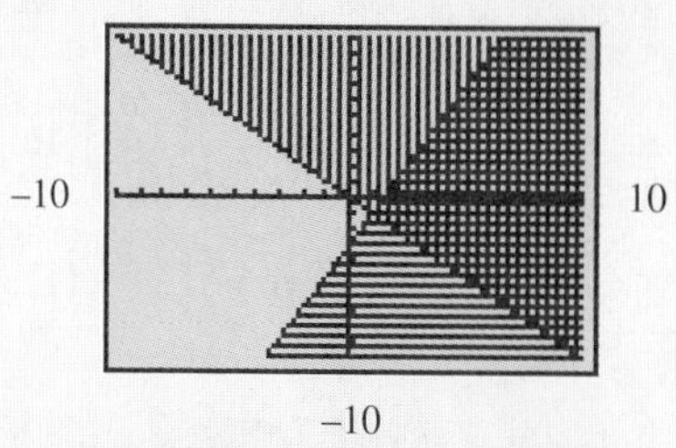

B.

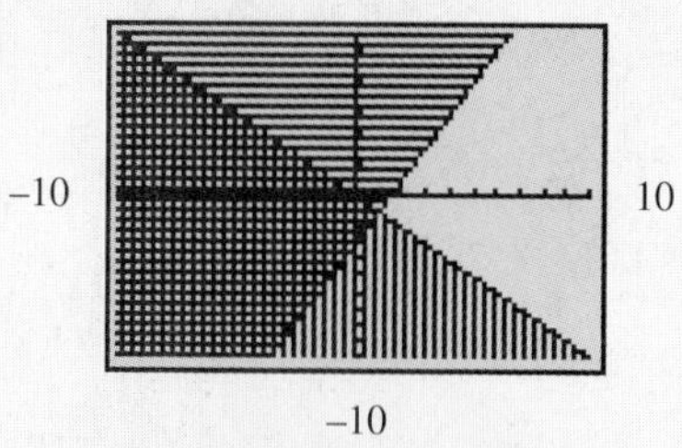

C.

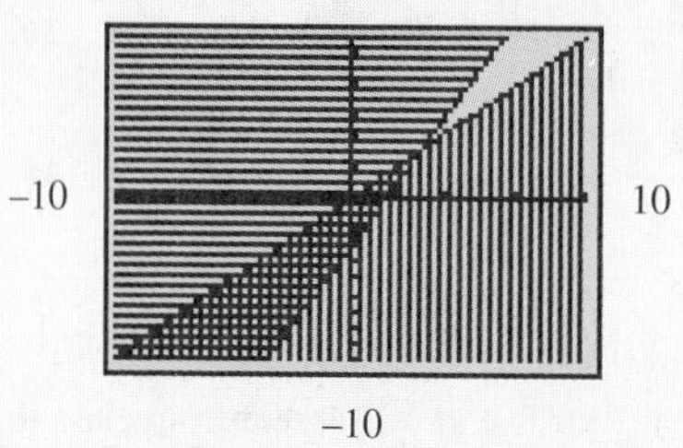

D.

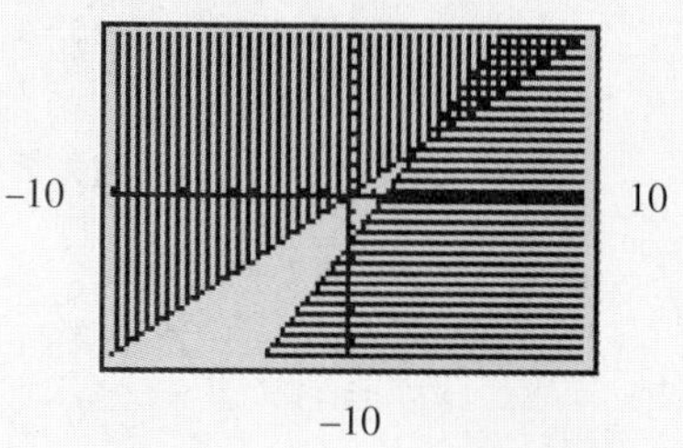

29. $y \geq x$
$y \leq 2x - 3$

30. $y \leq x$
$y \geq 2x - 3$

31. $y \geq -x$
$y \leq 2x - 3$

32. $y \leq -x$
$y \geq 2x - 3$

CHAPTER 8 GROUP ACTIVITY

Instructor's Note
Grouping: 2 students
Time: 20 minutes

In this activity students can work in pairs to solve systems of equations. One student can solve the systems related to the facts in part A, and the other student can solve the systems related to the facts in part B. Together they can answer the questions.

Top Concert Tours

Objective: Use systems of equations to analyze data.

In addition to movies and television as forms of entertainment, Americans continue to attend concerts. This activity will use systems of equations to analyze the top grossing concert tours of 1994 and 1997.

- In 1994 the top grossing North American concert tours were The Rolling Stones and Pink Floyd. Together they grossed \$224.7 million. The Rolling Stones grossed \$17.7 million more than Pink Floyd. They performed a total of 119 shows, and Pink Floyd performed one less show than The Rolling Stones. (*Source:* Pollstar.)
- In 1997 the top grossing North American concert tours were The Rolling Stones and U2. Together they grossed \$169.2 million. The Rolling Stones grossed \$9.4 million more than U2. They performed a total of 79 shows, and U2 performed 13 more shows than The Rolling Stones. (*Source:* Pollstar.)

Using the information given above, complete the chart. One student should complete the information for 1994 and the other student for 1997.

A. Write a system of equations to find the total gross for each group.

B. Write a system of equations to find the number of shows per year for each group.

A. Answers will vary.
B. Answers will vary.

C. See answers in the table.
D. 1. Answers will vary.
2. Rolling Stones in 1994
3. Rolling Stones in 1997

C. Use the information from Exercises A and B to find the gross per show.

Year	*Group*	*Total Gross*	*Shows per Year*	*Gross per Show*
1994	Rolling Stones	[$121.2 million]	[60]	[$2.02 million]
1994	Pink Floyd	[$103.5 million]	[59]	[$1.75 million]
1997	Rolling Stones	[$89.3 million]	[33]	[$2.71 million]
1997	U2	[$79.9 million]	[46]	[$1.74 million]

D. Once the chart is completed, compare the results. Answer these questions.

1. What differences do you notice between 1994 and 1997?
2. Which group had the highest total gross?
3. Which group had the highest gross per show?

CHAPTER 8 SUMMARY

KEY TERMS

8.1 **system of linear equations**
solution of a system
solution set of a system
inconsistent system
independent equations
dependent equations
consistent system

8.5 **system of linear inequalities**
solution set of a system of linear inequalities

TEST YOUR WORD POWER

See how well you have learned the vocabulary in this chapter. Answers, with examples, are given at the bottom of the page.

1. A **system of linear equations** consists of
(a) at least two linear equations with different variables
(b) two or more linear equations that have an infinite number of solutions
(c) two or more linear equations with the same variables
(d) two or more linear inequalities.

2. A **solution of a system** of linear equations is
(a) an ordered pair that makes one equation of the system true
(b) an ordered pair that makes all the equations of the system true at the same time
(c) any ordered pair that makes one or the other or both equations of the system true
(d) the set of values that make all the equations of the system false.

3. A **consistent system** is a system of equations
(a) with one solution
(b) with no solutions
(c) with an infinite number of solutions
(d) that have the same graph.

4. An **inconsistent system** is a system of equations
(a) with one solution
(b) with no solutions
(c) with an infinite number of solutions
(d) that have the same graph.

5. **Dependent equations**
(a) have different graphs
(b) have no solutions
(c) have one solution
(d) are different forms of the same equation.

Answers to Test Your Word Power

1. (c) *Example:* $2x + y = 7$, $3x - y = 3$ **2.** (b) *Example:* The ordered pair (2, 3) satisfies both equations of the above system, so it is a solution of the system. **3.** (a) *Example:* The above system is consistent. The graphs of the equations intersect at exactly one point, in this case the solution (2, 3). **4.** (b) *Example:* The equations of two parallel lines make up an inconsistent system; their graphs never intersect, so there is no solution to the system. **5.** (d) *Example:* The equations $4x - y = 8$ and $8x - 2y = 16$ are dependent because their graphs are the same line.

QUICK REVIEW

CONCEPTS	EXAMPLES
8.1 SOLVING SYSTEMS OF LINEAR EQUATIONS BY GRAPHING	
An ordered pair is a solution of a system if it makes all equations of the system true at the same time.	Is $(4, -1)$ a solution of the following system? $$x + y = 3$$ $$2x - y = 9$$ Yes, because $4 + (-1) = 3$, and $2(4) - (-1) = 9$ are both true.
If the graphs of the equations of a system are both sketched on the same axes, the points of intersection, if any, form the solution set of the system.	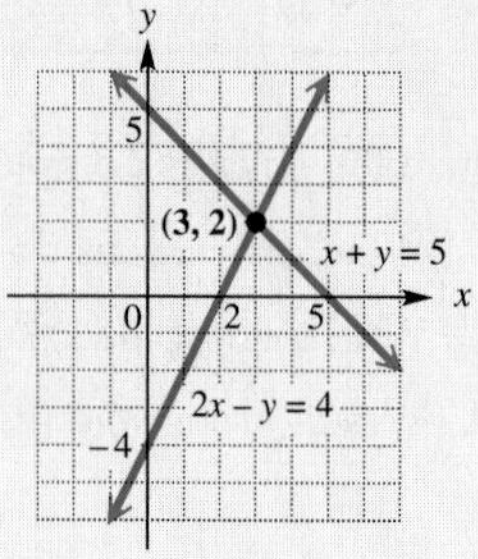 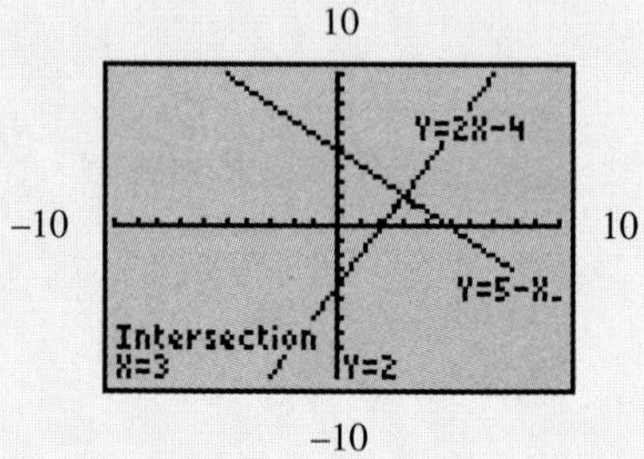
A graphing calculator can find the solution of a system by locating the point of intersection of the graphs.	$\{(3, 2)\}$ is the solution set of the system $$x + y = 5$$ $$2x - y = 4.$$
8.2 SOLVING SYSTEMS OF LINEAR EQUATIONS BY SUBSTITUTION	
Step 1 Solve one equation for one variable.	Solve by substitution. $$x + 2y = -5 \quad (1)$$ $$y = -2x - 1 \quad (2)$$ Equation (2) is already solved for y.
Step 2 Substitute for that variable in the other equation to get an equation in one variable.	Substitute $-2x - 1$ for y in equation (1). $$x + 2(-2x - 1) = -5$$
Step 3 Solve the equation from Step 2.	Solve to get $x = 1$.
Step 4 Substitute the result into the equation from Step 1 to get the value of the other variable.	To find y, let $x = 1$ in equation (2). $$y = -2(1) - 1 = -3$$
Step 5 Check. Write the solution set.	The solution, $(1, -3)$, checks so $\{(1, -3)\}$ is the solution set.
8.3 SOLVING SYSTEMS OF LINEAR EQUATIONS BY ELIMINATION	
Step 1 Write both equations in the form $$Ax + By = C.$$	Solve by elimination. $$x + 3y = 7$$ $$3x - y = 1$$
Step 2 Multiply one or both equations by appropriate numbers (if necessary) so that the coefficients of x (or y) are negatives of each other.	Multiply the first equation by -3 to eliminate the x-terms by addition.

CONCEPTS	EXAMPLES
Step 3 Add the equations to get an equation with only one variable.	$-3x - 9y = -21$ $3x - y = 1$ $-10y = -20$ Add.
Step 4 Solve the equation from Step 3.	$y = 2$ Divide by -10.
Step 5 Substitute the solution from Step 4 into either of the original equations to find the value of the remaining variable.	Substitute to get the value of x. $x + 3y = 7$ $x + 3(2) = 7$ Let $y = 2$. $x + 6 = 7$ Multiply. $x = 1$ Subtract 6.
Step 6 Write the solution set containing an ordered pair, and check the answer.	The solution, (1, 2), checks. The solution set is $\{(1, 2)\}$.
If the result of the addition step is a false statement, such as $0 = 4$, the graphs are parallel lines and *the solution set is* $\emptyset$.	$x - 2y = 6$ $-x + 2y = -2$ $0 = 4$ Solution set: $\emptyset$
If the result is a true statement, such as $0 = 0$, the graphs are the same line, and an *infinite number of ordered pairs are solutions.*	$x - 2y = 6$ $-x + 2y = -6$ $0 = 0$ Infinite number of solutions

8.4 APPLICATIONS OF LINEAR SYSTEMS

CONCEPTS	EXAMPLES
Use the six-step method.	The sum of two numbers is 30. Their difference is 6. Find the numbers.
Step 1 Choose a variable to represent each unknown value.	Let x represent one number. Let y represent the other number.
Step 2 Draw a figure or a diagram if it will help.	
Step 3 Translate the problem into a system of two equations using both variables.	$x + y = 30$ $x - y = 6$
Step 4 Solve the system.	$2x = 36$ Add. $x = 18$ Divide by 2. Let $x = 18$ in the top equation: $18 + y = 30$. Solve to get $y = 12$.
Step 5 Answer the question or questions asked.	The two numbers are 18 and 12.
Step 6 Check the solution in the words of the problem.	$18 + 12 = 30$ and $18 - 12 = 6$, so the solution checks.

8.5 SOLVING SYSTEMS OF LINEAR INEQUALITIES

CONCEPTS	EXAMPLES
To solve a system of linear inequalities, graph each inequality on the same axes. (This was explained in Section 7.2.) The solution set of the system is formed by the overlap of the regions of the two graphs.	The shaded region shows the solution set of the system $2x + 4y \geq 5$ $x \geq 1$.

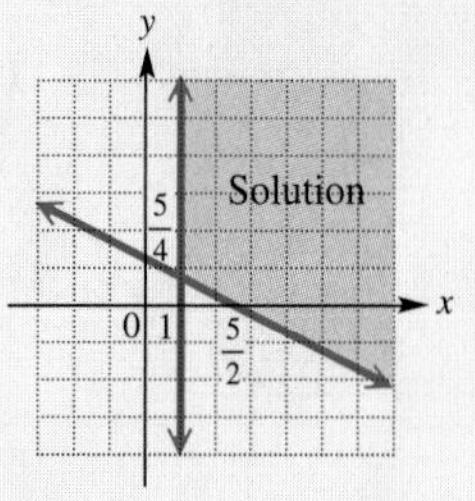

(continued)

CONCEPTS	EXAMPLES
Graphing calculators also can represent solutions of systems of inequalities.	The cross-hatched region shows the solution set of the system $y > 2x + 1$, $y < 3$. The boundary lines are not included.

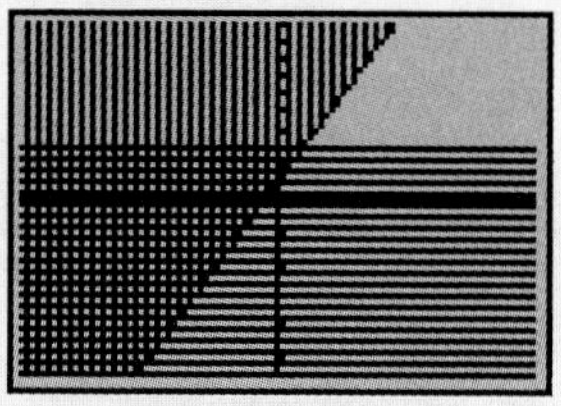

CHAPTER 8 REVIEW EXERCISES

1. yes 2. no 3. {(3, 1)} 4. ∅ 5. No, this is not correct. A false statement indicates that the solution set is ∅. 6. No, two lines cannot intersect in exactly three points. 7. {(2, 1)} 8. {(3, 5)} 9. infinite number of solutions 10. {(7, 1)} 11. His answer was incorrect since the system has infinitely many solutions (as indicated by the true statement 0 = 0). 12. It would be easiest to solve for x in the second equation because its coefficient is −1. No fractions would be involved. 13. (c)

[8.1] *Decide whether the given ordered pair is a solution of the given system.*

1. (3, 4)
$$4x - 2y = 4$$
$$5x + y = 19$$

2. (−5, 2)
$$x - 4y = -13$$
$$2x + 3y = 4$$

Solve each system by graphing.

3. $x + y = 4$
$2x - y = 5$

4. $2x + 4 = 2y$
$y - x = -3$

[8.2]

5. A student solves the system $\begin{aligned} 2x + y &= 6 \\ -2x - y &= 4 \end{aligned}$ and gets the equation $0 = 10$. The student gives the solution set as {(0, 10)}. Is this correct? Explain.

6. Can a system of two linear equations in two unknowns have exactly three solutions? Explain.

Solve each system by the substitution method.

7. $3x + y = 7$
$x = 2y$

8. $2x - 5y = -19$
$y = x + 2$

9. $5x + 15y = 30$
$x + 3y = 6$

10. $\frac{1}{3}x + \frac{1}{7}y = \frac{52}{21}$
$\frac{1}{2}x - \frac{1}{3}y = \frac{19}{6}$

11. After solving a system of linear equations by the substitution method, a student obtained the equation "0 = 0." He gave the solution set of the system as {(0, 0)}. Was his answer correct? Why or why not?

12. Suppose that you were asked to solve the system $\begin{aligned} 5x - 3y &= 7 \\ -x + 2y &= 4 \end{aligned}$ by substitution. Which variable in which equation would be easiest to solve for in your first step? Why?

[8.3]

13. Only one of the following systems does not require that we multiply one or both equations by a constant in order to solve the system by the elimination method. Which one is it?

(a) $-4x + 3y = 7$
$3x - 4y = 4$

(b) $5x + 8y = 13$
$12x + 24y = 36$

(c) $2x + 3y = 5$
$x - 3y = 12$

(d) $x + 2y = 9$
$3x - y = 6$

14. (a) 2 (b) 9 **15.** {(7, 1)} **16.** {(−4, 3)} **17.** infinite number of solutions **18.** ∅ **19.** {(−4, 1)} **20.** infinite number of solutions **21.** {(9, 2)} **22.** {(8, 9)} **23.** Answers will vary. **24.** In system B, the bottom equation is already solved for y. **25.** *How Stella Got Her Groove Back:* 782,699; *The Deep End of the Ocean:* 840,263 **26.** *Modern Maturity:* 20.5 million; *Reader's Digest:* 15.1 million **27.** Texas Commerce Tower: 75 stories; First Interstate Plaza: 71 stories **28.** Clinton: 379 votes; Dole: 159 votes **29.** length: 27 meters; width: 18 meters **30.** 13 twenties; 7 tens **31.** 25 pounds of \$1.30 candy; 75 pounds of \$.90 candy

14. For the system

$$\begin{aligned} 2x + 12y &= 7 \\ 3x + 4y &= 1, \end{aligned}$$

if we were to multiply the first equation by −3, by what number would we have to multiply the second equation in order to

(a) eliminate the x-terms when solving by the elimination method?

(b) eliminate the y-terms when solving by the elimination method?

Solve each system by the elimination method.

15. $2x - y = 13$, $x + y = 8$

16. $-4x + 3y = 25$, $6x - 5y = -39$

17. $3x - 4y = 9$, $6x - 8y = 18$

18. $2x + y = 3$, $-4x - 2y = 6$

[8.1–8.3] *Solve each system by any method.*

19. $2x + 3y = -5$, $3x + 4y = -8$

20. $6x - 9y = 0$, $2x - 3y = 0$

21. $2x + y - x = 3y + 5$, $y + 2 = x - 5$

22. $\dfrac{x}{2} + \dfrac{y}{3} = 7$, $\dfrac{x}{4} + \dfrac{2y}{3} = 8$

23. What are the three methods of solving systems discussed in this chapter? Choose one and discuss its drawbacks and advantages.

24. Why would it be easier to solve system B by the substitution method than system A?

A: $-5x + 6y = 7$, $2x + 5y = -5$ **B:** $2x + 9y = 13$, $y = 3x - 2$

[8.4] *Solve each problem by using a system of equations. Use the six-step method.*

25. A popular leisure activity of Americans is reading. In 1996, two popular fiction titles were *How Stella Got Her Groove Back* by Terry McMillan and *The Deep End of the Ocean* by Jacquelyn Mitchard. Together, these two titles sold 1,622,962 copies. The Mitchard book sold 57,564 more than the McMillan book. How many copies of each title were sold? (*Source: Publishers Weekly.*)

26. When people are not reading fiction during their leisure time, they are often reading magazines. Two of the most popular magazines in the United States are *Modern Maturity* and *Reader's Digest.* Together, the average total circulation for these two magazines during July–December 1996 was 35.6 million. *Reader's Digest* circulation was 5.4 million less than that of *Modern Maturity.* What were the circulation figures for each magazine? (*Source:* Audit Bureau of Circulations and Magazine Publishers of America.)

27. The two tallest buildings in Houston, Texas, are the Texas Commerce Tower and the First Interstate Plaza. The first of these is 4 stories taller than the second. Together the buildings have 146 stories. How many stories tall are the individual buildings? (*Source: The World Almanac and Book of Facts,* 1998.)

28. In the 1996 presidential election, Bill Clinton received 220 more electoral votes than Bob Dole. Together, the two candidates received a total of 538 votes. How many votes did each receive? (*Source: The World Almanac and Book of Facts,* 1998.)

29. The perimeter of a rectangle is 90 meters. Its length is $1\frac{1}{2}$ times its width. Find the length and width of the rectangle.

30. A cashier has 20 bills, all of which are \$10 or \$20 bills. The total value of the money is \$330. How many of each type does the cashier have?

31. Candy that sells for \$1.30 per pound is to be mixed with candy selling for \$.90 per pound to get 100 pounds of a mix that will sell for \$1 per pound. How much of each type should be used?

32. 60 liters of 40% solution; 30 liters of 70% solution
33. $7000 at 3%; $11,000 at 4%
34. apple: $.55; banana: $.45
35. plane: 250 miles per hour; wind: 20 miles per hour
36. 10,000 algebra books; 3000 calculus books

37.

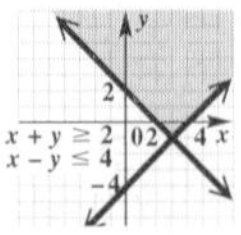

38.

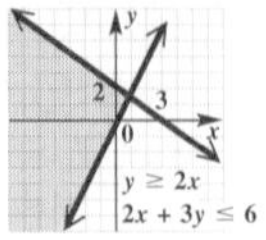

39.

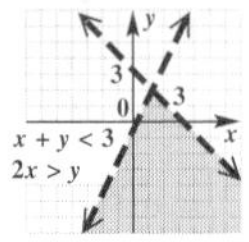

40.

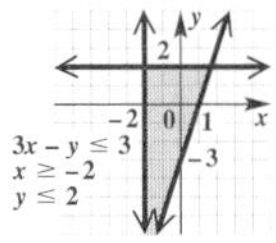

32. A 40% antifreeze solution is to be mixed with a 70% solution to get 90 liters of a 50% solution. How many liters of the 40% and 70% solutions will be needed?

Rate	*Number of Liters*	*Amount of Pure Antifreeze*
.40	x	
.70	y	
.50	90	

33. Ms. Cesar invested \$18,000. Part of it was invested at 3% annual simple interest and the rest was invested at 4%. Her interest income for the first year was \$650. How much did she invest at each rate?

Rate	*Amount of Principal*	*Interest*
.03	x	
.04	y	
Totals	\$18,000	

34. Della Dawkins can buy 6 apples and 5 bananas for \$5.55. She can also buy 12 apples and 2 bananas for \$7.50. Find the cost of one apple and the cost of one banana.

35. A certain plane flying with the wind travels 540 miles in 2 hours. Later, flying against the same wind, the plane travels 690 miles in 3 hours. Find the speed of the plane in still air and the speed of the wind.

36. A textbook author receives a \$2 royalty for each of his algebra books sold, and a \$3 royalty for each of his calculus books sold. During one royalty period, the two books together sold 13,000 copies and he received a total of \$29,000 in royalties. How many of each kind of book were sold during that period?

[8.5] *Graph the solution set for each system of linear inequalities.*

37. $x + y \geq 2$
$x - y \leq 4$

38. $y \geq 2x$
$2x + 3y \leq 6$

39. $x + y < 3$
$2x > y$

40. $3x - y \leq 3$
$x \geq -2$
$y \leq 2$

41. $\left\{\left(\frac{28}{5}, \frac{16}{5}\right)\right\}$ **42.** $\left\{\left(\frac{28}{5}, \frac{16}{5}\right)\right\}$
43. They are the same. It makes no difference which method we use.
44. $y_1 = 2x - 8$
45. $y_2 = \frac{12 - x}{2}$ or $y_2 = 6 - \frac{1}{2}x$
46. $\left\{\frac{28}{5}\right\}$; The solution is the x-value found in Exercises 41 and 42.
47. We get $\frac{16}{5} = \frac{16}{5}$. This value is the y-value found in Exercises 41 and 42. **48.** 2

RELATING CONCEPTS (EXERCISES 41–52)

Consider the system

$$2x - y = 8 \quad (1)$$
$$x + 2y = 12. \quad (2)$$

Work Exercises 41–52 in order.

41. Solve the system by the substitution method.

42. Solve the system by the elimination method.

43. Compare your answers in Exercises 41 and 42. What do you notice? Does it make a difference in the *method* of solution as to what answer you get?

44. Solve equation (1) for y, and call it y_1.

45. Solve equation (2) for y, and call it y_2.

46. Set the expressions for y_1 and y_2 equal to each other and solve the resulting equation. How does the solution compare to your answers in Exercises 41 and 42?

47. Substitute your solution from Exercise 46 back into the original equation you wrote. When you evaluate each side, you should get the same result. How does this result compare to your answers in Exercises 41 and 42?

48. What is the slope of the line in equation (1)?

49. $-\frac{1}{2}$

50. They are perpendicular.

51.

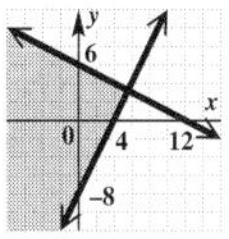

$x + 2y \le 12$
$2x - y \le 8$

52. $\left\{\left(\frac{28}{5}, \frac{16}{5}\right)\right\}$

53. $\{(4, 8)\}$ **54.** infinite number of solutions **55.** $\{(2, 0)\}$

56. $\{(-4, 15)\}$

57.

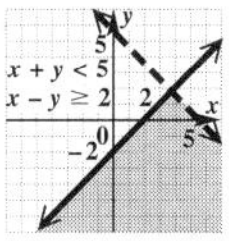

58.

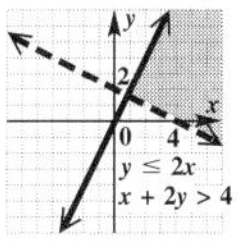

59. Great Smoky Mountains: 9.3 million; Rocky Mountain National Park: 2.9 million
60. 8 inches, 8 inches, 13 inches
61. slower car: 38 miles per hour; faster car: 68 miles per hour
62. 102 small bottles; 44 large bottles **63.** Yes. Let x represent each of the two equal side lengths. Then $x + 5$ is the length of the third side. The equation to solve is $x + x + (x + 5) = 29$, giving $x = 8$. The side lengths are 8, 8, and 13 inches.

RELATING CONCEPTS (EXERCISES 41–52) (CONTINUED)

49. What is the slope of the line in equation (2)?

50. Based on what you learned in Chapter 3, are the lines parallel, perpendicular, or neither?

51. Graph the system of linear inequalities

$$2x - y \le 8$$
$$x + 2y \le 12.$$

52. Look at the "corner point" of the shaded region in your answer to Exercise 51. What are its coordinates? (*Hint:* Look at your answers in Exercises 41 and 42.)

Did you make the connections between solving systems in two variables and solving and checking equations in one variable?

MIXED REVIEW EXERCISES

Solve each of the following using the methods of this chapter.

53. $\frac{2x}{3} + \frac{y}{4} = \frac{14}{3}$
$\frac{x}{2} + \frac{y}{12} = \frac{8}{3}$

54. $x + y = 2y + 6$
$y + 8 = x + 2$

55. $3x + 4y = 6$
$4x - 5y = 8$

56. $\frac{3x}{2} + \frac{y}{5} = -3$
$4x + \frac{y}{3} = -11$

57. $x + y < 5$
$x - y \ge 2$

58. $y \le 2x$
$x + 2y > 4$

59. In 1996, 12.2 million people visited the Great Smoky Mountains and Rocky Mountain National Park, two popular vacation destinations. The Great Smoky Mountains had 6.4 million more visitors than Rocky Mountain National Park. How many visitors did each of these destinations have? (*Source: The Wall Street Journal Almanac,* 1998.)

60. The perimeter of an isosceles triangle is 29 inches. One side of the triangle is 5 inches longer than each of the two equal sides. Find the lengths of the sides of the triangle.

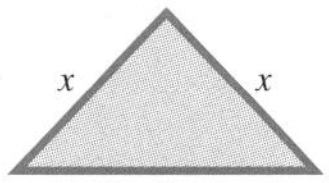

61. Two cars leave from the same place and travel in opposite directions. One car travels 30 miles per hour faster than the other. After $2\frac{1}{2}$ hours they are 265 miles apart. What are the rates of the cars?

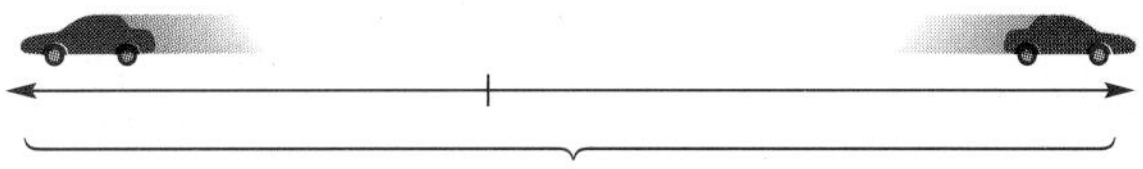

62. A hospital bought a total of 146 bottles of glucose solution. Small bottles cost \$2 each, and large ones cost \$3 each. The total cost was \$336. How many of each size bottle were bought?

63. Can the problem in Exercise 60 be worked using a single variable? If so, explain how it can be done, and then solve it.

64. Without actually graphing, determine which one of the following systems of inequalities has no solution.

(a) $x \geq 4$, $y \leq 3$ **(b)** $x + y > 4$, $x + y < 3$ **(c)** $x > 2$, $y < 1$ **(d)** $x + y > 4$, $x - y < 3$

64. (b)

CHAPTER 8 TEST

1. Decide whether each ordered pair is a solution of the system

$$\begin{aligned} 2x + y &= -3 \\ x - y &= -9. \end{aligned}$$

(a) $(1, -5)$ **(b)** $(1, 10)$ **(c)** $(-4, 5)$

2. Solve this system by graphing.

$$\begin{aligned} x + 2y &= 6 \\ -2x + y &= -7 \end{aligned}$$

Solve each system by substitution.

3. $2x + y = -4$, $x = y + 7$

4. $4x + 3y = -35$, $x + y = 0$

Solve each system by elimination.

5. $2x - y = 4$, $3x + y = 21$

6. $4x + 2y = 2$, $5x + 4y = 7$

7. $3x + 4y = 9$, $2x + 5y = 13$

8. $4x + 5y = 2$, $-8x - 10y = 6$

9. $6x - 5y = 0$, $-2x + 3y = 0$

10. $\frac{6}{5}x - \frac{1}{3}y = -20$, $-\frac{2}{3}x + \frac{1}{6}y = 11$

11. Solve the system by any method.

$$\begin{aligned} 4y &= -3x + 5 \\ 6x &= -8y + 10 \end{aligned}$$

12. Suppose that the graph of a system of two linear equations consists of lines that have the same slope but different y-intercepts. How many solutions does the system have?

Write a system of two equations and solve each problem. Use the six-step method.

13. The distance between Memphis and Atlanta is 300 miles less than the distance between Minneapolis and Houston. Together the two distances total 1042 miles. How far is it between Memphis and Atlanta? How far is it between Minneapolis and Houston?

14. In 1996, the two most popular amusement parks in the United States were Disneyland and the Magic Kingdom at Walt Disney World. Disneyland had 1.2 million more visitors than the Magic Kingdom, and together they had 28.8 million visitors. How many visitors did each park have? (*Source: The Wall Street Journal Almanac,* 1998.)

15. A 25% solution of alcohol is to be mixed with a 40% solution to get 50 liters of a final mixture that is 30% alcohol. How much of each of the original solutions should be used?

16. Two cars leave from Perham, Minnesota, and travel in the same direction. One car travels one and one third times as fast as the other. After 3 hours they are 45 miles apart. What are the speeds of the cars?

17. As of 1996, the two top-selling videos of all time were both Disney films: *The Lion King* and *Snow White. The Lion King* had sold 2 million more copies than *Snow White,* and together they had sold 58 million copies. How many copies had each title sold? (*Source:* Paul Kagan Associates, Inc.)

[8.1] 1. (a) no (b) no (c) yes 2. {(4, 1)} [8.2] 3. {(1, −6)} 4. {(−35, 35)} [8.3] 5. {(5, 6)} 6. {(−1, 3)} 7. {(−1, 3)} 8. ∅ 9. {(0, 0)} 10. {(−15, 6)} [8.1–8.3] 11. infinite number of solutions 12. It has no solution. [8.4] 13. Memphis and Atlanta: 371 miles; Minneapolis and Houston: 671 miles 14. Disneyland: 15.0 million; Magic Kingdom: 13.8 million 15. $33\frac{1}{3}$ liters of 25% solution; $16\frac{2}{3}$ liters of 40% solution 16. slower car: 45 miles per hour; faster car: 60 miles per hour 17. *The Lion King*: 30 million; *Snow White*: 28 million

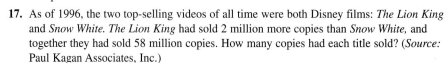

[8.5] 18.

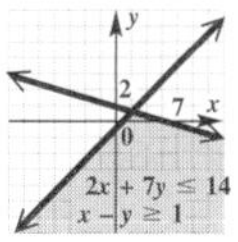

19.

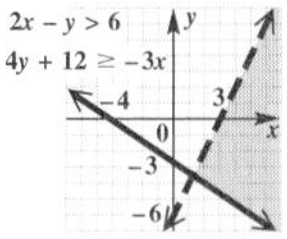

20. A

Graph the solution set of each system of inequalities.

18. $2x + 7y \le 14$
$x - y \ge 1$

19. $2x - y > 6$
$4y + 12 \ge -3x$

20. Which one of the following is a calculator representation of the graph of the system given here?

$$y \ge 3x + 5$$
$$y \le -2x - 1$$

A.

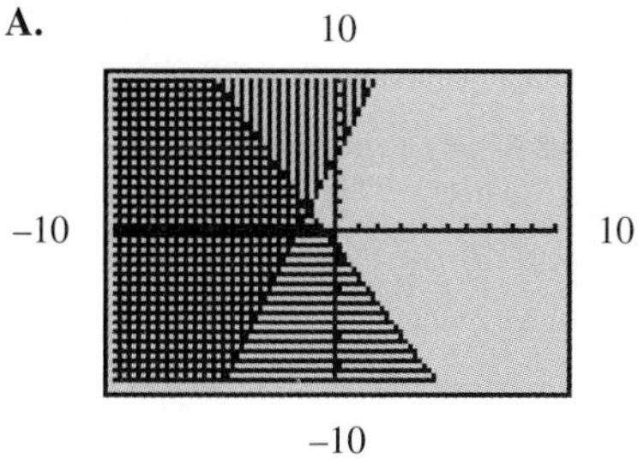

B.

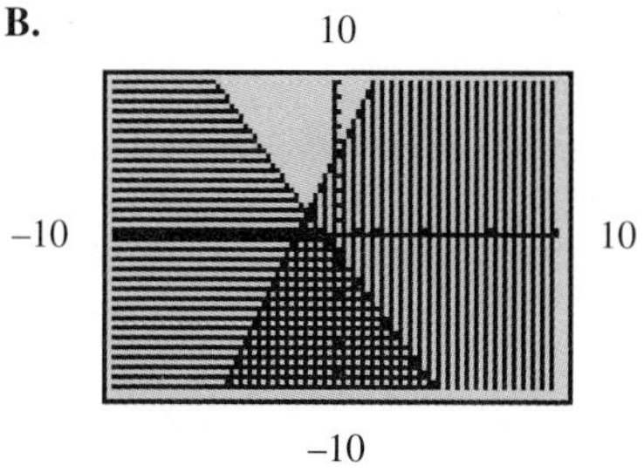

C.

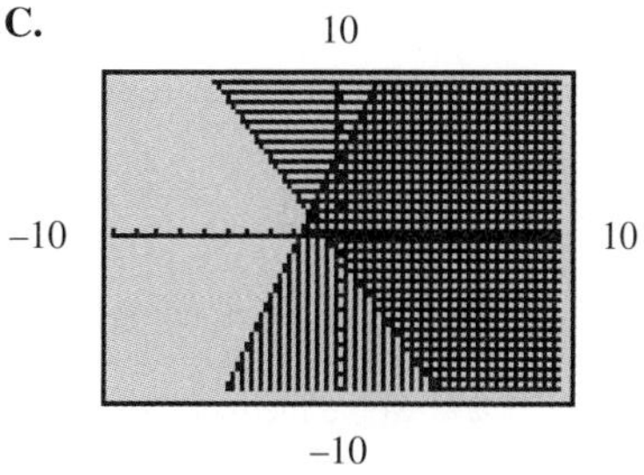

CUMULATIVE REVIEW EXERCISES CHAPTERS 1–8

[1.7] 1. −1, 1, −2, 2, −4, 4, −5, 5, −8, 8, −10, 10, −20, 20, −40, 40
[1.3] 2. 1
[1.7] 3. commutative property
4. distributive property
5. inverse property
[1.6] 6. 46
[2.2] 7. $\left\{-\frac{13}{11}\right\}$ **8.** $\left\{\frac{9}{11}\right\}$
[2.4] 9. width: $8\frac{1}{4}$ **inches; length:** $10\frac{3}{4}$ **inches**
[2.7] 10. $\left(-\frac{11}{2}, \infty\right)$
[3.1] 11.

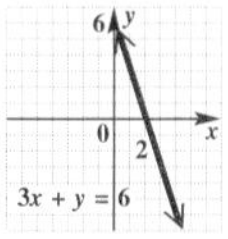

1. List all integer factors of 40.

2. Find the value of the expression if $x = 1$ and $y = 5$.

$$\frac{3x^2 + 2y^2}{10y + 3}$$

Name the property that justifies each statement.

3. $5 + (-4) = (-4) + 5$

4. $r(s - k) = rs - rk$

5. $-\frac{2}{3} + \frac{2}{3} = 0$

6. Evaluate $-2 + 6[3 - (4 - 9)]$.

Solve each linear equation.

7. $2 - 3(6x + 2) = 4(x + 1) + 18$

8. $\frac{3}{2}\left(\frac{1}{3}x + 4\right) = 6\left(\frac{1}{4} + x\right)$

9. No baseball fan should be without a copy of *The Sports Encyclopedia: Baseball 1997* by David S. Neft and Richard M. Cohen. Now in its 17th edition, it provides exhaustive statistics for professional baseball dating back to 1876. This book has a perimeter of 38 inches, and its width measures 2.5 inches less than its length. What are the dimensions of the book?

10. Solve the linear inequality $-8 < 2x + 3$.

11. Graph the linear equation $3x + y = 6$.

12. Find the slope of the line passing through the points $(-5, 6)$ and $(1, -2)$.

13. Find the slope of the line whose graph is perpendicular to that of $y = 4x - 3$.

Perform the indicated operations.

14. $-3(-5x^2 + 3x - 10) - (x^2 - 4x + 7)$

15. $(3x - 7)(2y + 4)$

16. $\dfrac{3k^3 + 17k^2 - 27k + 7}{k + 7}$

17. Write in scientific notation: 36,500,000,000.

18. Simplify and write the answer using only positive exponents: $\left(\dfrac{x^{-4}y^3}{x^2y^4}\right)^{-1}$.

Factor completely.

19. $10m^2 + 7mp - 12p^2$

20. $64t^2 - 48t + 9$

Solve each quadratic equation.

21. $6x^2 - 7x - 3 = 0$

22. $r^2 - 121 = 0$

23. For what value(s) of x is the rational expression $\dfrac{3x + 7}{2x^2 - 3x - 5}$ undefined?

Perform each operation and express each answer in lowest terms.

24. $\dfrac{-3x + 6}{2x + 4} - \dfrac{-3x - 8}{2x + 4}$

25. $\dfrac{16k^2 - 9}{8k + 6} \div \dfrac{16k^2 - 24k + 9}{6}$

26. Solve the equation $\dfrac{4}{x + 1} + \dfrac{3}{x - 2} = 4$.

27. Solve the inequality $x^2 - 5x - 6 \le 0$.

Write an equation for each line described. Express in the form $y = mx + b$.

28. Through $(2, -5)$ with slope 3

29. Through the points $(0, 4)$ and $(2, 4)$

30. (a) Write an equation of the vertical line through $(9, -2)$.
(b) Write an equation of the horizontal line through $(4, -1)$.

31. Refer to the accompanying graph. If 1992 is represented by $x = 0$ and 1996 is represented by $x = 4$, what is the equation of the line joining the tops of the bars for these two years? How is the slope of this line interpreted in the context of the data represented?

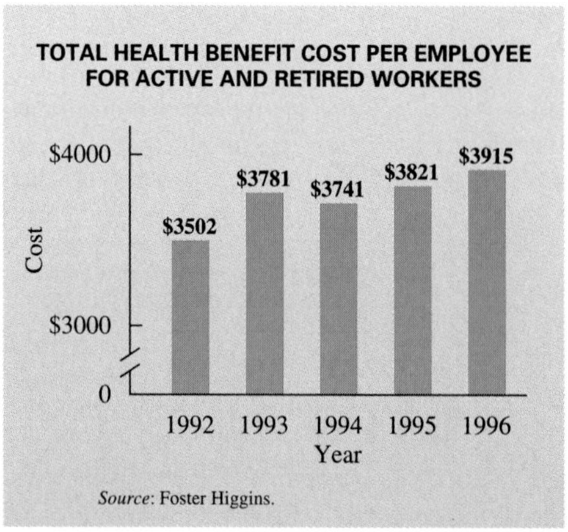

32. Graph the solution set of the linear inequality $x - 3y > 6$.

33. If $f(x) = 3x^2 - 4x + 5$, find $f(2)$.

[3.3] 12. $-\frac{4}{3}$ 13. $-\frac{1}{4}$
[4.1] 14. $14x^2 - 5x + 23$
[4.3] 15. $6xy + 12x - 14y - 28$
[4.6] 16. $3k^2 - 4k + 1$
[4.7] 17. 3.65×10^{10}
[4.5] 18. x^6y
[5.3] 19. $(5m - 4p)(2m + 3p)$
[5.4] 20. $(8t - 3)^2$
[5.5] 21. $\left\{-\frac{1}{3}, \frac{3}{2}\right\}$
22. $\{-11, 11\}$
[6.1] 23. $-1, \frac{5}{2}$
[6.4] 24. $\frac{7}{x + 2}$
[6.2] 25. $\frac{3}{4k - 3}$
[6.6] 26. $\left\{-\frac{1}{4}, 3\right\}$
[5.7] 27. $[-1, 6]$
[7.1] 28. $y = 3x - 11$
29. $y = 4$
30. (a) $x = 9$ (b) $y = -1$
31. $y = 103.25x + 3502$; The slope represents the average yearly increase in health benefit cost during the period.
[7.2] 32.

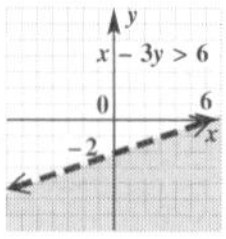

[7.3] 33. 9

[8.1–8.3] 34. {(−1, 6)}
35. {(3, −4)} 36. {(2, −1)}
[8.4] 37. 405 adults and 49 children
38. 19 inches, 19 inches, 15 inches
39. 4 girls and 3 boys
[8.5] 40. (b)

Solve each system by any method.

34. $2x - y = -8$
$x + 2y = 11$

35. $4x + 5y = -8$
$3x + 4y = -7$

36. $3x + 5y = 1$
$x = y + 3$

Use a system of equations to solve each problem.

37. Admission prices at a football game were \$6 for adults and \$2 for children. The total value of the tickets sold was \$2528, and 454 tickets were sold. How many adults and how many children attended the game?

Kind of Ticket	*Number Sold*	*Cost of Each (in dollars)*	*Total Value (in dollars)*
Adult	x	6	$6x$
Child	y		
Total	454		

38. The perimeter of a triangle is 53 inches. If two sides are of equal length, and the third side measures 4 inches less than each of the equal sides, what are the lengths of the three sides?

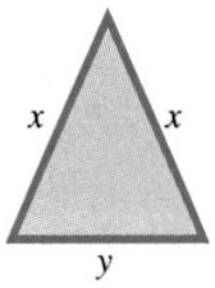

39. The Smith family is coming to visit, and no one knows how many children they have. Janet, one of the girls, says she has as many brothers as sisters; her brother Steve says he has twice as many sisters as brothers. How many boys and how many girls are in the family?

40. Which one of the following systems of linear inequalities is graphed in the figure?

(a) $x \le 3$, $y \le 1$ **(b)** $x \le 3$, $y \ge 1$

(c) $x \ge 3$, $y \le 1$ **(d)** $x \ge 3$, $y \ge 1$

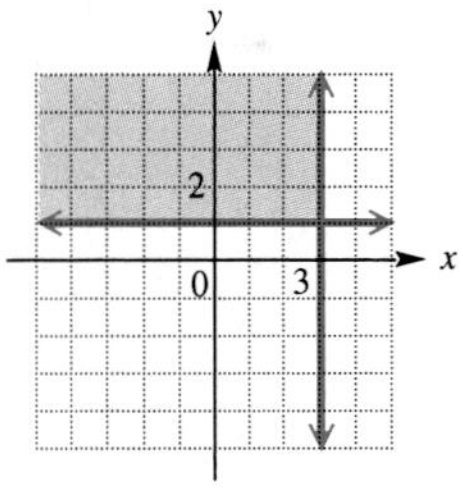

9 Roots and Radicals

Electronics

The electronics industry encompasses producing, selling, and servicing the computers, TVs, VCRs, CD players, and cell phones that pervade our lives. The industry boomed for many years, but recently industry sales have leveled off. The graph, which gives the percent change in orders from year to year, shows the turbulence in the electronics industry from 1983 to 1998.*

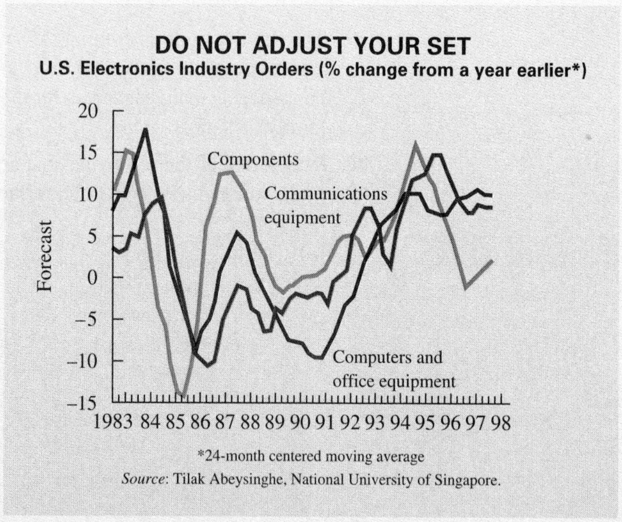

A major reason for the downturn in electronics business is the increase in the number of companies, particularly Asian companies, that have entered the industry. This large increase in supply has caused a glut of chips, with resulting price declines. This, in turn, has contributed to the collapse of export growth throughout Eastern Asia. In the exercises for Section 9.3, we examine data on U.S. exports and imports of electronics.

*"The silicon tigers' electric shocker," *The Economist,* November 9, 1996.

9.1 Evaluating Roots

OBJECTIVES

1. Find square roots.
2. Decide whether a given root is rational, irrational, or not a real number.
3. Find decimal approximations for irrational square roots.
4. Use the Pythagorean formula.
5. Use the distance formula.
6. Find higher roots.

FOR EXTRA HELP

SSG Sec. 9.1
SSM Sec. 9.1
Pass the Test Software
InterAct Math Tutorial Software
Video 14

In Section 1.2 we discussed the idea of the *square* of a number. Recall that squaring a number means multiplying the number by itself.

$$\text{If } a = 7, \text{ then } a^2 = 7 \cdot 7 = 49.$$
$$\text{If } a = -5, \text{ then } a^2 = (-5) \cdot (-5) = 25.$$
$$\text{If } a = -\frac{1}{2}, \text{ then } a^2 = \left(-\frac{1}{2}\right) \cdot \left(-\frac{1}{2}\right) = \frac{1}{4}.$$

In this chapter the opposite problem is considered.

$$\text{If } a^2 = 49, \text{ then } a = ?$$
$$\text{If } a^2 = 25, \text{ then } a = ?$$
$$\text{If } a^2 = \frac{1}{4}, \text{ then } a = ?$$

OBJECTIVE 1 Find square roots. To find a in the three statements above, we must find a number that when multiplied by itself results in the given number. The number a is called a **square root** of the number a^2.

CHALKBOARD EXAMPLE

Find all square roots of 64.

Answer: 8, −8

EXAMPLE 1 Finding All Square Roots of a Number

Find all square roots of 49.

To find a square root of 49, think of a number that when multiplied by itself gives 49. One square root is 7, since $7 \cdot 7 = 49$. Another square root of 49 is -7, since $(-7)(-7) = 49$. The number 49 has two square roots, 7 and -7; one is positive and one is negative.

TEACHING TIP Remind students that beginning algebra concentrates on the real numbers, which consist of the rationals and the irrationals. To this point, we have worked only with rational numbers. This chapter deals with irrational numbers as well.

Reals — Rationals / Irrationals

The positive square root of a number is written with the symbol $\sqrt{}$. For example, the positive square root of 49 is 7, written

$$\sqrt{49} = 7.$$

The symbol $-\sqrt{}$ is used for the negative square root of a number. For example, the negative square root of 49 is -7, written

$$-\sqrt{49} = -7.$$

Most calculators have a square root key, usually labeled $\boxed{\sqrt{x}}$, that allows us to find the square root of a number. For example, if we enter 49 and use the square root key, the display will show 7.

The symbol $\sqrt{}$ is called a **radical sign** and always represents the nonnegative square root. The number inside the radical sign is called the **radicand** and the entire expression, radical sign and radicand, is called a **radical.** An algebraic expression containing a radical is called a **radical expression.**

INSTRUCTOR'S RESOURCES

PTB Sec. 9.1 **ISM** Sec. 9.1
AB Sec. 9.1

TEST GENERATOR

WORLD WIDE WEB
www.LialAlgebra.com

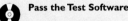

TEACHING TIP Mention that the statement $\sqrt{9} = \pm 3$ is **incorrect.** It says, in part, that a positive number equals a negative number.

Square Roots of a

If a is a positive real number,

$$\sqrt{a} \text{ is the positive square root of } a,$$
$$-\sqrt{a} \text{ is the negative square root of } a.$$

For nonnegative a,

$$\sqrt{a} \cdot \sqrt{a} = (\sqrt{a})^2 = a \quad \text{and} \quad -\sqrt{a} \cdot -\sqrt{a} = (-\sqrt{a})^2 = a.$$

Also, $\sqrt{0} = 0$.

CHALKBOARD EXAMPLE

Find the square root.

$$-\sqrt{225}$$

Answer: -15

EXAMPLE 2 Finding Square Roots

Find each square root.

(a) $\sqrt{144}$

The radical $\sqrt{144}$ represents the positive square root of 144. Think of a positive number whose square is 144.

$$12^2 = 144, \quad \text{so} \quad \sqrt{144} = 12.$$

(b) $-\sqrt{1024}$

This symbol represents the negative square root of 1024. A calculator with a square root key can be used to find $\sqrt{1024} = 32$. Then, $-\sqrt{1024} = -32$.

(c) $\sqrt{\dfrac{4}{9}} = \dfrac{2}{3}$ **(d)** $-\sqrt{\dfrac{16}{49}} = -\dfrac{4}{7}$

As shown in the definition above, when the square root of a positive real number is squared, the result is that positive real number. (Also, $(\sqrt{0})^2 = 0$.)

CHALKBOARD EXAMPLE

Find the square of the radical expression $\sqrt{2x^2 + 3}$.

Answer: $2x^2 + 3$

EXAMPLE 3 Squaring Radical Expressions

Find the *square* of each radical expression.

(a) $\sqrt{13}$

$(\sqrt{13})^2 = 13$ Definition of square root

(b) $-\sqrt{29}$

$(-\sqrt{29})^2 = 29$ The square of a *negative* number is positive.

(c) $\sqrt{p^2 + 1}$

$(\sqrt{p^2 + 1})^2 = p^2 + 1$

TEACHING TIP At this point, you may want to review the diagram of the real number system in Section 1.4 with students.

OBJECTIVE 2 Decide whether a given root is rational, irrational, or not a real number. All numbers with square roots that are rational are called **perfect squares.** For example, 100 and $\frac{25}{4}$ are perfect squares. A number that is not a perfect square has a square root that is not a rational number. For example, $\sqrt{5}$ is not a rational number because it cannot be written as the ratio of two integers. Its decimal neither terminates nor repeats. However, $\sqrt{5}$ is a real number and corresponds to a point on the number line. As mentioned in Chapter 1, a real number that is not rational is called an *irrational number.* The number $\sqrt{5}$ is irrational. Many square roots of integers are irrational.

If a is a positive number that is not a perfect square, then $\sqrt{a}$ is irrational.

Not every number has a *real number* square root. For example, there is no real number that can be squared to get -36. (The square of a real number can never be negative.) Because of this $\sqrt{-36}$ is not a real number. A calculator may show an error message in a case like this.

If a is a negative number, then $\sqrt{a}$ is not a real number.

CHALKBOARD EXAMPLE

Tell whether the square root is rational, irrational, or not a real number: $\sqrt{72}$.

Answer: irrational

EXAMPLE 4 Identifying Types of Square Roots

Tell whether each square root is rational, irrational, or not a real number.

(a) $\sqrt{17}$

Since 17 is a not a perfect square, $\sqrt{17}$ is irrational.

(b) $\sqrt{64}$

The number 64 is a perfect square, 8^2, so $\sqrt{64} = 8$, a rational number.

(c) $\sqrt{85}$ is irrational.

(d) $\sqrt{81}$ is rational ($\sqrt{81} = 9$).

(e) $\sqrt{-25}$

There is no real number whose square is -25. Therefore $\sqrt{-25}$ is not a real number.

NOTE: Not all irrational numbers are square roots of integers. For example, π (approximately 3.14159) is an irrational number that is not a square root of any integer.

OBJECTIVE 3 Find decimal approximations for irrational square roots. Even if a number is irrational, a decimal that approximates the number can be found by using a calculator. For example, a calculator shows that $\sqrt{10}$ is 3.16227766, although this is only a rational number *approximation* of $\sqrt{10}$.

CHALKBOARD EXAMPLE

Find a decimal approximation for the square root: $\sqrt{190}$.

Answer: 13.784

EXAMPLE 5 Approximating Irrational Square Roots

Find a decimal approximation for each square root. Round answers to the nearest thousandth.

(a) $\sqrt{11}$

Using the square root key of a calculator gives $\sqrt{11} \approx 3.31662479 \approx 3.317$ rounded to the nearest thousandth, where $\approx$ means "is approximately equal to."

(b) $\sqrt{39} \approx 6.245$

(c) $-\sqrt{745} \approx -27.295$

(d) $\sqrt{-180}$ is not a real number.

OBJECTIVE 4 **Use the Pythagorean formula.** One application of square roots uses the Pythagorean formula. Recall from Section 5.6 that by this formula, if c is the length of the hypotenuse of a right triangle, and a and b are the lengths of the two legs, then

$$c^2 = a^2 + b^2.$$

See Figure 1.

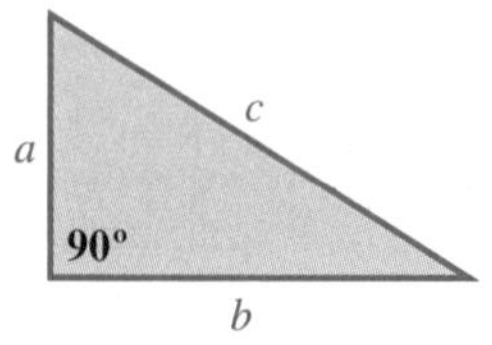

Figure 1

CHALKBOARD EXAMPLE

Find the unknown length of the third side in the right triangle with sides a, b, and c, where c is the hypotenuse.

$c = 11,\ a = 8$

Answer: $\sqrt{57} \approx 7.550$

EXAMPLE 6 Using the Pythagorean Formula

Find the unknown length of the third side of each right triangle with sides a, b, and c, where c is the hypotenuse.

(a) $a = 3, b = 4$

Use the formula to find c^2 first.

$$\begin{aligned} c^2 &= a^2 + b^2 \\ &= 3^2 + 4^2 && \text{Let } a = 3 \text{ and } b = 4. \\ &= 9 + 16 = 25 && \text{Square and add.} \end{aligned}$$

Now find the positive square root of 25 to get c.

$$c = \sqrt{25} = 5$$

(Although -5 is also a square root of 25, the length of a side of a triangle must be a positive number.)

(b) $c = 9, b = 5$

Substitute the given values in the formula $c^2 = a^2 + b^2$. Then solve for a^2.

$$\begin{aligned} 9^2 &= a^2 + 5^2 && \text{Let } c = 9 \text{ and } b = 5. \\ 81 &= a^2 + 25 && \text{Square.} \\ 56 &= a^2 && \text{Subtract 25.} \end{aligned}$$

Again, we want only the positive root $a = \sqrt{56} \approx 7.483$.

CAUTION

Be careful not to make the common mistake of thinking that $\sqrt{a^2 + b^2}$ equals $a + b$. As Example 6(a) shows,

$$\sqrt{9 + 16} = \sqrt{25} = 5 \neq \sqrt{9} + \sqrt{16} = 3 + 4,$$

so that, in general,

$$\sqrt{a^2 + b^2} \neq a + b.$$

CONNECTIONS

Pythagoras did not actually discover the Pythagorean formula. While he may have written the first proof, there is evidence that the Babylonians knew the concept quite well. The figure on the left illustrates the formula by using a tile pattern. In the figure, the side of the square along the hypotenuse measures 5 units, while the sides along the legs measure 3 and 4 units. If we let $a = 3$, $b = 4$, and $c = 5$, the equation of the Pythagorean formula is satisfied.

$$a^2 + b^2 = c^2$$

$$3^2 + 4^2 = 5^2 \quad ?$$

$$25 = 25 \quad \text{True}$$

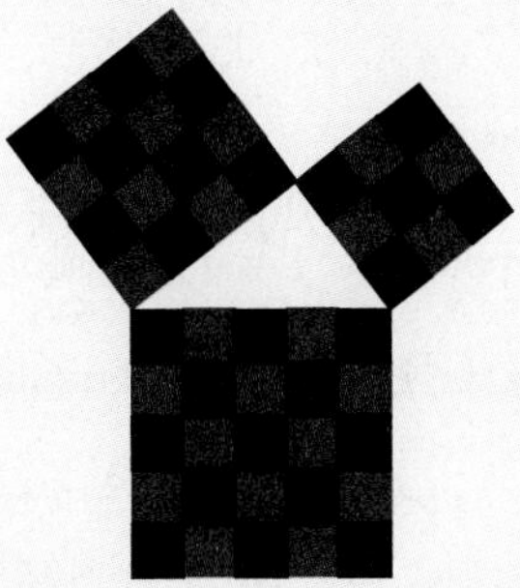

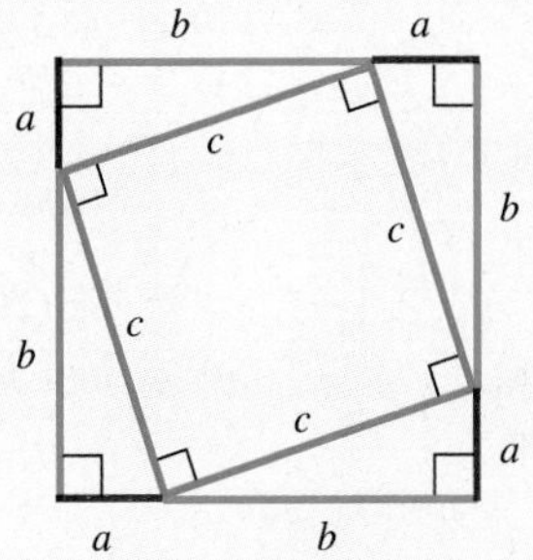

FOR DISCUSSION OR WRITING

The diagram on the right can be used to verify the Pythagorean formula. To do so, express the area of the figure in two ways: first, as the area of the large square, and then as the sum of the areas of the smaller square and the four right triangles. Finally, set the areas equal and simplify the equation.

The area of the large square is $(a + b)^2$ or $a^2 + 2ab + b^2$. The sum of the areas of the smaller square and the four right triangles is $c^2 + 2ab$. Set these equal to each other and subtract $2ab$ from both sides to get $a^2 + b^2 = c^2$.

PROBLEM SOLVING

The Pythagorean formula can be used to solve applied problems that involve right triangles. A good way to begin the solution is to sketch the triangle and label the three sides appropriately, using a variable as needed. Then use the Pythagorean formula to write an equation. This procedure is simply Steps 1–3 of the six-step problem-solving method given in Chapter 2, and used throughout the book. In Steps 4–6, we solve the equation, answer the question(s), and check the solution(s).

CHALKBOARD EXAMPLE

A rectangle has dimensions 5 feet by 12 feet. Find the length of its diagonal.

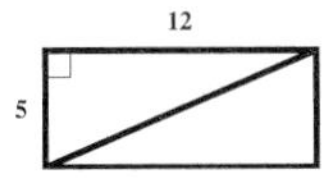

Answer: 13 feet

EXAMPLE 7 Solving an Application

A ladder 10 feet long leans against a wall. The foot of the ladder is 6 feet from the base of the wall. How high up the wall does the top of the ladder rest?

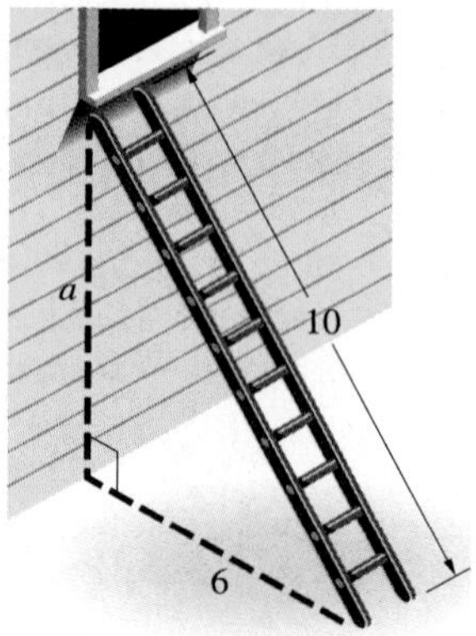

Figure 2

Steps 1 and 2 As shown in Figure 2, a right triangle is formed with the ladder as the hypotenuse. Let a represent the height of the top of the ladder.

Step 3 By the Pythagorean formula,

$$c^2 = a^2 + b^2.$$

$$10^2 = a^2 + 6^2 \qquad \text{Let } c = 10 \text{ and } b = 6.$$

$$100 = a^2 + 36 \qquad \text{Square.}$$

Step 4

$$64 = a^2 \qquad \text{Subtract 36.}$$

$$\sqrt{64} = a$$

$$a = 8 \qquad \sqrt{64} = 8$$

Step 5 Choose the positive square root of 64 since a represents a length.

Step 6 From Figure 2, we see that we must have

$$8^2 + 6^2 = 10^2 \qquad ?$$

$$64 + 36 = 100. \qquad \text{True}$$

The check shows that the top of the ladder rests 8 feet up the wall.

OBJECTIVE 5 Use the distance formula. The Pythagorean formula is used to develop another useful formula for finding the distance between two points on the plane. Figure 3 shows the two different points (x_1, y_1) and (x_2, y_2), as well as the point (x_2, y_1). The distance between (x_2, y_2) and (x_2, y_1) is $a = y_2 - y_1$, and the distance between (x_1, y_1) and (x_2, y_1) is $b = x_2 - x_1$. From the Pythagorean formula,

$$d^2 = (x_2 - x_1)^2 + (y_2 - y_1)^2.$$

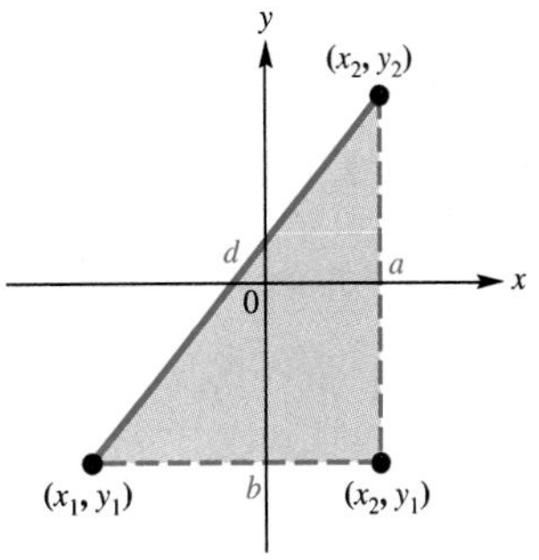

Figure 3

Taking the square root of each side, we get the **distance formula.**

Distance Formula

The distance between the points (x_1, y_1) and (x_2, y_2) is

$$d = \sqrt{(x_2 - x_1)^2 + (y_2 - y_1)^2}.$$

CHALKBOARD EXAMPLE

Find the distance between $(-6, 3)$ and $(-2, -4)$.

Answer: $\sqrt{65}$

EXAMPLE 8 Using the Distance Formula

Find the distance between $(-3, 4)$ and $(2, 5)$.

Use the distance formula. Choose $(x_1, y_1) = (-3, 4)$ and $(x_2, y_2) = (2, 5)$.

$$\begin{aligned} d &= \sqrt{(x_2 - x_1)^2 + (y_2 - y_1)^2} \\ &= \sqrt{(2 - (-3))^2 + (5 - 4)^2} \quad \text{Substitute.} \\ &= \sqrt{5^2 + 1^2} \\ &= \sqrt{26} \end{aligned}$$

OBJECTIVE 6 **Find higher roots.** Finding the square root of a number is the inverse (reverse) of squaring a number. In a similar way, there are inverses to finding the cube of a number, or finding the fourth power of a number. These inverses are the **cube root,** written $\sqrt[3]{a}$, and the **fourth root,** written $\sqrt[4]{a}$. Similar symbols are used for higher roots. In general we have the following.

The *n*th root of a is written $\sqrt[n]{a}$.

In $\sqrt[n]{a}$, the number n is the **index** or **order** of the radical. It would be possible to write $\sqrt[2]{a}$ instead of $\sqrt{a}$, but the simpler symbol $\sqrt{a}$ is customary since the square root is the most commonly used root. A calculator that has a key marked $\boxed{\sqrt[x]{y}}$ or $\boxed{x^y}$ can be used to find these roots. When working with cube roots or fourth roots, it is helpful to memorize the first few *perfect cubes* ($2^3 = 8$, $3^3 = 27$, and so on) and the first few perfect fourth powers.

CHALKBOARD EXAMPLE

Find the cube root: $\sqrt[3]{-27}$.

Answer: -3

EXAMPLE 9 Finding Cube Roots

Find each cube root.

(a) $\sqrt[3]{8}$

Look for a number that can be cubed to give 8. Since $2^3 = 8$, then $\sqrt[3]{8} = 2$.

(b) $\sqrt[3]{-8}$

$\sqrt[3]{-8} = -2$ because $(-2)^3 = -8$.

As Example 9 suggests, the cube root of a positive number is positive, and the cube root of a negative number is negative. *There is only one real number cube root for each real number.*

TEACHING TIP This concept will need to be clarified with several numerical examples.

When the index of the radical is even (square root, fourth root, and so on), the radicand must be nonnegative to get a real number root. Also, for even indexes the symbols $\sqrt{\ }$, $\sqrt[4]{\ }$, $\sqrt[6]{\ }$, and so on are used for the *nonnegative* roots, which are called **principal roots.** The symbols $-\sqrt{\ }$, $-\sqrt[4]{\ }$, $-\sqrt[6]{\ }$, and so on are used for the negative roots.

CHALKBOARD EXAMPLE

Find the root: $-\sqrt[6]{64}$.

Answer: -2

EXAMPLE 10 Finding Higher Roots

Find each root.

(a) $\sqrt[4]{16}$

$\sqrt[4]{16} = 2$ because 2 is positive and $2^4 = 16$.

(b) $-\sqrt[4]{16}$

From part (a), $\sqrt[4]{16} = 2$, so the negative root $-\sqrt[4]{16} = -2$.

(c) $\sqrt[4]{-16}$

To find the fourth root, the radicand must be nonnegative. There is no real number that equals $\sqrt[4]{-16}$.

(d) $\sqrt[3]{64} = 4$ since $4^3 = 64$.

(e) $-\sqrt[5]{32}$

First find $\sqrt[5]{32}$. The prime factorization of 32 as 2^5 shows that $\sqrt[5]{32} = 2$. If $\sqrt[5]{32} = 2$, then $-\sqrt[5]{32} = -2$.

9.1 EXERCISES

1. false; Zero has only one square root. 2. false; A negative number has no real square roots. 3. true 4. false; A positive number has just one real cube root. 5. true 6. true

Decide whether each statement is true or false. If false, tell why.

1. Every nonnegative number has two square roots.

2. A negative number has negative square roots.

3. Every positive number has two real square roots.

4. Every positive number has three real cube roots.

5. The cube root of every real number has the same sign as the number itself.

6. The positive square root of a positive number is its principal square root.

7. $-4, 4$ 8. $-3, 3$
9. $-12, 12$ 10. $-15, 15$
11. $-\frac{5}{14}, \frac{5}{14}$ 12. $-\frac{9}{20}, \frac{9}{20}$
13. $-30, 30$ 14. $-40, 40$
15. 7 16. 9 17. -11
18. 14 19. $-\frac{12}{11}$
20. $-\frac{7}{6}$ 21. not a real number
22. not a real number 23. 100
24. 36 25. 19 26. 99
27. $3x^2 + 4$ 28. $9y^2 + 3$
29. a must be positive.
30. a must be positive.
31. a must be negative.
32. a must be negative.
33. rational; 5
34. rational; 13
35. irrational; 5.385
36. irrational; 5.745
37. rational; -8
38. irrational; -22.361
39. not a real number
40. not a real number
41. The answer to Exercise 17 is the negative square root of a positive number. However, in Exercise 21, the square root of a negative number is not a real number.
42. The cube root of a negative number is negative, and the negative of the cube root of a positive number is negative, so both equal -2.
43. 23.896 44. 26.325
45. 28.249 46. 21.284
47. 1.985 48. .095
49. 5 50. 10 51. 17
52. 13 53. 3.606
54. 5.099 55. 2.289
56. 4.198 57. 5.074
58. 6.315 59. -4.431
60. -4.563 61. $c = 17$
62. $c = 26$ 63. $b = 8$
64. $a = 5$ 65. $c = 11.705$
66. $c = 15.811$

Find all square roots of each number. See Example 1.

7. 16 **8.** 9 **9.** 144 **10.** 225

11. $\frac{25}{196}$ **12.** $\frac{81}{400}$ **13.** 900 **14.** 1600

Find each square root that is a real number. See Examples 2 and 4(e).

15. $\sqrt{49}$ **16.** $\sqrt{81}$ **17.** $-\sqrt{121}$ **18.** $\sqrt{196}$

19. $-\sqrt{\frac{144}{121}}$ **20.** $-\sqrt{\frac{49}{36}}$ **21.** $\sqrt{-121}$ **22.** $\sqrt{-49}$

Find the square of each radical expression. See Example 3.

23. $\sqrt{100}$ **24.** $\sqrt{36}$ **25.** $-\sqrt{19}$

26. $-\sqrt{99}$ **27.** $\sqrt{3x^2 + 4}$ **28.** $\sqrt{9y^2 + 3}$

What must be true about the variable a for each statement to be true?

29. $\sqrt{a}$ represents a positive number. **30.** $-\sqrt{a}$ represents a negative number.

31. $\sqrt{a}$ is not a real number. **32.** $-\sqrt{a}$ is not a real number.

Write rational, irrational, *or* not a real number *for each number. If the number is rational, give its exact value. If the number is irrational, give a decimal approximation to the nearest thousandth. Use a calculator as necessary. See Examples 4 and 5.*

33. $\sqrt{25}$ **34.** $\sqrt{169}$ **35.** $\sqrt{29}$ **36.** $\sqrt{33}$

37. $-\sqrt{64}$ **38.** $-\sqrt{500}$ **39.** $\sqrt{-29}$ **40.** $\sqrt{-47}$

41. Explain why the answers to Exercises 17 and 21 are different.

42. Explain why $\sqrt[3]{-8}$ and $-\sqrt[3]{8}$ represent the same number.

Use a calculator with a square root key to find each root. Round to the nearest thousandth. See Example 5.

43. $\sqrt{571}$ **44.** $\sqrt{693}$ **45.** $\sqrt{798}$

46. $\sqrt{453}$ **47.** $\sqrt{3.94}$ **48.** $\sqrt{.00895}$

Find each square root. Use a calculator and round to the nearest thousandth, if necessary. (Hint: *First simplify the radicand to a single number.*)

49. $\sqrt{3^2 + 4^2}$ **50.** $\sqrt{6^2 + 8^2}$ **51.** $\sqrt{8^2 + 15^2}$

52. $\sqrt{5^2 + 12^2}$ **53.** $\sqrt{2^2 + 3^2}$ **54.** $\sqrt{(-1)^2 + 5^2}$

Use a calculator with a cube root key to find each root. Round to the nearest thousandth. (In Exercises 59 and 60, you may have to use the fact that if $a > 0$, $\sqrt[3]{-a} = -\sqrt[3]{a}$.)

55. $\sqrt[3]{12}$ **56.** $\sqrt[3]{74}$ **57.** $\sqrt[3]{130.6}$

58. $\sqrt[3]{251.8}$ **59.** $\sqrt[3]{-87}$ **60.** $\sqrt[3]{-95}$

Find the length of the unknown side of each right triangle with legs a and b and hypotenuse c. In Exercises 65 and 66, use a calculator and round to the nearest thousandth. See Example 6.

61. $a = 8, b = 15$ **62.** $a = 24, b = 10$ **63.** $a = 6, c = 10$

64. $b = 12, c = 13$ **65.** $a = 11, b = 4$ **66.** $a = 13, b = 9$

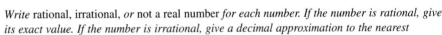

67. 24 centimeters
68. 41 meters **69.** 80 feet
70. 120 feet **71.** 195 miles
72. 18 feet

Use the Pythagorean formula to solve each problem. See Example 7.

67. The diagonal of a rectangle measures 25 centimeters. The width of the rectangle is 7 centimeters. Find the length of the rectangle.

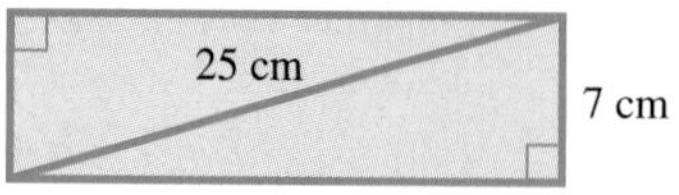

68. The length of a rectangle is 40 meters, and the width is 9 meters. Find the measure of the diagonal of the rectangle.

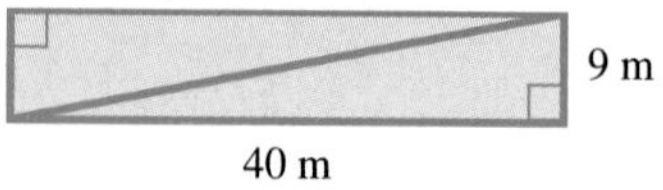

In Exercises 69–72, use these steps to solve each problem.

Step 1 *Let x represent the unknown length.*
Step 2 *Use the information in the problem to label the remaining sides of the triangle in the figure.*
Step 3 *Write an equation relating the three sides of the triangle.* (Hint: *Use the Pythagorean formula.*)
Step 4 *Solve the equation.*
Steps 5 and 6 *Give the answer; check it.*

69. Margaret is flying a kite on 100 feet of string. How high is it above her hand (vertically) if the horizontal distance between Margaret and the kite is 60 feet?

70. A guy wire is attached to the mast of a short-wave transmitting antenna. It is attached 96 feet above ground level. If the wire is staked to the ground 72 feet from the base of the mast, how long is the wire?

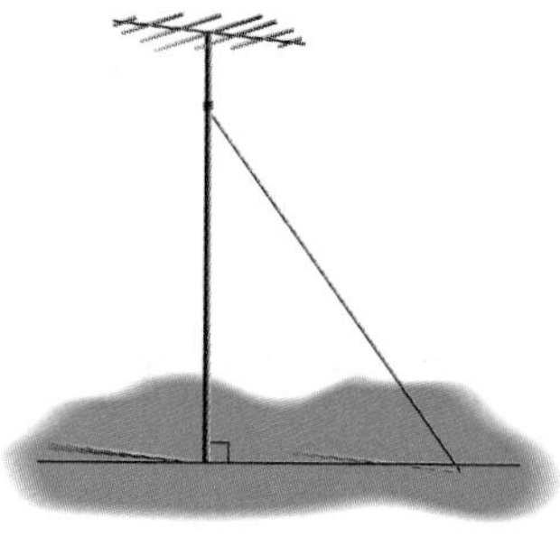

71. Two cars leave Tomball, Texas, at the same time. One travels north at 25 miles per hour and the other travels west at 60 miles per hour. How far apart are they after 3 hours? (*Hint:* Use $d = rt$ to find the distance traveled by each car.)

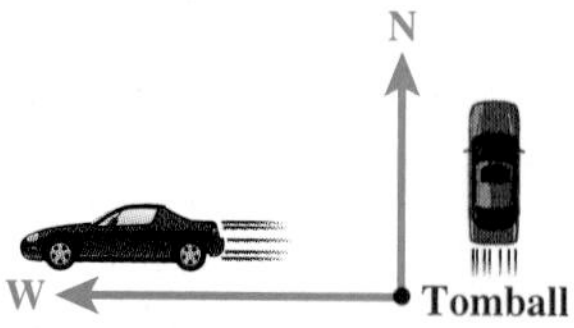

72. A boat is being pulled toward a dock with a rope attached at water level. When the boat is 24 feet from the dock, 30 feet of rope is extended. What is the height of the dock above the water?

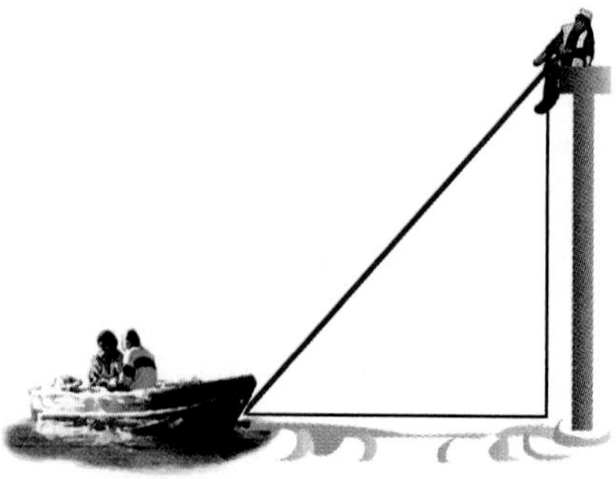

73. 9.434 **74.** 9.747
75. Answers will vary. For example, if $a = 2$ and $b = 7$, $\sqrt{a^2 + b^2} = \sqrt{53}$, while $a + b = 9$. $\sqrt{53} \neq 9$. If $a = 0$ and $b = 1$, then $\sqrt{a^2 + b^2}$ is equal to $a + b$ (which is not true in general).
76. The square root of a number, multiplied by itself, gives back the number, while the square of a number is the result of multiplying the number by itself. For example, $\sqrt{9} = 3$ gives 9 if we multiply $3 \cdot 3$, and $9^2 = 9 \cdot 9 = 81$. **77.** $\sqrt{29}$
78. 13 **79.** $\sqrt{2}$ **80.** $\sqrt{5}$
81. 10 **82.** 2 **83.** -3
84. -4 **85.** 5 **86.** 10
87. not a real number
88. not a real number **89.** -3
90. -4 **91.** 3 **92.** 10

In Exercises 73 and 74, round to the nearest thousandth.

73. What is the value of x in the figure?

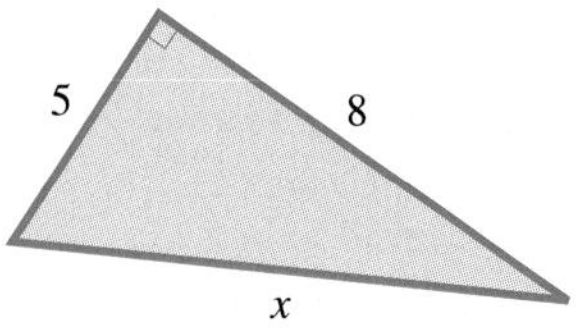

74. What is the value of y in the figure?

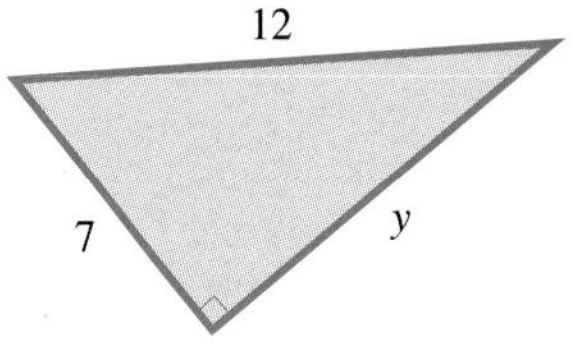

75. Use specific values for a and b different from those given in the "Caution" following Example 6 to show that $\sqrt{a^2 + b^2} \neq a + b$. Why would the values $a = 0$ and $b = 1$ *not* be satisfactory choices?

76. Explain how finding the square root of a number and squaring a number are related. Give examples.

Find the distance between each pair of points. See Example 8.

77. $(5, -3)$ and $(7, 2)$

78. $(2, 9)$ and $(-3, -3)$

79. $\left(-\frac{1}{4}, \frac{2}{3}\right)$ and $\left(\frac{3}{4}, -\frac{1}{3}\right)$

80. $\left(\frac{2}{5}, \frac{3}{2}\right)$ and $\left(-\frac{8}{5}, \frac{1}{2}\right)$

Find each root that is a real number. See Examples 9 and 10.

81. $\sqrt[3]{1000}$

82. $\sqrt[3]{8}$

83. $\sqrt[3]{-27}$

84. $\sqrt[3]{-64}$

85. $\sqrt[4]{625}$

86. $\sqrt[4]{10{,}000}$

87. $\sqrt[4]{-1}$

88. $\sqrt[4]{-625}$

TECHNOLOGY INSIGHTS (EXERCISES 89–92)

Exercises 89–92 illustrate the way one graphing calculator shows nth roots. Give each indicated root. Exercises 91 and 92 indicate 5th roots.

89.

90.

91.

92.

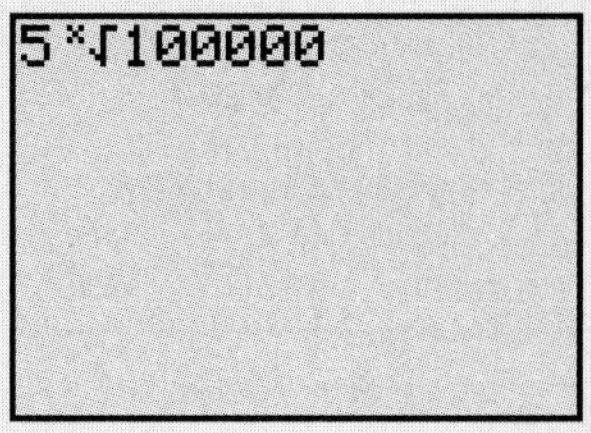

*These problems can be used as collaborative exercises with small groups of students.

93. c^2
94. $(b - a)^2$
95. $2ab$
96. $(b - a)^2 = a^2 - 2ab + b^2$
97. $c^2 = 2ab + (a^2 - 2ab + b^2)$
98. $c^2 = a^2 + b^2$

RELATING CONCEPTS* (EXERCISES 93–98)

One of the many proofs of the Pythagorean formula was given in the Connections box in this section. Here is another one, attributed to the Hindu mathematician Bhāskara.

Refer to the figures and ***work Exercises 93–98 in order.***

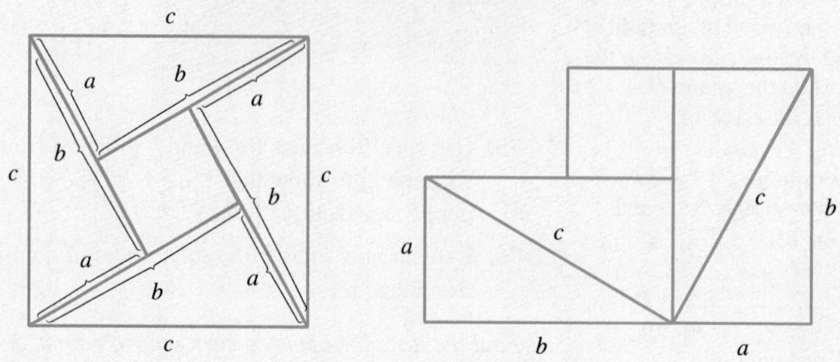

93. What is the area of the square on the left in terms of c?

94. What is the area of the small square in the middle of the figure on the left, in terms of $(b - a)$?

95. What is the sum of the areas of the two rectangles made up of triangles in the figure on the right?

96. What is the area of the small square in the figure on the right in terms of a and b?

97. The figure on the left is made up of the same square and triangles as the figure on the right. Write an equation setting the answer to Exercise 93 equal to the sum of the answers in Exercises 95 and 96.

98. Simplify the expressions you obtained in Exercise 97. What is your final result?

Did you make the connection that geometric figures can be used to prove algebraic formulas?

9.2 Multiplication and Division of Radicals

OBJECTIVES

1. Multiply radicals.
2. Simplify radicals using the product rule.
3. Simplify radical quotients using the quotient rule.
4. Use the product and quotient rules to simplify higher roots.

FOR EXTRA HELP

SSG Sec. 9.2
SSM Sec. 9.2
Pass the Test Software
InterAct Math Tutorial Software
Video 14

CONNECTIONS

The sixteenth century German radical symbol $\sqrt{\ }$ we use today is probably derived from the letter *R*. The radical symbol on the left below comes from the Latin word for root, *radix*. It was first used by Leonardo da Pisa (Fibonnaci) in 1220.

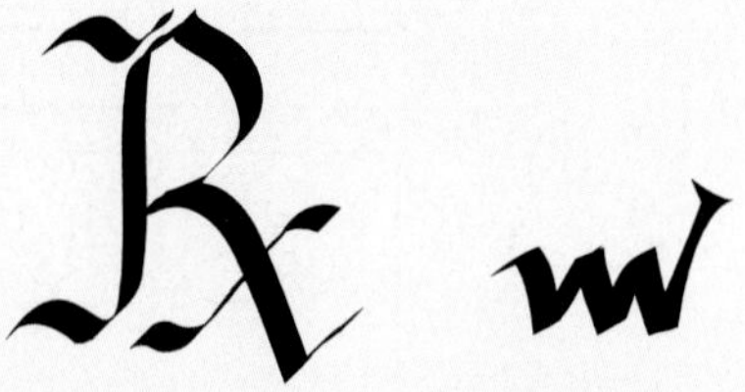

The cube root symbol shown on the right above was used by the German mathematician Christoff Rudolff in 1525. The symbol used today originated in the seventeenth century in France.

INSTRUCTOR'S RESOURCES

PTB Sec. 9.2 **ISM** Sec. 9.2
AB Sec. 9.2

TEST GENERATOR

WORLD WIDE WEB
www.LialAlgebra.com

1. x; $\sqrt{x}$ 2. The last part of it ($\surd$) is used as part of the radical symbol $\sqrt{\ }$.

CONNECTIONS (CONTINUED)

FOR DISCUSSION OR WRITING

1. In the radical sign shown on the left, the *R* referred to above is clear. What other letter do you think is part of the symbol? What would the equivalent be in our modern notation?
2. How is the cube root symbol on the right related to our modern radical sign?

OBJECTIVE 1 **Multiply radicals.** Several useful rules for finding products and quotients of radicals are developed in this section. To illustrate the rule for products, notice that

$$\sqrt{4} \cdot \sqrt{9} = 2 \cdot 3 = 6 \qquad \text{and} \qquad \sqrt{4 \cdot 9} = \sqrt{36} = 6,$$

showing that

$$\sqrt{4} \cdot \sqrt{9} = \sqrt{4 \cdot 9}.$$

This result is a particular case of the more general product rule for radicals.

Product Rule for Radicals

For nonnegative real numbers x and y,

$$\sqrt{x} \cdot \sqrt{y} = \sqrt{x \cdot y} \qquad \text{and} \qquad \sqrt{x \cdot y} = \sqrt{x} \cdot \sqrt{y}.$$

That is, the product of two radicals is the radical of the product.

CAUTION

In general, $\sqrt{x + y} \neq \sqrt{x} + \sqrt{y}$. To see why this is so, let $x = 16$ and $y = 9$.

$$\sqrt{16 + 9} = \sqrt{25} = 5$$

but

$$\sqrt{16} + \sqrt{9} = 4 + 3 = 7.$$

CHALKBOARD EXAMPLE

Find the product.

$$\sqrt{6} \cdot \sqrt{11}$$

Answer: $\sqrt{66}$

EXAMPLE 1 **Using the Product Rule to Multiply Radicals**

Use the product rule for radicals to find each product.

(a) $\sqrt{2} \cdot \sqrt{3} = \sqrt{2 \cdot 3} = \sqrt{6}$

(b) $\sqrt{7} \cdot \sqrt{5} = \sqrt{35}$

(c) $\sqrt{11} \cdot \sqrt{a} = \sqrt{11a}$ Assume $a \geq 0$.

OBJECTIVE 2 **Simplify radicals using the product rule.** A square root radical is **simplified** when no perfect square factor remains under the radical sign. This is accomplished by using the product rule as shown in Example 2.

CHALKBOARD EXAMPLE

Simplify the radical.

$$\sqrt{60}$$

Answer: $2\sqrt{15}$

EXAMPLE 2 Using the Product Rule to Simplify Radicals

Simplify each radical.

(a) $\sqrt{20}$

Since 20 has a perfect square factor of 4,

$$\begin{aligned}\sqrt{20} &= \sqrt{\mathbf{4} \cdot 5} && \text{4 is a perfect square.}\\ &= \sqrt{\mathbf{4}} \cdot \sqrt{5} && \text{Product rule}\\ &= \mathbf{2}\sqrt{5}. && \sqrt{4} = 2\end{aligned}$$

Thus, $\sqrt{20} = 2\sqrt{5}$. Since 5 has no perfect square factor other than 1, $2\sqrt{5}$ is called the **simplified form** of $\sqrt{20}$. Note that $2\sqrt{5}$ represents a product, where the factors are 2 and $\sqrt{5}$; also, $2\sqrt{5} \neq \sqrt{10}$.

(b) $\sqrt{72}$

Begin by looking for the largest perfect square that is a factor of 72. This number is 36, so

$$\begin{aligned}\sqrt{72} &= \sqrt{\mathbf{36} \cdot 2} && \text{36 is a perfect square.}\\ &= \sqrt{\mathbf{36}} \cdot \sqrt{2} && \text{Product rule}\\ &= \mathbf{6}\sqrt{2}. && \sqrt{36} = 6\end{aligned}$$

We could also factor 72 into its prime factors and look for pairs of like factors. Each pair of like factors produces a factor outside the radical in the simplified form.

$$\sqrt{72} = \sqrt{\mathbf{2} \cdot \mathbf{2} \cdot 2 \cdot \mathbf{3} \cdot \mathbf{3}} = 2 \cdot 3 \cdot \sqrt{2} = 6\sqrt{2}$$

In either case, we obtain $6\sqrt{2}$ as the simplified form of $\sqrt{72}$; our work is simpler, however, if we begin with the largest perfect square factor.

(c)
$$\begin{aligned}\sqrt{300} &= \sqrt{\mathbf{100} \cdot 3} && \text{100 is a perfect square.}\\ &= \sqrt{\mathbf{100}} \cdot \sqrt{3} && \text{Product rule}\\ &= \mathbf{10}\sqrt{3} && \sqrt{100} = 10\end{aligned}$$

(d) $\sqrt{15}$

The number 15 has no perfect square factor (except 1), so $\sqrt{15}$ cannot be simplified further.

TEACHING TIP In this example, again stress the use of the largest perfect square factor in 300. Although $300 = 4 \cdot 75$ or $300 = 25 \cdot 12$ could also be used, they would require additional simplifications.

Sometimes the product rule can be used to simplify a product, as Example 3 shows.

CHALKBOARD EXAMPLE

Find the product and simplify.

$$\sqrt{10} \cdot \sqrt{50}$$

Answer: $10\sqrt{5}$

EXAMPLE 3 Multiplying and Simplifying Radicals

Find each product and simplify.

(a)
$$\begin{aligned}\sqrt{9} \cdot \sqrt{75} &= 3\sqrt{75} && \sqrt{9} = 3\\ &= 3\sqrt{25 \cdot 3} && \text{25 is a perfect square.}\\ &= 3\sqrt{25} \cdot \sqrt{3} && \text{Product rule}\\ &= 3 \cdot 5\sqrt{3} && \sqrt{25} = 5\\ &= 15\sqrt{3} && \text{Multiply.}\end{aligned}$$

(b) $\sqrt{8} \cdot \sqrt{12} = \sqrt{8 \cdot 12}$ — Product rule

$= \sqrt{4 \cdot 2 \cdot 4 \cdot 3}$ — Factor; 4 is a perfect square.

$= \sqrt{4} \cdot \sqrt{4} \cdot \sqrt{2 \cdot 3}$ — Product rule

$= 2 \cdot 2 \cdot \sqrt{6}$ — $\sqrt{4} = 2$

$= 4\sqrt{6}$ — Multiply.

NOTE

Part (b) of Example 3 also could be simplified by using the product rule first to get

$$\begin{aligned} \sqrt{8} \cdot \sqrt{12} &= \sqrt{4 \cdot 2} \cdot \sqrt{4 \cdot 3} \\ &= 2\sqrt{2} \cdot 2\sqrt{3} \\ &= 2 \cdot 2 \cdot \sqrt{2} \cdot \sqrt{3} \\ &= 4\sqrt{6}. \end{aligned}$$

Both approaches are correct.

OBJECTIVE 3 Simplify radical quotients using the quotient rule. The *quotient rule for radicals* is very similar to the product rule. It, too, can be used either way.

Quotient Rule for Radicals

If x and y are nonnegative real numbers and $y \neq 0$,

$$\frac{\sqrt{x}}{\sqrt{y}} = \sqrt{\frac{x}{y}} \quad \text{and} \quad \sqrt{\frac{x}{y}} = \frac{\sqrt{x}}{\sqrt{y}}.$$

The quotient of the radicals is the radical of the quotient.

CHALKBOARD EXAMPLE

Simplify the radical.

$$\sqrt{\frac{10}{49}}$$

Answer: $\dfrac{\sqrt{10}}{7}$

EXAMPLE 4 Using the Quotient Rule to Simplify Radicals

Simplify each radical.

(a) $\sqrt{\dfrac{25}{9}} = \dfrac{\sqrt{25}}{\sqrt{9}} = \dfrac{5}{3}$ — Quotient rule

(b) $\dfrac{\sqrt{288}}{\sqrt{2}} = \sqrt{\dfrac{288}{2}} = \sqrt{144} = 12$ — Quotient rule

(c) $\sqrt{\dfrac{3}{4}} = \dfrac{\sqrt{3}}{\sqrt{4}} = \dfrac{\sqrt{3}}{2}$ — Quotient rule

CHALKBOARD EXAMPLE

Simplify $\dfrac{8\sqrt{50}}{4\sqrt{5}}$.

Answer: $2\sqrt{10}$

EXAMPLE 5 Using the Quotient Rule to Divide Radicals

Divide $27\sqrt{15}$ by $9\sqrt{3}$.

We use the quotient rule as follows.

$$\frac{27\sqrt{15}}{9\sqrt{3}} = \frac{27}{9} \cdot \frac{\sqrt{15}}{\sqrt{3}} = 3\sqrt{\frac{15}{3}} = 3\sqrt{5}$$

Some problems require both the product and quotient rules, as Example 6 shows.

CHALKBOARD EXAMPLE

Simplify $\sqrt{\frac{3}{8}} \cdot \sqrt{\frac{7}{2}}$.

Answer: $\frac{\sqrt{21}}{4}$

EXAMPLE 6 Using Both the Product and Quotient Rules

Simplify $\sqrt{\frac{3}{5}} \cdot \sqrt{\frac{4}{5}}$.

$$\sqrt{\frac{3}{5}} \cdot \sqrt{\frac{4}{5}} = \frac{\sqrt{3}}{\sqrt{5}} \cdot \frac{\sqrt{4}}{\sqrt{5}} \quad \text{Quotient rule}$$

$$= \frac{\sqrt{3} \cdot \sqrt{4}}{\sqrt{5} \cdot \sqrt{5}} \quad \text{Multiply fractions.}$$

$$= \frac{\sqrt{3} \cdot 2}{\sqrt{25}} \quad \text{Product rule; } \sqrt{4} = 2$$

$$= \frac{2\sqrt{3}}{5} \quad \sqrt{25} = 5$$

The product and quotient rules also apply when variables appear under the radical sign, as long as all the variables represent only nonnegative numbers. For example, $\sqrt{5^2} = 5$, but $\sqrt{(-5)^2} \neq -5$.

For a real number a, $\sqrt{a^2} = a$ only if a is nonnegative.

CHALKBOARD EXAMPLE

Simplify the radical. Assume all variables represent positive real numbers.

$\sqrt{100p^8}$

Answer: $10p^4$

EXAMPLE 7 Simplifying Radicals Involving Variables

Simplify each radical. Assume all variables represent positive real numbers.

(a) $\sqrt{25m^4} = \sqrt{25} \cdot \sqrt{m^4}$ Product rule

$= 5m^2$

(b) $\sqrt{64p^{10}} = 8p^5$ Product rule

(c) $\sqrt{r^9} = \sqrt{r^8 \cdot r}$

$= \sqrt{r^8} \cdot \sqrt{r} = r^4\sqrt{r}$ Product rule

(d) $\sqrt{\frac{5}{x^2}} = \frac{\sqrt{5}}{\sqrt{x^2}} = \frac{\sqrt{5}}{x}$ Quotient rule

TEACHING TIP Compare $\sqrt[3]{40}$ and $\sqrt{40}$.

OBJECTIVE 4 Use the product and quotient rules to simplify higher roots. The product and quotient rules for radicals also work for other roots, as shown in Example 8. To simplify cube roots, look for factors that are *perfect cubes.* A **perfect cube** is a number with a rational cube root. For example, $\sqrt[3]{64} = 4$, and since 4 is a rational number, 64 is a perfect cube. Higher roots are handled in a similar manner.

Properties of Radicals

For all real numbers where the indicated roots exist,

$$\sqrt[n]{x} \cdot \sqrt[n]{y} = \sqrt[n]{xy} \quad \text{and} \quad \frac{\sqrt[n]{x}}{\sqrt[n]{y}} = \sqrt[n]{\frac{x}{y}},\ y \neq 0.$$

CHALKBOARD EXAMPLE

Simplify the radical.

$$\sqrt[3]{-\frac{27}{64}}$$

Answer: $-\frac{3}{4}$

EXAMPLE 8 Simplifying Higher Roots

Simplify each radical.

(a) $\sqrt[3]{32} = \sqrt[3]{8 \cdot 4}$ 8 is a perfect cube.

$= \sqrt[3]{8} \cdot \sqrt[3]{4} = 2\sqrt[3]{4}$

(b) $\sqrt[3]{108} = \sqrt[3]{27 \cdot 4}$ 27 is a perfect cube.

$= \sqrt[3]{27} \cdot \sqrt[3]{4} = 3\sqrt[3]{4}$

(c) $\sqrt[4]{32} = \sqrt[4]{16} \cdot \sqrt[4]{2} = 2\sqrt[4]{2}$ 16 is a perfect fourth power.

(d) $\sqrt[3]{\frac{8}{125}} = \frac{\sqrt[3]{8}}{\sqrt[3]{125}} = \frac{2}{5}$

(e) $\sqrt[4]{\frac{16}{625}} = \frac{\sqrt[4]{16}}{\sqrt[4]{625}} = \frac{2}{5}$

(f) $\sqrt[3]{7} \cdot \sqrt[3]{49} = \sqrt[3]{7 \cdot 49} = \sqrt[3]{7 \cdot 7^2} = \sqrt[3]{7^3} = 7$

9.2 EXERCISES

Decide whether each statement is true or false for real numbers. If false, tell why.

1. $\sqrt{4} = \pm 2$ **2.** $\sqrt{(-6)^2} = -6$ **3.** $\sqrt{-6} \cdot \sqrt{6} = -6$

4. $\sqrt[3]{(-6)^3} = -6$ **5.** $\sqrt[3]{3} \cdot \sqrt[3]{2} = \sqrt[3]{6}$ **6.** $\sqrt[3]{4} \cdot \sqrt[3]{4} = 4$

7. $\sqrt{4} \cdot \sqrt{9} = \sqrt{4 \cdot 9}$ **8.** $\sqrt{4} + \sqrt{9} = \sqrt{4 + 9}$

Use the product rule for radicals to find each product. See Example 1.

9. $\sqrt{3} \cdot \sqrt{27}$ **10.** $\sqrt{2} \cdot \sqrt{8}$ **11.** $\sqrt{6} \cdot \sqrt{15}$

12. $\sqrt{10} \cdot \sqrt{15}$ **13.** $\sqrt{13} \cdot \sqrt{13}$ **14.** $\sqrt{17} \cdot \sqrt{17}$

15. $\sqrt{13} \cdot \sqrt{r}, r \ge 0$ **16.** $\sqrt{19} \cdot \sqrt{k}, k \ge 0$

17. Which one of the following radicals is simplified according to the guidelines of Objective 2?

(a) $\sqrt{47}$ **(b)** $\sqrt{45}$ **(c)** $\sqrt{48}$ **(d)** $\sqrt{44}$

18. If p is a prime number, is $\sqrt{p}$ in simplified form? Explain your answer.

Simplify each radical according to the method described in Objective 2. See Example 2.

19. $\sqrt{45}$ **20.** $\sqrt{56}$ **21.** $\sqrt{75}$ **22.** $\sqrt{18}$

23. $\sqrt{125}$ **24.** $\sqrt{80}$ **25.** $-\sqrt{700}$ **26.** $-\sqrt{600}$

27. $3\sqrt{27}$ **28.** $9\sqrt{8}$

Find each product and simplify. See Example 3.

29. $\sqrt{3} \cdot \sqrt{18}$ **30.** $\sqrt{3} \cdot \sqrt{21}$ **31.** $\sqrt{12} \cdot \sqrt{48}$

32. $\sqrt{50} \cdot \sqrt{72}$ **33.** $\sqrt{12} \cdot \sqrt{30}$ **34.** $\sqrt{30} \cdot \sqrt{24}$

35. In your own words, describe the product and quotient rules.

36. Simplify the radical $\sqrt{288}$ in two ways. First, factor 288 as $144 \cdot 2$ and then simplify completely. Second, factor 288 as $48 \cdot 6$ and then simplify completely. How do the answers compare? Make a conjecture concerning the quickest way to simplify such a radical.

Use the quotient rule and the product rule, as necessary, to simplify each radical expression. See Examples 4–6.

37. $\sqrt{\frac{16}{225}}$ **38.** $\sqrt{\frac{9}{100}}$ **39.** $\sqrt{\frac{7}{16}}$ **40.** $\sqrt{\frac{13}{25}}$

1. false; $\sqrt{4}$ represents only the principal (positive) square root. **2.** false; $\sqrt{(-6)^2} = \sqrt{36} = 6$. **3.** false; $\sqrt{-6}$ is not a real number. **4.** true **5.** true **6.** false; Three factors of $\sqrt[3]{4}$ are needed to get a product of 4. **7.** true **8.** false; $\sqrt{4} + \sqrt{9} = 2 + 3 = 5$ and $\sqrt{4 + 9} = \sqrt{13}$, so $\sqrt{4} + \sqrt{9} \ne \sqrt{4 + 9}$. **9.** 9 **10.** 4 **11.** $3\sqrt{10}$ **12.** $5\sqrt{6}$ **13.** 13 **14.** 17 **15.** $\sqrt{13r}$ **16.** $\sqrt{19k}$ **17.** (a) **18.** yes; Because a prime number cannot be factored (except as itself times 1), it cannot have any perfect square factors. **19.** $3\sqrt{5}$ **20.** $2\sqrt{14}$ **21.** $5\sqrt{3}$ **22.** $3\sqrt{2}$ **23.** $5\sqrt{5}$ **24.** $4\sqrt{5}$ **25.** $-10\sqrt{7}$ **26.** $-10\sqrt{6}$ **27.** $9\sqrt{3}$ **28.** $18\sqrt{2}$ **29.** $3\sqrt{6}$ **30.** $3\sqrt{7}$ **31.** 24 **32.** 60 **33.** $6\sqrt{10}$ **34.** $12\sqrt{5}$ **35.** The product rule says that the product of two radicals is the radical whose radicand is the product of their radicands. Similarly, by the quotient rule, the quotient of two radicals is the radical whose radicand is the quotient of their radicands. **36.** $\sqrt{288} = \sqrt{144 \cdot 2} = \sqrt{144} \cdot \sqrt{2} = 12\sqrt{2}$. Also, $\sqrt{288} = \sqrt{48 \cdot 6} = \sqrt{48} \cdot \sqrt{6} = \sqrt{16 \cdot 3} \cdot \sqrt{6} = \sqrt{16} \cdot \sqrt{3} \cdot \sqrt{6} = 4\sqrt{18} = 4\sqrt{9 \cdot 2} = 4 \cdot 3\sqrt{2} = 12\sqrt{2}$. Both methods give the same answer. The

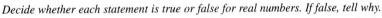

first method gets the answer with fewer steps, so it is better to use the largest perfect square factor first.

37. $\frac{4}{15}$ **38.** $\frac{3}{10}$ **39.** $\frac{\sqrt{7}}{4}$
40. $\frac{\sqrt{13}}{5}$ **41.** 5 **42.** 10
43. $\frac{25}{4}$ **44.** $\frac{64}{9}$ **45.** $6\sqrt{5}$
46. $25\sqrt{2}$ **47.** m **48.** k
49. y^2 **50.** s^2 **51.** $6z$
52. $7n$ **53.** $20x^3$ **54.** $30y^4$
55. $z^2\sqrt{z}$ **56.** $a^6\sqrt{a}$
57. x^3y^6 **58.** a^4b^5
59. $2\sqrt[3]{5}$ **60.** $2\sqrt[3]{6}$ **61.** $3\sqrt[3]{2}$
62. $4\sqrt[3]{3}$ **63.** $2\sqrt[4]{5}$ **64.** $3\sqrt[4]{3}$
65. $\frac{2}{3}$ **66.** $\frac{4}{5}$ **67.** $-\frac{6}{5}$
68. $-\frac{1}{4}$ **69.** 5 **70.** 4
71. $\sqrt[4]{12}$ **72.** $\sqrt[4]{28}$
73. $2x\sqrt[3]{4}$ **74.** $5p\sqrt[3]{25p}$

In Exercises 75 and 76, the number of displayed digits will vary among calculator models. Also, less sophisticated models may exhibit round-off error in the final decimal place.

75. (a) 4.472135955 (b) 4.472135955 (c) The numerical results are not a proof because both answers are approximations and they might differ if calculated to more decimal places.
76. $\sqrt{3} \times \sqrt{5} \approx 3.872983346$. The calculator should show the same digits for the approximation of $\sqrt{15}$, as a result of the product rule for radicals.
77. 6 centimeters
78. Each dimension is $4\sqrt[3]{2}$ feet.
79. 6 inches
80. It is doubled (to 12 inches).

41. $\sqrt{\frac{5}{7}} \cdot \sqrt{35}$ **42.** $\sqrt{\frac{10}{13}} \cdot \sqrt{130}$ **43.** $\sqrt{\frac{5}{2}} \cdot \sqrt{\frac{125}{8}}$

44. $\sqrt{\frac{8}{3}} \cdot \sqrt{\frac{512}{27}}$ **45.** $\frac{30\sqrt{10}}{5\sqrt{2}}$ **46.** $\frac{50\sqrt{20}}{2\sqrt{10}}$

Simplify each radical. Assume that all variables represent nonnegative real numbers. See Example 7.

47. $\sqrt{m^2}$ **48.** $\sqrt{k^2}$ **49.** $\sqrt{y^4}$ **50.** $\sqrt{s^4}$

51. $\sqrt{36z^2}$ **52.** $\sqrt{49n^2}$ **53.** $\sqrt{400x^6}$ **54.** $\sqrt{900y^8}$

55. $\sqrt{z^5}$ **56.** $\sqrt{a^{13}}$ **57.** $\sqrt{x^6y^{12}}$ **58.** $\sqrt{a^8b^{10}}$

Simplify each radical. See Example 8.

59. $\sqrt[3]{40}$ **60.** $\sqrt[3]{48}$ **61.** $\sqrt[3]{54}$ **62.** $\sqrt[3]{192}$

63. $\sqrt[4]{80}$ **64.** $\sqrt[4]{243}$ **65.** $\sqrt[3]{\frac{8}{27}}$ **66.** $\sqrt[3]{\frac{64}{125}}$

67. $\sqrt[3]{-\frac{216}{125}}$ **68.** $\sqrt[3]{-\frac{1}{64}}$ **69.** $\sqrt[3]{5} \cdot \sqrt[3]{25}$ **70.** $\sqrt[3]{4} \cdot \sqrt[3]{16}$

71. $\sqrt[4]{4} \cdot \sqrt[4]{3}$ **72.** $\sqrt[4]{7} \cdot \sqrt[4]{4}$ **73.** $\sqrt[3]{4x} \cdot \sqrt[3]{8x^2}$ **74.** $\sqrt[3]{25p} \cdot \sqrt[3]{125p^3}$

75. In Example 2(a) we showed *algebraically* that $\sqrt{20}$ is equal to $2\sqrt{5}$. To give *numerical support* to this result, use a calculator to do the following:
(a) Find a decimal approximation for $\sqrt{20}$ using your calculator. Record as many digits as the calculator shows.
(b) Find a decimal approximation for $\sqrt{5}$ using your calculator, and then multiply the result by 2. Record as many digits as the calculator shows.
(c) Your results in parts (a) and (b) should be the same. A mathematician would not accept this numerical exercise as *proof* that $\sqrt{20}$ is equal to $2\sqrt{5}$. Explain why.

76. On your calculator, multiply the approximations for $\sqrt{3}$ and $\sqrt{5}$. Now, predict what your calculator will show when you find an approximation for $\sqrt{15}$. What rule stated in this section justifies your answer?

The volume of a cube is found with the formula $V = s^3$, where s is the length of an edge of the cube. Use this information in Exercises 77 and 78.

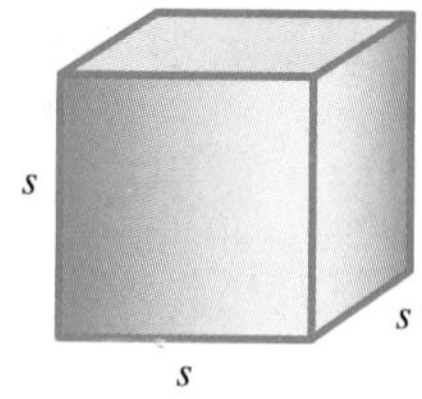

77. A container in the shape of a cube has a volume of 216 cubic centimeters. What is the depth of the container?

78. A cube-shaped box must be constructed to contain 128 cubic feet. What should the dimensions (height, width, and length) of the box be?

The volume of a sphere is found with the formula $V = \frac{4}{3}\pi r^3$, where r is the length of the radius of the sphere. Use this information in Exercises 79 and 80.

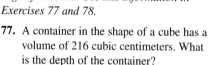

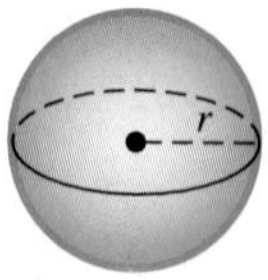

79. A ball in the shape of a sphere has a volume of 288π cubic inches. What is the radius of the ball?

▲ **80.** Suppose that the volume of the ball described in Exercise 79 is multiplied by 8. How is the radius affected?

81. The product rule for radicals requires that both a and b must be nonnegative. Otherwise $\sqrt{a}$ and $\sqrt{b}$ would not be real numbers (except when $a = b = 0$).
82. No, because all negative numbers have cube roots.

81. When we multiply two radicals with variables under the radical sign, such as $\sqrt{a} \cdot \sqrt{b} = \sqrt{ab}$, why is it important to know that both a and b represent nonnegative numbers?

82. Is it necessary to restrict k to a nonnegative number to say that $\sqrt[3]{k} \cdot \sqrt[3]{k} \cdot \sqrt[3]{k} = k$? Why?

RELATING CONCEPTS (EXERCISES 83–86)

An interesting way to represent the lengths corresponding to $\sqrt{2}$, $\sqrt{3}$, $\sqrt{4}$, $\sqrt{5}$, and so on is shown in the figure.

Work the following exercises in order.

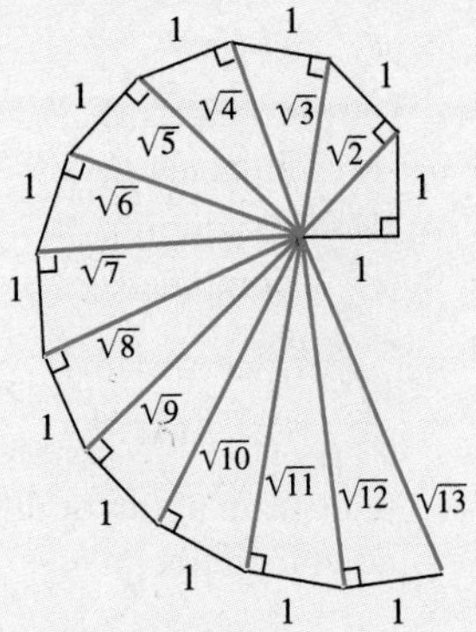

83. To verify the first length, we show, in the first triangle, that $1^2 + 1^2 = 2$, so the hypotenuse has length $\sqrt{2}$. Similarly, in the second triangle, $(\sqrt{2})^2 + 1^2 = 3$, so the hypotenuse has length $\sqrt{3}$, and so on. 84. $\sqrt{4}$, $\sqrt{9}$, and so on; $\sqrt{16} = 4$ and $\sqrt{25} = 5$
85. Look at the radicands: $9 - 4 = 5$, $16 - 9 = 7$, $25 - 16 = 9$, and so on. 86. The differences between consecutive whole number lengths increase by 2 each time, so we can predict that the next one will be $\sqrt{25 + 11} = \sqrt{36}$, and the one after that will be $\sqrt{36 + 13} = \sqrt{49}$.

83. Use the Pythagorean formula to verify the lengths in the figure.

84. Which of the lengths indicated as radicals equal whole numbers? If the figure is continued in the same way, what would the next two whole number lengths be?

85. Find the consecutive differences between the radicands in Exercise 84. (*Hint:* The first difference is $9 - 4 = 5$.)

86. Look for a pattern that determines where these whole number lengths occur. Use this pattern to predict where the next whole number length and the one after that will occur.

Did you make the connection between the lengths of the sides of the triangles and the Pythagorean formula?

9.3 Addition and Subtraction of Radicals

OBJECTIVES

1. Add and subtract radicals.
2. Simplify radical sums and differences.
3. Simplify radical sums involving multiplication.

FOR EXTRA HELP

SSG Sec. 9.3
SSM Sec. 9.3
Pass the Test Software
InterAct Math Tutorial Software
Video 14

OBJECTIVE 1 Add and subtract radicals. We add or subtract radicals by using the distributive property. For example,

$$8\sqrt{3} + 6\sqrt{3} = (8 + 6)\sqrt{3} \quad \text{Distributive property}$$
$$= 14\sqrt{3}.$$

Also,

$$2\sqrt{11} - 7\sqrt{11} = -5\sqrt{11}.$$

Like radicals are terms that have multiples of the *same root* of the *same number.* Only like radicals can be combined using the distributive property. In the example above, the like radicals are $2\sqrt{11}$ and $-7\sqrt{11}$. On the other hand, examples of *unlike radicals* are

$$2\sqrt{5} \quad \text{and} \quad 2\sqrt{3}, \qquad \text{Different radicands}$$

as well as

$$2\sqrt{3} \quad \text{and} \quad 2\sqrt[3]{3}. \qquad \text{Different indexes}$$

INSTRUCTOR'S RESOURCES

PTB Sec. 9.3 ISM Sec. 9.3
AB Sec. 9.3
TEST GENERATOR
WORLD WIDE WEB
www.LialAlgebra.com

CHALKBOARD EXAMPLE

Add $-4\sqrt{3} + 9\sqrt{3}$.

Answer: $5\sqrt{3}$

EXAMPLE 1 Adding and Subtracting Like Radicals

Add or subtract, as indicated.

(a) $3\sqrt{6} + 5\sqrt{6} = (3 + 5)\sqrt{6} = 8\sqrt{6}$ Distributive property

(b) $5\sqrt{10} - 7\sqrt{10} = (5 - 7)\sqrt{10} = -2\sqrt{10}$

(c) $\sqrt[3]{5} + \sqrt[3]{5} = 1\sqrt[3]{5} + 1\sqrt[3]{5} = (1 + 1)\sqrt[3]{5} = 2\sqrt[3]{5}$

(d) $\sqrt[4]{7} + 2\sqrt[4]{7} = 1\sqrt[4]{7} + 2\sqrt[4]{7} = 3\sqrt[4]{7}$

(e) $\sqrt{3} + \sqrt{13}$ cannot be added using the distributive property.

TEACHING TIP Also mention that $\sqrt{5} + \sqrt[3]{5}$ cannot be simplified.

CAUTION In general, $\sqrt{x} + \sqrt{y} \neq \sqrt{x + y}$. In Example 1(e), it would be **incorrect** to try to simplify $\sqrt{3} + \sqrt{13}$ as $\sqrt{3} + \sqrt{13} = \sqrt{16} = 4$. Only *like radicals* can be combined.

OBJECTIVE 2 Simplify radical sums and differences. Sometimes we must simplify one or more radicals in a sum or difference. Doing this may result in like radicals, which we can then add or subtract.

CHALKBOARD EXAMPLE

Subtract $5\sqrt{200} - 6\sqrt{18}$.

Answer: $32\sqrt{2}$

EXAMPLE 2 Adding and Subtracting Radicals That Require Simplification

Simplify as much as possible.

(a)
$$\begin{aligned} 3\sqrt{2} + \sqrt{8} &= 3\sqrt{2} + \sqrt{4 \cdot 2} && \text{Factor.} \\ &= 3\sqrt{2} + \sqrt{4} \cdot \sqrt{2} && \text{Product rule} \\ &= 3\sqrt{2} + 2\sqrt{2} && \sqrt{4} = 2 \\ &= 5\sqrt{2} && \text{Add like radicals.} \end{aligned}$$

(b)
$$\begin{aligned} \sqrt{18} - \sqrt{27} &= \sqrt{9 \cdot 2} - \sqrt{9 \cdot 3} && \text{Factor.} \\ &= \sqrt{9} \cdot \sqrt{2} - \sqrt{9} \cdot \sqrt{3} && \text{Product rule} \\ &= 3\sqrt{2} - 3\sqrt{3} && \sqrt{9} = 3 \end{aligned}$$

Since $\sqrt{2}$ and $\sqrt{3}$ are unlike radicals, this difference cannot be simplified further.

(c)
$$\begin{aligned} 2\sqrt{12} + 3\sqrt{75} &= 2(\sqrt{4} \cdot \sqrt{3}) + 3(\sqrt{25} \cdot \sqrt{3}) && \text{Product rule} \\ &= 2(2\sqrt{3}) + 3(5\sqrt{3}) && \sqrt{4} = 2 \text{ and } \sqrt{25} = 5 \\ &= 4\sqrt{3} + 15\sqrt{3} && \text{Multiply.} \\ &= 19\sqrt{3} && \text{Add like radicals.} \end{aligned}$$

(d)
$$\begin{aligned} 3\sqrt[3]{16} + 5\sqrt[3]{2} &= 3(\sqrt[3]{8} \cdot \sqrt[3]{2}) + 5\sqrt[3]{2} && \text{Product rule} \\ &= 3(2\sqrt[3]{2}) + 5\sqrt[3]{2} && \sqrt[3]{8} = 2 \\ &= 6\sqrt[3]{2} + 5\sqrt[3]{2} && \text{Multiply.} \\ &= 11\sqrt[3]{2} && \text{Add like radicals.} \end{aligned}$$

TEACHING TIP Warn students to avoid common errors when adding radicals. In particular, note that $3\sqrt{2} + \sqrt{8} \neq 3\sqrt{10}$.

OBJECTIVE 3 Simplify radical sums involving multiplication. Some radical expressions require both multiplication and addition (or subtraction). The order of operations presented in Chapter 1 still applies.

CHALKBOARD EXAMPLE

Multiply and combine terms. Assume all variables represent nonnegative real numbers.

$$\sqrt{3r} \cdot \sqrt{6} + \sqrt{8r}$$

Answer: $5\sqrt{2r}$

EXAMPLE 3 Simplifying Radical Sums Involving Multiplication

Simplify each radical expression. Assume that all variables represent nonnegative real numbers.

(a)

$\sqrt{5} \cdot \sqrt{15} + 4\sqrt{3} = \sqrt{5 \cdot 15} + 4\sqrt{3}$	Product rule
$= \sqrt{75} + 4\sqrt{3}$	Multiply.
$= \sqrt{25 \cdot 3} + 4\sqrt{3}$	25 is a perfect square.
$= \sqrt{25} \cdot \sqrt{3} + 4\sqrt{3}$	Product rule
$= 5\sqrt{3} + 4\sqrt{3}$	$\sqrt{25} = 5$
$= 9\sqrt{3}$	Add like radicals.

(b)

$\sqrt{2} \cdot \sqrt{6k} + \sqrt{27k} = \sqrt{12k} + \sqrt{27k}$	Product rule
$= \sqrt{4 \cdot 3k} + \sqrt{9 \cdot 3k}$	Factor.
$= \sqrt{4} \cdot \sqrt{3k} + \sqrt{9} \cdot \sqrt{3k}$	Product rule
$= 2\sqrt{3k} + 3\sqrt{3k}$	$\sqrt{4} = 2$ and $\sqrt{9} = 3$
$= 5\sqrt{3k}$	Add like radicals.

(c)

$\sqrt[3]{2} \cdot \sqrt[3]{16m^3} - \sqrt[3]{108m^3} = \sqrt[3]{32m^3} - \sqrt[3]{108m^3}$	Product rule
$= \sqrt[3]{(8m^3)4} - \sqrt[3]{(27m^3)4}$	Factor.
$= 2m\sqrt[3]{4} - 3m\sqrt[3]{4}$	$\sqrt[3]{8m^3} = 2m$ and $\sqrt[3]{27m^3} = 3m$
$= -m\sqrt[3]{4}$	Subtract like radicals.

CAUTION Remember that a sum or difference of radicals can be simplified only if the radicals are *like radicals.* For example, $2\sqrt{3} + 5\sqrt[3]{3}$ cannot be simplified further.

9.3 EXERCISES

1. distributive
2. root; radicand or number
3. radicands 4. roots or indexes
5. $-5\sqrt{7}$ 6. $-2\sqrt{2}$
7. $5\sqrt{17}$ 8. $6\sqrt{19}$
9. $5\sqrt{7}$ 10. $10\sqrt{14}$
11. $11\sqrt{5}$ 12. $20\sqrt{6}$
13. $15\sqrt{2}$ 14. $-2\sqrt{2}$
15. $-20\sqrt{2} + 6\sqrt{5}$
16. $-20\sqrt{3} + 6\sqrt{6}$

Fill in each blank with the correct response.

1. $5\sqrt{2} + 6\sqrt{2} = (5 + 6)\sqrt{2} = 11\sqrt{2}$ is an example of the _____ property.

2. Like radicals have the same _____ of the same _____.

3. $\sqrt{2} + 2\sqrt{3}$ cannot be simplified because the _____ are different.

4. $2\sqrt{3} + 4\sqrt[3]{3}$ cannot be simplified because the _____ are different.

Simplify and add or subtract wherever possible. See Examples 1 and 2.

5. $14\sqrt{7} - 19\sqrt{7}$

6. $16\sqrt{2} - 18\sqrt{2}$

7. $\sqrt{17} + 4\sqrt{17}$

8. $5\sqrt{19} + \sqrt{19}$

9. $6\sqrt{7} - \sqrt{7}$

10. $11\sqrt{14} - \sqrt{14}$

11. $\sqrt{45} + 4\sqrt{20}$

12. $\sqrt{24} + 6\sqrt{54}$

13. $5\sqrt{72} - 3\sqrt{50}$

14. $6\sqrt{18} - 5\sqrt{32}$

15. $-5\sqrt{32} + 2\sqrt{45}$

16. $-4\sqrt{75} + 3\sqrt{24}$

17. $17\sqrt{7}$ **18.** $-11\sqrt{11}$
19. $-16\sqrt{2} - 8\sqrt{3}$
20. $62\sqrt{2} - 12\sqrt{3}$
21. $20\sqrt{2} + 6\sqrt{3} - 15\sqrt{5}$
22. $18\sqrt{2} + 8\sqrt{3} + 12\sqrt{7}$
23. $4\sqrt{2}$ **24.** $5\sqrt{3}$
25. $2\sqrt{3} + 4\sqrt{3} = (2 + 4)\sqrt{3} = 6\sqrt{3}$ **26.** $\sqrt{3}$ and $\sqrt{7}$ are not *like radicals.* $\sqrt{3} + \sqrt{7} \approx 4.377802119$, while $\sqrt{10} \approx 3.16227766$. Therefore, $\sqrt{3} + \sqrt{7} \neq \sqrt{10}$.
27. $11\sqrt{3}$ **28.** $16\sqrt{5}$
29. $5\sqrt{x}$ **30.** $8\sqrt{a}$
31. $3x\sqrt{6}$ **32.** $11x\sqrt{3}$
33. 0 **34.** 0 **35.** $-20\sqrt{2k}$
36. $14\sqrt{3x}$ **37.** $42x\sqrt{5z}$
38. $19x\sqrt{3y}$ **39.** $-\sqrt[3]{2}$
40. $35\sqrt[3]{2}$ **41.** $6\sqrt[3]{p^2}$
42. $36k\sqrt[3]{2k}$ **43.** $21\sqrt[4]{m^3}$
44. $14m\sqrt[4]{m}$ **45.** Only like radicals can be added or subtracted. Thus, we can add $2\sqrt{3}$ and $5\sqrt{3}$ using the distributive property: $2\sqrt{3} + 5\sqrt{3} = (2 + 5)\sqrt{3} = 7\sqrt{3}$. We cannot subtract in the expression $2\sqrt{3} - 2\sqrt[3]{3}$, because $2\sqrt{3}$ and $2\sqrt[3]{3}$ are unlike radicals.
46. The variables may represent negative numbers, because negative numbers have cube roots.

17. $5\sqrt{7} - 3\sqrt{28} + 6\sqrt{63}$

18. $3\sqrt{11} + 5\sqrt{44} - 8\sqrt{99}$

19. $2\sqrt{8} - 5\sqrt{32} - 2\sqrt{48}$

20. $5\sqrt{72} - 3\sqrt{48} + 4\sqrt{128}$

21. $4\sqrt{50} + 3\sqrt{12} - 5\sqrt{45}$

22. $6\sqrt{18} + 2\sqrt{48} + 6\sqrt{28}$

23. $\frac{1}{4}\sqrt{288} + \frac{1}{6}\sqrt{72}$

24. $\frac{2}{3}\sqrt{27} + \frac{3}{4}\sqrt{48}$

25. The distributive property, which says $a(b + c) = ab + ac$ and $ba + ca = (b + c)a$, provides the justification for adding and subtracting like radicals. While we usually skip the step that indicates this property, we could not make the statement $2\sqrt{3} + 4\sqrt{3} = 6\sqrt{3}$ without it. Write an equation showing how the distributive property is actually used in this statement.

26. Refer to Example 1(e), and explain why $\sqrt{3} + \sqrt{7}$ cannot be further simplified. Confirm, by using calculator approximations, that $\sqrt{3} + \sqrt{7}$ is *not* equal to $\sqrt{10}$.

Perform the indicated operations. Assume that all variables represent nonnegative real numbers. See Example 3.

27. $\sqrt{6} \cdot \sqrt{2} + 9\sqrt{3}$

28. $4\sqrt{15} \cdot \sqrt{3} + 4\sqrt{5}$

29. $\sqrt{9x} + \sqrt{49x} - \sqrt{25x}$

30. $\sqrt{4a} - \sqrt{16a} + \sqrt{100a}$

31. $\sqrt{6x^2} + x\sqrt{24}$

32. $\sqrt{75x^2} + x\sqrt{108}$

33. $3\sqrt{8x^2} - 4x\sqrt{2} - x\sqrt{8}$

34. $\sqrt{2b^2} + 3b\sqrt{18} - b\sqrt{200}$

35. $-8\sqrt{32k} + 6\sqrt{8k}$

36. $4\sqrt{12x} + 2\sqrt{27x}$

37. $2\sqrt{125x^2z} + 8x\sqrt{80z}$

38. $\sqrt{48x^2y} + 5x\sqrt{27y}$

39. $4\sqrt[3]{16} - 3\sqrt[3]{54}$

40. $5\sqrt[3]{128} + 3\sqrt[3]{250}$

41. $6\sqrt[3]{8p^2} - 2\sqrt[3]{27p^2}$

42. $8k\sqrt[3]{54k} + 6\sqrt[3]{16k^4}$

43. $5\sqrt[4]{m^3} + 8\sqrt[4]{16m^3}$

44. $5\sqrt[4]{m^5} + 3\sqrt[4]{81m^5}$

45. Describe in your own words how to add and subtract radicals.

46. In the directions for Exercises 27–44, we made the assumption that all variables represent nonnegative real numbers. However, in Exercises 41 and 42, variables actually *may* represent negative numbers. Explain why this is so.

47. $-6x^2y$
48. $-6(p - 2q)^2(a + b)$
49. $-6a^2\sqrt{xy}$
50. The answers are alike because the numerical coefficient of the three answers is the same: -6. Also, the first variable factor is raised to the second power, and the second variable factor is raised to the first power. The answers are different because the variables are different: x and y, then $p - 2q$ and $a + b$, and then a and $\sqrt{xy}$.

RELATING CONCEPTS (EXERCISES 47–50)

Adding and subtracting like radicals is no different than adding and subtracting other like terms.

Work Exercises 47–50 in order.

47. Combine like terms: $5x^2y + 3x^2y - 14x^2y$.

48. Combine like terms: $5(p - 2q)^2(a + b) + 3(p - 2q)^2(a + b) - 14(p - 2q)^2(a + b)$.

49. Combine like radicals: $5a^2\sqrt{xy} + 3a^2\sqrt{xy} - 14a^2\sqrt{xy}$.

50. Compare your answers in Exercises 47–49. How are they alike? How are they different?

Did you make the connection between adding and subtracting like radicals and adding and subtracting like terms?

51. 9.220 **52.** 13 **53.** 5
54. 3.606

Perform the indicated operations following the order of operations we have used throughout the book. Use a calculator and round to the nearest thousandth, if necessary.

51. $\sqrt{(-3 - 6)^2 + (2 - 4)^2}$

52. $\sqrt{(-9 - 3)^2 + (3 - 8)^2}$

53. $\sqrt{(2 - (-2))^2 + (-1 - 2)^2}$

54. $\sqrt{(3 - 1)^2 + (2 - (-1))^2}$

55. $22\sqrt{2}$ **56.** $14\sqrt{5}$
57. (a) 1991 **(b)** 1997
58. (a) 1988; 1994
(b)

Year	1990	1993	1994	1996
Difference	−3.5	.2	−.1	−.4

The difference was positive (that is, exports exceeded imports) only in 1993.

Find the perimeter of each figure.

55.

56.

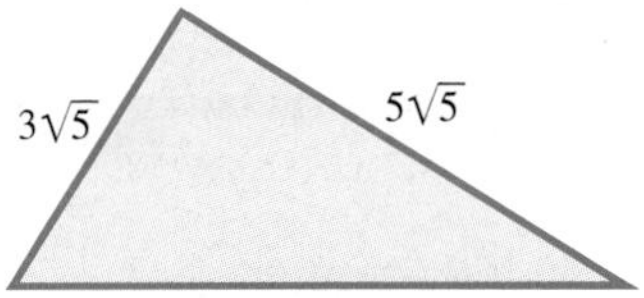

Work each problem.

57. The table shows the year in which U.S. exports of electronics were x billion dollars.

U.S. Exports

Billions of Dollars	*Year*
7.5	1990
19.6	1993
25.8	1994
36.4	1996

Source: U.S. Bureau of the Census; *U.S. Merchandise Trade,* series FT 900, December issue; and unpublished data.

The function defined by $y = 1.4\sqrt{x - 2.5} + 87.5$ gives a good approximation of the year y when exports were x billion dollars. Here $y = 90$ represents 1990, $y = 93$ represents 1993, and so on. Use the equation to find the years in which exports were

(a) \$10 billion

(b) \$50 billion.

58. U.S. imports of electronics for several years are shown in the table.

U.S. Imports

Billions of Dollars	*Year*
11.0	1990
19.4	1993
25.9	1994
36.8	1996

Source: U.S. Bureau of the Census; *U.S. Merchandise Trade,* series FT 900, December issue; and unpublished data.

The year when U.S. imports were x billion dollars is closely approximated by the function $y = 1.6\sqrt{x - 6} + 87$, where $y = 90$ represents 1990, and so on, as in Exercise 57.

(a) Use the equation to find the years when U.S. imports were \$7 billion; \$30 billion.

(b) Using the tables in Exercises 57 and 58, find the differences between U.S. exports and U.S. imports in the years listed in both tables. How do they compare?

9.4 Rationalizing the Denominator

OBJECTIVES

1. Rationalize denominators with square roots.
2. Write radicals in simplified form.
3. Rationalize denominators with cube roots.

FOR EXTRA HELP

SSG Sec. 9.4
SSM Sec. 9.4
Pass the Test Software
InterAct Math Tutorial Software
Video 14

CHALKBOARD EXAMPLE

Rationalize the denominator.

$$\frac{-12}{\sqrt{32}}$$

Answer: $\frac{-3\sqrt{2}}{2}$

TEACHING TIP Here is an alternative method:

$$\frac{12}{\sqrt{8}} \cdot \frac{\sqrt{2}}{\sqrt{2}} = \frac{12\sqrt{2}}{\sqrt{16}}$$
$$= \frac{12\sqrt{2}}{4} = 3\sqrt{2}.$$

OBJECTIVE 1 Rationalize denominators with square roots. Fractions are simplified by rewriting them without any radical expressions in the denominator. For example, the radical in the denominator of

$$\frac{\sqrt{3}}{\sqrt{2}}$$

can be eliminated by multiplying the numerator and the denominator by $\sqrt{2}$.

$$\frac{\sqrt{3}}{\sqrt{2}} = \frac{\sqrt{3} \cdot \sqrt{2}}{\sqrt{2} \cdot \sqrt{2}} = \frac{\sqrt{6}}{2} \qquad \text{Since } \sqrt{2} \cdot \sqrt{2} = 2$$

This process of changing the denominator from a radical (irrational number) to a rational number is called **rationalizing the denominator.** The value of the number is not changed; only the form of the number is changed, because the expression has been multiplied by 1 in the form $\frac{\sqrt{2}}{\sqrt{2}}$.

EXAMPLE 1 Rationalizing Denominators

Rationalize each denominator.

(a) $\frac{9}{\sqrt{6}}$

Multiply both numerator and denominator by $\sqrt{6}$.

$$\frac{9}{\sqrt{6}} = \frac{9 \cdot \sqrt{6}}{\sqrt{6} \cdot \sqrt{6}}$$
$$= \frac{9\sqrt{6}}{6} \qquad \sqrt{6} \cdot \sqrt{6} = 6$$
$$= \frac{3\sqrt{6}}{2} \qquad \text{Lowest terms}$$

(b) $\frac{12}{\sqrt{8}}$

The denominator here could be rationalized by multiplying by $\sqrt{8}$. However, the result can be found more directly by first simplifying the denominator.

$$\sqrt{8} = \sqrt{4} \cdot \sqrt{2} = 2\sqrt{2}$$

Then multiply numerator and denominator by $\sqrt{2}$.

$$\frac{12}{\sqrt{8}} = \frac{12}{2\sqrt{2}}$$
$$= \frac{12 \cdot \sqrt{2}}{2\sqrt{2} \cdot \sqrt{2}} \qquad \text{Multiply by } \frac{\sqrt{2}}{\sqrt{2}}.$$
$$= \frac{12\sqrt{2}}{2 \cdot 2} \qquad \sqrt{2} \cdot \sqrt{2} = 2$$
$$= \frac{12\sqrt{2}}{4} \qquad \text{Multiply.}$$
$$= 3\sqrt{2} \qquad \text{Lowest terms}$$

INSTRUCTOR'S RESOURCES

PTB Sec. 9.4 ISM Sec. 9.4
AB Sec. 9.4
TEST GENERATOR
WORLD WIDE WEB www.LialAlgebra.com

OBJECTIVE 2 **Write radicals in simplified form.** A radical is considered to be in simplified form if the following three conditions are met.

Simplified Form of a Radical

1. All nth power factors of the radicand of $\sqrt[n]{}$ are removed. (For example, $\sqrt{3^2} = 3$, $\sqrt[3]{5^3} = 5$, and so on.)
2. The radicand has no fractions.
3. No denominator contains a radical.

In the following examples, radicals are simplified according to these conditions.

CHALKBOARD EXAMPLE

Simplify $\sqrt{\frac{5}{18}}$ by rationalizing the denominator.

Answer: $\frac{\sqrt{10}}{6}$

EXAMPLE 2 Simplifying a Radical with a Fraction

Simplify $\sqrt{\frac{27}{5}}$ by rationalizing the denominator.

First use the quotient rule for radicals.

$$\sqrt{\frac{27}{5}} = \frac{\sqrt{27}}{\sqrt{5}}$$

Now multiply both numerator and denominator by $\sqrt{5}$.

$$\frac{\sqrt{27}}{\sqrt{5}} = \frac{\sqrt{27}\cdot\sqrt{5}}{\sqrt{5}\cdot\sqrt{5}} \qquad \text{Rationalize the denominator.}$$

$$= \frac{\sqrt{9\cdot 3}\cdot\sqrt{5}}{5} \qquad \sqrt{5}\cdot\sqrt{5} = 5$$

$$= \frac{\sqrt{9}\cdot\sqrt{3}\cdot\sqrt{5}}{5} \qquad \text{Product rule}$$

$$= \frac{3\cdot\sqrt{3\cdot 5}}{5} = \frac{3\sqrt{15}}{5} \qquad \text{Product rule}$$

CHALKBOARD EXAMPLE

Simplify $\sqrt{\frac{1}{2}}\cdot\sqrt{\frac{5}{6}}$.

Answer: $\frac{\sqrt{15}}{6}$

EXAMPLE 3 Simplifying a Product of Radicals

Simplify $\sqrt{\frac{5}{8}}\cdot\sqrt{\frac{1}{6}}$.

Use both the product and quotient rules.

$$\sqrt{\frac{5}{8}}\cdot\sqrt{\frac{1}{6}} = \sqrt{\frac{5}{8}\cdot\frac{1}{6}} \qquad \text{Product rule}$$

$$= \sqrt{\frac{5}{48}} \qquad \text{Multiply.}$$

$$= \frac{\sqrt{5}}{\sqrt{48}} \qquad \text{Quotient rule}$$

To rationalize the denominator, first simplify, then multiply the numerator and denominator by $\sqrt{3}$ as follows.

$$\frac{\sqrt{5}}{\sqrt{48}} = \frac{\sqrt{5}}{\sqrt{16}\cdot\sqrt{3}} \quad \text{Product rule}$$

$$= \frac{\sqrt{5}}{4\sqrt{3}} \quad \sqrt{16} = 4$$

$$= \frac{\sqrt{5}\cdot\sqrt{3}}{4\sqrt{3}\cdot\sqrt{3}} \quad \text{Rationalize the denominator.}$$

$$= \frac{\sqrt{15}}{4\cdot 3} \quad \text{Product rule}$$

$$= \frac{\sqrt{15}}{12} \quad \text{Multiply.}$$

CHALKBOARD EXAMPLE

Simplify $\frac{\sqrt{5p}}{\sqrt{q}}$. Assume that p and q are positive real numbers.

Answer: $\frac{\sqrt{5pq}}{q}$

EXAMPLE 4 Simplifying a Quotient of Radicals

Rationalize the denominator of $\frac{\sqrt{4x}}{\sqrt{y}}$. Assume that x and y represent positive real numbers.

Multiply numerator and denominator by $\sqrt{y}$.

$$\frac{\sqrt{4x}}{\sqrt{y}} = \frac{\sqrt{4x}\cdot\sqrt{y}}{\sqrt{y}\cdot\sqrt{y}} = \frac{\sqrt{4xy}}{y} = \frac{2\sqrt{xy}}{y}$$

OBJECTIVE 3 Rationalize denominators with cube roots. To rationalize a denominator with a cube root, change the radicand in the denominator to a perfect cube, as shown in the next example.

CHALKBOARD EXAMPLE

Rationalize the denominator.

$$\sqrt[3]{\frac{5}{6}}$$

Answer: $\frac{\sqrt[3]{180}}{6}$

TEACHING TIP Rationalize $\sqrt[3]{\frac{1}{2}}$ and $\sqrt[3]{\frac{1}{4}}$ and compare the number chosen in each case as the multiplier of the numerator and denominator.

EXAMPLE 5 Rationalizing a Denominator with a Cube Root

Rationalize each denominator.

(a) $\sqrt[3]{\frac{3}{2}}$

Begin by writing the expression as a quotient of radicals. Then, multiply the numerator and the denominator by enough factors of 2 to make the denominator a perfect cube. This will eliminate the radical in the denominator. Here, multiply by $\sqrt[3]{2^2}$.

$$\sqrt[3]{\frac{3}{2}} = \frac{\sqrt[3]{3}}{\sqrt[3]{2}} = \frac{\sqrt[3]{3}\cdot\sqrt[3]{2^2}}{\sqrt[3]{2}\cdot\sqrt[3]{2^2}} = \frac{\sqrt[3]{3\cdot 2^2}}{\sqrt[3]{2^3}} = \frac{\sqrt[3]{12}}{2} \quad \text{Since } \sqrt[3]{2^3} = \sqrt[3]{8} = 2$$

(b) $\frac{\sqrt[3]{3}}{\sqrt[3]{4}}$

Since $4\cdot 2 = 2^2\cdot 2 = 2^3$, multiply numerator and denominator by $\sqrt[3]{2}$.

$$\frac{\sqrt[3]{3}}{\sqrt[3]{4}} = \frac{\sqrt[3]{3}\cdot\sqrt[3]{2}}{\sqrt[3]{2\cdot 2}\cdot\sqrt[3]{2}} = \frac{\sqrt[3]{6}}{\sqrt[3]{2^3}} = \frac{\sqrt[3]{6}}{2}$$

CAUTION A common error in a problem like the one in Example 5(a) is to multiply by $\sqrt[3]{2}$ instead of $\sqrt[3]{2^2}$. Doing this would give a denominator of $\sqrt[3]{2} \cdot \sqrt[3]{2} = \sqrt[3]{4}$. Since 4 is not a perfect cube, the denominator is still not rationalized.

9.4 EXERCISES

1. radical 2. x^2 3. fraction 4. radical or irrational number 5. $4\sqrt{2}$ 6. $4\sqrt{3}$ 7. $\frac{-\sqrt{33}}{3}$ 8. $\frac{-\sqrt{65}}{5}$ 9. $\frac{7\sqrt{15}}{5}$ 10. $\frac{4\sqrt{30}}{5}$ 11. $\frac{\sqrt{30}}{2}$ 12. $\frac{3\sqrt{30}}{4}$ 13. $\frac{16\sqrt{3}}{9}$ 14. $4\sqrt{2}$ 15. $\frac{-3\sqrt{2}}{10}$ 16. $\frac{-\sqrt{3}}{3}$ 17. $\frac{21\sqrt{5}}{5}$ 18. $\frac{27\sqrt{2}}{8}$ 19. $\sqrt{3}$ 20. $\sqrt{2}$ 21. $\frac{\sqrt{2}}{2}$ 22. $\frac{\sqrt{3}}{3}$ 23. $\frac{\sqrt{65}}{5}$ 24. $\frac{\sqrt{187}}{11}$ 25. 1; identity property for multiplication 26. Both approximations are 3.674234614. 27. $\frac{\sqrt{21}}{3}$ 28. $\frac{\sqrt{57}}{3}$ 29. $\frac{3\sqrt{14}}{4}$ 30. $\frac{5\sqrt{3}}{12}$ 31. $\frac{1}{6}$ 32. $\frac{1}{4}$ 33. 1 34. 1 35. $\frac{\sqrt{7x}}{x}$ 36. $\frac{\sqrt{19y}}{y}$ 37. $\frac{2x\sqrt{xy}}{y}$ 38. $\frac{3t\sqrt{st}}{s}$ 39. $\frac{x\sqrt{3xy}}{y}$ 40. $\frac{t\sqrt{3pt}}{p}$ 41. $\frac{3ar^2\sqrt{7rt}}{7t}$ 42. $\frac{4xy\sqrt{13xz}}{13z}$

Fill in each blank with the correct response.

1. Rationalizing the denominator means to change the denominator from a(n) ______ to a rational number.
2. To simplify $\sqrt{x^2y}$, where $x \geq 0$, we must remove ______ from the radicand.
3. The expression $\sqrt{\frac{3m}{2}}$, where $m \geq 0$, is not simplified because the radical contains a(n) ______.
4. The expression $\frac{2x}{\sqrt{7}}$ is not simplified because the denominator is a(n) ______.

Rationalize each denominator. See Examples 1 and 2.

5. $\frac{8}{\sqrt{2}}$ 6. $\frac{12}{\sqrt{3}}$ 7. $\frac{-\sqrt{11}}{\sqrt{3}}$ 8. $\frac{-\sqrt{13}}{\sqrt{5}}$ 9. $\frac{7\sqrt{3}}{\sqrt{5}}$

10. $\frac{4\sqrt{6}}{\sqrt{5}}$ 11. $\frac{24\sqrt{10}}{16\sqrt{3}}$ 12. $\frac{18\sqrt{15}}{12\sqrt{2}}$ 13. $\frac{16}{\sqrt{27}}$ 14. $\frac{24}{\sqrt{18}}$

15. $\frac{-3}{\sqrt{50}}$ 16. $\frac{-5}{\sqrt{75}}$ 17. $\frac{63}{\sqrt{45}}$ 18. $\frac{27}{\sqrt{32}}$ 19. $\frac{\sqrt{24}}{\sqrt{8}}$

20. $\frac{\sqrt{36}}{\sqrt{18}}$ 21. $\sqrt{\frac{1}{2}}$ 22. $\sqrt{\frac{1}{3}}$ 23. $\sqrt{\frac{13}{5}}$ 24. $\sqrt{\frac{17}{11}}$

25. To rationalize the denominator of an expression such as $\frac{4}{\sqrt{3}}$, we multiply both the numerator and the denominator by $\sqrt{3}$. By what number are we actually multiplying the given expression, and what property of real numbers justifies the fact that our result is equal to the given expression?

26. In Example 1(a), we show algebraically that $\frac{9}{\sqrt{6}}$ is equal to $\frac{3\sqrt{6}}{2}$. Support this result numerically by finding the decimal approximation of $\frac{9}{\sqrt{6}}$ on your calculator, and then finding the decimal approximation of $\frac{3\sqrt{6}}{2}$. What do you notice?

Multiply and simplify each result. See Example 3.

27. $\sqrt{\frac{7}{13}} \cdot \sqrt{\frac{13}{3}}$ 28. $\sqrt{\frac{19}{20}} \cdot \sqrt{\frac{20}{3}}$ 29. $\sqrt{\frac{21}{7}} \cdot \sqrt{\frac{21}{8}}$ 30. $\sqrt{\frac{5}{8}} \cdot \sqrt{\frac{5}{6}}$

31. $\sqrt{\frac{1}{12}} \cdot \sqrt{\frac{1}{3}}$ 32. $\sqrt{\frac{1}{8}} \cdot \sqrt{\frac{1}{2}}$ 33. $\sqrt{\frac{2}{9}} \cdot \sqrt{\frac{9}{2}}$ 34. $\sqrt{\frac{4}{3}} \cdot \sqrt{\frac{3}{4}}$

Simplify each radical. Assume that all variables represent positive real numbers. See Example 4.

35. $\sqrt{\frac{7}{x}}$ 36. $\sqrt{\frac{19}{y}}$ 37. $\sqrt{\frac{4x^3}{y}}$ 38. $\sqrt{\frac{9t^3}{s}}$

39. $\sqrt{\frac{18x^3}{6y}}$ 40. $\sqrt{\frac{24t^3}{8p}}$ 41. $\sqrt{\frac{9a^2r^5}{7t}}$ 42. $\sqrt{\frac{16x^3y^2}{13z}}$

43. (b) 44. yes

45. $\frac{\sqrt[3]{12}}{2}$ 46. $\frac{\sqrt[3]{50}}{5}$

47. $\frac{\sqrt[3]{196}}{7}$ 48. $\frac{\sqrt[3]{4}}{2}$

49. $\frac{\sqrt[3]{6y}}{2y}$ 50. $\frac{\sqrt[3]{15x}}{5x}$

51. $\frac{\sqrt[3]{42mn^2}}{6n}$ 52. $\frac{\sqrt[3]{77pq^2}}{7q}$

53. (a) $\frac{9\sqrt{2}}{4}$ seconds
(b) 3.182 seconds

54. (a) $\frac{3\sqrt{15}}{2}$ kilometers per second
(b) 5.809 kilometers per second

43. Which one of the following would be an appropriate choice for multiplying the numerator and the denominator of $\frac{\sqrt[3]{2}}{\sqrt[3]{5}}$ in order to rationalize the denominator?

(a) $\sqrt[3]{5}$ **(b)** $\sqrt[3]{25}$ **(c)** $\sqrt[3]{2}$ **(d)** $\sqrt[3]{3}$

44. In Example 5(b), we multiply the numerator and denominator of $\frac{\sqrt[3]{3}}{\sqrt[3]{4}}$ by $\sqrt[3]{2}$ to rationalize the denominator. Suppose we had chosen to multiply by $\sqrt[3]{16}$ instead. Would we have obtained the correct answer after all simplifications were done?

Simplify. Rationalize each denominator. Assume that variables in each denominator are nonzero. See Example 5.

45. $\sqrt[3]{\frac{3}{2}}$ **46.** $\sqrt[3]{\frac{2}{5}}$ **47.** $\frac{\sqrt[3]{4}}{\sqrt[3]{7}}$ **48.** $\frac{\sqrt[3]{5}}{\sqrt[3]{10}}$

49. $\sqrt[3]{\frac{3}{4y^2}}$ **50.** $\sqrt[3]{\frac{3}{25x^2}}$ **51.** $\frac{\sqrt[3]{7m}}{\sqrt[3]{36n}}$ **52.** $\frac{\sqrt[3]{11p}}{\sqrt[3]{49q}}$

*In Exercises 53 and 54, **(a)** give the answer as a simplified radical and **(b)** use a calculator to give the answer correct to the nearest thousandth.*

53. The period p of a pendulum is the time it takes for it to swing from one extreme to the other and back again. The value of p in seconds is given by

$$p = k \cdot \sqrt{\frac{L}{g}},$$

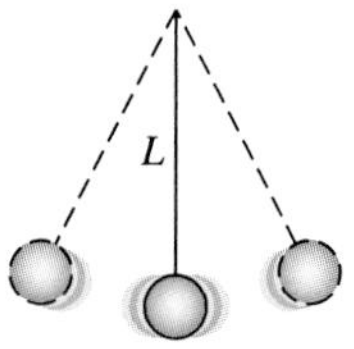

where L is the length of the pendulum, g is the acceleration due to gravity, and k is a constant. Find the period when $k = 6$, $L = 9$ feet, and $g = 32$ feet per second squared.

54. The velocity v of a meteorite approaching the earth is given by

$$v = \frac{k}{\sqrt{d}}$$

kilometers per second, where d is its distance from the center of the earth and k is a constant. What is the velocity of a meteorite that is 6000 kilometers away from the center of the earth, if $k = 450$?

9.5 Simplifying Radical Expressions

OBJECTIVES

1. Simplify products of radical expressions.
2. Simplify quotients of radical expressions.
3. Write radical expressions with quotients in lowest terms.

The conditions for which a radical is in simplest form were listed in the previous section. Below is a set of guidelines to follow when you are simplifying radical expressions. Although the guidelines are illustrated with square roots, they apply to higher roots as well.

Simplifying Radical Expressions

1. If a radical represents a rational number, use that rational number in place of the radical.

Examples: $\sqrt{49}$ is simplified by writing 7; $\sqrt{\frac{169}{9}}$ by writing $\frac{13}{3}$.

JOURNAL WRITING ▲ CHALLENGING SCIENTIFIC CALCULATOR GRAPHING CALCULATOR

INSTRUCTOR'S RESOURCES

PTB Sec. 9.5 **ISM** Sec. 9.5 **AB** Sec. 9.5 | **TEST GENERATOR** | **WORLD WIDE WEB** www.LialAlgebra.com

FOR EXTRA HELP

SSG Sec. 9.5
SSM Sec. 9.5
Pass the Test Software
InterAct Math Tutorial Software
Video 15

Simplifying Radical Expressions (continued)

2. If a radical expression contains products of radicals, use the product rule for radicals, $\sqrt{x} \cdot \sqrt{y} = \sqrt{xy}$, to get a single radical.

   Examples: $\sqrt{3} \cdot \sqrt{2}$ is simplified to $\sqrt{6}$; $\sqrt{5} \cdot \sqrt{x}$ to $\sqrt{5x}$.

3. If a radicand has a factor that is a perfect square, express the radical as the product of the positive square root of the perfect square and the remaining radical factor.

   Examples: $\sqrt{20}$ is simplified to $\sqrt{20} = \sqrt{4 \cdot 5} = \sqrt{4} \cdot \sqrt{5} = 2\sqrt{5}$;
   $\sqrt[3]{16} = \sqrt[3]{8 \cdot 2} = \sqrt[3]{8} \cdot \sqrt[3]{2} = 2\sqrt[3]{2}$.

4. If a radical expression contains sums or differences of radicals, use the distributive property to combine like radicals.

   Examples: $3\sqrt{2} + 4\sqrt{2} = 7\sqrt{2}$, but $3\sqrt{2} + 4\sqrt{3}$ cannot be further simplified.

5. Rationalize any denominator containing a radical.

   Examples: $\frac{5}{\sqrt{3}}$ is rationalized as $\frac{5}{\sqrt{3}} = \frac{5\sqrt{3}}{\sqrt{3} \cdot \sqrt{3}} = \frac{5\sqrt{3}}{3}$;
   $\sqrt{\frac{3}{2}} = \frac{\sqrt{3}}{\sqrt{2}} = \frac{\sqrt{3} \cdot \sqrt{2}}{\sqrt{2} \cdot \sqrt{2}} = \frac{\sqrt{6}}{2}$.

OBJECTIVE 1 Simplify products of radical expressions. Use the above guidelines to do this.

CHALKBOARD EXAMPLE

Find the product and simplify.

$(\sqrt{2} + 5\sqrt{3})(\sqrt{3} - 2\sqrt{2})$

Answer: $11 - 9\sqrt{6}$

EXAMPLE 1 Multiplying Radical Expressions

Find each product and simplify the answers.

(a) $\sqrt{5}(\sqrt{8} - \sqrt{32})$

Simplify inside the parentheses.

$$\sqrt{5}(\sqrt{8} - \sqrt{32}) = \sqrt{5}(2\sqrt{2} - 4\sqrt{2})$$
$$= \sqrt{5}(-2\sqrt{2}) \quad \text{Subtract like radicals.}$$
$$= -2\sqrt{5 \cdot 2} \quad \text{Product rule; commutative property}$$
$$= -2\sqrt{10} \quad \text{Multiply.}$$

(b) $(\sqrt{3} + 2\sqrt{5})(\sqrt{3} - 4\sqrt{5})$

The product of these sums of radicals can be found in the same way that we found the product of binomials in Chapter 4 using the FOIL method.

TEACHING TIP As with earlier binomial multiplication, encourage students to do the middle multiplication and addition steps mentally.

$$(\sqrt{3} + 2\sqrt{5})(\sqrt{3} - 4\sqrt{5})$$
$$\quad \text{F} \qquad \text{O} \qquad \text{I} \qquad \text{L}$$
$$= \sqrt{3} \cdot \sqrt{3} + \sqrt{3}(-4\sqrt{5}) + 2\sqrt{5} \cdot \sqrt{3} + 2\sqrt{5}(-4\sqrt{5})$$
$$= 3 - 4\sqrt{15} + 2\sqrt{15} - 8 \cdot 5 \quad \text{Product rule}$$
$$= 3 - 2\sqrt{15} - 40 \quad \text{Add like radicals.}$$
$$= -37 - 2\sqrt{15} \quad \text{Combine terms.}$$

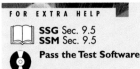

(c) $(\sqrt{3} + \sqrt{21})(\sqrt{3} - \sqrt{7})$

$$
\begin{aligned}
&(\sqrt{3} + \sqrt{21})(\sqrt{3} - \sqrt{7}) \\
&= \sqrt{3}(\sqrt{3}) + \sqrt{3}(-\sqrt{7}) + \sqrt{21}(\sqrt{3}) + \sqrt{21}(-\sqrt{7}) && \text{FOIL} \\
&= 3 - \sqrt{21} + \sqrt{63} - \sqrt{147} && \text{Product rule} \\
&= 3 - \sqrt{21} + \sqrt{9}\cdot\sqrt{7} - \sqrt{49}\cdot\sqrt{3} && \text{Simplify radicals.} \\
&= 3 - \sqrt{21} + 3\sqrt{7} - 7\sqrt{3} && \sqrt{9} = 3 \text{ and } \sqrt{49} = 7
\end{aligned}
$$

Since there are no like radicals, no terms may be combined.

Since radicals represent real numbers, the special products of binomials discussed in Chapter 4 can be used to find products of radicals. Example 2 uses the rule for the product that gives the difference of two squares,

$$(a + b)(a - b) = a^2 - b^2.$$

CHALKBOARD EXAMPLE

Find the product.

$$(\sqrt{3} - 2)(\sqrt{3} + 2)$$

Answer: -1

EXAMPLE 2 Using a Special Product with Radicals

Find each product.

(a) $(4 + \sqrt{3})(4 - \sqrt{3})$

Follow the pattern given above. Let $a = 4$ and $b = \sqrt{3}$.

$$
\begin{aligned}
(4 + \sqrt{3})(4 - \sqrt{3}) &= 4^2 - (\sqrt{3})^2 \\
&= 16 - 3 = 13 && 4^2 = 16 \text{ and } (\sqrt{3})^2 = 3
\end{aligned}
$$

(b)
$$
\begin{aligned}
(\sqrt{12} - \sqrt{6})(\sqrt{12} + \sqrt{6}) &= (\sqrt{12})^2 - (\sqrt{6})^2 \\
&= 12 - 6 && (\sqrt{12})^2 = 12 \text{ and } (\sqrt{6})^2 = 6 \\
&= 6
\end{aligned}
$$

Both products in Example 2 resulted in rational numbers. The pairs of expressions in those products, $4 + \sqrt{3}$ and $4 - \sqrt{3}$, and $\sqrt{12} - \sqrt{6}$ and $\sqrt{12} + \sqrt{6}$, are called **conjugates** of each other.

OBJECTIVE 2 Simplify quotients of radical expressions. Products of radicals similar to those in Example 2 can be used to rationalize the denominators in quotients with binomial denominators, such as

$$\frac{2}{4 - \sqrt{3}}.$$

By Example 2(a), if this denominator, $4 - \sqrt{3}$, is multiplied by $4 + \sqrt{3}$, then the product $(4 - \sqrt{3})(4 + \sqrt{3})$ is the rational number 13. Multiplying numerator and denominator by $4 + \sqrt{3}$ gives

$$
\begin{aligned}
\frac{2}{4 - \sqrt{3}} &= \frac{2(4 + \sqrt{3})}{(4 - \sqrt{3})(4 + \sqrt{3})} \\
&= \frac{2(4 + \sqrt{3})}{13}.
\end{aligned}
$$

The denominator now has been rationalized; it contains no radical signs.

Using Conjugates to Simplify a Radical Expression

To simplify a radical expression with two terms in the denominator, where at least one of those terms is a square root radical, multiply both the numerator and the denominator by the conjugate of the denominator.

CHALKBOARD EXAMPLE

Simplify by rationalizing the denominator.

$$\frac{3}{2-\sqrt{5}}$$

Answer: $-3(2+\sqrt{5})$

EXAMPLE 3 Using Conjugates to Rationalize a Denominator

Simplify by rationalizing the denominator.

(a) $\dfrac{7}{3+\sqrt{5}}$

We can eliminate the radical in the denominator by multiplying both numerator and denominator by $3-\sqrt{5}$, the conjugate of $3+\sqrt{5}$.

$$\frac{7}{3+\sqrt{5}} = \frac{7(3-\sqrt{5})}{(3+\sqrt{5})(3-\sqrt{5})} \qquad \text{Multiply by the conjugate.}$$

$$= \frac{7(3-\sqrt{5})}{3^2-(\sqrt{5})^2} \qquad (a+b)(a-b)=a^2-b^2$$

$$= \frac{7(3-\sqrt{5})}{9-5} \qquad 3^2 = 9 \text{ and } (\sqrt{5})^2 = 5$$

$$= \frac{7(3-\sqrt{5})}{4} \qquad \text{Subtract.}$$

(b) $\dfrac{6+\sqrt{2}}{\sqrt{2}-5}$

Multiply numerator and denominator by $\sqrt{2}+5$.

$$\frac{6+\sqrt{2}}{\sqrt{2}-5} = \frac{(6+\sqrt{2})(\sqrt{2}+5)}{(\sqrt{2}-5)(\sqrt{2}+5)}$$

$$= \frac{6\sqrt{2}+30+2+5\sqrt{2}}{2-25} \qquad \text{FOIL}$$

$$= \frac{11\sqrt{2}+32}{-23} \qquad \text{Combine terms.}$$

$$= -\frac{11\sqrt{2}+32}{23} \qquad \frac{a}{-b} = -\frac{a}{b}$$

CAUTION

Rationalizing the denominator in the two expressions

$$\frac{7}{\sqrt{x}+5} \quad \text{and} \quad \frac{7}{\sqrt{x}+\sqrt{5}}$$

involves different procedures. In the first denominator, which has one term, multiply both the numerator and the denominator by $\sqrt{x}+5$. In the second denominator, which has two terms, use $\sqrt{x}-\sqrt{5}$.

OBJECTIVE 3 Write radical expressions with quotients in lowest terms. The final example shows this.

CHALKBOARD EXAMPLE

Write $\frac{5\sqrt{3}-15}{10}$ in lowest terms.

Answer: $\frac{\sqrt{3}-3}{2}$

EXAMPLE 4 Writing a Radical Quotient in Lowest Terms

Write $\frac{3\sqrt{3}+15}{12}$ in lowest terms.

Factor the numerator and denominator, and then divide numerator and denominator by any common factors.

$$\frac{3\sqrt{3}+15}{12}=\frac{3(\sqrt{3}+5)}{3\cdot 4}=\frac{\sqrt{3}+5}{4}$$

This technique is used in Chapter 10.

CAUTION

A common error is to reduce an expression like the one in Example 4 incorrectly to lowest terms before factoring. For example,

$$\frac{4+8\sqrt{5}}{4}\neq 1+8\sqrt{5}.$$

The correct simplification is $1+2\sqrt{5}$. Why?

9.5 EXERCISES

1. 13 2. 1 3. 4 4. 8
5. 4 6. 6 7. 5 8. 0
9. $9\sqrt{5}$ 10. $14\sqrt{2}$
11. $16\sqrt{2}$ 12. $\sqrt{5}$
13. $\sqrt{15}-\sqrt{35}$
14. $\sqrt{70}+\sqrt{21}$
15. $2\sqrt{10}+30$
16. $42+12\sqrt{35}$ 17. $4\sqrt{7}$
18. $15\sqrt{2}$ 19. $57+23\sqrt{6}$
20. $48-20\sqrt{5}$
21. $81+14\sqrt{21}$
22. $30+18\sqrt{5}$
23. $37+12\sqrt{7}$
24. $105+40\sqrt{5}$ 25. 23
26. 4 27. 1 28. 1
29. $2\sqrt{3}-2+3\sqrt{2}-\sqrt{6}$
30. $3\sqrt{5}-\sqrt{15}+5\sqrt{3}-5$
31. $15\sqrt{2}-15$
32. $3\sqrt{2}+6\sqrt{3}-3-3\sqrt{6}$
33. $87+9\sqrt{21}$
34. $20+14\sqrt{5}$
35. Because multiplication must be performed before addition, it is incorrect to add -37 and -2. Since $-2\sqrt{15}$ cannot be simplified, the expression cannot be written in a simpler form, and the final answer is $-37-2\sqrt{15}$.

Based on the work so far, many simple operations involving radicals should now be performed mentally. In Exercises 1–8, perform the operations mentally, and write the answer without doing intermediate steps.

1. $\sqrt{49}+\sqrt{36}$
2. $\sqrt{100}-\sqrt{81}$
3. $\sqrt{2}\cdot\sqrt{8}$
4. $\sqrt{8}\cdot\sqrt{8}$
5. $\sqrt{2}(\sqrt{32}-\sqrt{8})$
6. $\sqrt{3}(\sqrt{27}-\sqrt{3})$
7. $\sqrt[3]{8}+\sqrt[3]{27}$
8. $\sqrt{4}-\sqrt[3]{64}+\sqrt[4]{16}$

Simplify each expression. Use the five guidelines given in the text. See Examples 1 and 2.

9. $3\sqrt{5}+2\sqrt{45}$
10. $2\sqrt{2}+4\sqrt{18}$
11. $8\sqrt{50}-4\sqrt{72}$
12. $4\sqrt{80}-5\sqrt{45}$
13. $\sqrt{5}(\sqrt{3}-\sqrt{7})$
14. $\sqrt{7}(\sqrt{10}+\sqrt{3})$
15. $2\sqrt{5}(\sqrt{2}+3\sqrt{5})$
16. $3\sqrt{7}(2\sqrt{7}+4\sqrt{5})$
17. $3\sqrt{14}\cdot\sqrt{2}-\sqrt{28}$
18. $7\sqrt{6}\cdot\sqrt{3}-2\sqrt{18}$
19. $(2\sqrt{6}+3)(3\sqrt{6}+7)$
20. $(4\sqrt{5}-2)(2\sqrt{5}-4)$
21. $(5\sqrt{7}-2\sqrt{3})(3\sqrt{7}+4\sqrt{3})$
22. $(2\sqrt{10}+5\sqrt{2})(3\sqrt{10}-3\sqrt{2})$
23. $(2\sqrt{7}+3)^2$
24. $(4\sqrt{5}+5)^2$
25. $(5-\sqrt{2})(5+\sqrt{2})$
26. $(3-\sqrt{5})(3+\sqrt{5})$
27. $(\sqrt{8}-\sqrt{7})(\sqrt{8}+\sqrt{7})$
28. $(\sqrt{12}-\sqrt{11})(\sqrt{12}+\sqrt{11})$
29. $(\sqrt{2}+\sqrt{3})(\sqrt{6}-\sqrt{2})$
30. $(\sqrt{3}+\sqrt{5})(\sqrt{15}-\sqrt{5})$
31. $(\sqrt{10}-\sqrt{5})(\sqrt{5}+\sqrt{20})$
32. $(\sqrt{6}-\sqrt{3})(\sqrt{3}+\sqrt{18})$
33. $(5\sqrt{7}-2\sqrt{3})(3\sqrt{7}+3\sqrt{3})$
34. $(2\sqrt{10}+5\sqrt{2})(3\sqrt{10}-4\sqrt{2})$
35. In Example 1(b), the original expression simplifies to $-37-2\sqrt{15}$. Students often try to simplify expressions like this by combining the -37 and the -2 to get $-39\sqrt{15}$, which is incorrect. Explain why.

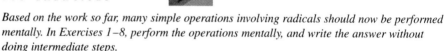

36. There would still be a radical in the denominator. You should multiply by $4 - \sqrt{3}$.
37. $\frac{3 - \sqrt{2}}{7}$ 38. $\frac{4 + \sqrt{3}}{13}$
39. $-4 - 2\sqrt{11}$ 40. $5 + \sqrt{6}$
41. $1 + \sqrt{2}$ 42. $\frac{\sqrt{7} + 1}{6}$
43. $-\sqrt{10} + \sqrt{15}$
44. $-\sqrt{6} + 3$ 45. $3 - \sqrt{3}$
46. $6 + 3\sqrt{2}$ 47. $2\sqrt{5} + \sqrt{15} + 4 + 2\sqrt{3}$
48. $\frac{4\sqrt{7} + \sqrt{35} + 12 + 3\sqrt{5}}{11}$
49. $\sqrt{11} - 2$ 50. $\sqrt{5} - 2$
51. $\frac{\sqrt{3} + 5}{8}$ 52. $\frac{\sqrt{6} + 6}{5}$
53. $\frac{6 - \sqrt{10}}{2}$ 54. $\frac{3 - 2\sqrt{2}}{4}$
55. $x\sqrt{30} + \sqrt{15x} + 6\sqrt{5x} + 3\sqrt{10}$
56. $2y\sqrt{5} - 5\sqrt{2y} - 2\sqrt{10y} + 10$
57. $6t - 3\sqrt{14t} + 2\sqrt{7t} - 7\sqrt{2}$
58. $2z - 2\sqrt{5z} - \sqrt{3z} + \sqrt{15}$
59. $m\sqrt{15} + \sqrt{10mn} - \sqrt{15mn} - n\sqrt{10}$
60. $2p\sqrt{2} + 6\sqrt{pk} - \sqrt{6pk} - 3k\sqrt{3}$
61. $2 - 3\sqrt[3]{4}$ 62. $4\sqrt[3]{25} - 5$
63. $12 + 10\sqrt[4]{8}$ 64. $36 - 18\sqrt[4]{3}$
65. $-1 + 3\sqrt[3]{2} - \sqrt[3]{4}$
66. $-17 - 4\sqrt[3]{9} + 5\sqrt[3]{3}$
67. 1 68. 6

69. $30 + 18x$ 70. They are not like terms. 71. $30 + 18\sqrt{5}$
72. They are not like radicals.
73. Make the first term $30x$, so that $30x + 18x = 48x$; make the first term $30\sqrt{5}$, so that $30\sqrt{5} + 18\sqrt{5} = 48\sqrt{5}$. 74. When combining like terms, we add (or subtract) the coefficients of the common factors of the terms: $2xy + 5xy = 7xy$. When combining like radicals, we add (or subtract) the coefficients of the common radical: $2\sqrt{ab} + 5\sqrt{ab} = 7\sqrt{ab}$.

36. If you try to rationalize the denominator of $\frac{2}{4 + \sqrt{3}}$ by multiplying the numerator and denominator by $4 + \sqrt{3}$, what problem arises? What should you multiply by?

Rationalize each denominator. See Example 3.

37. $\frac{1}{3 + \sqrt{2}}$ **38.** $\frac{1}{4 - \sqrt{3}}$ **39.** $\frac{14}{2 - \sqrt{11}}$ **40.** $\frac{19}{5 - \sqrt{6}}$

41. $\frac{\sqrt{2}}{2 - \sqrt{2}}$ **42.** $\frac{\sqrt{7}}{7 - \sqrt{7}}$ **43.** $\frac{\sqrt{5}}{\sqrt{2} + \sqrt{3}}$ **44.** $\frac{\sqrt{3}}{\sqrt{2} + \sqrt{3}}$

45. $\frac{\sqrt{12}}{\sqrt{3} + 1}$ **46.** $\frac{\sqrt{18}}{\sqrt{2} - 1}$ **47.** $\frac{\sqrt{5} + 2}{2 - \sqrt{3}}$ **48.** $\frac{\sqrt{7} + 3}{4 - \sqrt{5}}$

Write each quotient in lowest terms. See Example 4.

49. $\frac{6\sqrt{11} - 12}{6}$ **50.** $\frac{12\sqrt{5} - 24}{12}$ **51.** $\frac{2\sqrt{3} + 10}{16}$

52. $\frac{4\sqrt{6} + 24}{20}$ **53.** $\frac{12 - \sqrt{40}}{4}$ **54.** $\frac{9 - \sqrt{72}}{12}$

▲ *Simplify each radical expression. Assume all variables represent nonnegative real numbers.*

55. $(\sqrt{5x} + \sqrt{30})(\sqrt{6x} + \sqrt{3})$ **56.** $(\sqrt{10y} - \sqrt{20})(\sqrt{2y} - \sqrt{5})$

57. $(3\sqrt{t} + \sqrt{7})(2\sqrt{t} - \sqrt{14})$ **58.** $(2\sqrt{z} - \sqrt{3})(\sqrt{z} - \sqrt{5})$

59. $(\sqrt{3m} + \sqrt{2n})(\sqrt{5m} - \sqrt{5n})$ **60.** $(\sqrt{4p} - \sqrt{3k})(\sqrt{2p} + \sqrt{9k})$

61. $\sqrt[3]{4}(\sqrt[3]{2} - 3)$ **62.** $\sqrt[3]{5}(4\sqrt[3]{5} - \sqrt[3]{25})$

63. $2\sqrt[4]{2}(3\sqrt[4]{8} + 5\sqrt[4]{4})$ **64.** $6\sqrt[4]{9}(2\sqrt[4]{9} - \sqrt[4]{27})$

65. $(\sqrt[3]{2} - 1)(\sqrt[3]{4} + 3)$ **66.** $(\sqrt[3]{9} + 5)(\sqrt[3]{3} - 4)$

67. $(\sqrt[3]{5} - \sqrt[3]{4})(\sqrt[3]{25} + \sqrt[3]{20} + \sqrt[3]{16})$ **68.** $(\sqrt[3]{4} + \sqrt[3]{2})(\sqrt[3]{16} - \sqrt[3]{8} + \sqrt[3]{4})$

RELATING CONCEPTS (EXERCISES 69–74)

Work Exercises 69–74 in order. *They are designed to help you see why a common student error is indeed an error.*

69. Use the distributive property to write $6(5 + 3x)$ as a sum.

70. Your answer in Exercise 69 should be $30 + 18x$. Why can we not combine these two terms to get $48x$?

71. Repeat Exercise 22 from earlier in this exercise set.

72. Your answer in Exercise 71 should be $30 + 18\sqrt{5}$. Many students will, in error, try to combine these terms to get $48\sqrt{5}$. Why is this wrong?

73. Write the expression similar to $30 + 18x$ that simplifies to $48x$. Then write the expression similar to $30 + 18\sqrt{5}$ that simplifies to $48\sqrt{5}$.

74. Write a short explanation of the similarities between combining like terms and combining like radicals.

Did you make the connection that the procedure used in combining radical terms is the same as that used in combining variable terms?

75. **4 inches**

76. $\dfrac{\sqrt{AP} - P}{P}$; 8%

Solve each problem.

75. The radius of the circular top or bottom of a tin can with a surface area S and a height h is given by

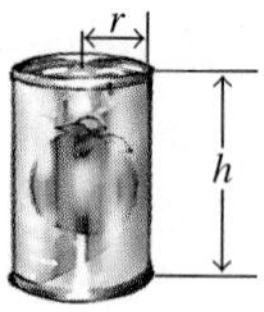

$$r = \frac{-h + \sqrt{h^2 + .64S}}{2}.$$

What radius should be used to make a can with a height of 12 inches and a surface area of 400 square inches?

▲ **76.** If an investment of P dollars grows to A dollars in two years, the annual rate of return on the investment is given by

$$r = \frac{\sqrt{A} - \sqrt{P}}{\sqrt{P}}.$$

First rationalize the denominator and then find the annual rate of return (as a percent) if \$50,000 increases to \$58,320.

9.6 Solving Equations with Radicals

OBJECTIVES

1. Solve equations with radicals.
2. Identify equations with no solutions.
3. Solve equations that require squaring a binomial.

FOR EXTRA HELP

SSG Sec. 9.6
SSM Sec. 9.6
Pass the Test Software
InterAct Math Tutorial Software
Video 15

CONNECTIONS

The most common formula for the area of a triangle is $A = \frac{1}{2}bh$, where b is the length of the base and h is the height. What if the height is not known? What if we know only the lengths of the sides? Another formula, known as *Heron's formula,* allows us to calculate the area of a triangle if we know the lengths of the sides a, b, and c. First let s equal the *semiperimeter,* which is one-half the perimeter.

$$s = \frac{1}{2}(a + b + c)$$

The area A is given by the formula

$$A = \sqrt{s(s - a)(s - b)(s - c)}.$$

For example, the familiar 3–4–5 right triangle has area

$$A = \frac{1}{2}(3)(4) = 6 \text{ square units,}$$

using the familiar formula. Using Heron's formula,

$$s = \frac{1}{2}(3 + 4 + 5) = 6.$$

Therefore,

$$\begin{aligned} A &= \sqrt{6(6 - 3)(6 - 4)(6 - 5)} \\ &= \sqrt{36} \\ &= 6 \end{aligned}$$

So $A = 6$ square units, as expected.

JOURNAL | WRITING | ▲ CHALLENGING | SCIENTIFIC CALCULATOR | GRAPHING CALCULATOR

INSTRUCTOR'S RESOURCES

PTB Sec. 9.6 **ISM** Sec. 9.6
AB Sec. 9.6
TEST GENERATOR
WORLD WIDE WEB www.LialAlgebra.com

CONNECTIONS (CONTINUED)

FOR DISCUSSION OR WRITING

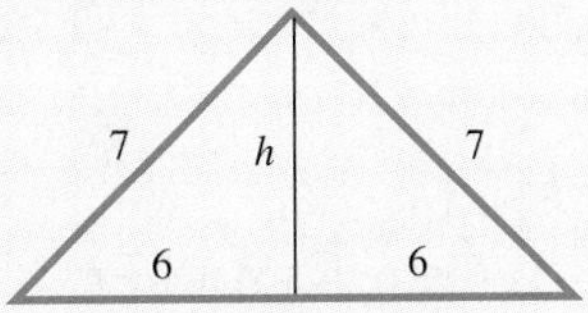

1. $6\sqrt{13} \approx 21.63$ (to the nearest hundredth)
2. $h = \sqrt{13}$; $6\sqrt{13} \approx 21.63$

1. Use Heron's formula to find the area of a triangle with sides 7, 7, and 12.
2. The area of this triangle can be found with the formula $A = \frac{1}{2}bh$ as follows. Divide the triangle into two equal triangles as shown in the figure. Use the Pythagorean formula to find h using one of the small triangles. Note that h is the altitude of the original triangle. Now find the area using the formula $A = \frac{1}{2}bh$. Which way do you prefer?

OBJECTIVE 1 **Solve equations with radicals.** The addition and multiplication properties of equality are not enough to solve an equation with radicals such as

$$\sqrt{x + 1} = 3.$$

Solving equations that have square roots requires a new property, the **squaring property.**

Squaring Property of Equality

If both sides of a given equation are squared, all solutions of the original equation are *among* the solutions of the squared equation.

TEACHING TIP Using the squaring property may produce a linear or a quadratic equation. Some review may be needed here as to the appropriate method to use to solve these equations.

CAUTION

Be very careful with the squaring property. Using this property can give a new equation with *more* solutions than the original equation. For example, starting with the equation $y = 4$ and squaring each side gives

$$y^2 = 4^2, \qquad \text{or} \qquad y^2 = 16.$$

This last equation, $y^2 = 16$, has *two* solutions, 4 or -4, while the original equation, $y = 4$, has only *one* solution, 4. Because of this possibility, checking is more than just a guard against algebraic errors when solving an equation with radicals. It is an essential part of the solution process. *All potential solutions from the squared equation must be checked in the original equation.*

CHALKBOARD EXAMPLE

Solve the equation.

$$\sqrt{9 - y} = 4$$

Answer: $\{-7\}$

EXAMPLE 1 **Using the Squaring Property of Equality**

Solve the equation $\sqrt{p + 1} = 3$.

Use the squaring property of equality to square both sides of the equation and then solve this new equation.

$$(\sqrt{p + 1})^2 = 3^2$$

$$p + 1 = 9 \qquad (\sqrt{p + 1})^2 = p + 1$$

$$p = 8 \qquad \text{Subtract 1.}$$

TEACHING TIP Emphasize that even though the algebraic work may be perfectly done, the answer produced may not make the original equation true.

Now check this answer in the original equation.

$$\sqrt{p+1} = 3$$
$$\sqrt{8+1} = 3 \quad ? \qquad \text{Let } p = 8.$$
$$\sqrt{9} = 3 \quad ?$$
$$3 = 3 \qquad \text{True}$$

Since this statement is true, the solution set of $\sqrt{p+1} = 3$ is $\{8\}$. In this case the squared equation had just one solution, which also satisfied the original equation.

CHALKBOARD EXAMPLE

Solve $\sqrt{3x+9} = 2\sqrt{x}$.

Answer: $\{9\}$

EXAMPLE 2 Using the Squaring Property with Radicals on Each Side

Solve $3\sqrt{x} = \sqrt{x+8}$.

Squaring both sides gives

$$(3\sqrt{x})^2 = (\sqrt{x+8})^2$$
$$3^2(\sqrt{x})^2 = (\sqrt{x+8})^2 \qquad (ab)^2 = a^2b^2$$
$$9x = x + 8 \qquad (\sqrt{x})^2 = x;\ (\sqrt{x+8})^2 = x + 8$$
$$8x = 8 \qquad \text{Subtract } x.$$
$$x = 1. \qquad \text{Divide by 8.}$$

Check this potential solution.

$$3\sqrt{x} = \sqrt{x+8}$$
$$3\sqrt{1} = \sqrt{1+8} \quad ? \qquad \text{Let } x = 1.$$
$$3(1) = \sqrt{9} \quad ? \qquad \sqrt{1} = 1$$
$$3 = 3 \qquad \text{True}$$

The check shows that the solution set of the given equation is $\{1\}$.

TEACHING TIP Distinguish between $(3\sqrt{x})^2$ and $(\sqrt{3x})^2$.

OBJECTIVE 2 **Identify equations with no solutions.** Not all equations with radicals have a solution, as shown in Examples 3 and 4.

CHALKBOARD EXAMPLE

Solve $\sqrt{y} + 4 = 0$.

Answer: $\emptyset$

EXAMPLE 3 Using the Squaring Property When One Side Is Negative

Solve the equation $\sqrt{y} = -3$.

Square both sides.

$$(\sqrt{y})^2 = (-3)^2$$
$$y = 9$$

Check this proposed answer in the original equation.

$$\sqrt{y} = -3$$
$$\sqrt{9} = -3 \quad ? \qquad \text{Let } y = 9.$$
$$3 = -3 \qquad \text{False}$$

Since the statement $3 = -3$ is false, the number 9 is not a solution of the given equation and is said to be *extraneous*. In fact, $\sqrt{y} = -3$ has no solution. Since $\sqrt{y}$ represents the *nonnegative* square root of y, we might have seen immediately that there is no solution. The solution set is $\emptyset$.

Use the following steps when solving an equation with radicals.

Solving an Equation with Radicals

Step 1 **Isolate the radical.** Arrange the terms so that the radical is alone on one side of the equation.

Step 2 **Square both sides.**

Step 3 **Combine like terms.**

Step 4 **Repeat Steps 1–3, if necessary.** If there is still a term with a radical, repeat Steps 1–3.

Step 5 **Solve the equation.** Find all potential solutions.

Step 6 **Check.** All potential solutions must be checked in the original equation.

CHALKBOARD EXAMPLE

Solve $m = \sqrt{m^2 - 4m - 16}$.

Answer: $\emptyset$

EXAMPLE 4 Using the Squaring Property with a Quadratic Expression

Solve $a = \sqrt{a^2 + 5a + 10}$.

Step 1 The radical is already alone on the right side of the equation.

Step 2 Square both sides.

$$a^2 = (\sqrt{a^2 + 5a + 10})^2$$

$$a^2 = a^2 + 5a + 10 \qquad (\sqrt{a^2 + 5a + 10})^2 = a^2 + 5a + 10$$

Step 3 $0 = 5a + 10$ Subtract a^2.

Step 4 This step is not needed.

Step 5 $a = -2$ Subtract 10; divide by 5.

Step 6 Check this proposed solution in the original equation.

$$a = \sqrt{a^2 + 5a + 10}$$

$$-2 = \sqrt{(-2)^2 + 5(-2) + 10} \quad ? \qquad \text{Let } a = -2.$$

$$-2 = \sqrt{4 - 10 + 10} \quad ? \qquad \text{Multiply.}$$

$$-2 = 2 \qquad \text{False}$$

Since $a = -2$ leads to a false result, the equation has no solution, and the solution set is $\emptyset$.

OBJECTIVE 3 Solve equations that require squaring a binomial. The next examples use the following facts from Section 4.4.

$$(a + b)^2 = a^2 + 2ab + b^2 \quad \text{and} \quad (a - b)^2 = a^2 - 2ab + b^2.$$

By the second pattern, for example,

$$(y - 3)^2 = y^2 - 2(y)(3) + (3)^2$$
$$= y^2 - 6y + 9.$$

CHALKBOARD EXAMPLE

Solve the equation $2u - 1 = \sqrt{10u + 9}$.

Answer: {4}

TEACHING TIP Watch for the error $(y - 3)^2 = y^2 + 9$.

EXAMPLE 5 Using the Squaring Property When One Side Has Two Terms

Solve the equation $\sqrt{2y - 3} = y - 3$.

Square each side, using the result above on the right side of the equation.

$$(\sqrt{2y - 3})^2 = (y - 3)^2$$
$$2y - 3 = y^2 - 6y + 9$$

This equation is quadratic because of the y^2-term. To solve it, as shown in Section 5.5, one side of the equation must be equal to 0. Subtract $2y$ and add 3, getting

$$0 = y^2 - 8y + 12.$$
$$0 = (y - 6)(y - 2) \qquad \text{Factor.}$$
$$y - 6 = 0 \quad \text{or} \quad y - 2 = 0 \qquad \text{Zero-factor property}$$
$$y = 6 \quad \text{or} \quad y = 2$$

Check both of these potential solutions in the original equation.

If $y = 6$,

$$\sqrt{2y - 3} = y - 3$$
$$\sqrt{2(6) - 3} = 6 - 3 \quad ?$$
$$\sqrt{12 - 3} = 3 \quad ?$$
$$\sqrt{9} = 3 \quad ?$$
$$3 = 3. \qquad \text{True}$$

If $y = 2$,

$$\sqrt{2y - 3} = y - 3$$
$$\sqrt{2(2) - 3} = 2 - 3 \quad ?$$
$$\sqrt{4 - 3} = -1 \quad ?$$
$$\sqrt{1} = -1 \quad ?$$
$$1 = -1. \qquad \text{False}$$

Only 6 is a solution of the equation, so the solution set is {6}.

Remember to use Step 1 and isolate the radical before squaring both sides. For example, suppose we want to solve $3\sqrt{x} - 1 = 2x$. If we skip Step 1 and square both sides, we get

$$(3\sqrt{x} - 1)^2 = (2x)^2$$
$$9x - 6\sqrt{x} + 1 = 4x^2,$$

a more complicated equation that still contains a radical. This is why we should begin by isolating the radical on one side of the equation. Example 6 shows how this is done.

CHALKBOARD EXAMPLE

Solve the equation

$\sqrt{z + 5} + 2 = z + 5$.

Answer: {−1}

EXAMPLE 6 Rewriting an Equation Before Using the Squaring Property

Solve the equation $3\sqrt{x} - 1 = 2x$.

Add 1 to both sides of the original equation.

$$3\sqrt{x} = 2x + 1$$
$$(3\sqrt{x})^2 = (2x + 1)^2 \qquad \text{Square both sides.}$$
$$9x = 4x^2 + 4x + 1$$
$$0 = 4x^2 - 5x + 1 \qquad \text{Subtract } 9x.$$
$$0 = (4x - 1)(x - 1) \qquad \text{Zero-factor property}$$
$$4x - 1 = 0 \quad \text{or} \quad x - 1 = 0$$
$$x = \frac{1}{4} \quad \text{or} \quad x = 1$$

Check both of these potential solutions in the original equation.

If $x = \frac{1}{4}$,		If $x = 1$,	
$3\sqrt{x} - 1 = 2x$		$3\sqrt{x} - 1 = 2x$	
$3\sqrt{\frac{1}{4}} - 1 = 2\left(\frac{1}{4}\right)$?		$3\sqrt{1} - 1 = 2(1)$?	
$\frac{3}{2} - 1 = \frac{1}{2}$.	True	$3 - 1 = 2$.	True

The solution set of the original equation is $\{\frac{1}{4}, 1\}$.

CAUTION

Errors often occur when each side of an equation is squared. For instance, in Example 6, when both sides of

$$3\sqrt{x} = 2x + 1$$

are squared, the *entire* binomial $2x + 1$ must be squared to get $4x^2 + 4x + 1$. It would be *incorrect* to square the $2x$ and the 1 separately to get $4x^2 + 1$.

Some equations with radicals require squaring twice, as in the next example.

CHALKBOARD EXAMPLE

Solve

$\sqrt{p+1} - \sqrt{p-4} = 1$.

Answer: $\{8\}$

EXAMPLE 7 Using the Squaring Property Twice

Solve $\sqrt{21 + x} = 3 + \sqrt{x}$.

Square both sides.

$$(\sqrt{21 + x})^2 = (3 + \sqrt{x})^2$$
$$21 + x = 9 + 6\sqrt{x} + x$$
$$12 = 6\sqrt{x} \qquad \text{Combine terms; simplify.}$$
$$2 = \sqrt{x}$$
$$4 = x. \qquad \text{Square both sides again.}$$

Check the potential solution.

If $x = 4$,

$$\sqrt{21 + x} = 3 + \sqrt{x} \qquad \text{Original equation}$$
$$\sqrt{21 + 4} = 3 + \sqrt{4} \quad ?$$
$$5 = 5. \qquad \text{True}$$

The solution set is $\{4\}$.

9.6 EXERCISES

1. $\{49\}$ 2. $\{100\}$ 3. $\{7\}$ 4. $\{18\}$ 5. $\{85\}$ 6. $\{21\}$ 7. $\{-45\}$ 8. $\{-16\}$ 9. $\left\{-\frac{3}{2}\right\}$ 10. $\left\{\frac{4}{5}\right\}$

Solve each equation. See Examples 1–4.

1. $\sqrt{x} = 7$

2. $\sqrt{k} = 10$

3. $\sqrt{y + 2} = 3$

4. $\sqrt{x + 7} = 5$

5. $\sqrt{r - 4} = 9$

6. $\sqrt{k - 12} = 3$

7. $\sqrt{4 - t} = 7$

8. $\sqrt{9 - s} = 5$

9. $\sqrt{2t + 3} = 0$

10. $\sqrt{5x - 4} = 0$

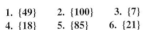

11. $\sqrt{3x - 8} = -2$

12. $\sqrt{6y + 4} = -3$

13. $\sqrt{m} - 4 = 7$

14. $\sqrt{t} + 3 = 10$

15. $\sqrt{10x - 8} = 3\sqrt{x}$

16. $\sqrt{17t - 4} = 4\sqrt{t}$

17. $5\sqrt{x} = \sqrt{10x + 15}$

18. $4\sqrt{y} = \sqrt{20y - 16}$

19. $\sqrt{3x - 5} = \sqrt{2x + 1}$

20. $\sqrt{5y + 2} = \sqrt{3y + 8}$

21. $k = \sqrt{k^2 - 5k - 15}$

22. $s = \sqrt{s^2 - 2s - 6}$

23. $7x = \sqrt{49x^2 + 2x - 10}$

24. $6m = \sqrt{36m^2 + 5m - 5}$

25. Explain how you can tell that the equation $\sqrt{x} = -8$ has no real number solution without performing any algebraic steps.

26. Explain why the equation $x^2 = 36$ has two real number solutions, while the equation $\sqrt{x} = 6$ has only one real number solution.

27. In your own words, explain how to solve a radical equation with a variable radical expression.

28. The first step in solving the equation $\sqrt{2x + 1} = x - 7$ is to square both sides of the equation. Errors often occur when the right side is squared incorrectly. Why is the square of the right side *not* equal to $x^2 + 49$? What is the square of the right side?

Solve each equation. Remember that $(a + b)^2 = a^2 + 2ab + b^2$. (Hint: *Be sure to get the radical term alone on one side of the equation before squaring.) See Examples 5 and 6.*

29. $\sqrt{2x + 1} = x - 7$

30. $\sqrt{3x + 3} = x - 5$

31. $\sqrt{3k + 10} + 5 = 2k$

32. $\sqrt{4t + 13} + 1 = 2t$

33. $\sqrt{5x + 1} - 1 = x$

34. $\sqrt{x + 1} - x = 1$

35. $\sqrt{6t + 7} + 3 = t + 5$

36. $\sqrt{10x + 24} = x + 4$

37. $x - 4 - \sqrt{2x} = 0$

38. $x - 3 - \sqrt{4x} = 0$

39. $\sqrt{x} + 6 = 2x$

40. $\sqrt{k} + 12 = k$

Solve each equation. You will need to square both sides twice. See Example 7.

41. $\sqrt{x + 1} - \sqrt{x - 4} = 1$

42. $\sqrt{2x + 3} + \sqrt{x + 1} = 1$

43. $\sqrt{x} = \sqrt{x - 5} + 1$

44. $\sqrt{2x} = \sqrt{x + 7} - 1$

45. What is wrong with the following "solution"?

$$\sqrt{3x - 6} + \sqrt{x + 2} = 12$$

$3x - 6 + x + 2 = 144$	Square both sides.
$4x - 4 = 144$	Combine terms.
$4x = 148$	Add 4 on both sides.
$x = 37$	Divide by 4.

46. What is wrong with the following "solution"?

$$-\sqrt{x - 1} = -4$$

$-(x - 1) = 16$	Square both sides.
$-x + 1 = 16$	Distributive property
$-x = 15$	Subtract 1 on each side.
$x = -15$	Multiply each side by -1.

Answers

11. $\emptyset$ 12. $\emptyset$ 13. {121} 14. {49} 15. {8} 16. {4} 17. {1} 18. {4} 19. {6} 20. {3} 21. $\emptyset$ 22. $\emptyset$ 23. {5} 24. {1} 25. Since $\sqrt{x}$ must be greater than or equal to zero for any replacement for x, it cannot equal -8, a negative number. 26. $x^2 = 36$ has two solutions because either -6 or 6 squared is 36. However, $\sqrt{x}$ designates the nonnegative square root of x, and there is only one number whose positive square root is 6, namely 36. 27. Arrange the equation so that one radical is alone on one side of the equation; then square both sides. Combine like terms. If there is still a radical term, repeat these steps. Solve the equation, and check all potential solutions in the original equation. 28. The square of the right side is $(x - 7)^2$, which equals $x^2 - 14x + 49$. 29. {12} 30. {11} 31. {5} 32. {3} 33. {0, 3} 34. {−1, 0} 35. {−1, 3} 36. {−2, 4} 37. {8} 38. {9} 39. {4} 40. {16} 41. {8} 42. {−1} 43. {9} 44. {2} 45. We cannot square term by term. The left side must be squared as a binomial in the first step. 46. When the left side is squared, the result should be $x - 1$, not $-(x - 1)$.

47. 158.6 feet **48.** 23.6 miles
49. (a) 70.5 miles per hour (b) 59.8 miles per hour (c) 53.9 miles per hour
50. (a) 149.4 miles (b) 163.7 miles (c) 189.0 miles

Solve each problem. Give answers to the nearest tenth.

▲ **47.** A surveyor wants to find the height of a building. At a point 110.0 feet from the base of the building he sights to the top of the building and finds the distance to be 193.0 feet. See the figure. How high is the building?

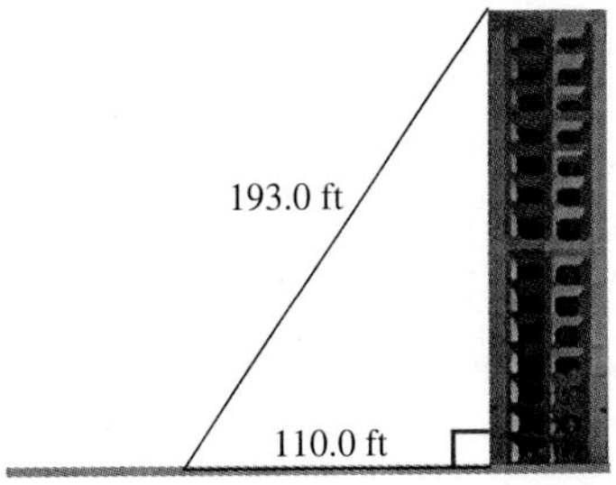

48. Two towns are separated by dense woods. To go from Town B to Town A, it is necessary to travel due west for 19.0 miles, then turn due north and travel for 14.0 miles. See the figure. How far apart are the towns?

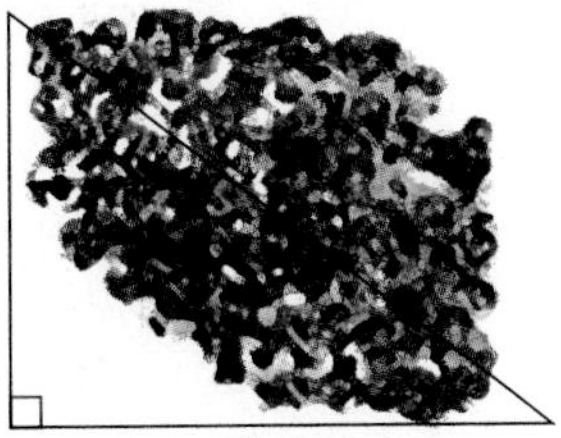

49. Police sometimes use the following procedure to estimate the speed at which a car was traveling at the time of an accident. A police officer drives the car involved in the accident under conditions similar to those during which the accident took place and then skids to a stop. If the car is driven at 30 miles per hour, then the speed at the time of the accident is given by

$$s = 30\sqrt{\frac{a}{p}},$$

where a is the length of the skid marks left at the time of the accident and p is the length of the skid marks in the police test. Find s for the following values of a and p.

(a) $a = 862$ feet; $p = 156$ feet
(b) $a = 382$ feet; $p = 96$ feet
(c) $a = 84$ feet; $p = 26$ feet

50. A formula for calculating the distance, d, one can see from an airplane to the horizon on a clear day is

$$d = 1.22\sqrt{x},$$

where x is the altitude of the plane in feet and d is given in miles. How far can one see to the horizon in a plane flying at the following altitudes? Round to the nearest tenth.

(a) 15,000 feet
(b) 18,000 feet
(c) 24,000 feet

51. (a) 1991 (b) 1997; Here, $f(3) \approx 97.5$. Thus 1997 is the year when about 3 million calls were made. (c) about 2.1 million 52. 6

Solve each problem.

51. In California, the number of 911 emergency calls has risen rapidly to 2.7 million, about a 9000% increase since 1985. The line graph shows the relationship between the number of calls in millions and the years from 1985 to 1997. This relationship is closely approximated by the function defined by the radical equation

$$y = \frac{1.76 + \sqrt{.0176 + .044x}}{.022}.$$

Here, y represents the year when x million calls were made. The years are coded so that 85 represents 1985, 95 represents 1995, and so on.

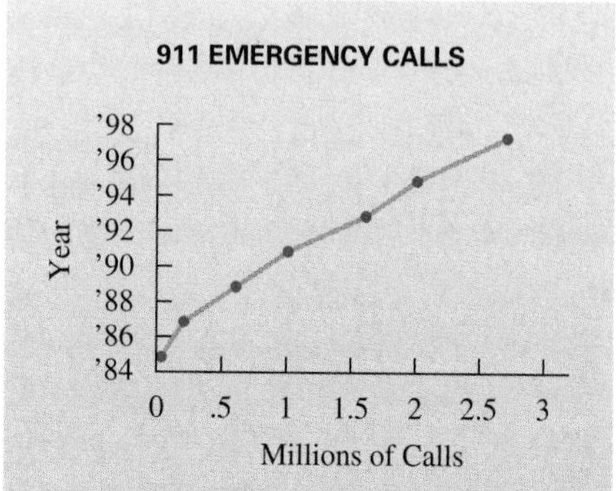

Source: California Highway Patrol.

(a) Find the year when the number of calls reached 1 million.

(b) If $y = f(x)$, find $f(3)$. Interpret your answer in the context of this problem.

(c) Use the equation to determine how many calls were made in 1995.

52. The five top-grossing films from 1995 ticket sales are listed here. The gross receipts y (in millions of dollars) and rank x are closely approximated by

$$x = \frac{9 + \sqrt{81 + 20(187 - y)}}{10}.$$

Note that the domain of x here must be restricted to whole numbers, since it includes possible ranks. Assuming the equation is still a good approximation, what rank would correspond to gross receipts of $60 million?

TOP-GROSSING FILMS

Rank/Film	*Gross (millions)*
1. *Toy Story*	$191.6
2. *Batman Forever*	184.0
3. *Apollo 13*	172.0
4. *Pocahontas*	141.6
5. *Ace Ventura: When Nature Calls*	108.3

Source: Exhibitor Relations Co., Inc.

53. 4 54. 3 55. -2
56. -3 57. -1

TECHNOLOGY INSIGHTS (EXERCISES 53-57)

In earlier chapters we showed how to find the solution of an equation with a graphing calculator by writing the equation with one side equal to 0, replacing 0 with y, then graphing the equation $y = f(x)$. Now we can use this method to solve equations with radicals. For example, to solve $\sqrt{3x + 10} = 4$, rewrite it as $\sqrt{3x + 10} - 4 = 0$, and then as $y = \sqrt{3x + 10} - 4$. The graph of y in the graphing calculator screen on the left shows that $y = 0$ when $x = 2$, so the solution is 2.

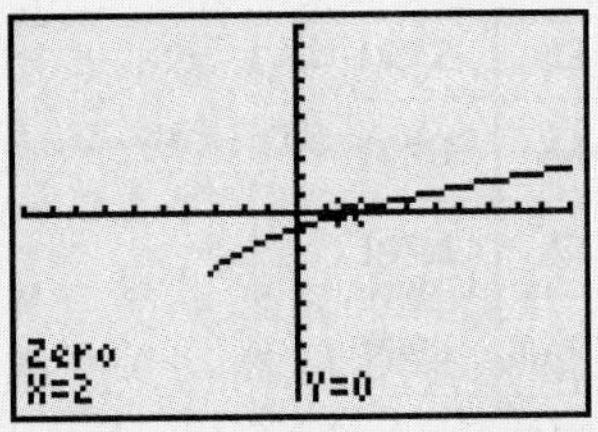

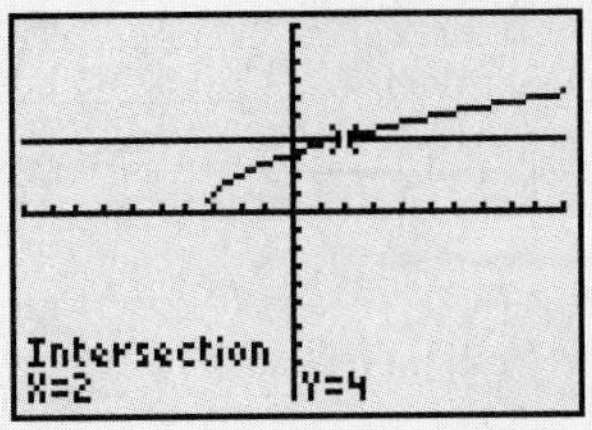

Another method, shown in Chapter 8, is to graph each side of the original equation separately, and look for the point of intersection. In the screen on the right, we have graphed $y_1 = \sqrt{3x + 10}$ and $y_2 = 4$. The value of x at the bottom of the screen again indicates that the solution is 2.

Use the information in each graph to determine the solution of the equation. Check each solution in the equation.

53. $\sqrt{5x - 4} = 4$

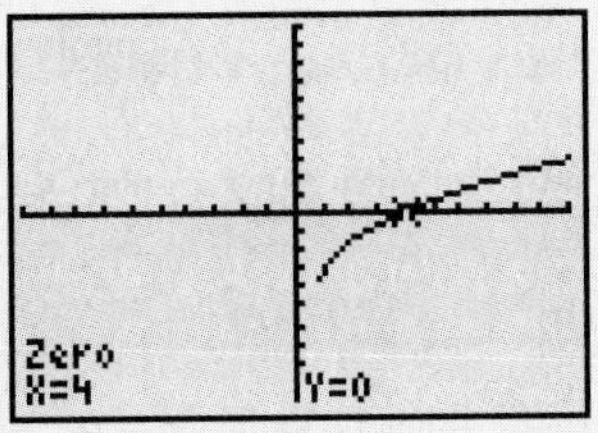

54. $\sqrt{2x + 6} = 2\sqrt{x}$

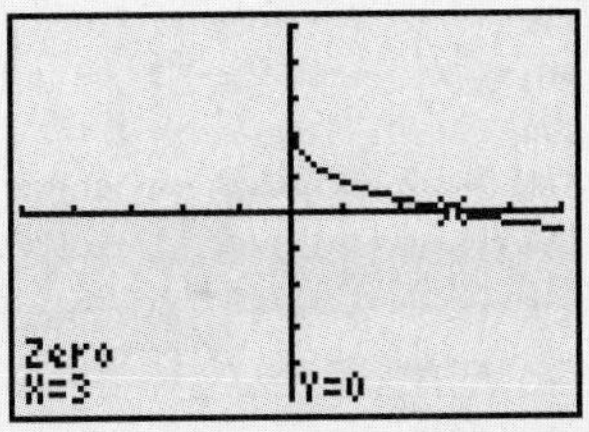

55. $\sqrt{1 - 3x} = \sqrt{5 - x}$

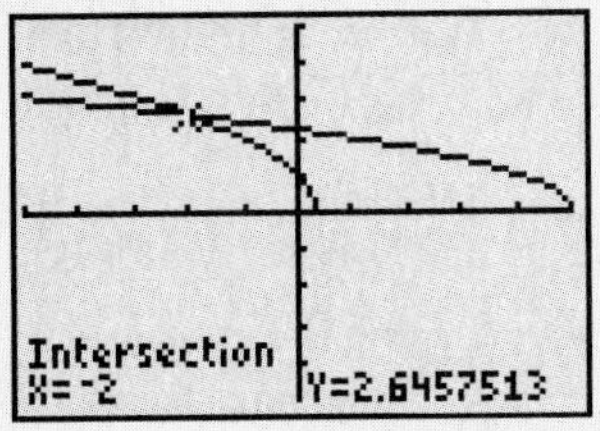

56. $\sqrt{2x^2 + 4x + 3} = -x$

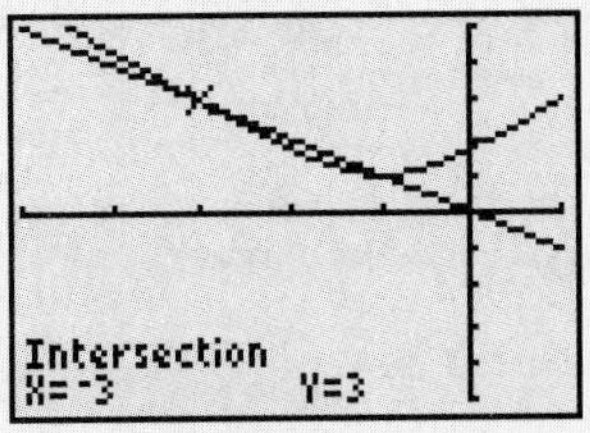

57. There are two solutions indicated in the screen for Exercise 56. They are both integers. Use the graph to determine the one not shown at the bottom of the screen.

9.7 Fractional Exponents

OBJECTIVES

1. Define and use $a^{1/n}$.
2. Define and use $a^{m/n}$.
3. Use rules for exponents with fractional exponents.
4. Use fractional exponents to simplify radicals.

FOR EXTRA HELP

SSG Sec. 9.7
SSM Sec. 9.7
Pass the Test Software
InterAct Math Tutorial Software
Video 15

TEACHING TIP Review the basic rules for exponents:

$$a^m a^n = a^{m+n}$$
$$\frac{a^m}{a^n} = a^{m-n}$$
$$(a^m)^n = a^{mn}.$$

OBJECTIVE 1 Define and use $a^{1/n}$. How should $5^{1/2}$ be defined? We want to define $5^{1/2}$ so that all the rules for exponents developed earlier in this book still hold. Then we should define $5^{1/2}$ so that

$$5^{1/2} \cdot 5^{1/2} = 5^{1/2+1/2} = 5^1 = 5.$$

This agrees with the product rule for exponents from Section 4.2. By definition,

$$(\sqrt{5})(\sqrt{5}) = 5.$$

Since both $5^{1/2} \cdot 5^{1/2}$ and $\sqrt{5} \cdot \sqrt{5}$ equal 5,

$5^{1/2}$ should equal $\sqrt{5}$.

Similarly,

$$5^{1/3} \cdot 5^{1/3} \cdot 5^{1/3} = 5^{1/3+1/3+1/3} = 5^{3/3} = 5,$$

and

$$\sqrt[3]{5} \cdot \sqrt[3]{5} \cdot \sqrt[3]{5} = \sqrt[3]{5^3} = 5,$$

so

$5^{1/3}$ should equal $\sqrt[3]{5}$.

These examples suggest the following definition.

$a^{1/n}$

If a is a nonnegative number and n is a positive integer,

$$a^{1/n} = \sqrt[n]{a}.$$

CHALKBOARD EXAMPLE

Simplify $1000^{1/3}$.

Answer: 10

EXAMPLE 1 Using the Definition of $a^{1/n}$

Simplify each expression by first writing it in radical form.

(a) $16^{1/2}$

By the definition above,

$$16^{1/2} = \sqrt{16} = 4.$$

(b) $27^{1/3} = \sqrt[3]{27} = 3$ **(c)** $64^{1/3} = \sqrt[3]{64} = 4$ **(d)** $64^{1/6} = \sqrt[6]{64} = 2$

OBJECTIVE 2 Define and use $a^{m/n}$. Now a more general exponential expression like $16^{3/4}$ can be defined. By the power rule, $(a^m)^n = a^{mn}$, so that

$$16^{3/4} = (16^{1/4})^3 = (\sqrt[4]{16})^3 = 2^3 = 8.$$

However, $16^{3/4}$ could also be written as

$$16^{3/4} = (16^3)^{1/4} = (4096)^{1/4} = \sqrt[4]{4096} = 8.$$

The expression can be evaluated either way to get the same answer. As the example suggests, taking the root first involves smaller numbers and is often easier. This example suggests the following definition for $a^{m/n}$.

TEACHING TIP Clarify this new idea with several numerical examples.

INSTRUCTOR'S RESOURCES

PTB Sec. 9.7 **ISM** Sec. 9.7
AB Sec. 9.7

TEST GENERATOR

WORLD WIDE WEB
www.LialAlgebra.com

$a^{m/n}$

If a is a nonnegative number and m and n are integers, with $n > 0$,

$$a^{m/n} = (a^{1/n})^m = (\sqrt[n]{a})^m.$$

CHALKBOARD EXAMPLE

Evaluate $-16^{3/4}$.

Answer: -8

EXAMPLE 2 Using the Definition of $a^{m/n}$

Evaluate each expression.

(a) $9^{3/2}$

Use the definition to write

$$9^{3/2} = (9^{1/2})^3 = 3^3 = 27.$$

(b) $64^{2/3} = (64^{1/3})^2 = 4^2 = 16$

(c) $-32^{4/5} = -(32^{1/5})^4 = -2^4 = -16$

Earlier, a^{-n} was defined as

$$a^{-n} = \frac{1}{a^n}$$

for nonzero numbers a and integers n. This same result applies for negative fractional exponents.

$a^{-m/n}$

If a is a positive number and m and n are integers, with $n > 0$,

$$a^{-m/n} = \frac{1}{a^{m/n}}.$$

CHALKBOARD EXAMPLE

Write the expression with a positive exponent and then evaluate.

$$36^{-3/2}$$

Answer: $\frac{1}{216}$

EXAMPLE 3 Using the Definition of $a^{-m/n}$

Write each expression with a positive exponent and then evaluate.

(a) $32^{-3/5} = \dfrac{1}{32^{3/5}} = \dfrac{1}{(32^{1/5})^3} = \dfrac{1}{2^3} = \dfrac{1}{8}$

(b) $27^{-4/3} = \dfrac{1}{27^{4/3}} = \dfrac{1}{(27^{1/3})^4} = \dfrac{1}{3^4} = \dfrac{1}{81}$

TEACHING TIP Emphasize that these rules are the same rules used previously with integer exponents. The rules do not change; only the type of numbers used as exponents changes.

CAUTION

In Example 3(b), a common mistake is to write $27^{-4/3} = -27^{3/4}$. **This is incorrect.** The negative exponent does not indicate a negative number. Also, the negative exponent indicates the reciprocal of the *base,* not the reciprocal of the *exponent.*

OBJECTIVE 3 Use rules for exponents with fractional exponents. All the rules for exponents given earlier still hold when the exponents are fractions. The next examples use these rules to simplify expressions with fractional exponents.

CHALKBOARD EXAMPLE

Simplify. Write the answer in exponential form with only positive exponents.

$$\frac{9^{2/3}}{9^{-1/3}}$$

Answer: 9

EXAMPLE 4 Using the Rules for Exponents with Fractional Exponents

Simplify each expression. Write each answer in exponential form with only positive exponents.

(a) $3^{2/3} \cdot 3^{5/3} = 3^{2/3+5/3} = 3^{7/3}$

(b) $\frac{5^{1/4}}{5^{3/4}} = 5^{1/4-3/4} = 5^{-2/4} = 5^{-1/2} = \frac{1}{5^{1/2}}$

(c) $(9^{1/4})^2 = 9^{2(1/4)} = 9^{2/4} = 9^{1/2} = \sqrt{9} = 3$

(d) $\frac{2^{1/2} \cdot 2^{-1}}{2^{-3/2}} = \frac{2^{1/2+(-1)}}{2^{-3/2}} = \frac{2^{-1/2}}{2^{-3/2}} = 2^{-1/2-(-3/2)} = 2^{2/2} = 2^1 = 2$

(e) $\left(\frac{9}{4}\right)^{5/2} = \frac{9^{5/2}}{4^{5/2}} = \frac{(9^{1/2})^5}{(4^{1/2})^5} = \frac{(\sqrt{9})^5}{(\sqrt{4})^5} = \frac{3^5}{2^5}$

CHALKBOARD EXAMPLE

Simplify. Write the answer in exponential form with only positive exponents. Assume that all variables represent positive numbers.

$$(a^{2/3}b^{1/3}c^2)^6$$

Answer: $a^4b^2c^{12}$

EXAMPLE 5 Using Fractional Exponents with Variables

Simplify each expression. Write each answer in exponential form with only positive exponents. Assume that all variables represent positive numbers.

(a) $m^{1/5} \cdot m^{3/5} = m^{1/5+3/5} = m^{4/5}$

(b) $\frac{p^{5/3}}{p^{4/3}} = p^{5/3-4/3} = p^{1/3}$

(c) $(x^2y^{1/2})^4 = (x^2)^4(y^{1/2})^4 = x^8y^2$

(d) $\left(\frac{z^{1/4}}{w^{1/3}}\right)^5 = \frac{(z^{1/4})^5}{(w^{1/3})^5} = \frac{z^{5/4}}{w^{5/3}}$

(e) $\frac{k^{2/3} \cdot k^{-1/3}}{k^{5/3}} = k^{2/3+(-1/3)-5/3} = k^{-4/3} = \frac{1}{k^{4/3}}$

CAUTION Errors often occur in problems like those in Examples 4 and 5 because students try to convert the expressions to radicals. Remember that the *rules of exponents* apply here.

OBJECTIVE 4 Use fractional exponents to simplify radicals. In fact, sometimes it is easier to simplify a radical by first writing it in exponential form.

CHALKBOARD EXAMPLE

Simplify the radical by first writing it in exponential form.

$$\sqrt[4]{12^2}$$

Answer: $2\sqrt{3}$

EXAMPLE 6 Simplifying Radicals by Using Rational Exponents

Simplify each radical by first writing it in exponential form.

(a) $\sqrt[6]{9^3} = (9^3)^{1/6} = 9^{3/6} = 9^{1/2} = \sqrt{9} = 3$

(b) $(\sqrt[4]{m})^2 = (m^{1/4})^2 = m^{2/4} = m^{1/2} = \sqrt{m}$

Here it is assumed that $m \geq 0$.

CONNECTIONS

Earlier in this chapter we used the rules for adding and multiplying polynomials with radical expressions. In this section we have treated radicals as exponential expressions and applied the rules for exponents. This should not be surprising since radicals are yet another way to represent real numbers, and all the rules developed in this book apply to any real number (with some exceptions). As we mentioned at the beginning of Chapter 4, polynomials are basic algebraic expressions. Our work with radicals just extends the kind of numbers that we use in polynomials.

CONNECTIONS (CONTINUED)

FOR DISCUSSION OR WRITING

Use the methods of Chapter 5 to factor the following expressions.

1. $x - 2\sqrt{x} - 3$

2. $x^2 - 10$

3. $x + 4\sqrt{x} + 4$

4. Divide the polynomial $x^2 + 3x + 5$ by $\sqrt{x}$, and simplify the result.

1. $(\sqrt{x} - 3)(\sqrt{x} + 1)$
2. $(x + \sqrt{10})(x - \sqrt{10})$
3. $(\sqrt{x} + 2)^2$
4. $x\sqrt{x} + 3\sqrt{x} + \dfrac{5\sqrt{x}}{x}$

9.7 EXERCISES

Decide which one of the four choices is not *equal to the given expression.*

1. $49^{1/2}$	**(a)** -7	**(b)** 7	**(c)** $\sqrt{49}$	**(d)** $49^{.5}$
2. $81^{1/2}$	**(a)** 9	**(b)** $\sqrt{81}$	**(c)** $81^{.5}$	**(d)** $\dfrac{81}{2}$
3. $-64^{1/3}$	**(a)** $-\sqrt{16}$	**(b)** -4	**(c)** 4	**(d)** $-\sqrt[3]{64}$
4. $-125^{1/3}$	**(a)** $-\sqrt{25}$	**(b)** -5	**(c)** $-\sqrt[3]{125}$	**(d)** 5

Simplify each expression by first writing it in radical form. See Examples 1–3.

5. $25^{1/2}$ **6.** $121^{1/2}$ **7.** $64^{1/3}$ **8.** $125^{1/3}$ **9.** $16^{1/4}$
10. $81^{1/4}$ **11.** $32^{1/5}$ **12.** $243^{1/5}$ **13.** $4^{3/2}$ **14.** $9^{5/2}$
15. $27^{2/3}$ **16.** $8^{5/3}$ **17.** $16^{3/4}$ **18.** $64^{5/3}$ **19.** $32^{2/5}$
20. $144^{3/2}$ **21.** $-8^{2/3}$ **22.** $-27^{5/3}$ **23.** $-64^{1/3}$ **24.** $-125^{5/3}$
25. $49^{-3/2}$ **26.** $9^{-5/2}$ **27.** $216^{-2/3}$ **28.** $27^{-4/3}$ **29.** $-16^{-5/4}$
30. $-81^{-3/4}$

Simplify each expression. Write answers in exponential form with only positive exponents. Assume that all variables represent positive numbers. See Examples 3–5.

31. $2^{1/2} \cdot 2^{5/2}$ **32.** $5^{2/3} \cdot 5^{4/3}$ **33.** $6^{1/4} \cdot 6^{-3/4}$ **34.** $12^{2/5} \cdot 12^{-1/5}$

35. $\dfrac{15^{3/4}}{15^{5/4}}$ **36.** $\dfrac{7^{3/5}}{7^{-1/5}}$ **37.** $\dfrac{11^{-2/7}}{11^{-3/7}}$ **38.** $\dfrac{4^{-2/3}}{4^{1/3}}$

39. $(8^{3/2})^2$ **40.** $(5^{2/5})^{10}$ **41.** $(6^{1/3})^{3/2}$ **42.** $(7^{2/5})^{5/3}$

43. $\left(\dfrac{25}{4}\right)^{3/2}$ **44.** $\left(\dfrac{8}{27}\right)^{2/3}$ **45.** $\dfrac{2^{2/5} \cdot 2^{-3/5}}{2^{7/5}}$ **46.** $\dfrac{3^{-3/4} \cdot 3^{5/4}}{3^{-1/4}}$

47. $\dfrac{6^{-2/9}}{6^{1/9} \cdot 6^{-5/9}}$ **48.** $\dfrac{8^{6/7}}{8^{2/7} \cdot 8^{-1/7}}$ **49.** $\dfrac{z^{2/3}}{z^{-1/3}}$ **50.** $\dfrac{r^{5/4}}{r^{3/4}}$

51. $(m^3n^{1/4})^{2/3}$ **52.** $(p^4q^{1/2})^{4/3}$ **53.** $\left(\dfrac{a^{1/2}}{b^{1/3}}\right)^{4/3}$ **54.** $\left(\dfrac{m^{2/3}}{n^{3/4}}\right)^{1/2}$

Simplify each radical by first writing it in exponential form. Give the answer as an integer or a radical in simplest form. Assume that all variables represent nonnegative numbers. See Example 6.

55. $\sqrt[6]{4^3}$ **56.** $\sqrt[9]{8^3}$ **57.** $\sqrt[8]{16^2}$ **58.** $\sqrt[9]{27^3}$
59. $\sqrt[4]{a^2}$ **60.** $\sqrt[9]{b^3}$ **61.** $\sqrt[6]{k^4}$ **62.** $\sqrt[8]{m^4}$

1. (a) 2. (d) 3. (c) 4. (d)
5. 5 6. 11 7. 4 8. 5
9. 2 10. 3 11. 2 12. 3
13. 8 14. 243 15. 9
16. 32 17. 8 18. 1024
19. 4 20. 1728 21. -4
22. -243 23. -4
24. -3125 25. $\dfrac{1}{343}$ 26. $\dfrac{1}{243}$
27. $\dfrac{1}{36}$ 28. $\dfrac{1}{81}$ 29. $-\dfrac{1}{32}$
30. $-\dfrac{1}{27}$ 31. 2^3 32. 5^2
33. $\dfrac{1}{6^{1/2}}$ 34. $12^{1/5}$ 35. $\dfrac{1}{15^{1/2}}$
36. $7^{4/5}$ 37. $11^{1/7}$ 38. $\dfrac{1}{4}$
39. 8^3 40. 5^4 41. $6^{1/2}$
42. $7^{2/3}$ 43. $\dfrac{5^3}{2^3}$ 44. $\dfrac{2^2}{3^2}$
45. $\dfrac{1}{2^{8/5}}$ 46. $3^{3/4}$ 47. $6^{2/9}$
48. $8^{5/7}$ 49. z 50. $r^{1/2}$
51. $m^2n^{1/6}$ 52. $p^{16/3}q^{2/3}$
53. $\dfrac{a^{2/3}}{b^{4/9}}$ 54. $\dfrac{m^{1/3}}{n^{3/8}}$ 55. 2
56. 2 57. 2 58. 3
59. $\sqrt{a}$ 60. $\sqrt[3]{b}$ 61. $\sqrt[3]{k^2}$
62. $\sqrt{m}$

63. $\sqrt{2} = 2^{1/2}$ and $\sqrt[3]{2} = 2^{1/3}$
64. $2^{1/2} \cdot 2^{1/3}$ 65. 6
66. $2^{3/6} \cdot 2^{2/6}$ 67. $2^{5/6}$
68. $\sqrt[6]{2^5}$ or $\sqrt[6]{32}$
69. 2 70. 3 71. 1.883
72. 1.361 73. 3.971
74. 2.508 75. 9.100
76. 13.967

RELATING CONCEPTS (EXERCISES 63–68)

The rules for multiplying and dividing radicals presented earlier in this chapter were stated for radicals having the same index. For example, we only multiplied or divided square roots, or multiplied or divided cube roots, and so on. Since we know how to write radicals with fractional exponents and from past work know how to add and subtract fractions with different denominators, we can now multiply and divide radicals having different indexes.

Work Exercises 63–68 in order, *to see how to multiply $\sqrt{2}$ by $\sqrt[3]{2}$.*

63. Write $\sqrt{2}$ and $\sqrt[3]{2}$ using fractional exponents.

64. Write the product $\sqrt{2} \cdot \sqrt[3]{2}$ using the expressions you found in Exercise 63.

65. What is the least common denominator of the fractional exponents in Exercise 64?

66. Repeat Exercise 64, but write the fractional exponents with the common denominator from Exercise 65.

67. Use the rule for multiplying exponential expressions with like bases to simplify the product in Exercise 66. (*Hint:* The base remains the same; do not multiply the two bases.)

68. Write the answer you obtained in Exercise 67 as a radical.

Did you make the connection that when radicals are written in exponential form, the rules for exponents may be used to multiply (or divide) unlike radical terms?

Use the exponential key on your calculator to find the following roots. For example, to find $\sqrt[5]{32}$, enter 32 and then raise to the 1/5 power. (The exponent 1/5 may be entered as .2 if you wish.) If the root is irrational, round it to the nearest thousandth.

69. $\sqrt[6]{64}$ **70.** $\sqrt[5]{243}$ **71.** $\sqrt[7]{84}$ **72.** $\sqrt[9]{16}$

73. $\sqrt[5]{987}$ **74.** $\sqrt[6]{249}$ **75.** $\sqrt[4]{19^3}$ **76.** $\sqrt[5]{27^4}$

TECHNOLOGY INSIGHTS (EXERCISES 77–80)

Each calculator screen gives the simplified real number value (or an approximation) of an exponential expression.

Use a scientific calculator to verify that the radical expression equals the exponential expression.

77.

78.

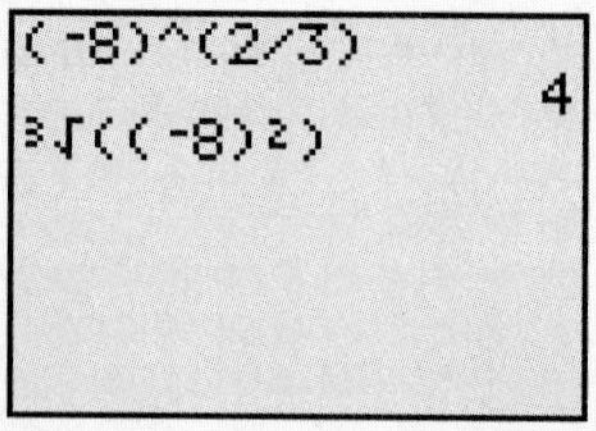

79.

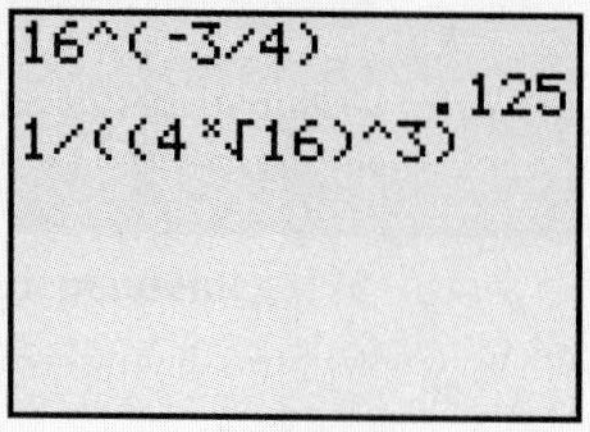

80.

81. (a) $d = 1.22x^{1/2}$
(b) 211.31 miles
82. (a) 57 species **(b)** 858 species **83.** Because $(7^{1/2})^2 = 7$ and $(\sqrt{7})^2 = 7$, they should be equal, so we define $7^{1/2}$ to be $\sqrt{7}$.
84. Because $(7^{1/3})^3 = 7$ and $(\sqrt[3]{7})^3 = 7$, they should be equal, so we define $7^{1/3}$ to be $\sqrt[3]{7}$.

Solve each problem.

▲ **81.** In Section 9.6, Exercise 50, we gave the formula $d = 1.22\sqrt{x}$ for calculating the distance in miles one can see from an airplane to the horizon on a clear day. Here, x is in feet.
(a) Write the formula using a fractional exponent.
(b) Find d to the nearest hundredth if the altitude x is 30,000 feet.

82. A biologist has shown that the number of different plant species S on a Galápagos Island is related to the area of the island, A (in miles), by

$$S = 28.6A^{1/3}.$$

How many plant species would exist on such an island with the following areas?
(a) 8 square miles **(b)** 27,000 square miles

83. Explain in your own words why $7^{1/2}$ is defined as $\sqrt{7}$.

84. Explain in your own words why $7^{1/3}$ is defined as $\sqrt[3]{7}$.

Instructor's Note
Grouping: 2 students
Time: 20 minutes
In this activity students are given various TV screen sizes and are asked to find missing dimensions. Advise them to round the answers up to the next whole number in part A in order to allow more accurate estimations in part B of the activity.

CHAPTER 9 GROUP ACTIVITY

Comparing Television Sizes

Objective: Use the Pythagorean formula to find the dimensions of different television sets.

Television sets are identified by the diagonal measurement of the viewing screen. For example, a 19-inch TV measures 19 inches from one corner of the viewing screen diagonally to the other corner.

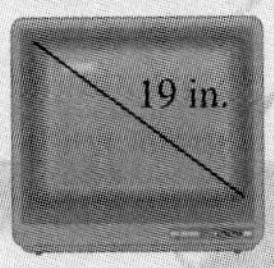

A. See table for answers.

A. The chart below gives some common TV sizes as well as their corresponding widths. Use the Pythagorean formula to find the heights of the viewing screens. Round up to the next whole number. (*Hint:* The TV size is the hypotenuse.)

TV Set	*TV Size*	*Width in Inches*	*Height in Inches*
A	13-inch	10	[9]
B	19-inch	15	[12]
C	25-inch	21	[14]
D	27-inch	22	[16]
E	32-inch	25	[20]
F	36-inch	30	[20]

Source: Data based on information from a Sears catalog.

(continued)

B. 1. set D **2.** set D **3.** set A

B. The dimensions in the table above correspond only to television viewing screens. Each television set also has an outer border of $1\frac{1}{2}$ inches as well as an additional 4 inches for a control panel at the bottom of the set. Also, it is recommended that a minimum of 2 inches of space be allowed above the set for ventilation.

Consider the following entertainment centers, with dimensions for television space as given below. Find the largest television set from the chart in part A that will fit in each entertainment center.

1. 27 inches by 29 inches
2. 28 inches by 25 inches
3. 20 inches by 20 inches

CHAPTER 9 SUMMARY

KEY TERMS

9.1 **square root**
radicand
radical
radical expression
perfect square
cube root
fourth root
index (order)
principal root

9.2 **perfect cube**

9.3 **like radicals**

9.4 **rationalizing the denominator**

9.5 **conjugate**

9.6 **squaring property**

NEW SYMBOLS

$\sqrt{\ }$	radical sign	$\sqrt[3]{a}$	cube root of a	$a^{1/m}$	mth root of a
$\approx$	is approximately equal to	$\sqrt[n]{a}$	nth root of a		

TEST YOUR WORD POWER

See how well you have learned the vocabulary in this chapter. Answers, with examples, are given at the bottom of the next page.

1. The **square root** of a number is
(a) the number raised to the second power
(b) the number under a radical sign
(c) a number that when multiplied by itself gives the original number
(d) the inverse of the number.

2. A **radical** is
(a) a symbol that indicates the nth root
(b) an algebraic expression containing a square root
(c) the positive nth root of a number
(d) a radical sign and the number or expression under it.

3. The **principal root** of a positive number with even index n is
(a) the positive nth root of the number
(b) the negative nth root of the number
(c) the square root of the number
(d) the cube root of the number.

4. Like radicals are
(a) radicals in simplest form
(b) algebraic expressions containing radicals
(c) multiples of the same root of the same number
(d) radicals with the same index.

TEST YOUR WORD POWER

5. Rationalizing the denominator is the process of
- **(a)** eliminating fractions from a radical expression
- **(b)** changing the denominator of a fraction from a radical to a rational number
- **(c)** clearing a radical expression of radicals
- **(d)** multiplying radical expressions.

6. The **conjugate** of $a + b$ is
- **(a)** $a - b$
- **(b)** $a \cdot b$
- **(c)** $a \div b$
- **(d)** $(a + b)^2$.

QUICK REVIEW

CONCEPTS	EXAMPLES
9.1 EVALUATING ROOTS	
If a is a positive real number, $\sqrt{a}$ is the positive square root of a;	$\sqrt{49} = 7$
$-\sqrt{a}$ is the negative square root of a; $\sqrt{0} = 0$.	$-\sqrt{81} = -9$
If a is a negative real number, $\sqrt{a}$ is not a real number.	$\sqrt{-25}$ is not a real number.
If a is a positive rational number, $\sqrt{a}$ is rational if a is a perfect square. $\sqrt{a}$ is irrational if a is not a perfect square.	$\sqrt{\frac{4}{9}}$, $\sqrt{16}$ are rational. $\sqrt{\frac{2}{3}}$, $\sqrt{21}$ are irrational.
Each real number has exactly one real cube root.	$\sqrt[3]{27} = 3$; $\sqrt[3]{-8} = -2$
9.2 MULTIPLICATION AND DIVISION OF RADICALS	
Product Rule for Radicals For nonnegative real numbers x and y, $\sqrt{x} \cdot \sqrt{y} = \sqrt{xy}$ and $\sqrt{xy} = \sqrt{x} \cdot \sqrt{y}$.	$\sqrt{5} \cdot \sqrt{7} = \sqrt{35}$ $\sqrt{8} \cdot \sqrt{2} = \sqrt{16} = 4$ $\sqrt{48} = \sqrt{16} \cdot \sqrt{3} = 4\sqrt{3}$
Quotient Rule for Radicals If x and y are nonnegative real numbers and y is not 0, $\frac{\sqrt{x}}{\sqrt{y}} = \sqrt{\frac{x}{y}}$ and $\sqrt{\frac{x}{y}} = \frac{\sqrt{x}}{\sqrt{y}}$.	$\sqrt{\frac{25}{64}} = \frac{\sqrt{25}}{\sqrt{64}} = \frac{5}{8}$ $\frac{\sqrt{8}}{\sqrt{2}} = \sqrt{\frac{8}{2}} = \sqrt{4} = 2$
If all indicated roots are real, $\sqrt[n]{x} \cdot \sqrt[n]{y} = \sqrt[n]{xy}$ and $\frac{\sqrt[n]{x}}{\sqrt[n]{y}} = \sqrt[n]{\frac{x}{y}}$, $y \neq 0$.	$\sqrt[3]{5} \cdot \sqrt[3]{3} = \sqrt[3]{15}$ $\frac{\sqrt[4]{12}}{\sqrt[4]{4}} = \sqrt[4]{\frac{12}{4}} = \sqrt[4]{3}$

Answers to Test Your Word Power

1. (c) *Examples:* 6 is a square root of 36 since $6^2 = 6 \cdot 6 = 36$; -6 is also a square root of 36. **2.** (d) *Examples:* $\sqrt{144}$, $\sqrt{4xy^2}$, and $\sqrt{4 + t^2}$
3. (a) *Examples:* $\sqrt{36} = 6$, $\sqrt[4]{81} = 3$, and $\sqrt[6]{64} = 2$ **4.** (c) *Examples:* $\sqrt{7}$ and $3\sqrt{7}$ are like radicals; so are $2\sqrt[3]{6k}$ and $5\sqrt[3]{6k}$.
5. (b) *Example:* To rationalize the denominator of $\frac{5}{\sqrt{3} + 1}$, multiply numerator and denominator by $\sqrt{3} - 1$ to get $\frac{5(\sqrt{3} - 1)}{2}$.
6. (a) *Example:* The conjugate of $\sqrt{3} + 1$ above is $\sqrt{3} - 1$.

CONCEPTS	EXAMPLES
9.3 ADDITION AND SUBTRACTION OF RADICALS	
Add and subtract like radicals by using the distributive property. Only like radicals can be combined in this way.	$2\sqrt{5} + 4\sqrt{5} = (2 + 4)\sqrt{5} = 6\sqrt{5}$ $\sqrt{8} + \sqrt{32} = 2\sqrt{2} + 4\sqrt{2} = 6\sqrt{2}$
9.4 RATIONALIZING THE DENOMINATOR	
The denominator of a radical can be rationalized by multiplying both the numerator and denominator by the same number.	$\frac{2}{\sqrt{3}} = \frac{2 \cdot \sqrt{3}}{\sqrt{3} \cdot \sqrt{3}} = \frac{2\sqrt{3}}{3}$ $\sqrt[3]{\frac{5}{6}} = \frac{\sqrt[3]{5}}{\sqrt[3]{6}} \cdot \frac{\sqrt[3]{6^2}}{\sqrt[3]{6^2}} = \frac{\sqrt[3]{180}}{6}$
9.5 SIMPLIFYING RADICAL EXPRESSIONS	
When appropriate, use the rules for adding and multiplying polynomials to simplify radical expressions.	$\sqrt{6}(\sqrt{5} - \sqrt{7}) = \sqrt{30} - \sqrt{42}$ $(\sqrt{5} - \sqrt{3})(\sqrt{5} + \sqrt{3}) = 5 - 3 = 2$
Any denominators with radicals should be rationalized.	$\frac{3}{\sqrt{6}} = \frac{3\sqrt{6}}{6} = \frac{\sqrt{6}}{2}$
If a radical expression contains two terms in the denominator and at least one of those terms is a radical, multiply both the numerator and the denominator by the conjugate of the denominator.	$\frac{6}{\sqrt{7} - \sqrt{2}} = \frac{6}{\sqrt{7} - \sqrt{2}} \cdot \frac{\sqrt{7} + \sqrt{2}}{\sqrt{7} + \sqrt{2}}$ $= \frac{6(\sqrt{7} + \sqrt{2})}{7 - 2}$ Multiply fractions. $= \frac{6(\sqrt{7} + \sqrt{2})}{5}$ Simplify.
9.6 SOLVING EQUATIONS WITH RADICALS	
Solving an Equation with Radicals	Solve $\sqrt{2x - 3} + x = 3$.
Step 1 Arrange the terms so that a radical is alone on one side of the equation.	$\sqrt{2x - 3} = 3 - x$ Isolate radical.
Step 2 Square each side. (By the squaring property of equality, all solutions of the original equation are *among* the solutions of the squared equation.)	$(\sqrt{2x - 3})^2 = (3 - x)^2$ Square. $2x - 3 = 9 - 6x + x^2$
Step 3 Combine like terms.	$0 = x^2 - 8x + 12$ Get one side equal to 0.
Step 4 If there is still a term with a radical, repeat Steps 1–3.	$0 = (x - 2)(x - 6)$ Factor. $x - 2 = 0$ or $x - 6 = 0$ Set each factor equal to 0.
Step 5 Solve the equation for potential solutions.	$x = 2$ or $x = 6$ Solve.
Step 6 Check all potential solutions from Step 5 in the original equation.	Verify that 2 is the only solution (6 is extraneous). Solution set: {2}
9.7 FRACTIONAL EXPONENTS	
Assume $a \geq 0$, m and n are integers, $n > 0$. $a^{1/n} = \sqrt[n]{a}$ $a^{m/n} = \sqrt[n]{a^m} = (\sqrt[n]{a})^m$ $a^{-m/n} = \frac{1}{a^{m/n}} \quad (a \neq 0)$	$8^{1/3} = \sqrt[3]{8} = 2$ $(81)^{3/4} = \sqrt[4]{81^3} = (\sqrt[4]{81})^3 = 3^3 = 27$ $36^{-3/2} = \frac{1}{36^{3/2}} = \frac{1}{(36^{1/2})^3} = \frac{1}{6^3} = \frac{1}{216}$

CHAPTER 9 REVIEW EXERCISES

1. $-7, 7$ 2. $-9, 9$ 3. $-14, 14$
4. $-11, 11$ 5. $-15, 15$
6. $-27, 27$ 7. 4 8. -6
9. 10 10. 3 11. not a real number 12. -65 13. $\frac{7}{6}$
14. $\frac{10}{9}$ 15. a must be negative.
16. 8 17. irrational; 4.796
18. rational; 13 19. rational; -5
20. not a real number 21. $5\sqrt{3}$
22. $-3\sqrt{3}$ 23. $4\sqrt{10}$
24. -5 25. 12 26. 18
27. $16\sqrt{6}$ 28. $25\sqrt{10}$
29. $-\frac{11}{20}$ 30. $\frac{\sqrt{3}}{7}$ 31. $\frac{\sqrt{7}}{13}$
32. $\frac{\sqrt{5}}{6}$ 33. $\frac{2}{15}$ 34. $3\sqrt{2}$
35. 8 36. $2\sqrt{2}$ 37. p
38. $\sqrt{km}$ 39. r^9 40. x^5y^8
41. $a^7b^{10}\sqrt{ab}$ 42. $11x^3y^5$
43. Yes, because both approximations are .7071067812.
44. $9\sqrt{2}$ 45. $21\sqrt{3}$
46. $12\sqrt{3}$ 47. 0 48. $3\sqrt{7}$
49. $2\sqrt{3} + 3\sqrt{10}$ 50. $2\sqrt{2}$
51. $6\sqrt{30}$ 52. $5\sqrt{x}$ 53. 0
54. $-m\sqrt{5}$ 55. $11k^2\sqrt{2n}$

[9.1] *Find all square roots of each number.*

1. 49 **2.** 81 **3.** 196 **4.** 121 **5.** 225 **6.** 729

Find each indicated root. If the root is not a real number, say so.

7. $\sqrt{16}$ **8.** $-\sqrt{36}$ **9.** $\sqrt[3]{1000}$ **10.** $\sqrt[4]{81}$

11. $\sqrt{-8100}$ **12.** $-\sqrt{4225}$ **13.** $\sqrt{\frac{49}{36}}$ **14.** $\sqrt{\frac{100}{81}}$

15. If $\sqrt{a}$ is not a real number, then what kind of number must a be?

16. Find the value of x in the figure.

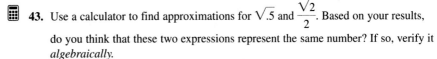

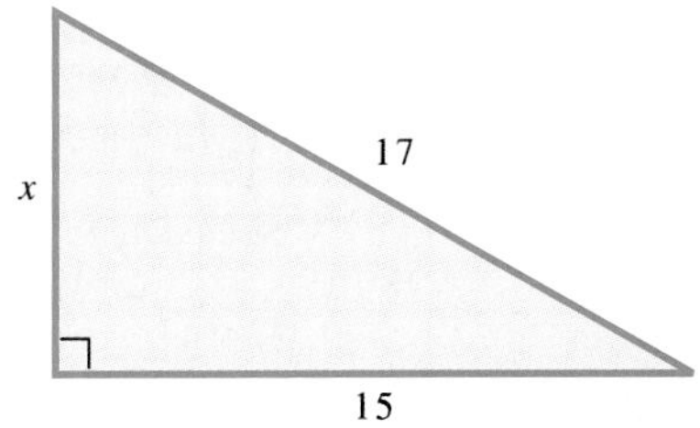

Determine whether each number is rational, irrational, *or* not a real number. *If the number is rational, give its exact value. If the number is irrational, give a decimal approximation rounded to the nearest thousandth.*

17. $\sqrt{23}$ **18.** $\sqrt{169}$ **19.** $-\sqrt{25}$ **20.** $\sqrt{-4}$

[9.2] *Use the product rule to simplify each expression.*

21. $\sqrt{5} \cdot \sqrt{15}$ **22.** $-\sqrt{27}$ **23.** $\sqrt{160}$ **24.** $\sqrt[3]{-125}$

25. $\sqrt[3]{1728}$ **26.** $\sqrt{12} \cdot \sqrt{27}$ **27.** $\sqrt{32} \cdot \sqrt{48}$ **28.** $\sqrt{50} \cdot \sqrt{125}$

Use the product rule, the quotient rule, or both to simplify each expression.

29. $-\sqrt{\frac{121}{400}}$ **30.** $\sqrt{\frac{3}{49}}$ **31.** $\sqrt{\frac{7}{169}}$ **32.** $\sqrt{\frac{1}{6}} \cdot \sqrt{\frac{5}{6}}$

33. $\sqrt{\frac{2}{5}} \cdot \sqrt{\frac{2}{45}}$ **34.** $\frac{3\sqrt{10}}{\sqrt{5}}$ **35.** $\frac{24\sqrt{12}}{6\sqrt{3}}$ **36.** $\frac{8\sqrt{150}}{4\sqrt{75}}$

Simplify each expression. Assume that all variables represent nonnegative real numbers.

37. $\sqrt{p} \cdot \sqrt{p}$ **38.** $\sqrt{k} \cdot \sqrt{m}$ **39.** $\sqrt{r^{18}}$

40. $\sqrt{x^{10}y^{16}}$ **41.** $\sqrt{a^{15}b^{21}}$ **42.** $\sqrt{121x^6y^{10}}$

43. Use a calculator to find approximations for $\sqrt{.5}$ and $\frac{\sqrt{2}}{2}$. Based on your results, do you think that these two expressions represent the same number? If so, verify it *algebraically.*

[9.3] *Simplify and combine terms where possible.*

44. $3\sqrt{2} + 6\sqrt{2}$ **45.** $3\sqrt{75} + 2\sqrt{27}$ **46.** $4\sqrt{12} + \sqrt{48}$

47. $4\sqrt{24} - 3\sqrt{54} + \sqrt{6}$ **48.** $2\sqrt{7} - 4\sqrt{28} + 3\sqrt{63}$ **49.** $\frac{2}{5}\sqrt{75} + \frac{3}{4}\sqrt{160}$

50. $\frac{1}{3}\sqrt{18} + \frac{1}{4}\sqrt{32}$ **51.** $\sqrt{15} \cdot \sqrt{2} + 5\sqrt{30}$

Simplify each expression. Assume that all variables represent nonnegative real numbers.

52. $\sqrt{4x} + \sqrt{36x} - \sqrt{9x}$ **53.** $\sqrt{16p} + 3\sqrt{p} - \sqrt{49p}$

54. $\sqrt{20m^2} - m\sqrt{45}$ **55.** $3k\sqrt{8k^2n} + 5k^2\sqrt{2n}$

56. $\frac{8\sqrt{10}}{5}$ 57. $\sqrt{5}$ 58. $\sqrt{6}$
59. $\frac{\sqrt{30}}{15}$ 60. $\frac{\sqrt{10}}{5}$ 61. $\sqrt{10}$
62. $\frac{\sqrt{42}}{21}$ 63. $\frac{r\sqrt{x}}{4x}$ 64. $\frac{\sqrt[3]{9}}{3}$
65. $\frac{\sqrt[3]{98}}{7}$

66. To show this, simplify the second expression using the guidelines given in Section 9.5.
$\sqrt{\frac{48}{128}} = \frac{\sqrt{48}}{\sqrt{128}} = \frac{\sqrt{16 \cdot 3}}{\sqrt{64 \cdot 2}} = \frac{4\sqrt{3}}{8\sqrt{2}} = \frac{4\sqrt{3} \cdot \sqrt{2}}{8\sqrt{2} \cdot \sqrt{2}} = \frac{4\sqrt{6}}{8 \cdot 2} = \frac{\sqrt{6}}{4}$

67. $-\sqrt{15} - 9$ 68. $3\sqrt{6} + 12$
69. $22 - 16\sqrt{3}$ 70. $179 + 20\sqrt{7}$
71. -2 72. -13
73. $-2 + \sqrt{5}$ 74. $\frac{-2\sqrt{2} - 6}{7}$
75. $\frac{-2 + 6\sqrt{2}}{17}$ 76. $\frac{-\sqrt{3} + 3}{2}$
77. $\frac{-\sqrt{10} + 3\sqrt{5} + \sqrt{2} - 3}{7}$
78. $\frac{2\sqrt{3} + 2 + 3\sqrt{2} + \sqrt{6}}{2}$
79. $\frac{3 + 2\sqrt{6}}{3}$ 80. $\frac{1 + 3\sqrt{7}}{4}$
81. $3 + 4\sqrt{3}$ 82. $\{25\}$
83. $\emptyset$ 84. $\{48\}$ 85. $\{1\}$
86. $\{2\}$ 87. $\{6\}$
88. $\{-2\}$ 89. $\{-2\}$
90. $\{4\}$ 91. (a) billions of dollars in exports (b) years (c) 1992 (d) Yes, because $13 billion is between $7.5 billion and $19.6 billion, so the year should be between 1990 and 1993.
(e) $31.2 billion; yes

[9.4] *Perform the indicated operations and write answers in simplest form. Assume that all variables represent positive real numbers.*

56. $\frac{8\sqrt{2}}{\sqrt{5}}$ **57.** $\frac{5}{\sqrt{5}}$ **58.** $\frac{12}{\sqrt{24}}$ **59.** $\frac{\sqrt{2}}{\sqrt{15}}$ **60.** $\sqrt{\frac{2}{5}}$

61. $\sqrt{\frac{5}{14}} \cdot \sqrt{28}$ **62.** $\sqrt{\frac{2}{7}} \cdot \sqrt{\frac{1}{3}}$ **63.** $\sqrt{\frac{r^2}{16x}}$ **64.** $\sqrt[3]{\frac{1}{3}}$ **65.** $\sqrt[3]{\frac{2}{7}}$

66. Explain how you would show, without using a calculator, that $\frac{\sqrt{6}}{4}$ and $\sqrt{\frac{48}{128}}$ represent the exact same number. Then actually perform the necessary steps.

[9.5] *Simplify each expression.*

67. $-\sqrt{3}(\sqrt{5} + \sqrt{27})$ **68.** $3\sqrt{2}(\sqrt{3} + 2\sqrt{2})$

69. $(2\sqrt{3} - 4)(5\sqrt{3} + 2)$ **70.** $(5\sqrt{7} + 2)^2$

71. $(\sqrt{5} - \sqrt{7})(\sqrt{5} + \sqrt{7})$ **72.** $(2\sqrt{3} + 5)(2\sqrt{3} - 5)$

Rationalize each denominator.

73. $\frac{1}{2 + \sqrt{5}}$ **74.** $\frac{2}{\sqrt{2} - 3}$ **75.** $\frac{\sqrt{8}}{\sqrt{2} + 6}$

76. $\frac{\sqrt{3}}{1 + \sqrt{3}}$ **77.** $\frac{\sqrt{5} - 1}{\sqrt{2} + 3}$ **78.** $\frac{2 + \sqrt{6}}{\sqrt{3} - 1}$

Write each quotient in lowest terms.

79. $\frac{15 + 10\sqrt{6}}{15}$ **80.** $\frac{3 + 9\sqrt{7}}{12}$ **81.** $\frac{6 + \sqrt{192}}{2}$

[9.6] *Solve each equation.*

82. $\sqrt{m} - 5 = 0$ **83.** $\sqrt{p} + 4 = 0$ **84.** $\sqrt{k + 1} = 7$

85. $\sqrt{5m + 4} = 3\sqrt{m}$ **86.** $\sqrt{2p + 3} = \sqrt{5p - 3}$ **87.** $\sqrt{4y + 1} = y - 1$

88. $\sqrt{-2k - 4} = k + 2$ **89.** $\sqrt{2 - x} + 3 = x + 7$ **90.** $\sqrt{x} - x + 2 = 0$

Work each problem.

91. Refer to Section 9.3, Exercise 57.
(a) Describe the elements in the domain.
(b) Describe the elements in the range.
(c) According to the equation there, $y = 1.4\sqrt{x - 2.5} + 87.5$, in what year were exports about $13 billion?
(d) Does your answer to part (c) seem reasonable, compared to the actual data repeated here? Explain.

U.S. Exports

Billions of Dollars (x)	Year (y)
7.5	1990
19.6	1993
25.8	1994
36.4	1996

(e) Use the equation from part (c) to find the exports in 1995. (*Hint:* Let $y = 95$ and solve for x.) Does your answer seem reasonable compared to the actual data in the table?

92. $f(x) = \dfrac{-100x^2 + 180x + 3740}{20}$

93. 9 94. −5 95. 7^3 or 343

96. $13^{7/5}$ 97. $x^{3/4}$ 98. 7

92. Refer to Section 9.6, Exercise 52, where $y = f(x)$ is equivalent to the equation

$$x = \frac{9 + \sqrt{81 + 20(187 - y)}}{10}.$$

Find $f(x)$. (*Hint:* Solve the equation for y, and replace y with $f(x)$.)

[9.7] *Simplify each expression. Assume that all variables represent positive real numbers.*

93. $81^{1/2}$

94. $-125^{1/3}$

95. $7^{2/3} \cdot 7^{7/3}$

96. $\dfrac{13^{4/5}}{13^{-3/5}}$

97. $\dfrac{x^{1/4} \cdot x^{5/4}}{x^{3/4}}$

98. $\sqrt[8]{49^4}$

MIXED REVIEW EXERCISES

99. 16 100. $11\sqrt{3}$

101. $\dfrac{5 - \sqrt{2}}{23}$ 102. $\dfrac{2\sqrt{10}}{5}$

103. $5y\sqrt{2}$ 104. −5

105. $-\sqrt{10} - 5\sqrt{15}$

106. $\dfrac{4r\sqrt{3rs}}{3s}$ 107. $\dfrac{2 + \sqrt{13}}{2}$

108. $-7\sqrt{2}$ 109. $7 - 2\sqrt{10}$

110. $166 + 2\sqrt{7}$ 111. −11

112. x^2 113. {7} 114. ∅

115. {8}

Simplify each expression if possible. Assume all variables represent positive real numbers.

99. $64^{2/3}$

100. $2\sqrt{27} + 3\sqrt{75} - \sqrt{300}$

101. $\dfrac{1}{5 + \sqrt{2}}$

102. $\sqrt{\dfrac{1}{3}} \cdot \sqrt{\dfrac{24}{5}}$

103. $\sqrt{50y^2}$

104. $\sqrt[3]{-125}$

105. $-\sqrt{5}(\sqrt{2} + \sqrt{75})$

106. $\sqrt{\dfrac{16r^3}{3s}}$

107. $\dfrac{12 + 6\sqrt{13}}{12}$

108. $-\sqrt{162} + \sqrt{8}$

109. $(\sqrt{5} - \sqrt{2})^2$

110. $(6\sqrt{7} + 2)(4\sqrt{7} - 1)$

111. $-\sqrt{121}$

112. $\dfrac{x^{8/3}}{x^{2/3}}$

Solve.

113. $\sqrt{x + 2} = x - 4$

114. $\sqrt{k} + 3 = 0$

115. $\sqrt{1 + 3t} - t = -3$

RELATING CONCEPTS (EXERCISES 116–120)

116. $\dfrac{10\sqrt{7} - 5\sqrt{7}}{-3\sqrt{14} - 2\sqrt{14}}$ or $\dfrac{5\sqrt{7} - 10\sqrt{7}}{2\sqrt{14} + 3\sqrt{14}}$

117. $-\dfrac{5\sqrt{7}}{5\sqrt{14}}$

118. $-\sqrt{\dfrac{1}{2}}$ 119. $-\dfrac{\sqrt{2}}{2}$

120. It falls from left to right.

In Chapter 3 we plotted points in the rectangular coordinate plane. In all cases our points had coordinates that were rational numbers. However, ordered pairs may have irrational coordinates as well.

Use your knowledge of the material in Chapters 3 and 9 to ***work Exercises 116–120 in order.***

Consider the points $A(2\sqrt{14}, 5\sqrt{7})$ *and* $B(-3\sqrt{14}, 10\sqrt{7})$.

116. Write an expression that represents the slope of the line containing points A and B. Do not simplify yet.

117. Simplify the numerator and the denominator in the expression from Exercise 116 by combining like radicals.

118. Write the fraction from Exercise 117 as the square root of a fraction in lowest terms.

119. Rationalize the denominator of the expression found in Exercise 118.

120. Based on your answer in Exercise 119, does line AB rise or fall from left to right?

Did you make the connection that the real number properties discussed up to this point apply to irrational numbers as well as rational numbers?

CHAPTER 9 TEST

[9.1] 1. $-14, 14$
2. (a) irrational **(b)** 11.916
[9.2–9.5] 3. 6
4. $-3\sqrt{3}$ **5.** $\frac{8\sqrt{2}}{5}$ **6.** $2\sqrt[3]{4}$
7. $4\sqrt{6}$ **8.** $9\sqrt{7}$ **9.** $-5\sqrt{3x}$
10. $2y\sqrt[3]{4x^2}$ **11.** 31
12. $6\sqrt{2} + 2 - 3\sqrt{14} - \sqrt{7}$
13. $11 + 2\sqrt{30}$
[9.1] 14. (a) $6\sqrt{2}$ inches
(b) 8.485 inches
[9.4] 15. $\frac{5\sqrt{14}}{7}$ **16.** $\frac{\sqrt{6x}}{3x}$
17. $-\sqrt[3]{2}$ **18.** $\frac{-12 - 3\sqrt{3}}{13}$
[9.6] 19. $\{3\}$ **20.** $\left\{\frac{1}{4}, 1\right\}$
[9.7] 21. 16 **22.** -25 **23.** 5
24. $\frac{1}{3}$ **[9.6] 25.** 12 is not a solution. A check shows that it does not satisfy the original equation.

On this test assume that all variables represent positive real numbers.

1. Find all square roots of 196.

2. Consider $\sqrt{142}$.
(a) Determine whether it is rational or irrational.
(b) Find a decimal approximation to the nearest thousandth.

3. Simplify $\sqrt[3]{216}$.

Simplify where possible.

4. $-\sqrt{27}$

5. $\sqrt{\frac{128}{25}}$

6. $\sqrt[3]{32}$

7. $\frac{20\sqrt{18}}{5\sqrt{3}}$

8. $3\sqrt{28} + \sqrt{63}$

9. $3\sqrt{27x} - 4\sqrt{48x} + 2\sqrt{3x}$

10. $\sqrt[3]{32x^2y^3}$

11. $(6 - \sqrt{5})(6 + \sqrt{5})$

12. $(2 - \sqrt{7})(3\sqrt{2} + 1)$

13. $(\sqrt{5} + \sqrt{6})^2$

14. The hypotenuse of a right triangle measures 9 inches, and one leg measures 3 inches. Find the measure of the other leg.
(a) Give its length in simplified radical form.
(b) Round the answer to the nearest thousandth.

Rationalize each denominator.

15. $\frac{5\sqrt{2}}{\sqrt{7}}$

16. $\sqrt{\frac{2}{3x}}$

17. $\frac{-2}{\sqrt[3]{4}}$

18. $\frac{-3}{4 - \sqrt{3}}$

Solve each equation.

19. $\sqrt{x + 1} = 5 - x$

20. $3\sqrt{x} - 1 = 2x$

Simplify each expression.

21. $8^{4/3}$

22. $-125^{2/3}$

23. $5^{3/4} \cdot 5^{1/4}$

24. $\frac{(3^{1/4})^3}{3^{7/4}}$

25. What is wrong with the following "solution"?

$$\sqrt{2x + 1} + 5 = 0$$
$$\sqrt{2x + 1} = -5 \quad \text{Subtract 5.}$$
$$2x + 1 = 25 \quad \text{Square both sides.}$$
$$2x = 24 \quad \text{Subtract 1.}$$
$$x = 12 \quad \text{Divide by 2.}$$

The solution set is $\{12\}$.

CUMULATIVE REVIEW EXERCISES CHAPTERS 1–9

[1.5–1.6] 1. 54 **2.** 6 **3.** 3
4. 18 **5.** 15 **6.** 4.223

Simplify each expression.

1. $3(6 + 7) + 6 \cdot 4 - 3^2$

2. $\frac{3(6 + 7) + 3}{2(4) - 1}$

3. $|-6| - |-3|$

4. $-9 + 14 + 11 + (-3 + 5)$

5. $13 - [-4 - (-2)]$

6. $-2.523 + 8.674 - 1.928$

[2.2] 7. $\{3\}$ **[2.7] 8.** $[-16, \infty)$
9. $(5, \infty)$ **[2.4] 10.** **207 cubic inches**
[3.2] 11.

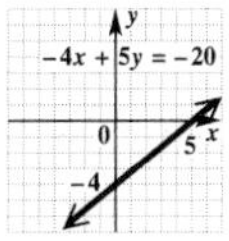

12.

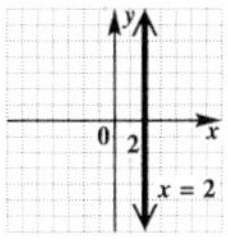

[3.3] 13. $-\frac{5}{6}$ **[4.2] 14.** $12x^{10}y^2$
[4.5] 15. $\frac{y^{15}}{5832}$
[4.1] 16. $3x^3 + 11x^2 - 13$
[4.6] 17. $4t^2 - 8t + 5$
[5.2–5.4] 18. $(m + 8)(m + 4)$
19. $(5t^2 + 6)(5t^2 - 6)$
20. $(6a + 5b)(2a - b)$
21. $(9z + 4)^2$ **[5.5] 22.** $\{3, 4\}$
23. $\{-2, -1\}$ **[6.1] 24.** $2, -7$
[6.2] 25. $\frac{x + 1}{x}$
26. $(t + 5)(t + 3)$
[6.4] 27. $\frac{y^2}{(y + 1)(y - 1)}$
28. $\frac{-2x - 14}{(x + 3)(x - 1)}$ **[6.5] 29.** -21
[7.2] 30.

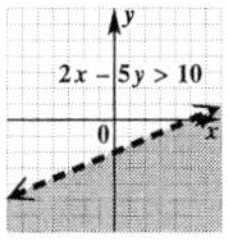

[8.1–8.3] 31. $\{(3, -7)\}$
32. **infinite number of solutions**
[8.4] 33. **from Chicago: 57 mph; from Des Moines: 50 mph**

Solve each equation or inequality.

7. $5(k - 4) - k = k - 11$

8. $-\frac{3}{4}y \leq 12$

9. $5z + 3 - 4 > 2z + 9 + z$

Solve the problem.

10. A Macintosh Powerbook 1400C computer has dimensions 9 inches by 11.5 inches by 2 inches. Use the formula $V = LWH$ to find its volume (in cubic inches).

Graph each equation in the rectangular coordinate plane.

11. $-4x + 5y = -20$

12. $x = 2$

13. Find the slope of the line through the points $(9, -2)$ and $(-3, 8)$.

Simplify and write each expression without negative exponents. Assume that variables represent positive real numbers.

14. $(3x^6)(2x^2y)^2$

15. $\left(\frac{3^2y^{-2}}{2^{-1}y^3}\right)^{-3}$

16. Subtract $7x^3 - 8x^2 + 4$ from $10x^3 + 3x^2 - 9$.

17. Divide: $\frac{8t^3 - 4t^2 - 14t + 15}{2t + 3}$.

Factor each polynomial completely.

18. $m^2 + 12m + 32$

19. $25t^4 - 36$

20. $12a^2 + 4ab - 5b^2$

21. $81z^2 + 72z + 16$

Solve each quadratic equation.

22. $x^2 - 7x = -12$

23. $(x + 4)(x - 1) = -6$

24. For what real number(s) is the expression $\frac{3}{x^2 + 5x - 14}$ undefined?

Perform the indicated operations. Express each answer in lowest terms.

25. $\frac{x^2 - 3x - 4}{x^2 + 3x} \cdot \frac{x^2 + 2x - 3}{x^2 - 5x + 4}$

26. $\frac{t^2 + 4t - 5}{t + 5} \div \frac{t - 1}{t^2 + 8t + 15}$

27. $\frac{y}{y^2 - 1} + \frac{y}{y + 1}$

28. $\frac{2}{x + 3} - \frac{4}{x - 1}$

29. Simplify the complex fraction: $\dfrac{\frac{2}{3} + \frac{1}{2}}{\frac{1}{9} - \frac{1}{6}}$.

30. Graph $2x - 5y > 10$.

Solve each system of equations.

31. $4x - y = 19$
$3x + 2y = -5$

32. $2x - y = 6$
$3y = 6x - 18$

Solve each problem.

33. Des Moines and Chicago are 321 miles apart. Two cars start from these cities traveling toward each other. They meet after 3 hours. The car from Chicago has an average speed 7 miles per hour faster than the other car. Find the average speed of each car.

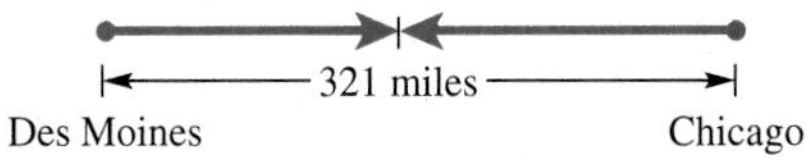

34. CNN: 67.8 million; ESPN: 67.9 million
[9.3–9.5] 35. $29\sqrt{3}$
36. $-\sqrt{3} + \sqrt{5}$
37. $10xy^2\sqrt{2y}$ [9.7] 38. 32
[9.5] 39. $21 - 5\sqrt{2}$
[9.6] 40. {16}

34. The top two cable television networks in 1995 were ESPN and CNN. ESPN had .1 million more subscribers than CNN. Together, the two networks had 135.7 million subscribers. How many subscribers did each network have?

Simplify each expression if possible. Assume all variables represent nonnegative real numbers.

35. $\sqrt{27} - 2\sqrt{12} + 6\sqrt{75}$

36. $\dfrac{2}{\sqrt{3} + \sqrt{5}}$

37. $\sqrt{200x^2y^5}$

38. $16^{5/4}$

39. $(3\sqrt{2} + 1)(4\sqrt{2} - 3)$

40. Solve the equation $\sqrt{x} + 2 = x - 10$.

Quadratic Equations 10

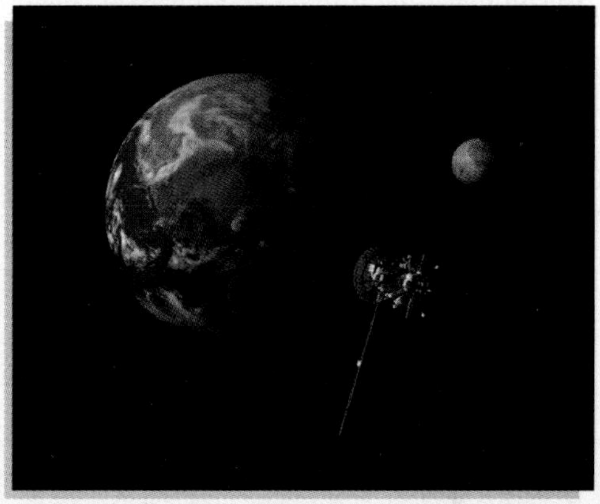

Astronomy

Quadratic equations and quadratic functions are important in the field of astronomy. When an object moves under the influence of a constant force, its path is *parabolic* (has the shape of a parabola), and its position is described by a *quadratic equation* $y = ax^2 + bx$. Here x is in seconds, y is in feet, and a and b are real numbers. This equation varies on the surface of different planets and moons. The table compares the heights of a toy rocket propelled upward at 60 miles per hour (88 feet per second) at an angle of 60° on Earth, our moon, and Mars (as x varies from 1 to 5 seconds).

Seconds	*Height (y, in feet)*		
(x)	*Earth*	*Moon*	*Mars*
1	1.724	1.726	1.724
2	3.431	3.449	3.442
3	5.121	5.170	5.154
4	6.795	6.888	6.860
5	8.453	8.604	8.560

Source: M. Zeilik, S. Gregory, and E. Smith, *Introductory Astronomy and Astrophysics*, Saunders College Publishers, 1992.

The familiar force referred to earlier is gravity. Gravitational force is different on Earth, our moon, and Mars. It is less powerful on our moon than on Earth, and therefore, with all other conditions being equal, the rocket will rise faster on the moon than on our home planet. How does the table support this statement? Look at the figures and surmise how gravity on Mars compares to gravity on Earth and on our moon. We examine these ideas in the exercises for Section 10.1.

10.1 Solving Quadratic Equations by the Square Root Property

OBJECTIVES

1. Solve equations of the form $x^2 =$ a number.
2. Solve equations of the form $(ax + b)^2 =$ a number.

FOR EXTRA HELP

SSG Sec. 10.1
SSM Sec. 10.1
Pass the Test Software
InterAct Math Tutorial Software
Video 16

TEACHING TIP Explain to students that until now, all quadratic equations have been carefully selected so that they could be solved by factoring. The methods in the following sections will enable us to solve *all* quadratic equations, including those that cannot be solved by factoring.

In Chapter 5 we solved quadratic equations by factoring. However, since not all quadratic equations can easily be solved by factoring, it is necessary to develop other methods. In this chapter we will do just that.

Recall that a *quadratic equation* is an equation that can be written in the form

$$ax^2 + bx + c = 0$$

for real numbers a, b, and c, with $a \neq 0$. To solve $x^2 + 4x + 3 = 0$ by the zero-factor property, begin by factoring the left side and then set each factor equal to zero.

$$x^2 + 4x + 3 = 0$$

$$(x + 3)(x + 1) = 0 \qquad \text{Factor.}$$

$$x + 3 = 0 \quad \text{or} \quad x + 1 = 0 \qquad \text{Zero-factor property}$$

$$x = -3 \quad \text{or} \quad x = -1$$

The solution set is $\{-3, -1\}$.

OBJECTIVE 1 Solve equations of the form $x^2 =$ a number. To solve equations such as $x^2 = 9$, factor as follows.

$$x^2 = 9$$

$$x^2 - 9 = 0 \qquad \text{Subtract 9.}$$

$$(x + 3)(x - 3) = 0 \qquad \text{Factor.}$$

$$x + 3 = 0 \quad \text{or} \quad x - 3 = 0 \qquad \text{Zero-factor property}$$

$$x = -3 \quad \text{or} \quad x = 3$$

The solution set is $\{-3, 3\}$.

This result is generalized as the **square root property of equations.**

Square Root Property of Equations

If k is a positive number and if $a^2 = k$, then

$$a = \sqrt{k} \quad \text{or} \quad a = -\sqrt{k},$$

and the solution set is $\{-\sqrt{k}, \sqrt{k}\}$.

NOTE When we solve an equation, we want to find *all* values of the variable that satisfy the equation. Therefore, we want both the positive and the negative square roots of k.

CHALKBOARD EXAMPLE

Solve the equation.

$$x^2 = 12$$

Answer: $\{2\sqrt{3}, -2\sqrt{3}\}$

EXAMPLE 1 Solving a Quadratic Equation of the Form $x^2 = k$

Solve each equation. Write radicals in simplified form.

(a) $x^2 = 16$

By the square root property, since $x^2 = 16$,

$$x = \sqrt{16} = 4 \quad \text{or} \quad x = -\sqrt{16} = -4.$$

An abbreviation for "$x = 4$ or $x = -4$" is written $x = \pm 4$ and is read "x equals positive or negative 4." The solution set is $\{-4, 4\}$. Check each solution by substituting back into the original equation.

INSTRUCTOR'S RESOURCES

PTB Sec. 10.1 ISM Sec. 10.1
AB Sec. 10.1

TEST GENERATOR

WORLD WIDE WEB
www.LialAlgebra.com

TEACHING TIP Emphasize that when the square root property is used, both the positive and the negative square roots must be given.

(b) $z^2 = 5$

The solutions are $z = \sqrt{5}$ or $z = -\sqrt{5}$, and the solution set may be written $\{\pm\sqrt{5}\}$.

(c) $m^2 = 8$

$$m^2 = 8$$

$$m = \sqrt{8} \quad \text{or} \quad m = -\sqrt{8} \qquad \text{Square root property}$$

$$m = 2\sqrt{2} \quad \text{or} \quad m = -2\sqrt{2} \qquad \text{Simplify } \sqrt{8}.$$

The solution set is $\{\pm 2\sqrt{2}\}$.

(d) $y^2 = -4$

Since -4 is a negative number and since the square of a real number cannot be negative, there is no real number solution for this equation. The solution set is $\emptyset$.

(e) $3x^2 + 5 = 11$

First solve the equation for x^2.

$$3x^2 + 5 = 11$$

$$3x^2 = 6 \qquad \text{Subtract 5.}$$

$$x^2 = 2 \qquad \text{Divide by 3.}$$

Now use the square root property to get the solution set $\{\pm\sqrt{2}\}$.

OBJECTIVE 2 **Solve equations of the form $(ax + b)^2$ = a number.** In each of the equations in Example 1, the exponent 2 appeared with a single variable as its base. The square root property of equations can be extended to solve equations where the base is a binomial, as shown in the next example.

CHALKBOARD EXAMPLE

Solve the equation.

$$(p - 4)^2 = 3$$

Answer: $\{4 + \sqrt{3}, 4 - \sqrt{3}\}$

EXAMPLE 2 **Solving a Quadratic Equation of the Form $(x + b)^2 = k$**

Solve each equation.

(a) $(x - 3)^2 = 16$

Apply the square root property, using $x - 3$ as the base.

$$(x - 3)^2 = 16$$

$$x - 3 = \sqrt{16} \quad \text{or} \quad x - 3 = -\sqrt{16}$$

$$x - 3 = 4 \quad \text{or} \quad x - 3 = -4 \qquad \sqrt{16} = 4$$

$$x = 7 \quad \text{or} \quad x = -1 \qquad \text{Add 3.}$$

Check both answers in the original equation.

$$(x - 3)^2 = 16 \qquad\qquad (x - 3)^2 = 16$$

$$(7 - 3)^2 = 16 \ ? \quad \text{Let } x = 7. \qquad (-1 - 3)^2 = 16 \ ? \quad \text{Let } x = -1.$$

$$4^2 = 16 \ ? \qquad\qquad (-4)^2 = 16 \ ?$$

$$16 = 16 \quad \text{True} \qquad\qquad 16 = 16 \quad \text{True}$$

The solution set is $\{7, -1\}$.

(b) $(x - 1)^2 = 6$

By the square root property,

$$x - 1 = \sqrt{6} \quad \text{or} \quad x - 1 = -\sqrt{6}$$

$$x = 1 + \sqrt{6} \quad \text{or} \quad x = 1 - \sqrt{6}.$$

TEACHING TIP In Example 3, point out that the common error of reducing the expression

$$\frac{2 + 3\sqrt{3}}{3} \quad \text{to} \quad 2 + \sqrt{3}$$

should be avoided. This is a good time to review the idea that only common factors may be divided out. Show why 3 is not a factor in the numerator.

Check: $(1 + \sqrt{6} - 1)^2 = (\sqrt{6})^2 = 6;$
$(1 - \sqrt{6} - 1)^2 = (-\sqrt{6})^2 = 6.$

The solution set is $\{1 + \sqrt{6}, 1 - \sqrt{6}\}$.

NOTE: The solutions in Example 2(b) may be written in abbreviated form as

$$1 \pm \sqrt{6}.$$

If they are written this way, keep in mind that *two* solutions are indicated, one with the + sign and the other with the − sign.

CHALKBOARD EXAMPLE

Solve the equation.

$$(5m + 1)^2 = 7$$

Answer: $\left\{\frac{-1 \pm \sqrt{7}}{5}\right\}$

EXAMPLE 3 Solving a Quadratic Equation of the Form $(ax + b)^2 = k$

Solve $(3r - 2)^2 = 27$.

$3r - 2 = \sqrt{27}$	or	$3r - 2 = -\sqrt{27}$	Square root property
$3r - 2 = 3\sqrt{3}$	or	$3r - 2 = -3\sqrt{3}$	$\sqrt{27} = \sqrt{9 \cdot 3} = 3\sqrt{3}$
$3r = 2 + 3\sqrt{3}$	or	$3r = 2 - 3\sqrt{3}$	Add 2.
$r = \frac{2 + 3\sqrt{3}}{3}$	or	$r = \frac{2 - 3\sqrt{3}}{3}$	Divide by 3.

The solution set is $\left\{\frac{2 \pm 3\sqrt{3}}{3}\right\}$.

CAUTION: The solutions in Example 3 are fractions that cannot be reduced, since 3 is *not* a common factor in the numerator.

CHALKBOARD EXAMPLE

Solve the equation.

$$(7z - 1)^2 = -1$$

Answer: $\emptyset$

EXAMPLE 4 Recognizing a Quadratic Equation with No Real Solution

Solve $(x + 3)^2 = -9$.

The square root of -9 is not a real number. The solution set is $\emptyset$.

10.1 EXERCISES

1. C **2.** D **3.** A **4.** E **5.** B **6.** true **7.** true **8.** false; If k is a positive integer that is not a perfect square, the solutions will be irrational. **9.** true **10.** false; For values of k that satisfy $0 \le k < 10$, there are real solutions. **11.** According to the square root property, -9 is also a solution, so her answer was not completely correct. The solution set is $\{\pm 9\}$. **12.** The example equation $x^2 = 21$ can be solved using the square root property. Because the square of x is 21, x can be either the positive or the negative square root of 21. The solution set is $\{\pm\sqrt{21}\}$.

Match each equation with the correct description of its solutions.

1. $x^2 = 0$	**A.** No real number solutions
2. $x^2 = 10$	**B.** Two integer solutions
3. $x^2 = -4$	**C.** One real solution
4. $x^2 = \frac{9}{16}$	**D.** Two irrational solutions
5. $x^2 = 9$	**E.** Two rational solutions that are not integers

Decide whether each statement is true or false. If it is false, tell why.

6. If k is a prime number, then $x^2 = k$ has two irrational solutions.

7. If k is a positive perfect square, then $x^2 = k$ has two rational solutions.

8. If k is a positive integer, then $x^2 = k$ must have two rational solutions.

9. If $-10 < k < 0$, then $x^2 = k$ has no real solutions.

10. If $-10 < k < 10$, then $x^2 = k$ has no real solutions.

11. When a student was asked to solve $x^2 = 81$, she wrote $\{9\}$ as her answer. Her teacher did not give her full credit. The student argued that because $9^2 = 81$, her answer had to be correct. Why was her answer not completely correct?

12. Explain the square root property for solving equations, and illustrate with an example.

Solve each equation by using the square root property. Express all radicals in simplest form. See Example 1.

13. $x^2 = 81$ **14.** $y^2 = 121$ **15.** $k^2 = 14$ **16.** $m^2 = 22$

17. $t^2 = 48$ **18.** $x^2 = 54$ **19.** $y^2 = -100$ **20.** $m^2 = -64$

21. $z^2 = 2.25$ **22.** $w^2 = 56.25$ **23.** $3x^2 - 8 = 64$ **24.** $2t^2 + 7 = 61$

Solve each equation by using the square root property. Express all radicals in simplest form. See Examples 2–4.

25. $(x - 3)^2 = 25$ **26.** $(y - 7)^2 = 16$ **27.** $(z + 5)^2 = -13$

28. $(m + 2)^2 = -17$ **29.** $(x - 8)^2 = 27$ **30.** $(y - 5)^2 = 40$

31. $(3k + 2)^2 = 49$ **32.** $(5t + 3)^2 = 36$ **33.** $(4x - 3)^2 = 9$

34. $(7y - 5)^2 = 25$ **35.** $(5 - 2x)^2 = 30$ **36.** $(3 - 2a)^2 = 70$

37. $(3k + 1)^2 = 18$ **38.** $(5z + 6)^2 = 75$ ▲ **39.** $\left(\frac{1}{2}x + 5\right)^2 = 12$

▲ **40.** $\left(\frac{1}{3}y + 4\right)^2 = 27$ ▲ **41.** $(4k - 1)^2 - 48 = 0$ ▲ **42.** $(2s - 5)^2 - 180 = 0$

43. Johnny solved the equation in Exercise 35 and wrote his answer as $\left\{\frac{5 + \sqrt{30}}{2}, \frac{5 - \sqrt{30}}{2}\right\}$.

Linda solved the same equation and wrote her answer as $\left\{\frac{-5 + \sqrt{30}}{-2}, \frac{-5 - \sqrt{30}}{-2}\right\}$. The teacher gave them both full credit. Explain why both students were correct, although their answers look different.

44. In the solutions found in Example 3 of this section, why is it not valid to reduce the answers by dividing out the threes in the numerator and denominator?

Use a calculator with a square root key to solve each equation. Round your answers to the nearest hundredth.

45. $(k + 2.14)^2 = 5.46$ **46.** $(r - 3.91)^2 = 9.28$

47. $(2.11p + 3.42)^2 = 9.58$ **48.** $(1.71m - 6.20)^2 = 5.41$

RELATING CONCEPTS* (EXERCISES 49–54)

In Section 5.4 we saw how certain trinomials can be factored as squares of binomials. *Use this idea and* ***work Exercises 49–54 in order,*** *considering the equation*

$$x^2 + 6x + 9 = 100.$$

49. Factor the left side of the equation as the square of a binomial, and write the resulting equation.

50. Write the equation from Exercise 49 as a compound statement using the word "or."

51. Solve the two individual equations from Exercise 50.

52. What is the solution set of the original equation?

53. Solve the equation $x^2 + 4x + 4 = 25$ using the method described in Exercises 49–52.

54. Solve the equation $4k^2 - 12k + 9 = 81$ using the method described in Exercises 49–52.

Did you make the connection that when the trinomial factors are the square of a binomial, the square root property of equations can be extended in scope?

---

13. $\{\pm 9\}$ **14.** $\{\pm 11\}$
15. $\{\pm\sqrt{14}\}$ **16.** $\{\pm\sqrt{22}\}$
17. $\{\pm 4\sqrt{3}\}$ **18.** $\{\pm 3\sqrt{6}\}$
19. $\emptyset$ **20.** $\emptyset$ **21.** $\{\pm 1.5\}$
22. $\{\pm 7.5\}$ **23.** $\{\pm 2\sqrt{6}\}$
24. $\{\pm 3\sqrt{3}\}$ **25.** $\{-2, 8\}$
26. $\{3, 11\}$ **27.** $\emptyset$ **28.** $\emptyset$
29. $\{8 \pm 3\sqrt{3}\}$ **30.** $\{5 \pm 2\sqrt{10}\}$
31. $\left\{-3, \frac{5}{3}\right\}$ **32.** $\left\{-\frac{9}{5}, \frac{3}{5}\right\}$
33. $\left\{0, \frac{3}{2}\right\}$ **34.** $\left\{0, \frac{10}{7}\right\}$
35. $\left\{\frac{5 \pm \sqrt{30}}{2}\right\}$
36. $\left\{\frac{3 \pm \sqrt{70}}{2}\right\}$
37. $\left\{\frac{-1 \pm 3\sqrt{2}}{3}\right\}$
38. $\left\{\frac{-6 \pm 5\sqrt{3}}{5}\right\}$
39. $\{-10 \pm 4\sqrt{3}\}$
40. $\{-12 \pm 9\sqrt{3}\}$
41. $\left\{\frac{1 \pm 4\sqrt{3}}{4}\right\}$
42. $\left\{\frac{5 \pm 6\sqrt{5}}{2}\right\}$
43. Johnny's first solution, $\frac{5 + \sqrt{30}}{2}$, is equivalent to Linda's second solution, $\frac{-5 - \sqrt{30}}{-2}$. This can be verified by multiplying $\frac{5 + \sqrt{30}}{2}$ by 1 in the form $\frac{-1}{-1}$. Similarly, Johnny's second solution is equivalent to Linda's first one.
44. 3 is not a common factor.
45. $\{-4.48, .20\}$ **46.** $\{.86, 6.96\}$
47. $\{-3.09, -.15\}$ **48.** $\{2.27, 4.99\}$

*These problems can be used as collaborative exercises with small groups of students.

49. $(x + 3)^2 = 100$
50. $x + 3 = -10$ or $x + 3 = 10$
51. $\{-13\}$; $\{7\}$ **52.** $\{-13, 7\}$
53. $\{-7, 3\}$ **54.** $\{-3, 6\}$

55. .983 foot
56. .993 foot
57. 3.442 feet
58. 3.4494 feet
59. about $\frac{1}{2}$ second 60. 9 feet
61. 9 inches 62. 3 feet

When an object moves under the influence of a constant force, its path is parabolic, and its position can be described by a quadratic equation $y = ax^2 + bx$, where x is in seconds, y is in feet, and a and b are real numbers. This equation varies on the surface of different celestial objects. Use this fact to work Exercises 55–58. (Source: *M. Zeilik, S. Gregory, and E. Smith,* Introductory Astronomy and Astrophysics, *Saunders College Publishers, 1992.)*

55. If an object is propelled upward on Earth at an angle of 45° with an initial velocity of 30 miles per hour, then the equation is $y = -.017x^2 + x$. Find the height of the object in feet after 1 second. (*Hint:* Let $x = 1$ and solve for y.)

56. If an object is propelled upward on Mars at an angle of 45° with an initial velocity of 30 miles per hour, then the equation is $y = -.007x^2 + x$. Find the height of the object in feet after 1 second.

57. If an object is propelled upward on Mars at an angle of 60° with an initial velocity of 60 miles per hour, then the equation is $y = -.003x^2 + 1.727x$. Find the height of the object in feet after 2 seconds.

58. If an object is propelled upward on the moon at an angle of 60° with an initial velocity of 60 miles per hour, then the equation is $y = -.0013x^2 + 1.7273x$. Find the height of the object in feet after 2 seconds.

Solve each problem.

59. One expert at marksmanship can hold a silver dollar at forehead level, drop it, draw his gun, and shoot the coin as it passes waist level. The distance traveled by a falling object is given by

$$d = 16t^2,$$

where d is the distance (in feet) the object falls in t seconds. If the coin falls about 4 feet, use the formula to estimate the time that elapses between the dropping of the coin and the shot.

60. The illumination produced by a light source depends on the distance from the source. For a particular light source, this relationship can be expressed as

$$d^2 = \frac{4050}{I},$$

where d is the distance from the source (in feet) and I is the amount of illumination in foot-candles. How far from the source is the illumination equal to 50 foot-candles?

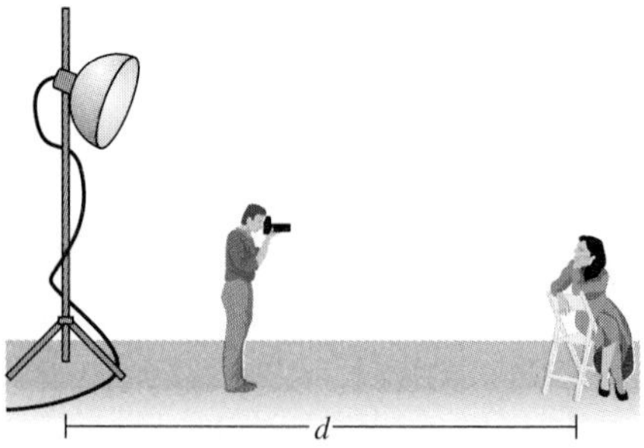

61. The area A of a circle with radius r is given by the formula

$$A = \pi r^2.$$

If a circle has area 81π square inches, what is its radius?

62. The surface area S of a sphere with radius r is given by the formula

$$S = 4\pi r^2.$$

If a sphere has surface area 36π square feet, what is its radius?

63. 5%
64. 7%

The amount A that P dollars invested at an annual rate of interest r will grow to in 2 years is

$$A = P(1 + r)^2.$$

63. At what interest rate will \$100 grow to \$110.25 in two years?

64. At what interest rate will \$500 grow to \$572.45 in two years?

10.2 Solving Quadratic Equations by Completing the Square

OBJECTIVES

1 Solve quadratic equations by completing the square when the coefficient of the squared term is 1.

2 Solve quadratic equations by completing the square when the coefficient of the squared term is not 1.

3 Simplify an equation before solving.

FOR EXTRA HELP

SSG Sec. 10.2
SSM Sec. 10.2
Pass the Test Software
InterAct Math Tutorial Software
Video 16

CHALKBOARD EXAMPLE

Solve $q^2 - 20q + 34 = 0$.

Answer: $\{10 \pm \sqrt{66}\}$

OBJECTIVE 1 Solve quadratic equations by completing the square when the coefficient of the squared term is 1. The methods studied so far are not enough to solve the equation

$$x^2 + 6x + 7 = 0.$$

If we could write the equation in the form $(x + b)^2 = k$, we could solve it with the square root property discussed in the previous section. To do that, we need to have a perfect square trinomial on one side of the equation. The next example shows how this is done.

EXAMPLE 1 Rewriting an Equation to Use the Square Root Property

Solve $x^2 + 6x + 7 = 0$.

Start by subtracting 7 from both sides of the equation to get $x^2 + 6x = -7$. To write $x^2 + 6x = -7$ in the form $(x + b)^2 = k$, the quantity on the left-hand side of $x^2 + 6x = -7$ must be made into a perfect square trinomial. The expression $x^2 + 6x + 9$ is a perfect square, since

$$x^2 + 6x + 9 = (x + 3)^2.$$

Therefore, if we add 9 to both sides of $x^2 + 6x = -7$, the equation will have a perfect square trinomial on one side, as needed.

$$x^2 + 6x + 9 = -7 + 9 \quad \text{Add 9 to both sides.}$$
$$(x + 3)^2 = 2 \quad \text{Factor; combine terms.}$$

Now use the square root property to complete the solution.

$$x + 3 = \sqrt{2} \quad \text{or} \quad x + 3 = -\sqrt{2}$$
$$x = -3 + \sqrt{2} \quad \text{or} \quad x = -3 - \sqrt{2}$$

The solution set is $\{-3 \pm \sqrt{2}\}$. Check this by substituting $-3 + \sqrt{2}$ and $-3 - \sqrt{2}$ for x in the original equation. ■

The process of changing the form of the equation in Example 1 from

$$x^2 + 6x + 7 = 0 \quad \text{to} \quad (x + 3)^2 = 2$$

is called **completing the square.** Completing the square changes only the form of the equation. To see this, multiply out the left side of $(x + 3)^2 = 2$ and combine terms. Then subtract 2 from both sides to see that the result is $x^2 + 6x + 7 = 0$.

Look again at the original equation

$$x^2 + 6x + 7 = 0$$

INSTRUCTOR'S RESOURCES

PTB Sec. 10.2 **ISM** Sec. 10.2
AB Sec. 10.2
TEST GENERATOR
WORLD WIDE WEB www.LialAlgebra.com

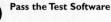

TEACHING TIP The method of completing the square is introduced here for the first time. Explain to students that it not only provides a method for solving quadratic equations, but is also used in other ways in algebra (finding the coordinates of the center of a circle, finding the vertex of a parabola, and so on).

CHALKBOARD EXAMPLE

Complete the square to solve $x^2 + 3x = 1$.

Answer: $\left\{\frac{-3 \pm \sqrt{13}}{2}\right\}$

in Example 1. Note that 9 is the square of half the coefficient of x, 6.

$$\frac{1}{2} \cdot 6 = 3 \qquad \text{and} \qquad 3^2 = 9$$

(↑ Coefficient of x)

So to complete the square in Example 1, 9 was added to each side.

EXAMPLE 2 Completing the Square to Solve a Quadratic Equation

Complete the square to solve $x^2 - 8x = 5$.

To complete the square on $x^2 - 8x$, take half the coefficient of x and square it.

$$\frac{1}{2}(-8) = -4 \qquad \text{and} \qquad (-4)^2 = 16$$

(↑ Coefficient of x)

Add the result, 16, to both sides of the equation.

$x^2 - 8x = 5$	Given equation
$x^2 - 8x + 16 = 5 + 16$	Add 16 to both sides.
$(x - 4)^2 = 21$	Factor the left side as the square of a binomial.

Now apply the square root property.

$x - 4 = \pm\sqrt{21}$	Square root property
$x = 4 \pm \sqrt{21}$	Add 4 to both sides.

A check indicates that the solution set is $\{4 \pm \sqrt{21}\}$.

To solve a quadratic equation by completing the square, follow the steps given below.

Solving a Quadratic Equation by Completing the Square

Step 1 **Be sure the squared term has coefficient 1.** If the coefficient of the squared term is 1, proceed to Step 2. If the coefficient of the squared term is not 1 but some other nonzero number a, divide both sides of the equation by a. This gives an equation that has 1 as coefficient of the squared term.

Step 2 **Put in correct form.** Make sure that all terms with variables are on one side of the equals sign and that all constants are on the other side.

Step 3 **Complete the square.** Take half the coefficient of the first-degree term, and square the result. Add the square to both sides of the equation. The side containing the variables now can be factored as a perfect square.

Step 4 **Solve.** Apply the square root property to solve the equation.

OBJECTIVE 2 Solve quadratic equations by completing the square when the coefficient of the squared term is not 1.

To complete the square, the coefficient of the squared term must be 1 (Step 1). The next example shows what to do when this coefficient is not 1.

CHALKBOARD EXAMPLE

Solve $4k^2 - 24k + 11 = 0$.

Answer: $\left\{\frac{11}{2}, \frac{1}{2}\right\}$

EXAMPLE 3 Solving a Quadratic Equation by Completing the Square

Solve $4y^2 + 24y - 13 = 0$.

Step 1 Divide each side by 4 so that the coefficient of y^2 is 1.

$$4y^2 + 24y - 13 = 0$$
$$y^2 + 6y - \frac{13}{4} = 0$$

Step 2 Add $\frac{13}{4}$ to each side to get the variable terms on the left and the constant on the right.

$$y^2 + 6y = \frac{13}{4}$$

TEACHING TIP Point out that if the solutions are rational numbers, as they are in Example 3, the equation can also be solved by factoring. Emphasize that the method of completing the square is a more powerful method than factoring because it allows us to solve *any* quadratic equation.

Step 3 To complete the square, take half the coefficient of y, $\frac{1}{2}(6) = 3$, and square the result: $3^2 = 9$. Add 9 to both sides of the equation, add on the right-hand side, and factor on the left.

$$y^2 + 6y + \mathbf{9} = \frac{13}{4} + \mathbf{9} \qquad \text{Add 9.}$$
$$y^2 + 6y + 9 = \frac{49}{4} \qquad \text{Add on the right.}$$
$$(y + 3)^2 = \frac{49}{4} \qquad \text{Factor on the left.}$$

Step 4 Use the square root property and solve for y.

$$y + 3 = \frac{7}{2} \quad \text{or} \quad y + 3 = -\frac{7}{2} \qquad \text{Square root property}$$
$$y = -3 + \frac{7}{2} \quad \text{or} \quad y = -3 - \frac{7}{2} \qquad \text{Add } -3.$$
$$y = \frac{1}{2} \quad \text{or} \quad y = -\frac{13}{2}$$

The solution set is $\left\{\frac{1}{2}, -\frac{13}{2}\right\}$. Check by substitution into the original equation.

CHALKBOARD EXAMPLE

Solve $5v^2 + 3v + 1 = 0$.

Answer: $\emptyset$

EXAMPLE 4 Solving a Quadratic Equation by Completing the Square

Solve $4p^2 + 8p + 5 = 0$.

First divide both sides by 4 to get a coefficient of 1 for the p^2-term (Step 1).

$$p^2 + 2p + \frac{5}{4} = 0$$

Add $-\frac{5}{4}$ to both sides (Step 2).

$$p^2 + 2p = -\frac{5}{4}$$

TEACHING TIP Perhaps some students have seen imaginary numbers and might question the result in Example 4. Explain that at this point we are only looking for real number solutions, and imaginary numbers will be discussed in Section 10.4.

The coefficient of p is 2. Take half of 2, square the result, and add it to both sides. The left-hand side can then be factored as a perfect square (Step 3).

$$p^2 + 2p + 1 = -\frac{5}{4} + 1 \qquad \left[\tfrac{1}{2}(2)\right]^2 = 1$$

$$(p + 1)^2 = -\frac{1}{4} \qquad \text{Factor; combine terms.}$$

The square root of $-\frac{1}{4}$ is not a real number, so the solution set is $\emptyset$.

OBJECTIVE 3 Simplify an equation before solving. Sometimes an equation must be simplified before completing the square. The next example illustrates this.

CHALKBOARD EXAMPLE

Solve the equation.

$$3p^2 + 7p = p^2 + p + 3$$

Answer: $\left\{\frac{-3 \pm \sqrt{15}}{2}\right\}$

EXAMPLE 5 Simplifying an Equation Before Completing the Square

Solve $(x + 3)(x - 1) = 2$.

$$(x + 3)(x - 1) = 2$$

$$x^2 + 2x - 3 = 2 \qquad \text{Use FOIL.}$$

$$x^2 + 2x = 5 \qquad \text{Add 3 to both sides.}$$

$$x^2 + 2x + 1 = 5 + 1 \qquad \text{Add 1 to get a perfect square on the left.}$$

$$(x + 1)^2 = 6 \qquad \text{Factor on the left; add on the right.}$$

$$x + 1 = \sqrt{6} \quad \text{or} \quad x + 1 = -\sqrt{6} \qquad \text{Square root property}$$

$$x = -1 + \sqrt{6} \quad \text{or} \quad x = -1 - \sqrt{6} \qquad \text{Add } -1.$$

The solution set is $\{-1 \pm \sqrt{6}\}$.

NOTE The solutions given in Example 5 are *exact*. In applications, decimal solutions are more appropriate. Using the square root key of a calculator, $\sqrt{6} \approx 2.449$. Evaluating the two solutions gives

$$x \approx 1.449 \quad \text{and} \quad x \approx -3.449.$$

CONNECTIONS

"Completing the square" has so far been applied only in an algebraic sense. We add an appropriate constant to a binomial to get a perfect square trinomial and then factor so that we can use the square root property. However, this procedure can literally be applied to a geometric figure so that it becomes a square. This gives a beautiful connection between algebra and geometry.

For example, to complete the square for $x^2 + 8x$, begin with a square having a side of length x. Add four rectangles of width 1 to the right side and to the bottom. To "complete the square" fill in the bottom right corner with 16 squares of area 1.

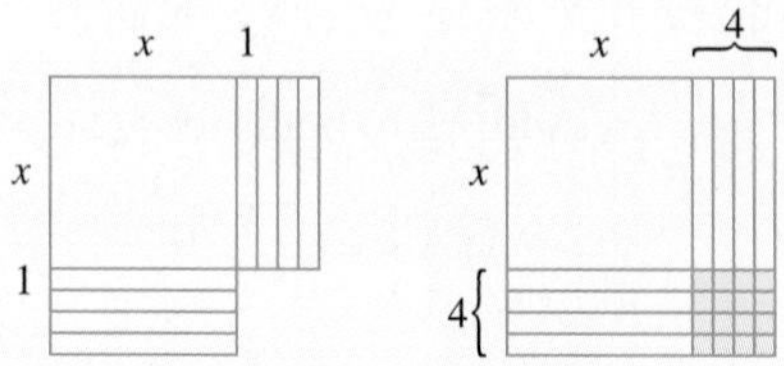

1. x^2 2. $x^2 + 8x$
3. $x^2 + 8x + 16$
4. It occurred when we added the 16 squares.

CONNECTIONS (CONTINUED)

FOR DISCUSSION OR WRITING

1. What is the area of the original square?
2. What is the area of the figure after the 8 rectangles are added?
3. What is the area of the figure after the 16 small squares are added?
4. At what point did we "complete the square"?

1. 16 2. 3 3. multiplying $(t + 2)(t - 5)$ to get $t^2 - 3t - 10$
4. it is not a quadratic equation because of the x^3-term
5. (d) 6. yes 7. 49
8. 81 9. $\frac{25}{4}$ 10. $\frac{81}{4}$
11. $\frac{1}{16}$ 12. $\frac{1}{36}$ 13. $\{1, 3\}$
14. $\{-2, 4\}$ 15. $\{-1 \pm \sqrt{6}\}$
16. $\{-2 \pm \sqrt{3}\}$ 17. $\{-3\}$
18. $\{4\}$ 19. $\left\{-\frac{3}{2}, \frac{1}{2}\right\}$
20. $\left\{-\frac{2}{3}, \frac{1}{3}\right\}$ 21. $\emptyset$ 22. $\emptyset$
23. $\left\{\frac{-7 \pm \sqrt{97}}{6}\right\}$
24. $\left\{\frac{-5 \pm \sqrt{33}}{4}\right\}$
25. $\{-4, 2\}$ 26. $\{-4, 10\}$
27. $\{1 \pm \sqrt{6}\}$ 28. $\{2 \pm \sqrt{3}\}$

10.2 EXERCISES

Fill in each blank with the correct response.

1. To solve the equation $x^2 - 8x = 4$ by completing the square, the first step is to add _____ to both sides of the equation.
2. To solve the equation $3x^2 - 8x - 1 = 0$ by completing the square, the first step is to divide both sides of the equation by _____.
3. To solve $(t + 2)(t - 5) = 18$ by completing the square, we should start by _______________.
4. It is not possible to solve $x^3 - x - 1 = 0$ by completing the square because _______________.
5. Which one of the following steps is an appropriate way to begin solving the quadratic equation
$$2x^2 - 4x = 9$$
by completing the square?
(a) Add 4 to both sides of the equation. (b) Factor the left side as $2x(x - 2)$.
(c) Factor the left side as $x(2x - 4)$. (d) Divide both sides by 2.
6. In Example 3 of Section 5.5, we solved the quadratic equation
$$4p^2 - 26p + 40 = 0$$
by factoring. If we were to solve by completing the square, would we get the same solution set, $\left\{\frac{5}{2}, 4\right\}$?

Find the constant that should be added to each expression to make it a perfect square. See Examples 1 and 2.

7. $y^2 + 14y$ 8. $z^2 + 18z$ 9. $k^2 - 5k$
10. $m^2 - 9m$ 11. $r^2 + \frac{1}{2}r$ 12. $s^2 - \frac{1}{3}s$

Solve each equation by completing the square. See Examples 1 and 2.

13. $x^2 - 4x = -3$ 14. $y^2 - 2y = 8$ 15. $x^2 + 2x - 5 = 0$
16. $r^2 + 4r + 1 = 0$ 17. $z^2 + 6z + 9 = 0$ 18. $k^2 - 8k + 16 = 0$

Solve each equation by completing the square. See Examples 3–5.

19. $4y^2 + 4y = 3$ 20. $9x^2 + 3x = 2$ 21. $2p^2 - 2p + 3 = 0$
22. $3q^2 - 3q + 4 = 0$ 23. $3k^2 + 7k = 4$ 24. $2k^2 + 5k = 1$
25. $(x + 3)(x - 1) = 5$ 26. $(y - 8)(y + 2) = 24$ 27. $-x^2 + 2x = -5$
28. $-x^2 + 4x = 1$

29. (a) $\left\{\frac{3 \pm 2\sqrt{6}}{3}\right\}$
(b) $\{-.633, 2.633\}$
30. (a) $\left\{\frac{1 \pm \sqrt{7}}{2}\right\}$
(b) $\{-.823, 1.823\}$
31. (a) $\{-2 \pm \sqrt{3}\}$
(b) $\{-3.732, -.268\}$
32. (a) $\{1 \pm \sqrt{5}\}$
(b) $\{-1.236, 3.236\}$
33. The student should have divided both sides of the equation by 2 as his first step. The correct solution set is $\{1, 4\}$.
34. 1; $\{\pm 2\}$
35. 75 feet by 100 feet
36. base: 10 meters; height: 3 meters
37. 1 second and 5 seconds
38. 1.3 seconds and 4.7 seconds
39. 3 seconds and 5 seconds
40. 8 seconds **41.** 8 miles

*Solve each equation by completing the square. Then **(a)** give the exact solutions and **(b)** give the solutions rounded to the nearest thousandth.*

29. $3r^2 - 2 = 6r + 3$

30. $4p + 3 = 2p^2 + 2p$

31. $(x + 1)(x + 3) = 2$

32. $(x - 3)(x + 1) = 1$

33. In using the method of completing the square to solve $2x^2 - 10x = -8$, a student began by adding the square of half the coefficient of x (that is, $\left[\frac{1}{2}(-10)\right]^2 = 25$) to both sides of the equation. He then encountered difficulty in his later steps. What was his error? Explain the steps needed to solve the problem, and give the solution set.

▲ **34.** The equation $x^4 - 2x^2 = 8$ can be solved by completing the square, even though it is not a quadratic equation. What number should be added to both sides of the equation so that the left side can be factored as the square of a binomial? Solve the equation for its real solutions.

Solve each problem.

35. A farmer has a rectangular cattle pen with perimeter 350 feet and area 7500 square feet. What are the dimensions of the pen? (*Hint:* Use the figure to set up the equation.)

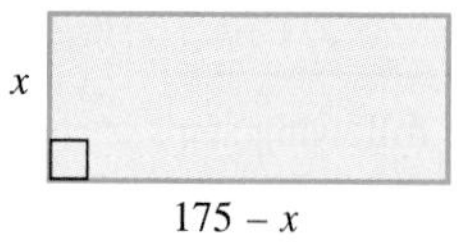

36. The base of a triangle measures 1 meter more than three times the height of the triangle. Its area is 15 square meters. Find the lengths of the base and the height.

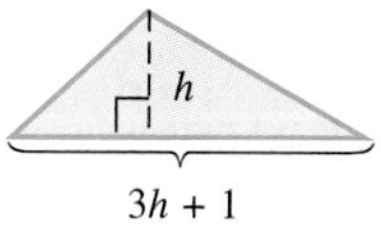

37. If an object is propelled upward from ground level on Earth with an initial velocity of 96 feet per second, its height, s (in feet), in t seconds is given by the formula $s = -16t^2 + 96t$. How long will it take for the object to be at a height of 80 feet? (*Hint:* Let $s = 80$.)

38. How long will it take the object described in Exercise 37 to be at a height of 100 feet? Round your answers to the nearest tenth.

39. If an object is propelled upward on the surface of Mars from ground level with an initial velocity of 104 feet per second, its height s (in feet) in t seconds is given by the formula $s = -13t^2 + 104t$. How long will it take for the object to be at a height of 195 feet?

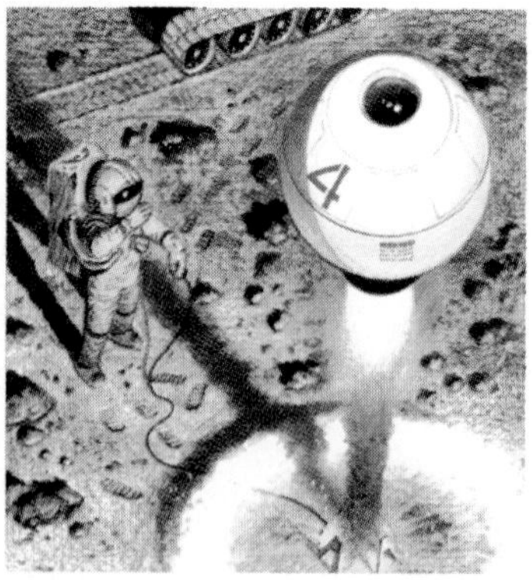

40. How long will it take the object in Exercise 39 to return to the surface? (*Hint:* When it returns to the surface, $s = 0$.)

41. Two cars travel at right angles to each other from an intersection until they are 17 miles apart. At that point one car has gone 7 miles farther than the other. How far did the slower car travel? (*Hint:* Use the Pythagorean formula.)

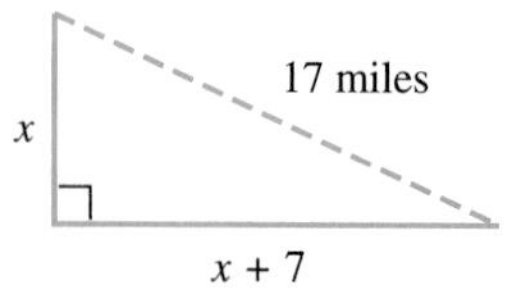

42. 8.7 hours

42. Two painters are painting a house in a development of new homes. One of the painters takes 2 hours longer to paint a house working alone than the other painter. When they do the job together, they can complete it in 4.8 hours. How long would it take the faster painter alone to paint the house? (Give your answer to the nearest tenth.)

10.3 Solving Quadratic Equations by the Quadratic Formula

OBJECTIVES

1. Identify the values of a, b, and c in a quadratic equation.
2. Use the quadratic formula to solve quadratic equations.
3. Solve quadratic equations with only one solution.
4. Solve quadratic equations with fractions.
5. Solve an applied problem that leads to a quadratic equation.

FOR EXTRA HELP

SSG Sec. 10.3
SSM Sec. 10.3
Pass the Test Software
InterAct Math Tutorial Software
Video 16

CHALKBOARD EXAMPLE

Write the equation in standard form, and then identify the values of a, b, and c.

$$3(m-2)(m+1)=7$$

Answer: $a=3$, $b=-3$, $c=-13$

TEACHING TIP Explain that in this section we will develop a formula for solving quadratic equations by using the method of completing the square for the general equation

$$ax^2+bx+c=0.$$

This may be the first time students have seen such a derivation. Explain that mathematical formulas do not just appear magically, but are derived from generalizations such as this one.

Completing the square can be used to solve any quadratic equation, but the method is tedious. In this section we complete the square on the general quadratic equation $ax^2+bx+c=0$ to get the *quadratic formula,* a formula that gives the solution for any quadratic equation. (Note that $a \neq 0$, or we would have a linear, not a quadratic, equation.)

OBJECTIVE 1 Identify the values of a, b, and c in a quadratic equation. The first step in solving a quadratic equation by this new method is to identify the values of a, b, and c in the standard form of the quadratic equation.

EXAMPLE 1 Determining Values of a, b, and c in a Quadratic Equation

For each of the following quadratic equations, write the equation in standard form if necessary, and then identify the values of a, b, and c.

(a) $2x^2+3x-5=0$

This equation is already in standard form. The values of a, b, and c are

$$a=2, \quad b=3, \quad \text{and} \quad c=-5.$$

(b) $-x^2+2=6x$

First rewrite the equation with 0 on the right side to match the standard form $ax^2+bx+c=0$.

$$-x^2+2=6x$$
$$-x^2-6x+2=0$$

Now identify $a=-1$, $b=-6$, and $c=2$. (Notice that the coefficient of x^2 is understood to be -1.)

(c) $(2x-7)(x+4)=-23$

$$(2x-7)(x+4)=-23$$
$$2x^2+x-28=-23 \quad \text{Use FOIL on the left.}$$
$$2x^2+x-5=0 \quad \text{Add 23 on each side.}$$

Now, identify the values: $a=2$, $b=1$, $c=-5$. ■

OBJECTIVE 2 Use the quadratic formula to solve quadratic equations. To develop the quadratic formula, we follow the steps for completing the square on $ax^2+bx+c=0$ given in the previous section. For comparison, we also show the corresponding steps for solving $2x^2+x-5=0$ (from Example 1(c)).

Step 1 Make the coefficient of the squared term equal to 1.

$2x^2+x-5=0$	$ax^2+bx+c=0$
$x^2+\frac{1}{2}x-\frac{5}{2}=0$ Divide by 2.	$x^2+\frac{b}{a}x+\frac{c}{a}=0$ Divide by a.

INSTRUCTOR'S RESOURCES

PTB Sec. 10.3 **ISM** Sec. 10.3
AB Sec. 10.3
TEST GENERATOR
WORLD WIDE WEB
www.LialAlgebra.com

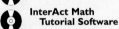

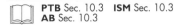

Step 2 Get the variable terms alone on the left side.

$x^2 + \frac{1}{2}x = \frac{5}{2}$ Add $\frac{5}{2}$.	$x^2 + \frac{b}{a}x = -\frac{c}{a}$ Subtract $\frac{c}{a}$.

Step 3 Add the square of half the coefficient of x to both sides, factor the left side, and combine terms on the right.

$x^2 + \frac{1}{2}x + \frac{1}{16} = \frac{5}{2} + \frac{1}{16}$ Add $\frac{1}{16}$.	$x^2 + \frac{b}{a}x + \frac{b^2}{4a^2} = -\frac{c}{a} + \frac{b^2}{4a^2}$ Add $\frac{b^2}{4a^2}$.
$\left(x + \frac{1}{4}\right)^2 = \frac{41}{16}$ Factor; add on right.	$\left(x + \frac{b}{2a}\right)^2 = \frac{b^2 - 4ac}{4a^2}$ Factor; add on right.

Step 4 Use the square root property to complete the solution.

$x + \frac{1}{4} = \pm\sqrt{\frac{41}{16}}$	$x + \frac{b}{2a} = \pm\sqrt{\frac{b^2 - 4ac}{4a^2}}$
$x + \frac{1}{4} = \pm\frac{\sqrt{41}}{4}$	$x = -\frac{b}{2a} \pm \frac{\sqrt{b^2 - 4ac}}{2a}$
$x = -\frac{1}{4} \pm \frac{\sqrt{41}}{4}$	$x = \frac{-b \pm \sqrt{b^2 - 4ac}}{2a}$
$x = \frac{-1 \pm \sqrt{41}}{4}$	

The final result in the column on the right is called the quadratic formula. *It is a key result that should be memorized.* Notice that there are two values, one for the + sign and one for the − sign.

Quadratic Formula

The solutions of the quadratic equation $ax^2 + bx + c = 0, a \neq 0$, are

$$x = \frac{-b + \sqrt{b^2 - 4ac}}{2a} \quad \text{and} \quad x = \frac{-b - \sqrt{b^2 - 4ac}}{2a}$$

or, in compact form,

$$x = \frac{-b \pm \sqrt{b^2 - 4ac}}{2a}.$$

TEACHING TIP Stress that the quadratic formula should be memorized, and that the values of *a*, *b*, and *c* are determined only after the equation is written in standard form.

CAUTION Notice that the fraction bar is under $-b$ as well as the radical. When using this formula, be sure to find the values of $-b \pm \sqrt{b^2 - 4ac}$ first, then divide those results by the value of $2a$.

CHALKBOARD EXAMPLE

Solve by using the quadratic formula.

$$2x^2 + x - 8 = 0$$

Answer: $\left\{\frac{-1 \pm \sqrt{65}}{4}\right\}$

EXAMPLE 2 Solving a Quadratic Equation by the Quadratic Formula

Solve $2x^2 + x - 5 = 0$ by the quadratic formula.

As found in Example 1(c), the values of a, b, and c are $a = 2$, $b = 1$, and $c = -5$. Substitute these numbers into the quadratic formula and simplify the result.

$$x = \frac{-b \pm \sqrt{b^2 - 4ac}}{2a}$$

$$x = \frac{-1 \pm \sqrt{(1)^2 - 4(2)(-5)}}{2(2)} \qquad \text{Let } a = 2,\ b = 1,\ c = -5.$$

$$x = \frac{-1 \pm \sqrt{1 + 40}}{4} \qquad \text{Perform the operations under the radical and in the denominator.}$$

$$x = \frac{-1 \pm \sqrt{41}}{4}$$

Notice that this result agrees with the result obtained in the left column when the quadratic formula was derived above. The solution set is $\left\{\frac{-1 \pm \sqrt{41}}{4}\right\}$.

CHALKBOARD EXAMPLE

Solve by using the quadratic formula.

$$-y^2 = 8y + 1$$

Answer: $\{-4 \pm \sqrt{15}\}$

EXAMPLE 3 Rewriting an Equation Before Using the Quadratic Formula

Solve $x^2 = 2x + 1$.

One side of the equation must be 0 before a, b, and c can be found. Subtract $2x$ and 1 from both sides of the equation to get

$$x^2 - 2x - 1 = 0.$$

Then $a = 1$, $b = -2$, and $c = -1$. Substitute these values into the quadratic formula.

$$x = \frac{-b \pm \sqrt{b^2 - 4ac}}{2a}$$

$$x = \frac{-(-2) \pm \sqrt{(-2)^2 - 4(1)(-1)}}{2(1)} \qquad \text{Let } a = 1,\ b = -2,\ c = -1.$$

$$x = \frac{2 \pm \sqrt{4 + 4}}{2} = \frac{2 \pm \sqrt{8}}{2}$$

$$x = \frac{2 \pm 2\sqrt{2}}{2} \qquad \sqrt{8} = \sqrt{4} \cdot \sqrt{2} = 2\sqrt{2}$$

Write the solutions in lowest terms by factoring $2 \pm 2\sqrt{2}$ as $2(1 \pm \sqrt{2})$ to get

$$x = \frac{2(1 \pm \sqrt{2})}{2} = 1 \pm \sqrt{2}.$$

The solution set is $\{1 \pm \sqrt{2}\}$.

OBJECTIVE 3 Solve quadratic equations with only one solution. The expression under the radical in the quadratic formula, $b^2 - 4ac$, is called the **discriminant.** The discriminant provides information about the number of real solutions of a quadratic equation. If the discriminant is positive, as in Examples 2 and 3, the equation has two real solutions. If it is zero, it has only one real solution (sometimes called a *double solution*). A negative discriminant indicates that there are no real solutions. The next example illustrates a quadratic equation with discriminant 0.

CHALKBOARD EXAMPLE

Solve $9x^2 = 42x - 49$.

Answer: $\left\{\frac{7}{3}\right\}$

EXAMPLE 4 Using the Quadratic Formula When There Is One Solution

Solve $4x^2 + 25 = 20x$.

Write the equation as $4x^2 - 20x + 25 = 0$. Here, $a = 4$, $b = -20$, and $c = 25$. By the quadratic formula,

$$x = \frac{-(-20) \pm \sqrt{(-20)^2 - 400}}{8} = \frac{20 \pm 0}{8} = \frac{5}{2}.$$

The solution set is $\left\{\frac{5}{2}\right\}$.

If the discriminant is 0, as in Example 4, then the quadratic trinomial is a perfect square, and the equation could have been solved by factoring. In Example 4, the expression $4x^2 - 20x + 25$ would be factored as $(2x - 5)^2$. Setting this equal to zero leads to the double solution $\frac{5}{2}$.

OBJECTIVE 4 Solve quadratic equations with fractions. It is usually easier to clear quadratic equations of fractions before solving them, as shown in the next example.

CHALKBOARD EXAMPLE

Solve by using the quadratic formula.

$$m^2 = \frac{2}{3}m + \frac{4}{9}$$

Answer: $\left\{\frac{1 \pm \sqrt{5}}{3}\right\}$

EXAMPLE 5 Solving a Quadratic Equation with Fractions

Solve the equation $\frac{1}{10}t^2 = \frac{2}{5}t - \frac{1}{2}$.

Eliminate the denominators by multiplying both sides of the equation by the least common denominator, 10.

$$10\left(\frac{1}{10}t^2\right) = 10\left(\frac{2}{5}t - \frac{1}{2}\right)$$

$$t^2 = 4t - 5 \quad \text{Distributive property}$$

$$t^2 - 4t + 5 = 0 \quad \text{Standard form}$$

TEACHING TIP Explain that the quadratic formula could be used with the fractions as coefficients, but using integer coefficients is much easier.

From this form identify $a = 1$, $b = -4$, and $c = 5$. Use the quadratic formula to complete the solution.

$$t = \frac{-(-4) \pm \sqrt{(-4)^2 - 4(1)(5)}}{2(1)} \quad \text{Substitute into the formula.}$$

$$t = \frac{4 \pm \sqrt{16 - 20}}{2} \quad \text{Perform the operations.}$$

$$t = \frac{4 \pm \sqrt{-4}}{2}$$

The discriminant -4 is less than 0. Because $\sqrt{-4}$ does not represent a real number, the solution set is $\emptyset$.

OBJECTIVE 5 **Solve an applied problem that leads to a quadratic equation.** The German astronomer Johannes Kepler (1571–1630) is responsible for discovering three laws of planetary motion. In 1619 he published his third law, which we examine in the next example. To understand this law, several terms must be defined. The *sidereal period* of a planet is the orbital period of a planet relative to the stars, or the time from one orbital point back to the same point. The *semimajor axis* of the elliptical orbit of a planet is half the length of the major (longer) axis of an ellipse. (See Figure 1.) One *astronomical unit* (AU) is the average distance between Earth and the sun.

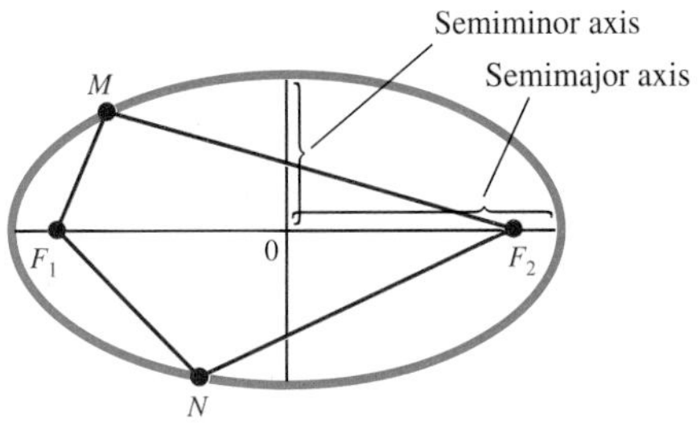

Figure 1

The orbit of Earth, as well as other planets in our solar system, takes the shape of an ellipse. In any ellipse, the sum of the distances of any point on the curve to each *focus* (labeled F_1 and F_2 in the figure) is constant. Thus $MF_1 + MF_2 = NF_1 + NF_2$, and so on for any point on the ellipse. The orbit of Earth has the sun at one focus.

CHALKBOARD EXAMPLE

Find P for Saturn, where $a = 9.54$.

Answer: $P = 29.47$

EXAMPLE 6 **Determining the Orbital Period P of Jupiter**

Kepler's third law of planetary motion states that the square of the sidereal period of a planet about the sun is proportional to the cube of its orbital semimajor axis. Relative to Earth's orbit, this harmonic law is usually expressed by the equation

$$P^2 = a^3,$$

where P is the planet's sidereal period in years and a is its orbital semimajor axis in AU. For Jupiter, $a = 5.20$. Find P. (*Source:* James B. Kaler, *Astronomy! A Brief Edition,* Addison Wesley, 1997.)

Substitute 5.20 for a in the formula to get

$$P^2 = (5.20)^3$$

$$P^2 = 140.608. \quad \text{Use a calculator.}$$

This is a quadratic equation in P. Apply the square root property. Using a calculator to find the positive square root of P to the nearest hundredth, we get

$$P = 11.86.$$

Thus, the sidereal period P for Jupiter is 11.86 years. (The negative square root, -11.86, is rejected since the period P must be a positive number.)

When we use a quadratic equation to solve an applied problem, it is important to check that the solution satisfies the physical requirements for the unknown. For example, it is impossible to have a negative length or a fractional number of people.

CHALKBOARD EXAMPLE

A ball is thrown vertically upward from the ground. Its distance in feet from the ground at t seconds is $s = -16t^2 + 64t$. At what times will the ball be 32 feet from the ground?

Answer: At .59 second and at 3.41 seconds

EXAMPLE 7 Solving an Applied Problem Using the Quadratic Formula

If an object is thrown upward on Earth from a height of 50 feet, with an initial velocity of 32 feet per second, then its height after t seconds is given by

$$h = -16t^2 + 32t + 50, \quad \text{where } h \text{ is in feet.}$$

After how many seconds will it reach a height of 30 feet?

We must find the value of t for which $h = 30$.

$$30 = -16t^2 + 32t + 50 \qquad \text{Let } h = 30.$$

$$16t^2 - 32t - 20 = 0 \qquad \text{Rewrite in standard form.}$$

$$4t^2 - 8t - 5 = 0 \qquad \text{Divide by 4.}$$

Use the quadratic formula, with $a = 4$, $b = -8$, and $c = -5$.

$$t = \frac{-b \pm \sqrt{b^2 - 4ac}}{2a}$$

$$t = \frac{-(-8) \pm \sqrt{(-8)^2 - 4(4)(-5)}}{2(4)}$$

$$t = \frac{8 \pm \sqrt{64 + 80}}{8}$$

$$t = \frac{8 \pm \sqrt{144}}{8} = \frac{8 \pm 12}{8}$$

The two solutions of the equation are

$$\frac{8 + 12}{8} = \frac{20}{8} = \frac{5}{2} \quad \text{and} \quad \frac{8 - 12}{8} = -\frac{4}{8} = -\frac{1}{2}.$$

Since t represents time, the solution $-\frac{1}{2}$ must be rejected, as time cannot be negative here. The object will reach a height of 30 feet after $\frac{5}{2}$, or 2.5, seconds.

10.3 EXERCISES

1. 4; 5; −9 **2.** 2 **3.** 2
4. 0; 1 **5.** $a = 3, b = -4, c = -2$
6. $a = 5, b = -3, c = 6$
7. $a = 3, b = 7, c = 0$
8. $a = 9, b = -8, c = 0$
9. $a = 1, b = 1, c = -12$
10. $a = 1, b = 12, c = 33$
11. $a = 9, b = 9, c = -26$
12. $a = 5, b = 14, c = -5$
13. If a were 0, the equation would be linear, not quadratic.

Fill in each blank with the correct response.

1. For the quadratic equation $4x^2 + 5x - 9 = 0$, the values of a, b, and c are respectively _____, _____, and _____.
2. To solve the equation $3x^2 - 5x = -2$ by the quadratic formula, the first step is to add _____ to both sides of the equation.
3. When using the quadratic formula, if the discriminant $b^2 - 4ac$ is positive, the equation has _____ real solution(s).
4. The discriminant for the equation $x^2 - 6x + 9 = 0$ is _____, and therefore the equation has _____ real solution(s).

Write each equation in the standard form $ax^2 + bx + c = 0$. Then identify the values of a, b, and c. Do not actually solve the equation. See Example 1.

5. $3x^2 = 4x + 2$
6. $5x^2 = 3x - 6$
7. $3x^2 = -7x$
8. $9x^2 = 8x$
9. $(x - 3)(x + 4) = 0$
10. $(x + 6)^2 = 3$
11. $9(x - 1)(x + 2) = 8$
12. $(3x - 1)(2x + 5) = x(x - 1)$
13. Why is the restriction $a \neq 0$ necessary in the definition of a quadratic equation?

14. In either case, the solution set is $\left\{\dfrac{-2 \pm \sqrt{10}}{2}\right\}$ or, equivalently, $\left\{\dfrac{2 \pm \sqrt{10}}{-2}\right\}$. **15.** No, because $2a$ should be the denominator for $-b$ as well. The correct formula is $x = \dfrac{-b \pm \sqrt{b^2 - 4ac}}{2a}$.

16. No, because the fraction bar should be under both $-b$ and $\sqrt{b^2 - 4ac}$. The term $2a$ should not be in the radicand. The correct formula is $x = \dfrac{-b \pm \sqrt{b^2 - 4ac}}{2a}$.

17. $\{-13, 1\}$ **18.** $\{-1, 9\}$

19. $\{2\}$ **20.** $\left\{-\dfrac{1}{3}\right\}$

21. $\left\{\dfrac{-6 \pm \sqrt{26}}{2}\right\}$

22. $\left\{\dfrac{-1 \pm \sqrt{21}}{10}\right\}$

23. $\left\{-1, \dfrac{5}{2}\right\}$ **24.** $\left\{-\dfrac{5}{2}, 6\right\}$

25. $\{-1, 0\}$ **26.** $\{0, 3\}$

27. $\left\{0, \dfrac{12}{7}\right\}$ **28.** $\left\{0, \dfrac{11}{9}\right\}$

29. $\{\pm 2\sqrt{6}\}$ **30.** $\{\pm 4\sqrt{6}\}$

31. $\left\{\pm\dfrac{2}{5}\right\}$ **32.** $\left\{\pm\dfrac{3}{4}\right\}$

33. $\left\{\dfrac{6 \pm 2\sqrt{6}}{3}\right\}$

34. $\left\{\dfrac{1 \pm \sqrt{13}}{4}\right\}$ **35.** $\emptyset$

36. $\emptyset$ **37.** $\emptyset$ **38.** $\emptyset$

39. $\left\{\dfrac{-5 \pm \sqrt{61}}{2}\right\}$

40. $\left\{\dfrac{-3 \pm \sqrt{57}}{4}\right\}$

41. (a) $\left\{\dfrac{-1 \pm \sqrt{11}}{2}\right\}$

(b) $\{-2.158, 1.158\}$

42. (a) $\left\{\dfrac{-1 \pm \sqrt{61}}{10}\right\}$

(b) $\{-.881, .681\}$

43. (a) $\left\{\dfrac{1 \pm \sqrt{5}}{2}\right\}$

(b) $\{-.618, 1.618\}$

44. (a) $\{2 \pm \sqrt{6}\}$

(b) $\{-.449, 4.449\}$

45. $\left\{-\dfrac{2}{3}, \dfrac{4}{3}\right\}$ **46.** $\left\{-1, \dfrac{5}{2}\right\}$

47. $\left\{\dfrac{-1 \pm \sqrt{73}}{6}\right\}$

14. To solve the quadratic equation $-2x^2 - 4x + 3 = 0$, we might choose to use $a = -2$, $b = -4$, and $c = 3$. On the other hand, we might decide to first multiply both sides by -1, obtaining the equation $2x^2 + 4x - 3 = 0$, and then use $a = 2$, $b = 4$, and $c = -3$. Show that in either case, we obtain the same solution set.

✍ **15.** A student writes the quadratic formula as $x = -b \pm \dfrac{\sqrt{b^2 - 4ac}}{2a}$. Is this correct? If not, explain the error, and give the correct formula.

✍ **16.** Another student writes the quadratic formula as $x = -b \pm \sqrt{\dfrac{b^2 - 4ac}{2a}}$. Is this correct? If not, explain the error, and give the correct formula.

Use the quadratic formula to solve each equation. Write all radicals in simplified form, and write all answers in lowest terms. See Examples 2–4.

17. $k^2 = -12k + 13$ **18.** $r^2 = 8r + 9$

19. $p^2 - 4p + 4 = 0$ **20.** $9x^2 + 6x + 1 = 0$

21. $2x^2 + 12x = -5$ **22.** $5m^2 + m = 1$

23. $2y^2 = 5 + 3y$ **24.** $2z^2 = 30 + 7z$

25. $6x^2 + 6x = 0$ **26.** $4n^2 - 12n = 0$

27. $7x^2 = 12x$ **28.** $9r^2 = 11r$

29. $x^2 - 24 = 0$ **30.** $z^2 - 96 = 0$

31. $25x^2 - 4 = 0$ **32.** $16x^2 - 9 = 0$

33. $3x^2 - 2x + 5 = 10x + 1$ **34.** $4x^2 - x + 4 = x + 7$

35. $-2x^2 = -3x + 2$ **36.** $-x^2 = -5x + 20$

37. $2x^2 + x + 5 = 0$ **38.** $3x^2 + 2x + 8 = 0$

39. $(x + 3)(x + 2) = 15$ **40.** $(2x + 1)(x + 1) = 7$

⊞ *Use the quadratic formula to solve each equation. (a) Give the solutions in* exact *form and* **(b)** *use a calculator to give the solutions correct to the nearest thousandth.*

41. $2x^2 + 2x = 5$ **42.** $5x^2 = 3 - x$

43. $x^2 = 1 + x$ **44.** $x^2 = 2 + 4x$

Use the quadratic formula to solve each equation. See Example 5.

45. $\dfrac{3}{2}k^2 - k - \dfrac{4}{3} = 0$ **46.** $\dfrac{2}{5}x^2 - \dfrac{3}{5}x - 1 = 0$

47. $\dfrac{1}{2}x^2 + \dfrac{1}{6}x = 1$ **48.** $\dfrac{2}{3}y^2 - \dfrac{4}{9}y = \dfrac{1}{3}$

49. $.5x^2 = x + .5$ **50.** $.25x^2 = -1.5x - 1$

51. $\dfrac{3}{8}x^2 - x + \dfrac{17}{24} = 0$ **52.** $\dfrac{1}{3}y^2 + \dfrac{8}{9}y + \dfrac{7}{9} = 0$

✍ **53.** If an applied problem leads to a quadratic equation, what must you be aware of after you have solved the equation?

54. Suppose that a problem asks you to find the length of a rectangle, and the problem leads to a quadratic equation. Which one of the following solutions to the equation cannot be an answer to the problem if L represents the length of the rectangle? Why?

(a) $L = 9$ **(b)** $L = 5\dfrac{1}{4}$ **(c)** $L = \dfrac{1 + \sqrt{5}}{2}$ **(d)** $L = \dfrac{1 - \sqrt{5}}{2}$

▲ **55.** Solve the formula $S = 2\pi rh + \pi r^2$ for r by first writing it in the form $ar^2 + br + c = 0$, and then using the quadratic formula.

▲ **56.** Solve the formula $V = \pi r^2h + \pi R^2h$ for r, using the method described in Exercise 55.

48. $\left\{\frac{2 \pm \sqrt{22}}{6}\right\}$
49. $\{1 \pm \sqrt{2}\}$
50. $\{-3 \pm \sqrt{5}\}$ **51.** $\emptyset$
52. $\emptyset$ **53.** The solution(s) must make sense in the original problem. (For example, a length cannot be negative.)
54. (d); Length cannot be negative.
55. $r = \frac{-\pi h \pm \sqrt{\pi^2h^2 + \pi S}}{\pi}$
56. $r = \pm\frac{\sqrt{(V - \pi hR^2)\pi h}}{\pi h}$
57. $P = .241$ (about 88 days)
58. $P = .615$ (about 224 days)
59. $P = 1.881$
60. $P = 4.610$
61. $P = 84.130$
62. $P = 165.139$
63. $1^2 = 1^3$, or $1 = 1$. This is a true statement.
64. 1 second, 3 seconds; Both answers are reasonable.
65. 4 seconds
66. 3.5 feet
67. approximately .2 second and 10.9 seconds
68. $\{16, -8\}$; Only 16 feet is a reasonable answer.
69. 30 units

*In each problem, find P, the planet's sidereal period in years, using Kepler's third law of planetary motion. The given value a is its orbital semimajor axis in AU. Refer to the formula in Example 6. (*Source: *James B. Kaler,* Astronomy! A Brief Edition, *Addison Wesley, 1997.)*

57. Mercury, $a = .387$

58. Venus, $a = .723$

59. Mars, $a = 1.524$

60. Ceres (the largest asteroid), $a = 2.77$

61. Uranus, $a = 19.2$

62. Neptune, $a = 30.1$

63. For Earth, $a = 1$ by definition and $P = 1$ year. Substitute these into the formula to show that Kepler's third law is valid for Earth.

Solve each problem. See Example 7.

64. The time t in seconds under certain conditions for a ball to be 48 feet in the air (on Earth) is given (approximately) by

$$48 = 64t - 16t^2.$$

Solve this equation for t. Are both answers reasonable?

65. A certain projectile is located $d = 2t^2 - 5t + 2$ feet from the ground after t seconds have elapsed. How many seconds will it take the projectile to be 14 feet from the ground?

66. A frog is sitting on a stump 3 feet above the ground. He hops off the stump and lands on the ground 4 feet away. During his leap, his height h is given by the equation $h = -.5x^2 + 1.25x + 3$, where x is the distance in feet from the base of the stump, and h is in feet. How far was the frog from the base of the stump when he was 1.25 feet above the ground?

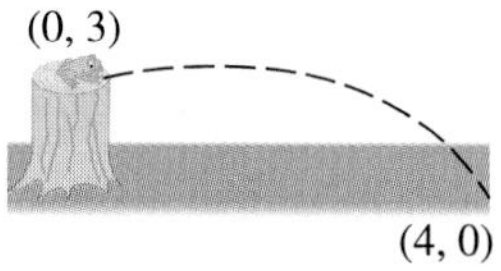

67. An astronaut on the moon throws a baseball upward. The height h of the ball, in feet, x seconds after he throws it, is given by the equation

$$h = -2.7x^2 + 30x + 6.5.$$

After how many seconds is the ball 12 feet above the moon's surface?

68. A rule for estimating the number of board feet of lumber that can be cut from a log depends on the diameter of the log. To find the diameter d required to get 9 board feet of lumber, we use the equation

$$\left(\frac{d - 4}{4}\right)^2 = 9.$$

Solve this equation for d. Are both answers reasonable?

69. An old Babylonian problem asks for the length of the side of a square, given that the area of the square minus the length of a side is 870. Find the length of the side. (*Source:* Howard Eves, *An Introduction to the History of Mathematics,* 6th Edition, Saunders College Publishing, 1990.)

70. It is the radicand in the quadratic formula.
71. (a) 225 (b) 169 (c) 4 (d) 121 72. Each is a perfect square. 73. (a) $(3x - 2)(6x + 1)$ (b) $(x + 2)(5x - 3)$ (c) $(8x + 1)(6x + 1)$ (d) $(x + 3)(x - 8)$
74. (a) 41 (b) −39 (c) 12 (d) 112 75. no
76. If the discriminant is a perfect square the trinomial is factorable. (a) yes (b) yes (c) no

RELATING CONCEPTS (EXERCISES 70–76)

In Chapter 5 we presented methods for factoring trinomials. Some trinomials cannot be factored using integer coefficients, however. There is a way to determine beforehand whether a trinomial of the form $ax^2 + bx + c$ can be factored using the discriminant.

Work Exercises 70–76 in order.

70. For the trinomial $ax^2 + bx + c$, the expression $b^2 - 4ac$ is called the discriminant. Where have you seen the discriminant before?

71. Each of the following trinomials is factorable. Find the discriminant for each one.
(a) $18x^2 - 9x - 2$ **(b)** $5x^2 + 7x - 6$
(c) $48x^2 + 14x + 1$ **(d)** $x^2 - 5x - 24$

72. What do you notice about the discriminants you found in Exercise 71.

73. Factor each of the trinomials in Exercise 71.

74. Each of the following trinomials is not factorable using the methods of Chapter 5. Find the discriminant for each one.
(a) $2x^2 + x - 5$ **(b)** $2x^2 + x + 5$ **(c)** $x^2 + 6x + 6$ **(d)** $3x^2 + 2x - 9$

75. Are any of the discriminants you found in Exercise 74 perfect squares?

76. Make a conjecture (an educated guess) concerning when a trinomial of the form $ax^2 + bx + c$ is factorable. Then use your conjecture to determine whether each trinomial is factorable. (Do not actually factor.)
(a) $42x^2 + 117x + 66$ **(b)** $99x^2 + 186x - 24$ **(c)** $58x^2 + 184x + 27$

Did you make the connection between the nature of the discriminant and whether a trinomial with that discriminant can be factored?

SUMMARY Exercises on Quadratic Equations

Four algebraic methods have now been introduced for solving quadratic equations written in the form $ax^2 + bx + c = 0$. The chart below shows some advantages and some disadvantages of each method.

Method	*Advantages*	*Disadvantages*
1. Factoring	It is usually the fastest method.	Not all equations can be solved by factoring. Some factorable polynomials are difficult to factor.
2. Square root property	It is the simplest method for solving equations of the form $(ax + b)^2$ = a number.	Few equations are given in this form.
3. Completing the square	It can always be used. (Also, the procedure is useful in other areas of mathematics.)	It requires more steps than other methods.
4. Quadratic formula	It can always be used.	It is more difficult than factoring because of the $\sqrt{b^2 - 4ac}$ expression.

SUMMARY EXERCISES

Solve each quadratic equation by the method of your choice.

1. $s^2 = 36$

2. $x^2 + 3x = -1$

3. $y^2 - \dfrac{100}{81} = 0$

4. $81t^2 = 49$

5. $z^2 - 4z + 3 = 0$

6. $w^2 + 3w + 2 = 0$

7. $z(z - 9) = -20$

8. $x^2 + 3x - 2 = 0$

9. $(3k - 2)^2 = 9$

10. $(2s - 1)^2 = 10$

11. $(x + 6)^2 = 121$

12. $(5k + 1)^2 = 36$

13. $(3r - 7)^2 = 24$

14. $(7p - 1)^2 = 32$

15. $(5x - 8)^2 = -6$

16. $2t^2 + 1 = t$

17. $-2x^2 = -3x - 2$

18. $-2x^2 + x = -1$

19. $8z^2 = 15 + 2z$

20. $3k^2 = 3 - 8k$

21. $0 = -x^2 + 2x + 1$

22. $3x^2 + 5x = -1$

23. $5y^2 - 22y = -8$

24. $y(y + 6) + 4 = 0$

25. $(x + 2)(x + 1) = 10$

26. $16x^2 + 40x + 25 = 0$

27. $4x^2 = -1 + 5x$

28. $2p^2 = 2p + 1$

29. $3m(3m + 4) = 7$

30. $5x - 1 + 4x^2 = 0$

31. $\dfrac{r^2}{2} + \dfrac{7r}{4} + \dfrac{11}{8} = 0$

32. $t(15t + 58) = -48$

33. $9k^2 = 16(3k + 4)$

34. $\dfrac{1}{5}x^2 + x + 1 = 0$

35. $y^2 - y + 3 = 0$

36. $4m^2 - 11m + 8 = -2$

37. $-3x^2 + 4x = -4$

38. $z^2 - \dfrac{5}{12}z = \dfrac{1}{6}$

39. $5k^2 + 19k = 2k + 12$

40. $\dfrac{1}{2}n^2 - n = \dfrac{15}{2}$

41. $k^2 - \dfrac{4}{15} = -\dfrac{4}{15}k$

42. If $D > 0$ and $\dfrac{5 + \sqrt{D}}{3}$ is a solution of $ax^2 + bx + c = 0$, what must be another solution of the equation?

43. How would you respond to this statement? "Since I know how to solve quadratic equations by the factoring method, there is no reason for me to learn any other method of solving quadratic equations."

44. How many real solutions are there for a quadratic equation that has a negative number as its radicand in the quadratic formula?

1. $\{\pm 6\}$ **2.** $\left\{\dfrac{-3 \pm \sqrt{5}}{2}\right\}$ **3.** $\left\{\pm\dfrac{10}{9}\right\}$ **4.** $\left\{\pm\dfrac{7}{9}\right\}$ **5.** $\{1, 3\}$ **6.** $\{-2, -1\}$ **7.** $\{4, 5\}$ **8.** $\left\{\dfrac{-3 \pm \sqrt{17}}{2}\right\}$ **9.** $\left\{-\dfrac{1}{3}, \dfrac{5}{3}\right\}$ **10.** $\left\{\dfrac{1 \pm \sqrt{10}}{2}\right\}$ **11.** $\{-17, 5\}$ **12.** $\left\{-\dfrac{7}{5}, 1\right\}$ **13.** $\left\{\dfrac{7 \pm 2\sqrt{6}}{3}\right\}$ **14.** $\left\{\dfrac{1 \pm 4\sqrt{2}}{7}\right\}$ **15.** $\emptyset$ **16.** $\emptyset$ **17.** $\left\{-\dfrac{1}{2}, 2\right\}$ **18.** $\left\{-\dfrac{1}{2}, 1\right\}$ **19.** $\left\{-\dfrac{5}{4}, \dfrac{3}{2}\right\}$ **20.** $\left\{-3, \dfrac{1}{3}\right\}$ **21.** $\{1 \pm \sqrt{2}\}$ **22.** $\left\{\dfrac{-5 \pm \sqrt{13}}{6}\right\}$ **23.** $\left\{\dfrac{2}{5}, 4\right\}$ **24.** $\{-3 \pm \sqrt{5}\}$ **25.** $\left\{\dfrac{-3 \pm \sqrt{41}}{2}\right\}$ **26.** $\left\{-\dfrac{5}{4}\right\}$ **27.** $\left\{\dfrac{1}{4}, 1\right\}$ **28.** $\left\{\dfrac{1 \pm \sqrt{3}}{2}\right\}$ **29.** $\left\{\dfrac{-2 \pm \sqrt{11}}{3}\right\}$ **30.** $\left\{\dfrac{-5 \pm \sqrt{41}}{8}\right\}$ **31.** $\left\{\dfrac{-7 \pm \sqrt{5}}{4}\right\}$ **32.** $\left\{-\dfrac{8}{3}, -\dfrac{6}{5}\right\}$ **33.** $\left\{\dfrac{8 \pm 8\sqrt{2}}{3}\right\}$ **34.** $\left\{\dfrac{-5 \pm \sqrt{5}}{2}\right\}$ **35.** $\emptyset$ **36.** $\emptyset$ **37.** $\left\{-\dfrac{2}{3}, 2\right\}$ **38.** $\left\{-\dfrac{1}{4}, \dfrac{2}{3}\right\}$ **39.** $\left\{-4, \dfrac{3}{5}\right\}$ **40.** $\{-3, 5\}$ **41.** $\left\{-\dfrac{2}{3}, \dfrac{2}{5}\right\}$ **42.** $\dfrac{5 - \sqrt{D}}{3}$ **43.** We need to know other methods because many quadratic equations cannot be solved by elementary factoring methods. **44.** There are no real solutions.

JOURNAL WRITING ▲ CHALLENGING SCIENTIFIC CALCULATOR GRAPHING CALCULATOR

10.4 Complex Numbers

OBJECTIVES

1. Write complex numbers as multiples of i.
2. Add and subtract complex numbers.
3. Multiply complex numbers.
4. Write complex number quotients in standard form.
5. Solve quadratic equations with complex number solutions.

FOR EXTRA HELP

SSG Sec. 10.4
SSM Sec. 10.4
Pass the Test Software
InterAct Math Tutorial Software
Video 17

As shown earlier in this chapter, some quadratic equations have no real number solutions. For example, the number

$$\frac{4 \pm \sqrt{-4}}{2},$$

which occurred in the solution of Example 5 in the previous section, is not a real number, because -4 appears as the radicand. For every quadratic equation to have a solution, we need a new set of numbers that includes the real numbers. This new set of numbers is defined using a new number i having the properties given below.

The Number i

$$i = \sqrt{-1} \quad \text{and} \quad i^2 = -1$$

OBJECTIVE 1 Write complex numbers as multiples of i. We can now write numbers like $\sqrt{-4}$, $\sqrt{-5}$, and $\sqrt{-8}$ as multiples of i, using a generalization of the product rule for radicals, as in the next example.

CHALKBOARD EXAMPLE

Write the number as a multiple of i.

$$\sqrt{-15}$$

Answer: $i\sqrt{15}$

EXAMPLE 1 Simplifying Square Roots of Negative Numbers

Write each number as a multiple of i.

(a) $\sqrt{-4} = \sqrt{-1 \cdot 4} = \sqrt{-1} \cdot \sqrt{4} = i\sqrt{4} = i \cdot 2 = 2i$

(b) $\sqrt{-5} = \sqrt{-1 \cdot 5} = \sqrt{-1} \cdot \sqrt{5} = i\sqrt{5}$

(c) $\sqrt{-8} = i\sqrt{8} = i \cdot 2 \cdot \sqrt{2} = 2i\sqrt{2}$

CAUTION It is easy to mistake $\sqrt{2}\,i$ for $\sqrt{2i}$, with the i under the radical. For this reason, it is customary to write the i factor first when it is multiplied by a radical. For example, we usually write $i\sqrt{2}$ rather than $\sqrt{2}\,i$.

Numbers that are nonzero multiples of i are *imaginary numbers.* The *complex numbers* include all real numbers and all imaginary numbers.

TEACHING TIP Show how this definition implies that real numbers are also complex numbers (by letting $b = 0$).

Complex Number

A **complex number** is a number of the form $a + bi$, where a and b are real numbers. If $b \neq 0$, $a + bi$ is also an **imaginary number.**

For example, the real number 2 is a complex number since it can be written as $2 + 0i$. Also, the imaginary number $3i = 0 + 3i$ is a complex number. Other complex numbers are

$$3 - 2i, \qquad 1 + i\sqrt{2}, \qquad \text{and} \qquad -5 + 4i.$$

In the complex number $a + bi$, a is called the **real part** and b (*not* bi) is called the **imaginary part.***

*Some texts *do* refer to bi as the imaginary part.

INSTRUCTOR'S RESOURCES

PTB Sec. 10.4 **ISM** Sec. 10.4 **AB** Sec. 10.4 | **TEST GENERATOR** | **WORLD WIDE WEB** www.LialAlgebra.com

A complex number written in the form $a + bi$ (or $a + ib$) is in **standard form.** Figure 2 shows the relationships among the various types of numbers discussed in this book. (Compare this figure to Figure 8 in Chapter 1.)

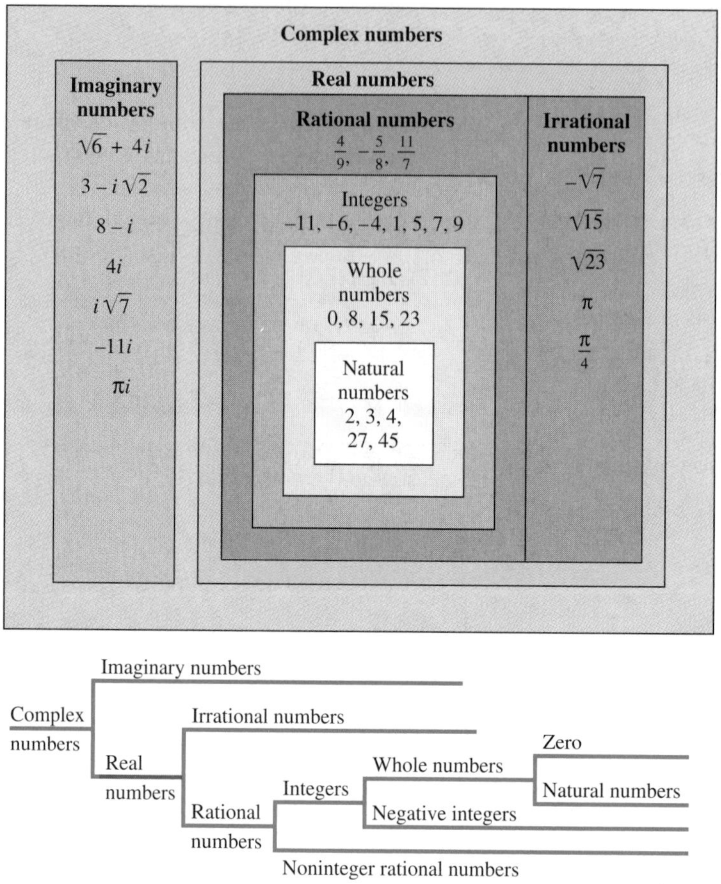

Figure 2

TEACHING TIP Explain that when we introduced radicals, we learned how to add, subtract, multiply, and divide with them. Now we introduce a new type of number, and similarly we must learn how to perform the four operations of arithmetic with these numbers.

OBJECTIVE 2 **Add and subtract complex numbers.** Adding and subtracting complex numbers is similar to adding and subtracting binomials.

Addition and Subtraction of Complex Numbers

1. To add complex numbers, add their real parts and add their imaginary parts.
2. To subtract complex numbers, change the number following the subtraction sign to its negative, and then add.

The properties of Section 1.7 (commutative, associative, etc.) also hold for operations with complex numbers.

CHALKBOARD EXAMPLE

Subtract $(-6 - i) - (-5 - 4i)$.

Answer: $-1 + 3i$

EXAMPLE 2 Adding and Subtracting Complex Numbers

Add or subtract.

(a) $(2 - 6i) + (7 + 4i) = (2 + 7) + (\mathbf{-6 + 4})i = 9 - 2i$

(b) $3i + (-2 - i) = -2 + (3 - 1)i = -2 + 2i$

(c) $(2 + 6i) - (-4 + i)$
Change $-4 + i$ to its negative, and then add.

$$\begin{aligned}(2 + 6i) - (-4 + i) &= (2 + 6i) + (4 - i) && -(-4 + i) = 4 - i \\ &= (2 + 4) + (6 - 1)i && \text{Commutative, associative, and distributive properties} \\ &= 6 + 5i\end{aligned}$$

(d) $(-1 + 2i) - 4 = (-1 - 4) + 2i = -5 + 2i$

TEACHING TIP Warn students that i is not equal to -1. This is a common misconception.

OBJECTIVE 3 Multiply complex numbers. We multiply complex numbers as we do polynomials. Since $i^2 = -1$ by definition, whenever i^2 appears, we replace it with -1.

CHALKBOARD EXAMPLE
Multiply $(1 - 5i)(3 + 7i)$.
Answer: $38 - 8i$

EXAMPLE 3 Multiplying Complex Numbers

Find the following products.

(a)
$$\begin{aligned}3i(2 - 5i) &= 6i - 15i^2 && \text{Distributive property} \\ &= 6i - 15(-1) && i^2 = -1 \\ &= 6i + 15 \\ &= 15 + 6i && \text{Commutative property}\end{aligned}$$

The last step gives the result in standard form.

(b) $(4 - 3i)(2 + 5i)$
Use FOIL.

$$\begin{aligned}(4 - 3i)(2 + 5i) &= 4(2) + 4(5i) + (-3i)(2) + (-3i)(5i) \\ &= 8 + 20i - 6i - 15i^2 \\ &= 8 + 14i - 15(-1) \\ &= 8 + 14i + 15 \\ &= 23 + 14i\end{aligned}$$

(c)
$$\begin{aligned}(1 + 2i)(1 - 2i) &= 1 - 2i + 2i - 4i^2 \\ &= 1 - 4(-1) \\ &= 1 + 4 \\ &= 5\end{aligned}$$

OBJECTIVE 4 Write complex number quotients in standard form. The quotient of two complex numbers is expressed in standard form by changing the denominator into a real number. For example, to write

$$\frac{8 + i}{1 + 2i}$$

in standard form, the denominator must be a real number. As seen in Example 3(c), the product $(1 + 2i)(1 - 2i)$ is 5, a real number. This suggests multiplying the numerator and denominator of the given quotient by $1 - 2i$ as follows.

$$\frac{8+i}{1+2i} = \frac{8+i}{1+2i} \cdot \frac{1-2i}{1-2i}$$

$$= \frac{8-16i+i-2i^2}{1-4i^2} \qquad \text{Multiply.}$$

$$= \frac{8-16i+i-2(-1)}{1-4(-1)} \qquad i^2 = -1$$

$$= \frac{10-15i}{5} \qquad \text{Combine terms.}$$

$$= \frac{5(2-3i)}{5} = 2-3i \qquad \text{Factor and write in standard form.}$$

Recall that this is the method used to rationalize some radical expressions in Chapter 9. The complex numbers $1 + 2i$ and $1 - 2i$ are *conjugates.* That is, the **conjugate** of the complex number $a + bi$ is $a - bi$. Multiplying the complex number $a + bi$ by its conjugate $a - bi$ gives the real number $a^2 + b^2$.

Product of Conjugates

$$(a + bi)(a - bi) = a^2 + b^2$$

The product of a complex number and its conjugate is the sum of the squares of the real and imaginary parts.

CHALKBOARD EXAMPLE

Write the quotient in standard form.

$$\frac{6+2i}{4-3i}$$

Answer: $\frac{18}{25} + \frac{26}{25}i$

EXAMPLE 4 Dividing Complex Numbers

Write the following quotients in standard form.

(a) $\frac{-4+i}{2-i}$

Multiply numerator and denominator by $2 + i$, the conjugate of the denominator.

$$\frac{-4+i}{2-i} \cdot \frac{2+i}{2+i} = \frac{-8-4i+2i+i^2}{4-i^2}$$

$$= \frac{-8-2i-1}{4-(-1)} \qquad i^2 = -1$$

$$= \frac{-9-2i}{5}$$

$$= -\frac{9}{5} - \frac{2}{5}i \qquad \text{Standard form}$$

(b) $\frac{3+i}{-i}$

Here, the conjugate of $0 - i$ is $0 + i$, or i.

$$\frac{3+i}{-i} \cdot \frac{i}{i} = \frac{3i+i^2}{-i^2}$$

$$= \frac{-1+3i}{-(-1)} \qquad i^2 = -1\text{; commutative property}$$

$$= -1 + 3i$$

1. $\langle -1, -4\rangle, \langle 0, 2\rangle, \langle -5, 0\rangle$
2. -1; 2

CONNECTIONS

The complex number $a + bi$ is also written with the notation $\langle a, b\rangle$. (Note the similarity to an ordered pair.) This notation suggests a way to graph complex numbers on a plane in a manner similar to the way we graph ordered pairs. For graphing complex numbers, the x-axis is called the *real axis* and the y-axis is called the *imaginary axis.* For example, we graph the complex number $2 + 3i$ or $\langle 2, 3\rangle$ just as we would the ordered pair $(2, 3)$, as shown in the figure. The figure also shows the graphs of the complex numbers $-1 - 4i$, $2i$, and -5.

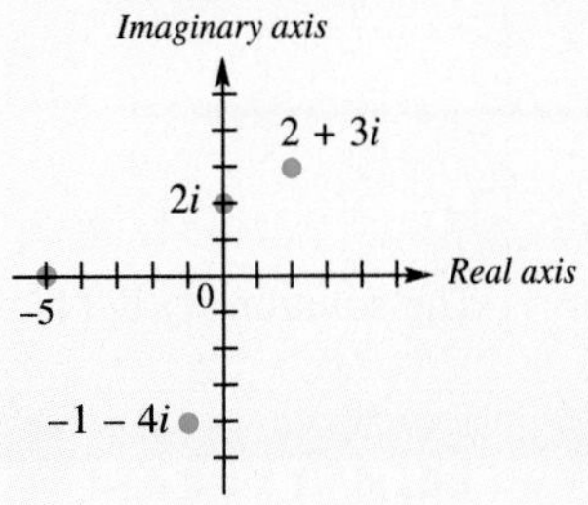

FOR DISCUSSION OR WRITING

1. Give the alternative notation for the last three complex numbers graphed above.
2. What is the real part of the complex number $-1 + 2i$? What is its imaginary part? Explain why we call the axes the real axis and the imaginary axis.

OBJECTIVE 5 Solve quadratic equations with complex number solutions. Quadratic equations that have no real solutions do have complex solutions, as shown in the next examples.

CHALKBOARD EXAMPLE

Solve $(x - 2)^2 = -64$.

Answer: $\{2 \pm 8i\}$

EXAMPLE 5 Solving a Quadratic Equation with Complex Solutions (Square Root Method)

Solve $(x + 3)^2 = -25$ for complex solutions.

Use the square root property.

$$(x + 3)^2 = -25$$

$$x + 3 = \sqrt{-25} \quad \text{or} \quad x + 3 = -\sqrt{-25}$$

Since $\sqrt{-25} = 5i$,

$$x + 3 = 5i \quad \text{or} \quad x + 3 = -5i$$

$$x = -3 + 5i \quad \text{or} \quad x = -3 - 5i.$$

The solution set is $\{-3 \pm 5i\}$.

CHALKBOARD EXAMPLE

Solve $x^2 - 2x = -26$.

Answer: $\{1 \pm 5i\}$

EXAMPLE 6 Solving a Quadratic Equation with Complex Solutions (Quadratic Formula)

Solve $2p^2 = 4p - 5$ for complex solutions.

Write the equation as $2p^2 - 4p + 5 = 0$. Then $a = 2$, $b = -4$, and $c = 5$. The solutions are

$$p = \frac{-(-4) \pm \sqrt{(-4)^2 - 4(2)(5)}}{2(2)}$$

$$= \frac{4 \pm \sqrt{16 - 40}}{4}$$

$$= \frac{4 \pm \sqrt{-24}}{4}.$$

Since $\sqrt{-24} = i\sqrt{24} = i \cdot \sqrt{4} \cdot \sqrt{6} = i \cdot 2 \cdot \sqrt{6} = 2i\sqrt{6}$,

$$p = \frac{4 \pm 2i\sqrt{6}}{4}$$

$$p = \frac{2(2 \pm i\sqrt{6})}{4} \quad \text{Factor out a 2.}$$

$$p = \frac{2 \pm i\sqrt{6}}{2} \quad \text{Lowest terms}$$

$$p = \frac{2}{2} \pm \frac{i\sqrt{6}}{2} \quad \text{Separate into real and imaginary parts.}$$

$$p = 1 \pm \frac{\sqrt{6}}{2}i. \quad \text{Standard form}$$

The solution set is $\left\{1 \pm \frac{\sqrt{6}}{2}i\right\}$.

10.4 EXERCISES

1. $3i$ 2. $6i$ 3. $2i\sqrt{5}$ 4. $3i\sqrt{3}$ 5. $3i\sqrt{2}$ 6. $5i\sqrt{2}$ 7. $5i\sqrt{5}$ 8. $7i\sqrt{2}$ 9. $5 + 3i$ 10. $11 + 3i$ 11. $6 - 9i$ 12. $-2 + 3i$ 13. $6 - 7i$ 14. $4 + 5i$ 15. $14 + 5i$ 16. $29 + 3i$ 17. $7 - 22i$ 18. $104 + 8i$ 19. 45 20. 125

Write each number as a multiple of i. See Example 1.

1. $\sqrt{-9}$ **2.** $\sqrt{-36}$ **3.** $\sqrt{-20}$ **4.** $\sqrt{-27}$

5. $\sqrt{-18}$ **6.** $\sqrt{-50}$ **7.** $\sqrt{-125}$ **8.** $\sqrt{-98}$

Add or subtract as indicated. See Example 2.

9. $(2 + 8i) + (3 - 5i)$ **10.** $(4 + 5i) + (7 - 2i)$

11. $(8 - 3i) - (2 + 6i)$ **12.** $(1 + i) - (3 - 2i)$

13. $(3 - 4i) + (6 - i) - (3 + 2i)$ **14.** $(5 + 8i) - (4 + 2i) + (3 - i)$

Find each product. See Example 3.

15. $(3 + 2i)(4 - i)$ **16.** $(9 - 2i)(3 + i)$ **17.** $(5 - 4i)(3 - 2i)$

18. $(10 + 6i)(8 - 4i)$ **19.** $(3 + 6i)(3 - 6i)$ **20.** $(11 - 2i)(11 + 2i)$

TECHNOLOGY INSIGHTS (EXERCISES 21-28)

Modern graphing calculators are capable of performing operations with complex numbers. The top screen shows how the TI-83 calculator can be set for complex mode ($a + bi$). The lower left screen shows how the square root of a negative number returns the product of a real number and i, and how addition and subtraction of complex numbers is accomplished. The lower right screen shows how multiplication is performed, how the calculator returns the real part of a complex number, and how the conjugate is given.

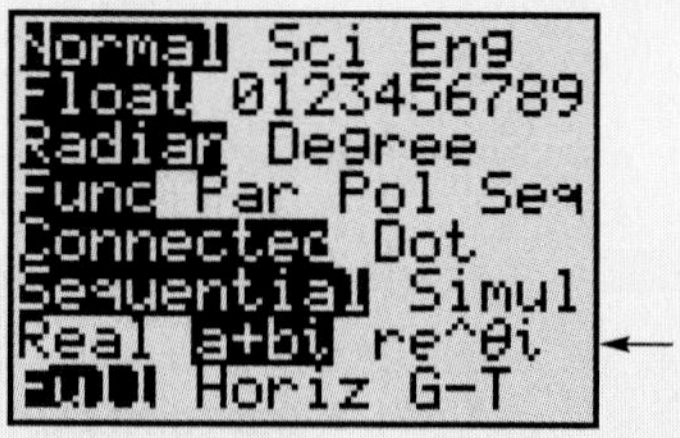

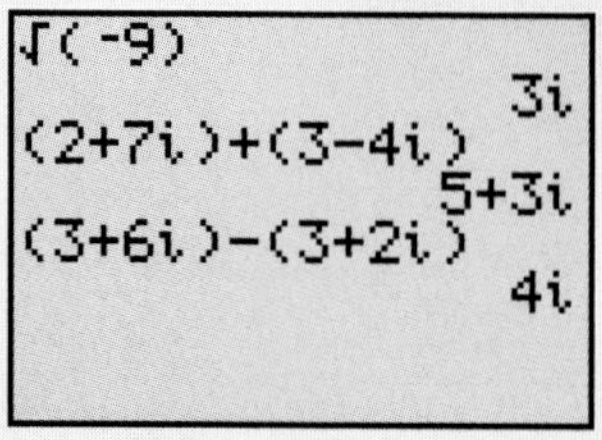

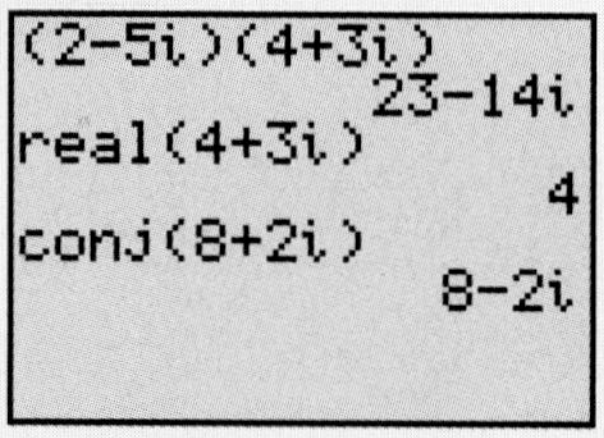

21. $18i$ 22. $16i$ 23. -4
24. $-9 + 10i$ 25. $6 + 4i$
26. 10 27. $4 - 3i$
28. $-5 + 12i$ 29. $3 - i$
30. $5 - i$ 31. $2 - 6i$
32. $3 - 2i$ 33. $-\frac{3}{25} + \frac{4}{25}i$
34. $-\frac{2}{5} - \frac{1}{5}i$
35. $-1 + 3i$ 36. The product of $-1 + 3i$ and $2 + 9i$ is $-29 - 3i$, which *is* the original dividend.
37. $-3 - 4i$ 38. The product of $-3 - 4i$ and i is $4 - 3i$, which *is* the original dividend. 39. Because $(3 - 2i)(4 + i) = 14 - 5i$ is true, the given statement is true.
40. The product of the quotient and the divisor must equal the dividend.
41. $\{-1 \pm 2i\}$ 42. $\{5 \pm 6i\}$
43. $\{3 \pm i\sqrt{5}\}$
44. $\{-6 \pm 2i\sqrt{3}\}$
45. $\left\{-\frac{2}{3} \pm i\sqrt{2}\right\}$

TECHNOLOGY INSIGHTS (EXERCISES 21-28) (CONTINUED)

Predict the display the calculator will give for each of the following screens.

21. `√(-169)+√(-25)`

22. `√(-36)+√(-100)`

23. `(5-i)+(-2+3i)-(7+2i)`

24. `3(1+2i)-4(3-i)`

(*Hint:* Use the order of operations.)

25. `conj(6-4i)`

26. `imag(5+10i)`

27. `(15-5i)/(3+i)`

28. `(2+3i)²`

Write each quotient in standard form. See Example 4.

29. $\frac{17 + i}{5 + 2i}$

30. $\frac{21 + i}{4 + i}$

31. $\frac{40}{2 + 6i}$

32. $\frac{13}{3 + 2i}$

33. $\frac{i}{4 - 3i}$

34. $\frac{-i}{1 + 2i}$

RELATING CONCEPTS (EXERCISES 35-40)

When you first divided whole numbers, you probably learned to check your work by multiplying your answer (the quotient) by the number doing the dividing (the divisor). For example,

$$\frac{2744}{28} = 98 \text{ is true, because } 98 \times 28 = 2744.$$

This same procedure works for real numbers other than whole numbers. Does it work for complex numbers?

(continued)

46. $\left\{\frac{1}{4} \pm \frac{\sqrt{5}}{2}i\right\}$ 47. $\{1 \pm i\}$

48. $\left\{-\frac{1}{2} \pm \frac{\sqrt{11}}{2}i\right\}$

49. $\left\{-\frac{3}{4} \pm \frac{\sqrt{31}}{4}i\right\}$

50. $\left\{\frac{1}{3} \pm \frac{2\sqrt{2}}{3}i\right\}$

51. $\left\{\frac{3}{2} \pm \frac{\sqrt{7}}{2}i\right\}$

52. $\left\{-\frac{1}{4} \pm \frac{\sqrt{23}}{4}i\right\}$

53. $\left\{\frac{1}{5} \pm \frac{\sqrt{14}}{5}i\right\}$

54. $\left\{-\frac{1}{6} \pm \frac{\sqrt{5}}{6}i\right\}$

55. $\left\{-\frac{1}{2} \pm \frac{\sqrt{13}}{2}i\right\}$

56. $\left\{-\frac{1}{4} \pm \frac{\sqrt{11}}{4}i\right\}$

57. $\left\{\frac{1}{2} \pm \frac{\sqrt{11}}{2}i\right\}$

58. $\left\{\frac{1}{4} \pm \frac{\sqrt{11}}{4}i\right\}$

59. If the discriminant $b^2 - 4ac$ is negative, the equation will have solutions that are not real. 60. conjugates 61. true 62. true 63. false; For example, $3 + 2i$ is a complex number but it is not real. 64. true 65. To add complex numbers, add the real parts to find the real part of the sum and add the imaginary parts to find the imaginary part of the sum. For example, $(5 + 2i) + (6 + 3i) = (5 + 6) + (2 + 3)i = 11 + 5i$. Subtraction is done similarly, but we subtract the real and imaginary parts instead: $(5 + 2i) - (6 + 3i) = (5 - 6) + (2 - 3)i = -1 - i$. To multiply complex numbers, use the FOIL method and replace i^2 with -1. Then combine terms. For example, $(5 + 2i)(6 + 3i) = 30 + 15i + 12i + 6i^2 = 30 + 15i + 12i + 6(-1) = 24 + 27i$. To divide complex numbers, multiply both the numerator and the denominator by the conjugate of the denominator. For example, $\frac{3 + i}{2 - i} = \frac{3 + i}{2 - i} \cdot \frac{2 + i}{2 + i} = \frac{6 + 5i + i^2}{4 - i^2} = \frac{5 + 5i}{5} = 1 + i$.

RELATING CONCEPTS (EXERCISES 35-40) (CONTINUED)

Work Exercises 35–40 in order *to see whether it does or not.*

35. Find the standard form of the quotient $\frac{-29 - 3i}{2 + 9i}$.

36. Multiply your answer from Exercise 35 by the divisor, $2 + 9i$. What is your answer? Is it equal to the original dividend (the numerator), $-29 - 3i$?

37. Find the standard form of the quotient $\frac{4 - 3i}{i}$.

38. Multiply your answer from Exercise 37 by the divisor, i. What is your answer? Is it equal to the original dividend?

39. Use the pattern established in Exercises 35–38 to determine whether the following is true: $\frac{14 - 5i}{4 + i} = 3 - 2i$.

40. State a rule, based on your observations, that tells whether the answer to a division problem involving complex numbers is correct.

Did you make the connection that checking complex number division is similar to checking real number division?

Solve each quadratic equation for complex solutions by the square root property. Write solutions in standard form. See Example 5.

41. $(a + 1)^2 = -4$
42. $(p - 5)^2 = -36$
43. $(k - 3)^2 = -5$
44. $(y + 6)^2 = -12$
45. $(3x + 2)^2 = -18$
46. $(4z - 1)^2 = -20$

Solve each quadratic equation for complex solutions by the quadratic formula. Write solutions in standard form. See Example 6.

47. $m^2 - 2m + 2 = 0$
48. $b^2 + b + 3 = 0$
49. $2r^2 + 3r + 5 = 0$
50. $3q^2 = 2q - 3$
51. $p^2 - 3p + 4 = 0$
52. $2a^2 = -a - 3$
53. $5x^2 + 3 = 2x$
54. $6y^2 + 2y + 1 = 0$
55. $2m^2 + 7 = -2m$
56. $4z^2 + 2z + 3 = 0$
57. $r^2 + 3 = r$
58. $4q^2 - 2q + 3 = 0$

Exercises 59–60 deal with quadratic equations having real number coefficients.

59. Suppose you are solving a quadratic equation by the quadratic formula. How can you tell, before completing the solution, whether the equation will have solutions that are not real numbers?

60. Refer to the solutions in Examples 5 and 6, and complete the following statement: If a quadratic equation has imaginary solutions, they are _____ of each other.

Answer true *or* false *to each of the following. If false, say why.*

61. Every real number is a complex number.

62. Every imaginary number is a complex number.

63. Every complex number is a real number.

64. Some complex numbers are imaginary.

65. Write a paragraph explaining how to add, subtract, multiply, and divide complex numbers. Give examples.

10.5 More on Graphing Quadratic Equations; Quadratic Functions

OBJECTIVES

1. Graph quadratic equations of the form $y = ax^2 + bx + c$ $(a \neq 0)$.
2. Use the vertex formula and then graph a parabola.
3. Use a graph to determine the number of real solutions of a quadratic equation.
4. Solve applications using quadratic functions.

FOR EXTRA HELP

SSG Sec. 10.5
SSM Sec. 10.5

Pass the Test Software

InterAct Math Tutorial Software

Video 17

In Section 4.1 we graphed the quadratic equation $y = x^2$. By plotting points, we obtained the graph of a parabola, shown again here in Figure 3.

x	y
3	9
2	4
1	1
0	0
−1	1
−2	4
−3	9

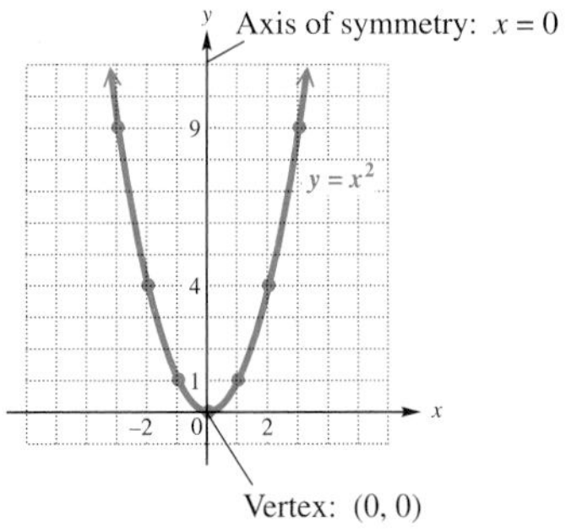

Figure 3

Recall that the lowest point on this graph is called the *vertex* of the parabola. (If the parabola opens downward, the vertex is the highest point.) The vertical line through the vertex is called the *axis,* or *axis of symmetry.* The two halves of the parabola are mirror images of each other across this axis.

We have also seen that other quadratic equations such as $y = -x^2 + 3$ and $y = (x + 2)^2$ also have parabolas as their graphs. See Figures 4 and 5.

x	y
2	−1
1	2
0	3
−1	2
−2	−1

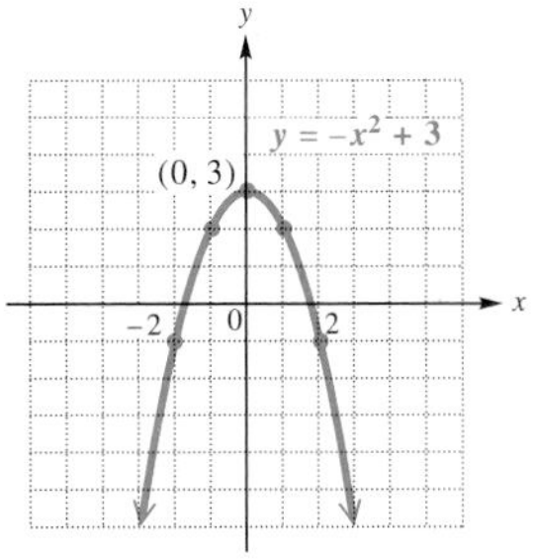

Figure 4

x	y
0	4
−1	1
−2	0
−3	1
−4	4

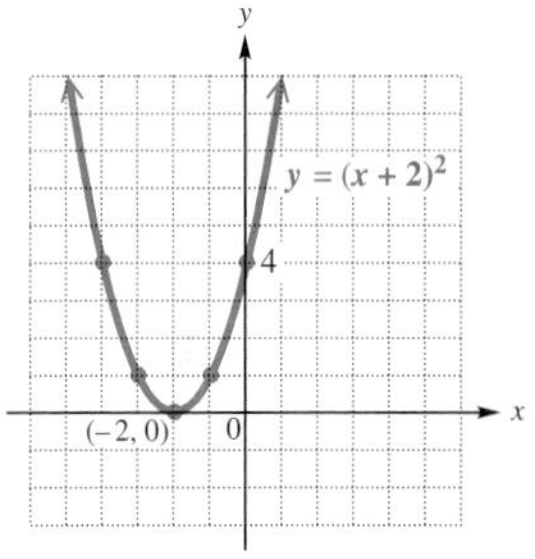

Figure 5

OBJECTIVE 1 **Graph quadratic equations of the form $y = ax^2 + bx + c$ $(a \neq 0)$.** Every equation of the form

$$y = ax^2 + bx + c,$$

with $a \neq 0$, has a graph that is a parabola. Because of its many useful properties, the parabola occurs frequently in real-life applications. For example, if an object is thrown into the air, the path that the object follows is a parabola (ignoring wind resistance). Cross sections of radar, spotlight, and telescope reflectors also form parabolas.

INSTRUCTOR'S RESOURCES

PTB Sec. 10.5 **ISM** Sec. 10.5
AB Sec. 10.5

TEST GENERATOR

WORLD WIDE WEB
www.LialAlgebra.com

When graphing parabolas, we are interested in finding the vertex, the x-intercept(s), if any, and the y-intercept. The first example shows how this is done.

CHALKBOARD EXAMPLE

Graph $y = x^2 + 2x - 8$.

Answer:

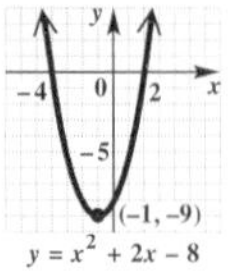

TEACHING TIP Explain that earlier in the course we solved linear equations in one variable of the form

$$Ax + B = C$$

and then graphed linear equations in two variables of the form

$$y = mx + b.$$

Now we are ready to do the same sort of thing for quadratic equations. We know how to solve

$$ax^2 + bx + c = 0$$

and now we will graph

$$y = ax^2 + bx + c.$$

EXAMPLE 1 Graphing a Parabola by Finding the Vertex and Intercepts

Graph $y = x^2 - 2x - 3$.

We want to find the vertex of the graph. Note in Figures 4 and 5 that the x-value of each vertex is exactly halfway between the x-intercepts. If a parabola has two x-intercepts this is always the case because of the symmetry of the figure. Therefore, begin by finding the x-intercepts. Let $y = 0$ in the equation and solve for x.

$$0 = x^2 - 2x - 3$$

$$0 = (x + 1)(x - 3) \qquad \text{Factor.}$$

$$x + 1 = 0 \quad \text{or} \quad x - 3 = 0 \qquad \text{Set each factor equal to 0.}$$

$$x = -1 \quad \text{or} \quad x = 3$$

There are two x-intercepts, $(-1, 0)$ and $(3, 0)$. Now find any y-intercepts. Substitute $x = 0$ in the equation.

$$y = 0^2 - 2(0) - 3 = -3$$

There is one y-intercept, $(0, -3)$.

As mentioned above, the x-value of the vertex is halfway between the x-values of the two x-intercepts. Thus, it is half their sum.

$$x = \frac{1}{2}(-1 + 3) = 1$$

Find the corresponding y-value by substituting 1 for x in the equation.

$$y = 1^2 - 2(1) - 3 = -4$$

The vertex is $(1, -4)$. The axis is the line $x = 1$. Plot the three intercepts and the vertex. Find additional ordered pairs as needed. For example, if $x = 2$,

$$y = 2^2 - 2(2) - 3 = -3,$$

leading to the ordered pair $(2, -3)$. A table of values with the ordered pairs we have found is shown with the graph in Figure 6.

	x	y
	-2	5
x-intercept	-1	0
y-intercept	0	-3
vertex	1	-4
	2	-3
x-intercept	3	0
	4	5

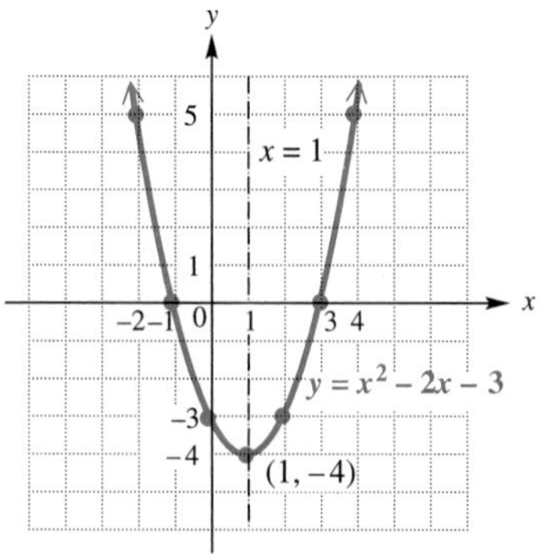

Figure 6

OBJECTIVE 2 **Use the vertex formula and then graph a parabola.** We can generalize from Example 1. The x-values of the x-intercepts for the equation $y = ax^2 + bx + c$, by the quadratic formula, are

$$x = \frac{-b + \sqrt{b^2 - 4ac}}{2a} \quad \text{and} \quad x = \frac{-b - \sqrt{b^2 - 4ac}}{2a}.$$

Thus, the x-value of the vertex is

$$x = \frac{1}{2}\left(\frac{-b + \sqrt{b^2 - 4ac}}{2a} + \frac{-b - \sqrt{b^2 - 4ac}}{2a}\right)$$

$$x = \frac{1}{2}\left(\frac{-b + \sqrt{b^2 - 4ac} - b - \sqrt{b^2 - 4ac}}{2a}\right)$$

$$x = \frac{1}{2}\left(\frac{-2b}{2a}\right) = -\frac{b}{2a}.$$

For the equation in Example 1, $y = x^2 - 2x - 3$, $a = 1$ and $b = -2$. Thus, the x-value of the vertex is

$$x = -\frac{b}{2a} = -\frac{-2}{2(1)} = 1,$$

which is the same x-value for the vertex we found in Example 1. (The x-value of the vertex is $x = -\frac{b}{2a}$ even if the graph has no x-intercepts.) A procedure for graphing quadratic equations follows.

Graphing the Parabola $y = ax^2 + bx + c$

Step 1 Find the y-intercept by letting $x = 0$.

Step 2 Find the x-intercepts, if any, by letting $y = 0$ and solving the equation.

Step 3 Find the vertex. Let $x = -\frac{b}{2a}$ and find the corresponding y-value by substituting for x in the equation.

Step 4 Plot the intercepts and the vertex.

Step 5 Find and plot additional ordered pairs near the vertex and intercepts as needed, using symmetry about the axis of the parabola.

CHALKBOARD EXAMPLE

Graph $y = -x^2 + 2x + 4$.

Answer:

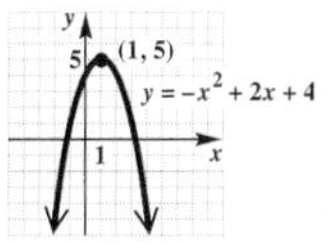

EXAMPLE 2 **Using the Steps to Graph a Parabola**

Graph $y = x^2 - 4x + 1$.

Step 1 To find the y-intercept, let $x = 0$ to get $y = 0^2 - 4(0) + 1 = 1$. The y-intercept is $(0, 1)$.

Step 2 To find the x-intercept(s), if any, let $y = 0$ and solve the equation

$$x^2 - 4x + 1 = 0.$$

Using the quadratic formula with $a = 1$, $b = -4$, and $c = 1$, we get

$$x = 2 \pm \sqrt{3}.$$

A calculator shows that the x-intercepts are $(3.7, 0)$ and $(.3, 0)$ to the nearest tenth.

TEACHING TIP You might want to explain a different method of finding the vertex, using completing the square.

$$y = x^2 - 4x + 1$$
$$y = (x^2 - 4x + 4) + 1 - 4$$
$$y = (x - 2)^2 - 3$$

The vertex is $(2, -3)$.

Step 3 The x-value of the vertex is

$$x = -\frac{b}{2a} = -\frac{-4}{2(1)} = 2.$$

The y-value of the vertex is

$$y = 2^2 - 4(2) + 1 = -3,$$

so the vertex is $(2, -3)$. The axis is the line $x = 2$.

Steps 4 and 5 A table of values of the points found so far, along with some others, is shown with the graph. Join these points with a smooth curve, as shown in Figure 7.

x	y
-1	6
0	1
$2 - \sqrt{3} \approx .3$	0
1	-2
2	-3
3	-2
$2 + \sqrt{3} \approx 3.7$	0
4	1
5	6

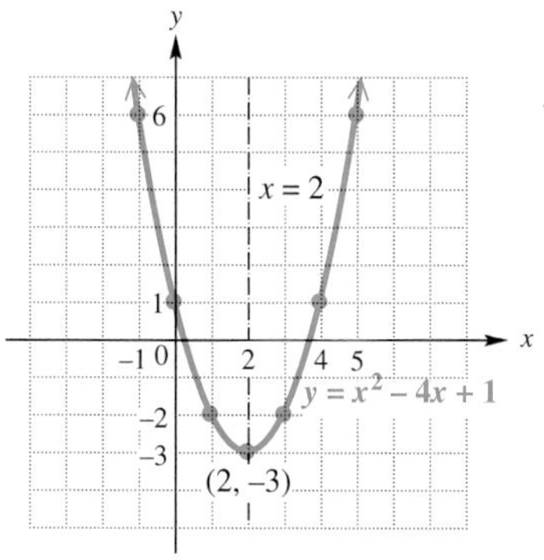

Figure 7

The maximum height is 8 feet at 16 feet from the initial point. The maximum distance is 32 feet.

CONNECTIONS

As shown in the figure, the trajectory of a shell fired from a cannon is a parabola. To reach the maximum range with a cannon, it is shown in calculus that the muzzle must be set at 45°. If the muzzle is elevated above 45°, the shell goes too high and falls too soon. If the muzzle is set below 45°, the shell is rapidly pulled to Earth by gravity.

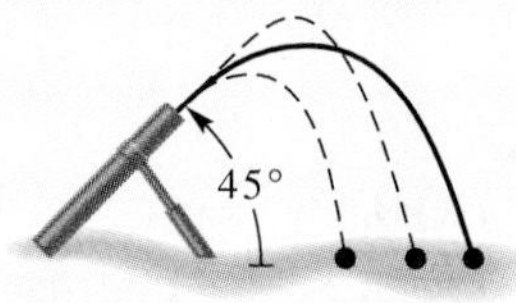

FOR DISCUSSION OR WRITING

If a shell is fired with an initial speed of 32 feet per second, and the muzzle is set at 45°, the equation of the parabolic trajectory is

$$y = x - \frac{1}{32}x^2,$$

where x is the horizontal distance in feet and y is the height in feet. At what distance is the maximum height attained and what is the maximum height? What is the maximum distance to where the shell strikes the ground?

OBJECTIVE 3 **Use a graph to determine the number of real solutions of a quadratic equation.** It can be verified by the vertical line test (Section 7.3) that the graph of an equation of the form $y = ax^2 + bx + c$ is the graph of a function. A function defined by an equation of the form $f(x) = ax^2 + bx + c$ $(a \neq 0)$ is called a **quadratic function.** The domain (possible x-values) of a quadratic function is all real numbers, or $(-\infty, \infty)$; the range (the resulting y-values) can be determined after the function is graphed. In Example 2, the domain is $(-\infty, \infty)$ and the range is $[-3, \infty)$.

Look again at Figure 7, the graph of $y = f(x) = x^2 - 4x + 1$. Recall that setting y equal to 0 gives the x-intercepts, where the x-values are

$$2 + \sqrt{3} \approx 3.7 \qquad \text{and} \qquad 2 - \sqrt{3} \approx .3.$$

The solutions of $0 = x^2 - 4x + 1$ are the x-values of the x-intercepts of the graph of the corresponding quadratic function.

Intercepts of the Graph of a Quadratic Function

The real number solutions of a quadratic equation $ax^2 + bx + c = 0$ are the x-values of the x-intercepts of the graph of the corresponding quadratic function $f(x) = ax^2 + bx + c$.

Since the graph of a quadratic function can intersect the x-axis in two, one, or no points, this result shows why some quadratic equations have two, some have one, and some have no real solutions.

CHALKBOARD EXAMPLE

Decide from the graph how many real number solutions the corresponding equation $x^2 - 4 = 0$ has.

$$f(x) = x^2 - 4$$

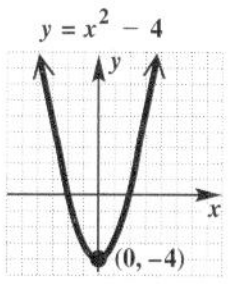

Answer: two real solutions

EXAMPLE 3 **Determining the Number of Real Solutions from a Graph**

(a) Figure 8 shows the graph of $f(x) = x^2 - 3$. The equation $0 = x^2 - 3$ has two real solutions, $\sqrt{3}$ and $-\sqrt{3}$, which correspond to the x-intercepts. The solution set is $\{\pm\sqrt{3}\}$.

(b) Figure 9 shows the graph of $f(x) = x^2 - 4x + 4$. The equation $0 = x^2 - 4x + 4$ has one real solution, 2, which corresponds to the x-intercept. The solution set is $\{2\}$.

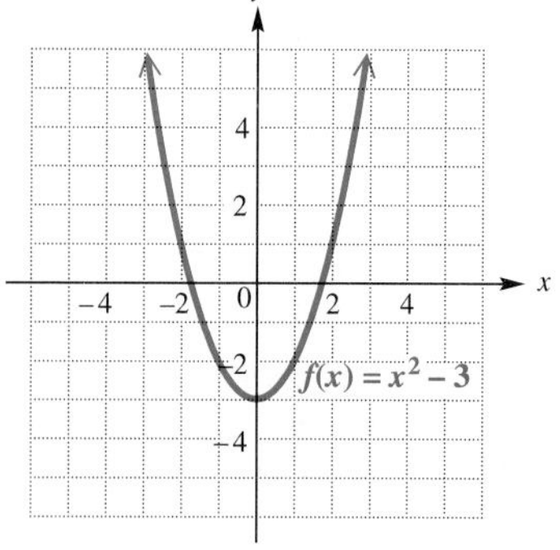

Figure 8

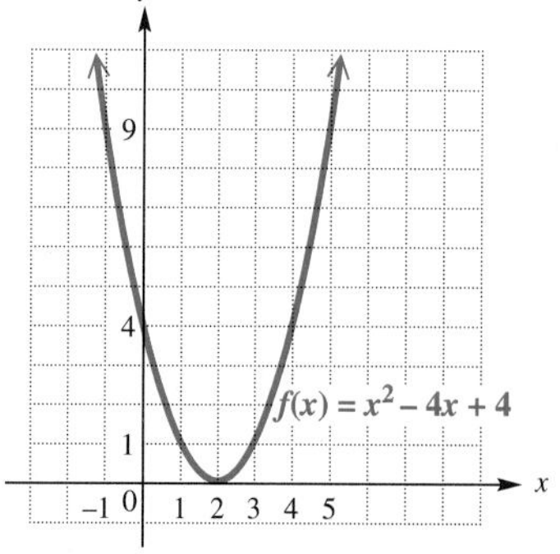

Figure 9

(c) Figure 10 shows the graph of $f(x) = x^2 + 2$. The equation $0 = x^2 + 2$ has no real solutions, since there are no x-intercepts. The solution set over the domain of real numbers is $\emptyset$. (The equation *does* have two imaginary solutions, $i\sqrt{2}$ and $-i\sqrt{2}$.)

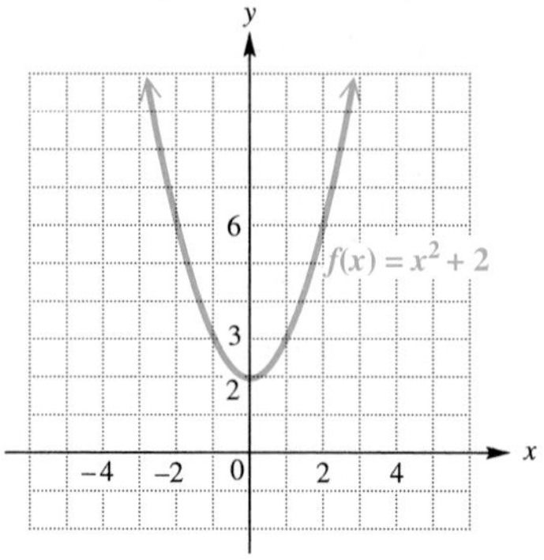

Figure 10

OBJECTIVE 4 Solve applications using quadratic functions. Because we can determine the coordinates of the vertex of the graph of a quadratic function, we are able to find the x-value that leads to the maximum or minimum y-value. This fact allows us to solve applications that lead to quadratic functions.

CHALKBOARD EXAMPLE

Repeat Example 4, but assume that the farmer has 400 feet of fencing.

Answer: 100 feet by 200 feet; 20,000 square feet

EXAMPLE 4 Solving a Problem Involving a Rectangular Region

A farmer wishes to enclose a rectangular region. He has 240 feet of fencing, and plans to use one side of his barn as part of the enclosure. See Figure 11. What dimensions should he use so that the enclosed region has maximum area? What will this maximum area be?

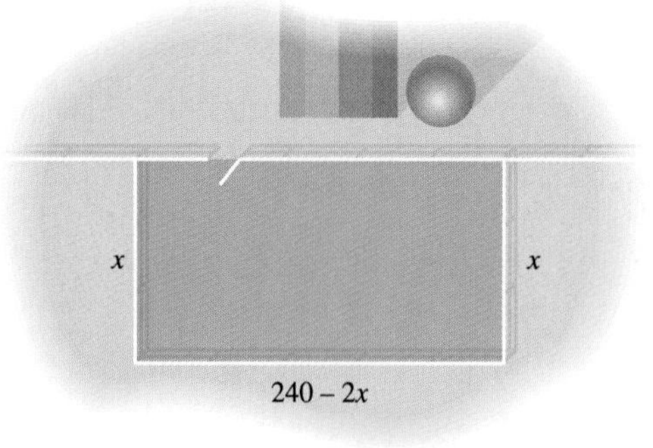

Figure 11

Let x represent the length of each of the two parallel sides of the enclosure. Then $240 - 2x$ represents the length of the third side formed from the fencing. Because area = length × width, the area of the region can be represented by the quadratic function

$$A(x) = x(240 - 2x)$$
$$A(x) = -2x^2 + 240x.$$

The graph of this function is a parabola that opens downward (because of the negative coefficient on x^2). The vertex is the highest point on the graph. Here, $a = -2$ and $b = 240$, so the x-coordinate of the vertex is

$$x = -\frac{b}{2a} = -\frac{240}{2(-2)} = 60.$$

So each of the two parallel sides of fencing should measure 60 feet. The remaining side will be $240 - 2(60) = 240 - 120 = 120$ feet long. The area of the region will be $60 \times 120 = 7200$ square feet. (This area can also be found by evaluating $A(60)$.)

Parabolic shapes are found all around us. Satellite dishes that deliver television signals are becoming more popular each year. Radio telescopes use parabolic reflectors to track incoming signals. The final example discusses how to describe a cross section of a parabolic dish using an equation.

CHALKBOARD EXAMPLE

Suppose that a radio telescope has a parabolic dish shape with a diameter of 350 feet and a depth of 48 feet. Find the equation of the graph of a cross section.

Answer: $y = \frac{48}{30{,}625}x^2$

EXAMPLE 5 Finding the Equation of a Parabolic Satellite Dish

The Parkes radio telescope has a parabolic dish shape with a diameter of 210 feet and a depth of 32 feet. (*Source:* J. Mar and H. Liebowitz, *Structure Technology for Large Radio and Radar Telescope Systems,* The MIT Press, Cambridge, MA, 1969.) Figure 12(a) shows a diagram of such a dish, and Figure 12(b) shows how a cross section of the dish can be modeled by a graph, with the vertex of the parabola at the origin of a coordinate system. Find the equation of this graph.

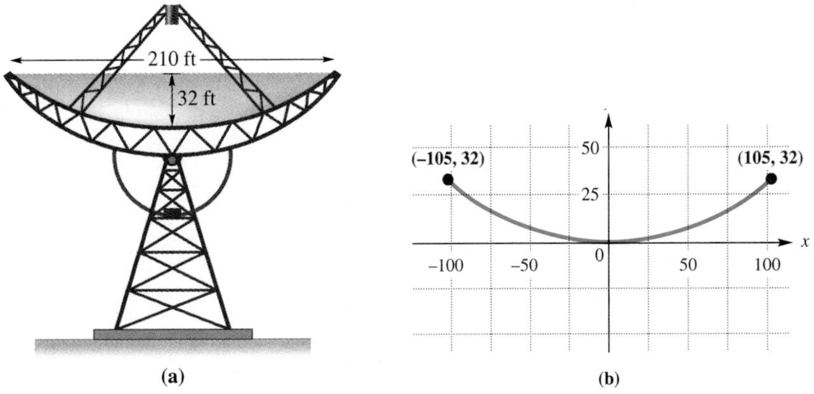

Figure 12

Because the vertex is at the origin, the equation will be of the form $y = ax^2$. As shown in Figure 12(b), one point on the graph has coordinates (105, 32). Letting $x = 105$ and $y = 32$, we can solve for a.

$y = ax^2$	General equation
$32 = a(105)^2$	Substitute for x and y.
$32 = 11{,}025a$	$105^2 = 11{,}025$
$a = \frac{32}{11{,}025}$	Divide by 11,025.

Thus the equation is $y = \frac{32}{11{,}025}x^2$.

1. vertex
2. axis (or axis of symmetry)
3. (2, −5) 4. 2
5. (−1, 2)

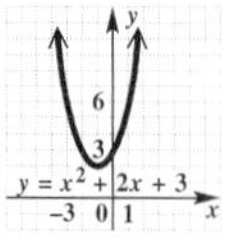

6. (2, −1)

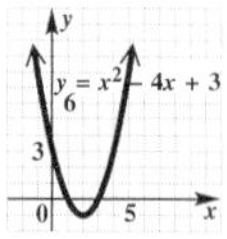

7. (−3, 0)

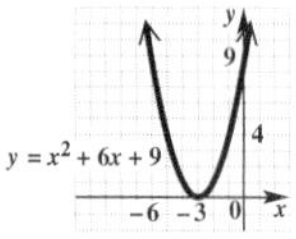

8. (4, 0)

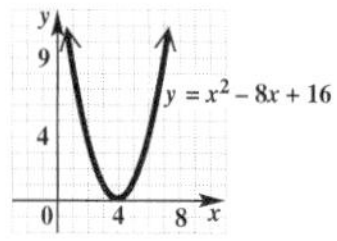

9. (3, 4)

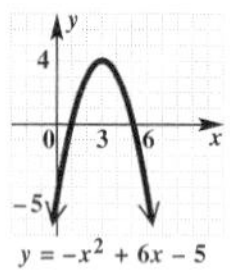

10. (−2, 1)

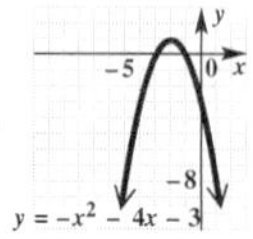

11. (2, 0)

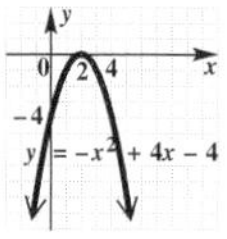

12. (−1, 0)

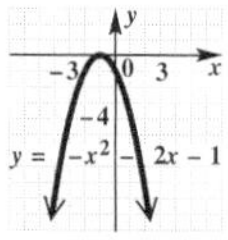

13. one real solution; {2}
14. one real solution; {−3}
15. two real solutions; {±2}
16. two real solutions; {−3, 1}
17. no real solutions; ∅
18. no real solutions; ∅

10.5 EXERCISES

Fill in each blank with the correct response.

1. The highest or lowest point on the graph of a quadratic function is called the _____ of the graph.
2. The vertical line through the vertex of the graph of a quadratic function is called its _____.
3. The vertex of the graph of $f(x) = 2x^2 - 8x + 3$ is _____.
4. If the vertex of the graph of $f(x) = ax^2 + bx + c$ is above the x-axis and the graph opens downward, then the equation $f(x) = 0$ has _____ real solution(s).

Sketch the graph of each equation and give the coordinates of the vertex. See Examples 1 and 2.

5. $y = x^2 + 2x + 3$
6. $y = x^2 - 4x + 3$
7. $y = x^2 + 6x + 9$
8. $y = x^2 - 8x + 16$
9. $y = -x^2 + 6x - 5$
10. $y = -x^2 - 4x - 3$
11. $y = -x^2 + 4x - 4$
12. $y = -x^2 - 2x - 1$

Decide from each graph how many real solutions $f(x) = 0$ has. Then give the solution set (of real solutions). See Example 3.

13.

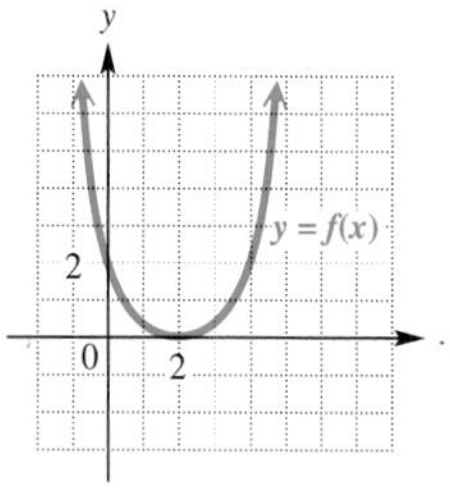

14.

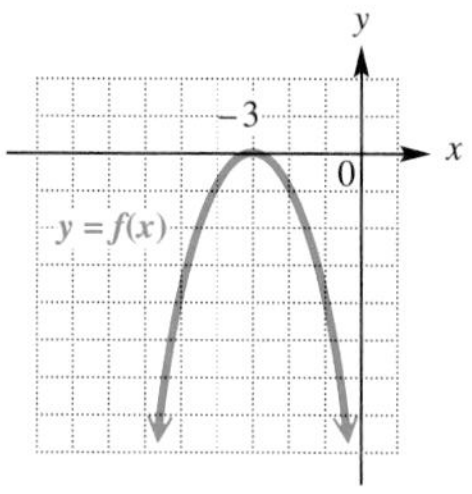

15.

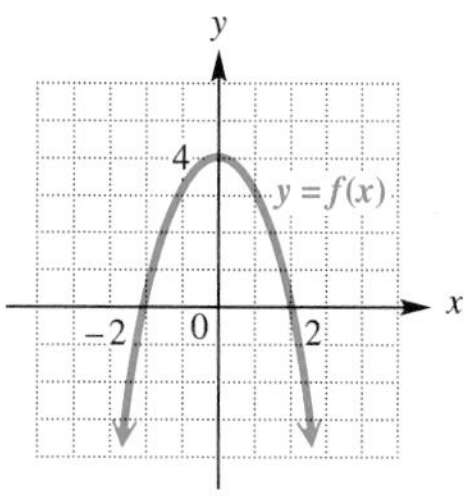

16.

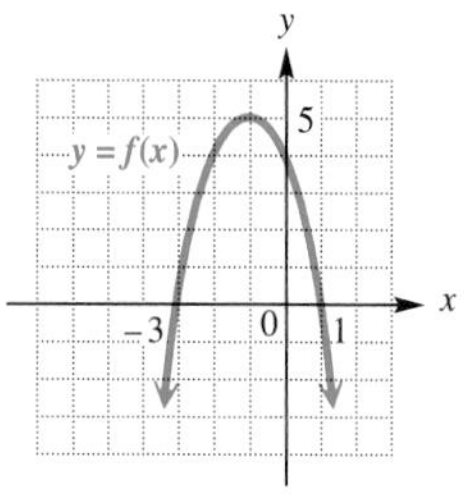

17.

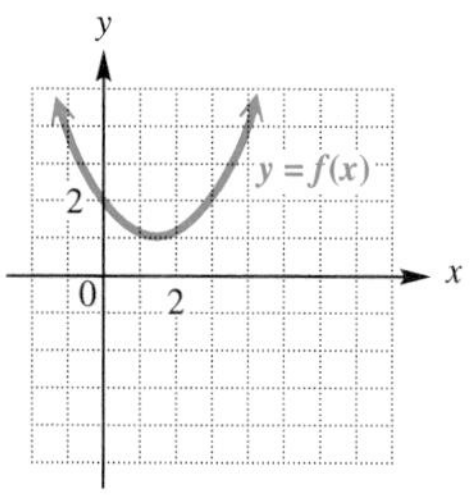

18.

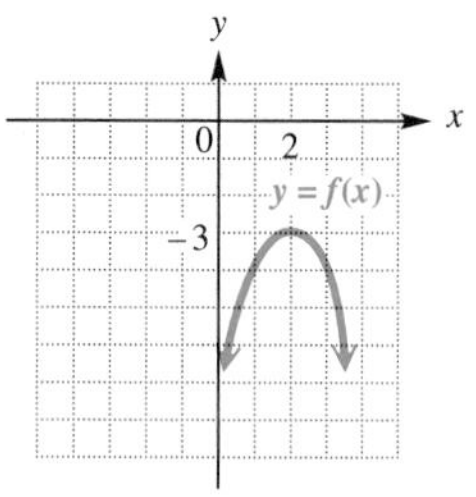

TECHNOLOGY INSIGHTS (EXERCISES 19–26)

A quadratic equation can be solved graphically using a graphing calculator. Assuming that the equation is in the form $ax^2 + bx + c = 0$, we enter $ax^2 + bx + c$ as Y_1, and then direct the calculator to find the x-intercepts of the graph. (These are also referred to as *zeros* of the function.) For example, to solve $x^2 - 5x - 6 = 0$ graphically, refer to the three screens on the next page. Notice that the displays at the bottoms of the lower two screens show the two solutions, −1 and 6.

19. $\{-2, 3\}$ 20. $\{-5, -1\}$

TECHNOLOGY INSIGHTS (EXERCISES 19–26) (CONTINUED)

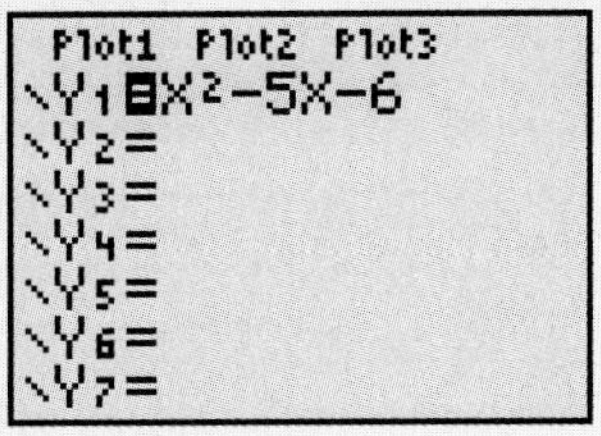

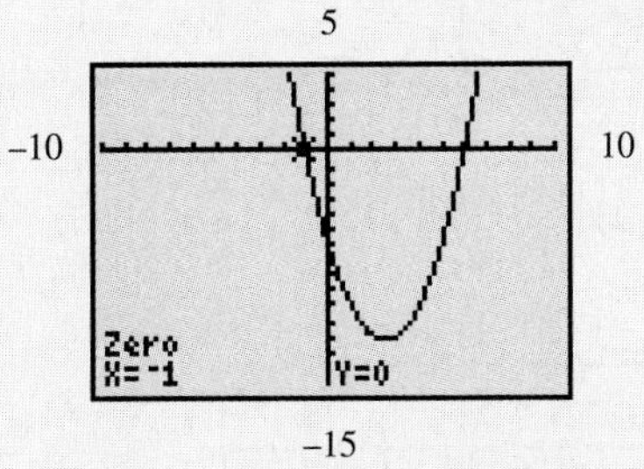

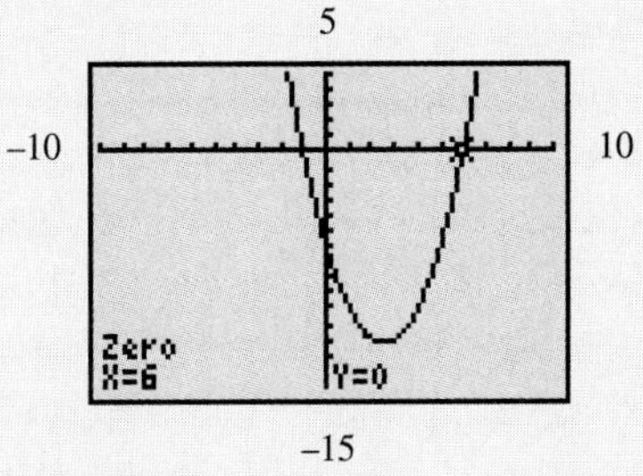

Determine the solution set of each quadratic equation by observing the corresponding screens. Then verify your answers by solving the quadratic equation using the method of your choice.

19. $x^2 - x - 6 = 0$

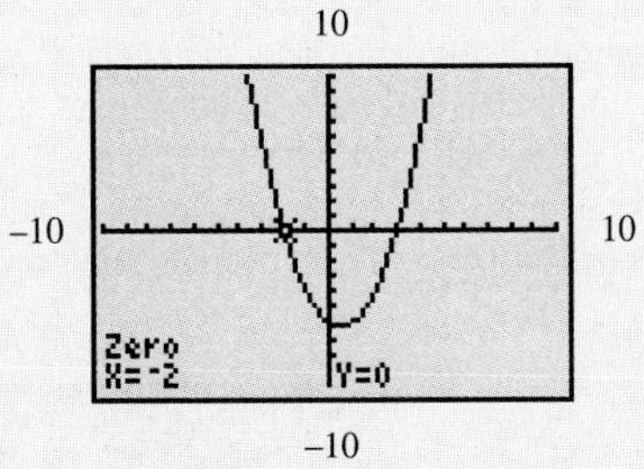

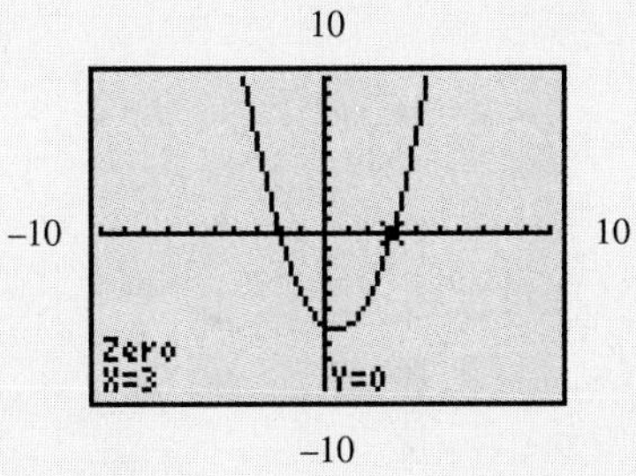

20. $x^2 + 6x + 5 = 0$

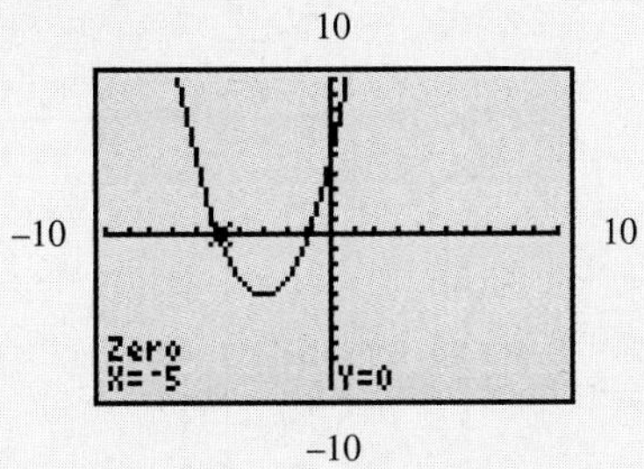

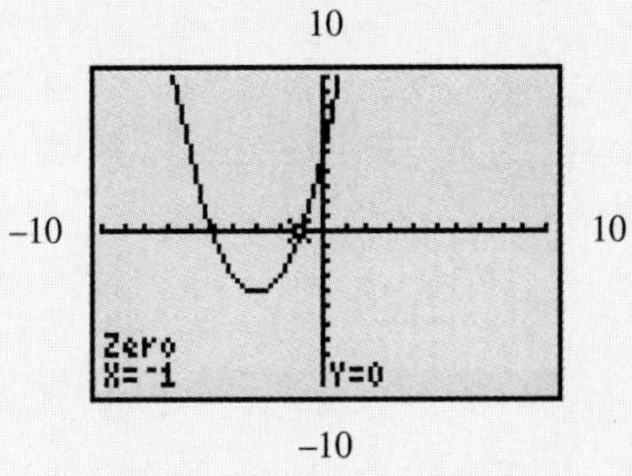

21. $\{-1, 1.5\}$ 22. $\{-.25, 3\}$

TECHNOLOGY INSIGHTS (EXERCISES 19–26) (CONTINUED)

21. $2x^2 - x - 3 = 0$

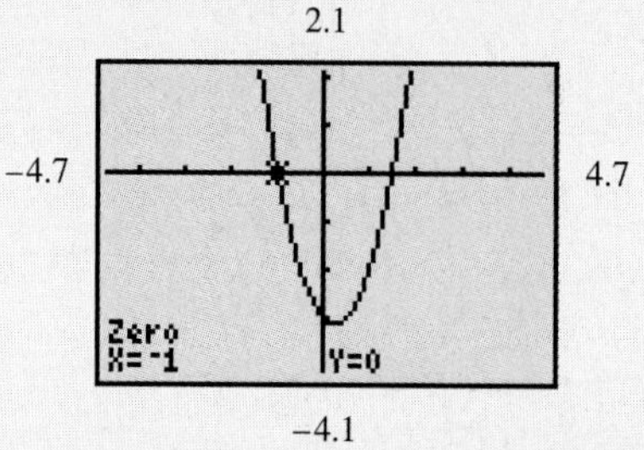

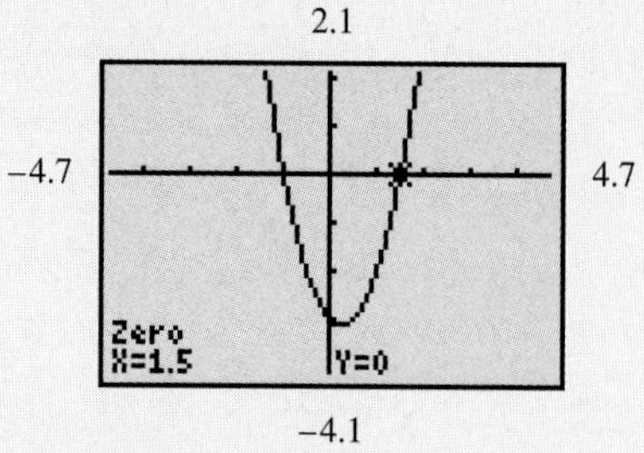

22. $4x^2 - 11x - 3 = 0$

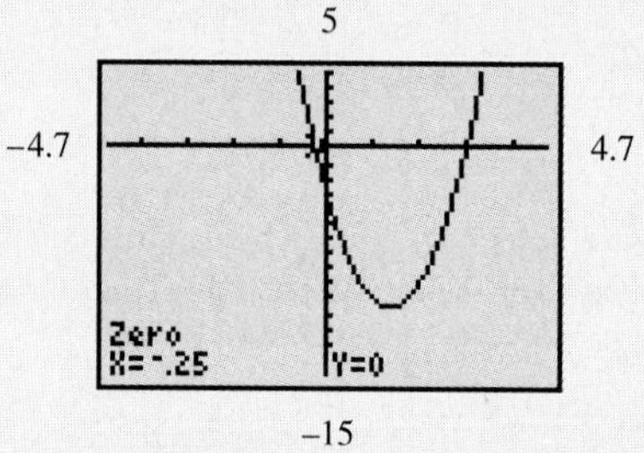

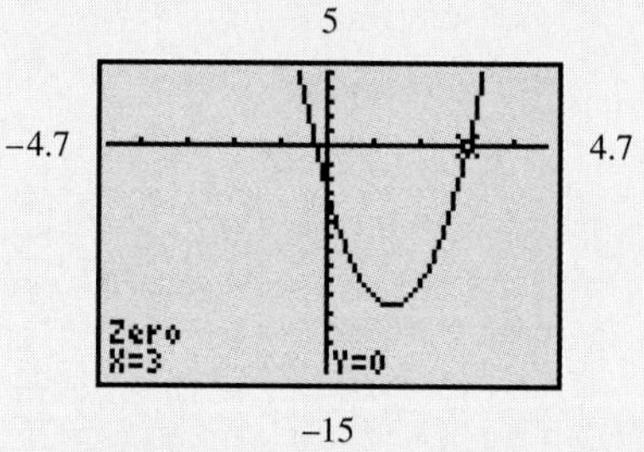

Another method of solving a quadratic equation graphically can be applied when the equation is not in standard form. For example, to solve $2x^2 = 9x + 5$, we graph the left side, $2x^2$, as Y_1 and the right side, $9x + 5$, as Y_2. Then we direct the calculator to find the coordinates of the points of intersection of the two graphs. As shown in the lower two screens below, the x-values of these points are $-.5$ and 5, giving the solution set $\{-.5, 5\}$.

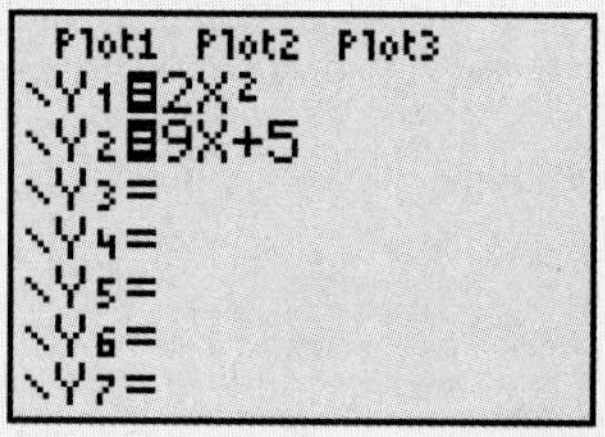

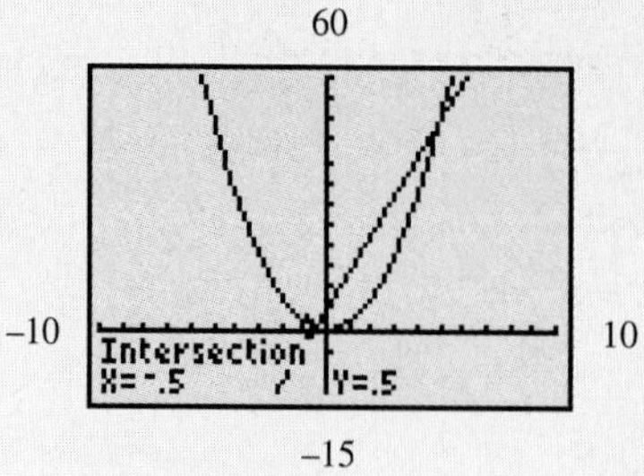

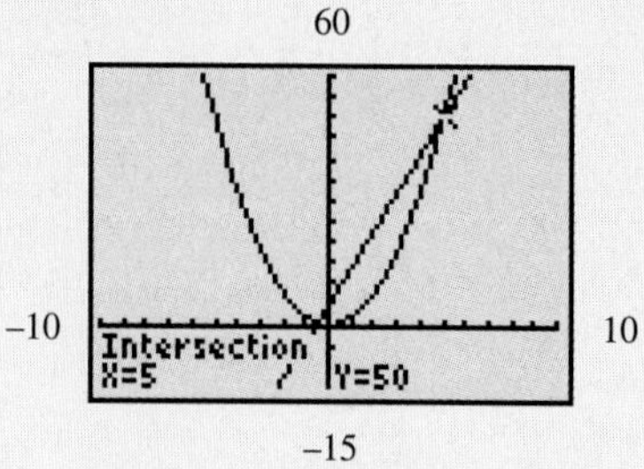

23. $\{-4, 2\}$ 24. $\{-2, 3\}$
25. (a) the point $(0, -2)$ (b) the points $(-1, 0)$ and $(2, 0)$
26. $y = .5^2 - .5 - 2 = .25 - .5 - 2 = -2.25$
In Exercises 27–32, the domain is first and the range is second.
27. $(-\infty, \infty)$; $[0, \infty)$
28. $(-\infty, \infty)$; $(-\infty, 0]$
29. $(-\infty, \infty)$; $(-\infty, 4]$
30. $(-\infty, \infty)$; $(-\infty, 5]$
31. $(-\infty, \infty)$; $[1, \infty)$
32. $(-\infty, \infty)$; $(-\infty, -3]$

TECHNOLOGY INSIGHTS (EXERCISES 19–26) (CONTINUED)

Determine the solution set of each quadratic equation by observing the corresponding screens. Then verify your answer by solving the quadratic equation using the method of your choice.

23. $x^2 = -2x + 8$

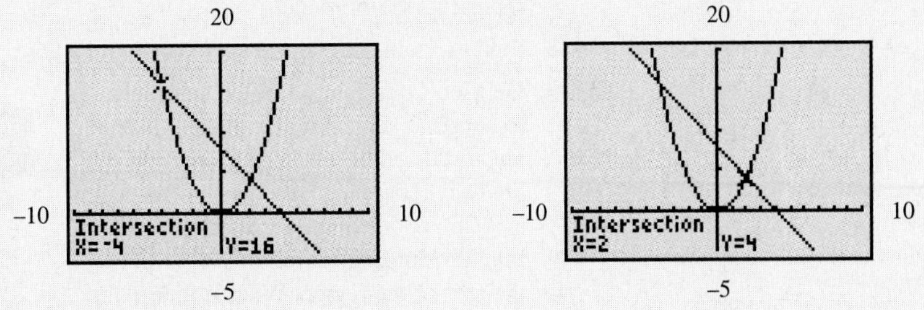

24. $x^2 = x + 6$

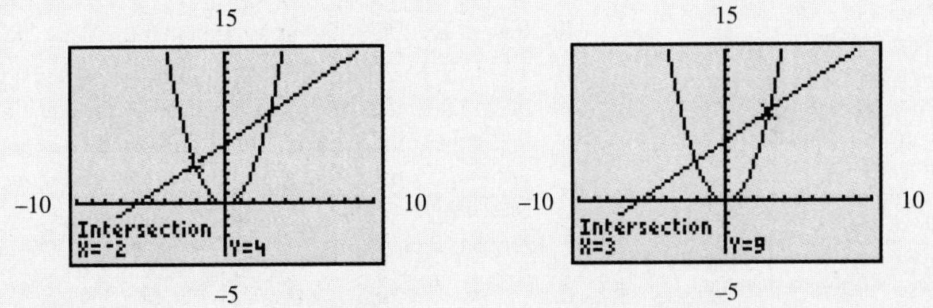

25. This table was generated by an equation of the form $Y_1 = ax^2 + bx + c$. Answer these questions by referring to the table.

(a) What is the y-intercept of the graph?

(b) What are the x-intercepts of the graph?

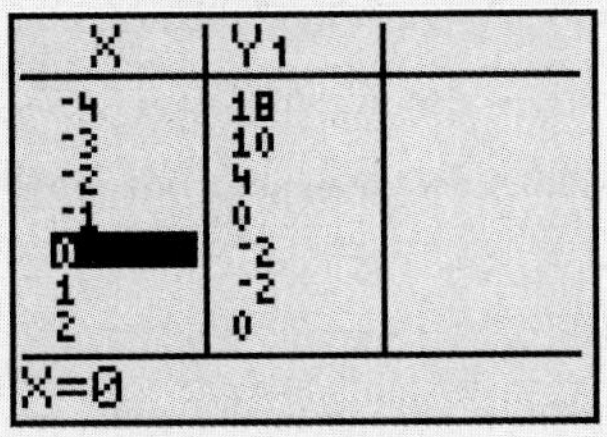

X	Y_1
-4	18
-3	10
-2	4
-1	0
0	-2
1	-2
2	0

X=0

26. The graph shown here is that of the equation from Exercise 25. The calculator has the capability of finding the vertex. If the parabola opens upward, as it does in this case, the vertex is a *minimum.* The equation is $y = x^2 - x - 2$. Show *algebraically* that the point $(.5, -2.25)$ lies on the graph, by letting $x = .5$ and solving for y.

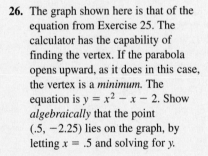

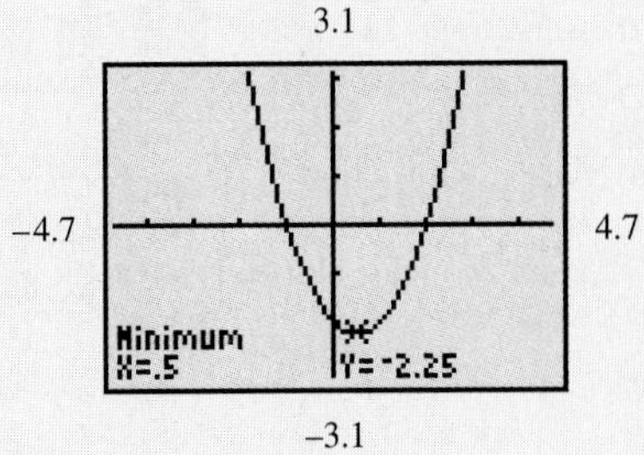

Find the domain and the range of each function graphed in the indicated exercise.

27. Exercise 13 **28.** Exercise 14 **29.** Exercise 15

30. Exercise 16 **31.** Exercise 17 **32.** Exercise 18

33. 3 **34.** 0 **35.** 21
36. 10 **37.** 40 and 40
38. 150 and 150
39. 320 feet by 640 feet; 204,800 square feet **40.** 750 feet by 1500 feet; 1,125,000 square feet
41. $y = \frac{11}{5625}x^2$
42. $y = \frac{1}{800}x^2$
43. In each case, there is a vertical "stretch" of the parabola. It becomes narrower as the coefficient gets larger. **44.** In each case, there is a vertical "shrink" of the parabola. It becomes wider as the coefficient gets smaller.
45. The graph of Y_2 is obtained by reflecting the graph of Y_1 across the x-axis.
46. When the coefficient of x^2 is negative, the parabola opens downward. **47.** By adding a positive constant k, the graph is shifted k units upward. By subtracting a positive constant k, the graph is shifted k units downward. **48.** Adding a positive constant k moves the graph k units to the left. Subtracting a positive constant k moves the graph k units to the right.
49. Begin by finding the y-intercept. To do this, let $x = 0$. In the given equation, we get $y = -5$. Next, find any x-intercepts by letting $y = 0$ and solving for x. In this example, $x^2 - 4x - 5 = 0$ leads to $(x - 5)(x + 1) = 0$, which gives $x = 5$ or $x = -1$. Next find the vertex: $x = -\frac{b}{2a} = -\frac{(-4)}{2(1)} = 2$, and $y = 2^2 - 4(2) - 5 = -9$. Plot these intercepts and vertex, plot any additional points as necessary, and join the points with a smooth parabolic curve.

Given $f(x) = 2x^2 - 5x + 3$, find each of the following.

33. $f(0)$ **34.** $f(1)$ **35.** $f(-2)$ **36.** $f(-1)$

Use a quadratic function to solve each problem. See Examples 4 and 5.

37. Find two numbers whose sum is 80 and whose product is a maximum. (*Hint:* Let x represent one of the numbers. Then $80 - x$ represents the other. A quadratic function represents their product.)

38. Find two numbers whose sum is 300 and whose product is a maximum.

39. Delgado Community College has plans to construct a rectangular parking lot on land bordered on one side by a highway. There are 1280 feet of fencing available to fence the three other sides. Find the dimensions that will maximize the area of the region. What is this area?

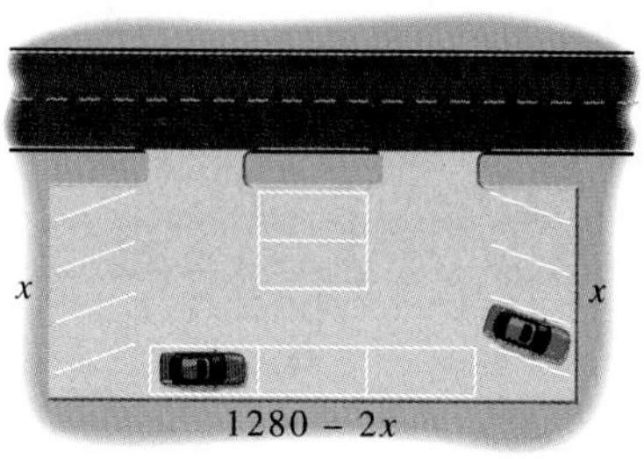

40. Repeat Exercise 39, but assume that there are 3000 feet of fencing available.

41. The U.S. Naval Research Laboratory designed a giant radio telescope that had a diameter of 300 feet and maximum depth of 44 feet. The graph depicts a cross section of this telescope. Find the equation of this parabola. (*Source:* J. Mar and H. Liebowitz, *Structure Technology for Large Radio and Radar Telescope Systems,* The MIT Press, Cambridge, MA, 1969.)

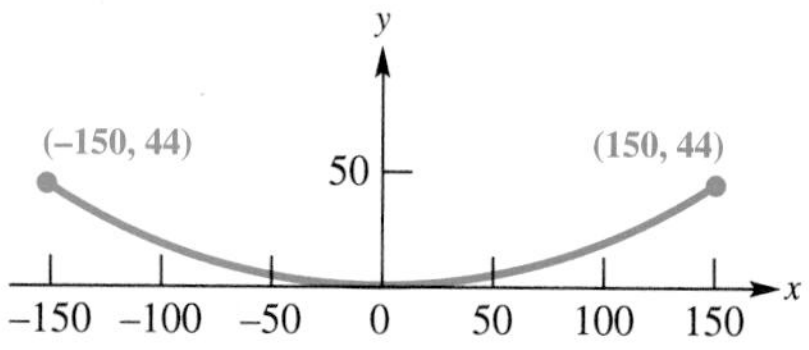

42. Suppose the telescope in Exercise 41 had a diameter of 400 feet and maximum depth of 50 feet. Find the equation of this parabola.

Exercises 43–48 use a graphing calculator to illustrate how the graph of $y = x^2$ can be transformed by using arithmetic operations.

43. In the standard viewing window of your calculator, graph the following one at a time, leaving the previous graphs on the screen as you move along.

$$Y_1 = x^2 \qquad Y_2 = 2x^2 \qquad Y_3 = 3x^2 \qquad Y_4 = 4x^2$$

Describe the effect the successive coefficients have on the parabola.

44. Repeat Exercise 43 for the following.

$$Y_1 = x^2 \qquad Y_2 = \frac{1}{2}x^2 \qquad Y_3 = \frac{1}{4}x^2 \qquad Y_4 = \frac{1}{8}x^2$$

45. Graph the pair of parabolas $Y_1 = x^2$ and $Y_2 = -x^2$ on the same screen. In your own words, describe how the graph of Y_2 can be obtained from the graph of Y_1.

50. 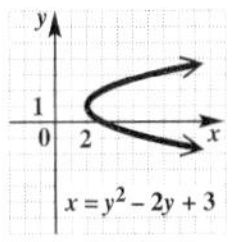

This is not the graph of a function because for many x-values, there are two y-values. The graph does not pass the vertical line test.

46. Graph $Y_1 = -x^2$, $Y_2 = -2x^2$, $Y_3 = -3x^2$ and $Y_4 = -4x^2$ on the same screen. Make a conjecture about what happens when the coefficient of x^2 is negative.

47. In the standard viewing window of your calculator, graph the following one at a time, leaving the previous graphs on the screen as you move along.

$$Y_1 = x^2 \qquad Y_2 = x^2 + 3 \qquad Y_3 = x^2 - 6$$

Describe the effect that adding or subtracting a constant has on the parabola.

48. Repeat Exercise 47 for the following.

$$Y_1 = x^2 \qquad Y_2 = (x + 3)^2 \qquad Y_3 = (x - 6)^2$$

49. Write an explanation of how to graph a quadratic equation such as $y = x^2 - 4x - 5$.

50. ▲ The equation $x = y^2 - 2y + 3$ has a parabola with a horizontal axis of symmetry as its graph. Sketch the graph of this equation by letting y take on the values -1, 0, 1, 2, and 3. Then write an explanation of why this is not the graph of a function.

CHAPTER 10 GROUP ACTIVITY

Instructor's Note
Grouping: 2 students
Time: 20 minutes

Most students will not be familiar with radio telescopes but some may be familiar with woks. This activity is a good one for answering the question "When will we ever use this?" You may want to ask the students what other parabolic shapes they have seen.

A. 1. No 2. It should more closely resemble the telescope. 3. The settings will vary but should be close to the dimensions of the telescope.

How Is a Radio Telescope Like a Wok?

Objective: Determine appropriate domains and ranges when graphing parabolas on a graphing calculator.

Example 5 of Section 10.5 explored the parabolic shape of a radio telescope. This activity will use that example and another to show the importance of setting appropriate domains and ranges on a graphing calculator. One student should write the answers to the questions while the other one does the graphing on a calculator. When you start part B, switch tasks.

A. In Example 5 of Section 10.5, an equation was found to represent the shape of the Parkes radio telescope. Using this equation, graph the parabola on a graphing calculator.

1. Use a standard viewing window, that is, Xmin $= -10$, Xmax $= 10$, Ymin $= -10$, Ymax $= 10$. Describe your graph. Does its shape look similar to the picture of the telescope in Section 10.5?
2. Change the domain and range settings to Xmin $= -50$, Xmax $= 50$, Ymin $= -5$, Ymax $= 50$. How does the graph look now? You may want to change Xscl and Yscl to 5.
3. Continue to adjust the domain and range settings until the calculator graph is close in shape to the picture of the telescope in Section 10.5. What are your settings?

B. A wok used in oriental cooking is also a parabola. Just as the shape of a parabola focuses radio waves from space, a wok focuses heat and oil for cooking.

(continued)

1. Find an equation that would model the cross section of a wok with diameter 14 inches and depth 4 inches.
2. Graph this equation using a standard viewing window on a calculator. Describe the graph of the equation. Describe the differences between this graph and the graph in part A, Exercise 1.
3. Adjust the settings for domain and range until the graph looks similar to the cross section of a wok.
4. How are a wok and a radio telescope alike?
5. How is the range (or domain) of a function different from the range (or domain) setting on a graphing calculator?

B. 1. $y = \frac{4}{49}x^2$ **2.** This graph is closer to the drawing than part A, Exercise 1. **3.** The best settings are Xmin $= -7$, Xmax $= 7$, Ymin $= 0$, Ymax $= 4$. **4.** They are both shaped like parabolic dishes. **5.** The range and domain of a graphing calculator determine the size of the coordinate system you will see on the screen. The range of a function is the set of all y-values for that function; the domain is the set of all x-values for the function.

CHAPTER 10 SUMMARY

KEY TERMS

10.1 quadratic equation
10.3 discriminant
10.4 complex number
imaginary number
real part
imaginary part
standard form (of a complex number)
conjugate (of a complex number)
10.5 parabola
vertex
axis (of symmetry)
quadratic function

NEW SYMBOLS

$\pm$ positive or negative (plus or minus)

i $i = \sqrt{-1}$ and $i^2 = -1$

TEST YOUR WORD POWER

See how well you have learned the vocabulary in this chapter. Answers, with examples, are given at the bottom of the page.

1. A **quadratic equation** is an equation that can be written in the form
   (a) $Ax + By = C$
   (b) $ax^2 + bx + c = 0$
   (c) $Ax + B = 0$
   (d) $y = mx + b$.

2. A **complex number** is defined as
   (a) a real number that includes a complex fraction
   (b) a nonzero multiple of i
   (c) a number of the form $a + bi$, where a and b are real numbers
   (d) the square root of -1.

3. An **imaginary number** is
   (a) a complex number $a + bi$ where $b \neq 0$
   (b) a number that does not exist
   (c) a complex number $a + bi$ where $b = 0$
   (d) any real number.

4. A **parabola** is the graph of
   (a) any equation in two variables
   (b) a linear equation
   (c) an equation of degree three
   (d) a quadratic equation.

5. The **vertex** of a parabola is
   (a) the point where the graph intersects the y-axis
   (b) the point where the graph intersects the x-axis
   (c) the lowest point on a parabola that opens up or the highest point on a parabola that opens down
   (d) the origin.

6. The **axis** of a vertical parabola is
   (a) either the x-axis or the y-axis
   (b) the vertical line through the vertex
   (c) the horizontal line through the vertex
   (d) the x-axis.

Answers to Test Your Word Power

1. (b) *Examples:* $z^2 + 6z + 9 = 0$, $y^2 - 2y = 8$, $(x + 3)(x - 1) = 5$ **2.** (c) *Examples:* -5 (or $-5 + 0i$), $7i$ (or $0 + 7i$), $\sqrt{2} - 4i$
3. (a) *Examples:* $2i$, $-13i$, $3 + i\sqrt{6}$ **4.** (d) *Examples:* See Figures 3–10 in Section 10.5. **5.** (c) *Example:* The graph of $y = (x + 3)^2$ has vertex $(-3, 0)$, which is the lowest point on the graph. **6.** (b) *Example:* The axis of $y = (x + 3)^2$ is the line $x = -3$.

QUICK REVIEW

CONCEPTS	EXAMPLES
10.1 SOLVING QUADRATIC EQUATIONS BY THE SQUARE ROOT PROPERTY	
Square Root Property of Equations If k is positive, and if $a^2 = k$, then $a = \sqrt{k}$ or $a = -\sqrt{k}$.	Solve $(2x + 1)^2 = 5$. $2x + 1 = \pm\sqrt{5}$ $2x = -1 \pm \sqrt{5}$ $x = \frac{-1 \pm \sqrt{5}}{2}$ Solution set: $\left\{\frac{-1 \pm \sqrt{5}}{2}\right\}$
10.2 SOLVING QUADRATIC EQUATIONS BY COMPLETING THE SQUARE	
Completing the Square **1.** If the coefficient of the squared term is 1, go to Step 2. If it is not 1, divide each side of the equation by this coefficient. **2.** Make sure that all variable terms are on one side of the equation and all constant terms are on the other. **3.** Take half the coefficient of x, square it, and add the square to each side of the equation. Factor the variable side and combine terms on the other. **4.** Use the square root property to solve the equation.	Solve $2x^2 + 4x - 1 = 0$. $x^2 + 2x - \frac{1}{2} = 0$ $x^2 + 2x = \frac{1}{2}$ $x^2 + 2x + 1 = \frac{1}{2} + 1$ $(x + 1)^2 = \frac{3}{2}$ $x + 1 = \pm\sqrt{\frac{3}{2}} = \pm\frac{\sqrt{6}}{2}$ $x = -1 \pm \frac{\sqrt{6}}{2}$ $x = \frac{-2 \pm \sqrt{6}}{2}$ Solution set: $\left\{\frac{-2 \pm \sqrt{6}}{2}\right\}$
10.3 SOLVING QUADRATIC EQUATIONS BY THE QUADRATIC FORMULA	
Quadratic Formula The solutions of $ax^2 + bx + c = 0$, $(a \neq 0)$, are $x = \frac{-b \pm \sqrt{b^2 - 4ac}}{2a}$. The discriminant for the quadratic equation is $b^2 - 4ac$. If $b^2 - 4ac > 0$, there are two real solutions. If $b^2 - 4ac = 0$, there is one real solution. If $b^2 - 4ac < 0$, there are no real solutions.	Solve $3x^2 - 4x - 2 = 0$. $x = \frac{-(-4) \pm \sqrt{(-4)^2 - 4(3)(-2)}}{2(3)}$ $a = 3$, $b = -4$, $c = -2$ $x = \frac{4 \pm \sqrt{16 + 24}}{6}$ $x = \frac{4 \pm \sqrt{40}}{6} = \frac{4 \pm 2\sqrt{10}}{6}$ Discriminant: 40 There are two real solutions. $x = \frac{2(2 \pm \sqrt{10})}{6} = \frac{2 \pm \sqrt{10}}{3}$ Solution set: $\left\{\frac{2 \pm \sqrt{10}}{3}\right\}$
10.4 COMPLEX NUMBERS	
The number $i = \sqrt{-1}$ and $i^2 = -1$. For the positive number b, $\sqrt{-b} = i\sqrt{b}$.	Simplify: $\sqrt{-19}$. $\sqrt{-19} = \sqrt{-1 \cdot 19} = i\sqrt{19}$

(continued)

CONCEPTS	EXAMPLES
Addition Add complex numbers by adding the real parts and adding the imaginary parts.	Add: $(3 + 6i) + (-9 + 2i)$. $(3 + 6i) + (-9 + 2i) = (3 - 9) + (6 + 2)i$ $= -6 + 8i$
Subtraction To subtract complex numbers, change the number following the subtraction sign to its negative and add.	Subtract: $(5 + 4i) - (2 - 4i)$. $(5 + 4i) - (2 - 4i) = (5 + 4i) + (-2 + 4i)$ $= (5 - 2) + (4 + 4)i$ $= 3 + 8i$
Multiplication Multiply complex numbers in the same way polynomials are multiplied. Replace i^2 with -1.	Multiply: $(7 + i)(3 - 4i)$. $(7 + i)(3 - 4i)$ $= 7(3) + 7(-4i) + i(3) + i(-4i)$ FOIL $= 21 - 28i + 3i - 4i^2$ $= 21 - 25i - 4(-1)$ $i^2 = -1$ $= 21 - 25i + 4$ $= 25 - 25i$
Division Divide complex numbers by multiplying the numerator and the denominator by the conjugate of the denominator.	Divide: $\frac{2}{6 + i}$. $\frac{2}{6+i} = \frac{2}{6+i} \cdot \frac{6-i}{6-i}$ $= \frac{2(6-i)}{36 - i^2}$ $= \frac{12 - 2i}{36 + 1}$ $= \frac{12 - 2i}{37}$ $= \frac{12}{37} - \frac{2}{37}i$ Standard form
Complex Solutions A quadratic equation may have nonreal, complex solutions. This occurs when the discriminant is negative. The quadratic formula will give complex solutions in such cases.	Solve for all complex solutions of $x^2 + x + 1 = 0$. Here, $a = 1$, $b = 1$, and $c = 1$. $x = \frac{-1 \pm \sqrt{1^2 - 4(1)(1)}}{2(1)}$ $x = \frac{-1 \pm \sqrt{1 - 4}}{2}$ $x = \frac{-1 \pm \sqrt{-3}}{2}$ Discriminant: -3 $x = \frac{-1 \pm i\sqrt{3}}{2}$ Two nonreal solutions Solution set: $\left\{-\frac{1}{2} \pm \frac{\sqrt{3}}{2}i\right\}$

10.5 MORE ON GRAPHING QUADRATIC EQUATIONS; QUADRATIC FUNCTIONS

Graphing $y = ax^2 + bx + c$ **1.** Find the y-intercept.	Graph $y = 2x^2 - 5x - 3$. $y = 2(0)^2 - 5(0) - 3 = -3$ The y-intercept is $(0, -3)$.

CONCEPTS	EXAMPLES
2. Find any x-intercepts.	$0 = 2x^2 - 5x - 3$ $0 = (2x + 1)(x - 3)$ $2x + 1 = 0$ or $x - 3 = 0$ $2x = -1$ or $x = 3$ $x = -\frac{1}{2}$ or $x = 3$ The x-intercepts are $\left(-\frac{1}{2}, 0\right)$ and $(3, 0)$.
3. Find the vertex: $x = -\frac{b}{2a}$; find y by substituting this value for x in the equation.	For the vertex: $x = -\frac{b}{2a} = -\frac{-5}{2(2)} = \frac{5}{4}$ $y = 2\left(\frac{5}{4}\right)^2 - 5\left(\frac{5}{4}\right) - 3$ $y = 2\left(\frac{25}{16}\right) - \frac{25}{4} - 3$ $y = \frac{25}{8} - \frac{50}{8} - \frac{24}{8} = -\frac{49}{8} = -6\frac{1}{8}.$ The vertex is $\left(\frac{5}{4}, -\frac{49}{8}\right)$.
4. Plot the intercepts and the vertex. **5.** Find and plot additional ordered pairs near the vertex and intercepts as needed.	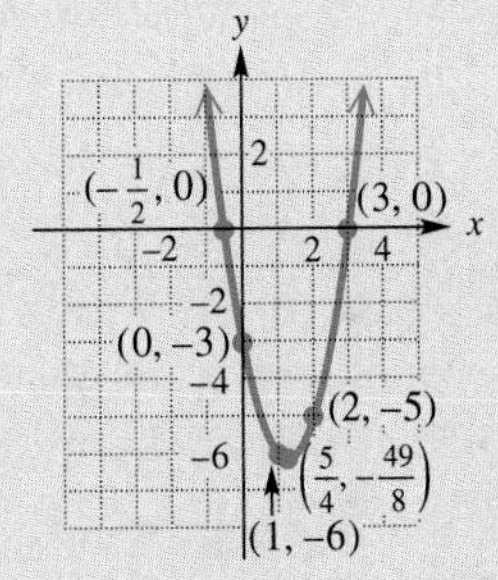
The number of real solutions of the equation $ax^2 + bx + c = 0$ can be determined from the number of x-intercepts of the graph of $y = ax^2 + bx + c.$	The figure shows that the equation $2x^2 - 5x - 3 = 0$ has two real solutions.

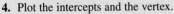

CHAPTER 10 REVIEW EXERCISES

1. $\{\pm 12\}$ 2. $\{\pm\sqrt{37}\}$
3. $\{\pm 8\sqrt{2}\}$ 4. $\{-7, 3\}$
5. $\{3 \pm \sqrt{10}\}$ 6. $\left\{\frac{-1 \pm \sqrt{14}}{2}\right\}$
7. $\emptyset$ 8. (d)

[10.1] *Solve each equation by using the square root property. Give only real number solutions. Express all radicals in simplest form.*

1. $y^2 = 144$ **2.** $x^2 = 37$ **3.** $m^2 = 128$ **4.** $(k + 2)^2 = 25$

5. $(r - 3)^2 = 10$ **6.** $(2p + 1)^2 = 14$ **7.** $(3k + 2)^2 = -3$

8. Which one of the following equations has two real solutions?
(a) $x^2 = 0$ **(b)** $x^2 = -4$ **(c)** $(x + 5)^2 = -16$ **(d)** $(x + 6)^2 = 25$

[10.2] *Solve each equation by completing the square. Give only real number solutions.*

9. $m^2 + 6m + 5 = 0$

10. $p^2 + 4p = 7$

11. $-x^2 + 5 = 2x$

12. $2y^2 - 3 = -8y$

13. $5k^2 - 3k - 2 = 0$

14. $(4a + 1)(a - 1) = -7$

Solve each problem.

15. If an object is thrown upward on Earth from a height of 50 feet, with an initial velocity of 32 feet per second, then its height after t seconds is given by $h = -16t^2 + 32t + 50$, where h is in feet. After how many seconds will it reach a height of 30 feet?

16. Find the lengths of the three sides of the right triangle shown.

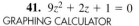

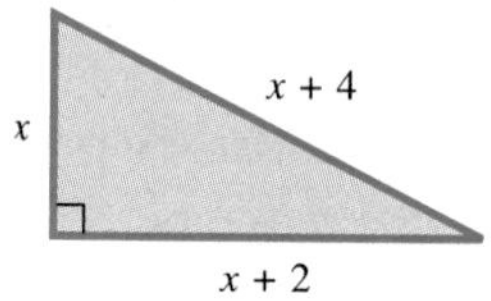

17. What must be added to $x^2 + kx$ to make it a perfect square?

[10.3]

18. Consider the equation $x^2 - 9 = 0$.
(a) Solve the equation by factoring.
(b) Solve the equation by the square root property.
(c) Solve the equation by the quadratic formula.
(d) Compare your answers. If a quadratic equation can be solved by both the factoring and the quadratic formula methods, should you always get the same results? Explain.

Solve each equation by using the quadratic formula. Give only real number solutions.

19. $x^2 - 2x - 4 = 0$

20. $3k^2 + 2k = -3$

21. $2p^2 + 8 = 4p + 11$

22. $-4x^2 + 7 = 2x$

23. $\frac{1}{4}p^2 = 2 - \frac{3}{4}p$

24. What is the value of the discriminant for $3x^2 - x - 2 = 0$? How many real solutions does this equation have?

[10.4]

25. Write an explanation of the method used to divide complex numbers.

26. Use the fact that $i^2 = -1$ to complete each of the following. Do them in order.
(a) Since $i^3 = i^2 \cdot i$, $i^3 =$ ______.
(b) Since $i^4 = i^3 \cdot i$, $i^4 =$ ______.
(c) Since $i^{48} = (i^4)^{12}$, $i^{48} =$ ______.

Perform the indicated operations.

27. $(3 + 5i) + (2 - 6i)$

28. $(-2 - 8i) - (4 - 3i)$

29. $(6 - 2i)(3 + i)$

30. $(2 + 3i)(2 - 3i)$

31. $\frac{1 + i}{1 - i}$

32. $\frac{5 + 6i}{2 + 3i}$

33. $\frac{1}{7 - i}$

34. What is the conjugate of the real number a?

35. Is it possible to multiply a complex number by its conjugate and get an imaginary product? Explain.

Find the complex solutions of each quadratic equation.

36. $(m + 2)^2 = -3$

37. $(3p - 2)^2 = -8$

38. $3k^2 = 2k - 1$

39. $h^2 + 3h = -8$

40. $4q^2 + 2 = 3q$

41. $9z^2 + 2z + 1 = 0$

9. $\{-5, -1\}$ **10.** $\{-2 \pm \sqrt{11}\}$
11. $\{-1 \pm \sqrt{6}\}$
12. $\left\{\frac{-4 \pm \sqrt{22}}{2}\right\}$ **13.** $\left\{-\frac{2}{5}, 1\right\}$
14. $\emptyset$ **15.** 2.5 seconds
16. 6, 8, 10 **17.** $\left(\frac{k}{2}\right)^2$ or $\frac{k^2}{4}$
18. (a) $\{\pm 3\}$ (b) $\{\pm 3\}$ (c) $\{\pm 3\}$ (d) We will always get the same results, no matter which method of solution is used. This is because there is only one solution set, and the solutions will always match. **19.** $\{1 \pm \sqrt{5}\}$
20. $\emptyset$ **21.** $\left\{\frac{2 \pm \sqrt{10}}{2}\right\}$
22. $\left\{\frac{-1 \pm \sqrt{29}}{4}\right\}$
23. $\left\{\frac{-3 \pm \sqrt{41}}{2}\right\}$
24. discriminant: 25; two real solutions **25.** Multiply both the numerator and the denominator by the conjugate of the denominator.
26. (a) $-i$ (b) 1 (c) 1
27. $5 - i$ **28.** $-6 - 5i$
29. 20 **30.** 13 **31.** i
32. $\frac{28}{13} - \frac{3}{13}i$ **33.** $\frac{7}{50} + \frac{1}{50}i$
34. a (the real number itself)
35. No, the product $(a + bi)(a - bi) = a^2 + b^2$ will always be the sum of the squares of two real numbers, which is a real number.
36. $\{-2 \pm i\sqrt{3}\}$
37. $\left\{\frac{2}{3} \pm \frac{2\sqrt{2}}{3}i\right\}$
38. $\left\{\frac{1}{3} \pm \frac{\sqrt{2}}{3}i\right\}$
39. $\left\{-\frac{3}{2} \pm \frac{\sqrt{23}}{2}i\right\}$
40. $\left\{\frac{3}{8} \pm \frac{\sqrt{23}}{8}i\right\}$
41. $\left\{-\frac{1}{9} \pm \frac{2\sqrt{2}}{9}i\right\}$

42. vertex: (1, 0)

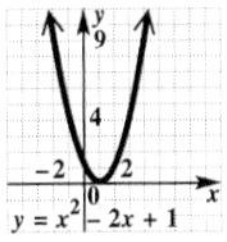

43. vertex: (1, 4)

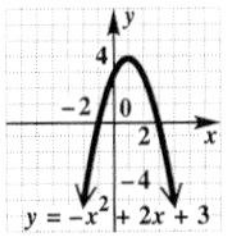

44. vertex: (−2, −2)

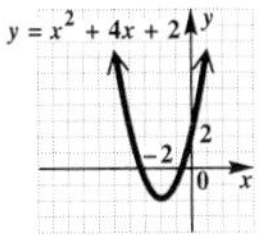

45. two; {±2}   46. one; {2}
47. two; {−1, 4}   48. none; ∅

49. none; ∅
50. one; {−3}

51. $\left\{-\frac{11}{2}, 5\right\}$ 52. $\left\{-\frac{11}{2}, \frac{9}{2}\right\}$

53. $\left\{\frac{-1 \pm \sqrt{21}}{2}\right\}$ 54. $\left\{-\frac{3}{2}, \frac{1}{3}\right\}$

55. $\left\{\frac{-5 \pm \sqrt{17}}{2}\right\}$

56. $\{-1 \pm \sqrt{3}\}$ 57. ∅

58. $\left\{\frac{9 \pm \sqrt{41}}{2}\right\}$ 59. $\left\{-\frac{5}{3}\right\}$

60. $\{-1 \pm 2\sqrt{2}\}$

61. $\{-2 \pm \sqrt{5}\}$ 62. $\{\pm 2\sqrt{2}\}$

[10.5] *Sketch the graph of each equation and identify the vertex.*

42. $y = x^2 - 2x + 1$ **43.** $y = -x^2 + 2x + 3$ **44.** $y = x^2 + 4x + 2$

Decide from the graph how many real number solutions the equation $f(x) = 0$ *has. Determine the solution set (of real solutions) for* $f(x) = 0$ *from the graph.*

45.

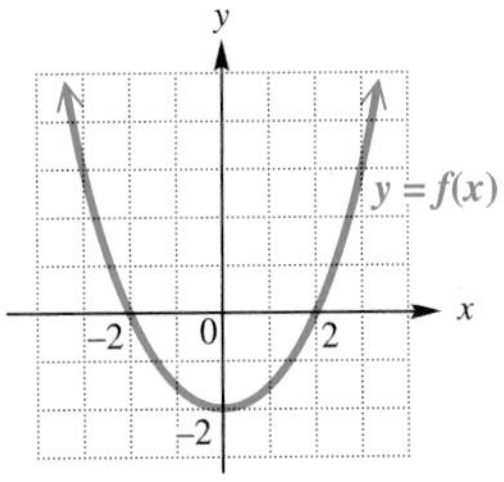

46.

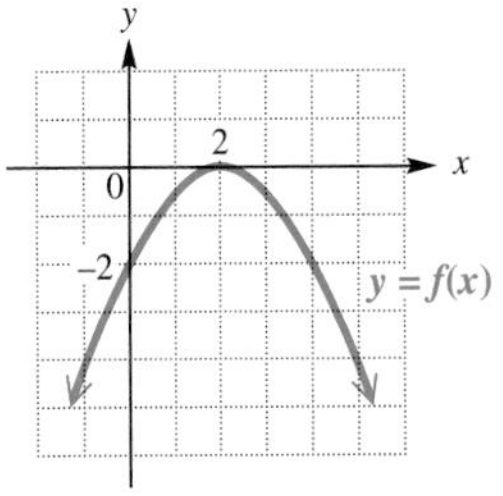

47.

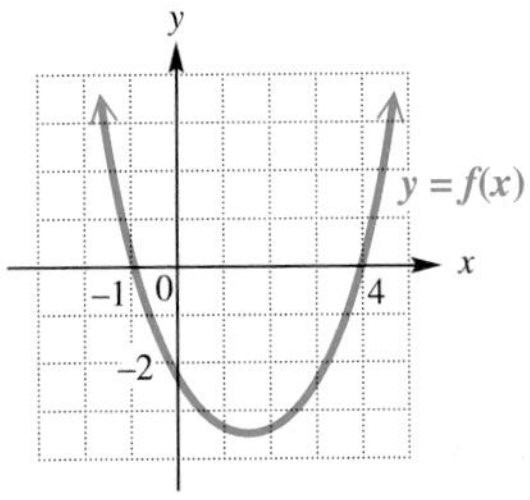

48.

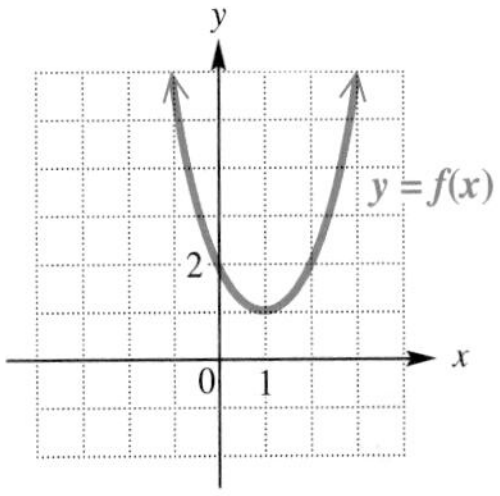

TECHNOLOGY INSIGHTS (EXERCISES 49 AND 50)

Follow the directions for Exercises 45–48.

49.

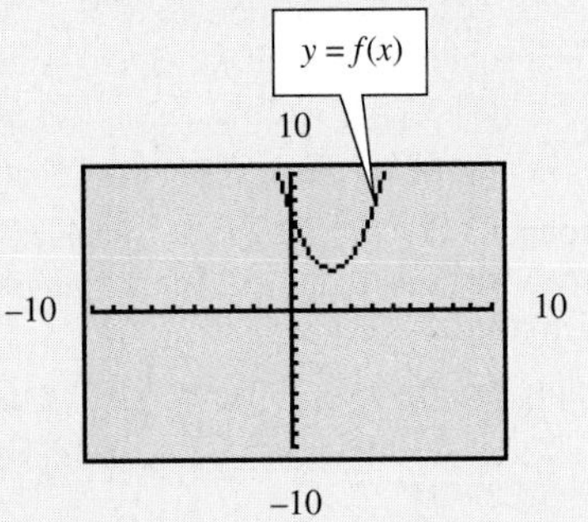

50.

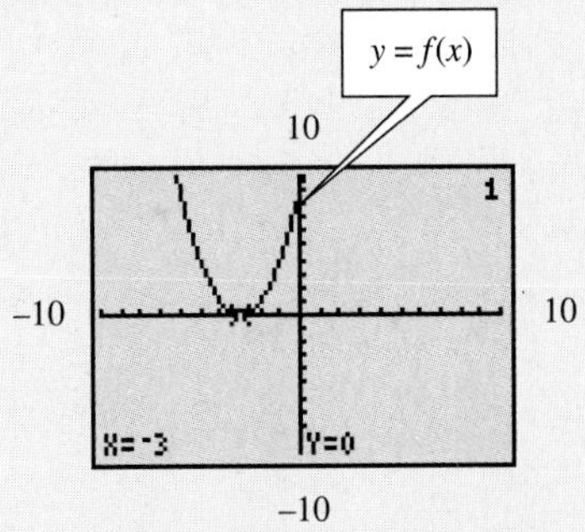

The parabola is tangent to the x-axis (that is, it touches at only one point).

MIXED REVIEW EXERCISES

Solve by any method. Give only real number solutions.

51. $(2t - 1)(t + 1) = 54$

52. $(2p + 1)^2 = 100$

53. $(k + 2)(k - 1) = 3$

54. $6t^2 + 7t - 3 = 0$

55. $2x^2 + 3x + 2 = x^2 - 2x$

56. $x^2 + 2x + 5 = 7$

57. $m^2 - 4m + 10 = 0$

58. $k^2 - 9k + 10 = 0$

59. $(3x + 5)^2 = 0$

60. $\frac{1}{2}r^2 = \frac{7}{2} - r$

61. $x^2 + 4x = 1$

62. $7x^2 - 8 = 5x^2 + 8$

63. approximately 424
64. 7.899, 8.899, 11.899

63. Becky and Brad are the owners of Cole's Baseball Cards. They have found that the price p, in dollars, of a particular Brad Radke baseball card depends on the demand d, in hundreds, for the card, according to the formula $p = (d - 2)^2$. What demand produces a price of $5 for the card?

64. Use a calculator and the Pythagorean formula to find the lengths of the sides of the triangle to the nearest thousandth.

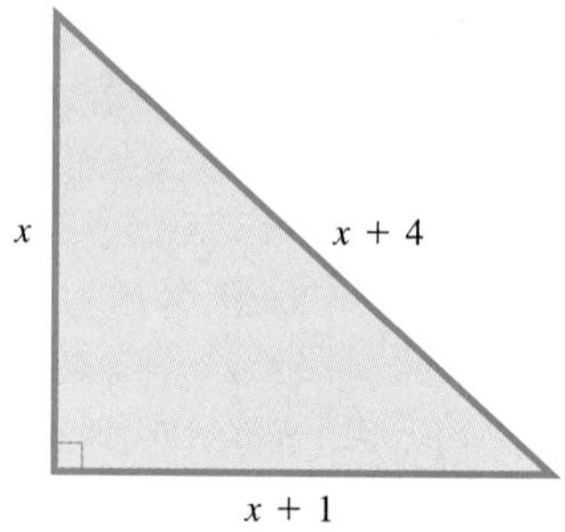

65. $\{1 \pm \sqrt{3}\}$
66. $(x - (1 + \sqrt{3}))(x - (1 - \sqrt{3}))$
67. $((x - 1) - \sqrt{3})((x - 1) + \sqrt{3})$
68. $(x - 1)^2 - \sqrt{3}^2 = x^2 - 2x + 1 - 3 = x^2 - 2x - 2$
69. They are both $x^2 - 2x - 2$.
70. $(x - (2 + \sqrt{5}))(x - (2 - \sqrt{5}))$ or $((x - 2) - \sqrt{5})((x - 2) + \sqrt{5})$

RELATING CONCEPTS (EXERCISES 65–70)

In courses such as Intermediate Algebra and College Algebra, we learn that if r and s are solutions of the equation $x^2 + bx + c = 0$, then $x^2 + bx + c$ factors as $(x - r)(x - s)$. For example, since 2 and 5 are solutions of $x^2 - 7x + 10 = 0$, we have

$$x^2 - 7x + 10 = (x - 2)(x - 5).$$

In Chapter 5 we learned various methods of factoring polynomials using integer coefficients. Now, with the property stated above, we can factor any trinomial of the form $x^2 + bx + c$.

Work Exercises 65–70 in order.

65. Solve the quadratic equation $x^2 - 2x - 2 = 0$ using the quadratic formula. It has two real solutions, both of which are irrational numbers.

66. Suppose that r and s represent your solutions from Exercise 65. Write the trinomial $x^2 - 2x - 2$ in the factored form $(x - r)(x - s)$. Use parentheses carefully.

67. Regroup the terms in the factors obtained in Exercise 66 so that in each factor, the first two terms are grouped. These "binomials within binomials" should be the same in each factor.

68. Multiply your factors in Exercise 67 by using the special product $(a + b)(a - b) = a^2 - b^2$. Then simplify by using the special product $(a + b)^2 = a^2 + 2ab + b^2$. Combine terms.

69. Compare your answer in Exercise 68 to the trinomial given in Exercise 66. They should be the same.

70. Use the method of Exercises 65–69 to factor $x^2 - 4x - 1$. (This process is called *factoring over the real numbers.*)

Did you make the connection between factors of a polynomial and solutions of the corresponding equation?

CHAPTER 10 TEST

[10.1] **1.** $\{\pm\sqrt{39}\}$ **2.** $\{-11, 5\}$
3. $\left\{\frac{-3 \pm 2\sqrt{6}}{4}\right\}$

*Items marked * require knowledge of complex numbers.*

Solve by using the square root property.

1. $x^2 = 39$ **2.** $(y + 3)^2 = 64$ **3.** $(4x + 3)^2 = 24$

[10.2] 4. $\{2 \pm \sqrt{10}\}$
5. $\left\{\frac{-6 \pm \sqrt{42}}{2}\right\}$
[10.3] 6. (a) 0 (b) one (a double solution) 7. $\left\{-3, \frac{1}{2}\right\}$
8. $\left\{\frac{3 \pm \sqrt{3}}{3}\right\}$ 9. $\left\{-1 \pm \frac{\sqrt{7}}{2}i\right\}$
10. $\left\{\frac{5 \pm \sqrt{13}}{6}\right\}$
[10.1–10.3] 11. $\{1 \pm \sqrt{2}\}$
12. $\left\{\frac{-1 \pm 3\sqrt{2}}{2}\right\}$
13. $\left\{\frac{11 \pm \sqrt{89}}{4}\right\}$
14. {5} 15. 2 seconds
16. 247.3 years
[10.4] 17. $-5 + 5i$
18. $-17 - 4i$ 19. 73
20. $2 - i$
[10.5]
21. vertex: (3, 0)

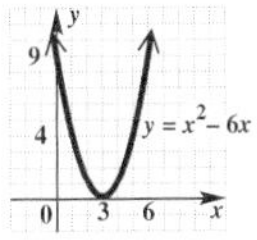

22. vertex: (−1, −3)

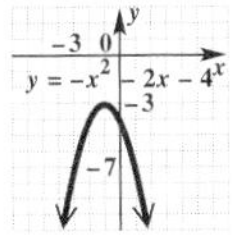

23. vertex: (−3, −2)

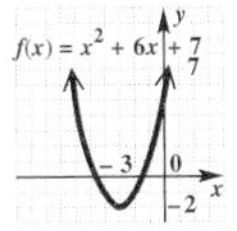

24. (a) two (b) $\{-3 \pm \sqrt{2}\}$
(c) $-3 - \sqrt{2} \approx -4.414$ and $-3 + \sqrt{2} \approx -1.586$
25. 200 and 200

Solve by completing the square.

4. $x^2 - 4x = 6$

5. $2x^2 + 12x - 3 = 0$

6. **(a)** Find the discriminant for the quadratic equation $4x^2 - 20x + 25 = 0$.
(b) Based on your answer to part (a), how many real solutions does the equation have?

Solve by the quadratic formula.

7. $2x^2 + 5x - 3 = 0$

8. $3w^2 + 2 = 6w$

***9.** $4x^2 + 8x + 11 = 0$

10. $t^2 - \frac{5}{3}t + \frac{1}{3} = 0$

Solve by the method of your choice.

11. $p^2 - 2p - 1 = 0$

12. $(2x + 1)^2 = 18$

13. $(x - 5)(2x - 1) = 1$

14. $t^2 + 25 = 10t$

Solve each problem.

15. If an object is propelled into the air from ground level on Earth with an initial velocity of 64 feet per second, its height s (in feet) after t seconds is given by the formula $s = -16t^2 + 64t$. After how many seconds will the object reach a height of 64 feet?

16. Use the formula from Kepler's third law of planetary motion, $P^2 = a^3$, to determine P, in years, for Pluto, if $a = 39.4$ AU.

Perform the indicated operations.

***17.** $(3 + i) + (-2 + 3i) - (6 - i)$

***18.** $(6 + 5i)(-2 + i)$

***19.** $(3 - 8i)(3 + 8i)$

***20.** $\frac{15 - 5i}{7 + i}$

Sketch each graph and identify the vertex.

21. $y = x^2 - 6x + 9$

22. $y = -x^2 - 2x - 4$

23. $f(x) = x^2 + 6x + 7$

24. Refer to the equation in Exercise 23, and do the following:
(a) Determine the number of real solutions of $x^2 + 6x + 7 = 0$ by looking at the graph.
(b) Use the quadratic formula to find the exact values of the real solutions. Give the solution set.
(c) Use a calculator to find approximations for the solutions. Round your answers to the nearest thousandth.

25. Find two numbers whose sum is 400 and whose product is a maximum.

CUMULATIVE REVIEW EXERCISES CHAPTERS 1–10

[1.2] 1. 15 [1.5] 2. −2
[1.6] 3. 5 [1.8] 4. $-r + 7$
5. $-2k$ 6. $19m - 17$
[2.1] 7. {18} [2.2] 8. {5}
9. {2}

***Note:* This cumulative review exercise set can be considered a final examination for the course.**

Perform the indicated operations.

1. $\frac{-4 \cdot 3^2 + 2 \cdot 3}{2 - 4 \cdot 1}$

2. $|-3| - |1 - 6|$

3. $-9 - (-8)(2) + 6 - (6 + 2)$

4. $-4r + 14 + 3r - 7$

5. $13k - 4k + k - 14k + 2k$

6. $5(4m - 2) - (m + 7)$

Solve each equation.

7. $x - 5 = 13$

8. $3k - 9k - 8k + 6 = -64$

9. $2(m - 1) - 6(3 - m) = -4$

[2.4] 10. $100°, 80°$

11. width: 50 feet; length: 94 feet

12. $L = \frac{P - 2W}{2}$ or $L = \frac{P}{2} - W$

[2.7] 13. $(-2, \infty)$

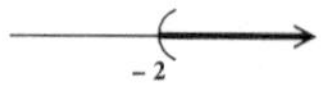

14. $(-\infty, 4]$

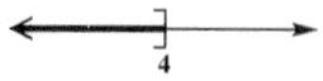

[3.2] 15.

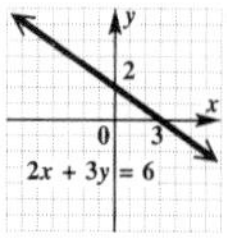

[3.3] 16. -1

[4.1] 17. $8x^5 - 17x^4 - x^2$

18.

x	y
-2	1
-1	-2
0	-3
1	-2
2	1

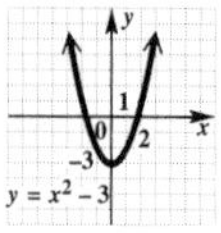

[4.5] 19. $\frac{x^4}{9}$ **20.** $\frac{b^{16}}{c^2}$

21. $\frac{27}{125}$

[4.3] 22. $2x^2 - 9x - 18$

23. $2x^4 + x^3 - 19x^2 + 2x + 20$

[4.6] 24. $3x^2 - 2x + 1$

[4.7] 25. (a) 6.35×10^9 (b) .00023

[5.1] 26. $16x^2(x - 3y)$

[5.3] 27. $(2a + 1)(a - 3)$

[5.4] 28. $(4x^2 + 1)(2x + 1)(2x - 1)$

29. $(5m - 2)^2$ **[5.5] 30.** $\{-9, 6\}$

[5.6] 31. $2\frac{1}{2}$ seconds

Solve each problem.

10. Find the measures of the marked angles.

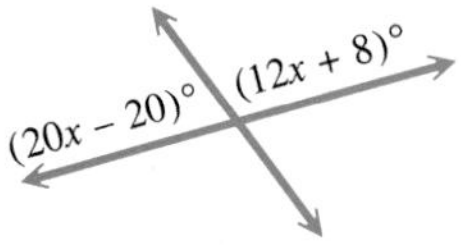

11. The perimeter of a basketball court is 288 feet. The width of the court is 44 feet less than the length. What are the dimensions of the court?

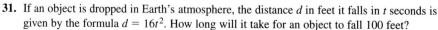

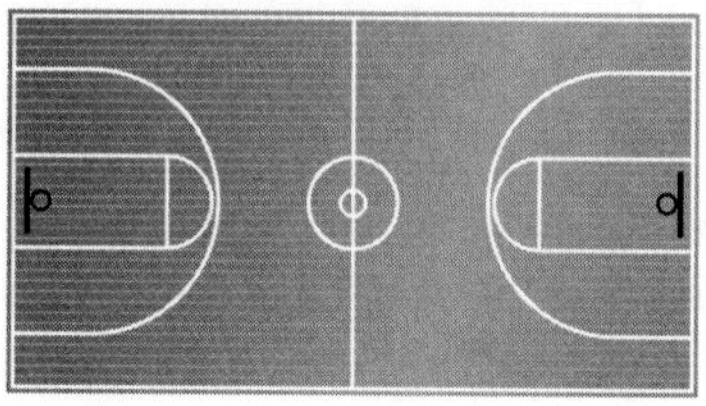

12. Solve the formula $P = 2L + 2W$ for L.

Solve each inequality and graph the solution set.

13. $-8m < 16$

14. $-9p + 2(8 - p) - 6 \geq 4p - 50$

15. Graph the equation $2x + 3y = 6$.

16. Find the slope of the line passing through the points $(5, -2)$ and $(1, 2)$.

17. Subtract: $(5x^5 - 9x^4 + 8x^2) - (9x^2 + 8x^4 - 3x^5)$.

18. Complete the table and graph $y = x^2 - 3$.

x	y
-2	
-1	
0	
1	
2	

Simplify each expression. Write answers with positive exponents.

19. $(3^2 \cdot x^{-4})^{-1}$

20. $\left(\frac{b^{-3}c^4}{b^5c^3}\right)^{-2}$

21. $\left(\frac{5}{3}\right)^{-3}$

Perform each indicated operation.

22. $(2x + 3)(x - 6)$

23. $(2x - 5)(x^3 + 3x^2 - 2x - 4)$

24. $\frac{3x^3 + 10x^2 - 7x + 4}{x + 4}$

25. **(a)** The number of possible hands in contract bridge is about 6,350,000,000. Write this number in scientific notation.

(b) The body of a 150-pound person contains about 2.3×10^{-4} pounds of copper. Write this number without using exponents.

Factor.

26. $16x^3 - 48x^2y$

27. $2a^2 - 5a - 3$

28. $16x^4 - 1$

29. $25m^2 - 20m + 4$

30. Solve by factoring: $x^2 + 3x - 54 = 0$.

31. If an object is dropped in Earth's atmosphere, the distance d in feet it falls in t seconds is given by the formula $d = 16t^2$. How long will it take for an object to fall 100 feet?

[5.7] 32. $(-\infty, -1] \cup [6, \infty)$

[6.1] 33. $\frac{x-3}{x-2}$; -2 and 2

[6.2] 34. $\frac{4}{5}$ [6.4] 35. $\frac{-k-1}{k(k-1)}$

36. $\frac{5a+2}{(a-2)^2(a+2)}$

[6.5] 37. $\frac{b+a}{b-a}$

[6.6] 38. $\left\{-\frac{15}{7}, 2\right\}$

[6.7] 39. 3 miles per hour

[7.1, 7.2] 40. (a) $-\frac{1}{3}$

(b) $y = -\frac{1}{3}x + 6$ (c) The line crosses the y-axis above the origin, indicating that $b > 0$, since $(0, b)$ is the y-intercept.

[7.1] 41. $2x - y = -3$

[7.2] 42.

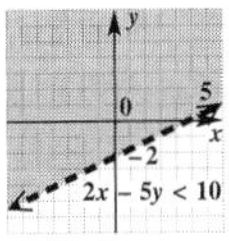

[7.3] 43. -18

[8.1–8.3] 44. $\{(-3, 2)\}$ 45. $\emptyset$

[8.4] 46. Krystalite: $39.99; Contempra II: $29.99

[8.5] 47.

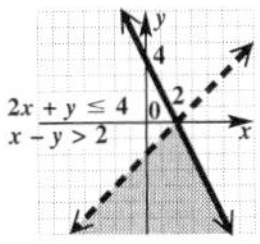

[9.1] 48. 10 [9.4] 49. $\frac{6\sqrt{30}}{5}$

32. Solve the quadratic inequality: $x^2 - 5x - 6 \geq 0$.

33. Write the rational expression $\frac{x^2 - x - 6}{x^2 - 4}$ in lowest terms. For what value(s) of x is the original rational expression undefined?

Perform each operation. Write the answer in lowest terms.

34. $\frac{2}{a-3} \div \frac{5}{2a-6}$

35. $\frac{1}{k} - \frac{2}{k-1}$

36. $\frac{2}{a^2-4} + \frac{3}{a^2-4a+4}$

37. $\dfrac{\frac{1}{a} + \frac{1}{b}}{\frac{1}{a} - \frac{1}{b}}$

38. Solve $\frac{1}{x+3} + \frac{1}{x} = \frac{7}{10}$.

39. Jake's boat goes 12 miles per hour. Find the speed of the current of the river if he can go 6 miles upstream in the same time that he can go 10 miles downstream.

40. Two views of the same line are shown, and the displays at the bottom indicate the coordinates of two points on the line.

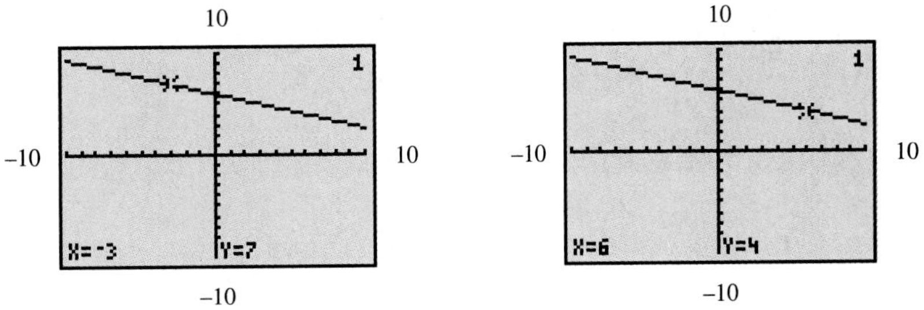

(a) Find the slope of the line.

(b) Find the equation of the line in the form $y = mx + b$.

(c) Explain why the graph indicates that when the equation is written in $y = mx + b$ form, the value of b must be positive.

41. Write an equation of a line with slope 2 and y-intercept $(0, 3)$. Give it in the form $Ax + By = C$.

42. Graph the inequality $2x - 5y < 10$.

43. If $f(x) = -3x^2 + x - 4$, find $f(-2)$.

Solve each system of equations.

44. $2x + y = -4$
$-3x + 2y = 13$

45. $3x - 5y = 8$
$-6x + 10y = 16$

46. Based on the prices in the 1998 Radio Shack catalogue, you can purchase 3 Krystalite phones and 2 Contempra II phones for $179.95. You can purchase 2 Krystalite phones and 3 Contempra II phones for $169.95. Find the price for a single phone of each model.

47. Graph the solution set of the system of inequalities.

$$2x + y \leq 4$$
$$x - y > 2$$

Simplify each expression as much as possible.

48. $\sqrt{100}$

49. $\frac{6\sqrt{6}}{\sqrt{5}}$

50. $\frac{\sqrt[3]{28}}{4}$ [9.3] 51. $4\sqrt{5}$
52. $-ab\sqrt[3]{2b}$ [9.6] 53. $\{7\}$
[9.7] 54. 4
[10.1] 55. $\left\{\frac{-2 \pm 2\sqrt{3}}{3}\right\}$
[10.2] 56. $\{-1 \pm \sqrt{6}\}$
[10.3] 57. $\left\{\frac{2 \pm \sqrt{10}}{2}\right\}$
[10.2, 10.3] 58. $\emptyset$ 59. 3, 4, 5
[10.4] 60. (a) $8i$ (b) $5 + 2i$
61. $\left\{-\frac{1}{2} \pm \frac{\sqrt{17}}{2}i\right\}$
[10.5] 62.

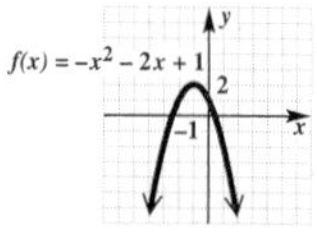

vertex: $(-1, 2)$; domain: $(-\infty, \infty)$; range: $(-\infty, 2]$ 63. positive

50. $\sqrt[3]{\frac{7}{16}}$

51. $3\sqrt{5} - 2\sqrt{20} + \sqrt{125}$

52. $\sqrt[3]{16a^3b^4} - \sqrt[3]{54a^3b^4}$

53. Solve $\sqrt{x + 2} = x - 4$.

54. Simplify $8^{2/3}$.

Solve each quadratic equation using the method indicated. Give only real solutions.

55. $(3x + 2)^2 = 12$ (square root property)

56. $-x^2 + 5 = 2x$ (completing the square)

57. $2x(x - 2) - 3 = 0$ (quadratic formula)

58. $(4x + 1)(x - 1) = -3$ (any method)

Solve each problem.

59. In a right triangle, the lengths of the sides are consecutive integers. Use the Pythagorean formula to find these lengths.

*__60.__ Write in standard form.

(a) $(-9 + 3i) + (4 + 2i) - (-5 - 3i)$ (b) $\frac{-17 - i}{-3 + i}$

*__61.__ Find the complex solutions of $2x^2 + 2x = -9$.

62. Graph the quadratic function $f(x) = -x^2 - 2x + 1$ and identify the vertex. Give the domain and the range.

63. If the graph of the parabola $y = ax^2 + bx + c$ opens upward and has two positive x-values for its x-intercepts, will the y-value for its y-intercept be positive or negative?

APPENDIX A REVIEW OF DECIMALS AND PERCENTS

OBJECTIVES

1. Add and subtract decimals.
2. Multiply and divide decimals.
3. Convert percents to decimals and decimals to percents.
4. Find percentages by multiplying.

FOR EXTRA HELP

SSG Ap. A
SSM Ap. A
Pass the Test Software
InterAct Math Tutorial Software
Video N/A

OBJECTIVE 1 Add and subtract decimals. A **decimal** is a number written with a decimal point, such as 4.2. The operations on decimals—addition, subtraction, multiplication, and division—are explained in the next examples.

EXAMPLE 1 Adding and Subtracting Decimals

Add or subtract as indicated.

(a) 6.92 + 14.8 + 3.217

Place the numbers in a column, with decimal points lined up, then add. If you like, attach zeros to make all the numbers the same length; this is a good way to avoid errors. For example,

Decimal points lined up

$$\begin{array}{r} 6.92 \\ 14.8 \\ +\ 3.217 \\ \hline 24.937 \end{array} \quad \text{or} \quad \begin{array}{r} 6.920 \\ 14.800 \\ +\ 3.217 \\ \hline 24.937. \end{array}$$

(b) 47.6 − 32.509

Write the numbers in a column, attaching zeros to 47.6.

$$\begin{array}{r} 47.6 \\ -32.509 \\ \hline \end{array} \quad \text{becomes} \quad \begin{array}{r} 47.600 \\ -32.509 \\ \hline 15.091 \end{array}$$

(c) 3 − .253

$$\begin{array}{r} 3.000 \\ -\ .253 \\ \hline 2.747 \end{array}$$

OBJECTIVE 2 Multiply and divide decimals. Multiplication and division of decimals are similar to the same operations with whole numbers.

EXAMPLE 2 Multiplying Decimals

Multiply.

(a) 29.3 × 4.52

Multiply as if the numbers were whole numbers. To find the number of decimal places in the answer, add the numbers of decimal places in the factors.

$$\begin{array}{rl} 29.3 & \text{1 decimal place} \\ \times 4.52 & \text{2 decimal places} \\ \hline 5\ 86 & \\ 14\ 6\ 5 & \\ 117\ 2 & \\ \hline 132.4\ 36 & \text{3 decimal places in answer} \end{array}$$

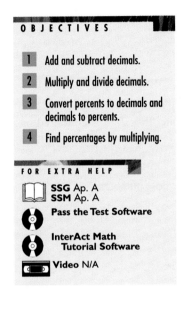

(b) 7.003×55.8

$$\begin{array}{r l}
7.003 & \text{3 decimal places} \\
\times\ 55.8 & \text{1 decimal place} \\
\hline
5\,602\,4 & \\
35\,015 & \\
350\,15 & \downarrow \\
\hline
390.767\,4 & \text{4 decimal places in answer}
\end{array}$$

EXAMPLE 3 Dividing Decimals

Divide: $279.45 \div 24.3$.

Move the decimal point in 24.3 one place to the right, to get 243. Move the decimal point the same number of places in 279.45. By doing this, 24.3 is converted into the whole number 243.

$$243.\overline{)2794.5}$$

Bring the decimal point straight up and divide as with whole numbers.

$$\begin{array}{r}
11.5 \\
243.\overline{)2794.5} \\
\underline{243} \\
364 \\
\underline{243} \\
121\,5 \\
\underline{121\,5} \\
0
\end{array}$$

OBJECTIVE 3 Convert percents to decimals and decimals to percents. One of the main uses of decimals is in percent problems. The word **percent** means "per one hundred." Percent is written with the sign %. One percent means "one per one hundred."

Percent

$$1\% = .01 \quad \text{or} \quad 1\% = \frac{1}{100}$$

EXAMPLE 4 Converting between Decimals and Percents

Convert.

(a) 75% to a decimal

Since $1\% = .01$,

$$75\% = 75 \cdot 1\% = 75 \cdot (.01) = .75.$$

The fraction form $1\% = \frac{1}{100}$ can also be used to convert 75% to a decimal.

$$75\% = 75 \cdot 1\% = 75 \cdot \left(\frac{1}{100}\right) = .75$$

(b) 2.63 to a percent

$$2.63 = 263 \cdot (.01) = 263 \cdot 1\% = 263\%$$

OBJECTIVE 4 Find percentages by multiplying. A part of a whole is called a **percentage.** For example, since 50% represents $\frac{50}{100} = \frac{1}{2}$ of a whole, 50% of 800 is half of 800, or 400. Multiply to find percentages, as in the next example.

EXAMPLE 5 Finding Percentages

Find the percentages.

(a) 15% of 600

The word *of* indicates multiplication here. For this reason, 15% of 600 is found by multiplying.

$$15\% \cdot 600 = (.15) \cdot 600 = 90$$

(b) 125% of 80

$$125\% \cdot 80 = (1.25) \cdot 80 = 100$$

(c) What percent of 52 is 7.8?

We can translate this sentence to symbols word by word.

What percent	of	52	is	7.8?
p	$\cdot$	52	=	7.8

Solve the equation.

$$52p = 7.8$$

$$p = .15 \quad \text{Divide both sides by 52.}$$

$$p = 15\% \quad \text{Change to percent.}$$

EXAMPLE 6 Using Percent in a Consumer Problem

A video movie with a regular price of $18 is on sale at 22% off. Find the amount of the discount.

The discount is 22% of $18. Change 22% to .22 and use the fact that *of* indicates multiplication.

$$22\% \cdot 18 = .22 \cdot 18 = 3.96$$

The discount is $3.96.

APPENDIX A EXERCISES

1. 117.385 **2.** 887.859 **3.** 13.21 **4.** 19.57 **5.** 150.49 **6.** 270.58 **7.** 96.101 **8.** 31.48

Perform the indicated operations. See Examples 1–3.

1. 14.23 + 9.81 + 74.63 + 18.715

2. 89.416 + 21.32 + 478.91 + 298.213

3. 19.74 − 6.53

4. 27.96 − 8.39

5. 219 − 68.51

6. 283 − 12.42

7.
$$\begin{array}{r} 48.96 \\ 37.421 \\ +\ 9.72 \\ \hline \end{array}$$

8.
$$\begin{array}{r} 9.71 \\ 4.8 \\ 3.6 \\ 5.2 \\ +8.17 \\ \hline \end{array}$$

9. 4.849 10. 14.218
11. 166.32 12. 43.01
13. 164.19 14. 80.0904
15. 1.344 16. 1655.9
17. 4.14 18. 8.21
19. 2.23 20. 6.14
21. 4800 22. 16.76
23. .53 24. .38
25. 1.29 26. 1.74
27. .96 28. .11
29. .009 30. .001
31. 80% 32. 75%
33. .7% 34. 140%
35. 67% 36. .3%
37. 12.5% 38. 98.3%
39. 109.2 40. 42
41. 238.92 42. 300
43. 5% 44. 65%
45. 110% 46. 120%
47. 25% 48. 2%
49. 148.44 50. 3.84
51. 7839.26 52. 7.50
53. 7.39% 54. 76.60%
55. $2760 56. $23,040
57. 805 miles 58. $96
59. 63 miles 60. $9240
61. $2400

9. $\begin{array}{r} 8.6 \\ -3.751 \\ \hline \end{array}$ **10.** $\begin{array}{r} 27.8 \\ -13.582 \\ \hline \end{array}$

11. 39.6×4.2 **12.** 18.7×2.3 **13.** 42.1×3.9 **14.** 19.63×4.08

15. $.042 \times 32$ **16.** 571×2.9 **17.** $24.84 \div 6$ **18.** $32.84 \div 4$

19. $7.6266 \div 3.42$ **20.** $14.9202 \div 2.43$ **21.** $2496 \div .52$ **22.** $.56984 \div .034$

Convert each percent to a decimal. See Example 4(a).

23. 53% **24.** 38% **25.** 129% **26.** 174%

27. 96% **28.** 11% **29.** .9% **30.** .1%

Convert each decimal to a percent. See Example 4(b).

31. .80 **32.** .75 **33.** .007 **34.** 1.4

35. .67 **36.** .003 **37.** .125 **38.** .983

Respond to each statement or question. Round your answer to the nearest hundredth if appropriate. See Example 5.

39. What is 14% of 780?

40. Find 12% of 350.

41. Find 22% of 1086.

42. What is 20% of 1500?

43. 4 is what percent of 80?

44. 1300 is what percent of 2000?

45. What percent of 5820 is 6402?

46. What percent of 75 is 90?

47. 121 is what percent of 484?

48. What percent of 3200 is 64?

49. Find 118% of 125.8.

50. Find 3% of 128.

51. What is 91.72% of 8546.95?

52. Find 12.741% of 58.902.

53. What percent of 198.72 is 14.68?

54. 586.3 is what percent of 765.4?

Solve each problem. See Example 6.

55. A retailer has $23,000 invested in her business. She finds that she is earning 12% per year on this investment. How much money is she earning per year?

56. Harley Dabler recently bought a duplex for $144,000. He expects to earn 16% per year on the purchase price. How many dollars per year will he earn?

57. For a recent tour of the eastern United States, a travel agent figured that the trip totaled 2300 miles, with 35% of the trip by air. How many miles of the trip were by air?

58. Capitol Savings Bank pays 3.2% interest per year. What is the annual interest on an account of $3000?

59. An ad for steel-belted radial tires promises 15% better mileage when the tires are used. Alexandria's Escort now goes 420 miles on a tank of gas. If she switched to the new tires, how many extra miles could she drive on a tank of gas?

60. A home worth $77,000 is located in an area where home prices are increasing at a rate of 12% per year. By how much would the value of this home increase in one year?

61. A family of four with a monthly income of $2000 spends 90% of its earnings and saves the rest. Find the *annual* savings of this family.

APPENDIX B SETS

OBJECTIVES

1. List the elements of a set.
2. Learn the vocabulary and symbols used to discuss sets.
3. Decide whether a set is finite or infinite.
4. Decide whether a given set is a subset of another set.
5. Find the complement of a set.
6. Find the union and the intersection of two sets.

FOR EXTRA HELP

SSG Ap. B
SSM Ap. B
Pass the Test Software
InterAct Math Tutorial Software
Video N/A

OBJECTIVE 1 List the elements of a set. A **set** is a collection of things. The objects in a set are called the **elements** of the set. A set is represented by listing its elements between **set braces,** { }. The order in which the elements of a set are listed is unimportant.

EXAMPLE 1 Listing the Elements of a Set

Represent the following sets by listing the elements.

(a) The set of states in the United States that border on the Pacific Ocean = {California, Oregon, Washington, Hawaii, Alaska}.

(b) The set of all counting numbers less than 6 = {1, 2, 3, 4, 5}. ■

OBJECTIVE 2 Learn the vocabulary and symbols used to discuss sets. Capital letters are used to name sets. To state that 5 is an element of

$$S = \{1, 2, 3, 4, 5\},$$

write $5 \in S$. The statement $6 \notin S$ means that 6 is not an element of S.

A set with no elements is called the **empty set,** or the **null set.** The symbols $\emptyset$ or { } are used for the empty set. If we let A be the set of all cats that fly, then A is the empty set.

$$A = \emptyset \qquad \text{or} \qquad A = \{\ \}$$

CAUTION Do not make the common error of writing the empty set as $\{\emptyset\}$.

In any discussion of sets, there is some set that includes all the elements under consideration. This set is called the **universal set** for that situation. For example, if the discussion is about presidents of the United States, then the set of all presidents of the United States is the universal set. The universal set is denoted U.

OBJECTIVE 3 Decide whether a set is finite or infinite. In Example 1, there are five elements in the set in part (a), and five in part (b). If the number of elements in a set is either 0 or a counting number, then the set is **finite.** On the other hand, the set of natural numbers, for example, is an **infinite** set, because there is no final number. We can list the elements of the set of natural numbers as

$$N = \{1, 2, 3, 4 \ldots\}$$

where the three dots indicate that the set continues indefinitely. Not all infinite sets can be listed in this way. For example, there is no way to list the elements in the set of all real numbers between 1 and 2.

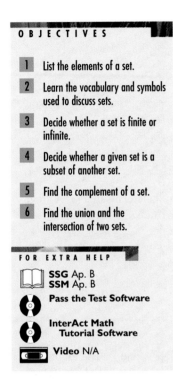

EXAMPLE 2 Distinguishing between Finite and Infinite Sets

List the elements of each set, if possible. Decide whether each set is finite or infinite.

(a) The set of all integers
One way to list the elements is $\{\ldots, -2, -1, 0, 1, 2, \ldots\}$. The set is infinite.

(b) The set of all natural numbers between 0 and 5
$\{1, 2, 3, 4\}$ The set is finite.

(c) The set of all irrational numbers
This is an infinite set whose elements cannot be listed.

Two sets are **equal** if they have exactly the same elements. Thus, the set of natural numbers and the set of positive integers are equal sets. Also, the sets

$$\{1, 2, 4, 7\} \quad \text{and} \quad \{4, 2, 7, 1\}$$

are equal. The order of the elements does not make a difference.

OBJECTIVE 4 Decide whether a given set is a subset of another set. If all elements of a set A are also elements of a new set B, then we say A is a **subset** of B, written $A \subseteq B$. We use the symbol $A \nsubseteq B$ to mean that A is not a subset of B.

EXAMPLE 3 Using Subset Notation

Let $A = \{1, 2, 3, 4\}$, $B = \{1, 4\}$, and $C = \{1\}$. Then $B \subseteq A$, $C \subseteq A$, and $C \subseteq B$, but $A \nsubseteq B$, $A \nsubseteq C$, and $B \nsubseteq C$.

The set $M = \{a, b\}$ has four subsets: $\{a, b\}$, $\{a\}$, $\{b\}$, and $\emptyset$. The empty set is defined to be a subset of any set. How many subsets does $N = \{a, b, c\}$ have? There is one subset with 3 elements: $\{a, b, c\}$. There are three subsets with 2 elements:

$$\{a, b\}, \quad \{a, c\}, \quad \text{and} \quad \{b, c\}.$$

There are three subsets with 1 element:

$$\{a\}, \quad \{b\}, \quad \text{and} \quad \{c\}.$$

There is one subset with 0 elements: $\emptyset$. Thus, set N has eight subsets.

The following generalization can be made.

Number of Subsets of a Set

A set with n elements has 2^n subsets.

To illustrate the relationships between sets, **Venn diagrams** are often used. A rectangle represents the universal set, U. The sets under discussion are represented by regions within the rectangle. The Venn diagram in Figure 1 shows that $B \subseteq A$.

OBJECTIVE 5 Find the complement of a set. For every set A, there is a set A', the **complement** of A, that contains all the elements of U that are not in A. The shaded region in the Venn diagram in Figure 2 represents A'.

$B \subseteq A$

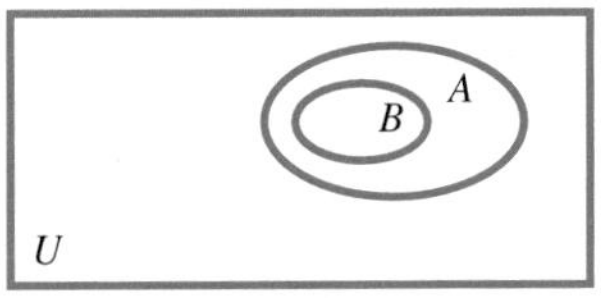

Figure 1

A' is in color.

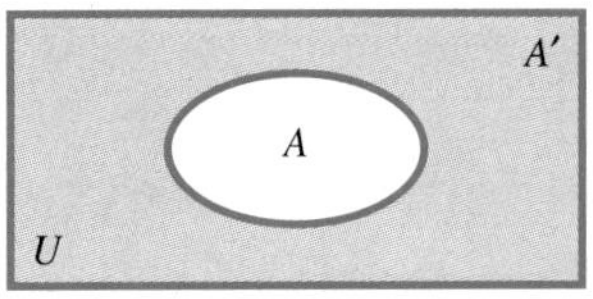

Figure 2

EXAMPLE 4 Determining the Complement of a Set

Given $U = \{a, b, c, d, e, f, g\}$, $A = \{a, b, c\}$, $B = \{a, d, f, g\}$, and $C = \{d, e\}$, find A', B', and C'.

(a) $A' = \{d, e, f, g\}$ **(b)** $B' = \{b, c, e\}$ **(c)** $C' = \{a, b, c, f, g\}$

OBJECTIVE 6 Find the union and the intersection of two sets. The **union** of two sets A and B, written $A \cup B$, is the set of all elements of A together with all elements of B. Thus, for the sets in Example 4,

$$A \cup B = \{a, b, c, d, f, g\}$$

and

$$A \cup C = \{a, b, c, d, e\}.$$

In Figure 3 the shaded region is the union of sets A and B.

$A \cup B$ is shaded.

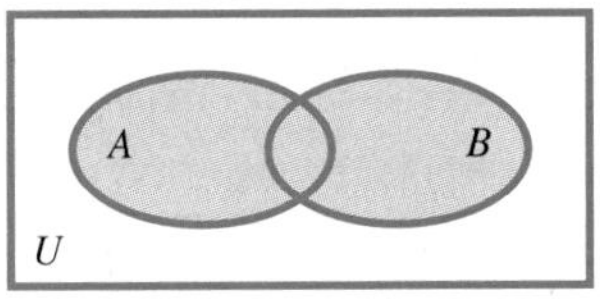

Figure 3

$A \cap B$ is shaded.

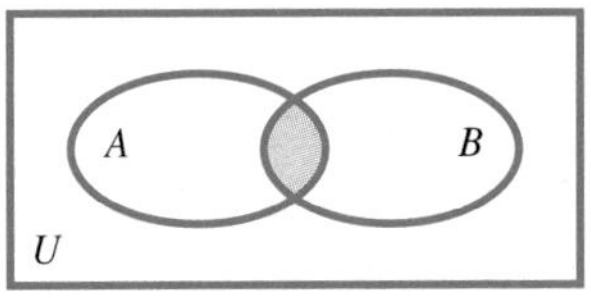

Figure 4

EXAMPLE 5 Finding the Union of Two Sets

If $M = \{2, 5, 7\}$ and $N = \{1, 2, 3, 4, 5\}$, then

$$M \cup N = \{1, 2, 3, 4, 5, 7\}.$$

The **intersection** of two sets A and B, written $A \cap B$, is the set of all elements that belong to both A and B. For example if,

$$A = \{\text{Jose, Ellen, Marge, Kevin}\}$$

and

$$B = \{\text{Jose, Patrick, Ellen, Sue}\},$$

then

$$A \cap B = \{\text{Jose, Ellen}\}.$$

The shaded region in Figure 4 represents the intersection of the two sets A and B.

EXAMPLE 6 Finding the Intersection of Two Sets

Suppose that $P = \{3, 9, 27\}$, $Q = \{2, 3, 10, 18, 27, 28\}$, and $R = \{2, 10, 28\}$.

(a) $P \cap Q = \{3, 27\}$ **(b)** $Q \cap R = \{2, 10, 28\} = R$ **(c)** $P \cap R = \emptyset$

Sets like P and R in Example 6 that have no elements in common are called **disjoint sets.** The Venn diagram in Figure 5 shows a pair of disjoint sets.

Disjoint sets; $A \cap B = \emptyset$

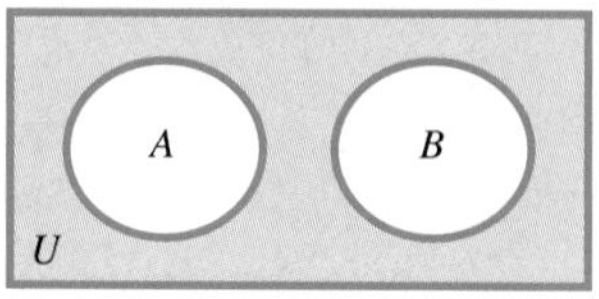

Figure 5

EXAMPLE 7 Using Set Operations

Let $U = \{2, 5, 7, 10, 14, 20\}$, $A = \{2, 10, 14, 20\}$, $B = \{5, 7\}$, and $C = \{2, 5, 7\}$. Find each of the following.

(a) $A \cup B = \{2, 5, 7, 10, 14, 20\} = U$

(b) $A \cap B = \emptyset$

(c) $B \cup C = \{2, 5, 7\} = C$

(d) $B \cap C = \{5, 7\} = B$

(e) $A' = \{5, 7\} = B$

APPENDIX B EXERCISES

List the elements of each set. See Examples 1 and 2.

1. The set of all natural numbers less than 8

2. The set of all integers between 4 and 10

3. The set of seasons

4. The set of months of the year

5. The set of women presidents of the United States

6. The set of all living humans who are more than 200 years old

7. The set of letters of the alphabet between K and M

8. The set of letters of the alphabet between D and H

9. The set of positive even integers

10. The set of all multiples of 5

11. Which of the sets described in Exercises 1–10 are infinite sets?

12. Which of the sets described in Exercises 1–10 are finite sets?

Tell whether each statement is true or false.

13. $5 \in \{1, 2, 5, 8\}$

14. $6 \in \{1, 2, 3, 4, 5\}$

15. $2 \in \{1, 3, 5, 7, 9\}$

16. $1 \in \{6, 2, 5, 1\}$

17. $7 \notin \{2, 4, 6, 8\}$

18. $7 \notin \{1, 3, 5, 7\}$

19. $\{2, 4, 9, 12, 13\} = \{13, 12, 9, 4, 2\}$

20. $\{7, 11, 4\} = \{7, 11, 4, 0\}$

1. {1, 2, 3, 4, 5, 6, 7}
2. {5, 6, 7, 8, 9}
3. {winter, spring, summer, fall}
4. {January, February, March, April, May, June, July, August, September, October, November, December} **5.** $\emptyset$ **6.** $\emptyset$
7. {L} **8.** {E, F, G}
9. {2, 4, 6, 8, 10, . . . }
10. { . . . , −15, −10, −5, 0, 5, 10, 15, . . . } **11.** The sets in Exercises 9 and 10 are infinite sets.
12. The sets in Exercises 1–8 are finite sets.
13. true **14.** false
15. false **16.** true
17. true **18.** false
19. true **20.** false

21. true **22.** false
23. true **24.** true
25. true **26.** false
27. false **28.** true
29. true **30.** false
31. true **32.** true
33. false **34.** true
35. true **36.** true
37. true **38.** false
39. false **40.** false
41. true **42.** true
43. false **44.** false
45. {g, h} **46.** {b, d, f, g, h}
47. {b, c, d, e, g, h}
48. {a, b, c, e, f, g, h}
49. {a, c, e} = *B*
50. {a, c, e} = *B* **51.** {d}
52. $\emptyset$ **53.** {a}
54. {a, b, c, d, e, f} = *A*
55. {a, c, d, e} **56.** {a, c, e, f}
57. {a, c, e, f} **58.** {a, d, f}
59. $\emptyset$ **60.** {a, c, e} = *B*
61. *B* and *D*; *C* and *D*

Let

$$A = \{1, 3, 4, 5, 7, 8\}$$
$$B = \{2, 4, 6, 8\}$$
$$C = \{1, 3, 5, 7\}$$
$$D = \{1, 2, 3\}$$
$$E = \{3, 7\}$$
$$U = \{1, 2, 3, 4, 5, 6, 7, 8, 9, 10\}.$$

Tell whether each statement is true or false. See Examples 3, 5, 6, and 7.

21. $A \subseteq U$ **22.** $D \subseteq A$ **23.** $\emptyset \subseteq A$ **24.** $\{1, 2\} \subseteq D$ **25.** $C \subseteq A$

26. $A \subseteq C$ **27.** $D \subseteq B$ **28.** $E \subseteq C$ **29.** $D \not\subseteq E$ **30.** $E \not\subseteq A$

31. There are exactly 4 subsets of *E*.

32. There are exactly 8 subsets of *D*.

33. There are exactly 12 subsets of *C*.

34. There are exactly 16 subsets of *B*.

35. $\{4, 6, 8, 12\} \cap \{6, 8, 14, 17\} = \{6, 8\}$

36. $\{2, 5, 9\} \cap \{1, 2, 3, 4, 5\} = \{2, 5\}$

37. $\{3, 1, 0\} \cap \{0, 2, 4\} = \{0\}$

38. $\{4, 2, 1\} \cap \{1, 2, 3, 4\} = \{1, 2, 3\}$

39. $\{3, 9, 12\} \cap \emptyset = \{3, 9, 12\}$

40. $\{3, 9, 12\} \cup \emptyset = \emptyset$

41. $\{4, 9, 11, 7, 3\} \cup \{1, 2, 3, 4, 5\} = \{1, 2, 3, 4, 5, 7, 9, 11\}$

42. $\{1, 2, 3\} \cup \{1, 2, 3\} = \{1, 2, 3\}$

43. $\{3, 5, 7, 9\} \cup \{4, 6, 8\} = \emptyset$

44. $\{5, 10, 15, 20\} \cup \{5, 15, 30\} = \{5, 15\}$

Let

$$U = \{a, b, c, d, e, f, g, h\}$$
$$A = \{a, b, c, d, e, f\}$$
$$B = \{a, c, e\}$$
$$C = \{a, f\}$$
$$D = \{d\}.$$

List the elements in each set. See Examples 4–7.

45. A' **46.** B' **47.** C' **48.** D'

49. $A \cap B$ **50.** $B \cap A$ **51.** $A \cap D$ **52.** $B \cap D$

53. $B \cap C$ **54.** $A \cup B$ **55.** $B \cup D$ **56.** $B \cup C$

57. $C \cup B$ **58.** $C \cup D$ **59.** $A \cap \emptyset$ **60.** $B \cup \emptyset$

61. Name every pair of disjoint sets among *A–D* above.

Answers to Selected Exercises

TO THE STUDENT

If you need further help with algebra, you may want to obtain a copy of the *Student's Solution Manual* that goes with this book. You also may want the *Student's Study Guide.* It has extra examples and exercises to complete, corresponding to each learning objective of the book. In addition, there is a practice test for each chapter. Your college bookstore either has these books or can order them for you.

In this section we provide the answers that we think most students will obtain when they work the exercises using the methods explained in the text. If your answer does not look exactly like the one given here, it is not necessarily wrong. In many cases there are equivalent forms of the answer that are correct. For example, if the answer section shows $\frac{3}{4}$ and your answer is .75, you have obtained the right answer but written it in a different (yet equivalent) form. Unless the directions specify otherwise, .75 is just as valid an answer as $\frac{3}{4}$.

In general, if your answer does not agree with the one given in the text, see whether it can be transformed into the other form. If it can, then it is the correct answer. If you still have doubts, talk with your instructor.

CHAPTER 1 THE REAL NUMBER SYSTEM

SECTION 1.1 (PAGE 9)

CONNECTIONS **Page 2:** Answers will vary.

EXERCISES **1.** true **3.** false; The fraction $\frac{17}{51}$ can be reduced to $\frac{1}{3}$. **5.** false; *Product* refers to multiplication, so the product of 8 and 2 is 16. **7.** prime **9.** composite; $2 \cdot 2 \cdot 2 \cdot 2 \cdot 2 \cdot 2$ **11.** composite; $2 \cdot 7 \cdot 13 \cdot 19$ **13.** neither **15.** composite; $2 \cdot 3 \cdot 5$ **17.** composite; $2 \cdot 2 \cdot 5 \cdot 5 \cdot 5$ **19.** composite; $2 \cdot 2 \cdot 31$ **21.** prime **23.** $\frac{1}{2}$ **25.** $\frac{5}{6}$ **27.** $\frac{1}{3}$ **29.** $\frac{6}{5}$ **31.** (c) **33.** $\frac{24}{35}$ **35.** $\frac{6}{25}$ **37.** $\frac{6}{5}$ or $1\frac{1}{5}$ **39.** $\frac{232}{15}$ or $15\frac{7}{15}$ **41.** $\frac{10}{3}$ or $3\frac{1}{3}$ **43.** 12 **45.** $\frac{1}{16}$ **47.** $\frac{84}{47}$ or $1\frac{37}{47}$ **51.** $\frac{2}{3}$ **53.** $\frac{8}{9}$ **55.** $\frac{27}{8}$ or $3\frac{3}{8}$ **57.** $\frac{17}{36}$ **59.** $\frac{11}{12}$ **61.** $\frac{4}{3}$ or $1\frac{1}{3}$ **63.** 6 cups **65.** 34 dollars **67.** $\frac{9}{16}$ inch **69.** $618\frac{3}{4}$ feet **71.** $5\frac{5}{24}$ inches **73.** $\frac{1}{3}$ cup **75.** 650 **77.** **(a)** Crum **(b)** Jordan **(c)** Jordan **(d)** Baldock **(e)** Tobin and Perry; $\frac{1}{2}$ **79.** (b)

SECTION 1.2 (PAGE 19)

EXERCISES **1.** false; $4 + 3(8 - 2) = 4 + 3 \cdot 6 = 4 + 18 = 22$. The common error leading to 42 is adding 4 to 3 and then multiplying by 6. One must follow the rules for order of operations. **3.** false; The correct interpretation is $4 = 16 - 12$.

5. 49 **7.** 144 **9.** 64 **11.** 1000 **13.** 81 **15.** 1024 **17.** $\frac{16}{81}$ **19.** .000064 **23.** 32 **25.** $\frac{49}{30}$ or $1\frac{19}{30}$ **27.** 12 **29.** 23.01 **31.** 95 **33.** 90 **35.** 14 **37.** 9 **41.** true **43.** false **45.** true **47.** true **49.** false **51.** false **53.** true **55.** $15 = 5 + 10$ **57.** $9 > 5 - 4$ **59.** $16 \neq 19$ **61.** $2 \leq 3$ **63.** Seven is less than nineteen. True **65.** Three is not equal to six. True **67.** Eight is greater than or equal to eleven. False **69.** Answers will vary. One example is $5 + 3 \geq 2 \cdot 2$. **71.** $30 > 5$ **73.** $3 \leq 12$ **75.** is younger than **77.** The inequality symbol $\geq$ implies a true statement if 12 equals 12 *or* if 12 is greater than 12. **79.** December 1996, January 1997, May 1997, June 1997, November 1997 **81.** **(a)** .7 **(b)** 1.3%

SECTION 1.3 (PAGE 25)

EXERCISES **1.** 10 **3.** $12 + x$; 21 **5.** no **7.** $2x^3 = 2 \cdot x \cdot x \cdot x$, while $2x \cdot 2x \cdot 2x = (2x)^3$. **9.** The exponent 2 applies only to its base, which is x. (The expression $(4x)^2$ would require multiplying 4 by $x = 3$ first.) **11.** (Answers will vary.) Two such pairs are $x = 0$, $y = 6$ and $x = 1$, $y = 4$. To determine them, choose a value for x, substitute it into the expression $2x + y$, and then subtract the value of $2x$ from 6. **13.** **(a)** 13 **(b)** 15 **15.** **(a)** 20 **(b)** 30 **17.** **(a)** 64 **(b)** 144 **19.** **(a)** $\frac{5}{3}$ **(b)** $\frac{7}{3}$ **21.** **(a)** $\frac{7}{8}$ **(b)** $\frac{13}{12}$ **23.** **(a)** 52 **(b)** 114 **25.** **(a)** 25.836 **(b)** 38.754 **27.** **(a)** 24 **(b)** 28 **29.** **(a)** 12 **(b)** 33 **31.** **(a)** 6 **(b)** $\frac{9}{5}$ **33.** **(a)** $\frac{4}{3}$ **(b)** $\frac{13}{6}$ **35.** **(a)** $\frac{2}{7}$ **(b)** $\frac{16}{27}$ **37.** **(a)** 12 **(b)** 55 **39.** **(a)** 1 **(b)** $\frac{28}{17}$ **41.** **(a)** 3.684 **(b)** 8.841 **43.** $12x$ **45.** $x + 7$ **47.** $x - 2$ **49.** $7 - x$ **51.** $x - 6$ **53.** $\frac{12}{x}$ **55.** $6(x - 4)$ **57.** No, it is a connective word that joins the two factors: the number and 6. **59.** yes **61.** no **63.** yes **65.** yes **67.** yes **69.** $x + 8 = 18$; 10 **71.** $16 - \frac{3}{4}x = 13$; 4 **73.** $2x + 1 = 5$; 2 **75.** $3x = 2x + 8$; 8 **77.** expression **79.** equation **81.** equation **83.** \$10.50; less by \$.33 **85.** \$12.10; less by \$.27

SECTION 1.4 (PAGE 34)

EXERCISES **1.** 4 **3.** 0 **5.** One example is $\sqrt{12}$. There are others. **7.** true **9.** true **11.** **(a)** 3, 7 **(b)** 0, 3, 7 **(c)** $-9, 0, 3, 7$ **(d)** $-9, -1\frac{1}{4}, -\frac{3}{5}, 0, 3, 5.9, 7$ **(e)** $-\sqrt{7}, \sqrt{5}$ **(f)** All are real numbers. **15.** 93,000 **17.** $-30°$ **19.** $-31{,}532$ **21.** -8 **23.** 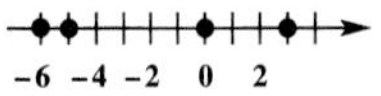

25. 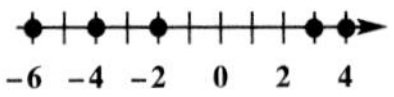

27. 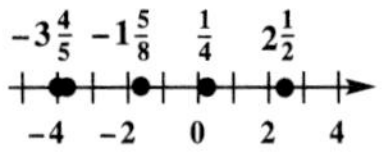

29. **(a)** A **(b)** A **(c)** B **(d)** B **31.** **(a)** 2 **(b)** 2 **33.** **(a)** -6 **(b)** 6 **35.** **(a)** -3 **(b)** 3 **37.** **(a)** 0 **(b)** 0 **39.** $a - b$ **41.** -12 **43.** -8 **45.** 3 **47.** $|-3|$ or 3 **49.** $-|-6|$ or -6 **51.** $|5 - 3|$ or 2 **53.** true **55.** true **57.** true **59.** false **61.** true **63.** false **65.** Softwood plywood from 1995 to 1996 represents the greatest drop. **67.** true

Answers will vary in Exercises 69–77. **69.** true: $a = 0$ or $b = 0$ or both $a = 0$ and $b = 0$; false: Choose any values for a and b so that neither a nor b is zero **71.** true: Choose a to be the opposite of b ($a = -b$); false: $a \neq -b$ **73.** $\frac{1}{2}, \frac{5}{8}, 1\frac{3}{4}$ **75.** $-3\frac{1}{2}, -\frac{2}{3}, \frac{3}{7}$ **77.** $\sqrt{5}, \pi, -\sqrt{3}$

SECTION 1.5 (PAGE 44)

EXERCISES **1.** negative **3.** -3; 5 **7.** 2 **9.** -3 **11.** -10 **13.** -13 **15.** -15.9 **17.** -1 **19.** 13 **21.** -3 **23.** -4 **25.** -10 **27.** -16 **29.** 11 **31.** 19 **33.** -4 **35.** 5 **37.** 0 **39.** $\frac{3}{4}$ **41.** -8 **43.** $\frac{15}{8}$ **45.** -6.3 **47.** -24 **49.** -16 **51.** no **53.** Answers will vary. One example is $-8 - (-2) = -6$. **55.** -1 **57.** $\frac{17}{9}$ or $1\frac{8}{9}$ **59.** $-5 + 12 + 6$; 13 **61.** $[-19 + (-4)] + 14$; -9 **63.** $[-4 + (-10)] + 12$; -2

65. $[8 + (-18)] + 4; -6$ **67.** $4 - (-8); 12$ **69.** $-2 - 8; -10$ **71.** $[9 + (-4)] - 7; -2$ **73.** $[8 - (-5)] - 12; 1$ **75.** \$13.2 billion **77.** −\$26.0 billion **79.** 50,395 feet **81.** 1345 feet **83. (a)** -10 **(b)** 5 **(c)** -12 **(d)** -31 **85.** 45°F **87.** −41°F **89.** 14,776 feet **91.** 365 pounds **93.** \$323.83

SECTION 1.6 (PAGE 57)

EXERCISES **1.** greater than 0 **3.** greater than 0 **5.** less than 0 **7.** greater than 0 **9.** equal to 0 **11.** 20 **13.** -28 **15.** 80 **17.** 0 **19.** $\frac{5}{6}$ **21.** $\frac{3}{2}$ **23.** $-32, -16, -8, -4, -2, -1, 1, 2, 4, 8, 16, 32$ **25.** $-40, -20, -10, -8, -5, -4, -2, -1, 1, 2, 4, 5, 8, 10, 20, 40$ **27.** $-31, -1, 1, 31$ **29.** -3 **31.** -2 **33.** 16 **35.** 0 **37.** 25.63 **39.** $\frac{3}{2}$ **43.** -11 **45.** -2 **47.** 35 **49.** 6 **51.** -18 **53.** 67 **55.** -8 **57.** 3 **59.** 7 **61.** 4 **63.** -3 **65.** First, substitute -3 for x and 4 for y to get $3(-3) + 2(4)$. Then perform the multiplications to get $-9 + 8$. Finally, add to get -1. **67.** 47 **69.** 72 **71.** $-\frac{78}{25}$ **73.** 0 **75.** -23 **77.** 2 **79.** $9 + (-9)(2); -9$ **81.** $-4 - 2(-1)(6); 8$ **83.** $(1.5)(-3.2) - 9; -13.8$ **85.** $12[9 - (-8)]; 204$ **87.** $\frac{-12}{-5 + (-1)}; 2$ **89.** $\frac{15 + (-3)}{4(-3)}; -1$ **91.** $\frac{\left(-\frac{1}{2}\right)\left(\frac{3}{4}\right)}{-\frac{2}{3}}; \frac{9}{16}$ **93.** $6x = -42; -7$ **95.** $\frac{x}{3} = -3; -9$ **97.** $x - 6 = 4; 10$ **99.** $x + 5 = -5; -10$ **101.** $8\frac{2}{5}$ **103.** 2 **105.** \$12.60 (to the nearest cent) **107.** 0 **109. (a)** 6 is divisible by 2. **(b)** 9 is not divisible by 2. **111. (a)** 64 is divisible by 4. **(b)** 35 is not divisible by 4. **113. (a)** 2 is divisible by 2 and $1 + 5 + 2 + 4 + 8 + 2 + 2 = 24$ is divisible by 3. **(b)** While 0 is divisible by 2, $2 + 8 + 7 + 3 + 5 + 9 + 0 = 34$ is not divisible by 3. **115. (a)** $4 + 1 + 1 + 4 + 1 + 0 + 7 = 18$ is divisible by 9. **(b)** $2 + 2 + 8 + 7 + 3 + 2 + 1 = 25$ is not divisible by 9.

SECTION 1.7 (PAGE 67)

EXERCISES **1.** B **3.** C **5.** B **7.** B **9.** G **11.** commutative property **13.** associative property **15.** associative property **17.** inverse property **19.** inverse property **21.** identity property **23.** commutative property **25.** distributive property **27.** identity property **29.** distributive property **31.** identity property **35.** $7 + r$ **37.** s **39.** $-6x + (-6) \cdot 7; -6x - 42$ **41.** $w + [5 + (-3)]; w + 2$ **43.** 11 **45.** 0 **47.** $-.38$ **49.** 1 **51.** Subtraction is not associative. **53.** The expression following the first equals sign should be $-3(4) -3(-6)$. The student forgot that 6 should be preceded by a − sign. The correct work is $-3(4 - 6) = -3(4) - 3(-6) = -12 + 18 = 6$. **55.** $(5 + 1)x; 6x$ **57.** $4t + 12$ **59.** $-8r - 24$ **61.** $-5y + 20$ **63.** $-16y - 20z$ **65.** $8(z + w)$ **67.** $7(2v + 5r)$ **69.** $24r + 32s - 40y$ **71.** $(1 + 1 + 1)q; 3q$ **73.** $(-5 + 1)x; -4x$ **75.** $-4t - 3m$ **77.** $5c + 4d$ **79.** $3q - 5r + 8s$ **81.** for example, "putting on your socks" and "putting on your shoes" **83.** 0 **84.** $-3(5) + (-3)(-5)$ **85.** -15 **86.** We must interpret $(-3)(-5)$ as 15, since it is the additive inverse of -15.

SECTION 1.8 (PAGE 73)

EXERCISES **1.** false **3.** true **5.** (c) **7.** (a) **9.** $4r + 11$ **11.** $5 + 2x - 6y$ **13.** $-7 + 3p$ **15.** -12 **17.** 5 **19.** 1 **21.** -1 **23.** 74 **25.** Answers will vary. For example, $-3x$ and $4x$. **27.** like **29.** unlike **31.** like **33.** unlike **35.** Apples and oranges are examples of unlike fruits, just like x and y are unlike terms. We cannot add x and y to get an expression any simpler than $x + y$; we cannot add, for example, 2 apples and 3 oranges to obtain 5 fruits that are all alike. **37.** $9k - 5$ **39.** $-\frac{1}{3}t - \frac{28}{3}$ **41.** $-4.1r + 5.6$ **43.** $-2y^2 + 3y^3$ **45.** $-19p + 16$ **47.** $-4y + 22$ **49.** $-16y + 63$ **51.** $4k - 7$ **53.** $-23.7y - 12.6$ **55.** $(x + 3) + 5x; 6x + 3$ **57.** $(13 + 6x) - (-7x); 13 + 13x$ **59.** $2(3x + 4) - (-4 + 6x); 12$ **61.** Wording will vary. One example is "the difference between 9 times a number and the sum of the number and 2." **63.** 2, 3, 4, 5 **64.** 1 **65. (a)** 1, 2, 3, 4 **(b)** 3, 4, 5, 6 **(c)** 4, 5, 6, 7 **66.** The value of $x + b$ also increases by 1 unit. **67. (a)** 2, 4, 6, 8 **(b)** 2, 5, 8, 11 **(c)** 2, 6, 10, 14 **68.** m **69. (a)** 7, 9, 11, 13 **(b)** 5, 8, 11, 14 **(c)** 1, 5, 9, 13 In comparison, we see that while the values themselves are different, the number of units of increase is the same as the corresponding parts of Exercise 67. **70.** m

CHAPTER 1 REVIEW EXERCISES (PAGE 79)

1. $\frac{3}{4}$ **3.** $\frac{9}{40}$ **5.** 625 **7.** .0000000032 **9.** 27 **11.** 39 **13.** true **15.** false **17.** $5 + 2 \neq 10$ **19.** 30 **21.** 14 **23.** $x + 6$ **25.** $6x - 9$ **27.** yes **29.** $2x - 6 = 10$; 8 **31.**

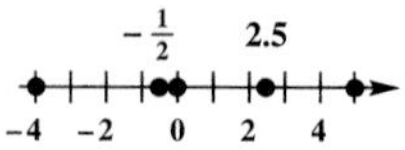

33. rational numbers, real numbers **35.** -10 **37.** $-\frac{3}{4}$ **39.** true **41.** true **43.** **(a)** 9 **(b)** 9 **45.** **(a)** -6 **(b)** 6 **47.** 12 **49.** -19 **51.** -6 **53.** -17 **55.** -21.8 **57.** -10 **59.** -11 **61.** 7 **63.** 10.31 **65.** 2 **67.** $(-31 + 12) + 19; 0$ **69.** $-4 - (-6); 2$ **71.** -2 **73.** \$26.25 **75.** $-\$29$ **77.** It gained 4 yards. **79.** 36 **81.** $\frac{1}{2}$ **83.** -20 **85.** -24 **87.** 4 **89.** $-\frac{3}{4}$ **91.** -1 **93.** 1 **95.** -18 **97.** 125 **99.** $-4(5) - 9; -29$ **101.** $\frac{12}{8 + (-4)}; 3$ **103.** $8x = -24; -3$ **105.** **(a)** \$61,374,559 **(b)** \$2,245,068 **107.** identity property **109.** inverse property **111.** associative property **113.** distributive property **115.** $(7 + 1)y; 8y$ **117.** $3(2s + 5y)$ **119.** $25 - (5 - 2) = 22$ and $(25 - 5) - 2 = 18$. Because different groupings lead to different results, we conclude that in general subtraction is not associative. **121.** $11m$ **123.** $16p^2 + 2p$ **125.** $-2m + 29$ **127.** 16 **129.** $\frac{8}{3}$ **131.** 2 **133.** $-1\frac{1}{2}$ **135.** $-\frac{28}{15}$ **137.** $8x^2 - 21y^2$ **139.** When dividing 0 *by* a nonzero number, the quotient will be 0. However, dividing a number *by* 0 is undefined. **141.** $5(x + 7)$; $5x + 35$ **143.** -79

CHAPTER 1 TEST (PAGE 85)

[1.1] **1.** $\frac{7}{11}$ **2.** $\frac{241}{120}$ **3.** $\frac{19}{18}$ **4.** **(a)** 492 million **(b)** 861 million [1.2] **5.** true [1.4] **6.**

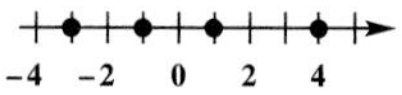

7. rational numbers, real numbers **8.** If -8 and -1 are both graphed on a number line, we see that the point for -8 is to the *left* of the point for -1. This indicates $-8 < -1$. **9.** $\frac{-6}{2 + (-8)}; 1$ [1.2] **10.** **(a)** Japanese, Vietnamese, Filipino, Hawaiian, Other Asian and Pacific Islander **(b)** Asian Indian, Chinese, Korean [1.1, 1.4–1.6] **11.** 4 **12.** $-2\frac{5}{6}$ **13.** 2 **14.** 6 **15.** 108 **16.** 3 **17.** $\frac{30}{7}$ [1.3, 1.5, 1.6] **18.** 6 **19.** 4 [1.4–1.6] **20.** -70 **21.** 3 [1.5, 1.6] **22.** **(a)** $-\$3100$ **(b)** \$3600 **(c)** \$6800 **(d)** \$4200 **23.** 15 **24.** **(a)** -14 **(b)** 55 **(c)** 31 [1.7] **25.** B **26.** D **27.** E **28.** A **29.** C **30.** distributive property **31.** **(a)** -18 **(b)** -18 **(c)** The distributive property assures us that the answers must be the same, because $a(b + c) = ab + ac$ for all a, b, c. [1.8] **32.** $21x$ **33.** $15x - 3$

CHAPTER 2 LINEAR EQUATIONS AND INEQUALITIES IN ONE VARIABLE

SECTION 2.1 (PAGE 95)

EXERCISES **1.** (a) and (c) **5.** $\{6\}$ **7.** $\{6.3\}$ **9.** $\{-6\}$ **11.** $\{-2\}$ **13.** $\{4\}$ **15.** $\{0\}$ **17.** $\{-2\}$ **19.** $\{-7\}$ **23.** $\{13\}$ **25.** $\{-4\}$ **27.** $\{0\}$ **29.** $\left\{\frac{7}{15}\right\}$ **31.** $\{7\}$ **33.** $\{-4\}$ **35.** $\{13\}$ **37.** $\{29\}$ **39.** $\{18\}$ **41.** If both sides of an equation are multiplied by 0, the resulting equation is $0 = 0$. This is true, but does not help to solve the equation. **43.** $\{6\}$ **45.** $\{-5\}$ **47.** $\left\{-\frac{18}{5}\right\}$ **49.** $\{12\}$ **51.** $\{0\}$ **53.** $\{-48\}$ **55.** $\{-12\}$

57. $\left\{\frac{4}{7}\right\}$ **59.** {40} **61.** {3} **63.** {7} **65.** {−35} **67.** $\left\{\frac{35}{2}\right\}$ **69.** $\left\{-\frac{27}{35}\right\}$ **71.** {−12.2} **73.** Answers will vary. For example, $\frac{3}{2}x = -6$. **75.** $3x = 2x + 17$; {17} **77.** $5x + 3x = 7x + 9$; {9} **79.** $\frac{x}{-5} = 2$; {−10}

SECTION 2.2 (PAGE 103)

EXERCISES **3.** {4} **5.** {−1} **7.** $\left\{-\frac{4}{5}\right\}$ **9.** {−6} **11.** $\emptyset$ **13.** {all real numbers} **15.** {1} **17.** $\emptyset$ **19.** No, it is incorrect to divide by a variable. If $-3x$ is added to both sides, the equation becomes $4x = 0$, so $x = 0$ and {0} is the correct solution set. **21.** Multiply both sides by the LCD of all fractions in the equation or by the power of 10 that makes all decimal numbers integers. **23.** {5} **25.** {0} **27.** $\left\{-\frac{7}{5}\right\}$ **29.** {120} **31.** {6} **33.** {15,000} **35.** 800 **36.** Yes, you will get $(100 \cdot 2) \cdot 4 = 200 \cdot 4 = 800$. This is a result of the associative property. **37.** No, because $(100a)(100b) = 10{,}000ab \neq 100ab$. **38.** The distributive property involves the operation of *addition* as well. **39.** Yes, the associative property of multiplication is used here. **40.** no **41.** {8} **43.** {0} **45.** {4} **47.** {20} **49.** {all real numbers} **51.** $\emptyset$ **53.** $11 - q$ **55.** $x + 7$ **57.** $a + 12$; $a - 5$ **59.** $\frac{t}{5}$

SECTION 2.3 (PAGE 112)

CONNECTIONS **Page 105:** Polya's Step 1 corresponds to our Steps 1 and 2. Polya's Step 2 corresponds to our Step 3. Polya's Step 3 corresponds to our Steps 4 and 5. Polya's Step 4 corresponds to our Step 6. Trial and error or guessing and checking fit into Polya's Step 2, devising a plan.

EXERCISES **1.** (c) **5.** 3 **7.** −4 **9.** 53 Republicans, 46 Democrats **11.** 6562 men, 3779 women **13.** Fernandez, 285; Irwin, 286 **15.** Airborne Express: 3; Federal Express: 9; United Parcel Service: 1 **17.** 36 million miles **19.** Smoltz: $253\frac{2}{3}$; Maddux: 245; Wohlers: $77\frac{1}{3}$ **21.** *A* and *B:* 40°; *C:* 100° **23.** exertion: 9443 calories; regulating body temperature: 1757 calories **25.** 18 prescriptions **27.** peanuts: $22\frac{1}{2}$ ounces; cashews: $4\frac{1}{2}$ ounces **29.** $k - m$ **31.** no **33.** $x - 1$ **35.** 80° **37.** 26° **39.** 55° **41.** 68, 69 **43.** 10, 12 **45.** 101, 102 **47.** 10, 11 **49.** 15, 17, 19 **51.** 1993: \$2.78 billion; 1994: \$3.33 billion; 1995: \$3.53 billion

SECTION 2.4 (PAGE 122)

EXERCISES **1.** The perimeter of a geometric figure is the sum of the lengths of the sides. **3.** area **5.** perimeter **7.** area **9.** area **11.** $P = 20$ **13.** $P = 24$ **15.** $A = 70$ **17.** $r = 40$ **19.** $I = 875$ **21.** $A = 91$ **23.** $r = 1.3$ **25.** $A = 452.16$ **27.** 4 **29.** $V = 384$ **31.** $V = 48$ **33.** $V = 904.32$ **35.** about 154,000 square feet **37.** 1978 feet **39.** perimeter: 172 inches; area: 1785 square inches **41.** $\frac{729}{32}$ or $22\frac{25}{32}$ cubic inches **43.** 23,800.10 square feet **45.** 107°, 73° **47.** 75°, 75° **49.** 139°, 139° **51.** $r = \frac{d}{t}$ **53.** $p = \frac{I}{rt}$ **55.** $a = P - b - c$ **57.** $b = \frac{2A}{h}$ **59.** $r = \frac{A - p}{pt}$ **61.** $h = \frac{V}{\pi r^2}$ **63.** $m = \frac{y - b}{x}$ **65. (a)** $P - 2L = 2W$ **(b)** $W = \frac{P - 2L}{2}$ **66. (a)** $\frac{P}{2} = L + W$ **(b)** $\frac{P}{2} - L = W$ **67. (a)** multiplication identity property **(b)** An expression divided by 1 is equal to itself. **(c)** rule for multiplication of fractions **(d)** rule for subtraction of fractions **68.** $\frac{5T}{4} + 1$

SECTION 2.5 (PAGE 131)

EXERCISES **1.** $\frac{5}{8}$ **3.** $\frac{1}{4}$ **5.** $\frac{2}{1}$ **7.** $\frac{3}{1}$ **9.** (d) **13.** true **15.** false **17.** true **19.** {35} **21.** {27} **23.** {−1} **25.** 6.875 ounces **27.** \$670.48 **29.** \$9.30 **31.** 4 feet **33.** 12,500 fish **35.** **(a)** $\frac{26}{100} = \frac{x}{350}$; \$91 million **(b)** \$112 million; \$11.2 million **(c)** \$119 million **37.** 30 count **39.** 31-ounce size **41.** 32-ounce size **43.** 4 **45.** 1 **47.** **(a)**

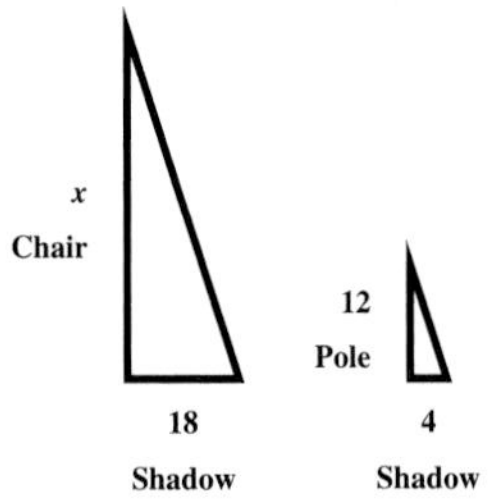

(b) 54 feet **49.** \$242 **51.** \$255 **53.** \$4850 **55.** 30

56. **(a)** $5x = 12$ **(b)** $\frac{12}{5}$ **57.** **(a)** $5x = 12$ **(b)** $\frac{12}{5}$ **58.** They are the same. Solving by cross products yields the same solution as multiplying by the least common denominator. **59.** \$40.32 **61.** $\frac{80}{3}$ or $26\frac{2}{3}$ inches **63.** approximately 2593 miles

SECTION 2.6 (PAGE 143)

EXERCISES **1.** 35 milliliters **3.** \$350 **5.** \$14.15 **7.** 3.334 hours **9.** 7.95 meters per second **11.** \$533 **13.** 284% **15.** **(a)** about \$785 million **(b)** about \$1726 million **(c)** about \$471 million **17.** 0 liters **19.** 160 gallons **21.** $53\frac{1}{3}$ kilograms **23.** 4 liters **25.** 25 milliliters **27.** (d) **29.** \$5000 at 3%; \$1000 at 5% **31.** \$40,000 at 3%; \$110,000 at 4% **33.** 25 fives **35.** 84 fives; 42 tens **37.** 20 pounds **39.** Yes. In Example 2, it would be reasonable to mix, say, $9\frac{1}{2}$ liters with 20 liters to get $29\frac{1}{2}$ liters of mixture. However, in Example 4, it is not possible to have a fractional part of a five- or ten-dollar bill. **40.** **(a)** $.05x + .10(3400 - x) = 290$ **(b)** 1000 nickels; 2400 dimes **41.** **(a)** $.05x + .10(3400 - x) = 290$ **(b)** \$1000 at 5%; \$2400 at 10% **42.** They are the same. **43.** No, you will get a different solution to the equation, but you *will* get the same answers to the problem. **45.** 530 miles **47.** No, since the rate is given in miles per *hour,* the time must be in hours. So the distance is $45 \times \frac{1}{2} = \frac{45}{2} = 22\frac{1}{2}$ miles. **49.** 10 hours **51.** northbound: 60 miles per hour; southbound: 80 miles per hour **53.** 5 hours **55.** single man: \$51.58; single woman: \$40.71 **57.** Business/Government: 6,100,000; K–12: 5,000,000; HED: 900,000; Homes: 9,800,000 **59.** 120 miles per hour

SECTION 2.7 (PAGE 158)

CONNECTIONS **Page 158:** The revenue is represented by $5x - 100$. The production cost is $125 + 4x$. The profit is represented by $R - C = (5x - 100) - (125 + 4x) = x - 225$. The solution of $x - 225 > 0$ is $x > 225$. In order to make a profit, more than 225 cassettes must be produced and sold.

EXERCISES **1.** Use a parenthesis if the symbol is $<$ or $>$. Use a square bracket if the symbol is $\leq$ or $\geq$. **3.** $x > -4$

5. $x \leq 4$ **7.** $(-\infty, 4]$ 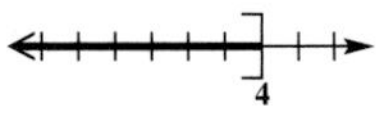

9. $(-\infty, -3)$ 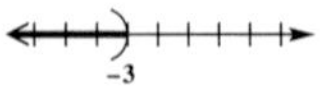

11. $(4, \infty)$ 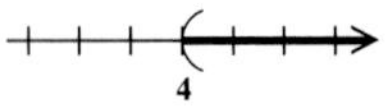

13. $[8, 10]$ 

15. $(0, 10]$ 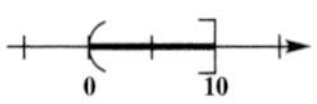

17. It would imply that $3 < -2$, a false statement.

19. $z \geq 1$ 

21. $k \geq 5$ 

23. $n < -11$ 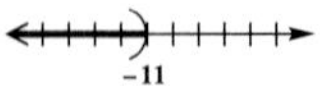

25. It must be reversed when multiplying or dividing by a negative number. **27.** His method is incorrect. He divided *by* the positive number 6. The sign of the number divided *into* does not matter. **29.** $(-\infty, 6)$ 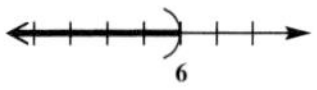

31. $[-10, \infty)$ 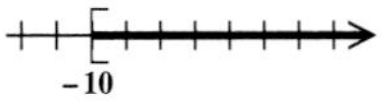

33. $(-\infty, -3)$ 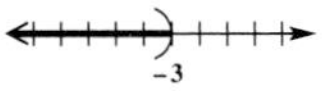

35. $(-\infty, 0]$ 

37. $(20, \infty)$ 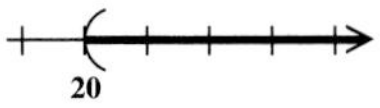

39. $[-3, \infty)$ 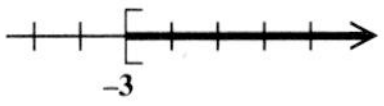

41. $[-5, \infty)$ 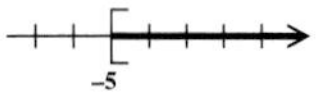

43. $(-\infty, 1)$ 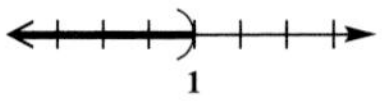

45. $(-\infty, 0]$ 

47. $[4, \infty)$ 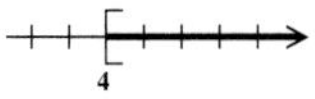

49. $(-\infty, 32)$ 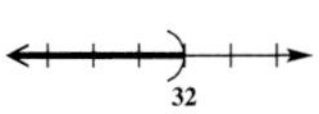

51. $\left[\frac{5}{12}, \infty\right)$ 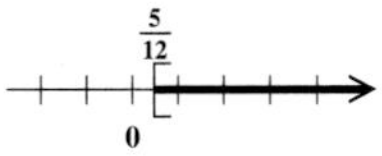

53. $(-21, \infty)$ 

55. $-1 < x < 2$ **57.** $-1 < x \leq 2$ **59.** $[-1, 6]$ 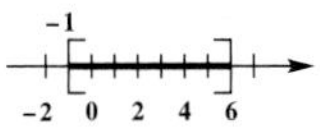

61. $\left(-\frac{11}{6}, -\frac{2}{3}\right)$ 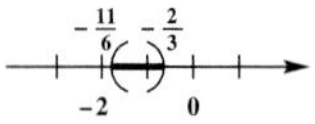

63. $(1, 3)$ 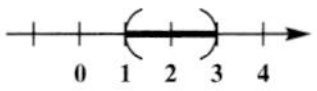

65. $[-26, 6]$ 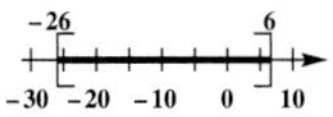

67. $[-3, 6]$ 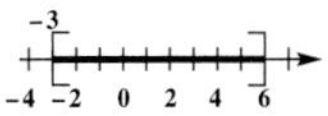

69. $\left[-\frac{24}{5}, 0\right]$ 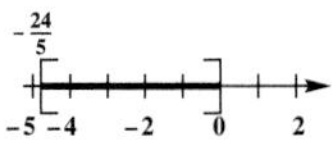

71. $\{4\}$ 

72. $(4, \infty)$ The solutions are all the numbers to the right of 4. 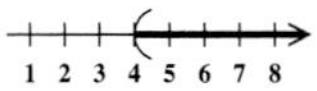

73. $(-\infty, 4)$ The solutions are all the numbers to the left of 4. 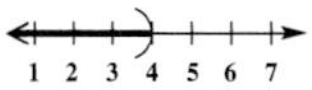

74. The graph would be all real numbers. **75.** The graph would be all real numbers. **76.** If a point on the number line satisfies an equation, the points on one side of that point will satisfy the corresponding less than inequality, and the points on the other side will satisfy the corresponding greater than inequality. **77.** 83 or greater **79.** It is never more than 86° Fahrenheit. **81.** $x \geq 500$ **83.** 32 or greater **85.** 15 minutes **87.** yes; yes **88.** no; no **89.** yes; no; no **90.** 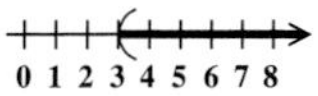

91. 

92. $(3, 8)$ **93.** 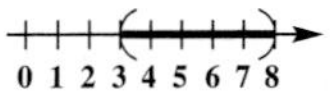

It is the graph of that interval (3, 8). **94.** $(-\infty, -4)$ 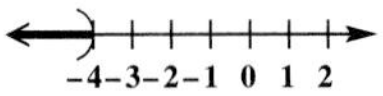

95. $\emptyset$ **96.** 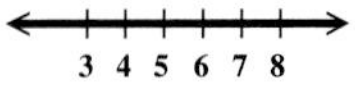

This graph includes both intervals graphed in Exercises 90 and 91. The result is all real numbers. **97.** $(-\infty, 2)$ 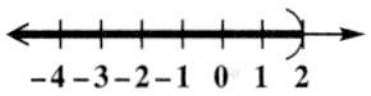

In Exercise 94 only the real numbers less than -4 were included, because both conditions had to be satisfied. Here all real numbers less than 2 satisfy at least one of the conditions. **98.** $(-\infty, -3) \cup (-2, \infty)$ 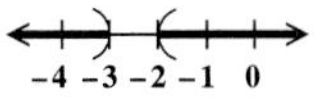

In Exercise 95, no real numbers satisfied both conditions. Here, the numbers in each interval satisfy at least one condition, so both intervals are in the solution set.

CHAPTER 2 REVIEW EXERCISES (PAGE 167)

1. $\{6\}$ **3.** $\{7\}$ **5.** $\{11\}$ **7.** $\{5\}$ **9.** $\{5\}$ **11.** $\left\{\frac{64}{5}\right\}$ **13.** {all real numbers} **15.** {all real numbers}
17. $\emptyset$ **19.** 112 nations **21.** Hawaii: 6425 square miles; Rhode Island: 1212 square miles **23.** 80° **25.** $A = 28$
27. $V = 904.32$ **29.** $h = \frac{2A}{b + B}$ **31.** 100°; 100° **33.** 42.2°; 92.8° **35.** $\frac{3}{2}$ **37.** $\frac{3}{4}$ **39.** $\left\{\frac{7}{2}\right\}$ **41.** $\left\{\frac{25}{19}\right\}$
43. .2 meter or 200 millimeters **45.** 375 kilometers **47.** 25.5-ounce size **49.** approximately $15.4 million
51. $5000 at 5%; $5000 at 6% **53.** 13 hours **55.** $\frac{5}{6}$ hour or 50 minutes **57.** 28 games **59.** Answers will vary.
61. $(-\infty, 7)$ 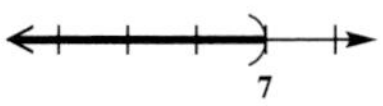

63. $[-3, \infty)$ 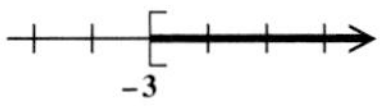

65. $[3, \infty)$ 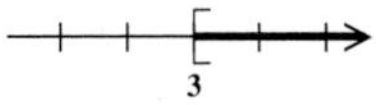

67. $(-\infty, -5)$ 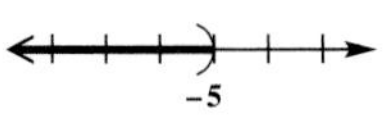

69. $\left[-2, \frac{3}{2}\right]$ 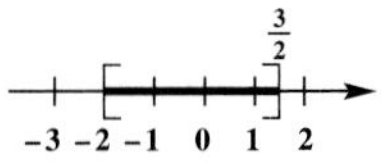

71. more than 2875 **73.** $\{7\}$

75. $(-\infty, 2)$ **77.** $\{70\}$ **79.** $\emptyset$ **81.** The first step is to distribute the negative sign over *both* terms in the parentheses, to get $3 - 8 - 4x$ on the left side of the equation. The student got $3 - 8 + 4x$ instead. The correct solution set is $\{-2\}$. **83.** Golden Gate Bridge: 4200 feet; Brooklyn Bridge: 1595 feet **85.** 64-ounce size **87.** faster train: 80 miles per hour; slower train: 50 miles per hour **89.** 8 centimeters **91.** No. Only equations that have one fractional term on each side can be solved directly by cross multiplication. This equation has two terms on the left.

CHAPTER 2 TEST (PAGE 171)

[2.1–2.2] **1.** $\{-6\}$ **2.** $\{21\}$ **3.** $\emptyset$ **4.** $\{30\}$ **5.** {all real numbers} [2.3] **6.** Hawaii: 4021 square miles; Maui: 728 square miles; Kauai: 551 square miles [2.4] **7.** **(a)** $W = \frac{P - 2L}{2}$ or $W = \frac{P}{2} - L$ **(b)** 18 **8.** 75°, 75° **9.** 50° [2.5] **10.** $\{-29\}$ **11.** 62.8 centimeters **12.** 2300 miles [2.6] **13.** \$8000 at 3%; \$14,000 at 4.5% **14.** 236% **15.** 4 hours [2.7] **16.** $(-\infty, 4]$ 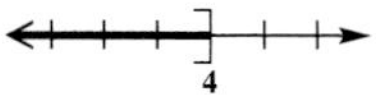

17. $(-2, 6]$ 

18. 2 is less than a number **19.** East 132, West 120 **20.** 26 points

CUMULATIVE REVIEW EXERCISES CHAPTERS 1–2 (PAGE 173)

[1.1] **1.** $\frac{3}{4}$ **2.** $\frac{37}{60}$ **3.** $\frac{48}{5}$ [1.2] **4.** $\frac{1}{2}x - 18$ **5.** $\frac{6}{x + 12} = 2$ [1.4] **6.** true [1.5–1.6] **7.** 11 **8.** -8 **9.** 28 [1.3] **10.** $-\frac{19}{3}$ [1.7] **11.** distributive property **12.** commutative property [1.8] **13.** $2k - 11$ [2.1–2.2] **14.** $\{-1\}$ **15.** $\{-1\}$ **16.** $\{-12\}$ [2.5] **17.** $\{26\}$ [2.4] **18.** $y = \frac{24 - 3x}{4}$ **19.** $n = \frac{A - P}{iP}$ [2.7] **20.** $(-\infty, 1]$ 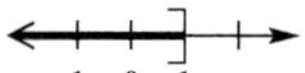

21. $(-1, 2]$ 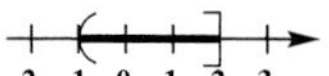

[2.3] **22.** 4 centimeters; 9 centimeters; 27 centimeters [2.4] **23.** 1029.92 feet [2.5] **24.** $\frac{25}{6}$ or $4\frac{1}{6}$ cups [2.6] **25.** 40 miles per hour; 60 miles per hour

CHAPTER 3 LINEAR EQUATIONS IN TWO VARIABLES

SECTION 3.1 (PAGE 183)

Exercises **1.** does; do not **3.** II **5.** 3 **7.** between 1990 and 1991; \$21,000 and between 1992 and 1993; \$122,000 **9.** 1993 **11.** (d) **13.** inkjet **15.** 1994–1995; 1993–1994 **17.** 4800; 3400; yes **19.** yes **21.** yes **23.** no **25.** yes **27.** yes **29.** no **33.** No, for two ordered pairs to be equal, the x-values must be equal, and the y-values must be equal. Here, $4 \neq -1$ and $-1 \neq 4$. **35.** 11 **37.** $-\frac{7}{2}$ **39.** -4 **41.** -5 **43.** Substituting $\frac{1}{3}$ for x in $y = 6x + 2$ gives $y = 6\left(\frac{1}{3}\right) + 2 = 2 + 2 = 4$. Because $6\left(\frac{1}{7}\right) = \frac{6}{7}$, calculating y requires working with fractions.
45. 4; 6; -6 **47.** 3; -5; -15 **49.** -9; -9; -9 **51.** -6; -6; -6 **53.–60.** 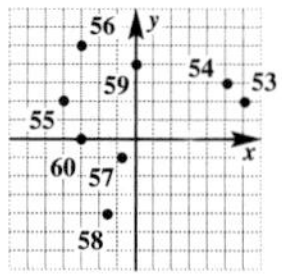

61. negative; negative **63.** positive; negative **65.** -3; 6; -2; 4 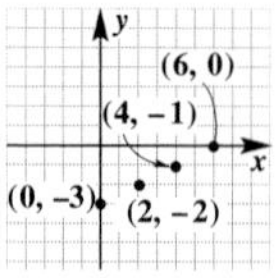

67. -3; 4; -6; $-\frac{4}{3}$ 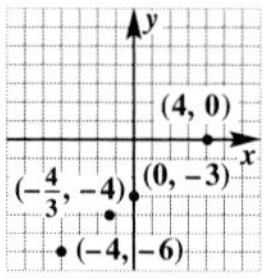

69. -4; -4; -4; -4 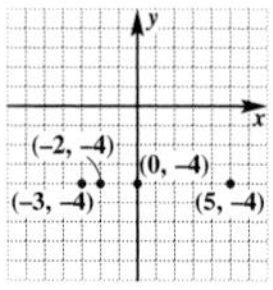

71. **(a)** (100, 5050) **(b)** (2000, 6000)

73. (a) Yes, there is an approximate linear relationship between the years since 1980 and the minutes beyond two hours to complete a game.

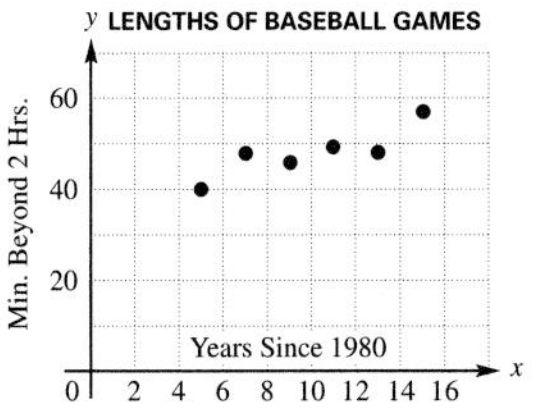

(b) Yes, a prediction could be made by determining the linear equation that describes the linear relationship. However, only input values for the years included in the given data, 1980–1996, should be used. **(c)** input: years since 1980; output: minutes beyond two hours

75. (a) (30, 133), (40, 126), (60, 112), (70, 105) **(b)** input: age; output: lower limit of the target heart rate zone
77. No. To go beyond the given data at either end assumes that the graph continues in the same way. This may not be true.

SECTION 3.2 (PAGE 196)

CONNECTIONS **Page 194:** **1.** $3x + 4 - 2x - 7 - 4x - 3 = 0$ **2.** $5x - 15 - 3(x - 2) = 0$

EXERCISES **1.** (c) **3.** (d) **5.** (b) **7.** 5; 5; 3

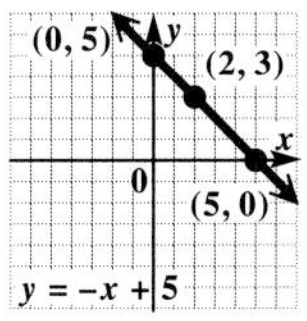

9. 1; 3; −1

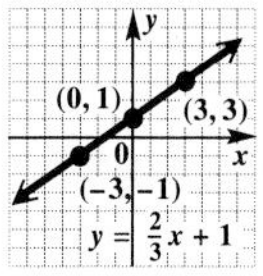

11. −6; −2; −5

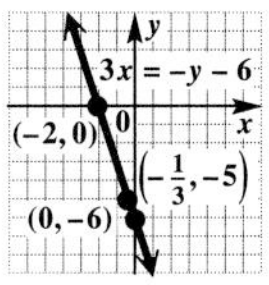

13. (12, 0); (0, −8) **15.** (0, 0); (0, 0) **17.** $y = 0$ **19.** Choose a value other than 0 for either x or y. For example, if $x = -5$, $y = 4$. **21.**

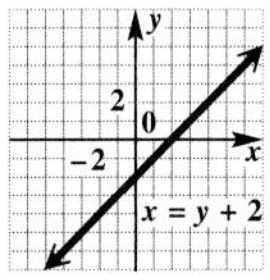

23.

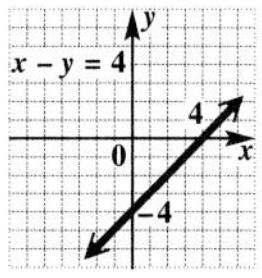

25.

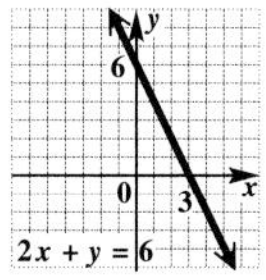

27.

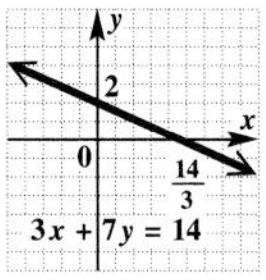

29.

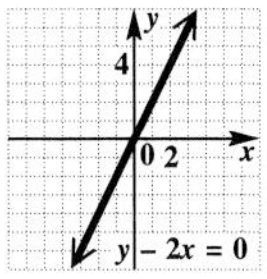

31.

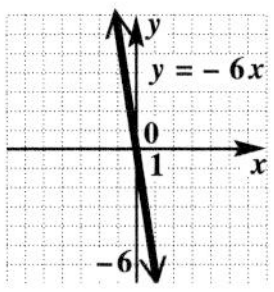

33.

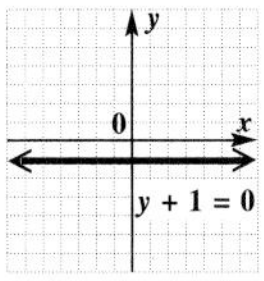

35.

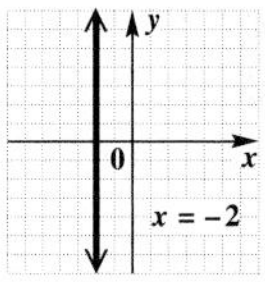

37. 2

39. 5 **41.** The y-value of each ordered pair gives the value of the expression in x for each x-value. The solution of the equation is the x-value that corresponds to $y = 0$, which is the x-intercept of the graph. **43. (a)** 1980: 52.2 thousand; 1985: 66.25 thousand; 1990: 80.3 thousand; 1994: 91.54 thousand **(b)** 1980: 54 thousand; 1985: 67 thousand; 1990: 77 thousand; 1994: 93 thousand **(c)** Yes, they are quite close. **45. (a)** 163.2 centimeters **(b)** 171 centimeters **(c)** 151.5 centimeters **(d)**

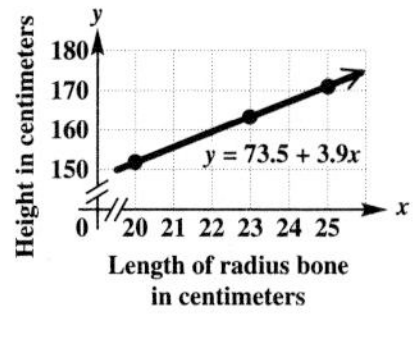

47. (a) \$1250 **(b)** \$1250 **(c)** \$1250 **(d)** \$6250 **49.** yes **51.** yes **53.** no **55.** yes

SECTION 3.3 (PAGE 207)

CONNECTIONS **Page 201:** Ski slopes, "grade" on a treadmill, and slope (or "fall") of a sewer pipe are some additional examples.

EXERCISES **1.** Rise is the vertical change between two different points on a line. Run is the horizontal change between two different points on a line. **3.** 4 **5.** $-\frac{1}{2}$ **7.** 0 **9.** Yes; it doesn't matter which point you start with. Both differences will be the negatives of the differences in Exercise 8, and the quotient will be the same.

In Exercises 11–13, sketches will vary.

11. The line must rise from left to right. **13.** The line must be horizontal. **15.** Because he found the difference $3 - 5 = -2$ in the numerator, he should have subtracted in the same order in the denominator to get $-1 - 2 = -3$. The correct slope is $\frac{-2}{-3} = \frac{2}{3}$. **17.** $\frac{5}{4}$ **19.** $\frac{3}{2}$ **21.** 0 **23.** undefined **25.** $-\frac{1}{2}$ **27.** 5 **29.** $\frac{1}{4}$ **31.** $\frac{3}{2}$ **33.** undefined **35.** **(a)** negative **(b)** zero **37.** **(a)** positive **(b)** negative **39.** **(a)** zero **(b)** negative **41.** (a) **43.** $-\frac{2}{5}$; $-\frac{2}{5}$; parallel **45.** $\frac{8}{9}$; $-\frac{4}{3}$; neither **47.** $\frac{3}{2}$; $-\frac{2}{3}$; perpendicular **49.** $\frac{3}{10}$ **51.** 232,000 **52.** positive; increased **53.** 232,000 students **54.** -2 **55.** negative; decreased **56.** 2 students per computer **57.** **(a)** Possible ordered pairs are (1991, 37.8), (1992, 38.6), (1993, 39.4), (1994, 40.2), (1995, 41.0), (1996, 41.8). The input numbers are 1991, 1992, 1993, 1994, 1995, 1996. The output numbers (in thousands) are 37.8, 38.6, 39.4, 40.2, 41.0, 41.8. **(b)** Yes, they are all within .4 thousand (400). **59.** The change for each year is .1 billion (or 100,000,000) square feet, so the graph is a straight line. **61.** 2 **63.** $\frac{2}{5}$ **65.** (0, 4)

CHAPTER 3 REVIEW EXERCISES (PAGE 217)

1. -1; 2; 1 **3.** 7; 7; 7 **5.** no **7.** I **9.** none **11.** I or III

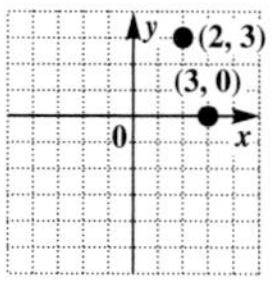

13. $\left(-\frac{5}{2}, 0\right)$; (0, 5)

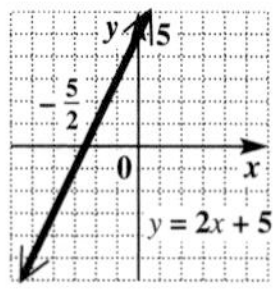

15. $(-4, 0)$; $(0, -2)$

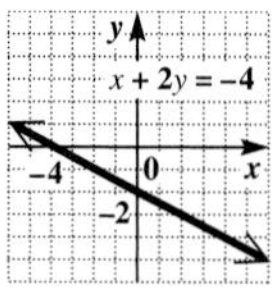

17. function **19.** function

21. $-\frac{1}{2}$ **23.** undefined **25.** $\frac{2}{3}$ **27.** $-\frac{4}{7}$ **29.** $\frac{1}{3}$ **31.** parallel **33.** neither **35.** 0; $\frac{8}{3}$; -9

37. $\left(-\frac{3}{2}, 0\right)$; (0, 3); 2 **39.** $\frac{1}{3}$ **41.** yes **43.**

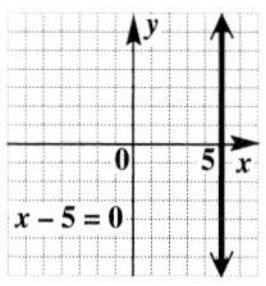

45. It is wise to plot three points before drawing the line, because if the three points do not lie in a line, then an error has been made and can be corrected.

CHAPTER 3 TEST (PAGE 219)

[3.1] **1.** (b) **2.** 16% **3.** -6; -10; -15 **4.** -12; -12; -12 **5.** no [3.2] **6.** To find the x-intercept, let $y = 0$, and to find the y-intercept, let $x = 0$. **7.** (2, 0); (0, 6)

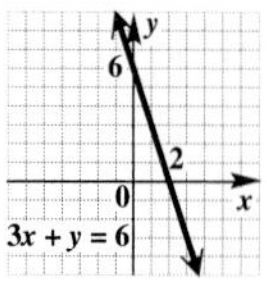

8. no x-intercept; $(0, -3)$

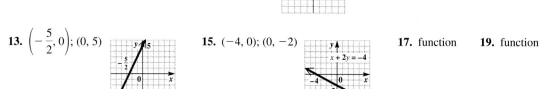

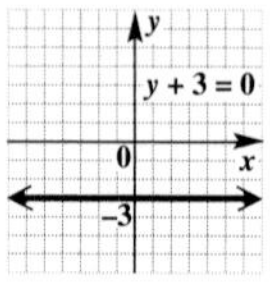

9. Yes, because each input determines exactly one output. **10.** The inputs are the years from 1990 to 1996, and the outputs are the corresponding Super Bowl advertising costs in thousands of dollars.

[3.3] **11.** $-\frac{8}{3}$ **12.** -2 **13.** $\frac{5}{2}$ **14.** $\frac{1}{2}$ **15.** -4 **16.** $\frac{1}{4}$

CUMULATIVE REVIEW EXERCISES CHAPTERS 1–3 (PAGE 220)

[1.1] **1.** $\frac{551}{40}$ or $13\frac{31}{40}$ **2.** $\frac{6}{5}$ [1.5] **3.** 7 [1.6] **4.** $\frac{73}{18}$ [1.1] **5.** $\frac{5}{12}$ [1.4] **6.** true [1.3] **7.** -134

[1.7] **8.** distributive property [1.8] **9.** $-p + 2$ [2.4] **10.** $h = \frac{3V}{\pi r^2}$ [2.2] **11.** $\{-1\}$ **12.** $\{2\}$

[2.2, 2.5] **13.** $\{-13\}$ [2.7] **14.** $(-\infty, 2]$ **15.** $\left(-\frac{8}{3}, -\frac{4}{3}\right)$ **16.** \$300 or more [2.4] **17.** 6 miles

[2.6] **18.** 10 liters [3.1] **19.**

x	y
12	89.45
28	81.95
36	78.20

20. \$7000; \$10,000 [3.2] **21.** x-intercept: (4, 0); y-intercept: (0, 6)

22.

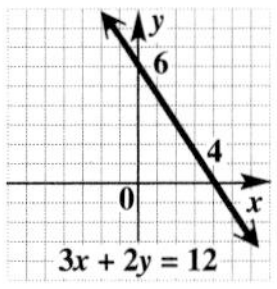

[3.3] **23.** $-\frac{3}{2}$ **24.** perpendicular **25.** Yes; (Taiwan, -6.6), (South Korea, -40.5); To determine the smallest and largest *change,* in the usual sense, we need to consider the absolute value of the changes.

CHAPTER 4 POLYNOMIALS AND EXPONENTS

SECTION 4.1 (PAGE 230)

CONNECTIONS **Page 226:** The polynomial gives the following: for 1996, 5730 million (greater than actual); for 1997, 5594 million (less than actual); for 1999, 5260 million (less than actual); for 2000, 5062 million (less than actual); for 2001, 4844 million (greater than actual).

EXERCISES **1.** 7; 5 **3.** 8 **5.** 26 **7.** 0 **9.** 1; 6 **11.** 1; 1 **13.** 2; -19, -1 **15.** 3; 1, 8, 5 **17.** $2m^5$ **19.** $-r^5$ **21.** cannot be simplified **23.** $-5x^5$ **25.** $5p^9 + 4p^7$ **27.** $-2y^2$ **29.** already simplified; 4; binomial **31.** $11m^4 - 7m^3 - 3m^2$; 4; trinomial **33.** x^4; 4; monomial **35.** 7; 0; monomial **37. (a)** 36 **(b)** -12 **39. (a)** 14 **(b)** -19 **41.** -3 **43.** 4; \$5.00 **44.** 6; \$27 **45.** 2.5; 130 **46.** 459.87 billion **47.** $5m^2 + 3m$ **49.** $4x^4 - 4x^2$ **51.** $\frac{7}{6}x^2 - \frac{2}{15}x + \frac{5}{6}$ **53.** $6m^3 + m^2 + 12m - 14$ **55.** $15m^3 - 13m^2 + 8m + 11$ **57.** Answers will vary. **59.** $5m^2 - 14m + 6$ **61.** $4x^3 + 2x^2 + 5x$ **63.** $-11y^4 + 8y^2 + y$ **65.** $a^4 - a^2 + 1$ **67.** $5m^2 + 8m - 10$ **69.** $-6x^2 - 12x + 12$ **71.** -10 **73.** $4b - 5c$ **75.** $6x - xy - 7$ **77.** $-3x^2y - 15xy - 3xy^2$ **79.** $8x^2 + 8x + 6$ **81. (a)** $23y + 5t$ **(b)** approximately 37.26°, 79.52°, and 63.22° **83.** $-6x^2 + 6x - 7$ **85.** $-7x - 1$ **87.** 0, -3, -4, -3, 0

89. 7, 1, -1, 1, 7

91. 0, 3, 4, 3, 0

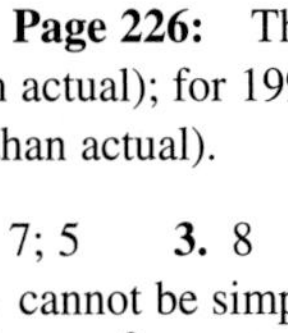

93. 4, 1, 0, 1, 4

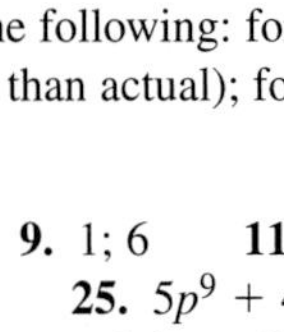

SECTION 4.2 (PAGE 239)

EXERCISES **1.** false **3.** false **5.** w^6 **7.** $\frac{1}{4^4}$ **9.** $(-7x)^4$ **11.** $\left(\frac{1}{2}\right)^6$ **13.** In $(-3)^4$, -3 is the base, while in -3^4, 3 is the base. $(-3)^4 = 81$, while $-3^4 = -81$. **15.** base: 3; exponent: 5; 243 **17.** base: -3; exponent: 5; -243 **19.** base: $-6x$; exponent: 4 **21.** base: x; exponent: 4 **23.** $5^2 + 5^3$ is a sum, not a product. $5^2 + 5^3 = 25 + 125 = 150$. **25.** 5^8 **27.** 4^{12} **29.** $(-7)^9$ **31.** t^{24} **33.** $-56r^7$ **35.** $42p^{10}$ **37.** The product rule applies only when the bases are the same. $3^2 \cdot 4^3 = 9 \cdot 64 = 576$. **39.** -32; -32; 1 **41.** 729; -729; 0 **43.** 4^6 **45.** t^{20} **47.** 7^3r^3 **49.** $5^5x^5y^5$ **51.** 5^{12} **53.** -8^{15} **55.** $8q^3r^3$ **57.** $\frac{1}{2^3}$ **59.** $\frac{a^3}{b^3}$ **61.** $\frac{9^8}{5^8}$ **63.** $\frac{5^5}{2^5}$ **65.** $\frac{9^5}{8^3}$ **67.** $2^{12}x^{12}$ **69.** -6^5p^5 **71.** $6^5x^{10}y^{15}$ **73.** x^{21} **75.** $2^2w^4x^{26}y^7$ **77.** $-r^{18}s^{17}$ **79.** $\frac{5^3a^6b^{15}}{c^{18}}$
81. This is incorrect. Using the product rule, it is simplified as follows: $(10^2)^3 = 10^{2 \cdot 3} = 10^6 = 1{,}000{,}000$. **83.** $12x^5$ **85.** $6p^7$ **87.** $125x^6$ **91.** \$304.16 **93.** \$1843.88

SECTION 4.3 (PAGE 246)

EXERCISES **1.** **(a)** B **(b)** D **(c)** A **(d)** C **3.** $40a^{14}$ **5.** $-6m^2 - 4m$ **7.** $24p - 18p^2 + 36p^4$ **9.** $-16z^2 - 24z^3 - 24z^4$ **11.** $14x^5y^3 + 21x^3y^2 - 28x^2y^2$ **13.** $12x^3 + 26x^2 + 10x + 1$ **15.** $20m^4 - m^3 - 8m^2 - 17m - 15$ **17.** $6x^6 - 3x^5 - 4x^4 + 4x^3 - 5x^2 + 8x - 3$ **19.** $5x^4 - 13x^3 + 20x^2 + 7x + 5$ **21.** $n^2 + n - 6$ **23.** $x^2 - 36$ **25.** $8r^2 - 10r - 3$ **27.** $9x^2 - 4$ **29.** $9q^2 + 6q + 1$ **31.** $3x^2 - 5xy - 2y^2$ **33.** $-3t^2 - 14t + 24$ **35.** $6y^5 - 21y^4 - 45y^3$ **37.** $36x^2 + 24x + 4$ **41.** $6p^2 - \frac{5}{2}pq - \frac{25}{12}q^2$ **43.** $2m^6 - 5m^3 - 12$ **45.** $2k^5 - 6k^3h^2 + k^2h^2 - 3h^4$ **47.** $6p^8 + 15p^7 + 12p^6 + 36p^5 + 15p^4$ **49.** $-24x^8 - 28x^7 + 32x^6 + 20x^5$ **51.** $14x + 49$ **53.** $\pi x^2 - 9$ **55.** $30x + 60$ **56.** $30x + 60 = 600$ **57.** 18 **58.** 10 yards by 60 yards **59.** \$2100 **60.** 140 yards **61.** \$1260 **62.** **(a)** $30kx + 60k$ (dollars) **(b)** $6rx + 32r$

SECTION 4.4 (PAGE 251)

EXERCISES **1.** **(a)** $4x^2$ **(b)** $12x$ **(c)** 9 **(d)** $4x^2 + 12x + 9$ **3.** $p^2 + 4p + 4$ **5.** $a^2 - 2ac + c^2$ **7.** $16x^2 - 24x + 9$ **9.** $64t^2 + 112st + 49s^2$ **11.** $25x^2 + 4xy + \frac{4}{25}y^2$ **13.** $4x^3 + 20x^2 + 25x$ **15.** $-16r^2 + 16r - 4$ **17.** **(a)** $49x^2$ **(b)** 0 **(c)** $-9y^2$ **(d)** $49x^2 - 9y^2$. Because 0 is the identity element for addition, it is not necessary to write "+ 0." **19.** $q^2 - 4$ **21.** $4w^2 - 25$ **23.** $100x^2 - 9y^2$ **25.** $4x^4 - 25$ **27.** $49x^2 - \frac{9}{49}$ **29.** $9p^3 - 49p$ **31.** $(a + b)^2$ **32.** a^2 **33.** $2ab$ **34.** b^2 **35.** $a^2 + 2ab + b^2$ **36.** They both represent the area of the entire large square. **37.** 1225 **38.** $30^2 + 2(30)(5) + 5^2$ **39.** 1225 **40.** They are equal. **41.** 9999 **43.** 39,999 **45.** $399\frac{3}{4}$ **47.** $m^3 - 15m^2 + 75m - 125$ **49.** $8a^3 + 12a^2 + 6a + 1$ **51.** $81r^4 - 216r^3t + 216r^2t^2 - 96rt^3 + 16t^4$ **53.** $x^2 + y^2$ is the sum of squares, while $(x + y)^2$ is the square of a sum. $(x + y)^2 = x^2 + 2xy + y^2$, and thus there is another term, $2xy$. **55.** $\frac{1}{2}m^2 - 2n^2$ **57.** $9a^2 - 4$ **59.** $\pi x^2 + 4\pi x + 4\pi$ **61.** $x^3 + 6x^2 + 12x + 8$

SECTION 4.5 (PAGE 259)

CONNECTIONS **Page 259:** (All are in millions of dollars.) 3501.09; 4792.64; 5607.39; 6560.65; 7675.96

EXERCISES **1.** false **3.** true **5.** false **7.** 1 **9.** 1 **11.** -1 **13.** 0 **15.** 0 **17.** 2 **19.** $\frac{1}{64}$ **21.** 16 **23.** $\frac{49}{36}$ **25.** $\frac{1}{81}$ **27.** $\frac{8}{15}$ **29.** 5^3 **31.** $\frac{1}{9}$ **33.** 5^2 **35.** x^{15} **37.** 6^3 **39.** $2r^4$ **41.** $\frac{5^2}{4^3}$

43. $\frac{p^5}{q^8}$ **45.** r^9 **47.** $\frac{x^5}{6}$ **49.** $a + b$ **51.** $(x + 2y)^2$ **53.** 1 **54.** $\frac{5^2}{5^2}$ **55.** 5^0 **56.** $1 = 5^0$; This supports the definition for 0 as an exponent. **57.** 7^3 or 343 **59.** $\frac{1}{x^2}$ **61.** $\frac{64x}{9}$ **63.** $\frac{x^2z^4}{y^2}$ **65.** $6x$ **67.** $\frac{1}{m^{10}n^5}$ **69.** $\frac{1}{xyz}$ **71.** x^3y^9 **73.** The student attempted to use the quotient rule with unequal bases. The correct way to simplify this expression is $\frac{16^3}{2^2} = \frac{(2^4)^3}{2^2} = \frac{2^{12}}{2^2} = 2^{10} = 1024$.

SECTION 4.6 (PAGE 266)

CONNECTIONS **Page 266:** **1.** -104 **2.** -104 **3.** They are both -104. **4.** The answers should agree.

EXERCISES **1.** $6x^2 + 8$; 2; $3x^2 + 4$ **3.** $3x^2 + 4$; 2 (These may be reversed.); $6x^2 + 8$ **5.** The first is a polynomial divided by a monomial, covered in Objective 1. This section does not cover dividing a monomial by a polynomial of several terms. **7.** $30x^3 - 10x + 5$ **9.** $4m^3 - 2m^2 + 1$ **11.** $4t^4 - 2t^2 + 2t$ **13.** $a^4 - a + \frac{2}{a}$ **15.** $4x^3 - 3x^2 + 2x$ **17.** $1 + 5x - 9x^2$ **19.** $\frac{12}{x} + 8 + 2x$ **21.** $\frac{4x^2}{3} + x + \frac{2}{3x}$ **23.** $9r^3 - 12r^2 + 2r + \frac{26}{3} - \frac{2}{3r}$ **25.** $-m^2 + 3m - \frac{4}{m}$ **27.** $4 - 3a + \frac{5}{a}$ **29.** $\frac{12}{x} - \frac{6}{x^2} + \frac{14}{x^3} - \frac{10}{x^4}$ **31.** $\frac{2}{3}x$ would not be an acceptable form, because $\frac{2}{3}x = \frac{2}{3} \cdot \frac{x}{1} = \frac{2x}{3}$, which is *not* equivalent to $\frac{2}{3x}$. **33.** $5x^3 + 4x^2 - 3x + 1$ **35.** $-63m^4 - 21m^3 - 35m^2 + 14m$ **37.** 1423 **38.** $(1 \times 10^3) + (4 \times 10^2) + (2 \times 10^1) + (3 \times 10^0)$ **39.** $x^3 + 4x^2 + 2x + 3$ **40.** They are similar in that the coefficients of powers of ten are equal to the coefficients of the powers of x. They are different in that one is a constant while the other is a polynomial. They are equal if $x = 10$ (the base of our decimal system). **41.** $x + 2$ **43.** $2y - 5$ **45.** $p - 4 + \frac{44}{p + 6}$ **47.** $r - 5$ **49.** $6m - 1$ **51.** $2a - 14 + \frac{74}{2a + 3}$ **53.** $4x^2 - 7x + 3$ **55.** $4k^3 - k + 2$ **57.** $5y^3 + 2y - 3$ **59.** $3k^2 + 2k - 2 + \frac{6}{k - 2}$ **61.** $2p^3 - 6p^2 + 7p - 4 + \frac{14}{3p + 1}$ **63.** $r^2 - 1 + \frac{4}{r^2 - 1}$ **65.** $y^2 - y + 1$ **67.** $a^2 + 1$ **69.** $x^2 - 4x + 2 + \frac{9x - 4}{x^2 + 3}$ **71.** $x^3 + 3x^2 - x + 5$ **73.** $\frac{3}{2}a - 10 + \frac{77}{2a + 6}$ **75.** The process stops when the degree of the remainder is less than the degree of the divisor. **77.** $x^2 + x - 3$ units **79.** $5x^2 - 11x + 14$ hours **81.** **(a)** is correct, **(b)** is incorrect. **82.** **(a)** is incorrect, **(b)** is correct. **83.** **(a)** is correct, **(b)** is incorrect. **84.** Because any power of 1 is 1, to evaluate a polynomial for 1 we simply add the coefficients. If the divisor is $x - 1$, this method does not apply, since $1 - 1 = 0$ and division by 0 is undefined.

SECTION 4.7 (PAGE 272)

CONNECTIONS **Page 270:** **1.** Move the decimal point 2 places to the right and use zeros as placeholders as necessary; use the same procedure, but move 4 places to the right. **2.** Move the decimal point 2 places to the left and use zeros as placeholders as necessary; use the same procedure, but move 4 places to the left. **3.** In 32,000, zero is used as a placeholder for the hundreds, tens, and units place values. In .00032, zero is used as a placeholder for the tenths, hundredths, and thousandths place values.

EXERCISES **1.** A **3.** C **5.** in scientific notation **7.** not in scientific notation; 5.6×10^6 **9.** not in scientific notation; 8×10^1 **11.** not in scientific notation; 4×10^{-3} **15.** 5.876×10^9 **17.** 8.235×10^4 **19.** 7×10^{-6} **21.** 2.03×10^{-3} **23.** 750,000 **25.** 5,677,000,000,000 **27.** 6.21 **29.** .00078 **31.** .000000005134 **33.** 600,000,000,000 **35.** 15,000,000 **37.** 60,000 **39.** .0003 **41.** 40 **43.** .000013 **45.** 4.7E−7 **47.** 2E7 **49.** 1E1 **51.** 1×10^{10} **53.** 2,000,000,000 **55.** $\$7.326 \times 10^9$ **57.** $\$3.0262 \times 10^{10}$ **59.** about 15,300 seconds **61.** **(a)** 1995: 1.063×10^{11}; 1996: 1.124×10^{11}; 1997: 1.250×10^{11} **(b)** 9.35×10^9

CHAPTER 4 REVIEW EXERCISES (PAGE 280)

1. $22m^2$; degree 2; monomial **3.** already in descending powers; degree 5; none of these **5.** $7r^4 - 4r^3 + 1$; degree 4; trinomial **7.** $a^3 + 4a^2$ **9.** $-13k^4 - 15k^2 + 18k$ **11.**

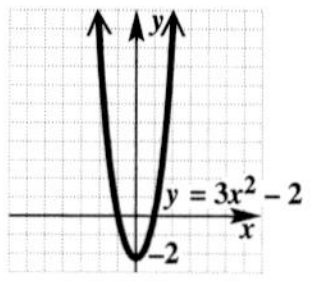

x	-2	-1	0	1	2
y	10	1	-2	1	10

13. 4^{11} **15.** $-72x^7$ **17.** 19^5x^5 **19.** $5p^4t^4$ **21.** $6^2x^{16}y^4z^{16}$ **23.** $a^3 - 2a^2 - 7a + 2$ **25.** $5p^5 - 2p^4 - 3p^3 + 25p^2 + 15p$ **27.** $6k^2 - 9k - 6$ **29.** $12k^2 - 32kq - 35q^2$ **31.** $2x^2 + x - 6$ **33.** $a^2 + 8a + 16$ **35.** $36m^2 - 25$ **37.** $r^3 + 6r^2 + 12r + 8$ **39.** **(a)** Answers will vary. For example, let $x = 1$ and $y = 2$. $(1 + 2)^2 \neq 1^2 + 2^2$, because $9 \neq 5$. **(b)** Answers will vary. For example, let $x = 1$ and $y = 2$. $(1 + 2)^3 \neq 1^3 + 2^3$, because $27 \neq 9$. **41.** In both cases, $x = 0$ and $y = 1$ lead to 1 on each side of the inequality. This would not be sufficient to show that *in general* the inequality is true. It would be necessary to choose other values of x and y. **43.** $\frac{4}{3}\pi(x + 1)^3$ or $\frac{4}{3}\pi x^3 + 4\pi x^2 + 4\pi x + \frac{4}{3}\pi$ cubic inches **45.** 0 **47.** $-\frac{1}{49}$ **49.** 5^8 **51.** $\frac{3}{4}$ **53.** x^2 **55.** $\frac{r^2}{6}$ **57.** $\frac{1}{a^3b^5}$ **59.** $\frac{-5y^2}{3}$ **61.** $-2m^2n + mn + \frac{6n^3}{5}$ **63.** This is not correct. The friend wrote the second term of the quotient as $-12x$ rather than $-2x$. Here is the correct method: $\frac{6x^2 - 12x}{6} = \frac{6x^2}{6} - \frac{12x}{6} = x^2 - 2x$. **65.** $2a^2 + 3a - 1 + \frac{6}{5a - 3}$ **67.** $m^2 + 4m - 2$ **69.** 2.8988×10^{10} **71.** 24,000 **73.** .000000897 **75.** 4,000,000 **77.** .000002 **79.** 4.2×10^{42} **81.** **(a)** 2.82×10^9 **(b)** 5.784×10^{11} **(c)** 1.77×10^{10} **83.** **(a)** 2 **(b)** 92 **(c)** $x^3 + x^2 - 10x - 6$ **(d)** Both results are 94. **84.** **(a)** 1 **(b)** 33 **(c)** $-x^3 - x^2 + 12x$ **(d)** Both results are -32. **85.** **(a)** -6 **(b)** 12 **(c)** $x^4 - 2x^3 - 14x^2 + 30x + 9$ **(d)** Both results are -72. **86.** **(a)** 3 **(b)** 183 **(c)** $x^2 + 4x + 1$ **(d)** Both results are 61. **87.** Answers will vary. **88.** We could not choose 3 because it would make the denominator equal to zero. It is the only invalid replacement for x. **89.** 2 **91.** $144a^2 - 1$ **93.** $\frac{1}{8^{12}}$ **95.** $\frac{2}{3m^3}$ **97.** r^{13} **99.** $-y^2 - 4y + 4$ **101.** $y^2 + 5y + 1$ **103.** $10p^2 - 3p - 5$ **105.** $49 - 28k + 4k^2$

CHAPTER 4 TEST (PAGE 284)

[4.1] **1.** $-7x^2 + 8x$; 2; binomial **2.** $4n^4 + 13n^3 - 10n^2$; 4; trinomial **3.** 4, -2, -4, -2, 4

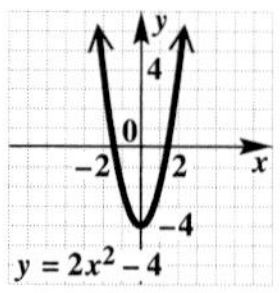

4. $-2y^2 - 9y + 17$ **5.** $-21a^3b^2 + 7ab^5 - 5a^2b^2$ **6.** $-12t^2 + 5t + 8$ [4.3] **7.** $-27x^5 + 18x^4 - 6x^3 + 3x^2$ **8.** $t^2 - 5t - 24$ **9.** $8x^2 + 2xy - 3y^2$ [4.4] **10.** $25x^2 - 20xy + 4y^2$ **11.** $100v^2 - 9w^2$ **12.** $2r^3 + r^2 - 16r + 15$ **13.** $9x^2 + 54x + 81$ [4.5] **14.** $\frac{1}{625}$ **15.** 2 **16.** $\frac{7}{12}$ [4.2] **17.** $9x^3y^5$ [4.5] **18.** 8^5 **19.** x^2y^6 **20.** Disagree, because $3^{-4} = \frac{1}{3^4} = \frac{1}{81}$, which is positive. [4.6] **21.** $4y^2 - 3y + 2 + \frac{5}{y}$ **22.** $-3xy^2 + 2x^3y^2 + 4y^2$ **23.** $3x^2 + 6x + 11 + \frac{26}{x - 2}$ [4.7] **24.** **(a)** 4.5×10^{10} **(b)** .0000036 **(c)** .00019 **25.** **(a)** 1.7×10^5; 1×10^3 **(b)** (more than) 1.7×10^8 pounds

CUMULATIVE REVIEW EXERCISES CHAPTERS 1–4 (PAGE 285)

[1.1] **1.** $\frac{7}{4}$ **2.** 5 **3.** $\frac{19}{24}$ **4.** $-\frac{1}{20}$ **5.** $31\frac{1}{4}$ cubic yards [1.5] **6.** \$1836 [1.6] **7.** 1, 3, 5, 9, 15, 45 [1.5] **8.** positive [1.6] **9.** -8 **10.** 24 **11.** $\frac{1}{2}$ **12.** -4 [1.7] **13.** associative property **14.** distributive property [1.5] **15. (a)** -175 **(b)** 575 [1.8] **16.** $-10x^2 + 21x - 29$ [2.1, 2.2] **17.** $\left\{\frac{13}{4}\right\}$ **18.** $\emptyset$

[2.4] **19.** $r = \frac{d}{t}$ [2.5] **20.** $\{-5\}$ [2.1, 2.2] **21.** $\{-12\}$ **22.** $\{20\}$ **23.** {all real numbers} [2.3] **24.** mouse: 160; elephant: 10 **25.** 4 **26.** 1995 and 1996: 22; 1994: 24 [2.7] **27.** $[10, \infty)$ **28.** $\left(-\infty, -\frac{14}{5}\right)$ **29.** $[-4, 2)$

[2.4] **30.** 11 feet and 22 feet [3.1] **31.**

x	0	-4	2	-2	4
y	2	0	3	1	4

[3.2] **32.**

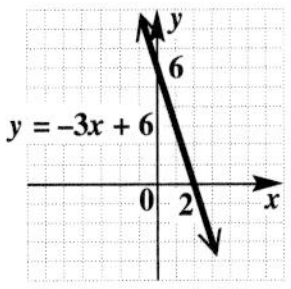

[4.1] **33.**

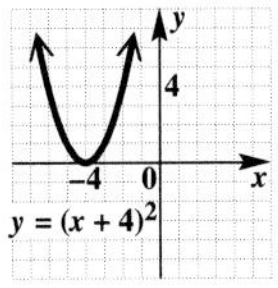

34. $11x^3 - 14x^2 - x + 14$ [4.3] **35.** $18x^7 - 54x^6 + 60x^5$ **36.** $63x^2 + 57x + 12$

[4.4] **37.** $25x^2 + 80x + 64$ [4.6] **38.** $2x^2 - 3x + 1$ **39.** $y^2 - 2y + 6$ [4.5] **40.** $\frac{5}{4}$ **41.** 2 **42.** 1

43. $\frac{2b}{a^{10}}$ [4.7] **44.** 3.45×10^4 **45.** .000000536

CHAPTER 5 FACTORING AND APPLICATIONS

SECTION 5.1 (PAGE 294)

EXERCISES **1.** no; 6 **3.** multiplication **5.** Answers will vary. One example is 5, 10, 15. **7.** 8 **9.** $10x^3$ **11.** $6m^3n^2$ **13.** xy^2 **15.** 6 **17.** factored **19.** not factored **21.** yes; x^3y^2 **23.** 2 **25.** x **27.** $3m^2$ **29.** $2z^4$ **31.** $2mn^4$ **33.** $-7x^3y^2$ **35.** $12(y - 2)$ **37.** $10a(a - 2)$ **39.** $5y^6(13y^4 + 7)$ **41.** no common factor (except 1) **43.** $8m^2n^2(n + 3)$ **45.** $13y^2(y^6 + 2y^2 - 3)$ **47.** $9qp^3(5q^3p^2 + 4p^3 + 9q)$ **49.** $a^3(a^2 + 2b^2 - 3a^2b^2 + 4ab^3)$ **51.** $(x + 2)(c - d)$ **53.** $(m + 2n)(m + n)$ **55.** not in factored form; $(7t + 4)(8 + x)$ **57.** in factored form **59.** not in factored form; $(y + 4)(18x^2 + 7)$ **61.** The quantities in parentheses are not the same, so there is no common factor in the two terms, $12k^3(s - 3)$ and $7(s + 3)$. **63.** $(p + 4)(p + 3)$ **65.** $(a - 2)(a + 5)$ **67.** $(z + 2)(7z - a)$ **69.** $(3r + 2y)(6r - x)$ **71.** $(a^2 + b^2)(3a + 2b)$ **73.** $(1 - a)(1 - b)$ **75.** $(4m - p^2)(4m^2 - p)$ **77.** $(5 - 2p)(m + 3)$ **79.** $(6r - y)(3r + 2y)$ **81.** $(a^5 - 3)(1 + 2b)$ **83.** commutative property and associative property **84.** $(2xy - 8x) + (-3y) + 12$; $2x$ **85.** $(2xy - 8x) + (-3y + 12)$; yes **86.** $2x(y - 4) - 3(y - 4)$ **87.** No, because it is the difference between two terms, $2x(y - 4)$ and $3(y - 4)$. **88.** $(2x - 3)(y - 4)$; yes **89. (a)** yes **(b)** When either one is multiplied out, the product is $1 - a + ab - b$.

SECTION 5.2 (PAGE 299)

EXERCISES **1.** a and b must have different signs, one positive and one negative. **5.** 1 and 48, -1 and -48, 2 and 24, -2 and -24, 3 and 16, -3 and -16, 4 and 12, -4 and -12, 6 and 8, -6 and -8; the pair with a sum of -19 is -3 and -16. **7.** 1 and -24; -1 and 24, 2 and -12, -2 and 12, 3 and -8, -3 and 8, 4 and -6, -4 and 6; the pair with a sum of -5 is 3 and -8. **9.** (c) **11.** $p + 6$ **13.** $x + 11$ **15.** $x - 8$ **17.** $y - 5$ **19.** $x + 11$ **21.** $y - 9$ **23.** 1 **24.** -1 **25.** The sums are opposites as well. **26.** The product is $x^2 - x - 12$. This is not correct because the middle term is incorrect. **27.** The product is $x^2 + x - 12$. This is correct because we obtain the exact trinomial we were given to factor. **28.** It is the opposite of what it should be. **29.** reverse the signs of the two second terms of the binomials **30.** $(x - 5)(x + 3)$ **31.** $(y + 8)(y + 1)$ **33.** $(b + 3)(b + 5)$ **35.** $(m + 5)(m - 4)$ **37.** $(y - 5)(y - 3)$ **39.** prime **41.** $(t - 4)^2$ or $(t - 4)(t - 4)$ **43.** $(r - 6)(r + 5)$ **45.** prime **47.** $(r + 2a)(r + a)$

49. $(t + 2z)(t - 3z)$ **51.** $(x + y)(x + 3y)$ **53.** $(v - 5w)(v - 6w)$ **55.** $4(x + 5)(x - 2)$ **57.** $2t(t + 1)(t + 3)$ **59.** $2x^4(x - 3)(x + 7)$ **61.** $mn(m - 6n)(m - 4n)$ **63.** The factored form $(2x + 4)(x - 3)$ is incorrect because $2x + 4$ has a common factor of 2, which must be factored out for the trinomial to be completely factored. **65.** $a^3(a + 4b)(a - b)$ **67.** $yz(y + 3z)(y - 2z)$ **69.** $z^8(z - 7y)(z + 3y)$ **71.** $(a + b)(x + 4)(x - 3)$ **73.** $(2p + q)(r - 9)(r - 3)$ **75.** $a^2 + 13a + 36$

SECTION 5.3 (PAGE 306)

EXERCISES **1.** 2; −21 **3.** $-6x$; $7x$ or $7x$; $-6x$ **5.** $x - 3$ **7.** $-4b$; $2a$ **9.** $-3x - 10$; -5; $+2$ or $-3x - 10$; $+2$; -5 **11.** (a) **13.** (a) **15.** The binomial $2x - 6$ cannot be a factor, because it has a common factor of 2, but the polynomial does not.

The order of the factors is irrelevant in Exercises 17–63.

17. $(3a + 7)(a + 1)$ **19.** $(4r - 3)(r + 1)$ **21.** $(3m - 1)(5m + 2)$ **23.** $(4m + 1)(2m - 3)$ **25.** $(4x + 3)(5x - 1)$ **27.** $(3m + 1)(7m + 2)$ **29.** $(4y - 1)(5y + 11)$ **31.** $(2b + 1)(3b + 2)$ **33.** $3(4x - 1)(2x - 3)$ **35.** $q(5m + 2)(8m - 3)$ **37.** $2m(m - 4)(m + 5)$ **39.** $3n^2(5n - 3)(n - 2)$ **41.** $3x^3(2x + 5)(3x - 5)$ **43.** $y^2(5x - 4)(3x + 1)$ **45.** $(3p + 4q)(4p - 3q)$ **47.** $(5a + 2b)(5a + 3b)$ **49.** $(3a - 5b)(2a + b)$ **51.** $m^4n(3m + 2n)(2m + n)$ **53.** $(5 - x)(1 - x)$ **55.** $(4 + 3x)(4 + x)$ **57.** $-5x(2x + 7)(x - 4)$ **59.** $-1(x + 7)(x - 3)$ **61.** $-1(3x + 4)(x - 1)$ **63.** $-1(a + 2b)(2a + b)$ **65.** Yes, $(x + 7)(3 - x)$ is equivalent to $-1(x + 7)(x - 3)$, because $-1(x - 3) = -x + 3 = 3 - x$. **67.** $(m + 1)^3(5q - 2)(5q + 1)$ **69.** $(r + 3)^3(5x + 2y)(3x - 8y)$ **71.** −4, 4 **73.** −11, −7, 7, 11 **75.** $5 \cdot 7$ **76.** $(-5)(-7)$ **77.** The product of $3x - 4$ and $2x - 1$ is indeed $6x^2 - 11x + 4$. **78.** The product of $4 - 3x$ and $1 - 2x$ is indeed $6x^2 - 11x + 4$. **79.** The factors in Exercise 78 are the opposites of the factors in Exercise 77. **80.** $(3 - 7t)(5 - 2t)$ **81.** $-a$; $-b$ **82.** $-P$; $-Q$

SECTION 5.4 (PAGE 314)

EXERCISES **1.** 1; 4; 9; 16; 25; 36; 49; 64; 81; 100; 121; 144; 169; 196; 225; 256; 289; 324; 361; 400 **3.** 1; 8; 27; 64; 125; 216; 343; 512; 729; 1000 **5.** **(a)** both of these **(b)** a perfect cube **(c)** a perfect square **(d)** a perfect square **7.** $(y + 5)(y - 5)$ **9.** $(3r + 2)(3r - 2)$ **11.** $\left(6m + \frac{4}{5}\right)\left(6m - \frac{4}{5}\right)$ **13.** $4(3x + 2)(3x - 2)$ **15.** $(14p + 15)(14p - 15)$ **17.** $(4r + 5a)(4r - 5a)$ **19.** prime **21.** $(p^2 + 7)(p^2 - 7)$ **23.** $(x^2 + 1)(x + 1)(x - 1)$ **25.** $(p^2 + 16)(p + 4)(p - 4)$ **27.** The teacher was justified, because it was not factored completely; $x^2 - 9$ can be factored as $(x + 3)(x - 3)$. The complete factored form is $(x^2 + 9)(x + 3)(x - 3)$. **29.** $(w + 1)^2$ **31.** $(x - 4)^2$ **33.** $\left(t + \frac{1}{2}\right)^2$ **35.** $(x - .5)^2$ **37.** $2(x + 6)^2$ **39.** $(4x - 5)^2$ **41.** $(7x - 2y)^2$ **43.** $(8x + 3y)^2$ **45.** $-2(5h - 2y)^2$ **47.** $(a + 1)(a^2 - a + 1)$ **49.** $(a - 1)(a^2 + a + 1)$ **51.** $(p + q)(p^2 - pq + q^2)$ **53.** $(3x - 1)(9x^2 + 3x + 1)$ **55.** $(2p + 9q)(4p^2 - 18pq + 81q^2)$ **57.** $(y - 2x)(y^2 + 2yx + 4x^2)$ **59.** $(3a - 4b)(9a^2 + 12ab + 16b^2)$ **61.** $(5t + 2s)(25t^2 - 10ts + 4s^2)$ **64.** $(5x - 2)(2x + 3)$ **65.** $5x - 2$ **66.** Yes. If $10x^2 + 11x - 6$ factors as $(5x - 2)(2x + 3)$, then when $10x^2 + 11x - 6$ is divided by $2x + 3$, the quotient should be $5x - 2$. **67.** $x^2 + x + 1$; $(x - 1)(x^2 + x + 1)$ **69.** $-2b(3a^2 + b^2)$ **71.** $3(r - k)(1 + r + k)$ **72.** $(x + 1)^2 - 2^2$ or $(x + 1)^2 - 4$ **73.** $[(x + 1) - 2][(x + 1) + 2]$ **74.** $(x - 1)(x + 3)$ **75.** $x^2 + 2x - 3$ **76.** $(x - 1)(x + 3)$ **77.** The results are the same. (Preference is a matter of individual choice.) **79.** 10 **81.** 9

SUMMARY: EXERCISES ON FACTORING (PAGE 317)

EXERCISES **1.** $(a - 6)(a + 2)$ **3.** $6(y - 2)(y + 1)$ **5.** $6(a + 2b + 3c)$ **7.** $(p - 11)(p - 6)$ **9.** $(5z - 6)(2z + 1)$ **11.** $(m + n + 5)(m - n)$ **13.** $8a^3(a - 3)(a + 2)$ **15.** $(z - 5a)(z + 2a)$ **17.** $(x - 5)(x - 4)$ **19.** $(3n - 2)(2n - 5)$ **21.** $4(4x + 5)$ **23.** $(3y - 4)(2y + 1)$ **25.** $(6z + 1)(z + 5)$ **27.** $(2k - 3)^2$ **29.** $6(3m + 2z)(3m - 2z)$ **31.** $(3k - 2)(k + 2)$ **33.** $7k(2k + 5)(k - 2)$ **35.** $(y^2 + 4)(y + 2)(y - 2)$ **37.** $8m(1 - 2m)$ **39.** $(z - 2)(z^2 + 2z + 4)$ **41.** prime **43.** $8m^3(4m^6 + 2m^2 + 3)$ **45.** $(4r + 3m)^2$ **47.** $(5h + 7g)(3h - 2g)$ **49.** $(k - 5)(k - 6)$ **51.** $3k(k - 5)(k + 1)$ **53.** $(10p + 3)(100p^2 - 30p + 9)$ **55.** $(2 + m)(3 + p)$ **57.** $(4z - 1)^2$ **59.** $3(6m - 1)^2$ **61.** prime **63.** $8z(4z - 1)(z + 2)$ **65.** $(4 + m)(5 + 3n)$ **67.** $2(3a - 1)(a + 2)$ **69.** $(a - b)(a^2 + ab + b^2 + 2)$ **71.** $(8m - 5n)^2$ **73.** $(4k - 3h)(2k + h)$ **75.** $(m + 2)(m^2 + m + 1)$ **77.** $(5y - 6z)(2y + z)$ **79.** $(8a - b)(a + 3b)$ **81.** $(x^3 - 1)(x^3 + 1)$

82. $(x-1)(x^2+x+1)(x+1)(x^2-x+1)$ **83.** $(x^2-1)(x^4+x^2+1)$ **84.** $(x-1)(x+1)(x^4+x^2+1)$ **85.** The result in Exercise 82 is completely factored. **86.** Show that $x^4+x^2+1=(x^2+x+1)(x^2-x+1)$. **87.** difference of squares **88.** $(x-3)(x^2+3x+9)(x+3)(x^2-3x+9)$

SECTION 5.5 (PAGE 323)

CONNECTIONS **Page 322:** **1.** 4 seconds **2.** 64 feet

EXERCISES **1.** ax^2+bx+c **3.** factoring **5.** $\left\{-\frac{8}{3}, -7\right\}$ **7.** $\{0, -4\}$ **9.** To solve $2x(3x-4)=0$, set each *variable* factor equal to 0 to get $x=0$ or $3x-4=0$. Then solve the second equation to get $x=\frac{4}{3}$. **11.** Because $(x-9)^2=(x-9)(x-9)=0$ leads to two solutions of 9, we call 9 a double solution. **13.** $\{-2, -1\}$ **15.** $\{1, 2\}$ **17.** $\{-8, 3\}$ **19.** $\{-1, 3\}$ **21.** $\{-2, -1\}$ **23.** $\{-4\}$ **25.** $\left\{-2, \frac{1}{3}\right\}$ **27.** $\left\{-\frac{4}{3}, \frac{1}{2}\right\}$ **29.** $\left\{-\frac{2}{3}\right\}$ **31.** $\{-3, 3\}$ **33.** $\left\{-\frac{7}{4}, \frac{7}{4}\right\}$ **35.** $\{-11, 11\}$ **37.** Another solution is -11. **39.** $\{0, 7\}$ **41.** $\left\{0, \frac{1}{2}\right\}$ **43.** $\{2, 5\}$ **45.** $\left\{-4, \frac{1}{2}\right\}$ **47.** $\left\{-12, \frac{11}{2}\right\}$ **49.** $\{-1, 3\}$ **51.** $\left\{-\frac{5}{2}, \frac{1}{3}, 5\right\}$ **53.** $\left\{-\frac{7}{2}, -3, 1\right\}$ **55.** $\left\{-\frac{7}{3}, 0, \frac{7}{3}\right\}$ **57.** $\{-2, 0, 4\}$ **59.** $\{-5, 0, 4\}$ **61.** $\{-3, 0, 5\}$ **63.** $\left\{-\frac{4}{3}, -1, \frac{1}{2}\right\}$ **65.** $\left\{-\frac{2}{3}, 4\right\}$ **67.** To use the zero-factor property, one side of the equation must be 0. Before solving this equation, multiply the factors on the left side, and subtract 1 from both sides so the right side is 0. **69.** $\{-.5, .1\}$ **71.** $\{-1.1, -2.5\}$

SECTION 5.6 (PAGE 330)

CONNECTIONS **Page 329:** The diagonal of the floor should be $\sqrt{208} \approx 14.4$ ft, or about 14 feet, 5 inches. The carpenter is off by 3 inches and must correct the error to avoid major construction problems.

EXERCISES **1.** a variable; the unknown **3.** an equation **5. (a)** $80=(x+8)(x-8)$ **(b)** 12 **(c)** length: 20 units; width: 4 units **7. (a)** $60=\frac{1}{2}(3x+6)(x+5)$ **(b)** 3 **(c)** base: 15 units; height: 8 units **9.** length: 7 inches; width: 4 inches **11.** length: 11 inches; width: 8 inches **13.** base: 12 inches; height: 5 inches **15.** height: 13 inches; width: 10 inches **17.** 12 centimeters **19.** 8 feet **21.** 6 meters **23.** 1 second **25.** 3 seconds **27.** 1 second and 3 seconds **29.** 4 seconds **31.** 9 centimeters **33.** 256 feet **35.** 5.7 seconds **37.** 8.0 seconds **39.** $-3, -2$ or 4, 5 **41.** 7, 9, 11 **43.** $-2, 0, 2$ or 6, 8, 10 **45. (a)** 1990 **(b)** 8.8 billion passengers; This approximation is .2 billion less than the value in the table. **(c)** 25 **(d)** 8.0 billion passengers; This is correct to the nearest one tenth billion. **(e)** 1995 **47. (a)** approximately 17,000 injuries **(b)** approximately .7 or 7 to 10; This indicates that in 1998, on the average, 1.4 people were injured per accident. **49.** c^2 **50.** b^2 **51.** a^2 **52.** $a^2+b^2=c^2$; This is the Pythagorean formula.

SECTION 5.7 (PAGE 339)

CONNECTIONS **Page 339:** The profit is found by subtracting cost from revenue, represented by $R-C$. If $R-C=x^2-x-12$, the solution of $R-C>0$ is $x<-3$ or $x>4$. Only $x>4$ makes sense, since x must be positive in the context of the problem.

EXERCISES **1.** (c) **3. (a)** true **(b)** true **(c)** false **(d)** true **5.** $(-3, 3)$ 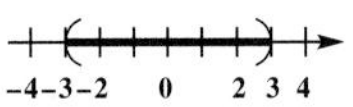

7. $(-\infty, -6] \cup [7, \infty)$ 

9. $(-\infty, -3) \cup (-2, \infty)$ 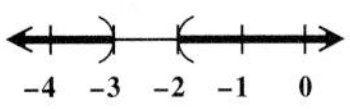

11. $[-1, 5]$ 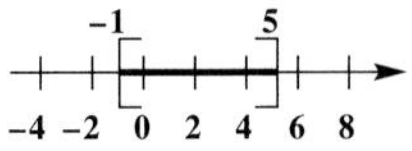

13. $\left(-1, \frac{2}{5}\right)$ 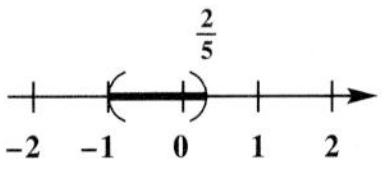

15. $\left(-\frac{1}{2}, \frac{4}{3}\right)$ 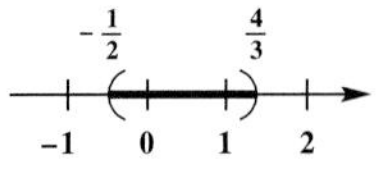

17. (1, 6) 

19. $\left(-\infty, -\frac{1}{2}\right) \cup \left(\frac{1}{3}, \infty\right)$ 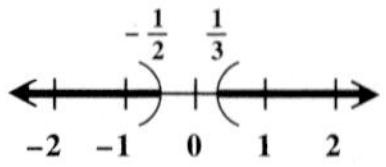

21. $\left(-\frac{2}{3}, -\frac{1}{4}\right)$ 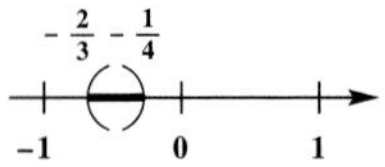

23. $(-\infty, -2) \cup (2, \infty)$ 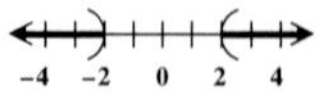

25. $(-\infty, -4) \cup (4, \infty)$ 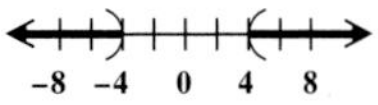

27. $[-1.5, .8]$ **29.** $(-\infty, 1.5] \cup [3.5, \infty)$

31. $\left[-2, \frac{1}{3}\right] \cup [4, \infty)$ 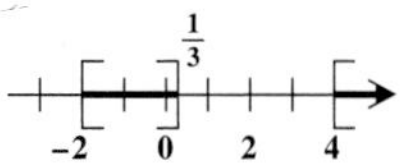

33. $(-\infty, -1) \cup (2, 4)$ 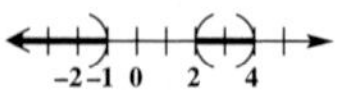

35. 2 seconds and 14 seconds **37.** between 0 and 2 seconds or between 14 and 16 seconds

CHAPTER 5 REVIEW EXERCISES (PAGE 345)

1. $7(t + 2)$ **3.** $(x - 4)(2y + 3)$ **5.** $(x + 3)(x + 2)$ **7.** $(q + 9)(q - 3)$ **9.** $(r + 8s)(r - 12s)$ **11.** $8p(p + 2)(p - 5)$ **13.** $p^5(p - 2q)(p + q)$ **15.** r and $6r$, $2r$ and $3r$ **17.** $(2k - 1)(k - 2)$ **19.** $(3r + 2)(2r - 3)$ **21.** $(v + 3)(8v - 7)$ **23.** $-3(x + 2)(2x - 5)$ **25.** (b) **27.** $(n + 7)(n - 7)$ **29.** $(7y + 5w)(7y - 5w)$ **31.** prime **33.** $(3t - 7)^2$ **35.** $(5k + 4x)(25k^2 - 20kx + 16x^2)$ **37.** $\{-3, -1\}$ **39.** $\{3, 5\}$ **41.** $\left\{-\frac{8}{9}, \frac{8}{9}\right\}$ **43.** $\{-1, 6\}$ **45.** $\{6\}$ **47.** length: 10 meters; width: 4 meters **49.** length: 6 meters; height: 5 meters **51.** 112 feet **53.** 256 feet **55.** 2 inches **57.** **(a)** \$17.4 million **(b)** −\$48.6 million **(c)** The equation is based on the data for 1994 to 1996. To use it to predict much beyond 1996 may lead to incorrect results, since it is possible that the conditions the equation was based on will change.

59. $(-\infty, -4] \cup \left[\frac{1}{2}, \infty\right)$ **61.** $(-\infty, -4] \cup \left[\frac{3}{2}, \infty\right)$ **63.** $(-\infty, -5) \cup (7, \infty)$ **65.** $(3k + 5)(k + 2)$ **67.** $(y^2 + 25)(y + 5)(y - 5)$ **69.** $8abc(3b^2c - 7ac^2 + 9ab)$ **71.** $6xyz(2xz^2 + 2y - 5x^2yz^3)$ **73.** $(2r + 3q)(6r - 5q)$ **75.** $(7t + 4)^2$ **77.** $\{-5, 2\}$ **79.** **(a)** 417,000 vehicles **(b)** The estimate may be unreliable because the conditions that prevailed in the years 1995–1997 may have changed, causing either a greater increase or a greater decrease in the numbers of alternative-fueled vehicles. **81.** width: 10 meters; length: 17 meters **83.** 34 miles **85.** 25 miles

CHAPTER 5 TEST (PAGE 349)

[5.1–5.4] **1.** (d) **2.** $m^2n(2mn + 3m - 5n)$ **3.** $(x + 3)(x - 8)$ **4.** $(2x + 3)(x - 1)$ **5.** $(5z - 1)(2z - 3)$ **6.** prime **7.** prime **8.** $(2 - a)(6 + b)$ **9.** $(3y + 8)(3y - 8)$ **10.** $(2x - 7y)^2$ **11.** $-2(x + 1)^2$ **12.** $3t^2(2t + 9)(t - 4)$ **13.** $(r - 5)(r^2 + 5r + 25)$ **14.** $8(k + 2)(k^2 - 2k + 4)$ **15.** The product $(p + 3)(p + 3) = p^2 + 6p + 9$, which does not equal $p^2 + 9$. The binomial $p^2 + 9$ is a prime polynomial. [5.5] **16.** $\left\{6, \frac{1}{2}\right\}$ **17.** $\left\{-\frac{2}{5}, \frac{2}{5}\right\}$ **18.** $\{10\}$ **19.** $\{-3, 0, 3\}$ [5.6] **20.** $1\frac{1}{2}$ seconds and $4\frac{1}{2}$ seconds **21.** **(a)** $3x - 7$ **(b)** $2x - 1$ **(c)** 17 feet **22.** July 1, 1995: 341,000; July 1, 1996: 368,000 [5.7] **23.** $\left(-\frac{5}{2}, \frac{1}{3}\right)$ 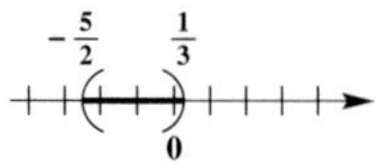

24. $(-\infty, -4] \cup [6, \infty)$ 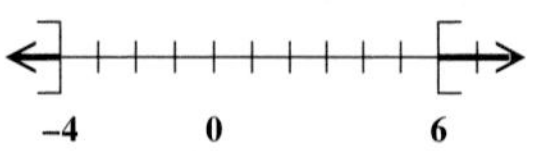

[5.5] **25.** Another solution is $-\frac{2}{3}$.

CUMULATIVE REVIEW EXERCISES CHAPTERS 1–5 (PAGE 351)

[2.2] **1.** $\{0\}$ **2.** $\{.05\}$ **3.** $\{6\}$ [2.4] **4.** $P = \frac{A}{1 + rt}$ [2.3] **5.** 110° and 70° **6.** exports: \$741 million; imports: \$426 million [2.5] **7.** 35 pounds [3.1] **8.** **(a)** negative; positive **(b)** negative; negative **9.** (2, 27,000), (5, 63,000) [3.2] **10.** x-intercept: $\left(-\frac{1}{4}, 0\right)$; y-intercept: (0, 3)

11. 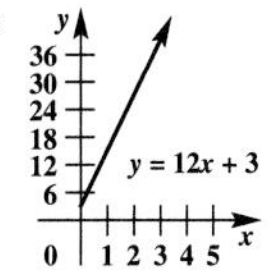

[3.3] **12.** 209.2; A slope of 209.2 means that the number of radio stations increased on the average by about 209 per year.

[4.2, 4.5] **13.** 4 **14.** $\frac{16}{9}$ **15.** 256 **16.** $\frac{1}{p^2}$ [4.1] **17.** $-4k^2 - 4k + 8$
[4.3] **18.** $6m^8 - 15m^6 + 3m^4$ **19.** $3y^3 + 8y^2 + 12y - 5$ [4.4] **20.** $4p^2 - 9q^2$ [4.6] **21.** $4x^3 + 6x^2 - 3x + 10$
22. $6p^2 + 7p + 1 + \frac{7}{2p - 2}$ [5.2–5.3] **23.** $(2a - 1)(a + 4)$ **24.** $(2m + 3)(5m + 2)$ **25.** $(5x + 3y)(3x - 2y)$
[5.4] **26.** $(3x + 1)^2$ **27.** $-2(4t + 7z)^2$ **28.** $(5r + 9t)(5r - 9t)$ [5.1] **29.** $25(4x^2 + 1)$
[5.3] **30.** $2pq(3p + 1)(p + 1)$ [5.1] **31.** $(2x + y)(a - b)$ [5.5] **32.** $\left\{\frac{3}{2}, -2, 6\right\}$ **33.** $\left\{-\frac{2}{3}, \frac{1}{2}\right\}$
[5.7] **34.** $(-\infty, -2] \cup \left[\frac{3}{2}, \infty\right)$ 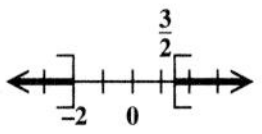

[5.6] **35.** $-4, -2$ or $8, 10$ **36.** 5 meters, 12 meters, 13 meters

CHAPTER 6 RATIONAL EXPRESSIONS

SECTION 6.1 (PAGE 360)

CONNECTIONS **Page 360:** **1.** $3x^2 + 11x + 8$ cannot be factored, so this quotient cannot be reduced. By long division the quotient is $3x + 5 + \frac{-2}{x + 2}$. **2.** The numerator factors as $(x - 2)(x^2 + 2x + 4)$, so by reducing, the quotient is $x - 2$. Long division gives the same quotient.

EXERCISES **1. (a)** 3; -5 **(b)** q; -1 **5.** 0 **7.** $-\frac{5}{3}$ **9.** $-3, 2$ **11.** never undefined **13. (a)** 1 **(b)** $\frac{17}{12}$ **15. (a)** 0 **(b)** $-\frac{10}{3}$ **17. (a)** $\frac{9}{5}$ **(b)** undefined **19. (a)** $\frac{2}{7}$ **(b)** $\frac{13}{3}$ **21.** Any number divided by itself is 1, *provided the number is not 0.* This expression is equal to $\frac{1}{x + 2}$ for all values of x except -2 and 2. **23.** $3r^2$ **25.** $\frac{2}{5}$ **27.** $\frac{x - 1}{x + 1}$ **29.** $\frac{7}{5}$ **31.** $m - n$ **33.** $\frac{3(2m + 1)}{4}$ **35.** $\frac{3m}{5}$ **37.** $\frac{3r - 2s}{3}$ **39.** $\frac{z - 3}{z + 5}$ **41.** $k - 3$ **43.** $\frac{x + 1}{x - 1}$ **45.** -1 **47.** $-(m + 1)$ **49.** -1 **51.** already in lowest terms **53.** $x^2 + 3$ **Answers may vary in Exercises 55–59.** **55.** $\frac{-(x + 4)}{x - 3}, \frac{-x - 4}{x - 3}, \frac{x + 4}{-(x - 3)}, \frac{x + 4}{-x + 3}$ **57.** $\frac{-(2x - 3)}{x + 3}, \frac{-2x + 3}{x + 3}, \frac{2x - 3}{-(x + 3)}, \frac{2x - 3}{-x - 3}$ **59.** $-\frac{3x - 1}{5x - 6}, \frac{-(3x - 1)}{5x - 6}, -\frac{-3x + 1}{-5x + 6}, \frac{3x - 1}{-5x + 6}$ **61.** $\frac{m + n}{2}$ **63.** $-\frac{b^2 + ba + a^2}{a + b}$ **65.** $\frac{z + 3}{z}$ **67.** $x + 3$ **69.** $x + 5$ **71.** $x - 3$

SECTION 6.2 (PAGE 367)

EXERCISES **1. (a)** B **(b)** D **(c)** C **(d)** A **3.** $\frac{3a}{2}$ **5.** $-\frac{4x^4}{3}$ **7.** $\frac{2}{c + d}$ **9.** 5 **11.** $-\frac{3}{2t^4}$ **13.** $\frac{1}{4}$ **15.** -4 causes the denominator in the first fraction to equal 0. **16.** -5 causes the denominator in the second fraction to equal 0. **17.** -7 causes the numerator in the divisor to equal 0, meaning that we would be dividing by 0. This is undefined. **18.** We *are* allowed to divide 0 by a nonzero number. **21.** $\frac{10}{9}$ **23.** $-\frac{3}{4}$ **25.** $-\frac{9}{2}$ **27.** $\frac{p + 4}{p + 2}$

29. $\frac{(k-1)^2}{(k+1)(2k-1)}$ 31. $\frac{4k-1}{3k-2}$ 33. $\frac{m+4p}{m+p}$ 35. $\frac{m+6}{m+3}$ 37. $\frac{y+3}{y+4}$ 39. $\frac{m}{m+5}$ 41. $\frac{r+6s}{r+s}$
43. $\frac{(q-3)^2(q+2)^2}{q+1}$ 45. $\frac{x+10}{10}$ 47. $\frac{3-a-b}{2a-b}$ 49. $-\frac{(x+y)^2(x^2-xy+y^2)}{3y(y-x)(x-y)}$ or $\frac{(x+y)^2(x^2-xy+y^2)}{3y(x-y)^2}$
51. $\frac{5xy^2}{4q}$

SECTION 6.3 (PAGE 373)

EXERCISES **1.** (c) **3.** (c) **5.** 60 **7.** 1800 **9.** x^5 **11.** $30p$ **13.** $180y^4$ **15.** $15a^5b^3$ **17.** $12p(p-2)$
19. $2^3 \cdot 3 \cdot 5$ **20.** $(t+4)^3(t-3)(t+8)$ **21.** The similarity is that 2 is replaced by $t+4$, 3 is replaced by $t-3$, and 5 is replaced by $t+8$. **22.** The procedure used is the same. The only difference is that for algebraic fractions, the factors may contain variables, while in a common fraction, the factors are constants. **23.** $18(r-2)$ **25.** $12p(p+5)^2$
27. $8(y+2)(y+1)$ **29.** $m-3$ or $3-m$ **31.** $p-q$ or $q-p$ **33.** $a(a+6)(a-3)$
35. $(k+3)(k-5)(k+7)(k+8)$ **37.** Yes, because $(2x-5)^2=(5-2x)^2$. **39.** 7 **40.** 1 **41.** identity property of multiplication **42.** 7 **43.** 1 **44.** identity property of multiplication **45.** $\frac{60m^2k^3}{32k^4}$ **47.** $\frac{57z}{6z-18}$
49. $\frac{-4a}{18a-36}$ **51.** $\frac{6(k+1)}{k(k-4)(k+1)}$ **53.** $\frac{36r(r+1)}{(r-3)(r+2)(r+1)}$ **55.** $\frac{ab(a+2b)}{2a^3b+a^2b^2-ab^3}$
57. $\frac{(t-r)(4r-t)}{t^3-r^3}$ **59.** $\frac{2y(z-y)(y-z)}{y^4-z^3y}$ or $\frac{-2y(y-z)^2}{y^4-z^3y}$

SECTION 6.4 (PAGE 381)

EXERCISES **1.** $\frac{5}{4}$ **2.** $\frac{-5}{-7+3}$ **3.** 1 **4.** $\frac{-5}{-4}=\frac{5}{4}$; The two answers are equal. **5.** Jack's answer was correct. His answer can be obtained by multiplying Jill's correct answer by $\frac{-1}{-1}=1$, the identity element for multiplication.
6. Changing the sign of each term in the fraction is a way of multiplying by $\frac{-1}{-1}$ or 1, the identity element for multiplication.
7. Putting a negative sign in front of a fraction gives the opposite of the fraction. Changing the signs in either the numerator or the denominator also gives the opposite. Thus we have multiplied by $(-1)(-1)=1$, the identity element for multiplication.
8. **(a)** not equivalent **(b)** equivalent **(c)** equivalent **(d)** not equivalent **(e)** not equivalent **(f)** not equivalent
9. $\frac{11}{m}$ **11.** b **13.** x **15.** $y-6$ **19.** $\frac{3z+5}{15}$ **21.** $\frac{10-7r}{14}$ **23.** $\frac{-3x-2}{4x}$ **25.** $\frac{x+1}{2}$ **27.** $\frac{5x+9}{6x}$
29. $\frac{7-6p}{3p^2}$ **31.** $\frac{x+8}{x+2}$ **33.** $\frac{3}{t}$ **35.** $m-2$ or $2-m$ **37.** $\frac{-2}{x-5}$ or $\frac{2}{5-x}$ **39.** -4 **41.** $\frac{-5}{x-y^2}$ or $\frac{5}{y^2-x}$
43. $\frac{x+y}{5x-3y}$ or $\frac{-x-y}{3y-5x}$ **45.** $\frac{-6}{4p-5}$ or $\frac{6}{5-4p}$ **47.** $\frac{-(m+n)}{2(m-n)}$ **49.** $\frac{-x^2+6x+11}{(x+3)(x-3)(x+1)}$
51. $\frac{-5q^2-13q+7}{(3q-2)(q+4)(2q-3)}$ **53.** $\frac{9r+2}{r(r+2)(r-1)}$ **55.** $\frac{2x^2+6xy+8y^2}{(x+y)(x+y)(x+3y)}$ or $\frac{2x^2+6xy+8y^2}{(x+y)^2(x+3y)}$
57. $\frac{15r^2+10ry-y^2}{(3r+2y)(6r-y)(6r+y)}$ **59.** $\frac{2k^2-10k+6}{(k-3)(k-1)^2}$ **61.** $\frac{7k^2+31k+92}{(k-4)(k+4)^2}$ **63.** **(a)** $\frac{9k^2+6k+26}{5(3k+1)}$ **(b)** $\frac{1}{4}$

SECTION 6.5 (PAGE 388)

CONNECTIONS **Page 388:** 1.413793103; 2

EXERCISES **1.** **(a)** 6; $\frac{1}{6}$ **(b)** 12; $\frac{3}{4}$ **(c)** $\frac{1}{6} \div \frac{3}{4}$ **(d)** $\frac{2}{9}$ **3.** Choice (d) is correct, because every sign has been changed in the fraction. **7.** -6 **9.** $\frac{1}{xy}$ **11.** $\frac{2a^2b}{3}$ **13.** $\frac{m(m+2)}{3(m-4)}$ **15.** $\frac{2}{x}$ **17.** $\frac{8}{x}$ **19.** $\frac{a^2-5}{a^2+1}$ **21.** $\frac{31}{50}$

23. $\frac{y^2 + x^2}{xy(y - x)}$ **25.** $\frac{40 - 12p}{85p}$ **27.** $\frac{5y - 2x}{3 + 4xy}$ **29.** $\frac{a - 2}{2a}$ **31.** $\frac{z - 5}{4}$ **33.** $\frac{-m}{m + 2}$ **35.** $\frac{3m(m - 3)}{(m - 1)(m - 8)}$

37. division **39.** $\frac{\frac{3}{8} + \frac{5}{6}}{2}$ **40.** $\frac{29}{48}$ **41.** $\frac{29}{48}$ **42.** Answers will vary. **43.** $\frac{5}{3}$ **45.** $\frac{13}{2}$ **47.** $\frac{19r}{15}$

SECTION 6.6 (PAGE 396)

EXERCISES **1.** expression; $\frac{43}{40}x$ **3.** equation; $\left\{\frac{40}{43}\right\}$ **5.** expression; $-\frac{1}{10}y$ **7.** equation; $\{-10\}$ **9.** $-2, 0$

11. $-3, 4, -\frac{1}{2}$ **13.** $-9, 1, -2, 2$ **17.** $\left\{\frac{1}{4}\right\}$ **19.** $\left\{-\frac{3}{4}\right\}$ **21.** $\{-15\}$ **23.** $\{7\}$ **25.** $\{-15\}$ **27.** $\{-5\}$

29. $\{-6\}$ **31.** $\emptyset$ **33.** $\{5\}$ **35.** $\{4\}$ **37.** $\{1\}$ **39.** $\{4\}$ **41.** $\{5\}$ **43.** $\{-2, 12\}$ **45.** $\emptyset$ **47.** $\{3\}$

49. $\{3\}$ **51.** $\left\{-\frac{1}{5}, 3\right\}$ **53.** $\left\{-\frac{1}{2}, 5\right\}$ **55.** $\{3\}$ **57.** $\left\{-\frac{1}{3}, 3\right\}$ **59.** $\left\{-6, \frac{1}{2}\right\}$ **61.** $\{6\}$

63. Transform the equation so that the terms with k are on one side and the remaining term is on the other. **65.** $F = \frac{ma}{k}$

67. $a = \frac{kF}{m}$ **69.** $R = \frac{E - Ir}{I}$ or $R = \frac{E}{I} - r$ **71.** $A = \frac{h(B + b)}{2}$ **73.** $a = \frac{2S - ndL}{nd}$ or $a = \frac{2S}{nd} - L$

75. $y = \frac{xz}{x + z}$ **77.** $z = \frac{3y}{5 - 9xy}$ or $z = \frac{-3y}{9xy - 5}$ **79.** The solution set is $\{-5\}$. The number 3 must be rejected.

80. The simplified form is $x + 5$. **(a)** It is the same as the actual solution. **(b)** It is the same as the rejected solution.

81. The solution set is $\{-3\}$. The number 1 must be rejected. **82.** The simplified form is $\frac{x + 3}{2(x + 1)}$. **(a)** It is the same as the actual solution. **(b)** It is the same as the rejected solution. **83.** Answers will vary. **84.** If we transform so that 0 is on one side, then perform the operation(s) and *reduce to lowest terms* before solving, no rejected values will appear.

85. $-\frac{3}{10}$ **87.** An error message would occur. **89.** An error message would occur.

SUMMARY: EXERCISES ON OPERATIONS AND EQUATIONS WITH RATIONAL EXPRESSIONS (PAGE 401)

1. operation; $\frac{10}{p}$ **3.** operation; $\frac{1}{2x^2(x + 2)}$ **5.** operation; $\frac{y + 2}{y - 1}$ **7.** equation; $\{39\}$ **9.** operation; $\frac{13}{3(p + 2)}$

11. equation; $\left\{\frac{1}{7}, 2\right\}$ **13.** operation; $\frac{7}{12z}$ **15.** operation; $\frac{3m + 5}{(m + 3)(m + 2)(m + 1)}$ **17.** equation; $\emptyset$

19. operation; $\frac{t + 2}{2(2t + 1)}$

SECTION 6.7 (PAGE 409)

CONNECTIONS **Page 405:** 54.5 miles per hour; 57.4 miles per hour; yes; More time is spent at the slower speed, and thus the average speed is less than the average of the two speeds.

EXERCISES **1. (a)** an amount **(b)** $5 + x$ **(c)** $\frac{5 + x}{6} = \frac{13}{3}$ **3.** $\frac{12}{18}$ **5.** $\frac{1386}{97}$ **7.** 1989: 33,579 (thousand); 1990: 34,203 (thousand) **9.** female: 144,000; male: 576,000 **11.** 24.15 kilometers per hour **13.** 3.429 hours

15. 7.91 meters per second **17.** $\frac{500}{x - 10} = \frac{600}{x + 10}$ **19.** $\frac{D}{R} = \frac{d}{r}$ **21.** 8 miles per hour **23.** $18\frac{1}{2}$ miles per hour

25. N'Deti: 12.02 miles per hour; McDermott: 8.78 miles per hour **27.** $\frac{1}{10}$ job per hour **29.** $\frac{1}{8}x + \frac{1}{6}x = 1$ or $\frac{1}{8} + \frac{1}{6} = \frac{1}{x}$ **31.** $4\frac{4}{17}$ hours **33.** $5\frac{5}{11}$ hours **35.** 3 hours **37.** $2\frac{7}{10}$ hours **39.** $9\frac{1}{11}$ minutes **41.** direct

43. direct **45.** direct **47.** inverse **49.** 9 **51.** $\frac{16}{5}$ **53.** $\frac{4}{9}$ **55.** increases **57.** $106\frac{2}{3}$ miles per hour **59.** 25 kilograms per hour **61.** 20 pounds per square foot **63.** 15 feet **65.** 144 feet **67.** direct **69.** inverse **71.** approximately 2364 in 1995 and 716 in 1985 (The actual numbers were 2361 and 719.)

CHAPTER 6 REVIEW EXERCISES (PAGE 421)

1. 3 **3.** $-5, -\frac{2}{3}$ **5. (a)** $\frac{11}{8}$ **(b)** $\frac{13}{22}$ **7.** $\frac{b}{3a}$ **9.** $\frac{-(2x+3)}{2}$

Answers may vary in Exercise 11.

11. $\frac{-(4x-9)}{2x+3}, \frac{-4x+9}{2x+3}, \frac{4x-9}{-(2x+3)}, \frac{4x-9}{-2x-3}$ **13.** $\frac{72}{p}$ **15.** $\frac{5}{8}$ **17.** $\frac{3a-1}{a+5}$ **19.** $\frac{p+5}{p+1}$ **21.** $108y^4$ **23.** $\frac{15a}{10a^4}$ **25.** $\frac{15y}{50-10y}$ **27.** $\frac{15}{x}$ **29.** $\frac{4k-45}{k(k-5)}$ **31.** $\frac{-2-3m}{6}$ **33.** $\frac{7a+6b}{(a-2b)(a+2b)}$ **35.** $\frac{5z-16}{z(z+6)(z-2)}$ **37. (a)** $\frac{a}{b}$ **(b)** $\frac{a}{b}$ **(c)** Answers will vary. **39.** $\frac{4(y-3)}{y+3}$ **41.** $\frac{xw+1}{xw-1}$ **43.** It would cause the first and third denominators to equal 0. **45.** $\emptyset$ **47.** $t = \frac{Ry}{m}$ **49.** $m = \frac{4+p^2q}{3p^2}$ **51.** $\frac{2}{6}$ **53.** $3\frac{1}{13}$ hours **55.** inverse **57.** 4 centimeters **59. (a)** -3 **(b)** -1 **(c)** $-3, -1$ **60.** $\frac{15}{2x}$ **61.** If $x = 0$, the divisor R is equal to 0, and division by 0 is undefined. **62.** $(x+3)(x+1)$ **63.** $\frac{7}{x+1}$ **64.** $\frac{11x+21}{4x}$ **65.** $\emptyset$ **66.** We know that -3 is not allowed because P and R are undefined for $x = -3$. **67.** Rate is equal to distance divided by time. Here, distance is 6 miles and time is $x + 3$ minutes, so rate $= \frac{6}{x+3}$, which is the expression for P. **68.** $\frac{6}{5}, \frac{5}{2}$ **69.** $\frac{(5+2x-2y)(x+y)}{(3x+3y-2)(x-y)}$ **71.** $8p^2$ **73.** 3 **75.** $r = \frac{3kz}{5k-z}$ or $r = \frac{-3kz}{z-5k}$ **77.** approximately 7443 for hearts and 3722 for livers (The actual numbers were 7467 and 3698.) **79.** $\frac{36}{5}$

CHAPTER 6 TEST (PAGE 425)

[6.1] **1.** $-2, 4$ **2. (a)** $\frac{11}{6}$ **(b)** undefined **3.** (Answers may vary.) $\frac{-(6x-5)}{2x+3}, \frac{-6x+5}{2x+3}, \frac{6x-5}{-(2x+3)}, \frac{6x-5}{-2x-3}$ **4.** $-3x^2y^3$ **5.** $\frac{3a+2}{a-1}$ [6.2] **6.** $\frac{25}{27}$ **7.** $\frac{3k-2}{3k+2}$ **8.** $\frac{a-1}{a+4}$ [6.3] **9.** $150p^5$ **10.** $(2r+3)(r+2)(r-5)$ **11.** $\frac{240p^2}{64p^3}$ **12.** $\frac{21}{42m-84}$ [6.4] **13.** 2 **14.** $\frac{-14}{5(y+2)}$ **15.** $\frac{-x^2+x+1}{3-x}$ or $\frac{x^2-x-1}{x-3}$ **16.** $\frac{-m^2+7m+2}{(2m+1)(m-5)(m-1)}$ [6.5] **17.** $\frac{2k}{3p}$ **18.** $\frac{-2-x}{4+x}$ [6.6] **19.** $-1, 4$ **20.** $\left\{-\frac{1}{2}\right\}$ **21.** $D = \frac{dF-k}{F}$ or $D = \frac{k-dF}{-F}$ [6.7] **22.** 3 miles per hour **23.** $2\frac{2}{9}$ hours **24.** 27 days

CUMULATIVE REVIEW EXERCISES CHAPTERS 1–6 (PAGE 426)

[1.2, 1.5, 1.6] **1.** 2 [2.2] **2.** $\{17\}$ [2.4] **3.** $b = \frac{2A}{h}$ [2.5] **4.** $\left\{-\frac{2}{7}\right\}$ [2.7] **5.** $[-8, \infty)$ **6.** $(4, \infty)$

[3.1] **7. (a)** $(-3, 0)$ **(b)** $(0, -4)$ [3.2] **8.**

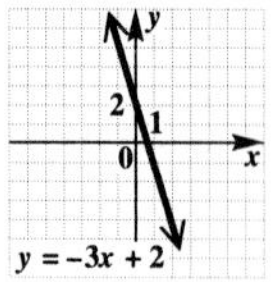

[4.1] **9.**

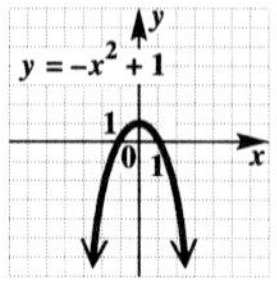

[4.2, 4.5] **10.** $\frac{1}{2^4x^7}$ **11.** $\frac{1}{m^6}$ **12.** $\frac{q}{4p^2}$ [4.1] **13.** $k^2 + 2k + 1$ [4.2] **14.** $72x^6y^7$ [4.4] **15.** $4a^2 - 4ab + b^2$

[4.3] **16.** $3y^3 + 8y^2 + 12y - 5$ [4.6] **17.** $6p^2 + 7p + 1 + \frac{3}{p - 1}$ [4.7] **18.** 1.4×10^5 seconds

[5.3] **19.** $(4t + 3v)(2t + v)$ **20.** prime [5.4] **21.** $(4x^2 + 1)(2x + 1)(2x - 1)$ [5.5] **22.** $\{-3, 5\}$

23. $\left\{5, -\frac{1}{2}, \frac{2}{3}\right\}$ [5.6] **24.** -2 or -1 **25.** 6 meters **26.** 30 (The maximum percent is 30%.) [6.1] **27.** (a)

28. (d) [6.4] **29.** $\frac{4}{q}$ **30.** $\frac{3r + 28}{7r}$ **31.** $\frac{7}{15(q - 4)}$ **32.** $\frac{-k - 5}{k(k + 1)(k - 1)}$ [6.2] **33.** $\frac{7(2z + 1)}{24}$

[6.5] **34.** $\frac{195}{29}$ [6.6] **35.** 4, 0 **36.** $\left\{\frac{21}{2}\right\}$ **37.** $\{-2, 1\}$ [6.7] **38.** 150 miles **39.** $1\frac{1}{5}$ hours **40.** 20 pounds

CHAPTER 7 EQUATIONS OF LINES, INEQUALITIES, AND FUNCTIONS

SECTION 7.1 (PAGE 435)

CONNECTIONS **Page 429:** **1.** \$3500, \$5000; \$35,000, \$50,000 **2.** The loss in value each year; the depreciation when the item is brand new ($D = 0$)

EXERCISES **1.** D **3.** B **5.** The slope m of a vertical line is undefined, so it is not possible to write the equation of the line in the form $y = mx + b$. **7.** $y = 3x - 3$ **9.** $y = -x + 3$ **11.** $y = 4x - 3$ **13.** $y = 3$

15.
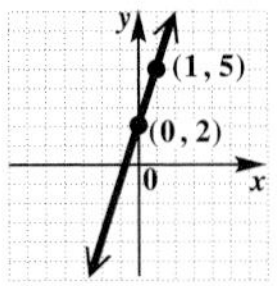

17.
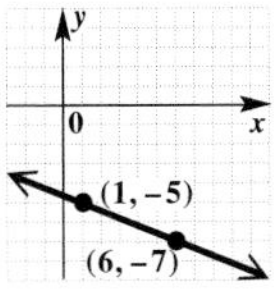

19.
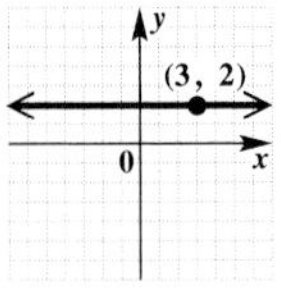

21.
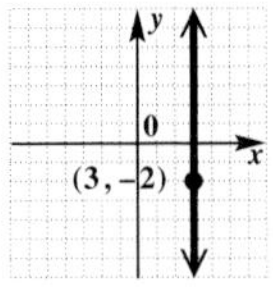

23. the y-axis

25. $y = 2x - 7$ **27.** $y = \frac{2}{3}x + \frac{19}{3}$ **29.** $y = -\frac{4}{5}x + \frac{9}{5}$ **31.** $4 = -3(-2) + b$ **32.** $b = -2$

33. $y = -3x - 2$ **34.** The equations are the same. **35.** $y = x$ (There are other forms as well.) **37.** $y = x - 3$

39. $y = -\frac{5}{7}x - \frac{54}{7}$ **41.** $y = -\frac{2}{3}x - 2$ **43.** $y = \frac{1}{3}x + \frac{4}{3}$ **45.** **(a)** (5, 42), (15, 61), (25, 76)

(b) yes
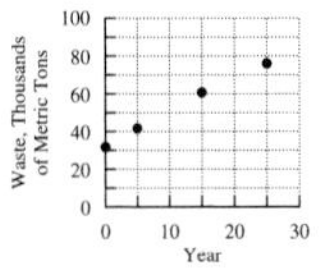

(c) $y = 1.76x + 32$ **(d)** $y = 49.6$, so there will be about 49,600 metric tons in 2005. **47.** **(a)** \$400 **(b)** \$.25 **(c)** $y = .25x + 400$ **(d)** \$425 **(e)** 1500 **49.** **(a)** (1, 822.3), (2, 774.9) **(b)** $y = -47.4x + 869.7$ **(c)** The slope represents the change in sales from 1996 to 1997. The negative slope indicates that sales *decreased.* **51.** $y = -3x + 6$

53. $Y_1 = \frac{3}{4}x + 1$ **55.** (0, 32); (100, 212) **56.** $\frac{9}{5}$ **57.** $F - 32 = \frac{9}{5}(C - 0)$ or $F - 212 = \frac{9}{5}(C - 100)$

58. $F = \frac{9}{5}C + 32$ **59.** $C = \frac{5}{9}(F - 32)$ or $C = \frac{5}{9}F - \frac{160}{9}$ **60.** 86° **61.** 10° **62.** $-40°$

SECTION 7.2 (PAGE 445)

CONNECTIONS **Page 444:** **1.** Answers will vary. **2. (a)** $(5, \infty)$ **(b)** $(-\infty, 5)$ **(c)** $\left[3, \frac{11}{3}\right)$

EXERCISES **1.** $>$ **3.** $\geq$ **5.** $\leq$ **7.** false **9.** true **11.**

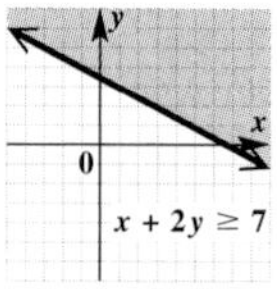

13.

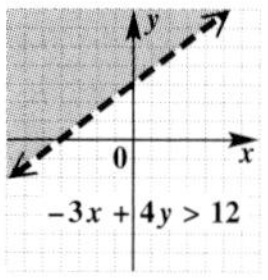

15.

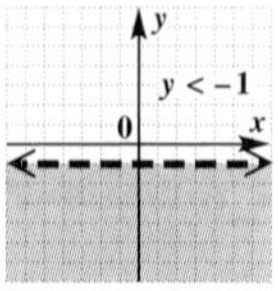

19.

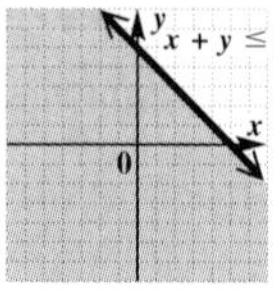

21.

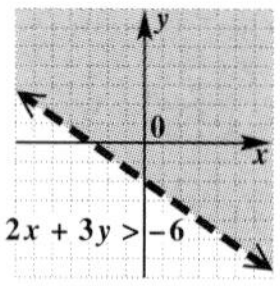

23.

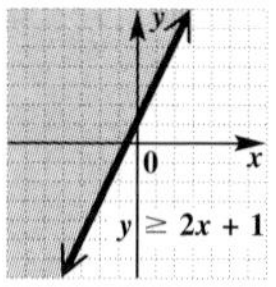

25.

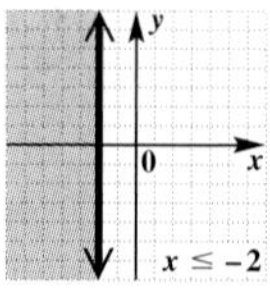

27.

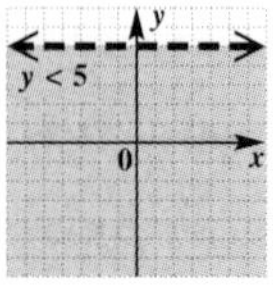

29.

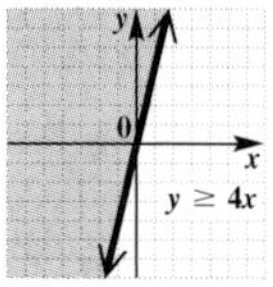

31. Every point in quadrant IV has a positive x-value and a negative y-value. Substituting into $y > x$ would imply that a negative number is greater than a positive number, which is always false. Thus, the graph of $y > x$ cannot lie in quadrant IV. **33.** A **35.** C

In part (b) of Exercises 37 and 39, other answers are possible.

37. (a)

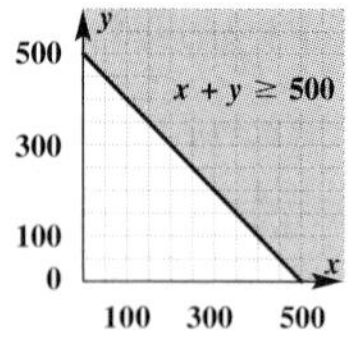

(b) (500, 0), (200, 400) **39. (a)**

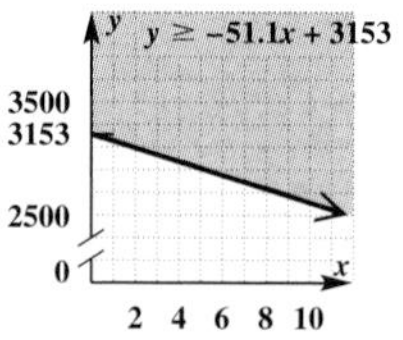

(b) (0, 3153), (2, 3050.8), (10, 2642)

41. (a) $\{-2\}$ **(b)** $(-2, \infty)$ **(c)** $(-\infty, -2)$ **43. (a)** $\{-4\}$ **(b)** $(-\infty, -4)$ **(c)** $(-4, \infty)$

SECTION 7.3 (PAGE 454)

EXERCISES **1.** 3; 3; (1, 3) **3.** 5; 5; (3, 5) **5.** The graph consists of the four points (0, 2), (1, 3), (2, 4), and (3, 5). **7.** not a function; domain: $\{-4, -2, 0\}$; range: $\{3, 1, 5, -8\}$ **9.** function; domain: {A, B, C, D, E}; range: {2, 3, 6, 4} **11.** function **13.** not a function **15.** function **17.** function **19.** function **21.** not a function **23.** domain: $(-\infty, \infty)$; range: $[2, \infty)$ **25.** domain: $(-\infty, \infty)$; range: $(-\infty, \infty)$ **27.** domain: $[0, \infty)$; range: $[0, \infty)$ **29.** (2, 4) **30.** $(-1, -4)$ **31.** $\frac{8}{3}$ **32.** $f(x) = \frac{8}{3}x - \frac{4}{3}$ **33. (a)** 11 **(b)** 3 **(c)** -9 **35. (a)** 4 **(b)** 2 **(c)** 14 **37. (a)** 2 **(b)** 0 **(c)** 3 **39. (a)** 4 **(b)** 2 **41.** {(1970, 9.6), (1980, 14.1), (1990, 19.8), (1997, 25.8)}; yes **43.** $g(1980) = 14.1$; $g(1990) = 19.8$ **45.** For the year 2000, the function predicts 29.3 million foreign-born residents in the United States. **47.** $y = -\frac{14}{3}x + 9506$ **49.** $y = -4x + 8176$ **51.** 4 **53.** 1 **55.** 1 **57.** $y = x + 1$

CHAPTER 7 REVIEW EXERCISES (PAGE 461)

1. $y = -x + \frac{2}{3}$ **3.** $y = x - 7$ **5.** $y = -\frac{3}{4}x - \frac{1}{4}$ **7.** $y = 1$ **9.**

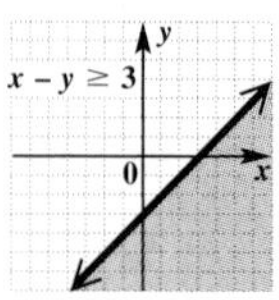

11.

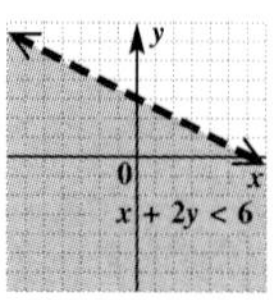

13. 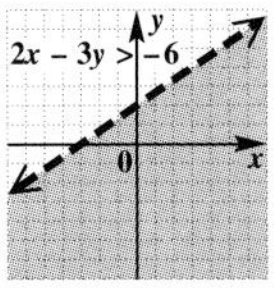

15. not a function; domain: $\{-2, 0, 2\}$; range: $\{4, 8, 5, 3\}$ **17.** not a function **19.** function

21. not a function **23.** domain: $(-\infty, \infty)$; range: $[1, \infty)$ **25.** **(a)** 8 **(b)** -1 **27.** **(a)** 5 **(b)** 2 **29.** $y = -1$
31. $y = -3x + 30$ **33.** 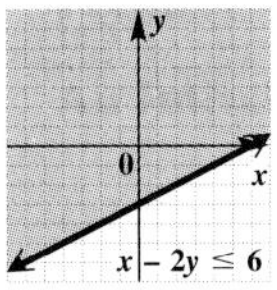

35. 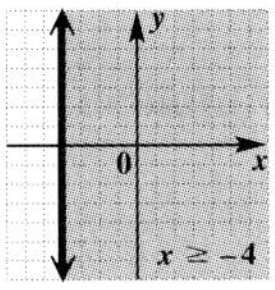

37. Either a $>$ inequality or a $<$ inequality has a dashed boundary line to indicate that the points on the line are not included in the solution set. **38.** 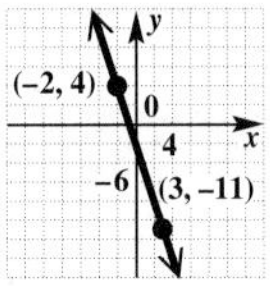

39. -3 **40.** $y = -3x - 2$ **41.** $\left(-\frac{2}{3}, 0\right)$, $(0, -2)$ **42.** 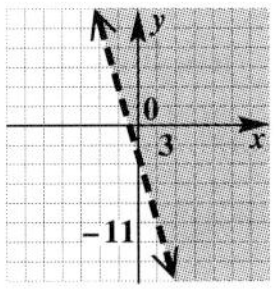

43. -32 **44.** domain: $(-\infty, \infty)$; range: $(-\infty, \infty)$ **45.** **(a)** \$2.910 billion; 1996 **(b)** domain: $\{1994, 1995, 1996, 1997\}$; range: $\{2.677, 2.910, 3.297, 3.561\}$ **(c)** $y = .3255x + 2.5845$; yes **(d)** \$3.2355 billion; Yes, the result is reasonably close.

CHAPTER 7 TEST (PAGE 464)

[7.1] **1.** $y = 2x + 6$ **2.** $y = \frac{5}{2}x - 4$ **3.** $y = \frac{1}{2}x + 4$ **4.** $2x + 3y = 15$ [7.2] **5.** 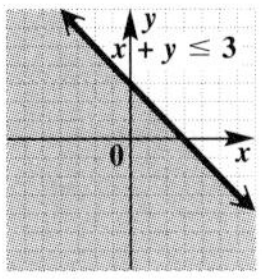

6. 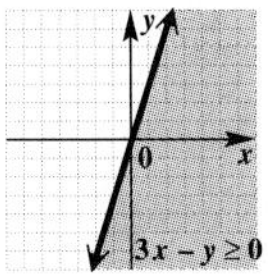

[7.3] **7.** 20 **8.** not a function **9.** function; domain: $(-\infty, \infty)$; range: $[2, \infty)$

10. function; domain: $(-\infty, \infty)$; range: $(-\infty, \infty)$ **11.** not a function **12.** Every vertical line intersects the graph (a line) in only one point. **13.** yes; \$4459 billion or \$4,459,000,000,000 **14.** $x = 1996$ **15.** 234.6 (billion per year)
16. Consumer expenditures are increasing by \$234.6 billion each year.

CUMULATIVE REVIEW EXERCISES CHAPTERS 1–7 (PAGE 465)

[2.2] **1.** $\{-65\}$ [2.4] **2.** $t = \frac{A - p}{pr}$ [5.5] **3.** $\left\{-1, -\frac{1}{7}\right\}$ [6.6] **4.** $\{3\}$ **5.** $\{5\}$

[2.7] **6.** $(-2.6, \infty)$ 

7. $(0, \infty)$ 

8. $(-\infty, -4]$ 

[4.1–4.5] **9.** $\frac{1}{x^2y}$ **10.** $\frac{y^7}{x^{13}z^2}$ **11.** $\frac{m^6}{8n^9}$ **12.** $2x^2 - 4x + 38$ **13.** $15x^2 + 7xy - 2y^2$ **14.** $x^3 + 8y^3$
[4.6] **15.** $m^2 - 2m + 3$ [5.1–5.4] **16.** $(y + 6k)(y - 2k)$ **17.** $(3x^2 - 5y)(3x^2 + 5y)$ **18.** $5x^2(5x - 13y)(5x - 3y)$
19. $(f + 10)^2$ **20.** prime [6.4] **21.** 1 **22.** $\frac{6x + 22}{(x + 1)(x + 3)}$ or $\frac{2(3x + 11)}{(x + 1)(x + 3)}$ [6.2] **23.** $\frac{4(x - 5)}{3(x + 5)}$

24. $\dfrac{x+1}{x}$ **25.** $\dfrac{(x+3)^2}{3x}$ **26.** $\dfrac{4xy^4}{z^2}$ [6.5] **27.** $\dfrac{5}{8}$ **28.** 6 [3.3] **29.** $-\dfrac{4}{3}$ **30.** 0

[7.1] **31.** $y = -4x + 15$ **32.** $y = 4x$ [3.2] **33.**

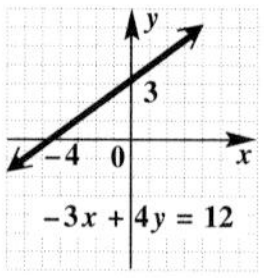

[7.2] **34.**

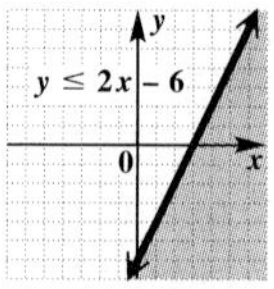

35.

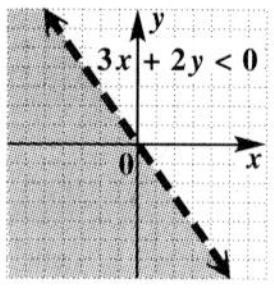

[2.3] **36.** corporate income taxes: $171.8 billion; individual income taxes: $656.4 billion [2.6] **37.** 15°, 35°, 130°
[5.6] **38.** 7 inches [2.5] **39.** 14 [6.7] **40.** 1 hour

CHAPTER 8 LINEAR SYSTEMS

SECTION 8.1 (PAGE 473)

EXERCISES **1. (a)** B **(b)** C **(c)** D **(d)** A **3.** yes **5.** no **7.** yes **9.** yes **11.** no **13.** (a); The ordered pair solution must be in quadrant II, and $(-4, -4)$ is in quadrant III. **15.** $\{(4, 2)\}$

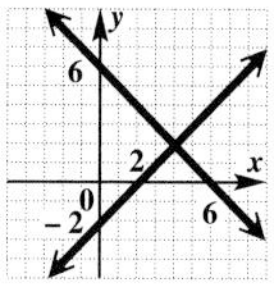

17. $\{(0, 4)\}$

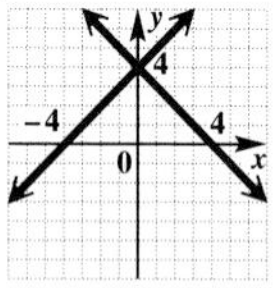

19. $\{(4, -1)\}$

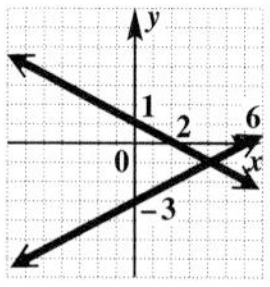

In Exercises 21–29, we do not show the graphs.

21. $\{(1, 3)\}$ **23.** $\{(0, 2)\}$ **25.** $\emptyset$ (inconsistent system) **27.** infinite number of solutions (dependent equations) **29.** $\{(4, -3)\}$ **33.** Yes, it is possible. For example, the system

$$\begin{aligned} x + y &= 5 \\ x - y &= -1 \\ 2x - y &= 1 \end{aligned}$$

has the single solution (2, 3).

35. about 350 million for each format **37.** 200 million **39.** between 1984 and 1986, and between 1988 and 1990 **41. (a)** neither **(b)** intersecting lines **(c)** one solution **43. (a)** dependent **(b)** one line **(c)** infinite number of solutions **45. (a)** inconsistent **(b)** parallel lines **(c)** no solution **47.** 40 **49.** (40, 30) **51.** B **53.** A **55.** $\{(-1, 5)\}$ **57.** $\{2\}$ **58.** 5 **59.** $\{(2, 5)\}$

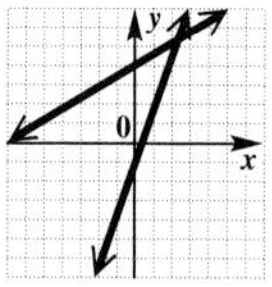

60. The x-coordinate, 2, is equal to the solution of the equation. **61.** The y-coordinate, 5, is equal to the value we obtained on both sides when checking. **62.** 5; 3; 5

SECTION 8.2 (PAGE 481)

EXERCISES **1.** No, it is not correct, because the solution set is $\{(3, 0)\}$. The y-value in the ordered pair must also be determined. **3.** $\{(3, 9)\}$ **5.** $\{(7, 3)\}$ **7.** $\{(0, 5)\}$ **9.** $\{(-4, 8)\}$ **11.** $\{(3, -2)\}$ **13.** infinite number of solutions **15.** $\left\{\left(\dfrac{1}{3}, -\dfrac{1}{2}\right)\right\}$ **17.** $\emptyset$ **19.** infinite number of solutions **21.** The first student had less work to do, because the coefficient of y in the first equation is -1. The second student had to divide by 2, introducing fractions into the expression for x. **23.** $\{(2, -3)\}$ **25.** $\{(3, 2)\}$ **27.** $\{(-2, 1)\}$ **29.** 1993 **31.** To find the total cost, multiply the number of bicycles (x) by the cost per bicycle ($400), and add the fixed cost ($5000). Thus, $y_1 = 400x + 5000$ gives this total cost (in dollars).

32. $y_2 = 600x$ **33.** $y_1 = 400x + 5000$; solution set: $\{(25, 15{,}000)\}$
$y_2 = 600x$ **34.** 25; 15,000; 15,000

35. $\{(2, 4)\}$

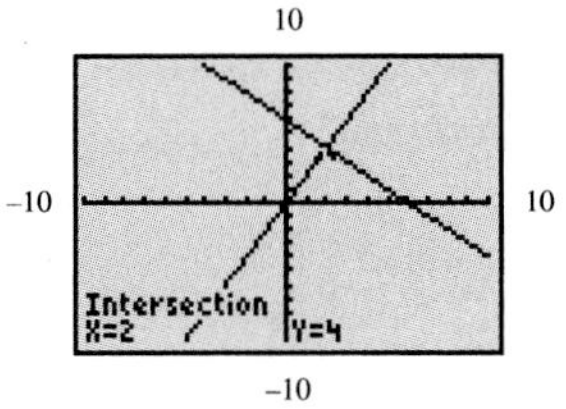

37. $\{(1, 5)\}$

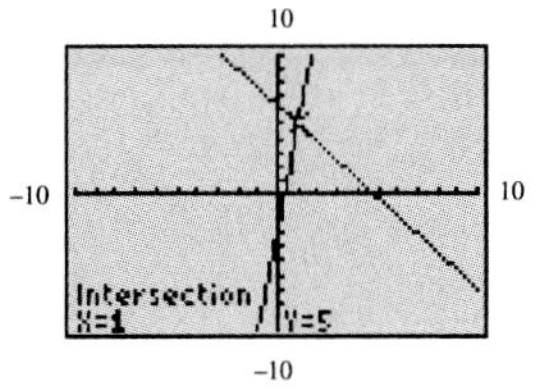

39. $\{(5, -3)\}$; The equations to input are $y_1 = \dfrac{5 - 4x}{5}$ and $y_2 = \dfrac{1 - 2x}{3}$.

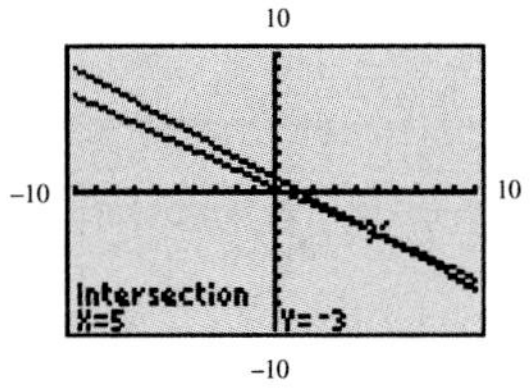

41. Adjust the viewing window so that it does appear.

SECTION 8.3 (PAGE 489)

EXERCISES **1.** true **3.** true **5.** $\{(4, 6)\}$ **7.** $\{(-1, -3)\}$ **9.** $\{(-2, 3)\}$ **11.** $\left\{\left(-\frac{2}{3}, \frac{17}{2}\right)\right\}$ **13.** $\{(3, -6)\}$ **15.** $\{(7, 4)\}$ **17.** $\{(0, 3)\}$ **19.** $\{(3, 0)\}$ **21.** $\left\{\left(-\frac{32}{23}, -\frac{17}{23}\right)\right\}$ **25.** $\{(-3, 4)\}$ **27.** infinite number of solutions **29.** $\{(0, 6)\}$ **31.** $\emptyset$ **33.** **(a)** $\{(1, 4)\}$ **(b)** $\{(1, 4)\}$ **(c)** Answers will vary. **35.** Yes, they should both get the same answer, since both procedures are mathematically valid. **37.** $\{(0, 3)\}$ **39.** $\{(24, -12)\}$ **41.** $\{(3, 2)\}$ **43.** $1141 = 1991a + b$ **44.** $1339 = 1996a + b$ **45.** $1991a + b = 1141$
$1996a + b = 1339$; solution set: $\{(39.6, -77{,}702.6)\}$
46. $y = 39.6x - 77{,}702.6$ **47.** 1220.2 (million); This is slightly less than the actual figure.
48. It is not realistic to expect the data to lie in a perfectly straight line; as a result, the quantity obtained from an equation determined in this way will probably be "off" a bit. One cannot put too much faith in models such as this one, because not all data points are linear in nature.

SECTION 8.4 (PAGE 496)

EXERCISES **1.** (d) **3.** (b) **5.** (c) **7.** the second number; $x - y = 48$; The two numbers are 73 and 25. **9.** Boyz II Men: 134; Bruce Springsteen & the E St. Band: 40 **11.** Terminal Tower: 708 feet; Society Center: 948 feet **13.** 46 ones; 28 tens **15.** 2 copies of *Godzilla*; 5 Aerosmith compact discs **17.** \$2500 at 4%; \$5000 at 5% **19.** Japan: \$17.19; Switzerland: \$13.15 **21.** 80 liters of 40% solution; 40 liters of 70% solution **23.** 30 pounds at \$6 per pound; 60 pounds at \$3 per pound **25.** 60 barrels at \$40 per barrel; 40 barrels at \$60 per barrel **27.** boat: 10 miles per hour; current: 2 miles per hour **29.** plane: 470 miles per hour; wind: 30 miles per hour **31.** car leaving Cincinnati: 55 miles per hour; car leaving Toledo: 70 miles per hour **33.** Roberto: 3 miles per hour; Juana: 2.5 miles per hour

SECTION 8.5 (PAGE 504)

EXERCISES **1.** C **3.** B **5.**

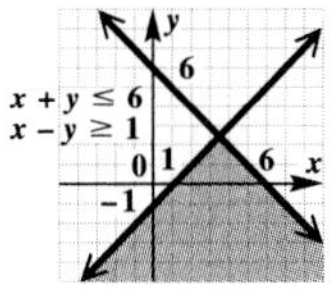

7.

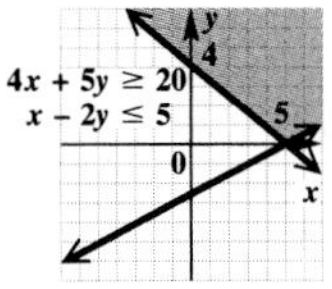

9.

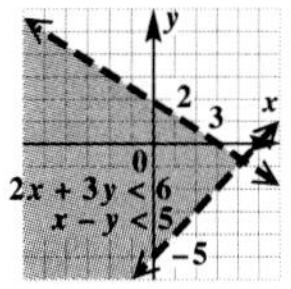

11. 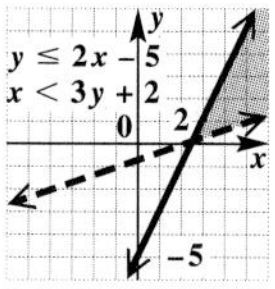

13. 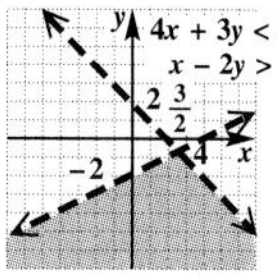

15. 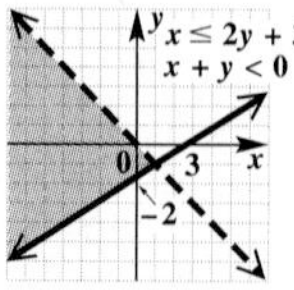

17. 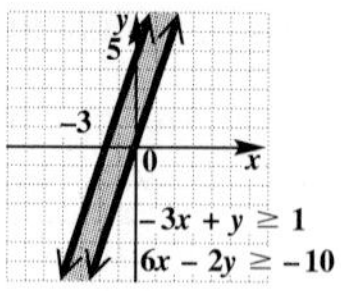

19. 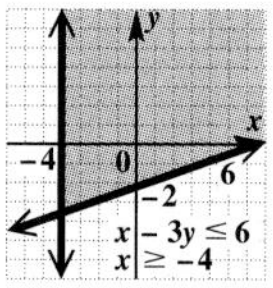

21. 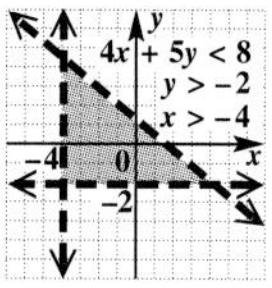

23. 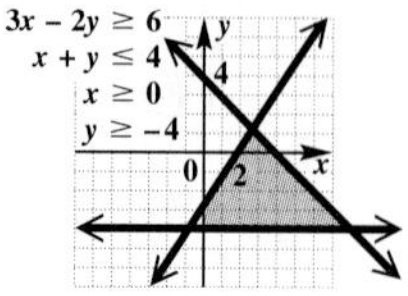

25. (4, 0), (4, 5), (9, 5)

27. (−3, 3), (5, 3), (5, −5) 29. D 31. A

CHAPTER 8 REVIEW EXERCISES (PAGE 510)

1. yes **3.** $\{(3, 1)\}$ **5.** No, this is not correct. A false statement indicates that the solution set is $\emptyset$. **7.** $\{(2, 1)\}$ **9.** infinite number of solutions **11.** His answer was incorrect since the system has infinitely many solutions (as indicated by the true statement $0 = 0$). **13.** (c) **15.** $\{(7, 1)\}$ **17.** infinite number of solutions **19.** $\{(-4, 1)\}$ **21.** $\{(9, 2)\}$ **25.** *How Stella Got Her Groove Back*: 782,699; *The Deep End of the Ocean*: 840,263 **27.** Texas Commerce Tower: 75 stories; First Interstate Plaza: 71 stories **29.** length: 27 meters; width: 18 meters **31.** 25 pounds of \$1.30 candy; 75 pounds of \$.90 candy **33.** \$7000 at 3%; \$11,000 at 4% **35.** plane: 250 miles per hour; wind: 20 miles per hour

37. 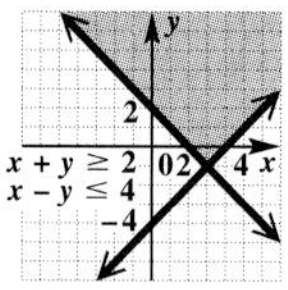

39. 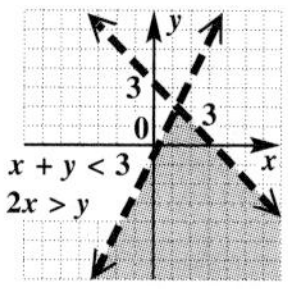

41. $\left\{\left(\frac{28}{5}, \frac{16}{5}\right)\right\}$ **42.** $\left\{\left(\frac{28}{5}, \frac{16}{5}\right)\right\}$ **43.** They are the same. It makes no difference which method we use. **44.** $y_1 = 2x - 8$ **45.** $y_2 = \frac{12 - x}{2}$ or $y_2 = 6 - \frac{1}{2}x$ **46.** $\left\{\frac{28}{5}\right\}$; The solution is the x-value found in Exercises 41 and 42. **47.** We get $\frac{16}{5} = \frac{16}{5}$. This value is the y-value found in Exercises 41 and 42. **48.** 2 **49.** $-\frac{1}{2}$ **50.** They are perpendicular.

51. 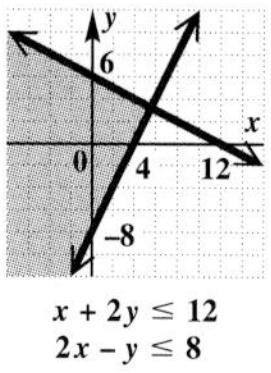

52. $\left\{\left(\frac{28}{5}, \frac{16}{5}\right)\right\}$ **53.** $\{(4, 8)\}$

55. $\{(2, 0)\}$

57. 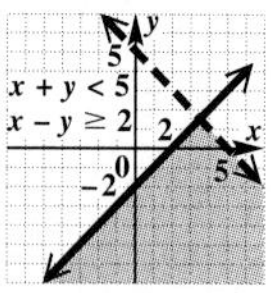

59. Great Smoky Mountains: 9.3 million; Rocky Mountain National Park: 2.9 million

61. slower car: 38 miles per hour; faster car: 68 miles per hour **63.** Yes. Let x represent each of the two equal side lengths. Then $x + 5$ is the length of the third side. The equation to solve is $x + x + (x + 5) = 29$, giving $x = 8$. The side lengths are 8, 8, and 13 inches.

CHAPTER 8 TEST (PAGE 514)

[8.1] **1. (a)** no **(b)** no **(c)** yes **2.** $\{(4, 1)\}$ [8.2] **3.** $\{(1, -6)\}$ **4.** $\{(-35, 35)\}$ [8.3] **5.** $\{(5, 6)\}$ **6.** $\{(-1, 3)\}$ **7.** $\{(-1, 3)\}$ **8.** $\emptyset$ **9.** $\{(0, 0)\}$ **10.** $\{(-15, 6)\}$ [8.1–8.3] **11.** infinite number of solutions **12.** It has no solution. [8.4] **13.** Memphis and Atlanta: 371 miles; Minneapolis and Houston: 671 miles

14. Disneyland: 15.0 million; Magic Kingdom: 13.8 million **15.** $33\frac{1}{3}$ liters of 25% solution; $16\frac{2}{3}$ liters of 40% solution

16. slower car: 45 miles per hour; faster car: 60 miles per hour **17.** *The Lion King*: 30 million; *Snow White*: 28 million

[8.5] **18.**

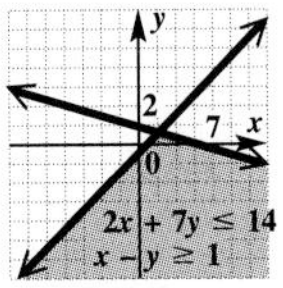

19.

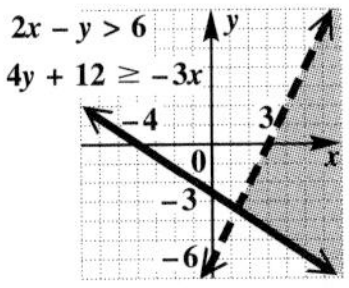

20. A

CUMULATIVE REVIEW EXERCISES CHAPTERS 1–8 (PAGE 515)

[1.7] **1.** $-1, 1, -2, 2, -4, 4, -5, 5, -8, 8, -10, 10, -20, 20, -40, 40$ [1.3] **2.** 1 [1.7] **3.** commutative property

4. distributive property **5.** inverse property [1.6] **6.** 46 [2.2] **7.** $\left\{-\frac{13}{11}\right\}$ **8.** $\left\{\frac{9}{11}\right\}$

[2.4] **9.** width: $8\frac{1}{4}$ inches; length: $10\frac{3}{4}$ inches [2.7] **10.** $\left(-\frac{11}{2}, \infty\right)$ [3.1] **11.**

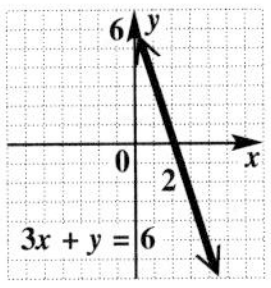

[3.3] **12.** $-\frac{4}{3}$

13. $-\frac{1}{4}$ [4.1] **14.** $14x^2 - 5x + 23$ [4.3] **15.** $6xy + 12x - 14y - 28$ [4.6] **16.** $3k^2 - 4k + 1$

[4.7] **17.** 3.65×10^{10} [4.5] **18.** x^6y [5.3] **19.** $(5m - 4p)(2m + 3p)$ [5.4] **20.** $(8t - 3)^2$

[5.5] **21.** $\left\{-\frac{1}{3}, \frac{3}{2}\right\}$ **22.** $\{-11, 11\}$ [6.1] **23.** $-1, \frac{5}{2}$ [6.4] **24.** $\frac{7}{x + 2}$ [6.2] **25.** $\frac{3}{4k - 3}$

[6.6] **26.** $\left\{-\frac{1}{4}, 3\right\}$ [5.7] **27.** $[-1, 6]$ [7.1] **28.** $y = 3x - 11$ **29.** $y = 4$ **30.** **(a)** $x = 9$ **(b)** $y = -1$

31. $y = 103.25x + 3502$; The slope represents the average yearly increase in health benefit cost during the period.

[7.2] **32.**

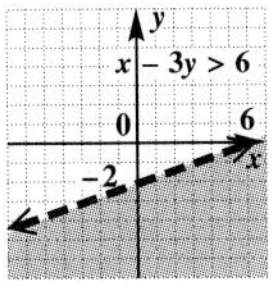

[7.3] **33.** 9 [8.1–8.3] **34.** $\{(-1, 6)\}$ **35.** $\{(3, -4)\}$ **36.** $\{(2, -1)\}$

[8.4] **37.** 405 adults and 49 children **38.** 19 inches, 19 inches, 15 inches **39.** 4 girls and 3 boys [8.5] **40.** (b)

CHAPTER 9 ROOTS AND RADICALS

SECTION 9.1 (PAGE 526)

CONNECTIONS **Page 523:** The area of the large square is $(a + b)^2$ or $a^2 + 2ab + b^2$. The sum of the areas of the smaller square and the four right triangles is $c^2 + 2ab$. Set these equal to each other and subtract $2ab$ from both sides to get $a^2 + b^2 = c^2$.

EXERCISES **1.** false; Zero has only one square root. **3.** true **5.** true **7.** $-4, 4$ **9.** $-12, 12$ **11.** $-\frac{5}{14}, \frac{5}{14}$

13. $-30, 30$ **15.** 7 **17.** -11 **19.** $-\frac{12}{11}$ **21.** not a real number **23.** 100 **25.** 19 **27.** $3x^2 + 4$

29. a must be positive. **31.** a must be negative. **33.** rational; 5 **35.** irrational; 5.385 **37.** rational; -8 **39.** not a real number **41.** The answer to Exercise 17 is the negative square root of a positive number. However, in Exercise 21, the square root of a negative number is not a real number. **43.** 23.896 **45.** 28.249 **47.** 1.985 **49.** 5 **51.** 17 **53.** 3.606 **55.** 2.289 **57.** 5.074 **59.** -4.431 **61.** $c = 17$ **63.** $b = 8$ **65.** $c = 11.705$ **67.** 24 centimeters **69.** 80 feet **71.** 195 miles **73.** 9.434 **75.** Answers will vary. For example, if $a = 2$ and

$b = 7$, $\sqrt{a^2 + b^2} = \sqrt{53}$, while $a + b = 9$. $\sqrt{53} \neq 9$. If $a = 0$ and $b = 1$, then $\sqrt{a^2 + b^2}$ is equal to $a + b$ (which is not true in general). **77.** $\sqrt{29}$ **79.** $\sqrt{2}$ **81.** 10 **83.** -3 **85.** 5 **87.** not a real number **89.** -3 **91.** 3 **93.** c^2 **94.** $(b - a)^2$ **95.** $2ab$ **96.** $(b - a)^2 = a^2 - 2ab + b^2$ **97.** $c^2 = 2ab + (a^2 - 2ab + b^2)$ **98.** $c^2 = a^2 + b^2$

SECTION 9.2 (PAGE 535)

CONNECTIONS **Page 530:** **1.** x; $\sqrt{x}$ **2.** The last part of it ($\sqrt{}$) is used as part of the radical symbol $\sqrt{}$.

EXERCISES **1.** false; $\sqrt{4}$ represents only the principal (positive) square root. **3.** false; $\sqrt{-6}$ is not a real number. **5.** true **7.** true **9.** 9 **11.** $3\sqrt{10}$ **13.** 13 **15.** $\sqrt{13r}$ **17.** (a) **19.** $3\sqrt{5}$ **21.** $5\sqrt{3}$ **23.** $5\sqrt{5}$ **25.** $-10\sqrt{7}$ **27.** $9\sqrt{3}$ **29.** $3\sqrt{6}$ **31.** 24 **33.** $6\sqrt{10}$ **37.** $\frac{4}{15}$ **39.** $\frac{\sqrt{7}}{4}$ **41.** 5 **43.** $\frac{25}{4}$ **45.** $6\sqrt{5}$ **47.** m **49.** y^2 **51.** $6z$ **53.** $20x^3$ **55.** $z^2\sqrt{z}$ **57.** x^3y^6 **59.** $2\sqrt[3]{5}$ **61.** $3\sqrt[3]{2}$ **63.** $2\sqrt[4]{5}$ **65.** $\frac{2}{3}$ **67.** $-\frac{6}{5}$ **69.** 5 **71.** $\sqrt[4]{12}$ **73.** $2x\sqrt[3]{4}$

In Exercise 75, the number of displayed digits will vary among calculator models. Also, less sophisticated models may exhibit round-off error in the final decimal place.

75. **(a)** 4.472135955 **(b)** 4.472135955 **(c)** The numerical results are not a proof because both answers are approximations and they might differ if calculated to more decimal places.

77. 6 centimeters **79.** 6 inches **81.** The product rule for radicals requires that both a and b must be nonnegative. Otherwise $\sqrt{a}$ and $\sqrt{b}$ would not be real numbers (except when $a = b = 0$). **83.** To verify the first length, we show, in the first triangle, that $1^2 + 1^2 = 2$, so the hypotenuse has length $\sqrt{2}$. Similarly, in the second triangle, $(\sqrt{2})^2 + 1^2 = 3$, so the hypotenuse has length $\sqrt{3}$, and so on. **84.** $\sqrt{4}$, $\sqrt{9}$, and so on; $\sqrt{16} = 4$ and $\sqrt{25} = 5$ **85.** Look at the radicands: $9 - 4 = 5$, $16 - 9 = 7$, $25 - 16 = 9$, and so on. **86.** The differences between consecutive whole number lengths increase by 2 each time, so we can predict that the next one will be $\sqrt{25 + 11} = \sqrt{36}$, and the one after that will be $\sqrt{36 + 13} = \sqrt{49}$.

SECTION 9.3 (PAGE 539)

EXERCISES **1.** distributive **3.** radicands **5.** $-5\sqrt{7}$ **7.** $5\sqrt{17}$ **9.** $5\sqrt{7}$ **11.** $11\sqrt{5}$ **13.** $15\sqrt{2}$ **15.** $-20\sqrt{2} + 6\sqrt{5}$ **17.** $17\sqrt{7}$ **19.** $-16\sqrt{2} - 8\sqrt{3}$ **21.** $20\sqrt{2} + 6\sqrt{3} - 15\sqrt{5}$ **23.** $4\sqrt{2}$ **25.** $2\sqrt{3} + 4\sqrt{3} = (2 + 4)\sqrt{3} = 6\sqrt{3}$ **27.** $11\sqrt{3}$ **29.** $5\sqrt{x}$ **31.** $3x\sqrt{6}$ **33.** 0 **35.** $-20\sqrt{2k}$ **37.** $42x\sqrt{5z}$ **39.** $-\sqrt[3]{2}$ **41.** $6\sqrt[3]{p^2}$ **43.** $21\sqrt[4]{m^3}$ **47.** $-6x^2y$ **48.** $-6(p - 2q)^2(a + b)$ **49.** $-6a^2\sqrt{xy}$ **50.** The answers are alike because the numerical coefficient of the three answers is the same: -6. Also, the first variable factor is raised to the second power, and the second variable factor is raised to the first power. The answers are different because the variables are different: x and y, then $p - 2q$ and $a + b$, and then a and $\sqrt{xy}$. **51.** 9.220 **53.** 5 **55.** $22\sqrt{2}$ **57.** **(a)** 1991 **(b)** 1997

SECTION 9.4 (PAGE 545)

EXERCISES **1.** radical **3.** fraction **5.** $4\sqrt{2}$ **7.** $\frac{-\sqrt{33}}{3}$ **9.** $\frac{7\sqrt{15}}{5}$ **11.** $\frac{\sqrt{30}}{2}$ **13.** $\frac{16\sqrt{3}}{9}$ **15.** $\frac{-3\sqrt{2}}{10}$ **17.** $\frac{21\sqrt{5}}{5}$ **19.** $\sqrt{3}$ **21.** $\frac{\sqrt{2}}{2}$ **23.** $\frac{\sqrt{65}}{5}$ **25.** 1; identity property for multiplication **27.** $\frac{\sqrt{21}}{3}$ **29.** $\frac{3\sqrt{14}}{4}$ **31.** $\frac{1}{6}$ **33.** 1 **35.** $\frac{\sqrt{7x}}{x}$ **37.** $\frac{2x\sqrt{xy}}{y}$ **39.** $\frac{x\sqrt{3xy}}{y}$ **41.** $\frac{3ar^2\sqrt{7rt}}{7t}$ **43.** (b) **45.** $\frac{\sqrt[3]{12}}{2}$ **47.** $\frac{\sqrt[3]{196}}{7}$ **49.** $\frac{\sqrt[3]{6y}}{2y}$ **51.** $\frac{\sqrt[3]{42mn^2}}{6n}$ **53.** **(a)** $\frac{9\sqrt{2}}{4}$ seconds **(b)** 3.182 seconds

SECTION 9.5 (PAGE 550)

EXERCISES **1.** 13 **3.** 4 **5.** 4 **7.** 5 **9.** $9\sqrt{5}$ **11.** $16\sqrt{2}$ **13.** $\sqrt{15}-\sqrt{35}$ **15.** $2\sqrt{10}+30$ **17.** $4\sqrt{7}$ **19.** $57+23\sqrt{6}$ **21.** $81+14\sqrt{21}$ **23.** $37+12\sqrt{7}$ **25.** 23 **27.** 1 **29.** $2\sqrt{3}-2+3\sqrt{2}-\sqrt{6}$ **31.** $15\sqrt{2}-15$ **33.** $87+9\sqrt{21}$ **35.** Because multiplication must be performed before addition, it is incorrect to add -37 and -2. Since $-2\sqrt{15}$ cannot be simplified, the expression cannot be written in a simpler form, and the final answer is $-37-2\sqrt{15}$. **37.** $\frac{3-\sqrt{2}}{7}$ **39.** $-4-2\sqrt{11}$ **41.** $1+\sqrt{2}$ **43.** $-\sqrt{10}+\sqrt{15}$ **45.** $3-\sqrt{3}$ **47.** $2\sqrt{5}+\sqrt{15}+4+2\sqrt{3}$ **49.** $\sqrt{11}-2$ **51.** $\frac{\sqrt{3}+5}{8}$ **53.** $\frac{6-\sqrt{10}}{2}$ **55.** $x\sqrt{30}+\sqrt{15x}+6\sqrt{5x}+3\sqrt{10}$ **57.** $6t-3\sqrt{14t}+2\sqrt{7t}-7\sqrt{2}$ **59.** $m\sqrt{15}+\sqrt{10mn}-\sqrt{15mn}-n\sqrt{10}$ **61.** $2-3\sqrt[3]{4}$ **63.** $12+10\sqrt[4]{8}$ **65.** $-1+3\sqrt[3]{2}-\sqrt[3]{4}$ **67.** 1 **69.** $30+18x$ **70.** They are not like terms. **71.** $30+18\sqrt{5}$ **72.** They are not like radicals. **73.** Make the first term $30x$, so that $30x+18x=48x$; make the first term $30\sqrt{5}$, so that $30\sqrt{5}+18\sqrt{5}=48\sqrt{5}$. **74.** When combining like terms, we add (or subtract) the coefficients of the common factors of the terms: $2xy+5xy=7xy$. When combining like radicals, we add (or subtract) the coefficients of the common radical: $2\sqrt{ab}+5\sqrt{ab}=7\sqrt{ab}$. **75.** 4 inches **76.** $\frac{\sqrt{AP}-P}{P}$; 8%

SECTION 9.6 (PAGE 557)

CONNECTIONS **Page 552:** **1.** $6\sqrt{13}\approx 21.63$ (to the nearest hundredth) **2.** $h=\sqrt{13}$; $6\sqrt{13}\approx 21.63$

EXERCISES **1.** {49} **3.** {7} **5.** {85} **7.** {−45} **9.** $\left\{-\frac{3}{2}\right\}$ **11.** ∅ **13.** {121} **15.** {8} **17.** {1} **19.** {6} **21.** ∅ **23.** {5} **25.** Since $\sqrt{x}$ must be greater than or equal to zero for any replacement for x, it cannot equal -8, a negative number. **29.** {12} **31.** {5} **33.** {0, 3} **35.** {−1, 3} **37.** {8} **39.** {4} **41.** {8} **43.** {9} **45.** We cannot square term by term. The left side must be squared as a binomial in the first step. **47.** 158.6 feet **49.** **(a)** 70.5 miles per hour **(b)** 59.8 miles per hour **(c)** 53.9 miles per hour **51.** **(a)** 1991 **(b)** 1997; Here, $f(3)\approx 97.5$. Thus, 1997 is the year when about 3 million calls were made. **(c)** about 2.1 million **53.** 4 **55.** −2 **57.** −1

SECTION 9.7 (PAGE 565)

CONNECTIONS **Page 564:** **1.** $(\sqrt{x}-3)(\sqrt{x}+1)$ **2.** $(x+\sqrt{10})(x-\sqrt{10})$ **3.** $(\sqrt{x}+2)^2$ **4.** $x\sqrt{x}+3\sqrt{x}+\frac{5\sqrt{x}}{x}$

EXERCISES **1.** (a) **3.** (c) **5.** 5 **7.** 4 **9.** 2 **11.** 2 **13.** 8 **15.** 9 **17.** 8 **19.** 4 **21.** −4 **23.** −4 **25.** $\frac{1}{343}$ **27.** $\frac{1}{36}$ **29.** $-\frac{1}{32}$ **31.** 2^3 **33.** $\frac{1}{6^{1/2}}$ **35.** $\frac{1}{15^{1/2}}$ **37.** $11^{1/7}$ **39.** 8^3 **41.** $6^{1/2}$ **43.** $\frac{5^3}{2^3}$ **45.** $\frac{1}{2^{8/5}}$ **47.** $6^{2/9}$ **49.** z **51.** $m^2n^{1/6}$ **53.** $\frac{a^{2/3}}{b^{4/9}}$ **55.** 2 **57.** 2 **59.** $\sqrt{a}$ **61.** $\sqrt[3]{k^2}$ **63.** $\sqrt{2}=2^{1/2}$ and $\sqrt[3]{2}=2^{1/3}$ **64.** $2^{1/2}\cdot 2^{1/3}$ **65.** 6 **66.** $2^{3/6}\cdot 2^{2/6}$ **67.** $2^{5/6}$ **68.** $\sqrt[6]{2^5}$ or $\sqrt[6]{32}$ **69.** 2 **71.** 1.883 **73.** 3.971 **75.** 9.100 **81.** **(a)** $d=1.22x^{1/2}$ **(b)** 211.31 miles **83.** Because $(7^{1/2})^2=7$ and $(\sqrt{7})^2=7$, they should be equal, so we define $7^{1/2}$ to be $\sqrt{7}$.

CHAPTER 9 REVIEW EXERCISES (PAGE 571)

1. −7, 7 **3.** −14, 14 **5.** −15, 15 **7.** 4 **9.** 10 **11.** not a real number **13.** $\frac{7}{6}$ **15.** a must be negative. **17.** irrational; 4.796 **19.** rational; −5 **21.** $5\sqrt{3}$ **23.** $4\sqrt{10}$ **25.** 12 **27.** $16\sqrt{6}$ **29.** $-\frac{11}{20}$ **31.** $\frac{\sqrt{7}}{13}$

33. $\frac{2}{15}$ **35.** 8 **37.** p **39.** r^9 **41.** $a^7b^{10}\sqrt{ab}$ **43.** Yes, because both approximations are .7071067812.

45. $21\sqrt{3}$ **47.** 0 **49.** $2\sqrt{3} + 3\sqrt{10}$ **51.** $6\sqrt{30}$ **53.** 0 **55.** $11k^2\sqrt{2n}$ **57.** $\sqrt{5}$ **59.** $\frac{\sqrt{30}}{15}$

61. $\sqrt{10}$ **63.** $\frac{r\sqrt{x}}{4x}$ **65.** $\frac{\sqrt[3]{98}}{7}$ **67.** $-\sqrt{15} - 9$ **69.** $22 - 16\sqrt{3}$ **71.** -2 **73.** $-2 + \sqrt{5}$

75. $\frac{-2 + 6\sqrt{2}}{17}$ **77.** $\frac{-\sqrt{10} + 3\sqrt{5} + \sqrt{2} - 3}{7}$ **79.** $\frac{3 + 2\sqrt{6}}{3}$ **81.** $3 + 4\sqrt{3}$ **83.** $\emptyset$ **85.** $\{1\}$ **87.** $\{6\}$

89. $\{-2\}$ **91.** **(a)** billions of dollars in exports **(b)** years **(c)** 1992 **(d)** Yes, because \$13 billion is between \$7.5 billion and \$19.6 billion, so the year should be between 1990 and 1993. **(e)** \$31.2 billion; yes **93.** 9 **95.** 7^3 or 343

97. $x^{3/4}$ **99.** 16 **101.** $\frac{5 - \sqrt{2}}{23}$ **103.** $5y\sqrt{2}$ **105.** $-\sqrt{10} - 5\sqrt{15}$ **107.** $\frac{2 + \sqrt{13}}{2}$ **109.** $7 - 2\sqrt{10}$

111. -11 **113.** $\{7\}$ **115.** $\{8\}$ **117.** $-\frac{5\sqrt{7}}{5\sqrt{14}}$ **119.** $-\frac{\sqrt{2}}{2}$

CHAPTER 9 TEST (PAGE 574)

[9.1] **1.** $-14, 14$ **2.** **(a)** irrational **(b)** 11.916 [9.2–9.5] **3.** 6 **4.** $-3\sqrt{3}$ **5.** $\frac{8\sqrt{2}}{5}$ **6.** $2\sqrt[3]{4}$ **7.** $4\sqrt{6}$ **8.** $9\sqrt{7}$ **9.** $-5\sqrt{3x}$ **10.** $2y\sqrt[3]{4x^2}$ **11.** 31 **12.** $6\sqrt{2} + 2 - 3\sqrt{14} - \sqrt{7}$ **13.** $11 + 2\sqrt{30}$

[9.1] **14.** **(a)** $6\sqrt{2}$ inches **(b)** 8.485 inches [9.4] **15.** $\frac{5\sqrt{14}}{7}$ **16.** $\frac{\sqrt{6x}}{3x}$ **17.** $-\sqrt[3]{2}$ **18.** $\frac{-12 - 3\sqrt{3}}{13}$

[9.6] **19.** $\{3\}$ **20.** $\left\{\frac{1}{4}, 1\right\}$ [9.7] **21.** 16 **22.** -25 **23.** 5 **24.** $\frac{1}{3}$ [9.6] **25.** 12 is not a solution. A check shows that it does not satisfy the original equation.

CUMULATIVE REVIEW EXERCISES CHAPTERS 1–9 (PAGE 574)

[1.5–1.6] **1.** 54 **2.** 6 **3.** 3 **4.** 18 **5.** 15 **6.** 4.223 [2.2] **7.** $\{3\}$ [2.7] **8.** $[-16, \infty)$ **9.** $(5, \infty)$

[2.4] **10.** 207 cubic inches [3.2] **11.**

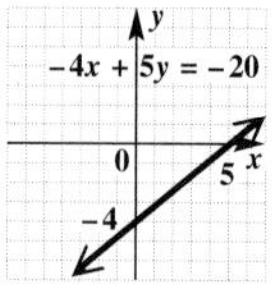

12.

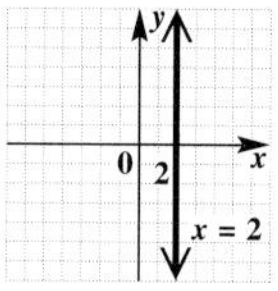

[3.3] **13.** $-\frac{5}{6}$ [4.2] **14.** $12x^{10}y^2$

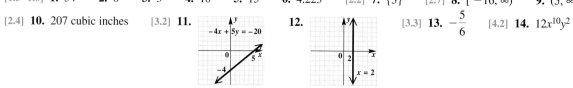

[4.5] **15.** $\frac{y^{15}}{5832}$ [4.1] **16.** $3x^3 + 11x^2 - 13$ [4.6] **17.** $4t^2 - 8t + 5$ [5.2–5.4] **18.** $(m + 8)(m + 4)$ **19.** $(5t^2 + 6)(5t^2 - 6)$ **20.** $(6a + 5b)(2a - b)$ **21.** $(9z + 4)^2$ [5.5] **22.** $\{3, 4\}$ **23.** $\{-2, -1\}$ [6.1] **24.** $2, -7$

[6.2] **25.** $\frac{x + 1}{x}$ **26.** $(t + 5)(t + 3)$ [6.4] **27.** $\frac{y^2}{(y + 1)(y - 1)}$ **28.** $\frac{-2x - 14}{(x + 3)(x - 1)}$ [6.5] **29.** -21

[7.2] **30.**

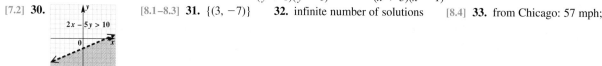

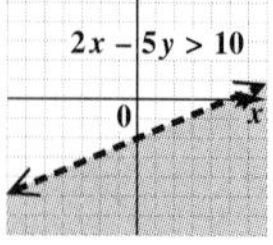

[8.1–8.3] **31.** $\{(3, -7)\}$ **32.** infinite number of solutions [8.4] **33.** from Chicago: 57 mph; from Des Moines: 50 mph **34.** CNN: 67.8 million; ESPN: 67.9 million [9.3–9.5] **35.** $29\sqrt{3}$ **36.** $-\sqrt{3} + \sqrt{5}$ **37.** $10xy^2\sqrt{2y}$ [9.7] **38.** 32 [9.5] **39.** $21 - 5\sqrt{2}$ [9.6] **40.** $\{16\}$

CHAPTER 10 QUADRATIC EQUATIONS

SECTION 10.1 (PAGE 580)

EXERCISES **1.** C **3.** A **5.** B **7.** true **9.** true **11.** According to the square root property, -9 is also a solution, so her answer was not completely correct. The solution set is $\{\pm 9\}$. **13.** $\{\pm 9\}$ **15.** $\{\pm\sqrt{14}\}$ **17.** $\{\pm 4\sqrt{3}\}$ **19.** $\emptyset$ **21.** $\{\pm 1.5\}$ **23.** $\{\pm 2\sqrt{6}\}$ **25.** $\{-2, 8\}$ **27.** $\emptyset$ **29.** $\{8 \pm 3\sqrt{3}\}$ **31.** $\left\{-3, \frac{5}{3}\right\}$ **33.** $\left\{0, \frac{3}{2}\right\}$ **35.** $\left\{\frac{5 \pm \sqrt{30}}{2}\right\}$ **37.** $\left\{\frac{-1 \pm 3\sqrt{2}}{3}\right\}$ **39.** $\{-10 \pm 4\sqrt{3}\}$ **41.** $\left\{\frac{1 \pm 4\sqrt{3}}{4}\right\}$ **43.** Johnny's first solution, $\frac{5+\sqrt{30}}{2}$, is equivalent to Linda's second solution, $\frac{-5-\sqrt{30}}{-2}$. This can be verified by multiplying $\frac{5+\sqrt{30}}{2}$ by 1 in the form $\frac{-1}{-1}$. Similarly, Johnny's second solution is equivalent to Linda's first one. **45.** $\{-4.48, .20\}$ **47.** $\{-3.09, -.15\}$ **49.** $(x+3)^2 = 100$ **50.** $x + 3 = -10$ or $x + 3 = 10$ **51.** $\{-13\}$; $\{7\}$ **52.** $\{-13, 7\}$ **53.** $\{-7, 3\}$ **54.** $\{-3, 6\}$ **55.** .983 foot **57.** 3.442 feet **59.** about $\frac{1}{2}$ second **61.** 9 inches **63.** 5%

SECTION 10.2 (PAGE 587)

CONNECTIONS **Page 586: 1.** x^2 **2.** $x^2 + 8x$ **3.** $x^2 + 8x + 16$ **4.** It occurred when we added the 16 squares.

EXERCISES **1.** 16 **3.** multiplying $(t+2)(t-5)$ to get $t^2 - 3t - 10$ **5.** (d) **7.** 49 **9.** $\frac{25}{4}$ **11.** $\frac{1}{16}$ **13.** $\{1, 3\}$ **15.** $\{-1 \pm \sqrt{6}\}$ **17.** $\{-3\}$ **19.** $\left\{-\frac{3}{2}, \frac{1}{2}\right\}$ **21.** $\emptyset$ **23.** $\left\{\frac{-7 \pm \sqrt{97}}{6}\right\}$ **25.** $\{-4, 2\}$ **27.** $\{1 \pm \sqrt{6}\}$ **29. (a)** $\left\{\frac{3 \pm 2\sqrt{6}}{3}\right\}$ **(b)** $\{-.633, 2.633\}$ **31. (a)** $\{-2 \pm \sqrt{3}\}$ **(b)** $\{-3.732, -.268\}$ **35.** 75 feet by 100 feet **37.** 1 second and 5 seconds **39.** 3 seconds and 5 seconds **41.** 8 miles

SECTION 10.3 (PAGE 594)

EXERCISES **1.** 4; 5; -9 **3.** 2 **5.** $a = 3, b = -4, c = -2$ **7.** $a = 3, b = 7, c = 0$ **9.** $a = 1, b = 1, c = -12$ **11.** $a = 9, b = 9, c = -26$ **13.** If a were 0, the equation would be linear, not quadratic. **15.** No, because $2a$ should be the denominator for $-b$ as well. The correct formula is $x = \frac{-b \pm \sqrt{b^2 - 4ac}}{2a}$. **17.** $\{-13, 1\}$ **19.** $\{2\}$ **21.** $\left\{\frac{-6 \pm \sqrt{26}}{2}\right\}$ **23.** $\left\{-1, \frac{5}{2}\right\}$ **25.** $\{-1, 0\}$ **27.** $\left\{0, \frac{12}{7}\right\}$ **29.** $\{\pm 2\sqrt{6}\}$ **31.** $\left\{\pm\frac{2}{5}\right\}$ **33.** $\left\{\frac{6 \pm 2\sqrt{6}}{3}\right\}$ **35.** $\emptyset$ **37.** $\emptyset$ **39.** $\left\{\frac{-5 \pm \sqrt{61}}{2}\right\}$ **41. (a)** $\left\{\frac{-1 \pm \sqrt{11}}{2}\right\}$ **(b)** $\{-2.158, 1.158\}$ **43. (a)** $\left\{\frac{1 \pm \sqrt{5}}{2}\right\}$ **(b)** $\{-.618, 1.618\}$ **45.** $\left\{-\frac{2}{3}, \frac{4}{3}\right\}$ **47.** $\left\{\frac{-1 \pm \sqrt{73}}{6}\right\}$ **49.** $\{1 \pm \sqrt{2}\}$ **51.** $\emptyset$ **53.** The solution(s) must make sense in the original problem. (For example, a length cannot be negative.) **55.** $r = \frac{-\pi h \pm \sqrt{\pi^2 h^2 + \pi S}}{\pi}$ **57.** $P = .241$ (about 88 days) **59.** $P = 1.881$ **61.** $P = 84.130$ **63.** $1^2 = 1^3$, or $1 = 1$. This is a true statement. **65.** 4 seconds **67.** approximately .2 second and 10.9 seconds **69.** 30 units **70.** It is the radicand in the quadratic formula. **71. (a)** 225 **(b)** 169 **(c)** 4 **(d)** 121 **72.** Each is a perfect square. **73. (a)** $(3x-2)(6x+1)$ **(b)** $(x+2)(5x-3)$ **(c)** $(8x+1)(6x+1)$ **(d)** $(x+3)(x-8)$ **74. (a)** 41 **(b)** -39 **(c)** 12 **(d)** 112 **75.** no **76.** If the discriminant is a perfect square the trinomial is factorable. **(a)** yes **(b)** yes **(c)** no

SUMMARY: EXERCISES ON QUADRATIC EQUATIONS (PAGE 598)

1. $\{\pm 6\}$ **3.** $\left\{\pm\frac{10}{9}\right\}$ **5.** $\{1, 3\}$ **7.** $\{4, 5\}$ **9.** $\left\{-\frac{1}{3}, \frac{5}{3}\right\}$ **11.** $\{-17, 5\}$ **13.** $\left\{\frac{7 \pm 2\sqrt{6}}{3}\right\}$ **15.** $\emptyset$
17. $\left\{-\frac{1}{2}, 2\right\}$ **19.** $\left\{-\frac{5}{4}, \frac{3}{2}\right\}$ **21.** $\{1 \pm \sqrt{2}\}$ **23.** $\left\{\frac{2}{5}, 4\right\}$ **25.** $\left\{\frac{-3 \pm \sqrt{41}}{2}\right\}$ **27.** $\left\{\frac{1}{4}, 1\right\}$
29. $\left\{\frac{-2 \pm \sqrt{11}}{3}\right\}$ **31.** $\left\{\frac{-7 \pm \sqrt{5}}{4}\right\}$ **33.** $\left\{\frac{8 \pm 8\sqrt{2}}{3}\right\}$ **35.** $\emptyset$ **37.** $\left\{-\frac{2}{3}, 2\right\}$ **39.** $\left\{-4, \frac{3}{5}\right\}$
41. $\left\{-\frac{2}{3}, \frac{2}{5}\right\}$ **43.** We need to know other methods because many quadratic equations cannot be solved by elementary factoring methods.

SECTION 10.4 (PAGE 604)

CONNECTIONS **Page 603:** **1.** $\langle -1, -4\rangle, \langle 0, 2\rangle, \langle -5, 0\rangle$ **2.** $-1; 2$

EXERCISES **1.** $3i$ **3.** $2i\sqrt{5}$ **5.** $3i\sqrt{2}$ **7.** $5i\sqrt{5}$ **9.** $5 + 3i$ **11.** $6 - 9i$ **13.** $6 - 7i$ **15.** $14 + 5i$
17. $7 - 22i$ **19.** 45 **21.** $18i$ **23.** -4 **25.** $6 + 4i$ **27.** $4 - 3i$ **29.** $3 - i$ **31.** $2 - 6i$
33. $-\frac{3}{25} + \frac{4}{25}i$ **35.** $-1 + 3i$ **36.** The product of $-1 + 3i$ and $2 + 9i$ is $-29 - 3i$, which *is* the original dividend.
37. $-3 - 4i$ **38.** The product of $-3 - 4i$ and i is $4 - 3i$, which *is* the original dividend. **39.** Because $(3 - 2i)(4 + i) = 14 - 5i$ is true, the given statement is true. **40.** The product of the quotient and the divisor must equal the dividend. **41.** $\{-1 \pm 2i\}$ **43.** $\{3 \pm i\sqrt{5}\}$ **45.** $\left\{-\frac{2}{3} \pm i\sqrt{2}\right\}$ **47.** $\{1 \pm i\}$ **49.** $\left\{-\frac{3}{4} \pm \frac{\sqrt{31}}{4}i\right\}$
51. $\left\{\frac{3}{2} \pm \frac{\sqrt{7}}{2}i\right\}$ **53.** $\left\{\frac{1}{5} \pm \frac{\sqrt{14}}{5}i\right\}$ **55.** $\left\{-\frac{1}{2} \pm \frac{\sqrt{13}}{2}i\right\}$ **57.** $\left\{\frac{1}{2} \pm \frac{\sqrt{11}}{2}i\right\}$ **59.** If the discriminant $b^2 - 4ac$ is negative, the equation will have solutions that are not real. **61.** true **63.** false; For example, $3 + 2i$ is a complex number but it is not real.

SECTION 10.5 (PAGE 614)

CONNECTIONS **Page 610:** The maximum height is 8 feet at 16 feet from the initial point. The maximum distance is 32 feet.

EXERCISES **1.** vertex **3.** $(2, -5)$ **5.** $(-1, 2)$

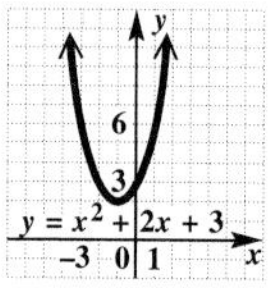

7. $(-3, 0)$

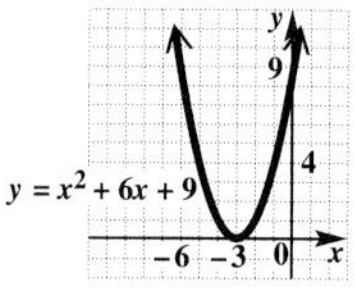

9. $(3, 4)$

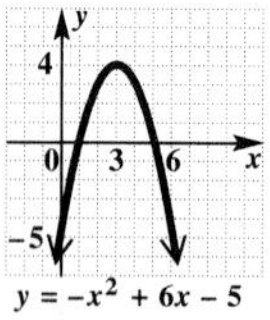

11. $(2, 0)$

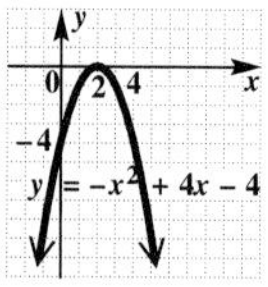

13. one real solution; $\{2\}$ **15.** two real solutions; $\{\pm 2\}$

17. no real solutions; $\emptyset$ **19.** $\{-2, 3\}$ **21.** $\{-1, 1.5\}$ **23.** $\{-4, 2\}$ **25.** **(a)** the point $(0, -2)$ **(b)** the points $(-1, 0)$ and $(2, 0)$

In Exercises 27–31, the domain is first and the range is second.
27. $(-\infty, \infty)$; $[0, \infty)$ **29.** $(-\infty, \infty)$; $(-\infty, 4]$ **31.** $(-\infty, \infty)$; $[1, \infty)$ **33.** 3 **35.** 21 **37.** 40 and 40
39. 320 feet by 640 feet; 204,800 square feet **41.** $y = \frac{11}{5625}x^2$ **43.** In each case, there is a vertical "stretch" of the parabola. It becomes narrower as the coefficient gets larger. **45.** The graph of Y_2 is obtained by reflecting the graph of Y_1 across the x-axis. **47.** By adding a positive constant k, the graph is shifted k units upward. By subtracting a positive constant k, the graph is shifted k units downward.

CHAPTER 10 REVIEW EXERCISES (PAGE 623)

1. $\{\pm 12\}$ **3.** $\{\pm 8\sqrt{2}\}$ **5.** $\{3 \pm \sqrt{10}\}$ **7.** $\emptyset$ **9.** $\{-5, -1\}$ **11.** $\{-1 \pm \sqrt{6}\}$ **13.** $\left\{-\frac{2}{5}, 1\right\}$ **15.** 2.5 seconds **17.** $\left(\frac{k}{2}\right)^2$ or $\frac{k^2}{4}$ **19.** $\{1 \pm \sqrt{5}\}$ **21.** $\left\{\frac{2 \pm \sqrt{10}}{2}\right\}$ **23.** $\left\{\frac{-3 \pm \sqrt{41}}{2}\right\}$ **25.** Multiply both the numerator and the denominator by the conjugate of the denominator. **27.** $5 - i$ **29.** 20 **31.** i **33.** $\frac{7}{50} + \frac{1}{50}i$ **35.** No, the product $(a + bi)(a - bi) = a^2 + b^2$ will always be the sum of the squares of two real numbers, which is a real number. **37.** $\left\{\frac{2}{3} \pm \frac{2\sqrt{2}}{3}i\right\}$ **39.** $\left\{-\frac{3}{2} \pm \frac{\sqrt{23}}{2}i\right\}$ **41.** $\left\{-\frac{1}{9} \pm \frac{2\sqrt{2}}{9}i\right\}$ **43.** vertex: $(1, 4)$

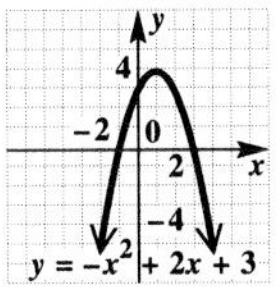

45. two; $\{\pm 2\}$ **47.** two; $\{-1, 4\}$ **49.** none; $\emptyset$ **51.** $\left\{-\frac{11}{2}, 5\right\}$ **53.** $\left\{\frac{-1 \pm \sqrt{21}}{2}\right\}$ **55.** $\left\{\frac{-5 \pm \sqrt{17}}{2}\right\}$ **57.** $\emptyset$ **59.** $\left\{-\frac{5}{3}\right\}$ **61.** $\{-2 \pm \sqrt{5}\}$ **63.** approximately 424 **65.** $\{1 \pm \sqrt{3}\}$ **66.** $(x - (1 + \sqrt{3})) \cdot (x - (1 - \sqrt{3}))$ **67.** $((x - 1) - \sqrt{3})((x - 1) + \sqrt{3})$ **68.** $(x - 1)^2 - \sqrt{3}^2 = x^2 - 2x + 1 - 3 = x^2 - 2x - 2$ **69.** They are both $x^2 - 2x - 2$. **70.** $(x - (2 + \sqrt{5}))(x - (2 - \sqrt{5}))$ or $((x - 2) - \sqrt{5})((x - 2) + \sqrt{5})$

CHAPTER 10 TEST (PAGE 626)

[10.1] **1.** $\{\pm\sqrt{39}\}$ **2.** $\{-11, 5\}$ **3.** $\left\{\frac{-3 \pm 2\sqrt{6}}{4}\right\}$ [10.2] **4.** $\{2 \pm \sqrt{10}\}$ **5.** $\left\{\frac{-6 \pm \sqrt{42}}{2}\right\}$ [10.3] **6.** **(a)** 0 **(b)** one (a double solution) **7.** $\left\{-3, \frac{1}{2}\right\}$ **8.** $\left\{\frac{3 \pm \sqrt{3}}{3}\right\}$ **9.** $\left\{-1 \pm \frac{\sqrt{7}}{2}i\right\}$ **10.** $\left\{\frac{5 \pm \sqrt{13}}{6}\right\}$ [10.1–10.3] **11.** $\{1 \pm \sqrt{2}\}$ **12.** $\left\{\frac{-1 \pm 3\sqrt{2}}{2}\right\}$ **13.** $\left\{\frac{11 \pm \sqrt{89}}{4}\right\}$ **14.** $\{5\}$ **15.** 2 seconds **16.** 247.3 years [10.4] **17.** $-5 + 5i$ **18.** $-17 - 4i$ **19.** 73 **20.** $2 - i$ [10.5] **21.** vertex: $(3, 0)$

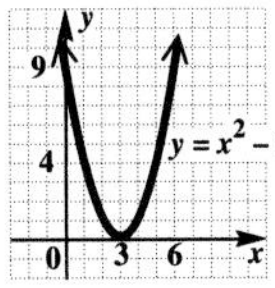

22. vertex: $(-1, -3)$

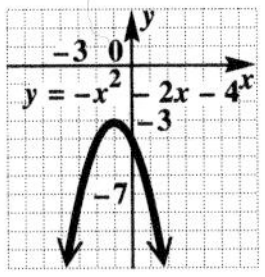

23. vertex: $(-3, -2)$

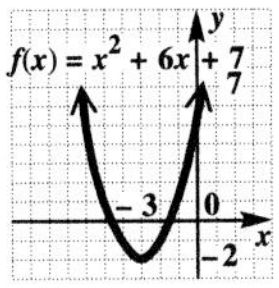

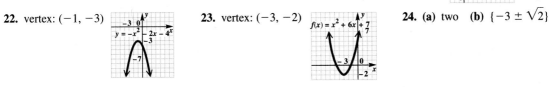

24. **(a)** two **(b)** $\{-3 \pm \sqrt{2}\}$ **(c)** $-3 - \sqrt{2} \approx -4.414$ and $-3 + \sqrt{2} \approx -1.586$ **25.** 200 and 200

CUMULATIVE REVIEW EXERCISES CHAPTERS 1–10 (PAGE 627)

[1.2] **1.** 15 [1.5] **2.** -2 [1.6] **3.** 5 [1.8] **4.** $-r + 7$ **5.** $-2k$ **6.** $19m - 17$ [2.1] **7.** $\{18\}$ [2.2] **8.** $\{5\}$ **9.** $\{2\}$ [2.4] **10.** 100°, 80° **11.** width: 50 feet; length: 94 feet **12.** $L = \frac{P - 2W}{2}$ or $L = \frac{P}{2} - W$ [2.7] **13.** $(-2, \infty)$

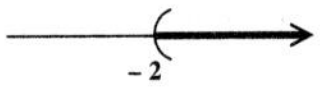

14. $(-\infty, 4]$

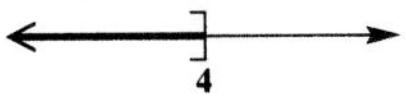

[3.2] **15.**

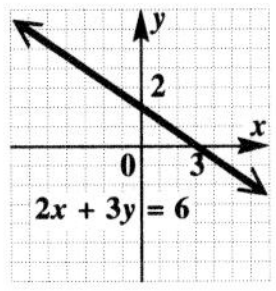

[3.3] **16.** -1

[4.1] **17.** $8x^5 - 17x^4 - x^2$ **18.**

x	y
-2	1
-1	-2
0	-3
1	-2
2	1

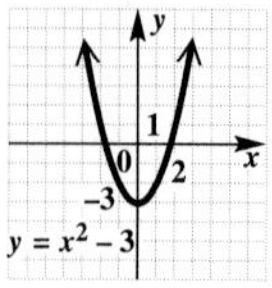

[4.5] **19.** $\frac{x^4}{9}$ **20.** $\frac{b^{16}}{c^2}$ **21.** $\frac{27}{125}$

[4.3] **22.** $2x^2 - 9x - 18$ **23.** $2x^4 + x^3 - 19x^2 + 2x + 20$ [4.6] **24.** $3x^2 - 2x + 1$ [4.7] **25. (a)** 6.35×10^9 **(b)** .00023 [5.1] **26.** $16x^2(x - 3y)$ [5.3] **27.** $(2a + 1)(a - 3)$ [5.4] **28.** $(4x^2 + 1)(2x + 1)(2x - 1)$ **29.** $(5m - 2)^2$ [5.5] **30.** $\{-9, 6\}$ [5.6] **31.** $2\frac{1}{2}$ seconds [5.7] **32.** $(-\infty, -1] \cup [6, \infty)$

[6.1] **33.** $\frac{x - 3}{x - 2}$; -2 and 2 [6.2] **34.** $\frac{4}{5}$ [6.4] **35.** $\frac{-k - 1}{k(k - 1)}$ **36.** $\frac{5a + 2}{(a - 2)^2(a + 2)}$ [6.5] **37.** $\frac{b + a}{b - a}$

[6.6] **38.** $\left\{-\frac{15}{7}, 2\right\}$ [6.7] **39.** 3 miles per hour [7.1, 7.2] **40. (a)** $-\frac{1}{3}$ **(b)** $y = -\frac{1}{3}x + 6$ **(c)** The line crosses the y-axis above the origin, indicating that $b > 0$, since $(0, b)$ is the y-intercept. [7.1] **41.** $2x - y = -3$

[7.2] **42.**

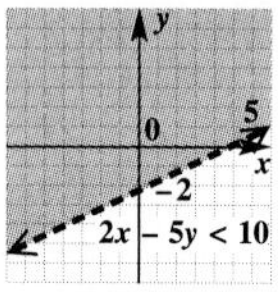

[7.3] **43.** -18 [8.1–8.3] **44.** $\{(-3, 2)\}$ **45.** $\emptyset$ [8.4] **46.** Krystalite: $39.99; Contempra II: $29.99 [8.5] **47.**

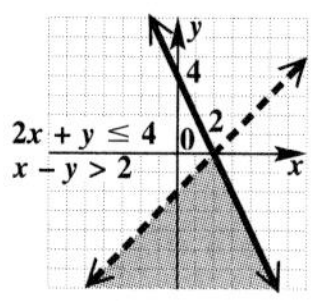

[9.1] **48.** 10 [9.4] **49.** $\frac{6\sqrt{30}}{5}$ **50.** $\frac{\sqrt[3]{28}}{4}$ [9.3] **51.** $4\sqrt{5}$

52. $-ab\sqrt[3]{2b}$ [9.6] **53.** $\{7\}$ [9.7] **54.** 4 [10.1] **55.** $\left\{\frac{-2 \pm 2\sqrt{3}}{3}\right\}$ [10.2] **56.** $\{-1 \pm \sqrt{6}\}$

[10.3] **57.** $\left\{\frac{2 \pm \sqrt{10}}{2}\right\}$ [10.2, 10.3] **58.** $\emptyset$ **59.** 3, 4, 5 [10.4] **60. (a)** $8i$ **(b)** $5 + 2i$

61. $\left\{-\frac{1}{2} \pm \frac{\sqrt{17}}{2}i\right\}$ [10.5] **62.**

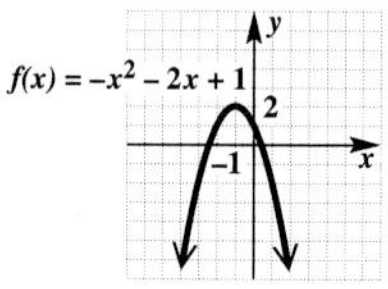

vertex: $(-1, 2)$; domain: $(-\infty, \infty)$; range: $(-\infty, 2]$ **63.** positive

APPENDIX A (PAGE 633)

1. 117.385 **3.** 13.21 **5.** 150.49 **7.** 96.101 **9.** 4.849 **11.** 166.32 **13.** 164.19 **15.** 1.344 **17.** 4.14 **19.** 2.23 **21.** 4800 **23.** .53 **25.** 1.29 **27.** .96 **29.** .009 **31.** 80% **33.** .7% **35.** 67% **37.** 12.5% **39.** 109.2 **41.** 238.92 **43.** 5% **45.** 110% **47.** 25% **49.** 148.44 **51.** 7839.26 **53.** 7.39% **55.** $2760 **57.** 805 miles **59.** 63 miles **61.** $2400

APPENDIX B (PAGE 638)

1. {1, 2, 3, 4, 5, 6, 7} **3.** {winter, spring, summer, fall} **5.** $\emptyset$ **7.** {L} **9.** {2, 4, 6, 8, 10, . . . } **11.** The sets in Exercises 9 and 10 are infinite sets. **13.** true **15.** false **17.** true **19.** true **21.** true **23.** true **25.** true **27.** false **29.** true **31.** true **33.** false **35.** true **37.** true **39.** false **41.** true **43.** false **45.** {g, h} **47.** {b, c, d, e, g, h} **49.** {a, c, e} = B **51.** {d} **53.** {a} **55.** {a, c, d, e} **57.** {a, c, e, f} **59.** $\emptyset$ **61.** B and D; C and D

Index

A

B

C

D

G

H

I

K

L

M

N

O

P

S

T

U

V

W

X

Y

Z

Index of Applications

Astronomy/Aerospace

Biology

Business

Chemistry

Construction

Consumer Applications

Economics

Education

Environment

FINANCE

GEOMETRY

GOVERNMENT

HEALTH/LIFE SCIENCES

LABOR

MISCELLANEOUS

PHYSICS

Sports/Entertainment

Statistics/Demographics

Transportation